Birkhäuser

# Contemporary Mathematicians

Joseph P.S. Kung

Editor

For further volumes:
http://www.springer.com/series/4817

Patrizia Pucci • Vicentiu D. Radulescu
Hans Weinberger
Editors

# James Serrin. Selected Papers

## Volume 2

Birkhäuser

*Editors*
Patrizia Pucci
Università di Perugia
Perugia, Italy

Vicentiu D. Radulescu
King Abdulaziz University
Jeddah, Saudi Arabia

Hans Weinberger
University of Minnesota
Minneapolis, USA

ISBN 978-3-0348-0686-2 ISBN 978-3-0348-0847-7 (eBook)
DOI 10.1007/978-3-0348-0847-7
Springer Basel Heidelberg New York Dordrecht London

Library of Congress Control Number: 2014939300

Mathematics Subject Classification (2010): 26, 34, 35, 76



Printed on acid-free paper

Springer Basel is part of Springer Science+Business Media (www.birkhauser-science.com)

# Table of Contents

## Volume 2

# Introduction

This volume is part of a two-volumes set of the Selected Works of James Serrin. The purpose of this set is to make available to a wider audience some of the main mathematical contributions of James Serrin. We feel that this is useful, because some of his papers are not as well known as they should be, while others are not easily available since they were published in volumes with only limited circulation.

The first volume is composed of the basic contributions to mathematics made by Professor Serrin, while the second volume is devoted to works which are more directly related to applications in the physical sciences. Each volume is grouped into a large variety of fields, and into topics within the fields, as shown in the table of contents. A special feature of the volumes is that, in most cases, the papers in a particular topic are preceded by a commentary by an expert on the subject. These commentaries are intended to indicate how the papers are related to other work, including future work, on this topic. We wish to express our gratitude and thanks to all the authors of these commentaries.

In order to give an overview of a few very long papers without overburdening the volumes, Professor Serrin chose to omit some details of the proofs in the reprinted versions. Such omissions are clearly marked in the articles.

The table of contents assigns capital Roman numerals to the fields, and, if necessary, Arabic numerals to the topics within a field. The field and topic numbers appear before the commentary, if any, and the reprinted articles on this topic. The papers within a topic are designated by lower case Roman numerals. A commentary on a topic uses the lower case numerals to refer to the papers in this topic, which appear just below the commentary.

We also draw the reader's attention to the final Section VIII in Volume 2, containing various essays of James Serrin. These include assessments of the work of *Jean Leray*, *Eberhard Hopf* and *Clifford Truesdell*, and some further comments on special topics, in particular an essay on the position of applied mathematics in scientific thought. In addition, the two papers in Section I.3 constitute a discussion of elliptic partial differential equations in the second half of the previous century, while the paper I.2 (ii) contains a historical essay on the origin of the minimal surface equation (see pages 491–493 of the article), and the paper I.6 (ii) includes an essay on the history of *the Liouville theorem* for entire solutions of elliptic partial differential equations (see pages 139–141 of the article).

A complete list of the publications of James Serrin appears at the end of each volume. Articles in this list which are reprinted in these volumes are indicated by boldface item numbers.

The articles in these volumes are reprinted with the kind permission of the copyright holders, who are indicated at the tops of the articles.

P. Pucci, V. Rădulescu, H. Weinberger

James Serrin, Minneapolis, MN, 1990.

# II. Evolution Equations

# II.1. Ordinary Differential Equations – Stability

David Russell and Roberto Triggiani

**Mathematics Subject Classification (2010).** 34D20, 34D05, 34A34, 34C99, 34C15.

**Keywords.** Asymptotic stability; nonlinear oscillator; nonlinear ordinary differential system.

This article is concerned with joint work of Pucci and Serrin in the area of asymptotic stability properties of certain nonlinear second-order differential equations, which is displayed in in the papers [i], [ii], and [iii] below. However, the authors feel obliged to comment also on a fourth companion paper [1] by the same researchers, whose results are closely related to the first three.

**A Dynamical Model.** The papers [i], [ii], and [1] all study a very general second-order (in time $t$) $N$-dimensional system of quasi-variational ordinary differential equations (ODEs) of the form

$$(\nabla G(u,\, u'))' - \nabla_u G(u,\, u') + \nabla_u F(t,\, u) = Q(t,\, u,\, u'),\ {}' = \frac{d}{dt},\ t \in J = [T, \infty), \qquad (1)$$

where $u \equiv 0$ is a solution. It is assumed that $F(t,\, u)$ is a $\mathcal{C}^2$ scalar-valued function. Here $u$ is the unknown $N$-vector, and $\nabla G(u,p)$ means $\nabla_p G(u,p)$.

The function $F$ represents a restoring potential and $Q$ a general nonlinear damping term with $(Q(t,\, u,\, p),\, p) \le 0$. An important special case of (1), obtained when $G(u,\, p) = \frac{1}{2}\|p\|^2$, $F(t,u) \equiv F(u)$, $Q(t,\, u,\, p) = -A(t,\, u,\, p)\, p$ for some $N \times N$ matrix function $A(t,\, u,\, p)$, is

$$u'' + A(t,\, u,\, u')\, u' + \nabla_u F(u) = 0, \qquad (2)$$

which is still a large and important class of problems.

From a mechanical point of view, system (1) can be considered as the governing law of a holonomic dynamical system having $N$ degrees of freedom and subject to nonlinear damping.

**Goal.** The common objective of all four of the works [i], [ii], [iii], and [1] is to provide various forms of precise and general sufficient conditions, as well as necessary conditions, for the asymptotic stability of the trivial solution. That is, the equilibrium solution $u(t) \equiv 0$ is a *global attractor* in the sense that any bounded solution $u(t)$ defined on an interval $J = [T, \infty)$ satisfies $u(t), u'(t) \to 0$ as $t \to \infty$ in the case of [i], [ii], and [1] for (1), and $u(t), u'(t)/q(t) \to 0$, $t \to \infty$ in the case of [iii] for $(1)'$. (For some results these asymptotic properties are shown to be true for all unbounded solutions as well.)

The delicacy of the problem is already evident from classical results concerning the linear scalar case of (2), i.e., with $u$ scalar, $A(t) = a(t)$ and $F(u) = \frac{1}{2}u^2$. Then, if $a(t)$


DOI 10.1007/978-3-0348-0847-7_1, © Springer Basel 2014

is small or, more precisely, if $a \in L^1(J)$, resonant behavior results in solutions which are oscillatory and do not tend to a limit as $t \to \infty$ (cf. Levin–Nohel, 1950). On the other hand, if $a(t)$ is sufficiently large, there again exist solutions not tending to zero as $t \to \infty$ (Ballieu–Pfeiffer, 1978). In both cases the rest state loses asymptotic stability.

**Results of [iii].** The paper [iii] considers the more specialized version of (1), which we denote henceforth by $(1)'$, in which $G(u, p)$ is strictly convex and homogeneous of degree $m > 1$ in the variable $p$, and $F(t, u)$ has the form $q(t)^m F(u)$. It is assumed that $(u, \nabla_u F(u)) \geq 0$ for all $u$, and that $Q(t, u, u')$ satisfies certain growth conditions.

By employing the well-known Liouville transformation $t \to y = y(t) = \int_T^t q(s)\, ds$, $t \in J$, $(1)'$ is reduced to an equation of the form (1) with $t$ replaced by $y$, $u'$ by $\dot{u} = du/dy$, $Q(t, u, u')$ by a corresponding $\tilde{Q}(y, u, \dot{u})$, and $t \in J$ by a corresponding $y \in \tilde{J}$. The results of [i] and [1] are thus transferred into the result of [iii]. These results include Theorem 1, p. 215, and its variation, Theorem 2, p. 217, both in §3 of the paper. They may be illustrated by applying them to the classical scalar case:

$$u'' + h(t)\, u' + q^2(t)\, f(u) = 0, \quad t \in J, \tag{3}$$

with $m = 2$. The result applied to this problem yields Theorem B, p. 209: assume there exists a function $k(t)$ which is of bounded variation on $J$ and satisfies certain conditions involving $q$ and $A = h + q'/q$. Also assume that $uf(u) > 0$ for $u \neq 0$. Then every bounded solution of (3) tends to zero as $t \to \infty$. Moreover, this conclusion holds for *all* solutions of (3) if the integral of $f(u)$ diverges at both $+\infty$ and $-\infty$. Even these specialized conclusions considerably extend classical results.

**Results of [i] and [1].** To explain the results in their clearest form, we first concentrate on the simplified canonical model (2) with the basic assumptions

$$(A(t, u, p), p) \geq k_0 \, \|A(t, u, p)\, p\| \, \|p\| , \quad k_0 > 0, \tag{4}$$

$$\sigma(t) \, \|p\|^\mu \leq \|A(t, u, p)\, p\| \leq \delta(t) \, \|p\|^\mu , \quad \mu > 0, \tag{5}$$

and $(f(u), u) > 0$ for $u \neq 0$. Then with precise assumptions on $\sigma(t)$, $\delta(t)$, [i] provides in Theorem A, p. 277, both sufficient and necessary conditions for the rest state to be a global attractor. To illustrate: if, in addition to some suitable smoothness conditions, the product $\sigma\,\delta$ is bounded on $J$, then a sufficient condition is $\sigma \notin L^1(J)$, while $\delta \notin L^1(J)$ is necessary. Alternatively, if $1/(\sigma\,\delta)$ is bounded on $J$, then the condition $1/\delta \notin L^1(J)$ is sufficient while $1/\sigma \notin L^1(J)$ is necessary. When the hypotheses on $\sigma$, $\delta$ in Theorem A are not satisfied, [i] provides a more general Theorem B, p. 278, wherein three different conditions are given, each sufficient for the rest state to be a global attractor. One such condition is expressed solely in terms of the functions $\sigma$, $\delta$ in (5); the other two also require a postulated auxiliary function $k(t)$ of bounded variation satisfying some additional conditions involving $\sigma, \delta$. An additional Theorem C, p. 278, of negative type states that if $\delta \in L^1(J)$, then there is no nonzero solution of (2) which approaches a limit in $R^N$ as $t \to \infty$. This result nicely complements a prior conclusion obtained by the authors ca. 1992, recalled here as Theorem D: if $1/\sigma \in L^1(J)$, then every bounded solution of (2) approaches a limit as $t \to \infty$ and the set of such attainable limits is dense in $R^N$. The combination of the above results provides necessary <u>and</u> sufficient conditions. These are illustrated by the effectively scalar case $A(t) = t^\alpha \,(\log t)^\beta \; I$. First, the rest state of (2) is a global attractor in case either $|\alpha| < 1$, or $\alpha = 1$ and $\beta \leq 1$, or $\alpha = -1$ and $\beta \geq -1$. Next, when $\alpha = -1$ and $\beta > -1$, or when $\alpha > 1$, each solution of (2) approaches a limit and the set of attainable limits is all of $R^N$. Finally, if $\alpha = -1$ and $\beta < -1$ or if $\alpha < -1$, no nonzero solution of (2) can approach a limit as $t \to \infty$.

In [i], the analysis of asymptotic stability of the trivial solution is mainly focused on the case $\delta(t) \to 0$ as $t \to \infty$. In contrast, the contemporaneous and independent work [1] provides a precise complementary condition on $\delta(t)$, allowing it to be unbounded as $t \to \infty$ in a definite sense.

Regarding the *general model* (1), four main results where the rest state is a global attractor are given, two in §3 and two in §4. These employ very general and technical hypotheses; in particular the function $G(u, \cdot)$ is assumed convex throughout $R^N$ and for given $U > 0$, $p_0 > 0$ one has $(\nabla_u G(u,p), u) \leq c\,(\nabla_p G(u,p), p)$ for some constant $c$ and $\|u\| < V$, $\|p\| \geq p_0$. A series of useful and enlightening corollaries follows, illustrating mathematically challenging special cases. Finally, §5 provides several necessary conditions for $u \equiv 0$ to be a global attractor for (1).

**Basic approach in [i] and** [1] **(hence [iii]).** The above results greatly extend, even in the scalar case, numerous classical works through a *unified theory* based on a *single proof technique*. The basic approach to asymptotic stability properties of the solution $u \equiv 0$ of (1) depends on the construction of an appropriate Lyapunov function for that system. This construction is, in turn, based on the general theory of variational identities (cf. (2.8), p. 283), introduced by the same authors in [2]. This generalizes the well-known Pokozhaev, or Rellich, identity for elliptic problems. In general, (1) involves coupling of the second derivatives in each component equation of the system. This makes the determination of an appropriate Lyapunov function far from transparent and causes additional problems when $G$ is not suitably smooth. By introducing a Lyapunov function directly in terms of a variational identity, these difficulties are avoided. The choice of the appropriate test function in the main variational identity is given at the bottom of p. 285.

**Results of [ii].** The paper [ii] studies the *same* oscillatory-type equations given by (1) and pursues the *same* goal of establishing the rest state as a global attractor. The *novelty* here, with respect to [i], [1] (hence [iii]), lies in the authors' focus on the concept of *intermittent damping*, previously introduced in the literature for the scalar case. This consists in imposing on the damping term $Q$ certain restrictions, or controls, on non-overlapping intervals $I_n = [a_n, b_n] \subset J = [T, \infty)$, with $a_n \to \infty$, while on the gaps between these intervals no restrictions are imposed. Alternatively, the damping term $Q$ may be assumed to be bounded away from zero and otherwise uncontrolled. The intervals $I_n$ may be arbitrarily widely spread and may have arbitrarily long gaps between them. The general and technical results of [ii] for (1) are expressed in particularly clean form when specialized to the simplified canonical model (2). Let the matrix $A$ be bounded and uniformly positive definite for $t \in I = \bigcup_1^\infty I_n$ and $(u,p)$ in any compact set of $R^N \times R^N$ – no other restrictions other than non-negativity are imposed on $A$ in the set $J/I$. Then the following, rather unexpected, result holds: if $\sum_1^\infty |I_n|^3 = \infty$, then $u = 0$ is a global attractor for (2). Moreover, the exponent 3 is best possible in the sense that, without further restrictions, no larger exponent yields this general conclusion.

Regarding the *general model* (1), with $G(u, \cdot)$ strictly convex in $R^N$ for all $u \in R^N$, the main results under appropriate technical assumptions are given in Theorem A, p. 818, and Theorem B, p. 818, in §2. Theorem A refers to the case where the damping magnitude $\|Q\|$ is controlled from below; Theorem B, instead, covers the complementary situation in which the damping magnitude is bounded away from zero when $\|u\|$ and $\|p\|$ are bounded away from zero (in the same way as the setting of [1] is complementary to the setting of [i]). In turn, Theorems A and B form the basis, in their respective settings, for the most general results of [ii] – Theorem 1, p. 821, and Theorem 2, p. 823 – under technical assumptions. These, in turn, are illustrated by a set of more "digestible" corollaries. A final §6 is devoted to showing, via examples, that the exponents in the paper (in particular the exponent "3" in $\sum_1^\infty |I_n|^3 = \infty$) are best possible.

**Importance and influence of [i], [ii], [iii], and [1] in later developments.** Cumulatively, this body of work may be characterized by two main features. First, in the area of ODEs, it vastly improves, and combines under a unifying proof technique, numerous classical results of the literature even in the scalar case (cf. works by Artstein–Infante; Ballieu–Pfeiffer; Nakao; Salvadori; Sansone; Smith; Thurston–Wang; Levin–Nohel; Hotnemi–Totik; Cesari; Prudi; Li; Nieto, etc.). Moreover, as a consequence, this work has stimulated subsequent research developments in further directions. ODEs with complex-valued coefficients or with random coefficients; or with "on–off" damping feedback. In addition, this work impacted – both with respect to the subsequent investigations of the authors and of other researchers – the area of abstract wave equations, concrete instances of hyperbolic PDEs with on–off and positive–negative or intermittently controlled feedback damping, anisotropic Kirchhoff systems and even in the area of elliptic systems, e.g., existence of non-trivial entire radial solutions.

While time obscures many mathematical contributions, those discussed here seem destined to have permanent value and importance.

## References

[1] Pucci, P., and J. Serrin: *Precise damping conditions for global asymptotic stability for nonlinear second order systems, II*, J. Diff. Eqns. **113** (1994), pp. 505–534.

[2] ——: *A general variational identity*, Indiana Univ. Math. J. **35** (1986), pp. 681–703.

David Russell
Dept. Math., Virginia Tech
Blacksburg, VA, USA
e-mail: `russell@math.vt.edu`

Roberto Triggiani
Department of Mathematics Sciences
The University of Memphis
Memphis, TN 38152, USA
e-mail: `rtrggani@memphis.edu`

*Acta Math.* 170 (1993), 275–307. **Reprint II.1.[i]**


# Precise damping conditions for global asymptotic stability for nonlinear second order systems

by

PATRIZIA PUCCI
*Università di Perugia*
*Perugia, Italy*

and

JAMES SERRIN
*University of Minnesota*
*Minneapolis, MN, U.S.A.*

## 1. Introduction

The global asymptotic stability of the rest point for nonlinear equations has been treated by Levin and Nohel, Salvadori, Thurston and Wong, Artstein and Infante, and Ballieu and Peiffer. These studies have been generalized to scalar variational problems in [8] and to variational systems in [9]. Here we shall unify and further extend this work, by obtaining necessary and sufficient conditions for the asymptotic stability of solutions of quasi-variational systems in terms of the damping functions of the systems treated.

Throughout the paper we thus consider vector unknowns $u: J\to\mathbf{R}^N$, where $J$ is a half open interval of the form $[R,\infty)$. The typical system which we shall study then has the form

$$(\nabla\mathcal{L}(t,u,u'))'-\nabla_u\mathcal{L}(t,u,u')=Q(t,u,u'), \tag{1.1}$$

where $\mathcal{L}(t,u,p)=G(u,p)-F(t,u)$ and where

$$G:\mathbf{R}^N\times\mathbf{R}^N\to\mathbf{R},\quad F:J\times\mathbf{R}^N\to\mathbf{R},\quad Q:J\times\mathbf{R}^N\times\mathbf{R}^N\to\mathbf{R}^N$$

are given continuously differentiable functions. The most important of the conditions which will be imposed on (1.1) are that

$$G(u,\cdot)\text{ is strictly convex in }\mathbf{R}^N;\quad G(u,0)=0,\quad \nabla G(u,0)=0, \tag{1.2}$$

$$(\nabla_u F(t,u),u)>0\quad\text{for }u\neq 0;\quad F(t,0)=0, \tag{1.3}$$

$$(Q(t,u,p),p)\leqslant 0. \tag{1.4}$$

Here $(\cdot,\cdot)$ denotes the inner product in $\mathbf{R}^N$ and

$$\nabla=\left(\frac{\partial}{\partial p_1},\dots,\frac{\partial}{\partial p_N}\right),\quad \nabla_u=\left(\frac{\partial}{\partial u_1},\dots,\frac{\partial}{\partial u_N}\right).$$

The function $F$ represents a restoring potential, this being analytically described by (1.3), while $Q$ represents a general nonlinear damping, expressed by (1.4). In Section 2 we shall give a complete set of hypotheses, while explicit examples are given at the end of the introduction and also in references [4] and [9].

Since $\nabla G(u,0)=\nabla_u G(u,0)=\nabla_u F(t,0)=Q(t,u,0)=0$ it is clear that the rest state $u\equiv 0$ is a solution of (1.1). We shall say that the rest state is *a global attractor* for the system if any bounded solution $u$, defined on some interval $J$, has the property

$$u(t),\ u'(t)\to 0 \quad \text{as } t\to\infty.$$

This concept of asymptotic stability is defined with respect to *bounded* solutions rather than with respect to *all* solutions. The motivation for this point of view is that the concepts of boundedness and of stability are essentially different and accordingly should be treated separately (for the boundedness of solutions see [2], [3], [10] and [13]).

Our approach depends on the construction of an appropriate Liapunov function for the system (1.1), based on the general theory of variational identities introduced in [7]. In general (1.1) involves coupling of the second derivatives in each equation of the system. This makes the determination of an appropriate Liapunov function far from transparent and causes additional problems when $G$ is not suitably smooth or presents other singularities. By introducing a Liapunov function directly in terms of a variational identity, these difficulties are avoided.

To present our results, we consider first the important special case of (1.1) given by

$$u''+A(t,u,u')u'+f(u)=0, \tag{1.5}$$

for which the conclusions can be stated in simple form. Here $A$ is a continuous $N\times N$ damping matrix for which there exist non-negative measurable functions $\sigma$, $\delta$: $J\to\mathbf{R}$ such that

$$(A(t,u,p)p,p)\geqslant \text{Pos. Const.}\,|A(t,u,p)p|\cdot|p| \tag{1.6}$$

and

$$\sigma(t)|p|\leqslant|A(t,u,p)p|\leqslant\delta(t)|p| \tag{1.7}$$

hold for all $t\in J$ and $u,p\in\mathbf{R}^N$. Furthermore $f$ is a continuous gradient of a scalar potential $F$, with

$$(f(u),u)>0 \quad \text{for } u\neq 0.$$

The system (1.5) is naturally identified with (1.1) by the choice $\mathcal{L}(t,u,p)=\frac{1}{2}|p|^2-F(u)$ and $Q(t,u,p)=-A(t,u,p)p$. The scalar case of (1.5) was the main object of the studies [1], [2], [5], [12] and [13].

Levin and Nohel have shown for the linear scalar case of (1.5), i.e. when $A=a(t)$ and $f(u)=u$, that if $a(t)$ is small, or more precisely if it is in $L^1(J)$, then there exist solutions which are oscillatory and moreover do not tend to a limit as $t\to\infty$. On the other hand, Ballieu and Peiffer have shown that when $a(t)$ is sufficiently large then again there exist solutions which do not tend to zero as $t\to\infty$ (Artstein and Infante have given a specific example of this behavior).

The delicacy of the problem is evident from these results: both for small $a$, due to resonant behavior, and for large $a$, due to overdamping, the rest state loses asymptotic stability. Moreover the question of asymptotic stability is separate from the question of oscillation of solutions, as is clear from the example of Bessel's equation

$$u''+\frac{\beta}{t}u'+u=0,$$

for which the rest state is a global attractor for all $\beta>0$, but whose solutions are oscillatory when $\beta<1$ and monotone when $\beta\geqslant 1$. Reflecting this situation, and as a corollary of our general results for the system (1.1), the following conclusions hold for the system (1.5).

THEOREM A. (i) *Suppose* $\sigma\delta$ *is bounded on* $J$. *If* $\sigma$ *is continuous and of bounded variation, or if* $\log\sigma$ *is uniformly Lipschitz continuous, then a sufficient condition for the rest state of* (1.5) *to be a global attractor is that*

$$\sigma\notin L^1(J),$$

*while a necessary condition is that*

$$\delta\notin L^1(J).$$

(ii) *Suppose* $1/\sigma\delta$ *is bounded on* $J$. *If* $1/\delta$ *is continuous and of bounded variation, or if* $1/\delta$ *is absolutely continuous and* $|(1/\delta)'|\leqslant\text{Const.}\sqrt{\sigma/\delta}$, *then a sufficient condition for the rest state to be a global attractor is that*

$$\frac{1}{\delta}\notin L^1(J),$$

*while a necessary condition is that*

$$\frac{1}{\sigma}\notin L^1(J).$$

When $\sigma,\delta$ fail to satisfy the hypotheses of Theorem A, for example when the product $\sigma\delta$ is at the same time neither bounded from zero nor bounded above, the situation is more delicate. In this case we have the following more general conditions for the stability of solutions of (1.5).

THEOREM B. *The rest state of* (1.5) *is a global attractor if one of the following conditions* (a), (b), (c) *is satisfied.*

(a) *There exist positive numbers c and $\eta$ such that*

$$t\sigma(t) \geqslant c, \qquad \int_R^t \delta(s)s^{\eta-2}\,ds \leqslant \text{Const.}\, t^{\eta}.$$

(b) *There exists a non-negative continuous function $k=k(t)$ of bounded variation on $J$ such that*

$$k \notin L^1(J), \quad k(t) \leqslant \text{Const.}\, \sigma(t) \quad \text{in } J, \tag{1.8}$$

*and either*

$$\text{(i) } \delta k \in L^\infty(J) \quad \text{or} \quad \text{(ii) } \delta k^2 \in L^1(J).$$

(c) *There exists a non-negative bounded absolutely continuous function $k=k(t)$ satisfying* (1.8) *and* (i) *or* (ii), *such that*

$$|k'| \leqslant \text{Const.}\sqrt{k\sigma} \quad \text{a.e. in } J.$$

Sufficiency for the first part of Theorem A follows from (b) and (c) with $k=\sigma$, and for the second part from (b) and (c) with $k=1/\delta$. Necessity for the first part of Theorem A comes from Corollary 1 of Section 5, and for the second part from Theorem D below.

Part (a) of Theorem B is the case $\mu=1$ of Corollary 5 in Section 4. Parts (b) and (c) are Corollary 1 in Section 4, with $\mu=1$ and with the respective decompositions $d=\delta$, $e=0$ for (i) and $d=0$, $e=\delta$ for (ii). Case (b) then corresponds to the application of Theorem 4.1, and (c) to the application of Theorem 4.2 with the parameter values $m=2$, $\nu=1$, $\lambda=\frac{1}{2}$.

Theorems A and B are new in both the scalar and vector case of (1.5). When (1.5) is scalar, several subcases of Theorems A and B are already known: in particular, Smith [12] obtained, in the linear case, the first part of Theorem A when $\sigma=\delta$ and $\sigma$ is decreasing; Artstein and Infante found (a) when $\sigma(t)$=Constant and $\eta=2$; and Ballieu and Peiffer obtained (b) in the following special cases: when $\sigma(t)$=Constant and $\delta k \in L^\infty(J)$, and when $\delta(t)$=Constant and $k$ is decreasing. Finally, Salvadori obtained asymptotic stability for the vector system (1.5) when $\sigma$ and $\delta$ are bounded both from zero and above, a special case of (a).

From Section 5 we also obtain

THEOREM C. *Let $\delta \in L^1(J)$. Then there are no solutions of* (1.5) *except $u\equiv 0$ which approach a limit in $\mathbf{R}^N$ as $t\to\infty$.*

Theorem C is complemented by the following result (cf. [10], Section 4).

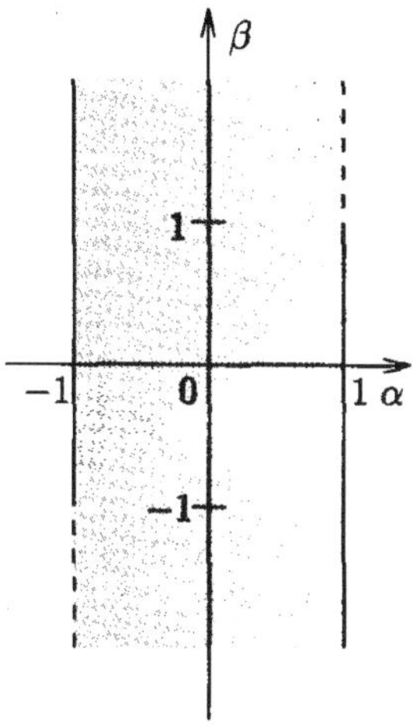


Fig. 1. Asymptotic stability for the system (1.5) when $A=t^{\alpha}(\log t)^{\beta}I$. The rest state is a global attractor if and only if $(\alpha,\beta)$ is in the shaded region.

THEOREM D. *Let* $1/\sigma\in L^1(J)$. *Then every bounded solution of* (1.5) *approaches a limit as* $t\to\infty$, *and the set of attainable limits is dense in* $\mathbf{R}^N$.

The matrix $A=t^{\alpha}(\log t)^{\beta}I$ serves as an example to illustrate these results. For this case the rest state of (1.5) is a global attractor if and only if either $|\alpha|<1$; or $\alpha=1$, $\beta\leqslant 1$; or $\alpha=-1$, $\beta\geqslant -1$. When $\alpha=1$ and $\beta>1$ or when $\alpha>1$ each solution of (1.5) approaches a limit, and the set of attainable limits is $\mathbf{R}^N$. On the other hand, if $\alpha=-1$ and $\beta<-1$ or if $\alpha<-1$ no solution of (1.5), with the exception of $u\equiv 0$, can approach a limit as $t\to\infty$. See Figure 1.

A direct generalization of condition (1.7), for which results similar to Theorems A–D also hold is

$$\sigma(t)|p|^{\mu}\leqslant |A(t,u,p)p|\leqslant \delta(t)|p|^{\mu} \tag{1.9}$$

where $\mu>0$, see particularly the corollaries in Section 4.

For the general system (1.1) our results are entirely analogous to those stated above for (1.5). These further results, which are the main purpose of the paper, are contained in the theorems of Sections 3 and 4, in a series of corollaries to these theorems, and in Section 5.

That a set of corollaries is needed for an adequate statement of the conclusions is due, we believe, to the fact that no single useful condition can encompass the general theory of asymptotic stability. On the other hand, the corollaries present a *unified* theory of asymptotic stability, based on a single proof technique.

There are a number of situations which can be represented by the system (1.1).

19–935202 *Acta Mathematica* 170. Imprimé le 30 juin 1993

When $G(p)=|p|^m/m$, $m>1$, the corresponding system is

$$|u'|^{m-4}\left[|u'|^2\delta_{ij}+(m-2)u_i'u_j'\right]u_j''+f_i(t,u)=Q_i(t,u,u'),\quad f=\nabla F, \tag{1.10}$$

where $\delta_{ij}$ denotes the Kronecker symbol and the repeated index $j$ is to be summed from 1 to $N$; note that when $m\neq 2$ this system is singular at points where $u'=0$. A similar example is $G(p)=\sum_{i=1}^N |p_i|^m/m$, $m>1$, the related system then being

$$|u_i'|^{m-2}u_i''+f_i(t,u)=Q_i(t,u,u'). \tag{1.11}$$

The second derivatives are uncoupled here, but again when $m\neq 2$ the system is singular if at least one $u_i'=0$. If $m=2$ both (1.10) and (1.11) reduce to the important form

$$u''+f(t,u)=Q(t,u,u').$$

When $G(p)=\sqrt{1+|p|^2}-1$, that is, when $G$ is the mean curvature operator, the system (1.1) becomes

$$(1+|u'|^2)^{-3/2}[(1+|u'|^2)\delta_{ij}-u_j'u_j']u_j''+f_i(t,u)=Q_i(t,u,u'). \tag{1.12}$$

The system (1.1) also arises as the *radial* version of the partial differential system in $\mathbf{R}^n$

$$\operatorname{div}\nabla G(Du)+f(t,u)=0,\quad t=|x|,$$

where $Du$ denotes the Jacobian matrix $(\partial u_i/\partial x_j)$, $i=1,...,N$, $j=1,...,n$ and where $G(Du)$ has the special form $\overline{G}(|\nabla u_1|,...,|\nabla u_N|)$. To place this in the context of (1.1) we take $G(p)=\overline{G}(|p_1|,...,|p_N|)$ and $Q=-(n-1)\nabla G/t$.

The general system (1.1) may be considered as the motion equation for a holonomic dynamical system with $N$ degrees of freedom, whose Lagrangian $\mathcal{L}$ is defined by an action energy $T=G(u,p)$ and a potential $U=F(t,u)$, and whose dynamics are governed by a general nonlinear damping term $Q=Q(t,u,p)$. Here the variables $u_i$ represent appropriate Lagrangian coordinates.

A final example worth noting arises from the Euler–Lagrange system for extremals of the variational integral

$$\int_J g(t)\mathcal{L}(t,u,u')\,dt,$$

where $\mathcal{L}$ has the form given earlier and $g\colon J\to\mathbf{R}$ is continuously differentiable, positive and non-decreasing. Extremals for this functional satisfy (1.1) with

$$Q(t,u,p)=-\frac{g'(t)}{g(t)}\nabla G(u,p).$$

Here the condition $(Q,p)\leqslant 0$ is a consequence of the fact that $g(t)>0$, $g'(t)\geqslant 0$ together with (1.2), namely $(\nabla G(u,p),p)\geqslant 0$. For further generalizations to variational problems we refer to the work of Leoni, Manfredini and Pucci [4].

It is also possible to treat the non-homogeneous analogue of (1.1), namely

$$(\nabla G(u,u'))'-\nabla_u G(u,u')+f(t,u)=Q(t,u,u')+e(t),$$

and the corresponding analogue of (1.5),

$$u''+A(t,u,u')u'+f(u)=e(t).$$

In the final section of the paper we show that the same asymptotic stability results hold for these cases as for homogeneous systems, provided the non-homogeneous term can be decomposed in the form

$$e(t)=e_1(t)+e_2(t),$$

where $e_1\in L^1(J)$ and $e_2$ is of bounded variation on $J$ with $e_2(t)\to 0$ as $t\to\infty$.

The system (1.10) is further discussed at the end of Section 5. Various other examples and applications which illustrate our results have been given in [1], [4], [5] and [9].

The following section contains preliminary material, including the statement of the important identity (2.8) for solutions of (1.1). In Sections 3 and 4 we present our main sufficiency results for global asymptotic stability. Section 3 covers functions $G$ which are merely assumed to be strictly convex, while in Section 4 we allow $G$ to have a more specialized structure, of the type possessed in particular by the systems (1.10)–(1.12). Section 5 contains necessary conditions for global asymptotic stability. Finally, Theorem 5.1 is a backward uniqueness result of interest in itself.

## 2. Preliminaries

We consider vector solutions $u=(u_1,\dots,u_N)$ of the quasi-variational ordinary differential system

$$(\nabla G(u,u'))'-\nabla_u G(u,u')+f(t,u)=Q(t,u,u'),\quad J=[R,\infty), \tag{2.1}$$

where $\nabla$ denotes the gradient operator with respect to the variable $p$ and

$$f(t,u)=\nabla_u F(t,u).$$

It will be supposed throughout the paper that

$$G\in C^1(\mathbf{R}^N\times\mathbf{R}^N;\mathbf{R}),\quad F\in C^1(J\times\mathbf{R}^N;\mathbf{R}),\quad Q\in C(J\times\mathbf{R}^N\times\mathbf{R}^N;\mathbf{R}^N),$$

and also, except in Section 5, that the following natural conditions hold:

($H_1$) $G(u,\cdot)$ is strictly convex in $\mathbf{R}^N$ for all $u\in\mathbf{R}^N$, with $G(u,0)=0$ and $\nabla G(u,0)=0$. Finally, for every $U>0$ and $p_0>0$ there exists a non-negative constant such that

$$(\nabla_u G(u,p),u)\leqslant \text{Const.}\,(\nabla G(u,p),p) \quad \text{for all } |u|\leqslant U \text{ and } |p|\geqslant p_0. \tag{2.2}$$

($H_2$) $F(t,0)=0$. For all $u_0, U$ with $0<u_0\leqslant U$ there exist a constant $\varkappa>0$ and a non-negative function $\psi\in L^1(J)$ such that

$$(f(t,u),u)\geqslant\varkappa \quad \text{when } t\in J \text{ and } |u|\in[u_0,U], \tag{2.3}$$

$$F_t(t,u)\leqslant\psi(t) \quad \text{when } t\in J \text{ (a.e) and } |u|\leqslant U. \tag{2.4}$$

($H_3$) $(Q(t,u,p),p)\leqslant 0$ for all $t\in J$, $u\in\mathbf{R}^N$ and $p\in\mathbf{R}^N$.

A condition of the type (2.2) first appears in [4]; of course this condition automatically holds in the important case when $G$ is independent of $u$. Moreover (2.2) is implied by the somewhat simpler condition $(\nabla_u G(u,p),u)\leqslant\text{Const.}\,G(u,p)$, since for any function satisfying ($H_1$) one has $0\leqslant G(u,p)\leqslant(\nabla G(u,p),p)$. A simple example of this type occurs when $G(u,p)=g(u)\overline{G}(u,p)$ with $g(u)>0$ in $\mathbf{R}^N$ and $\overline{G}$ satisfying ($H_1$).

Condition (2.3) implies that $(f(t,u),u)>0$ and $F(t,u)>0$ for $t\in J$ and $u\neq 0$. If $F$ does not depend on $t$, then (2.3) follows from the condition $(f(u),u)>0$ for $u\neq 0$, while (2.4) is irrelevant. Finally, when $N=1$ any continuous function $f$ is of gradient type, with $F(t,u)=\int_0^u f(t,s)\,ds$.

When $f$ has the form $f(t,u)=l(t)\chi(u)$ with $l\colon J\to\mathbf{R}^+$ and $\chi\colon\mathbf{R}^N\to\mathbf{R}^N$, then (2.3) is equivalent to the simple condition

$$l(t)\geqslant\text{Const.}>0 \quad \text{for } t\in J, \quad (\chi(u),u)>0 \quad \text{for } u\neq 0.$$

Property (2.4) then holds if and only if $(l')^+\in L^1(J)$.

In what follows, we consider (weak) solutions of (2.1) on $J$, namely vector functions $u\colon J\to\mathbf{R}^N$ of class $C^1$, for which

$$\nabla G(u(t),u'(t))\in C^1(J;\mathbf{R}^N) \tag{2.5}$$

and which satisfy the system (2.1) in $J$. Conditions (2.3) and ($H_3$) imply that $f(t,0)=Q(t,u,0)=0$. Moreover from ($H_1$) it follows that $\nabla G(u,0)=\nabla_u G(u,0)=0$ so that $u\equiv 0$ is a solution of (2.1) on $J$.

Let $H(u,\cdot)$ be the Legendre transform of $G(u,\cdot)$, namely

$$H(u,p)=(\nabla G(u,p),p)-G(u,p). \tag{2.6}$$

If $u=u(t)$ is a weak solution of (2.1) on $J$, then we have

$$\{H(u,u')+F(t,u)\}'=(Q(t,u,u'),u')+F_t(t,u). \tag{2.7}$$

Under the minimal conditions which we are assuming here, namely that $G$ is of class $C^1$ and is strictly convex in the variable $p$, this identity is proved in a recent paper of the authors, *On the derivation of Hamilton's equations*; see also [4]. (It is worth noting as well that (2.7) is directly and easily verified for weak solutions $u$ which are also assumed to be of class $C^2(J,\mathbf{R}^N)$; this additional degree of smoothness is *not* always satisfied however, e.g. when $G(p)=|p|^m/m$, $m>2$.)

In view of (2.5) and (2.7) the following identity holds for solutions of (2.1) and for any pair of scalar functions $\varphi,\omega\in C^1(J;\mathbf{R})$,

$$\begin{aligned}\{\omega[H(u,u')+F(t,u)]&+\varphi(\nabla G(u,u'),u)\}'\\ =\omega F_t(t,u)&+\omega'[H(u,u')+F(t,u)]-\varphi(f(t,u),u)\\ &+\varphi[(\nabla G(u,u'),u')+(\nabla_u G(u,u'),u)]+\varphi'(\nabla G(u,u'),u)\\ &+\omega(Q(t,u,u'),u')+\varphi(Q(t,u,u'),u).\end{aligned} \tag{2.8}$$

This formula was originally discovered as a special case of the main identity in [7], see particularly the remark on p. 685. Note that (2.8) reduces to (2.7) when $\omega=1$ and $\varphi=0$.

The identity (2.8) can also be usefully written in the notation of analytical dynamics, by putting

$$v=\nabla\mathcal{L}=\nabla G$$

and

$$\mathcal{H}=(p,v)-\mathcal{L}=H+F,$$

so that $v$ is the conjugate variable to $p$ and $\mathcal{H}$ is the Hamiltonian associated with the Lagrangian $\mathcal{L}$. Then it becomes

$$\{\omega\mathcal{H}+\varphi(u,v)\}'=\omega'\mathcal{H}-\omega\mathcal{L}_t+(\varphi u,\nabla_u\mathcal{L})+((\varphi u)',v)+(\omega u'+\varphi u,Q).$$

In the sequel we shall use the properties that $H(u,0)=0$ and $H(u,p)>0$ for $p\neq0$. The first follows from the condition $G(u,0)=0$ and the latter is a standard consequence of the strict convexity of $G(u,\cdot)$. Moreover $\nabla G(u,p)\neq0$ for $p\neq0$ since $\nabla G(u,0)=0$. Another consequence of these facts is the following useful

LEMMA 2.1. *For every $u$, $p\in\mathbf{R}^N$ we have*

$$|\nabla G(u,p)|\leqslant H(u,p)+\max_{|v|=1}G(u,v). \tag{2.9}$$

*Proof.* Fix $u\in\mathbf{R}^N$. Obviously (2.9) is valid when $p=0$. Take $p\neq 0$ so that

$$v=\frac{\nabla G(u,p)}{|\nabla G(u,p)|}$$

is well defined. Then from the convexity condition

$$G(u,p)-G(u,v)\leqslant(\nabla G(u,p),p-v)$$

we immediately obtain, using (2.6),

$$|\nabla G(u,p)|=(\nabla G(u,p),v)\leqslant(\nabla G(u,p),p)-G(u,p)+G(u,v)=H(u,p)+G(u,v),$$

which yields (2.9).

## 3. Sufficient conditions for stability, Part 1

The purpose of this section is to study the asymptotic stability of bounded solutions of (2.1). We shall always assume, without further mention, that $(\mathrm{H}_1)$–$(\mathrm{H}_3)$ are valid. Moreover, for this section we suppose the additional hypotheses:

$(\mathrm{C}_1)$ For every $U>0$ there exist a measurable control set $I\subset J$, two non-negative measurable damping functions $\sigma,\delta\colon I\to\mathbf{R}$, and two continuous increasing weight functions $\phi,\varrho\colon\mathbf{R}_0^+\to\mathbf{R}_0^+$ with $\phi(0)=\varrho(0)=0$, such that

$$|(Q(t,u,p),p)|\geqslant\sigma(t)\phi(|p|)|p| \tag{3.1}$$

for $t\in I$, $|u|\leqslant U$ and $p\in\mathbf{R}^N$; and

$$(Q(t,u,p),u)\leqslant\delta(t)\varrho(|p|) \tag{3.2}$$

for $t\in I$, $|u|\leqslant U$ and all sufficiently small $p\in\mathbf{R}^N$.

$(\mathrm{C}_2)$ For every $U>0$ there exists a positive constant $\gamma\geqslant 1$ such that

$$|Q(t,u,p)|\cdot|p|\leqslant\gamma|(Q(t,u,p),p)|\quad\text{for all } t\in I,\ |u|\leqslant U \text{ and } p\in\mathbf{R}^N.$$

Although $I$, $\sigma$, $\delta$ and $\phi$, $\varrho$, $\gamma$ may depend on $U$, for simplicity we do not specifically indicate this dependence.

Conditions $(\mathrm{C}_1)$ and $(\mathrm{C}_2)$ are automatically satisfied, for example, by the damping function $Q=-a(t)b(|p|)p$ when $a,b\geqslant 0$ and $sb(s)$ is increasing. Note that $(\mathrm{C}_2)$ also is satisfied automatically when $N=1$ and indeed whenever $-Q$ and $p$ have the same direction, with $\gamma=1$.

THEOREM 3.1. *Suppose that for every $U>0$ there is a continuous function $k$ of bounded variation on $J$ such that*

$$k \notin L^1(J), \quad k(t)=0 \quad \text{for } t\in J\setminus I \tag{3.3}$$

*and*

$$0 \leqslant k(t) \leqslant \beta\sigma(t) \quad \text{for } t\in I \tag{3.4}$$

*for some positive constant $\beta$. Assume also that there exists $M>0$ such that*

$$\int_{(R,t)\cap I} \delta(s)k(s)\exp\left(-\int_s^t k(r)\,dr\right)ds \leqslant M \quad \text{for } t\in J. \tag{3.5}$$

*Then the rest state is a global attractor for the system* (2.1).

*Proof.* We first give the proof when the BV function $k$ is of class $C^1(J)$.

Let $u$ be a solution of (2.1) with $|u(t)|\leqslant L$ on $J$ for some $L>0$. We take $U=L$ and let $\psi$, $\sigma$, $\delta$, etc., be the associated functions and constants in $(H_1)$–$(H_3)$ and $(C_1)$–$(C_2)$, and similarly let $k$ be the associated function in the theorem.

By (2.7) the expression

$$H(u,u')+F(t,u)-\int_R^t (Q(s,u,u'),u')\,ds-\int_R^t F_t(s,u)\,ds \tag{3.6}$$

is constant. Since the first three terms in (3.6) are non-negative, it follows from (2.4) that

$$(Q(t,u(t),u'(t)),u'(t))\in L^1(J). \tag{3.7}$$

Therefore also $F_t(t,u(t))\in L^1(J)$ and in turn there exists a number $l\geqslant 0$ such that

$$H(u,u')+F(t,u)\to l \quad \text{as } t\to\infty. \tag{3.8}$$

First assume $l=0$. Since $H(u,p)>0$ for $p\neq 0$, and since (again by the strict convexity of the function $G(u,\cdot)$)

$$\inf\{H(u,p):|u|\leqslant L, |p|\geqslant 1\}>0,$$

it follows easily that $u'(t)\to 0$ as $t\to\infty$. Moreover by (2.3), see Lemma 2 of [9], there is a positive function $\omega_L$ defined on $(0,L]$ such that $F(t,u)\geqslant\omega_L(|u|)$ for any $t\in J$. Hence also $u(t)\to 0$ as $t\to\infty$, completing the proof when $l=0$.

Now assume for contradiction that $l>0$ in (3.8). The proof relies on the main identity (2.8) with the choice $\varphi=\alpha\omega'$, where $\omega$ is the positive non-decreasing function defined by

$$\omega(t)=\exp\left(\int_R^t k(s)\,ds\right) \tag{3.9}$$

and $\alpha$ is a positive number which will be determined later. Then, recalling that $f=\nabla_u F$, the identity (2.8) reduces to

$$\begin{aligned}&\{\omega[H(u,u')+F(t,u)+\alpha k(\nabla G(u,u'),u)]\}'\\&\quad=\omega F_t(t,u)+\omega'[H(u,u')+F(t,u)-\alpha(f(t,u),u)]\\&\qquad+\alpha\omega'[(\nabla G(u,u'),u')+(\nabla_u G(u,u'),u)]+\alpha\omega''(\nabla G(u,u'),u)\\&\qquad+\omega(Q(t,u,u'),u')+\alpha\omega'(Q(t,u,u'),u),\end{aligned}\tag{3.10}$$

where

$$\omega'=k\omega,\quad \omega''=(k^2+k')\omega.\tag{3.11}$$

By (3.8) the function $H(u,u')$ is bounded on $J$. Hence by Lemma 2.1 also $|\nabla G(u,u')|$ is bounded, say

$$|\nabla G(u,u')|\leqslant C\quad\text{for } t\in J.\tag{3.12}$$

We take $R_1$ so large that

$$H(u,u')+F(t,u)\geqslant\tfrac{3}{4}l\quad\text{on } J_1\quad\text{and}\quad\int_{J_1}\psi(s)\,ds\leqslant\tfrac{1}{4}l,\tag{3.13}$$

where $J_1=[R_1,\infty)$.

Let $\mathcal{R}(t)$ denote the right hand side of the identity (3.10). Our purpose, and the principal effort in the proof, will be to obtain an appropriate estimate for $\mathcal{R}(t)$ when $t\in J_1$ and $\alpha$ is suitably small.

First, since $k(t)=k'(t)=0$ on $J\backslash I$ it follows from (3.10), (3.11), (2.4) and ($H_3$) that

$$\mathcal{R}(t)\leqslant\omega\psi(t),\quad t\in J_1\backslash I\quad\text{(a.e.)}.\tag{3.14}$$

On the remaining set $I'=I\cap J_1$ we partition $\mathcal{R}(t)$ in the form

$$\frac{1}{\omega}\mathcal{R}=F_t+k(H+F)+\sum_1^5\mathcal{R}_i,$$

where

$$\begin{aligned}\mathcal{R}_1&=\tfrac{1}{5}(Q,u')-\alpha k(f,u)\\\mathcal{R}_2&=\tfrac{1}{5}(Q,u')+\alpha k\{(\nabla G,u')+(\nabla_u G,u)\}\\\mathcal{R}_3&=\tfrac{1}{5}(Q,u')+\alpha k^2(\nabla G,u)\\\mathcal{R}_4&=\tfrac{1}{5}(Q,u')+\alpha k'(\nabla G,u)\\\mathcal{R}_5&=\tfrac{1}{5}(Q,u')+\alpha k(Q,u),\end{aligned}$$

the notation being obvious, and $u$, $u'$ clearly standing for $u(t)$, $u'(t)$. The important term $(Q, u')$ lies behind the desired estimates; accordingly this term is partitioned into each of the expressions $\mathcal{R}_1$ through $\mathcal{R}_5$.

Again $F_t \leqslant \psi$ from (2.4), while from (3.8)

$$|H+F-l| = \varepsilon(t) \to 0 \quad \text{as } t \to \infty. \tag{3.15}$$

The remaining terms will be treated one-by-one.

*1.* We assert that $\mathcal{R}_1 \leqslant -\alpha\varkappa k$ for some constant $\varkappa > 0$, provided that $\alpha$ is suitably small. To see this, first let $p_1 > 0$ be fixed so that $H(u,p) \leqslant \frac{1}{4}l$ when $|u| \leqslant L$ and $|p| \leqslant p_1$. Clearly this can be done since $H$ is continuous and $H(u,0)=0$. Then by (3.13) we have

$$F(t,u(t)) \geqslant \tfrac{1}{2}l \quad \text{in } I_1 = \{t \in I' : |u'(t)| \leqslant p_1\}.$$

On the other hand, by (2.4) and (3.13),

$$\begin{aligned} F(t,u(t)) &= F(R_1,u(t)) + \int_{R_1}^{t} F_t(s,u(t))\,ds \\ &\leqslant F(R_1,u(t)) + \int_{R_1}^{\infty} \psi(s)\,ds \leqslant F(R_1,u(t)) + \tfrac{1}{4}l \end{aligned}$$

for $t \in J_1$. Thus

$$F(R_1,u(t)) \geqslant \tfrac{1}{4}l \quad \text{in } I_1.$$

Since $F(R_1,0)=0$ it follows that there exists a number $u_0 > 0$ such that $|u(t)| \geqslant u_0$ for $t \in I_1$. Denote by $\varkappa = \varkappa(u_0, L) > 0$ the constant given in (2.3) corresponding to $u_0$ and $U = L$, so that now

$$(f(t,u(t)), u(t)) \geqslant \varkappa > 0 \quad \text{for } t \in I_1. \tag{3.16}$$

Since $(Q, u') \leqslant 0$ by ($H_3$), this gives the required estimate when $t \in I_1$.

In the remaining set $I' \setminus I_1$ we have $|u'(t)| > p_1$. Hence by (3.1) and (3.4)

$$|(Q,u')| \geqslant \sigma\phi(|u'|)\cdot|u'| \geqslant \frac{k}{\beta}\phi(|u'|)\cdot|u'| \geqslant 5\alpha\varkappa k \quad \text{in } I' \setminus I_1,$$

provided that

$$\alpha \leqslant \frac{p_1}{5\beta\varkappa}\phi(p_1); \tag{3.17}$$

here $\phi(|u'|) \geqslant \phi(p_1) > 0$ since $\phi(0)=0$ and $\phi$ is increasing. Since $(f(t,u),u) \geqslant 0$ by (2.3), the required estimate is now valid for all $t \in I'$, provided of course (3.17) is satisfied.

*2.* We claim that $\mathcal{R}_2 \leqslant \frac{1}{8}\alpha k\varkappa$ on $I'$, provided $\alpha$ is suitably small. Indeed

$$(\nabla G, u') + (\nabla_u G, u) \leqslant \tfrac{1}{8}\varkappa$$

if $|u'|$ is suitably small, say $|u'|\leqslant p_2$, since $\nabla G(u,0)=\nabla_u G(u,0)=0$ and the solution is bounded. Hence in this case the claim holds. Otherwise, if $|u'|>p_2$, then by (2.2) there is a constant $b>0$ such that

$$(\nabla G,u')+(\nabla_u G,u)\leqslant(1+b)\cdot(\nabla G,u').$$

Moreover by (3.4) and (3.12) we have

$$\begin{aligned}k(\nabla G,u')\leqslant kC|u'|\leqslant \beta C\sigma|u'|&\leqslant\frac{\beta C\sigma\phi(|u'|)\cdot|u'|}{\phi(p_2)}\\&=C_2\sigma\phi(|u'|)\cdot|u'|\leqslant C_2|(Q,u')|\end{aligned}$$

by (3.1), where $C_2$=Constant$>0$. Hence if

$$\alpha\leqslant\frac{1}{5C_2(1+b)}\tag{3.18}$$

then $\mathcal{R}_2\leqslant 0$ when $|u'|>p_2$, completing step *2*.

*3.* Again $\mathcal{R}_3\leqslant\frac{1}{8}\alpha k\varkappa$ on $I'$. This is treated exactly as in step *2*, the only exception being that (3.18) is replaced by a corresponding bound

$$\alpha\leqslant\frac{1}{5C_3},\tag{3.19}$$

where $C_3=\beta CL\sup k/p_3\phi(p_3)$ and $p_3$ is an appropriate positive constant analogous to $p_2$ in step *2*. Here we use the condition that $k$ is bounded, since $\mathcal{R}_3$ includes the factor $k^2$ rather than $k$.

*4.* By $(H_3)$ and (3.12) obviously $\mathcal{R}_4\leqslant\alpha CL|k'|$ in $I'$.

*5.* We claim that $\mathcal{R}_5\leqslant\varepsilon_5(\alpha)\alpha\delta k$, where $\varepsilon_5(\alpha)\to 0$ as $\alpha\to 0$. Let

$$I_5=\{t\in I':|u'(t)|\geqslant\Lambda_5\alpha\},\quad \Lambda_5=5L\gamma\sup_{t\in J}k(t).$$

In this set, by $(C_2)$,

$$\alpha k(Q,u)\leqslant\alpha Lk|Q|\leqslant\frac{1}{5\gamma}|Q|\cdot|u'|\leqslant\frac{1}{5}|(Q,u')|$$

so that $\mathcal{R}_5\leqslant 0$ in $I_5$. In $I'\backslash I_5$ we have $|u'(t)|<\Lambda_5\alpha$. Hence

$$\mathcal{R}_5\leqslant\alpha k\delta\varrho(|u'(t)|)\leqslant\varepsilon_5(\alpha)\alpha\delta k$$

by (3.2), where $\varepsilon_5(\alpha)=\varrho(\Lambda_5\alpha)$. Thus the claim is proved for all $t\in I'$.

Combining the previous estimates, including (3.14) and (3.15), now yields, for almost all $t \in J_1$ and all $\alpha > 0$ satisfying (3.17), (3.18) and (3.19)

$$\mathcal{R} \leqslant \omega\{\psi + \alpha CL|k'| + \varepsilon_5(\alpha)\alpha\delta k\} + \omega'\{l + \varepsilon - \alpha\varkappa + 2\cdot\tfrac{1}{8}\alpha\varkappa\}, \tag{3.20}$$

where we introduce the agreement $\delta k = 0$ on $J \setminus I$. Recall that $k = 0$ on $J \setminus I$ by assumption $(3.3)_2$.

Now fix $\alpha > 0$ so small that (3.17), (3.18) and (3.19) hold and also so that

$$\varepsilon_5(\alpha) \leqslant \frac{\varkappa}{8M}, \tag{3.21}$$

where $M$ is the constant given in (3.5). Moreover take $R_2 \geqslant R_1$ such that

$$\varepsilon = \varepsilon(t) \leqslant \tfrac{1}{8}\alpha\varkappa \quad \text{for all } t \geqslant R_2 \tag{3.22}$$

and

$$\int_{R_2}^{\infty} \psi(s)\, ds \leqslant \alpha\frac{\varkappa}{8}, \quad \int_{R_2}^{\infty} |k'(s)|\, ds \leqslant \frac{\varkappa}{8CL}. \tag{3.23}$$

Condition $(3.23)_2$ can be attained since $k \in \mathrm{BV}(J)$ and so $k' \in L^1(J)$. Then from (3.10), (3.20) and (3.22) we see that the function

$$\begin{aligned}\Psi(t) = \omega(t)\Big\{ & H(u,u') + F(t,u) + \alpha k(\nabla G(u,u'),u) - \Big(l + \alpha\frac{\varkappa}{8}\Big) + \alpha\varkappa - \alpha\frac{\varkappa}{4} \\ & - \frac{1}{\omega(t)}\int_{R_2}^{t} \psi(s)\omega(s)\, ds - \frac{\alpha CL}{\omega(t)}\int_{R_2}^{t} |k'(s)|\omega(s)\, ds \\ & - \frac{\varepsilon_5(\alpha)\alpha}{\omega(t)}\int_{R_2}^{t} \delta(s)k(s)\omega(s)\, ds\Big\}\end{aligned} \tag{3.24}$$

is decreasing in $J_2 = [R_2, \infty)$.

We claim that there exists a sequence $(t_n)$ with $t_n \nearrow \infty$ such that

$$k(t_n)\nabla G(u(t_n), u'(t_n)) \to 0 \quad \text{as } n \to \infty. \tag{3.25}$$

By (3.12) and the boundedness of $k$, this follows immediately if either $\liminf_{t\to\infty} k(t) = 0$ or $\liminf_{t\to\infty} |\nabla G(u(t), u'(t))| = 0$. But if neither of these occurs, then for some sufficiently large constant $\bar{t}$ there holds

$$k(t) \geqslant k_0 > 0 \quad \text{and} \quad |\nabla G(u,u')| \geqslant C_0 > 0 \quad \text{for } t \geqslant \bar{t}.$$

In turn, by $(3.3)_2$ we must have $I \supset [\bar{t}, \infty)$ and so using (3.4) and the strict convexity of $G(u,\cdot)$, see $(\mathrm{H}_1)$, we have

$$\sigma(t) \geqslant \frac{k_0}{\beta} \quad \text{and} \quad |u'(t)| \geqslant p_0 > 0 \quad \text{for } t \geqslant \bar{t}.$$

Hence by (3.1) we obtain

$$|(Q,u')|\geqslant\sigma\phi(|u'|)\cdot|u'|\geqslant\frac{k_0p_0\phi(p_0)}{\beta}>0\quad\text{in }[\bar{t},\infty),$$

which contradicts (3.7).

Now, along the sequence $(t_n)$, by (3.5), (3.15), (3.21)–(3.24) there results

$$\Psi(t_n)\geqslant\omega(t_n)\Big\{\Big(l-\alpha\frac{\varkappa}{8}\Big)-\alpha Lk(t_n)|\nabla G(u(t_n),u'(t_n))|-\Big(l+\alpha\frac{\varkappa}{8}\Big)+\alpha\varkappa-5\alpha\frac{\varkappa}{8}\Big\}.$$

But then using (3.25) it follows that $\Psi(t_n)\geqslant\frac{1}{16}\alpha\varkappa\omega(t_n)$ for $n$ sufficiently large. Hence $\Psi(t_n)\to\infty$ as $n\to\infty$ by (3.9) and $(3.3)_1$, which contradicts the fact that $\Psi$ is decreasing in $J_2$. Therefore the case $l\neq0$ cannot occur in (3.8), and the proof is complete when $k\in C^1(J)$.

We now turn to the general case when $k\in\mathrm{BV}(J)$. The proof is based on the following lemma, whose demonstration will be given in the Appendix.

LEMMA A. *Let $k$ be a non-negative continuous function of bounded variation on $J$. Then for every constant $\theta>1$ there exists a function $\bar{k}\in C^1(J)$ and an open set $E\subset J$ such that*

$$\text{(i) }\theta k\geqslant\bar{k}\geqslant\begin{cases}k & \text{in } J\setminus E\\ 0 & \text{in } E\end{cases};\quad\text{(ii) }\operatorname{Var}\bar{k}\leqslant\theta\operatorname{Var}k;\quad\text{(iii) }\int_E k\leqslant1.$$

Now fix $\theta>1$ and let $\bar{k}$ and $E$ be as given in the lemma. We assert that $\bar{k}$ satisfies (3.3)–(3.5) with $\beta$ replaced by $\bar{\beta}=\theta\beta$. Indeed by (i) it is clear that $(3.3)_2$ is satisfied. Also $(3.3)_1$ holds thanks to (i) and (iii). Similarly, in view of (i) and (iii),

$$\int_R^t\delta(s)\bar{k}(s)\exp\Big(-\int_s^t\bar{k}(r)\,dr\Big)\,ds\leqslant\theta e\int_R^t\delta(s)k(s)\exp\Big(-\int_s^t k(r)\,dr\Big)\,ds,$$

and so (3.5) holds. This completes the proof.

The condition that $k\in\mathrm{BV}(J)$ in Theorem 3.1 can be replaced by an interesting alternative hypothesis.

THEOREM 3.2. *Suppose that the control set $I$ in $(\mathrm{C}_1)$ is the entire interval $J$ and that for every $U>0$ there is a bounded function $k$ on $J$ such that*

$$k\notin L^1(J),\tag{3.3$'$}$$

$$0<k(t)\leqslant\beta\sigma(t)\ \text{ in } J,\quad\log k\in\mathrm{Lip}(J),\tag{3.4$'$}$$

*for some constant* $\beta>0$. *Assume also that* (3.5) *holds. Then the rest state is a global attractor for the system* (2.1).

*Proof.* Clearly $k\in\mathrm{Lip}(J)$, so $k\in\mathrm{AC}(J)$ with $|k'|\leqslant\mathrm{Const.}\,k$ a.e. in $J$. The previous proof now applies word-for-word, except for step *4*. For this step, however, we have the alternative estimate

$$\alpha k'(\nabla G,u)\leqslant\mathrm{Const.}\,\alpha k|\nabla G|\quad\text{a.e. in } J_1.$$

Therefore, using the argument of steps *2* and *3*, we easily derive that for $\alpha$ suitably small

$$\mathcal{R}_4\leqslant\tfrac{1}{8}\alpha k\varkappa\quad\text{a.e. in } J_1.$$

In turn the second term in the first braces of (3.20) should be deleted, while at the same time an extra term $\frac{1}{8}\alpha\varkappa$ should be inserted in the second braces. The corresponding function $\Psi$ again is decreasing in $J_2$. The rest of the proof follows exactly as before.

Theorem 3.1 has several interesting corollaries. Corresponding results can also be obtained from Theorem 3.2, but we leave their statement to the interested reader.

COROLLARY 1. *Suppose that the hypotheses of Theorem* 3.1 *are satisfied, except in place of* (3.5) *we assume that the function* $\delta$ *allows a decomposition* $\delta=d+e$ *with*

$$d\in L^\infty(I)\quad\textit{and}\quad ek\in L^1(I). \tag{3.5$'$}$$

*Then the rest state is a global attractor.*

*Proof.* Clearly

$$\int_R^t k(s)\exp\left(-\int_s^t k(r)\,dr\right)ds=1-\exp\left(-\int_R^t k(r)\,dr\right)\leqslant 1.$$

Hence (3.5)′ implies (3.5), with $M=\|d\|_{L^\infty(I)}+\|ek\|_{L^1(I)}$, and Theorem 3.1 can be applied.

COROLLARY 2. *Define* $\bar\sigma(t)=\sigma(t)$ *for* $t\in I$ *and* $\bar\sigma(t)=0$ *for* $t\in J\setminus I$, *for any* $\sigma$ *given in* ($C_1$). *Assume also that for every* $U>0$,

$$\sigma\notin L^1(I),\quad \delta\in L^\infty(I),\quad \bar\sigma\in\mathrm{CBV}(J).$$

*Then the rest state is a global attractor.*

*Proof.* Take $k(t)=\bar\sigma(t)$. Then (3.3) and (3.4) hold (with $\beta=1$), while (3.5)′ is satisfied with $d=\delta$, $e=0$. Thus Corollary 1 can be applied.

COROLLARY 3. *Condition* (3.5) *in Theorem* 3.1 *can be replaced by*

$$\liminf_{t\to\infty}\int_{(R,t)\cap I}\delta(s)k(s)\exp\left(-\int_s^t k(r)\,dr\right)ds=M<\infty, \tag{3.5)''$$

*provided that* $k(t)\to 0$ *as* $t\to\infty$.

*Proof.* We choose the sequence $(t_n)$ in the proof of Theorem 3.1 so that

$$\int_{(R,t_n)\cap I}\delta(s)k(s)\exp\left(-\int_s^{t_n} k(r)\,dr\right)ds\leqslant 2M.$$

Then (3.25) holds as before, by (3.12) and the assumption that $k(t)\to 0$ as $t\to\infty$. The rest of the proof is the same as that in Theorem 3.1.

COROLLARY 4. *Suppose that the control set* $I$ *in* $(C_1)$ *is the entire interval* $J$, *and that for every* $U>0$ *there exist positive numbers* $c$ *and* $\eta$ *such that*

$$t\sigma(t)\geqslant c,\quad \liminf_{t\to\infty}\int_R^t\left(\frac{s}{t}\right)^{\eta}\delta(s)\,\frac{ds}{s}=M'<\infty. \tag{3.26}$$

*Then the rest state is a global attractor.*

*Proof.* Taking $k(t)=\eta/t$, and $\beta=\eta/c$, we immediately derive all the assumptions of Corollary 3, with $M=M'\eta$ in (3.5)″.

The case $\eta=1$ in Corollary 4 is particularly interesting. Moreover, the result of Corollary 4 is in a sense best possible. For example, $(3.26)_2$ is satisfied when $\delta$ is bounded but fails when $\delta(t)\geqslant t^{\varepsilon}$ for some $\varepsilon>0$, while there exist equations with $\delta(t)\geqslant t^{\varepsilon}$ for which the rest state is *not* a global attractor (see [9], Section 5).

*Remarks.* Condition (2.4) can be weakened to

$$F_t(t,u)\leqslant\psi(t)\{1+F(t,u)\},\quad |u|\leqslant U. \tag{3.27}$$

Indeed if this condition holds, then by integration

$$\log\{1+F(t,u)\}\leqslant\int_R^\infty\psi(s)\,ds+\log\{1+F(R,u)\}\leqslant\text{Const.}$$

Hence $1+F(t,u)$ is also bounded above for $t\in J$ and $|u|\leqslant U$ and we get $F_t(t,u)\leqslant$ Const. $\psi(t)$, which is essentially condition (2.4). That is (3.27) is in fact equivalent to (2.4).

The preliminary change of variable $t \mapsto \tau = \tau(t) = \int_R^t q(s)\,ds$, with $q>0$ and with $q \notin L^1(J)$, can occasionally be helpful in applications, particularly when properties (2.3), (2.4) are not satisfied. For example (see [2], p. 327) the system

$$u'' + q^2(t) f(u) = Q(t, u, u')$$

transforms to

$$\ddot{u} + f(u) = \frac{1}{q^2(t(\tau))}\{Q(t(\tau), u, q\dot{u}) - q'(t(\tau))\dot{u}\}.$$

Various results which can be obtained in this connection will be treated in a paper to appear later.

## 4. Sufficient conditions for stability, Part 2

We suppose as before that conditions $(H_1)$–$(H_3)$ and $(C_2)$ are satisfied. However, in place of assumption $(C_1)$ we shall require the following more specific hypothesis in which the functions $\phi$, $\varrho$ are assumed to be algebraic.

$(C_1)'$ There exist exponents $0<\mu\leqslant\nu$ such that assumption $(C_1)$ is satisfied with the specific functions

$$\phi(s) = \min\{1, s^\nu\}, \quad \varrho(s) = s^\mu. \tag{4.1}$$

The first main result of the section is the following

THEOREM 4.1. *Suppose that for every $U>0$ there is a continuous function $k$ of bounded variation on $J$ such that*

$$k \notin L^1(J), \quad k(t) = 0 \quad \text{for } t \in J\backslash I \tag{4.2}$$

*and for some $\beta>0$*

$$0 \leqslant k(t) \leqslant \beta\sigma(t) \quad \text{for } t \in I. \tag{4.3}$$

*Assume also that there exists $M>0$ for which*

$$\int_{(R,t)\cap I} \delta(s) k^{\mu+1}(s) \exp\left(-\int_s^t k(r)\,dr\right) ds \leqslant M \quad \text{for } t \in J. \tag{4.4}$$

*Then the rest state is a global attractor for the system* (2.1).

*Proof.* The first part of the proof is the same as in Theorem 3.1, while steps *1–4* are also essentially as before.

For step *5* the only change is in the estimate for the set $I'\backslash I_5$. We assert that on this set there holds

$$\alpha k(Q,u) \leqslant -\alpha^{1/2}L\gamma(Q,u')+\alpha^{1+\mu/2}\delta k^{\mu+1}. \tag{4.5}$$

Since this is obvious if $u'=0$, we may suppose $u'\neq 0$. Then by Young's inequality

$$\alpha k \leqslant \alpha^{1/2}\{|u'|+|u'|^{-\mu}(\alpha^{1/2}k)^{\mu+1}\}$$

and so by (3.2) with $\varrho(s)=s^{\mu}$, see $(\mathrm{C}_1)'$,

$$\alpha k(Q,u) \leqslant \alpha^{1/2}L|Q|\cdot|u'|+\alpha^{1+\mu/2}\delta k^{\mu+1}.$$

This gives the required result (4.5) in view of $(\mathrm{C}_2)$ and $(\mathrm{H}_3)$. To apply (3.2) the derivative $|u'(t)|$ must be sufficiently small, but this is certainly true in $I'\backslash I_5$ provided $\alpha$ is small enough.

In view of (4.5) we thus have

$$\mathcal{R}_5 \leqslant \left(\tfrac{1}{5}-\alpha^{1/2}L\gamma\right)(Q,u')+\alpha^{1+\mu/2}\delta k^{\mu+1} \quad \text{in } I'\backslash I_5.$$

Taking $\alpha\leqslant(5L\gamma)^{-2}$ therefore gives

$$\mathcal{R}_5 \leqslant \alpha^{1+\mu/2}\delta k^{1+\mu} \quad \text{in } I'\backslash I_5. \tag{4.6}$$

Since we have already shown that $\mathcal{R}_5\leqslant 0$ in $I_5$, the estimate (4.6) then holds in all of $I'$.

The rest of the proof follows exactly that for Theorem 3.1, except that in (3.20) the term $\varepsilon_5(\alpha)\alpha\delta k$ should be replaced by $\alpha^{1+\mu/2}\delta k^{1+\mu}$. Of course in the estimates for the corresponding function $\Psi$ we use (4.4) instead of (3.5).

As in Section 3, the condition that $k\in\mathrm{BV}(J)$ can be replaced by other hypotheses. It is clear, for example, that we could suppose $\log k\in\mathrm{Lip}(J)$ in parallel with the result of Theorem 3.2. Another alternative is given by the following theorem, in which we assume that:

For every $U>0$ there exists a positive constant $\Theta=\Theta(U)>0$ and an exponent $m>1$ such that

$$|\nabla G(u,p)| \leqslant \Theta|p|^{m-1} \quad \text{when } |u|\leqslant U \text{ and } |p|\leqslant 1. \tag{4.7}$$

*Remark.* Obviously (4.7) holds when $G(p)=|p|^m/m$, with $\Theta=1$. It is also satisfied when $G(p)=\sqrt{1+|p|^2}-1$, with $\Theta=1$ and $m=2$. Another example, in which $G$ depends explicitly on $u$, is $G(u,p)=(1+|u|^a)|p|^m/m$ where $a>0$, $m>1$. In general $m$ can be thought of as being independent of $U$, as in the examples above. However, this is not strictly necessary, as the function $G(u,p)=|p|^{2+|u|^2}$ shows.

THEOREM 4.2. *Suppose that for every $U>0$ there is a bounded absolutely continuous function $k$ on $J$ such that* (4.2)–(4.4) *are satisfied. Assume also that*

$$|k'|\leqslant B\sigma^{\lambda}k^{1-\lambda} \quad a.e.\ on\ I, \tag{4.8}$$

*where $B$ is a positive constant and*

$$\lambda=\begin{cases}\dfrac{m-1}{\nu+1}, & if\ \nu>m-2\\ 1, & if\ \nu\leqslant m-2\ \ (and\ m>2).\end{cases} \tag{4.9}$$

*Then the rest state is a global attractor for the system* (2.1).

Observe that the two values of $\lambda$ in (4.9) coincide when $\nu=m-2$. In case $\nu=m-1$ we can write $\lambda=1/q$ and $1-\lambda=1/m$, so $q$ is the Hölder conjugate of $m$. Finally, the limiting case $m\to 1$, $\lambda\to 0$ corresponds to the condition $\log k\in\mathrm{Lip}(J)$ noted above.

*Proof of Theorem* 4.2. The previous proof applies essentially word-for-word, except for step *4*. Here, in $I'$ we see by (4.8) that

$$\begin{aligned}\alpha k'(\nabla G,u)&\leqslant \alpha BL\sigma^{\lambda}k^{1-\lambda}|\nabla G|\\ &\leqslant \alpha BL\sigma\begin{cases}\Theta(k/\sigma)^{1-\lambda}|u'|^{m-1}, & \text{if } |u'|<1\\ \beta^{1-\lambda}C, & \text{if } |u'|\geqslant 1,\end{cases}\end{aligned} \tag{4.10}$$

by (4.7) and by (3.12), (4.3). We now proceed somewhat as in steps *2* and *3* of the proof of Theorem 3.1. First, if $|u'(t)|\geqslant 1$ then from ($H_3$), (3.1) and $(4.1)_1$ we have $(Q,u')\leqslant-\sigma$. Thus by $(4.10)_2$ if

$$\alpha\leqslant\frac{\beta^{\lambda-1}}{5BCL},$$

then $\mathcal{R}_4\leqslant 0$ at such values of $t$. Second, if $|u'(t)|<1$ but $\nu\leqslant m-2$ (so that $m>2$ and $\lambda=1$), then by $(4.10)_1$

$$\alpha k'(\nabla G,u)\leqslant\alpha BL\Theta\sigma|u'|^{m-1}\leqslant\alpha BL\Theta\sigma|u'|^{\nu+1}$$

while by ($H_3$), (3.1) and $(4.1)_1$ once more, $(Q,u')\leqslant-\sigma|u'|^{\nu+1}$. Hence if

$$\alpha\leqslant(5B\Theta L)^{-1},$$

then $\mathcal{R}_4\leqslant 0$ also for the second set of values of $t$.

In the remaining case we have $\nu>m-2$ and $|u'(t)|<1$. Let

$$I_4'=\{t\in I':\Lambda_4(k/\sigma)^{1-\lambda}\alpha\leqslant|u'(t)|^{\nu-m+2},\ |u'(t)|<1\}$$
$$I_4''=\{t\in I':|u'(t)|^{\nu-m+2}<\Lambda_4(k/\sigma)^{1-\lambda}\alpha,\ |u'(t)|<1\},$$

where $\Lambda_4 = 5BL\Theta$. In $I_4'$ it is easy to see that $\mathcal{R}_4 \leqslant 0$, using the usual estimates for the term $(Q, u')$. On the other hand, in $I_4''$ we have

$$|u'(t)| < [\Lambda_4(k/\sigma)^{1-\lambda}\alpha]^{1/(\nu-m+2)} \quad (\text{since } \nu - m + 2 > 0).$$

Consequently in this set, by $(4.10)_1$ and $(\mathrm{H}_3)$,

$$\mathcal{R}_4 \leqslant \alpha BL\Theta\sigma^{\lambda}k^{1-\lambda}[\Lambda_4(k/\sigma)^{1-\lambda}\alpha]^{\lambda'} = \text{Const.}\, \alpha^{1+\lambda'}k,$$

where $\lambda' = (m-1)/(\nu - m + 2) = \lambda/(1-\lambda) > 0$. By choosing $\alpha$ sufficiently small we then get

$$\mathcal{R}_4 \leqslant \tfrac{1}{8}\alpha\varkappa k$$

as in steps *2* and *3*. Hence in (3.20) the second term on the right side is dropped and another term $\frac{1}{8}\alpha\varkappa k$ is added in the second braces.

The rest of the proof is essentially the same as for Theorem 4.1.

*Remark.* While Theorem 4.2 requires the full strength of assumption $(\mathrm{C}_1)'$ as well as condition (4.7), the result of Theorem 4.1 actually uses only the second part of $(\mathrm{C}_1)'$, i.e. the condition $\varrho(s) = s^{\mu}$.

The following corollaries are helpful in the application of Theorems 4.1 and 4.2 to specific systems.

COROLLARY 1. *Suppose that the hypotheses of Theorem* 4.1 *or of Theorem* 4.2 *are satisfied, except that in place of* (4.4) *we assume the function* $\delta$ *has a decomposition* $\delta = d + e$ *with*

$$dk^{\mu} \in L^{\infty}(I) \quad \textit{and} \quad ek^{\mu+1} \in L^1(I). \tag{4.4$'$}$$

*Then the rest state is a global attractor.*

*Proof.* It is easily seen, as in the proof of Corollary 1 of Theorem 3.1, that (4.4)$'$ implies (4.4). Hence Theorem 4.1 and Theorem 4.2 can be applied.

COROLLARY 2. *Define* $\bar\sigma(t) = \sigma(t)$ *for* $t \in I$ *and* $\bar\sigma(t) = 0$ *for* $t \in J \setminus I$, *for any* $\sigma$ *given in* $(\mathrm{C}_1)'$. *Assume also that for every* $U > 0$

$$\sigma \notin L^1(I), \quad \delta\sigma^{\mu} \in L^{\infty}(I), \quad \bar\sigma \in \mathrm{CBV}(J).$$

*Then the rest state is a global attractor.*

*Proof.* Take $k = \bar\sigma$ and apply Corollary 1 in the case of Theorem 4.1.

COROLLARY 3. *Assume $I=J$ and that for every $U>0$*

$$\sigma \notin L^1(J), \quad \delta\sigma^\mu \in L^\infty(J), \quad \log\sigma \in \mathrm{Lip}(J).$$

*Then the rest state is a global attractor.*

*Proof.* Take $k=\sigma$ and apply Corollary 1 in the case of Theorem 4.2. (Note that $\sigma$ is Lipschitz continuous, and so also absolutely continuous as is required for Theorem 4.2.)

COROLLARY 4. *Assume $I=J$. Suppose for every $U>0$ that $\delta$ admits the decomposition $\delta=d+e$, with $d>0$, and*

$$\frac{1}{d} \notin L^{1/\mu}(J), \quad \frac{1}{d}\left(1+\frac{1}{\sigma^\mu}\right) \in L^\infty(J), \quad \frac{e}{d} \in L^1(J).$$

*Suppose also that*

$$d^{-1/\mu} \in \mathrm{CBV}(J) \tag{4.11}$$

*or that* (4.7) *holds and*

$$d^{-1/\mu} \in \mathrm{AC}(J), \quad |(d^{-1/\mu})'| \leqslant \mathrm{Const.}\,(d^{-1/\mu})^{1-\lambda}\sigma^\lambda. \tag{4.12}$$

*Then the rest state is a global attractor.*

*Proof.* Take $k=d^{-1/\mu}$ so that (4.2)–(4.4)$'$ are satisfied and $k$ is bounded. If (4.11) holds, then we apply Corollary 1 in the case of Theorem 4.1. If (4.12) holds then (4.8) is satisfied and Corollary 1 in the case of Theorem 4.2 can be applied.

COROLLARY 5. *Suppose $I=J$ and that for every $U>0$ there exist positive numbers $c$ and $\eta$ such that*

$$t\sigma(t) \geqslant c, \quad \liminf_{t\to\infty} \int_R^t \left(\frac{s}{t}\right)^\eta \delta(s)\,\frac{ds}{s^{\mu+1}} < \infty.$$

*Then the rest state is a global attractor.*

The proof is parallel to that of Corollaries 3 and 4 in Section 3. Just as for Corollary 4 in Section 3, the second condition of Corollary 5 is in a sense best possible. That is, it is satisfied when

$$\delta(t) \leqslant \mathrm{Const.}\, t^\mu$$

but fails when $\delta(t) \geqslant \mathrm{Const.}\, t^{\mu+\varepsilon}$ for some $\varepsilon>0$, while there are equations with $\delta(t) \geqslant \mathrm{Const.}\, t^{\mu+\varepsilon}$ for which the rest state is not stable.

*Remarks.* The last condition in Corollary 5 is obviously satisfied if $\delta(t)/t^{\mu+1} \in L^1(J)$. Similarly the last condition in Corollary 4 of Section 3 holds if $\delta(t)/t \in L^1(J)$.

For the system (1.5), the last condition of Corollary 5 becomes, since $\mu=1$,

$$\liminf_{t\to\infty} \int_R^t \left(\frac{s}{t}\right)^{\eta} \delta(s)\, \frac{ds}{s^2} < \infty.$$

Taking $\eta=2$ yields

$$\liminf_{t\to\infty} \frac{1}{t^2} \int_R^t \delta(s)\, ds < \infty. \tag{4.13}$$

For the scalar case of (1.5), and under the assumption $\sigma(t)\geqslant$Constant$>0$, condition (4.13) is essentially due to Artstein and Infante [1].

There are several special cases of Corollaries 2–4 which are of interest, e.g. when $\delta\leqslant$Const.$\,\sigma$ and when $\mu=m-1$. We leave these to the reader.

## 5. Necessary conditions for stability

In this section we shall obtain several necessary conditions for global asymptotic stability of the rest state $u\equiv 0$ for the system (2.1).

We shall not require the full strength of the hypotheses ($H_1$)–($H_3$). In fact, only the following standing assumptions will be made, beyond the basic continuity and smoothness conditions stated at the outset of Section 2,

$$G(u,0)=0, \quad F(t,0)=0.$$

Recall finally that $H(u,p)=(\nabla G(u,p),p)-G(u,p)$, so in particular $H(u,0)=0$ for all $u$.

THEOREM 5.1. *Let $u$ be a solution of* (2.1) *on $J$ such that*

$$u(t)\to 0 \quad \text{and} \quad u'(t)\to 0 \quad \text{as } t\to\infty.$$

*Suppose that for every $t\in J$ and for all $u$ and $p$ sufficiently small,*

$$H(u,p)>0, \quad p\neq 0, \tag{5.1}$$

$$F(t,u)\geqslant 0, \quad 0\leqslant F_t(t,u)\leqslant \psi(t) \quad \text{with } \psi\in L^1(J), \tag{5.2}$$

$$0\leqslant -(Q(t,u,p),p)\leqslant \hat\delta(t)H(u,p), \tag{5.3}$$

*where*

$$\hat\delta\in L^1(J). \tag{5.4}$$

*Then $u\equiv 0$ in $J$.*

*Proof.* Take $R_1\geqslant R$ such that $|u(t)|$ and $|u'(t)|$ are small enough in $J_1=[R_1,\infty)$ to allow the application of (5.1)–(5.3) along the solution.

By (2.7) the expression

$$H(u,u')+F(t,u)-\int_{R_1}^{t}(Q(s,u,u'),u')\,ds-\int_{R_1}^{t}F_t(s,u)\,ds \tag{5.5}$$

is constant. Since the first three terms are non-negative by (5.1), $(5.2)_1$ and $(5.3)_1$, from $(5.2)_2$ it follows that

$$(Q(t,u,u'),u')\in L^1(J_1)$$

and so

$$H(u,u')+F(t,u)=l-\int_{t}^{\infty}(Q(s,u,u'),u')\,ds-\int_{t}^{\infty}F_t(s,u)\,ds \tag{5.6}$$

for some $l$.

We claim that $F(t,u(t))\to 0$ as $t\to\infty$. Indeed by $(5.2)_2$ we obtain

$$F(t,u(t))\leqslant F(r,u(t))+\int_{r}^{t}\psi(s)\,ds, \quad R_1\leqslant r\leqslant t.$$

Since $u(t)\to 0$, this gives by $(5.2)_1$ and the fact that $F(t,0)=0$

$$0\leqslant \limsup_{t\to\infty} F(t,u(t))\leqslant \int_{r}^{\infty}\psi(s)\,ds.$$

The claim is proved by letting $r\to\infty$.

Now let $t\to\infty$ in (5.6) to obtain $l=0$. Hence in turn, because of (5.3) and $(5.2)_2$

$$\begin{aligned} H(u,u')+F(t,u)&=-\int_{t}^{\infty}(Q(s,u,u'),u')\,ds-\int_{t}^{\infty}F_t(s,u)\,ds\\ &\leqslant \int_{t}^{\infty}\hat{\delta}(s)H(u,u')\,ds=X(t). \end{aligned} \tag{5.7}$$

Since $H(u,u')\to 0$ as $t\to\infty$ and since $\hat{\delta}\in L^1(J)$, it is clear that $X(t)\to 0$ as $t\to\infty$. By (5.7) and $(5.2)_1$

$$X'=-\hat{\delta}H\geqslant -\hat{\delta}X,$$

so that for any $t\in J_1$ and $r>t$

$$X(t)\leqslant X(r)\exp\int_{t}^{r}\hat{\delta}(s)\,ds.$$

Letting $r\to\infty$ and using (5.4) again, we obtain $X(t)\leqslant 0$ for all $t\in J_1$.

By (5.1), $(5.2)_1$ and (5.7) it follows that $H(u(t),u'(t))\equiv 0$ on $J_1$. Hence $u'(t)\equiv 0$ on $J_1$ by (5.1), and so $u(t)\equiv 0$ on $J_1$ since $u(t)\to 0$ as $t\to\infty$. In particular $u(R_1)=u'(R_1)=0$. By repetition of the previous argument one finds that the subset of $J$ where $u=0$ is both open and closed in $J$ and so is all of $J$. This completes the proof.

COROLLARY 1. *Suppose that* (5.1)–(5.3) *hold. Then a necessary condition for the rest state to be a global attractor for* (2.1) *is that*

$$\hat{\delta} \notin L^1(J).$$

In fact if $\hat{\delta} \in L^1(J)$ then by Theorem 5.1 there are no solutions whatsoever, except the trivial one $u \equiv 0$, which approach zero together with their derivatives at infinity.

*Remarks.* To prove Corollary 1 it would in fact be enough to find just one solution (or one bounded solution) which does not tend to zero as $t \to \infty$. Hence Theorem 5.1 is actually a much stronger result than simply that the rest state is not a global attractor.

The hypotheses (5.1) and (5.3) of Theorem 5.1 hold for the function $G(p) = |p|^m/m$, $m > 1$, and for $Q$ satisfying $(H_3)$ and (3.2) of $(C_1)'$ in the stronger form

$$|Q(t,u,p)| \leqslant \delta(t) \cdot |p|^\mu \quad \text{for } t \in J,\ |u| \leqslant U \text{ and } |p| \leqslant q, \tag{3.2$'$}$$

with $\mu \geqslant m-1$. Indeed $H(p) = (m-1)|p|^m/m$, giving (5.1), while also for $t \in J$, $|u| \leqslant U$ and $|p| \leqslant q \leqslant 1$

$$-(Q,p) \leqslant |Q| \cdot |p| \leqslant \delta |p|^{1+\mu} \leqslant \hat{\delta} H, \quad \text{with } \hat{\delta} = \frac{m\delta}{m-1}.$$

Hence (5.3) holds and $\hat{\delta} \in L^1(J)$ if and only if $\delta \in L^1(J)$. The simplest example of this behavior is the system (1.5) discussed in the introduction.

The same ideas apply to the mean curvature operator $G(p) = \sqrt{1+|p|^2} - 1$. Here $H(p) = 1 - 1/\sqrt{1+|p|^2}$, giving (5.1) immediately. Moreover for $Q$ satisfying $(H_3)$ and $(3.2)'$ with $\mu = 1$, we have $-(Q,p) \leqslant 4\delta H$ when $|p| \leqslant q \leqslant 1$. Here it is essential to restrict to small $p$ to obtain the validity of (5.3). Clearly a wide class of functions $G$ and $Q$ can be treated in the same way.

Finally (5.2) is a consequence of $(H_2)$ if we add the assumption $F_t(t,u) \geqslant 0$. Of course, if $F$ is independent of $t$, only $(5.2)_1$ is required.

As a special example, consider the strongly nonlinear systems (1.10) and (1.11) in the introduction, with $F = F(u)$, $(f(u),u) > 0$, and

$$\sigma(t) = t^\alpha (\log t)^\beta, \quad \delta(t) = \text{Pos. Const.}\, \sigma(t), \tag{5.8}$$

in $(C_1)'$; of course here we replace (3.2) by $(3.2)'$. If $\alpha < 0$ then from Corollary 2 of Section 4 we see that the rest state is a global attractor if

$$\alpha > -1 \quad \text{or} \quad \alpha = -1,\ \beta \geqslant -1. \tag{5.9}$$

On the other hand, by Corollary 1 of Theorem 5.1, condition (5.9) is also necessary for the rest state to be a global attractor provided $\mu \geqslant m-1$. It is doubtful that (5.9) is a necessary condition when $\mu < m-1$.

Next suppose $\alpha > 0$, but that otherwise $(C_1)'$ holds with the conditions given above. Then by Corollary 4 of Section 4 (with $d=\delta$, $e=0$ and (4.11) holding) the rest state is a global attractor if

$$\alpha < \mu \quad \text{or} \quad \alpha = \mu, \ \beta \leqslant \mu.$$

This condition is also necessary, as shown in [10, Theorem 4.4]; that is, if $1/\sigma \in L^{1/\mu}(J)$ then the rest state cannot be a global attractor. Finally, it is easy to check that when $\alpha = 0$ the rest state is always a global attractor.

When $N=1$ the condition $u'(t) \to 0$ as $t \to \infty$ in Theorem 5.1 can be omitted and the proof simplified. For this improvement we require the further structural condition that $H(0,p)$ be strictly increasing for $p>0$ and strictly decreasing for $p<0$. (This is a consequence of the strict convexity of $G(0,p)$.)

THEOREM 5.1′. *Assume* $N=1$. *Let the hypotheses of Theorem* 5.1 *hold, with the exception that the condition* $u'(t) \to 0$ *as* $t \to \infty$ *is omitted. Then* $u \equiv 0$ *in* $J$.

*Proof.* The argument is the same as before, except that now we obtain $l=0$ in a different way and without using the limit assumption on $u'$.

Suppose then that $N=1$. In the same way as before $F(t,u(t)) \to 0$ as $t \to \infty$. Then by (5.6) it follows that $H(u(t),u'(t)) \to l$ as $t \to \infty$. Since $H(0,p)$ is strictly increasing for $p>0$ and strictly decreasing for $p<0$, it is clear that

$$u'(t) \to p_\infty \text{ (possibly infinite)} \quad \text{as } t \to \infty$$

(note that for any $t_0 \geqslant R$ the values of $u'$ for $t \geqslant t_0$ form an interval of $\mathbf{R}$). This is obviously impossible when $u(t) \to 0$ unless $p_\infty = 0$. Thus $l=0$ and the previous proof applies without further changes.

A result corresponding to Theorem 5.1′ can also be proved for vector solutions. We indicate how this can be accomplished for the system

$$u'' + f(u) = Q(t,u,u'),$$

when $F$ and $Q$ satisfy the conditions (5.2)–(5.4) and $Q$ obeys $(C_2)$. Thus suppose that $u(t) \to 0$ as $t \to \infty$. By Theorem 5.1 it is enough to prove that also $u'(t) \to 0$ as $t \to \infty$ since (5.1) obviously holds. From (2.7) and (5.3) we obtain $\{H(u') + F(u)\}' \leqslant 0$ so that $H(u') = \frac{1}{2}|u'|^2 \to \text{limit} = l$. If $l=0$ we are done, so suppose that $l>0$. Then

$$u'(t) - u'(r) = \int_r^t Q(s,u,u')\,ds - \int_r^t f(u)\,ds.$$

As in the first part of the proof of Theorem 5.1, $(Q,u')\in L^1(J)$. But by $(C_2)$, for $t$ sufficiently large,

$$|Q|\leqslant\frac{|Q|\cdot|u'|}{\frac{1}{2}\sqrt{2l}}\leqslant\frac{2\gamma}{\sqrt{2l}}|(Q,u')|.$$

Hence also $Q(t,u,u')\in L^1(J)$. On the other hand $f(u(t))\to f(0)=0$ as $t\to\infty$, because $u(t)\to 0$. Consequently, for $t\in(r,r+1]$ and $r\to\infty$ we have

$$u'(t)-u'(r)\to 0.$$

Then from the relation $|u'|^2\to 2l\neq 0$ there results

$$|u(r+1)-u(r)|=\left|\int_r^{r+1}u'(t)\,dt\right|\geqslant|u'(r)|-\int_r^{r+1}|u'(t)-u'(r)|\,dt\geqslant\tfrac{1}{2}\sqrt{2l}$$

for $r$ large, which contradicts the fact that $u(t)\to 0$ as $t\to\infty$.

The same idea can be extended to strictly convex action energies $G(p)$, for which $H(p)\to\infty$ as $|p|\to\infty$. Indeed from $H(u')\to l$ it is clear that $|u'(t)|\geqslant\text{Const.}>0$ for large $t$ (when $l>0$). Thus $Q\in L^1(J)$ as above, and in turn $\nabla G(u'(t))-\nabla G(u'(r))\to 0$. Since $u'$ is bounded by the assumption on $H$, it now follows that $u'(t)-u'(r)\to 0$, and the contradiction is obtained as before.

## 6. The non-homogeneous case

In this section we consider the non-homogeneous analogue of the system (2.1), namely

$$(\nabla G(u,u'))'-\nabla_u G(u,u')+f(t,u)=Q(t,u,u')+e(t),\quad J=[R,\infty). \tag{6.1}$$

Under the conditions (a) or (b) in the following lemma, we shall show that the asymptotic stability of the rest state $u\equiv 0$ for (6.1) is essentially the same as for the system (2.1).

LEMMA. *Suppose $F(t,u)\geqslant 0$ in $J\times\mathbf{R}^N$, that* (2.4) *and* $(5.3)_1$ *hold, and that either*

(a) $e'\in L^1(J)$, $e(t)\to 0$ *as* $t\to\infty$,

*or*

(b) $e\in L^1(J)$ *and* $|p|\leqslant\text{Const.}\{1+H(u,p)\}$ *for* $u,p\in\mathbf{R}^N$.

*Then the quantity $H(u,u')+F(t,u)$ tends to a non-negative limit as $t\to\infty$ along any bounded solution of* (6.1).

*Proof.* From (6.1) we obtain, corresponding to (2.7) for the system (2.1),

$$\{H(u,u')+F(t,u)\}'=(Q(t,u,u'),u')+F_t(t,u)+(e,u').$$

In case (a) we write

$$(e,u')=(e,u)'-(e',u).$$

Then by (2.4), $(5.3)_1$ and the fact that $|u|\leqslant L$, say, we have

$$\left\{H(u,u')+F(t,u)+\int_t^\infty \psi(s)\,ds-(e,u)+L\int_t^\infty |e'(s)|\,ds\right\}'\leqslant 0$$

and the result follows at once. In case (b)

$$\{H(u,u')+F(t,u)\}'\leqslant \psi(t)+\text{Const.}\,|e(t)|\{1+H(u,u')\}.$$

Thus by Gronwall's inequality and the fact that $F(t,u)\geqslant 0$, we find that $H(u,u')$ is bounded. In turn

$$\{H(u,u')+F(t,u)\}'\leqslant \psi(t)+\text{Const.}\,|e(t)|$$

and again the conclusion follows.

We can now give the main result of the section.

THEOREM 6.1. *Let the hypotheses of Theorems* 3.1, 3.2, 4.1, *or* 4.2 *hold, and suppose also that e satisfies either condition* (a) *or condition* (b) *of the lemma.*

*Then the rest state is a global attractor for the system* (6.1).

*Proof.* We show that the proofs of Theorems 3.1, 3.2, 4.1, and 4.2 are essentially unchanged if either (a) or (b) holds.

Consider first case (a). We use the variational identity for (6.1) corresponding to the identity (3.10) for the problem (2.1). The new identity in fact differs from (3.10) only by the addition, to the left hand side of (3.10), of the term $-(e,u)\omega$ and, to the right hand side, of the quantity

$$(\alpha-1)(e,u)\omega'-(e',u)\omega,$$

see [7]. Clearly

$$|(\alpha-1)(e,u)\omega'|\leqslant \text{Const.}\,|e|\omega',$$

and since $e(t)\to 0$ as $t\to\infty$ one has

$$|(\alpha-1)(e,u)\omega'|\leqslant \frac{\alpha\varkappa}{8}\omega'$$

for large $t$. Similarly

$$|(e',u)\omega|\leqslant L|e'|\omega,\quad |e'|\in L^1(J),$$

and this term can be treated almost exactly the same as the term Const. $|k'|\omega$ in step *4* of Theorem 3.1.

21 – 935202 *Acta Mathematica* 170. Imprimé le 30 juin 1993

Next if (b) holds we write the additional term on the left hand side of the identity, namely $\{-(e,u)\omega\}'$, in the form

$$-(e,u)\omega'-(e',u)\omega-(e,u')\omega.$$

These terms cancel several of the additional terms on the right hand side, leaving only the following additional terms on the right:

$$\alpha(e,u)\omega'+(e,u')\omega.$$

Since $u$ and $u'$ are bounded in case (b)—recall that $|p|\leqslant\text{Const.}\{1+H(u,p)\}$ and that, by the lemma, $H(u,u')$ is bounded—and since

$$\omega'=k\omega\leqslant\text{Const.}\,\omega,$$

the additional terms are bounded by

$$\text{Const.}\,|e|\omega.$$

Using the fact that $e\in L^1$, this term can also be treated as in step *4* of Theorem 3.1. This completes the proof.

*Remarks.* The condition $|p|\leqslant\text{Const.}\{1+H(u,p)\}$ in case (b) is satisfied when $G(p)=|p|^m/m$, $m>1$, and in particular for the system (1.5). In the scalar case, Levin and Nohel had found the condition $e\in L^1(J)$, $e(t)\to 0$ as $t\to\infty$, whereas here for case (b) only $e\in L^1(J)$ is needed.

It is, finally, worth noting that hypothesis (b) can be generalized without difficulty to the case when the non-homogeneous term $e$ in (6.1) depends not only on $t$ but also on $u$ and $p$, provided we assume that

$$|e(t,u,p)|\leqslant\psi(t)\frac{H(u,p)+F(t,u)+1}{|p|+1} \tag{6.2}$$

where $\psi\in L^1(J)$.

## Appendix

We prove the lemma used in the proof of Theorem 3.1.

*Proof of Lemma* A. If $k>0$ in $J$ we take $E=\varnothing$, and define $\bar{\bar{k}}$ to be a continuous piecewise linear function with vertices on the graph of the function $\frac{1}{2}(\theta+1)k$, such that

$$k<\bar{\bar{k}}<\theta k,\quad t\in J.$$

(For this construction it is necessary that $k$ be continuous; in particular on any compact set $k$ then has a positive lower bound and the graphs of $k$, $\frac{1}{2}(\theta+1)k$ and $\theta k$ are separated by a positive distance.)

Rounding the corners of the graph of $\bar{\bar{k}}$ (say with small circular arcs) then produces the desired function $\bar{k}$, since clearly

$$\operatorname{Var}\bar{k} \leqslant \operatorname{Var}\bar{\bar{k}} \leqslant \operatorname{Var}\left(\tfrac{1}{2}(\theta+1)k\right) \leqslant \theta \operatorname{Var} k. \tag{A1}$$

When the set $C$ of zeros of $k$ is non-empty, the construction is slightly more delicate. Clearly $C$ is closed since $k$ is continuous. Hence $J\backslash C$ is a relatively open set, $O$, and we can write

$$O = \bigcup I_j,$$

where $I_j$, $j=1,2,\ldots$, are disjoint open intervals in $J$, with $I_1=[R,a)$ if $k(R)>0$.

Let $J_1$, $J_2$, ... be a sequence of disjoint intervals such that $|J_n|=1$ and $J=\bigcup J_n$. It is not difficult to see that there exists a subsequence of intervals $I_i'$ drawn from $(I_j)$ such that

(1) only a finite number of intervals $I_i'$ intersect any $J_n$,

(2) $\int_{J_n\backslash O'} k \leqslant 2^{-(n+1)}$, where $O'=\bigcup I_i'$.

To define $E$, we first choose a closed subinterval $G_i$ in each interval $I_i'$, with the property that

$$\int_{S_i} k \leqslant 2^{-(i+1)}, \quad S_i = I_i'\backslash G_i, \quad i=1,2,\ldots,$$

and put $E=\operatorname{Int} J\backslash\bigcup G_i$ (here $E$ is open, since $\bigcup G_i$ is certainly closed; if $I_1=[R,a)$ we take $I_1'=I_1$ and $G_1$ in the form $[R,b)$, $b<a$). Clearly

$$E=(C\cup O''\cup S)\backslash\{R\}, \quad \text{where } O''=O\backslash O' \text{ and } S=\bigcup S_i.$$

Since $\int_{O''} k \leqslant \frac{1}{2}$ by (2), it follows that

$$\int_E k = \int_{O''} k + \int_S k \leqslant 1$$

and (iii) is verified.

Next, define $\bar{\bar{k}}$ to be a continuous piecewise linear function on $J$, all of whose vertices lie in $O'=\bigcup I_i'$, such that:

> In $G_i$, the vertices of $\bar{\bar{k}}$ lie on the graph of $\frac{1}{2}(\theta+1)k$ and $k<\bar{\bar{k}}<\theta k$ (recall that $k>0$ on the closed set $G_i$),
>
> $\bar{\bar{k}}$ has exactly one vertex $(c_i, 0)$ in each $S_i$, $i=1,2,\ldots$, and $0\leqslant\bar{\bar{k}}<\theta k$ in $S_i$.

Clearly $0\leqslant\bar{\bar{k}}\leqslant\theta k$ in $E$. Finally, we obtain $\bar{k}$ by rounding the corners $\bar{\bar{k}}$ as before (obviously $\bar{\bar{k}}$ has at most a finite number of corners in any bounded interval of $J$). Conditions (i) and (ii) now follow as in the first part of the proof.

*Acknowledgment.* The first author is a member of *Gruppo Nazionale di Analisi Funzionale e sue Applicazioni* of the *Consiglio Nazionale delle Ricerche*, Rome, Italy. This research has been partly supported by the Italian *Ministero della Università e della Ricerca Scientifica e Tecnologica.*

## References

[1] ARTSTEIN, Z. & INFANTE, E. F., On the asymptotic stability oscillators with unbounded damping. *Quart. Appl. Math.*, 34 (1976), 195–199.

[2] BALLIEU, R. J. & PEIFFER, K., Attractivity of the origin for the equation $\ddot{x}+f(t,x,\dot{x})|\dot{x}|^{\alpha}\dot{x}+g(x)=0$. *J. Math. Anal. Appl.*, 65 (1978), 321–332.

[3] BURTON, T. A., On the equation $u''+f(x)h(u')u'+g(u)=e(t)$. *Ann. Mat. Pura Appl.*, 85 (1970), 227–285.

[4] LEONI, G., MANFREDINI, M. & PUCCI, P., Stability properties for solutions of general Euler–Lagrange systems. *Differential Integral Equations*, 5 (1992), 537–552.

[5] LEVIN, J. J. & NOHEL, J. A., Global asymptotic stability for nonlinear systems of differential equations and applications to reactor dynamics. *Archive Rational Mech. Anal.*, 5 (1960), 194–211.

[6] NAKAO, M., Asymptotic stability for some nonlinear evolution equations of second order with unbounded dissipative terms. *J. Differential Equations*, 30 (1978), 54–63.

[7] PUCCI, P. & SERRIN, J., A general variational identity. *Indiana Univ. Math. J.*, 35 (1986), 681–703.

[8] — Continuation and limit properties for solutions of strongly nonlinear second order differential equations. *Asymptotic Anal.*, 4 (1991), 97–160.

[9] — Global asymptotic stability for strongly nonlinear second order systems, in *Proc. Conf. on Nonlinear Diffusion Equations and Their Equilibrium States* (N. G. Lloyd, W.-M. Ni, L. A. Peletier and J. Serrin, eds.), pp. 437–449. Birkhäuser, Boston–Basel–Berlin, 1992.

[10] — Continuation and limit behavior for damped quasi-variational systems, in *Proc. Conf. on Degenerate Diffusions, 1992* (W.-M. Ni, L. A. Peletier and J. L. Vazquez, eds.). The IMA Volumes and its Applications, 47. To appear.

[11] SALVADORI, L., Famiglie ad un parametro di funzioni di Liapunov nello studio della stabilità. *Symposia Math.*, 6 (1971), 309–330. Istituto Nazionale di Alta Matematica, Roma.

[12] SMITH, R. A., Asymptotic stability of $x''+a(t)x'+x=0$. *Quart. J. Math. Oxford*, 12 (1961), 123–126.

[13] THURSTON, L. H. & WONG, J. S. W., On global asymptotic stability of certain second order differential equations with integrable forcing terms. *SIAM J. Appl. Math.*, 24 (1973), 50–61.

PRECISE DAMPING CONDITIONS FOR GLOBAL ASYMPTOTIC STABILITY 307

PATRIZIA PUCCI
Dipartimento di Matematica
Università di Perugia
Via Vanvitelli, 1
I-06100 Perugia
Italy

JAMES SERRIN
Department of Mathematics
University of Minnesota
Minneapolis, MN 55455
U.S.A.

*Received February 28, 1991*
*Received in revised form January 28, 1992*

*SIAM J. Math. Anal.* 25 (1994), 815–835. **Reprint II.1.[ii]**


SIAM J. MATH. ANAL.
Vol. 25, No. 3, pp. 815–835, May 1994

001

# ASYMPTOTIC STABILITY FOR INTERMITTENTLY CONTROLLED NONLINEAR OSCILLATORS*

PATRIZIA PUCCI† AND JAMES SERRIN‡

**Abstract.** The authors prove a number of asymptotic stability theorems for intermittently damped quasi-variational systems, extending and generalizing previous work on the subject.

**Key words.** global asymptotic stability, intermittent damping, control set

**AMS subject classifications.** 34DXX, 35A15

**1. Introduction.** The problem of global asymptotic stability of solutions of second order equations with intermittent damping has been studied by Smith, Thurston and Wong, Artstein and Infante, and Hatvani and Totik. In this paper we give various generalizations of this work and extensions to quasi-variational systems.

As in our earlier work [5], [7] on asymptotic stability, we consider vector unknowns $u : J \to \mathbb{R}^N$ and systems having the general form

$$(\nabla\mathcal{L}(t,u,u'))' - \nabla_u\mathcal{L}(t,u,u') = Q(t,u,u'), \qquad t \in J, \tag{1.1}$$

where $J$ is a half open interval of the form $[T,\infty)$ and $\mathcal{L}(t,u,p) = G(u,p) - F(t,u)$, and where $G$, $F$, $Q$ are given continuously differentiable functions. The most important of the conditions which will be imposed on (1.1) are that

$$G(u,\cdot) \text{ is strictly convex in } \mathbb{R}^N; \qquad G(u,0) = 0, \quad \nabla G(u,0) = 0, \tag{1.2}$$

$$\big(\nabla_u F(t,u), u\big) > 0 \quad \text{for } u \neq 0; \qquad F(t,u) = 0, \tag{1.3}$$

$$\big(Q(t,u,p), p\big) \leq 0. \tag{1.4}$$

Here $(\cdot,\cdot)$ denotes the inner product in $\mathbb{R}^N$ and

$$\nabla = \nabla_p = \left(\frac{\partial}{\partial p_1}, \ldots, \frac{\partial}{\partial p_N}\right), \qquad \nabla_u = \left(\frac{\partial}{\partial u_1}, \ldots, \frac{\partial}{\partial u_N}\right).$$

The function $F$ represents a restoring potential and $Q$ a general nonlinear damping, expressed by (1.4). In §2 we shall give a complete set of hypotheses, while explicit examples are given in [5] and [7].

Since $\nabla G(u,0) = \nabla_u G(u,0) = \nabla_u F(t,0) = Q(t,u,0) = 0$ it is clear that the rest state $u = 0$ is a solution of (1.1). This state is said to be *a global attractor* for the system if any bounded solution $u$, defined on some interval $J$, has the property

$$u(t),\ u'(t) \to 0 \quad \text{as } t \to \infty.$$

By the concept of *intermittent damping* we mean that certain restrictions or controls are placed on the damping term on a sequence of nonoverlapping intervals $I_n = [a_n, b_n]$ of $J$, with $a_n \to \infty$; on the other hand, in the gaps *between* these intervals either no restrictions are imposed or, alternatively, the damping is assumed

* Received by the editors November 23, 1992; accepted for publication April 8, 1993. This research has been partially supported by the Italian Ministero della Universitá e della Ricerca Scientifica e Tecnologica.

† Dipartimento di Matematica, Università di Perugia, 06123 Perugia, Italy.

‡ Department of Mathematics, University of Minnesota, Minneapolis, Minnesota 55455.

to be bounded from zero but to be otherwise uncontrolled. We emphasize that the intervals $I_n$ may be arbitrarily widely spaced, leaving gaps between them that can be as long as one wishes.

Our purpose is to show that under appropriate conditions on the measures $|I_n|$ and on the damping term $Q(t,u,p)$ for $t \in \cup_1^\infty I_n$, the rest state $u = 0$ becomes a global attractor for (1.1).

From a mechanical point of view the system (1.1) can be considered as the governing law of a holonomic dynamical system, having $N$ degrees of freedom and subject to nonlinear damping. The notion of intermittent damping then occurs if the system is positively damped in the time intervals $I_n$, but has its damping either *switched off* or *unrestricted* at other times. The system is oscillatory when no damping is present, because $(f(t,u),u) > 0$ for $u \neq 0$; that is, it is not possible to have any solution, other than the trivial one $u = 0$, approaching a limit as $t \to \infty$ (see [5, §5] for a more complete discussion). From this point of view the question we consider is whether the damping which occurs on the time intervals $I_n$ is sufficient to drive the solution to its rest state as $t \to \infty$. The following example provides a specific illustration of this situation in perhaps its simplest form.

Consider the system

$$u'' + A(t,u,u')u' + f(u) = 0, \tag{1.5}$$

where $A$ is a continuous $N \times N$ nonnegative definite matrix and $f(u) = \nabla_u F(u)$. This system arises from (1.1) by the specializations

$$G(p) = \tfrac{1}{2}|p|^2, \qquad Q(t,u,p) = -A(t,u,p)p.$$

We suppose that $(f(u),u) > 0$ for $u \neq 0$, and that $A$ is bounded and uniformly positive definite for $t \in I = \cup_1^\infty I_n$ and $(u,p)$ in any given compact set of $\mathbb{R}^N \times \mathbb{R}^N$; no restrictions, however, other than nonnegativity, are placed on $A$ in the set $J \setminus I$. Then the following rather unexpected result holds:

*If the measures of the intervals $I_n$ satisfy*

$$\sum_1^\infty |I_n|^3 = \infty, \tag{1.6}$$

*then $u = 0$ is a global attractor for* (1.5).

*The exponent 3 is the best possible one: that is, without further restrictions no smaller exponent can yield the general conclusion.*

A stronger result is valid if the damping matrix $A$ has the decomposition

$$A(t,u,p) = \beta(t,u,p)I + B(t,u,p), \tag{1.7}$$

where $B(t,u,p)$ is bounded and nonnegative definite for $t \in J$ and $(u,p)$ in any compact set of $\mathbb{R}^N \times \mathbb{R}^N$; the coefficient $\beta(t,u,p)$ is such that for every compact set $K$ of $\mathbb{R}^N \times \mathbb{R}^N$ there exist positive constants $\beta_1$, $\beta_2$ such that

$$\beta(t,u,p) \geq \beta_1 \quad \text{in } J \times K, \tag{1.8}$$

$$\beta(t,u,p) \leq \beta_2 \quad \text{in } I \times K. \tag{1.9}$$

*Then $u = 0$ is a global attractor for* (1.5) *provided that*

$$\sum_1^\infty |I_n|^2 = \infty. \tag{1.10}$$

*Again the exponent is best possible.*

The above results are special cases, respectively, of Corollaries 3 and 4 in §3; see the comments at the end of §3. Indeed, in those results the damping need not even be bounded on $I$ but only have a controlled $L^1$ norm. Moreover, the constant $\beta_1$ in (1.8) can be replaced by a nonnegative measurable function $\hat{\sigma}$ satisfying a *positive mean value criterion*; see condition (2.13) below.

References [1]–[3], [8], and [9] treat the case $N = 1$ of (1.5); moreover, in [8] the coefficient $A$ is independent of $u, u'$ and $f(u)$ is linear. Our results are improvements of the corresponding ones in these papers, even when restricted to the cases treated there.

In §2 we present the setting of the paper and state two important preliminary theorems upon which our further results are based. The main results for the system (1.1) are given in §3 and proved in §§4 and 5. In §6 we present specific examples, showing that the exponents 2 and 3 in the above results are best possible.

**2. Preliminaries.** We consider vector solutions $u = (u_1, \dots, u_N)$ of the quasi-variational ordinary differential system

$$(2.1) \qquad \big(\nabla G(u,u')\big)' - \nabla_u G(u,u') + f(t,u) = Q(t,u,u'), \qquad t \in J = [T,\infty),$$

where $\nabla$ denotes the gradient operator with respect to the variable $p$ and

$$f(t,u) = \nabla_u F(t,u).$$

It will be supposed throughout the paper that

$$G \in C^1(\mathbb{R}^N \times \mathbb{R}^N; \mathbb{R}), \quad F \in C^1(J \times \mathbb{R}^N; \mathbb{R}), \quad Q \in C(J \times \mathbb{R}^N \times \mathbb{R}^N; \mathbb{R}^N),$$

and also that the following natural conditions hold.

(H$_1$) $G(u,\cdot)$ is strictly convex in $\mathbb{R}^N$ for all $u \in \mathbb{R}^N$; with $G(u,0) = 0$ and $\nabla G(u,0) = 0$. For all $U > 0$ there exists a positive constant $\Theta = \Theta(U)$ and an exponent $m > 1$ independent of $U$ such that

$$(2.2) \qquad |\nabla G(u,p)| \le \Theta |p|^{m-1} \quad \text{for all } |u| \le U \text{ and } |p| \le 1.$$

(H$_2$) $F(t,0) = 0$ for all $t \in J$. For all $u_0, U$ with $0 < u_0 \le U$ there exists a constant $\kappa > 0$ and a nonnegative function $\psi \in L^1(J)$ such that

$$(2.3) \qquad \big(f(t,u),u\big) \ge \kappa \quad \text{when } t \in J \quad \text{and} \quad |u| \in [u_0, U],$$

$$(2.4) \qquad |F_t(t,u)| \le \psi(t) \quad \text{when } t \in J(\text{a.e.}) \quad \text{and} \quad |u| \le U.$$

(H$_3$) $\big(Q(t,u,p),p\big) \le 0$ for all $t \in J$, $u \in \mathbb{R}^N$ and $p \in \mathbb{R}^N$.

If $F$ does not depend on $t$, then (2.3) follows from the condition $\big(f(u),u\big) > 0$ for $u \ne 0$, while (2.4) is irrelevant. Finally, when $N = 1$ any function $f$ is of gradient type, with $F(t,u) = \int_0^u f(t,s)\,ds$.

Obviously (H$_1$) is satisfied by any strictly convex homogeneous function $G = G(p)$ of degree $m > 1$, and in particular by the model function $G(p) = |p|^m/m$, $m > 1$; another example is $G(p) = \sqrt{1+|p|^2} - 1$, with $m = 2$. The system (1.5) arises when $G(p) = \frac{1}{2}|p|^2$, with the corresponding exponent $m = 2$.

The next hypothesis places in evidence the concept of a control set $I \subset J$ where the damping term $Q$ is subject to restrictions.

($H_4$) For all $U > 0$ there exists a measurable control set $I \subset J$ and a number $\gamma \geq 1$ such that

$$|Q(t,u,p)| \cdot |p| \leq \gamma \left|\left(Q(t,u,p),p\right)\right| \quad \text{for all } t \in I, |u| \leq U \text{ and } p \in \mathbb{R}^N. \tag{2.5}$$

Moreover, there exists a positive measurable damping function $\delta : I \to \mathbb{R}$ and numbers $\mu$, $q > 0$ such that

$$\left(Q(t,u,p),u\right) \leq \delta(t)\,|p|^\mu \quad \text{for } t \in I, |u| \leq U \text{ and } |p| \leq q. \tag{2.6}$$

Although $I$, $\delta$ and $\gamma$, $\mu$, $q$ may depend on $U$, for simplicity we do not specifically indicate this dependence. When $N = 1$ condition (2.5) holds automatically with $\gamma = 1$.

In [5] we considered the asymptotic stability of the system (2.1) when the damping magnitude $|Q|$ is controlled from below, but not bounded away from zero. Specifically the following condition was required:

For every $U > 0$ there exist a nonnegative measurable damping control $\sigma : I \to \mathbb{R}$ and an exponent $\nu > 0$ such that

$$|Q(t,u,p)| \geq \sigma(t) \min\{1,\ |p|^\nu\} \quad \text{for all } t \in I, |u| \leq U \text{ and } p \in \mathbb{R}^N. \tag{2.7}$$

A further technical hypothesis concerning the function $G$ was also assumed:

For every $U > 0$ and $p_0 > 0$ there is a constant such that

$$\left(\nabla_u G(u,p),u\right) \leq \text{Const.} \left(\nabla G(u,p),p\right) \quad \text{whenever } |u| \leq U \text{ and } |p| \geq p_0. \tag{2.8}$$

Note that (2.8) holds whenever $G(u,p) = g(u)\overline{G}(p)$, with $g(u) > 0$ a smooth function in $\mathbb{R}^N$ and $\overline{G}$ satisfying ($H_1$).

Under the natural assumptions ($H_1$)–($H_4$), together with (2.7) and (2.8), the following result is valid; see [5, Thm. 4.2] and the modified version of this result proved in §3.2 of [4]. This will be the basis for the first main theorem in §3. In its statement we agree that the function $\delta k$ is extended to all of $J$ by the definition $\delta(t)k(t) = 0$ for $t \in J \setminus I$.

THEOREM A. *Assume that for every $U > 0$ there exists a bounded absolutely continuous function $k$ on $J$ such that*

$$k \notin L^1(J), \qquad k = 0 \quad \text{on } J \setminus I, \tag{2.9}$$

$$0 \leq k \leq \text{Const. } \sigma \text{ on } I, \qquad |k'| \leq \text{Const. } \sigma^\lambda k^{1-\lambda} \quad \text{a.e. on } I, \tag{2.10}$$

*where*

$$\lambda = \begin{cases} \dfrac{m-1}{\nu+1} & \text{if } \nu > m-2, \\ 1 & \text{if } \nu \leq m-2 \quad (\text{and } m > 2). \end{cases} \tag{2.11}$$

*Suppose furthermore that there exists a constant $M > 0$ for which*

$$\int_T^t \delta(s) k^{\mu+1}(s)\,ds \leq M \int_T^t k(s)\,ds, \qquad t \in J. \tag{2.12}$$

*Then the rest state $u = 0$ is a global attractor for the system* (2.1).

In [7] we also studied asymptotic stability for the complementary situation in which the damping magnitude $|Q|$ is bounded from zero when $|u|$ and $|p|$ are bounded from zero. In particular, the following condition was assumed:

There is (i) a continuous function $\varphi : \mathbb{R}^N \times \mathbb{R}^N \to [0, \infty)$ with

$$\varphi(u,p) > 0 \quad \text{when } u \neq 0 \text{ and } p \neq 0;$$

and (ii) a measurable function $\hat{\sigma} : J \to [0, \infty)$ and a positive function $a$ on $(0,1)$ satisfying

$$\int_L \hat{\sigma}(t)\, dt \geq a(|L|) > 0 \quad \text{for all intervals } L \subset J \text{ with } |L| \in (0,1);$$

such that for all $U > 0$ there holds

$$\big|(Q(t,u,p),p)\big| \geq \hat{\sigma}(t)\, \varphi(u,p) \quad \text{for all } t \in J, |u| \leq U \text{ and } |p| \leq q, \tag{2.13}$$

where $q = q(U) > 0$ is given in ($H_4$).

This is in fact condition ($H_3$) of [7], in the weaker version involving (ii) which was given in §7 of [7]. (In this context, condition (ii) was first introduced by Hatvani [2].) If $Q(t,u,p) = \hat{\sigma}(t)\, \hat{Q}(u,p)$ and $\big(\hat{Q}(u,p),p\big) < 0$ for $u,p \neq 0$, it is easy to see that (2.13) is satisfied.

Two further technical hypotheses were introduced in [7]; they are required only when $N > 1$, though in fact the second, ($V_2$), automatically holds when $N = 1$ with $\varepsilon(p) = 0$, $g(u) = \frac{1}{2}u^2$ and $C = 0$ in view of ($H_1$) and ($H_3$); see also [7, Lemma 2.1].

($V_1$) For all $U > 0$ and $p_0 > 0$ there is a nonnegative measurable function $h \notin L^1(J)$ such that

$$\big|(Q(t,u,p),p)\big| \geq h(t) \quad \text{for all } t \in J, |u| \leq U \text{ and } |p| \geq p_0.$$

($V_2$) For all $U > 0$ there exists a continuous function $\varepsilon(p)$ with $\varepsilon(0) = 0$, such that

$$\big(Q(t,u,p),u\big) \leq \varepsilon(p)$$

when $t \in J$, $|u| \leq U$, $|p| \leq q$ and $\big(\nabla G(u,p),u\big) \geq 0$. Moreover, there exists a $C^1$ function $g(u)$ and a constant $C \geq 0$ such that

$$\frac{(\nabla_u g(u),p)}{|p|} - \frac{(\nabla G(u,p),u)}{|\nabla G(u,p)|} \leq C\,\frac{\varphi(u,p)}{|p|},$$

when $|u| \leq U$, $|p| \leq q$ and $(\nabla G(u,p),u) < 0$. Again $q$ and $\varphi$ are given in (2.13).

It is worth noting that ($V_2$) is satisfied if the vectors $p$, $\nabla G(u,p)$ and $-Q(t,u,p)$ all have the same direction when $p \neq 0$.

Again under the natural hypotheses ($H_1$)–($H_4$), and also assuming ($V_1$)–($V_2$) and (2.13), we have the following result; see [7, Thm. 2], its extension in §7 of [7], and the modified version of this result proved in §3.1 of [4].

THEOREM B. *Suppose that for all $U > 0$ there is a bounded absolutely continuous function $k$ on $J$ such that* (2.9), (2.12), *and*

$$|k'| \leq \begin{cases} \text{Const. } k^{2-m}, & 1 < m < 2, \\ \text{Const.}, & m \geq 2, \end{cases} \qquad \textit{a.e. in } J, \tag{2.14}$$

*are satisfied. Then the rest state is a global attractor for the system* (2.1).

Theorem B will be the basis for the second main theorem in §3. Now let

$$H(u,p) = \big(\nabla G(u,p),p\big) - G(u,p)$$

be the Legendre transform in the variable $p$ of the action function $G(u,p)$. The following observation shows that, when

$$H(u,p) \to \infty \quad \text{as } |p| \to \infty \tag{2.15}$$

uniformly for $u$ in compact subsets of $\mathbb{R}^N$, then several of the earlier hypotheses can be weakened, while (2.8) can be omitted entirely.

We first recall that solutions of (2.1) have the property that

$$H(u(t),u'(t)) + F(t,u(t)) \to \text{limit} \quad \text{as } t \to \infty.$$

see [5, (3.7)] or [7, Lemma 5.1(i)]. Hence in turn, since $F(t,u) \geq 0$ by ($H_2$), the function

$$H(u(t),u'(t))$$

is bounded along any solution $u = u(t)$, $t \in J$. Thus by (2.15), *for any bounded solution of* (2.1) *the function* $u'(t)$ *is also bounded on* $J$.

It follows that in applying the hypotheses of Theorems A and B for any given bounded solution of (2.1), *one can restrict consideration to compact subsets of vectors* $(u,p)$ *in* $\mathbb{R}^N \times \mathbb{R}^N$. In particular, when (2.15) holds, the condition (2.5) can be weakened to the following form:

For every compact set $K$ in $\mathbb{R}^N \times \mathbb{R}^N$ there exists a measurable control set $I \subset J$ and a number $\gamma \geq 1$ such that

$$|Q(t,u,p)| \cdot |p| \leq \gamma \big|\big(Q(t,u,p),p\big)\big| \quad \text{for all } t \in I \text{ and } (u,p) \in K. \tag{2.5$'$}$$

Analogous restatements of (2.7) and ($V_1$) also hold when (2.15) is assumed. Finally, (2.8) is automatically satisfied when $|u \leq U$ and $p_o \leq |p| \leq P$, with the constant depending only on $U$, $P$ and $p_o$. Thus (2.8) can be omitted if (2.15) holds.

We conclude the section with a useful estimate.

LEMMA. *Let* (2.5)–(2.7) *hold. Then for all* $U > 0$ *there is a positive constant* $c = c(U)$ *such that*

$$\delta(t) \geq c\,\sigma(t) \quad \textit{for } t \in I. \tag{2.16}$$

*If* (2.5), (2.6), *and* (2.13) *hold, then for all* $U > 0$ *and* $\vartheta \in (0,1)$ *there is a positive constant* $d = d(U,\vartheta)$ *such that*

$$\frac{1}{|L|}\int_L \delta(t)\,dt \geq d \quad \textit{for all intervals } L \subset J \textit{ with } |L| \geq \vartheta. \tag{2.17}$$

*Proof.* Fix $U > 0$. By (2.5) and (2.6), with $u = -p$ and $|p| = \min\{1U,\ q\} = r > 0$, we get

$$\delta(t)\,|p|^\mu \geq -\big(Q(t,-p,p),p\big) \geq |Q(t,-p,p)| \cdot |p|/\gamma.$$

On the other hand, by (2.7) and the fact that $|p| \leq \min\{1,U\}$,

$$|Q(t,-p,p)| \geq \sigma(t)\,|p|^\nu,$$

proving (2.16) with $c = r^{\nu+1-\mu}/\gamma$.

Next by (2.13)

$$|Q(t,-p,p)|\cdot|p| \geq \hat{\sigma}(t)\,\varphi(-p,p),$$

so that $\delta(t) \geq \hat{d}\,\hat{\sigma}(t)$ in $J$, with $\hat{d} = \max\{\varphi(-p,p)\ :\ |p| = r\}\cdot r^{-\mu}/\gamma$. In turn,

$$\frac{1}{|L|}\int_L \delta(t)\,dt \geq \frac{\hat{d}}{|L|}\int_L \hat{\sigma}(t)\,dt \geq \frac{\hat{d}}{2\vartheta}\,a(\vartheta) = d > 0$$

by application of inequality (7.2) of [7] with $\lambda = \vartheta$. This completes the proof.

**3. Main results.** Here we state our main theorems and related consequences. *It is assumed throughout, without further comment, that the conditions* ($H_1$)–($H_4$) *are satisfied.*

Let $(I_n)_n$ be a sequence of nonoverlapping intervals $I_n = [a_n, b_n]$ of $J$ with $a_n < b_n$ and, for the results of this section, $a_n \to \infty$. Let the control set in ($H_3$)–($H_4$) have the form $I = \cup_1^\infty I_n$, and introduce the notation

$$d_n = \frac{1}{|I_n|}\int_{I_n} \delta(t)\,dt \quad \text{(possibly } \infty\text{)},$$

which will be used throughout the paper.

Our first result is based on Theorem A of §2. Conditions (2.7) and (2.8) are of course required here, and also for the corresponding corollaries.

THEOREM 1. *Suppose that for every* $U > 0$ *there are positive constants* $A, B$ *such that*

$$\text{(3.1)}\qquad \sum_1^\infty \sigma_n\cdot\min\left\{|I_n|^q,\ \frac{A}{B+x_n}|I_n|\right\} = \infty, \qquad q = \begin{cases} \dfrac{m+\nu}{m-1}, & \nu > m-2,\\ 2, & \nu \leq m-2,\end{cases}$$

*where*

$$\sigma_n = \inf_{I_n}\sigma(t) \quad \textit{and} \quad x_n = \sigma_n\, d_n^{\ell}, \qquad \ell = 1/\mu.$$

*Then* $u = 0$ *is a global attractor for* (2.1).

The proof of Theorem 1 is given in §4. In the case $N = 1$, $G(p) = p^2/2$, $Q = -a(t)\,p$ and $f(u) = u$, Smith [8] obtained the weaker result that $u = 0$ is a global attractor when

$$\sum_1^\infty \sigma_n|I_n|\cdot\min\left\{|I_n|^2,\ \frac{1}{(1+\Delta_n)^2}\right\} = \infty, \qquad \Delta_n = \max_{I_n} a(t);$$

in particular, for this case

$$m = 2, \quad \mu = \nu = 1, \quad q = 3, \quad \sigma(t) = a(t), \quad \delta(t) = Ua(t),$$

so that by taking $A = B = U$ in (3.1) we get

$$\frac{A}{B+x_n} = \frac{1}{1+\sigma_n d_n/U} \geq \frac{1}{1+\Delta_n^2} \geq \frac{1}{(1+\Delta_n)^2}.$$

Several special cases of Theorem 1 are of particular importance. In stating the results, we recall throughout that the hypotheses are understood to apply for each fixed $U > 0$.

COROLLARY 1. *Suppose that*

$$\sup_n x_n < \infty. \tag{3.2}$$

*Then $u = 0$ is a global attractor for* (2.1) *if*

$$\sum_1^\infty \sigma_n \min\{|I_n|^q,\ |I_n|\} = \infty. \tag{3.3}$$

*Proof.* Condition (3.1) with $A = 1+\sup_n x_n$ and $B = 1$ follows at once from (3.2), (3.3).

In view of (2.16) it is clear that (3.2) holds whenever $\delta$ is bounded on $I = \cup_1^\infty I_n$. From Corollary 1 we also get the following consequence:

*Suppose that*

$$\sup_n x_n < \infty, \qquad \inf_n |I_n| > 0.$$

*Then $u = 0$ is a global attractor if $\sum_1^\infty \sigma_n = \infty$.*

A related result, applying however only for the scalar case of (1.5), appears in [3, Cor. 4.2].

COROLLARY 2. *Suppose that*

$$\inf_n x_n > 0.$$

*Then $u = 0$ is a global attractor for* (2.1) *if*

$$\sum_1^\infty d_n^{-\ell} \min\{|I_n|^q, |I_n|\} = \infty, \quad \textit{where } \ell = 1/\mu. \tag{3.4}$$

*Proof.* Condition (3.1) with $A = 1 + \inf_n x_n$, $B = 1$ follows easily from (3.4) together with the relations

$$\sigma_n = x_n d_n^{-\ell} \geq x\, d_n^{-\ell}, \qquad \frac{\sigma_n A}{1+x_n} = \frac{x_n(1+x)}{1+x_n} d_n^{-\ell} \geq x\, d_n^{-\ell},$$

where $x = \inf x_n > 0$.

COROLLARY 3. *Suppose that*

$$\inf_I \sigma(t) > 0, \qquad \sup_n d_n < \infty. \tag{3.5}$$

*Then $u = 0$ is a global attractor for* (2.1) *if*

$$\sum_1^\infty |I_n|^q = \infty. \tag{3.6}$$

*Proof.* This is an immediate consequence of Corollary 1 or 2. For example, (3.5) implies that

$$\inf_n x_n \geq c^\ell \left(\inf_n \sigma_n\right)^{1+\ell} > 0 \quad \text{and} \quad \inf_n d_n^{-\ell} \geq \left(\sup_n d_n\right)^{-\ell} > 0,$$

where $c$ is the constant in (2.16). Hence by Corollary 2 the rest state is a global attractor provided that

$$\sum_1^\infty \min\{|I_n|^q, |I_n|\} = \infty.$$

But this series diverges if and only if (3.6) diverges (since $q > 1$).

Our second main result is based on Theorem B. In this case conditions (2.13) and ($V_1$)–($V_2$) are required (instead of (2.7)–(2.8)). We recall again that the control set has the form $I = \cup_1^\infty I_n$.

THEOREM 2. *Suppose that for every $U > 0$ there exists a positive constant $A$ such that*

$$\sum_1^\infty \min\left\{|I_n|^{\bar q}, \frac{A}{d_n^\ell}|I_n|\right\} = \infty, \qquad \bar q = \begin{cases} \dfrac{m}{m-1}, & 1 < m \le 2, \\ 2, & m \ge 2. \end{cases} \tag{3.7}$$

*Then $u = 0$ is a global attractor for* (2.1).

The proof of Theorem 2 is given in §5. The hypotheses of Theorem 2 hold for the system (1.5) when $A$ has the decomposition (1.7), see the comments at the end of the section.

Theorem 2 has the following consequences.

COROLLARY 4. *Suppose that*

$$\sup_n d_n < \infty.$$

*Then $u = 0$ is a global attractor for* (2.1) *if*

$$\sum_1^\infty |I_n|^{\bar q} = \infty. \tag{3.8}$$

*Proof.* Taking $A = (\sup_n d_n)^\ell$, we see that the series in (3.7) is greater than

$$\sum_1^\infty \min\left\{|I_n|^{\bar q}, |I_n|\right\}.$$

This diverges if and only if $\sum_1^\infty |I_n|^{\bar q}$ diverges (since $\bar q > 1$).

For the canonical case $m = 2$, $\mu = 1$, $\nu = 1$, the exponents $q$ and $\bar q$ in Corollaries 3 and 4 have the respective values 3 and 2, these values being the best possible as shown in §6.

COROLLARY 5. *Suppose there is a positive constant such that*

$$d_n \ge \text{Const.} \begin{cases} |I_n|^{-\mu} & \text{if } m \ge 2, \\ |I_n|^{-\mu/(m-1)} & \text{if } 1 < m \le 2. \end{cases} \tag{3.9}$$

*Then $u = 0$ is a global attractor for* (2.1) *if*

$$\sum_1^\infty |I_n|\, d_n^{-\ell} = \infty. \tag{3.10}$$

*Proof.* Let the constant in (3.9) be denoted by $D$, and choose $A = D^\ell$ in (3.7). Then the second term in braces in (3.7) is less than the first, so that (3.10) implies (3.7).

COROLLARY 6 (Criterion of Thurston–Wong type). *Let* $\inf_n |I_n| > 0$. *Then* $u = 0$ *is a global attractor for* (2.1) *if* (3.10) *is satisfied.*

*Proof.* Let $U > 0$ be fixed as usual. From (2.17) with $d = d(U, \vartheta) > 0$, where $\vartheta = \min\{\frac{1}{2}, \inf_n |I_n|\}$, we have $d_n \ge d > 0$ for all $n$, and in turn (3.9) obviously holds because $\inf_n |I_n| > 0$.

Thurston and Wong discovered the special case of Corollary 6 when $N = 1$, $|I_n| = 1$, $G(p) = p^2/2$, $Q(t,u,p) = -a(t,u,p)\,p$, and $f$ is independent of $t$. Their assumptions imply $\mu = \ell = 1$, in which case (3.10) takes exactly their form $\sum_1^\infty (\int_{I_n} \delta(t)\,dt)^{-1} = \infty$.

Artstein and Infante [1, cond. (2.7)] showed for the same case that $u = 0$ is a global attractor provided

$$\frac{1}{K^2}\sum_1^K d_n \le B$$

for some constant $B$ independent of $K$. In fact, more generally, without any restrictions on the measures of $I_n$, and whatever the value of $\mu$, condition (3.10) is implied by

$$\frac{1}{K^{\mu+1}}\sum_1^K c_n \le B, \quad \text{where } c_n = \frac{1}{|I_n|^{\mu+1}}\int_{I_n} \delta(t)\,dt = \frac{d_n}{|I_n|^\mu}. \tag{3.11}$$

To see this, note that for any positive integers $0 < L < K$ we have, by Hölder's inequality,

$$K - L = \sum_L^K 1 \le \left(\sum_L^K c_n\right)^{1/(\mu+1)} \left(\sum_L^K c_n^{-1/\mu}\right)^{\mu/(\mu+1)},$$

so that in turn

$$\sum_L^K c_n^{-\ell} \ge \left[\frac{1}{(K-L)^{\mu+1}}\sum_L^K c_n\right]^{-1/\mu}. \tag{3.12}$$

But, by (3.11), if $K = 2L$ then

$$\frac{1}{(K-L)^{\mu+1}}\sum_L^K c_n \le 2^{\mu+1}B.$$

Hence from (3.12) it follows that

$$\sum_1^{2^\nu} c_n^{-\ell} \ge \frac{\nu}{2(2B)^\ell}, \qquad \nu = 1, 2, \ldots.$$

Thus the series (3.10) diverges. We have proved the following result.

COROLLARY 7 (Criterion of Artstein–Infante type). *Suppose that* $\inf_n |I_n| > 0$, *or more generally that* (3.9) *holds. Then* $u = 0$ *is a global attractor for* (2.1) *if* (3.11) *is satisfied.*

In essentially the same way, condition (3.4) in Corollary 2 can also be deduced from the Artstein–Infante type condition (3.11), provided $\inf_n |I_n| > 0$.

*Remark*. When $\inf_n |I_n| > 0$ the criteria (3.4) of Corollary 2 and (3.10) of Corollary 6 are equivalent. Since by (2.16) the condition $\inf_n x_n > 0$ holds whenever $\inf_I \sigma(t) > 0$, one can see a connection between the hypotheses of these corollaries. On the other hand, the assumptions of Theorem 2 are different enough from those of Theorem 1 that the corollaries are not directly comparable.

**The system (1.5).** We show that Corollaries 3 and 4 apply to the system (1.5). For Corollary 3 the hypotheses of Theorem A must be verified, on the basis of the assumptions immediately following (1.5).

Fix a compact subset $K$ of $\mathbb{R}^N \times \mathbb{R}^N$, and let $\alpha > 0$ be such that

$$\big(A(t,u,p)p,p\big) \geq \alpha\,|p|^2 \quad \text{for } (t,u,p) \in I \times K;$$

also denote by $\|A\|$, the $L^\infty$ norm of $A$ on $I \times K$. Then one easily sees that (2.5) holds in $I \times K$ with $\gamma = \alpha/\|A\|$, that (2.6) is satisfied with $\delta(t) = U\,\|A\| = \text{Const.}$ and $\mu = 1$, and that (2.7) is verified with $\sigma(t) = \alpha$, $\nu = 1$. Finally, taking into account the observations just before the lemma of §2, together with the fact that (2.15) holds since $H(u,p) = \frac{1}{2}|p|^2$, we see that Theorem A is applicable to (1.5), with $m = 2$.

In turn, since $\nu = 1$, we get $q = (m+\nu)/(m-1) = 3$ in (3.1). Moreover, $\inf_I \sigma(t) = \alpha > 0$ and $\sup_n d_n = U\,\|A\| < \infty$, so (3.5) is satisfied. Corollary 3 then gives the criterion (1.6).

For Corollary 4 the hypotheses of Theorem B must be verified on the basis of the assumptions (1.7)–(1.9). Again fix a compact set $K$ of $\mathbb{R}^N \times \mathbb{R}^N$. Then one easily sees that

$$\big(A(t,u,p)p,p\big) = \beta(t,u,p)\,|p|^2 + \big(B(t,u,p)p,p\big) \geq \beta_1\,|p|^2 \quad \text{on } J \times K,$$

by (1.8) and the fact that $B$ is nonnegative definite. Hence we can take $\hat{\sigma}(t) = 1$ and $\varphi(u,p) = \beta_1\,|p|^2$ in (2.13). Also $\|A\| \leq \beta_2 + \|B\| < \infty$ by (1.9). Thus (2.5) holds in $I \times K$ with $\gamma = \beta_1/\|A\|$. As before we can take $\delta(t) = U\,\|A\|$ and $\mu = 1$.

Next $(V_1)$ is satisfied with $h(t) = \beta_1 p_0^2$. Finally, for $(V_2)$,

$$\big(Q(t,u,p),u\big) = -\beta(t,u,p)\,(p,u) - \big(B(t,u,p)p,u\big) \leq U\,\|B\|\,|p|,$$

when $|u| \leq U$ and $(p,u) \geq 0$. This gives the first part of $(V_2)$ with $\varepsilon(p) = U\,\|B\|\,|p|$. The second part of $(V_2)$ is automatic for (1.5) with $g(u) = \frac{1}{2}u^2$ and $C = 0$. Again taking into account the observations just before the lemma of §2, we see that Theorem B is applicable.

As before $\sup_n d_n = U\,\|A\| < \infty$. Corollary 4 then gives the criterion (1.10), since $m = 2$ and $\bar{q} = 2$ by (3.7).

**4. Proof of Theorem 1.** Recall that $I = \cup_1^\infty I_n$ where $I_n = [a_n,\ b_n]$. We begin with a simple lemma.

LEMMA. *Let $k$ be a nonnegative measurable function such that $k(t) = 0$ for $t \in J \setminus I$ and for which*

$$d_n\, k_n^\mu \leq M_1 \tag{4.1}$$

*and*

$$\int_{I_n} k(t)dt \geq \frac{1}{M_2}\,|I_n|\,k_n, \tag{4.2}$$

*where $k_n = \sup_{I_n} k(t)$ and $M_1$, $M_2$ are positive constants.*

*Then $k$ satisfies condition* (2.12) *with $M = M_1 M_2$.*

*Proof.* We have

$$\begin{aligned}\int_{I_n} \delta(t) k^{\mu+1}(t)dt &\le k_n^{\mu+1} \int_{I_n} \delta(t)dt = k_n |I_n|\, k_n^{\mu} d_n \\ &\le k_n |I_n| M_1 \quad \text{by } (4.1) \\ &\le M_1 M_2 \int_{I_n} k(t)dt \quad \text{by } (4.2).\end{aligned}$$

Condition (2.12) now follows by summation over $n$ and the fact that $R = 0$ on $J/I$.

*Proof of Theorem* 1. Recall that $x_n = \sigma_n\, d_n^{\ell}$, $\ell = 1/\mu$, $q = (\nu + m)/(m-1)$ if $\nu > m-2$ and $q = 2$ if $\nu \le m-2$.

We now construct a bounded piecewise smooth function $k = k(t)$ satisfying the assumptions $(2.9)_2$ and (2.10) of Theorem A. (Of course the functions $\sigma, \delta$ in Theorem A depend on $U$, so also $k$ depends on $U$. As in Section 2 we do not specifically indicate this dependence.) In particular, let $k = 0$ on $J \setminus I$. To obtain $k$ on the intervals $I_n$, we separately consider the two subcases

$$\text{(i)} \qquad |I_n|^{q-1} \le \frac{A}{B + x_n}$$

and

$$\text{(ii)} \qquad |I_n|^{q-1} > \frac{A}{B + x_n}.$$

*Subcase* (i). Let $I_n = [a_n, b_n]$ and put

$$k(t) = \begin{cases} C\,\sigma_n (t - a_n)^{q-1}, & a_n \le t \le \frac{1}{2}(a_n + b_n), \\ C\,\sigma_n (b_n - t)^{q-1}, & \frac{1}{2}(a_n + b_n) \le t \le b_n, \end{cases}$$

where $C = 2^{q-1} B/A$. Then $(2.10)_2$ holds on $I_n$ with the Const. $= 2\,(q-1)\,(B/A)^{\lambda}$, independent of $n$; note that the exponent $\lambda$ is defined in (2.11), and that $\lambda = 1/(q-1)$ by virtue of $(3.1)_2$. Next, letting $k_n = \max_{I_n} k(t)$ as in the lemma, we have

$$(4.3) \qquad k_n = k\left(\frac{a_n + b_n}{2}\right) = C\,\sigma_n \left(\frac{|I_n|}{2}\right)^{q-1} \le 2^{1-q} C\,\sigma_n \frac{A}{B + x_n} = \frac{B\sigma_n}{B + x_n}$$

by (i) and the choice of $C$. In turn, since $x_n = \sigma_n\, d_n^{\ell} \ge c^{\ell}\, \sigma_n^{\ell+1}$ by (2.16), there follows

$$(4.4) \qquad k_n \le \frac{B\sigma_n}{B + c^{\ell} {\sigma_n}^{\ell+1}} \le D,$$

where $D$ is a constant depending only on $B$, $c$, and $\ell$. Hence $k$ is uniformly bounded on intervals $I_n$ of type (i), with the bound independent of $n$. Moreover, again from (4.3), we have $k(t) \le \sigma_n \le \sigma(t)$, so $(2.10)_1$ holds on these intervals with Const. $= 1$.

*Subcase* (ii). Put

$$k(t) = \begin{cases} C\,\sigma_n (t - a_n)^{q-1}, & a_n \le t \le t_n, \\ \dfrac{B\sigma_n}{B + x_n}, & t_n < t < \bar{t}_n, \\ C\,\sigma_n (b_n - t)^{q-1}, & \bar{t}_n \le t \le b_n, \end{cases}$$

where $t_n$ and $\bar{t}_n$ are chosen so that $k$ is continuous on $I_n$. This can be done because of condition (ii).

As before $k$ satisfies $(2.10)_2$ on $I_n$ with the same Const. $= 2(q-1)(B/A)^\lambda$, since $k' = 0$ on $(t_n, \bar{t}_n)$. Moreover, we have

$$k_n = \frac{B\sigma_n}{B + x_n} \le D, \tag{4.5}$$

as in (4.4). Therefore $(2.10)_1$ is satisfied and $k$ is uniformly bounded on intervals $I_n$ of type (ii).

We next show that $k$ satisfies conditions (4.1) and (4.2) of the lemma. Indeed by (4.3) and (4.5), for each $n$,

$$d_n k_n^\mu \le d_n \left(\frac{B\,\sigma_n}{B + x_n}\right)^\mu = B^\mu \left(\frac{x_n}{B + x_n}\right)^\mu \le B^\mu = M_1.$$

Thus (4.1) is verified.

In case (i) an easy calculation and the use of (4.3) gives

$$\int_{I_n} k(t)\,dt = \frac{2}{q}\,C\,\sigma_n \left(\frac{|I_n|}{2}\right)^q = \frac{1}{q}\,|I_n|\,k_n,$$

while in case (ii)

$$\begin{aligned} \int_{I_n} k(t)\,dt &= \frac{(t_n - a_n)}{q}\,k(t_n) + (\bar{t}_n - t_n)\frac{B\,\sigma_n}{B + x_n} + \frac{(b_n - \bar{t}_n)}{q}\,k(\bar{t}_n) \\ &= \left\{\frac{1}{q}(t_n - a_n) + (\bar{t}_n - t_n) + \frac{1}{q}(b_n - \bar{t}_n)\right\}\frac{B\,\sigma_n}{B + x_n} \\ &> \frac{1}{q}\,|I_n|\,k_n, \end{aligned} \tag{4.6}$$

since $q > 1$. Hence (4.2) holds with $M_2 = q$.

The lemma now shows that condition (2.12) of Theorem A is satisfied. It remains only to verify the hypothesis $(2.9)_1$ to finish the proof. We already know that in case (i)

$$\int_{I_n} k(t)\,dt = \frac{2}{q}\,C\,\sigma_n \left(\frac{|I_n|}{2}\right)^q = \frac{1}{q}\frac{B}{A}\,\sigma_n\,|I_n|^q$$

by the choice of $C$, while in case (ii) by (4.5) and (4.6),

$$\int_{I_n} k(t)\,dt > \frac{1}{q}\,\sigma_n\,|I_n|\,\frac{B}{B + x_n}.$$

Consequently for each $n$,

$$\int_{I_n} k(t)\,dt \ge \frac{1}{q}\frac{B}{A}\,\sigma_n\,\min\left\{|I_n|^q,\ \frac{A}{B + x_n}\,|I_n|\right\}.$$

Since

$$\int_J k(t)\,dt = \sum_1^\infty \int_{I_n} k(t)\,dt,$$

condition (3.1) now shows that $k \notin L^1(J)$, as required in $(2.9)_1$.

This completes the proof.

**5. Proof of Theorem 2.** As in the proof of Theorem 1, we shall construct a bounded piecewise smooth function $k = k(t)$ satisfying the assumptions (2.9) and (2.14) of Theorem B. (As before, $k$ depends on $U$ in what follows.) In particular, we define $k$ to be 0 on $J \setminus \cup_1^\infty I_n$ and, to obtain $k$ on the intervals $I_n$, consider separately the two cases:

$$\text{(i)} \qquad |I_n|^{\bar{q}-1} \le \frac{A}{d_n^\ell}$$

and

$$\text{(ii)} \qquad |I_n|^{\bar{q}-1} > \frac{A}{d_n^\ell}.$$

*Case* (i). Put

$$k(t) = \begin{cases} C(t-a_n)^{\bar{q}-1}, & a_n \le t \le \frac{1}{2}(a_n+b_n), \\ C(b_n-t)^{\bar{q}-1}, & \frac{1}{2}(a_n+b_n) \le t \le b_n, \end{cases}$$

where $C = 2^{\bar{q}-1}/A$. Then recalling the definition of $\bar{q}$ in (3.7), we see that (2.14) holds on $I_n$ with Const. $= (\bar{q}-1)C^{m-1}$ when $1 < m < 2$ and Const. $= C$ when $m \ge 2$.

Next

$$k_n = k\left(\tfrac{1}{2}(a_n+b_n)\right) = C\left(\tfrac{1}{2}|I_n|\right)^{\bar{q}-1}. \tag{5.1}$$

By (i) and the choice of $C$ we obtain

$$k_n \le d_n^{-\ell} \quad \text{for each } n. \tag{5.2}$$

Now applying (2.17) in the lemma of §2 with $d = d(U, 1/2)$, we have $d_n \ge d$ whenever $|I_n| \ge 1/2$, so that $k_n \le d^{-\ell}$ for such $I_n$'s. On the other hand, when $|I_n| \le 1/2$ we derive from (5.1) that $k_n \le 2^{1-\bar{q}}/A$. Hence $k$ is bounded on each $I_n$ of type (i), uniformly in $n$. Thus $k$ is bounded on $J$, with

$$k(t) \le \max\{d^{-\ell},\ 2^{1-\bar{q}}/A\}, \qquad t \in J.$$

*Case* (ii). Put

$$k(t) = \begin{cases} C(t-a_n)^{\bar{q}-1}, & a_n \le t \le t_n, \\ d_n^{-\ell}, & t_n < t < \bar{t}_n, \\ C(b_n-t)^{\bar{q}-1}, & \bar{t}_n \le t \le b_n, \end{cases}$$

where $t_n$ and $\bar{t}_n$ are chosen so that $k$ is continuous on $I_n$. This can be done in virtue of (ii). As in case (i), we see that (2.14) is satisfied and that $k$ is bounded on $I_n$ uniformly in $n$, namely

$$k_n = d_n^{-\ell} \le \max\{d^{-\ell},\ 2^{1-\bar{q}}/A\},$$

since by (2.17) we have $d_n \ge d = d(U, 1/2)$ when $|I_n| \ge 1/2$, while by (ii) on the other hand, $d_n^{-\ell} \le 2^{1-\bar{q}}/A$ when $|I_n| \le 1/2$.

We next show that $k$ satisfies conditions (4.1) and (4.2) of the lemma in §4. Indeed for each $n$,

$$d_n k_n^\mu \leq d_n (d_n^{-\ell})^\mu = 1 = M_1,$$

by (5.2) in case (i) and by the definition of $k$ in case (ii). Thus (4.1) is verified.

In case (i) by (5.1)

$$\int_{I_n} k(t)\, dt = 2\frac{C}{\bar{q}}\left(\frac{1}{2}|I_n|\right)^{\bar{q}} = \frac{1}{\bar{q}}\,|I_n|\,k_n,$$

while in case (ii), as in the proof of Theorem 1,

$$\int_{I_n} k(t)\, dt > \frac{1}{\bar{q}}\,|I_n|\,k_n.$$

Hence (4.2) holds with $M_2 = \bar{q}$.

The lemma now shows that condition (2.12) is satisfied, so that to apply Theorem B it remains only to verify $(2.9)_1$. Arguing as in the proof of Theorem 1, we obtain

$$\int_{I_n} k(t)\, dt \geq \frac{1}{\bar{q}A}\min\left\{|I_n|^{\bar{q}},\ \frac{A}{d_n^\ell}|I_n|\right\}.$$

Consequently (3.7) implies that $k \notin L^1(J)$, and this completes the proof.

**6. Examples.** The purpose of this section is to show that the exponents which appear in conditions (1.6) and (1.10) are best possible.

Consider the linear equation

$$u'' + a(t)u' + u = 0, \qquad t \in J = [0,\infty), \tag{6.1}$$

where $a\colon J \to \mathbb{R}$ is an *on–off* damping function of the form

$$a(t) = \begin{cases} 0 & \text{in } J \setminus \bigcup_1^\infty I_n, \\ 2 & \text{in } \bigcup_1^\infty I_n. \end{cases} \tag{6.2}$$

The following result shows that the exponent 3 in (1.6) is best possible.

PROPOSITION 1. *Let $\epsilon \in (0,2]$ be fixed and let $(\lambda_n)_n$ be a sequence of positive real numbers such that*

$$\sum_1^\infty \lambda_n^{3-\epsilon} = \infty, \qquad \sum_1^\infty \lambda_n^3 < \infty. \tag{6.3}$$

*Then there exists a sequence of disjoint intervals $I_n = [a_n,\ a_n + \lambda_n]$, with $a_n \to \infty$, such that $u = 0$ is not a global attractor for (6.1) with the damping (6.2).*

*Remarks.* Since the damping function (6.2) is not continuous the corresponding solutions of (6.1) must be sought in the class $C^1(J)$. Of course, a smoothing procedure will obviously yield a corresponding result for (6.1) with continuous damping.

From the proof it will be clear that the sequence $(I_n)_n$ can be chosen to have arbitrarily large gaps, i.e., with $a_{n+1} - a_n$ unbounded.

*Proof.* We place the interval $I_1$ arbitrarily in $J = [0, \infty)$, and recursively determine the location of the successive intervals $I_n$ for $n \geq 2$. In particular we shall impose the Cauchy conditions

$$u(a_n) = A_n, \qquad u'(a_n) = 0, \qquad A_1 \neq 0 \tag{6.4}$$

in order to construct a bounded solution $u$ of (6.1)–(6.2) that does not approach 0 as $t \to \infty$. The values $A_n$ will be recursively determined, along with the location of the intervals $I_n$.

From (6.4) and (6.2) it is clear that

$$u(t) = A_n(1 + t - a_n)e^{a_n - t} \quad \text{for } t \in I_n. \tag{6.5}$$

On the other hand, in the intervals $(a_n + \lambda_n,\ a_{n+1})$ between the sets $I_n$ and $I_{n+1}$, we have

$$u(t) = \varphi_n(t) = B_n \cos(t + \theta_n) \tag{6.6}$$

for some constants $B_n$ and $\theta_n$, again by (6.2). Clearly (6.6) should join smoothly with (6.5) at the point $a_n + \lambda_n$ and also satisfy the conditions $\varphi_n(a_{n+1}) = A_{n+1}$ and $\varphi_n'(a_{n+1}) = 0$. These latter conditions take the specific form

$$B_n \cos(a_{n+1} + \theta_n) = A_{n+1}, \qquad B_n \sin(a_{n+1} + \theta_n) = 0,$$

so that $B_n^2 = A_{n+1}^2$. The former conditions are

$$B_n \cos(a_n + \lambda_n + \theta_n) = A_n(1 + \lambda_n)e^{-\lambda_n},$$

$$B_n \sin(a_n + \lambda_n + \theta_n) = A_n \lambda_n e^{-\lambda_n}.$$

Squaring and adding gives

$$A_{n+1}^2 = A_n^2\{(1 + \lambda_n)^2 + \lambda_n^2\}e^{-2\lambda_n} \equiv A_n^2 \Phi(\lambda_n). \tag{6.7}$$

This determines $A_{n+1}^2$ in terms of $A_n^2$ and $\lambda_n$. Because $\theta_n$ is not yet chosen, it is clear that $B_n$ and $A_{n+1}$ are so far determined only up to their signs. We can choose the sign of $B_n$ as we wish, say sign $B_n =$ sign $A_{n+1}$. Then $\cos(a_{n+1} + \theta_n) = 1$, so without loss of generality $\theta_n = -a_{n+1}$. In turn

$$\tan(a_{n+1} - a_n - \lambda_n) = -\frac{\lambda_n}{1 + \lambda_n},$$

which determines $a_{n+1}$ modulo $\pi$; indeed, if sign $A_{n+1} =$ sign $A_n$, then one sees that

$$a_{n+1} - a_n - \lambda_n \in \left(\tfrac{7}{4}\pi,\ 2\pi\right) \qquad \text{modulo } 2\pi,$$

while if sign $A_{n+1} = -$sign $A_n$, then

$$a_{n+1} - a_n - \lambda_n \in \left(\tfrac{3}{4}\pi,\ \pi\right) \qquad \text{modulo } 2\pi.$$

Clearly $a_{n+1} - a_n$ can be arbitrarily large, though not arbitrarily small; in fact since $\lambda_n \to 0$, there is a sequence $(k_n)_n$ of positive integers such that $\lim_n (a_{n+1} - a_n - k_n\pi) = 0$.

An easy calculation shows that

$$1 - \tfrac{4}{3}x^3 < \Phi(x) < 1 \quad \text{for } x > 0.$$

Hence $|A_{n+1}| < |A_n|$ by (6.7), so that

$$\limsup_{t\to\infty} |u(t)|^2 = \lim_n A_n^2 = A_1^2 \prod_1^\infty \Phi(\lambda_n).$$

We can assume without loss of generality that $\lambda_n^3 < 3/4$ for all $n$, in virtue of $(6.3)_2$. Therefore,

$$\limsup_{t\to\infty} |u(t)|^2 > A_1^2 \prod_1^\infty (1 - \tfrac{4}{3}\lambda_n^3) > 0,$$

where the last inequality is equivalent to $\sum_1^\infty \lambda_n^3 < \infty$. This completes the demonstration.

The proof sharply brings out the role of the exponent 3 in condition (1.6). Figure 1 shows a typical graph of $u$ with all the $A_n > 0$ and with varying spacing between the $I_n$.

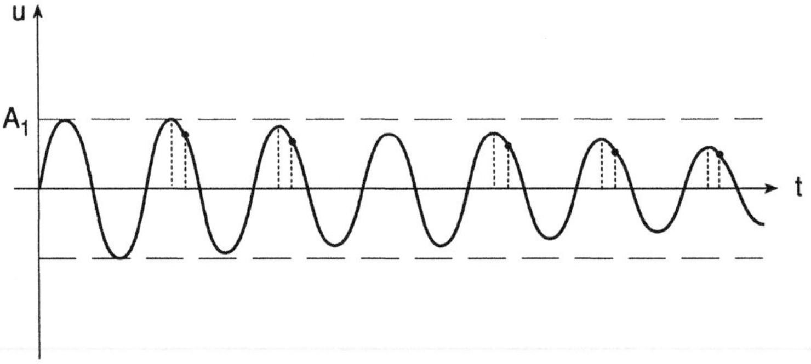


FIG. 1. *The solution u of Proposition* 1.

In Fig. 2 the heavy curve is the solution (6.5). The light curve is one arch of the cosine wave $\varphi_n$ defined by (6.6), whose amplitude is $A_{n+1}$. The dashed curve is one arch of the cosine wave $\varphi_{n-1}$, whose amplitude is $A_n$.

The next result shows that (1.6) is *not* necessary for the global stability of the rest state of (6.1) when

$$\inf_n \sigma_n > 0 \quad \text{and} \quad \sup_n d_n < \infty.$$

It also indicates the extreme delicacy of the situation when one has *on–off* damping, that is, the exact switching times can be of great importance.

PROPOSITION 2. *Under the hypotheses of Proposition* 1 *there also exists a sequence of disjoint intervals* $I_n = [a_n, a_n + \lambda_n]$, *with* $a_{n+1} - a_n > \pi$, *such that* $u = 0$ *is a global attractor for* (6.1) *with the damping* (6.2).

*Proof.* We place $I_1$ arbitrarily. To construct the remaining intervals, choose some sequence $(k_n)_n$ of positive integers and define

$$a_{n+1} = a_n + \lambda_n + k_n\pi.$$

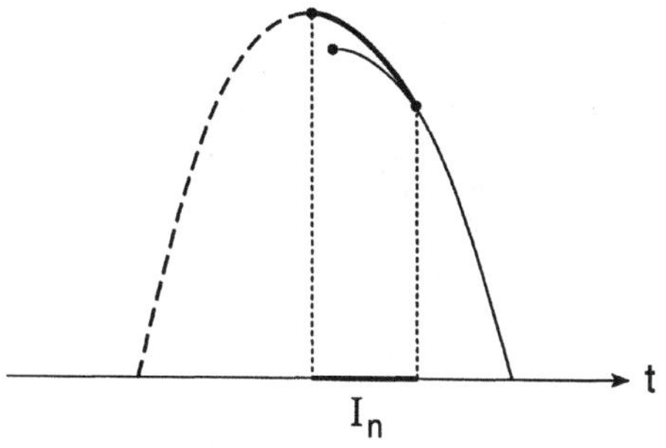


FIG. 2. *Behavior of u near an interval $I_n$.*

Now let $u$ be any solution of (6.1)–(6.2) on $J$. Since (6.6) holds on the intervals $(a_n + \lambda_n, a_{n+1})$, whose lengths are multiples of $\pi$, it is evident that

$$(6.8)\qquad \begin{aligned} u(a_{n+1}) &= (-1)^{k_n}\, u(a_n+\lambda_n),\\ u'(a_{n+1}) &= (-1)^{k_n}\, u'(a_n+\lambda_n). \end{aligned}$$

Now, for $\tau \in [0,\infty)$, we put

$$v(\tau) = (-1)^{j_n}\, u(a_n + \tau - \ell_n) \quad \text{if } \tau \in [\ell_n, \ell_{n+1}), \quad n = 1, 2, \ldots,$$

where $j_n = \sum_1^{n-1} k_i$ and $\ell_n = \sum_1^{n-1} \lambda_i$ when $n \geq 2$, and $j_1 = \ell_1 = 0$. By $(6.3)_1$ it is clear that $\ell_n \to \infty$ as $n \to \infty$, so $v$ is well defined. Also $v$ is of class $C^1$ in view of (6.8).

Directly from the definition of $v$ we see that $v$ is a solution of

$$(6.9)\qquad v'' + 2v' + v = 0 \quad \text{in } [0,\infty),$$

so that $v$ is certainly of class $C^2$. The equation (6.9) has characteristic values $r_1 = r_2 = -1$; hence $v(\tau) \to 0$ and $v'(\tau) \to 0$ as $\tau \to \infty$.

From this we get

$$\sup_{I_n} |u(t)| \to 0, \quad \sup_{I_n} |u'(t)| \to 0 \quad \text{as } n \to \infty;$$

it then follows at once that the coefficients $B_n$ in (6.6) also approach 0 as $n \to \infty$. This completes the proof.

*Remark.* Similar conclusions can be obtained for the equation

$$u'' + a(t)\,|u'|^{\nu-1}u' + u = 0, \qquad \nu > 0,$$

but we shall not pursue this here.

To investigate the exponent 2 in condition (1.10), we consider the equation (6.1) with the singular *on–off* damping

$$(6.10)\qquad a(t) = \begin{cases} \infty & \text{in } J \setminus \bigcup_1^\infty I_n, \\ 2 & \text{in } \bigcup_1^\infty I_n. \end{cases}$$

Of course $a$ is neither continuous nor has its values in the reals, so that equation (6.1) on $J \setminus \cup_1^\infty I_n$ must therefore be interpreted in terms of a family of equations in which the damping uniformly approaches $\infty$ on any compact (or even any bounded) subset of $J \setminus \cup_1^\infty I_n$. The corresponding solutions approach constants on the open intervals between the $I_n$'s. In fact, the appropriate interpretation of a solution of the initial value problem $u(a_n + \lambda_n) = \alpha$, $u'(a_n + \lambda_n) = \beta$ on the interval $J_n = (a_n + \lambda_n,\ a_{n+1}]$ is $u(t) \equiv \alpha$. That is, if $u_M$ is the solution of the initial value problem

$$\begin{cases} u'' + Mu' + u = 0, \\ u_M(a_n + \lambda_n) = \alpha, \qquad u_M'(a_n + \lambda_n) = \beta \end{cases}$$

on the interval $\overline{J_n}$, then $u_M(t) \to \alpha$ as $M \to \infty$ uniformly on $\overline{J_n}$, while $u_M'(t) \to 0$ uniformly on any compact subset of $J_n$. Accordingly, solutions of (6.1), (6.10) are to be interpreted as functions of class

$$C(J) \cap C^1(J \setminus \{a_n + \lambda_n,\ n = 1, 2, \ldots\})$$

which satisfy (6.1) with $a(t) = 2$ on each interval $[a_n,\ a_n + \lambda_n)$ and which are constant on each interval $\overline{J_n}$.

PROPOSITION 3. *Let $(I_n)_n$ be a sequence of disjoint intervals $I_n = [a_n,\ a_n + \lambda_n]$. Then $u = 0$ is a global attractor for* (6.1) *with the damping* (6.10) *if and only if*

$$\sum_1^\infty |I_n|^2 = \infty.$$

*Proof.* If $\sum_1^\infty |I_n|^2 = \infty$, then $a_n \to \infty$ and by Corollary 4 the rest state $u = 0$ is a global attractor.

Now assume that $\sum_1^\infty |I_n|^2 < \infty$ and, without loss of generality, that $\lambda_n^2 < 2$. For simplicity we first consider the case

$$a_n \nearrow \infty \quad \text{as } n \to \infty. \tag{6.11}$$

Then every solution of (6.1), (6.10) on $J$ has the form

$$u(t) = \begin{cases} A_1 & \text{if } t \in [0, a_1], \\ A_n(1 + t - a_n)e^{a_n - t} & \text{if } t \in I_n, \\ A_{n+1} & \text{if } t \in J_n. \end{cases}$$

Furthermore, by continuity at the point $a_n + \lambda_n$ we have the recursive formula

$$A_{n+1} = A_n(1 + \lambda_n)e^{-\lambda_n}.$$

Obviously,

$$1 - \tfrac{1}{2}x^2 < (1 + x)e^{-x} < 1 \quad \text{for } x > 0.$$

It follows that $(|A_n|)_n$ is decreasing and so also $|u(t)|$ is decreasing on $[0, \infty)$. Thus

$$\lim_{t \to \infty} |u(t)| = |A_1| \cdot \prod_1^\infty (1 + \lambda_n)e^{-\lambda_n}.$$

Then if $A_1 \neq 0$

$$\lim_{t\to\infty} |u(t)| > |A_1| \cdot \prod_1^\infty (1 - \tfrac{1}{2}\lambda_n^2) > 0.$$

Consequently every solution, except the trivial one ($A_1 = 0$), approaches a nonzero limit at $\infty$. In particular, $u = 0$ is not a global attractor for (6.1), (6.10).

If (6.11) fails, then $a_n \nearrow$ finite $a$. The previous proof shows that $u(t) \to u_0 \neq 0$ as $t \nearrow a$, so that in turn

$$u(t) \equiv u_0 \quad \text{for } t \geq a;$$

the solution of course need not be smooth at the point $t = a$. This completes the proof.

The condition $\delta \notin L^1(J)$ is known to be a *necessary condition* for the rest state $u = 0$ to be a global attractor for the system (1.5); see [5, Cor. 1, §5]. (The notation used in §5 of [5] gives $\hat{\delta} = 2\delta/U$ and $\hat{\delta} \notin L^1(J)$: the multiplicative constant $2/U$ clearly does not affect the conclusion.) This result holds in particular for the linear equation

$$u'' + A(t)u' + u = 0, \tag{6.12}$$

where of course the necessary condition becomes $A \notin L^1(J)$.

On the other hand, *the condition $\sigma \notin L^1(J)$ alone is not sufficient for the rest state to be a global attractor*, even for equation (6.12), where $\sigma = \delta/U$, and even in the representative case in which $A$ is bounded.

*Proof.* We must show that there exist equations of the form (6.12) with $A \in L^\infty(J)\backslash L^1(J)$ such that zero is not a global attractor. Indeed, consider equation (6.12) with $A = a$, and $a$ given by (6.2). Moreover, suppose the intervals $I_n$ in (6.2) satisfy (6.3). By $(6.3)_2$, eventually $|I_n| \leq 1$, say for all $n \geq K$, so that

$$\sum_K^\infty |I_n| \geq \sum_K^\infty |I_n|^{3-\varepsilon} = \infty$$

by $(6.3)_1$. Hence by (6.2)

$$\int_J A(t)dt = 2\sum_1^\infty |I_n| = \infty,$$

so $A \in L^\infty(J)\backslash L^1(J)$. On the other hand, by Propositon 1, if the intervals $I_n$ are located properly, then $u = 0$ is *not* a global attractor.

A related result was shown in [6, Thm. 4.4], namely, that the condition $1/\sigma \notin L^1(J)$ is a *necessary condition* for the rest state to be a global attractor for (1.5). (In the notation of Theorem 4.4 of [6] the function here called $\sigma$ is there called $\delta$; moreover, in the application of Theorem 4.4 to (1.5) we have $q = k = 2$, $C = 0$, $\psi = 0$ and of course $H = |p|^2/2$ and $Q = -A(t, u, p)p$.)

On the other hand, *the condition $1/\delta \notin L^1(J)$ alone is not sufficient for the rest state to be a global attractor*, even for (6.12) and even when $1/A \in L^\infty(J)\backslash L^1(J)$.

*Proof.* We must show that there exist equations of the form (6.12) with $1/A \in L^\infty(J)\backslash L^1(J)$ such that zero is not a global attractor. Indeed consider (6.12) with

$A = a$, and $a$ given by (6.10). Moreover, suppose the intervals $I_n$ in (6.10) satisfy

$$\sum_{1}^{\infty} |I_n| = \infty, \qquad \sum_{1}^{\infty} |I_n|^2 < \infty.$$

Then

$$\int_J \frac{dt}{A(t)} = \frac{1}{2}\sum_{1}^{\infty} |I_n| = \infty,$$

so $1/A \in L^\infty(J)\backslash L^1(J)$. On the other hand, by Proposition 3 the rest state is *not* a global attractor.

**Acknowledgment.** The first author is a member of *Gruppo Nazionale di Analisi Funzionale e sue Applicazioni* of the *Consiglio Nazionale delle Ricerche.*

REFERENCES

[1] Z. ARTSTEIN AND E. F. INFANTE, *On the asymptotic stability of oscillators with unbounded damping*, Quart. Appl. Math., 35 (1976), pp. 195–199.

[2] L. HATVANI, *Nonlinear oscillation with large damping*, Dynamical Systems Appl., 1994, to appear.

[3] L. HATVANI AND V. TOTIK, *Asymptotic stability of the equilibrium of the damped oscillator*, Differential Integral Equations, 6 (1993), pp. 835–848.

[4] G. LEONI, *A note on a theorem of Pucci and Serrin*, J. Differential Equations, 1994 to appear.

[5] P. PUCCI AND J. SERRIN, *Precise damping conditions for global asymptotic stability for nonlinear second order systems*, Acta Math., 170 (1993), pp. 275–307.

[6] ———, *Continuation and limit behavior for damped quasi-variational systems*, in Degenerate Diffusions, IMA Vols. in Math. Appl. 47, W.-M. Ni, L. A. Peletier and J. L. Vazquez, eds., Springer-Verlag, New York, 1993, pp. 157–173.

[7] ———, *Precise damping conditions for global asymptotic stability for nonlinear second order systems*, II, J. Differential Equations, 1994, to appear.

[8] R. A. SMITH, *Asymptotic stability of $x'' + a(t)x' + x = 0$*, Quart. J. Math. Oxford, 12 (1961), pp. 123–126.

[9] L. H. THURSTON AND J. W. WONG, *On global asymptotic stability of certain second order differential equations with integrable forcing terms*, SIAM J. Appl. Math., 24 (1973), pp. 50–61.

*Arch. Ration. Mech. Anal.* 132 (1995), 207–232. **Reprint II.1.[iii]**


Arch. Rational Mech. Anal. 132 (1995) 207–232. © Springer-Verlag 1995

# *Asymptotic Stability for Ordinary Differential Systems with Time-Dependent Restoring Potentials*

PATRIZIA PUCCI & JAMES SERRIN

## 1. Introduction

Asymptotic stability for the classical second-order ordinary differential equation

$$u'' + q^2(t)u = 0, \quad t \in [T, \infty), \tag{1.1}$$

where $q(t) > 0$, has been widely studied in the literature. Important results are due to SANSONE, WIMAN and BALLIEU & PEIFFER, and a good bibilography is given by CESARI.

In particular, it was shown by BALLIEU & PEIFFER that if there exists a non-increasing continuous function $k$ such that

(i) $$q'(t) \geqq k(t)q^2(t) > 0 \quad \text{for } t \in [T, \infty),$$

(ii) $$\int^{\infty} k(t)q(t)\,dt = \infty,$$

(iii) $$q'/q^2 \text{ is bounded in } [T, \infty),$$

then every solution $u$ of (1.1) tends to zero as $t \to \infty$. SANSONE earlier had obtained the special case $k = \text{Const.}/q^2$, but without requiring condition (iii).

Here, as a consequence of our general conclusions we establish the following theorem, containing both the results of SANSONE [14, page 67] and of BALLIEU & PEIFFER [1, Corollary 11].

**Theorem A.** *Suppose that there exists a non-negative continuous function $k$ of bounded variation on $[T, \infty)$, such that*

(i) $$q'(t) \geqq k(t)q^2(t) > 0 \quad \textit{for } t \in [T, \infty),$$

(ii) $$\int^{\infty} k(t)q(t)\,dt = \infty,$$

(iii) $$\liminf_{t\to\infty} \int^{t} k^2(s)\frac{q'(s)}{q(s)}\,ds \Bigg/ \int^{t} k(s)q(s)\,ds < \infty.$$

*Then every solution of (1.1) tends to zero as $t \to \infty$.*

The cases $q(t) = t^\beta$ and $q(t) = (\log t)^\beta$, $\beta > 0$, are covered by Theorem A as well as by BALLIEU & PEIFFER, if we take respectively $k(t) = \beta t^{-\beta-1}$ and $k(t) = \beta/t(\log t)^{\beta+1}$. Condition (iii) of Theorem A is implied by (iii) of the theorem of BALLIEU & PEIFFER, and even by the weaker pointwise condition, $kq'/q^2$ bounded on $[T, \infty)$. The hypotheses (i) and (ii), on the other hand, show that $q'/q \notin L^1[T, \infty)$. This is in fact a *necessary condition* for $u = 0$ to be an attractor as $t \to \infty$, as follows by applying Theorem 5 of [10] to the transformed system (2.9).

By taking $k(t) = \text{Const.}/tq(t)$ we obtain the following striking, and apparently new, result. (In this case $k$ is clearly continuous and of bounded variation on $[T, \infty)$ since $q$ is certainly non-decreasing.)

**Corollary A1.** *Suppose that*

$$tq'(t) \geqq \text{Const. } q(t) \quad \textit{for } t \in [T, \infty).$$

*Then every solution of* (1.1) *tends to zero as* $t \to \infty$.

A variant of Corollary A1 is the following slightly more general result.

**Corollary A2.** *Suppose that there exists a non-negative continuous function* $\psi$ *of bounded variation on* $[T, \infty)$, *such that*

$$q' \geqq \psi q \textit{ in } [T, \infty), \quad \int^{\infty} \psi(t)\, dt = \infty.$$

*Then every solution of* (1.1) *tends to zero as* $t \to \infty$.

The important advantage of Corollary A2 over the result of BALLIEU & PEIFFER is that no upper bound on the growth rate of $q$ is required. The theorem of SANSONE, moreover, arises as the special case $\psi(t) = \text{Const.}/q(t)$.

In all the above results necessarily $q(t) \nearrow \infty$ as $t \to \infty$, so that under the given assumptions all solutions are oscillatory with frequency approaching $\infty$ and amplitude approaching zero as $t \to \infty$. The case $q(t) \nearrow \infty$ as $t \to \infty$ was treated also by PRODI, who showed under this assumption alone that there exists *at least one solution* of (1.1) which tends to zero as $t \to \infty$. PRODI's result, like Corollary A2, places no upper bound on the growth rate of $q$; the advantage of Corollary A2 is that *every solution* must tend to zero as $t \to \infty$, though of course this is bought at the expense of the stronger condition $q' \geqq \psi q$.

Consider now the equation

$$u'' + h(t)u' + q^2(t) f(u) = 0, \quad t \in [T, \infty), \tag{1.2}$$

where $f, h\colon \mathbb{R} \to \mathbb{R}$ are continuous functions, with $f$ satisfying the restoring condition

$$uf(u) > 0 \quad \text{for } u \neq 0. \tag{1.3}$$

Equation (1.2), which has similar but more delicate behavior than (1.1), has been studied by SALVADORI and, in the linear case, by BALLIEU & PEIFFER and by CORNE. Here we have the following conclusion.

**Theorem B.** *Let $A = h + q'/q$. Suppose that there exists a non-negative continuous function $k$ of bounded variation on $[T, \infty)$, such that*

(i) $$A \geqq kq \quad \text{in } [T, \infty),$$

(ii) $$\int^{\infty} k(t)q(t)\,dt = \infty,$$

(iii) $$\liminf_{t \to \infty} \int^{t} k^2(s)A(s)\,ds \Big/ \int^{t} k(s)q(s)\,ds < \infty.$$

*Then every bounded solution of* (1.2) *tends to zero as $t \to \infty$.*

*If $f^+, f^- \notin L^1(\mathbb{R})$, then the conclusion holds for all solutions of* (1.2).

Theorem A is the special case of Theorem B when $h = 0$ and $f(u) = u$. Theorem B also includes Corollary 8 of [1] by taking $k = 1/F$, in their notation.

An important example of (1.2) is the radial Matukuma equation,

$$u'' + \frac{n-1}{t}u' + \frac{u|u|^{P-1}}{1+t^2} = 0, \quad n > 2, P > 1. \tag{1.4}$$

It is easy to see that the hypotheses of Theorem B are satisfied with $k(t) = \rho$, where $\rho \in (0, n-2)$. Hence *all solutions of* (1.4) *approach zero as $t \to \infty$*. For positive solutions on $[0, \infty)$ of (1.4) this is a well-known result; see for example [6, 7] and the references quoted therein.

If the term $1 + t^2$ in (1.4) is replaced by $(1 + t^2)^\gamma$, then by the same proof, solutions necessarily approach zero when $\gamma \leqq 1$ (take $k(t) = \rho t^{\gamma - 1}$). On the other hand, if $\gamma > 1$, there exist solutions which tend to finite non-zero limits as $t \to \infty$. Thus the Matukuma equation is a borderline case for asymptotic stability of the rest state. In fact a more precise borderline is provided by the function $(1 + t^2)\log(1 + t^2)$, as will be seen in Theorems 7 and 8 of Section 4.

Finally, Theorem B has the following corollaries corresponding to those of Theorem A. (Throughout the paper, the notation Const. denotes a generic positive constant.)

**Corollary B1.** *Let $h \in L^1[T, \infty)$. Suppose that*

$$tA(t) \geqq \text{Const.} \quad \text{for } t \in [T, \infty).$$

*Then every bounded solution of* (1.2) *tends to zero as $t \to \infty$.*

**Corollary B2.** *Let $h \in L^1[T, \infty)$. Assume that there exists a non-negative continuous function $\psi$ of bounded variation on $[T, \infty)$, such that*

$$A \geqq \psi \text{ in } [T, \infty), \quad \int^{\infty} \psi(t)\,dt = \infty.$$

*Then every bounded solution of* (1.2) *tends to zero as $t \to \infty$.*

To prove these results, put $E(t) = \exp \int^t h(s)\,ds$ and observe that $A = (qE)'/qE$, that $qE$ is non-decreasing since $A \geqq 0$, and that $E$ is of bounded variation on $[T, \infty)$ since $h \in L^1[T, \infty)$. One can then proceed as for Corollaries A1 and A2.

In both corollaries, if $q$ is non-decreasing (or more generally if $1/q$ is of bounded variation), then the condition $h \in L^1[T, \infty)$ can be replaced by the weaker hypothesis $\sup\{\int^t h(s)\,ds \colon t \in [T, \infty)\} < \infty$; since it is no longer necessary to have $E$ of bounded variation. This remark applies in particular when $h \leqq 0$, and also, more interestingly, when $h$ takes both positive and negative values, e.g., $h(t) = t^{-1} \sin t$.

When $\int^t h(s)\,ds$ is unbounded above, the situation is somewhat more complicated; see Theorems 5 and 6 below. We give one representative corollary which is applicable in this case.

**Corollary B3.** *Let $q'(t) \geqq 0$ for $t \in [T, \infty)$. Suppose there exists a non-negative continuous function $\psi$ of bounded variation on $[T, \infty)$, such that*

$$A \geqq \psi, \quad \psi h \leqq \text{Const. } q^2 \quad \text{in } [T, \infty),$$

*and $\int^\infty \psi(t)\,dt = \infty$. Then every bounded solution of (1.2) tends to zero as $t \to \infty$.*

We now turn to the main purpose of the paper, to develop a fairly general asymptotic stability theory for ordinary differential systems, containing results of the type discussed above as special cases. We include, as well, nonlinear mechanical systems of the sort introduced by SALVADORI [13] and also systems arising from degenerate variational problems.

For the rest of the paper, therefore, we consider vector unknowns $u \colon J \to \mathbb{R}^N$, where $J$ is a half open interval of the type $[T, \infty)$, and systems of the form

$$(\nabla_p G(u, u'))' - \nabla_u G(u, u') + q^m(t) f(u) = Q(t, u, u'), \quad t \in J. \tag{1.5}$$

The most important assumptions on (1.5) are that $G = G(u, p)$ be strictly convex and homogeneous of degree $m > 1$ in the variable $p \in \mathbb{R}^N$, that $f = \nabla_u F$ be a continuous function of gradient type satisfying $(f(u), u) > 0$ for $u \neq 0$, and that $Q$ be a continuous vector function satisfying the growth conditions $(H_4)$ below.

There are a number of interesting examples which can be represented by the system (1.5). The case studied in [13] is given by $G(u,p) = \frac{1}{2}(M(u)p, p)$ and $Q(t, u, p) = -L(t, u)p$, where $L$ is a continuous, bounded and uniformly positive-definite matrix and $M$ is a symmetric positive-definite matrix of class $C^1$.

When $G(p) = \sum_{i=1}^N |p_i|^m/m$, $m > 1$, the system (1.5) reduces to

$$(|u_i'|^{m-2} u_i')' + q^m(t) f_i(u) = Q_i(t, u, u'), \quad i = 1, \ldots, N, \tag{1.6}$$

which is singular when $m \neq 2$ at points where at least one $u_i' = 0$. If $m = 2$, the system (1.6) takes the familiar form

$$u'' + q^2(t) f(u) = Q(t, u, u'), \tag{1.7}$$

and, when $Q = 0$,

$$u'' + q^2(t) f(u) = 0. \tag{1.8}$$

For this system, in which of course $(f(u), u) > 0$ for $u \neq 0$ and $f$ is of gradient type, $f = \nabla_u F$, we have the following results corresponding to Theorem A and its corollaries.

**Theorem C.** *Assume that the hypotheses of Theorem A, or of Corollary A1 or A2, are satisfied. Then every bounded solution of the system* (1.8) *tends to zero as* $t \to \infty$.

*If additionally* $F(u) \to \infty$ *as* $|u| \to \infty$, *then all solutions of* (1.8) *approach zero as* $t \to \infty$.

A final example worth noting is the Euler-Lagrange system for extremals of the variational integral

$$\int_J g(t)\{G(u,u') - q^m(t)F(u)\}\,dt,$$

where $G, F$ and $q$ are of the type discussed above and $g: J \to \mathbb{R}$ is a positive function of class $C^1$. Extremals for this functional satisfy (1.5) with

$$Q(t,u,p) = -\frac{g'(t)}{g(t)}\,\nabla_p G(u,p). \tag{1.9}$$

The paper is organized as follows. Section 2 contains the formulation of the problem and preliminary results, including a boundedness theorem for solutions of (1.5) which is of independent interest. In Section 3 we give our main theorems. Theorem 1 deals with the general system (1.5), while Theorem 2 treats the variational subcase (1.9). Theorems A–C above are special cases of Theorem 2.

Section 4 includes applications to holonomic mechanical systems with $N$ degrees of freedom and strongly time-dependent restoring potentials, to generalized Matukuma equations, and to the canonical equation (1.2). These applications — *which are a central part of the paper* — are consequences of the principal Theorems 5–8 of Section 4. Theorems 5 and 6 in particular bear the same relation to Theorem 1 as the corollaries above bear to Theorems A and B. Finally Theorems 7 and 8 are concerned with equation (1.2) in the special case $h(t) = (n-1)/t$, corresponding to the radial Matukuma equation (1.4) and to related problems for partial differential equations.

The last part of Section 4 treats the case when the restoring potential $q^m(t)F(u)$ for (1.5) is replaced by the more general form $\Phi(t,u)$; see (4.31).

## 2. Formulation of the problem

We consider the system (1.5) under the following specific hypotheses, which are assumed to hold throughout the paper unless otherwise mentioned.

($H_1$) $G \in C^1(\mathbb{R}^N \times \mathbb{R}^N; \mathbb{R})$; $G(u,\cdot)$ is strictly convex and homogeneous of degree $m > 1$; for every $U > 0$ and $p_0 > 0$ there is a positive constant such that

$$(\nabla_u G(u,p), u) \leqq \text{Const. } G(u,p) \quad \text{for all } |u| \leqq U \text{ and } |p| \geqq p_0.$$

$(H_2)$ There exists $F \in C^1(\mathbb{R}^N; \mathbb{R})$ such that $f = \nabla_u F$ and

$$(f(u), u) > 0 \quad \text{for } u \neq 0. \tag{2.1}$$

$(H_3)$
$$q \in C^1(J; \mathbb{R}^+), \quad q \notin L^1(J).$$

$(H_4)$ For every $U > 0$ there exist two measurable functions $\sigma, \delta$ defined on $J$, such that for all $t \in J$, $|u| \leqq U$ and $p \in \mathbb{R}^N$,

$$(Q(t,u,p), p) \leqq -\sigma(t)|p|^m, \tag{2.2}$$

$$|Q(t,u,p)| \leqq \delta(t)|p|^{m-1}. \tag{2.3}$$

Conditions (2.2) and (2.3) imply that $\delta \geqq \sigma$ and $\delta \geqq 0$. Note also that $\sigma$ is allowed to take positive, negative and zero values.

The convexity and homogeneity of $G(u,\cdot)$ immediately give the following properties. For every $\lambda > 0$ and all $u, p \in \mathbb{R}^N$

$$G(u, \lambda p) = \lambda^m G(u,p), \quad \nabla G(u, \lambda p) = \lambda^{m-1} \nabla G(u,p), \tag{2.4}$$

and for every $U > 0$ there exist positive constants, $c, c_1, c_2$ such that

$$|\nabla G(u,p)| \leqq c|p|^{m-1}, \tag{2.5}$$

$$c_1|p|^m \leqq mG(u,p) = (\nabla G(u,p), p) \leqq c_2|p|^m \tag{2.6}$$

for all $|u| \leqq U$ and $p \in \mathbb{R}^N$. (From here on we write $\nabla$ rather than $\nabla_p$.)

From these properties we also obtain

$$G(u,0) = 0, \quad \nabla G(u,0) = 0, \quad \nabla_u G(u,0) = 0, \tag{2.7}$$

while from (2.1) and (2.3) it follows that $f(0) = 0$, $Q(t,u,0) = 0$. Hence $u = 0$ is a solution of (1.5).

Our proofs below are based on the well-known Liouville transformation, namely, the change of independent variables $t \mapsto y$ given by

$$y = y(t) = \int_T^t q(s)\,ds, \quad t \in J. \tag{2.8}$$

By $(H_3)$ we have $y'(t) = q(t) > 0$ and $y(t) \nearrow \infty$ as $t \to \infty$, so that $J = [T, \infty)$ is diffeomorphically mapped into $\tilde{J} = [0, \infty)$. After a short calculation using the homogeneity of $G(u,\cdot)$, the system (1.5) is transformed into

$$(\nabla G(u,\dot{u}))^{\dot{}} - \nabla_u G(u,\dot{u}) + f(u) = \tilde{Q}(y,u,\dot{u}), \quad y \in \tilde{J}, \tag{2.9}$$

where $\dot{} = d/dy$ and, for $(y,u,v) \in \tilde{J} \times \mathbb{R}^N \times \mathbb{R}^N$,

$$\tilde{Q}(y,u,v) = \frac{1}{q}\left\{\frac{Q(t,u,qv)}{q^{m-1}} - (m-1)\frac{q'}{q}\nabla G(u,v)\right\}. \tag{2.10}$$

The new system (2.9) has exactly the form (1.5), except that the term $q^m(t)f(u)$ is replaced simply by $f(u)$ and $Q(t,u,u')$ by $\tilde{Q}(y,u,\dot{u})$. For this system the hypotheses $(H_1)$–$(H_3)$ still hold. The appropriate conditions on $\tilde{Q}$ replacing (2.2) and (2.3) are given by the following

**Lemma 1.** *For every* $U > 0$

$$(\tilde{Q}(y,u,v),v) \leqq -\frac{S(t)}{q(t)}|v|^m, \quad |\tilde{Q}(y,u,v)| \leqq \frac{D(t)}{q(t)}|v|^{m-1} \tag{2.11}$$

*for all* $y \in \tilde{J}$, $|u| \leqq U$ *and* $v \in \mathbb{R}^N$, *with*

$$S(t) = \begin{cases} \sigma(t) + c_1(m-1)\dfrac{q'(t)}{q(t)} & \text{if } q'(t) \geqq 0, \\ \sigma(t) + c_2(m-1)\dfrac{q'(t)}{q(t)} & \text{if } q'(t) \leqq 0, \end{cases} \tag{2.12}$$

$$D(t) = \delta(t) + c(m-1)\frac{|q'(t)|}{q(t)}, \tag{2.13}$$

*where* $\sigma, \delta$ *are the functions in* (2.2) *and* (2.3), *while* $c, c_1, c_2$ *are the constants in* (2.5) *and* (2.6). *The quantities* $\sigma, \delta, c, c_1$ *and* $c_2$ *may depend on* $U$.

**Proof.** By (2.10), (2.2) and (2.6) we have

$$\begin{aligned}(\tilde{Q}(y,u,v),v) &= \frac{1}{q}\left\{\frac{(Q(t,u,qv),qv)}{q^m} - (m-1)\frac{q'}{q}(\nabla G(u,v),v)\right\} \\ &\leqq \frac{1}{q}\left\{-\frac{\sigma(t)|qv|^m}{q^m} - (m-1)\frac{q'}{q}(\nabla G(u,v),v)\right\} \\ &\leqq \begin{cases} -\dfrac{|v|^m}{q}\left[\sigma(t) + c_1(m-1)\dfrac{q'}{q}\right] & \text{if } q'(t) \geqq 0, \\ -\dfrac{|v|^m}{q}\left[\sigma(t) + c_2(m-1)\dfrac{q'}{q}\right] & \text{if } q'(t) \leqq 0, \end{cases}\end{aligned} \tag{2.14}$$

proving $(2.11)_1$. The inequality $(2.11)_2$ is obtained in the same way, using (2.3) and (2.5).

In the variational case, when

$$Q(t,u,p) = -\frac{g'(t)}{g(t)}\nabla G(u,p), \tag{2.15}$$

the above result can be improved. Setting $h = g'/g$, and

$$A = h + (m-1)\frac{q'}{q},$$

we get

**Lemma 2.** *Let* $Q$ *have the special form* (2.15). *Then for every* $U > 0$,

$$(\tilde{Q}(y,u,v),v) \leqq -\tilde{\sigma}(y)|v|^m, \quad |\tilde{Q}(y,u,v)| \leqq c\frac{|A(t)|}{q(t)}|v|^{m-1} \tag{2.16}$$

*for all* $y \in \tilde{J}$, $|u| \leqq U$ *and* $v \in \mathbb{R}^N$, *where*

$$\tilde{\sigma}(y) = \begin{cases} c_1 \dfrac{A(t)}{q(t)} & \text{if } A(t) \geqq 0, \\ c_2 \dfrac{A(t)}{q(t)} & \text{if } A(t) < 0, \end{cases}$$

*and* $c, c_1, c_2$ *are given by* (2.5) *and* (2.6).

**Proof.** From (2.10) and (2.15) we obtain

$$(\tilde{Q}(y,u,v),v) = \frac{1}{q}\left\{h\frac{(\nabla G(u,qv),qv)}{q^m} + (m-1)\frac{q'}{q}(\nabla G(u,v),v)\right\}$$

$$= -\frac{1}{q}\left\{h + (m-1)\frac{q'}{q}\right\}(\nabla G(u,v),v) \tag{2.17}$$

by $(2.4)_2$. Condition $(2.16)_1$ is now a consequence of (2.6). The estimate $(2.16)_2$ follows similarly using (2.5) and (2.10).

We conclude Section 2 with a useful boundedness theorem of independent interest.

**Proposition.** *Suppose* $S(t) \geqq 0$ *in* $J$, *or* $A(t) \geqq 0$ *for the variational case. Assume also that*

$$F(u) \to \infty \quad as\ |u| \to \infty. \tag{2.18}$$

*Then every solution of* (1.5) *is bounded on* $J$.

**Proof.** Let $u = u(t)$ be a solution of (1.5) in $J$. Then $u(y)$ is a solution of (2.9) in $\tilde{J}$. By Theorem 8 of [12] we have the Hamilton identity

$$\{H(u,\dot{u}) + F(u)\}^{\cdot} = (\tilde{Q}(y,u,\dot{u}),\dot{u}),$$

where $H(u,v)$ is the Legendre transform of $G(u,v)$ in the variable $v$, namely,

$$H(u,v) = (\nabla G(u,v),v) - G(u,v).$$

By $(2.11)_1$, or $(2.16)_1$, and by assumption, we get

$$\{H(u,\dot{u}) + F(u)\}^{\cdot} \leqq 0 \quad \text{in } \tilde{J},$$

so that $H(u,\dot{u}) + F(u)$ is bounded above. Since $H(u,v) \geqq 0$ by $(H_1)$, it follows that $F(u)$ is bounded above in $\tilde{J}$. Hence (2.18) implies that $u(y)$ is bounded in $\tilde{J}$, which completes the proof.

A similar idea appears in [1, Theorem 4]. We have used this proposition in obtaining the final conclusions of Theorem B and C, as well as for Theorem A and the Matukuma equation (for the latter cases $F(u) = \int_0^u f(s)\,ds = u^2/2$ and $|u|^{p+1}/(p+1)$, respectively).

### 3. Main results

This section contains our main results and their proofs. Applications to special cases will be given in the next section.

We say that the forcing vector $Q = Q(t, u, p)$ in (1.5) is *tame* if for every $U > 0$ there exists a constant $\gamma \geqq 1$ such that

$$|Q(t,u,p)| \cdot |p| \leqq \gamma |(Q(t,u,p),p)|$$

for all $t \in J$, $|u| \leqq U$ and $p \in \mathbb{R}^N$. A similar definition applies for the forcing vector $\tilde{Q} = \tilde{Q}(y,u,v)$ in (2.9), (2.10).

Since $|(Q,p)|/|Q| \cdot |p|$ is exactly the cosine of the angle between the vectors $Q = Q(t,u,p)$ and $-p$, the condition that $Q$ be tame can be expressed geometrically by saying that the vectors $Q$ and $-p$ are *bounded* away from orthogonality.

We note that when $N = 1$, both $Q$ and $\tilde{Q}$ are tame, with $\gamma = \tilde{\gamma} = 1$. In Theorems 3 and 4 below we present other natural sufficient conditions guaranteeing that $\tilde{Q}$ is tame.

We can now state our first main result.

**Theorem 1.** *Let $\tilde{Q}$ be tame. Assume that for every $U > 0$ there exists a non-negative continuous function $k = k(t)$ of bounded variation on $J$, such that*

$$S \geqq kq \quad \text{in } J; \tag{3.1}$$

$$kq \notin L^1(J); \tag{3.2}$$

$$\liminf_{t\to\infty} \int^t k^m(s)D(s)\,ds \Big/ \int^t k(s)q(s)\,ds < \infty, \tag{3.3}$$

*where $S$ and $D$ are the functions defined in* (2.12) *and* (2.13). *Then every bounded solution of* (1.5) *tends to zero as $t \to \infty$.*

**Proof.** We apply Theorem 4.1 of [11] to the transformed system (2.9). First, the basic assumptions $(H_1)$–$(H_3)$ of Section 2 of [11] are satisfied, as direct consequences of $(H_1)$, $(H_2)$, $(2.11)_1$ and (3.1).

Next, condition $(C_1)'$ of [11] can be written, for (2.9), in the following form: for every $U > 0$ there exist two non-negative measurable functions $\tilde{\sigma}, \tilde{\delta} : \tilde{J} \to \mathbb{R}$ such that, for all $y \in \tilde{J}$, $|u| \leqq U$ and $v \in \mathbb{R}^N$,

$$(\tilde{Q}(y,u,v),v) \leqq -\tilde{\sigma}(y)|v|^m, \quad |\tilde{Q}(y,u,v)| \leqq \tilde{\delta}(y)|v|^{m-1}.$$

[Here $\mu = \nu = m - 1$.]

By (2.11) this condition is satisfied for (2.9) with

$$\tilde{\sigma}(y) = \frac{S(t)}{q(t)}, \quad \tilde{\delta}(y) = \frac{D(t)}{q(t)}, \tag{3.4}$$

where $y$ and $t$ and related by (2.8).

Condition $(C_2)$ of [11] is exactly the assumption that $\tilde{Q}$ be tame.

The basic conditions $(H_1)$–$(H_3)$, $(C_1)'$ and $(C_2)$ of [11] are therefore satisfied for (2.9). It remains to check the specific hypotheses of Theorem 4.1, in Leoni's version in which (4.4) of [11] is replaced by (3.2) of [5].

For the system (2.9) these hypotheses take the form: *for every $U > 0$ there exists a continuous function $\tilde{k}$ of bounded variation on $\tilde{J}$ such that*

$$\tilde{k} \notin L^1(\tilde{J}); \tag{3.5}$$

$$0 \leqq \tilde{k}(y) \leqq \tilde{\sigma}(y), \quad y \in \tilde{J}; \tag{3.6}$$

$$\liminf_{y \to \infty} \int^y \tilde{k}^m(s)\tilde{\delta}(s)\, ds \bigg/ \int^y \tilde{k}(s)\, ds < \infty. \tag{3.7}$$

Now we make the choice

$$\tilde{k}(y) = k(t),$$

with $y$ and $t$ related as usual by (2.8). Then (3.5) holds in view of (3.2) and the differential relation $dy = q(t)\,dt$. Condition (3.6) follows directly from (3.1) and $(3.4)_1$. Finally (3.7) is satisfied in view of $(3.4)_2$, (3.3) and, again, the relation $dy = q(t)\,dt$.

This being shown, Theorem 4.1 implies that every bounded solution $u$ of (2.9) satisfies

$$u(y) \to 0 \quad \text{and} \quad \dot{u}(y) \to 0 \quad \text{as } y \to \infty.$$

Therefore, returning to the variable $t$, we see that

$$u(t) \to 0 \quad \text{and} \quad u'(t)/q(t) \to 0 \quad \text{as } t \to \infty.$$

This completes the proof.

*Remarks.* 1. In general, one cannot expect that $u'(t) \to 0$ as $t \to \infty$, at least when $q(t) \to \infty$ as $t \to \infty$. A particular example illustrating this fact was obtained by WIMAN, who showed for equation (1.1) in the special case $q(t) = t^\beta$, $\beta > -1$, that both functions

$$t^{\beta/2}u(t) \quad \text{and} \quad t^{-\beta/2}u'(t)$$

tend to finite limits as $t \to \infty$. A more general conclusion of the same sort is given in [10], Theorem 7.

2. Condition (3.1) is actually required only for all sufficiently large $t$, because the conclusion of Theorem 1 refers to asymptotic behavior at $\infty$. A similar remark naturally applies in corresponding places throughout the paper (for instance, for conditions (i) in Theorems A–C of the Introduction).

**Corollary.** *Let $\delta \in L^1(J)$ and assume $\tilde{Q}$ is tame. Suppose that for every $U > 0$ there exists a non-negative continuous function $\psi$ of bounded variation on $J$, such that*

$$S \geqq \psi \text{ in } J, \quad \psi \notin L^1(J).$$

*Then every bounded solution of* (1.5) *tends to zero as $t \to \infty$.*

**Proof.** Put $E(t) = \exp(n \int^t \delta(s)\,ds)$, where $n = (2c + c_2)/cc_2(m-1)$, and choose $k = \psi/q$. Then

$$\frac{(qE)'}{qE} = n\delta + \frac{q'}{q} = \begin{cases} \dfrac{2\delta}{c_2(m-1)} + \dfrac{1}{c(m-1)}\left[\delta + c(m-1)\dfrac{|q'|}{q}\right] & \text{if } q' \geqq 0 \\ \dfrac{2}{c_2(m-1)}\left[\delta + c_2(m-1)\dfrac{q'}{q}\right] + \dfrac{1}{c(m-1)}\left[\delta + c(m-1)\dfrac{|q'|}{q}\right] & \text{if } q' < 0 \end{cases}$$

$$\geqq \frac{D}{c(m-1)},$$

since $\delta \geqq \max\{0, \sigma\}$ and $S \geqq \psi \geqq 0$. Consequently in (3.3) we have

$$\int^t k^m D\,ds \leqq c(m-1) \int^t k^m \frac{(qE)'}{qE}\,ds = c(m-1) \int^t (\psi E)^m \frac{(qE)'}{(qE)^{m+1}}\,ds.$$

Here $\psi E$ is bounded on $J$ since $\delta \in L^1(J)$ and $\psi$ is of bounded variation on $J$. Therefore

$$\int^t k^m D\,ds \leqq \text{const.}\left[-\frac{1}{(qE)^m}\right]^t \leqq \text{Const.},$$

because $(qE)' \geqq 0$ on $J$. Hence (3.1)–(3.3) are satisfied.

Finally, $k = \psi/q = \psi(qE)^{-1}E$, so that $k$, being the product of three $BV$ functions, is itself of bounded variation. Consequently Theorem 1 applies and the proof is complete.

Our second main result concerns the variational case of (1.5), when $Q$ is of the form (2.15), that is,

$$Q(t, u, p) = -h(t)\nabla G(u, p), \quad h \in C(J). \tag{3.8}$$

Here we define

$$A(t) = h(t) + (m-1)\frac{q'(t)}{q(t)}, \quad t \in J. \tag{3.9}$$

**Theorem 2.** *Suppose that there exists a non-negative continuous function $k$ of bounded variation on $J$, such that*

$$A \geqq kq \quad \text{in } J; \tag{3.10}$$

$$kq \notin L^1(J), \tag{3.11}$$

$$\liminf_{t\to\infty} \int^t k^m(s)A(s)\,ds \Big/ \int^t k(s)q(s)\,ds < \infty. \tag{3.12}$$

*Then every bounded solution of* (1.5), (3.8) *tends to zero as $t \to \infty$.*

**Proof.** This is the same as for Theorem 1, with the exception that (2.11) is replaced by (2.16), and the tameness of $\tilde{Q}$ must be verified directly.

Noting that $A(t) \geqq 0$ in $J$ by (3.10), the relations (3.4) become for the present case

$$\tilde{\sigma}(y) = c_1 \frac{A(t)}{q(t)}, \quad \tilde{\delta}(y) = c \frac{A(t)}{q(t)}, \tag{3.13}$$

where $y$ and $t$ are related by (2.8), and $c, c_1$ are the constants in (2.5), (2.6). Therefore, if we choose $\tilde{k}(y) = k(t)$ and use (3.13), the conditions (3.5)–(3.7) hold in view of (3.10)–(3.12).

It remains to see that $\tilde{Q}$ is tame. Fix $U > 0$. By (2.17), $(2.6)_1$ and (2.5) we have

$$|(\tilde{Q}(y,u,v),v)| = \frac{A(t)}{q(t)} (\nabla G(u,v),v) \geqq \frac{c_1}{c} \frac{A(t)}{q(t)} |\nabla G(u,v)| \cdot |v| = \frac{c_1}{c} |\tilde{Q}(y,u,v)| \cdot |v|,$$

by (2.10), (3.8) and (3.9). Hence $\tilde{Q}$ is tame, with $\tilde{\gamma} = c/c_1$. This completes the proof of the theorem.

Theorem B of the Introduction is the special case of Theorem 2 when $N = 1$, $m = 2$ and $G(u,p) = |p|^2/2$. Theorem C is the special case when $G(u,p) = |p|^2/2$ and $h(t) = 0$ in (3.8).

## *Appendix. Tameness Conditions*

We have observed that $\tilde{Q}$ is tame when $N = 1$ and also in the variational case. The following results apply when these latter conditions are not satisfied.

**Theorem 3.** *Suppose that $q'(t) \geqq 0$ in $J$ and $(Q(t,u,p),p) \leqq 0$ in $J \times \mathbb{R}^N \times \mathbb{R}^N$. Then if $Q$ is tame, also $\tilde{Q}$ is tame.*

**Proof.** Fix $U > 0$. By (2.10) and the hypotheses of the theorem we have

$$\begin{aligned} |(\tilde{Q}(y,u,v),v)| &= \frac{1}{q}\left\{\frac{|Q(t,u,qv),qv)|}{q^m} + (m-1)\frac{q'}{q}(\nabla G(u,v),v)\right\} \\ &\geqq \frac{1}{q}\left\{\frac{1}{\gamma}\frac{|Q(t,u,qv)|\cdot|qv|}{q^m} + \frac{c_1}{c}(m-1)\frac{q'}{q}|\nabla G(u,v)|\cdot|v|\right\} \\ &= \frac{|v|}{q}\left\{\frac{1}{\gamma}\frac{|Q(t,u,qv)|}{q^{m-1}} + \frac{c_1}{c}(m-1)\frac{q'}{q}|\nabla G(u,v)|\right\} \end{aligned}$$

by (2.5) and (2.6). Now putting $\tilde{\gamma} = \max\{\gamma, c/c_1\}$ we obtain

$$\begin{aligned} |(\tilde{Q}(y,u,v),v)| &\geqq \frac{1}{\tilde{\gamma}}\frac{|v|}{q}\left\{\frac{|Q(t,u,qv)|}{q^{m-1}} + (m-1)\frac{q'}{q}|\nabla G(u,v)|\right\} \\ &\geqq \frac{1}{\tilde{\gamma}}|\tilde{Q}(y,u,v)|\cdot|v| \end{aligned} \tag{3.14}$$

by (2.10). Thus $\tilde{Q}$ is tame.

The next result will not be used in the later parts of the paper, but is presented for its general interest.

**Theorem 4.** *Suppose for every $U > 0$ that there exists numbers*

$$\theta_1 < c_1(m-1), \quad \theta_2 > c_2(m-1) \tag{3.15}$$

*such that the function*

$$\tilde{S}(t) = \begin{cases} \sigma(t) + \theta_1 \dfrac{q'(t)}{q(t)} & \text{if } q'(t) \geqq 0, \\ \sigma(t) + \theta_2 \dfrac{q'(t)}{q(t)} & \text{if } q'(t) \leqq 0, \end{cases} \tag{3.16}$$

*is non-negative in $J$; here $\sigma$ is the function given in (2.2). Then $\tilde{Q}$ is tame if $Q$ is tame.*

**Proof.** Fix $U > 0$ and $y \in \tilde{J}$, $|u| \leqq U$, $v \in \mathbb{R}^N$. Let $t \in J$ correspond to $y$ through (2.8), and define

$$p = qv, \quad \text{where } q = q(t).$$

We distinguish four cases.

1. $q'(t) \geqq 0, (Q(t,u,p),p) \leqq 0$. By the proof of Theorem 3 (see (3.14)) we get

$$|\tilde{Q}(y,u,v)| \cdot |v| \leqq \gamma_1 |(\tilde{Q}(y,u,v),v)|,$$

where $\gamma_1 = \max\{\gamma, c/c_1\}$.

2. $q'(t) \geqq 0, (Q(t,u,p),p) > 0$. Here by (2.10), (2.2), (2.6) and the fact that $\tilde{S}(t) \geqq 0$ we have

$$\begin{aligned} (\tilde{Q}(y,u,v),v) &= \frac{1}{q}\left\{\frac{(Q(t,u,p),p)}{q^m} - (m-1)\frac{q'}{q}(\nabla G(u,v),v)\right\} \\ &\leqq \frac{1}{q}\left\{-\sigma(t)|v|^m - c_1(m-1)\frac{q'}{q}|v|^m\right\} \\ &\leqq \frac{q'}{q^2}\{\theta_1 - c_1(m-1)\}|v|^m < 0, \end{aligned} \tag{3.17}$$

by $(3.15)_1$. Hence

$$|(\tilde{Q}(y,u,v))| \geqq \frac{q'}{q^2}\{c_1(m-1) - \theta_1\}|v|^m.$$

On the other hand, because $Q$ is tame, we obtain from (2.5), (2.2) and the condition $(Q(t,u,p),p) > 0$ that

$$\begin{aligned} |\tilde{Q}(y,u,v)| \cdot |v| &\leqq \frac{1}{q}\left\{\frac{|Q(t,u,p)|}{q^m} + (m-1)\frac{q'}{q}|\nabla G(u,v)|\right\}|v| \\ &\leqq \frac{1}{q}\left\{\gamma\frac{(Q(t,u,p),p)}{q^m} + c(m-1)\frac{q'}{q}|v|^m\right\} \\ &\leqq \frac{|v|^m}{q}\left\{-\gamma\sigma(t) + c(m-1)\frac{q'}{q}\right\} \\ &\leqq \frac{q'}{q}\{\gamma\theta_1 + c(m-1)\}|v|^m, \end{aligned} \tag{3.18}$$

since $\tilde{S}(t) \geqq 0$. Choose

$$\gamma_2 = \frac{\gamma\theta_1 + c(m-1)}{c_1(m-1) - \theta_1}, \tag{3.19}$$

so that $\gamma_2 > c/c_1$. Then by (3.17) and (3.18) we obtain

$$|\tilde{Q}(y,u,v)| \cdot |v| \leqq \gamma_2 |(\tilde{Q}(y,u,v),v)|. \tag{3.20}$$

3. $q'(t) < 0$, $(Q(t,u,p),p) < 0$. By (2.10), (2.2), (2.6), and since $\tilde{S}(t) \geqq 0$, we have

$$\begin{aligned}(\tilde{Q}(y,u,v),v) &\leqq \frac{1}{q}\left\{-\sigma(t)|v|^m + (m-1)\frac{|q'|}{q}c_2|v|^m\right\}\\ &\leqq \frac{|q'|}{q^2}\{-\theta_2 + c_2(m-1)\} < 0\end{aligned}$$

by $(3.15)_2$. Hence

$$|(\tilde{Q}(y,u,v),v)| \geqq -\frac{1}{q}\left\{\frac{(Q(t,u,p),p)}{q^m} + c_2(m-1)\frac{|q'|}{q}|v|^m\right\}. \tag{3.21}$$

On the other hand, since $Q$ is tame, we find from (2.5) and the condition $(Q(t,u,p),p) < 0$ that

$$\begin{aligned}|\tilde{Q}(y,u,v)| \cdot |v| &\leqq \frac{|v|}{q}\left\{\frac{|Q(t,u,p)|}{q^m} + (m-1)\frac{|q'|}{q} \cdot |\nabla G(u,v)|\right\}\\ &\leqq \frac{1}{q}\left\{-\frac{\gamma}{q^m}(Q(t,u,p),p) + c(m-1)\frac{|q'|}{q}|v|^m\right\}.\end{aligned} \tag{3.22}$$

Choose now

$$\gamma_3 = \frac{\theta_2\gamma + c(m-1)}{\theta_2 - c_2(m-1)},$$

so that $\gamma_3 > \gamma$ and

$$(\gamma_3 - \gamma)\theta_2 = (c_2\gamma_3 + c)(m-1). \tag{3.23}$$

We assert that

$$|\tilde{Q}(y,u,v)| \cdot |v| \leqq \gamma_3 |(\tilde{Q}(y,u,v),v)|. \tag{3.24}$$

Indeed, by (3.21) and (3.22) it is enough to show that

$$-\gamma\frac{(Q(t,u,p),p)}{q^m} + c(m-1)\frac{|q'|}{q}|v|^m \leqq -\gamma_3\left\{\frac{(Q(t,u,p),p)}{q^m} + c_2(m-1)\frac{|q'|}{q}|v|^m\right\},$$

namely,

$$-(\gamma_3 - \gamma)\frac{(Q(t,u,p),p)}{q^m} \geqq (\gamma_3 c_2 + c)(m-1)\frac{|q'|}{q}. \tag{3.25}$$

However, by (2.2) and the fact that $\tilde{S}(t) \geqq 0$ we have

$$-(Q(t,u,p),p) \geqq \sigma(t) q^m |v|^m \geqq \theta_2 q^m |v|^m \frac{|q'|}{q};$$

thus (3.25) follows from (3.23). This completes the proof of (3.24).

4. $q'(t) < 0$, $(Q(t,u,p),p) \geqq 0$. This case cannot occur. Indeed, since $\tilde{S}(t) \geqq 0$ by assumption, there results $\sigma(t) \geqq \theta_2 |q'(t)|/q(t) > 0$ and consequently $(Q(t,u,p),p) < 0$ by (2.2).

The assertion of the theorem is now proved by choosing $\tilde{\gamma} = \max\{\gamma_2, \gamma_3\}$.

The condition $\tilde{S}(t) \geqq 0$ in $J$ should be compared with the related hypothesis of Theorem 1:

$$S(t) \geqq k(t) q(t) > 0 \quad \text{in } J.$$

In fact the following result holds.

**Corollary.** *Suppose that*

$$S(t) \geqq \text{Const.} \frac{|q'(t)|}{q(t)} \quad \textit{for all } t \in J. \tag{3.26}$$

*Then $\tilde{Q}$ is tame if $Q$ is tame.*

**Proof.** By (3.16) and (2.12) we have

$$\begin{aligned}\tilde{S}(t) &= S(t) - \frac{|q'(t)|}{q(t)} \begin{cases} c_1(m-1) - \theta_1 & \text{if } q'(t) \geqq 0 \\ \theta_2 - c_2(m-1) & \text{if } q'(t) < 0 \end{cases} \\ &\geqq S(t) \begin{cases} 1 - \kappa[c_1(m-1) - \theta_1] & \text{if } q'(t) \geqq 0 \\ 1 - \kappa[\theta_2 - c_2(m-1)] & \text{if } q'(t) < 0, \end{cases}\end{aligned}$$

where $1/\kappa$ is the constant in (3.26). Now it is enough to choose

$$\theta_1 \in (c_1(m-1) - 1/\kappa, c_1(m-1)), \quad \theta_2 \in (c_2(m-1), c_2(m-1) + 1/\kappa).$$

Then (3.15) is satisfied and $\tilde{S}(t) > 0$ in $J$. Hence Theorem 4 can be applied and the proof is complete.

## 4. Applications and extensions

### *4.1. Holonomic mechanical systems*

A well-known example studied by Salvadori [13] is the holonomic bilaterally constrained mechanical system

$$(\nabla G(u,u'))' - \nabla_u G(u,u') + q^2(t) f(u) = -L(t,u)u', \quad t \in J, \tag{4.1}$$

where

$$G(u,p) = \tfrac{1}{2}(M(u)p,p).$$

Here $L$ is a continuous, bounded and uniformly positive-definite $N \times N$ matrix, $M$ is a symmetric positive definite $N \times N$ matrix of class $C^1$, and $q$ is a positive non-decreasing function of class $C^1$; see [13, Sections 3.2 and 4.5]. The function $f$ is assumed to satisfy hypothesis $(H_2)$ in Section 2.

Under these conditions SALVADORI showed that, if also

$$\frac{q'}{q^2} \text{ is bounded in } J, \tag{4.2}$$

*then any bounded solution of* (4.1) *approaches zero as* $t \to \infty$.

[SALVADORI actually considered only uniform asymptotic stability, rather than asymptotic stability of bounded solutions. For simplicity we have restated his theorem in the context of our earlier results.]

Here we present a generalization of SALVADORI's theorem, in which condition (4.2) is *not* required.

**Criterion of Salvadori-type.** *Let $L, M, f$ and $q$ satisfy the conditions stated above, except* (4.2). *Then every bounded solution of* (4.1) *tends to zero as* $t \to \infty$.

This criterion is a consequence of the following theorem for the system (1.5) under the natural hypotheses $(H_1)$–$(H_4)$; in particular the condition $q'(t) \geqq 0$ in $J$ is not assumed.

**Theorem 5.** *Let $\tilde{Q}$ be tame. Assume that for every $U > 0$ there exists a non-negative $C^1$ function $\psi$ such that*

$$\psi \notin L^1(J), \quad \frac{\psi}{q} \in BV(J), \tag{4.3}$$

$$S \geqq \psi \quad \text{in } J, \tag{4.4}$$

$$\liminf_{t\to\infty} \int^t \left[\frac{\psi(s)}{q(s)}\right]^m \delta(s)\,ds \Big/ \int^t \psi(s)\,ds < \infty, \tag{4.5}$$

*and either*

$$q' \leqq \text{Const.}\,\frac{q^{m+1}}{\psi^{m-1}} \quad \text{or} \quad \psi' \leqq \text{Const.}\,\frac{q^m}{\psi^{m-2}} \quad \text{in } J, \tag{4.6}$$

*where $S$ is defined by* (2.12), *and $\sigma, \delta$ are the functions given in* $(H_4)$. *Then every bounded solution of* (1.5) *tends to zero as* $t \to \infty$.

**Proof.** Take $k = \psi/q$ in Theorem 1, so that $k \in CBV(J)$. Then (3.1) follows from (4.4), and (3.2) from $(4.3)_1$.

As shown in the proof of the corollary of Theorem 1 and using the same notation,

$$D \leqq c(m-1)\left[n\delta + \frac{q'(t)}{q(t)}\right].$$

Consequently

$$\int^t k^m(s)D(s)\,ds \leqq c(m-1)\left\{n\int^t \left[\frac{\psi(s)}{q(s)}\right]^m \delta(s)\,ds + \int^t \left[\frac{\psi(s)}{q(s)}\right]^m \frac{q'(s)}{q(s)}\,ds\right\}.$$

If $(4.6)_1$ holds, then clearly

$$\int^t k^m(s)D(s)\,ds \leqq \text{Const.}\left\{\int^t \left[\frac{\psi(s)}{q(s)}\right]^m \delta(s)\,ds + \int^t \psi(s)\,ds\right\}$$

and (3.3) therefore follows from (4.5). If $(4.6)_2$ holds, then by integration by parts

$$\int^t \psi^m(s)\frac{q'(s)}{q^{m+1}(s)}\,ds = -\frac{1}{m}\left[\frac{\psi(s)}{q(s)}\right]^m\Bigg|^t + \int^t \left[\frac{\psi(s)}{q(s)}\right]^m \frac{\psi'(s)}{\psi(s)}\,ds$$

$$\leqq \text{Const.}\left(1 + \int^t \psi(s)\,ds\right).$$

Combining the two previous inequalities now yields

$$\int^t k^m(s)D(s)\,ds \leqq \text{Const.}\left(\int^t \left[\frac{\psi(s)}{q(s)}\right]^m \delta(s)\,ds + 1 + \int^t \psi(s)\,ds\right),$$

so that (3.3) is again satisfied, by (4.5) and $(4.3)_1$.

Thus the conditions of Theorem 1 hold, and all bounded solutions of (1.5) tend to zero as $t \to \infty$. This completes the proof.

**Theorem 6.** *If $1/q$ is of bounded variation on $J$, then Theorem 5 remains valid when $(4.3)_2$ and (4.6) are replaced by $\psi \in BV(J)$.*

**Proof.** Take $k = \psi/q$ as in the proof of Theorem 5. Now we have

$$\int^t \left[\frac{\psi(s)}{q(s)}\right]^m \frac{q'(s)}{q(s)}\,ds \leqq \text{Const.}\int^t \frac{|q'(s)|}{q^2(s)}\,ds,$$

since $\psi$ and $1/q$ are bounded. The last integral, however, is convergent since $1/q$ is of bounded variation on $J$. It now follows, as in the proof of Theorem 5, that

$$\int^t k^m(s)D(s)\,ds \leqq \text{Const.}\left(\int^t \left[\frac{\psi(s)}{q(s)}\right]^m \delta(s)\,ds + 1\right)$$

and (3.3) then follows from (4.5) and $(4.3)_1$. Finally, (3.1) and (3.2) are just (4.4) and $(4.3)_1$, while $k$ is of bounded variation since both $\psi$ and $1/q$ are. This completes the proof.

Corollary A2 is the special case of Theorem 6 when $\sigma = \delta = 0$, since then (4.4) implies $q' \geqq 0$ in $J$, so that $1/q$ is of bounded variation.

**Corollary.** *Let the hypotheses of Theorem 5 or Theorem 6 be satisfied, except that (4.5) is replaced by*

$$\delta \leqq \text{Const.}\,\frac{q^m}{\psi^{m-1}} \quad \text{on } J. \tag{4.7}$$

*Then every bounded solution of* (1.5) *tends to zero as* $t \to \infty$.

**Proof.** The pointwise condition (4.7) obviously implies (4.5). Consequently Theorem 5 or Theorem 6 can be applied.

We now prove the Salvadori criterion stated above. First, since $Q(t,u,p) = -L(t,u)p$ and $L$ is bounded and uniformly positive definite, we get

$$|(Q(t,u,p),p)| = (L(t,u)p,p) \geqq b|p|^2 \geqq \frac{b}{l}|Q(t,u,p)|\cdot|p|,$$

where $b = \text{Const.} > 0$ and $l = \sup\{\|L(t,u)\| : t \in J,\ u \in \mathbb{R}^N\}$. Thus $Q$ is tame with $\gamma = l/b$.

We can now apply Theorem 3, since $q'(t) \geqq 0$ by assumption and $(Q(t,u,p),p) = -(L(t,u)p,p) \leqq 0$. Hence $\tilde{Q}$ is tame in (4.1).

In view of the regularity and positive definiteness of $M$, it is clear that condition $(H_1)$ holds for (4.1) with $m = 2$, while conditions $(H_2)$ and $(H_3)$ have already been directly assumed. Finally $(H_4)$ is satisfied with $\sigma(t) = \text{Const.} = b$ and $\delta(t) = \text{Const.} = l$.

Take $\psi(t) = 1/t$ in the last corollary. Then $(4.3)_{1,2}$ are satisfied since $\psi(t)/q(t) = 1/tq(t)$ is decreasing ($q'(t) \geqq 0$ in $J$). Similarly $S \geqq \sigma$, so (4.4) and $(4.6)_2$ hold ($\psi' < 0$). Finally, condition (4.7) is obvious since $q(t) \geqq \text{Const.} > 0$. Consequently the previous corollary applies and the proof is complete.

When $N = 1$, or more generally whenever $Q$ is tame in (4.1) — as, for example, when the matrix $L$ is a multiple of the identity — it is possible to weaken significantly the assumptions on $L$ in the Salvadori criterion. In particular it is enough to require merely that

$$(L(t,u)p,p) \geqq \frac{b}{t}, \quad \|L(t,u)\| \leqq lt^{m-1}q^m(t)$$

for $t \in J$, $|u| \leqq U$ and $|p| = 1$, where $b$ and $l$ are positive constants. Indeed in this case we can take $\sigma(t) = b/t$ and $\delta(t) = lt^{m-1}q^m(t)$ in $(H_4)$, so that the corollary above again applies with $\psi(t) = 1/t$.

### *4.2. Radial Matukuma-type equations*

We consider here a generalized version of (1.4),

$$u'' + \frac{n-1}{t}u' + q^2(t)f(u) = 0, \quad t > 0, \tag{4.8}$$

where $N = 1$, $n > 1$, and $f, q$ are assumed to satisfy conditions $(H_2)$, $(H_3)$ with $J = [T, \infty)$, $T > 0$.

**Theorem 7.** *Suppose that there exists a non-negative $C^1$ function $\psi$ such that*

$$\psi \notin L^1(J), \quad t\psi \in L^\infty(J), \quad \frac{\psi}{q} \in BV(J), \tag{4.9}$$

$$\psi'(t) \leqq \text{Const. } q^2(t), \quad tq'(t) + \theta q(t) \geqq 0 \quad \textit{for } \theta \in (0, n-1), \tag{4.10}$$

$$\liminf_{t\to\infty} \int^t \left[\frac{\psi(s)}{q(s)}\right]^2 \frac{ds}{s} \Big/ \int^t \psi(s)\,ds < \infty. \tag{4.11}$$

*Then every (ultimately) bounded solution of* (4.8) *tends to zero as $t \to \infty$.*

**Proof.** We apply Theorem 5 with $N = 1$, $G(p) = |p|^2/2$, $m = 2$, and $\sigma(t) = \delta(t) = (n-1)/t$. Then (2.6) holds with $c_1 = c_2 = 1$, and (2.12) becomes

$$S(t) = \frac{n-1}{t} + \frac{q'(t)}{q(t)}, \quad t \in J.$$

Consequently (4.4) follows from $(4.10)_2$ and $(4.9)_2$. Also $(4.6)_2$ is satisfied in view of $(4.10)_1$. Finally $(4.3)_{1,2}$ hold and (4.5) is a consequence of (4.11). This completes the proof.

*Remarks.* The same result applies to the $m$-Laplace equation

$$(|u'|^{m-2}u')' + \frac{n-1}{t}|u'|^{m-2}u' + q^m(t)f(u) = 0, \quad m > 1, \tag{4.12}$$

provided $(4.10)_1$ is replaced by $\psi' \leqq \text{Const. } q^m/\psi^{m-2}$, the $\theta$ condition in $(4.10)_2$ is replaced by $\theta \in (0, (n-1)/(m-1))$, and the exponent 2 in (4.11) is replaced by $m$.

The condition $\theta \in (0, n-1)$ in $(4.10)_2$ cannot be replaced by $\theta \in (0, n-1]$. Indeed, if $n = 2$, $q(t) = 1/t$, then $(4.10)_2$ holds with $\theta = 1 = n - 1$, but the transformed equation (2.9) takes the form $\ddot{u} + f(u) = 0$, and has periodic solutions.

*Examples.* 1. $\psi(t) = \text{Const.}/t$. Then $(4.9)_{1,2}$ and $(4.10)_1$ hold automatically, while condition (4.11) becomes

$$\liminf_{t\to\infty} \frac{1}{\log t} \int^t \left[\frac{\psi(s)}{q(s)}\right]^2 \frac{ds}{s} < \infty.$$

Since $\psi/q$ is assumed to be of bounded variation on $J$, see $(4.9)_3$, it follows that $\psi/q$ is bounded; thus (4.11) is also automatically satisfied.

The specific Matukuma function $q(t) = 1/\sqrt{1+t^2}$ is covered by this example, with $\theta = 1$ in $(4.10)_2$.

2. The case $\psi(t) = \text{Const.}/t^\alpha$, $\alpha > 1$, fails since $\psi \in L^1(J)$.

3. $\psi(t) = \text{Const.}/t \log t$. Again $(4.9)_{1,2}$ and $(4.10)_1$ hold, while (4.11) becomes

$$\liminf_{t\to\infty} \frac{1}{\log\log t} \int^t \left[\frac{\psi(s)}{q(s)}\right]^2 \frac{ds}{s} < \infty.$$

To investigate this condition, we make the principal assumption

$$tq(t) \geqq \frac{\text{Const.}}{\log^\alpha t}, \quad \alpha \in (0, 1] \tag{4.13}$$

(when $\alpha > 1$, condition $(4.9)_3$ fails). Then

$$\int^t \left[\frac{\psi(s)}{q(s)}\right]^2 \frac{ds}{s} \leqq \text{Const.} \int^t (\log s)^{2(\alpha-1)}\, ds = \begin{cases} -(\log t)^{2\alpha-1} + \text{Const.} & \text{if } \alpha < \frac{1}{2}, \\ \log\log t + \text{Const.} & \text{if } \alpha = \frac{1}{2}, \\ (\log t)^{2\alpha-1} + \text{Const.} & \text{if } \alpha > \frac{1}{2}. \end{cases}$$

Hence (4.11) is satisfied if and only if $\alpha \leqq \frac{1}{2}$.

4. $\psi(t) = \text{Const.}\, q(t)/(\log t)^{1-\alpha}$, with $q$ obeying (4.13). Here $(4.9)_1$ and $(4.9)_3$ are satisfied, while $(4.9)_2$, $(4.10)_1$ respectively become

$$tq(t) \leqq \text{Const.}\, (\log t)^{1-\alpha}, \qquad q'(t) \leqq \text{Const.}\, q^2(t)(\log t)^{1-\alpha}. \tag{4.14}$$

Finally from (4.13)

$$\int^t \left[\frac{\psi(s)}{q(s)}\right]^2 \frac{ds}{s} \Big/ \int^t \psi(s)\, ds \leqq \frac{\text{Const.}}{\log\log t} \int^t \frac{ds}{s(\log s)^{2(1-\alpha)}}.$$

Therefore condition (4.11) again holds if and only if $\alpha \leqq \frac{1}{2}$.

The function $q^2(t) = 1/(1+t^2)\log(1+t^2)$ noted in the introduction is covered by both cases 3 and 4, with $\theta \in (1, n-1)$, $n > 2$, in $(4.10)_2$.

Comparing examples 3 and 4, we see that under the assumption (4.13) *both* $(4.10)_2$ *and* $\alpha \leqq \frac{1}{2}$ *are needed.* On the other hand, in example 3 one also requires $(4.9)_3$, that is

$$\frac{1}{tq(t)\log t} \in BV(J), \tag{4.15}$$

while example 4 needs additionally (4.14) instead of (4.15). The conditions (4.14) and (4.15) are independent, so cases 3 and 4 provide different sufficient conditions for asymptotic stability.

### *4.3. Non-stability for Matukuma type equations*

The case $\alpha = \frac{1}{2}$ of (4.13), namely, the condition

$$q(t) \geqq \frac{\text{Const.}}{t\sqrt{\log t}} \quad \text{in } J,$$

effectively defines the borderline case for the validity of Theorem 7, as shown by the following

**Theorem 8.** *Suppose that* $n > 2$ *and that*

$$\int^{\infty} sq^2(s)\,ds < \infty. \tag{4.16}$$

*Then there exists a (bounded) solution* $u$ *of* (4.8) *on* $J = [T, \infty) \subset (0, \infty)$ *such that*

$$u(t) \to \text{limit} > 0 \quad \text{as } t \to \infty. \tag{4.17}$$

*The set of attainable limits in* (4.17) *is dense in* $\mathbb{R}$.

**Proof.** Consider equation (4.8) subject to initial conditions

$$u(T_1) = a, \quad u'(T_1) = 0, \tag{4.18}$$

where $a \in \mathbb{R}$ and where $T_1 \geqq T > 0$ will be chosen later. By standard theory this problem has a local solution, and by Theorem 3 of [10] it can be continued to all of $J$, as required. The hypotheses of [10] are satisfied with $g(t) = t^{n-1}$ and $F(t, u)$ replaced by $q^2(t)F(u)$; moreover, the four conditions (i)–(ii)′ on page 106 of [10] clearly hold since $F(u) \geqq 0$ and $q'/q \in L^1_{\text{loc}}(J)$.

We now show that this solution has a limit as $t \to \infty$. First, integration of (4.8) gives

$$u'(t) = -\frac{1}{t^{n-1}} \int_{T_1}^{t} s^{n-1} q^2(s) f(u(s))\,ds, \quad t \in [T_1, \infty), \tag{4.19}$$

and in turn

$$u(t) = a - \int_{T_1}^{t} \frac{1}{r^{n-1}} \int_{T_1}^{r} s^{n-1} q^2(s) f(u(s))\, ds\, dr$$

$$= a - \int_{T_1}^{t} s^{n-1} q^2(s) f(u(s)) \int_{s}^{t} \frac{dr}{r^{n-1}}\, ds \tag{4.20}$$

$$= a - \frac{1}{n-2} \int_{T_1}^{t} s^{n-1} q^2(s) f(u(s)) [s^{2-n} - t^{2-n}]\, ds$$

by Fubini's theorem.

Assume $a > 0$ without loss of generality, and fix $\varepsilon \in (0, a/2)$. Denote by $B > 0$ the maximum value of $f(u)$ for $u \in [a - \varepsilon, a]$. Now choose $T_1 = T_1(\varepsilon) \geqq T$, so that

$$\frac{B}{n-2} \int_{T_1}^{\infty} s q^2(s)\, ds \leqq \varepsilon. \tag{4.21}$$

This can be done by virtue of (4.16).

We assert that

$$u(t) \geqq a - \varepsilon \quad \text{for } t \geqq T_1. \tag{4.22}$$

To see this, let $[T_1, T_2)$, $T_2 \leqq \infty$, be the maximal half-open interval for which (4.22) holds. Clearly $T_2 > T_1$ by (4.18) and (4.19). If $T_2 < \infty$, then by (4.20) and (4.21)

$$u(T_2) > a - \frac{B}{n-2} \int_{T_1}^{T_2} s q^2(s)\, ds \geqq a - \varepsilon.$$

This contradicts the maximality of $T_2$. Hence $T_2 = \infty$ and (4.22) is proved.

From (4.22), (4.19) and the fact that $a - \varepsilon > 0$, we get $u'(t) < 0$ for $t > T_1$. Therefore $u(t) \searrow \text{limit} \in [a - \varepsilon, a)$ as $t \to \infty$. Since $\varepsilon > 0$ and $a > 0$ are arbitrary, we have thus obtained the existence of a family of solutions on $J$ of (4.8) having a dense set of limits as $t \to \infty$. This completes the proof.

**Corollary.** *Under the hypotheses of Theorem 8 the rest state $u = 0$ is not an attractor for bounded solutions of* (4.8).

*Remarks.* 1. It is evident that conditions (4.9)–(4.11) are incompatible with (4.16), since the conclusions of Theorems 6 and 7 cannot hold simultaneously. The incompatibility of $(4.9)_1$, (4.11) and (4.16) also follows directly from the inequality

$$\left[ \int^{t} \left[ \frac{\psi(s)}{q(s)} \right]^2 \frac{ds}{s} \Big/ \int^{t} \psi(s)\, ds \right] \cdot \left[ \int^{t} s q^2(s)\, ds \right] \geqq \int^{t} \psi(s)\, ds, \quad t \in J. \tag{4.23}$$

[To prove (4.23), we have

$$\int^t \psi(s)\,ds \leqq \left[\int^t \left[\frac{\psi(s)}{q(s)}\right]^2 \frac{ds}{s}\right]^{1/2} \cdot \left[\int^t sq^2(s)\,ds\right]^{1/2}$$

by Hölder's inequality.]

2. If the function $q$ is assumed to be locally Hölder continuous on $[0, \infty)$, as in the case, for example, for the Matukuma equation, then one can obtain a family of positive regular solutions of (4.8), which are defined in the entire interval $[0, \infty)$ and tend to non-zero limits as $t \to \infty$, provided that (4.16) holds and the function $f = f(u)$ is monotone in $[0, \infty)$; see [8] and [15]. The set of limits $l$ which are attainable in this way include all sufficiently small $l > 0$ when $f$ is non-decreasing, or all sufficiently large $l$ when $f$ is non-increasing.

### *4.4. A canonical example*

Consider the equation

$$u'' + h(t)u' + q^2(t) f(u) = 0, \quad t \in J, \tag{4.24}$$

where $J = [T, \infty)$, $T > 0$, and where

$$th(t) \geqq bt^\alpha, \quad b = \text{constant} > 0, \quad \alpha = \text{constant} \geqq 0, \tag{4.25}$$

$$tq(t) \quad \text{is non-decreasing in } J. \tag{4.26}$$

For the linear case of (4.24), see [1, page 326], and [4] with $th(t) = bt^\alpha$, $tq(t) = \sqrt{c t^\beta}$, where $b, c$ are positive constants.

Here we apply the Corollary of Theorems 5 and 6, with $\psi(t) = \text{Const.}/t$. Then (4.3) holds by (4.26). Next by (4.25) we can take $\sigma(t) = \text{Const. } t^{\alpha-1}$ in (2.2), so that

$$S(t) = \sigma(t) + \frac{q'(t)}{q(t)} = bt^{\alpha-1} + \frac{[tq(t)]'}{tq(t)} - \frac{1}{t} \geqq bt^{\alpha-1} - \frac{1}{t}$$

by (2.12), (4.26), and the fact that $m = 2$, $c_1 = c_2 = 1$. Thus (4.4) is satisfied provided $\alpha > 0$. Moreover $(4.6)_2$ is obviously valid.

It remains to check condition (4.7), namely, since we can take $\delta = h$ in (2.3),

$$h(t) \leqq \text{Const. } tq^2(t) \quad \text{in } J. \tag{4.27}$$

Combining (4.25) and (4.27) then gives the condition

$$bt^\alpha \leqq th(t) \leqq \text{Const. } [tq(t)]^2. \tag{4.28}$$

It follows now that *if* (4.26), (4.28) *hold and* $\alpha > 0$, *then the Corollary of Theorems 5 and 6 applies and every bounded solution of* (4.24) *tends to zero as* $t \to \infty$.

Note that (4.28) implies in particular that

$$tq(t) \geqq \text{Const. } t^{\alpha/2}. \tag{4.29}$$

For the case treated in [4], namely, $q(t) = \sqrt{c t^{\beta-1}}$, this gives specifically

$$\alpha \leqq 2\beta, \tag{4.30}$$

a condition already found for asymptotic stability in [4].

If (4.26) is strengthened to

$$q \quad \text{is non-decreasing in } J,$$

then it is clear that

$$S(t) \geqq \text{Const. } t^{\alpha-1},$$

and so (4.4) then holds also when $\alpha = 0$.

The case $q(t) = \sqrt{c t^{\beta-1}}$, $\beta > 0$, is of particular interest. Here (4.29) reduces exactly to (4.30), and (4.26) certainly holds. Hence *every bounded solution of* (4.24) *tends to zero as* $t \to \infty$, *provided* $0 < \alpha \leqq 2\beta$ *and* $bt^{\alpha} \leqq th(t) \leqq \text{Const. } t^{2\beta}$.

Corollary 8 of [1] applies to this case *whenever additionally* $\alpha \geqq \beta$, that is, in view of (4.30), whenever $\beta \leqq \alpha \leqq 2\beta$, $\alpha > 0$, and (4.28) holds. On the other hand, when $\alpha < \beta$, Corollary 8 of [1] is not generally applicable, and indeed fails whenever $\liminf_{t \to \infty} t^{1-\alpha} h(t) < \infty$.

A second special case of interest occurs when the damping function $h$ in (4.24) is bounded, and bounded from zero. In this case we can take $\alpha = 1$ in (4.25), so that (4.26) and (4.28) reduce to the simple conditions

$$tq(t) \quad \text{non-decreasing in } J, \quad tq^2(t) \geqq \text{Const.}$$

### *4.5. More general systems*

Extensions to the system

$$(\nabla G(u, u'))' - \nabla_u G(u, u') + \nabla_u \Phi(t, u) = Q(t, u, u'), \quad t \in J, \tag{4.31}$$

are also possible, provided the function $\Phi(t, u)$ has asymptotic behavior corresponding to $q^m(t) F(u)$, as in the case of the system (1.5).

Under the change of variables (2.8) the system (4.31) becomes

$$(\nabla G(u, \dot{u}))^{\cdot} - \nabla_u G(u, \dot{u}) + \frac{\nabla_u \Phi(t, u)}{q^m(t)} = \tilde{Q}(y, u, \dot{u}), \quad y \in \tilde{J}, \tag{4.32}$$

where $\tilde{Q}$ is given by (2.10). In order to apply Theorem 4.1 of [11] to the system (4.32), as in the proof of Theorem 1 above, we must verify, for the function

$$\tilde{F}(y, u) = \frac{\Phi(t, u)}{q^m(t)}, \quad t = t(y),$$

the hypothesis $(H_2)$ of [11]. It is easy to see that this requires exactly the following condition:

$(\tilde{H}_2)$ $\Phi \in C^1(J \times \mathbb{R}^N; \mathbb{R})$ and $\Phi(t,0) = 0$. For all $u_0$, $U$ with $0 < u_0 \leqq U$ there exist a constant $\kappa > 0$ and a non-negative function $\varphi \in L^1(J)$ such that

$$(\nabla_u \Phi(t,u), u) \geqq \kappa q^m(t) \quad \text{whenever } t \in J \text{ and } |u| \in [u_0, U],$$

$$\Phi_t(t,u) - m \frac{q'(t)}{q(t)} \Phi(t,u) \leqq q^m(t) \varphi(t) \quad \text{whenever } t \in J \text{ and } |u| \leqq U.$$

The assumption $(\tilde{H}_2)$ being made, all the previous conclusions carry over without change to the system (4.31).

*Note added in proof.* The present work can be refined by taking into account a forthcoming paper of the authors, *Stability for Abstract Evolution Equations*, to appear in *Festschrift in honor of Carlo Pucci*, 1995.

To illustrate the type of improvements which can be given, we note the following sharper version of Corollary A1:

*Suppose that $q'(t) > 0$ for all t sufficiently large, and that*

$$\liminf_{t \to \infty} \frac{1}{t^2} \int^t \frac{q(s)}{q'(s)} ds < \infty.$$

*Then every solution of* (1.1) *tends to zero as $t \to \infty$.*

A full discussion will appear in a later paper.

*Acknowledgement.* P. PUCCI is a member of *Gruppo Nazionale di Analisi Funzionale e sue Applicazioni* of the *Consiglio Nazionale delle Ricerche*. This research has been partly supported by the Italian *Ministero della Università e della Ricerca Scientifica e Tecnologica*.

## References

1. BALLIEU, R. J. & PEIFFER, K., Attractivity of the origin for the equation $\ddot{x} + f(t,x,\dot{x})|\dot{x}|^\alpha \dot{x} + g(x) = 0$, *J. Math. Anal. Appl.* **65** (1978), 321–332.
2. CESARI, L., *Asymptotic Behavior and Stability Problems in Ordinary Differential Equations*, 3rd ed., Springer-Verlag, Berlin, 1970.
3. CORNE, J. L., On the asymptotic stability, *Ann. Soc. Sci. Bruxelles Sér. I* **87** (1973), 217–235.
4. HARRIS, W. A., PUCCI, P. & SERRIN, J., Asymptotic behavior of solutions of a nonstandard second order differential equation, *Diff. and Int. Eqs.* **6** (1993), 1201–1215.
5. LEONI, G., A note on a theorem of Pucci and Serrin, *J. Diff. Equations* **113** (1994), 535–542.
6. LI, Y., On the positive solutions of the Matukuma equation, *Duke Math. J.* **70** (1993), 575–589.

7. LI, Y. & NI, W.-M., On conformal scalar curvature equations in $\mathbb{R}^n$, *Duke Math. J.* **57** (1988), 895–924.
8. NAITO, M., A note on bounded positive entire solutions of semilinear elliptic equations, *Hiroshima Math. J.* **14** (1984), 211–214.
9. PRODI, G., Un'osservazione sugl'integrali dell'equazione $\ddot{y} + A(x)y = 0$ nel caso $A(x) \to +\infty$ per $x \to \infty$, *Rend. Accad. Naz. Lincei* **8** (1950), 462–464.
10. PUCCI, P. & SERRIN, J., Continuation and limit properties for solutions of strongly nonlinear second order differential equations, *Asymptotic Anal.* **4** (1991), 97–160.
11. PUCCI, P. & SERRIN, J., Precise damping conditions for global asymptotic stability for nonlinear second order systems, *Acta Math.* **170** (1993), 275–307.
12. PUCCI, P. & SERRIN, J., On the derivation of Hamilton's equations, *Arch. Rational Mech. Anal.* **125** (1994), 297–310.
13. SALVADORI, L., Famiglie ad un parametro di funzioni di Liapunov nello studio della stabilità, *Ist. Naz. di Alta Matematica, Symposia Matematica* **4** (1971), 307–330.
14. SANSONE, G., *Equazioni differenziali nel campo reale,* Vol II, 2nd edition, Zanichelli Editore, Bologna, Italy, 1949.
15. USAMI, H., On bounded positive entire solutions of semilinear elliptic equations, *Funkc. Ekv.* **29** (1986), 189–195.
16. WIMAN, A., Über eine Stabilitätsfrage in der Theorie der linearen Differentialgleichungen, *Acta Math.* **66** (1936), 121–145.

Dipartimento di Matematica
Università di Perugia
Via Vanvitelli, 1
I-06123 Perugia

and

Department of Mathematics
University of Minnesota
Minneapolis, Minnesota

*(Accepted November 7, 1994)*

# II.2. Ordinary Differential Equations – Asymptotic Behavior

Henghui Zou

**Mathematics Subject Classification (2010).** 34D05, 34C11, 34A34, 49J45.

**Keywords.** Asymptotic properties, nonlinear systems.

The following two articles are concerned with the question of continuation, boundedness and asymptotic behavior for solutions of the ordinary differential system

$$(A(t, u, u')P(t, u, u'))' + Q(t, u, u') = 0, \tag{1}$$

where $A$, $P$ and $Q$ are suitable real functions, and the solution $u : J \to \mathbb{R}^N$ is vector valued ($J = (a, b)$ is an open interval of the real line).

In the comprehensive 64-page paper [i] below, Pucci and Serrin consider the scalar case $N = 1$ of (1). It is assumed that (1) takes the form

$$(A(u')u')' + \delta(t)A(u')u' + f(t, u) = 0. \tag{2}$$

Observing the variational structure of (2), the authors establish a key variational identity for all solutions. Utilizing this identity, they obtain a number of continuation and boundedness properties, summarized in Theorems 1–4. When $J = \mathbb{R}^+$, they also prove that all bounded solutions necessarily decay at infinity, that is,

$$u(t) = o(1), \qquad u'(t) = o(1) \tag{3}$$

as $t \to \infty$, see Theorem 5.

The general structure of (2) enables it to cover a wide class of equations of interest. A canonical prototype is the radial version of the quasi-linear partial differential equation

$$\operatorname{div}\{A(|\nabla u|)\nabla u\} + f(r, u) = 0 \quad \text{in} \quad \mathbb{R}^n, \tag{4}$$

where $r = |x|$ is the radius from some given origin. Writing $u(x) = u(r)$ with some abuse of notation, (4) becomes (2) with $t = r$, $J = \mathbb{R}^+$ and $\delta(r) = (n-1)/r$. (Note that $\beta = \beta' = n - 1$). Needless to say, the topic of asymptotic behavior at infinity has been a central issue in studying (4). For instance, the Bessel equation and the Lane–Emden–Fowler equation are classical examples with a linear or semi-linear structure, and the topic of asymptotic behavior of their solutions has been extensively studied, see for example [1, 2, 3, 7] and the references therein.

In particular, as a classical quasi-linear example of (4), Pucci and Serrin systematically and extensively study the asymptotic behavior at infinity of radial solutions of the

P. Pucci et al. (eds.), *James Serrin. Selected Papers: Volume 2*, Contemporary Mathematicians,
DOI 10.1007/978-3-0348-0847-7_2, © Springer Basel 2014

$m$-Laplace equation ($m > 1$), that is, equation (2) with

$$A(t) = t^{m-2}, \quad t > 0.$$

First, they strengthen the conclusion (3) by showing that solutions of (2) have the following algebraic decay,

$$u(r) = O(r^{-\lambda}), \qquad u(r) = O(r^{-\mu}), \tag{5}$$

where $\lambda$, $\mu$ are positive constants, see Theorem 6. Moreover, under appropriate additional conditions on the asymptotic behavior of the functions $\delta(r)$ and $F(r, u)$, where

$$F(r,u) = \int_0^u f(r,t)dt,$$

they find explicit values for $\lambda$, $\mu$, see Theorems 7, 8 and the remarks at the end of Section 7. Even more, they further establish that the following limits exist and are finite,

$$\lim_{r\to\infty} r^{\lambda}u(r) = \text{Limit}, \qquad \lim_{r\to\infty} r^{\mu}u'(r) = \text{Limit}, \tag{6}$$

see Theorems 9–11.

Related to the conclusions above, the paper [i] also obtains sufficient conditions for classifying monotone and oscillatory solutions, respectively. Furthermore, for oscillatory solutions they prove that there exists $L > 0$ such that there necessarily holds

$$\lim_{j\to\infty} (r_{j+1}^{\tau} - r_{j+1}^{\tau}) = 2\tau L, \qquad \lim_{j\to\infty} (\rho_j^{\tau} - r_{j+1}^{\tau}) = \tau L, \tag{7}$$

where $r_j$ are the successive zeros and $\rho_j$ the interlaced critical points of $u$, see Theorems 12–15.

The above results for the $m$-Laplace equation generalize earlier results of Fowler [1, 2], Levin & Nohel [4] and others. In some cases the results have been extended to other equations (e.g., the mean-curvature equation), see the detailed results of Sections 12 and 13.

A later paper [5] generalizes [i] to quasi-variational systems rather than single equations. To discuss the results of this paper would go beyond the limits of this commentary, though the generality of the results makes it equally comprehensive as the paper [i] itself.

Paper [ii] deals with equation (1) with $J = \mathbb{R}^N$, where it is assumed that the matrix function $A$ has an inverse $A^{-1}$ and the vector-valued function $P$ has an inverse $\Phi$. In the special case when

$$N = 1, \qquad A(t,u,u') = a(t), \qquad P(t,u,u') = u', \qquad Q(t,u,u') = q(t)u,$$

Wintner [6] had already shown in 1948 that every solution of (1) has a finite limit at infinity and, in fact, the set of attainable limits as $t \to \infty$ is $\mathbf{R}$.

Let $\Psi$ be the function defined either by

$$\Psi[u,w](t) = -A^{-1}(t,u,\Phi(t,u,w)) \int_T^t Q(r,u,\Phi(r,u,w))dr,$$

where $T > 0$ is an appropriate constant, or by

$$\Psi[u,w](t) = A^{-1}(t,u,\Phi(t,u,w)) \int_t^{\infty} Q(r,u,\Phi(r,u,w))dr.$$

Writing (1) in the following integral forms,

$$\begin{aligned}(F_1(u,w))(t) &= u_0 - \int_t^\infty \Phi(s,u,\Psi[u,w](s))ds \\ (F_2(u,w))(t) &= \Psi[u,w](t),\end{aligned} \tag{8}$$

Serrin and Zou [ii] are able to show that (8) is solvable for every $u_0 \in \mathbb{R}^N$, under suitable conditions on the function $\Psi$. In turn, for every $u_0 \in \mathbb{R}^N$, (1) has a solution $u$ such that

$$\lim_{t\to\infty} u(t) = u_0.$$

Finally, it is proved that every bounded solution of (1) has a finite limit at infinity, see Theorems A, B and C. This remarkable result extends the earlier results of Wintner to general quasi-linear systems.

In a later article [8], the commentator generalizes the ideas of [ii] to derive a sufficient condition on the damping coefficient matrix in order to obtain the results of [ii] for over-damped systems. (See [8], Theorem A.) This enables him also to show instability of the rest point for such systems, see [8], Theorem 2.

## References

[1] Fowler, R.H., *The form near infinity of real continuous solutions of a certain differential equation of second order*, Quart. J. Math, Cambridge Series, **45** (1914), pp. 289–350.

[2] Fowler, R.H., *Further studies of Emden's and similar differential equations*, Quart. J. Math, Oxford Series, **2** (1931), pp. 259–298.

[3] Lebedev, N.N., *Special fuctions & their applications*, Dove, 1972.

[4] Levin, J.J. and J.A. Nohel, *Global asymptotic stability for non-linear systems of differential equations and applications to reactor dynamics*, Arch. Ration. Mech. Anal., **5** (1960), pp. 194–211.

[5] Pucci, P. and J. Serrin, *Continuation and limit behavior for damped quasi-variational systems*, IMA Volumes in Mathematics and its Applications, **47** (1993), pp. 137–153.

[6] Winter, A., *Asymptotic integration of the adiabetic oscillator in its hyperbolic range*, Duke Math. J., **15** (1948), pp. 55–67.

[7] Yanagida, E. and S. Yotsutani, *Classification of the structure of positive radial solutions to* $\Delta u + K(|x|)u^p = 0$ *in* $\mathbb{R}^n$, Arch. Ration. Mech. Anal., **124** (1993), pp. 239–259.

[8] Zou, H., *Asymptotic integration of over damped second order systems and applications to instability*, Nonlinear Analysis, **23** (1994), pp. 215–224.

Henghui Zou
Department of Mathematics
University of Alabama at Birmingham
Birmingham, AL 35294, USA
e-mail: `zou@uab.edu`

*Asymptotic Anal.* 4 (1991), 97–160. **Reprint II.2.[i]**


Asymptotic Analysis 4 (1991) 97–160 97
North-Holland

# Continuation and limit properties for solutions of strongly nonlinear second order differential equations

Patrizia Pucci
*Dipartimento di Matematica 'G. Vitali', Università di Modena, Italy*

James Serrin
*Department of Mathematics, University of Minnesota, Minneapolis, MN, USA*

Received 4 April 1990

*Abstract*

Pucci, P. and J. Serrin, Continuation and limit properties for solutions of strongly nonlinear second order differential equations, Asymptotic Analysis 4 (1991) 97–160.

This paper deals with continuation, boundedness and asymptotic behavior of solutions of second order ordinary differential equations of variational type. The equations considered are nonlinear both in the solution variable and in its derivative. For equations arising from the degenerate Laplace operator, precise asymptotic behavior is established, along with various monotonicity and oscillation properties. Similar results are obtained also for other classes of nonlinear second order differential operators. The results of the paper also include generalizations of theorems of Fowler, Levin and Nohel, and Artstein and Infante. The main tool of the paper is a variational identity which supplies a useful new family of Liapunov functions for nonlinear equations.

## 1. Introduction

The purpose of this paper is to study continuation, boundedness, and asymptotic behavior properties for solutions of nonlinear ordinary differential equations of second order, in which the nonlinearity occurs both in the solution variable and its derivative. The typical equation we consider has the form

$$(A(u')u')' + \delta(r)A(u')u' + f(r, u) = 0, \tag{1.1}$$

where the solution $u = u(r)$ is defined on some interval $I$ of the real line, the function $A = A(p)$ determines the nonlinearity of the equation in the derivative $u'$, and $f = f(r, u)$ is a continuous function containing the nonlinearity in $u$. Equation (1.1) can be also written in the equivalent form

$$(g(r)A(u')u')' + g(r)f(r, u) = 0, \tag{1.1$'$}$$

where $g(r) = \exp\int_{r_0}^{r}\delta(s)\,ds$.

0921-7134/91/$03.50 © 1991 – Elsevier Science Publishers B.V. (North-Holland)

An immediate and elementary example of (1.1) is the Bessel equation

$$u'' + \frac{1}{r}u' + \left(1 - \frac{\mu^2}{r^2}\right)u = 0, \tag{1.2}$$

in which $A(p) \equiv 1$ and $f(r, u)$ is linear in $u$. More generally, (1.1) includes the Lane–Emden equation of astrophysics

$$u'' + \frac{2}{r}u' + u^{q-1} = 0, \quad u > 0, \ q > 2,$$

and various generalizations of this equation for composite stellar densities; the Emden–Fowler equation, either in the form given by Fowler

$$u'' + r^{\nu}u|u|^{q-2} = 0, \quad \nu \in \mathbb{R}, \ q \geqslant 2,$$

or its forerunner

$$u'' + \frac{n-1}{r}u' + u|u|^{q-2} = 0, \quad n \neq 2, \ q \geqslant 2;$$

and the Haraux–Weissler equation [6]

$$u'' + \left(\frac{n-1}{r} + \frac{r}{2}\right)u' + \frac{l}{2}u + u|u|^{q-2} = 0, \quad n > 1, \ l > 0, \ q \geqslant 2.$$

Another interesting example occurs when $A(p) = |p|^{m-2}$, $m > 1$, for which (1.1) has the form

$$\left(|u'|^{m-2}u'\right)' + \delta(r)|u'|^{m-2}u' + f(r, u) = 0. \tag{1.3}$$

In this case the defining function $A(p)$ is usually called the degenerate Laplace operator, or the 'm-Laplacian'. This equation of course has a strongly singular behavior when $u' = 0$.

Equation (1.1), when $A(p)$ depends only on $|p|$ and $\delta(r) = (n-1)/r$, can also be considered as the radial version of the partial differential equation in $\mathbb{R}^n$.

$$\operatorname{div}\{A(|\mathrm{D}u|)\mathrm{D}u\} + f(r, u) = 0,$$

an equation whose study has been initiated in [5] and in [9]. An important case occurs here when $A$ is the mean curvature operator, that is $A(p) = 1/\sqrt{1 + |p|^2}$.

From a more abstract point of view, perhaps the most fruitful way to visualize (1.1) is as the Euler–Lagrange equation for extremals of the variational problem

$$\delta \int_J g(r)[G(u') - F(r, u)] \, \mathrm{d}r = 0. \tag{1.4}$$

Here $J$ is an open interval of the real line, $g = g(r)$ is a positive, differentiable function on $J$ with

$$g(r) = \exp\int_{r_0}^{r} \delta(s) \, \mathrm{d}s, \quad r_0 \in J, \quad \delta(r) = g'(r)/g(r),$$

the function $G = G(p)$ is related to $A$ by

$$G(p) = \int_0^p tA(t) \, \mathrm{d}t, \quad \text{and} \quad F(r, u) = \int_0^u f(r, t) \, \mathrm{d}t.$$

That (1.1) or (1.4) be regular is expressed by the condition that $pA(p)$ be a strictly increasing continuous function on $\mathbb{R}$ which vanishes at $p=0$, or equivalently that $G$ be a strictly convex, continuously differentiable function on $\mathbb{R}$ with $\mathrm{d}G/\mathrm{d}p=0$ at $p=0$. The strict convexity of $G$ is itself equivalent to the strong Weierstrass condition for the integrand in (1.4).

In the main part of the paper we shall frequently frame our hypotheses in terms of the variational problem (1.4); the reader may easily interpret them directly for the differential equation (1.1). We shall suppose always that (1.1) and (1.4) are regular.

The principal conclusions of Part I of the paper are that under quite weak conditions on the functions $F$ and $G$ any extremal of (1.4) can be extended over the full interval $J$, and that under slightly stronger hypotheses on $g$ and $F$, including the condition $\delta(r)\geqslant 0$, both $u$ and $u'$ are bounded as $r\to\infty$ when $J=\mathbb{R}^+$. Finally, again in the case $J=\mathbb{R}^+$, if $F(r, u)$ satisfies the further principal condition $u\,f(r, u)>0$ for $u\neq 0$, then, for any solution $u$ which is bounded as $r\to\infty$, both $u$ and $u'$ tend to zero as $r\to\infty$. (For results in case the principal condition fails, cf. for example [5].)

The conclusions can be illustrated most simply for the special case $A(p)=|p|^{m-2}$, $m>1$, with the function $F$ independent of $r$, that is, $F=F(u)$. Then equation (1.3) becomes

$$(|u'|^{m-2}u')'+\delta(r)|u'|^{m-2}u'+f(u)=0 \tag{1.5}$$

and $G(p)=|p|^m/m$. By Theorem 3, if $F(u)\geqslant -M$ then all solutions of (1.5) can be extended to the full interval $J$ in which $g(r)>0$. By Theorem 4, if $J=\mathbb{R}^+$ and $\delta(r)\geqslant 0$ for $r$ suitably large, and if furthermore

$$F(u)=\int_0^u f(t)\,\mathrm{d}t\to\infty$$

when $u\to\pm\infty$, then all solutions of (1.5) are bounded as $r\to\infty$. From Theorem 5, if there are positive constants $\beta$, $\beta'$ such that

$$\beta\leqslant r\delta(r)\leqslant\beta' r^m \quad \text{as } r\to\infty, \tag{1.6}$$

and if

$$uf(u)>0 \quad \text{for } u\neq 0,$$

then all solutions $u$ which are bounded as $r\to\infty$ satisfy

$$u(r),\ u'(r)\to 0 \quad \text{as } r\to\infty. \tag{1.7}$$

Simple examples show that this conclusion may fail if the upper bound $\beta' r^m$ in (1.6) is replaced by $\beta' r^{m+\epsilon}$ with $\epsilon>0$. As a case in point, the function

$$u(r)=\exp\left(\frac{m-1}{\epsilon r^{\epsilon/(m-1)}}\right)$$

is a bounded solution on $[1,\infty)$ of equation (1.5) with

$$\delta(r)=r^{m-1+\epsilon}\left(1+\frac{m-1+\epsilon}{r^{m+\epsilon}}+\frac{m-1}{r^{m+\epsilon m/(m-1)}}\right), \qquad f(u)=u|u|^{m-2},$$

yet $u(r)\to 1$ as $r\to\infty$ (see also Artstein and Infante [1]). Similarly, the lower bound in (1.6) is also essentially best possible; it cannot for example be weakened even to $\delta\in L^1[1,\infty)$, as is shown by an example of Levin and Nohel ([7, Remark 2.1]).

Theorem 5 is closely related to Theorem 1 of Levin and Nohel and to the results of Artstein and Infante, and indeed provides a considerable generalization of this work; see the remarks and Corollary following Theorem 5 in Section 5.

When the function $f(u)$ behaves algebraically as $u \to 0$, the conclusion (1.7) can be further refined and discussed in detail. This is the main goal of Part II of the paper. Suppose in particular that (1.6) holds in the form

$$\beta \leqslant r\delta(r) \leqslant \beta' \quad \text{as } r \to \infty, \quad \beta,\, \beta' \in \mathbb{R}, \tag{1.8}$$

and that there exists an exponent $q > 1$ such that

$$F(u) \geqslant \text{Pos. Const.}\ |u|^q, \qquad (\operatorname{sign}\beta)(uf(u) - qF(u)) \geqslant 0 \quad \text{if } \beta \neq 0, \tag{1.9}$$

for all $u$, conditions which substantially generalize the specific power-law behavior $f(u) = u|u|^{q-2}$, $\delta(r) = \beta/r$, studied by Fowler in [3,4]. Then the following series of conclusions hold for solutions $u$ which are defined as $r \to \infty$ (see Sections 7–12 with the parameter $\nu = 0$).

**1**. *Subcritical conditions.* Assume

$$q \geqslant m, \qquad q(\beta + 1 - m) < m(\beta + 1). \tag{1.10}$$

Then

$$u(r) = \mathrm{O}(r^{-\lambda_1}), \quad u'(r) = \mathrm{O}(r^{-\lambda_2}) \qquad \text{as } r \to \infty,$$

where

$$\lambda_1 = \frac{m\beta}{m + (m-1)q}, \qquad \lambda_2 = \frac{q\beta}{m + (m-1)q}.$$

In particular, if $\beta > 0$ then $u$ decays algebraically to zero as $r \to \infty$. See Theorem 7.

**2**. *Subcritical oscillations.* In the previous case, suppose that $\lambda_1 \neq 0$, that

$$u(r) \neq \mathrm{o}(r^{-\lambda_1}) \quad \text{as } r \to \infty,$$

and that $(1.9)_1$ holds in the stronger form

$$uf(u) \geqslant \text{Pos. Const.}\ |u|^q. \tag*{$(1.9)_1'$}$$

Then $u$ is oscillatory and there exist positive numbers $A$, $B$ such that along the solution $u = u(r)$ there holds

$$r^{\lambda_1}\left(\frac{m-1}{m}|u'|^m + \binom{A}{B}|u|^q\right) \to \text{Pos. Limit} \quad \text{as } r \to \infty, \tag{1.11}$$

where $\binom{A}{B}$ denotes the functions with values $A$ when $u \geqslant 0$ and $B$ when $u < 0$. Moreover, the successive zeros $r_j$ of $u$ and the interlaced critical points $\rho_j$ of $u$, $r_j < \rho_j < r_{j+1}$, possess the property

$$r_{j+2}^{\tau} - r_j^{\tau} \to 2\gamma, \quad \rho_{j+1}^{\tau} - \rho_j^{\tau} \to \gamma \qquad \text{as } j \to \infty,$$

where

$$\tau = \frac{m(\beta+1) - q(\beta + 1 - m)}{m + (m-1)q} \quad (>0)$$

and $\gamma$ is a positive number depending only on $q$, $m$, $\beta$, $A$, $B$ and the limit value in (1.11). See the corollary to Theorem 7 and Theorem 15.

The case $\beta < 0$, $q > m$ is particularly remarkable, for then $\lambda_1 < 0$, $\tau > 1$: the oscillations not only become infinitely large in amplitude as $r \to \infty$ but also the spacing between successive zeros approaches zero. If one should attempt to graph this behavior for large $r$ the result would be, assuming a finite thickness for the graph curve, a solid black region expanding infinitely in size as $r$ approaches infinity. If $m = 2$, $q = 4$, for example, the enveloped region has asymptotically straight boundaries when $\beta = -3$, is asymptotically convex when $-3 < \beta < 0$, and is asymptotically 'concave' when $\beta < -3$. Moreover when $\beta = -3$ we have $\tau = 2$, so that the 'period' $r_{j+2} - r_j \approx \text{const.}(r_{j+2} + r_j)^{-1} \approx O(r_j^{-1})$. An explicit equation having just this behavior is, for example,

$$u'' - \frac{3}{r}u' + u^3 = 0,$$

entirely innocuous in appearance.

**3.** *Supercritical conditions.* Assume

$$q > m, \qquad q(\beta + 1 - m) > m(\beta + 1). \tag{1.12}$$

Then as $r \to \infty$,

$$u(r) = O(r^{-m/(q-m)}), \qquad u'(r) = O(r^{-q/(q-m)}).$$

If $(1.12)_2$ holds with equality, then as $r \to \infty$

$$u(r) = O\big((r/\log r)^{-m/(q-m)}\big), \qquad u'(r) = O\big((r/\log r)^{-q/(q-m)}\big).$$

See Theorem 8 and the following remarks.

**4.** *Monotonic solutions.* If $u$ tends monotonically to zero as $r \to \infty$ and $(1.9)'_1$ holds, then necessarily

$$\beta' > m - 1, \qquad q(\beta' + 1 - m) > m\beta', \tag{1.13}$$

and

$$u(r) = O(r^{-m/(q-m)}) \quad \text{as } r \to \infty.$$

See Theorem 12.

**5.** Conversely, suppose

$$q > m, \quad \beta > m - 1, \qquad uf(u) \leqslant \text{Pos. Const. } |u|^q,$$

and $u(r) = o(r^{-m/(q-m)})$ as $r \to \infty$. Then $u$ approaches zero monotonically. If in addition $q(\beta + 1 - m) > m\beta$ we have

$$r^{\lambda_4}u(r) \to \text{Limit}, \quad r^{\lambda_4 + 1}u'(r) \to \text{Limit} \qquad \text{as } r \to \infty,$$

where $\lambda_4 = (\beta + 1 - m)/(m - 1)$. See Theorem 13.

**6.** Suppose that $q > m$ and that in place of (1.8) we have

$$r\delta(r) \leqslant \beta' r^{\eta}, \quad \eta > 0; \qquad r\delta(r) \to \infty \quad \text{as } r \to \infty.$$

Then as $r \to \infty$,

$$u(r) = O(r^{-(m-\eta)/(q-m)}), \qquad u'(r) = O(r^{-(m-\eta)q/(q-m)m}).$$

See Theorem 16.III, and note that we can take $\beta$ arbitrarily large.

**7.** Suppose $1 < q < m$. If $q(\beta + 1 - m) < m\beta$ then the results of **1** and **2** hold. On the other hand, if $q(\beta + 1 - m) > m\beta$ then

$$r^{\lambda_4} u(r) \to \text{Limit}, \quad r^{\lambda_4 + 1} u'(r) \to \text{Limit} \qquad \text{as } r \to \infty,$$

see Theorems 10 and 11.I.

In the conclusions **1**–**7** the various conditions placed on the function $f(u)$ are satisfied, e.g., by polynomials $f(u) = cu|u|^{q-2} + du|u|^{s-2}$, where $c$, $d$ are positive constants and $s \geqslant q > 1$ when $\beta > 0$, while $q \geqslant s > 1$ when $\beta < 0$. Other examples which show that the conclusions **1**–**7** are essentially best possible are given in Sections 7 and 9.

For functions $F(r, u)$ depending explicitly on $r$, as for example in the Emden–Fowler equation, we obtain corresponding conclusions under further conditions on the behavior of the function $\partial F/\partial r$ as $r \to \infty$. The case when $(1.9)_1$ is replaced by

$$F(r, u) \geqslant \text{Pos. Const. } r^{\nu} |u|^q, \quad \nu \in \mathbb{R},$$

is particularly important and accordingly is treated in detail in Theorems 7–16. Here, when $\nu + m > 0$, the results are analogous to those given above, while when $\nu + m < 0$ several new facets appear which require separate treatment, cf. Theorems 9, 10, 11. The variety of the results which are contained in Theorems 7–11 are illustrated in an efficient, if somewhat oversimplified, form in Figs. 1–5 of Section 9.

The operator $A(p) = |p|^{m-2}$ in the above examples was introduced principally in order to make the statements more transparent. The main conclusions however continue to hold for much wider classes of operators $A(p)$, including in particular the mean curvature operator $A(p) = 1/\sqrt{1 + p^2}$ where the appropriate value to be assigned to the parameter $m$ is 2. For specific results in this case, see Sections 3, 4, 5 and 13.

When $\beta = n - 1$, as is the case for radial solutions of partial differential equations, the exponent ranges (1.10) and (1.12) become

$$q(n - m) < nm \quad \text{and} \quad q(n - m) > nm,$$

with the critical value $q = nm/(n - m)$ as expected (see [9]). The exponents $\lambda_1$ and $\lambda_2$ then have the form

$$\lambda_1 = (n - 1)\frac{m}{m + (m - 1)q}, \qquad \lambda_2 = (n - 1)\frac{q}{m + (m - 1)q},$$

(note that $\lambda_2$ is in general different from $\lambda_1 + 1$). For radial solutions we also have $\beta' = n - 1$ so condition (1.13) becomes

$$n > m, \qquad q(n - m) > (n - 1)m,$$

a result first obtained in [9].

The following section is devoted to preliminary material, including the statement of the important identity (2.9) for extremals of (1.4). This identity, which is itself obtained from the general theory introduced in [10], provides in essence a powerful family of Liapunov functions for the study of the asymptotic behavior of extremals. Section 3 is concerned with continuation of solutions, Section 4 with boundedness, and Section 5 with the decay of solutions. Sections 6 through 12 treat specifically the algebraic behavior of solutions when $A(p) = |p|^{m-2}$, $G(p) = |p|^m/m$. In Section 12 we introduce weaker versions of the hypotheses (1.8) and (1.9) under which our results, and thus also the conclusions above, continue to hold. Finally in Section 13 we extend the results of Sections 6–12 to more general operators $A = A(p)$.

We point out that the decomposition (5.4), (5.5) used in the proof of Theorem 5 appears earlier in [8], as well as the motivation for Theorem 6. The specific contents of the paper are as follows.

**Part I. General Theory**



**Part II. Special Operators**



## Part I. General Theory

### 2. Preliminary results

Throughout the paper we consider extremals of the variational problem

$$\delta \int_J g(r)[G(u') - F(r, u)] \, dr = 0, \tag{2.1}$$

where $J$ is an open interval of $\mathbb{R}$, say $J = (a_0, a_1)$ with $a_0 = -\infty$ and $a_1 = +\infty$ allowed. We suppose always, and without further comment, that

(a) $g \in C^1(J)$, $\quad g(r) > 0$ for all $r \in J$;

(b) $G \in C^1(\mathbb{R})$, $\quad G(p) > 0$ for all $p \neq 0$, $G(0) = 0$, and $G$ strictly convex in $\mathbb{R}$;

(c) $F \in C^1(J \times \mathbb{R})$, $\quad F(r, 0) = 0$ for all $r \in J$.

The first parts of assumptions (a), (b), (c) are just the condition that the integrand $\mathscr{F}(r, u, p)$ in (2.1) be of class $C^1$ in $J \times \mathbb{R} \times \mathbb{R}$.

The fact that $G$ is strictly convex shows that the associated variational problem is weakly elliptic. Clearly the derivative $G_p$ of $G$ is strictly increasing in $\mathbb{R}$, so that $s = G_p(p)$ is invertible; we denote its inverse function by $p = G_p^{-1}(s)$. Of course $G_p(0) = 0$.

Now let $H = H(p)$ be the Legendre transform of $G = G(p)$, namely the continuous function defined by

$$H(p) = pG_p(p) - G(p) \quad \text{for every } p \in \mathbb{R}. \tag{2.2}$$

**Lemma 1.** *We have*

$$H(p) = \int_0^{G_p(p)} G_p^{-1}(s)\, \mathrm{d}s \quad \textit{for all } p \in \mathbb{R}. \tag{2.3}$$

*In turn,* $H(p) > 0$ *for* $p \neq 0$ *and* (sign $p$)$H(p)$ *is a strictly increasing function in* $\mathbb{R}$. *Moreover* $H(p)/p \to 0$ *as* $p \to 0$.

**Proof.** By Riemann–Stieltjes integration

$$H(p) = pG_p(p) - \int_0^p G_p(t)\, \mathrm{d}t = \int_0^p t\, \mathrm{d}G_p(t) = \int_0^{G_p(p)} G_p^{-1}(s)\, \mathrm{d}s. \quad \square$$

**Remark.** If $H$, the energy function of the problem, is considered to be the fundamental element of the theory, rather than $G$, then we must assume that $H$ is continuous, that (sign $p$)$H(p)$ is strictly increasing in $\mathbb{R}$, that $H(p)/p \to 0$ as $p \to 0$, and that the integral

$$\int_0^p \frac{H(t)}{t^2}\, \mathrm{d}t$$

is convergent. In this case $G(p) = p\int_0^p (H(t)/t^2)\, \mathrm{d}t$, as follows directly from (2.2).

**Lemma 2.** *For every* $(u, p) \in \mathbb{R}^2$ *we have*

$$G(p) \leqslant pG_p(p), \qquad H(p) \leqslant pG_p(p), \qquad |uG_p(p)| \leqslant G(\sigma u) + H(p), \tag{2.4}$$

*where* $\sigma =$ sign $(pu)$.

**Proof.** The first two inequalities follow at once from assumption (b) and Lemma 1. To prove the third one we first assume that $u, p \geqslant 0$. Then by Young's inequality we find that

$$uG_p(p) \leqslant \int_0^u G_p(t)\, \mathrm{d}t + \int_0^{G_p(p)} G_p^{-1}(s)\, \mathrm{d}s = G(u) + H(p),$$

taking (2.3) into account. The alternative case $u, p < 0$ can be treated similarly; the remaining cases are easily derived from the two already obtained. $\square$

We now turn to the variational equation associated with problem (2.1). Define

$$f(r, u) = F_u(r, u) \quad \text{for every } (r, u) \in J \times \mathbb{R},$$

and

$$A(p) = G_p(p)/p \quad \text{for all } p \neq 0.$$

Clearly $p\,A(p)$ can be extended by continuity to be zero at $p=0$. Also of course $A(p)>0$ for all $p\neq 0$.

An *extremal* for (2.1) is by definition a $C^1$ function $u$ defined on some subinterval $I$ of $J$, such that

$$G_p(u'(r)) = A(u'(r))u'(r) \in C^1(I) \tag{2.5}$$

and for which the corresponding (weak) Euler–Lagrange equation is satisfied on $I$, namely

$$(g\,A(u')u')' + g f(r,u) = 0, \tag{2.6}$$

or equivalently

$$g(r)A(u'(r))u'(r) + \int_{r_0}^{r} g(t)f(t,u(t))\,\mathrm{d}t = \text{Constant},$$

where $r_0$ is any fixed point of $I$. In the special case when $g(r)=r^{n-1}$, $n>1$, and $G(p)=\frac{1}{2}p^2$, we have $A(p)=1$ and (2.6) reduces to the radial Laplace equation

$$u'' + \frac{n-1}{r}u' + f(r,u) = 0,$$

in which $n$ can be thought of as a real parameter.

As noted in the introduction, if the function $A=A(p)$ in (2.6) is considered to be the fundamental element of the theory, rather than $G$, then we have to define

$$G(p) = \int_0^p tA(t)\,\mathrm{d}t \quad \text{for every } p\in\mathbb{R}.$$

In this case (b) is equivalent to the assumption that $p\,A(p)$ be continuous and strictly increasing in $\mathbb{R}$.

Rather surprisingly, in spite of the fact that neither $u'(r)$ nor $H(p)$ need be separately differentiable, the composite function $H(u'(r))$ is differentiable in $I$ along an extremal of (2.1). Indeed by (2.3), (2.5) and (2.6) we have for every $r\in I$

$$\begin{aligned}(H(u'(r)))' &= u'(r)\frac{\mathrm{d}}{\mathrm{d}r}G_p(u'(r)) = u'(r)\left[-\frac{g'(r)}{g(r)}G_p(u'(r)) - f(r,u(r))\right]\\ &= -\frac{g'(r)}{g(r)}A(u'(r))u'^2(r) - u'(r)f(r,u(r)).\end{aligned}$$

Therefore, along an extremal $u$ of (2.1) in $I$ we have the identity

$$(H(u'(r)) + F(r,u(r)))' = -\frac{g'(r)}{g(r)}A(u'(r))u'^2(r) + F_r(r,u(r)).$$

For simplicity in what follows we shall frequently use the common notation of ordinary differential equations, where $u=u(r)$ and $u'=u'(r)$ denote the solution and its derivative. Thus the above formula can be written simply as

$$(H(u') + F(r,u))' = -\frac{g'}{g}A(u')u'^2 + F_r(r,u). \tag{2.7}$$

Since along an extremal $u$ of (2.1) the function $H(u')$ is differentiable with respect to $r$, the main identity of [10, Proposition 1] holds even for $C^1$ extremals of (2.1), yielding the following useful result.

*Let $u$ be an extremal of problem* (2.1) *and let $a$, $h$ be scalar functions of class $C^1(I)$. Then the following identity holds in $I$,*

$$\{gh[H(u')+F(r,u)]+agA(u')uu'\}'$$
$$=g\Big[\Big(h'+h\frac{g'}{g}\Big)F(r,u)+hF_r(r,u)-auf(r,u)$$
$$-\Big(h'+h\frac{g'}{g}\Big)G(u')+(a+h')A(u')u'^2+a'A(u')uu'\Big]. \tag{2.8}$$

*In the particular case in which $a(r)=\alpha r^{k-1}/g(r)$ and $h(r)=r^k/g(r)$, where $\alpha$ and $k$ are constants, the identity* (2.8) *reduces to*

$$\Big\{r^k\Big[H(u')+F(r,u)+\frac{\alpha}{r}A(u')uu'\Big]\Big\}'$$
$$=r^{k-1}\Big[kF(r,u)+rF_r(r,u)-\alpha uf(r,u)$$
$$-kG(u')+\Big(\alpha+k-r\frac{g'}{g}\Big)A(u')u'^2+\frac{\alpha}{r}\Big(k-1-r\frac{g'}{g}\Big)A(u')uu'\Big]. \tag{2.9}$$

*The identity* (2.7) *is the special case of* (2.9) *where $k=\alpha=0$.*

The question of existence of extremals of (2.1) has been avoided in the above discussion, since it is not in itself relevant to the conclusions of the paper. It is nevertheless worth noting that this question is settled by the results of [9], where it is (in essence) shown that the initial value problem

$$u(r_1)=u_1,\qquad u'(r_1)=u_1'$$

for (2.6) always has a solution in some neighborhood of $r_1$. An outline of the proof is indicated in Section 3 below, as part of the demonstration of Theorem 1.

## 3. Continuation properties for extremals

In this section we shall treat nonlinearities $F$ which satisfy growth conditions of the following type:

(i) there is a sequence $(r_j)$ with $r_j \nearrow a_1$ such that $F(r_j,\cdot)$ is bounded below for all $j$;
(ii) there is a nonnegative function $\psi\in L^1_{loc}(J)$ such that $F_r(r,u)\leqslant(1+|F(r,u)|)\psi(r)$ a.e. in $J$ and for all $u\in\mathbb{R}$;

(i)′ there is a sequence $(\rho_j)$ with $\rho_j\searrow a_0$ such that $F(\rho_j,\cdot)$ is bounded below for all $j$;
(ii)′ there is a nonnegative function $\psi\in L^1_{loc}(J)$ such that $F_r(r,u)\geqslant-(1+|F(r,u)|)\psi(r)$ a.e. in $J$ and for all $u\in\mathbb{R}$.

In the important special case when $f(r, u) = \rho(r)\phi(u)$ condition (c) of Section 2 is satisfied when $\rho \in C^1(J)$ and $\phi \in C(\mathbb{R})$, and the hypotheses (i)–(ii)′ when $\rho(r)$ is positive and $\Phi(u) = \int_0^u \phi(t)\,dt$ is bounded below in $\mathbb{R}$. Indeed since $F(r, u) = \rho(r)\Phi(u)$ it is enough to take $\psi(r) = \rho'(r)/\rho(r)$.

The following preliminary result will be useful throughout the section.

**Lemma.** *Suppose* (i) *and* (ii) *hold. Then the function $F$ is bounded below in $I \times \mathbb{R}$, for any interval $I$ whose closure is in $J$.*

**Proof.** Let $I = [b_0, b_1]$ be a compact subinterval of $J$. Consider any fixed value $u = u_0$ and any fixed $\bar{r}$ in $I$. We shall show that there exists a real number $M$, depending only on $I$, such that $F(\bar{r}, u_0) \geqslant M$, which will complete the proof.

If $F(\bar{r}, u_0) \geqslant 0$ then $M = 0$ satisfies the claim. Hence suppose $F(\bar{r}, u_0) < 0$. Let $b = \inf_{r_j > b_1} r_j$. We consider two cases:

$$F(r, u_0) < 0 \quad \text{for all } r \in [\bar{r}, b), \tag{3.1}$$

$$\text{there is a point } \bar{b} \in [\bar{r}, b) \text{ such that } F(r, u_0) < 0 \quad \text{for } r \in [\bar{r}, \bar{b}),\ F(\bar{b}, u_0) = 0. \tag{3.2}$$

Obviously one of these must occur.

Consider case (3.1). Then for $r \in [\bar{r}, b]$ we have by (ii):

$$F_r(r, u_0) \leqslant (1 - F(r, u_0))\psi(r).$$

Integrating from $\bar{r}$ to $b$, we get

$$\begin{aligned} F(\bar{r}, u_0) &\geqslant 1 - (1 - F(b, u_0)) \exp\int_{\bar{r}}^{b} \psi(s)\,ds \\ &\geqslant 1 - (1 - F(b, u_0)) \exp\int_{b_0}^{b} \psi(s)\,ds. \end{aligned}$$

Clearly $b$ depends only on the interval $I$, the value $F(b, u_0)$ is bounded below by a number independent of $u_0$ by virtue of (i), and the exponential is bounded since $\psi \in L^1_{\text{loc}}(J)$. Hence $F(\bar{r}, u_0)$ is bounded below by a number depending only on $I$.

Next consider case (3.2). By the same procedure, integrating from $\bar{r}$ to $\bar{b}$ we obtain

$$F(\bar{r}, u_0) \geqslant 1 - \exp\int_{\bar{r}}^{\bar{b}} \psi(s)\,ds \geqslant 1 - \exp\int_{b_0}^{b} \psi(s)\,ds,$$

a lower bound depending only on $I$. This completes the proof. □

The conclusion of the lemma continues to hold when (i), (ii) are replaced by (i)′, (ii)′, the argument requiring only obvious changes.

Our first result gives a simple continuation property of solutions to the right endpoint $a_1$ of $J$. We recall that the functions $g$, $G$, $F$ always satisfy conditions (a), (b), (c) of Section 2.

**Theorem 1.** *Assume that g is an increasing function in J, that is $\delta(r) \geqslant 0$. Suppose also that conditions* (i) *and* (ii) *hold and that the Legendre transform of G satisfies*

$$H(p) \to \infty \quad \text{as } |p| \to \infty. \tag{3.3}$$

*Then any extremal of problem* (2.1) *can be continued to the right endpoint of J.*

**Proof.** Let $u$ be an extremal of problem (2.1) in some domain $I$ and denote by $r_0$ any fixed point of $I$. By (2.7), for any $r \in I$ we have

$$(H(u') + F(r, u))' = -\frac{g'}{g} A(u')u'^2 + F_r(r, u). \tag{3.4}$$

Thanks to assumptions (a), (ii) and the fact that $g$ is increasing, from (3.4) we see that

$$(H(u') + F(r, u))' \leqslant F_r(r, u) \leqslant (1 + |F(r, u)|)\psi(r) \quad \text{a.e. in } I. \tag{3.5}$$

Now, assume for contradiction that the extremal $u$ cannot be continued beyond some point $r_1$ in $J$, $r_0 < r_1 < a_1$.

Integrating (3.5) on $[r_0, r]$, with $r_0 < r < r_1$, we have

$$H(u') + F(r, u) \leqslant \int_{r_0}^{r} (1 + |F(s, u(s))|)\psi(s) \, ds + c_0, \tag{3.6}$$

where $c_0 = H(u'(r_0)) + F(r_0, u(r_0))$.

We now assert that there is a positive number $k$ such that

$$|F(r, u(r))| \leqslant k \quad \text{for each } r \in [r_0, r_1). \tag{3.7}$$

Indeed, from (3.6) and the condition $\psi \in L^1_{\mathrm{loc}}(J)$, it follows that for every $r \in [r_0, r_1)$

$$|F(r, u)| \leqslant \int_{r_0}^{r} |F(s, u(s))| \psi(s) \, ds + c_1,$$

where $c_1 = c_0 + \int_{r_0}^{r_1} \psi(s) \, ds - 2\min\{0, M\}$ and $M$ is a lower bound for $F$ in $[r_0, r_1] \times \mathbb{R}$. It is now immediate from Gronwall's inequality that (3.7) holds with $k = c_1 \exp \int_{r_0}^{r_1} \psi(s) \, ds$. From (3.6) and (3.7) also $H(u')$ is bounded in $[r_0, r_1)$. Hence by Lemma 1 and (3.3) there is a constant $P > 0$ such that

$$|u'(r)| \leqslant P \quad \text{for all } r \in [r_0, r_1).$$

Consequently $u$ is Lipschitz continuous in $[r_0, r_1)$ and so $u(r) \to u_1$ as $r \nearrow r_1$.

Now, from (3.5) and (3.7) we also have

$$(H(u') + F(r, u))' \leqslant (1 + k)\psi(r) \quad \text{a.e. in } (r_0, r_1).$$

Therefore, by assumption (ii),

$$H(u') + F(r, u) - (1 + k)\int_{r_0}^{r} \psi(s) \, ds$$

is decreasing in $[r_0, r_1)$ with

$$F(r, u(r)) \to F(r_1, u_1) \quad \text{and} \quad \int_{r_0}^{r} \psi(s) \, ds \to \int_{r_0}^{r_1} \psi(s) \, ds \qquad \text{as } r \nearrow r_1.$$

Thus the function $H(u'(r))$ also tends to a finite limit as $r \nearrow r_1$, and by Lemma 1 and (3.3) this means that

$$u'(r) \to u_1' \quad \text{as } r \nearrow r_1,$$

for some finite number $u_1'$.

We now use the existence theorem given in [9], Proposition 1, to show that the Cauchy problem

$$\left(g\, G_p(u')\right)' + g f(r, u) = 0, \quad u(r_1) = u_1, \ u'(r_1) = u_1',$$

admits a $C^1$ solution $\bar{u}$ in some interval $[r_1, R]$, $r_1 < R < a_1$. To do this requires only that we consider the operator

$$T[u](r) = u_1 + \int_{r_1}^{r} G_p^{-1}\left( -\int_{r_1}^{t} \frac{g(s)}{g(t)} f(s, u(s))\, \mathrm{d}s + \frac{g(r_1)}{g(t)} G_p(u_1') \right) \mathrm{d}t \tag{3.8}$$

defined on the set

$$C = \{ u \in C[r_1, R] : \|u(\cdot) - u_1\|_\infty \leqslant 1 \},$$

where $C[r_1, R]$ denotes the space of continuous functions on $[r_1, R]$ endowed with the uniform norm $\|\cdot\|_\infty$. As in [9], the operator $T$ is compact and continuous on $C$, with $T(C) \subset C$, provided that $R - r_1$ is sufficiently small, and the existence of $\bar{u}$ is proved.

It now follows that $u$ on $[r_0, r_1)$ can be continued as an extremal to $[r_0, R]$ by putting

$$u(r) \equiv \begin{cases} u(r), & r_0 \leqslant r < r_1, \\ \bar{u}(r), & r_1 \leqslant r \leqslant R. \end{cases}$$

This contradiction completes the proof. □

In the next theorem we establish continuation of extremals without the assumption that $g$ is increasing.

**Theorem 2.** *Suppose that the conditions* (i) *and* (ii) *hold. Assume also that there is a positive number $\theta$ such that*

$$G(p) \leqslant \theta H(p) \quad \textit{for all } p \in \mathbb{R}. \tag{3.9}$$

*Then any extremal of problem* (2.1) *can be continued to the right endpoint of $J$.*

**Proof.** Let $u$ be an extremal of problem (2.1) and let $r_0$ be a point of its domain $I$. Assume for contradiction that $u$ cannot be continued beyond some point $r_1$, $r_0 < r_1 < a_1$. Integrating (3.4) on $[r_0, r]$, with $r_0 < r < r_1$, and setting $c_0 = H(u'(r_0)) + F(r_0, u(r_0))$ as before, we obtain

$$H(u') + F(r, u) = -\int_{r_0}^{r} \frac{g'}{g} A(u') u'^2\, \mathrm{d}s + \int_{r_0}^{r} F_r(s, u)\, \mathrm{d}s + c_0.$$

By (2.2) and (3.9) we have $p^2A(p) \leqslant (1+\theta)H(p)$ for every $p \in \mathbb{R}$. Let $c$ be such that $|g'(r)/g(r)| \leqslant c$ for $r \in [r_0, r_1]$. Then, using assumptions (i) and (ii), we see that for every $r \in [r_0, r_1)$,

$$H(u') + |F(r, u)| \leqslant c(1+\theta)\int_{r_0}^{r} H(u')\,\mathrm{d}s + \int_{r_0}^{r} |F(s, u)|\,\psi(s)\,\mathrm{d}s + c_1$$

$$\leqslant \int_{r_0}^{r} (H(u') + |F(s, u)|)\phi(s)\,\mathrm{d}s + c_1,$$

where $c_1 = c_0 + \int_{r_0}^{r_1}\psi(s)\,\mathrm{d}s - 2\min\{0, M\}$, $M$ is a lower bound for $F$ in $[r_0, r_1] \times \mathbb{R}$, and $\phi(r) = c(1+\theta) + \psi(r)$, so that $\phi \in L^1_{\mathrm{loc}}(J)$. It is now a consequence of Gronwall's inequality that $H(u') + |F(r, u)|$ is bounded in $[r_0, r_1)$, say by $c_1 \exp \int_{r_0}^{r_1}\phi(s)\,\mathrm{d}s$. Thus, since (3.9) implies (3.3), there is a positive number $P$ such that $|u'(r)| \leqslant P$ for every $r \in [r_0, r_1)$.

The rest of the proof is now essentially the same as in Theorem 1, that is the extremal $u$ can here be continued to some interval $[r_0, R]$, with $r_1 < R < a_1$. This contradiction completes the proof. □

Continuation of extremals to the left endpoint $a_0$ of $J$ can be obtained as in Theorems 1 and 2, using (i)′, (ii)′ instead of (i), (ii). Regarding continuation of extremals over the entire interval $J$ where $g > 0$, we state the following

**Theorem 3.** *Suppose that* (i), (i)′, (ii) *and* (ii)′ *hold and also that* (3.9) *is satisfied. Then any extremal $u$ of* (2.1) *can be continued to all of $J$.*

**Proof.** Continuation to $a_1$ follows from Theorem 2. To prove continuation to $a_0$ one can apply the proof of Theorem 2 with obvious modifications, and in particular the use of (i)′, (ii)′ rather than (i), (ii). □

**Remark.** The continuation property of Theorem 1 can be obtained under weaker conditions on $F$ if (3.3) is replaced by

$$H(p) \geqslant \text{Pos. Const. } |p| \quad \text{for all } p \text{ sufficiently large.} \tag{3.3$'$}$$

In particular, in this case it is enough if $F$ has the form

$$F(r, u) = F_1(r, u) + F_2(r, u), \tag{3.10}$$

where $F_1$ satisfies conditions (i)–(ii) and $F_2$ is such that there exists a nonnegative function $\psi_2 \in L^1_{\mathrm{loc}}(J)$ for which

$$\left|\frac{\partial F_2}{\partial u}(r, u)\right| \leqslant \psi_2(r) \quad \text{a.e. in } J \text{ and for all } u \in \mathbb{R}.$$

The proof is essentially the same as before, except that (3.5) is replaced by

$$(H(u') + F_1(r, u))' \leqslant F_{1r}(r, u) - F_{2u}(r, u)u',$$

which leads in turn to

$$H(u') + |F_1(r, u)| \leqslant \text{Constant},$$

instead of (3.7). Thus $|u'(r)|$ is bounded in $[r_0, r_1)$ and the conclusion is then obtained exactly as before. The same remark holds for Theorems 2 and 3, since (3.9) implies (3.3)′.

Condition (3.10) is satisfied in particular when $F(r, u) = F_1(r, u) - e(r)u$ and $e \in L^1_{\text{loc}}(J)$. In this case equation (1.1) takes the form

$$(A(u')u')' + \delta(r)A(u')u' + f_1(r, u) = e(r);$$

cf. also the corollary to Theorem 5 in Section 5.

### Examples

1. The degenerate Laplace operator $G(p) = |p|^m/m$, $m > 1$, has

$$G_p(p) = p|p|^{m-2}, \quad A(p) = |p|^{m-2} \quad \text{and} \quad H(p) = \frac{m-1}{m}|p|^m.$$

Hence (3.9) holds with $\theta = m - 1$.

2. The mean curvature operator $G(p) = \sqrt{1+p^2} - 1$ has

$$G_p(p) = \frac{p}{\sqrt{1+p^2}}, \quad A(p) = \frac{1}{\sqrt{1+p^2}} \quad \text{and} \quad H(p) = 1 - \frac{1}{\sqrt{1+p^2}}.$$

Thus both (3.3) and (3.9) fail.

3. The function $G(p) = p \tan^{-1} p - \frac{1}{2}\log(1+p^2)$ has

$$G_p(p) = \tan^{-1} p, \quad G_{pp}(p) = 1/(1+p^2) > 0,$$

$$A(p) = \frac{1}{p}\tan^{-1} p \quad \text{and} \quad H(p) = \tfrac{1}{2}\log(1+p^2).$$

Here (3.3) holds while obviously (3.9) fails for any $\theta > 0$. Therefore Theorem 1 can be applied, but both Theorems 2 and 3 are inapplicable.

### 4. Boundedness of extremals

Here we discuss the boundedness of extremals on $J$. The results vary slightly depending on whether we consider boundedness as $r \to a_1$, the right endpoint of $J$, or as $r \to a_0$, the left endpoint. For simplicity, we shall present two results for the case $r \to a_1$, leaving other examples to the reader.

**Theorem 4.** *Let $u$ be an extremal of problem* (2.1) *whose domain $I$ contains $I_1 = [r_0, a_1)$. Assume that there exists a sequence $(r_j)$ with $r_j \nearrow a_1$, such that $F(r_j, \cdot)$ is bounded below uniformly in $j$ and*

$$F(r_j, u) \to \infty \quad \text{as } |u| \to \infty \text{ uniformly in } j. \tag{4.1}$$

*Suppose also that* (ii) *holds with $\psi \in L^1(I_1)$ and that $g$ is increasing. Then $u$ is bounded in $I_1$.*

**Proof.** The proof is the same as for Theorem 1, with $r_1$ taken to be $a_1$. We find as there, by using the Gronwall inequality, that

$$|F(r, u)| \leqslant c_1 \exp \int_{I_1} \psi(s)\,\mathrm{d}s \quad \text{for every } r \in I_1,$$

where $c_1 = c_0 + \int_{I_1} \psi(s)\,\mathrm{d}s - 2\min\{0, M\}$, and $M$ is a lower bound for $F$ in $I_1 \times \mathbb{R}$. Existence of this bound follows exactly as in the proof of the lemma in Section 3. In the same way (4.1) and (ii) imply that

$$F(r, u) \to \infty \quad \text{as } |u| \to \infty, \text{ uniformly in } r \in I_1.$$

It now follows at once that $u$ is bounded as $r$ tends to $a_1$, completing the proof. □

**Remarks.** Theorem 4 can be applied even when $G(p) = \sqrt{1+p^2} - 1$, in which case equation (2.6) becomes the mean curvature equation

$$\left( g(r) \frac{u'}{\sqrt{1+u'^2}} \right)' + g(r) f(r, u) = 0.$$

Here the associated function $H$ does not satisfy either (3.3) or (3.9).

If hypothesis (4.1) fails then the conclusion need not hold, as shown by the simple example $u'' + u'/r = 0$ with $a_1 = +\infty$ and $u(r) = \log r$.

We next give an alternative version of Theorem 4 for $F$ depending only on $u$ and for the special case when the extremal $u$ is oscillatory as $r \to a_1$, that is, possesses an infinite number of zeros which accumulate at $a_1$ (the usual case is $a_1 = +\infty$).

**Theorem 4′.** *Let $u$ be an extremal of problem* (2.1) *whose domain contains $I_1 = [r_0, a_1)$ and which is oscillatory as $r \to a_1$. Assume also that $g$ is increasing and that the function $F$ given in* (c) *depends only on $u$ and satisfies the assumption*

$$\begin{aligned} F(u) &< \lim_{t \to \infty} F(t) && \text{for } u \geqslant 0, \\ F(u) &< \lim_{t \to -\infty} F(t) && \text{for } u \leqslant 0. \end{aligned} \tag{4.2}$$

*Then $u$ is bounded in $I_1$.*

**Proof.** Since $F_r(r, u) \equiv 0$ and $g'(r) \geqslant 0$, from (3.4) we have

$$(H(u') + F(u))' \leqslant 0 \quad \text{for all } r \in I_1.$$

Therefore the function $H(u') + F(u)$ is decreasing in $I_1$ and so is bounded above. We denote by $u_{\text{crit}}$ the values of $u$ at local maxima and local minima in $I_1$. Thus $F(u_{\text{crit}})$ is decreasing and by the growth condition (4.2) this shows that $|u_{\text{crit}}|$ is bounded. Hence $u$ is bounded in $I_1$. □

**Remark.** If $H$ satisfies (3.3), then, under the assumptions of either Theorem 4 or Theorem 4′, the extremal $u$ and its derivative $u'$ are bounded in $I_1$.

If moreover the interval $J$ is bounded on the right, that is $a_1 < \infty$, then clearly $u$ is Lipschitz continuous in $I_1$, and so has a finite limit as $r \to a_1$. Thus the extremal $u$ can be extended to a function of class $C(\bar{I}_1)$. The derivative $u'$ need not approach a limit unless further assumptions are made concerning the continuity of $F$ at $r = a_1$. If, however, we assume also that $F \in C(\bar{J} \times \mathbb{R})$ then by the arguments in the proof of Theorem 1 we see that $u'$ also approaches a finite limit as $r \to a_1$. The condition $F \in C(\bar{J} \times \mathbb{R})$ holds in particular when $f(r, u) = \rho(r)\phi(u)$ and $\rho(r) \to$ limit as $r \to a_1$.

In both Theorems 4 and 4′ it is evident that $g$ is bounded from zero near the endpoint $a_1$. Some condition of this sort is in fact essential for the boundedness of extremals, as is clear even from the case of Bessel's equation, where $g(r) = r^2 \to 0$ and the corresponding solutions may become unbounded as $r \to 0$.

## 5. Global asymptotic stability of extremals

In the preceding section we noted, for intervals $J$ bounded on the right, that extremals can be continuously extended, under certain conditions, to $\bar{I}_1 \ni a_1$.

When $J$ is unbounded on the right the situation is more delicate. The purpose of this section is to present several results in this direction, under mild assumptions on the functions $g = g(r)$ and $F = F(r, u)$. The following properties of $F$ will be crucial in the discussion. Without loss of generality we assume from here on that $J = \mathbb{R}^+$.

(P) There exists a number $R > 0$ such that for each pair of values $u_0$, $U$, with $0 < u_0 \leqslant U$, there is a constant $\kappa > 0$ for which $uf(r, u) \geqslant \kappa$ when $r \geqslant R$ and $|u| \in [u_0, U]$.

(P)′ For every $U > 0$ there is a nonnegative function $\psi \in L^1[R, \infty)$ such that $F_r(r, u) \leqslant \psi(r)$ for almost all $r \in [R, \infty)$ and all $|u| \leqslant U$.

In the important case when $f(r, u) = \rho(r)\phi(u)$, it is easy to see that property (P) is equivalent to the simple condition

$$\rho(r) \geqslant \text{Const.} > 0 \quad \text{for } r \geqslant R; \qquad u\phi(u) > 0 \quad \text{for } u \neq 0.$$

Moreover, condition (P)′ holds when $\rho' \in L^1[R, \infty)$. Of course when $f$ does not depend on $r$ property (P)′ is irrelevant.

Condition (P) implies $uf(r, u) > 0$ for $r \geqslant R$ and $u \neq 0$, and so in turn $F(r, u) > 0$ for those values of $r$ and $u$. The following lemma makes this statement more precise.

**Lemma.** *Suppose that $F = F(r, u)$ has property* (P). *Then for every $U > 0$ there is a positive increasing function $\omega = \omega(t)$, $0 < t \leqslant U$, such that*

$$F(r, u) \geqslant \omega(|u|) \quad \textit{for } r \geqslant R \textit{ and } 0 < |u| \leqslant U.$$

**Proof.** Let $\kappa(u_0, U)$ be the value of $\kappa$ in (P) corresponding to the interval $[u_0, U]$. If this function is not already increasing in the variable $u_0$, we replace it by the new function

$$\bar{\kappa}(u_0, U) = \sup_{0 < t \leqslant u_0} \kappa(t, U),$$

which is positive and increasing and for which (P) is also satisfied.

Now for any $r \geqslant R$ and $u \in (0, U]$ we have

$$F(r, u) = \int_0^u f(r, t)\,\mathrm{d}t \geqslant \int_{u/2}^u \frac{\bar{\kappa}(\frac{1}{2}u, U)}{u}\,\mathrm{d}t = \tfrac{1}{2}\bar{\kappa}(\tfrac{1}{2}u, U).$$

Similarly $F(r, u) \geqslant \frac{1}{2}\bar{\kappa}(\frac{1}{2}|u|, U)$ when $u \in [-U, 0)$. The proof is thus completed by taking $\omega(t) = \frac{1}{2}\bar{\kappa}(\frac{1}{2}t, U)$, $0 < t \leqslant U$. □

We recall the relation $\delta(r) = g'(r)/g(r)$. Our main theorem is now

**Theorem 5.** *Let $u$ be an extremal of problem* (2.1), *bounded in* $[r_0, \infty)$ *for some* $r_0 > 0$. *Suppose that there are positive numbers $\beta$ and $\beta'$ such that*

$$\beta \leqslant r\delta(r) \leqslant \beta' r \quad \textit{for every } r \in [r_0, \infty). \tag{5.1}$$

*Assume also that conditions* (P) *and* (P)$'$ *hold.*

*Then both $u$ and $u'$ tend to zero as $r \to \infty$.*

Condition (5.1) can be significantly weakened if the operator $A$ satisfies a mild algebraic growth condition as $p \to 0$, see the remarks at the end of the proof.

**Proof of Theorem 5.** Let $u$ be an extremal of problem (2.1), with $|u(r)| \leqslant L$ for every $r \in [r_0, \infty)$. Without loss of generality we shall assume that the number $R$ given in (P) is less than $r_0$, and that $L > 0$. Let $\psi$ denote the function given in (P)$'$ corresponding to $U = L$. From (2.7), (5.1) and (P)$'$ we have $(H(u') + F(r, u))' \leqslant \psi(r)$ a.e. in $[R, \infty)$. Therefore,

$$H(u') + F(r, u) + \int_r^\infty \psi(s)\,\mathrm{d}s$$

is decreasing in $[R, \infty)$. Hence, since $H$ and $F$ are nonnegative, there is a number $l \geqslant 0$ such that

$$H(u') + F(r, u) \to l \quad \text{as } r \to \infty. \tag{5.2}$$

If $l = 0$, then by Lemma 1 of Section 2 and the lemma above we immediately derive that $u'(r) \to 0$ and $u(r) \to 0$ as $r \to \infty$, completing the proof.

Let us therefore assume for contradiction that $l > 0$ in (5.2). We choose $r_1 \geqslant r_0$ so that

$$\tfrac{3}{4}l \leqslant H(u') + F(r, u) \leqslant 2l \quad \text{for } r \geqslant r_1 \quad \text{and} \quad \int_{r_1}^\infty \psi(s)\,\mathrm{d}s \leqslant \tfrac{1}{8}l. \tag{5.3}$$

Consider the two disjoint subsets $I_1$ and $I_2$ of $[r_1, \infty)$ such that

$$F(r, u(r)) \leqslant \tfrac{1}{4}l \quad \text{whenever } r \in I_1, \tag{5.4}$$

$$F(r, u(r)) > \tfrac{1}{4}l \quad \text{whenever } r \in I_2. \tag{5.5}$$

By (5.3) and Lemma 1 there is a positive number $p_0$ such that

$$H(u'(r)) \geqslant \tfrac{1}{2}l \quad \text{and} \quad |u'(r)| \geqslant p_0 > 0 \qquad \text{for all } r \in I_1. \tag{5.6}$$

We claim moreover that there is a positive number $u_0$ such that

$$|u(r)| \geqslant u_0 > 0 \quad \text{for every } r \in I_2. \tag{5.7}$$

Indeed, by (5.5), (P)′ and (5.3) we have for $r \in I_2$

$$\tfrac{1}{4}l < F(r, u(r)) = F(r_1, u(r)) + \int_{r_1}^{r} F_r(s, u(r))\, ds \leqslant F(r_1, u(r)) + \int_{r_1}^{\infty} \psi(s)\, ds$$

$$\leqslant F(r_1, u(r)) + \tfrac{1}{8}l; \tag{5.8}$$

in other words $F(r_1, u(r)) > l/8$ for every $r \in I_2$. The claim now follows, since $F$ is continuous in the variable $u$ and $F(r_1, 0) = 0$ by (c).

We shall show that for a suitable choice of the constants $\alpha$ and $k$ the right-hand side of the main identity (2.9) is less than $r^k F_r(r, u)$ for all $r$ sufficiently large. We take in particular

$$0 < \alpha \leqslant \tfrac{1}{4}\beta, \qquad 0 < k \leqslant \tfrac{1}{4}\beta. \tag{5.9}$$

It is convenient in what follows to express the right-hand side $\mathscr{R} = \mathscr{R}(r)$ of (2.9) in the form

$$\mathscr{R} = r^{k-1}\Gamma(r) + r^k F_r(r, u),$$

so that what we must show is that $\Gamma(r) < 0$ for $r$ sufficiently large. Moreover, we shall write $\xi = \xi(r) = rg'(r)/g(r) = r\delta(r)$. For later use, note by (5.9) and (5.1) that

$$\frac{\xi - k + 1}{\xi - k - \alpha} \leqslant 2\frac{1+\beta}{\beta} \quad \text{for } r \in [r_0, \infty). \tag{5.10}$$

We now divide the discussion according to whether $r \in I_1$ or $r \in I_2$.

*Case* $r \in I_1$. By (b) and the fact that $uf(r, u) > 0$ we have

$$\Gamma \leqslant kF(r, u) - (\xi - \alpha - k)A(u')u'^2 - \frac{\alpha}{r}(\xi - k + 1)A(u')uu'$$

$$\leqslant kF(r, u) - (\xi - \alpha - k)A(u')|u'|\left(|u'| - \frac{2\alpha}{r} \cdot \frac{1+\beta}{\beta}|u|\right)$$

by (5.10). We now take $r_2 \geqslant r_1$ so that $r_2 \geqslant (1+\beta)L/p_0$. Hence for $r \geqslant r_2$ in $I_1$ we have by $(5.6)_2$ and (5.9)

$$|u'| - \frac{2\alpha}{r} \cdot \frac{1+\beta}{\beta}|u| \geqslant \tfrac{1}{2}|u'|,$$

and

$$\Gamma \leqslant kF(r, u) - \tfrac{1}{4}\beta A(u')u'^2 \leqslant \tfrac{1}{4}\beta\big(F(r, u) - H(u')\big)$$

by (5.9) and $(2.4)_2$. Hence $\Gamma(r) < 0$ for any $r \in I_1 \cap (r_2, \infty)$ by (5.4) and $(5.6)_1$. In other words, taking (2.9) and (P)′ into account, we have shown that

$$\left\{r^k\left(H(u') + F(r, u) + \frac{\alpha}{r}A(u')uu'\right)\right\}' < r^k\psi(r) \quad \text{for almost all } r \in I_1 \cap (r_2, \infty). \tag{5.11}$$

*Case* $r \in I_2$. By (5.3), (P) and (5.9), when $r \in I_2$ there holds

$$\Gamma \leqslant 2kl - \alpha\kappa - (\xi - \alpha - k)A(u')u'^2 - \frac{\alpha}{r}(\xi - k + 1)A(u')uu', \tag{5.12}$$

where $\kappa = \kappa(u_0, L)$ is the number of property (P) corresponding to $u_0$ given in (5.7) and to $U = L$. As in the previous case, if

$$|u'| \geqslant \frac{2\alpha}{r} \cdot \frac{1+\beta}{\beta} L,$$

we get

$$\Gamma \leqslant 2kl - \alpha\kappa.$$

On the other hand, if

$$|u'| < \frac{2\alpha}{r} \cdot \frac{1+\beta}{\beta} L, \tag{5.13}$$

then from (5.12)

$$\Gamma \leqslant 2kl - \alpha\kappa + \frac{\alpha}{r}(1+\beta' r) LA(u')u'. \tag{5.14}$$

Thus, if $r$ is suitably large, say $r \geqslant r_3$, we get from (5.13), (5.14) and (2.5)

$$\Gamma < 2kl - \tfrac{1}{2}\alpha\kappa.$$

Hence for $r \geqslant r_3$ in $I_2$ and for $k \leqslant \alpha\kappa/4l$ we have $\Gamma(r) < 0$.

In conclusion, recalling (5.11), we have shown that

$$\left\{ r^k \left( H(u') + F(r, u) + \frac{\alpha}{r} A(u') uu' \right) \right\}' < r^k \psi(r) \quad \text{for almost all } r > r_4,$$

where $r_4 = \max\{r_1, r_2, r_3\}$. Consequently the function

$$\Psi(r) = r^k \left( H(u') + F(r, u) + \frac{\alpha}{r} A(u') uu' - \frac{1}{r^k} \int_{r_4}^{r} s^k \psi(s)\, \mathrm{d}s \right)$$

is decreasing in $[r_4, \infty)$. Moreover

$$\int_{r_4}^{r} \left(\frac{s}{r}\right)^k \psi(s)\, \mathrm{d}s \leqslant \int_{r_1}^{\infty} \psi(s)\, \mathrm{d}s \leqslant \tfrac{1}{8} l$$

by (5.3), while also by (5.9) and $(2.4)_3$ we have

$$\frac{\alpha}{r} |A(u') uu'| = \frac{\alpha}{r} |uG_p(u')| \leqslant \frac{\alpha}{r}(G(\sigma u) + H(u')) \leqslant \frac{\alpha}{r}(G(L) + G(-L) + 2l),$$

thanks to (5.3). Hence for all sufficiently large $r$

$$\Psi(r) \geqslant r^k \left( \frac{3}{4} l - \frac{l}{8} - \frac{l}{8} \right) = \frac{l}{2} r^k,$$

which contradicts the fact that $\Psi$ is decreasing in $[r_4, \infty)$ and completes the proof of the theorem. □

**Remarks.** If for some $m > 1$ we have

$$A(p) \leqslant \text{Const.}\ |p|^{m-2}, \quad p \neq 0, \tag{5.15}$$

then condition (5.1) can be improved to

$$\beta \leqslant r\delta(r) \leqslant \beta' r^m. \tag{5.1$'$}$$

For example, in equation (1.5) we can allow $\delta(r) = r^{m-1}$ and still retain the conclusion $u(r)$, $u'(r) \to 0$ as $r \to \infty$. The proof is essentially the same as before, except that (5.14) now becomes

$$\Gamma \leqslant 2kl - \alpha\kappa + \text{Const.}\frac{\alpha}{r}(1 + \beta' r^m) L |u'|^{m-1}$$

$$\leqslant 2kl - \alpha\kappa + \text{Const. } \alpha^m L^m \left(\frac{1+\beta}{\beta}\right)^{m-1} \frac{1 + \beta' r^m}{r^m},$$

by (5.13). Thus if $\alpha$ is suitably small we again obtain

$$\Gamma < 2kl - \tfrac{1}{2}\alpha\kappa,$$

and the rest of the proof is unchanged.

A further refinement of the proof (details omitted) shows, even more, that (5.1) can be weakened to the form

$$r\delta(r) \geqslant \beta, \qquad r^{-k}\int_{r_0}^{r} s^{k-1}\delta(s)\, ds \leqslant \text{Constant}$$

for some number $k$ with $0 < k < \beta$ and all $r \geqslant r_0$. Similarly, if (5.15) holds then it is enough if

$$r\delta(r) \geqslant \beta, \qquad r^{-k}\int_{r_0}^{r} s^{k-m}\delta(s)\, ds \leqslant \text{Constant}$$

for some number $k$ with $0 < k < \beta$ and all $r \geqslant r_0$. If equality holds in (5.15), then $0 < k < m\beta/(m-1)$. If $\beta > m$, or if $A(p) = |p|^{m-2}$, $p \neq 0$, and $\beta > m - 1$, then one can specifically choose $k = m$ to obtain the condition

$$r^{-m}\int_{r_0}^{r} \delta(s)\, ds \leqslant \text{Constant},$$

discovered by Artstein and Infante [1] in the case when $m = 2$, $\delta(r) \geqslant$ Pos. Constant and $F$ is independent of $r$.

**Corollary to Theorem 5.** *Let $u$ be a bounded solution of the equation*

$$(A(u')u')' + h(r, u, u')A(u')u' + f(r, u) = e(r) \quad \textit{on } [r_0, \infty). \tag{5.16}$$

*Suppose that conditions* (P) *and* (P)$'$ *are satisfied by the functions $f = f(r, u)$ and $F(r, u) = \int_0^u f(r, t)\, dt$, and that for some positive constants $\beta$, $\beta'$:*

$$\beta \leqslant rh(r, u, p) \leqslant \beta' r \quad \textit{for every } r \geqslant r_0 \textit{ and all } u \in \mathbb{R},\ p \in \mathbb{R}. \tag{5.17}$$

*Finally, assume that either*

$$e(r) \to 0 \quad \textit{as } r \to \infty, \qquad e' \in C[r_0, \infty) \cap L^1[r_0, \infty), \tag{5.18}$$

*or*

$$e \in L^1[r_0, \infty), \qquad H(p) \geqslant \text{Pos. Const. } |p| \quad \textit{when } |p| \geqslant p_1 \tag{5.19}$$

*for some number $p_1 \geqslant 0$. Then*

$$u(r) \to 0, \quad u'(r) \to 0 \qquad \textit{as } r \to \infty.$$

*In case* (5.19) *the condition that u is bounded can be replaced by*

$$F(r, u) \to \infty \quad as\ |u| \to \infty,\ uniformly\ in\ r, \tag{5.20}$$

*provided that* (P)′ *holds for all* $u \in \mathbb{R}$.

**Proof.** Corresponding to the given solution $u = u(r)$, we define

$$\delta(r) \equiv h(r, u(r), u'(r)), \quad r \geqslant r_0.$$

Then by (5.17) the condition (5.1) is satisfied. Now set

$$\hat{f}(r, u) = f(r, u) - e(r) \quad \text{for } r \geqslant r_0 \text{ and } \quad u \in \mathbb{R},$$

so that in turn

$$u\hat{f}(r, u) = uf(r, u) - e(r)u \quad \text{and} \quad \hat{F}(r, u) = F(r, u) - e(r)u.$$

Suppose now that (5.18) is satisfied. Then, as in the first part of the proof of Theorem 5, we see that

$$H(u') + \hat{F}(r, u) + \int_r^\infty (\psi(s) + |e'(s)|L)\, \mathrm{d}s$$

is decreasing in $[R, \infty)$. Hence (5.2) holds in view of (5.18). Thus we are done if $l = 0$.

Therefore, as before, assume for contradiction that $l > 0$. We follow the proof of Theorem 5 with $F$ replaced by $\hat{F}$ and with the right hand side $\mathscr{R} = \mathscr{R}(r)$ of (2.9) in the form:

$$\mathscr{R}(r) = r^{k-1}\hat{\Gamma}(r) + r^k\hat{F}_r(r, u),$$

where

$$\hat{\Gamma}(r) = \Gamma(r) + (\alpha - k)e(r)u.$$

Then as before there is an $r_4 > 0$ such that $\hat{\Gamma}(r) < 0$ for all $r \geqslant r_4$, since $|e(r)u(r)| \leqslant |e(r)|L \to 0$ as $r \to \infty$. Consequently the function

$$\hat{\Psi}(r) = r^k\left\{H(u') + F(r, u) + \frac{\alpha}{r}A(u')uu' - e(r)u \right.$$
$$\left. - \frac{1}{r^k}\int_{r_4}^r s^k(\psi(s) + |e'(s)|L)\, \mathrm{d}s\right\}$$

is decreasing in $[r_4, \infty)$, which yields the same contradiction as before.

Next suppose that (5.19) is satisfied. We show first that $u'$ is bounded in $[r_0, \infty)$. By the calculation used in Section 2 for the proof of (2.7) we find

$$\mathscr{L}' = (H(u') + F(r, u))' = -\delta(r)A(u')u'^2 + F_r(r, u) + e(r)u'$$
$$\leqslant \psi(r) + |e(r)|\,|u'|,$$

a.e. in $[R, \infty)$ by (P)′. In turn by (5.19)

$$\mathscr{L}' \leqslant \psi(r) + p_1|e(r)| + \text{Const.}\ |e(r)|\mathscr{L}.$$

The Gronwall inequality implies that there exists $M > 0$ such that

$$\mathscr{L}(r) \leqslant M \quad \text{on } [R, \infty). \tag{5.21}$$

Hence $H(u')$ is bounded and $u'$ must be bounded by $(5.19)_2$, say $|u'(r)| \leqslant L'$.

As in the first part of the proof of Theorem 5 we then see that

$$H(u') + F(r, u) + \int_r^\infty (\psi(s) + |e(s)|L')\,\mathrm{d}s$$

is decreasing in $[R, \infty)$. Hence (5.2) holds in view of $(5.19)_1$ and the proof is complete when $l = 0$. If $l > 0$ we continue as in the proof of Theorem 5. Again as in Section 2 we derive after some calculation

$$\left\{r^k\left(H(u') + F(r, u) + \frac{\alpha}{r}A(u')uu'\right)\right\}'$$
$$= r^{k-1}\Gamma(r) + r^k\left\{F_r(r, u) + e(r)\left(u' + \frac{\alpha u}{r}\right)\right\}.$$

The rest of the proof in case (5.19) is the same as for Theorem 5 since

$$e\left(u' + \frac{\alpha u}{r}\right) \in L^1[r_0, \infty)$$

by $(5.19)_1$ and the boundedness of $u$ and $u'$.

To obtain the final part of the corollary note that (5.21) implies also that $F(r, u)$ is bounded along the extremal. Therefore condition (5.20) gives the required boundedness of $u$, and the proof proceeds unchanged. □

The case (5.19–20) of the preceding corollary generalizes Theorem 1 of [7]. In particular the latter result applies when $A \equiv 1$ (so $H(p) = \frac{1}{2}p^2$) and when $h = h(r, u, p)$ is bounded above and below by positive constants. In [7] the upper bound is stated only for bounded ranges of $u$ and $p$, but the a priori boundedness of $u = u(r)$ and $u' = u'(r)$ has been established here in the last part of the proof of the corollary using (5.21) together with $(5.19)_2$ and (5.20). (Note also that (5.20) corresponds to (1.7) in [7].) Finally, in [7] the function $f$ is assumed independent of $r$ and Lipschitz continuous in $u$, and $e$ is supposed to be bounded and of class $L^1[r_0, \infty)$, while we have assumed only that $e$ is of class $L^1[r_0, \infty)$.

It almost goes without saying that condition (5.17) in the Corollary can be weakened in the directions indicated in the remarks following the proof of Theorem 5. In particular, in case $A \equiv 1$ the upper bound for $rh$ in (5.17) can be weakened to $\beta' r^2$.

## Part II. Special Operators

### 6. The degenerate Laplacian

The special case of the degenerate Laplace operator, when $G(p) = |p|^m/m$ and $A(p) = |p|^{m-2}$, $m > 1$, is of particular interest both as a special case and also because it illustrates an alternative approach to the earlier theorems, allowing for stronger dependence on $r$ of the function $F = F(r, u)$ than we have previously treated. In addition, for functions $F = F(r, u)$

with approximately power-law growth for large $r$ and small $u$, say Pos. Const. $r^\nu|u|^q \leqslant F(r, u) \leqslant$ Pos. Const. $r^\nu|u|^q$, $q>1$, we can obtain sharp estimates for the asymptotic behavior of solutions as $r \to \infty$.

Problem (2.1) here becomes

$$\delta\int_{\mathbb{R}^+} g(r)\left[\frac{1}{m}|u'|^m - F(r, u)\right] dr = 0, \quad m>1, \tag{6.1}$$

where we take $J=\mathbb{R}^+$. The corresponding differential equation (2.6) for this variational problem is

$$(g|u'|^{m-2}u')' + gf(r, u) = 0, \tag{6.1$'$}$$

or alternatively, written out in full,

$$(m-1)|u'|^{m-2}u'' + \delta(r)|u'|^{m-2}u' + f(r, u) = 0, \tag{6.1$''$}$$

where $\delta(r) = g'(r)/g(r)$ and $f(r, u) = \partial F(r, u)/\partial u$. The assumptions (a) and (c) of Section 2 are of course maintained throughout; assumption (b) is automatic since we treat the special operator $G(p) = |p|^m/m$, $m>1$.

As noted in Section 2, equation (6.1)$'$ is to be interpreted in the weak form that $|u'|^{m-2}u'$ is of class $C^1$ on any domain of interest; in turn equation (6.1)$''$ applies rigorously only when $u' \neq 0$ or $m=2$.

Throughout Part II we shall state our theorems in terms of extremals of (6.1). Naturally they apply in equal measure to solutions of (6.1)$'$ and, if $u$ is of class $C^2$, to solutions of (6.1)$''$.

The main result of this section requires the following properties.

(Q) There are numbers $R>0$, $C>0$, $q>1$, $\nu \in \mathbb{R}$ such that

$$F(r, u) \geqslant Cr^\nu|u|^q \quad \text{for every } r \geqslant R \text{ and } u \in \mathbb{R}.$$

(Q)$'$ There are numbers $R>0$ and $\alpha$, $\kappa \in \mathbb{R}$ such that for each $r \geqslant R$ and $u \in \mathbb{R}$

$$\kappa F(r, u) + rF_r(r, u) \leqslant \alpha u f(r, u).$$

The condition $q>1$ is used to guarantee that $f(r, u)$ is continuous as $u \to 0$, a necessity as long as one maintains $F \in C^1(\mathbb{R}^+ \times \mathbb{R})$.

When $f(r, u)$ is independent of $r$, equations (6.1)$'$ and (6.1)$''$ reduce to (1.5). The conclusions **1**–**7** described in the introduction are then obtained by specialization of the theorems below to the case when $\nu = 0$.

**Theorem 6.** *Let $u$ be an extremal of problem* (6.1) *whose domain contains $I_1 = [R, \infty)$. Suppose that there are real numbers $\beta$ and $\beta'$ such that*

$$\beta \leqslant r\delta(r) \leqslant \beta' \quad \textit{for } r \in I_1, \tag{6.2}$$

*and that* (Q), (Q)$'$ *hold with*

$$q \geqslant m, \qquad \nu + m > 0, \qquad \kappa + \nu > 0, \qquad \beta + \frac{m-1}{m}\nu > \alpha. \tag{6.3}$$

*Then $u$ decays algebraically to zero as $r \to \infty$.*

**Proof.** The principal identity (2.9) for the degenerate Laplace operator takes the form

$$\left\{r^k\left[\frac{m-1}{m}|u'|^m+F(r,u)+\frac{\alpha}{r}uu'|u'|^{m-2}\right]\right\}'$$
$$=r^{k-1}\left[kF(r,u)+rF_r(r,u)-\alpha uf(r,u)-\left(\xi-\alpha-\frac{m-1}{m}k\right)|u'|^m\right.$$
$$\left.-\frac{\alpha}{r}(\xi+1-k)uu'|u'|^{m-2}\right], \tag{6.4}$$

where $\xi=\xi(r)=rg'(r)/g(r)=r\delta(r)$ and $\alpha$ is the number in (Q)$'$. We choose $k\in\mathbb{R}$ such that

$$k\leqslant\kappa,\qquad \frac{m-1}{m}k\leqslant\beta-\alpha. \tag{6.5}$$

Thus, by (Q)$'$ and the fact that $F(r,u)\geqslant 0$, we have

$$kF(r,u)+rF_r(r,u)-\alpha uf(r,u)\leqslant 0\quad \text{for } r\geqslant R \text{ and } u\in\mathbb{R}.$$

Let us define

$$\theta=1-\frac{1}{m}+\frac{1}{q}.$$

Then by Young's inequality

$$|u|\cdot|u'|^{m-1}\leqslant r^{-\nu/q}\left\{\frac{1}{q\theta}(r^\nu|u|^q)^\theta+\frac{m-1}{m\theta}|u'|^{m\theta}\right\}$$
$$\leqslant r^{-\nu/q}\left\{(r^\nu|u|^q)^\theta+|u'|^{m\theta}\right\}.$$

Next, since $x^\theta+y^\theta\leqslant 2^{1-\theta}(x+y)^\theta$ if $x\geqslant 0$, $y\geqslant 0$ and $0<\theta\leqslant 1$, we obtain

$$|u|\cdot|u'|^{m-1}\leqslant 2^{1-\theta}r^{-\nu/q}\{r^\nu|u|^q+|u'|^m\}^\theta. \tag{6.6}$$

Note also by (6.2),

$$|\xi+1-k|\leqslant 1+|\beta|+|\beta'|+|k|=C_1\quad\text{for } r\geqslant R.$$

Thus by (6.6) the right hand side $\mathscr{R}=\mathscr{R}(r)$ of (6.4) satisfies

$$\mathscr{R}\leqslant 2^{1-\theta}|\alpha|C_1r^{k-2-\nu/q}\{r^\nu|u|^q+|u'|^m\}^\theta, \tag{6.7}$$

provided that $r\geqslant R$.

From here on the proof is slightly different depending on whether $q=m$ or $q>m$. We treat first the case $q>m$, observing that here $0<\theta<1$. Then, by using Young's inequality twice more, there results

$$\frac{|\alpha u|}{r}\cdot|u'|^{m-1}\leqslant\frac{1}{2}\left\{\frac{1}{m}\frac{|2\alpha u|^m}{r^m}+\frac{m-1}{m}|u'|^m\right\}$$
$$\leqslant\frac{1}{2}\left\{\frac{m-1}{m}|u'|^m+Cr^\nu|u|^q\right\}+C_2r^{-\sigma},\quad r\geqslant R,$$

where $C_2$ is an appropriate constant depending only on $\alpha$, $m$, $q$ and $C$, and

$$\sigma=m\frac{q+\nu}{q-m}.$$

Thus using (Q) the quantity $\mathscr{L}=\mathscr{L}(r)$ in braces on the left hand side of (6.4) satisfies

$$\frac{r^k}{2}\left(\frac{m-1}{m}|u'|^m + Cr^\nu |u|^q\right) \leqslant \mathscr{L} + C_2 r^{k-\sigma} \equiv \hat{\mathscr{L}} \quad \text{for } r \geqslant R. \tag{6.8}$$

Consequently by (6.7) and (6.8),

$$\mathscr{R} \leqslant C_3 r^{k(1-\theta)-2-\nu/q} \hat{\mathscr{L}}^\theta = C_3 r^{-\mu-1} \hat{\mathscr{L}}^\theta \quad \text{for } r \geqslant R, \tag{6.9}$$

where $C_3$ depends only on $\alpha$, $m$, $q$, $C$ and $C_1$, and

$$\mu m = (k+\nu)\left(\frac{m}{q}-1\right)+\nu+m.$$

Now by integration of (6.4) on $[r_1, r]$ and use of (6.8), we obtain

$$\hat{\mathscr{L}} \leqslant C_2 r^{k-\sigma} + \mathscr{L}(r_1) + C_3 \int_{r_1}^{r} s^{-\mu-1}\left(\hat{\mathscr{L}}(s)\right)^\theta \, \mathrm{d}s, \quad r \geqslant r_1 \geqslant R. \tag{6.10}$$

In what follows we shall assume that $k \leqslant \sigma$, that is

$$k \leqslant m\frac{q+\nu}{q-m}. \tag{6.11}$$

Hence,

$$\hat{\mathscr{L}} \leqslant C_4 + C_3 \int_{r_1}^{r} s^{-\mu-1}\left(\hat{\mathscr{L}}(s)\right)^\theta \, \mathrm{d}s, \tag{6.12}$$

where

$$C_4 = C_2 r_1^{k-\sigma} + \mathscr{L}(r_1).$$

The integral inequality (6.12) can be integrated by setting

$$Y(r) = C_4 + C_3 \int_{r_1}^{r} s^{-\mu-1}\left(\hat{\mathscr{L}}(s)\right)^\theta \, \mathrm{d}s,$$

so that

$$Y' \leqslant C_3 r^{-\mu-1} Y^\theta, \quad r \geqslant r_1. \tag{6.13}$$

Dividing each side of (6.13) by $Y^\theta$ and integrating over $[r_1, r]$ yields

$$\frac{1}{1-\theta}\left(Y(r)^{1-\theta} - Y(r_1)^{1-\theta}\right) \leqslant \frac{C_3}{\mu}\left(r_1^{-\mu} - r^{-\mu}\right), \tag{6.14}$$

provided $\mu \neq 0$.

To complete the proof we now make the final choice of $k$ so that

$$\mu > 0. \tag{6.15}$$

To see that this can be done, consistent with the earlier restrictions (6.5) and (6.11), it is enough to verify the following conditions

$$\begin{aligned} &k+\nu < q\frac{\nu+m}{q-m}, && [\text{see (6.11) and (6.15)}], \\ &k+\nu \leqslant \kappa+\nu, && [\text{see (6.5)}], \\ &k+\nu \leqslant \frac{m}{m-1}\left(\beta+\frac{m-1}{m}\nu-\alpha\right), && [\text{see (6.5)}]. \end{aligned} \tag{6.16}$$

It now follows from (6.14) that $Y = Y(r)$ is bounded on $[r_1, \infty)$, say $Y(r) \leqslant M$. Then (6.8) and (6.12) give

$$\frac{r^k}{2}\left(\frac{m-1}{m}|u'|^m + Cr^\nu |u|^q\right) \leqslant \hat{\mathscr{L}} \leqslant M, \tag{6.17}$$

whence in particular

$$|u(r)| \leqslant \left(\frac{2M}{C}\right)^{1/q} r^{-(k+\nu)/q} \quad \text{for } r \geqslant r_1.$$

By (6.3) condition (6.16) can be met with

$$k + \nu > 0.$$

This completes the proof when $q > m$.

When $q = m$ the condition (6.8) clearly holds with $C_2 = 0$ and $r_1$ suitably large, since $\nu + m > 0$. The integral inequality (6.12) can be treated as before (of course now with $\theta = 1$), or one can use the Gronwall inequality directly, to show that

$$\log \frac{Y(r)}{Y(r_1)} \leqslant \frac{C_3}{\mu}(r_1^{-\mu} - r^{-\mu}), \qquad \mu m = \nu + m. \tag{6.14$'$}$$

The rest of the proof is unchanged, except that the first inequality of (6.16) is no longer needed. This completes the proof of Theorem 6. □

**Corollary.** *Suppose that* (6.2) *holds and that for every* $r \geqslant R$ *and* $u \in \mathbb{R}$

$$rF_r(r, u) \leqslant \nu F(r, u), \qquad F(r, u) \geqslant r^\nu \Phi(u), \qquad uF_u(r, u) \geqslant \hat{\Phi}(u) F(r, u), \tag{6.18}$$

*where* $\Phi$, $\hat{\Phi}$ *are continuous functions on* $\mathbb{R}$ *such that*

$$\begin{aligned} &\Phi(u) > 0 \quad \textit{for } u \neq 0, \qquad \hat{\Phi}(u) > 0 \quad \textit{for } u \in \mathbb{R}, \\ &\Phi(u) > \text{Pos. Const.}\ |u|^p \quad \textit{for } u \textit{ small and } p > 0, \\ &\Phi(u) \to \infty \quad \textit{if } |u| \to \infty, \end{aligned} \tag{6.19}$$

*and* $\nu$ *is a real number such that*

$$\nu + m > 0, \qquad m\beta + (m-1)\nu \geqslant 0. \tag{6.20}$$

*Then any extremal* $u$ *of* (6.1) *on* $I_1$ *decays algebraically to zero as* $r \to \infty$.

**Proof.** Choose $\alpha = 0$, $k = -\nu$ in (6.4). Then by $(6.2)_1$, $(6.18)_1$ and $(6.20)_2$ we obtain along $u = u(r)$

$$\left\{r^{-\nu}\left[\frac{m-1}{m}|u'|^m + F(r, u)\right]\right\}' \leqslant 0.$$

Hence $r^{-\nu}F(r, u)$ is bounded, and in turn by $(6.18)_2$ also $\Phi(u)$ is bounded. By the final condition on $\Phi$ it now follows that $u$ is bounded, say $|u(r)| \leqslant U$ for large $r$.

We now apply Theorem 6. It is enough to show that (Q) and (Q)$'$ hold for $|u| \leqslant U$ and for some constants $C > 0$ and $\alpha$, $\kappa$ satisfying (6.3). First, for some sufficiently large $q > m$ it is clear that

$$\Phi(u) > |u|^q \quad \text{for } |u| < \tfrac{1}{2}.$$

Let $c$ be such that $\Phi(u) \geqslant c > 0$ for $\frac{1}{2} \leqslant |u| \leqslant U$. Then obviously (Q) holds for $|u| \leqslant U$ with

$$C = \min\{1, c|U|^{-q}\}.$$

Also $\hat{\Phi}(u) \geqslant \hat{c} > 0$ for $|u| \leqslant U$. Hence, if $\kappa + \nu > 0$ then

$$\kappa F(r, u) + rF_r(r, u) \leqslant (\kappa + \nu)F(r, u) \leqslant \frac{\kappa + \nu}{\hat{c}} uF_u(r, u)$$

for every $r \geqslant R$ and $|u| \leqslant U$. We now choose $\alpha = (\kappa + \nu)/\hat{c}$, and $\kappa$ so that

$$0 < \kappa + \nu < \hat{c}\left(\beta + \frac{m-1}{m}\nu\right),$$

which can surely be done in view of $(6.20)_2$. Thus both (Q)′ and (6.3) are satisfied, and the result of the corollary follows by application of Theorem 6. □

**Remarks.** A particular case when the corollary applies is

$$F(r, u) = r^\nu \Phi(u),$$

where $\nu$, $\Phi$ satisfy the conditions (6.19), (6.20), and also

$$u\Phi'(u) > 0 \quad \text{for } u \neq 0 \quad \text{and} \quad \liminf_{u \to 0} \frac{u\Phi'(u)}{\Phi(u)} > 0.$$

If $\nu \leqslant 0$ in Theorem 6 or its corollary, then also $u' \to 0$ algebraically as $r \to \infty$. Indeed, one sees easily that

$$u'(r) = O(r^{-k/m}) = O(r^{-(k+\nu)/m + \nu/m}),$$

proving the assertion.

If condition (6.3) is dropped then one can no longer meet (6.16) with a choice of $k$ such that $k + \nu > 0$. In such a case the proof shows that the growth rate of $u(r)$ as $r \to \infty$ cannot be *more* than algebraic.

## 7. Asymptotic estimates for the degenerate Laplacian

The decay exponent $(k + \nu)/q$ obtained for extremals $u$ in the previous section can be estimated quite accurately in many cases, including for example $F(r, u) = Cr^\nu |u|^q$. In particular, in the next two theorems we shall consider functions $g$ satisfying (6.2) and $F$ verifying (Q), for which also

$$rF_r(r, u) \leqslant \nu F(r, u), \tag{7.1}$$

and

$$\begin{aligned} uF_u(r, u) &\geqslant qF(r, u) \quad \text{if } \beta + \frac{m-1}{m}\nu > 0 \\ uF_u(r, u) &\leqslant qF(r, u) \quad \text{if } \beta + \frac{m-1}{m}\nu < 0 \end{aligned} \tag{7.2}$$

for all $r \geqslant R$ and every $u \in \mathbb{R}$. The numbers $\lambda_1$, $\lambda_2$, $\lambda_3$, defined by

$$\lambda_1 = \frac{m\beta + (m-1)\nu}{m + (m-1)q}, \qquad \lambda_2 = \frac{\beta q - \nu}{m + (m-1)q}, \qquad \lambda_3 = \frac{\nu + m}{q - m} \quad \text{if } q \neq m,$$

play an important role in what follows. We observe in the case $q > m$ that

$$\lambda_1 < \lambda_3 \quad \text{if and only if} \quad (q-m)(\beta+1-m) < m(\nu+m).$$

When $q - m \neq 0$ the crossover between the two cases $\lambda_1 < \lambda_3$ and $\lambda_1 > \lambda_3$ occurs when $(q-m)(\beta+1-m) = m(\nu+m)$, that is when

$$q = q_1 = m\frac{\beta+1+\nu}{\beta+1-m}, \qquad \beta+1-m \neq 0.$$

Our conclusions are quite different depending on whether $\lambda_1 < \lambda_3$ or $\lambda_1 > \lambda_3$. The former case is called the subcritical case, the latter the supercritical case. The value $q_1$ is the critical exponent for problem (6.1).

Theorem 7 below treats the subcritical case, while Theorem 8 deals with the supercritical one.

**Theorem 7.** *Suppose that the conditions* (Q), (6.2), (7.1) *and* (7.2) *are satisfied and assume*

$$q \geqslant m, \qquad (q-m)(\beta+1-m) < m(\nu+m). \tag{7.3}$$

*Let $u$ be an extremal of problem* (6.1) *on $I_1 = [R, \infty)$. Then*

$$u(r) = O(r^{-\lambda_1}), \quad u'(r) = O(r^{-\lambda_2}) \qquad \text{as } r \to \infty, \tag{7.4}$$

*and*

$$r^{q\lambda_1 - \nu}\left(\frac{m-1}{m}|u'|^m + F(r, u)\right) \to \text{Limit} \quad \text{as } r \to \infty. \tag{7.5}$$

*If $u$ is oscillatory as $r \to \infty$, then along the critical points of $u$*

$$r^{q\lambda_1 - \nu}F(r, u) \to \text{Limit}.$$

**Proof.** Put

$$\alpha = \lambda_1, \qquad \kappa = \alpha q - \nu. \tag{7.6}$$

By (7.1) and (7.2) for $r \geqslant R$ and $u \in \mathbb{R}$ we have

$$\kappa F(r, u) + rF_r(r, u) \leqslant \alpha q F(r, u) \leqslant \alpha u f(r, u),$$

since

$$\operatorname{sign} \alpha = \operatorname{sign} \lambda_1 = \operatorname{sign}\left(\beta + \frac{m-1}{m}\nu\right).$$

Therefore condition (Q)$'$ is satisfied. We now proceed as in the proof of Theorem 6. As there, it is convenient to divide the discussion into the two cases $q > m$ and $q = m$. When $q > m$, we obtain (6.17) for any $k$ satisfying (6.16). With the choice of $\alpha$ and $\kappa$ given in (7.6), and since $\lambda_1 < \lambda_3$ by (7.3), condition (6.16) reduces simply to $k + \nu \leqslant q\lambda_1$. Hence we can choose

$$k = q\lambda_1 - \nu. \tag{7.7}$$

Now by (6.17) it follows that $r^{k+\nu}|u|^q$ and $r^k|u'|^m$ are bounded. The choice (7.7) now gives (7.4) when $q > m$.

To obtain (7.5), we return to (6.4) and observe by (6.9) and (6.17) that the right hand side $\mathscr{R}$ satisfies

$$\mathscr{R} \leqslant C_3 r^{-\mu-1}M^\theta = C_5 r^{-\mu-1}.$$

Hence (6.4) yields the inequality

$$\mathscr{L}' \leqslant C_5 r^{-\mu-1}.$$

With the choice (7.7) of $k$ we have explicitly

$$\mu m = (q-m)(\lambda_3 - \lambda_1) > 0,$$

which in turn implies

$$\mathscr{L} + \frac{C_5}{\mu} r^{-\mu} \searrow \text{Limit} \quad (\text{or } -\infty). \tag{7.8}$$

On the other hand, by (7.4) we have

$$\begin{aligned} |\alpha r^{k-1} u u' |u'|^{m-2}| &= |\alpha| r^{k-1} |u| \cdot |u'|^{m-1} \\ &\leqslant \text{Const. } r^{k-1} \cdot r^{-\lambda_1} \cdot r^{-(m-1)\lambda_2} = \text{Const. } r^{-\mu}. \end{aligned}$$

It now follows from (7.8) and the definition of $\mathscr{L}$ that

$$r^k \left( \frac{m-1}{m} |u'|^m + F(r, u) \right) \to \text{Limit},$$

completing the proof of (7.5) when $q > m$. If $q = m$, the hypothesis (7.3) implies $\nu + m > 0$. Thus the previous calculations continue to apply, except that (6.17) is obtained from (6.14)$'$.

The final statement of the theorem follows at once from (7.5). □

In Theorem 7 the exponent $\lambda_1$ may have any sign. When it is positive, the solution $u = u(r)$ necessarily decays to zero as $r \to \infty$, only if

$$\beta + \frac{m-1}{m} \nu > 0.$$

In this regard it is interesting to consider the example

$$u'' + \frac{\beta}{r} u' + \left(1 + \gamma r^{-(\nu+2)}\right) r^{\nu} u = 0, \tag{7.9}$$

where $\gamma = (2\beta + \nu)(2\beta - \nu - 4)/16$ and $\nu + 2 \neq 0$. Here $q = m = 2$ of course. This has the general solution

$$u(r) = r^{-(2\beta+\nu)/4} \left[ a \sin\left( \frac{2}{\nu+2} r^{(\nu+2)/2} \right) + b \cos\left( \frac{2}{\nu+2} r^{(\nu+2)/2} \right) \right], \tag{7.10}$$

which becomes unbounded as $r \to \infty$ when $\beta + \nu/2 < 0$, and which tends to zero only when $\beta + \nu/2 > 0$. Equation (7.9) satisfies the hypotheses of Theorem 7 when $0 < \nu + 2 < 2|\beta - 1|$. We find then from Theorem 7

$$\lambda_1 = \tfrac{1}{4}(2\beta + \nu), \qquad \lambda_2 = \tfrac{1}{4}(2\beta - \nu),$$

so that (7.10) is in agreement with (7.4) and (7.5). There are corresponding examples when $q \neq m$ and when $\nu + m = 0$.

Conditions (7.1) and (7.2)$_1$ imply that when $u \neq 0$

$$\frac{\partial}{\partial r} \left( \frac{F(r, u)}{r^{\nu}} \right) \leqslant 0, \qquad \operatorname{sign} u \, \frac{\partial}{\partial u} \left( \frac{F(r, u)}{|u|^q} \right) \geqslant 0, \tag{7.11}$$

whence also

$$\frac{\partial}{\partial r}\left(\frac{F(r, u)}{r^{\nu}|u|^{q}}\right) \leqslant 0, \qquad \operatorname{sign} u \, \frac{\partial}{\partial u}\left(\frac{F(r, u)}{r^{\nu}|u|^{q}}\right) \geqslant 0. \tag{7.12}$$

With the help of these inequalities we have the following

**Lemma.** *Let* (7.1) *and* $(7.2)_1$ *hold with* $F(r, u) \geqslant 0$ *in* $\mathbb{R}^+ \times \mathbb{R}$. *Then there exist nonnegative constants* $C_{\pm}$ *such that*

$$\lim_{r \to \infty,\, u \to \pm 0} \frac{F(r, u)}{r^{\nu}|u|^{q}} = C_{\pm}.$$

**Proof.** It is enough to prove the lemma as $u \to +0$. If for contradiction the limit does not exist, then there would be monotone sequences $r_j \nearrow \infty$, $u_j \searrow 0$ and $\bar{r}_j \nearrow \infty$, $\bar{u}_j \searrow 0$ such that

$$\lim_j \frac{F(r_j, u_j)}{r_j^{\nu}|u_j|^{q}} = C_1, \qquad \lim_j \frac{F(\bar{r}_j, \bar{u}_j)}{\bar{r}_j^{\nu}|\bar{u}_j|^{q}} = C_2,$$

with, say, $0 \leqslant C_1 < C_2 \leqslant \infty$.

Fix $(\bar{r}, \bar{u})$, $\bar{r} \geqslant R$, $\bar{u} > 0$. Consider the path $\Gamma$ in the $(r, u)$ plane consisting of straight segments joining the successive vertices $(\bar{r}, \bar{u})$, $(\bar{r}_k, \bar{u}_k)$, $(\bar{r}_{k+1}, \bar{u}_{k+1}), \ldots$, where $k$ is chosen so large that $\bar{r}_k > \bar{r}$, $\bar{u}_k < \bar{u}$. Then by (7.12) the function $F(r, u)/r^{\nu}|u|^{q}$ is decreasing along $\Gamma$. Hence

$$\frac{F(\bar{r}, \bar{u})}{\bar{r}^{\nu}|\bar{u}|^{q}} \geqslant C_2,$$

(so $C_2 < \infty$). Since $(\bar{r}, \bar{u})$ is arbitrary we find that $C_1 \geqslant C_2$. This contradiction proves the lemma. □

If (7.1) and $(7.2)_2$ hold, with $F(r, u) \geqslant 0$ on $\mathbb{R}^+ \times \mathbb{R}$, then similarly

$$\lim_{r \to \infty,\, u \to \pm\infty} \frac{F(r, u)}{r^{\nu}|u|^{q}} = D_{\pm},$$

where $D_{\pm} \geqslant 0$.

A further consequence of these results in the estimate

$$F(r, u) \leqslant \text{Const. } r^{\nu} \times \begin{cases} |u|^{q}, & |u| \leqslant M, \quad \lambda_1 > 0, \\ |u|, & |u| \leqslant M, \quad \lambda_1 = 0, \\ 1 + |u|^{q}, & u \in \mathbb{R}, \quad \lambda_1 < 0, \end{cases} \tag{Q$''$}$$

provided (7.1) and (7.2) are satisfied and $r \geqslant R$. Indeed, by (7.12)

$$\frac{F(r, u)}{r^{\nu}|u|^{q}} \leqslant \frac{\max\{F(R, -M), F(R, M)\}}{R^{\nu}M} = \text{Const.} \quad \text{for } r \geqslant R, \ |u| \leqslant M, \ \lambda_1 > 0,$$

and analogously

$$\frac{F(r, u)}{r^{\nu}|u|^{q}} \leqslant \text{Const.} \quad \text{for } r \geqslant R, \ |u| \geqslant M, \ \lambda_1 < 0,$$

where $M$ is any fixed positive number. It remains only to prove that

$$F(r, u) \leqslant \text{Const. } r^{\nu}|u| \quad \text{for } r \geqslant R, \ |u| \leqslant M.$$

But this follows directly from $(7.11)_1$ together with the estimate $F(R, u) \leqslant \text{Const.}\,|u|$ for $|u| \leqslant M$ (see, Section 2, assumption (c)).

If these results are used in (7.5) we obtain the following.

**Corollary to Theorem 7.** *Under the hypotheses of Theorem 7, if $\lambda_1 \neq 0$, there exist constants $A$, $B$ such that*

$$r^{q\lambda_1-\nu}\left(\frac{m-1}{m}|u'|^m + \binom{A}{B} r^{\nu}|u|^q\right) \to \text{Limit} \quad \textit{as } r \to \infty, \tag{7.13}$$

*where $\binom{A}{B}$ is the function whose value is $A$ when $u \geqslant 0$ and $B$ when $u < 0$.*

We turn next to the supercritical case.

**Theorem 8.** *Suppose that conditions* (Q), (6.2), (7.1) *and* (7.2) *hold, and that*

$$q > m, \qquad (q-m)(\beta+1-m) > m(\nu+m). \tag{7.14}$$

*Let $u$ be an extremal of* (6.1) *on $I_1 = [R, \infty)$. Then*

$$u(r) = O(r^{-\lambda_3}), \quad u'(r) = O(r^{-\lambda_3-1}) \qquad \textit{as } r \to \infty. \tag{7.15}$$

*The order behavior in* (7.15) *can be taken independent of the extremal, depending only on $\beta$, $m$, $\nu$, $q$ and $C$.*

**Proof.** With the choice of $\alpha$ and $\kappa$ given by (7.6) the condition (Q)′ is satisfied, as shown in the first part of the proof of Theorem 7. Thus as before we can apply the proof technique of Theorem 6. The conditions (6.16) for the choice of $k$ now become simply $k + \nu < q\lambda_3$ since by (7.14) we have $\lambda_3 < \lambda_1$. We shall nevertheless choose $k$ so that

$$q\lambda_3 - \nu < k \leqslant q\lambda_1 - \nu, \tag{7.16}$$

in other words, while satisfying (6.5) we also allow $k > \sigma(= q\lambda_3 - \nu)$ in the final part of the proof of Theorem 6. Thus (6.10) now gives, not (6.12), but

$$\hat{\mathscr{L}} \leqslant C_2 r^{k-\sigma} + \mathscr{L}(r_1) + C_3\int_{r_1}^{r} s^{-\mu-1}(\hat{\mathscr{L}}(s))^{\theta}\,\mathrm{d}s. \tag{6.12$'$}$$

Next, setting

$$\hat{Y}(r) = \mathscr{L}(r_1) + C_3\int_{r_1}^{r} s^{-\mu-1}(\hat{\mathscr{L}}(s))^{\theta}\,\mathrm{d}s,$$

(6.12)′ becomes, after some calculation,

$$\hat{Y}' \leqslant C_3 r^{k-\sigma-1}\left(C_2 + r^{\sigma-k}\hat{Y}\right)^{\theta}. \tag{6.13$'$}$$

To solve (6.13)′, put $Z = r^{\sigma-k}\hat{Y}$ to obtain

$$Z' \leqslant \frac{1}{r}\left\{(\sigma-k)Z + C_3(C_2+Z)^{\theta}\right\}.$$

Clearly $Z' \leqslant 0$ for $Z$ sufficiently large, since $k - \sigma > 0$ and $\theta < 1$. Hence $Z$ is bounded, with the bound independent of $u$. It follows in turn that

$$\hat{\mathscr{L}} \leqslant \text{Const. } r^{k-\sigma},$$

from which the theorem results by use of (6.8) and the relation $\sigma = q\lambda_3 - \nu$. □

When $\nu + m > 0$ the estimate (7.15) is best possible, as is shown by the example

$$r\delta(r) = \beta, \quad F(r, u) = C\, r^{\nu} |u|^{q}, \qquad r \in \mathbb{R}^{+},\ u \in \mathbb{R}.$$

Indeed in this case a direct calculation from (6.1)″ shows that (6.1) has a unique extremal of the form

$$u(r) = \text{Pos. Const. } r^{-\lambda_3} \tag{7.17}$$

if and only if

$$(q - m)(\beta + 1 - m) > (m - 1)(\nu + m). \tag{7.18}$$

This condition is clearly satisfied when (7.14) holds. When $\nu + m < 0$ we find similarly that there is an extremal of the form (7.17) if and only if

$$(q - m)(\beta + 1 - m) < (m - 1)(\nu + m). \tag{7.19}$$

When $\beta < m - 1$ this condition and (7.14) can be simultaneously satisfied, for $q$ in a bounded interval of values. Thus, for this range, (7.15) is again best possible. On the other hand, whenever (7.14) holds but (7.19) fails, the conclusion (7.15) can be improved (see the next Theorem 9).

**Remarks.** 1. Clearly (7.1) and (7.2) are satisfied by any function of the form $F(r, u) = \rho(r)\Phi(u)$ provided that $\Phi(0) = 0$, $\Phi(u) > 0$ for $u \neq 0$ and

$$\operatorname{sign} \lambda_1 \{ u\Phi'(u) - q\Phi(u) \} \geqslant 0 \quad \text{for } u \in \mathbb{R}, \qquad \nu\rho(r) \geqslant r\rho'(r) \quad \text{for } r \geqslant R.$$

2. In the radial case when $g(r) = r^{n-1}$, $n > 1$, we have $\beta = n - 1$. The exponents $\lambda_1$, $\lambda_3$ are then

$$\lambda_1 = \frac{m(n-1) + (m-1)\nu}{m + (m-1)q} \quad \text{and} \quad \lambda_3 = \frac{\nu + m}{q - m}.$$

When $q > m$ the inequalities $\lambda_1 < \lambda_3$ and $\lambda_1 > \lambda_3$ occur respectively if and only if $q$ is sub- or super-critical; that is

$$q(n - m) < m(n + \nu) \quad \text{or} \quad q(n - m) > m(n + \nu).$$

For the Laplace operator, and when $\nu = 0$, we find, even more specifically,

$$\lambda_1 = (n-1)\frac{2}{q+2} \quad \text{and} \quad \lambda_3 = \frac{2}{q-2}.$$

These are the well-known values essentially due to Fowler [4].

3. In the above discussion of the ranges of the exponent $q$ we have avoided the critical case

$$q = q_1 = m\frac{\beta + 1 + \nu}{\beta + 1 - m}, \quad \beta \neq m - 1.$$

For this particular value of $q$, however, and with $k$ chosen to be $q\lambda_1 - \nu$, we get $\lambda_1 = \lambda_3$, $k = \sigma$ and $\mu = 0$. Let us suppose also $q > m$. In this case the principal inequality (6.13), or (6.13)′, becomes $Y' \leqslant C_3 r^{-1} Y^{\theta}$, which introduces an additional logarithmic term in (7.15). This then takes the form

$$u(r) = O\left(r^{-\lambda_3}(\log r)^{m/(q-m)}\right), \quad u'(r) = O\left(r^{-\lambda_3 - 1}(\log r)^{q/(q-m)}\right) \qquad \text{as } r \to \infty.$$

4. We note that if the extremal $u$ is assumed to be bounded, then conditions (Q), (7.1) and (7.2) need hold only in the range of $u$.

## 8. Four lemmas

At the end of the previous section, we noted that when the condition (7.14) holds but (7.19) fails one can expect to improve the conclusion of Theorem 8. In Section 9 we shall take up this situation in more detail; the present section is devoted to some preliminary results which are convenient to group together here.

**Lemma 1.** *Let $u$ be an extremal of* (6.1) *on $I_1 = [R, \infty)$. Suppose conditions* (Q), $(6.2)_1$ *and* (7.1) *hold. Then*

(i) *if $m\beta + (m-1)\nu \geqslant 0$, we have*

$$u(r) = O(1), \quad u'(r) = O(r^{\nu/m}) \qquad \text{as } r \to \infty;$$

(ii) *if $m\beta + (m-1)\nu < 0$, then along the extremal $u$ we have*

$$r^{m\beta/(m-1)}\left(\frac{m-1}{m}|u'|^m + F(r, u)\right) \searrow \text{Limit}$$

*and*

$$u(r) = O\left(r^{-(m\beta+(m-1)\nu)/(m-1)q}\right), \quad u'(r) = O\left(r^{-\beta/(m-1)}\right) \qquad \text{as } r \to \infty.$$

**Proof.** Choose $\alpha = 0$ in (6.4). By (7.1)

$$kF(r, u) + rF_r(r, u) \leqslant (k+\nu)F(r, u).$$

Moreover by $(6.2)_1$ we have

$$\xi - \frac{m-1}{m}k = r\delta(r) - \frac{m-1}{m}k \geqslant 0$$

provided that $k + \nu \leqslant (m\beta + (m-1)\nu)/(m-1)$.

(i) Suppose $m\beta + (m-1)\nu \geqslant 0$. Then taking $k = -\nu$ we see that the right hand side of (6.4) is non-positive. In turn, with the help of condition (Q),

$$r^{-\nu}\left\{\frac{m-1}{m}|u'|^m + Cr^{\nu}|u|^q\right\} \leqslant \text{Constant} \quad \text{as } r \to \infty,$$

and so

$$u(r) = O(1), \quad u'(r) = O(r^{\nu/m}) \qquad \text{as } r \to \infty.$$

(ii) Suppose $m\beta + (m-1)\nu < 0$. Then taking $k = m\beta/(m-1)$ it is clear that $k + \nu < 0$. The result now follows exactly as in case (i). □

In what follows it is convenient to set

$$\lambda_4 = \frac{\beta+1-m}{m-1}, \qquad \lambda_4' = \frac{\beta'+1-m}{m-1}. \tag{8.1}$$

Recall also that

$$\lambda_3 = \frac{\nu+m}{q-m} \quad \text{if } q \neq m.$$

**Lemma 2.** *Suppose that u is extremal of* (6.1) *on* $I_1 = [R, \infty)$ *satisfying, for some* $\lambda > 0$,

$$u(r) = \mathrm{O}(r^{-\lambda}) \quad \text{as } r \to \infty. \tag{8.2}$$

*Assume that*

$$\begin{aligned} &|f(r, u)| \leqslant Cr^{\nu}|u|^{q-1} \quad \text{for } r \geqslant R,\ |u| \leqslant U, \\ &r^{-\beta}g(r) \leqslant \text{Const. } s^{-\beta}g(s) \quad \text{for } R \leqslant r < s, \end{aligned} \tag{8.3}$$

*where C, U are positive numbers and* $q > 1$.

*If any one of the following conditions is satisfied*:

(i) $q > m$, $\quad \lambda_3 < \lambda \leqslant \lambda_4$,

(ii) $q = m$, $\quad \nu + m < 0$, $\quad \lambda \leqslant \lambda_4$,

(iii) $q < m$, $\quad \lambda \leqslant \lambda_4 < \lambda_3$,

*then*

$$u(r) = \mathrm{O}(r^{-\lambda_4}), \quad u'(r) = \mathrm{O}(r^{-\lambda_4 - 1}) \qquad \text{as } r \to \infty.$$

*Furthermore, if*

(iv) $q < m$, $\quad \lambda < \lambda_3 < \lambda_4$,

*then*

$$u(r) = \mathrm{O}(r^{-\lambda_3}), \quad u'(r) = \mathrm{O}(r^{-\lambda_3 - 1}) \qquad \text{as } r \to \infty.$$

**Proof.** We follow an idea of Egnell [2]. By (6.1)′ together with the fact that $u(r) \to 0$ as $r \to \infty$ since $\lambda > 0$,

$$\begin{aligned} |u(r)| &= \left| \lim_{r_1 \to \infty} \int_r^{r_1} u'(s)\, \mathrm{d}s \right| \\ &\leqslant \int_r^{\infty} \left| \frac{1}{g(s)} \int_R^s g(t) f(t, u(t))\, \mathrm{d}t + \frac{\text{Const.}}{g(s)} \right|^{1/(m-1)} \mathrm{d}s. \end{aligned} \tag{8.4}$$

Thus for all suitably large $r$, by (8.2) and (8.3)

$$\begin{aligned} |u(r)| &\leqslant \text{Const.} \int_r^{\infty} \left\{ \int_R^s \left( \frac{t}{s} \right)^{\beta} t^{\nu - \lambda(q-1)}\, \mathrm{d}t + s^{-\beta} \right\}^{1/(m-1)} \mathrm{d}s \\ &= \text{Const.} \int_r^{\infty} \left\{ \frac{s^{\nu+1-\lambda(q-1)}}{\sigma_1} + \left( 1 - \frac{R^{\sigma_1}}{\sigma_1} \right) s^{-\beta} \right\}^{1/(m-1)} \mathrm{d}s, \end{aligned} \tag{8.5}$$

when $\sigma_1 \equiv \beta + \nu + 1 - \lambda(q-1) \neq 0$. We can in fact always suppose $\sigma_1 \neq 0$ in this integration, and similarly in later formulas, by slightly reducing the value of $\lambda$ in the original estimate (8.2).

To carry out the outer integration, observe that the exponents in (8.5) satisfy

$$\frac{\beta}{m-1} > 1, \qquad \frac{\lambda(q-1)-(\nu+1)}{m-1} > 1;$$

the first of these is an immediate consequence of the relation $\lambda_4 \geqslant \lambda > 0$, which holds in all cases of the lemma; the second similarly is easily checked in all cases. Thus from (8.5), in all cases of the lemma,

$$|u(r)| \leqslant \frac{\text{Const.}}{|\sigma_1|^{1/(m-1)}} \left\{ \frac{r^{(\nu+m-\lambda(q-1))/(m-1)}}{\lambda(q-1)-(\nu+m)} + \frac{r^{-\lambda_4}}{\lambda_4} \right\}.$$

An easy iteration of this procedure leads by induction, after $n$ steps, to

$$u(r) = \begin{cases} O(r^{-\lambda_3-(\lambda-\lambda_3)\zeta^n} + r^{-\lambda_4}) & q \neq m, \quad \text{cases (i), (iii), (iv)}, \\ O(r^{-\lambda+n(\nu+m)/(m-1)} + r^{-\lambda_4}), & q = m, \quad \text{case (ii)}, \end{cases} \tag{8.6}$$

where $\zeta = (q-1)/(m-1)$. The validity of the iteration process relies on the fact that at each step the outer integral in (8.5) is convergent, that is, for $n = 1, 2, \ldots$,

$$\lambda_n = \begin{cases} \lambda_3 + (\lambda-\lambda_3)\zeta^n \geqslant \dfrac{(\lambda-\lambda_3)(q-m)}{m-1} > 0, & \text{cases (i), (iii), (iv)}, \\ \lambda - n\dfrac{\nu+m}{m-1} \geqslant -\dfrac{\nu+m}{m-1} > 0, & \text{case (ii)}. \end{cases}$$

With the exponents decreased by 1, the estimate (8.6) also holds for $u'(r)$ as $r \to \infty$.

In the first case of the lemma we have $\zeta > 1$, so that for $n$ suitably large

$$\lambda_n = \lambda_3 + (\lambda-\lambda_3)\zeta^n \geqslant \lambda_4.$$

Hence

$$u(r) = O(r^{-\lambda_4}), \quad u'(r) = O(r^{-\lambda_4-1}) \qquad \text{as } r \to \infty,$$

as required. In the second case, for $n$ suitably large

$$\lambda_n = \lambda - n\frac{\nu+m}{m-1} \geqslant \lambda_4,$$

and the result follows as before. Next $\zeta < 1$ and $\lambda < \lambda_3$ in the third case, so that again for $n$ large

$$\lambda_n = \lambda_3 + (\lambda-\lambda_3)\zeta^n \geqslant \lambda_4.$$

Finally, if $q < m$ and $\lambda < \lambda_3 < \lambda_4$ a careful repetition of the iteration process shows that in fact, for all $n = 1, 2, \ldots$,

$$|u(r)| \leqslant A_n r^{-\lambda_3+(\lambda_3-\lambda)\zeta^n},$$

where $A_n$ satisfies the recursive relation $A_n \leqslant AA_{n-1}^{\zeta}$ and $A$ is a constant independent of $n$; here we have strongly used the relations

$$\lambda_n \geqslant \frac{(\lambda-\lambda_3)(q-m)}{m-1} > 0,$$

$$\sigma_n = \beta + \nu + 1 - \lambda_n(q-1) = (m-1)(\lambda_4-\lambda_3) + (q-1)(\lambda_3-\lambda)\zeta^n$$
$$\geqslant (m-1)(\lambda_4-\lambda_3) > 0.$$

It follows now that

$$|u(r)| \leqslant A^{1+\zeta+\dots+\zeta^{n-1}} A_0^{\zeta^n} r^{-\lambda_3+(\lambda_3-\lambda)\zeta^n}$$
$$= A^{1/(1-\zeta)}\left(A^{1/(\zeta-1)} A_0 r^{\lambda_3-\lambda}\right)^{\zeta^n} r^{-\lambda_3}. \tag{8.7}$$

Letting $n \to \infty$ and noting that $\zeta < 1$ now completes the proof. □

If $\lambda_3 = \lambda_4$ in case (iv) of Lemma 2 then $\sigma_n = (q-1)(\lambda_3 - \lambda)\zeta^n$ and the iteration relation becomes

$$A_n \leqslant A A_{n-1}^{\zeta} \zeta^{-n/(m-1)},$$

and in turn

$$|u(r)| \leqslant \text{Const. } r^{(\lambda_3-\lambda)\zeta^n} r^{-\lambda_3} \zeta^{-n/(m-q)}, \quad n = 1, 2, \dots.$$

For fixed $r$, choose $n$ so that $\zeta^{1-n} < \log r \leqslant \zeta^{-n}$. Then

$$|u(r)| \leqslant \text{Const. } (\log r)^{1/(m-q)} r^{-\lambda_3}.$$

**Lemma 3.** *Suppose that* (8.3) *holds and u is an extremal of* (6.1) *on* $I_1 = [R, \infty)$ *which tends to a finite, non-zero limit* $u_0$ *as* $r \to \infty$. *Then as* $r \to \infty$

$$u(r) - u_0 = \begin{cases} \mathrm{O}(r^{-\lambda_4}), & \text{if } \beta + \nu + 1 < 0, \\ \mathrm{O}\left(r^{-\lambda_4}(\log r)^{1/(m-1)}\right), & \text{if } \beta + \nu + 1 = 0, \\ \mathrm{O}\left(r^{(\nu+m)/(m-1)}\right), & \text{if } \beta + \nu + 1 > 0. \end{cases}$$

This result is clearly of interest only if $\beta > m - 1$ and $\nu + m < 0$, and therefore we assume that these conditions are satisfied in the proof.

**Proof.** As in the proof of Lemma 2, we find

$$|u(r) - u_0| \leqslant \text{Const.} \int_r^\infty \left\{ \int_R^s \left(\frac{t}{s}\right)^\beta t^\nu \, \mathrm{d}t + s^{-\beta} \right\}^{1/(m-1)} \mathrm{d}s.$$

If $\beta + \nu + 1 < 0$ the inner integral is convergent and the result follows at once. When $\beta + \nu + 1 = 0$ we get

$$|u(r) - u_0| \leqslant \text{Const.} \int_r^\infty \left(s^{-\beta} \log s\right)^{1/(m-1)} \mathrm{d}s = \mathrm{O}\left(r^{-\lambda_4}(\log r)^{1/(m-1)}\right),$$

by L'Hôpital's rule. Finally if $\beta + \nu + 1 > 0$ then

$$|u(r) - u_0| \leqslant \text{Const.} \int_r^\infty s^{(\nu+1)/(m-1)} \, \mathrm{d}s = \text{Const. } r^{(\nu+m)/(m-1)},$$

completing the proof. □

**Lemma 4.** *Suppose that u is an extremal of* (6.1) *on* $I_1 = [R, \infty)$ *satisfying*

$$r^{\lambda_4'} u(r) \to 0 \quad \text{as } r \to \infty \quad \left(\text{if } \lambda_4' = 0 \text{ replace } r^{\lambda_4'} \text{ by } 1/\log r\right). \tag{8.8}$$

*Assume in addition*

$$\begin{aligned} &|f(r,u)| \leqslant Cr^{\nu}|u|^{q-1} \quad \text{for } r \geqslant R,\ u \in \mathbb{R}, \\ &r^{-\beta'}g(r) \leqslant \text{Const. } s^{-\beta'}g(s) \quad \text{for } R \leqslant s < r, \end{aligned} \tag{8.9}$$

*where $C$, $R$ are positive numbers and $q > 1$.*

*If either of the following conditions is satisfied*:

(i) $q > m, \quad \nu + m < 0, \quad \lambda_4' > \lambda_3,$

(ii) $q \leqslant m, \quad \nu + m < 0,$

*then $u$ tends to a finite limit as $r \to \infty$. On the other hand, if*

(iii) $q < m, \quad \nu + m > 0,$

*then*

$$u(r) = O(r^{-\lambda_3}), \quad u'(r) = O(r^{-\lambda_3 - 1}) \qquad \text{as } r \to \infty.$$

**Proof.** We assert first that

$$u'(r) = o(r^{-\beta'/(m-1)}), \tag{8.10}$$

on some sequence of values of $r$ tending to $\infty$. Indeed, if (8.10) fails then

$$|u'(r)| \geqslant \text{Pos. Const. } r^{-\beta'/(m-1)} \quad \text{as } r \to \infty. \tag{8.11}$$

Therefore, for every $r \geqslant R$

$$|u(r)| = \begin{cases} \displaystyle\int_r^{\infty} |u'(s)|\,\mathrm{d}s, & \lambda_4' > 0, \\ \displaystyle\int_R^{r} |u'(s)|\,\mathrm{d}s + \text{Const.}, & \lambda_4' \leqslant 0, \end{cases}$$

where, when $\lambda_4' > 0$ we use that fact that $u(r) \to 0$ as $r \to \infty$. Now using (8.11) again, if $\lambda_4' \neq 0$ then $r^{\lambda_4'}|u(r)| \geqslant$ Pos. Const., violating (8.8). Similarly, if $\lambda_4' = 0$, then $(\log r)^{-1}|u(r)| \geqslant$ Pos. Const., again contradicting (8.8).

From $(8.9)_2$ and (8.10) we get

$$g(r)|u'(r)|^{m-1} = O(r^{\beta'})o(r^{-\beta'}) \to 0$$

on some sequence of values of $r \to \infty$. Therefore from (6.1)′, as in Lemma 2,

$$|u(r)| \leqslant \int_R^r \left\{ \frac{1}{g(s)} \int_s^{\infty} |g(t) f(t, u(t))|\,\mathrm{d}t \right\}^{1/(m-1)} \mathrm{d}s + \text{Const.} \tag{8.12}$$

We now consider each of the three cases of the lemma separately. In case (i)

$$\beta' + \nu + 1 - \lambda_4'(q-1) = (q-m)(\lambda_3 - \lambda_4') < 0,$$

so, as in Lemma 2, the inner integral can be estimated by means of (8.8) and (8.9). Therefore by integration and then iteration $n$ times, as in the proof of Lemma 2,

$$u(r) = O(r^{-\lambda_3 - (\lambda_4' - \lambda_3)\zeta^n} + 1).$$

Hence, for large enough $n$, it is apparent that $u$ is bounded and $u'$ integrable on $I_1$, as required.

If (ii) holds, if $\lambda'_4 < 0$ and also $q < m$, then since $\lambda_3 > 0$ the inner integral can be estimated as before, and iteration leads to

$$u(r) = O(r^{-\lambda_3 + (\lambda_3 - \lambda'_4)\zeta''} + 1).$$

Thus ultimately $u$ is bounded and $u'$ integrable on $I_1$. If $\lambda'_4 = 0$ a logarithmic factor occurs in the iteration procedure, but otherwise the proof is unchanged. Finally, when $q = m$ the same argument gives

$$u(r) = O(r^{-\lambda'_4 + n(\nu+m)/(m-1)} + 1),$$

and the result follows as before.

If (iii) is satisfied and $\lambda'_4 \geqslant \lambda_3$, then by (8.8) there is nothing to prove. Otherwise, if $\lambda'_4 < \lambda_3$ the result follows as in the final part of the proof of Lemma 2 (replace $\lambda$ by $\lambda'_4$). □

## 9. Further results. The case $1 < q < m$

With the lemmas of Section 8 in hand, we can now extend the results of Theorem 8 to further parameter values and can also treat the previously omitted range $1 < q < m$ of the exponent $q$.

In particular, instead of condition (7.14) of Theorem 8 we shall assume that

$$q \geqslant m, \qquad \nu + m < 0, \qquad (q-m)(\beta + 1 - m) > (m-1)(\nu + m) \tag{9.1}$$

is satisfied. Note that (9.1) is incompatible with (7.4); Theorem 7 is optimal as it stands.

**Theorem 9.** *Suppose that conditions* (Q), (Q)″, $(6.2)_1$, (7.1) *and* (9.1) *hold. Let $u$ be an extremal of* (6.1) *on $I_1 = [R, \infty)$. Then*

I. *If $\beta < m - 1$ we have*

$$r^{\lambda_4} u(r) \to \text{Limit}, \quad r^{\lambda_4 + 1} u'(r) \to \text{Limit} \qquad \text{as } r \to \infty.$$

II. *If $\beta = m - 1$ then*

$$u(r)/\log r \to \text{Limit}, \quad r u'(r) \to \text{Limit} \qquad \text{as } r \to \infty.$$

III. *If $\beta > m - 1$ then $u(r)$ tends to a finite limit $u_0$ as $r \to \infty$. If in addition $(8.3)_1$ holds, the asymptotic behavior of $u(r)$ is given by Lemma 3 when $u_0 \neq 0$, and by*

$$r^{\lambda_4} u(r) \to \text{Limit}, \quad r^{\lambda_4 + 1} u'(r) \to \text{Limit} \qquad \text{as } r \to \infty$$

*when $u_0 = 0$.*

IV. *Suppose that* (8.8) *and* (8.9) *are satisfied. Then $u$ tends to a finite limit $u_0$ as $r \to \infty$. If $u_0 = 0$ then $u(r) \equiv 0$ for large $r$.*

**Proof.** In order not to interrupt the later development, we note here that if $(6.2)_1$ holds then

$$r^{-\beta} g(r) \leqslant s^{-\beta} g(s), \quad R \leqslant r < s, \tag{9.2}$$

as follows from the relation

$$g(r) = \exp \int_R^r \delta(t)\, dt.$$

Now suppose $\beta \leqslant m-1$. Then from

$$m\beta + (m-1)\nu = m(\beta + 1 - m) + (m-1)(\nu + m) \tag{9.3}$$

we get $m\beta + (m-1)\nu < 0$ by $(9.1)_2$. From Lemma 1 (ii) and $(8.1)_1$ we now find

$$u'(r) = O(r^{-\beta/(m-1)}) = O(r^{-\lambda_4 - 1}) \quad \text{as } r \to \infty,$$

and then by integration $u(r) = O(r^{-\lambda_4})$ as $r \to \infty$. Hence from (Q)$''$, along the extremal $u$,

$$r^{m\beta/(m-1)} F(r, u) \leqslant \text{Const. } r^{m\beta/(m-1)} r^{\nu} r^{-q\lambda_4}$$

$$= \text{Const. } r^{\nu + m - (q-m)\lambda_4} \to 0 \quad \text{as } r \to \infty,$$

by $(9.1)_3$. Thus, again by Lemma 1 (ii),

$$r^{m\beta/(m-1)} |u'(r)|^m \to \text{Limit},$$

that is $r^{\lambda_4 + 1} u'(r) \to$ Limit, which proves I and II.

Next suppose $\beta > m-1$. Since also $\nu + m < 0$ we see that in both cases of Lemma 1 the function $u'$ is integrable on $I_1$. Hence $u(r)$ tends to a finite limit $u_0$ as $r \to \infty$. If $u_0 \neq 0$ the asymptotic behavior of $u - u_0$ follows from Lemma 3. Thus suppose $u_0 = 0$. There are two cases:

*Case* 1. $m\beta + (m-1)\nu < 0$. The result follows as in case I, since by (Q)$''$ and the fact that $u(r) \to 0$ as $r \to \infty$, we have

$$r^{m\beta/(m-1)} F(r, u) \leqslant \text{Const. } r^{(m\beta + (m-1)\nu)/(m-1)} \to 0 \quad \text{as } r \to \infty.$$

*Case* 2. $m\beta + (m-1)\nu \geqslant 0$. By Lemma 1 (i) as $r \to \infty$

$$u(r) = \int_r^{\infty} u'(s)\, ds = O(r^{(\nu + m)/m}),$$

that is, the hypotheses of Lemma 2 hold with

$$\lambda = -\frac{\nu + m}{m} > 0.$$

Note that $\lambda \leqslant \lambda_4$ in view of (9.3), since here $m\beta + (m-1)\nu \geqslant 0$.

If $q = m$, then by Lemma 2(ii)

$$u(r) = O(r^{-\lambda_4}), \quad u'(r) = O(r^{-\lambda_4 - 1}) \qquad \text{as } r \to \infty.$$

When $q > m$ then $\lambda > \lambda_3$ since $\nu + m < 0$, and the same order behavior is now obtained from Lemma 2(i). Now consider (6.4) with $\alpha = 0$, $k = m\beta/(m-1)$, namely

$$\left\{ r^k \left( \frac{m-1}{m} |u'|^m + F(r, u) \right) \right\}' = r^{k-1} \left\{ kF + rF_r - \left( \xi - \frac{m-1}{m} k \right) |u'|^m \right\}$$

$$\leqslant r^{k-1}(k + \nu) F(r, u) \tag{9.4}$$

by (7.1) and $(6.2)_1$. In case $m\beta + (m-1)\nu = 0$ we have $k + \nu = 0$ and the right hand side of the inequality of (9.4) is zero. If $m\beta + (m-1)\nu > 0$ then from (Q)$''$

$$r^k F(r, u) \leqslant \text{Const. } r^{k+\nu} |u|^q \leqslant \text{Const. } r^{k + \nu - q\lambda_4} = \text{Const. } r^{-\mu_1},$$

where $\mu_1 = (q-m)\lambda_4 - (\nu+m) > 0$ by (9.1). It follows that, in both cases $m\beta + (m-1)\nu = 0$ and $m\beta + (m-1)\nu > 0$, along the extremal $u$ we have

$$r^k\left(\frac{m-1}{m}\,|u'|^m + F(r,u)\right) \to \text{Limit} \quad \text{as } r \to \infty.$$

Also in both cases $r^k F(r,u) \to 0$ as $r \to \infty$ (using (Q)$''$ again) and the proof of III is complete.

To obtain the first part of IV we note by (8.9) and (9.2) that necessarily $\beta' \geqslant \beta$. Then from $(9.1)_3$ follows $\lambda'_4 \geqslant \lambda_4 > \lambda_3$ when $q > m$. This being shown, the first part of IV is then an immediate consequence of Lemma 4(i) when $q > m$ and of Lemma 4(ii) when $q = m$. To obtain the last part we apply the iteration of Lemma 4 again, but now beginning with

$$|u(r)| \leqslant \int_r^\infty \left\{\frac{1}{g(s)}\int_s^\infty |g(t) f(t, u(t))|\,dt\right\}^{1/(m-1)} ds, \tag{9.5}$$

instead of (8.12), since $u_0 = 0$. We now proceed recursively, as before dividing the discussion into several cases.

*Case* 1. $q > m$ and $\lambda'_4 \geqslant 0$. Since $r^{\lambda'_4} u(r) \to 0$ as $r \to \infty$ we can argue as in the proof of Lemma 2, starting with (9.5) and taking $\lambda = \lambda'_4$. There follows, in analogy with (8.7),

$$|u(r)| \leqslant A_0 r^{-\lambda'_4}\left(A_0 A^{1/(\zeta-1)} r^{\lambda_3 - \lambda'_4}\right)^{\zeta^n - 1}, \tag{9.6}$$

where $A$ and $A_0$ are constants independent of $n$. Clearly $\lambda_3 - \lambda'_4 < 0$, so

$$A_0 A^{1/(\zeta-1)} r^{\lambda_3 - \lambda'_4} < 1$$

for $r$ sufficiently large, say $r \geqslant r_1$. Thus letting $n \to \infty$ in (9.6) and recalling that $\zeta > 1$, we obtain

$$u(r) \equiv 0 \quad \text{for } r \geqslant r_1.$$

*Case* 2. $q > m$ and $\lambda'_4 < 0$. The argument is the same as for case 1, except that in the iteration of (9.5) we replace $\lambda = \lambda'_4$ by $\lambda = 0$, which can surely be done since $u(r) \to u_0 = 0$ as $r \to \infty$. Then (9.6) becomes

$$|u(r)| \leqslant A_0\left(A_0 A^{1/(\zeta-1)} r^{\lambda_3}\right)^{\zeta^n - 1}. \tag{9.7}$$

Here $\lambda_3 < \lambda'_4 < 0$. Again letting $n \to \infty$ gives $u(r) \equiv 0$ for sufficiently large $r$.

*Case* 3. $q = m$. The same conclusion follows from the iteration formulas

$$|u(r)| \leqslant \begin{cases} A_0 r^{-\lambda'_4}\left(A r^{(\nu+m)/(m-1)}\right)^n & \text{if } \lambda'_4 \geqslant 0 \\ A_0\left(A r^{(\nu+m)/(m-1)}\right)^n & \text{if } \lambda'_4 < 0. \end{cases} \qquad \square$$

**Remark.** When $q > m$ the condition $(9.1)_3$ is equivalent to $\lambda_3 < \lambda_4$. Thus it is evident that Theorem 9 gives a better result than Theorem 8 when both (7.14) and (9.1) apply. If $(9.1)_3$ fails and $q > m$ then $\lambda_4 \leqslant \lambda_3 < 0$. In this case Theorem 8 is clearly the appropriate result.

In general the hypotheses of Theorem 9 imply that the extremal is monotone; see the remarks after the proofs of Theorems 13 and 14.

We consider finally the case $1 < q < m$.

**Theorem 10.** *Let u be an extremal of* (6.1) *on* $I_1 = [R, \infty)$. *Suppose* $1 < q < m$ *and that conditions* (Q), (6.2), (7.1) *and* (7.2) *hold. If*

$$\nu + m > 0, \qquad (q-m)(\beta+1-m) < (m-1)(\nu+m), \tag{9.8}$$

*then the conclusions of Theorem 7 and its corollary are valid.*

**Proof.** The proof relies on the argument already used for Theorem 7, which in turn is based on the calculations given in the proof of Theorem 6. For the present case, when $1 < q < m$, only small modifications are necessary. In particular, using Young's inequality as in the proof of Theorem 6, we can replace (6.6) by

$$|u| \cdot |u'|^{m-1} \leqslant r^{-\nu/m} |u|^{(m-q)/m} (r^\nu |u|^q + |u'|^m). \tag{9.9}$$

If $m\beta + (m-1)\nu \geqslant 0$ then $u(r) = O(1)$ as $r \to \infty$ by Lemma 1(i), that is,

$$|u(r)| \leqslant L \quad \text{for } r \geqslant R. \tag{9.10}$$

On the other hand, if $m\beta + (m-1)\nu < 0$, then by Lemma 1(ii),

$$|u(r)| \leqslant L\, r^{-(m\beta+(m-1)\nu)/(m-1)q} \quad \text{for } r \geqslant R, \tag{9.11}$$

where $L$ is an appropriate positive constant.

Inserting (9.10) and (9.11) into (9.9) yields

$$|u| \cdot |u'|^{m-1} \leqslant L^{(m-q)/m} r^{-\mu+1} (r^\nu |u|^q + |u'|^m),$$

where

$$\mu = \begin{cases} \dfrac{\nu+m}{m} & \text{if } m\beta + (m-1)\nu \geqslant 0, \\[2ex] \dfrac{m-q}{q}(\lambda_4 - \lambda_3) & \text{if } m\beta + (m-1)\nu < 0. \end{cases}$$

In turn, see (6.7), there results

$$\mathscr{R} \leqslant |\alpha| C_1 L^{(m-q)/m} r^{k-\mu-1} (r^\nu |u|^q + |u'|^m).$$

Here $\mu > 0$ by $(9.8)_1$ and $(9.8)_2$, respectively.

Now using Young's inequality once more, we obtain

$$\begin{aligned} \frac{|\alpha u|}{r} \cdot |u'|^{m-1} &\leqslant \frac{1}{2}\left\{ \frac{1}{m} \frac{|2\alpha u|^m}{r^m} + \frac{m-1}{m} |u'|^m \right\} \\ &\leqslant \frac{1}{2}\left\{ \frac{m-1}{m} |u'|^m + C r^\nu |u|^q \right\} \end{aligned}$$

for $r$ sufficiently large, thanks to $(9.8)_1$. Thus one can take $\theta = 1$, $C_2 = 0$, and follow the proof of Theorem 6 for the case $q = m$ to obtain (6.14)′ and the corresponding (6.17). In turn $r^{k+\nu} |u|^q$ and $r^k |u'|^m$ are bounded on $I_1$.

We can now continue as in the proof of Theorem 7, choosing $\alpha$, $\kappa$ as in (7.6) and $k$ as in (7.7). This proves (7.4), and (7.5) then follows at once. □

**Theorem 11.** *Let u be an extremal of* (6.1) *on* $[R, \infty)$. *Suppose* $1 < q < m$ *and that conditions* (Q), (Q)″, $(6.2)_1$, (7.1) *are satisfied.*

I. *If* $\beta < m-1$, $(q-m)(\beta+1-m) > (m-1)(\nu+m)$, *then*

$$r^{\lambda_4}u(r) \to \text{Limit}, \quad r^{\lambda_4+1}u'(r) \to \text{Limit} \qquad \text{as } r \to \infty.$$

II. *If* $\beta = m-1$, $\nu + m < 0$, *then*

$$u(r)/\log r \to \text{Limit}, \quad ru'(r) \to \text{Limit} \qquad \text{as } r \to \infty.$$

III. *If* $\beta > m-1$, $\nu + m < 0$, *then* $u(r)$ *tends to a finite limit* $u_0$ *as* $r \to \infty$. *If in addition* $(8.3)_1$ *holds, then the behavior at* $\infty$ *of* $u(r)$ *is given by Lemma* 3 *when* $u_0 \neq 0$, *and by*

$$\begin{aligned} &r^{\lambda_4}u(r) \to \text{Limit}, && r^{\lambda_4+1}u'(r) \to \text{Limit}, && \text{for } \lambda_3 > \lambda_4,\\ &u(r) = O(r^{-\lambda_3}), && u'(r) = O(r^{-\lambda_3-1}), && \text{for } \lambda_3 < \lambda_4, \end{aligned}$$

*when* $u_0 = 0$.

IV. *If* (8.8) *and* (8.9) *are satisfied, then u tends to a finite limit* $u_0$ *as* $r \to \infty$ *when* $\nu + m < 0$, *and satisfies*

$$u(r) = O(r^{-\lambda_3}), \quad u'(r) = O(r^{-\lambda_3-1}) \qquad \text{as } r \to \infty,$$

*when* $\nu + m > 0$.

**Proof.** I. Here

$$\begin{aligned} m\beta + (m-1)\nu &= m(\beta+1-m) + (m-1)(\nu+m)\\ &< q(\beta+1-m) < 0 \end{aligned} \tag{9.12}$$

by the assumption. Lemma 1(ii) now yields

$$u(r) = O(r^{-\lambda_4}), \quad u'(r) = O(r^{-\lambda_4-1}) \qquad \text{as } r \to \infty.$$

Since $\lambda_3 > \lambda_4$ the argument then proceeds exactly as in the proof of Theorem 9.I.

II. Since $\beta = m-1$, $\nu + m < 0$ we must have $m\beta + (m-1)\nu < 0$ as in (9.12), and so can proceed as in the proof of Theorem 9.II.

III. The assumption implies that in both cases of Lemma 1 the derivative $u'$ is integrable on $I_1$. Thus $u(r)$ tends to a finite limit $u_0$ as $r \to \infty$. If $u_0 \neq 0$ we apply Lemma 3. If $u_0 = 0$ we use Lemma 2, there being two cases:

*Case* 1. $m\beta + (m-1)\nu < 0$. Here, using (9.12)

$$\lambda_3 - \lambda_4 = \frac{m-1}{q-m}\{m\beta + (m-1)\nu - q(\beta+1-m)\} > 0.$$

The conclusion now follows almost exactly as in case III of Theorem 9.

*Case* 2. $m\beta + (m-1)\nu \geqslant 0$. Then by integration of Lemma 1(i) we get (as in the proof of Theorem 9)

$$u(r) = O(r^{-\lambda}), \qquad \lambda = -\frac{\nu+m}{m} > 0.$$

Here $\lambda < \lambda_3$ and $\lambda \leqslant \lambda_4$. If $\lambda_4 < \lambda_4$ the conclusion is obtained from Lemma 2(iv). If $\lambda_3 > \lambda_4$ then

$$u(r) = O(r^{-\lambda_4}), \quad u'(r) = O(r^{-\lambda_4 - 1}) \qquad \text{as } r \to \infty,$$

from Lemma 2(iii). By treating (6.4) with $\alpha = 0$, $k = m\beta/(m-1)$ as in the proof of Theorem 9.III, we obtain the remaining part of III.

IV. The final assertion is an immediate consequence of Lemma 4(ii)–(iii). □

The conclusions of Theorems 9–11 are essentially best possible. This is shown both by the examples of the previous section and the following:

The function $u(r) = r^{-\lambda_4}$ satisfies (6.1)″ when

$$f(r, u) = r^{\nu} u |u|^{q-2}, \qquad r\delta(r) = \beta + \lambda_4 |\lambda_4|^{-m} r^{(q-m)(\lambda_3 - \lambda_4)},$$
$$\beta \neq m - 1, \qquad (q-m)(\lambda_3 - \lambda_4) < 0.$$

The function $u(r) = 1 + cr^{(\nu+m)/(m-1)}$ satisfies (6.1)″ when

$$f(r, u) = r^{\nu} u |u|^{q-2}, \qquad r\delta(r) = \beta - (\beta + \nu + 1)\{1 - (1 + cr^{\alpha})^{q-1}\},$$
$$\alpha = \frac{\nu + m}{m - 1} < 0, \qquad \beta + 1 + \nu \neq 0, \qquad c = -\frac{\beta + \nu + 1}{\alpha |\beta + \nu + 1|^{m/(m-1)}}.$$

The function $u(r) = 1 + r^{-\lambda_4}$, satisfies (6.1)″ when

$$f(r, u) = r^{\nu} u |u|^{q-2}, \qquad r\delta(r) = \beta + \lambda_4^{1-m} r^{\beta+\nu+1} (1 + r^{-\lambda_4})^{q-1},$$
$$\nu + m < 0, \qquad \beta + 1 - m > 0, \qquad \beta + \nu + 1 < 0.$$

**Recapitulation**

The large number of results contained in Theorems 7–11 for various ranges of the parameters $\nu$, $m$, $\beta$, $q$ can be efficiently organized, even if somewhat oversimplified, by giving a graphical description of the maximal exponent $\lambda$ for which the condition

$$u(r) = O(r^{-\lambda}) \quad \text{as } r \to \infty \tag{9.13}$$

can be expected to hold. That is, the exponent $\lambda$ in the following figures is such that (9.13) holds for all extremals of (6.1), but at the same time cannot be increased (in view of explicit examples). We consider specifically the following parameter ranges:

Fig. 1. $\nu + m > 0$, $\beta + 1 - m > 0$;
Fig. 2. $\nu + m > 0$, $\beta + 1 - m \leqslant 0$, $\beta + 1 + \nu \geqslant 0$;
Fig. 3. $\nu + m \geqslant 0$, $\beta + 1 - m < 0$, $\beta + 1 + \nu < 0$;
Fig. 4. $\nu + m < 0$, $\beta + 1 - m < 0$;
Fig. 5. $\nu + m \leqslant 0$, $\beta + 1 - m \geqslant 0$;

which mutually exhaust all possible values for the parameters $\nu$, $m$, $\beta$. In each of these ranges $\lambda$

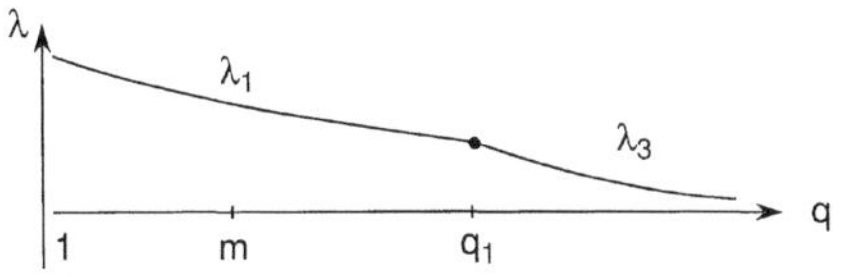


Fig. 1. $\nu + m > 0$, $\beta + 1 - m > 0$. Theorem 7 for $m \leqslant q < q_1$; Theorem 8 for $q > q_1$; Theorem 10 for $1 < q < m$. Here $q_1 > m$.

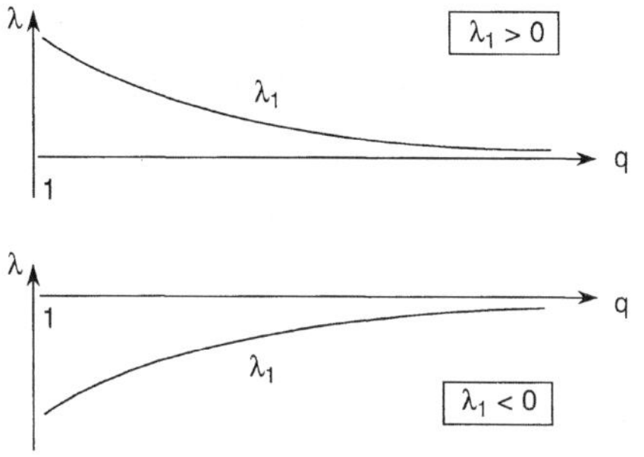


Fig. 2. $\nu + m > 0$, $\beta + 1 - m \leqslant 0$, $\beta + 1 + \nu \geqslant 0$. Theorem 7 for $q \geqslant m$; Theorem 10 for $1 < q < m$.

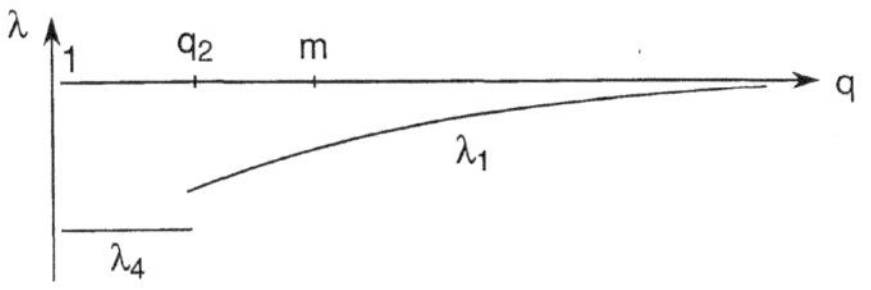


Fig. 3. $\nu + m \geqslant 0$, $\beta + 1 - m < 0$, $\beta + 1 + \nu < 0$. Theorem 7 for $q \geqslant m$, $q \neq q_2$; Theorem 10 for $q_2 < q < m$; Theorem 11–I for $1 < q < q_2$. Here $1 < q_2 \leqslant m$.

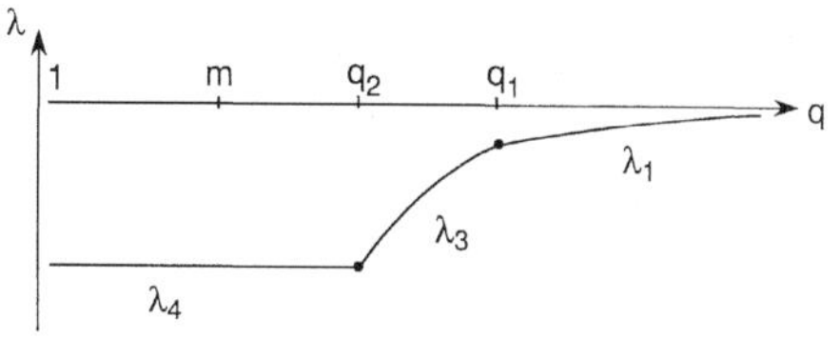


Fig. 4. $\nu + m < 0$, $\beta + 1 - m < 0$. Theorem 7 for $q > q_1$; Theorem 8 for $q_2 \leqslant q < q_1$; Theorem 9–I for $m \leqslant q < q_2$; Theorem 11–I for $1 < q < m$. Here $m < q_2 < q_1$.

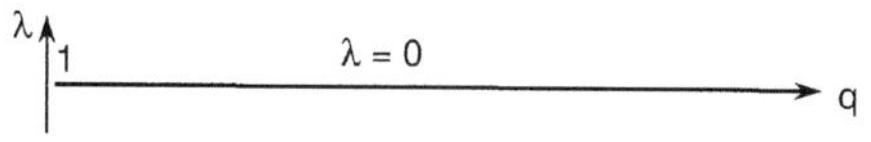


Fig. 5. $\nu + m \leqslant 0$, $\beta + 1 - m \geqslant 0$. Lemma 1(i) when $\nu + m = 0$; Theorems 9–II and 11–II when $\nu + m < 0$, $\beta + 1 - m = 0$ (here $u(r)/\log r \to$ Limit); Theorems 9–III and 11–III when $\nu + m < 0$, $\beta + 1 - m > 0$.

is indicated as a function of the remaining, and in a sense definitive, parameter $q$. In Figs. 1, 3 and 4 the numbers $q_1$ and $q_2$ are given by

$$q_1 - m = m\frac{\nu + m}{\beta + 1 - m}, \qquad q_2 - m = (m-1)\frac{\nu + m}{\beta + 1 - m}.$$

## 10. Monotone solutions

The asymptotic behavior obtained in Sections 7 and 9 can be improved in certain respects without adding essentially to the conditions on the function $F$.

**Theorem 12.** *Suppose that*

$$uf(r,u) \geqslant Cr^{\nu}|u|^{q} \quad \text{for all } r \geqslant R \text{ and } |u| \leqslant U,$$
$$r^{-\beta'}g(r) \leqslant \text{Const. } s^{-\beta'}g(s) \quad \text{for } R \leqslant s < r,$$

*where $C$, $R$, $U$ are given positive numbers and $\nu + m > 0$, $q > 1$. Let $u$ be a one-signed extremal of* (6.1) *on $I_1 = [R, \infty)$ which tends to zero as $r \to \infty$. Then necessarily*

$$\beta' > m-1, \qquad (q-m)(\beta'+1-m) > (m-1)(\nu+m) \tag{10.1}$$

*and $u$ is ultimately strictly monotone with*

$$u(r) = O(r^{-\lambda_3}) \quad \text{as } r \to \infty, \qquad \lambda_3 = \frac{\nu+m}{q-m}. \tag{10.2}$$

**Remarks.** A particular consequence of this result is that if the exponent $q$ in Theorem 7 or Theorem 10 satisfies

$$q - m \leqslant (m-1)\frac{\nu+m}{\beta'+1-m},$$

then the extremal $u$ must necessarily be oscillatory.

It is clear that Theorem 12 can also be formulated equivalently as a result about solutions of the differential inequality

$$(|u'|^{m-2}u')' + \delta(r)|u'|^{m-2}u' + Cr^{\nu}|u|^{q-2}u \leqslant 0$$

when $u > 0$ (in case $u < 0$ the direction of the differential inequality should be reversed). A similar remark holds for the following Theorems 13 and 14.

**Proof of Theorem 12.** We follow the procedure given in [9], Theorems 2.0, 2.1 and 2.2. First, as in Theorem 2.0 we see that when $\beta' \leqslant m-1$ there can be no one-signed extremals tending to zero as $r \to \infty$. Hence necessarily $\beta' > m-1$. Moreover, as is obvious from equation (6.1)$'$ or as shown in [9], the extremal is necessarily strictly monotone.

Next, the argument in the proof of Theorem 2.1 yields

$$|u(r)| \geqslant \text{Pos. Const. } r^{-(\beta'+1-m)/(m-1)} \tag{10.3}$$

for all $r$ sufficiently large.

Finally, we apply the proof method of [9], Theorem 2.2, with obvious changes to account for the fact that now $m \neq 2$, $\nu \neq 0$ and $g(r) \neq r^{n-1}$. The procedure is straightforward, except that the integral on line 3 of page 236 now takes the form

$$\int_R^r g(s)s^{\nu}\,\mathrm{d}s.$$

This term however can easily be estimated if we first divide by $g(r)$, thus

$$\frac{1}{g(r)}\int_R^r g(s)s^{\nu}\,\mathrm{d}s \geqslant \text{Const.}\int_R^r \left(\frac{s}{r}\right)^{\beta'} s^{\nu}\,\mathrm{d}s$$
$$= \frac{\text{Const.}}{\beta'+1+\nu} r^{1+\nu}\left(1 - \left(\frac{R}{r}\right)^{\beta'+1+\nu}\right), \quad r > R,$$

the constant being positive. Continuing as in [9], this now leads to the relation

$$u(r) = \mathrm{O}(r^{-(\nu+m)/(q-m)}) \tag{10.4}$$

when $q > m$ and

$$u(r) = \mathrm{O}(\exp(-Cr^{(\nu+m)/(m-1)})), \tag{10.5}$$

when $q = m$. The assumption $q < m$, on the other hand, leads to an impossibility (again see [9, p. 236]). The estimate (10.5) stands in contradiction to (10.3) so necessarily $q > m$. Moreover (10.4) contradicts (10.3) unless

$$q - m \geqslant (m-1)\frac{\nu + m}{\beta' + 1 - m}.$$

The case of equality cannot occur here, this being shown mutatis mutandis as in the final part of the proof of Theorem 2.2 ([9], pp. 236–7). This completes the proof. □

If a solution of (6.1) decays even slightly faster than the rate $\mathrm{O}(r^{-\lambda_3})$ as $r \to \infty$, then it will generally decay at a considerably faster rate, and, even more, its precise asymptotic behavior can be ascertained. In particular the following result holds.

**Theorem 13.** *Suppose that*

$$uf(r, u) \leqslant Cr^{\nu}|u|^q, \quad \beta \leqslant r\delta(r) + r\psi(r) \qquad \text{for all } r \geqslant R, \ |u| \leqslant U, \tag{10.6}$$

*where* $C$, $R$, $U$ *are given positive numbers, the function* $\psi$ *satisfies*

$$\psi(r) = \mathrm{o}(r^{-1}) \quad \text{as } r \to \infty, \ \psi \in L^1[R, \infty), \tag{10.6$'$}$$

*and*

$$q > m, \qquad \nu + m > 0, \qquad \beta > m - 1. \tag{10.7}$$

*If* $u$ *is an extremal of* (6.1) *on* $I_1 = [R, \infty)$ *such that*

$$u(r) = \mathrm{o}(r^{-\lambda_3}) \quad \text{as } r \to \infty, \tag{10.8}$$

*then* $u$ *is ultimately monotone and in addition*

$$u(r) = \mathrm{O}(r^{-\lambda_4}), \quad u'(r) = \mathrm{O}(r^{-\lambda_4 - 1}) \qquad \text{as } r \to \infty, \tag{10.9}$$

*where* $\lambda_4$ *is defined in* (8.1).

*Moreover if* (7.1) *holds, if* $|F(r, u)| \leqslant Cr^{\nu}|u|^q$ *for* $r \geqslant R$, $|u| \leqslant U$, *and also*

$$(q - m)(\beta + 1 - m) > (m - 1)(\nu + m), \tag{10.10}$$

*then*

$$r^{\lambda_4}u(r) \to \text{Limit}, \quad r^{\lambda_4+1}u'(r) \to \text{Limit} \qquad \text{as } r \to \infty. \tag{10.11}$$

As a corollary of Theorems 12 and 13 suppose that $u$ is a non-trivial monotone solution of (6.1) and that

$$\delta(r) - \beta/r \in L^1[R, \infty), \qquad \text{Pos. Const. } r^{\nu}|u|^q \leqslant uf(r, u) \leqslant \text{Pos. Const. } r^{\nu}|u|^q,$$

where

$$\nu + m > 0, \qquad q > 1.$$

Then Theorem 12 implies (10.1) with $\beta' = \beta$, and in turn (10.7) and (10.10) are satisfied. By (10.2) and (10.3)

$$\text{Pos. Const. } r^{-\lambda_4} \leqslant |u(r)| \leqslant \text{Pos. Const. } r^{-\lambda_3}.$$

By Theorem 13, if also

$$u(r) = \mathrm{o}(r^{-\lambda_3}), \qquad r\delta(r) - \beta = \mathrm{o}(1),$$

then

$$r^{\lambda_4}u(r) \to \text{Non-zero Limit}.$$

This provides a precise classification of the limit behavior of monotone solutions.

**Proof of Theorem 13.** It can be assumed that the solution $u$ is not identically zero for large $r$.

We introduce the new variable $\phi = \phi(r)$ by means of

$$\phi(r) = r^{\omega}u(r),$$

where the exponent $\omega$ is chosen to satisfy

$$0 < \omega < \min\left\{\lambda_3, \frac{\beta + 1 - m}{m}\right\} \tag{10.12}$$

(a similar change of variables occurs in [9]). By direct calculation from (6.1)″, at all points where $u' \neq 0$ we have

$$\left|\phi' - \omega\frac{\phi}{r}\right|^{m-2}\left\{(m-1)\phi'' + [\xi(r) - 2\omega(m-1)]\frac{\phi'}{r} + \omega[(m-1)(\omega+1) - \xi(r)]\frac{\phi}{r^2}\right\}$$
$$+ r^{(m-1)\omega}f(r, u) = 0,$$

where $\xi(r) = r\delta(r)$. We assert that, for sufficiently large $r$, necessarily $\phi'' > 0$ at a positive critical point of $\phi$, namely where $\phi > 0$, $\phi' = 0$. Indeed, when $\phi' = 0$, $\phi > 0$ there holds

$$(m-1)\phi'' = \omega[\xi(r) - (m-1)(\omega+1)]\frac{\phi}{r^2} - r^{(m-1)\omega}f(r, u)\left(\omega\frac{\phi}{r}\right)^{2-m}$$
$$\geqslant \omega\left\{[\beta + 1 - m - (m-1)\omega - r\psi(r)] - C\omega^{1-m}r^{\nu+m}u^{q-m}\right\}\frac{\phi}{r^2},$$

by (10.6). By (10.8) we have $r^{\nu+m}u^{q-m}(r) \to 0$ as $r \to \infty$, and by $(10.6)_1'$ also $r\psi(r) \to 0$ as $r \to \infty$. Hence by (10.12) for suitably large $r$ the right hand side is positive and $\phi'' > 0$. A similar argument shows that $\phi'' < 0$ at a critical point with negative critical value.

This result shows that $\phi$ must ultimately be monotone, with $\phi' \neq 0$. We may thus suppose without loss of generality that ultimately $\phi > 0$, and in consequence $u > 0$. Moreover, since $\omega < \lambda_3$,

$$\phi(r) = r^{\omega}u(r) = \mathrm{o}(r^{\omega - \lambda_3}),$$

by (10.8), so that $\phi(r) \to 0$ as $r \to \infty$. It follows that the case $\phi' > 0$ is impossible. Hence $\phi' < 0$ for large $r$, and in turn

$$u'(r) = r^{-\omega-1}(r\phi'(r) - \omega\phi(r)) < 0.$$

Thus $u$ is ultimately monotone.

We next prove (10.9). If $\lambda_4 \leqslant \lambda_3$ there is nothing to show, so we can assume $\lambda_4 > \lambda_3$. Now, from (6.4) with $k = m(1+\omega)$, $\alpha = \omega$ and $d > 0$, there follows without difficulty

$$\left\{ r^k \left[ \frac{m-1}{m} |u'|^m + \frac{\omega}{r} uu' |u'|^{m-2} + d\left(\frac{u}{r}\right)^m \right] \right\}'$$

$$= r^{k-1} \left\{ -(ru' + \omega u) \left[ f(r, u) + \frac{md}{r} \left(\frac{u}{r}\right)^{m-1} \right] - \left( \xi - \omega - \frac{m-1}{m} k \right) |u'|^m - \frac{\omega}{r} (\xi + 1 - k) uu' |u'|^{m-2} \right\}$$

$$= r^{k-2} (ru' + \omega u) \left\{ -rf(r, u) + md\left(\frac{u}{r}\right)^{m-1} - (\xi + 1 - k) u' |u'|^{m-2} \right\}. \tag{10.13}$$

In this relation, by (10.6), (10.6)′ and (10.12),

$$\xi + 1 - k \geqslant \beta + 1 - m(1+\omega) - r\psi(r) > 0 \tag{10.14}$$

for all sufficiently large $r$. Moreover, also for suitably large $r$,

$$-rf(r, u) + md\left(\frac{u}{r}\right)^{m-1} > 0 \tag{10.15}$$

by $(10.6)_1$ and (10.8). Finally, as noted earlier,

$$ru' + \omega u = r^{1-\omega}\phi' < 0, \quad u > 0, \ u' < 0 \tag{10.16}$$

again for large $r$. Let $r_1$ be chosen so that the three conditions (10.14), (10.15) and (10.16) all hold for $r \geqslant r_1$.

Define

$$\bar{\mathscr{L}}(r) = r^k \left\{ \frac{m-1}{m} |u'|^m + \frac{\omega}{r} uu' |u'|^{m-2} + d\left(\frac{u}{r}\right)^m \right\}.$$

By (10.14), (10.15) and (10.16) the right hand side of (10.13) is negative for $r \geqslant r_1$, and hence $\bar{\mathscr{L}}$ is decreasing. If we set $x = -ru'/u$, then

$$x > \omega \ (>0),$$

by (10.16). Therefore

$$\bar{\mathscr{L}}(r) = \left( \frac{m-1}{m} x^m - \omega x^{m-1} + d \right) \phi^m > \left( -\frac{\omega^m}{m} + d \right) \phi^m,$$

and so

$$\lim_{r\to\infty} \bar{\mathscr{L}}(r) \geqslant \left( -\frac{\omega^m}{m} + d \right) \lim_{r\to\infty} \phi^m(r) = 0.$$

Consequently $\bar{\mathscr{L}}(r) > 0$ because $\bar{\mathscr{L}}$ is decreasing; in turn

$$\frac{m-1}{m} x^m - \omega x^{m-1} + d > 0 \quad \text{for } r \geqslant r_1.$$

A simple analysis of this condition on the set $x > \omega$ shows that in fact $x > \lambda$, where $\lambda$ depends only on $\omega$, $m$, $d$, and moreover increases to $m\omega/(m-1)$ as $d \to 0$. It follows that

$$ru' + \lambda u < 0 \quad \text{for } r \geqslant r_1,$$

and by integration $r^\lambda u$ is decreasing. Consequently

$$u(r) = O(r^{-\lambda}) \quad \text{as } r \to \infty. \tag{10.17}$$

Since $\omega$ is chosen to satisfy (10.12) and since $\lambda$ can be taken arbitrarily near $m\omega/(m-1)$, one can then allow in (10.17) any exponent $\lambda$ with

$$\lambda_3 < \lambda < \min\left\{\frac{m}{m-1}\lambda_3,\ \lambda_4\right\}.$$

We now use Lemma 2(i) of Section 8 to conclude the validity of (10.9). To justify this, it must be shown first that condition $(8.3)_2$ is valid, and second that $(8.3)_1$ can be replaced in the present circumstance by the weaker condition $(10.6)_1$.

To obtain $(8.3)_2$, however, one has for $r < s$,

$$\frac{g(r)}{g(s)} = \exp\int_s^r \delta(t)\,\mathrm{d}t \leqslant \exp\int_s^r\left(\frac{\beta}{t} - \psi(t)\right)\mathrm{d}t$$
$$= \left(\frac{r}{s}\right)^\beta \exp\int_s^r \psi(t)\,\mathrm{d}t \leqslant \text{Const.}\left(\frac{r}{s}\right)^\beta$$

by $(10.6)_2$ and $(10.6)_2'$, as required. Next, to show that $(10.6)_1$ is sufficient for the validity of Lemma 2(i), recall that now $u > 0$, $u' < 0$ for $r \geqslant r_1$. Hence from (6.1)$'$ follows

$$u(r) = \int_r^\infty\left(\frac{1}{g(s)}\int_{r_1}^s g(t)f(t,\, u(t))\,\mathrm{d}t + \frac{\text{Const.}}{g(s)}\right)^{1/(m-1)}\mathrm{d}s \quad \text{for } r \geqslant r_1 \tag{10.18}$$

rather than (8.4) as earlier, the quantity in parentheses being positive. But (10.18), $(10.6)_1$ and $(8.3)_2$ imply (8.5) and the rest of the proof of Lemma 2(i) carries through unchanged.

The final part of the theorem is proved almost exactly as in Theorem 9.III, where (9.4) is applied. In particular, with $\alpha = 0$, $k = m\beta/(m-1)$, we have in the present case, along $u$,

$$\mathscr{L}' = \left\{r^k\left(\frac{m-1}{m}|u'|^m + F(r,\, u)\right)\right\}' = r^{k-1}\left\{kF + rF_r - \left(\xi - \frac{m-1}{m}k\right)|u'|^m\right\}$$
$$\leqslant r^{k-1}\left(k + \nu - \frac{m}{m-1}r\psi(r)\right)F + \frac{m}{m-1}\psi(r)\mathscr{L},$$

by (7.1) and $(10.6)_2$. From the choice of $k$ and (10.7) it follows that

$$k + \nu - \frac{m}{m-1}r\psi(r) > m + \nu - \frac{m}{m-1}r\psi(r) > 0$$

for sufficiently large $r$; see $(10.6)_1'$. Therefore by (10.9),

$$\mathscr{L}' \leqslant r^{k-1}(k+\nu)Cr^\nu|u|^q + \frac{m}{m-1}\psi(r)\mathscr{L}$$
$$\leqslant \text{Const.}\ r^{-\mu_1-1} + \frac{m}{m-1}\psi(r)\mathscr{L}$$

for sufficiently large $r$, where $\mu_1 = (q-m)\lambda_4 - (\nu+m) > 0$ by (10.10). From Gronwall's inequality we then derive that $\mathscr{L}$ is bounded above and hence

$$\mathscr{L}' \leqslant \text{Const.}(r^{-\mu_1-1} + \psi(r)),$$

that is, as $r \to \infty$

$$\mathscr{L} + \text{Const.}\left(r^{-\mu_1} - \int_R^r \psi(s)\,\mathrm{d}s\right) \searrow \text{Limit} \quad (\text{or } -\infty).$$

By $(10.6)_2'$ and the fact that $\mu_1 > 0$ this gives

$$\mathscr{L} = r^k\left(\frac{m-1}{m}|u'|^m + F(r,u)\right) \to \text{Limit} \quad (\text{or } -\infty).$$

From (10.9) and the growth condition on $F$ we obtain that $r^k F(r,u) = O(r^{-\mu_1}) \to 0$ as $r \to \infty$. Therefore the limit is finite and the required result follows at once. □

Monotonicity theorems corresponding to Theorems 12 and 13 can also be obtained for the range of parameter values

$$q \geqslant m, \qquad \beta < m-1, \qquad \nu + m < 0,$$

relying on similar methods. We omit statement of these results.

The technique used in the first part of the proof can be formalized to give the following general monotonicity result.

**Theorem 14.** *Let $u$ be an extremal of problem* (6.1) *on $I_1 = [R, \infty)$. Suppose that $q \geqslant m$ and*

$$0 \leqslant uf(r,u) \leqslant Cr^\nu |u|^q \quad \text{for } r \geqslant R, \ u \in \mathbb{R},$$

*where $C$ and $R$ are positive constants. If $q > m$ there exists a constant $\kappa > 0$ such that for larger $r$ either*

$$|r\delta(r) + 1 - m| \geqslant \kappa > 0; \qquad |u(r)| < Dr^{-\lambda_3}, \tag{10.19}$$

*or*

$$r\delta(r) + 1 - m \leqslant \frac{m-1-\kappa}{\log r}; \qquad |u(r)| < Dr^{-\lambda_3}(\log r)^{-m/(q-m)}, \tag{10.20}$$

*where*

$$D = (\kappa/mC^m)^{m/(q-m)},$$

*then $u$ is ultimately monotone for large $r$. If $q = m$ the order behavior of $u$ in* (10.19), (10.20) *should be replaced by the condition $\nu + m < 0$.*

**Proof.** We assume first that $q > m$. If (10.19) is satisfied, the argument given in the first part of the proof of Theorem 13 applies almost verbatim—with the choices $\omega = \kappa/m$ if $r\delta(r) \geqslant m - 1 + \kappa$ and $\omega = -\kappa/m$ if $r\delta(r) \leqslant m - 1 - \kappa$—to show that $\phi$ is ultimately monotone and so $u$ ultimately one-signed (or identically zero). It follows in turn from (6.1)′ and the condition $uf(r,u) \geqslant 0$ that $g|u'|^{m-2}u'$ is ultimately of one sign, completing the proof when (10.19) is satisfied.

If (10.20) holds we set

$$\phi(r) = (\log r)^{-\omega} u(r), \qquad \omega = \kappa/m.$$

Then at a critical point of $\phi$ with positive critical value we have from (6.1)″

$$\begin{aligned}(m-1)\phi'' &\geqslant \omega\{(m-1)(1-\omega) + (m-1-r\delta(r))\log r \\ &\qquad - C\omega^{1-m} r^{\nu+m}(\log r)^m u^{q-m}\}\frac{\phi}{(r\log r)^2} \\ &\geqslant \left(\frac{\kappa}{m}\right)^2\{1 - D^{m-q} r^{\nu+m}(\log r)^m u^{q-m}\}\frac{\phi}{(r\log r)^2}\end{aligned}$$

by $(10.20)_1$, which is positive for large $r$ by $(10.20)_2$. Hence $\phi$ is ultimately monotone, and the rest of the proof is the same as before.

The case $q=m$ and $\nu+m<0$ can be treated by the same calculations, except that it is not necessary to place any restriction on the order behavior of $u$ at $\infty$. □

In Theorem 9 the condition (9.1) implies that $\lambda_3<0$ and $\lambda_3<\lambda_4$ when $q>m$. Since (10.19) is satisfied in Theorem 9.III, the extremal $u$ must be monotone. Moreover, in case I, if $r\delta(r)\leqslant\beta'<m-1$, then condition (10.19) is satisfied and hence the extremal $u$ is again monotone; in case II if $\beta=\beta'=m-1$ then (10.20) is verified and $u$ is monotone.

## 11. Oscillatory solutions

Here we study the relation (7.13) in the subcritical case, with $A=B$ for simplicity.

**Theorem 15.** *Let $u=u(r)$ be an extremal of* (6.1) *on $I_1=[R,\infty)$ such that* (7.13) *holds, with a positive limit and with $A=B>0$. Assume also that* (6.2) *and* $(7.3)_2$ *are satisfied and that*

$$uf(r,u)\geqslant Cr^{\nu}|u|^{q}\quad \text{for } r\geqslant R,\ u\in\mathbb{R}. \tag{11.1}$$

*Then $u$ is necessarily oscillatory. The successive zeros $r_j$ of $u$ and the interlaced critical points $\rho_j$, $r_j<\rho_j<r_{j+1}$, $j=1,2,\ldots$, possess the property*

$$r_{j+1}^{\tau}-r_j^{\tau}\to 2\tau c,\quad \rho_j^{\tau}-r_j^{\tau}\to\tau c\qquad \text{as } j\to\infty, \tag{11.2}$$

*where $c$ is positive and*

$$m\tau=\nu+m-(q-m)\lambda_1.$$

**Proof.** Here we may write (7.13) in the form

$$r^{q\lambda_1-\nu}\left(\frac{m-1}{m}|u'|^{m}+Ar^{\nu}|u|^{q}\right)\to Al^{q}\quad \text{as } r\to\infty,$$

with $l>0$. Since $\tau>0$ by $(7.3)_2$, the change of variables

$$u(r)=r^{-\lambda_1}v(x),\qquad x=\frac{1}{\tau}r^{\tau}, \tag{11.3}$$

reduces this to

$$\frac{m-1}{m}\left|\dot v-\lambda_1 r^{-\tau}v\right|^{m}+A|v|^{q}\to Al^{q}\quad \text{as } x\to\infty, \tag{11.4}$$

where $\dot{}\ =d/dx$. Now $r^{-\tau}v(x)\to 0$ as $x\to\infty$, because $v$ is bounded by (11.4). Since also $\dot v-\lambda_1 r^{-\tau}v$ is bounded for all $x$ large, this yields easily that

$$\frac{m-1}{m}|\dot v|^{m}+A|v|^{q}\to Al^{q}\quad \text{as } x\to\infty. \tag{11.5}$$

We now show that $u$ is oscillatory. We claim first that either

$$v(x)\to\pm l\quad \text{as } x\to\infty, \tag{11.6}$$

or else

$$\liminf_{x\to\infty} v(x) = -l \quad \text{and} \quad \limsup_{x\to\infty} v(x) = l. \tag{11.7}$$

To see this, note that if $v(x) \to L$ as $x \to \infty$, then $|\dot{v}(x)| \to$ Limit, which must be zero and so $L = \pm l$, namely in this case (11.6) holds. Otherwise, if $v$ does not possess a limit at $\infty$, then

$$l' = \liminf_{x\to\infty} v(x) < \limsup_{x\to\infty} v(x) = l'',$$

with $l'$ and $l''$ finite since $v$ is bounded. Thus there exist sequences $(\xi_j')$ and $(\xi_j'')$ tending to $\infty$ such that

$$\lim_j v(\xi_j') = l', \qquad \dot{v}(\xi_j') = 0, \quad j = 1, 2, \ldots,$$

$$\lim_j v(\xi_j'') = l'', \qquad \dot{v}(\xi_j'') = 0, \quad j = 1, 2, \ldots.$$

But then, by (11.5) again, we get $l' = -l$, $l'' = l$. Hence (11.7) holds and the claim is proved.

We next show that (11.6) is impossible. To this end, observe from (6.1)$''$ and the change of variables (11.3) that $v$ satisfies the equation

$$(m-1)\ddot{v} + (\xi - 2\lambda_1 - 1 + \tau)\frac{\dot{v}}{\tau x} - (\xi - \lambda_1 - 1)\lambda_1 \frac{v}{\tau^2 x^2}$$

$$+ \left|\dot{v} - \lambda_1 \frac{v}{\tau x}\right|^{2-m} v|v|^{q-2} \frac{uf(r, u)}{r^\nu |u|^q} = 0 \tag{11.8}$$

at any point where $v \neq 0$ and $\dot{v} \neq \lambda_1 v/\tau x$ (so $u \neq 0$, $u' \neq 0$). By (11.6) there is an $x_1$ such that $v(x) \neq 0$ for $x \geqslant x_1$; additionally $u'$ can have at most one zero in $[r(x_1), \infty)$ by virtue of (6.1)$'$ and (11.1). Hence (11.8) is valid for $x \geqslant x_1$ except possibly at one point $\hat{x}$.

Now by (11.8) and (11.1)

$$(m-1)\ddot{v} + (\xi - 2\lambda_1 - 1 + \tau)\frac{\dot{v}}{\tau x} - (\xi - \lambda_1 - 1)\lambda_1 \frac{v}{\tau^2 x^2}$$

$$+ C\left|\dot{v} - \lambda_1 \frac{v}{\tau x}\right|^{2-m} v|v|^{q-2} \leqslant 0 \quad \text{if } v > 0 \text{ and } x \neq \hat{x}, \tag{11.9}$$

the inequality sign being reversed if $v < 0$. Here $|v(x)| \to l$ and $x \to \infty$ by (11.6). In turn, $v$ has at most one critical point for $x$ sufficiently large, since at any point where $\dot{v} = 0$ we would have

$$(\operatorname{sign} v)(m-1)\ddot{v} \leqslant (\xi - \lambda_1 - 1)\lambda_1 \frac{|v|}{\tau^2 x^2} - C\left|\lambda_1 \frac{v}{\tau x}\right|^{2-m} |v|^{q-1}$$

$$\leqslant -C\left|\frac{\lambda_1}{\tau}\right|^{2-m} \frac{|v|}{x^2}\{|v|^{q-m} x^m - \text{Const.}\} < 0, \tag{11.10}$$

if $x$ is large enough.

We now integrate (11.9) from $x_2$ to $\bar{x}$, where $x_2 \geqslant x_1$ is such that $\dot{v} \neq 0$ for $x \geqslant x_2$. Thus, by (6.2)

$$\int_{x_2}^{\bar{x}} C\left|\dot{v} - \lambda_1 \frac{v}{\tau x}\right|^{2-m} |v|^{q-1}\, dx$$

$$\leqslant (m-1)|\dot{v}(\bar{x}) - \dot{v}(x_2)| + \text{Const.} \int_{x_2}^{\bar{x}} \frac{|\dot{v}|}{x}\, dx + \text{Const.} \int_{x_2}^{\bar{x}} \frac{|v|}{x^2}\, dx.$$

Here

$$\int_{x_2}^{\bar{x}} \frac{|\dot{v}|}{x}\,\mathrm{d}x = \pm\int_{x_2}^{\bar{x}} \frac{\dot{v}}{x}\,\mathrm{d}x = \pm\left(\left[\frac{v}{x}\right]_{x_2}^{\bar{x}} + \int_{x_2}^{\bar{x}} \frac{v}{x^2}\,\mathrm{d}x\right) \leqslant \text{Const.},$$

the constant being independent of $\bar{x}$. Hence, recalling that $|v(x)| \to l$, $\dot{v}(x) \to 0$ as $x \to \infty$, there results by (11.5)

$$I(\bar{x}) = \int_{x_2}^{\bar{x}} \left|\dot{v} - \lambda_1 \frac{v}{\tau x}\right|^{2-m} \mathrm{d}x \leqslant \text{Const.} \tag{11.11}$$

Since the constant is independent of $\bar{x}$, this is impossible when $m \geqslant 2$.

If $1 < m < 2$

$$I(\bar{x}) \geqslant \int_{x_2}^{\bar{x}} \left|\dot{v} - \lambda_1 \frac{v}{\tau x}\right| \mathrm{d}x,$$

provided $|\dot{v} - \lambda_1 v/\tau x| < 1$ for $x \geqslant x_2$, which we can certainly assume without loss of generality. In turn

$$I(\bar{x}) \geqslant \int_{x_2}^{\bar{x}} \frac{|\lambda_1 v|}{\tau x}\,\mathrm{d}x - \int_{x_2}^{\bar{x}} |\dot{v}|\,\mathrm{d}x \to \infty \quad \text{as } r \to \infty.$$

This contradicts (11.11), so (11.6) cannot occur.

It follows that (11.7) holds. Consequently $v$ is oscillatory and in turn so is $u$, completing the first part of the proof.

It remains to show (11.2). By (11.7) there is a sequence $(x_j)$ tending to $\infty$ such that

$$\lim_j v(x_j) = l, \qquad \dot{v}(x_j) = 0, \quad j = 1, 2, \ldots.$$

Fix $\varepsilon \in (0, \frac{1}{2}l)$ and let $j_\varepsilon$ be such that

$$|v(x_j) - l| < \varepsilon \quad \text{for every } j \geqslant j_\varepsilon.$$

For fixed $j \geqslant j_\varepsilon$ let $x_j'$, $x_j''$ be numbers defined by the conditions

$$x_j' < x_j < x_j'', \qquad |v(x) - l| < \varepsilon \quad \text{for } x \in (x_j', x_j''), \qquad v(x_j') = v(x_j'') = l - \varepsilon. \tag{11.12}$$

We claim that if $j_\varepsilon$ is chosen even larger, if necessary, then

$$0 < x_j'' - x_j' \leqslant \text{Const. } \varepsilon^{(m-1)/m}, \tag{11.13}$$

where the constant is independent of $\varepsilon$.

We integrate (11.9) on $[x_j', x_j'']$, and estimate the various terms which arise exactly as in the earlier part of the proof. Here, as before, $u'$ and $\dot{v}$ can be zero at most once in the interval of integration. Moreover, using (11.4), (11.5) and (11.12) we obtain

$$|\dot{v}(x)|, \left|\dot{v} - \lambda_1 \frac{v}{\tau x}\right| \leqslant \text{Const. } \varepsilon^{1/m} \quad \text{for } x \in [x_j', x_j''],$$

with $j_\varepsilon$ possibly larger, if necessary. Hence there results for $m \geqslant 2$

$$\text{Pos. Const. } \varepsilon^{(2-m)/m}(x_j'' - x_j') \leqslant \text{Const.}\left(\varepsilon^{1/m} + \frac{l}{x_{j-1}}\right),$$

which is the required result (11.13), provided $j_\varepsilon$ is chosen so that also

$$x_j' \geqslant x_{j-1} \geqslant \varepsilon^{-1/m} \quad \text{for every } j \geqslant j_\varepsilon.$$

When $1 < m < 2$ the integration procedure yields the weaker condition

$$\text{Pos. Const.} \log\left(\frac{x_j''}{x_j'}\right) \leqslant \text{Const.}\left(\varepsilon^{1/m} + \frac{l}{x_{j-1}}\right),$$

from which we can derive only that

$$1 \leqslant \frac{x_j''}{x_j'} \leqslant 1 + \text{Const. } \varepsilon^{1/m}, \tag{11.14}$$

the constant being independent of $\varepsilon$ and $j$. To obtain (11.13) in this case an additional procedure is required.

Integrating (6.1)′ on $[x_j', x_j'']$, we find

$$g(r)u'(r)\,|u'(r)|^{m-2}\Big]_{x_j'}^{x_j''} = -\int_{x_j'}^{x_j''} g(s)f(s, u(s))s^{1-\tau}\,\mathrm{d}x, \tag{11.15}$$

where $r$ and $s$ are relate to $x$ by (11.3). By (11.3), (11.4) and the condition $(11.12)_3$

$$u'(r_j') = (r_j')^{\tau-\lambda_1-1}\left(\dot{v}(r_j') - \frac{\lambda_1 v(r_j')}{\tau x_j'}\right)$$

$$\geqslant \text{Pos. Const. } (r_j')^{\tau-\lambda_1-1}\varepsilon^{1/m}$$

for $j_\varepsilon$ larger if necessary. Similarly,

$$u'(r_j'') \leqslant -\text{Pos. Const. } (r_j'')^{\tau-\lambda_1-1}\varepsilon^{1/m}.$$

Hence from (11.15),

$$\text{Pos. Const.}\left\{1 + \left(\frac{r_j'}{r_j''}\right)^{(\tau-\lambda_1-1)(m-1)}\frac{g(r_j')}{g(r_j'')}\right\}\varepsilon^{m/(m-1)}$$

$$\geqslant \frac{1}{(r_j'')^{(\tau-\lambda_1-1)(m-1)}g(r_j'')}\int_{x_j'}^{x_j''} g(s)f(s, u(s))s^{1-\tau}\,\mathrm{d}x$$

$$\geqslant \text{Pos. Const.}\frac{\tilde{s}_j^{\,\nu-\lambda_1(q-1)+1-\tau}}{(r_j'')^{(\tau-\lambda_1-1)(m-1)}}(x_j'' - x_j'), \tag{11.16}$$

by (11.1) and $(6.2)_2$, where $\tilde{s}_j \in (r_j', r_j'')$. By direct calculation and the choice of $\tau$ one has

$$\nu - \lambda_1(q-1) + 1 - \tau = (\tau - \lambda_1 - 1)(m-1).$$

Calling this exponent $\gamma$ for convenience, and using $(6.2)_1$, we can rewrite (11.16) in the form

$$x_j'' - x_j' \leqslant \text{Pos. Const.}\left(\frac{\tilde{s}}{r_j''}\right)^{-\gamma}\left\{1 + \left(\frac{r_j'}{r_j''}\right)^{\gamma+\beta}\right\}\varepsilon^{m/(m-1)}.$$

The required result (11.13) now follows from (11.14).

To complete the proof of (11.2), consider the initial value problem

$$\frac{m-1}{m}|\dot{w}|^m + A|w|^q = Al^q, \quad \dot{w} > 0, \qquad w(x_j') = l - \varepsilon. \tag{11.17}$$

This has the solution $\bar{w} = \bar{w}(x)$, in inverse terms,

$$x - x_j' = \left(\frac{m-1}{mA}\right)^{1/m} \int_{l-\varepsilon}^{\bar{w}} \frac{\mathrm{d}t}{(l^q - |t|^q)^{1/m}}$$

as long as $-l < \bar{w}(x) < l$. In particular $\bar{w}$ is defined in the interval

$$(x_j' + a(\varepsilon) - 2c,\ x_j' + a(\varepsilon)), \quad \text{where} \quad c = \left(\frac{m-1}{mA}\right)^{1/m} \int_0^l \frac{\mathrm{d}w}{(l^q - |w|^q)^{1/m}},$$

and $a(\varepsilon)$ is determined by $\bar{w}(x_j' + a(\varepsilon)) = l$. Clearly the integral defining $c$ is convergent since $m > 1$, $q > 1$. Also $a(\varepsilon) \to 0$ as $\varepsilon \to 0$. By the continuous dependence of solutions of the initial value problem on the equation, it follows that

$$|v(x) - \bar{w}(x)| \leqslant \varepsilon \quad \text{for } x \in (x_j' + 2a(\varepsilon) - 2c,\ x_j'),$$

provided $j_\varepsilon$ is taken even larger, if necessary. Let $y_j$ be the first zero of $v$ to the left of $x_j$, namely the number such that $v(x) > 0$ for $x \in (y_j, x_j]$ and $v(y_j) = 0$. Obviously $\bar{w}$ vanishes at $y_j' = x_j' + a(\varepsilon) - c$ and

$$|y_j' - y_j| \leqslant b(\varepsilon) \quad \text{where } b(\varepsilon) = \left(\frac{m-1}{mA}\right)^{1/m} \int_0^\varepsilon \frac{\mathrm{d}w}{(l^q - |w|^q)^{1/m}}.$$

With the help of (11.12) it now follows that

$$\begin{aligned} |x_j - y_j - c| &\leqslant x_j'' - x_j' + a(\varepsilon) + b(\varepsilon) \\ &\leqslant \text{Const. } \varepsilon^{(m-1)/m} + a(\varepsilon) + b(\varepsilon), \end{aligned} \tag{11.18}$$

using (11.13). Similarly, letting $y_{j+1}$ be the first zero of $v$ to the right of $x_j$, we get

$$|y_{j+1} - x_j - c| \leqslant \text{Const. } \varepsilon^{(m-1)/m} + a(\varepsilon) + b(\varepsilon).$$

Thus the nearest zeros on each side of the critical point $x_j$ are separated from $x_j$ by at least $c/2$. Between these zeros $v$ has only the single critical point $x_j$, for $x_j$ is the unique critical point of $v$ in $(x_j', x_j'')$, while $v$ has no critical points at all in the intervals $[y_j, x_j')$ and $(x_j'', y_{j+1}]$ by (11.5). In turn, the sequence $(x_j)$ can be taken to include all the positive critical points of $v$ for $x \geqslant x_{j_\varepsilon}$, namely where $v > 0$, $\dot{v} = 0$.

The negative critical points of $v$ of course have an analogous behavior, and can similarly be enumerated. With this proved, it now is obvious that the positive and negative critical points separate each other, and that the successive zeros $y_j$ of $v$ are such that

$$y_{j+1} - y_j \to 2c \quad \text{as } j \to \infty.$$

Finally, by (11.3),

$$\frac{1}{\tau}(r_{j+1}^\tau - r_j^\tau) \to 2c \quad \text{as } j \to \infty,$$

proving $(11.2)_1$. The remaining condition $(11.2)_2$ follows in the same way, using (11.18) and the fact that $x(\rho_j) \in (x_j', x_j'')$. This completes the proof of the theorem. □

The relation of the zero separation $2\tau c$ to the value of $l$ is of some interest. Indeed, one sees at once that

$$c = \left(\frac{m-1}{mA}\right)^{1/m} l^{1-q/m} \int_0^1 \frac{\mathrm{d}t}{(1-|t|^q)^{1/m}} = K(m, q, A) l^{1-q/m}.$$

In particular $c \to \infty$ as $l \to 0$ when $q > m$, while $c$ is a fixed constant independent of $l$ when $q = m$ and $c \to 0$ as $l \to 0$ when $q < m$.

Finally we observe that if the number $\lambda_1$ in Theorem 15 is taken as in Section 7, namely $\lambda_1 = (m\beta + (m-1)\nu)/(m + (m-1)q)$, then

$$\tau = \frac{m}{m + (m-1)q}\left[m(\nu + m) - (q-m)(\beta + 1 - m)\right].$$

The required assumption $(7.3)_2$, that is $\tau > 0$, is evidently equivalent to $q$ being subcritical.

In general, if the limit in (7.5) or in (7.13) is zero, one may expect that the extremal $u$ would be monotone (its zeros are infinitely distantly spaced). To prove this, however, would require sharper asymptotics on $F = F(r, u)$ than we have already stated.

## 12. Weaker hypotheses

Conditions (7.1) and (7.2) are more restrictive than necessary for the results of Theorems 7 and 8, at least if one is given a priori that $u(r) \to 0$ as $r \to \infty$ (or if this can be proved independently, as in Theorem 5).

Indeed in this case, and assuming also that

$$\beta + \frac{m-1}{m}\nu > 0, \qquad \nu + m > 0, \tag{12.1}$$

Theorems 7 and 8 continue to hold even if one weakens (7.1), $(7.2)_1$ to the form

$$\begin{aligned} \nu F(r, u) &\geqslant r F_r(r, u) - \left\{h\left(\frac{1}{r}\right) + h(u)\right\} F(r, u) \\ q F(r, u) &\leqslant u F_u(r, u) + \left\{h\left(\frac{1}{r}\right) + h(u)\right\} F(r, u) \end{aligned} \tag{12.2}$$

for all sufficiently small $u$ and sufficiently large $r$, where $h = h(\tau)$ is a positive, increasing function defined for $\tau \in (0, 1)$ and such that

$$\int_0^1 \frac{h(\tau)}{\tau}\, \mathrm{d}\tau < \infty. \tag{12.3}$$

To prove Theorems 7 and 8 under the weaker hypothesis (12.2), though with the added assumption that $u(r) \to 0$ as $r \to \infty$, we first observe that necessarily $h(\tau) \to 0$ as $\tau \to 0$. Hence the conditions (12.2) together with the condition $u(r) \to 0$ as $r \to \infty$ imply that (7.1) and $(7.2)_1$ hold *for the given solution* with any slightly smaller $q$ and slightly larger $\nu$. Thus, appropriate values $\alpha$, $\kappa$ can be found so that both (6.3) and condition (Q)′ hold (the argument being the same as at the beginning of the proof of Theorem 7). In turn Theorem 6 (or its corollary in case the slightly smaller value of $q$ is less than $m$) can be applied to show that $u$ decays algebraically to zero as $r \to \infty$.

Next, using (12.2) and the values of $\alpha$ and $\kappa$ actually given in (7.6), one finds the revised estimate in the proof of Theorem 6

$$kF(r, u) + rF_r(r, u) - \alpha u f(r, u) \leqslant (1+\alpha)\{h(r^{-1}) + h(u)\} F(r, u).$$

Consequently in (6.12) and in (6.12)′ there will be an additional term

$$\text{Const.} \int_{r_1}^{r} \tau^{-1} \hat{h}(\tau) \hat{\mathscr{L}}(\tau)\, \mathrm{d}\tau \tag{12.4}$$

on the right hand side, where

$$\hat{h}(\tau) = h(\tau^{-1}) + h(\tau^{-\varepsilon}), \quad \varepsilon > 0,$$

since $u$ decays algebraically to zero. Clearly

$$\int_1^{\infty} \frac{\hat{h}(\tau)}{\tau}\, \mathrm{d}\tau < \infty$$

in view of (12.3). The proofs are then completed in essentially the same ways as before.

**Remarks.**

1. Just as one can improve upon conditions (7.1), $(7.2)_1$, it is also possible to improve on (6.2). The same procedures indeed allow for the weakened lower estimate

$$\beta \leqslant r\delta(r) + r\psi(r), \quad \psi \in L^1[R, \infty).$$

The upper estimate in (6.2) also can be improved similarly, but the best results require a different approach which we give below.

2. As an example of a nonlinearity $f$ for which one can show that $u(r) \to 0$ as $r \to \infty$ and also for which $(12.2)_2$ but not $(7.2)_1$ is applicable, consider the function

$$f(u) = u|u|^{q-2} - cu|u|^{s-1} + u|u|^{t-2},$$

where $1 < q < s < t$ and $c$ is positive and small. For this function one clearly has $F(u) \to \infty$ as $|u| \to \infty$ so that any solution $u$ is bounded as $r \to \infty$ (see Theorem 4). Also $uf(u) = |u|^q - c|u|^s + |u|^t > 0$ when $c$ is suitably small, so Theorem 5 implies that $u(r) \to 0$ as $r \to \infty$.

Finally, we have $F(u) \geqslant |u|^q/2q$ for $c$ suitably small, while

$$qF(u) - uf(u) = c\frac{s-q}{s}|u|^s - \frac{t-q}{t}|u|^t,$$

the right hand side being positive for all sufficiently small $u$. Thus $(7.2)_1$ does not hold, but (12.2) does, with $\nu = 0$ and $h(\tau) = 2cq(s-q)\tau^{s-q}/s$.

In the final part of this section we consider the situation when the upper estimate of (6.2) is weakened. In particular, we suppose only the condition

$$\beta \leqslant r\delta(r) \leqslant \beta' r^{\eta}, \quad \eta > 0. \tag*{(6.2)′}$$

In this case the exponent corresponding to $\lambda_3$ becomes

$$\lambda_5 = \frac{\nu + m - \eta}{q - m} \quad \text{if } q \neq m.$$

As before, the results are different depending on whether $\lambda_1 < \lambda_5$ or $\lambda_1 > \lambda_5$, with the critical exponent $q_1$ now being defined by

$$q_1 - m = m\frac{\nu + m - \eta}{\beta + 1 - m + (m-1)\eta/m}.$$

The following generalization of Theorems 7, 8 and 10 then holds.

**Theorem 16.** *Let u be an extremal of problem* (6.1) *on* $I_1 = [R, \infty)$, *and suppose that* (Q), (6.2)′, (7.1) *ad* (7.2) *are satisfied.*

I. *If* $q \geqslant m$ *and*

$$(q-m)\left(\beta + 1 - m + \frac{m-1}{m}\eta\right) < m(\nu + m - \eta) \tag{12.5}$$

(*subcritical case*), *then for any* $\varepsilon > 0$,

$$u(r) = \mathrm{O}(r^{-\lambda_1+\varepsilon}), \quad u'(r) = \mathrm{O}(r^{-\lambda_2+\varepsilon}) \qquad \text{as } r \to \infty. \tag{12.6}$$

II. *If* $1 < q < m$, $\nu + m - \eta > 0$ *and*

$$(q-m)\left(\beta + 1 - m + \frac{m-1}{m}\eta\right) < (m-1)(\nu + m - \eta), \tag{12.7}$$

*then* (12.6) *continues to hold.*

III. *If* $q > m$ *and*

$$(q-m)\left(\beta + 1 - m + \frac{m-1}{m}\eta\right) > m(\nu + m - \eta) \tag{12.5$'$}$$

(*supercritical case*), *then*

$$u(r) = \mathrm{O}(r^{-\lambda_5}), \quad u'(r) = \mathrm{O}(r^{-\lambda_5 - 1 + \eta/m}) \qquad \text{as } r \to \infty. \tag{12.8}$$

**Proof.** I. The principal difference, and the only part of the proof which we indicate in detail, concerns the estimation of the right hand side $\mathscr{R}$ of (6.4). Here we fix $k < q\lambda_1 - \nu$, $\alpha = \lambda_1$. Then the term $\xi - \alpha - (m-1)k/m$ in (6.4) is bounded below by $\xi - \beta + \bar{\varepsilon}$, where $\bar{\varepsilon}$ can be taken without loss of generality to be between 0 and 1. Then following the proof of Theorem 6, the inequality (6.7) can be replaced by

$$\begin{aligned}
\mathscr{R} &\leqslant -r^{k-1}\left\{(\xi - \beta + \bar{\varepsilon})\,|u'|^m + \frac{\alpha}{r}(\xi + 1 - k)uu'\,|u'|^{m-2}\right\} \\
&\leqslant \text{Const. } r^{k-1}(\xi - \beta + \bar{\varepsilon})\left|\frac{\xi + 1 - k}{\xi - \beta + \bar{\varepsilon}}\frac{\alpha}{r}\right|^m |u|^m \\
&\leqslant \text{Const. } r^{k-1+\eta-m}\,|u|^m \\
&\leqslant \text{Const. } r^{k-1+\eta-m} r^{-\nu\bar{\theta}}(r^\nu |u|^q + |u'|^m)^{\bar{\theta}},
\end{aligned} \tag{12.9}$$

where $\tilde{\theta} = m/q \leqslant 1$. Hence, continuing as in the proof of Theorem 6, instead of (6.10) we obtain

$$\hat{\mathscr{L}} \leqslant C_2 r^{k-\sigma} + \mathscr{L}(r_1) + C_3 \int_{r_1}^{r} s^{-\tilde{\mu}-1} \big(\hat{\mathscr{L}}(s)\big)^{\tilde{\theta}} \, \mathrm{d}s, \tag{12.10}$$

where

$$\tilde{\mu} = (k+\nu)\left(\frac{m}{q} - 1\right) + \nu + m - \eta > 0,$$

by (12.5) and the choice of $k$.

We have $k < \sigma (= q\lambda_3 - \nu)$ in view of the relation $\lambda_1 < \lambda_5 < \lambda_3$, and the rest of the proof is essentially the same as for Theorem 7. The occurence of the exponent $\varepsilon$ in (12.6) is due to the fact that $k$ can be chosen arbitrarily near $q\lambda_1 - \nu$, but not equal to this value.

II. The estimate (12.9) for $\mathscr{R}$, when $1 < q < m$, becomes

$$\begin{aligned} \mathscr{R} &\leqslant \text{Const. } r^{k-1+\eta-m-\nu} \, |u|^{m-q} r^{\nu} \, |u|^{q} \\ &\leqslant \text{Const. } r^{-1-(\nu+m-\eta)} \, |u|^{m-q} \hat{\mathscr{L}}. \end{aligned}$$

We now use Lemma 1 in Section 8 to estimate the term $|u|^{m-q}$. If $m\beta + (m-1)\nu \geqslant 0$ then $u$ is bounded and so

$$\mathscr{R} \leqslant \text{Const. } r^{-\tilde{\mu}-1} \hat{\mathscr{L}},$$

with $\tilde{\mu} = \nu + m - \eta > 0$. Thus $\hat{\mathscr{L}}$ is bounded by Gronwall's inequality and we are done.

On the other hand, if $m\beta + (m-1)\nu < 0$ then

$$u(r) = \mathrm{O}\big(r^{-(m\beta + (m-1)\nu)/(m-1)q}\big) \quad \text{as } r \to \infty,$$

by Lemma 1(ii). Hence

$$r^{-(\nu+m-\eta)} \, |u(r)|^{m-q} = \mathrm{O}(r^{-\tilde{\mu}}),$$

where

$$\tilde{\mu} = \frac{m}{q}\left\{\nu + m - \eta - \frac{q-m}{m-1}\left(\beta + 1 - m + \frac{m-1}{m}\eta\right)\right\} > 0,$$

by (12.7). Thus again $\hat{\mathscr{L}}$ is bounded by Gronwall's inequality, as required.

III. We now choose $k$ in (12.9) so that

$$q\lambda_5 < k + \nu < q \min\{\lambda_1, \lambda_3\},$$

which can surely be done since $\lambda_5 < \lambda_1$ by (12.5)$'$ and of course also $\lambda_5 < \lambda_3$. Then $k < \sigma (= q\lambda_3 - \nu)$ and (12.10) holds with

$$\begin{aligned} \tilde{\mu} &= (k+\nu)\left(\frac{m}{q} - 1\right) + \nu + m - \eta \\ &< (m-q)\lambda_5 + \nu + m - \eta = 0. \end{aligned}$$

One can now follow the proof of Theorem 6 from (6.10) onward to obtain (6.14) with $\theta$, $\mu$ replaced by $\tilde{\theta}$, $\tilde{\mu}$. Since $\tilde{\mu} < 0$, this yields

$$Y(r) \leqslant M r^{-\tilde{\mu}/(1-\tilde{\theta})} = M r^{k+\nu-q\lambda_5}.$$

The required estimate (12.8) now follows at once. □

The case $1 < q \leqslant m$ in Theorem 16 is somewhat unusual, for in this situation the additional hypothesis $r\delta(r) \to \infty$ as $r \to \infty$ yields further information about the asymptotic behavior. Indeed we can then take $\beta$ arbitrarily large in (6.2)′, whence (12.6) yields

$$u(r),\ u'(r) = O(r^{-\lambda}) \quad \text{as } r \to \infty$$

for all $\lambda > 0$. More precise results are also possible here, see for example [11], in which the case $q = m = 2$, $r\delta(r) = \beta' r^{\eta}$ is considered.

## 13. Asymptotic estimates for other operators

The procedures of Sections 6–12 apply not only to the degenerate Laplace operator, but in fact much more generally. By examining the successive parts of the proofs given there one sees that the specific nature of the operator $A(p) = |p|^{m-2}$ was used only in the following ways:

(1) To show that in going from (2.9) to (6.4)

$$-kG(u') + (\alpha + k - \xi)A(u')u'^2 \leqslant \left(\frac{m-1}{m}k + \alpha - \xi\right)A(u')u'^2.$$

(2) To obtain

$$|A(u')u'| \leqslant \text{Const.}\ |u'|^{m-1},$$

used in the demonstration of (6.6) and (6.8), and also in the proof of Theorem 12.

(3) To obtain

$$\text{Const.}\ |u'|^{m} \leqslant H(u'),$$

also used in the proof of (6.8).

(4) To obtain the specific result (7.5), rather than simply

$$r^{q\lambda_1 - \nu}(H(u') + F(r, u)) \to \text{Limit}. \tag{7.5$'$}$$

It is apparent that (1) remains valid for a general function $G$ provided that either

$$k \geqslant 0 \quad \text{and} \quad H(p) \leqslant (m-1)G(p), \quad m > 1,$$

or

$$k \leqslant 0 \quad \text{and} \quad H(p) \geqslant (m-1)G(p), \quad m > 1.$$

In the proof of Theorem 6, in particular, it is clear from (6.16) that $k + \nu$ must in general be chosen small and positive. Hence we must take $k > 0$ when $\nu \leqslant 0$, while if $\nu > 0$ we can always choose $k \leqslant 0$.

In Theorem 7

$$k = q\lambda_1 - \nu = \frac{m(\beta q - \nu)}{m + (m-1)q}.$$

Hence $k \geqslant 0$ when $\nu \leqslant \beta q$, and $k \leqslant 0$ when $\nu \geqslant \beta q$.

For Theorem 8

$$q\lambda_1 - \nu \geqslant k > q\lambda_3 - \nu = m\frac{q+\nu}{q-m}.$$

We can now state the results corresponding to Theorems 6–9 for extremals $u$ of the general problem (2.1), rather that of the special problem (6.1) corresponding to the degenerate Laplace operator.

**Theorem 17.** *Suppose that for some number $m>1$ we have*

$$H(p) \leqslant (m-1)G(p), \tag{13.1}$$

$$H(p) \geqslant \text{Pos. Const.}\ |p|^m, \qquad A(p) \leqslant \text{Pos. Const.}\ |p|^{m-2} \tag{13.2}$$

*for all $p \neq 0$ contained in some interval $T$ of $\mathbb{R}$.*

*Let $u$ be an extremal on $I_1=[R,\infty)$ of problem* (2.1), *rather than of* (6.1), *such that $u'(r)\in T$ for all $r$ sufficiently large. Then*

*Theorems 6 and its corollary hold for $u$ when $\nu \leqslant 0$;*

*Theorem 7 holds for $u$ with (7.5)′ replacing (7.5), provided that $\nu \leqslant \beta q$;*

*The corollary to Theorem 7 holds for $u$ provided that $\nu < \beta q$ and*

$$|p|^{-m}H(p) \to \text{Pos. Limit} \quad \text{as } p\to 0; \tag{13.3}$$

*Theorem 8 holds when $\nu \leqslant \beta q$;*

*Theorem 9 holds provided that $\beta>0$ and (13.3) is satisfied.*

**Proof.** Under the given assumptions we can take $k \geqslant 0$ in the first four cases, so that (1) holds. Clearly (2) and (3) follow from (13.2). Thus Theorem 6 and its corollary hold, using (13.2) moreover to derive (6.6) and (6.8). Next, we obtain (7.5)′ as in the proof of Theorem 7. To get (7.4) it is enough to use (7.5)′ and $(13.2)_1$.

We next show (7.13) when (13.3) is valid. By rescaling $F$ and $G$ it may be assumed without loss of generality that

$$|p|^{-m}H(p) \to \frac{m-1}{m} \quad \text{as } p\to 0.$$

But

$$r^{q\lambda_1-\nu}H(u') = r^{q\lambda_1-\nu}\left\{\frac{m-1}{m}|u'|^m + |u'|^m\left(\frac{H(u')}{|u'|^m} - \frac{m-1}{m}\right)\right\}.$$

From $(7.4)_2$ it follows that $u'(r)\to 0$ and $r\to\infty$, since $\nu<\beta q$. Thus

$$r^{q\lambda_1-\nu}|u'|^m\left(\frac{H(u')}{|u'|^m} - \frac{m-1}{m}\right) \to 0 \quad \text{as } r\to\infty.$$

To prove (7.15), we first find $\hat{\mathscr{L}}(r) \leqslant \text{Const.}\ r^{k-\sigma}$, as in the proof of Theorem 8. Then (7.15) follows from (6.8) and $(13.2)_1$.

It remains finally to consider Theorem 9 when $\beta>0$. Here $\nu+m<0$ by (9.1) and so for Lemma 1(i) we have $k=-\nu>0$, while for Lemma 1(ii) similarly $k=m\beta/(m-1)>0$. The validity of Lemmas 2–4 of Section 8 for general operators requires that

$$A(p) \geqslant \text{Pos. Const.}\ |p|^{m-2} \quad \text{for } p\neq 0 \text{ small},$$

which follows directly from (2.2) and $(13.2)_1$. The proof of Theorem 9 now carries over without essential change. Condition (13.3) in particular is used here to show that

$$r^{m\beta/(m-1)}|u'(r)|^m \to \text{Limit} \quad \text{as } r\to\infty,$$

since Lemma 1(ii) for the general problem (2.1) yields only that

$$r^{m\beta/(m-1)}(H(u') + F(r, u)) \to \text{Limit} \quad \text{as } r \to \infty.$$

**Theorem 18.** *Suppose that for some number $m > 1$ we have*

$$H(p) \geqslant (m-1)G(p), \tag{13.4}$$

$$H(p) \geqslant \text{Pos. Const.} \mid p \mid^m, \qquad A(p) \leqslant \text{Const.} \mid p \mid^{m-2} \tag{13.5}$$

*for all $p \neq 0$ contained in some interval $T$ of $\mathbb{R}$.*

*Let $u$ be an extremal on $I_1 = [R, \infty)$ of problem* (2.1), *rather than of* (6.1), *such that $u'(r) \in T$ for all $r$ sufficiently larger. Then*

*Theorems* 6 *and its corollary hold for $u$ when $\nu > 0$;*

*Theorem* 7 *holds for $u$ with* (7.5)′ *replacing* (7.5), *provided that $\nu \geqslant \beta q$;*

*Theorem* 8 *holds if $q + \nu < 0$.*

The proof is obvious since it follows exactly the lines of the preceding discussion.

**Theorem 19.** *Suppose that for some $m > 1$*

$$A(p) \leqslant \text{Const.} \mid p \mid^{m-2}$$

*for all $p \neq 0$ contained in some interval $T$ of $\mathbb{R}$. Then Theorem* 12 *holds for any extremal $u$ of* (2.1) *provided that $u'(r) \in T$ for all $r$ sufficiently large.*

The proof is almost exactly the same as before. Extensions of Theorems 10, 11, 13, 14, 15 and 16 are also possible, but can be left to the interested researcher.

As an example of the results above, consider the mean curvature operator

$$G(p) = \sqrt{1 + p^2} - 1.$$

In this case (13.1) holds for $m = 2$ as one easily checks. On the other hand (13.2) requires that the interval $T$ should be bounded. Finally (13.3) is valid with the limit $1/2$. It follows that Theorem 17 may be applied to the mean curvature operator when the extremal $u$ is such that $u'(r)$ is bounded as $r \to \infty$. It is interesting that (7.4), (7.5) and (7.15) are then exactly the same as for the Laplace operator.

Theorem 18 on the other hand does not directly apply to the mean curvature operator since (13.5) requires $m = 2$ while (13.4) fails for this value of $m$ and for any interval $T$. A similar case is the third example at the end of Section 3, for which (13.1), (13.2) hold with $m = 2$, while (13.4) and (13.5) again fail.

## Acknowledgment

The paper was written partly at the University of Modena and partly at the University of Minnesota. The first author wishes to thank the University of Minnesota for its hospitality, and the second author similarly expresses his thanks to the University of Modena and the Consiglio Nazionale delle Ricerche, Rome, for their support.

## References

[1] Z. Artstein and E.F. Infante, On the asymptotic stability oscillators with unbounded damping, *Q. Appl. Math. 34* (1976) 195–199.

[2] H. Egnell, Existence results for some quasilinear elliptic equations, in: H. Berestycki, J.M. Coron and I. Ekeland, eds., *Variational Methods, Progress in Nonlinear Differential Equations,* vol. 4 (Birkhäuser, Boston, 1990) 61–76.

[3] R.H. Fowler, The form near infinity of real continuous solutions of a certain differential equation of second order, *Q. J. Math. (Cambridge Series) 45* (1914) 289–350.

[4] R.H. Fowler, Further studies of Emden's and similar differential equations, *Q. J. Math. (Oxford Series) 2* (1931) 259–288.

[5] B. Franchi, E. Lanconelli and J. Serrin, Existence and uniqueness of nonnegative solutions of quasilinear equations in $\mathbb{R}^n$, in: *Diffusion Equations and Their Equilibrium States*, MSRI Publ., Vol. 12 (Springer, New York, 1988) 293–300.

[6] A. Haraux and F.B. Weissler, Non-uniqueness of a semilinear initial value problem, *Indiana Univ. Math. J. 31* (1982) 167–189.

[7] J.J. Levin and J.A. Nohel, Global asymptotic stability for nonlinear systems of differential equations and applications to reactor dynamics, *Arch. Ration. Mech. Anal. 5* (1960) 194–211.

[8] J.B. McLeod, W.-M. Ni and J. Serrin, Unpublished notes.

[9] W.-M. Ni and J. Serrin, Existence and non-existence theorems for ground states of quasilinear partial differential equations. The anomalous case, *Accad. Naz. Lincei, Convegni dei Lincei 77* (1986) 231–257.

[10] P. Pucci and J. Serrin, A general variational identity, *Indiana Univ. Math. J. 35* (1986) 681–703.

[11] P. Pucci and J. Serrin, Asymptotic properties for solutions of strongly nonlinear second order differential equations, *Rend. Sem. Mat. Univ. Politec. Torino, Fascicolo Speciale* (1989) 121–129.

*Am. J. Math.* 116 (1994), no. 5, 1241–1264. **Reprint II.2.[ii]**


# ASYMPTOTIC INTEGRATION OF SECOND ORDER SYSTEMS

By James Serrin and Henghui Zou

*Dedicated to Philip Hartman and Aurel Wintner*

---

**1. Introduction.** In an elegant paper, Aurel Wintner considered the linear differential equation

$$(p(t)u')' + q(t)u = 0 \tag{1.1}$$

and proved that every solution of (1.1) tends to a finite limit as $t \to \infty$, and at least one solution has a nonzero limit, provided $p(t) \neq 0$, $q(t) \not\equiv 0$ and

$$\int_0^\infty \frac{1}{|p(t)|} \int_0^t |q(s)| ds dt < \infty, \tag{1.2}$$

see also [7]. In particular, the rest point of (1.1) cannot be a global attractor if (1.2) holds - even more, since (1.1) is linear and homogeneous, every point $u_0 \in \mathbf{R}^N$ is the limit of some solution of (1.1) as $t \to \infty$.

In this paper, we shall investigate similar problems for the more general nonlinear system

$$(P(t, u, u')A(t, u, u'))' + Q(t, u, u') = 0 \tag{1.3}$$

and also for the system

$$(A(t, u, u'))' + D(t, u, u')A(t, u, u') + Q(t, u, u') = 0, \tag{1.3$'$}$$

where $u : J \to \mathbf{R}^N$ and $J$ is a half open interval of the form $[T, \infty)$. We suppose throughout that

$$A(t, u, v),\ Q(t, u, v) : J \times \mathbf{R}^N \times \mathbf{R}^N \to \mathbf{R}^N$$

and

$$P(t, u, v),\ D(t, u, v) : J \times \mathbf{R}^N \times \mathbf{R}^N \to \mathrm{Lin}(\mathbf{R}^N, \mathbf{R}^N)$$

---

Manuscript received December 2, 1991.
*American Journal of Mathematics* 116 (1994), 1241–1264.

are continuous with $A(t,u,0) = 0$. It is assumed moreover that for fixed $t$ and $u$ the mapping $A(t,u,\cdot) : \mathbf{R}^N \to \mathbf{R}^N$ is invertible and that the inverse matrix $P^{-1}$ of $P$ exists for all values of $t$, $u$ and $v$. We note that a solution of (1.3) or (1.3)′ need not be of class $C^2$. Indeed, for the system (1.3), we require only that $u \in C^1(J)$ and $P(t,u,u')A(t,u,u') \in C^1(J)$, while for (1.3)′, correspondingly, the conditions $u \in C^1(J)$ and $A(t,u,u') \in C^1(J)$ are enough.

We shall extend Wintner's results to (1.3) and (1.3)′, and in particular prove that under appropriate conditions generalizing (1.2) the set of attainable limits as $t \to \infty$ of solutions of (1.3) and of (1.3)′ is exactly $\mathbf{R}^N$, as in the linear case, and that all bounded solutions approach a finite limit. Moreover, if $Q$ satisfies additional growth conditions, then all solutions are bounded.

An important special case of (1.3) and (1.3)′ occurs when $A(t,u,v) \equiv v$. Then (1.3) becomes

$$(P(t,u,u')u')' + Q(t,u,u') = 0, \tag{1.4}$$

the most immediate generalization of Wintner's case to nonlinear systems, while (1.3)′ becomes

$$u'' + D(t,u,u')u' + Q(t,u,u') = 0. \tag{1.4$'$}$$

To illustrate our results, it is convenient to state them first for the prototype special case (1.4). The main results for the general system (1.3) are then presented in Sections 2, 3, and 4, and for (1.3)′ and (1.4)′ in Section 6 and in Corollaries 4, 5 at the end of the introduction. A number of examples are given in Sections 5 and 7.

Let $K$ be a set, $Y$ a Banach space, and $C = C(J \to Y)$ the space of all continuous $Y$-valued functions on $J$. In what follows we shall say that a functional $G : K \to C$ is *uniformly integrable* on $K$ if there exists a function $\psi(t) \in L^1(J)$ such that

$$||G[x](t)||_Y \le \psi(t)$$

for all $x \in K$ and $t \in J$. Similarly, we say that $G$ is *equicontinuous* (as a function of $t$) on $K$ if for any $\epsilon > 0$ there exists $\delta = \delta(\epsilon) > 0$ such that for all $x \in K$.

$$||G[x](t') - G[x](t'')||_Y < \epsilon \qquad \text{when } |t' - t''| < \delta.$$

*Remark.* If the functional $G[x]$ is conventional, that is, if $K = C$ and

$$G[x](t) = f(t, x(t)) : C \to C$$

for some function $f$, then the uniform integrability condition is equivalent to the requirement that $||f(t, x(t))||_Y \leq \psi(t)$ for all $x \in K$, while equicontinuity on $K$ is expressed by

$$\|f(t', x(t')) - f(t'', x(t''))\|_Y < \epsilon \qquad \text{when } |t' - t''| < \delta.$$

In the sequel we shall consider mainly the case $Y = \mathbf{R}^N$. Of course, then $C = C(J \to \mathbf{R}^N)$, a notation we shall maintain throughout the introduction and for the major part of the paper, until Section 6. For $u, v \in C$ and $t \in J$, we define two mappings $\Psi_1, \Psi_2 : C \times C \to C$ by

$$\Psi_1[u, v](t) = -P^{-1}(t, u(t), v(t)) \int_T^t Q(s, u(s), v(s)) \tag{1.5}$$

and

$$\Psi_2[u, v](t) = P^{-1}(t, u(t), v(t)) \int_t^\infty Q(s, u(s), v(s)). \tag{1.6}$$

Of course $\Psi_2$ is well-defined only if the integrand $Q(s, u(s), v(s))$ is of class $L^1$; thus whenever the functional $\Psi_2$ is under consideration it will be tacitly assumed that this condition holds. When $P$ and $Q$ are independent of $v$, we shall continue to use the notation (1.5) and (1.6), but with $[u, v]$ replaced by $[u]$.

Let $M > 0$ be fixed, and define

$$B_M = \{u \in C : ||u||_J \leq M\}, \qquad ||u||_J = \sup_{t \in J} |u(t)|.$$

Our main results for the system (1.4) are then as follows.

THEOREM A. *Let $P$ and $Q$ be independent of $v$. Suppose that either $\Psi_1[u]$ or $\Psi_2[u]$ is uniformly integrable on the set $K = B_M$, for all fixed $M > 0$. Then the set of attainable limits as $t \to \infty$ of the solutions of* (1.4) *is exactly $\mathbf{R}^N$.*

When either $P$ or $Q$ depends on $v$, a result corresponding to Theorem A holds under additional assumptions.

THEOREM B. *Suppose that for all fixed $M > 0$, there exists a constant $\epsilon > 0$ such that at least one of the functionals $\Psi_1[u, v]$ and $\Psi_2[u, v]$ is uniformly integrable and equicontinuous on $B_M \times B_\epsilon$. Then the set of attainable limits as $t \to \infty$ of solutions of* (1.4) *is exactly $\mathbf{R}^N$.*

Uniform integrability hypotheses of the type used in Theorems A and B also imply boundedness and limit theorems for solutions of (1.4). In particular we have the following result.

THEOREM C. *Suppose that both $P^{-1}(\cdot, u, v)$ and $\Psi_1[u, v]$ are uniformly integrable on $B_M \times C$ for all fixed $M > 0$. Then every bounded solution of (1.4) tends to a finite limit.*

*If, additionally, the function*

$$\frac{\Psi_1[u, v](t)}{1 + \|u\|_{[T,t]}}$$

*is uniformly integrable on $C \times C$, then every solution of (1.4) tends to a limit.*

Theorem A is a special case of Theorem 1 in the following section. Theorem B follows from Theorem 2 and the remark after the proof of Theorem 2. Theorem C is the special case $m = 2$ of Theorem 4.

Theorems A, B, C extend Wintner's results to nonlinear systems, see in particular Theorems 6 and 7 in Section 5. Related, overlapping, results for quasi-variational systems are given in [6].

COROLLARY 1. *Let $P$ and $Q$ be independent of $v$. Suppose for each $M > 0$ that there exists $\lambda(s, t)$ such that either*

(i) $|P^{-1}(t, u)Q(s, u)| \le \lambda(s, t)$ *when* $|u| \le M$, $T \le s \le t$,

$$\int_T^\infty \int_T^t \lambda(s, t)dsdt < \infty;$$

*or*

(ii) $|P^{-1}(t, u)Q(s, u)| \le \lambda(s, t)$ *when* $|u| \le M$, $T \le t \le s$,

$$\int_T^\infty \int_t^\infty \lambda(s, t)dsdt < \infty.$$

*Then the set of attainable limits as $t \to \infty$ of solutions of (1.4) is exactly $\mathbf{R}^N$.*

This is an immediate consequence of Theorem A, since the assumptions of the corollary imply that either $\Psi_1[u]$ or $\Psi_2[u]$ is uniformly integrable on $B_M$ for all fixed $M > 0$.

There are cases when Theorem A applies but the corollary fails. For example, let $N = 1$ and take $f$ to be any continuous function on $\mathbf{R}$. Put

$$P(t, u) = t^{3/2}, \qquad Q(t, u) = \sin t + f(u)/t$$

with $T = 1$. Clearly $\Psi_1[u]$ is uniformly integrable on $B_M$ for all fixed $M > 0$ and hence Theorem A holds. On the other hand, one easily verifies that

$$|P^{-1}(t, u)Q(s, u)| \ge |\sin(s)|t^{-3/2} - Cs^{-1}t^{-3/2}$$

and that

$$\int_1^\infty \int_1^t \left[|\sin(s)|t^{-3/2} - Cs^{-1}t^{-3/2}\right] = \infty,$$

where $C$ is a constant depending on $M$. Hence, obviously, the assumptions in Corollary 1 cannot hold.

COROLLARY 2. *Let $P$ and $Q$ be independent of $v$. Suppose for each $M > 0$ that there exist $r(t)$ and $q(t)$ such that*

$$|P^{-1}(t,u)| \le r(t), \qquad |Q(t,u)| \le q(t) \quad \text{when } |u| \le M,\ t \ge T,$$

*and that either*

$$\int_T^\infty \int_T^t r(t)q(s)dsdt < \infty \tag{1.7}$$

*or*

$$\int_T^\infty \int_t^\infty r(t)q(s)dsdt < \infty. \tag{1.8}$$

*Then the set of attainable limits as $t \to \infty$ of solutions of (1.4) is exactly $\mathbf{R}^N$.*

Assumption (1.7) is a direct generalization of Wintner's condition (1.2). Note, however, that for *systems* the result of Corollary 2 is strictly weaker than that of Corollary 1, as shown in the example at the end of the introduction.

When $Q$ depends on $v$, an analogue of Corollary 1 remains true. The following result is a consequence of the corollary at the end of Section 3.

COROLLARY 3. *Let $P$ be independent of $v$. Suppose for each $M > 0$ that there exist $\epsilon > 0$, $L > 0$, and $\mu(s)$ such that*

$$|P^{-1}(t,u)| \le L, \qquad |Q(s,u,v)| \le \mu(s),$$

*for $|u| \le M$, $|v| \le \epsilon$ and that*

$$\int_T^\infty s\mu(s)ds < \infty.$$

*Then the set of attainable limits as $t \to \infty$ of solutions of (1.4) is exactly $\mathbf{R}^N$.*

The case $P = I$ of this result is due to Hale, Onichic and Hartman [3], [4]; see also Hartman [5].

Finally, we give two applications of our results to the system (1.4)$'$.

COROLLARY 4. *Suppose for each $M > 0$ that there exist constants $\epsilon > 0$, $L > 0$, and a function $\mu(s)$ such that*

$$\left|\int_T^t D(s, u(s), v(s))ds\right| \leq L, \qquad |Q(s,u,v)| \leq \mu(s), \tag{1.9}$$

*for $|u| \leq M$, $|v| \leq \epsilon$ and that*

$$\int_T^\infty s\mu(s)ds < \infty. \tag{1.10}$$

*Then the set of attainable limits as $t \to \infty$ of solutions of* (1.4)′ *is exactly* $\mathbf{R}^N$.

Corollary 4 is a consequence of Theorem 2* and the remark at the end of Section 6. In (1.9)-(1.10) both the damping matrix $D$ and the forcing term $Q$ are (integrally) small for large $t$. A complementary result holds when the damping matrix $D$ is large. For simplicity we give this result for the case when $D$ and $Q$ are independent of $v$; the proof is in Part 4 of Section 7.

COROLLARY 5. *Let $D$ be independent of $v$, and $Q = f(u)$. Suppose for each $M > 0$ that there exists a positive function $\sigma(t)$ such that*

$$D(t,u) \geq \sigma(t)I, \qquad \text{when } |u| \leq M,$$

*where $I$ is the $N \times N$ identity matrix and $\sigma(t)$ satisfies*

$$1/\sigma \in L^1(J), \qquad (1/\sigma(t))' \to 0 \quad \text{as } t \to \infty.$$

*Then the set of attainable limits as $t \to \infty$ of solutions of (1.4)′ is exactly* $\mathbf{R}^N$.

*Example.* Let $N = 2$ and in (1.4) put $P(t,u) = P(t)$, $Q(t,u) = Q(t)f(u)$ where

$$P^{-1}(t) = \begin{pmatrix} 1+t^{-4} & 2 \\ 1 & 2 \end{pmatrix}, \qquad Q(t) = \begin{pmatrix} 2t \\ -t \end{pmatrix}, \qquad f : \mathbf{R}^2 \to \mathbf{R}^1$$

Clearly

$$|P^{-1}(t)Q(s)| = \left|\begin{pmatrix} 2s/t^4 \\ 0 \end{pmatrix}\right| \leq 2st^{-4}$$

the first part of Corollary 1 applies with $\lambda(s,t)$ =Const $st^{-4}$. On the other hand,

$$r(t) \geq |P^{-1}(t)| = 2, \qquad q(t) \geq |Q(t)| = 2t,$$

and the assumptions in Corollary 2 fail.

**2. Main Results, I.** In this section we investigate the system (1.3) when $P$ and $Q$ are independent of $v$. We begin with some basic notation and assumptions.

($A_1$) For all fixed $(t,u) \in J \times \mathbf{R}^N$, the function $A(t,u,\cdot)$ is one-to-one from $\mathbf{R}^N$ to an open set $\Omega = \Omega_{t,u} = A(t,u,\mathbf{R}^N) \subset \mathbf{R}^N$. Moreover $A(t,u,0) = 0$.

($A_2$) There exists an exponent $m > 1$ such that $|v|^{m-1} \leq \sigma(u)|A(t,u,v)|$ for $t \in J$ and $v \in \mathbf{R}^N$, where $\sigma \in C(\mathbf{R}^N \to \mathbf{R}^+)$.

Condition ($A_2$) implies that the set $\Omega$ in ($A_1$) is all of $\mathbf{R}^N$, see the lemma below. A weaker condition than ($A_2$) is the following

($A_2$)′ There exist $m > 1$ and $R > 0$ such that $|v|^{m-1} \leq \sigma(u)|A(t,u,v)|$ for $t \in J$, $u \in \mathbf{R}^N$ and $v \in B(R) \subset \mathbf{R}^N$, where $B(R)$ is the ball of radius $R$ in $\mathbf{R}^N$ centered at the origin.

LEMMA. *If* ($A_1$) *and* ($A_2$) *hold, then* $\Omega = \mathbf{R}^N$. *If* ($A_1$) *and* ($A_2$)′ *hold, then* $\Omega$ *contains the ball* $B(\rho) \subset \mathbf{R}^N$, *where* $\rho = R^{m-1}/\sigma(u)$.

*Proof.* Since $A$ is continuous, the set $\Omega$ is simply connected. By ($A_2$)′, we have $B(\rho) \subset A(t,u,B(R)) \subset \Omega$, where $\rho = R^{m-1}/\sigma(u)$. The proof is completed by observing that ($A_2$) is the special case $R = \infty$.

For fixed $(t,u) \in J \times \mathbf{R}^N$ we denote the inverse of $A(t,u,\cdot)$ by $\Phi(t,u,\cdot)$, if it exists. Obviously $\Phi(t,u,v)$ is continuous in all its arguments. As in the introduction, for $u \in C = C(J \to \mathbf{R}^N)$ we introduce the functionals (from $C$ into $C$)

$$\Psi_1[u](t) = -P^{-1}(t,u(t)) \int_T^t Q(s,u(s)) \tag{2.1}$$

and

$$\Psi_2[u](t) = P^{-1}(t,u(t)) \int_t^\infty Q(s,u(s)). \tag{2.2}$$

For the statement of the following results it is useful to generalize the definition of uniform integrability given in Section 1 to the case where $\psi \in L^k(J)$ rather than $L^1(J)$, using for this situation the terminology that $G$ is 'uniformly $L^k$-integrable' on $K$.

In refering to an 'attainable limit' of a solution it is tacitly understood that the limit value is finite (that is, in $\mathbf{R}^N$).

THEOREM 1. *Suppose that* $(A_1)$ *and* $(A_2)$ *are satisfied. Assume that for all fixed* $M > 0$ *either* $\Psi_1[u]$ *or* $\Psi_2[u]$ *is uniformly* $L^{1/(m-1)}$*-integrable on* $B_M$. *Then the set of attainable limits of solutions of* (1.3) *is equal to* $\mathbf{R}^N$.

*Proof.* We shall only prove the theorem for $\Psi_1[u]$, the other case being essentially the same. Moreover for simplicity in printing we write $\Psi[u]=\Psi_1[u]$. For fixed $u_0 \in \mathbf{R}^N$ and for $T_0 \in J$, put $I = [T_0, \infty)$, $C(I) = C(I \to \mathbf{R}^N)$ and

$$X = X_{u_0} = \{u \in C(I) \mid \|u(t) - u_0\|_I \leq 1 \},$$

where $T_0$ will be determined later. Consider the mapping $F : X \to C(I)$ given by

$$(Fu)(t) = u_0 - \int_t^\infty \Phi(s, u(s), \Psi[\bar{u}](s)), \qquad t \in I, \tag{2.3}$$

where $\bar{u} \in C$ is defined by

$$\bar{u}(t) = \begin{cases} u(T_0), & T \leq t < T_0 \\ u(t), & t \geq T_0. \end{cases}$$

We claim that $F$ is well-defined and compact, and that $F(X) \subset X$ if $T_0$ is sufficiently large. The proof is divided into three steps.

*Step* 1. $F$ is well-defined on $X$. First, $\Phi(t, u(t), \Psi[\bar{u}](t))$ is well-defined on $X$ and continuous for $t \in I$.

Next we show that $\Phi(t, u(t), \Psi[\bar{u}](t))$ is integrable on $I$, whence $F$ is well-defined. Indeed, for $u \in X$ we have from $(A_2)$

$$|v|^{m-1} \leq C_0 |A(t, \bar{u}(t), v)| \quad \text{for } t \in J, \ v \in \mathbf{R}^N,$$

with the constant $C_0$ depending only on $R_0 = 1 + |u_0|$. Replacing $v$ by $\Phi(t, \bar{u}(t), v)$ yields

$$|\Phi(t, \bar{u}(t), v)| \leq C_1 |v|^{1/(m-1)} \qquad (C_1 = C_0^{1/(m-1)}), \tag{2.4}$$

and in particular

$$|\Phi(t, \bar{u}(t), \Psi[\bar{u}](t))| \leq C_1 |\Psi[\bar{u}](t)|^{1/(m-1)}, \qquad t \in J. \tag{2.5}$$

Then since $\bar{u} \in B_{R_0}$ and $\Psi[u]$ is uniformly $L^{1/(m-1)}$-integrable on $B_M$ for all $M > 0$, it follows that

$$|\Psi[\bar{u}](t)| \leq \psi(t), \qquad t \in J, \tag{2.6}$$

where $\psi \in L^{1/(m-1)}(J)$. In turn, by (2.5) and (2.6), we see that $\Phi(t, \bar{u}(t), \Psi[\bar{u}](t))$ is uniformly integrable on $B_{R_0}$. This proves that $F$ is well-defined on $X$, since in the integrand of (2.3) we have $s \in I$ and

$$\Phi(s, u(s), \Psi[\bar{u}](s)) \equiv \Phi(s, \bar{u}(s), \Psi[\bar{u}](s))) \in L^1(I).$$

Finally, making use of the fact that $\Phi$ is uniformly integrable, it is easy to check that the mapping $F$ is continuous on $X$.

*Step* 2. $F : X \to X$. We need to show that, for some $T_0$,

$$||Fu - u_0||_I \leq 1.$$

But this is a direct consequence of (2.5) and (2.6), that is

$$||Fu - u_0||_I = \sup_{T_0 \leq t < \infty} \left| \int_t^\infty \Phi(s, u(s), \Psi[\bar{u}](s)) \right| \leq C_1 |\psi|_{L^{1/(m-1)}(I)}^{1/(m-1)} \leq 1,$$

provided $T_0$ is sufficiently large.

*Step* 3. $F$ is compact from $X$ to $X$. Let $u_n \in X$, $n = 1, 2, \ldots$, and put

$$z_n = Fu_n, \qquad n = 1, 2, \ldots \tag{2.8}$$

It is clear, by Step 2, that $z_n \in X$ and so

$$||z_n||_I \leq R_0 = |u_0| + 1, \qquad n = 1, 2, \ldots. \tag{2.9}$$

Let $S$ denote the closed interval $[\tan^{-1} T_0, \pi/2]$, and define the functions $\tilde{z}_n : S \to \mathbf{R}^N$ by

$$\tilde{z}_n(\tau) = \begin{cases} z_n(\tan\tau), & \tau \in S\backslash\{\pi/2\} \\ u_0, & \tau = \pi/2. \end{cases}$$

Clearly $\tilde{z}_n$ is uniformly bounded in $C(S)$, by (2.9). Also put

$$\varphi(\tau) = C_1 \int_{T_0}^{\tan\tau} |\psi(s)|^{1/(m-1)} ds, \quad \tau \in S$$

(so $\varphi$ is continuous on $S$).

Then by (2.3), (2.5) and (2.6)

$$|\tilde{z}_n(\tau'') - \tilde{z}_n(\tau')| \leq |\varphi(\tau'') - \varphi(\tau')| \leq \omega(|\tau'' - \tau'|)$$

where $\omega$ is a modulus of continuity for $\varphi$. It now follows from the Arzelá-Ascoli theorem that the family $\{\tilde{z}_n(\tau)\}$ is compact in $C(S)$. In turn, $\{z_n(t)\}$ is compact in $C(I)$, that is for some sequence of values $n$ we have

$$\|z_n - z\|_I \to 0$$

for some $z \in X$ (recall that $X$ is closed and $z_n \in X$).

This proves that $F$ maps $X$ into a relatively compact subset of $X$. By definition, $F: X \to X$ is a compact continuous map.

The theorem now follows at once. Obviously $X$ is convex, so by the Schauder fixed point theorem $F$ has a fixed point $x$ in $X$, that is,

$$Fx = x. \tag{2.10}$$

It is clear that $x \in C^1(I)$ and satisfies (1.3) on $I$. Indeed by (2.3) one has

$$x(t) = u_0 - \int_t^\infty \Phi(s, x(s), \Psi[\bar{x}](s)), \qquad t \in I.$$

It follows that

$$x'(t) = \Phi(t, x(t), \Psi[\bar{x}](t)), \qquad t \in I,$$

that is,

$$A(t, x(t), x'(t)) = \Psi[\bar{x}](t).$$

Therefore by (2.1)

$$P(t, x(t))A(t, x(t), x'(t)) = P(t, x(t))\Psi[\bar{x}](t) = -\int_T^t Q(s, \bar{x}(s)).$$

Differentiation now yields

$$[P(t, x(t))A(t, x(t), x'(t))]' = -Q(t, \bar{x}(t)) = -Q(t, x(t)), \qquad t \geq T_0,$$

as required. Moreover, from (2.3) one has

$$\lim_{t\to\infty} x(t) = u_0. \tag{2.11}$$

Since $u_0$ is an arbitrary point in $\mathbf{R}^N$, this completes the proof of the theorem.

Corresponding to condition $(A_2)'$, we have the following

THEOREM 1$'$. *Suppose that* $(A_1)$ *and* $(A_2)'$ *are satisfied. Assume also for all fixed* $M > 0$ *that either* $\Psi_1[u]$ *or* $\Psi_2[u]$ *is uniformly* $L^{1/(m-1)}$*-integrable on* $B_M$*, with the corresponding functions* $\psi(t)$ *satisfying*

$$\lim_{t\to\infty} \psi(t) = 0. \tag{2.12}$$

*Then the set of attainable limits of solutions of* (1.3) *is equal to* $\mathbf{R}^N$.

*Proof.* The argument is the same as before, except for the first step. In this case the inverse function $\Phi(t, u(t), \cdot)$ is defined only on a subset $\Omega = \Omega_{t,u(t)}$ of $\mathbf{R}^N$. By the lemma, for $t \in J$ and $|u| \leq R_0$, there exists $\rho > 0$ such that $B(\rho) \subset \Omega$. On the other hand, from (2.6) we have, for all $u \in X$,

$$|\Psi[\bar{u}](t)| \leq \psi(t) \leq \sup_I \psi(t), \quad t \in I.$$

In turn, if $T_0$ is fixed suitably large, it follows from (2.12) that

$$|\Psi[\bar{u}](t)| \leq \rho, \quad t \in I.$$

Hence $\Phi(t, u(t), \Psi[\bar{u}](t))$ is well-defined for $t \in I$, $u \in X$. Then, as in the remaining part of the proof of Step 1, we see from (2.5), (2.6) and $(A_2)'$ that $F$ is well-defined on $X$.

**3. Main Results, II.** In the last section we proved that every target $u_0 \in \mathbf{R}^N$ is attainable by solutions of (1.3) provided both $P$ and $Q$ are independent of $v$. If either $P$ or $Q$ depends on $v$, Theorems 1 and 1$'$ will fail because the mapping $F$ is not well-defined. It can however be modified so that the preceding arguments apply.

We introduce the mappings $\Psi_3, \Psi_4 : C \times C \to C$ by

$$\Psi_3[u, w](t) = -P^{-1}(t, u, \Phi(t, u, w)) \int_T^t Q(s, u, \Phi(s, u, w)), \tag{3.1}$$

$$\Psi_4[u, w](t) = P^{-1}(t, u, \Phi(t, u, w)) \int_t^\infty Q(s, u, \Phi(s, u, w)). \tag{3.2}$$

Obviously $\Psi_4$ is well-defined only when $Q(t, u(t), \Phi(t, u(t), w(t)))$ belongs to $L^1(J)$. Accordingly this condition is assumed in the following results.

THEOREM 2. *Suppose that* $(A_1)$ *and* $(A_2)$ *are satisfied. Assume that for all fixed* $M > 0$ *there exists* $\epsilon > 0$ *such that at least one of the functionals* $\Psi_3[u, w]$ *and*

*$\Psi_4[u,w]$ is uniformly $L^{1/(m-1)}$-integrable and equicontinuous on $B_M \times B_\epsilon$. Then the set of attainable limits of solutions of* (1.3) *is equal to* $\mathbf{R}^N$.

*Proof.* We prove the theorem only for $\Psi_4$. The case for $\Psi_3$ is similar and is left to the reader; see also the proof in Section 2. We modify the proof of Theorem 1 by first introducing a larger system and then applying the fixed point argument to this system. Put $I = [T_0, \infty)$,

$$W = \{w \in C(I) \mid \|w(t)\| \leq \epsilon\},$$

and $Z = X \times W$, where $X$ is as before. For fixed $u_0 \in \mathbf{R}^N$, consider the mapping $F(u,w) = (F_1, F_2)$ on $Z$ given by

$$\left\{ \begin{array}{l} (F_1(u,w))(t) = u_0 - \int_t^\infty \Phi(s,u,\Psi_4[u,w](s)) \\ (F_2(u,w))(t) = \Psi_4[u,w](t) \end{array} \right. \qquad t \in I. \tag{3.3}$$

Applying the argument of Theorem 1, and fixing $T_0$ suitably large, one may readily verify that $F_1$ maps $Z$ to a relatively compact subset of $X$.

Similarly, we claim that $F_2$ maps $Z$ to a relatively compact subset of $W$. To see this, first observe that there exists $\psi \in L^{1/(m-1)}(J)$ such that

$$|\Psi_4[u,w](t)| \leq \psi(t), \quad t \in I, \tag{3.4}$$

for all $(u,w) \in Z$. Next we show that

$$\Psi_4[u,w](t) \to 0 \quad \text{as } t \to \infty \quad \text{uniformly on } Z. \tag{3.5}$$

Otherwise there exists a sequence $\{u_k, w_k\}$ in $Z$, and a sequence $\{t_k\}$ in $I$ with $t_k \to \infty$, such that

$$\Psi_4[u_k, w_k](t_k) \geq \epsilon_0$$

for some $\epsilon_0 > 0$. But by the equicontinuity of $\Psi_4$ (see Section 1) there exists $\delta_0 > 0$ such that

$$\Psi_4[u,w](t) \geq \epsilon_0/2 \quad \text{for } t \in I,\ |t - t_k| \leq \delta_0,$$

and all $k$. This contradicts (3.4), proving (3.5).

By (3.5) it is clear that $F_2$ maps $Z$ to $W$, provided $T_0$ is chosen so large that $|\Psi_4[u,w](t)| \leq \epsilon$ for $t \in I$.

Now define

$$\Psi_4[u,w](\infty) = 0 \tag{3.6}$$

and consider the family of functions

$$\Psi_4[u, w](\tan \tau) : S \to \mathbf{R}^N,$$

where $(u, w) \in Z$. By (3.5), (3.6) and the equicontinuity assumption on $\Psi_4$ it is easy to see that this family is equicontinuous on $Z$. Applying the Arzela-Ascoli theorem in the same way as before, now proves the claim. Therefore $F$ is a compact mapping from $Z$ to $Z$, and $F$ has a fixed point $(u, w)$ in $Z$.

It remains to verify that $(u, w)$ is a solution of (1.3) having the limit $u_0$ at infinity. In particular, we have

$$F(u, w) = (u, w).$$

It then follows from (3.3) that

$$w(t) = \Psi_4[u, w](t), \qquad u'(t) = \Phi(t, u(t), \Psi_4[u, w](t)) = \Phi(t, u(t), w(t)).$$

Hence

$$w = A(t, u(t), u'(t)), \qquad P(t, u, u')w(t) = \int_t^\infty Q(t, u, u').$$

Therefore

$$[P(t, u, u')A(t, u(t), u'(t))]' = (P(t, u, u')w)' = -Q(t, u, u'),$$

as required.

*Remark.* Note that $w = v$, $\Psi_3[u, w] = \Psi_1[u, v]$, $\Psi_4[u, w] = \Psi_2[u, v]$ when $A(t, u, v) = v$.

The equicontinuity condition on $\Psi$ in Theorem 2 may not always be easy to verify, since it involves the equicontinuity of $P^{-1}$. If $P$ is independent of $v$, then we can further modify $F$ to avoid this problem. We define

$$\Psi_5[u, w](t) = -P^{-1}(t, u) \int_T^t Q(s, u, \Phi(s, u, P^{-1}(s, u)w)),$$

$$\Psi_6[u, w](t) = P^{-1}(t, u) \int_t^\infty Q(s, u, \Phi(s, u, P^{-1}(s, u)w)).$$

The following result then holds.

THEOREM 3. *Let P be independent of v. Suppose that* $(A_1)$ *and* $(A_2)$ *are satisfied. Assume that for all fixed* $M > 0$ *there exist* $\epsilon > 0$ *and* $L > 0$ *such that*

(i) $|P^{-1}(t,u)| \leq L$ *when* $|u| \leq M$,

(ii) $Q(t,u,v)$ *is uniformly integrable on* $B_M \times B_\epsilon$,

(iii) $\Psi_5[u,w]$ *or* $\Psi_6[u,w]$ *is uniformly* $L^{1/(m-1)}$*-integrable on* $B_M \times B_\epsilon$.

*Then the set of attainable limits of solutions of* (1.3) *is equal to* $\mathbf{R}^N$.

The arguments in Theorem 2 can be carried over essentially unchanged. For instance, for the case of $\Psi_6[u,w]$ we modify $F$ to the form

$$\begin{cases} (F_1(u,w))(t) = u_0 - \int_t^\infty \Phi(s,u(s),\Psi_6[u,w](s)) \\ (F_2(u,w))(t) = P(t,u(t))\Psi_6[u,w](t). \end{cases} \tag{3.7}$$

The proof in Theorem 2 then applies directly, once we prove the family of functions

$$F_2(u,w)(t) = \int_t^\infty Q(s,u,\Phi(s,u,P^{-1}(s,u)w))ds$$

is equicontinuous on Z, and

$$F_2(u,w)(t) \to 0 \quad \text{as} \quad t \to \infty \quad \text{uniformly on Z.}$$

To do this, it is necessary to redefine the set $W$, replacing $\epsilon$ by $\min(\epsilon, \epsilon^{m-1}/LC_0)$. Then by (i), (ii) and (2.4) one easily obtains the required behavior.

It remains only to show that a fixed point of (3.7) is a solution of (1.3). But for any such point $(u,w)$ we have

$$\begin{cases} w(t) = P(t,u(t))\Psi_6[u,w](t), \\ u'(t) = \Phi(t,u(t),\Psi_6[u,w](t)) = \Phi(t,u(t),P^{-1}(t,u(t))w(t)), \end{cases}$$

so that by the definition of $\Phi$ and $\Psi_6$

$$w(t) = P(t,u(t))A(t,u(t),u'(t)) = \int_t^\infty Q(s,u(s),u'(s)),$$

as required.

COROLLARY. *In Theorem 3 we can replace (ii) and (iii) by*

$$|Q(s,u,v)| \leq \mu(s), \qquad \text{for } |u| \leq M,\ |v| \leq \epsilon \tag{3.8}$$

*where*

$$\int_T^\infty \left(\int_t^\infty \mu(s)ds\right)^{1/(m-1)} dt < \infty. \tag{3.9}$$

If $m = 2$ the condition (3.9) is equivalent to $\int_T^\infty s\mu(s)ds < \infty$.

**4. Boundness and Limit Properties.** In Theorems 1, 2 and 3 we showed that for each $u_0 \in \mathbf{R}^N$ there exists a solution with the limit $u_0$ at infinity. However, does every solution tend to a limit?

The hypotheses of Theorem 1 are enough to assure a partial answer, that is, every *bounded* solution of (1.3) tends to a limit; see Theorem 4. On the other hand, the assumptions of Theorem 2 assure that every solution $u$ of (1.3) tends to a certain generalized limit even when the solution is not bounded, see Theorem 5.

We show finally that under appropriate further conditions every solution of (1.3) tends to a finite limit at infinity.

For simplicity, throughout this section it will be assumed that the following strengthened version of $(A_2)$ is satisfied, in which the function $\sigma(u)$ is relpaced by a constant.

$(A_2)''$ The inverse $\Phi(t, u, \cdot)$ of $A(t, u, \cdot)$ is well defined for all $t \in J$ and all $u \in \mathbf{R}^N$. Moreover there exists a constant $c > 0$ such that for $t \in J$ and all $u, v \in \mathbf{R}^N$,

$$\Phi(t, u, v) \le c|v|^{1/(m-1)} \quad \text{for } t \in J \text{ and } u, v \in \mathbf{R}^N.$$

Corresponding to Theorem 1, the following result holds.

THEOREM 4. *For all fixed $M > 0$, suppose that $P^{-1}(\cdot, u, v)$ and $\Psi_1[u, v]$ are uniformly $L^{1/(m-1)}$-integrable on $B_M \times C$. Then every bounded solution of (1.3) tends to a limit as $t \to \infty$.*

*If in addition the function*

$$\frac{\Psi_1[u, v](t)}{1 + \|u\|_{[T,t]}^{m-1}} \tag{4.1}$$

*is uniformly $L^{1/(m-1)}$-integrable on $C \times C$, then every solution of (1.3) is necessarily bounded.*

*Remark.* The first part of this theorem requires only $(A_2)$, not $(A_2)''$.

*Proof.* For any solution $u = u(t)$ of (1.3) there holds

$$A(t, u, u'(t)) = \Psi_1[u, u'](t) + P^{-1}(t, u(t), u'(t))A(T, u(T), u'(T)). \tag{4.2}$$

Assume that $u$ is bounded. Since both $P^{-1}(\cdot, u, v)$ and $\Psi_1[u, v]$ are uniformly $L^{1/(m-1)}$-integrable on $B_M \times C$ for all fixed $M > 0$, it follows from $(A_2)$ that the right hand side of (4.2) belongs to $L^{1/(m-1)}(J)$. Hence $u'$ belongs to $L^1(J)$. Therefore $u$ tends to a finite limit at infinity.

Next, consider an arbitrary solution $u$. For any $t_1 > t \geq T$, we see from (4.2) that $u$ satisfies the integral equation

$$u(t_1) - u(t) = \int_t^{t_1} \Phi\left(s, u(s), \Psi_1[u, u'](s) P^{-1}(s, u(s), u'(s)) A(T, u(T), u'(T))\right). \tag{4.3}$$

By $(A_2)''$

$$\begin{aligned} &\left|\int_t^{t_1} \Phi\left(s, u(s), \Psi_1[u, u'](s) + P^{-1}(s, u(s), u'(s)) A(T, u(T), u'(T))\right)\right| \\ &\quad \leq c \int_t^{t_1} \left|\Psi_1[u, u'](s) + P^{-1}(s, u(s), u'(s)) A(T, u(T), u'(T))\right|^{1/(m-1)} \\ &\quad \leq c \int_t^{t_1} |\Psi_1[u, u'](s)|^{1/(m-1)} + c(T) \int_t^{t_1} |P^{-1}(s, u(s), u'(s))|^{1/(m-1)}. \end{aligned} \tag{4.4}$$

Put

$$M_1 = \sup_{t \leq s < t_1} |u(s)| = ||u||_{[t, t_1]}.$$

Thanks to (4.1),

$$\int_t^{t_1} |\Psi_1[u, u'](s)|^{1/(m-1)} \leq (1 + M_1) ||\psi||^{1/(m-1)}_{L^{1/(m-1)}[t, \infty)}. \tag{4.5}$$

Combining (4.3)-(4.5) yields

$$M_1 \leq c[(1 + M_1) + c(T)] ||\psi||^{1/(m-1)}_{L^{1/(m-1)}[t, \infty)} + |u(t)|.$$

Since $\psi \in L^{1/(m-1)}$, we may choose $t$ so large that

$$M_1 \leq \frac{1}{2}(1 + M_1 + c(T)) + |u(t)| \qquad \text{for } t_1 > t. \tag{4.6}$$

It follows from (4.6) that $u$ is bounded on $J$, which proves the theorem.

THEOREM 5. *Suppose that $Q(\cdot, u, v)$ is uniformly integrable on $C \times C$. Also assume that the matrix $P(t, u, v)$ tends to a limit $P_0$ (independent of $u, v$) as $t$ tends to infinity, where $P_0$ is invertible. Then there exists a vector $w_0 \in \mathbf{R}^N$ such that*

$$\lim_{t \to \infty} A(t, u(t), u'(t)) = w_0.$$

*Furthermore, if* $w_0 = 0$ *then* $u$ *tends to a finite limit* $u_0$ *provided that* $\Psi_2[u, v]$ *is uniformly* $L^{1/(m-1)}$*-integrable on* $C \times C$.

*Proof.* This is similar to the demonstration of Theorem 4. For any solution $u = u(t)$ we have for $t_1 > t \geq T$

$$(4.7) \qquad P(t, u(t), u'(t))A(t, u, u'(t)) = P(t_1, u(t_1), u'(t_1))A(t_1, u(t_1), u'(t_1)) - \int_t^{t_1} Q(s, u, u').$$

Letting $t_1 \to \infty$ shows that

$$\bar{w}_0 = \lim_{t_1 \to \infty} P(t_1, u(t_1), u'(t_1))A(t_1, u(t_1), u'(t_1))$$

exists, e.g. the left hand side of (4.7) is constant and the integral converges by assumption. Hence

$$(4.8) \quad A(t, u(t), u'(t)) = P^{-1}(t, u(t), u'(t))\bar{w}_0 - P^{-1}(t, u(t), u'(t)) \int_t^{\infty} Q(s, u, u').$$

Now let $t \to \infty$ to get

$$\lim_{t \to \infty} A(t, u(t), u'(t)) = P_0^{-1}\bar{w}_0 = w_0.$$

This proves the first part of the theorem.

Finally suppose $w_0 = 0$. Then by (4.8), since $\bar{w}_0 = P_0 w_0 = 0$,

$$A(t, u(t), u'(t)) = -P^{-1}(t, u(t), u'(t)) \int_t^{\infty} Q(s, u, u') = -\Psi_2[u, u'](t).$$

Therefore $u'(t) \in L^1(J)$ as in Theorem 4, so $u(t) \to u_0$ as $t \to \infty$.

**5. Linear and Semilinear Systems.** Consider the special case when (1.3) has the linear form

$$(5.1) \qquad (p(t)u')' + q(t)u = 0,$$

where $p(t)$ and $q(t)$ are $N \times N$ matrices with $p(t)$ invertible for all $t \in J$.

Correspondinglywe have $A \equiv v$, and the functionals $\Psi_1$ and $\Psi_2$ are given by

(5.2)
$$\Psi_1[u](t) = -p^{-1}(t) \int_T^t q(s)u(s),$$

(5.3)
$$\Psi_2[u](t) = p^{-1}(t) \int_t^\infty q(s)u(s).$$

Parallel to Theorems 1, 4 and 5, the following results now hold.

THEOREM 6. *Suppose that either*

(5.4)
$$\int_T^\infty \int_T^t \left| p^{-1}(t)q(s) \right| dsdt < \infty$$

*or*

(5.5)
$$\int_T^\infty \int_T^t \left| p^{-1}(s)q(t) \right| dsdt < \infty.$$

*Then the set of attainable limits as $t \to \infty$ of solutions of* (5.1) *is exactly* $\mathbf{R}^N$.

THEOREM 7. *Suppose that* (5.4) *is satisfied, and that $|p^{-1}(t)|$ is in $L^1(J)$. Then every solution of* (5.1) *tends to a finite limit as $t \to \infty$.*

Theorem 6 follows directly from Corollary 1. To prove Theorem 7, it is enough to observe that the assumptions in Theorem 4 are satisfied.

When $N = 1$ and $q \not\equiv 0$, condition (5.4) implies that $p^{-1}(t) = 1/p(t)$ is integrable. Thus Wintner's theorem is a special case of Theorems 6 and 7.

The semilinear system

(5.6)
$$(p(t)u')' + q(t)f(u) = 0$$

where $f \in C(\mathbf{R}^N \to \mathbf{R}^N)$ can be treated in exactly the same way, since $|f(u(s))|$ is uniformly bounded when $|u(s)| \leq M$. This gives

THEOREM 8. *Theorem* 6 *remains valid for the system* (5.6). *Similarly, Theorem* 7 *continues to hold for* (5.6) *provided that*

$$|f(u)| \leq \text{Const}\,(1 + |u|).$$

**6. Functional Differential Equations.** The previous results extend naturally to functional differential systems, since their proofs actually involve only the behavior of the *functionals* $\Psi_1$ through $\Psi_6$. Let $Y$ be a real Banach space and let $J = [T, \infty)$ be as before. Denote by $C = C(J \to Y)$ the space of all continuous $Y$-valued functions on $J$ and by $\mathrm{Lin}(Y, Y)$ the space of all continuous linear transformations from $Y$ to $Y$. Consider the system

$$(P[u,u'](t)A(t,u,u'))' + Q[u,u'](t) = 0 \tag{6.1}$$

where $u: J \to Y$. We suppose throughout that

$$A(t,u,v) : J \times Y \times Y \to Y, \qquad A(t,u,0) = 0,$$

$$Q[u,v] : C \times C \to C, \quad . \quad P[u,v] : C \times C \to \mathrm{Lin}(Y,Y).$$

It is assumed moreover that the inverse $P^{-1}$ of $P$ exists for all $u$, $v$ and all $t \in J$.

Defining $\Psi_1, \ldots, \Psi_6$ as before, but with the obvious interpretation of $P$ and $Q$, we may carry out the preceding arguments without changing the proofs in any significant way. This leads to the following results (here $B_M$ is of course defined with respect to the norm $||u||_J = \sup_{t \in J} ||u(t)||_Y$).

THEOREM 1*. *Let $P$ and $Q$ be independent of $v$, and suppose that* $(A_1)$ *and* $(A_2)$ *are satisfied. Assume that for all fixed $M > 0$ either $\Psi_1[u]$ or $\Psi_2[u]$ is uniformly $L^{1/(m-1)}$-integrable on $B_M$. Then the set of attainable limits of solutions of* (6.1) *is equal to $Y$.*

THEOREM 2*. *Suppose that* $(A_1)$ *and* $(A_2)$ *are satisfied, and that for all fixed $M > 0$, there exists $\epsilon > 0$ such that at least one of the functionals $\Psi_3[u,w]$ and $\Psi_4[u,w]$ is uniformly $L^{1/(m-1)}$-integrable and equicontinuous on $B_M \times B_\epsilon$. Then the set of attainable limits of solutions of* (6.1) *is equal to $Y$.*

THEOREM 3*. *Let $P$ be independent of $v$, and suppose that* $(A_1)$ *and* $(A_2)$ *are satisfied. Assume that for all fixed $M > 0$ there exists $\epsilon > 0$ such that*

(i) *$P^{-1}[u]$ is uniformly bounded on $B_M$,*

(ii) *$Q[u,v]$ is uniformly integrable on $B_M \times B_\epsilon$,*

(iii) *$\Psi_5[u,w]$ or $\Psi_6[u,w]$ is uniformly $L^{1/(m-1)}$-integrable on $B_M \times B_\epsilon$.*

*Then the set of attainable limits of solutions of* (6.1) *is equal to $Y$.*

THEOREM 4*. *For all fixed $M > 0$, suppose that $P^{-1}[u,v]$ and $\Psi_1[u,v]$ are uniformly $L^{1/(m-1)}$-integrable on $B_M \times C$. Then every bounded solution of (6.1) tends to a finite limit.*

*If in addition the function*

$$\frac{\Psi_1[u,v](t)}{1 + ||u||_{[T,t]}^{m-1}}$$

*is uniformly $L^{1/(m-1)}$-integrable on $C \times C$, then every solution of (6.1) is necessarily bounded.*

THEOREM 5*. *Suppose that $Q[u, v]$ is uniformly integrable on $C \times C$. Assume also that for all $u, v \in C$ the mapping $P[u, v]$ tends to a limit $P_0$ (independent of $u, v$) as $t \to \infty$, where $P_0$ is invertible. Then there exists a vector $w_0 \in Y$ such that*

$$\lim_{t\to\infty} A(t, u(t), u'(t)) = w_0.$$

*Furthermore, if $w_0 = 0$ then $u$ tends to a finite limit $u_0$ provided that $\Psi_2$ is uniformly $L^{1/(m-1)}$-integrable on $C \times C$.*

The above results apply at once to the system $(1.3)'$,

$$(A(t, u, u'))' + D(t, u, u')A(t, u, u') + Q(t, u, u') = 0.$$

Here $Y = \mathbf{R}^N$ and $C = C(J \to \mathbf{R}^N)$. For $u, v \in C$, we define the $N \times N$ matrix $P[u, v]$ by

$$P[u, v](t) = \exp \int_T^t D(s, u(s), v(s)) \tag{6.2}$$

and the map $Q[u, v] : C \times C \to C$ by

$$Q[u, v](t) = P[u, v](t)Q(t, u(t), v(t))$$

(note that this definition of $Q[u, v]$ is not the expected one, but nevertheless should not cause confusion). Then $(1.3)'$ takes the form (6.1), and Theorems 1*–5* carry over directly.

*Remarks.* As a special case of importance, suppose that for all fixed $M > 0$ there exist positive constants $\epsilon$ and $L$ such that $|\int_T^t D(s, u(s), v(s))| \le L$ and $Q(t, u, v)$ is uniformly integrable on $B_M \times B_\epsilon$. Then both $P$ and $P^{-1}$ are uniformly bounded, and $\Psi_4[u, w]$ is equicontinuous on $B_M \times B_\epsilon$. In turn, in Theorem 2* one can drop the explicit assumption that $\Psi_4[u, w]$ be equicontinuous.

If the damping matrix $D$ is sufficiently large, then also $\Psi_3[u, w]$ is equicontinuous on $B_M \times B_\epsilon$, see [9].

**7. Further Examples.** (1) Consider the Emden-Fowler type equation

$$u'' + t^{-k}f(u) = 0, \qquad k > 2, \tag{7.1}$$

where $f : \mathbf{R} \to \mathbf{R}$ is continuous. Here it is easy to verify that (5.5) is satisfied, so that every target $u_0 \in \mathbf{R}$ is attainable by solutions of (7.1); see [1], where this result is crucial to the argument.

In the following examples, we need a simple lemma.

LEMMA. *Let $k > 0$, and suppose $\sigma \in C(J \to \mathbf{R}^+)$ satisfies*

$$1/\sigma \in L^k(J), \qquad (1/\sigma(t))' \to 0 \quad \textit{as } t \to \infty. \tag{7.2}$$

*Then*

$$\sigma(t)\frac{\int_T^t \exp\left(\int_T^s \sigma(r)\right)}{\exp\left(\int_T^t \sigma(s)\right)} \to 1 \quad \text{as } t \to \infty. \tag{7.3}$$

*Proof.* By Hölder's inequality and (7.2)

$$\int_T^t \sigma(s) \geq \left[\int_T^t 1\right]^{(k+1)/k} \Big/ \left[\int_T^t \sigma^{-k}(s)\right]^{1/k} \to \infty \quad \text{as } t \to \infty$$

and so

$$\exp \int_T^t \sigma(s) \to \infty \quad \text{as } t \to \infty.$$

Even more, we claim that

$$\frac{\exp\left(\int_T^t \sigma(s)\right)}{\sigma(t)} \to \infty \quad \text{as } t \to \infty.$$

Indeed, by Jensen's inequality

$$\exp \int_T^t \sigma(s) \geq \int_T^t \exp(\sigma(s)).$$

Obviously $e^x \geq x^2$ for $x \geq 0$, so

$$\exp \int_T^t \sigma(s) \geq \int_T^t \exp(\sigma(s)) \geq \int_T^t \sigma^2(s) \geq \int_T^t M(s)|\sigma'(s)|,$$

where

$$M(t) = \frac{\sigma^2(t)}{|\sigma'(t)|} \to \infty \quad \text{as } t \to \infty$$

($M(t) = \infty$ if $\sigma'(t) = 0$), which implies the claim. Now by L'Hospital's rule

$$\lim_{t\to\infty} \frac{\int_T^t \exp\left(\int_T^s \sigma(r)\right)}{\exp\left(\int_T^t \sigma(s)\right)/\sigma(t)} = \lim_{t\to\infty} \frac{\exp\left(\int_T^t \sigma(s)\right)}{\exp\left(\int_T^t \sigma(s)\right) + \exp\left(\int_T^t \sigma(s)\right)(1/\sigma(t))'}$$
$$= \lim_{t\to\infty} \frac{1}{1+(1/\sigma(t))'} = 1.$$

(2) Consider the degenerate equation

$$(|u'|^{m-2}u')' + \sigma(t)|u'|^{m-2}u' + f(u) = 0, \qquad m > 1,\ N = 1. \tag{7.4}$$

If condition (7.2) is satisfied with $k = 1/(m-1)$, then the set of attainable limits as $t \to \infty$ of solutions of (7.4) is exactly **R**. Moreover, every bounded solution tends to a finite limit as $t \to \infty$.

To prove this, note that here $A(t, u, v) = |v|^{m-2}v$, so ($A_1$) and ($A_2$) are satisfied. Equation (7.4) then takes the form (1.3) with

$$P(t) = \exp \int_T^t \sigma(s), \qquad Q(t, u) = P(t)f(u) = f(u)\ \exp \int_T^t \sigma(s). \tag{7.5}$$

Put

$$\Psi[u](t) = -P^{-1}(t) \int_T^t Q(s, u(s)) = -\exp\left(-\int_T^t \sigma(s)\right) \int_T^t f(u(s))\ \exp \int_T^s \sigma(r).$$

It is easy to see that

$$|\Psi[u](t)| \le C \exp\left(-\int_T^t \sigma(s)\right) \int_T^t \exp \int_T^s \sigma(r)$$

if $||u||_J \le M$, where $C$ is a constant depending on $M$. On the other hand, by the lemma we have

$$\exp\left(-\int_T^t \sigma(s)\right) \int_T^t \exp \int_T^s \sigma(r) \le c/\sigma(t)$$

for some $c > 0$. It follows that $\Psi[u](t)$ and $P^{-1}(t)$ are uniformly $L^{1/(m-1)}$-integrable for $||u||_J \le M$, with $\psi(t) = cC/\sigma(t)$. Therefore Theorem 1 implies the first part of the claim, while the second comes from Theorem 4.

(3) Consider the mean curvature equation

$$\left(\frac{u'}{\sqrt{1+u'^2}}\right)' + \sigma(t)\frac{u'}{\sqrt{1+u'^2}} + f(u) = 0. \tag{7.6}$$

If condition (7.2) is satisfied with $k = 1$, and if $1/\sigma(t) \to 0$ as $t \to \infty$, then the conclusions of the previous example remain true.

Here $A(t, u, v) = v/\sqrt{1+v^2}$, and (7.6) takes the form (1.3) if $P$ and $Q$ are defined by (7.5). Then for the first conclusion it suffices to show that (2.12) holds (we now use Theorem 1′ rather than Theorem 1 since in the present case $(A_2)'$ but not $(A_2)$ is satisfied), while the second part follows from Theorem 4 directly. However, (2.12) follows at once from the calculation in the preceding section and the fact that $1/\sigma(t) \to 0$ as $t \to \infty$.

*Note.* An alternate and more delicate argument shows that the required conclusions hold even without the condition that $1/\sigma(t) \to 0$ as $t \to \infty$, see also [9].

(4) Finally, consider the system

$$u'' + D(t,u)u' + f(u) = 0 \tag{7.7}$$

with

$$D(t,u) \geq \sigma(t)I, \tag{7.8}$$

where $I$ is the $N \times N$ identity matrix and $\sigma$ satisfies (7.2) with $k = 1$. Then again the conclusions in the second example hold.

In this case $A(t, u, v) = v$ and we use Theorems 1* and 4*. Corresponding to Part 2 above, we have

$$P[u](t) = \exp \int_T^t D(s, u(s)),$$

$$Q[u](t) = P[u](t)f(u(t)) = \left(\exp \int_T^t D(s, u(s))\right) f(u(t)),$$

$$\begin{aligned}\Psi[u](t) &= -P^{-1}[u](t) \int_T^t Q[u](s) \\ &= -\exp\left(-\int_T^t D(s, u(s))\right) \int_T^t \left(\exp \int_T^s D(r, u(r))\right) f(u(s)).\end{aligned}$$

Clearly

$$|\Psi[u](t)| \leq C \int_T^t \left|\exp\left(-\int_s^t D(r, u(r))\right)\right|$$

if $||u||_J \leq M$, where $C$ is a constant depending on $M$. But by (7.8)

$$\left|\exp\left(-\int_s^t D(r,u(r))\right)\right| \leq \left|\exp\left(-\int_s^t \sigma(r)\right)\right|. \tag{7.9}$$

Hence, by the lemma and (7.9), we obtain

$$|\Psi[u](t)| \leq M \int_T^t \left|\exp\left(-\int_s^t \sigma(r)\right)\right|$$

so that $\Psi[u]$ is uniformly $L^1$-integrable for $||u||_J \leq M$, and we are done.

The system

$$(|u'|^{m-2}u')' + D(t,u)|u'|^{m-2}u' + f(u) = 0, \qquad m > 1$$

can of course be treated in exactly the same way.

Examples 2–4 are special cases of a large class of over-damped second order systems which allow similar conclusions, see [6], Section 4, and [9].

SCHOOL OF MATHEMATICS, UNIVERSITY OF MINNESSOTA, MINNEAPOLIS, MN 55455

SCHOOL OF MATHEMATICS, UNIVERSITY OF ALABAMA AT BIRMINGHAM, BIRMINGHAM, AL 35294

---

REFERENCES

[1] F. V. Atkinson and L. A. Peletier, Ground states of $-\Delta u = f(u)$ and the Emden-Fowler equation, *Arch. Rational Mech. Anal.* 93 (1986), 103–127.

[2] M. Cecchi, M. Marini and G. Villari, Integral criteria for a classification of solutions of linear differential equations, *J. Differential Equations* (to appear).

[3] J. Hale and N. Onuchic, On the asymptotic behavior of solutions of a class of differential equations, *Contributions to Differential Equations*, vol. 1, 1963, pp. 61–75.

[4] P. Hartman, *Ordinary Differential Equations*, 2nd ed., Wiley, New York.

[5] P. Hartman and N. Onuchic, On the asymptotic integration of ordinary differential equations, *Pacific J. Math.* 13 (1963), 1193–1207.

[6] P. Pucci and J. Serrin, Continuation and limit behavior for damped quasi-variational systems, *Proceedings on Degenerate Diffusion Equations*, Institute for Mathematics and its Applications.

[7] H. Weyl, Über gewöhnliche lineare Differentialgleichungen mit singulären Stellen und ihre Eigenfunktionen, *Nachrichten von der Königlichen Gesellschaft der Wissenschaften zu Göttingen* (1909), 37–63.

[8] A. Wintner, Asymptotic integration of the adiabetic oscillator in its hyperbolic range, *Duke Math. J.* 15 (1948), 55–67.

[9] H. Zou, Asymptotic integration of over damped second order systems and applications to instability, (to appear 1994).

# II.3. Parabolic Equations – Local Behavior of Solutions

Gary M. Lieberman

**Mathematics Subject Classification (2010).** 35K10, 35K59.

**Keywords.** Quasilinear parabolic equations, Harnack inequalities.

The paper [i] below, Local behavior of solutions of quasilinear parabolic equations by Aronson and Serrin, first appeared during a time of great activity on regularity theory for second-order elliptic and parabolic equations. In the late 1950s, De Giorgi proved that weak solutions of linear second-order elliptic equations are Hölder continuous, and Nash proved (via very different methods) the corresponding result for elliptic and parabolic equations. De Giorgi's method was then used in the early 1960s by Ladyzhenskaya and Ural′tseva to obtain the Hölder continuity of weak solutions for a large class of quasilinear elliptic equations and for a smaller class of quasilinear parabolic equations. Further, Moser showed that a Harnack inequality was true for positive weak solutions of linear elliptic equations and for positive weak solutions of linear parabolic equations; he also used his Harnack inequality to give a different proof of Hölder continuity of such solutions. In 1964, Serrin extended Moser's results to the same class of quasilinear elliptic equations studied by Ladyzhenskaya and Ural′tseva. In retrospect, it seems obvious that a similar extension should be available for parabolic equations, but that extension took two more years to be published in [i], and a major reason that the parabolic extension now seems obvious is the publication of [i]. On the other hand, some of the ideas in this paper were simultaneously considered by several other authors: notably A.V. Ivanov [4], M. Kurihara [5], and N.S. Trudinger [7]. Two elements make [i] especially noteworthy compared to these three papers. The first is the timing: [i] appeared in January, 1967 while the others appeared later in 1967 ([4] and [5]) and in 1968 ([7]); preliminary versions of these works are cited in [1] but these are the final published versions. The second is the thoroughness of [i]. It contains complete proofs of all results with full details, especially concerning the assumed regularity of the weak solution and of the coefficients in the problem; it also discusses several consequences of the Harnack inequality, such as asymptotic behavior of solutions in unbounded domains. As a consequence, [i] is quoted much more frequently than the other three papers combined.

Even today, more than forty years after its publication, [i] is the basis for ongoing research. Its primary use today is as a reference for Harnack inequalities for quasilinear parabolic equations of a particular type, but it is also a springboard for modern investigations into Harnack inequalities for more general parabolic equations. In particular,

P. Pucci et al. (eds.), *James Serrin. Selected Papers: Volume 2*, Contemporary Mathematicians, DOI 10.1007/978-3-0348-0847-7_3, 

DiBenedetto, Gianazza, and Vespri [2, 3] have proved analogs of the Aronson–Serrin–Harnack inequality for a class of parabolic equations essentially as broad as the class of elliptic equations studied by Ladyzhenskaya and Ural′tseva. Although the earlier works of Hadamard and Pini are quoted in more detail by these three authors, [i] plays an important role in their considerations because it is the first work to take advantage of nonlinear information about the equations. Liskevich and Skrypnik [6] look at another related class of quasilinear equations in which the coefficients lie in a somewhat larger class than that of [i].

## References


[1] D.G. Aronson and James Serrin, *Local Behavior of Solutions of Quasilinear Parabolic Equations.* Arch. Rational Mech. Anal. **25** (1966), 81–122.

[2] E. DiBenedetto, U. Gianazza, and V. Vespri, *Harnack estimates for quasi-linear degenerate parabolic differential equations.* Acta Math. **200** (2008), 181–209.

[3] E. DiBenedetto, U. Gianazza, and V. Vespri, *Harnack type estimates and Hölder continuity for non-negative solutions to certain sub-critically singular parabolic partial differential equations.* Manuscripta Math. **131** (2010), 231–245.

[4] A.V. Ivanov, *The Harnack inequality for generalized solutions of second order quasilinear parabolic equations.* (Russian) Trudy Mat. Inst. Steklov. **102** (1967), 51–84.

[5] M. Kurihara, *On a Harnack inequality for nonlinear parabolic equations.* Publ. Res. Inst. Math. Sci. Ser. A **3** (1967/1968), 211–241.

[6] V. Liskevich and I.L. Skrypnik, *Harnack inequality and continuity of solutions to quasilinear degenerate parabolic equations with coefficients from Kato-type classes.* J. Differential Equations **247** (2009), 2740–2777.

[7] N.S. Trudinger, *Pointwise estimates and quasilinear parabolic equations.* Comm. Pure Appl. Math. **21** (1968), 205–226.



Gary M. Lieberman
Department of Mathematics
Iowa State University
Ames, IA 50011, USA
e-mail: lieb@iastate.edu

*Arch. Ration. Mech. Anal.* 25 (1967), 81–122.
Reprint II.3.[i]


# *Local Behavior of Solutions of Quasilinear Parabolic Equations*

D. G. Aronson & James Serrin

## Introduction

The main results in this paper are three theorems concerning the behavior of solutions of a class of quasilinear partial differential equations of parabolic type. Specifically, we prove a global maximum principle, a theorem on local boundedness of solutions, and a Harnack type inequality. All of these results depend on certain fundamental inequalities which are derived from the weak form of the differential equation. Moreover, in each case the proof depends on the iterative techniques introduced by Moser and developed in papers by Serrin and Aronson & Serrin. The above results have many applications, of which we shall consider two main types, the interior Hölder continuity of solutions and the growth properties of non-negative solutions near certain parts of the boundary of their domain of definition.

Let $x=(x_1, \ldots, x_n)$ denote points in $n$-dimensional Euclidian space $E^n$ $(n \geqq 1)$ and $t$ denote points on the real line. Let $\Omega$ be a bounded domain in $E^n$, and consider the space-time cylinder $Q=\Omega \times (0, T)$ for some fixed $T>0$. For $(x, t) \in Q$ we treat the second order quasilinear equation

$$u_t = \operatorname{div} \mathscr{A}(x, t, u, u_x) + \mathscr{B}(x, t, u, u_x) \tag{1}$$

where $\mathscr{A}=(\mathscr{A}_1, \ldots, \mathscr{A}_n)$ is a given vector function of $(x, t, u, u_x)$, $\mathscr{B}$ is a given scalar function of the same variables, and $u_x=(\partial u/\partial x_1, \ldots, \partial u/\partial x_n)$ denotes the spatial gradient of the dependent variable $u=u(x, t)$. Also here $\operatorname{div}\mathscr{A}$ refers to the divergence of the vector $\mathscr{A}(x, t, u(x, t), u_x(x, t))$ with respect to the variables $(x_1, \ldots, x_n)$. The structure of (1) is determined by the functions $\mathscr{A}(x, t, u, p)$ and $\mathscr{B}(x, t, u, p)$. We assume that they are defined and measurable for all $(x, t) \in Q$ and for all values of $u$ and $p$. In addition, $\mathscr{A}$ and $\mathscr{B}$ will be required to satisfy certain inequalities whose description follows.

A function $w=w(x, t)$ which is defined and measurable on $Q$ will be said to belong to class $L^{p,q}(Q)$ if the iterated integral

$$\|w\|_{p,q} = \left\{ \int_0^T \left( \int_\Omega |w|^p \, dx \right)^{q/p} dt \right\}^{1/q}$$

is finite. Here $p, q$ may be any real numbers $\geqq 1$; and with the obvious use of $L^\infty$ norms rather than integrals we can also allow $p$ or $q$ to have the value $\infty$.

Throughout the paper we shall then assume that the functions $\mathscr{A}$ and $\mathscr{B}$ satisfy inequalities of the form

$$
\begin{aligned}
&p\cdot\mathscr{A}(x,t,u,p)\geqq a\,|p|^2-b^2u^2-f^2\\
(2)\qquad &|\mathscr{B}(x,t,u,p)|\leqq c\,|p|+d\,|u|+g\\
&|\mathscr{A}(x,t,u,p)|\leqq \bar{a}\,|p|+e\,|u|+h\,.
\end{aligned}
$$

Here $\bar{a}$ and $a$ are positive constants, while the coefficients $b, c, \ldots, h$ are non-negative functions of $(x, t)$ *each contained in some space* $L^{p,q}(Q)$, *where* $p$ *and* $q$ *are non-negative real numbers (possibly different from coefficient to coefficient) such that*

$$(3)\qquad p>2 \quad and \quad \frac{n}{2p}+\frac{1}{q}<\frac{1}{2} \quad for \quad b,c,e,f,h$$

*and*

$$(4)\qquad p>1 \quad and \quad \frac{n}{2p}+\frac{1}{q}<1 \quad for \quad d,g.$$

The norms of the functions $b, c, \ldots, h$ in their respective spaces will be denoted simply by $\|b\|, \|c\|, \ldots, \|h\|$.

It is easily verified that the linear equation

$$(5)\qquad u_t=\{A_{ij}(x,t)\,u_{x_i}+A_j(x,t)\,u+F_j(x,t)\}_{x_j}+B_j(x,t)\,u_{x_j}+C(x,t)\,u+G(x,t)$$

satisfies the preceding hypotheses if (i) the $A_{ij}$ are bounded and measurable on $Q$ and there exists a constant $\nu>0$ such that $A_{ij}(x,t)\,\xi_i\xi_j\geqq\nu|\xi|^2$ for almost all $(x,t)\,\varepsilon Q$ and all $\xi\,\varepsilon\,E^n$; (ii) the coefficients $A_j$, $B_j$ and $F_j$ each belong to some space $L^{p,q}(Q)$ with $p$, $q$ satisfying (3); and (iii) the coefficients $C$ and $G$ each belong to some space $L^{p,q}(Q)$ with $p$, $q$ satisfying (4).

Without further hypotheses on $\mathscr{A}$ and $\mathscr{B}$ it is not possible, in general, to speak of a classical solution of equation (1), and it is correspondingly necessary to introduce the notion of a generalized solution. Let $u=u(x,t)$ be a function which is locally of class $L^{2,\infty}$ in $Q$ and possesses a strong derivative $u_x$ which is locally of class $L^{2,2}$ in $Q$. Then $u$ will be called a *weak solution* of equation (1) in $Q$ if

$$(6)\qquad \iint_Q\{-u\,\varphi_t+\varphi_x\cdot\mathscr{A}(x,t,u,u_x)\}\,dx\,dt=\iint_Q\varphi\,\mathscr{B}(x,t,u,u_x)\,dx\,dt$$

for any continuously differentiable function $\varphi=\varphi(x,t)$ having compact support in $Q$. It would be simpler to consider a less general class of weak solutions, namely those for which both $u_x$ and $u_t$ are locally of class $L^{2,2}$ in $Q$. However, from the point of view of existence theory the latter class is not as natural as the former. Thus even in the linear case some smoothness of the coefficients must be assumed in order to prove the existence of solutions having a strong time derivative. Moreover, in applications of the results of this paper it is often necessary to extend the domain of definition of a given solution. Regardless of whether the solution in question is classical or weak the most natural extension process leads in general to a weak solution of the type considered here. An example of such an extension can be found in reference [2].

Theorem 1, which is proved in section 3, states that if $u$ is a weak solution of (1) such that $u \leqq M$ on the parabolic boundary of $Q$, then $u(x, t) \leqq M + Ck$ almost everywhere in $Q$. Here $k = (\|b\| + \|d\|)\,|M| + (\|f\| + \|g\|)$ and $C$ depends only on $T$, $|\Omega|$ and the structure of (1). (By the *structure* of (1) we mean the numbers $a$, $\bar{a}$ and $n$, the various exponents $p$ and $q$, and the norms of the functions $b, c, \ldots, h$.) Moreover, for Theorem 1 the last of the structural inequalities (2) is irrelevant, and $C$ is independent of $\bar{a}$, $e, f, g$ and $h$. Theorem 1 extends to nonlinear equations results for linear equations proved earlier by LADYŽENSKAJA & URAL'CEVA [10], GUGLIELMINO [4], ARONSON [1], IVANOV [6], and IVANOV, LADYŽENSKAJA, TRESKUNOV & URAL'CEVA [7]. [In references [6] and [7] the coefficients $B_j$ of equation (5) are permitted to belong to a space $L^{p,q}(Q)$ with

$$\frac{n}{2p} + \frac{1}{q} = \frac{1}{2} \qquad (n \geqq 2).$$

The same result can be obtained by a slight modification of the methods used here. However, in the interest of simplicity and a uniform exposition we omit such considerations and restrict ourselves to a single set of exponent conditions which suffice for all our results. Thus our exponent conditions are slightly stronger than necessary, in that occasionally the equality sign could be used while only the inequality sign is indicated.]

A maximum principle for more strongly nonlinear equations was proved by ARONSON & SERRIN [3] using techniques which are essentially the same as those employed here. The result proved there overlaps the present maximum principle only in the case in which all of the coefficients in the structural inequalities (2) are constant. Finally, we remark that the restriction of the domain in Theorem 1 to a space-time cylinder is not essential; the reader can easily formulate maximum principles for more general domains.

In the absence of information about the behavior of a solution $u$ on the parabolic boundary of $Q$ it is still possible to show that $u$ is locally bounded. Theorem 2, which is proved in section 4, states that in any proper subdomain of $Q$ the modulus of $u$ has a bound determined by the structure of (1), the geometry of the domains, and the $L^{2,2}(Q)$ norm of $u$. This generalizes earlier results for linear equations in references [6, 7, 9, 12]. A by-product of the proof of Theorem 2 is a local bound for the $L^{2,2}$ norm of $u_x$.

Sections 5 and 6 are devoted to the proof of a Harnack inequality for non-negative weak solutions of (1). This theorem includes as special cases the Harnack inequality proved by MOSER [12] for solutions of

$$u_t - \{A_{ij}(x, t)\, u_{x_i}\}_{x_j} = 0, \tag{7}$$

ARONSON'S [2] extension of MOSER'S result, and the Harnack inequalities for the equation of heat conduction due to HADAMARD [5] and PINI [15]. It does not, however, contain the very sharp result obtained by RESCH [16] for non-negative solutions of the heat conduction equation defined in an infinite strip. In case everything is independent of $t$ the present result reduces to the Harnack inequality for quasilinear elliptic equations proved by SERRIN [17]. The demonstration of the Harnack inequality given here follows closely the outline of MOSER'S proof, and specifically makes essential use of his remarkable Main Lemma.

As MOSER observed in reference [*12*], the Harnack inequality and the local boundedness theorem can be used to prove the interior Hölder continuity of weak solutions of (7), and to estimate the growth as $|x| \to \infty$ of non-negative solutions of (7) defined in a strip $E^n \times (0, T)$. These results are generalized to solutions of (1) in section 7 of this paper. Interior Hölder continuity has, of course, been proved by a number of other authors by various methods. In this connection we mention the pioneering work of NASH [*13*] as well as the papers [*1, 6, 7, 9, 10, 14*].

In section 7 the Harnack inequality is also used to generalize two results stated by NASH on the behavior of non-negative solutions of (7) as $t \to 0$. Specifically, for a non-negative solution $u$ of (1) we give a condition which guarantees that $t^{n/2} u$ is bounded above for small $t$, and another sufficient condition for $u$ to be bounded below by a function of the form

$$\text{Const}\, t^{-n/2} \exp(-\text{const}\, |x|^2 t^{-1})$$

for small $t$. In a forthcoming paper these results will be used to derive upper and lower bounds for the fundamental solution of the linear equation (5).

The proofs of the maximum principle, the local boundedness theorem, and the Harnack inequality are based on certain inequalities which are derived from (6) by taking powers of the solution as test functions. Since weak solutions $u$ are not required to possess a derivative with respect to $t$, it is clear that there are technical difficulties which must be overcome in this process. Further difficulties arise from the fact that even with the existence of strong derivatives with respect to both $x$ and $t$ the derivatives of powers of $u$ do not, in general, belong to appropriate function spaces unless we know *a priori* that $u$ is bounded. To avoid making the preliminary sections too lengthy we have divided the derivation of the basic inequalities into two parts. In section 2 these inequalities are derived under the temporary assumption that $u$ has strong derivatives $u_x$, $u_t$ which are locally of class $L^{2,2}$ in $Q$. This derivation is similar to the corresponding one given by SERRIN in the elliptic case. At the end of the paper, in section 9, we remove the temporary assumption and show that the basic inequalities are valid for the general class of weak solutions defined above. For the reader who is interested only in classical solutions, section 9 is irrelevant. Section 1 is devoted to certain preliminary lemmas concerning function spaces and the calculus of strong derivatives. Of these, the interpolation Lemma 3 is the main result, and indeed has some independent interest.

While the manuscript for this paper was being prepared, we learned of independent but related work [*20*] of TRUDINGER. Although TRUDINGER'S paper is mainly devoted to the Dirichlet problem for quasilinear elliptic equations, he also obtained a Harnack principle for *bounded solutions* of divergence structure parabolic equations. Since the solutions he deals with are bounded, the structure assumed in [*20*] differs somewhat from the structure assumed here. He notes, however, that a result similar to our Theorem 3 can be derived by his procedure. In section 8 of this paper we indicate how TRUDINGER'S result can be obtained using the methods employed here.

*Note.* An abstract announcing the Harnack inequality for quasilinear parabolic equations (Theorem 3) was submitted to the American Mathematical Society in

1965, and appeared in the Notices of the Society, vol. **13** (1966). After the paper was completed we learned of similar work of M. KURIHARA in Japan and A.V. IVANOV in Leningrad. The work of KURIHARA appears in Funkcialaj Ekvacioj (in Japanese), vol. **17**, pp. 3–57 and vol. **18**, pp. 3–32; his result corresponds to the special case of Theorem 3 where $b, c, e, f, h$ are in $L^{2p,\infty}$ and $d, g$ are in $L^{p,\infty}$ with $p=\max(3, n)$, and his proof like ours is based on that given originally by MOSER.

In his Leningrad dissertation (1966), IVANOV obtained a maximum principle and a local boundedness theorem for weak solutions of quasilinear equations of essentially the type considered here. Moreover, he determined a Harnack inequality corresponding to the special case $n\geqq 2$, $q=\infty$ of Theorem 3. In a talk at the International Congress of Mathematicians (Moscow), IVANOV indicated that he has eliminated the restriction $q=\infty$ and now has the Harnack inequality in exactly the form given here, though still retaining the restriction $n\geqq 2$.

Because of the overlap between our work and that of KURIHARA, IVANOV, and TRUDINGER, it is important to emphasize the independence of all these researches, and the fact that they were all undertaken at roughly the same time.

## 1. Preliminaries

In this section we state and prove several preliminary lemmas which will be needed in our subsequent work.

**Lemma 1.** *If $w$ is contained in $L^{q,q_1}\cap L^{r,r_1}$, then $w$ is contained in $L^{p,p_1}$, where*

$$\frac{1}{p}=\frac{\lambda}{q}+\frac{\mu}{r},\quad \frac{1}{p_1}=\frac{\lambda}{q_1}+\frac{\mu}{r_1}\qquad (\lambda, \mu\geqq 0, \lambda+\mu=1).$$

*Moreover*

$$\|w\|_{p,p_1}\leqq \|w\|^{\lambda}_{q,q_1}\,\|w\|^{\mu}_{r,r_1}.$$

The proof is an easy exercise in the application of Hölder's inequality. Our remaining results have to do with the calculus of strongly differentiable functions. Suppose $w=w(x,t)$ is locally summable in $Q$. We say that $w$ is *strongly differentiable* with respect to $x$ if there exists a locally summable vector function, denoted by $w_x(x,t)$, such that

$$\iint w\,\varphi_x\,dx\,dt=-\iint \varphi\,w_x\,dx\,dt$$

for all continuously differentiable functions $\varphi=\varphi(x,t)$ having compact support in $Q$. The function $w_x$ is called the strong derivative of $w$. A similar definition applies for differentiation with respect to $t$.

**Lemma 2.** (Sobolev's inequality). *Suppose $n\geqq 2$. Let $w$ have strong derivatives $w_x$ which belong to $L^{2,2}(Q)$. Suppose also that $|\Omega|=1$ and that $w$ has compact support as a function of $x$. Then*

$$\|w\|_{2^*,2}\leqq K\,\|w_x\|_{2,2}$$

*where $2^*=2n/(n-2)$ when $n>2$, and $2^*$ is any finite number when $n=2$. The constant $K$ depends only on the dimension $n$ (if $n=2$, then $K$ depends on the choice of $2^*$).*

**Proof.** It is clear from the definition of strong derivatives that $w(x,t)$, considered as a function of $x$ for fixed $t$, is strongly differentiable in $x$ for almost all values of $t$. Hence by Sobolev's inequality (see [19])

$$\Big(\int_\Omega |w|^{2^*}\,dx\Big)^{1/2^*} \leqq K\Big(\int_\Omega |w_x|^2\,dx\Big)^{1/2}.$$

The required result follows by integrating both sides with respect to $t$.

**Lemma 3.** *Let $w$ be contained in $L^{2,\infty}(Q)$ and have strong derivatives $w_x$ belonging to $L^{2,2}(Q)$. Suppose that $|\Omega|=1$ and that $w$ has compact support in $\Omega$ as a function of $x$. Then $w$ is in $L^{2p',2q'}$ for all exponents pairs $(p',q')$ whose Hölder conjugates $(p,q)$ satisfy*

$$\frac{n}{2p}+\frac{1}{q}\leqq 1 \qquad (<1 \text{ if } n=2).$$

*Moreover*

$$\|w\|^2_{2p',2q'} \leqq K\,T^\theta\{\|w\|^2_{2,\infty}+\|w_x\|^2_{2,2}\}$$

*where*

$$\theta = 1-\frac{1}{q}-\frac{n}{2p}$$

*and $K$ depends only on $n$. (If $n=2$ then $T^\theta$ should be replaced by $T^{\theta/2}$, and $K$ depends on $\theta$.)*

**Proof.** Assume $n\geqq 2$. Let $r'$, $s$ be real numbers $\geqq 1$. Then by Lemma 1

$$\|w\|^2_{2p',2q'} \leqq T^\nu\,\|w\|^2_{2p',2r'} \leqq T^\nu\,\|w\|^\lambda_{s,2}\,\|w\|^{2-\lambda}_{2,\infty}$$

provided that $\nu\geqq 0$, $0\leqq\lambda\leqq 2$, and

$$\frac{1}{q'}=\frac{1}{r'}+\nu, \qquad \frac{1}{p'}=\frac{\lambda}{s}+\frac{2-\lambda}{2}, \qquad \frac{1}{r'}=\frac{\lambda}{2}.$$

These relations imply

$$\lambda=\frac{2s}{(s-2)\,p} \quad\text{and}\quad \nu=1-\frac{1}{q}-\frac{s}{(s-2)\,p}.$$

If $n>2$ we choose $s=2^*=2n/(n-2)$, whence $\lambda=n/p$ and $\nu=\theta=1-1/q-n/2p$. The required conclusion then follows from Young's inequality and Lemma 2. When $n=2$ we choose $s=2+4/\theta p$, whence

$$\lambda=\frac{1}{p}+\frac{1}{q'}, \qquad \nu=\frac{\theta}{2}=\frac{1}{2}\Big(1-\frac{1}{p}-\frac{1}{q}\Big).$$

The conclusion then follows exactly as before.

The case $n=1$ needs a special treatment. Consider a strongly differentiable function $w$ of the real variable $x$, $0<x<d$. Then obviously

$$|w(x)|^2 \leqq \frac{2}{d}\int_0^d |w|^2\,dx + 2d\int_0^d |w_x|^2\,dx,$$

and by Minkowski's inequality

$$\Big(\int_0^d |w|^{2s}\,dx\Big)^{\frac{1}{s}} \leqq 2d^{\frac{1}{s}-1}\int_0^d |w|^2\,dx + 2d^{\frac{1}{s}+1}\int_0^d |w_x|^2\,dx, \quad (1\leqq s\leqq\infty).$$

If now $w$ is defined over the entire real axis, it follows by summation of this result over all intervals $(md, (m+1)\,d)$ that

$$\left(\int |w|^{2s}\,dx\right)^{\frac{1}{s}} \leqq 2d^{\frac{1}{s}-1} \int |w|^2\,dx + 2d^{\frac{1}{s}+1} \int |w_x|^2\,dx .$$

Now set $d^2 = \int |w|^2\,dx / \int |w_x|^2\,dx$, whence

$$(8) \qquad \left(\int |w|^{2s}\,dx\right)^{\frac{1}{s}} \leqq 4\left(\int |w|^2\,dx\right)^{\frac{1}{2}\left(1+\frac{1}{s}\right)} \left(\int |w_x|^2\,dx\right)^{\frac{1}{2}\left(1-\frac{1}{s}\right)}.$$

Considering $w$ as a function of the two variables $x, t$ we obtain easily from (8)

$$\|w\|^2_{2s,\,4s'} \leqq 4\,\|w\|^{1+1/s}_{2,\,\infty}\,\|w_x\|^{1-1/s}_{2,\,2}$$

where $s'$ is the Hölder conjugate of $s$. Consequently

$$\|w\|^2_{2p',\,2q'} \leqq T^{\theta}\,\|w\|^2_{2p',\,4p} \leqq 4\,T^{\theta}\,\|w\|^{2-1/p}_{2,\,\infty}\,\|w_x\|^{1/p}_{2,\,2} \leqq 4\,T^{\theta}\left\{\|w\|^2_{2,\,\infty} + \|w_x\|^2_{2,\,2}\right\}$$

by Young's inequality. This completes the proof of Lemma 3. [It should be noted that the condition $|\Omega|=1$ was required only when $n=2$; for simplicity in stating Lemma 3, however, we have not distinguished this fact.]

For convenience in what follows we introduce the

**Definition.** *If a function $w(x,t)$ is in $L^{2p',2q'}(Q)$ for all exponents $(p',q')$ satisfying the conditions of Lemma 3, we say that $w$ is in the space $\mathscr{U}(Q)$.*

In terms of this definition, Lemma 3 can be restated as follows: *Let $w$ have compact support in $\Omega$. If $w$ is in $L^{2,\infty}(Q)$ and has strong derivatives $w_x$ in $L^{2,2}(Q)$ then $w$ is in $\mathscr{U}(Q)$.* Now suppose that $w$ belongs to $L^{2,2}(Q)$ and has strong derivatives $w_x$, $w_t$ in $L^{2,2}(Q)$. In view of the elementary estimate

$$\|w\|^2_{2,\,\infty} \leqq 2\,T^{-1}\,\|w\|^2_{2,\,2} + 2\,T\,\|w_t\|^2_{2,\,2}$$

it follows that $w$ is contained in $L^{2,\infty}(Q)$. Thus we have the

**Corollary.** *Let $w$ belong to $L^{2,2}(Q)$ and have strong derivatives $w_x$ and $w_t$ belonging to $L^{2,2}(Q)$. Suppose further that $w$ has compact support in $\Omega$. Then $w$ is in $\mathscr{U}(Q)$.*

We shall often have to deal with composite functions of the form $\mathscr{G}(u(x,t))$, where $u$ is strongly differentiable. The next two lemmas provide a calculus for such functions (for simplicity the lemmas are stated for functions $u(x)$ rather $u(x,t)$, though of course this is no essential restriction).

**Lemma 4.** *Let $\mathscr{G}(u)$ be a piecewise smooth function of the real variable $u$, $-\infty < u < \infty$, with corners at the points $\ell_1, \ldots, \ell_N$. Suppose also that $\mathscr{G}$ satisfies a uniform Lipschitz condition. Then if $u(x)$ is a strongly differentiable function, the composite function $\mathscr{G}(u(x))$ is also strongly differentiable, and*

$$\mathscr{G}(u)_x = \mathscr{G}'(u)\,u_x$$

*where $\mathscr{G}'(u)$ is set equal to $\mathscr{G}'(u+0)$ when $u=\ell_1, \ldots, \ell_N$.*

This result is proved in [*19*, pp. 48–49]. Note that we are able to specify $\mathscr{G}'$ at the points $\ell_1, \ldots, \ell_N$ because of the well-known fact that the strong derivatives of a function are zero almost everywhere on any set where the function is constant.

**Lemma 5.** *Let $\mathscr{G}(u)$ be a continuously differentiable function of the real variable $u$, $-\infty<u<\infty$, such that $\mathscr{G}'(u)$ is uniformly Lipschitz continuous. Let $u(x)$ be a strongly differentiable function whose derivatives are locally in $L^2$. Then the composite function $\mathscr{G}(u(x))$ is also strongly differentiable, and $\mathscr{G}(u)_x=\mathscr{G}'(u)\,u_x$.*

The proof is a straightforward exercise, in which $u$ is first approximated by its integral average $u_h$ and then $h$ is allowed to tend to zero. It almost goes without saying that both Lemmas 4 and 5 remain valid without the uniformity assumption, provided the function $u(x)$ is essentially bounded.

The following lemma is a version of the fundamental theorem of integral calculus. In stating it we use the symbol $\Gamma$ to denote the set $\{\partial\Omega\times(0,T)\}\cup\{\Omega\times(t=0)\}$ forming the lower and lateral boundaries of the cylinder $Q$. The set $\Gamma$ is often called the parabolic boundary of $Q$.

**Lemma 6.** *Suppose that $w$ is strongly differentiable with respect to $t$ in the cylinder $Q$ and vanishes in a neighborhood of the boundary set $\Gamma$. Then*

$$\int_0^\tau\int_\Omega w_t\,dx\,dt=\int_\Omega w|_{t=\tau}\,dx$$

*for almost all values of $t$ in the interval* $(0,T)$.

**Proof.** We first show that

$$\int_\Omega w(x,t)\,dx$$

is a strongly differentiable function of $t$. Indeed let $\psi=\psi(t)$ be a smooth test function vanishing near $t=0$ and $t=T$. Then

$$\int_0^T\psi_t\Big(\int_\Omega w\,dx\Big)dt=\iint_Q\psi_t w\,dx\,dt=-\iint_Q\psi w_t\,dx\,dt=-\int_0^T\psi\Big(\int_\Omega w_t\,dx\Big)dt.$$

According to the definition of strong derivatives we have thus shown

$$\Big(\int_\Omega w\,dx\Big)_t=\int_\Omega w_t\,dx.$$

Since strongly differentiable functions of a single variable are equal (almost everywhere) to the integrals of their derivatives, the lemma is proved.

## 2. Fundamental Inequalities

In the following sections we shall derive various bounds for solutions of equation (1), and use these bounds to study the local behavior of solutions. The material of the present section is basic to this purpose.

Concerning the coefficients $b$ through $h$ in the structural inequalities (2), we recall that each of them is contained in some space $L^{p,q}(Q)$ where $p$ and $q$ are non-negative real numbers such that

$$p>2 \quad\text{and}\quad \frac{n}{2p}+\frac{1}{q}<\frac{1}{2} \quad\text{for}\quad b,c,e,f,h$$

and

$$p>1 \quad\text{and}\quad \frac{n}{2p}+\frac{1}{q}<1 \quad\text{for}\quad d,g.$$

In view of these conditions it is clear the there exists a positive number $\theta$ such that

$$p \geqq \frac{2}{1-\theta} \quad \text{and} \quad \frac{n}{2p} + \frac{1}{q} \leqq \frac{1-\theta}{2} \quad \text{for} \quad b, c, e, f, h$$

and

$$p \geqq \frac{1}{1-\theta} \quad \text{and} \quad \frac{n}{2p} + \frac{1}{q} \leqq 1-\theta \quad \text{for} \quad d, g.$$

The number $\theta$ will play an important role in the sequel. Indeed the quantitative structure of equation (1) is entirely determined by the coefficients $a$ and $\bar{a}$, the values of the norms $\|b\|$ through $\|h\|$ in their respective spaces, and the numbers $\theta$ and $n$. *We shall say that a constant depends on the structure of equation* (1) *if it is determined by these quantities (and is uniformly bounded whenever these quantities are).*

Now let $u = u(x, t)$ be a weak solution of (1) in $Q$. Then $u$ has a strong derivative $u_x$ which is locally of class $L^{2,2}$ in $Q$, that is, $u_x$ is contained in $L^{2,2}(Q')$ for each compact subcylinder $Q'$ of $Q$. As already explained in the introduction, in the present section we make the temporary assumption that $u$ also has a strong derivative $u_t$ which is locally of class $L^{2,2}$ in $Q$. The weak form (6) of equation (1) can then be replaced by

$$\iint_Q \{\varphi u_t + \varphi_x \cdot \mathscr{A}(x, t, u, u_x)\} \, dx \, dt = \iint_Q \varphi \mathscr{B}(x, t, u, u_x) \, dx \, dt, \tag{9}$$

valid for any continuously differentiable function $\varphi = \varphi(x, t)$ having compact support in $Q$.

Since $u_x$ and $u_t$ are locally in $L^{2,2}$ it follows from the Corollary to Lemma 3 that $u$ is locally in the space $\mathscr{U}(Q)$. This being the case, using (2) and the assumptions stated above on the coefficients $b$ through $h$, we see that

$$\mathscr{A}(x, t, u, u_x) \quad \text{is locally in } L^{2,2}$$

while

$$\mathscr{B}(x, t, u, u_x) \quad \text{is locally in the dual of } \mathscr{U}(Q).$$

[We illustrate by considering the term $d(x, t)\,|u|$ in the majorant of $\mathscr{B}$. If $\varphi$ is in $\mathscr{U}(Q)$ then

$$\|d\,u\,\varphi\|_{1,1} \leqq \|d\|_{p,q} \|u\|_{2p', 2q'} \|\varphi\|_{2p', 2q'} < \infty,$$

the middle integral being finite since $u \in \mathscr{U}(Q)$.] It follows therefore that *if $u$ is a solution of* (1), *then* (9) *holds not only for continuously differentiable functions $\varphi$, but in fact for any $\varphi$ in $\mathscr{U}(Q)$ with strong derivatives $\varphi_x$ in $L^{2,2}$ and with compact support in $Q$.* This remark will be of considerable importance later.

The fundamental estimates of this section will be obtained by using powers of $u$ as test functions in the weak integral relation (9). This computation would be relatively easy to perform if we knew *a priori* that $u$ is bounded, since positive powers of a bounded strongly differentiable function are clearly also strongly differentiable in the same Lebesgue class (see the previous section). However, we do not have an *a priori* bound on $u$, nor is it reasonable to postulate one. Accordingly we shall have to overcome certain technical difficulties, for which purpose it is convenient to employ a device already introduced by one of us in a similar context.

Let $\kappa$ be a fixed positive constant (except $\kappa=0$ if $f$, $g$, $h$ are zero) and let $\ell$ be another constant greater than $\kappa$. Set

$$\bar{u}=\max(0,u)+\kappa$$

and define the function

$$\mathscr{G}(u)=\begin{cases}\bar{u}^{\beta}-\kappa^{\beta} & \text{for } -\infty<u\leqq \ell-\kappa\\ \ell^{\beta-1}\bar{u}-\kappa^{\beta} & \text{for } \ell-\kappa\leqq u<\infty\end{cases}\tag{10}$$

where $\beta$ is a fixed exponent $\geqq 1$. It is clear that $\mathscr{G}$ is a non-negative piecewise smooth function of $u$ for $-\infty<u<\infty$, with corners at $u=0$ and $\ell-\kappa$. The function $\bar{u}^{\beta}-\kappa^{\beta}$ is essentially the function we want. To use it at this stage of the argument, however, it is necessary to linearize it for large values of $u$ as in (10). Also define

$$\mathscr{H}(u)=\begin{cases}\dfrac{1}{\beta+1}\{\bar{u}^{\beta+1}-(\beta+1)\bar{u}\,\kappa^{\beta}+\beta\,\kappa^{\beta+1}\} & \text{for } -\infty<u\leqq \ell-\kappa\\ \dfrac{1}{2}\ell^{\beta-1}\bar{u}^{2}-\kappa^{\beta}\bar{u}+\dfrac{1}{\beta+1}\left(\beta\,\kappa^{\beta+1}-\dfrac{\beta-1}{2}\ell^{\beta+1}\right) & \text{for } \ell-\kappa\leqq u<\infty.\end{cases}$$

It is easily verified that $\mathscr{H}$ is a non-negative continuously differentiable function for $-\infty<u<\infty$ and that $\mathscr{H}'(u)=\mathscr{G}(u)$.

Now let $\eta=\eta(x,t)$ be a piecewise smooth non-negative function vanishing in a neighborhood of the parabolic boundary set $\Gamma$ (see section 1). Also let $\chi(t,\tau)$ be the characteristic function of the open interval $(0,\tau)$, where $\tau$ is a fixed number less than $T$. We set

$$\varphi(x,t)=\eta^{2}\mathscr{G}(u)\,\chi(t,\tau).\tag{11}$$

Since $\eta^{2}\chi(t,\tau)$ has compact support in $Q$ while $\mathscr{G}(u)$ is linear in $u$ for large values of its argument, it is evident that $\varphi$ is in $\mathscr{U}(Q)$. Moreover, in view of Lemma 4, $\varphi$ is strongly differentiable with respect to $x$ and

$$\varphi_x=\begin{cases}\eta^{2}\mathscr{G}'(u)\,u_x+2\eta\,\eta_x\mathscr{G}(u) & 0<t<\tau\\ 0 & \tau\leqq t.\end{cases}$$

Clearly $\varphi_x$ is in $L^{2,2}(Q)$. Altogether, therefore, $\varphi$ is admissible as a test function in (9).

Observe that $\mathscr{H}(u)$ satisfies the hypothesis of Lemma 5. Hence $\mathscr{H}(u)_t=\mathscr{G}(u)\,u_t$ and

$$\eta^{2}\mathscr{G}(u)\,u_t=(\eta^{2}\mathscr{H}(u))_t-2\eta\,\eta_t\mathscr{H}(u).$$

Since $\eta^{2}\mathscr{H}(u)$ satisfies the hypothesis of Lemma 6, we have then

$$\iint_Q \varphi\,u_t\,dx\,dt=\int_0^{\tau}\int_{\Omega}\eta^{2}\mathscr{G}u_t\,dx\,dt=\int_{\Omega}\eta^{2}\mathscr{H}|_{t=\tau}\,dx-\int_0^{\tau}\int_{\Omega}2\eta\,\eta_t\mathscr{H}\,dx\,dt.$$

Thus, with $\varphi$ as test function, (9) becomes

$$\int_{\Omega}\eta^{2}\mathscr{H}(u)|_{t=\tau}\,dx+\iint_Q\{\varphi_x\cdot\mathscr{A}-\varphi\,\mathscr{B}\}\,dx\,dt=\int_0^{\tau}\int_{\Omega}2\eta\,\eta_t\mathscr{H}(u)\,dx\,dt,\tag{12}$$

valid for almost all values $\tau$ in $(0, T)$. In order to estimate the second integral in (12) we shall require some preliminary calculations.

First, on the set where $\varphi>0$ we have necessarily $u>0$, and consequently $u_x=\bar{u}_x$ and $|u|<\bar{u}$. Therefore on this set, using the structure hypotheses (2),

$$\begin{aligned}\varphi_x\cdot\mathscr{A}-\varphi\mathscr{B}\geqq\eta^2\mathscr{G}'\{a|\bar{u}_x|^2-b^2\bar{u}^2-f^2\}-\\ -2\eta|\eta_x|\mathscr{G}\{\bar{a}|\bar{u}_x|+e\bar{u}+h\}+\eta^2\mathscr{G}\{c|\bar{u}_x|+d\bar{u}+g\}.\end{aligned} \tag{13}$$

Now observe that

$$\mathscr{G}^2\leqq\bar{u}^{\beta+1}\mathscr{G}'. \tag{14}$$

Thus

$$2\bar{a}\eta\mathscr{G}|\eta_x|\,|\bar{u}_x|\leqq\frac{a}{4}\eta^2\mathscr{G}'|\bar{u}_x|^2+\frac{4\bar{a}^2}{a}|\eta_x|^2\bar{u}^{\beta+1}$$

and

$$c\mathscr{G}|\bar{u}_x|\leqq\frac{a}{4}\mathscr{G}'|\bar{u}_x|^2+\frac{1}{a}c^2\bar{u}^{\beta+1}.$$

Also since $\mathscr{G}\leqq\bar{u}^{\beta}$ and $\kappa\leqq\bar{u}$ we have

$$\mathscr{G}\{d\bar{u}+g\}\leqq\left(d+\frac{g}{\kappa}\right)\bar{u}^{\beta+1},\qquad \mathscr{G}\{e\bar{u}+h\}\leqq\left(e+\frac{h}{\kappa}\right)\bar{u}^{\beta+1}.$$

Combining the preceding inequalities yields then (since $\mathscr{G}'\leqq\beta\bar{u}^{\beta-1}$)

$$\frac{a}{2}\eta^2\mathscr{G}'|\bar{u}_x|^2\leqq\varphi_x\cdot\mathscr{A}-\varphi\mathscr{B}+(F\eta^2+2G\eta|\eta_x|+H|\eta_x|^2)\,\bar{u}^{\beta+1} \tag{15}$$

where

$$F=\beta\left(b^2+\frac{f^2}{\kappa^2}\right)+\left(d+\frac{g}{\kappa}\right)+\frac{c^2}{a},\qquad G=e+\frac{h}{\kappa},\qquad H=\frac{4\bar{a}^2}{a}. \tag{16}$$

Inequality (15) has been proved to hold only in the set where $\varphi>0$. When $\varphi=0$ and $t<\tau$, we have either $\eta=0$ or $u\leqq0$ and $\bar{u}_x=0$. In either case (15) clearly remains valid; thus in fact, (15) holds throughout $\Omega\times(0,\tau)$.

Integrating (15) over $\Omega\times(0,\tau)$ and adding the result to (12), we get (since $\mathscr{H}\leqq(\beta+1)^{-1}\bar{u}^{\beta+1}$)

$$\begin{aligned}&\int_\Omega\eta^2\mathscr{H}|_{t=\tau}dx+\frac{a}{2}\iint\eta^2\mathscr{G}'|\bar{u}_x|^2dx\,dt\\ &\quad\leqq\iint(F\eta^2+2G\eta|\eta_x|+H|\eta_x|^2)\,\bar{u}^{\beta+1}dx\,dt+\frac{2}{\beta+1}\iint\eta|\eta_t|\,\bar{u}^{\beta+1}dx\,dt,\end{aligned}$$

the integrations extending over $\Omega\times(0,\tau)$.

At this stage it is possible to let the parameter $\ell$ tend to infinity. Clearly $\mathscr{G}'$ and $\mathscr{H}$ are non-decreasing functions of $\ell$, with

$$\lim\mathscr{H}=\frac{1}{\beta+1}\{\bar{u}^{\beta+1}-(\beta+1)\kappa^\beta\bar{u}+\beta\kappa^{\beta+1}\},\qquad \lim\mathscr{G}'|\bar{u}_x|^2=\beta\bar{u}^{\beta-1}|\bar{u}_x|^2.$$

Applying the monotone convergence theorem*, we thus obtain for almost all values $\tau$ in $(0, T)$

$$\text{(I)}\quad \begin{aligned} &\frac{1}{\beta+1}\int_\Omega \eta^2\{\bar{u}^{\beta+1}-(b+1)\kappa^\beta\bar{u}+\beta\kappa^{\beta+1}\}_{t=\tau}\,dx+\frac{a\beta}{2}\iint \eta^2\bar{u}^{\beta-1}|\bar{u}_x|^2\,dx\,dt\\ &\qquad\leqq \iint \mathscr{F}\bar{u}^{\beta+1}\,dx\,dt+\frac{2}{\beta+1}\iint \eta|\eta_t|\bar{u}^{\beta+1}\,dx\,dt,\end{aligned}$$

the integrations extending over $\Omega\times(0,\tau)$. Here

$$\mathscr{F}=F\eta^2+2G\eta|\eta_x|+H|\eta_x|^2 \tag{17}$$

and $F$, $G$, $H$ are defined by (16).

Inequality (I) is fundamental to our subsequent work. Because of the particular method of derivation it clearly holds without the assumption that either side of the inequality is finite. Of course, if the right hand side of (I) is finite then so is the left hand side; the iterative arguments which we shall use in the next sections are designed to exploit just this fact.

A variant of (I) arises if in place of the test function (11) we use

$$\varphi(x,t)=\eta^2\mathscr{G}(u)\chi(t,\tau_1,\tau_2) \tag{18}$$

where $\chi(t,\tau_1,\tau_2)$ denotes the characteristic function of the open interval $(\tau_1,\tau_2)$, with $0<\tau_1<\tau_2<T$. The preceding calculation evidently then leads to the inequality

$$\text{(II)}\quad \begin{aligned} &\frac{1}{\beta+1}\int_\Omega \eta^2\{\bar{u}^{\beta+1}-(\beta+1)\kappa^\beta\bar{u}+\beta\kappa^{\beta+1}\}_{t=\tau_1}^{t=\tau_2}\,dx+\frac{a\beta}{2}\iint \eta^2\bar{u}^{\beta-1}|\bar{u}_x|^2\,dx\,dt\\ &\qquad\leqq \iint \mathscr{F}\bar{u}^{\beta+1}\,dx\,dt+\frac{2}{\beta+1}\iint \eta|\eta_t|\bar{u}^{\beta+1}\,dx\,dt,\end{aligned}$$

valid for almost all $\tau_1$, $\tau_2$ in $(0, T)$. Here the $(x, t)$ integrations extend over the set $\Omega\times(\tau_1,\tau_2)$, and $\mathscr{F}$ is defined by (17).

In section 5 another variant of (I) will be required. Suppose in fact that $u$ is *non-negative* and *locally bounded* in $Q$. For arbitrary $\varepsilon>0$ set

$$\bar{u}=u+\kappa+\varepsilon$$

and define the function $\mathscr{G}(u)=\bar{u}^\beta$, where $\beta$ is a fixed exponent different from zero. We also put

$$\mathscr{H}(u)=\begin{cases}\dfrac{1}{\beta+1}\bar{u}^{\beta+1} & \text{if } \beta\neq-1\\ \log\bar{u} & \text{if } \beta=-1,\end{cases} \tag{19}$$

so that $\mathscr{H}'(u)=\mathscr{G}(u)$.

Now let $\chi(t,\tau_1,\tau_2)$ be as before and set

$$\varphi(x,t)=\eta^2\mathscr{G}(u)\chi(t,\tau_1,\tau_2).$$

Since $\bar{u}$ is locally bounded, it is evident that $\varphi$ is admissible as a test function in (9).

* It is convenient to use a denumerable sequence of values of $t$ in order to avoid difficulties with the fact that the integral $\int_\Omega \eta^2\mathscr{H}\,dx$ is only defined for almost all $t$.

The resulting calculation is very similar to the one just completed, the main difference being that $\mathscr{G}'<0$ when $\beta<0$. Thus (13) must be replaced by

$$\text{(13')}\quad \begin{aligned} \operatorname{sign}\beta\,(\varphi_x\cdot\mathscr{A}-\varphi\,\mathscr{B}) &\geqq \eta^2\,|\mathscr{G}'|\,\{a\,|\bar{u}_x|^2-b^2\bar{u}^2-f^2\} \\ &\quad -2\eta\,\mathscr{G}\,|\eta_x|\,\{\bar{a}\,|\bar{u}_x|+e\,\bar{u}+h\}-\eta^2\mathscr{G}\{c\,|\bar{u}_x|+d\,\bar{u}+g\}, \end{aligned}$$

and (14) by

$$\text{(14')}\quad \mathscr{G}^2=\bar{u}^{\beta+1}\mathscr{G}'/\beta.$$

We obtain thereby the following inequality, valid for almost all $\tau_1$, $\tau_2$ in $(0, T)$,

$$\text{(III)}\quad \begin{aligned} &\operatorname{sign}\beta\left\{\int_\Omega \eta^2\,\mathscr{H}\,\Big|_{t=\tau_1}^{t=\tau_2}dx+\frac{a\,\beta}{2}\iint\eta^2\,\bar{u}^{\beta-1}\,|\bar{u}_x|^2\,dx\,dt\right\} \\ &\qquad\leqq\iint\mathscr{F}_1\,\bar{u}^{\beta+1}\,dx\,dt+2\iint\eta\,|\eta_t|\,|\mathscr{H}|\,dx\,dt, \end{aligned}$$

the integrations extending over $\Omega\times(\tau_1,\tau_2)$. Here $\mathscr{H}$ is defined by (19), while

$$\mathscr{F}_1=F_1\,\eta^2+2G_1\,\eta\,|\eta_x|+H_1\,|\eta_x|^2,$$

and

$$F_1=|\beta|\left(b^2+\frac{f^2}{\kappa^2}\right)+\left(d+\frac{g}{\kappa}\right)+\frac{c^2}{a\,|\beta|},\qquad G_1=e+\frac{h}{\kappa},\qquad H_1=\frac{4\bar{a}^2}{a\,|\beta|}.$$

We observe that both sides of (III) are finite because of the assumption that $u$ is locally bounded.

## 3. A Maximum Principle

Here we shall prove a maximum principle for weak solutions of (1) in a bounded cylinder $Q=\Omega\times(0,T)$. The demonstration will be carried out in two stages, first an estimate for the maximum of $u$ in terms of a certain norm of $u$, and then an estimate for this norm in terms of other quantities. The class of weak solutions treated here has been defined strictly in terms of local properties. In particular, there is no restriction on the behavior of a weak solution in the neighborhood of the parabolic boundary $\Gamma=\{\partial\Omega\times[0,T)\}\cup\{\Omega\times(t=0)\}$ of $Q$. It is clear the in order to obtain a maximum principle it will be necessary to make some assumption on the behavior of $u$ near $\Gamma$.

We shall say that $u\leqq M$ on the boundary set $\Gamma$ of $Q$ if for every $\varepsilon>0$ there exists a neighborhood of $\Gamma$ in which $u\leqq M+\varepsilon$.

**Theorem 1.** *Let $u$ be a weak solution of* (1) *in $Q$ such that $u\leqq M$ on $\Gamma$. Then almost everywhere in $Q$*

$$u(x,t)\leqq M+Ck,$$

*where $C$ depends only on $T$, $|\Omega|$ and the structure of* (1), *while*

$$k=(\|b\|+\|d\|)\,|M|+(\|f\|+\|g\|).$$

The particular functional dependence of $C$ on the structure of (1) can be expressed by writing

$$C=C(a,\|b\|,\|c\|,\|d\|,\theta,n,T,|\Omega|).$$

Note that this is independent of $\bar{a}$, $e, f, g$ and $h$. The detailed dependence of $C$ on the various parameters listed will become apparent during the proof. Moreover the structural inequality $|\mathscr{A}| \leqq \bar{a}\,|p| + e\,|u| + h$ is not used explicitly here. All that is really needed is that the term $\varphi_x \cdot \mathscr{A}$ be integrable for $\varphi_x$ belonging to $L^{2,2}$ locally in $Q$.

**Proof.** We assume temporarily that $|\Omega| = T = 1$ and $M < 0$. At the conclusion of the proof these normalizations will be removed.

Our first step involves showing that, under the normalizations here assumed,

$$u(x,t) \leqq C_1 \{\|\tilde{u}\|_{2,\infty} + \|\tilde{u}_x\|_{2,2} + \kappa\}, \tag{20}$$

where $\tilde{u} = \max(0, u)$ and $\kappa = \|f\| + \|g\|$. The constant $C_1$, as well as later constants $C_2$, $C_3$, etc., depends only on the norms $\|b\|$, $\|c\|$ and $\|d\|$, the coefficient $a$, the dimension $n$, and $\theta$.

We set $\bar{u} = \max(0, u) + \kappa$ and apply the considerations of the previous section. In this process the compact support factor $\eta^2$ can be omitted from (11) since $\mathscr{G}$ already vanishes identically in a neighborhood of $\Gamma$. We thus obtain from the fundamental inequality (I)

$$\begin{aligned} \frac{1}{\beta+1} \int_{\Omega} \{\bar{u}^{\beta+1} - (\beta+1)\,\kappa^{\beta}\,\bar{u} + \beta\,\kappa^{\beta+1}\}_{t=\tau}\, dx + \frac{a\beta}{2} \iint \bar{u}^{\beta+1}\, |\bar{u}_x|^2\, dx\, dt \\ \leqq \iint F\, \bar{u}^{\beta+1}\, dx\, dt, \end{aligned} \tag{21}$$

where

$$F = \beta\left(b^2 + \frac{f^2}{\kappa^2}\right) + \left(d + \frac{g}{\kappa}\right) + \frac{c^2}{a}$$

and where the integrations extend over $\Omega \times (0, \tau)$.

Write $r = (\beta+1)/2$ and $v = \bar{u}^r$. Then $|v_x|^2 = r^2\, \bar{u}^{\beta-1}\, |\bar{u}_x|^2$. Since the first integral in (21) is non-negative, we obtain (letting $\tau \to 1$)

$$a\,\|v_x\|_{2,2}^2 \leqq \frac{2r^2}{\beta}\, \|F v^2\|_{1,1}\,.$$

At this stage it is convenient to introduce the special norm

$$|||v||| = \sup \|v\|_{2p', 2q'}$$

where the supremum is taken over all exponent pairs $(p', q')$ whose Hölder conjugates $(p, q)$ satisfy

$$\frac{n}{2p} + \frac{1}{q} \leqq 1 - \theta \quad \text{and} \quad p \geqq \frac{1}{1-\theta}, \tag{22}$$

$\theta$ being the constant introduced at the beginning of section 2. In particular, using Hölder's inequality and the hypothesis on the coefficients in the structural inequalities (2), it is clear that

$$\frac{1}{\beta}\, \|F v^2\|_{1,1} \leqq (\|b\|^2 + a^{-1}\, \|c\|^2 + \|d\| + 1)\, |||v|||^2 = C_2\, |||v|||^2 \tag{23}$$

(recall that $\beta \geqq 1$). Inserting this estimate into the earlier inequality for $\|v_x\|_{2,2}$ yields

(24) $$a\,\|v_x\|_{2,2}^2 \leqq 2\,C_2\, r^2\, |||v|||^2.$$

This being shown, we return once more to (21). By Young's inequality

$$\bar{u}^{\beta+1} - (\beta+1)\,\kappa^\beta\,\bar{u} + \beta\,\kappa^{\beta+1} \geqq \tfrac{1}{2}\,\bar{u}^{\beta+1} - \beta\,\kappa^{\beta+1}.$$

Consequently, since the second integral in (21) is non-negative, we have

$$\frac{1}{2(\beta+1)} \int_\Omega v^2|_{t=\tau}\, dx \leqq \beta\, C_2\, |||v|||^2 + \frac{\beta}{\beta+1}\, |||v|||^2,$$

valid for almost all values $\tau$ in $(0, T)$. It follows that

(25) $$\|v\|_{2,\infty}^2 \leqq C_3\, r^2\, |||v|||^2$$

where $C_3 = 8\,C_2 + 2$.

Now let $\sigma = 1 + 2\theta/n$ (if $n=2$, let $\sigma = 1 + \theta/2$). We assert that

(26) $$|||v^\sigma|||^{2/\sigma} \leqq 2(K+1)\,\{\|v\|_{2,\infty}^2 + \|v_x\|_{2,2}^2\}.$$

To see this, let $(p', q')$ be an exponent pair such that (22) holds. Then

$$\|v^\sigma\|_{2p',\,2q'}^{1/\sigma} = \|v\|_{2\sigma p',\,2\sigma q'}.$$

Setting $\bar{p}' = \sigma p'$, $\bar{q}' = \sigma q'$, a short calculation shows that the exponent pair $(\bar{p}, \bar{q})$ fulfills the conditions of Lemma 3 and the assertion follows immediately (here we use the fact that $v - \kappa^r$ has compact support in $\Omega$ as a function of $x$, and also the obvious relation $v \geqq \kappa^r$).

Combining (24), (25), (26), there results

$$|||v^\sigma|||^{2/\sigma} \leqq C_4\, r^2\, |||v|||^2,$$

when $C_4 = 4(K+1)\,\{1 + (a^{-1}+4)\,C_2\}$. For $m = 0, 1, 2, \ldots$, put

$$\varphi_m = |||v|||^{2/r}, \qquad (v = \bar{u}^r,\ r = \sigma^m).$$

The preceding inequality can then be rewritten

$$\varphi_{m+1} \leqq C_4^{1/\sigma^m}\, \sigma^{2m/\sigma^m}\, \varphi_m.$$

By iteration we obtain

(27) $$\varphi_{m+1} \leqq C_4^{\Sigma \sigma^{-j}}\, \sigma^{2\Sigma j \sigma^{-j}}\, \varphi_0,$$

the sums running from $j=0$ to $j=m$. Clearly $\sum \sigma^{-j} < \infty$ and $\sum j\sigma^{-j} < \infty$. Thus we can let $m$ tend to infinity in (27) to get

$$\max \bar{u}^2 \leqq C_5\, \varphi_0.$$

Since

$$\varphi_0 = \sup \|\bar{u}\|_{2p',\,2q'}^2 \leqq 2(K+1)\,\{\|\tilde{u}\|_{2,\infty}^2 + \|\tilde{u}_x\|_{2,2}^2 + \kappa^2\},$$

the demonstration of (20) is completed. Clearly $C_1 = \{2(K+1)\,C_5\}^{1/2}$.

The second stage of the proof is concerned with estimation of the norms $\|\tilde{u}\|_{2,\infty}$ and $\|\tilde{u}_x\|_{2,2}$. To this end, set $\beta=1$ in the fundamental estimate (II). Since the cut-off function $\eta^2$ can be dropped in the present case, we thus obtain

$$\tfrac{1}{2}\int_\Omega \tilde{u}^2\Big|_{t=\tau_1}^{t=\tau_2} dx+\frac{a}{2}\iint|\tilde{u}_x|^2\,dx\,dt\leqq\iint F\,\bar{u}^2\,dx\,dt \tag{28}$$

where

$$F=\left(b^2+\frac{f^2}{\kappa^2}\right)+\left(d+\frac{g}{\kappa}\right)+\frac{c^2}{a},$$

the integrations extending over the set $\Omega\times(\tau_1,\tau_2)$. Note here that $\bar{u}=\tilde{u}+\kappa$.

Now by Hölder's inequality and Lemma 3

$$\iint F\,\bar{u}^2\,dx\,dt\leqq 2(K+1)\,C_2(\tau_2-\tau_1)^\theta\{\|\tilde{u}\|_{2,\infty}^2+\|\tilde{u}_x\|_{2,2}^2+\kappa^2\},$$

where $\theta$ is the constant introduced earlier in the proof* and the norms on the right side are taken over the interval $(\tau_1,\tau_2)$. Now consider $\tau_2=t$ as a variable over the time interval $I$:

$$\tau_1\leqq t\leqq\tau_1+\mu$$

where $\mu=\{\operatorname{Min}(a,1)/8(K+1)\,C_2\}^{1/\theta}$, it being assumed that (28) is meaningful for the given $\tau_1$. Then setting

$$X(t)=\int_\Omega \tilde{u}^2\,dx$$

we obtain for almost all $t$ in $I$

$$X(t)+\frac{a}{2}\iint|\tilde{u}_x|^2\,dx\,dt\leqq\frac{\operatorname{Min}(a,1)}{2}\quad(\|\tilde{u}\|_{2,\infty}^2+\kappa^2)+X(\tau_1), \tag{29}$$

the integral extending over the set $\Omega\times(\tau_1,t)$ while the norm is taken over $\Omega\times I$.

By disregarding the integral on the left hand side we obtain at once

$$\operatorname{ess\,sup}_I X(t)\leqq\kappa^2+2X(\tau_1).$$

We can obviously iterate this inequality from one time interval $(\tau_1,\tau_1+\mu)$ to the next one $(\tau_1+\mu,\tau_1+2\mu)$, and so forth. Since $X(0)=0$ by hypothesis, it follows that

$$X(t)\leqq 2^{1+t/\mu}\,\kappa^2,$$

that is, with $t=1$,

$$\|\tilde{u}\|_{2,\infty}^2\leqq 2^{1+1/\mu}\,\kappa^2. \tag{30}$$

Returning to (29) and disregarding the first term we find

$$\iint|\tilde{u}_x|^2\,dx\,dt\leqq 2\kappa^2+2(1+a^{-1})\,X(\tau_1),$$

whence by iteration

$$\|\tilde{u}_x\|_{2,2}^2\leqq 2^{2+1/\mu}(1+a^{-1})\,\kappa^2. \tag{31}$$

* If $n=2$ then $(\tau_2-\tau_1)^\theta$ should be replaced by $(\tau_2-\tau_1)^{\theta/2}$, with corresponding changes later in the proof.

Inserting (30) and (31) into (20) shows that

$$u(x,t) \leqq C\kappa, \tag{32}$$

completing the proof of Theorem 1 for the special normalization introduced at the outset.

We now consider the case where $u \leqq M$ on $\Gamma$, still retaining the restrictions $|\Omega| = T = 1$. Let $\delta > 0$ be arbitrary, and put $U = u - M - \delta$. Then $U < 0$ on $\Gamma$. Since $u$ is a solution of (1), we have

$$U_t = \operatorname{div} \tilde{\mathscr{A}}(x,t,U,U_x) + \tilde{\mathscr{B}}(x,t,U,U_x)$$

where

$$\tilde{\mathscr{A}}(x,t,U,P) = \mathscr{A}(x,t,U+M+\delta,P) \text{ and } \tilde{\mathscr{B}}(x,t,U,P) = \mathscr{B}(x,t,U+M+\delta,P).$$

In view of (2)

$$P \cdot \tilde{\mathscr{A}}(x,t,U,P) \geqq a|P|^2 - 2b^2U^2 - 2b^2(M+\delta)^2 - f^2$$

and

$$|\tilde{\mathscr{B}}(x,t,U,P)| \leqq c|P| + d|U| + d|M+\delta| + g.$$

Thus we can apply (32) to obtain $U \leqq C'\kappa'$, where

$$C' = \sqrt{2}\,C \quad \text{and} \quad \kappa' = (\|b\| + \|d\|)|M+\delta| + (\|f\| + \|g\|).$$

Returning to the original dependent variable yields (since we can now let $\delta \to 0$)

$$u(x,t) \leqq M + C'k. \tag{33}$$

Finally suppose that $|\Omega|$ and $T$ are no longer normalized to have the value 1. Introduce new variables $t' = T^{-1}t$ and $x'_\ell = |\Omega|^{-1/n} x_\ell$, $\ell = 1, \ldots, n$. In the new variables $u$ satisfies an equation of the form (1) in a cylinder $Q' = \Omega' \times (0,1)$ with $|\Omega'| = 1$. Moreover, for the new equation the coefficients $a$ through $h$ in (2) are respectively replaced by

$$aT|\Omega|^{-2/n}, \quad \bar{a}T|\Omega|^{-2/n}, \quad bT^{1/2}, \quad cT|\Omega|^{-1/n}, \quad dT,$$
$$eT|\Omega|^{-1/n}, \quad fT^{1/2}, \quad gT, \quad \text{and} \quad hT|\Omega|^{-1/n}.$$

Let the new coefficients be denoted by $a'$ through $h'$, and write $\|\cdot\|'$ for norms over the cylinder $Q'$. Then we have

$$\begin{aligned}
(a', \bar{a}') &= T|\Omega|^{-2/n}(a, \bar{a}) \\
(\|b'\|', \|f'\|') &= T^{1/2-1/q}|\Omega|^{-1/p}(\|b\|, \|f\|) \\
(\|c'\|', \|e'\|', \|h'\|') &= T^{1-1/q}|\Omega|^{-1/n-1/p}(\|c\|, \|e\|, \|h\|) \\
(\|d'\|', \|g'\|') &= T^{1-1/q}|\Omega|^{-1/p}(\|d\|, \|g\|).
\end{aligned} \tag{34}$$

In the new variables, relation (33) evidently holds in the same general form, with $C$ now depending also on $T$ and $|\Omega|$. This completes the proof of Theorem 1.

Under the hypothesis of Theorem 1 a minimum principle holds equally well. Thus if $u \geqq M$ on $\Gamma$, then in $Q$

$$u(x,t) \geqq M - Ck$$

where $C$ and $k$ are as in Theorem 1.

It should also be observed that the maximum principle of Theorem 1 remains true if we replace equation (1) by the differential inequality

$$u_t \leqq \operatorname{div} \mathscr{A}(x, t, u, u_x) + \mathscr{B}(x, t, u, u_x).$$

In fact, in this case the fundamental estimates (I) and (II) clearly continue to hold, and the remaining part of the proof is exactly the same.

Finally, the method of proof indicates that the same results can be obtained for a wider class of coefficients $b$ through $g$ than is here assumed. In particular, for the proof to be valid it is only necessary that the coefficients be such that (23) holds. Thus, for example, it is sufficient that the quantity

$$b^2 + c^2 + d + f^2 + g$$

be a linear combination of functions, each of which is in $L^{p,q}$ for some exponent pair satisfying

$$\frac{n}{2p} + \frac{1}{q} < 1, \qquad p \geqq 1.$$

Clearly this requirement is weaker than the conditions originally posed.

## 4. Local Boundedness

In this section we consider the problem of bounding a weak solution $u$ when we do not have any information about its behavior at the boundary of $Q$. The main result is a bound for $u$ in a subcylinder in terms of the norm $\|u\|_{2,2}$ over the full cylinder. This is obtained by an iteration argument similar to the one employed in proving Theorem 1.

Let $(\bar{x}, \bar{t})$ be a fixed point in the basic set $Q = \Omega \times (0, T)$. We denote by $R(\rho)$ the open cube in $E^n$ of edge length $\rho$ centered at $\bar{x}$, and define

$$Q(\rho) = R(\rho) \times (\bar{t} - \rho^2, \bar{t}).$$

The symbol $\|\cdot\|_{p,q,\rho}$ will be used to denote the $L^{p,q}$ norm of a function over the cylinder $Q(\rho)$.

**Theorem 2.** *Let $u$ be a weak solution of* (1) *in $Q$. Suppose that the set $Q(3\rho)$ is contained in $Q$. Then almost everywhere in $Q(\rho)$ we have*

$$|u(x, t)| \leqq C\left(\rho^{-(n+2)/2} \|u\|_{2,2,3\rho} + \rho^{\theta} k\right)$$

*where $C$ is a constant depending only on $\rho$ and the structure of* (1), *and*

$$k = \|f\| + \|g\| + \|h\|.$$

*In particular, weak solutions of* (1) *must be locally essentially bounded.*

**Proof.** We may clearly suppose that $Q \equiv Q(3\rho)$. Moreover, after an appropriate normalization there is no loss of generality in assuming $\rho = 1/3$; cf. the transformation formulae at the end of the preceding section.

With this normalization understood, the proof is now based on the fundamental estimate (I). If we set $v = \bar{u}^r$ where $\bar{u} = \max(0, u) + k$ and $r = (\beta + 1)/2$, this inequality takes the form

$$\frac{1}{\beta+1}\int \eta^2\{\bar{u}^{\beta+1}-(\beta+1)\,k^\beta\,\bar{u}+\beta k^{\beta+1}\}_{t=\tau}\,dx+\frac{a\beta}{2r^2}\iint \eta^2\,|v_x|^2\,dx\,dt \tag{35}$$
$$\leqq \|(\mathscr{F}+\eta\,|\eta_t|)\,v^2\|_{1,1}$$

(here $\|\cdot\|_{p,q}$ denotes the $L^{p,q}$ norm taken over the full cylinder $Q=Q(1)$).

Let $\ell$ and $\ell'$ be real numbers satisfying $1/3<\ell'<\ell\leqq 2/3$. We choose the function $\eta$ so that $\eta\equiv 1$ in $Q(\ell')$ and $\eta\equiv 0$ outside $Q(\ell)$. We can do this, moreover, in such a way that $0\leqq\eta\leqq 1$ and $|\eta_x|$, $|\eta_t|\leqq 2(\ell-\ell')^{-1}$. As in the proof of Theorem 1 it is convenient to introduce a special norm over $Q(\ell)$ defined by

$$|||\cdot|||_\ell=\sup\,\|\cdot\|_{2p',\,2q',\,\ell}$$

where the supremum is taken over all exponent pairs $(p', q')$ whose Hölder conjugates $(p, q)$ satisfy (22). Then by Hölder's inequality we find (since $\beta\geqq 1$)

$$\frac{1}{\beta}\,\|(\mathscr{F}+\eta\,|\eta_t|)\,v^2\|_{1,1}\leqq C_1\,|||\eta v|||_\ell^2+C_2(\ell-\ell')^{-2}\,\|v\|_{2,2,\ell}^2 \tag{36}$$

where

$$C_1=\|b\|^2+a^{-1}\|c\|^2+\|d\|+\|e\|^2+1$$

and

$$C_2=9+16\,a^{-1}\,\bar{a}^2$$

(we use here a new sequence of constants $C_1$, $C_2$, etc., depending only on the structure of (1)). Inequalities (35) and (36) are basic for the remainder of the proof.

The integral on the left hand side of (35) is non-negative. Disregarding it we obtain as in the proof of Theorem 1,

$$a\,\|\eta v_x\|_{2,2}^2\leqq C_3(\ell-\ell')^{-2}\,r^2\,|||v|||^2$$

where $C_3=C_1+2C_2$. Similarly disregarding the norm on the left hand side of (35), we also obtain as in the proof of Theorem 1,

$$\|\eta v\|_{2,\infty}^2\leqq(4C_3+1)(\ell-\ell')^{-2}\,r^2\,|||v|||_\ell^2.$$

Now by Lemma 3, again as in the proof of Theorem 1,

$$\begin{aligned} |||(\eta v)^\sigma|||_\ell^{2/\sigma}&\leqq K\{\|\eta v\|_{2,\infty}^2+\|(\eta v)_x\|_{2,2}^2\}\\ &\leqq 2K\{\|\eta v\|_{2,\infty}^2+\|\eta v_x\|_{2,2}^2+4(\ell-\ell')^{-2}\,\|v\|_{2,2,\ell}^2\}\\ &\leqq C_4(\ell-\ell')^{-2}\,r^2\,|||v|||_\ell^2, \end{aligned} \tag{37}$$

the last step following from the estimates above. Here $C_4=2K\{5+(a^{-1}+4)\,C_3\}$ and $\sigma=1+2\theta/n$ (if $n=2$, then $\sigma=1+\theta/2$).

For $m=0, 1, 2, \ldots$ we put

$$\ell=\ell_m=3^{-1}(1+2^{-m}),\qquad \ell'=\ell'_m=3^{-1}(1+2^{-m-1})$$

and

$$\varphi_m=|||v|||_\ell^{2/r}\qquad (v=\bar{u}^r,\ r=\sigma^m).$$

A straightforward calculation based on (37) and the fact that $\eta \equiv 1$ in $Q(\ell')$ shows that

$$\varphi_{m+1} \leqq (36\, C_4)^{1/\sigma^m} (2\sigma)^{2\, m/\sigma^m}\, \varphi_m\, .$$

This can be iterated from $m=0$ to $m=\infty$, yielding

$$\max_{Q(\frac{1}{3})} \bar{u}^2 \leqq C_5\, \varphi_0 = C_5\, |||\bar{u}|||^2_{(2/3)}\, . \tag{38}$$

This much of the proof already establishes that solutions of (1) must be locally bounded functions.

To complete the proof we must estimate $|||\bar{u}|||_{(2/3)}$ in terms of $\|u\|_{2,2}$ and $k$. To this end, put $\beta=1$ in the fundamental inequality (II). Then we get

$$\int \eta^2\, \tilde{u}^2 \big|_{t=\tau_1}^{t=\tau_2} dx + a \iint \eta^2 |\tilde{u}_x|^2\, dx\, dt \leqq 2 \iint (\mathscr{F} + \eta\, |\eta_t|)\, \bar{u}^2\, dx\, dt \tag{39}$$

where $\tilde{u} = \max(0, u) = \bar{u} - k$. Here the $(x, t)$ integrations are to be taken only over the time interval $(\tau_1, \tau_2)$, that is, over the set $R(1) \times (\tau_1, \tau_2)$.

Now choose $\eta$ to be identically one in $Q(2/3)$ and to vanish in a neighborhood of the lower and lateral boundaries of $Q(1)$. We can do this in such a way that $0 \leqq \eta \leqq 1$ and $|\eta_x|$, $|\eta_t| \leqq 8$. Then by Hölder's inequality and Lemma 3

$$\begin{aligned} \|(\mathscr{F} + \eta\, |\eta_t|)\, \bar{u}^2\|_{1,1} &\leqq C_1 |||\eta\, \bar{u}|||^2 + 2^4\, C_2\, \|\bar{u}\|^2_{2,2} \\ &\leqq K\, C_1 (\tau_2 - \tau_1)^\theta \{\|\eta\, \bar{u}\|^2_{2,\infty} + \|(\eta\, \bar{u})_x\|^2_{2,2}\} + 2^4\, C_2\, \|\bar{u}\|^2_{2,2} \\ &\leqq 2K\, C_1 (\tau_2 - \tau_1)^\theta \{\|\eta\, \tilde{u}\|^2_{2,\infty} + \|\eta\, \tilde{u}_x\|^2_{2,2} + k^2 + 2^6\, \|\bar{u}\|^2_{2,2}\} + 2^4\, C_2\, \|\bar{u}\|^2_{2,2}\, . \end{aligned} \tag{40}$$

Here the norms are to be taken over the time interval $(\tau_1, \tau_2)$, and, if $n=2$, then $\theta$ is to be replaced by $\theta/2$.

Now return to inequality (39) and consider $\tau_2 = t$ as a variable over the time interval $I$:

$$\tau_1 \leqq t \leqq \tau_1 + \mu$$

where $\mu = \{\mathrm{Min}(a, 1)/8KC_1\}^{1/\theta}$, it being assumed that (39) is meaningful for the given $\tau_1$. Then setting

$$X(t) = \int \eta^2\, \tilde{u}^2\, dx$$

we obtain from (39) and (40) the following estimate, valid for almost all $t$ in $I$,

$$\begin{aligned} X(t) + \frac{a}{2} \iint \eta^2\, |\tilde{u}_x|^2\, dx\, dt \leqq & \frac{\mathrm{Min}(a, 1)}{2} (\|\eta\, \tilde{u}\|^2_{2,\infty} + k^2 + 2^6\, \|\bar{u}\|^2_{2,2}) + \\ & + 2^5\, C_2\, \|\bar{u}\|^2_{2,2} + X(\tau_1)\, . \end{aligned} \tag{41}$$

Here the integral extends over the set $R(1) \times (\tau_1, \tau_2)$ while the norms are over $R(1) \times I$. Inequality (41) can be solved in exactly the same way as the corresponding inequality (29) in the proof of Theorem 1. We obtain in particular, over the full space $Q(1)$,

$$\|\eta\, \tilde{u}\|^2_{2,\infty} \leqq 2^{1+1/\mu} \{k^2 + 2^6 (1 + C_2)\, \|\bar{u}\|^2_{2,2}\}$$

and

$$\|\eta\, \tilde{u}_x\|^2_{2,2} \leqq 2^{2+1/\mu} (1 + a^{-1})\, \{k^2 + 2^6 (1 + C_2)\, \|\bar{u}\|^2_{2,2}\}\, .$$

The norm $|||\bar{u}|||_{(2/3)}$ in (38) can now be estimated with the help of Lemma 3. Using the fact that $\eta \equiv 1$ in $Q(2/3)$ we have

$$|||\bar{u}|||^2_{(2/3)} \leqq |||\eta \bar{u}|||^2 \leqq 2K\{\|\eta \bar{u}\|^2_{2,\infty} + \|\eta \bar{u}_x\|^2_{2,2} + k^2 + 2^6 \|\bar{u}\|^2_{2,2}\}$$
$$\leqq C_6\{k^2 + \|\bar{u}\|^2_{2,2}\}$$

where $C_6 = 2^{10+1/\mu} K(1+a^{-1})(1+C_2)$. Combining this result with (38) gives finally

$$\max_{Q(\frac{1}{3})} u \leqq C(\|u\|_{2,2} + k)$$

where $C = \sqrt{C_5 C_6}$. Since a similar result can be obtained for $-u$, the proof is completed.

It may be noted that the constant $C$, though of googol-like magnitude [*8*], depends only on the norms of $\|b\|$ through $\|e\|$, the structure constant $\theta$, the coefficients $a$ and $\bar{a}$, and the dimension $n$. Moreover, as in the remark at the end of section 3, we observe that Theorem 2 also holds for a wider class of coefficients than is here assumed.

In the proof of Theorem 2 we have derived a local bound for the $L^{2,2}$ norm of $u_x$. Since this result is of interest we state it formally as

**Theorem 2′.** *Under the hypothesis of Theorem 2*

$$\|u_x\|_{2,2,\rho} \leqq C\rho^{-1}(\|u\|_{2,2,3\rho} + \rho^{\theta + (n+2)/2} k),$$

*where $C$ depends only on $T$ and the structure of* (1).

Theorem 1 gives a global bound for a solution of (1) which is known to be bounded on $\Gamma$, while Theorem 2 gives a local bound in the absence of any information about boundary behavior. The proofs of these results are quite similar, the essential difference being the choice of the cut-off function $\eta$. By the same techniques one can prove semiglobal bounds for solutions which are known to be bounded on some part of the boundary set $\Gamma$. For example, in treating the Cauchy problem for (1) in the strip $E^n \times (0, T)$, only a bound for the solution $u$ on the hyperplane $t=0$ would in general be known. However, if $u$ belongs to $L^{2,2}$ in a cylinder $R(3\rho) \times (0, T)$ then $u$ is bounded on $R(\rho) \times [0, T)$, where the bound is determined by the bound for $u$ on $R(3\rho) \times (t=0)$, the $L^{2,2}$ norm of $u$ in $R(3\rho) \times (0, T)$, and the structure of (1). We leave the proof of this statement to the reader, and only remark that here $\eta$ is chosen to have compact support as a function of $x$ but not as a function of $t$.

By a slight modification of the technique employed here it is possible to prove local boundedness of a weak solution of (1) with a condition on the structure of $\mathscr{A}$ which is less stringent then the corresponding condition in Theorem 2. Specifically, in the last of the structural inequalities (2) the coefficient $\bar{a}$ can be allowed to be an element of $L^{p,q}(Q)$ with exponents $(p,q)$ satisfying (3), while the coefficients $e$ and $h$ may each belong to some space $L^{p,q}(Q)$ with exponents satisfying (4) (instead of (3)). In this case it is no longer true that $\mathscr{A}(x, t, u, u_x)$ is automatically in $L^{2,2}$ locally when $u$ is a weak solution of (1). Hence we add this requirement to the definition of a weak solution. The coefficients $\bar{a}$, $e$ and $h$ enter in the proof of the estimate (38) only via the terms $2G\eta\,|\eta_x|$ and $H|\eta_x|^2$ in $\mathscr{F}$. In deriving the intermediate estimate (36) for $\|\mathscr{F}v^2\|_{1,1}$ we have used the fact that

$$2G\eta|\eta_x|+H|\eta_x|^2 \leqq \left(e^2+\frac{h^2}{k^2}\right)\eta^2+\left(2+\frac{4\bar{a}^2}{a}\right)|\eta_x|^2.$$

Under the new hypothesis, however, one has the alternate estimate

$$\|(2G\eta|\eta_x|+H|\eta_x|^2)v^2\|_{1,1} \leqq \left\{4(\ell-\ell')^{-1}(\|e\|+1)+\frac{16}{a}(\ell-\ell')^{-2}\|\bar{a}\|^2\right\}|||v|||_\ell^2,$$

and the proof of (38) can be carried out exactly as before. Although the new structure suffices for local boundedness, it is not sufficiently strong for the remaining statements of Theorem 2.

## 5. The Harnack Inequality

We shall now derive an analogue of the well-known Harnack inequality for non-negative harmonic functions. Roughly speaking, this inequality for harmonic functions gives a bound for the maximum over an interior subset of the domain of definition in terms of a minimum on the same subset. For parabolic equations the result is similar except that now the subsets must be separated by a non-empty time interval. The reader is referred to the introduction of MOSER's paper for further discussion of this point. The proof which we shall present below is similar in outline to the proof for the special case $u_t=\{A_{ij}u_{x_i}\}_{x_j}$ given by MOSER in reference [12].

Let $(\bar{x},\bar{t})$ be a fixed point in the basic set $Q$. As in section 4 let

$$Q(\rho)=R(\rho)\times(\bar{t}-\rho^2,\bar{t}).$$

We also set

$$Q^*(\rho)=R(\rho)\times(\bar{t}-8\rho^2,\bar{t}-7\rho^2),$$

that is, $Q^*(\rho)$ is the cylinder $Q(\rho)$ translated downwards a distance $7\rho^2$. Then the following theorem of Harnack type holds.

**Theorem 3.** *Let $u$ be a non-negative weak solution of* (1) *in $Q$. Suppose that the set $Q(3\rho)$ is contained in $Q$. Then*

$$\max_{Q^*(\rho)} u \leqq C \min_{Q(\rho)} (u+\rho^\theta k)$$

*where $C$ is a constant depending only on $\rho$ and the structure of* (1), *and* $k=\|f\|+\|g\|+\|h\|$.

The symbols max and min here of course stand for the essential maximum and the essential minimum. In view of Theorem 2 both numbers are finite. Moreover, we note that $C$ is independent of $\|f\|$, $\|g\|$, and $\|h\|$.

**Proof of Theorem 3.** As in the proof of Theorem 2 there is clearly no loss of generality in assuming that $\rho=1/3$ and

$$Q=Q(3\rho)=R(1)\times(0,1).$$

For arbitrary $\varepsilon>0$ we define $\bar{u}=u+k+\varepsilon$. In order to prove the theorem it is enough to show that

$$\max_{Q^*(\frac{1}{3})} \bar{u} \leqq C \min_{Q(\frac{1}{3})} \bar{u}.$$

The proof is based on the fundamental estimate (III).

Suppose to begin with that $\beta \neq -1$. Then if $v=\bar{u}^r$ where $r=(\beta+1)/2$, this inequality takes the form

$$\operatorname{sign}\beta \left\{\frac{1}{\beta+1} \int \eta^2 v^2 |_{t=\tau_1}^{t=\tau_2} dx + \frac{a\beta}{2r^2} \iint \eta^2 |v_x|^2 \, dx\, dt\right\} \tag{42}$$
$$\leqq \|(\mathscr{F}_1 + 2|\beta+1|^{-1} \eta |\eta_t|) v^2\|_{1,1}.$$

Here the $(x,t)$ integration extends over the set $R(1)\times(\tau_1, \tau_2)$, and the quantity $\mathscr{F}_1$ is defined by

$$\mathscr{F}_1 = F_1 \eta^2 + 2 G_1 \eta |\eta_x| + H_1 |\eta_x|^2$$

where

$$F_1 = |\beta| \left(b^2 + \frac{f^2}{k^2}\right) + \left(d + \frac{g}{k}\right) + \frac{c^2}{a|\beta|}, \quad G_1 = e + \frac{h}{k}, \quad H_1 = \frac{4\bar{a}^2}{a|\beta|}.$$

The further discussion of (42) divides naturally into two parts, when $\beta>-1$ and when $\beta<-1$.

I. $\beta>-1$. Let $s$ be a real number, $1/3 \leqq s \leqq 1/2$, and define

$$S(s) = R(s) \times \{\tfrac{1}{6}(1-s), \tfrac{1}{6}(1+s)\}.$$

Notice particularly that $S(\frac{1}{3}) = R(\frac{1}{3}) \times (\frac{1}{9}, \frac{2}{9}) = Q^*(\frac{1}{3})$, while $S(\frac{1}{2}) = R(\frac{1}{2}) \times (\frac{1}{12}, \frac{1}{4})$; see Figure 1.

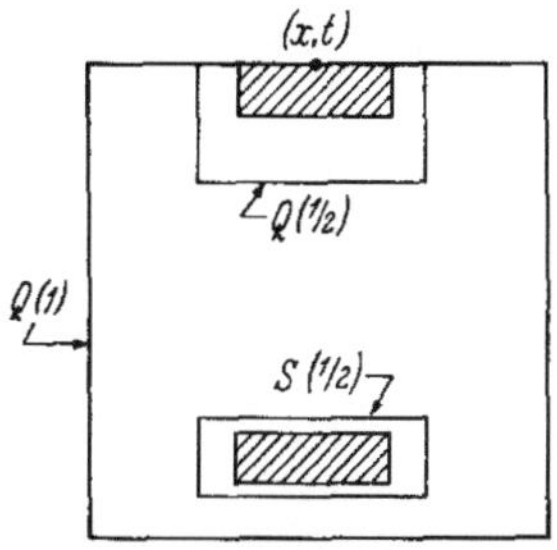


Fig. 1. Diagram illustrating the domain $Q(1)=R(1)\times(0,1)$. The upper and lower shaded regions are, respectively, $Q(\frac{1}{3})$ and $S(\frac{1}{3})=Q^*(\frac{1}{3})$

Let $\ell$ and $\ell'$ be real numbers satisfying $\frac{1}{3}<\ell'<\ell\leqq\frac{1}{2}$. We choose the function $\eta$ so that $\eta\equiv 1$ in $S(\ell')$ and $\eta\equiv 0$ outside $S(\ell)$. This can be done, moreover, so that $0\leqq\eta\leqq 1$ and $|\eta_x|\leqq 2(\ell-\ell')^{-1}$, $|\eta_t|\leqq 6(\ell-\ell')^{-1}$. Then, in an obvious notation,

$$\|(\mathscr{F}_1 + 2|\beta+1|^{-1}\eta|\eta_t|)v^2\|_{1,1} \leqq \Lambda C_1 (\ell-\ell')^{-2} |||v|||_\ell^2 \tag{43}$$

where

$$\Lambda = \operatorname{Max}\left(|\beta|, |\beta|^{-1}, |1+\beta|^{-1}\right),$$
$$C_1 = \|b\|^2 + a^{-1}\|c\|^2 + \|d\| + \|e\|^2 + 16a^{-1}\bar{a}^2 + 12.$$

Suppose $\beta>0$. We set $\tau_1=\frac{1}{12}$ and $\tau_2=\tau$. Then as in the previous proofs we obtain

$$a\,\|\eta\, v_x\|^2_{2,2}\leqq 2C_1(1+\beta^{-2})(\ell-\ell')^{-2}\,r^2\,|||v|||^2_\ell$$

and

$$\|\eta\, v\|^2_{2,\infty}\leqq 4C_1(1+\beta^{-1})(\ell-\ell')^{-2}\,r^2\,|||v|||^2_\ell.$$

On the other hand, when $-1<\beta<0$, we set $\tau_1=\tau$ and $\tau_2=\frac{1}{4}$. Then we find

$$a\,\|\eta\, v_x\|^2_{2,2}\leqq C_1\,\beta^{-2}(\ell-\ell')^{-2}\,|||v|||^2_\ell$$

and

$$\|\eta\, v\|^2_{2,\infty}\leqq C_1\,|\beta|^{-1}(\ell-\ell')^{-2}\,|||v|||^2_\ell.$$

These inequalities can be combined into a single pair, valid for any $\beta>-1$ other than $\beta=0$, namely

$$a\,\|\eta\, v_x\|^2_{2,2}\leqq 2C_1(1+\beta^{-2})(\ell-\ell')^{-2}(1+r^2)\,|||v|||^2_\ell$$

and

$$\|\eta\, v\|^2_{2,\infty}\leqq 4C_1(1+|\beta|^{-1})(\ell-\ell')^{-2}(1+r^2)\,|||v|||^2_\ell.$$

As in the previous estimate (37), these results yield

$$|||(\eta\, v)^\sigma|||^{2/\sigma}\leqq C_2(1+\beta^{-2})(\ell-\ell')^{-2}(1+r^2)\,|||v|||^2_\ell \tag{44}$$

where $C_2=4K\{2+(4+a^{-1})C_1\}$ and $\sigma=1+2\theta/n$ (if $n=2$ then $\sigma=1+\theta/2$).

We wish to write the last inequality in a form suitable for iteration, beginning with a small positive value $r=r_0$ and setting generally

$$r=r_m=r_0\,\sigma^m,\qquad m=0,1,2,\ldots.$$

In order that (44) be applicable at each stage, the successive iterates $r_m$ must avoid the point $r=\frac{1}{2}$ where $\beta=0$. In order to accomplish this in a definite way we shall suppose that $r_0$ has the form

$$r_0=\frac{\sigma^{-\nu}}{1+\sigma} \tag{45}$$

for some integer $\nu\geqq 0$. The point $r=\frac{1}{2}$ then lies midway between $r_\nu$ and $r_{\nu+1}$, and at all stages of the iteration we have

$$|\beta|^{-1}\leqq(\sigma+1)/(\sigma-1).$$

In addition, we also have

$$1+r^2\leqq(1+r_0^2)(r/r_0)^2\leqq 2(r/r_0)^2.$$

Thus for our purposes (44) can be written simply

$$|||(\eta\, v)^\sigma|||^{2/\sigma}\leqq C_3(\ell-\ell')^{-2}(r/r_0)^2\,|||v|||^2_\ell$$

where $C_3=4C_2(1+\sigma^2)\,(\sigma-1)^{-2}$.

Now put

$$\ell=\ell_m=3^{-1}(1+2^{-m-1}),\qquad \ell'=\ell'_m=3^{-1}(1+2^{-m-2})$$

and

$$\varphi_m=|||v|||_\ell^{2r_0/r}\qquad (v=\bar{u}^r,\ r=r_0\,\sigma^m).$$

By the usual iteration procedure, it then follows that

(46) $$\max_{Q^*(\frac{1}{3})} \bar{u}^{r_0} \leqq C_4 \varphi_0^{\frac{1}{2}} = C_4 \,|||\bar{u}^{r_0}|||_{S(\frac{1}{2})}\,.$$

Here $r_0$ is any number of the form (45), and $C_4$ depends only on the structure of (1). Later in the proof we shall fix $r_0$ in a definite way.

It is worth pausing here to notice that (46) provides an estimate for the maximum of $\bar{u}$ over $Q^*(\frac{1}{3})$ in terms of a certain norm over the set $S(\frac{1}{2})$. In the next stage of the proof we shall obtain an estimate for the minimum of $\bar{u}$ over $Q(\frac{1}{3})$ in terms of another norm on the set $Q(\frac{1}{2})$.

II. $\beta < -1$. Let $\ell$ and $\ell'$ be real numbers satisfying $\frac{1}{3} < \ell' < \ell \leqq \frac{1}{2}$, as before. We now choose the test function $\eta$ in (42) in a different manner, so that $\eta \equiv 1$ in $Q(\ell')$ and $\eta \equiv 0$ outside $Q(\ell)$; cf. Fig. 1. This can be done with $0 \leqq \eta \leqq 1$ and $|\eta_x|$, $|\eta_t| \leqq 2(\ell - \ell')^{-1}$. Then

$$\|(\mathscr{F}_1 + 2|\beta+1|^{-1} \eta\, |\eta_t|)\, v^2\|_{1,1} \leqq \Lambda\, C_1 (\ell-\ell')^{-2}\, |||v|||_\ell^2$$

where $\Lambda$ and $C_1$ are as before, and $|||v|||_\ell$ denotes the special norm now taken over the set $Q(\ell)$.

We take $\tau_1 = \frac{1}{2}$ and $\tau_2 = \tau$. Then by the argument already employed in part I of the proof, one finds

$$a\, \|\eta\, v_x\|_{2,2}^2 \leqq 2 C_1 (\ell-\ell')^{-2} (1+r^2)\, |||v|||_\ell^2$$

and

$$\|\eta\, v\|_{2,\infty}^2 \leqq 5 C_1 (\ell-\ell')^{-2} (1+r^2)\, |||v|||_\ell^2 .$$

Consequently

$$|||(\eta\, v)^\sigma|||^{2/\sigma} \leqq C_2 (\ell-\ell')^{-2} (1+r^2)\, |||v|||_\ell^2,$$

valid for all $r<0$.

In order to iterate this result, we take $\ell$ and $\ell'$ as before and

$$\varphi_m = |||v|||_\ell^{-2r_0/r} \qquad (v = \bar{u}^r,\ r = -r_0 \sigma^m)\,.$$

Replacing $r$ by $-r$ and $\bar{u}$ by $1/\bar{u}$, we obtain from the preceding iteration procedure

$$\max_{Q(\frac{1}{3})} (1/\bar{u})^{r_0} \leqq C_4\, |||(1/\bar{u})^{r_0}|||_{Q(\frac{1}{2})},$$

whence

(47) $$|||(1/\bar{u})^{r_0}|||_{Q(\frac{1}{2})}^{-1} \leqq C_4 \min_{Q(\frac{1}{3})} \bar{u}^{r_0}.$$

In view of (46) and (47) it is clear that the remaining task is to establish an estimate of the form

(48) $$|||\bar{u}^{r_0}|||_{S(\frac{1}{2})}\, |||(1/\bar{u})^{r_0}|||_{Q(\frac{1}{2})} \leqq C'.$$

We now turn our attention to this result.

In outline, our procedure will be to reduce the proof of (48) to the demonstration of a lemma. The lemma will then be proved in the following section. We first require some notation.

Let $D$ denote the set $R(\frac{1}{2})\times(0,1)$. By $Q^-(\ell)$, $Q^+(\ell)$ we denote pairs of rectangles which lie in $D$ and are obtained from the fixed pair

$$R(\tfrac{1}{2})\times(0,\tfrac{1}{2}),\qquad R(\tfrac{1}{2})\times(\tfrac{1}{2},1)$$

by the transformations

$$t\to\ell^2 t+\text{constant},\qquad x\to\ell x+\text{constant}. \tag{49}$$

We also introduce the function

$$\psi(w)=\begin{cases}\sqrt{w} & \text{when } w>0\\ 0 & \text{when } w\leqq 0.\end{cases}$$

Then the following result holds.

**Lemma.** *Let $u$ be a non-negative weak solution of* (1) *in the cylinder* $R(1)\times(0,1)$. *Then for any pair of cylinders $Q^+$, $Q^-$ defined as above, we have*

$$\iint_{Q^-}\iint_{Q^+}\psi\left(\log\frac{\bar{u}(y,s)}{\bar{u}(x,t)}\right)dy\,ds\,dx\,dt\leqq B\ell^{2n+4} \tag{50}$$

*where $B$ depends only on the structure of* (1).

Inequality (48) is a consequence of this lemma, via a remarkable theorem due to MOSER [*12*, p. 106]. Let $D^-$ and $D^+$ denote the respective rectangles

$$R(\tfrac{1}{2})\times(0,\tfrac{1}{4})\quad\text{and}\quad R(\tfrac{1}{2})\times(\tfrac{3}{4},1).$$

MOSER's theorem states that *if $\bar{u}$ is a non-negative measurable function satisfying* (50) *for all pairs of cylinders $Q^+$, $Q^-$ in $D$ then*

$$\iint_{D^-}\bar{u}^{\lambda/B^2}dy\,ds\cdot\iint_{D^+}(1/\bar{u})^{\lambda/B^2}dx\,dt\leqq\mu \tag{51}$$

*where $\lambda$ and $\mu$ are positive constants depending only on $n$.* [Actually MOSER's formulation is slightly different; the present version is obtained by means of the transformation $f=(-\log\bar{u})/B^2$ and an inessential change of linear dimension.]

We now choose the constant $r_0$ to be the unique number of the form (45) lying in the interval

$$\left(\frac{\lambda\theta}{2\sigma B^2},\frac{\lambda\theta}{2B^2}\right).$$

Now it is easily checked that if $(p',q')$ is a pair of exponents whose Hölder conjugates satisfy (22), then $p',q'\leqq 1/\theta$. Moreover,

$$S(\tfrac{1}{2})\subset D^-\quad\text{and}\quad Q(\tfrac{1}{2})\subset D^+.$$

Hence

$$|||\bar{u}^{r_0}|||_{S(\frac{1}{2})}\leqq\|\bar{u}^{r_0}\|_{2/\theta,2/\theta,S(\frac{1}{2})}\leqq\Big(\iint_{D^-}\bar{u}^{\lambda/B^2}dy\,ds\Big)^{r_0B^2/\lambda}$$

and

$$|||(1/\bar{u})^{r_0}|||_{Q(\frac{1}{2})}\leqq\Big(\iint_{D^+}(1/\bar{u})^{\lambda/B^2}dx\,dt\Big)^{r_0B^2/\lambda};$$

here we have repeatedly used the fact that increasing the exponent increases the $L^p$ norm when the domain measure does not exceed unity. Applying MOSER's theorem we therefore obtain

$$|||\bar{u}^{r_0}|||_{S(\frac{1}{2})}\,|||(1/\bar{u})^{r_0}|||_{Q(\frac{1}{2})} \leqq \mu^{r_0 B^2/\lambda}.$$

Thus (48) is established, with $C' = \mu^{r_0 B^2/\lambda}$.

The proof of Theorem 3 is now easily completed. Combining (46), (47), and (48) there results

$$\max_{Q^*(\frac{1}{2})} \bar{u} \leqq C_5 \min_{Q(\frac{1}{2})} \bar{u}$$

where

$$C_5 = (C' C_4^2)^{1/r_0} \leqq \{\mu C_4^{(4/\theta + 8/n)}\}^{B^2/\lambda}.$$

It remains only to demonstrate the basic lemma above.

## 6. Completion of the Proof

Here we shall prove the lemma of the preceding section. Any pair of rectangles $Q^-$, $Q^+$ of the type described in the statement of the lemma can be mapped into the pair $R(\frac{1}{2})\times(0, \frac{1}{2})$, $R(\frac{1}{2})\times(\frac{1}{2}, 1)$ by a transformation of the form (49) with $\ell \leqq 1$. It follows from the relations (34) that the coefficients $a$ and $\bar{a}$ are invariant under all such transformations. Moreover, since $\ell \leqq 1$ the norms of the coefficients $b$ through $h$ are not increased. Hence there is no loss of generality in assuming that $\ell = 1$ and that $Q^-$, $Q^+$ is the particular pair $R(\frac{1}{2})\times(0, \frac{1}{2})$, $R(\frac{1}{2})\times(\frac{1}{2}, 1)$.

The proof is based on the fundamental estimate (III) in the case (so far not considered) when $\beta = -1$. If we set $v = -\log \bar{u}$, this inequality takes the form

$$(52)\qquad \int \eta^2 v|_{t=\tau_1}^{t=\tau_2} dx + \frac{a}{2}\iint \eta^2 |v_x|^2 \,dx\,dt \leqq \iint \mathscr{F}_1 \,dx\,dt + 2\iint \eta\,|\eta_t|\,|v|\,dx\,dt$$

where the $(x, t)$ integrations extend over the set $R(1)\times(\tau_1, \tau_2)$.

We choose the cut-off function $\eta$ in the form

$$\eta(x, t) = \zeta(x)\,\omega(t)$$

where $\zeta$ is identically one in $R(\frac{1}{2})$ and vanishes outside $R(\frac{5}{6})$, while $\omega$ is identically one for $t \geqq \tau_1$ and vanishes for $t \leqq \tau_1/2$. We can do this, moreover, in such a way that $0 \leqq \zeta \leqq 1$, $|\zeta_x| \leqq 6$, and $\zeta$ has convex level surfaces (say cubes concentric with $R(\frac{1}{2})$). Then the second integral on the right hand side of (52) vanishes, while also

$$(53)\qquad \|\mathscr{F}_1\|_{1,1} \leqq C_1 + 64(1 + 2a^{-1}\bar{a}^2) = C_6.$$

Now define

$$V(t) = \frac{\int \zeta^2 v(x, t)\,dx}{\int \zeta^2\,dx}, \qquad 0 < t < 1.$$

By Lemma 3 of reference [12] one has*

$$\iint \zeta^2 (v - V)^2 \,dx\,dt \leqq \frac{\omega_n n^{1+n/2}}{2\int \zeta^2\,dx} \iint \zeta^2 |v_x|^2 \,dx\,dt.$$

* The constant in the statement of this lemma in [12] seems to be in error. In particular, in the final equation on page 133 the $z$-integration must be taken over a set larger than the support of $p$. To correct the difficulty, it is enough to replace the integral $\int_{p>0} dx$ by $\omega_n B^n$, where $\omega_n$ is the volume of the unit ball in $E^n$.

Since $\int \zeta^2 dx \geqq |R(\frac{1}{2})| = 2^{-n}$ we obtain from (52)

$$V(\tau_2) - V(\tau_1) + A \iint (v-V)^2 dx\, dt \leqq 2^n \iint \mathscr{F}_1\, dx\, dt,$$

the integral on the left extending over $R(\frac{1}{2}) \times (\tau_1, \tau_2)$ and that on the right over $R(1) \times (\tau_1, \tau_2)$. Here $A = a/\omega_n n^{1+n/2}$.

As a by-product of the derivation of the basic inequality (III) in section 2 it is not hard to see that $V(t)$ can be written as an integral. Consequently $V(t)$ is essentially absolutely continuous, and therefore strongly differentiable. By modifying $V$ on a set of measure zero we can suppose moreover that $V$ is continuous. If we now divide both sides of the last inequality by $\tau_2 - \tau_1$ and let $\tau_2 \to \tau_1 = t$ there results

$$\frac{dV}{dt} + A \int (v-V)^2\, dx \leqq B(t) \tag{53}$$

almost everywhere in (0, 1). Here

$$B(t) = 2^n \int_{R(1)} \mathscr{F}_1\, dx.$$

Now apply Lemma 7 (see the appendix to this section) to the differential inequality (53). Taking $\alpha = 2$ and $\gamma = \frac{1}{2}$ and considering the time interval $[\frac{1}{2}, 1)$, we get

$$\iint_{Q^+} \psi(v-m)\, dx\, dt \leqq \frac{1}{A} + \frac{2 + \sqrt{2^n C_6}}{2^{n+1}}, \qquad (m = V(\tfrac{1}{2})).$$

Similarly, applying the corollary of Lemma 7 in the time interval $(0, \frac{1}{2}]$ yields

$$\iint_{Q^-} \psi(m-v)\, dx\, dt \leqq \frac{1}{A} + \frac{2 + \sqrt{2^n C_6}}{2^{n+1}}, \qquad (m = V(\tfrac{1}{2})).$$

It is easily verified that $\psi(w_1 + w_2) \leqq \psi(w_1) + \psi(w_2)$. Therefore from the two previous inequalities it follows that

$$\begin{aligned}
&\iint_{Q^-} \iint_{Q^+} \psi\{v(x,t) - v(y,s)\}\, dy\, ds\, dx\, dt \\
&\qquad \leqq 2^{-n-1} \Big\{ \iint_{Q^+} \psi(v-m)\, dx\, dt + \iint_{Q^-} \psi(m-v)\, dy\, ds \Big\} \\
&\qquad \leqq 2^{-n} \{ A^{-1} + 2^{-n} + \sqrt{2^{-n} C_6} \} = B,
\end{aligned}$$

completing the proof of the lemma, since $v = -\log \bar{u}$.

## Appendix

This appendix contains the proof of a lemma which was used earlier in this section.

**Lemma 7.** *Let $V = V(t)$ be a strongly differentiable continuous function on $[a, b)$, and let $v = v(x, t)$ be bounded and measurable on the cylinder $Q = \Omega \times (a, b)$. Suppose that*

$$\frac{dV}{dt} + A \int |v - V|^\alpha\, dx \leqq B(t)$$

*where* $A>0$ *and* $\alpha>1$ *are constants and* $B(t)$ *is a non-negative element of* $L^1(a, b)$. *Then for any constant* $\gamma$ *satisfying* $0<\gamma<\alpha-1$ *we have*

$$\iint_{v>c} (v-c)^\gamma \, dx\, dt \leqq \frac{\gamma A^{-1}}{(\alpha-1)(\alpha-1-\gamma)} + \left\{2+\left(\int_a^b B\, dt\right)^\gamma\right\} |Q|$$

*where* $c=V(a)$. (*When* $\gamma>1$ *the right hand side should be multiplied by* $2^{\gamma-1}$.)

In our application of this result the constants $\alpha$ and $\gamma$ have the special values 2 and $\frac{1}{2}$ respectively. The general result, however, is interesting and worthwile. The idea for the proof is due to MOSER [*12*, pp. 122–123].

**Proof.** Define

$$W(t)=V(t)-\int_a^t B(\tau)\, d\tau, \qquad w(x,t)=v(x,t)-\int_a^t B(\tau)\, d\tau$$

whence

(54) $$\frac{dW}{dt}+A\int |w-W|^\alpha \, dx \leqq 0.$$

Clearly $dW/dt \leqq 0$ and $W(a)=c$. Hence $W \leqq c$ everywhere in $(a, b)$.

Let $s$ be a real number larger than $c$. For fixed $t$ we consider the set of points $x$ in $\Omega$ where $w(x, t)>s$, and denote its measure by $m(s, t)$. Using (54) it follows that

$$\frac{dW}{dt}+A(s-W)^\alpha\, m(s,t) \leqq 0.$$

Dividing by $(s-W)^\alpha$ and integrating from $t=a$ to $t=b$ yields

(55) $$A\, m(s) \leqq \frac{1}{(\alpha-1)(s-c)^{\alpha-1}} \qquad (s>c)$$

where

$$m(s)=\int_a^b m(s,t)\, dt.$$

We note particularly that $m(s)$ is the $(n+1)$ – dimensional measure of the set of points in $Q$ where $w>s$.

It is convenient to introduce the function

$$\Psi(u)=\begin{cases} u^\gamma & \text{when } u>0 \\ 0 & \text{when } u\leqq 0, \end{cases}$$

and to let $I$ denote the integral

$$\iint_{v>c} (v-c)^\gamma \, dx\, dt.$$

Then

$$I=\iint \Psi(v-c)\, dx\, dt = \iint \Psi\left(w-c+\int_a^t B\, dt\right) dx\, dt.$$

Now $\Psi(u_1+u_2)\leqq\Psi(u_1)+\Psi(u_2)$, (if $\gamma>1$ a factor of $2^{\gamma-1}$ should be inserted on the right hand side). Hence

$$I\leqq\iint\Psi(w-c)\,dx\,dt+\Psi\left(\int_a^b B\,dt\right)\cdot|Q|$$
$$\leqq\iint_{w>c+1}(w-c)^\gamma\,dx\,dt+\left\{1+\Psi\left(\int_a^b B\,dt\right)\right\}|Q|.$$

Also, since $w$ is bounded,

$$\iint_{w>c+1}(w-c)^\gamma\,dx\,dt=-\int_{c+1}^{\infty}(s-c)^\gamma\,dm(s)$$
$$=m(c+1)+\gamma\int_{c+1}^{\infty}(s-c)^{\gamma-1}\,m(s)\,ds$$
$$\leqq|Q|+\gamma[(\alpha-1)(\alpha-\gamma-1)A]^{-1}$$

using (55). This completes the proof.

**Corollary.** *Let the hypothesis of Lemma 7 be satisfied, except let $V$ be strongly differentiable and continuous on $(a, b]$. Then for any constant $\gamma$ satisfying $0<\gamma<\alpha-1$ we have*

$$\iint_{v<d}(d-v)^\gamma\,dx\,dt\leqq\frac{\gamma A^{-1}}{(\alpha-1)(\alpha-1-\gamma)}+\left\{2+\left(\int_a^b B\,dt\right)^\gamma\right\}|Q|$$

*where $d=V(b)$. (When $\gamma>1$ the right hand side should be multiplied by $2^{\gamma-1}$.)*

**Proof.** This follows from Lemma 7 by making the transformations $V\to-V$, $v\to-v$, and $t\to b+a-t$.

It is not hard to show that both the lemma and the corollary fail to hold for $\gamma=\alpha-1$.

## 7. Consequences of the Harnack inequality

Theorem 3 can be used to obtain estimates for the uniform Hölder continuity of solutions of (1); and once we know that solutions of (1) are necessarily continuous, it is then possible to rephrase Theorem 3 in several more convenient and useful forms.

Let us write $x'=(x,t)$, $y'=(y,s)$, etc. to denote points in space-time. It is then convenient to introduce a pseudo-distance according to the definition

$$|x'|^2=\begin{cases}\max(x_i^2,\ -t/4) & \text{for}\quad t\leqq 0\\ \infty & \text{for}\quad t>0.\end{cases}$$

Thus the set $|y'-x'|<\rho$ for fixed $x'$ is the cylinder $|x_i-y_i|<\rho$, $t-4\rho^2<s\leqq t$. The basic result on the Hölder continuity of weak solutions of (1) can now be stated as follows.

**Theorem 4.** *Suppose that $u$ is a weak solution of (1) in $Q$. Then $u$ is (essentially) Hölder continuous in $Q$. Moreover, if $|u|\leqq L$ and $x'$, $y'$ are points of $Q$ with $s\leqq t$ then*

$$|u(y')-u(x')|\leqq H(L+k)\left(\frac{|y'-x'|}{R}\right)^\alpha$$

*where $H$ and $\alpha$ are positive constants depending only on the structure of* (1), $k=\|f\|+\|g\|+\|h\|$, *and $R$ is the pseudo-distance from $x'$ to the boundary of $Q$, (or $R=1$, if this is smaller).*

**Proof.** Since solutions of (1) are locally bounded functions, it is clearly enough to prove the second part of the theorem. Thus we may suppose that $|u|\leqq L$ in $Q$.

Let $(\bar{x}, \bar{t})$ be a fixed point in $Q$, and let the associated cylinder $Q(\rho)$ be contained in $Q$ for all $\rho\leqq\sigma$. Let $M$ and $m$ be respectively the essential maximum and the essential minimum of $u$ in $Q(\rho)$. The functions $M-u$ and $u-m$ are then non-negative in $Q(\rho)$. Let $w$ denote either $M-u$ or $u-m$. Then since $M$ and $m$ do not exceed $L$ in absolute value, it is easily verified that $w$ is a weak solution of the equation

$$w_t=\operatorname{div}\bar{\mathscr{A}}(x, t, w, w_x)+\bar{\mathscr{B}}(x, t, w, w_x),$$

where

$$p\cdot\bar{\mathscr{A}}\geqq a\,|p|^2-\bar{b}^2 w^2-\bar{f}^2, \qquad \bar{b}=\sqrt{2}\,b, \qquad \bar{f}=\sqrt{2}\,b\,L+f,$$

and so forth. Thus we may apply Theorem 3 to $w$ in the cylinder $Q(\rho)$.

Letting $M_+$ and $m_+$ denote respectively the maximum and minimum of $u$ in $Q(\rho/3)$, and similarly letting $M^*$ and $m^*$ denote the maximum and minimum of $u$ in $Q^*(\rho/3)$, we obtain

$$M-m^*\leqq C(M-M_+ +\rho^\theta\,\bar{k})$$

$$M^*-m\leqq C(m_+ -m+\rho^\theta\,\bar{k}).$$

Here $C$ depends only on $\sigma$ and the structure of (1), while

$$\bar{k}=\|\bar{f}\|+\|\bar{g}\|+\|\bar{h}\|.$$

Adding the two preceding inequalities yields (since $M^*-m^*\geqq 0$)

$$M_+ - m_+\leqq\frac{C-1}{C}\,(M-m)+2\rho^\theta\,\bar{k}. \tag{56}$$

Setting

$$\vartheta=\frac{C-1}{C}, \qquad \tau=\frac{2C\bar{k}}{C-1},$$

and denoting the oscillation $M-m$ of $u$ in $Q(\rho)$ by $\omega(\rho)$, we may write (56) in the form

$$\omega(\rho/3)\leqq\vartheta\{\omega(\rho)+\tau\,\rho^\theta\}.$$

We wish to iterate this last relation to successively smaller values of $\rho$, beginning with $\rho=\sigma$. For this purpose, it is first convenient to define constants $\lambda$ and $\alpha$ by means of

$$\vartheta=3^{-\lambda}, \qquad \alpha=\operatorname{Min}(\lambda, \theta/2).$$

Then

$$\omega(\rho/3)\leqq 3^{-\alpha}\{\omega(\rho)+\tau\,\rho^\theta\},$$

and by iteration

$$\begin{aligned}\omega(3^{-\nu}\sigma)&\leqq 3^{-\alpha\nu}\{\omega(\sigma)+[1+3^{\alpha-\theta}+\cdots+3^{(\nu-1)(\alpha-\theta)}]\,\tau\,\sigma^\theta\}\\ &\leqq 3^{-\alpha\nu}\{\omega(\sigma)+(1-3^{-\theta/2})^{-1}\,\tau\,\sigma^\theta\}, \qquad (\nu=1,2,\ldots).\end{aligned}$$

Since the oscillation of $u$ on smaller and smaller cylinders $Q(\rho)$ thus tends to zero, it follows already that $u$ is (essentially) Hölder continuous. We suppose from here on that $u$ denotes the continuous representative.

Now consider two points $x'$ and $y'$ in $Q$, with $s \leqq t$. We identify $x'$ with $(\bar{x}, \bar{t})$, and observe that the cylinder $Q(\sigma)$, $\sigma = 2R$, is contained in $Q$.

Since $R$ is less than or equal to one, the preceding constants $C$, $\vartheta$, and $\alpha$ depend only on the structure of (1). Moreover,

$$\omega(\sigma) \leqq 2L, \qquad \bar{k} \leqq (\sqrt{2}\,\|b\| + \|d\| + \|e\|)\, L + k.$$

Hence

$$\omega(3^{-\nu}\sigma) \leqq H(L+k)\, 3^{-\alpha(\nu+1)} \qquad (\nu = 0, 1, 2, \ldots) \tag{57}$$

where $H$ and $\alpha$ depend only on the structure of (1), and are independent of $f$, $g$, and $h$. There are now two cases.

(i) $|y' - x'| < R$. Then for some integer $\nu$ we have

$$3^{-(\nu+1)} R \leqq |y' - x'| < 3^{-\nu} R,$$

and $y'$ is contained in the closure of $Q(3^{-\nu}\sigma)$. Consequently from (57)

$$|u(y') - u(x')| \leqq \omega(3^{-\nu}\sigma) \leqq H(L+k) \left(\frac{|y' - x'|}{R}\right)^{\alpha}.$$

(ii) $|y' - x'| \geqq R$. Then since $|u| \leqq L$ and $H > 2$ we have

$$|u(y') - u(x')| \leqq 2L \leqq H(L+k) \left(\frac{|y' - x'|}{R}\right)^{\alpha}.$$

This completes the proof of Theorem 4.

Since every weak solution of (1) is locally continuous it is meaningful to consider the value of a solution at a point. Using Theorem 3 we can derive a pointwise Harnack inequality which, at the same time, makes explicit the dependence of the Harnack constant on the domains.

**Theorem 5.** *Let $u$ be a non-negative weak solution of* (1) *in $Q = \Omega \times (0, T)$. Suppose $\Omega'$ is a convex subdomain of $\Omega$ which has a positive distance $\delta$ from the boundary of $\Omega$. Then for all $x$, $y$ in $\Omega'$ and all $s$, $t$ satisfying $0 < s < t < T$ we have*

$$u(y, s) + k \leqq \{u(x, t) + k\} \exp C \left(\frac{|x-y|^2}{t-s} + \frac{t-s}{R} + 1\right),$$

*where $C$ depends only on the structure of* (1), *$k = \|f\| + \|g\| + \|h\|$, and $R = \mathrm{Min}\,(1, s, \delta^2)$.*

**Proof.** Let $(\xi, \tau)$ be a fixed point in $Q$. If $r$ is sufficiently small, say $r \leqq r_0$, the cylinder

$$|x_i - \xi_i| < \tfrac{3}{2} r, \qquad -r^2 < t - \tau \leqq 8 r^2$$

is contained in $Q$. Then by Theorem 3, assuming $r_0 \leqq 1$,

$$u(\xi, \tau) + k \leqq \gamma \{u(x, \tau + 8 r^2) + k\}$$

for $|x_i-\xi_i|<\frac{1}{2}r$, where $\gamma$ depends only on the structure of (1). This may be rewritten

(58) $$u(\xi,\tau)+k\leqq\gamma\{u(x,t)+k\},$$

valid for all points $(x, t)$ satisfying

(59) $$|x-\xi|^2\leqq\frac{t-\tau}{32},\quad 0\leqq t-\tau\leqq 8r^2,\quad r\leqq r_0.$$

Now connect the points $(y, s)$ and $(x, t)$ by a straight line. Let $N$ be a positive integer and consider the points $(x^\nu, t^\nu)$, where

$$x^\nu=y+\frac{\nu}{N}(x-y),\quad t^\nu=s+\frac{\nu}{N}(t-s),\quad \nu=0,1,\dots,N.$$

Since $\Omega'$ is convex it is clear that each $x^\nu$ is in $\Omega'$. The rectangles

$$|x_i-x_i^\nu|<\tfrac{3}{2}r,\quad -r^2<t-t^\nu\leqq 8r^2,\quad \nu=0,1,\dots,N-1,$$

are then contained in $Q$ when $r\leqq r_0=\sqrt{(t-s)/8N}$, provided that also

(60) $$r_0^2\leqq s,\quad \tfrac{3}{2}\sqrt{n}\,r_0\leqq\delta.$$

According to (58), therefore,

(61) $$u(x^\nu,t^\nu)+k\leqq\gamma\{u(x^{\nu+1},t^{\nu+1})+k\},\quad \nu=0,1,\dots,N-1,$$

provided that the points $(x^{\nu+1}, t^{\nu+1})$ satisfy the condition (59), that is

(62) $$|x^{\nu+1}-x^\nu|^2=\frac{|x-y|^2}{N^2}\leqq\frac{t-s}{32N}.$$

Choose $N$ to be the smallest integer which satisfies both

$$N\geqq 32\frac{|x-y|^2}{t-s}\quad\text{and}\quad N\geqq n\frac{t-s}{R}.$$

Then (60), (62), and $r_0\leqq 1$ are verified, and it follows by iteration of (61) that

$$u(y,s)+k\leqq\gamma^N\{u(x,t)+k\}\leqq\{u(x,r)+k\}\exp C\left(\frac{|x-y|^2}{t-s}+\frac{t-s}{R}+1\right),$$

where $C$ depends only on the structure of (1).

A slight variant of the above argument gives a somewhat neater but less explicit statement of the theorem. If $N$ is chosen to be the smallest integer which satisfies both

$$N\geqq 32\frac{|x-y|^2}{t-s}\quad\text{and}\quad N\geqq\left(\frac{1}{8}+\frac{T}{8}+\frac{9nT}{32\delta^2}\right)\frac{t}{s}$$

then (60), (62), and $r_0\leqq 1$ are again verified. Iterating (61) it follows that

(63) $$u(y,s)+k\leqq\{u(x,t)+k\}\exp C'\left(\frac{|x-y|^2}{t-s}+\frac{t}{s}\right),$$

where now $C'$ depends on the structure of (1) and the geometric quantities $\delta$ and $T$.

Theorem 5 is of course valid when $u$ is a non-negative weak solution of (1) in a strip $S=E^n\times(0,T)$. The value of $\delta$ is then immaterial and the result holds with $R=\mathrm{Min}(1,s)$. There is one drawback here, however, namely the requirement that the norms $\|b\|$ through $\|h\|$ be finite over the entire strip. Thus the result as stated does not apply to a solution in $S$ when, for example, any of the quantities $b$ through $h$ is constant. This defect can be removed by a simple expedient.

Let $\mathscr{S}$ denote the set of all cylinders of the form

$$Q(\sigma), \quad \sigma=\mathrm{Min}(1,\sqrt{T}),$$

contained in $S$. *We say that a function $w=w(x,t)$ defined on $S$ belongs to the class $L^{p,q}(\mathscr{S})$ if*

$$\sup \|w\|_{p,q}<\infty$$

*where the norms are taken over cylinders in the family $\mathscr{S}$.*

Now consider equation (1) in $S$, where the coefficients in the structural inequalities are contained in the respective classes $L^{p,q}(\mathscr{S})$ rather than $L^{p,q}(Q)$ as in section 2. The proof of Theorem 5 then carries through essentially unchanged, yielding the following result applicable in particular when the coefficients in (2) are constant.

**Theorem 5′.** *Let $u$ be a non-negative weak solution of* (1) *in $S$, where the coefficients $b$ through $h$ in* (2) *are contained in the classes $L^{p,q}(\mathscr{S})$. Then for all points $(x,t)$ and $(y,s)$ in $S$ with $0<s<t<T$ we have*

$$u(y,s)+k\leqq\{u(x,t)+k\}\exp C\left(\frac{|x-y|^2}{t-s}+\frac{t}{s}\right).$$

*Here $k=\sup\|f\|+\sup\|g\|+\sup\|h\|$ and $C$ depends only on the structure of* (1) *and on $T$.*

[By a further argument (cf. [*12*], pp. 111–112) it is possible to replace the term $t/s$ in Theorem 5′ by $\log(t/s)+1$. We leave the details to the reader. Even with this improvement the result falls short of what one hopes to get, namely, the assertion of Theorem 5′ with the exponential factor replaced by

$$\left(\frac{t}{s}\right)^{n/2}\exp C\left(\frac{|x-y|^2}{t-s}+1\right).$$

For a non-negative solution of the equation of heat conduction in $S$, RESCH [*16*] has established the sharp result

$$u(y,s)\leqq u(x,t)\left(\frac{t}{s}\right)^{n/2}\exp\frac{|x-y|^2}{4(t-s)}$$

valid for all points $(x,t)$ and $(y,s)$ in $S$ with $0<s<t<T$].

According to the preceding result, a non-negative weak solution $u=u(x,t)$ of (1) cannot grow more rapidly than $e^{\mathrm{const}\,|x|^2}$ as $|x|\to\infty$ for any fixed $t$ in $(0,T)$. On the other hand, the results thus far do not give very precise information about the behavior of $u$ as $t\to 0$. In order to derive such information we have found it necessary to impose certain additional restrictions on the solution. Among various conditions which present themselves here, we consider one which can be shown to be satisfied by the Green's functions and fundamental solutions of linear equations of the form (5).

**Theorem 6.** *Let u be a non-negative weak solution of* (1) *in Q. If*

$$M = \sup_{0<t<T} \int_\Omega u(x,t)\,dx < \infty,$$

*then for x in any compact subset $\Omega'$ of $\Omega$ and t sufficiently small we have*

$$u(x,t) \leqq C M t^{-n/2} + C k t^{\theta/2}$$

*where $k = \|f\| + \|g\| + \|h\|$ and C depends only on the structure of* (1).

**Proof.** Let $(\xi, \tau)$ be a fixed point in $\Omega' \times (0, t)$. If $\tau$ is sufficiently small, the cylinder

$$|x_i - \xi_i| < \tfrac{3}{2}\sqrt{\tau}, \qquad 0<t<9\tau$$

is contained in $Q$. Then by Theorem 3, assuming $\tau \leqq 1$

$$u(\xi,\tau) \leqq C \min_{Q^+}(u + \tau^{\theta/2} k)$$

where $C$ depends only on the structure of (1), and $Q^+$ denotes the set

$$|x_i - \xi_i| < \tfrac{1}{2}\sqrt{\tau}, \qquad 8\tau < t < 9\tau.$$

The assertion of the theorem now follows from the fact that

$$\min_{Q^+} u \leqq \frac{1}{|Q^+|} \iint_{Q^+} u\,dx\,dt \leqq M \tau^{-n/2}.$$

A similar result, which need not be stated explicitly here, clearly holds for solutions defined in the strip $S$ when the coefficients in (2) are contained in the classes $L^{p,q}(\mathscr{S})$. This latter result generalizes a theorem of Nash concerning the growth of the fundamental solution of the linear equation (7) as $t \to 0$. Indeed, if $g(x, t; y, 0)$ denotes the fundamental solution of (7) with singularity at the point $(y, 0)$ then

$$\int_{E^n} g(x,t;y,0)\,dx = 1$$

for all $t > 0$.

According to Theorem 6 a non-negative solution of (1) belonging to $L^{1,\infty}(Q)$ cannot tend to infinity more rapidly than $t^{-n/2}$ as $t \to 0$. It is natural to seek conditions which assure that the solution will actually tend to infinity at precisely that rate. In the next theorem we give a sufficient condition for this, a condition which can be shown to be satisfied by the Green's function and fundamental solution of the linear equation (5).

**Theorem 7.** *Let u be a non-negative solution of* (1) *in Q. Suppose that $x=0$ is contained in $\Omega$, and that for some $\alpha > 0$*

$$\liminf \int_{|x|^2 < \alpha t} u(x,t)\,dx > 0 \quad as \quad t \to 0.$$

*Let $\Omega'$ be a convex subdomain of $\Omega$ which has a positive distance $\delta$ from the boundary of $\Omega$ and which contains $x=0$ in its interior. Then there exists a constant $\mathscr{C} > 0$ such* *that for all $t$ sufficiently small and all $x$ in $\Omega'$*

$$u(x,t)+k \geqq \mathscr{C}\, t^{-n/2} \exp(-2C\,|x|^2/t)$$

*where $C$ depends only on $\delta$, $T$ and the structure of* (1).

We note that after a suitable translation of axes this result would apply equally well to any point $x=x_0$ contained in $\Omega$.

**Proof of Theorem 7.** Let $y$ satisfy $|y|^2<\alpha t$. Then for $t$ sufficiently small we have, by (63),

$$u(y,t/2) \leqq \{u(0,t)+k\} \exp 2C(\alpha+1). \tag{64}$$

Clearly for some $y_0$ in $|y|^2<\alpha t$ we have

$$u(y_0,t/2)=\frac{1}{\omega_n(\alpha t)^{n/2}} \int\limits_{|y|^2<\alpha t} u(y,t/2)\,dy.$$

In view of our hypotheses, putting $y=y_0$ in (64) therefore yields

$$u(0,t)+k \geqq \mathscr{C}'\, t^{-n/2}$$

if $t$ is sufficiently small, say $t \leqq t_0$. On the other hand, a second application of (63) gives

$$u(0,t/2)+k \leqq \{u(x,t)+k\} \exp 2C\left(\frac{|x|^2}{t}+1\right),$$

valid for all $x$ in $\Omega'$. Combining the last two inequalities we obtain

$$u(x,t)+k \geqq \mathscr{C}\, t^{-n/2} \exp(-2C\,|x|^2/t)$$

for all $x$ in $\Omega'$ and $t \leqq 2t_0$, where $C$ depends only on $\delta$, $T$ and the structure of (1).

Suppose next that $u$ is a non-negative solution of (1) in the strip $S$, where the coefficients in (2) are contained in the classes $L^{p,q}(\mathscr{S})$. If $u$ satisfies the conditions of Theorem 7 it is clear that the steps of the preceding proof apply equally well, using Theorem 5′ in place of (63). The conclusion of Theorem 7 is therefore valid in this case also, and even without the requirement that $t$ be small (since the previous restriction on $t$ comes from the fact that $|y|^2<\alpha t$ must be contained in $\Omega'$). By keeping track of the constants in the proof it is easily seen that the following result holds.

**Theorem 7′.** *Let $u$ be a non-negative weak solution of* (1) *in the strip $S$, where the coefficients in* (2) *are contained in the classes $L^{p,q}(\mathscr{S})$. Suppose that for some $\alpha>0$ we have*

$$\mathscr{M}=\inf_{0<t<T} \int\limits_{|x|^2<\alpha t} u(x,t)\,dx>0.$$

*Then there exist positive constants $C_1$ and $C_2$ such that*

$$u(x,t)+k \geqq C_1\, t^{-n/2} \exp(-C_2\,|x|^2/t)$$

*in $S$. Here $C_1$ depends only on $\alpha$, $\mathscr{M}$, $n$, $T$ and the structure of* (1), *while $C_2$ depends only on $T$ and the structure of* (1).

## 8. The Harnack Principle for Bounded Solutions

LADYŽENSKAJA and URAL'CEVA have shown that *bounded weak solutions* of parabolic divergence structure equations are Hölder continuous under conditions weaker than those so far assumed here. Moreover, under similar assumptions TRUDINGER *[20]* has recently proved a Harnack principle for solutions of (1). In this section we shall show that these results can also be obtained using the methods of the present paper, and in fact can be somewhat improved insofar as the class of weak solutions in question is concerned. In addition, our exponent conditions are slightly better than those required earlier.

In particular, we consider equation (1) with the following structure inequalities,

$$\begin{aligned} p\cdot\mathscr{A} &\geqq a\,|p|^2-f^2 \\ |\mathscr{A}| &\leqq \bar{a}\,|p|+h \qquad (65)\\ |\mathscr{B}| &\leqq \bar{a}\,|p|^2+g\,. \end{aligned}$$

Here the coefficients $a$ and $\bar{a}$ are assumed to be positive functions of $u$, while $f$, $g$, and $h$ depend additionally on $x$ and $t$. For each real number $L>0$ it is supposed that $a^{-1}$ and $\bar{a}$ are bounded whenever $|u|\leqq L$, and similarly that $f$, $g$, $h$ are in the Lebesgue classes of section 2 whenever $u=u(x,t)$ is a measurable function such that $|u(x,t)|\leqq L$. The corresponding bounds and norms will be denoted by

$$1/a_L, \quad \bar{a}_L \quad \text{and} \quad \|f\|_L,\ \|g\|_L,\ \|h\|_L.$$

We note further that the structure constant $\theta$ associated with the Lebesgue classes of $f$, $g$, $h$ (see section 2) may also depend on $L$, though this dependence need not be explicitly indicated.

With the notation of section 5 the following version of the Harnack principle then holds.

**Theorem 8.** *Let $u$ be a non-negative weak solution of* (1) *in $Q$, and let $\mathscr{A}$ and $\mathscr{B}$ have the structure indicated above. Suppose that $Q$ contains the set $Q(3\rho)$ and that $|u|\leqq L$ there. Then*

$$\max_{Q^*(\rho)} u \leqq C_L \min_{Q(\rho)} (u+\rho^\theta k)$$

*where $k=\|f\|_L+\|g\|_L+\|h\|_L$. The coefficient $C_L$ depends only on $\rho$, $L$, and the structure constants $n$, $\theta$, $a_L$, $\bar{a}_L$, and $\|f\|_L$.*

**Proof.** We follow closely the proof of Theorem 3, the main difference in the argument being an alternate version of the fundamental estimate (III).

To obtain this alternate estimate it is convenient to proceed as in the final paragraphs of section 2. Here we take $\bar{u}=u+k+\varepsilon$ and

$$\mathscr{G}(u)=\bar{u}^\beta e^{(\operatorname{sign}\beta)\sigma u}, \qquad \beta\neq 0, \quad \sigma>0,$$

(the use of an exponential test function appears in the work of LADYŽENSKAYA and URAL'CEVA and also in *[20]*). Then setting

$$\varphi=\eta^2\,\mathscr{G}(u)\,\chi(t,\tau_1,\tau_2)$$

in the weak form of (1) we find in place of (13′) the relation

$$\operatorname{sign}\beta\,(\varphi_x\cdot\mathscr{A}-\varphi\,\mathscr{B})\geqq\eta^2\,|\mathscr{G}'|\,\{a\,|\bar{u}_x|^2-f^2\}-$$
$$-2\eta\,|\eta_x|\,\mathscr{G}\{\bar{a}\,|\bar{u}_x|+h\}-\eta^2\,\mathscr{G}\{\bar{a}\,|\bar{u}_x|^2+g\}.$$

If now $\sigma=\bar{a}_L/a_L$, the right hand side of this inequality is bounded below by

$$\{\tfrac{1}{2}a\,|\beta|\,\bar{u}^{\beta-1}\,|\eta\,\bar{u}_x|^2-\mathscr{F}_2\,\bar{u}^{\beta+1}\}\,E$$

where

$$E=e^{(\operatorname{sign}\beta)\,\sigma u},\qquad \mathscr{F}_2=F_2\,\eta^2+2\eta\,|\eta_x|\,G_2+|\eta_x|^2\,H_2,$$

$$F_2=|\beta|\frac{f^2}{k^2}+(\sigma f)\,\frac{f}{k}+\frac{g}{k},\qquad G_2=\frac{h}{k},\qquad H_2=\frac{2\bar{a}^2}{a\,|\beta|}.$$

Thus if $\mathscr{H}(u)$ is a primitive of $\mathscr{G}(u)$ we obtain as in section 2,

$$\text{(III}'\text{)}\qquad \operatorname{sign}\beta\left\{\int_\Omega \eta^2\,\mathscr{H}\big|_{t=\tau_1}^{t=\tau_2}\,dx+\frac{a\,\beta}{2}\iint\eta^2\,\bar{u}^{\beta-1}\,|\bar{u}_x|^2\,E\,dx\,dt\right\}$$
$$\leqq\iint\mathscr{F}_2\,\bar{u}^{\beta+1}\,E\,dx\,dt+2\iint\eta\,|\eta_t|\,|\mathscr{H}|\,dx\,dt,$$

the integrations extending over $\Omega\times(\tau_1,\tau_2)$. For our purposes we shall fix $\mathscr{H}$ by means of the relation

$$\mathscr{H}(u)=\int_{u_0}^{u}\mathscr{G}(u)\,du,\quad\text{where}\quad u_0=\begin{cases}0 & \text{if } \beta\geqq-1\\ L & \text{if } \beta<-1.\end{cases}$$

Suppose now that $\beta\neq 0,\ -1$. Then it is easy to check that

$$\frac{\bar{u}^{\beta+1}-(k+\varepsilon)^{\beta+1}}{\beta+1}\leqq\ \mathscr{H}(u)\leqq e^{\sigma L}\,\frac{\bar{u}^{\beta+1}}{\beta+1}\qquad\text{for } \beta>0$$

$$e^{-\sigma L}\,\frac{\bar{u}^{\beta+1}-(k+\varepsilon)^{\beta+1}}{\beta+1}\leqq\ \mathscr{H}(u)\leqq\frac{\bar{u}^{\beta+1}}{\beta+1}\qquad\text{for } -1<\beta<0$$

$$e^{-\sigma L}\,\frac{\bar{u}^{\beta+1}-(k+L+\varepsilon)^{\beta+1}}{|\beta+1|}\leqq-\mathscr{H}(u)\leqq\frac{\bar{u}^{\beta+1}}{|\beta+1|}\qquad\text{for } \beta<-1.$$

Defining $v=\bar{u}^r$ where $r=(\beta+1)/2$, the proofs of parts I and II in section 5 carry over intact to the present situation except for some obvious algebraic changes. These changes affect the functional dependence of the various constants (as indicated in our statement of Theorem 8) but leave the form of relations (46) and (47) unvaried.

This being the case, it therefore remains only to prove an inequality of the form (48), for the conclusion of Theorem 8 will then follow exactly as in the proof of Theorem 3. To obtain the required inequality it is enough (by a simple change of variables) to consider the case $\ell=1$. We take $\beta=-1$ in (III′). Setting $v=-\mathscr{H}(u)$ this inequality reduces to the form

$$\int_\Omega\eta^2\,v\big|_{t=\tau_1}^{t=\tau_2}\,dx+\frac{a}{2}\iint\eta^2\,|v_x|^2\,dx\,dt\leqq\iint\mathscr{F}_2\,dx\,dt+2\iint\eta\,|\eta_t|\cdot|v|\,dx\,dt,$$

which is exactly the same as (52). Consequently as in section 6 we find

$$\iint_{Q^-}\iint_{Q^+} \Psi\{\mathscr{H}[u(y,s)]-\mathscr{H}[u(x,t)]\}\,dy\,ds\,dx\,dt \leqq \text{Constant}.$$

In view of the definition of $\mathscr{H}(u)$ and the fact that $\beta=-1$ one easily checks that

$$\Psi\{\mathscr{H}[u(y,s)]-\mathscr{H}[u(x,t)]\} \geqq e^{-\sigma L/2}\,\Psi\left\{\log\frac{\bar{u}(y,s)}{\bar{u}(x,t)}\right\}.$$

This proves the desired inequality, and consequently completes the demonstration of Theorem 8.

In view of Theorem 8, we can carry over the conclusions of Theorem 4 and 5 to bounded weak solutions of (1) under the structure conditions (65). For completeness we give a formal statement of the Hölder continuity theorem.

**Theorem 9.** *Suppose that $u$ is a locally bounded weak solution of* (1) *in $Q$, and let $\mathscr{A}$ and $\mathscr{B}$ have the structure indicated above. Then $u$ is (essentially) Hölder continuous in $Q$.*

As pointed out earlier, this result was derived first (under somewhat stronger assumptions) by LADYŽENSKAJA and URAL'CEVA.

## 9. General Weak Solutions

In the introduction, a weak solution of (1) in a domain $Q=\Omega\times(0,T)$ was defined to be a function $u(x,t)$ which is locally of class $L^{2,\infty}(Q)$ and has strong derivatives $u_x$ which are locally in $L^{2,2}(Q)$, and is such that

$$u_t=\operatorname{div}\mathscr{A}+\mathscr{B}$$

in the sense of distribution theory. It was asserted that all of the results of the paper hold for weak solutions of this type. However, in section 2 we derived the basic inequalities (I), (II) and (III) with the additional assumption that $u$ has a strong derivative $u_t$ belonging locally to class $L^{2,2}$ in $Q$. It is the purpose of this section to show that the basic inequalities remain valid without this extra assumption. Since these inequalities were the basis for our further theorems, it will thus be shown that these results hold equally for weak solutions which do not possess a strong derivative with respect to $t$.*

Let $K(x,t)$ be a non-negative smooth function defined on $E^n\times E^1$, with support in $|x|^2+t^2\leqq h^2$ and such that

$$\iint K(x,t)\,dx\,dt=1,\qquad K(x,t)\leqq \text{const}/h^{n+1}.$$

Also let $Q_h$ denote the set of points in the basic cylinder $Q$ which are further than $h$ from the boundary of $Q$. If $w=w(x,t)$ is locally of class $L^{1,1}$ in $Q$ we define its regularization $w_h$ by means of the formula

$$w_h=w_h(x,t)=\iint K(x-y,t-s)\,w(y,s)\,dy\,ds,\qquad (x,t)\in Q_h.$$

Evidently $w_h$ is of class $C^1$ in $Q_h$.

* In section 8 we require another inequality of the same type as (I), (II) and (III). That this latter inequality also holds without the assumption of a strong $t$ derivative is easily checked using the methods of this section.

Since $u$ satisfies (1) in the sense of distribution theory, it follows that in $Q_h$

$$u_{ht} = \operatorname{div} \mathscr{A}_h + \mathscr{B}_h$$

(take $\varphi = K(y-x, s-t)$ in (6)). Thus if $\Phi$ is any continuously differentiable function with compact support in $Q_h$ we have

$$\iint \{\Phi u_{ht} + \Phi_x \cdot \mathscr{A}_h\} \, dx \, dt = \iint \Phi \mathscr{B}_h \, dx \, dt. \tag{66}$$

The relation takes the place of the earlier identity (9), which of course does not hold under the present weak hypothesis on the solution.

We can now proceed as in section 2. Define

$$\Phi(x, t) = \eta^2 \mathscr{G}(u_h) \chi(t, \tau)$$

where $h$ is small enough so that $Q_h$ contains the (closed) support of $\eta^2 \chi$. Then inserting $\Phi$ into (66) yields the identity

$$\int_\Omega \eta^2 \mathscr{H}(u_h)|_{t=\tau} \, dt + \iint_Q \{\Phi_x \cdot \mathscr{A}_h - \Phi \mathscr{B}_h\} \, dx \, dt = \int_0^\tau \int_\Omega 2\eta \eta_t \mathscr{H}(u_h) \, dx \, dt. \tag{67}$$

We wish to let $h$ tend to zero in this identity. For this the following lemmas are required.

**Lemma a).** *Let $w$ be locally in $L^{p,q}(Q)$ where $1 \leq p < \infty$, $1 \leq q < \infty$. Then $\| w_h - w \|_{p,q} \to 0$ as $h \to 0$, the norm being taken over any compact subset of $Q$.*

**Proof.** By a double application of the integral form of Minkowski's inequality we find

$$\| w_h - w \|_{p,q} \leq \max_{|y|^2 + s^2 \leq h^2} \{\int (\int |w(x+y, t+s) - w(x, t)|^p \, dx)^{q/p} \, dt\}^{1/q}.$$

The required conclusion then follows from the translation continuity of the norm $\| \cdot \|_{p,q}$.

**Lemma b).** *Let $w$ be locally in $L^{p,\infty}(Q)$, where $1 \leq p < \infty$. Then for any compact subset $\Omega'$ of $\Omega$ we have*

$$\int_{\Omega'} |w_h(x, t) - w(x, t)|^p \, dx \to 0 \quad as \quad h \to 0$$

*for almost all $t$. Moreover $\| w_h \|_{p,\infty}$ is uniformly bounded on any compact cylinder in $Q$ as $h$ tends to zero.*

**Proof.** The second part of the lemma is a simple consequence of Minkowski's inequality. The first part of the lemma requires Minkowski's inequality together with the usual Lebesgue set construction and the fact that there exists a denumerable dense family of continuous functions in $L^p(\Omega')$. We omit the details.

When the conclusions of Lemma b) are fulfilled we say that $w_h$ tends to $w$ locally in $L^{p,\infty}(Q)$, or simply that $w_h$ tends to $w$ locally in $L^{p,\infty}$.

From Lemma a) it follows directly that $\mathscr{A}_h$ tends to $\mathscr{A}$ and $u_{hx}$ tends to $u_x$ locally in $L^{2,2}$. From Lemma b) we have that $u_h$ tends to $u$ locally in $L^{2,\infty}$.

Now the function $\mathscr{G}(u)$ is uniformly Lipschitz continuous, hence $|\mathscr{G}(u_h) - \mathscr{G}(u)| \leq \text{Const.} \, |u_h - u|$. Consequently $\mathscr{G}(u_h)$ tends to $\mathscr{G}(u)$ locally in $L^{2,\infty}$. Put

$$\varphi(x, t) = \eta^2 \mathscr{G}(u) \chi(t, \tau).$$

Then $\Phi(x, t)$ evidently tends to $\varphi(x, t)$ in $L^{2,\infty}(Q)$. Furthermore, since

$$\Phi_x = \begin{cases} \eta^2 \mathscr{G}'(u_h) u_{hx} + 2\eta \eta_x \mathscr{G}(u_h) & 0<t<\tau \\ 0 & \tau \leqq t, \end{cases}$$

it is not difficult to see that $\Phi_x$ tends to $\varphi_x$ in $L^{2,2}(Q)$. [Indeed, after a straightforward calculation we find

$$|\Phi_x - \varphi_x|^2 \leqq \text{Const}\, \eta^2(\{\mathscr{G}'(u_h) - \mathscr{G}'(u)\}^2 |u_x|^2 + |u_{hx} - u_x|^2 + |u_h - u|^2).$$

Integrating over $\Omega \times (0, \tau)$, and using the dominated convergence theorem together with the fact that $u_h \to u$ almost everywhere, the required conclusion is easily obtained.]

Next, since $\mathscr{H}'(u) \leqq \text{Const.}\, |u|$, it follows that $\mathscr{H}(u_h)$ tends to $\mathscr{H}(u)$ locally in $L^{1,\infty}$.

In order to calculate the limit in (67) as $h$ tends to zero, it now remains only to consider the term $\Phi \mathscr{B}_h$. A particular difficulty here is that the function $\mathscr{B}$ is not in any single class $L^{p,q}$. However, we can write $\mathscr{B}$ in the form

$$\mathscr{B} = \frac{c|u_x|}{B}\mathscr{B} + \frac{d|u|}{B}\mathscr{B} + \frac{g}{h}\mathscr{B}, \qquad \text{where } B = c|u_x| + d|u| + g.$$

Thus, in view of (2), $\mathscr{B}$ is a sum of three terms each of which is locally in some class $L^{(2p')', (2q')'}$ where $p$ and $q$ satisfy

$$\frac{n}{2p} + \frac{1}{q} < 1. \tag{68}$$

On the other hand, by Lemma 3 and what has already been proved it is clear that $\Phi$ tends to $\varphi$ in any space $L^{2p', 2q'}(Q)$ where $p$ and $q$ satisfy (68).* Consequently we find that $\Phi \mathscr{B}_h$ tends to $\varphi \mathscr{B}$ in $L^{1,1}(Q)$, as desired.

The results of the four preceding paragraphs allow us to send $h$ to zero in (67), with the result that

$$\int_\Omega \eta^2 \mathscr{H}(u)|_{t=\tau}\, dx + \iint_Q \{\varphi_x \cdot \mathscr{A} - \varphi \mathscr{B}\}\, dx = \int_0^\tau \int_\Omega 2\eta \eta_t \mathscr{H}(u)\, dx\, dt, \tag{69}$$

valid for almost all $\tau$ in $(0, T)$.

Since (69) is precisely the same as (12) and since the existence of $u_t$ is irrelevant in the derivation of (I) from (12) the calculations of section 2 apply from here on. This proves that (I) holds for the general weak solutions of this section. The estimates (II) and (III) obviously may be established by the same process. We have therefore proved that all results of the paper, the maximum principle, local boundedness of solutions, the Harnack principle, Hölder continuity of solutions, and so forth, hold without the assumption that the weak solution $u$ has a time derivative.

* A formal proof goes as follows. By Lemma 3, or more precisely, a slight modification of Lemma 3, we have

$$\|\Phi - \varphi\|^2_{2p', 2q'} \leqq \text{Const}\{\|\Phi - \varphi\|^2_{2,\rho} + \|\Phi_x - \varphi_x\|^2_{2,2}\},$$

where $\rho$ is some finite number. Both terms on the right hand side tend to zero (the first by the bounded convergence theorem and the fact that $\Phi$ tends to $\varphi$ in $L^{2,\infty}(Q)$), and the assertion is proved.

*Note.* This research was partially supported by the United States Air Force Office of Scientific Research under Contract No. AF-AFOSR 883-65. The second author was also partly supported by the National Science Foundation under a grant to the Department of Mechanics, Johns Hopkins University.

## References

[*1*] ARONSON, D. G., On the Green's function for second order parabolic differential equations with discontinuous coefficients. Bulletin of the American Mathematical Society **69**, 841-847 (1963).

[*2*] ARONSON, D. G., Uniqueness of positive weak solutions of second order parabolic equations. Annales Polonici Mathematici **16**, 285-303 (1965).

[*3*] ARONSON, D. G., & JAMES SERRIN, A maximum principle for nonlinear parabolic equations. Annali Scuola Normale Superiore di Pisa (to appear).

[*4*] GUGLIELMINO, F., Sulla regolarizzazione delle soluzioni deboli dei problemi al contorno per operatori parabolici. Ricerche di Matematica **12**, 44-66, 140-150 (1963).

[*5*] HADAMARD, J., Extension à l'équation de la chaleur d' un théorème de A. Harnack. Rendiconti del Circolo Matematico di Palermo, series 2, **3**, 337-346 (1954).

[*6*] IVANOV, A. V., A priori estimates for solutions of linear equations of second order of elliptic and parabolic type. Doklady Akademii Nauk S.S.S.R. **161**, 1270-1273 (1965).

[*7*] IVANOV, A. V., O. A. LADYŽENSKAJA, A. L. TRESKUNOV, & N. N. URAL'CEVA, Certain properties of generalized solutions of parabolic equations of second order. Doklady Akademii Nauk S.S.S.R. **168**, 17-20 (1966).

[*8*] KASNER, E., & J. NEWMAN, Mathematics and the Imagination. New York: Simon and Schuster 1940.

[*9*] KRUŽKOV, S. N., A priori estimates and certain properties of solutions of elliptic and parabolic equations. Matematičeskii Sbornik **65**, 522-570 (1964).

[*10*] LADYŽENSKAJA, O. A., & N. N. URAL'CEVA, A boundary value problem for linear and quasilinear parabolic equations I, II, III. Izvestija Akademii Nauk S.S.S.R., Serija Matematičeskaja **26**, 5-52 (1962); **26**, 753-780 (1962); **27**, 161-240 (1963).

[*11*] MOSER, J., On Harnack's theorem for elliptic differential equations. Communications on Pure and Applied Mathematics **14**, 577-591 (1961).

[*12*] MOSER, J., A Harnack inequality for parabolic differential equations. Communications on Pure and Applied Mathematics **17**, 101-134 (1964); Correction, ibid. **20**, 231-236 (1967).

[*13*] NASH, J., Continuity of solutions of parabolic and elliptic equations. American Journal of Mathematics **80**, 931-954 (1958).

[*14*] OLEINIK, O. A., & S. N. KRUŽKOV, Quasilinear second order parabolic equations with many independent variables. Russian Mathematical Surveys **16**, 106-146 (1961).

[*15*] PINI, B., Sulla soluzione generalizate di Wiener per il primo problema di valori al contorno nel caso parabolico. Rendiconti del Seminario Matematico dell' Università di Padova **23**, 422-434 (1954).

[*16*] RESCH, D., Temperature bounds on an infinite rod. Proceedings of the American Mathematical Society **3**, 632-634 (1952).

[*17*] SERRIN, J., Local behavior of solutions of quasi-linear equations. Acta Mathematica **111**, 247-302 (1964).

[*18*] SERRIN, J., Isolated singularities of solutions of quasi-linear equations. Acta Mathematica **113**, 219-240 (1965).

[*19*] SERRIN, J., Introduction to Differentiation Theory. Lecture notes, University of Minnesota, School of Mathematics 1965.

[*20*] TRUDINGER, N. S., Quasilinear elliptic partial differential equations in *n* variables. Stanford University, Department of Mathematics 1966.

University of Minnesota
Minneapolis, Minnesota

*(Received November 15, 1966)*

# II.4 Wave Operators – Part I

Howard A. Levine

**Mathematics Subject Classification (2010).** 34G20, 35L70, 35B35.

**Keywords.** Nonlinear second-order hyperbolic equations, qualitative properties of solutions.

## 1. Introduction

It is a pleasure and an honor to contribute some remarks to this volume. James Serrin has been a mathematical idol of mine as well as a good friend for many years.

There can be no doubt about the importance and influence of Professor Serrin's work on the profession. An informal comparison of his citation record with other influential individuals in partial differential equations reveals him to be one of the leaders of the field in the last sixty years. He is also listed in the ISI Web of Science as among most highly cited mathematicians in the world.

The two papers of Professor Serrin and his coauthors [8, 12] or [i, ii] below, we discuss here are concerned with the long time asymptotic behavior of solutions of damped wave equations when they exist on a half-line or when local solutions fail to exist on a half-line (global non-existence or, less accurately, although more colorfully, "blow-up in finite time" of solutions). According to Mathscinet, the eleven items on Professor Serrin's cv, [5]–[8], [10]–[16], that fall into these two topics constitute 10% of the total number of his citations although they represent 6% of his total number of publications.

The papers [8, 12] are concerned with generalizations to partial differential equations of properties that have long known for solutions of the damped and forced harmonic oscillator equation: $y''(t) + \gamma y'(t) + \alpha^2 y(t) + f(t, y) = 0$ where $y(t)$ is the displacement of the oscillator from rest at time $t$, $\gamma, \alpha$ are positive constants and $f(t, y)$ is some forcing function on the system. The scientific literature is rich with examples of equations in this form. The classical example is the displacement of a mass suspended from a spring where the system's motion is resisted by a dashpot. Using the nomenclature from physics, the term $\gamma y'(t)$ is called the damping term, the term $\alpha^2 y(t)$ is the restoring force (Hooke's Law) while the source term, $f(t, y)$, denotes the sum of all remaining external forces that act on the system. The critical property here is that when there are no external forces $(f(t, y) \equiv 0)$, all solutions of this ordinary differential equation converge asymptotically and exponentially rapidly to the solution $y_0(t) \equiv 0$. The exponential rate of convergence does not depend on the initial displacement or on the initial velocity $(y(0), y'(0))$.

P. Pucci et al. (eds.), *James Serrin. Selected Papers: Volume 2*, Contemporary Mathematicians, DOI 10.1007/978-3-0348-0847-7_4, © Springer Basel 2014

If the forcing term depends nonlinearly on the displacement, the long time behavior can change radically. To see this, consider the choices $f(t, y) = y^3$ and $f(t, y) = -y^3$. In the former case, all solutions exist on $J = [0, \infty)$ and decay asymptotically to the null solution. In the latter case, there are two more stationary solutions, namely $y_{\pm} = \pm\alpha$. In this case it is known that if $y'(0)$ is sufficiently small and $|y(0)| < \alpha/\sqrt{3}$, the corresponding solutions converges to the null solution. If $|y(0)| > \alpha/\sqrt{3}$, the solution can only exist on a finite interval $[0, T)$ and $t = T$ is a vertical asymptote for $y(\cdot)$, i.e., $\lim_{t\uparrow T^-} |y(t)| = +\infty$. The physical meaning in the case of a mass suspended from a spring is that the spring will either exceed its elastic limit or perhaps break in finite time if the displacement is sufficiently large.

In [12] the issue of asymptotic stability is taken up for the system of partial differential equations

$$u_{tt} + Q(t, x, u, u_t) - \Delta u + f(x, u) = 0. \tag{1}$$

Here $u, Q, f$ take values in $R^N$, $x \in \Omega$, a bounded open subset of $R^n$ and $t \in J$. For this to be a dissipative system of wave equations, one requires that $(Q(\cdot, \cdot, \cdot, \eta), \eta) \geq 0$ for $\eta \in R^N$ where $(,)$ denotes the scalar product there.The forcing term $f(\cdot, \cdot)$ is to be a restoring force, i.e., $(f(x, u), u) > 0$ when $u \neq 0$ and is derivable from a scalar function (a potential) $F$, i.e., $f(x, u) = \nabla_u F(x, u)$. The authors follow the time evolution of appropriately defined weak solutions by means of the associated energy

$$E(t) \equiv \frac{1}{2}\int_\Omega |u_t(t, x)|^2\, dx + \frac{1}{2}\int_\Omega |\nabla_x u(t, x)|^2\, dx + \int_\Omega F(x, u(t, x))\, dx. \tag{2}$$

If $F(x, 0) \equiv 0$, the potential function is given by the line integral

$$F(x, u) = \int_0^1 ((f(x, tu), u)dt.$$

Then $E(t)$ does not change sign and vanishes when and only when $u = u_t \equiv 0$. It can therefore be used to define a norm on an appropriate function space. This functional is used to define asymptotic stability of the null solution in the mean and establish, under rather general conditions on $f$ and $Q$, sufficient conditions for which asymptotic stability in the mean holds. The asymptotic stability is global in the sense that it holds for all (appropriately defined weak) solutions that exist locally in the norm induced by the energy functional. The variety of operators that can be used in place of the Laplacian $(-\Delta)$ is quite impressive. They include the degenerate the polyharmonic operator $(-\Delta)^l$ and nonlinear operators $-\sum_{ij=1}^n \partial_i(A_{ij}(x, \nabla u))$ even when the matrix $A_{ij}$ is degenerate as in the case of the $l$ Laplacian, $-\nabla \cdot (|\nabla u|^{l-1}\nabla u)$. As a consequence of this work, in [2], the authors applied the ideas of [12] to an abstract evolutionary equation of the form $(P(u_t))_t + A(u) + \rho(t)A(u_t) + Q(t, u_t) + F(u) = 0$ and provided further interesting examples. See also [17].

The case for which one can have blow-up in finite time (global nonexistence in some norm) for equations such as (1) also has a long history, most of which will not be cited here. However, we mention two papers, [3, 4]. The first of these treated abstract damped nonlinear wave equations of the form

$$Pu_{tt} + Qu_t + Au + F(u) = 0 \tag{3}$$

where $P, Q, A$ are positive linear operators defined on some dense subset of a Hilbert Space and $F$ is a nonlinear function also defined on that dense domain and is derivable from a potential functional.[1, 2] A simple example of such an equation is a that of a damped wave equation for the motion of a vibrating string with a potentially nonuniform density and nonconstant damping term forced by a nonlinear function of the displacement that can change sign. For example, $\rho(x)u_{tt} + q(x)u_t - u_{xx} - |u|^{p-1}u = 0$ for $p > 1$ on an interval $(0, 1)$ with homogeneous Dirichlet boundary conditions. Here $\rho$ is continuous and positive and $q \geq 0$ on $[0, 1]$. The corresponding energy functional is then

$$E(t) = \frac{1}{2}\int_0^1 \rho(x)u_t^2\,dx + \frac{1}{2}\int_0^1 u_x^2\,dx - \frac{1}{p+1}\int_0^1 |u|^{p+1}\,dx. \tag{4}$$

Unlike the functional in (2), the energy can no longer be used to define a norm. Indeed, if the initial data $u(0, x), u_t(0, x)$ is such that $E(0) < 0$, the solution cannot exist globally on $J \times (0, 1)$. In particular, as for the ordinary differential equation case, global asymptotic stability of the null solution is not possible. However, for a sufficiently small ball about $(0, 0)$, one can establish local asymptotic stability using potential well arguments. Thus linear damping cannot prevent nonlinearities from causing finite time blow-up. One is then led to ask: Can nonlinear damping inhibit blow up? A simple example from ordinary differential equations gives us a hint to the answer.

Suppose that $p, m > 1$. Consider the initial value problem for $|u_t|^{m-1}u_t - |u|^{p-1}u = 0$ on $J$ with $u(0) > 0$. With this initial condition, $u_t(0) > 0$ also. The exact solution is given by

$$u(t) \equiv u^0(t) = \begin{cases} \{(1-p/m)t + u(0)^{1-p/m}\}^{1/(1-p/m)}, & \text{if } p < m; \\ u(0)\exp(t), & \text{if } p = m; \\ \{u(0)^{1-p/m} - (p/m-1)t\}^{-1/(p/m-1)}, & \text{if } p > m \end{cases}$$

for as long as it exists. If $p \leq m$, the solution exists on $J$, decaying to zero if $p < m$. If $p > m$ the solution blows up in finite time. Using the method of matched asymptotic expansions ( [18], pp. 177–204) these same properties can be shown to be true for the singularly perturbed equation $\epsilon\, u_{tt}^\epsilon + |u_t^\epsilon|^{m-1}u_t^\epsilon - |u^\epsilon|^{p-1}u^\epsilon = 0$ where $\epsilon$ is small and positive. (This follows from the O'Malley/Hoppenstead inner-outer solution construction, [18], since on any compact interval where $u^0$ exists, the coefficient $|u_t^0|^{m-1}$ is bounded away from zero.)

This example provides strong motivation for the following result of [3]. It is stated here for the case of nonlinear damping in one space dimension, namely

$$u_{tt} + |u_t|^{m-1}u_t - u_{xx} - |u|^{p-1}u = 0 \tag{5}$$

on a bounded interval (say $[0, 1]$) with Dirichlet boundary conditions. It was shown that if $p \leq m$, all solutions of the initial value problem exist on $J$ while if $p > m$, solutions with sufficiently large initial data cannot be global. Here "sufficiently large" means that $E(0)$ is sufficiently negative where $E(t)$ is given by (4) with $\rho \equiv 1$. (As stated it also holds in two space dimensions. When $n \geq 3$, it is also required that $1 < p < n/(n-2)$.)

[1] Abstract equations of the type (3) are not just mathematical toys. These equations, whether linear or not in any of its terms, are known as equations of Sobolev type. We refer the reader to [1] for a comprehensive list of physical examples of such (nonlinear) equations and a detailed study/review of finite time blow up properties for them.

[2] Weak formulations of solutions of abstract solutions of equations such as (3) involve an additional energy inequality hypothesis. The primary reason for this is that one cannot assume that the weak time derivative, $u_t$, can be used an admissible test function in the distribution equation that replaces (3). Such side hypotheses are not unusual in the literature. Similar extra, physically motivated, conditions on weak solutions of first-order nonlinear hyperbolic equations include the entropy condition or solutions that are defined as vanishing viscosity limits of semilinear parabolic equations.

The papers [5, 7, 8, 11] provide a number of interesting developments and extensions of this result. For example, in [8] abstract equations of the form:

$$(P(u_t))_t + Q(t, u, u_t) + A(u) + F(u) = 0$$

where $P, Q, A$ can also be nonlinear in $u$ or $u_t$ are treated. There it is shown under certain very general structure conditions on these nonlinearities that when the initial energy is negative, (not just sufficiently negative) corresponding solutions cannot be global. This result includes the equation studied in [3]. In [11], Pucci and Serrin showed that, at least for equations of the form $Pu_{tt} + Q(t)u_t + A(t, u) + F(t, u) = 0$, some local (in time) solutions with small, positive initial energies will fail to be continuable to global solutions on the half-line. Motivated by this result, it was later shown in [9, 19] for equations of the form $u_{tt} + |u_t|^{m-1}u_t - \Delta u - |u|^{p-1}u = 0$ on the time-space-cylinder $J \times R^n$ that when $m < p < n/(n-2)$ there are choices of compact support initial data with arbitrarily large positive initial energies for which finite time blow up occurs. The finite speed of propagation principle permits the same result when $R^n$ is replaced by a bounded domain and the solution vanishes on the boundary.

Professor Serrin's contributions to these papers are insightful, concise and illuminating. They form a valuable part of our understanding of the effect of nonlinearities on the behavior of solutions of continuous dynamical systems in partial differential equations.

## References

[1] Al'shin, A,B., Korpusov, M.O., and Sveshnikov, A.G., Blow-up in Nonlinear Sobolev Type Equations, De Gruyter, Berlin/New York, 2011.

[2] Boccuto, A. and Vitillaro, E., Asymptotic stability for abstract evolution equations and applications to partial differential systems, *Rend. Circ. Mat. Palermo*, **47(2)**, (1998), 25–48.

[3] Georgiev, V. and Todorova, G., Existence of a solution of the wave equation with nonlinear damping and source terms, *JDE*, **109 (2)**, (1994), 295–308.

[4] Levine, H. A., Some Additional Remarks on the Nonexistence of Global Solutions to Nonlinear Wave Equations, *SIMA*, **4**, (1974), 138–146.

[5] Levine, H.A., Park, S.R., Serrin, J., Global existence and global nonexistence of solutions of the Cauchy problem for a nonlinearly damped wave equation, *J. Math. Anal. Appl.*, **228 (1)**, (1998), 181–205.

[6] Levine, H.A., Park, S.R., Serrin, J., Global existence and nonexistence theorems for quasilinear evolution equations of formally parabolic type, *JDE*, **150** (1998), 212–229.

[7] Levine, H.A., Serrin, J., Global nonexistence theorems for quasilinear evolution equations with dissipation. *Arch. Rational Mech. Anal.*, **137 (4)**, (1997), 341–361.

[8] Levine, H.A., Pucci, P., and Serrin, J., Some remarks on global nonexistence for nonautonomous abstract evolution equations, *Contemporary Mathematics*, **208**, Amer. Math. Soc., Providence, RI, (1997) 253–263.

[9] Levine, H.A. and Todorova, G., Blow up of solutions of the Cauchy problem for a wave equation with nonlinear damping and source terms and positive initial energy, *PAMS*, **129 (3)**, (2001), 793–805.

[10] Pucci, P., and Serrin, J., Local asymptotic stability for dissipative wave systems, *Israel J. Math.*, **104**, (1998), 29–50.

[11] Pucci, P., Serrin, J., Global nonexistence for abstract evolution equations with positive initial energy, *JDE*, **150 (1)**, (1998), 203–214.

[12] Pucci, P., and Serrin, J., Asymptotic stability for nonautonomous dissipative wave equations, *Comm. Pure. Appl. Math.*, **46**, (1996), 177–216.

[13] Pucci, P., and Serrin, J., Asymptotic stability for intermittently controlled nonlinear oscillators, *SIAM J. Math. Anal.*, **25 (3)**, (1994) 815–835.

[14] Pucci, P., and Serrin, J., Precise damping conditions for global asymptotic stability for nonlinear second order systems. II, *JDE*, **113 (2)**, (1994), 505–534.

[15] Pucci, P., and Serrin, J., Precise damping conditions for global asymptotic stability for nonlinear second order systems, *Acta Math.*, **170 (2)**, (1993), 275–307.

[16] Serrin, J., Todorova, G., and Vitillaro, E., Existence for a nonlinear wave equation with damping and source terms, *Diff. Int. Eqns.*, **16**, (2003), 13–50.

[17] Salvatori, M.C. and Vitillaro, E., Decay for the solutions of nonlinear abstract damped evolution equations with applications to partial and ordinary differential systems, *Diff. Int. Eqns.*, **11 (2)**, (1998), 223–262.

[18] Smith, D.R., Singular-Perturbation Theory, Cambridge Univ. Press, Cambridge, 1985.

[19] Todorova, G., and Vitillaro, E., Blow-up for nonlinear dissipative wave equations in $R^n$, *JMAA*, **303 (1)**, (2005), 242–257.

Howard A. Levine
Department of Mathematics
Iowa State University
Ames, Iowa, 50011, USA
e-mail: halevine@iastate.edu

*Proc. Am. Math. Soc.* 5 (1954), 307–308. **Reprint II.4.[i]**


# A NOTE ON THE WAVE EQUATION

J. B. SERRIN

It is well known that if $u$ is a regular harmonic function and if $E_1$ and $E_2$ are confocal solid ellipsoids contained in the region where $u$ is defined, then the mean value of $u$ over $E_1$ is equal to the mean value of $u$ over $E_2$.[1] In this note we shall give a generalization of this result for solutions $u(x, t) = u(x_1, \cdots, x_n, t)$ of the wave equation

$$\sum_{i=1}^{n} u_{x_i x_i} = u_{tt}.$$

Let $P$ and $Q$ be two points in $(xt)$ space, and let $\Gamma_P$ and $\Gamma_Q$ be the corresponding characteristic cones. We suppose that

$$k(P, Q) \equiv (t_P - t_Q)^2 - \sum (x_P - x_Q)^2 > 0$$

where $(x_P, t_P)$ and $(x_Q, t_Q)$ are the coordinates of $P$ and $Q$ respectively. It follows that the intersecting nappes of $\Gamma_P$ and $\Gamma_Q$ bound a finite region $R$. We shall say that an ellipsoid $E$ is inscribed in $R$ if $E$ is contained in $R$ and if for any two points $P'$, $Q'$ such that $k(P', Q') < k(P, Q)$, the corresponding region $R'$ does not contain $E$.

Now note that there exists a Lorentz transformation[2] $\mathcal{L}$ making the line $PQ$ a time axis, and finally let $L$ denote the plane of intersection of $\Gamma_P$ and $\Gamma_Q$. Then we have the following theorem.

*If $E_1$ and $E_2$ are solid ellipsoids inscribed in $R$, then the mean value of $u(x, t)$ over $E_1$ is equal to the mean value of $u(x, t)$ over $E_2$, provided that the cross sections of $E_1$ and $E_2$ by the plane $L$ are transformed into confocal ellipsoids by $\mathcal{L}$.*[3]

Since the Lorentz transformation $\mathcal{L}$ clearly leaves the "inscribed" relation invariant, it follows that the theorem may be reduced to the special case when $PQ$ is a time axis, and $\mathcal{L}$ is the identity. In this case one readily verifies that $PQ$ is also a principal axis of $E_1$ and $E_2$.

By taking a suitable coordinate system we may therefore assume in our proof that the principal axes of $E_1$ and $E_2$ coincide with the coordinate axes, and take $P$ and $Q$ to be $(0, \cdots, 0, \tau)$ and

Received by the editors May 25, 1953 and, in revised form, July 17, 1953.

[1] L. Asgeirsson, Math. Ann. vol. 113 (1936) pp. 321–346, or Courant-Hilbert, *Methoden der mathematischen Physik*, vol. II, pp. 417–424.

[2] I.e. a linear transformation of $(xt)$ leaving the form $t^2 - \sum x_i^2$, and hence the wave equation invariant.

[3] In other words the cross sections of $E_1$ and $E_2$ by $L$ are confocal ellipsoids in the metric $d(P, Q) = (-k(P, Q))^{1/2}$.

$(0, \cdots, 0, -\tau)$ respectively. Then $L$ becomes the plane $t=0$, and $E_1$ and $E_2$ may be represented by

$$\sum_{i=1}^{n} \frac{x_i^2}{a_i^2} + \frac{t^2}{b^2-1} \leqq r^2 \quad \text{and} \quad \sum_{i=1}^{n} \frac{x_i^2}{a_i^2-1} + \frac{t^2}{b^2} \leqq r^2,$$

where

$$a_1 \geqq a_2 \geqq \cdots \geqq a_n > 1, \; b > 1, \quad \text{and} \quad r = \tau(a_1^2 + b^2 - 1)^{-1/2}.$$

The last relations guarantee that both ellipsoids are inscribed in $R$.

Now consider the function

$$v(\xi, \eta) = v(\xi_1, \cdots, \xi_{n+1}, \eta_1, \cdots, \eta_{n+1}) = u(x_1, \cdots, x_n, t),$$

where

$$x_i = a_i\xi_i + (a_i^2 - 1)^{1/2}\eta_i, \qquad i = 1, \cdots, n,$$
$$t = b\eta_{n+1} + (b^2 - 1)^{1/2}\xi_{n+1}.$$

Then

$$\sum_{i=1}^{n+1} v_{\xi_i\xi_i} = \sum_{i=1}^{n+1} v_{\eta_i\eta_i}.$$

From the Asgeirsson mean value theorem,[1] we obtain

$$\int\cdots\int_{\sum_{i=1}^{n+1}\xi_i^2\leqq r^2} v(\xi, 0)d\xi_1 \cdots d\xi_{n+1}$$
$$= \int\cdots\int_{\sum_{i=1}^{n+1}\eta_i^2\leqq r^2} v(0, \eta)d\eta_1 \cdots d\eta_{n+1},$$

or, expressed in terms of the $(xt)$ variables,

$$\frac{\int\cdots\int_{E_1} u(x, t)dx_1 \cdots dx_n dt}{\int\cdots\int_{E_1} dx_1 \cdots dx_n dt} = \frac{\int\cdots\int_{E_2} u(x, t)dx_1 \cdots dx_n dt}{\int\cdots\int_{E_2} dx_1 \cdots dx_n dt} \quad \text{Q.E.D.}$$

One may of course find analogous results for equations

$$\sum_{i=1}^{n} u_{x_ix_i} = \sum_{i=1}^{m} u_{y_iy_i},$$

but the interpretation is no longer so apparent.

MASSACHUSETTS INSTITUTE OF TECHNOLOGY

*Commun. Pure Appl. Math.* 49 (1996), 177–216.


**Reprint II.4.[ii]**

# Asymptotic Stability for Nonautonomous Dissipative Wave Systems

PATRIZIA PUCCI
*University of Perugia, Italy*

AND

JAMES SERRIN
*University of Minnesota*

**Abstract**

We study the problem of asymptotic stability, as time tends to infinity, of solutions of dissipative wave systems, governed by time-dependent nonlinear damping forces and subject to the action of strongly nonlinear potential energies. In the final part of the paper we also consider the case of polyharmonic wave operators and higher-order dissipation terms. 

## 1. Introduction

The problem of asymptotic stability for second-order ordinary differential equations is well-known in the literature, even in the nonautonomous case. Recently, several authors have extended part of this work to abstract evolutionary equations, as well as to the case of damped wave equations [6], [8], [9], [11]. Here we present an approach that systematically treats this class of problems. We shall concentrate our attention particularly on nonlinear wave systems with nonautonomous and unbounded dissipation.

Consider specifically the nonlinear damped wave system with Dirichlet data

$$(1.1)\qquad \begin{gathered} u_{tt} - \Delta u + Q(t,x,u,u_t) + f(x,u) = 0, \qquad (t,x) \in J \times \Omega, \\ u(t,x) = 0, \qquad (t,x) \in J \times \partial\Omega, \end{gathered}$$

where $J = [0,\infty)$ and $\Omega$ is a bounded open subset of $\mathbb{R}^n$. The values of $u$ are taken in $\mathbb{R}^N$, $N \geqq 1$, and

$$Q \in C(J \times \Omega \times \mathbb{R}^N \times \mathbb{R}^N \to \mathbb{R}^N), \qquad f \in C(\Omega \times \mathbb{R}^N \to \mathbb{R}^N).$$

The function $Q$ represents a *nonlinear damping*, so that

$$(1.2)\qquad \big(Q(t,x,u,v), v\big) \geqq 0 \quad \text{for all arguments } t,x,u,v,$$

where $(\cdot,\cdot)$ is the inner product in $\mathbb{R}^N$; moreover, $f$ is a *restoring force* derivable from a potential $F$. That is,

$$(1.3)\qquad \big(f(x,u), u\big) \geqq 0$$

and

$$f(x,u) = \frac{\partial F}{\partial u}(x,u), \tag{1.4}$$

where $F \in C^1(\Omega \times \mathbb{R}^N \to \mathbb{R})$ and $F(x,0) = 0$. The last condition is a normalization that can be assumed without loss of generality.

Problem (1.1) is well-known in a variety of models in mathematical physics, for example, elastic vibrations in a dissipative medium, the telegraphic equation, and the damped Klein-Gordon equation. The setting for all these models, it should be noted, is the scalar case $N = 1$, where the existence of a potential $F$ is redundant because one can simply integrate the force $f(x,u)$ with respect to $u$.

A canonical example of the type of functions $f$ and $Q$ that we contemplate here is given by the system

$$u_{tt} - \Delta u + A(t)|u_t|^{m-2}u_t + V(x)|u|^{p-2}u = 0, \tag{1.5}$$

where

$$1 < p < \infty, \qquad 2 \leqq m \leqq \max\{p,\ 2n/(n-2)\}$$

($2 \leqq m < \infty$ if $n = 1, 2$) and where $V(x)$ is a continuous function on $\overline{\Omega}$ and $A(t)$ an element of $\mathbb{R}^{N \times N}$, continuous on $J$. In order that (1.2) and (1.3) should hold, we assume

$$(A(t)v, v) \geqq 0, \qquad V(x) > 0, \quad \text{for all arguments } t,\ x,\ v.$$

The corresponding potential $F$ in (1.4) is clearly given by $F(x,u) = V(x)\,|u|^p/p$. Even for systems such as this, and even in the scalar case $N = 1$, our results improve upon earlier work. See, for example, [11], where it was required that $N = 1$, $A(t) = h(t) \geqq h_0 = \text{const} > 0$ and

$$2 \leqq m, \qquad p \leqq 2\frac{n-1}{n-2},$$

or [9], where asymptotic stability was obtained for a larger range of exponents than in [11]; that is,

$$2 \leqq m, \qquad p \leqq \frac{2n}{n-2},$$

but restricted to autonomous equations only (see Section 5).

In the context of problem (1.1) the question of asymptotic stability is best considered by means of the natural energy associated with solutions of the system, namely,

$$Eu(t) = \frac{1}{2}\int_\Omega \{|u_t(t,x)|^2 + |Du(t,x)|^2\}dx + \int_\Omega F(x, u(t,x))dx.$$

In particular, the rest field $u(t,x) \equiv 0$ will be called *asymptotically stable in the mean*, or simply *asymptotically stable*, if and only if

$$\lim_{t\to\infty} Eu(t) = 0 \quad \text{for all solutions } u = u(t,x) \text{ of (1.1).}$$

In this formulation we have tacitly assumed that the solutions in question are classical, but for an adequate and useful theory one must actually consider solutions in a wider class of functions. Indeed, one of the goals of the paper is to formulate a rational definition of solution for (1.1) that is independent of detailed properties of the functions $f$ and $Q$, and at the same time provides a useful framework for the study of asymptotic stability. This we do in Section 2.

Sections 3 and 4 are devoted to our two main asymptotic stability results, Theorems 3.1 and 4.1. Both of these are based on the a priori existence of a suitable auxiliary function $k = k(t)$. Section 5 presents specific applications in which the function $k$ is chosen in explicit form. Further applications to more general problems than (1.1) are given in Sections 6 and 7. In Section 6 we treat systems in which the Laplace operator $-\Delta$ is replaced by the polyharmonic operator $(-\Delta)^L$, $L \geqq 1$, by the degenerate Laplace operator $-\Delta_s$, $s > 1$, and by the general quasilinear operator $-\operatorname{div} A(x, Du)$. The last class contains in particular the mean curvature case $A(x, Du) = Du/\sqrt{1 + |Du|^2}$. Finally, the systems discussed in Section 7 specifically involve higher-order dissipation terms of the type introduced by [1], [2], [4], [5], [18].

The primary consideration of the paper is the asymptotic stability of solutions of (1.1). Accordingly, we shall not be concerned with the problem of existence of solutions. This problem has a large literature (see, e.g., [6], [7], [10], and the references cited therein), and has also been treated recently by Zhu [19] under hypotheses closely related to, but somewhat stronger than, (H1)–(H3) below. The reader is referred to this literature for further details.

**The Linear Case**

An important special case of (1.5) occurs when $p = m = 2$ and $A(t) = a(t)t^\alpha$, $\alpha \in \mathbb{R}$, so that (1.5) takes the form

$$u_{tt} - \Delta u + a(t)t^\alpha u_t + V(x)u = 0 \tag{1.6}$$

(if $\alpha < 0$ we restrict the interval $J$ to the set $t \geqq 1$). For simplicity we also suppose $N = 1$, and assume that

$$1/C \leqq a(t) \leqq C \quad \text{for } t \in J,$$

where $C$ is a positive constant.

The behavior of solutions as $t \to \infty$ depends crucially on the parameter $\alpha$. We show in Section 5 that *if* $|\alpha| \leqq 1$, *then the rest field is asymptotically stable*. On the other hand, when $\alpha < -1$ there exist oscillatory solutions that *do not* approach zero when $t \to \infty$, whereas if $\alpha > 1$ there exist solutions that approach *nonzero* functions $\psi = \psi(x)$ as $t \to \infty$.

A similar situation occurs when there are higher-order dissipation terms present in the equation, for example, in the case

$$u_{tt} - \Delta\big(u + b(t)t^{\beta}u_t\big) + a(t)t^{\alpha}u_t + V(x)u = 0, \tag{1.7}$$

where $a$, $V$, and $\alpha$ are as before, and

$$1/D \leqq b(t) \leqq D \quad \text{on } J,$$

with $D > 0$ constant. In Section 7 we prove the following result.

*The rest field is asymptotically stable for equation* (1.7) *provided that*

$$\alpha \leqq 1, \quad |\beta| \leqq 1 \qquad or \qquad \beta \leqq 1, \quad |\alpha| \leqq 1. \tag{1.8}$$

*Moreover, when $\alpha < 1$ and $\beta < 1$ there exist oscillatory solutions that do not approach zero as $t \to \infty$; on the other hand, if either $\alpha > 1$ or $\beta > 1$ there are again solutions that approach nonzero functions $\psi = \psi(x)$ as $t \to \infty$.*

## 2. Definition of Solutions

It is axiomatic that the notion of a classical solution of (1.1) is inadequate not only for showing the existence of solutions, but also for maintaining a reasonable generality of treatment. On the other hand, in research primarily devoted to the existence of solutions of (1.1) it has been necessary to assume special conditions on the functions $f$ and $Q$ in order for the existence process to be carried out successfully. This leaves open whether one can frame a definition of solution that is suitably general for existence purposes, independent of detailed properties of $f$ and $Q$, and adequate for the investigation of asymptotic stability.

To provide such a definition it is convenient first to introduce the elementary bracket pairing in $\Omega \subset \mathbb{R}^n$:

$$\langle \varphi, \psi\rangle = \langle \varphi, \psi\rangle(t) = \int_\Omega (\varphi, \psi)dx \quad \text{where } \varphi, \psi : J \times \Omega \to \mathbb{R}^N.$$

This is a well-defined real function of time for all $t \in J$ such that $(\varphi, \psi) \in L^1(\Omega)$. We write for simplicity

$$L^p = \left[L^p(\Omega)\right]^N, \qquad X = \left[W_0^{1,2}(\Omega)\right]^N,$$

where $p > 1$; these spaces are endowed, respectively, with the natural norms

$$\|\varphi\|_{L^p} = \left(\int_\Omega |\varphi|^p\, dx\right)^{1/p},$$

$$\|\varphi\|_X = \|D\varphi\|^2 = \int_\Omega |D\varphi|^2 dx = \int_\Omega \sum_{i=1}^n |D_i\varphi|^2\, dx.$$

When $p = 2$ it is convenient to use simply $\|\varphi\|$ instead of $\|\varphi\|_{L^2}$. We also put

$$\langle D\varphi, D\psi\rangle = \int_\Omega \sum_{i=1}^n (D_i\varphi, D_i\psi) dx \quad \text{for all } \varphi, \psi : J \to X,$$

so that $\langle D\varphi, D\psi\rangle : J \to \mathbb{R}$. Of course, $\langle D\varphi, D\varphi\rangle = \|D\varphi\|^2$.

Now define $K' = C(J \to X) \cap C^1(J \to L^2)$ and

$$K = \{\phi \in K' : E\phi \text{ is locally bounded on } J\},$$

where $E\phi$ is the *total energy of the field* $\phi$; that is,

$$E\phi = E\phi(t) = \frac{1}{2}\|\phi'\|^2 + \frac{1}{2}\|D\phi\|^2 + \mathscr{F}\phi,$$

and $\mathscr{F}\phi$, the *potential energy of the field*, is given by

$$\mathscr{F}\phi = \mathscr{F}\phi(t) = \int_\Omega F(x, \phi(t,x))dx.$$

The quantity $\phi'$ in $E\phi$ can be written more explicitly as $\phi'(t)$, where $\phi'(t) = (d\phi/dt)(t) \in L^2$ for each $t \in J$. Similarly, again because $\phi \in K$, we have $D\phi = D\phi(t) \in L^2$ for each $t \in J$. Clearly, $\|\phi'\|$ and $\|D\phi\|$ are then continuous functions of $t$.

In writing $E\phi$ and $\mathscr{F}\phi$ we make the tacit agreement that $\mathscr{F}\phi$ is *well-defined*, namely, that $F(\cdot, \phi(t,\cdot)) \in L^1(\Omega)$ for all $t \in J$. An equivalent definition for $K$ is clearly given by

$$K = \{\phi \in K' : \mathscr{F}\phi \text{ is locally bounded on } J\}.$$

Our motivation for introducing the set $K$ is that solutions of (1.1) should, whatever else, be sought in a function space for which the total energy is well-defined and bounded on any finite time interval, and $K$ has essentially just this property.

The definition of $K$, moreover, applies without reference to the restoring force condition (1.3). Indeed, *throughout this section we shall not make use of* (1.3), so that the definition of solution given below applies equally whether or not $f$ represents a restoring force. Of course, $f$ must still be derivable from a potential as in (1.4).

We can now give our principal definition: A *strong solution of* (1.1) is a function $u \in K$ satisfying the following *two conditions*:

(A) *Conservation law*

$$\text{(i)} \qquad \langle Q(t,\cdot,u,u'),u'\rangle \in L^1_{\text{loc}}(J),$$

$$\text{(ii)} \qquad Eu\big]_0^t = -\int_0^t \langle Q(s,\cdot,u,u'),u'\rangle ds \quad \text{for all } t \in J.$$

(B) *Distribution identity*

$$\langle u',\phi\rangle\big]_0^t = \int_0^t \{\langle u',\phi'\rangle - \langle Du, D\phi\rangle - \langle Q(s,\cdot,u,u'),\phi\rangle - \langle f(\cdot,u),\phi\rangle\}ds$$

for all $t \in J$ and $\phi \in K$.

This definition is justified by the fact that any function $u \in C^2(J\times\Omega)\cap C(J\times\overline{\Omega})$ that has zero boundary values and satisfies (A) and (B) is a classical solution; conversely, any classical solution $u \in C^1(J \times \overline{\Omega})$ satisfies (A) and (B) provided appropriate structure conditions are assumed. See the comments at the end of the section.

We emphasize that *condition* (A) *is an essential attribute of solutions*. Indeed, standard existence theorems for (1.1) in the literature always yield solutions satisfying *both* (A) and (B). On the other hand, (A) is not implied by (B) alone, even if the integrability condition (A)(i) is assumed a priori; see [17].

Some further remarks are in order concerning the individual terms in (A) and (B). Let $u \in K$. Then for each $t \in J$ the terms $Q(t,\cdot,u,u')$ and $u'$ are defined a.e. in $\Omega$ with values in $\mathbb{R}^N$. Hence $\langle Q,u'\rangle$ is a well-defined function of time with values in the nonnegative extended reals; see (1.2). The pairing $\langle Q,u'\rangle$ in fact represents the *damping rate* for the system (1.1), so that (A)(i) expresses the natural requirement that the total damping over any finite time interval should be finite.

A remaining issue is to determine a category of functions $f$ and $Q$ for which the pairings $\langle Q,\phi\rangle$ and $\langle f,\phi\rangle$ are well-defined along with their time integrals. In particular, it must be shown that

$$\text{(2.1)} \qquad \langle f(\cdot,u),\phi\rangle,\ \langle Q(t,\cdot,u,u'),\phi\rangle \in L^1_{\text{loc}}(J)$$

so that the right-hand integral in the identity (B) will be meaningful.[1]

To obtain (2.1), observe first that if $u$, $\phi \in K$, then

$$\text{(2.2)} \qquad u,\ \phi \in C(J \to L^r),$$

[1]Condition (2.1) could in fact be introduced as an integral part of condition (B) in the definition of a strong solution. We do not follow this approach, however, because the hypotheses required for the proof of (2.1)—see (H1) through (H3) below—are natural and elementary, as well as essentially necessary for further development of the theory.

where $r = 2n/(n-2)$ is the Sobolev exponent for the space $X$ ($r = \infty$ if $n = 1$; $2 < r < \infty$ if $n = 2$, because $\Omega$ is bounded). We now make the following natural hypotheses on $f$ and $Q$, in the principal case $n \geqq 2$:

(H1) There exists an exponent $p > 1$ such that

$$\text{(a)} \qquad |f(x,u)| \leqq \text{const}\,[1 + |u|^{p-1}].$$

Moreover, if $n \geqq 3$ and $p > r$, there are constants $\kappa > 0$ and $\kappa_1 \geqq 0$ for which

$$\text{(b)} \qquad (f(x,u),u) \geqq \kappa|u|^p - \kappa_1|u|.$$

$$\text{(H2)} \qquad |Q(t,x,u,v)| \leqq \delta(t)\,[1 + |u|^{q-1} + |v|^{q-1}],$$

where $\delta \in L^1_{\text{loc}}(J)$ and

$$q = \max\{p,\ r\} \quad (2 < q < \infty \text{ if } n = 2).$$

(H3) $Q$ is *tame*; that is, there exists $\gamma \geqq 1$ such that

$$|Q(t,x,u,v)| \cdot |v| \leqq \gamma\big(Q(t,x,u,v),v\big) \quad (\text{automatic if } N = 1).$$

Note that each of these conditions is assumed to hold for all values of the given arguments, namely, $t \in J$, $x \in \Omega$, $u \in \mathbb{R}^N$, and $v \in \mathbb{R}^N$. The case $n = 1$ is treated separately at the end of the section, along with the related question of classical solutions.

It is worth observing that the homogeneous function $f(u) = |u|^{p-2}u$ satisfies (H1), whereas the function $Q(v) = |v|^{m-2}v$ satisfies (H2) and (H3) if $1 < m \leqq q$.

The following lemmas will be used to show that (2.1) is satisfied. The next result shows in particular that (2.2) can be improved when the exponent $p$ in (H1) is greater than $r$.

LEMMA 2.1. *Let* (H1) *hold. If $n \geqq 3$, then $\phi(t) \in L^q$ for all $t \in J$ and $\phi \in K$, and*

$$\text{(2.3)} \qquad \|\phi\|_{L^q} \quad \textit{is locally bounded on } J.$$

*If $n = 2$, then $\phi(t) \in L^r$ for all $\phi \in K$ and $r \in (2,\infty)$; moreover, $\|\phi\|_{L^r}$ is locally bounded on $J$.*

Proof: In view of (2.2) it is enough to consider the case $n \geqq 3$, $p > r$, in which case $q = p$. Now by (1.4) we can write

$$\text{(2.4)} \qquad F(x,u) = \int_0^1 (f(x,\tau u),u)\,d\tau, \qquad u \in \mathbb{R}^N,$$

so by the second condition of (H1)

$$F(x,u) \geqq \int_0^1 (\kappa\,|u|^p \tau^{p-1} - \kappa_1 |u|)d\tau = \frac{\kappa}{p}\,|u|^p - \kappa_1 |u|.$$

Consequently, for $\phi \in K$ we have

$$\|\phi\|_{L^p}^p \leqq \frac{p}{\kappa}\mathscr{F}\phi + \frac{p\kappa_1}{\kappa}\|\phi\|_{L^1}, \quad t \in J.$$

This proves the result, because both $F(x,\phi)$ and $|\phi|$ are integrable on $\Omega$, and $\mathscr{F}\phi$ and $\|\phi\|_{L^1}$ are locally bounded on $J$; see the definition of $K$.

LEMMA 2.2. *Let* (H1) *hold. Then*

$$|\mathscr{F}\phi| \leqq \operatorname{const}\left(\|\phi\|_{L^1} + \|\phi\|_{L^p}^p\right), \tag{2.5}$$

*for all $\phi \in K$, and*

$$|\langle f(\cdot,u),\phi\rangle| \leqq \operatorname{const}\left(\|\phi\|_{L^1} + \|u\|_{L^p}^{p-1}\cdot\|\phi\|_{L^p}\right) \quad \textit{for } u,\phi \in K. \tag{2.6}$$

Proof: From (2.4) and the first part of (H1), we have

$$|F(x,u)| \leqq \operatorname{const}(|u| + |u|^p), \qquad u \in \mathbb{R}^N, \tag{2.7}$$

and (2.5) follows at once. Condition (2.6) is equally clear, in view of Hölder's inequality.

Lemma 2.1 and (2.6) show that $\langle f(\cdot,u),\phi\rangle$ is locally bounded on $J$ whenever $u$, $\phi \in K$, so that this term in (B) is well-defined.

LEMMA 2.3. *Let $u \in K$ satisfy* (A)(i), *and suppose that* (H1) *through* (H3) *hold. Then $\langle Q(t,\cdot,u,u'),\phi(t)\rangle$ exists for all $\phi \in K$ and almost all $t \in J$, and*

$$\left|\langle Q(t,\cdot,u,u'),\phi(t)\rangle\right| \leqq 2\{\delta(t)\}^{1/q}\cdot\left\{c\delta(t) + \gamma\langle Q(t,\cdot,u,u'),u'\rangle\right\}^{1/q'}\cdot\|\phi(t)\|_{L^q},$$

*where $q'$ is the Hölder conjugate of $q$, and $c = c(t)$ is a locally bounded function on $J$.*

Proof: Write

$$\langle Q(t,\cdot,u,u'),\phi\rangle = \left(\int_{\Omega_1} + \int_{\Omega_2}\right)(Q,\phi)dx, \qquad t \in J, \tag{2.8}$$

where

$$\Omega_1 = \Omega_1(t) = \{x \in \Omega : |u'(t)(x)|^{q-1} \geqq 1 + |u(t)(x)|^{q-1}\},$$
$$\Omega_2 = \Omega_2(t) = \{x \in \Omega : |u'(t)(x)|^{q-1} < 1 + |u(t)(x)|^{q-1}\}.$$

Here $u(t)(x)$ denotes the value of the function $u(t,\cdot)$ at the point $x \in \Omega$, and similarly for $u'(t)(x)$. By Schwarz's inequality and (H2), for all $t \in J$ and $x \in \Omega_1$, we get

$$\begin{aligned}|(Q,\phi)| &\leqq |Q|^{1/q} \cdot |Q|^{1/q'} \cdot |\phi| \leqq [2\delta(t)|u'|^{q-1}]^{1/q} \cdot |Q|^{1/q'} \cdot |\phi| \\ &= \{2\delta(t)\}^{1/q} \cdot (|Q| \cdot |u'|)^{1/q'} \cdot |\phi|.\end{aligned}$$

Then from Hölder's inequality and (H3)

$$\int_{\Omega_1} |(Q,\phi)|\, dx \leqq \{2\delta(t)\}^{1/q} \cdot \left(\gamma \int_\Omega (Q,u')dx\right)^{1/q'} \cdot \|\phi\|_{L^q}. \tag{2.9}$$

Also, by (H2),

$$\begin{aligned}\int_{\Omega_2} |(Q,\phi)|\, dx &\leqq \left(\int_{\Omega_2} |Q|^{q'} dx\right)^{1/q'} \left(\int_{\Omega_2} |\phi|^q dx\right)^{1/q} \\ &\leqq 2\,\delta(t) \cdot \left(2 \int_\Omega (1 + |u|^q) dx\right)^{1/q'} \cdot \|\phi\|_{L^q},\end{aligned} \tag{2.10}$$

where in the last line we have used that $(a+b)^p \leqq 2^{p-1}(a^p + b^p)$ for all $a, b \geqq 0$ and $p > 1$.

Now by (2.3) both $\|u\|_{L^q}$ and $\|\phi\|_{L^q}$ are locally bounded on $J$, and by (H2) and (A)(i) the terms $\delta(t)$ and $\langle Q, u'\rangle$ are in $L^1_{\text{loc}}(J)$. Hence Hölder's inequality implies that the right side of (2.9) is in $L^1_{\text{loc}}(J)$, and the same is true for the right side of (2.10). Then by (2.8) it follows that $\langle Q, \phi\rangle$ exists for almost all $t \in J$, is in $L^1_{\text{loc}}(J)$, and satisfies the required estimate—with, for example, $c(t) = 4\left(|\Omega| + \|u(t)\|^q_{L^q}\right)$. This completes the proof.

The second part of (2.1) is of course proved as well, and the distribution identity (B) is well-defined, as required.

The next lemma gives an alternative definition for the set $K$, which further clarifies the meaning of this set.

LEMMA 2.4. *Let* (H1) *be satisfied. Then, for* $n \geqq 3$,

$$K \equiv \{\phi \in K' : \phi(t) \in L^q \textit{ for all } t \in J, \textit{ and } \|\phi\|_{L^q} \textit{ is locally bounded on } J\}.$$

Proof: Let $\hat{K}$ denote the set on the right side above. By Lemma 2.1, clearly $K \subset \hat{K}$. On the other hand, if $\phi \in \hat{K}$, then by the argument of Lemma 2.2 (see (2.7)) we get $F(\cdot, \phi(t, \cdot)) \in L^1(\Omega)$ for all $t \in J$, so that $\mathscr{F}\phi$ is well-defined on $J$. In turn, using (2.5) and the fact that $p \leqq q$, we infer that $\mathscr{F}\phi$ is locally bounded on $J$, so that $\phi \in K$. This completes the proof.

COROLLARY 2.1. *When $n = 2$, or when $p \leqq r$ in* (H1), *we have $K \equiv K'$.*

LEMMA 2.5. *If* (1.3) *holds then $\mathscr{F}\phi \geqq 0$ and $E\phi \geqq 0$ for all $\phi \in K$ and $t \in J$.*

This follows at once from (2.4).

*Remark* 2.1. Concerning the case $n \geqq 3$, $p > r$ in hypothesis (H1), it is important to observe that a general existence theorem for such rapidly growing potentials is somewhat unlikely. At the same time, there do exist special solutions in such cases; for example, the equation

$$u_{tt} - \Delta u + (1 + \mu + |u_t|^{p-2})u_t + |u|^{p-2}u = 0$$

where $\mu$ is an eigenvalue of the Laplacian and has the explicit solution $e^{-t}\varphi(x)$ provided that $\varphi$ is an eigenfunction corresponding to $\mu$. Thus results for $p > r$ still remain of interest for the problem of asymptotic stability, which justifies the relatively small additional effort involved in treating this situation.

A final comment concerns the terminology *strong solution* in the definition (A), (B). Our purpose is to emphasize the fact that in (B) the test space for the functions $\phi$ is precisely the set $K$ in which solutions reside, whereas in contrast the typical test space for distribution (*weak*) solutions is the smaller set $C_c^\infty(J \times \Omega)$. In general, it seems impossible to obtain stability results for solutions of the latter type, a fact that motivates our choice of the test space $K$ and the distinction made between weak and strong solutions.

**2.1. The Case $n = 1$**

When $n = 1$ we can again take $K \equiv K'$. Moreover, the hypothesis (H1) can be weakened to the form

$$\text{(H1)}' \qquad |f(x, u)| \leqq g(u), \qquad g \in C(\mathbb{R}^N \to \mathbb{R}_0^+),$$

for all $(x, u) \in \Omega \times \mathbb{R}^N$.

LEMMA 2.2′. *Let* (H1)′ *hold. Then for all $u$, $\phi \in K$, and $t \in J$*

$$|\langle f(\cdot, u), \phi \rangle| \leqq c_1 \|\phi\|_{L^\infty},$$

*where*

$$c_1 = c_1(t) = |\Omega| \sup_{w \in B(t)} g(w),$$

*and* $B(t) = \{w \in \mathbb{R}^N : |w| \leqq \|u(t)\|_{L^\infty}\}$.

Proof: Because $n = 1$ and $\phi(t) \in X$ for all $t \in J$, we get

$$\|\phi(t)\|_{L^\infty} \leqq \sqrt{|\Omega|/2}\, \|D\phi(t)\| \in C(J),$$

with a similar estimate for $\|u(t)\|_{L^\infty}$. Thus, as in the proof of Lemma 2.2, the conclusions follow from (H1)$'$. Note in particular that $c_1$ is locally bounded on $J$, thus proving the first part of (2.1).

*Remark* 2.2. Condition (H1)$'$ is automatic if $f \in C(\overline{\Omega} \times \mathbb{R}^N \to \mathbb{R}^N)$, or if $f$ does not depend on $x$.

Hypothesis (H2) can similarly be weakened to the form

$$\text{(H2)}' \qquad |Q(t,x,u,v)| \leqq \delta(t)g(u) \quad \text{for } |v| \leqq 1,$$

where $\delta \in L^1_{\text{loc}}(J)$ and $g \in C(\mathbb{R}^N \to \mathbb{R}_0^+)$, as in (H1)$'$.

LEMMA 2.3$'$. *Let $u \in K$ satisfy* (A)(i), *and suppose that* (H2)$'$ *and* (H3) *hold. Then $\langle Q(t,\cdot,u,u'),\phi(t)\rangle$ exists for all $\phi \in K$ and almost all $t \in J$, and*

$$|\langle Q(t,\cdot,u,u'),\phi(t)\rangle| \leqq \{c_1\delta(t) + \gamma\langle Q(t,\cdot,u,u'),u'\rangle\} \cdot \|\phi(t)\|_{L^\infty},$$

*where $c_1 = c_1(t)$ is defined in Lemma 2.2$'$.*

The proof is essentially the same as for Lemma 2.3. The second part of (2.1) also follows at once, because both $\delta$ and $\langle Q, u'\rangle$ are in $L^1_{\text{loc}}(J)$.

### 2.2. Classical Solutions

Although classical solutions of (1.1) are unlikely except in special circumstances, it is nevertheless worth commenting on their treatment within the present context.

A classical solution is a function $u = u(t,x) \in C^2(J \times \Omega) \cap C(J \times \overline{\Omega})$ satisfying (1.1) in the natural sense. Let

$$K^* = \{\phi \in C^1(J \times \overline{\Omega}) : \phi(t,x) = 0 \text{ when } x \in \partial\Omega\}.$$

Then we have the following result.

*Suppose that* (H1)$'$ *and* (H2)$'$ *hold, and that the boundary $\partial\Omega$ is suitably smooth, say of class $C^1$. Then a classical solution $u \in K^*$ satisfies* (A)*; moreover,* (B) *holds for any $\phi \in K^*$.*

To see this, let $u \in K^*$ be a classical solution of (1.1). Then from (H2)′ and the identity $u_t(t,x) = 0$ on $J \times \partial\Omega$, it follows that (A)(i) is satisfied. Moreover, if we multiply (1.1) by $u_t$, integrate over $[0,t] \times \Omega_d$, where $\{\Omega_d\}$ is an appropriate nested family of domains in $\Omega$, and carry out the natural integrations by parts followed by the limiting process $\Omega_d \to \Omega$, then with the help of (H1)′ we get (A)(ii). Similarly, if we multiply (1.1) by $\phi \in K^*$, then the same procedure yields the distribution identity (B).

## 3. Asymptotic Stability, Part I

In this and the following section we give our main results on the global asymptotic stability of solutions of (1.1). We recall that a solution of (1.1) is a function $u \in K$ satisfying the requirements (A) and (B) in the previous section.

We retain the principal conditions (1.2)–(1.4) together with hypotheses (H1) and (H3). In addition, the following stronger version of (H2) will be required:

$$\text{(AS)} \qquad \sigma(t)\omega(|v|) \leqq |Q(t,x,u,v)| \leqq \delta(t)\,[|v|^{m-1} + |v|^{q-1}],$$

where $q$ and $\delta$ are defined in condition (H2), $m$ is an exponent satisfying $1 < m \leqq q$,

$$1/\sigma \in L^{m-1}_{\text{loc}}(J), \qquad \sigma \geqq 0 \quad \text{in } J,$$

and $\omega$ is a continuous real function on $[0,\infty)$ such that

$$\omega(0) = 0, \qquad \omega(\tau) > 0 \quad \text{for } 0 < \tau < 1, \qquad \omega(\tau) = \tau \quad \text{for } \tau \geqq 1.$$

Here it is tacitly assumed that $n \geqq 2$; the case $n = 1$ is treated briefly at the end of the section.

The system (1.5) clearly satisfies condition (AS). More explicitly, let $h(t)$ be the least eigenvalue of the symmetric part of the coefficient matrix $A(t)$ and let $H(t) = |A(t)|$, with $h(t) \geqq 0$ by assumption. Then for $t \in J$ and $v \in \mathbb{R}^N$ we have

$$h(t)\,|v|^2 \leqq (A(t)v,\, v), \qquad |A(t)v| \leqq H(t)\,|v|.$$

In turn, because $Q(t,v) = A(t)\,|v|^{m-2}v$ and $2 \leqq m \leqq q$, it follows that (AS) is satisfied with

$$\sigma(t) = h(t), \qquad \delta(t) = H(t), \qquad \omega(\tau) = \min\{\tau^{m-1},\ \tau\}.$$

Moreover, because $h(t)|Q| \cdot |v| \leqq H(t)\,(Q,\, v)$, it is apparent that (H3) holds for (1.5) if and only if there exists a constant $\gamma \geqq 1$ such that

$$H(t) \leqq \gamma\, h(t), \qquad t \in J$$

(weak uniform definiteness of $A(t)$). Finally, condition (H1) is satisfied when $p \leqq r$; when $p > r$ it is clearly necessary to assume also that $V(x) > 0$ for $x \in \overline{\Omega}$.

Another case, similar to (1.5), occurs when $N \geqq 2$ and $f$ is derivable from the potential $F(x,u) = |u|^{p-2}(V(x)u,u)$, where $V(x)$ is a symmetric nonnegative definite element of $\mathbb{R}^{N\times N}$ for every $x \in \overline{\Omega}$. In this case $(f,u) = (\partial F/\partial u, u) = pF \geqq 0$, whereas (H1) is also satisfied when $p \leqq r$, or when $p > r$ and $V(x)$ is positive definite in $\overline{\Omega}$. (The case (1.5) occurs exactly when $V(x)$ is a multiple of the identity matrix.)

We now turn to the first main result of the paper.

THEOREM 3.1. *Let* (H1), (H3), *and* (AS) *hold. Suppose there exists a nonnegative continuous function $k$ of bounded variation on $J$, such that*

$$k \notin L^1(J), \tag{3.1}$$

$$\liminf_{t\to\infty} \int_0^t (\delta + \sigma^{1-m})k^m ds \bigg/ \left(\int_0^t k ds\right)^m < \infty. \tag{3.2}$$

*Then along any strong solution $u$ of* (1.1) *we have*

$$\lim_{t\to\infty} Eu(t) = 0. \tag{3.3}$$

The integral condition (3.2) prevents the damping term $Q$ from being either too small (*underdamping*), or too large (*overdamping*) as $t \to \infty$. In Section 5 we present a number of corollaries that make this remark more explicit.

The proof of Theorem 3.1 is based on several preliminary lemmas, with the use of arguments related to those in [14].

In what follows $u = u(t) \in K$ denotes a given solution of (1.1) in $J$, in the sense of Section 2.

LEMMA 3.1. *The total energy $Eu$ is a Lyapunov function, that is,*

$$Eu \geqq 0 \quad \text{and} \quad Eu \text{ is decreasing on } J. \tag{3.4}$$

*Moreover,*

$$\|u'\|,\ \|Du\| \in L^\infty(J), \qquad \|u\|,\ \|u\|_{L^p},\ \|u\|_{L^r} \in L^\infty(J), \tag{3.5}$$

*and*

$$\langle Q(t,\cdot,u,u'),u'\rangle \in L^1(J). \tag{3.6}$$

Proof: Condition $(3.4)_1$ follows from Lemma 2.5, and $(3.4)_2$ from (A) and (1.2). In turn, $Eu \in L^\infty(J)$. Consequently, each of the terms $\|u'\|$, $\|Du\|$, and $\mathcal{F}u$ making up $Eu$ must be bounded, because each is nonnegative. Then, from the argument of Lemma 2.1, we find that $\|u\|_{L^p}$ is bounded, and from Sobolev, that

$\|u\|_{L^r}$ is bounded. Finally $\|u\|$ is bounded as a consequence of Hölder's inequality and the boundedness of $\Omega$.

Again, with the use of (1.2) and the fact that $Eu$ is bounded on $J$, it is clear from (A)(ii) that (3.6) is valid.

LEMMA 3.2. *Suppose that* (AS) *and* (H3) *hold. Then, for any* $T \in J$ *and* $t \geqq T$, *we have*

$$\int_T^t k\,|\langle Q(s,\cdot,u,u'),u\rangle|ds \leqq \varepsilon(T)\left[1+\left(\int_0^t \delta k^m ds\right)^{1/m}\right], \tag{3.7}$$

*where* $\varepsilon(T) \to 0$ *as* $T \to \infty$.

Proof: Define, for any fixed $s \in J$,

$$\Omega_1 = \Omega_1(s) = \{x \in \Omega : |u'(s)(x)| \leqq 1\},$$
$$\Omega_2 = \Omega_2(s) = \{x \in \Omega : |u'(s)(x)| > 1\}.$$

Then, as in the demonstration of Lemma 2.3 (see (2.9)), we get by the right-hand inequality of (AS) and by (H3),

$$k\left|\int_{\Omega_1}(Q(s,\cdot,u,u'),u)\,dx\right| \leqq (2\delta k^m)^{1/m}\left(\gamma\int_\Omega (Q(s,\cdot,u,u'),u')\,dx\right)^{1/m'}\|u(s)\|_{L^m}$$

and

$$k\left|\int_{\Omega_2}(Q(s,\cdot,u,u'),u)\,dx\right| \leqq (2\delta k^q)^{1/q}\left(\gamma\int_\Omega (Q(s,\cdot,u,u'),u')\,dx\right)^{1/q'}\|u(s)\|_{L^q}.$$

Consequently, if we integrate with respect to $s$ from $T$ to $t$, we find by Hölder's inequality

$$\begin{aligned}\int_T^t & k\,|\langle Q(s,\cdot,u,u'),u\rangle|ds \\ &\leqq \left(2\gamma^{m-1}\int_T^t \delta k^m ds\right)^{1/m}\left(\int_T^t \langle Q(s,\cdot,u,u'),u'\rangle ds\right)^{1/m'}\sup_{[T,t]}\|u(s)\|_{L^m} \\ &\quad+\left(2\gamma^{q-1}\int_T^t \delta k^q ds\right)^{1/q}\left(\int_T^t \langle Q(s,\cdot,u,u'),u'\rangle ds\right)^{1/q'}\sup_{[T,t]}\|u(s)\|_{L^q}.\end{aligned}$$

Because $k$ is of bounded variation on $J$, we have $k(t) \leqq \text{const} = k_0$ on $J$. Moreover $1 < m \leqq q$, so that by Hölder's inequality

$$\|u(s)\|_{L^m} \leqq |\Omega|^{1/m-1/q}\|u(s)\|_{L^q},$$

and by Young's inequality

$$\left(2\gamma^{q-1}\int_T^t \delta k^q ds\right)^{1/q} \leqq \gamma k_0 + \left(2\gamma^{m-1}\int_T^t \delta k^m ds\right)^{1/m}.$$

Thus,

$$\int_T^t k\,|\langle Q(s,\cdot,u,u'),u\rangle|ds \leqq \text{const}\left[\gamma k_0 + \left(\gamma^{m-1}\int_T^t \delta k^m ds\right)^{1/m}\right]$$

$$\cdot\left\{\left(\int_T^t \langle Q(s,\cdot,u,u'),u'\rangle ds\right)^{1/m'} + \left(\int_T^t \langle Q(s,\cdot,u,u'),u'\rangle ds\right)^{1/q'}\right\}$$

$$\cdot \sup_{[T,t]} \|u(s)\|_{L^q},$$

where the constant depends only on $m$, $q$, and $|\Omega|$. The proof is now complete, because $\|u\|_{L^q} \in L^\infty(J)$ and $\langle Q(s,\cdot,u,u'),u'\rangle \in L^1(J)$ by Lemma 3.1.

LEMMA 3.3. *Suppose that* (AS) *and* (H3) *hold. Let $\vartheta$ be a given positive constant. Then there exists $C(\vartheta) > 0$ such that for all $t \geqq T \geqq 0$*

$$\int_T^t k\|u'\|^2 ds \leqq \vartheta\int_T^t k ds + \varepsilon(T)C(\vartheta)\left(\int_T^t \sigma^{1-m}k^m ds\right)^{1/m}, \tag{3.8}$$

*where $\varepsilon(T) \to 0$ as $T \to \infty$.*

Proof: For fixed $s \in J$, define

$$\Omega_1 = \Omega_1(s) = \left\{x \in \Omega : |u'(s)(x)| \leqq \sqrt{\vartheta/|\Omega|}\right\},$$

$$\Omega_2 = \Omega_2(s) = \left\{x \in \Omega : |u'(s)(x)| > \sqrt{\vartheta/|\Omega|}\right\}.$$

Clearly

$$\int_{\Omega_1} |u'|^2 dx \leqq \vartheta. \tag{3.9}$$

On the other hand,

$$\int_{\Omega_2} |u'|^2 dx = \int_{\Omega_2} \frac{|u'|}{\omega(|u'|)}\cdot\omega(|u'|)\,|u'|dx, \tag{3.10}$$

where $\omega$ is the function appearing on the left-hand side of (AS). Now

$$(3.11)\qquad \sup_{\tau \geqq \sqrt{\vartheta/|\Omega|}} \frac{\tau}{\omega(\tau)} = \sup_{\sqrt{\vartheta/|\Omega|} \leqq \tau \leqq 1} \frac{\tau}{\omega(\tau)} = \Lambda, \qquad \Lambda = \Lambda(\vartheta) \geqq 1,$$

because $\omega$ is continuous and positive for $\tau > 0$, and $\omega(\tau) = \tau$ for $\tau \geqq 1$. From (3.10), (3.11), and $(AS)_1$, we obtain

$$\int_{\Omega_2} |u'|^2 dx \leqq \Lambda(\vartheta) \int_{\Omega_2} \frac{|Q(s,\cdot,u,u')| \cdot |u'|}{\sigma(s)} dx \leqq \frac{\gamma\Lambda(\vartheta)}{\sigma(s)} \int_\Omega (Q(s,\cdot,u,u'),u')dx$$

by (H3). In turn, because $\|u'\| \leqq$ const in $J$,

$$\begin{aligned} k\int_{\Omega_2} |u'|^2 dx &\leqq \text{const}\, k \left( \int_{\Omega_2} |u'|^2 dx \right)^{1/m'} \\ &\leqq \text{const}\, [\gamma\Lambda(\vartheta)]^{1/m'} (\sigma^{1-m}k^m)^{1/m} \langle Q(s,\cdot,u,u'),u'\rangle^{1/m'}, \end{aligned}$$

so that (3.9) yields

$$k\|u'(s)\|^2 \leqq \vartheta k + \text{const}\, [\gamma\Lambda(\vartheta)]^{1/m'} (\sigma^{1-m}k^m)^{1/m} \langle Q(s,\cdot,u,u'),u'\rangle^{1/m'}.$$

The required result now follows by integration from $T$ to $t$ and another use of Hölder's inequality; in particular, we can take $C(\vartheta) = [\Lambda(\vartheta)]^{1/m'}$, and

$$\varepsilon(T) = \text{const} \left( \int_T^\infty \langle Q(s,\cdot,u,u'),u'\rangle ds \right)^{1/m'} \to 0 \quad \text{as } T \to \infty$$

by (3.6).

Note that $C(\vartheta) \to \infty$ as $\vartheta \to 0$, whenever $\omega(\tau) = o(\tau)$ as $\tau \to 0$, as can be seen from the calculation (3.11).

By (3.4) it is clear that there exists $\ell \geqq 0$ such that

$$(3.12)\qquad \lim_{t\to\infty} Eu(t) = \ell.$$

LEMMA 3.4. *Suppose $\ell > 0$ and that* (H1) *holds. Then there exists $\alpha = \alpha(\ell) > 0$ such that*

$$(3.13)\qquad \|u'\|^2 + \|Du\|^2 + \langle f(\cdot,u),u\rangle \geqq \alpha \quad \textit{on } J.$$

Before giving the proof, it is worth observing that (3.13) follows almost trivially in the special case of the system (1.5). Indeed, for this case

$$\langle f,u\rangle = p\mathscr{F}u = \int_\Omega |u|^{p-2}(V(x)u,u)dx,$$

so that (3.13) is an immediate consequence of (3.12), with $\alpha = \min\{2, p\}\,\ell$.

Proof of Lemma 3.4: Because $Eu(t) \geqq \ell$ for all $t \in J$, it follows that

$$\|u'\|^2 + \|Du\|^2 \geqq 2(\ell - \mathscr{F}u) \quad \text{on } J.$$

Let

$$J_1 = \{t \in J : \mathscr{F}u(t) \leqq \ell/2\} \quad J_2 = \{t \in J : \mathscr{F}u(t) > \ell/2\}.$$

For $t \in J_1$

$$\|u'(t)\|^2 + \|Du(t)\|^2 \geqq \ell.$$

Next, consider $t \in J_2$. There are two cases, depending on whether $p \leqq r$ or $p > r$. In the first case, by (2.5),

$$\frac{1}{2}\ell < \mathscr{F}u(t) \leqq C_1(\|u(t)\|_{L^1} + \|u(t)\|_{L^p}^p). \tag{3.14}$$

With the use of the Sobolev theorem, we have $\|u(t)\|_{L^r} \leqq C\,\|Du(t)\|$, and in turn

$$\begin{aligned} &\|u(t)\|_{L^1} \leqq |\Omega|^{1-1/r}\|u(t),\|_{L^r} \leqq C\,|\Omega|^{1-1/r}\|Du(t)\|, \\ &\|u(t)\|_{L^p} \leqq C\,|\Omega|^{1/p-1/r}\|Du(t)\|. \end{aligned} \tag{3.15}$$

Then, by (3.14) and (3.15),

$$\frac{1}{2}\ell < C_2(\|Du\| + \|Du\|^p)$$

for an appropriate constant $C_2$. It follows that

$$\|Du\| \geqq \min\{1, \ell/4C_2\}.$$

Because $\langle f, u\rangle \geqq 0$, this proves (3.13) for the first case, with

$$\alpha = \min\{1, \ell, (\ell/4C_2)^2\}.$$

Next, if $p > r$, we have by (2.5) again

$$\begin{aligned} \frac{1}{2}\ell &< \mathscr{F}u(t) \leqq C_1(\|u(t)\|_{L^1} + \|u(t)\|_{L^p}^p) \\ &\leqq C_1\left\{\|u(t)\|_{L^1} + \frac{1}{\kappa}(\langle f,u\rangle + \kappa_1\|u(t)\|_{L^1})\right\}, \end{aligned} \tag{3.16}$$

where the last step follows from the second part of (H1). Then, from $(3.15)_1$ and (3.16), there results

$$\frac{1}{2}\ell < C_3(\|Du\| + \langle f,u\rangle), \qquad t \in J_2,$$

for an appropriate constant $C_3$. This implies either

$$\langle f,u\rangle \geqq \ell/4C_3 \quad \text{or} \quad \|Du\| \geqq \ell/4C_3,$$

and therefore (3.13) holds with

$$\alpha = \min\left\{\ell,\ \ell/4C_3,\ (\ell/4C_3)^2\right\},$$

which completes the proof.

Proof of Theorem 3.1: First we treat the simpler case in which $k$ is not only $CBV(J)$, but also of class $C^1(J)$. Suppose for contradiction that $\ell > 0$ in (3.12). Define a second Lyapunov function by

$$V(t) = k(t)\langle u,u'\rangle = \langle u',\phi\rangle, \qquad \phi = k(t)u. \tag{3.17}$$

We note that $\phi \in K$. Indeed, because $k \in C^1(J)$ and $u \in K$, we have

$$\phi' = (k\,u)' = k'u + ku',$$

so that obviously $\phi \in K'$. Moreover, Lemma 3.1 implies $\phi = ku \in L^q$, with $\|\phi\|_{L^q}$ locally bounded on $J$. Consequently $\phi \in K$ by Lemma 2.4, as asserted. It now follows, from the distribution identity (B) in Section 2, that for any $t \geqq T \geqq 0$

$$\begin{aligned} V(s)\big]_T^t = \int_T^t \{k'\langle u,u'\rangle + 2k\|u'\|^2 \\ - k(\|u'\|^2 + \|Du\|^2 + \langle f(\cdot,u),u\rangle) - k\langle Q(s,\cdot,u,u'),u\rangle\}ds. \end{aligned} \tag{3.18}$$

Note that an additional term $k\|u'\|^2$ has been added and subtracted inside the integral. (It is exactly here that we use the fact that $u$ is a *strong* solution of (1.1); otherwise, if $u$ is only a weak solution, then $\phi = ku$ need not be in the test space $C_c^\infty(J \times \Omega)$ and (3.18) can no longer be inferred.)

We now estimate the right-hand side of (3.18). First

$$|\langle u,u'\rangle| \leqq \|u\| \cdot \|u'\| \leqq \text{const} \quad \text{in } J,$$

by (3.5). Applying Lemmas 3.2 through 3.4 to the remaining terms, we then obtain

$$\begin{aligned} V(s)\big]_T^t \leqq \text{const}\int_T^\infty |k'|ds + 2\vartheta\int_T^t kds + 2\varepsilon(T)C(\vartheta)\left(\int_0^t \sigma^{1-m}k^m ds\right)^{1/m} \\ - \alpha\int_T^t kds + \varepsilon(T)\left\{1 + \left(\int_0^t \delta k^m ds\right)^{1/m}\right\}. \end{aligned}$$

Choose $\vartheta = \vartheta(\ell) = \alpha/4$ and fix $T > 0$ sufficiently large so that

$$\text{const} \int_T^\infty |k'| ds \leqq \frac{1}{2}, \qquad \varepsilon(T) \leqq \frac{1}{2}.$$

Then

$$V(s)]_T^t \leqq 1 - \frac{\alpha}{2} \int_T^t kds$$
$$+ \varepsilon(T) \left\{ 2C(\vartheta) \left( \int_0^t \sigma^{1-m} k^m ds \right)^{1/m} + \left( \int_0^t \delta k^m ds \right)^{1/m} \right\}.$$

By (3.2) there is a sequence $t_i \nearrow \infty$ and a number $M > 0$ such that

$$\int_0^{t_i} (\delta + \sigma^{1-m}) k^m \, ds \leqq \left( M \int_0^{t_i} k \, ds \right)^m .$$

Consequently, for $t_i \geqq T$,

$$V(t_i) - V(T) \leqq 1 - \frac{\alpha}{2} \int_T^{t_i} k \, ds + \varepsilon(T) [2C(\vartheta) + 1] M \int_0^{t_i} k \, ds. \tag{3.19}$$

We now take $T$ even larger, if necessary, so that

$$\varepsilon(T)[2C(\vartheta) + 1]M \leqq \alpha/4,$$

which gives, for all $t_i \geqq T$,

$$V(t_i) \leqq V(T) + 1 + \frac{\alpha}{4} \int_0^T k \, ds - \frac{\alpha}{4} \int_T^{t_i} k \, ds.$$

Thus by (3.1) we get

$$\lim_{i \to \infty} V(t_i) = -\infty.$$

On the other hand, if we recall that $k$ is bounded,

$$|V(t)| \leqq \sup_J k \cdot \|u(t)\| \cdot \|u'(t)\| \leqq \text{const} \quad \text{for } t \in J,$$

by (3.5). This contradiction completes the first part of the proof.

For the general case, in which $k \notin C^1(J)$, the proof is based on [14], p. 290, lemma A. We take $\theta = 2$ in this lemma and let $\bar{k} \in C^1(J)$ and $E \subset J$ be an open set as it is there. Namely, it is such that

$$\text{(i)} \quad 2k \geqq \bar{k} \geqq \begin{cases} k, & \text{in } J \setminus E, \\ 0, & \text{in } E, \end{cases} \qquad \text{(ii)} \quad \operatorname{Var} \bar{k} \leqq 2 \operatorname{Var} k,$$

$$\text{(iii)} \quad \int_E k \, ds \leqq 1.$$

Clearly, $\bar{k} \in CBV(J)$ by (ii).

We assert that $\bar{k}$ satisfies (3.1) and (3.2). Indeed, let $T_1$ be such that $\int_0^{T_1} k\,ds \geqq 2$, and consider values $t \geqq T_1$. Then, by (i) and (iii)

$$(3.20)\qquad \int_0^t \bar{k}\,ds \geqq \int_{[0,t]\setminus E} k\,ds \geqq \int_0^t k\,ds - \int_E k\,ds \geqq \int_0^t k\,ds - 1 \geqq \frac{1}{2}\int_0^t k\,ds.$$

Hence $\bar{k}$ satisfies (3.1). Moreover, by (i) and (3.20),

$$\frac{\int_0^t(\delta + \sigma^{1-m})\bar{k}^m ds}{\left(\int_0^t \bar{k}\,ds\right)^m} \leqq 4^m \frac{\int_0^t(\delta + \sigma^{1-m})k^m ds}{\left(\int_0^t k\,ds\right)^m}, \qquad t \geqq T_1,$$

which shows that $\bar{k}$ also satisfies (3.2).

The general case is therefore reduced to the situation when $k$ is smooth, and the proof is complete.

*Remark* 3.1. As in Section 2, our hypotheses can be weakened in case $n = 1$. Specifically, in this case (H1) can be replaced by (H1)′ and (AS) by

$$(\mathrm{AS})' \qquad \sigma(t)\omega(|v|) \leqq |Q(t,x,u,v)| \leqq \delta(t)g(u)\,|v|^{m-1},$$

where

$$1 < m < \infty, \qquad g \in C(\mathbb{R}^N \to \mathbb{R}_0^+).$$

$\sigma$, $\delta$, $\omega$ are the same as in (AS), and the right-hand inequality of (AS)′ need hold only for $|v| \leqq 1$.

The proof of asymptotic stability is the same as before, taking into account the fact that

$$\|u(t)\|_{L^\infty} \leqq \sqrt{|\Omega|/2}\,\|Du(t)\| \leqq \text{const} \quad \text{in } J;$$

see (3.5).

Similarly, Theorem 3.1 continues to apply in the case of classical solutions $u \in K^*$, provided the boundary of $\Omega$ is smooth. The proof is essentially the same as before, once we observe that (A) is satisfied, that (B) holds with $K$ replaced by the subspace $K^*$, and that $\phi = ku$ is an admissible test function in $K^*$.

## 4. Asymptotic Stability, Part II

We suppose as before that conditions (H1), (H3), and (AS) are satisfied. At the same time, we introduce the following more specific behavior for the function $\omega$ in (AS)

$$(4.1)\qquad \omega(\tau) = \tau^\nu, \quad 0 \leqq \tau \leqq 1; \qquad \omega(\tau) = \tau, \quad \tau \geqq 1,$$

where $\nu \geqq m-1$. Recall also that we have defined $q = \max\{p, r\}$ when $n \geqq 3$.

THEOREM 4.1. *Suppose there exists a bounded, nonnegative, absolutely continuous function k on J such that* (3.1) *and* (3.2) *are satisfied and also*

$$|k'| \leqq \text{const}\, \sigma^{\lambda} k^{1-\lambda}, \qquad \textit{a.e. in } J, \tag{4.2}$$

*where*

$$0 < \lambda \leqq \begin{cases} \min\left\{\dfrac{q-1}{q}, \dfrac{1}{\nu+1}\right\}, & \textit{if } n \geqq 3, \\ \dfrac{1}{\nu+1}, & \textit{if } n = 2. \end{cases} \tag{4.3}$$

*Then* (3.3) *holds for any strong solution u of* (1.1).

Proof: We give only the case $n \geqq 3$; when $n = 2$ the idea is the same. Because $\nu$ in (4.1) can always be made larger, we can clearly suppose without loss of generality that (4.3) has the specific form

$$\lambda = 1/(\nu+1) \leqq 1 - 1/q. \tag{4.4}$$

The proof of Theorem 3.1 now applies almost word for word, except for the estimation of the term

$$\int_T^t k'\langle u, u'\rangle\, ds$$

in (3.18); here it should be noted that (3.18) holds even though $k'$ is only assumed to be integrable, as one can see by an easy approximation procedure. Now for any $s \in J$, introduce the sets

$$\Omega_1 = \Omega_1(s) = \{x \in \Omega : |u'(s)(x)| \leqq 1\},$$
$$\Omega_2 = \Omega_2(s) = \{x \in \Omega : |u'(s)(x)| > 1\}.$$

Then from (4.2)

$$\begin{aligned}\int_{\Omega_1} k'(u, u')dx &\leqq \text{const} \int_{\Omega_1} \sigma^{\lambda} k^{1-\lambda} |u| \cdot |u'|\, dx \\ &\leqq \text{const}\, k^{1-\lambda} \left(\int_{\Omega_1} |u|^{1/(1-\lambda)} dx\right)^{1-\lambda} \left(\int_{\Omega_1} \sigma |u'|^{1/\lambda} dx\right)^{\lambda} \\ &\leqq \text{const}\, |\Omega|^{1-\lambda-1/q} k^{1-\lambda} \|u\|_{L^q} \left(\int_{\Omega_1} \sigma |u'|^{\nu+1} dx\right)^{\lambda},\end{aligned}$$

where by (4.4) we have $1-\lambda-1/q \geqq 0$, as required for the application of Hölder's inequality in the third step. Use of (AS), (H3), (4.1), and (3.5) now gives

$$\int_{\Omega_1} k'(u,u')dx \leqq \operatorname{const} k^{1-\lambda}\left(\int_{\Omega}(Q(s,\cdot,u,u'),u')dx\right)^{\lambda}. \tag{4.5}$$

Next, because $q' < 2$ and $|u'(s)| \geqq 1$ in $\Omega_2$,

$$\begin{aligned}\int_{\Omega_2} k'(u,u')dx &\leqq \operatorname{const}\int_{\Omega_2}\sigma^{\lambda}k^{1-\lambda}|u|\cdot|u'|^{2/q'}dx\\ &\leqq \operatorname{const} k^{1-\lambda}\left(\int_{\Omega_2}\sigma|u'|^2dx\right)^{\lambda}\left(\int_{\Omega_2}|u|^q dx\right)^{1/q}\\ &\quad\cdot\left(\int_{\Omega_2}|u'|^2dx\right)^{1-\lambda-1/q}\\ &\leqq \operatorname{const} k^{1-\lambda}|\langle Q(s,\cdot,u,u'),u'\rangle|^{\lambda},\end{aligned} \tag{4.6}$$

again by (AS), (H3), and (3.5).

From (4.5), (4.6), and integration from $T$ to $t \geqq T$, we now find

$$\begin{aligned}\int_T^t k'\langle u,u'\rangle ds &\leqq \operatorname{const}\left(\int_T^t k\,ds\right)^{1-\lambda}\cdot\left(\int_T^t\langle Q(s,\cdot,u,u'),u'\rangle ds\right)^{\lambda}\\ &\leqq \operatorname{const}\left\{1+\int_T^t k\,ds\right\}\left(\int_T^t\langle Q(s,\cdot,u,u'),u'\rangle ds\right)^{\lambda},\end{aligned}$$

by Young's inequality. Finally, taking $T$ sufficiently large yields the estimate

$$\int_T^t k'\langle u,u'\rangle ds \leqq \frac{1}{2}+\frac{\alpha}{8}\int_T^t k\,ds,$$

where $\alpha = \alpha(\ell)$ is defined in Lemma 3.4.

In conclusion, as in the proof of Theorem 3.1, we find in place of (3.19),

$$V(t_i)-V(T) \leqq 1-\frac{\alpha}{8}\int_T^{t_i} k\,ds + \varepsilon(T)[2C(\vartheta)+1]M\int_T^{t_i} k\,ds,$$

and this produces the same contradiction as before.

**Remarks**

1. When $n = 1$, we can replace (H1) by (H1)′ and (AS) by (AS)′.

2. If $f \equiv 0$ in (1.1), then (H1)(b) fails. In this case only (H1)(a) is true, so that it is necessary to take $q = r$ in (H2). Therefore $K = K'$, the result of Lemma 2.1 requires no proof, and the final part of the proof of Lemma 3.4 can be omitted.

3. One can combine Theorems 3.1 and 4.1 by assuming that $k$ has the decomposition

$$k = k_1 + k_2,$$

where $k_1 \in CBV(J)$ and $k_2 \in AC(J) \cap L^\infty(J)$ with

$$|k_2'| \leqq \text{const}\, \sigma^\lambda k^{1-\lambda}, \quad \text{a.e. in } J;$$

see also [14], section 4, where a similar result was given for ordinary differential systems.

4. Professor Herbert Koch has suggested that, in the definition of a solution of (1.1) given in Section 2, condition (A)(ii) can be weakened by replacing the equality sign by the inequality sign ($\leqq$). In this case, Theorem 3.1 (and later results) continue to hold, but with (3.3) modified to the weaker conclusion

$$\liminf_{t\to\infty} Eu(t) = 0. \tag{3.3$'$}$$

Of course, if it is known a priori that $Eu(t)$ has a limit as $t \to \infty$, then we recover the original conclusion (3.3) as before.

5. Condition (AS) need not hold in the entire interval $J = [0, \infty)$, but in fact can be restricted to a measurable *control subset* $I \subset J$, see [14], section 3. Theorems 3.1 and 4.1 then need to be revised only in two places. First, (3.1) should be modified to the form

$$k \notin L^1(J), \qquad k = 0 \quad \text{in } J \setminus I, \tag{3.1$'$}$$

and second, (3.2) should be replaced by

$$\liminf_{t\to\infty} \int_{[0,t]\cap I} (\delta + \sigma^{1-m}) k^m ds \Bigg/ \left(\int_0^t k ds\right)^m < \infty. \tag{3.2$'$}$$

The proofs are almost exactly the same.

The importance of control subsets in studying asymptotic stability is illustrated in [15], in the context of ordinary differential systems.

6. The previous results carry over almost without change to hyperbolic systems that arise when the second derivative term $u_{tt}$ in (1.1) is replaced by $P(x)u_{tt}$, where $P(x)$ is an element of $\mathbb{R}^{N\times N}$, is continuous in $\overline{\Omega}$, and is Hermitian and nonnegative definite.

7. We note finally that similar methods can be applied to parabolic systems of the form, for example,

$$A(t)u_t = \Delta u - f(x,u),$$

or even

$$A(t)|u_t|^{m-2}u_t = \Delta u - f(x,u), \qquad m > 1;$$

see [16].

## 5. Applications

With appropriate choices of the auxiliary function $k$ we can obtain a number of useful special cases of Theorems 3.1 and 4.1. It is assumed throughout that (H1), (H3), and (AS) are satisfied.

COROLLARY 5.1. *Suppose*

$$\liminf_{t\to\infty} \frac{1}{t^m}\int_0^t (\delta + \sigma^{1-m})ds < \infty. \tag{5.1}$$

*Then* (3.3) *holds for any strong solution u of* (1.1).

Proof: Take $k = 1$ in Theorem 3.1.

For the system (1.5) in particular, Corollary 5.1 shows that the rest state is asymptotically stable provided that

$$\liminf_{t\to\infty} \frac{1}{t^m}\int_0^t [H(s) + h(s)^{1-m}]ds < \infty, \tag{5.2}$$

and

$$H(t) \leqq \gamma\, h(t), \qquad \gamma = \text{const} \geqq 1, \tag{5.3}$$

where $h(t)$ is the least eigenvalue of the symmetric part of the matrix $A(t)$, and $H(t) = |A(t)|$; see the discussion preceding Theorem 3.1 in Section 3. The relation (5.3) is automatically satisfied, it should be noted, when (1.5) is either scalar ($N = 1$) or autonomous with $h(t) = \text{const} > 0$.

If (5.3) fails, that is, if $A(t)$ is not weakly uniformly definite, then (H3) cannot be used for the system (1.5). Hence the criterion (5.2) for asymptotic stability no longer applies. On the other hand, in work to be published elsewhere we show that, even when condition (5.3) is not imposed, a more general criterion for the asymptotic stability of the rest state of (1.5) remains applicable, namely,

$$\liminf_{t\to\infty} \frac{1}{t^m}\int_0^t \frac{H^m + 1}{h^{m-1}}ds < \infty.$$

We note that an interesting special case of Corollary 5.1 was obtained earlier by Nakao [11], p. 62. In his work $N = 1$, and the damping term $Q(t, x, u, v)$ was assumed to have the separated form $h(t)\rho(x, v)$. Furthermore, he requires (in our notation) the following stronger versions of (H1) and (AS):

$$0 \leqq f(x,u)\,u \leqq \text{const}\,|u|^{1+r/2}, \qquad h(t) \geqq \text{const} = h_0 > 0,^2$$
$$C_1|v|^m \leqq \rho(x,v)\,v \leqq C_2|v|^m, \qquad m \in [2, 1 + r/2].$$

Under these assumptions we can take $p = 1 + r/2$ in (H1) and $q = m$, $\sigma(t) = C_1 h_0$, $\delta(t) = C_2 h(t)$, and $\omega(\tau) = \min\{\tau^{m-1},\ \tau\}$ in (AS), in which case (5.1) reduces to

$$\liminf_{t\to\infty} \frac{1}{t^m} \int_0^t h(s)ds < \infty,$$

which is exactly condition (1.3) in [11], p. 57.

Asymptotic stability for *autonomous equations* of the form (1.1) has also been studied by Marcati [9], pp. 244 and 245, under the following versions of conditions (H1) and (AS):

$$0 \leqq f(x,u)\,u \leqq \text{const}\,(|u|^2 + |u|^r),$$
$$C_1|v|^\mu \leqq Q(x,u,v)\,v \leqq C_2(|v|^2 + |v|^\mu),$$

where $2 \leqq \mu \leqq r$. His conclusion, that under these conditions the rest state of (1.1) is asymptotically stable, is likewise an immediate consequence of Corollary 5.1, with $\sigma$ and $\delta$ constant, $m = 2$, and $\omega(\tau) = \min\{\tau^{\mu-1},\ \tau\}$. For other related work in the autonomous case, see [12].

In addition to the case $k \equiv 1$, which yields Corollary 5.1, other choices for $k$ are also of interest. In the following two corollaries we indicate some of these possibilities.

COROLLARY 5.2. *Assume that*

$$\liminf_{t\to\infty} \frac{1}{\log^m t} \int_1^t (\delta + \sigma^{1-m}) \frac{ds}{s^m} < \infty.$$

*Then* (3.3) *holds for any strong solution u of* (1.1).

[2]Nakao [11], p. 62, actually writes only the weaker condition $0 \leqq f(x,u)u \leqq \text{const}(1 + |u|^{1+r/2})$, but this seems inaccurate. First, it is too weak in itself to obtain his earlier hypothesis ($A_1$). Second, he later adds continuity hypotheses that would imply (at the least) that

$$0 \leqq f(x,u)u \leqq \text{const}\,|u|\,(1 + |u|^{r/2}),$$

whereas in an earlier paper [10], which he refers to in the same context, it is explicitly required that $|f(x,u)| \leqq \text{const}\,|u|^{r/2}$, which is equivalent to the condition given here.

Proof: Take $k(t) = \min\{1, 1/t\}$ in Theorem 3.1.

It is possible to state a series of further corollaries, corresponding to the functions

$$k(t) = \begin{cases} 1/2\log 2, & t \leqq 2, \\ 1/t\log t, & t \geqq 2, \end{cases} \qquad k(t) = \begin{cases} 1/3\log 3 \cdot \log\log 3, & t \leqq 3, \\ 1/t\log t \cdot \log\log t, & t \geqq 3, \end{cases}$$

and so forth, but we shall not pursue this here.

COROLLARY 5.3. *Suppose that any one of the following conditions is satisfied:*

(a) $$\sigma \in CBV(J); \qquad \int_0^\infty \sigma\, dt = \infty; \qquad \sigma^{m-1}\delta \in L^\infty(J);$$

(b) $$\delta^{1/(1-m)} \in CBV(J); \qquad \int_0^\infty \delta^{1/(1-m)} dt = \infty; \qquad 1/\sigma^{m-1}\delta \in L^\infty(J);$$

(c) $$(\delta + \sigma^{1-m})^{1/(1-m)} \in CBV(J); \qquad \int_0^\infty (\delta + \sigma^{1-m})^{1/(1-m)} dt = \infty.$$

*Then* (3.3) *holds for any strong solution u of* (1.1).

Proof: We take, respectively,

$$k = \sigma; \quad k = \delta^{1/(1-m)}; \quad k = (\delta + \sigma^{1-m})^{1/(1-m)}; \tag{5.4}$$

and find in each case that (3.1) and (3.2) are satisfied. Consequently, the assertion follows from Theorem 3.1.

*Remark.* The choices

$$k = \sigma; \qquad k = \delta^{1/(1-m)}$$

can also be used in Theorem 4.1, in which case the results of Corollary 5.3(a) and 5.3(b) continue to hold, but with the bounded variation conditions in (a) and (b) replaced by the corresponding relations (4.2), namely,

$$|\sigma'| \leqq \text{const}\, \sigma \quad \text{for (a)},$$

and, with $\lambda = \min\{1/q', 1/(\nu + 1)\}$,

$$|\delta'| \leqq \text{const}\, \delta(\sigma^{m-1}\delta)^{\lambda/(m-1)} \quad \text{for (b)}.$$

**The Linear Case**

Consider the problem

$$(5.5)\qquad \begin{cases} u_{tt} - \Delta u + a(t)t^{\alpha}u_t + V(x)u = 0, & \text{in } J \times \Omega, \\ u(t,x) = 0, & \text{on } J \times \partial\Omega, \end{cases}$$

where $J = [1, \infty)$, $\Omega$ is a bounded open subset of $\mathbb{R}^n$, $N = 1$ for simplicity, $V$ is continuous in $\overline{\Omega}$, and $\alpha \in \mathbb{R}$. Assume, moreover, that there is a number $C > 0$ such that

$$(5.6)\qquad 1/C \leqq a(t) \leqq C, \quad \text{in } J.$$

Here

$$Q(t,x,u,v) = a(t)t^{\alpha}v, \qquad F(x,u) = \frac{1}{2}V(x)u^2,$$

so that (H1) is satisfied, and (AS) holds with

$$\delta(t) = 1/\sigma(t) = C\,t^{|\alpha|}, \qquad m = 2, \qquad \omega(s) = s.$$

Consider first the case $|\alpha| \leqq 1$. Then

$$\delta + 1/\sigma = 2C\,t^{|\alpha|}.$$

Hence, by Corollary 5.1 or by Corollary 5.2, every strong solution $u$ of (5.5) has the property that

$$\lim_{t\to\infty} Eu(t) = 0.$$

When $|\alpha| > 1$, Corollaries 5.1 and 5.2 do not apply. In fact, if $\alpha < -1$ there exist oscillating solutions of (5.5) that *do not* approach zero as $t \to \infty$. To show this, consider solutions of (5.5) having the separated form

$$(5.7)\qquad u(t,x) = w(t)\varphi(x),$$

where $\varphi = \varphi_k$ is an eigenfunction of $-\Delta + V(x)$ in $\Omega$, with Dirichlet boundary conditions, having the corresponding eigenvalue $\mu = \mu_k > 0$. An easy calculation shows that $w$ is a solution of the ordinary differential equation

$$(5.8)\qquad w'' + a(t)t^{\alpha}w' + \mu w = 0, \qquad t \in J.$$

This equation satisfies hypotheses (5.1)–(5.4) of theorem 5.1′ in [14], in particular with $H(u,p) = \frac{1}{2}p^2$ and $\hat{\delta}(t) = 2a(t)t^{\alpha}$. Clearly $\hat{\delta} \in L^1(J)$ by (5.6). Therefore the only solution of (5.8) that approaches zero at $\infty$ is $w \equiv 0$.

This being shown, it is easy to argue that all nontrivial solutions of (5.8) are oscillatory, with amplitude approaching a nonzero limit as $t \to \infty$. It may be noted that the behavior of solutions of (5.8) when $\alpha < -1$ is then essentially the same as for the *wave equation* itself in a bounded domain with zero boundary data.

When $\alpha > 1$, solutions of (5.5) again do not in general approach zero as $t \to \infty$, though their behavior is quite different from the case $\alpha < -1$. We say that a function

$$\psi = \psi(x) \in Y \equiv \operatorname{span}\{\varphi_k\}_{k=1}^{\infty}$$

is *attainable* if there exists a solution $u \in K$ of (5.5) such that

$$\lim_{t\to\infty} \|u(t) - \psi\|_{L^2} = 0. \tag{5.9}$$

THEOREM 5.1. *Every function $\psi \in Y$ is attainable for problem* (5.5). *In turn, the set of attainable functions is dense in $L^2$.*

Proof: We first show that every eigenfunction $\varphi_k$ of $-\Delta + V(x)$ in $\Omega$, with eigenvalue $\mu_k > 0$, is attainable. For this purpose consider the function

$$u_k(t,x) = w_k(t)\varphi_k(x),$$

which satisfies (5.5) if and only if $w_k$ is a solution of (5.8) with $\mu = \mu_k$. By [13], theorem 4.4, because

$$1/a(t)t^{\alpha} \in L^1(J)$$

by (5.6), it follows that the set of attainable limits at $\infty$ of solutions of (5.8) is dense in $\mathbb{R}$. (Note that all solutions of (5.8) can be continued over all of $J$ because of linearity.) On the other hand, again because (5.8) is linear, the set of attainable limits for (5.8) must in fact be all of $\mathbb{R}$. Hence for an appropriate solution $w_k \not\equiv 0$ we get

$$\lim_{t\to\infty} \|u_k(t) - \varphi_k\|_{L^2} = 0.$$

Finally, again from the linearity of (5.5), we obtain (5.9) for every $\psi \in Y$.

It is interesting to note that when $\alpha > 1$ the specific function

$$u(t,x) = \left(1 + \frac{\mu}{(\alpha-1)t^{\alpha-1}}\right)\varphi(x)$$

solves (5.5) corresponding to the choice

$$a(t) = a_{\mu}(t) = 1 + \frac{\mu}{(\alpha-1)t^{\alpha-1}} + \frac{\alpha}{t^{\alpha+1}}.$$

Note that (5.6) is then satisfied with $C = (\alpha^2 - 1 + \mu)/(\alpha - 1) > 1$. Here it is obvious that

$$\lim_{t\to\infty} \|u(t) - \varphi\|_{L^2} = 0.$$

We can also treat the problem

$$(5.10)\qquad \begin{aligned} &u_{tt} - \Delta u + a(t)t^{\alpha}(\log t)^{\beta}u_t + V(x)u = 0, \quad \text{in } J \times \Omega, \\ &u(t,x) = 0, \quad \text{on } J \times \partial\Omega. \end{aligned}$$

The results are essentially the same, namely,

(a) If $|\alpha| < 1$; or if $\alpha = 1$, $\beta \leqq 1$; or if $\alpha = -1$, $\beta \geqq -1$, then $Eu(t) \to 0$ as $t \to \infty$ for every strong solution $u$ of (5.10).

(b) If $\alpha = -1$, $\beta < -1$; or if $\alpha < -1$, then there exist oscillating solutions of (5.10) that do not approach zero as $t \to \infty$.

(c) If $\alpha = 1$, $\beta > 1$; or if $\alpha > 1$, then every function $\psi \in Y$ is attainable at $\infty$, and in turn the set of attainable functions is dense in $L^2$.

## 6. Other Operators

As noted in the introduction, the Laplace operator $-\Delta$ in (1.1) can be replaced by the polyharmonic operator $(-\Delta)^L$, $L \geqq 1$, by the degenerate Laplace operator $-\Delta_s$, or, more generally, by

$$-\operatorname{div} A(x, Du) = -\sum_{i=1}^{n} D_i A_{ij}(x, Du), \qquad j = 1, \ldots, N,$$

where $u = (u_1, \ldots, u_N)$ and $A \in C^1(\Omega \times \mathbb{R}^{n\times N} \to \mathbb{R}^{n\times N})$. We treat each of these cases briefly.

### 6.1. The Polyharmonic Operator

Consider the problem

$$(6.1)\qquad \begin{cases} u_{tt} + (-\Delta)^L u + Q(t, x, u, u_t) + f(x, u) = 0, & \text{in } J \times \Omega, \\ u(t,x) = 0,\ Du(t,x) = 0,\ \ldots,\ D^{2(L-1)}u(t,x) = 0, & \text{on } J \times \partial\Omega. \end{cases}$$

Define

$$X = \left[W_0^{L,2}(\Omega)\right]^N, \qquad K' = C(J \to X) \cap C^1(J \to L^2),$$

$$\mathscr{D}_L\varphi = \begin{cases} D\Delta^j\varphi & \text{if } L = 2j+1, \\ \Delta^j\varphi & \text{if } L = 2j, \end{cases}$$

$$K = \{\phi \in K' : \mathscr{F}\phi \text{ is well-defined and locally bounded on } J\},$$

where $\mathscr{F}\phi$ is as in Section 2, and the total energy of $\phi$ is given by

$$E\phi = E\phi(t) = \frac{1}{2}\|\phi'\|^2 + \frac{1}{2}\|\mathscr{D}_L\phi\|^2 + \mathscr{F}\phi.$$

By a *strong solution* of (6.1) we mean a function $u \in K$ satisfying the following two conditions:

(A) *Conservation law*

$$\text{(i)} \qquad \langle Q(t,\cdot,u,u'),u'\rangle \in L^1_{\text{loc}}(J),$$

$$\text{(ii)} \qquad Eu]_0^t = -\int_0^t \langle Q(s,\cdot,u,u'),u'\rangle ds \quad \text{for all } t \in J.$$

(B) *Distribution identity*

$$\langle u',\phi\rangle]_0^t = \int_0^t \{\langle u',\phi'\rangle - \langle \mathscr{D}_L u, \mathscr{D}_L\phi\rangle - \langle Q(s,\cdot,u,u'),\phi\rangle - \langle f(\cdot,u),\phi\rangle\} ds$$

for all $t \in J$ and $\phi \in K$.

By the argument of Section 2, this definition is meaningful under hypotheses (H1) through (H3) provided $r$ now denotes the Sobolev exponent for the space $W_0^{L,2}(\Omega)$; that is,

$$r = \frac{2n}{n-2L}, \qquad n > 2L \tag{6.2}$$

($r = \infty$ if $n = 1,\dots,2L-1$; $2 < r < \infty$ if $n = 2L$, because $\Omega$ is bounded).

Theorems 3.1 and 4.1 carry over equally, the only change being that $r$ is given by (6.2).

It almost goes without saying that (H1), (H2), and (AS) can be replaced by (H1)$'$, (H2)$'$, and (AS)$'$ when $n = 1,\dots,2L-1$.

### 6.2. The Degenerate Laplace Operator

Consider the problem

$$\begin{cases} u_{tt} - \operatorname{div}(|Du|^{s-2}Du) + Q(t,x,u,u_t) + f(x,u) = 0, & \text{in } J\times\Omega, \\ u(t,x) = 0, & \text{on } J \times \partial\Omega, \end{cases} \tag{6.3}$$

where $s > 1$. Define $X = [W_0^{1,s}(\Omega)]^N$, $K' = C(J \to X) \cap C^1(J \to L^2)$, and

$$K = \{\phi \in K' : \mathscr{F}\phi \text{ is well-defined and locally bounded on } J\}.$$

The total energy is now given by

$$E\phi = E\phi(t) = \frac{1}{2}\|\phi'\|^2 + \frac{1}{s}\|D\phi\|^s_{L^s} + \mathscr{F}\phi,$$

while the distribution identity becomes

$$\langle u', \phi\rangle]_0^t = \int_0^t \{\langle u', \phi'\rangle - \langle |Du|^{s-2}Du, D\phi\rangle - \langle Q(\tau, \cdot, u, u'), \phi\rangle - \langle f(\cdot, u), \phi\rangle\} d\tau$$

for all $t \in J$ and $\phi \in K$. The corresponding conditions (A) and (B) are then meaningful under assumptions (H1) through (H3), again provided that $r$ is the Sobolev number for the space $[W_0^{1,s}(\Omega)]^N$, namely,

$$r = \frac{ns}{n-s}, \qquad n > s \tag{6.4}$$

($r = \infty$ if $n < s$; $2 < r < \infty$ if $n = s$). Theorems 3.1 and 4.1 carry over equally, *mutatis mutandis*.

### 6.3. The Operator div $A(x, Du)$

In order to formulate a conservation law for this case, it is necessary to assume that $A = A(x, w)$ is derivable from a potential, namely, that

$$A(x, w) = \frac{\partial G}{\partial w}(x, w), \tag{6.5}$$

where $G \in C^2(\Omega \times \mathbb{R}^{n\times N} \to \mathbb{R})$ and $G(x, 0) = 0$. Note, however, that for strong solutions there is no need to assume either A of class $C^1$ or $G$ of class $C^2$. The total energy now has the form

$$Eu = Eu(t) = \frac{1}{2}\|u'\|^2 + \|G(x, Du)\|_{L^1} + \mathcal{F}u. \tag{6.6}$$

For an adequate definition of a solution of the Dirichlet problem, we suppose that

$$|A(x, w)| \leqq \text{const}\,(1 + |w|^{s-1}), \qquad s \geqq 1, \tag{6.7}$$

for all $(x, w) \in \Omega \times \mathbb{R}^{n\times N}$. We can then take $X = [W_0^{1,s}(\Omega)]^N$ and $K$ in the usual way. Relation (6.5) yields the formula

$$G(x, w) = \int_0^1 (A(x, \tau w), w)\, d\tau, \tag{6.8}$$

(cf. (2.4)), and so by (6.7)

$$|G(x, w)| \leqq \text{const}\,(|w| + |w|^s). \tag{6.9}$$

Thus the energy $Eu$ is meaningfully defined. Similarly, the distribution identity,

$$\langle u',\phi\rangle]_0^t = \int_0^t \{\langle u',\phi'\rangle - \langle A(\cdot,Du),D\phi\rangle - \langle Q(\tau,\cdot,u,u'),\phi\rangle - \langle f(\cdot,u),\phi\rangle\}d\tau$$

for all $t \in J$ and $\phi \in K$, is well-defined in view of (6.7) and the assumptions (H1) through (H3), where $r$ is again given by (6.4).

Finally, in order to carry out the proofs for asymptotic stability, in particular the preliminary results given in Lemmas 3.1 and 3.4, we must assume $A$ to be weakly coercive, in the sense that for all $w_0 > 0$ there exists a constant $C_1 = C_1(w_0) > 0$ such that

$$(6.10) \qquad (A(x,w),w) \geqq C_1\,|w|^s \quad \text{for all } (x,w) \in \Omega \times \mathbb{R}^{n\times N} \text{ with } |w| \geqq w_0.$$

To apply the condition (6.10), we require the following abstract result.

LEMMA 6.1. *Let $Y$ be a real Banach space and $\Psi$ a continuous real function on $\Omega \times Y$, having the property that for every $y_0 > 0$ there exists a constant $C_1 = C_1(y_0) > 0$ such that*

$$(6.11) \qquad \Psi(x,y) \geqq C_1\,|y|_Y^s, \qquad s \geqq 1,$$

*for all $(x,y) \in \Omega \times Y$ with $|y|_Y \geqq y_0$.*

*Assume that $w \in L^s(\Omega \to Y)$ with $\|w\|_{L^s} \geqq c > 0$. Then*

$$(6.12) \qquad \int_\Omega \Psi(x,w(x))\,dx \geqq C_2\|w\|_{L^s}^s > 0,$$

*where $C_2$ depends only on $c$, $s$, and $|\Omega|$.*

Proof: Let $y_0$ be a positive constant to be chosen later, and define

$$\Omega_0 = \{x \in \Omega : |w(x)|_Y \geqq y_0\}.$$

Then, because $\Psi(x,y)$ is obviously nonnegative in view of (6.11), we have

$$\begin{aligned}\int_\Omega \Psi(x,w(x))\,dx &\geqq \int_{\Omega_0} \Psi(x,w(x))\,dx \geqq C_1(y_0)\int_{\Omega_0} |w(x)|_Y^s\,dx \\ &\geqq C_1(y_0)\{\|w\|_{L^s}^s - y_0^s|\Omega\setminus\Omega_0|\} \\ &\geqq \tfrac{1}{2}C_1(y_0)\|w\|_{L^s}^s + \tfrac{1}{2}C_1(y_0)\{c^s - 2y_0^s|\Omega|\},\end{aligned}$$

because $\|w\|_{L^s}^s \geqq c$. Now fix

$$y_0 = \bar{y}_0 = c/(2|\Omega|)^{1/s}$$

to obtain (6.12) with $C_2 = \frac{1}{2}C_1(\bar{y}_0)$, as required.

The proof of the analogue of Lemma 3.4 can now be carried out by following essentially the ideas in Section 3. In correspondence to (3.13) it must be shown that

$$\|u'\|^2 + \langle A(\cdot, Du), Du\rangle + \langle f(\cdot, u), u\rangle \geqq \alpha \quad \text{on } J, \tag{6.13}$$

provided that $Eu(t) \geqq \ell > 0$. In the set $J_1$ we have

$$\|u'(t)\|^2 + 2\|G(\cdot, Du)\|_{L^1} \geqq \ell.$$

If $\|u'(t)\|^2 \geqq \ell/2$ we get (6.13) with $\alpha = \ell/2$. On the other hand, if we have $\|G(\cdot, Du)\|_{L^1} \geqq \ell/4$, then by (6.9) there results

$$\text{const}\left(\|Du\|_{L^1} + \|Du\|_{L^s}^s\right) \geqq \ell/4,$$

and in turn, then,

$$\|Du\|_{L^s} \geqq c = c(\ell, |\Omega|) > 0.$$

Hence by (6.10) and Lemma 6.1, with $Y = \mathbb{R}^{n\times N}$ and $\Psi(x, w) = \big(A(x, w), w\big)$, we get

$$\langle A(\cdot, Du), Du\rangle \geqq C_3 = C_2(c, |\Omega|)c^s > 0,$$

with $c = c(\ell, |\Omega|)$. Thus (6.13) holds for $\alpha = C_3$.

For $t \in J_2$ we follow the proof of Lemma 3.4 essentially word for word, with the exception that $\|Du\|$ is replaced by $\|Du\|_{L^s}$, and that Lemma 6.1 is again applied at the final step of the discussion of each of the cases $p \leqq r$ and $p > r$. Thus (6.13) is proved for all $t \in J$.

To obtain the analogue of Lemma 3.1 it remains only to prove

$$\|Du\|_{L^s} \in L^\infty(J). \tag{6.14}$$

However, with the use of (6.8) and (6.10), it is easy to show that

$$G(x, w) \geqq \tilde{C}_1\,|w|^s \quad \text{for all } (x, w) \in \Omega \times \mathbb{R}^{n\times N} \text{ with } |w| \geqq w_0, \tag{6.15}$$

where $\tilde{C}_1 = C_1(\frac{1}{2}w_0)/2s$. Condition (6.14) then follows at once from Lemma 6.1 and the fact that $Eu$, and hence $\|G(\cdot, Du)\|_{L^1}$, is bounded on $J$.

The remaining arguments of Sections 3 and 4 can be repeated essentially unchanged, and therefore the main Theorems 3.1 and 4.1 continue to hold for the operator $\operatorname{div} A(x, Du)$, of course with the energy given by (6.6).

We note finally that the principal conclusion $Eu(t) \to 0$ as $t \to \infty$ also implies that $\|Du\|_{L^s} \to 0$ as $t \to \infty$, this again being a consequence of (6.15) and Lemma 6.1.

Conditions (6.7) and (6.10) are satisfied for the canonical function

$$A(w) = (1 + |w|^a)^b |w|^{c-2} w, \quad a \geqq 0, \quad s = ab + c \geqq 1, \quad c > 1.$$

This example includes the degenerate Laplace operator when $a = b = 0$ and $c = s > 1$. A second case of importance occurs when $a = c = 2$, $b = -1/2$ and $s = 1$, yielding the mean curvature operator $A(w) = w/\sqrt{1+|w|^2}$.

## 7. Nonautonomous Higher-Order Damping

Consider the problem

$$
\left\{
\begin{array}{ll}
u_{tt} - \Delta(u + \rho(t)u_t) + Q(t,x,u,u_t) + f(x,u) = 0, & \text{in } J \times \Omega, \\
u(t,x) = 0, & \text{on } J \times \partial\Omega,
\end{array}
\right. \tag{7.1}
$$

where $\rho \in L^1_{\text{loc}}(J \to \mathbb{R}^+_0)$. Here we take

$$X = \left[W_0^{1,2}(\Omega)\right]^N, \qquad K' = C^1(J \to X),$$

and $K$ in the usual way. Moreover, again,

$$Eu = Eu(t) = \frac{1}{2}\|u'\|^2 + \frac{1}{2}\|Du\|^2 + \mathcal{F}u.$$

By a *strong solution* of (7.1) we mean a function $u \in K$ satisfying

(A) *Conservation law*

(i) $$\langle Q(t,\cdot,u,u'),u'\rangle \in L^1_{\text{loc}}(J),$$

(ii) $$Eu\big]_0^t = -\int_0^t \{\langle Q(s,\cdot,u,u'),u'\rangle + \rho\|Du'\|^2\}ds \quad \text{for all } t \in J.$$

(B) *Distribution identity*

$$\langle u',\phi\rangle\big]_0^t = \int_0^t \{\langle u',\phi'\rangle - \langle Du,D\phi\rangle - \rho(s)\langle Du',D\phi\rangle - \langle Q(s,\cdot,u,u'),\phi\rangle - \langle f(\cdot,u),\phi\rangle\}ds$$

for all $t \in J$ and $\phi \in K$.

As before, it is easy to see that this definition is meaningful when hypotheses (H1) through (H3) hold. We now turn to the asymptotic stability of solutions of (7.1). We must amplify the discussion already given in Section 3 to take into account the term $\rho(t)\langle Du',D\phi\rangle$ in (B).

THEOREM 7.1. *Let* (H1), (H3), *and* (AS) *hold. Suppose there exists* $k \in CBV(J \to \mathbb{R}_0^+)$ *such that*

$$k \notin L^1(J), \tag{7.2}$$

$$\liminf_{t\to\infty} \frac{\left(\int_0^t \rho k^2 ds\right)^{1/2} + \left(\int_0^t (\delta + \sigma^{1-m}) k^m ds\right)^{1/m}}{\int_0^t k ds} < \infty. \tag{7.3}$$

*Then along any strong solution u of* (7.1) *there holds*

$$\lim_{t\to\infty} Eu(t) = 0. \tag{7.4}$$

Before proving this result we require two further lemmas.

LEMMA 7.1. *In addition to the conclusions* (3.4) through (3.6) *of Lemma 3.1, we also have*

$$\rho(t)\|Du'(t)\|^2 \in L^1(J). \tag{7.5}$$

The proof is the same as that of Lemma 3.1, with (3.6) and (7.5) deduced in exactly the same way as (3.6) before.

LEMMA 7.2. *For all* $t \geqq T \geqq 0$ *we have*

$$\int_T^t k\rho\langle Du, Du'\rangle ds \leqq \varepsilon(T)\left(\int_T^t \rho k^2 \, ds\right)^{1/2}, \tag{7.6}$$

*where* $\varepsilon(T) \to 0$ *as* $T \to \infty$.

Proof: By Schwarz's inequality and (3.5)

$$\langle Du, Du'\rangle \leqq \text{const}\, \|Du'\|.$$

Hence, by integration from $T$ to $t$, and another use of Schwarz's inequality,

$$\begin{aligned}\int_T^t k\rho\langle Du, Du'\rangle ds &\leqq \text{const}\left(\int_T^t \rho k^2 ds\right)^{1/2}\left(\int_T^t \rho\|Du'\|^2 ds\right)^{1/2}\\ &\leqq \varepsilon(T)\left(\int_T^t \rho k^2 ds\right)^{1/2},\end{aligned}$$

where $\varepsilon(T) \to 0$ as $T \to \infty$ by (7.5).

Proof of Theorem 7.1: As in the proof of Theorem 3.1, we first treat the simpler case in which $k$ is also of class $C^1(J)$. Suppose for contradiction that $\ell > 0$ in (3.12).

Consider the Lyapunov function

$$V(t) = \langle u', \phi\rangle, \qquad \phi = k(t)u \in K.$$

Then by the distribution identity (B) above, for any $t \geqq T \geqq 0$, we have

$$V(s)]_T^t = \int_T^t \{k'\langle u, u'\rangle + 2k\|u'\|^2 - k(\|u'\|^2 + \|Du\|^2 + \langle f(\cdot, u), u\rangle) \\ - k\rho\langle Du, Du'\rangle - k\langle Q(s, \cdot, u, u'), u\rangle\} ds.$$

If we apply Lemmas 3.2 through 3.4, which continue to hold in the present case, as one easily sees, together with Lemmas 7.1 and 7.2, there results

$$\begin{aligned} V(s)]_T^t \leqq \operatorname{const} \int_T^\infty |k'| ds + 2\vartheta \int_T^t k\, ds \\ - \alpha \int_T^t k\, ds + 2\varepsilon(T)C(\vartheta)\left(\int_0^t \sigma^{1-m} k^m\, ds\right)^{1/m} \\ + \varepsilon(T)\left\{1 + \left(\int_0^t \rho k^2\, ds\right)^{1/2} + \left(\int_0^t \delta k^m\, ds\right)^{1/m}\right\}. \end{aligned} \tag{7.7}$$

The rest of the proof is the same as for Theorem 3.1, except of course (7.3) is used instead of (3.2).

We also have a related result of independent interest, valid under a much weaker form of (AS); that is,

$$|Q(t, x, u, v)| \leqq \delta(t)[|v|^{m-1} + |v|^{q-1}], \tag{AS1}$$

where $m$, $q$, and $\delta$ are as in (AS); see Section 3.

THEOREM 7.2. *Assume that* (H1), (H3), *and* (AS1) *are satisfied, and suppose there exists* $k \in CBV(J \to \mathbb{R}_0^+)$ *such that*

$$k \notin L^1(J), \qquad k(t) \leqq \operatorname{const} \rho(t) \quad \textit{for all sufficiently large } t \in J, \tag{7.8}$$

$$\liminf_{t\to\infty} \frac{\left(\int_0^t \rho k^2\, ds\right)^{1/2} + \left(\int_0^t \delta k^m\, ds\right)^{1/m}}{\int_0^t k\, ds} < \infty. \tag{7.9}$$

*Then* (7.4) *holds.*

To begin with, observe that Lemmas 7.1, 7.2 and 3.2, 3.4 continue to hold when (AS) is replaced by (AS1). Lemma 3.3 is no longer valid, but in its place we use the following.

LEMMA 7.3. *Let* $(7.8)_2$ *be satisfied. Then, for all* $t \geqq T \geqq 0$,

$$\int_T^t k\,\|u'\|^2\,ds \leqq \varepsilon(T),$$

*where* $\varepsilon(T) \to 0$ *as* $T \to \infty$.

Proof: By Poincaré's inequality and the fact that $u' \in C(J \to X)$, we have

$$\|u'\| \leqq \text{const}\,\|Du'\|.$$

Consequently, for all $t \geqq T \geqq 0$, with $T$ sufficiently large,

$$\int_T^t k\,\|u'\|^2 ds \leqq \text{const} \int_T^t \rho\,\|Du'\|^2 ds$$

by $(7.8)_2$. The assertion now follows from (7.5).

Proof of Theorem 7.2: We argue as in the proof of Theorem 7.1, until (7.7). By Lemmas 3.2, 3.4, and 7.1–7.3, we obtain, instead of (7.7),

$$V(s)\big]_T^t \leqq \text{const} \int_T^\infty |k'|ds - \alpha \int_T^t k\,ds + \varepsilon(T)\left\{3 + \left(\int_0^t \rho k^2\,ds\right)^{1/2} + \left(\int_0^t \delta k^m\,ds\right)^{1/m}\right\}.$$

The proof is now completed as before.

*Remark* 7.1. In case $n = 1$ we can allow slightly weaker hypotheses of the type indicated in Sections 2 and 3. We leave the details to the reader.

Theorems 7.1 and 7.2 have a number of corollaries corresponding to those of Section 5.

COROLLARY 7.1. *Suppose*

$$\text{(7.10)} \qquad \liminf_{t\to\infty}\left\{\left(\frac{1}{t^2}\int_0^t \rho\,ds\right)^{1/2} + \left(\frac{1}{t^m}\int_0^t (\delta + \sigma^{1-m})ds\right)^{1/m}\right\} < \infty.$$

*Then* (7.4) *holds.*

*If $\rho(t) \geqq \rho_0 > 0$ for all $t \in J$ sufficiently large, then the term $\sigma^{1-m}$ can be dropped in* (7.10).

Proof: Take $k = 1$ in Theorems 7.1 and 7.2.

COROLLARY 7.2. *Suppose*

$$(7.11)\quad \liminf_{t\to\infty}\left\{\left(\frac{1}{\log^2 t}\int_1^t \rho\frac{ds}{s^2}\right)^{1/2}+\left(\frac{1}{\log^m t}\int_1^t(\delta+\sigma^{1-m})\frac{ds}{s^m}\right)^{1/m}\right\}<\infty.$$

*Then* (7.4) *holds.*

*If $\rho(t) \geqq \rho_0/t$ for all $t \in J$ sufficiently large, then the term $\sigma^{1-m}$ can be dropped in* (7.11).

Proof: Take $k(t) = \min\{1, 1/t\}$ in Theorems 7.1 and 7.2.

A result corresponding to Corollary 5.3 can also be given. We leave the statement to the reader.

The condition that $k$ be of bounded variation in Theorems 7.1 and 7.2 can be replaced by a differential relation corresponding to the results in Section 4. Exactly as there, we get the following analogue of Theorem 4.1.

THEOREM 7.3. *Let the assumptions of Theorem 7.1 hold, except that $k$ is bounded and absolutely continuous rather than of bounded variation. Suppose also that the function $\omega$ in* (AS) *satisfies* (4.1), *and that $k$ obeys* (4.2) *and* (4.3). *Then* (7.4) *is valid.*

THEOREM 7.4. *Let the assumptions of Theorem 7.1 hold, except that $k$ is bounded and absolutely continuous rather than of bounded variation. If also*

$$(7.12)\qquad |k'| \leqq \text{const}\sqrt{\rho k}\quad a.e.\ in\ J,$$

*then* (7.4) *holds.*

Proof: Again it is only necessary to estimate the term $k'\langle u, u'\rangle$ in the identity (3.18). We have

$$|k'\langle u,u'\rangle| \leqq \text{const}\sqrt{\rho k}\,\|u\|\cdot\|u'\| \leqq \text{const}\sqrt{\rho k}\,\|u\|\cdot\|Du'\|$$

by the Poincaré–Sobolev inequality. Then by (3.5) and Hölder's inequality

$$\int_T^t |k'\langle u,u'\rangle|ds \leqq \text{const}\left(\int_T^t k ds\right)^{1/2}\left(\int_T^t \rho\,\|Du'\|^2 ds\right)^{1/2}$$
$$\leqq \varepsilon(T)\left\{1+\int_T^t k\,ds\right\}.$$

The remainder of the proof is the same as before.

THEOREM 7.5. *Let the assumptions of Theorem 7.2 hold, except that $k$ is bounded and absolutely continuous rather than of bounded variation. If* (7.12) *is satisfied, then* (7.4) *holds.*

The proof is the same as that of Theorem 7.2, except that the term $k'\langle u, u'\rangle$ is estimated as in the proof of Theorem 7.4.

**The Linear Case**

Consider the problem

$$(7.13)\qquad \begin{cases} u_{tt} - \Delta(u + b(t)t^{\beta}u_t) + a(t)t^{\alpha}u_t + V(x)u = 0, & \text{in } J \times \Omega, \\ u(t,x) = 0, & \text{on } J \times \partial\Omega, \end{cases}$$

where $N = 1$, $J = [1, \infty)$, $\Omega$ is a bounded open subset of $\mathbb{R}^n$, $V$ is a continuous function on $\overline{\Omega}$, and $\alpha$, $\beta \in \mathbb{R}$. We assume that

$$1/C \leqq a(t) \leqq C, \qquad 1/D \leqq b(t) \leqq D, \quad \text{in } J,$$

where $C$, $D > 0$ are constants. Clearly (H1) is satisfied, and (AS) holds with $m = 2$, $\omega(s) \equiv s$, $\rho(t) = b(t)t^{\beta}$, and appropriate $\delta$ and $\sigma$.

*Then strong solutions of* (7.13) *are asymptotically stable, that is,* (7.4) *holds, whenever either*

$$\text{(i)}\quad \alpha \leqq 1 \textit{ and } |\beta| \leqq 1 \qquad \textit{or} \qquad \text{(ii)}\quad \beta \leqq 1 \textit{ and } |\alpha| \leqq 1.$$

To see this, in case (i) use the second part of Corollary 7.2 with $m = 2$, $\delta(t) = Ct$, and $(1/Dt) \leqq \rho(t) \leqq Dt$. Similarly, in case (ii) we use the first part of Corollary 7.1 with $m = 2$, $\delta(t) = Ct$, $\sigma(t) = 1/Ct$, and $\rho(t) \leqq Dt$.

Note in fact that for the result (i) the condition $a(t) \geqq 1/C$ is not needed — merely that $a(t) \geqq 0$; similarly, for (ii) the condition $b(t) \geqq 1/D$ is not necessary.

Exactly as in the case of problem (5.5), when $\alpha < -1$ and $\beta < -1$, there exist oscillating solutions of (7.13) that do not approach zero as $t \to \infty$. Indeed, it is enough to consider solutions having the separated form (5.7) and to observe that the function $w$ then is a solution of the linear ordinary differential equation

$$w'' + [a(t)t^{\alpha} + \mu b(t)t^{\beta}]w' + \mu w = 0.$$

Similarly, when either $\alpha > 1$ or $\beta > 1$, the conclusion of Theorem 5.4 remains valid for (7.13), with essentially the same proof.

**Acknowledgment.** P. Pucci is a member of Gruppo Nazionale di Analisi Funzionale e sue Applicazioni of the Consiglio Nazionale delle Ricerche. This research has been partly supported by the Italian Ministero della Università e della Ricerca Scientifica e Tecnologica.

## Bibliography


[1] Ball, J. M., *Stability theory for an extensible beam*, J. Differential Equations 14, 1973, pp. 399–418.

[2] Cannarsa, P., Da Prato, G., and Zolesio, J. P., *The damped wave equation in a moving domain*, J. Differential Equations 85, 1990, pp. 1–16.

[3] Caughen, J. K., and Ellison, J., *Existence, uniqueness and stability of solutions of a class of nonlinear differential equations*, J. Math. Anal. Appl. 51, 1975, pp. 1–32.

[4] Ebihara, Y., *On some nonlinear evolution equations with strong dissipation*, J. Differential Equations 30, 1978, pp. 149–164.

[5] Hale, J. K., *Asymptotic behavior of dissipative systems*, Mathematical Surveys and Monographs No. 25, American Mathematical Society, Providence, RI, 1988.

[6] Haraux, A., *Recent results on semilinear hyperbolic problems in bounded domains*, pp. 118–126, in: *Partial Differential Equations*, Lecture Notes in Mathematics Vol. 1324, Springer-Verlag, Berlin, 1988.

[7] Lions, J. L., and Strauss, W. A., *On some nonlinear evolution equations*, Bull. Soc. Math. France 93, 1965, 43–96.

[8] Marcati, P., *Decay and stability for nonlinear hyperbolic equations*, J. Differential Equations 155, 1984, pp. 30–58.

[9] Marcati, P., *Stability for second order abstract evolution equations*, Nonlinear Anal. 18, 1984, pp. 237–252.

[10] Nakao, M., *Bounded, periodic or almost periodic solutions of nonlinear hyperbolic partial differential equations*, J. Differential Equations 23, 1977, pp. 368–386.

[11] Nakao, M., *Asymptotic stability for some nonlinear evolution equations of second order with unbounded dissipative terms*, J. Differential Equations 30, 1978, pp. 54–63.

[12] Nakao, M., *Energy decay for the wave equation with a nonlinear weak dissipation*, Differential Integral Equations 18, 1995, pp. 681–688.

[13] Pucci, P., and Serrin, J., *Continuation and limit behavior for damped quasi–variational systems*, pp. 157-173 in: *Degenerate Diffusions*, W.–M. Ni et al., eds., IMA Vol. Math. Appl. Vol. 47, Springer-Verlag, New York, 1993.

[14] Pucci, P., and Serrin, J., *Precise damping conditions for global asymptotic stability for nonlinear second order systems*, Acta Math. 170, 1993, pp. 275–307.

[15] Pucci, P., and Serrin, J., *Asymptotic stability for intermittently controlled nonlinear oscillators*, SIAM J. Math. Anal. 25, 1994, pp. 815–835.

[16] Pucci, P., and Serrin, J., *Asymptotic stability for nonlinear parabolic systems*, in: *Proc. Conference, Energy Methods in Continuum Mechanics*, Kluwer, Dordrecht, in press.

[17] Strauss, W. A., *On continuity of functions with values in various Banach spaces*, Pacific J. Math. 19, 1966, pp. 543–551.

[18] Webb, G., *Existence and asymptotic behavior for a strongly damped nonlinear wave equation*, Canad. J. Math. 32, 1980, pp. 631–643.

[19] Zhu, X., Thesis, University of Minnesota, 1996.



PATRIZIA PUCCI
Dipartimento di Matematica
Universitá di Perugia
Via Vanvitelli, 1
06123, Perugia, Italy
E-mail: pucci@unipg.it

JAMES SERRIN
University of Minnesota
School of Mathematics
127 Vincent Hall
206 Church St. S.E.
Minneapolis, MN 55455
E-mail: serrin@math.umn.edu


Received December 1994.

*Contemp. Math.* 208 (1997), 253–263.


**Reprint II.4.[iii]**

©Contemporary Mathematics
Volume 208, 1997

# Some Remarks on Global Nonexistence for Nonautonomous Abstract Evolution Equations

*Howard A. Levine   Patrizia Pucci   James Serrin*

Abstract. In this paper we consider the problem of non-continuation of solutions of the initial value problem for abstract evolution equations of the form

$$Pu_{tt} + Q(t)u_t + A(t,u) = F(t,u), \quad t \in J = [0,\infty)$$

where $P$ and $Q$ are linear self-adjoint operators, and $A(t,u)$ and $F(t,u)$ are respectively a linear operator in $u$ (typically of differential type) and a nonlinear driving force. We also consider the case in which the linear dissipation, $Q(t)u_t$ is replaced by a nonlinear dissipation, $Q(t,u,u_t)$. The results extend earlier works of Levine and Levine and Serrin.

## §1. Introduction.

In a series of papers [1–3] by the first author the problem of non–continuation was studied for abstract evolution equations of the type

$$\text{(1.1)} \qquad Pu_{tt} + Qu_t + A(t,u) = F(u), \quad t \in J = [0,\infty),$$

where $P$ and $Q$ are linear self-adjoint operators, and $A(t,u)$ and $F(u)$ are respectively a linear operator in $u$ (typically of differential type) and a nonlinear driving force. In [4] the operator $A$ was allowed some degree of nonlinearity, but at the expense of dropping dependence on $t$.

In a recent paper [5] this work was extended to evolution equations of the form

$$\text{(1.2)} \qquad (P(u_t))_t + Q(t,u,u_t) + A(u) = F(u),$$

where all the operators are allowed to be nonlinear, though with $A$ independent of time.

The purpose of the present paper is to generalize the above results, and to some extent to unify them, by admitting *the same degree of nonlinearity* for the operators $A$ and $F$ in both (1.1) and (1.2), by letting both $A$ *and* $F$ depend on time, and finally by allowing time–dependent operators $Q$ in (1.1).

Our results are established in two theorems, the first (Section 2) corresponding to the ideas and methods of [1–3], and the second (Section 3) to those of [5]. We end with the presentation of various examples, of interest in themselves, illustrating the range of our conclusions when interpreted in concrete cases. In this respect it almost goes without saying that the results of both Sections 2 and 3 apply to the wave operator $u_{tt} - \Delta u$, or its degenerate counterpart $u_{tt}-\text{div}(|Du|^{s-2}Du)$, and to the corresponding parabolic operators $u_t - \Delta u$ and $u_t-\text{div}(|Du|^{s-2}Du)$.

While the results of Section 2 are, in some ways, more special than those of Section 3, we note in Section 4 several situations of interest covered by Theorem 1 but not by Theorem 2.

**§2. The case with linear dissipation.**

Let $X$ be a Banach space, and $X'$ its dual space. If $x \in X$ and $x' \in X'$, we shall write $\langle x', x\rangle_X$ to denote the natural pairing of $x$ and $x'$, that is $\langle x', x\rangle_X = x'(x)$.

Let $V$ be a Hilbert space. An operator $P : V \to V'$ will be called *symmetric* if

$$\langle Pv, w\rangle_V = \langle Pw, v\rangle_V \quad \text{for all } v, w \in V,$$

and *non-negative definite* if

$$\langle Pv, v\rangle_V \geq 0 \quad \text{for all } v \in V.$$

It is easy to check that a symmetric operator must be linear and, moreover, continuous by the uniform boundedness theorem.

In this section we study the evolution equation

$$(2.1) \qquad Pu_{tt} + Q(t)u_t + A(t, u) = F(t, u), \qquad t \in J,$$

where $P$ is symmetric and non-negative definite from $V$ into $V'$. We suppose that the *dissipation operator* $Q(t)$ is, for each $t \in J$, symmetric and non-negative definite from an appropriate Hilbert space $Y$ into its dual $Y'$. In addition, assume $Q \in C(J \to B(Y, Y'))$, that is $\langle Q(\cdot)v, w\rangle_Y : J \to \mathbb{R}$ is continuous for each $v, w \in Y$. Note that $P \equiv 0$ and $Q \equiv 0$ are specifically allowed.

Finally, the operators $A$ and $F$ are given by

$$A : J \times W \to W', \qquad F : J \times X \to X',$$

with $W$, $X$ Banach spaces and $W'$, $X'$ their duals. In order to define the energy $\mathcal{E}u$ of a solution of (2.1), see below, it is necessary that there exist $C^1$ potentials

$$\mathcal{A} : J \times W \to \mathbb{R}, \qquad \mathcal{F} : J \times X \to \mathbb{R},$$

such that for each fixed $t$ the operators $A$ and $F$ are the Fréchet derivatives with respect to $u$ of $\mathcal{A}$ and $\mathcal{F}$, respectively; by normalization we can take $\mathcal{A}(t, 0) \equiv 0$, $\mathcal{F}(t, 0) \equiv 0$.

Now suppose that there is given a nontrivial subspace $G$ of $V$, $W$, $X$ and $Y$ – not necessarily closed. Let

$$K = \{\varphi : J \to G \mid \varphi \in C(J \to W) \cap C(J \to X) \cap C^1(J \to V) \cap AC(J \to Y)\}.$$

We say that $u$ is a (*strong*) *solution* of (2.1) if

(a) $u \in K$;

(b) Distribution Identity:

$$\langle Pu_t(\tau), \varphi(\tau)\rangle_V \Big|_0^t = \int_0^t \{\langle Pu_t(\tau), \varphi_t(\tau)\rangle_V - \langle Q(\tau)u_t(\tau), \varphi(\tau)\rangle_Y - \langle A(\tau, u(\tau)), \varphi(\tau)\rangle_W + \langle F(\tau, u(\tau)), \varphi(\tau)\rangle_X\} d\tau$$

for all $t \in J$ and $\varphi \in K$;

(c) Energy Conservation:

$$\mathcal{E}u(t) - \mathcal{E}u(0) \leq -\int_0^t \{\langle Q(\tau)u_t(\tau), u_t(\tau)\rangle_Y - \mathcal{A}_t(\tau, u(\tau)) + \mathcal{F}_t(\tau, u(\tau))\}d\tau,$$

where

$$\mathcal{E}u(t) = \tfrac{1}{2}\langle Pu_t(t), u_t(t)\rangle_V + \mathcal{A}(t, u(t)) - \mathcal{F}(t, u(t)), \quad t \in J, \tag{2.2}$$

is the total energy of $u$.

For our first result we assume even more that $Q \in C^1(J \to B(Y, Y'))$ with $Q_t(t) : Y \to Y'$ being non-positive definite and (necessarily) symmetric for all $t \in J$.

**Theorem 1.** *Suppose there is a constant $q > 2$ such that, for all $(t, u) \in J \times G$ for which $\mathcal{A}(t, u) < \mathcal{F}(t, u)$, we have*

$$\langle A(t, u), u\rangle_W - \langle F(t, u), u\rangle_X \leq q\{\mathcal{A}(t, u) - \mathcal{F}(t, u)\} \tag{2.3}$$

*and*

$$\mathcal{A}_t(t, u) - \mathcal{F}_t(t, u) \leq 0. \tag{2.4}$$

*Then there is no solution $u$ of* (2.1) *on $J$ such that $\mathcal{E}u(0) < 0$.*

*Proof.* We first show that, along any solution $u$ of (2.1) with $\mathcal{E}u(0) < 0$, we have

$$\mathcal{A}(t, u(t)) - \mathcal{F}(t, u(t)) < 0 \tag{2.5}$$

and

$$\mathcal{E}u(t) - \mathcal{E}u(0) \leq -\int_0^t \langle Q(\tau)u_t(\tau), u_t(\tau)\rangle_Y d\tau \tag{2.6}$$

for all $t \in J$. Indeed, since $\mathcal{E}u(0) < 0$, by (2.2) and the fact that $P$ is a non-negative definite operator, we obtain

$$\mathcal{A}(0, u(0)) - \mathcal{F}(0, u(0)) < 0.$$

If (2.5) fails to hold for all $t \in J$, then there exists a first value $T > 0$ such that

$$\mathcal{A}(T, u(T)) - \mathcal{F}(T, u(T)) = 0.$$

Consequently by (2.4)

$$\mathcal{A}_t(t, u(t)) - \mathcal{F}_t(t, u(t)) \leq 0 \quad \text{for all } t \in [0, T).$$

Then by the energy conservation law (c) we get, since $Q(t)$ is a non-negative definite operator,

$$\mathcal{E}u(t) \leq \mathcal{E}u(0) < 0 \quad \text{in } [0, T].$$

Hence by (2.2)

$$\mathcal{A}(T, u(T)) - \mathcal{F}(T, u(T)) < 0,$$

which is the required contradiction. Consequently (2.5) holds, and in turn by (c) and (2.4) also (2.6) is verified.

The inequalities (2.5) and (2.6) being shown, we now define, corresponding to any solution $u$ of (2.1) on $J$,

(2.7)
$$\mathcal{I}(t) = \langle Pu(t), u(t)\rangle_V + \int_0^t \{\langle Q(\tau)u(\tau), u(\tau)\rangle_Y + (\tau - t)\langle Q_t(\tau)u(\tau), u(\tau)\rangle_Y\}d\tau + (T_0 - t)\langle Q(0)u(0), u(0)\rangle_Y + \beta(t+t_0)^2,$$

where $t_0$, $T_0$, $\beta$ are positive constants which will be fixed later (see [3, p.145]). Then one finds, from the assumption that $P$, $Q(t)$, $Q_t(t)$ are linear, continuous and symmetric,

$$\begin{aligned}\mathcal{I}'(t) &= 2\langle Pu(t), u_t(t)\rangle_V + \langle Q(t)u(t), u(t)\rangle_Y - \langle Q(0)u(0), u(0)\rangle_Y \\ &\quad - \int_0^t \langle Q_t(\tau)u(\tau), u(\tau)\rangle_Y + 2\beta(t+t_0) \\ &= 2\langle Pu(t), u_t(t)\rangle_V + 2\int_0^t \langle Q(\tau)u(\tau), u_t(\tau)\rangle_Y d\tau + 2\beta(t+t_0).\end{aligned}$$

With the help of the distribution identity (b), with $\varphi = u \in K$, it follows next that

$$\begin{aligned}\tfrac{1}{2}\mathcal{I}''(t) &= \{\langle Pu_t(t), u_t(t)\rangle_V - \langle Q(t)u_t(t), u(t)\rangle_Y \\ &\qquad - \langle A(t,u(t)), u(t)\rangle_W + \langle F(t,u(t)), u(t)\rangle_X\} \\ &\qquad + \langle Q(t)u(t), u_t(t)\rangle_Y + \beta \\ &= \langle Pu_t(t), u_t(t)\rangle_V - \langle A(t,u(t)), u(t)\rangle_W + \langle F(t,u(t)), u(t)\rangle_X + \beta.\end{aligned}$$

This may be simplified by using (2.3) and (2.2), namely

$$\begin{aligned}\tfrac{1}{2}\mathcal{I}''(t) &\geq \langle Pu_t(t), u_t(t)\rangle_V - q\{\mathcal{A}(t,u(t)) - \mathcal{F}(t,u(t))\} + \beta \\ &= \left(1 + \frac{q}{2}\right)\langle Pu_t(t), u_t(t)\rangle_V - q\mathcal{E}u(t) + \beta.\end{aligned}$$

Now assume for contradiction that $u$ has the property $\mathcal{E}u(0) < 0$. Using (2.6) to eliminate $\mathcal{E}u(t)$ from the previous inequality and choosing $\beta = 2|\mathcal{E}u(0)|$, we obtain the crucial estimate

$$\mathcal{I}''(t) \geq (q+2)\{\langle Pu_t(t), u_t(t)\rangle_V + \beta\} + 2q\int_0^t \langle Q(\tau)u_t(\tau), u_t(\tau)\rangle_Y d\tau. \tag{2.8}$$

Take $t_0$ so large that $\mathcal{I}'(0) = 2\langle Pu(0), u_t(0)\rangle_V + 2\beta t_0 > 0$. Then, using the fact that $P$, $Q(t)$ are non-negative definite yields

$$\mathcal{I}'', \mathcal{I}', \mathcal{I} > 0 \quad \text{on } J.$$

We assert that

$$\mathcal{I}\mathcal{I}'' - \alpha\mathcal{I}'^2 \geq 0 \quad \text{on } [0, T_0], \tag{2.9}$$

where $\alpha = (q+2)/4$. Indeed, put

$$\mathbb{A} = \langle Pu(t), u(t)\rangle_V + \int_0^t \langle Q(\tau)u(\tau), u(\tau)\rangle_Y d\tau + \beta(t+t_0)^2,$$

$$\mathbb{C} = \langle Pu_t(t), u_t(t)\rangle_V + \int_0^t \langle Q(\tau)u_t(\tau), u_t(\tau)\rangle_Y d\tau + \beta,$$

and $\mathbb{B} = \frac{1}{2}\mathcal{I}'$. Since $Q(t)$ is non-negative definite and $Q_t(t)$ non-positive definite for each $t \in J$, we see that

$$\mathbb{A} \leq \mathcal{I} \quad \text{on } [0, T_0]. \tag{2.10}$$

Moreover, by (2.8) and the fact that $2q > q+2$,

$$\mathbb{C} \leq \mathcal{I}''/(q+2) \quad \text{on } J. \tag{2.11}$$

Now observe that, for all $(\xi, \eta) \in \mathbb{R}^2$ and $t \in J$,

$$\begin{aligned}\mathbb{A}\xi^2 + 2\mathbb{B}\xi\eta + \mathbb{C}\eta^2 = {} & \langle \xi Pu(t) + \eta Pu_t(t), \xi u(t) + \eta u_t(t)\rangle_V \\ & + \int_0^t \langle \xi Q(\tau)u(\tau) + \eta Q(\tau)u_t(\tau), \xi u(\tau) + \eta u_t(\tau)\rangle_Y d\tau \\ & + \beta\{(t+t_0)\xi + \eta\}^2 \geq 0,\end{aligned}$$

because $P$, $Q(t)$ are linear, symmetric and non-negative definite. Thus $\mathbb{AC} - \mathbb{B}^2 \geq 0$. In turn (2.9) is valid by virtue of (2.10), (2.11) and the fact that $\mathbb{A}$, $\mathbb{C} > 0$.

Of course $\alpha > 1$ since $q > 2$ by assumption. The inequality (2.9) can be written as $(\mathcal{I}^{-\alpha}\mathcal{I}')' \geq 0$, so

$$\frac{\mathcal{I}'(t)}{\mathcal{I}^\alpha(t)} \geq \frac{\mathcal{I}'(0)}{\mathcal{I}^\alpha(0)} > 0 \quad \text{for } t \in [0, T_0].$$

This is a Riccati inequality with blow-up time

$$T < \frac{1}{\alpha - 1}\frac{\mathcal{I}(0)}{\mathcal{I}'(0)}.$$

Consequently, if $T_0$ is chosen as the right hand side of the above inequality, we have a contradiction. In fact, since $\mathcal{I}(0)$ depends on $T_0$, this gives an easily solved linear equation for $T_0$, the solution being *positive* for all $t_0$ large enough, e.g., whenever

$$\beta t_0 > \frac{2}{q-2}\langle Q(0)u(0), u(0)\rangle_Y - \langle Pu(0), u_t(0)\rangle_V.$$

(An optimal choice for $t_0$, to minimize $T_0$, is easily determined, but is unnecessary for our purposes.) This completes the proof of the theorem.

**Corollary.** *Assume that $\mathcal{A}(t,u)\geq 0$ on the entire space $J\times G$. Then the assertion of Theorem 1 remains true if (2.3) is replaced by the property*

(2.12) $(t,u)\in J\times G$ *and* $\mathcal{F}(t,u)>0$ *implies* $\mathcal{F}_t(t,u)\geq 0$ *and* $(\mathcal{A}/\mathcal{F})_t(t,u)\leq 0$.

*Proof.* We show that (2.12) implies (2.4). Thus let $(t,u)\in J\times G$ be such that $\mathcal{A}(t,u)<\mathcal{F}(t,u)$. Then $\mathcal{F}(t,u)>0$ and so (2.12) can be applied, that is

$$\mathcal{A}_t(t,u)-\mathcal{F}_t(t,u)\leq \frac{\mathcal{F}_t(t,u)}{\mathcal{F}(t,u)}\{\mathcal{A}(t,u)-\mathcal{F}(t,u)\}\leq 0.$$

**Special Case.** $\mathcal{A}(t,u)=a(t)\hat{\mathcal{A}}(u)$ and $\mathcal{F}(t,u)=b(t)\hat{\mathcal{F}}(u)$. If $a\geq 0$ and $b>0$ on $J$, and $\hat{\mathcal{A}}\geq 0$ on $G$, then (2.12) holds provided $a/b$ is non–increasing and $b$ is non–decreasing on $J$.

**Remark.** The special case of (2.1) when $F=F(u)$, $Q(t)\equiv 0$, $A(t,u)=A(t)u$ is a linear non–negative definite symmetric operator in $W$, with $W$ a Hilbert space, was treated in [2]. In this case, condition (2.4) is implied by condition (A-iii) of Theorem 1 of [2]. Indeed, we clearly have

$$\mathcal{A}(t,u)=\int_0^1\langle\tau A(t)u,u\rangle_W d\tau\geq 0$$

and moreover

$$\mathcal{A}_t(t,u)=\int_0^1\langle\tau A_t(t)u,u\rangle_W d\tau\leq 0$$

by (A-iii). Hence $\mathcal{A}_t(t,u)-\mathcal{F}_t(t,u)=\mathcal{A}_t(t,u)\leq 0$ for all $(t,u)\in J\times W$, as required.

A similar remark applies to the parabolic subcase of (2.1) treated in Theorem 1 of [1], for which $P\equiv 0$ and the corresponding space $V$ is no longer needed.

## §3. The case with general dissipation.

Here we study the more general evolution equation

$$Pu_{tt}+Q(t,u,u_t)+A(t,u)=F(t,u),\qquad t\in J, \tag{3.1}$$

where $P$, $A$, $F$ are as in Section 2, while $Q$ is a nonlinear damping operator. We introduce the set

$$S=J\times X\times Y$$

and suppose $Q\in C(S\to X')$. Here $Y$ no longer need be a Hilbert space.

By a (*strong*) *solution* of (3.1) we mean a function $u$ such that

(a) $u\in K$;

(b) Distribution Identity:

$$\langle Pu_t(\tau),\varphi(\tau)\rangle_V\Big|_0^t=\int_0^t\{\langle Pu_t(\tau),\varphi_t(\tau)\rangle_V-\langle Q(\tau,u(\tau),u_t(\tau)),\varphi(\tau)\rangle_X$$
$$-\langle A(\tau,u(\tau)),\varphi(\tau)\rangle_W+\langle F(\tau,u(\tau)),\varphi(\tau)\rangle_X\}d\tau$$

for all $t \in J$ and $\varphi \in K$;
(c) Energy Conservation:

$$\mathcal{E}u(t) - \mathcal{E}u(0) \leq - \int_0^t \{\mathcal{D}(\tau, u(\tau), u_t(\tau)) - \mathcal{A}_t(\tau, u(\tau)) + \mathcal{F}_t(\tau, u(\tau))\} d\tau,$$

where $\mathcal{E}u$ is defined as in (2.2).

The function $\mathcal{D} = \mathcal{D}(t, u, v) \in C(S \to \mathbb{R}_0^+)$ is the *dissipation rate*. Clearly it has the property that

$$\mathcal{D}(\cdot, u(\cdot), u_t(\cdot)) \in L^1_{\text{loc}}(J) \tag{3.2}$$

along any solution $u$ of (3.1).

We assume that *there is an exponent $m > 1$, a real constant $\kappa$ and a positive function $\delta = \delta(t)$ of class of $C^1(J)$ such that*

$$\|Q(t, u, v)\|_{X'} \leq [\delta(t) \cdot \|u\|_X^\kappa]^{1/m} [\mathcal{D}(t, u, v)]^{1/m'} \quad \text{on } S,$$

*where $m'$ is the Hőlder conjugate of $m$.*

For a complete discussion of this abstract formulation, see [5] and also [6, 7].

We now present a generalization of Theorem 1 of [5], covering the case when $A$ and $F$ depend on $t$ as well as $u$. In this result, condition (2.3) is replaced by the following three growth conditions:

*There exists $q > 0$ such that*

$$\langle A(t, u), u \rangle_W \leq q\mathcal{A}(t, u) \quad \text{for all } (t, u) \in J \times G. \tag{3.3}$$

*For all $\varepsilon > 0$ there are there are positive constants $c_1 = c_1(\varepsilon)$, $c_2 = c_2(\varepsilon)$ and an exponent $p > 1$ such that*

$$c_1\mathcal{F}(t, u) \leq c_2\|u\|_X^p \leq \langle F(t, u), u \rangle_X - q\mathcal{F}(t, u) \tag{3.4}$$

*whenever $(t, u) \in J \times G$ and $\mathcal{F}(t, u) \geq \varepsilon$.*

*There exists a positive constant $c_3$ such that*

$$\|Pv\|_{V'}^2 \leq c_3 \langle Pv, v \rangle_V \qquad \text{for all } v \in V. \tag{3.5}$$

Finally, we suppose that the space $X$ is continuously embedded in $V$.

**Theorem 2.** *Suppose $\delta'(t) = o(\delta(t))$ as $t \to \infty$. Assume that $p > \max\{2, m + \kappa\}$ and*

$$\min\{1, \delta^{(1+\theta)/(1-m)}\} \notin L^1(J),$$

*for some constant $\theta$, with $0 < \theta < \theta_0$, where*

$$\theta_0 = \begin{cases} \dfrac{p - m - \kappa}{pm - (p - m - \kappa)} & \text{if } \dfrac{\kappa}{p} + \dfrac{m}{2} \geq 1 \\[2ex] \dfrac{p - 2}{2p - (p - 2)} & \text{if } \dfrac{\kappa}{p} + \dfrac{m}{2} < 1. \end{cases}$$

*Then no solution $u$ of* (3.1) *can exist on $J$ whenever* (2.4) *holds and* $\mathcal{E}u(0) < 0$.

*Proof.* We first observe, in view of (2.4), that along any solution $u$ of (3.1) the conditions (2.5) and

$$\mathcal{E}u(t) - \mathcal{E}u(0) \leq - \int_0^t \mathcal{D}(\tau, u(\tau), u_t(\tau))d\tau \tag{3.6}$$

hold on $J$, see the proof of Theorem 1 above. The proof of the theorem therefore reduces exactly to that of Theorem 1 of [5], or more precisely to the case where $k(t) = 1$, $\rho = \delta^{1/(m-1)}$ on $J$, and $P$ is linear – these latter specializations are not essential, but have been made only for the sake of simplicity. Note also that the Remark at the end of Section 3 of [5] has been used to treat the case when $\kappa$ is not zero.

## §4. Examples.

A number of concrete examples relative to linear operators $A$ were given in Section III of [2], to which we refer the reader. Example VI of [2, p.16] in particular deserves special mention since for precision the space $Y$ as well as $W$ must be chosen as $H_0^1(\Omega)$.

Other concrete operators $A(u)$, not appearing in [2], are noted in [4] and in Section 6 of [6], notably the degenerate Laplacian $-\text{div}(|Du|^{p-2}Du)$ and the polyharmonic operator $(-\Delta)^L$, where $L \geq 1$ is an integer, and still further examples are given in Section 4 of [5].

We conclude with a discussion of several model cases, the first being

$$u_{tt} - \Delta u = f(t, x, u), \quad x \in \Omega, \ t \in J, \tag{4.1}$$

where

$$f(t, x, u) = g(t, x)u + c|u|^{p-2}u, \quad c > 0, \ p > 2.$$

For the conclusions below we ask that

$$\frac{\partial g}{\partial t} \geq 0 \quad \text{on } J \times \Omega, \qquad \|g(t,\cdot)\|_{L^{p/(p-2)}} \leq c_0 \quad \text{for all } t \in J. \tag{4.2}$$

The natural spaces associated to (4.1) are $V = L^2(\Omega)$, $W = H_0^1(\Omega)$, $X = L^p(\Omega)$ and $G = L^p(\Omega) \cap H_0^1(\Omega)$. (The remaining spaces $S$ and $Y$ are unneeded, since we have omitted the damping terms from the equation. Note moreover that in applications of Theorem 2 the domain $\Omega$ must be bounded, in order that $X = L^p(\Omega)$ be embedded in $V = L^2(\Omega)$.)

The operator $P$ corresponding to (4.1) is given by $\langle Pv, w\rangle_V = (v, w)_{L^2}$; clearly $P$ is symmetric and positive definite, and satisfies (3.5). Moreover $A(u) = -\Delta u$, so

$$\langle A(u), u\rangle_{H_0^1} = \|u\|_{H_0^1}^2, \qquad \mathcal{A}(u) = \frac{1}{2}\|u\|_{H_0^1}^2$$

and $q\mathcal{A}(u) \geq \langle A(u), u\rangle_{H_0^1}$ whenever $q \geq 2$. On the other hand,

$$\mathcal{F}(t, u) = \frac{1}{2}\int_\Omega g(t, x)|u|^2 dx + \frac{c}{p}\|u\|_{L^p}^p. \tag{4.3}$$

Hence, (2.4) is satisfied on $J \times G$ by $(4.2)_1$ – we assume suitable regularity of $g = g(t,x)$ so that $\mathcal{F}_t(t,u)$ can be calculated on $J \times G$ by differentiation under the integral sign in (4.3).

Moreover,

$$\langle F(t,u), u\rangle_{L^p} = \int_\Omega g(t,x)|u|^2 dx + c\|u\|^p_{L^p}.$$

Therefore (2.3) is verified on $J \times G$ and Theorem 1 can be applied, *provided that*

$$\left(\frac{q}{2}-1\right)\int_\Omega g(t,x)|u|^2 dx \le c\left(1-\frac{q}{p}\right)\|u\|^p_{L^p}, \tag{4.4}$$

with $2 < q \le p$. This obviously occurs whenever $g(t,x) \le 0$ a.e. in $J \times \Omega$, but fails when $g(t,x) > 0$ on $J \times \Omega$ and $|u(x)|$ is sufficiently small.

Nevertheless non–continuation still occurs. To see this, observe from (2.2) and (3.6) that necessarily, for all solutions $u$ of (2.1) satisfying $\mathcal{E}u(0) < 0$, we have

$$\mathcal{F}(t,u(t)) = \tfrac{1}{2}\|u_t(t)\|^2_{L^2} + \tfrac{1}{2}\|u(t)\|^2_{H^1_0} - \mathcal{E}u(t) \ge -\mathcal{E}u(t) \ge -\mathcal{E}u(0) = |\mathcal{E}u(0)|, \quad t \in J.$$

Thus we only need to verify (4.4) whenever $(t,u) \in J \times G$ is such that

$$\mathcal{F}(t,u) \ge |\mathcal{E}u(0)| = \varepsilon, \qquad \text{that is} \qquad \frac{1}{2}\int_\Omega g(t,x)|u|^2 dx + \frac{c}{p}\|u\|^p_{L^p} \ge \varepsilon.$$

In particular, this implies by Hőlder's inequality and $(4.2)_2$

$$\frac{c_0}{2}\|u\|^2_{L^p} + \frac{c}{p}\|u\|^p_{L^p} \ge \varepsilon.$$

Consequently we need to check (4.4) only when $\|u\|_{L^p} \ge \lambda(\varepsilon) = \lambda > 0$, where

$$\frac{c_0}{2}\lambda^2 + \frac{c}{p}\lambda^p = \varepsilon.$$

In fact, rather than (4.4), consider the stronger inequality

$$\left(\frac{q}{2}-1\right)c_0\|u\|^2_{L^p} \le c\left(1-\frac{q}{p}\right)\|u\|^p_{L^p};$$

this however holds for $\|u(t)\|_{L^p} \ge \lambda$ provided that

$$c\left(1-\frac{q}{p}\right)\lambda^{p-2} - \left(\frac{q}{2}-1\right)c_0 \ge 0.$$

Thus it is enough if $q$ obeys

$$2 < q \le \frac{c\lambda^{p-2}+c_0}{2c\lambda^{p-2}+pc_0}\,2p = q_1;$$

note that $q_1 > 2$ since $p > 2$ by assumption. Of course $q_1 = p$ when $c_0 = 0$, while $q_1 < p$ when $c_0 > 0$ and $p > 2$.

This example can also be treated by means of Theorem 2. With the above calculations in mind, we take $q = 2$. Then (2.4) is verified as above, while (3.4) becomes

$$c_1\left(\frac{1}{2}\int_\Omega g(t,x)|u|^2dx + \frac{c}{p}\|u\|^p_{L^p}\right) \le c_2\|u\|^p_{L^p} \le c\left(1-\frac{2}{p}\right)\|u\|^p_{L^p}.$$

This holds if $c_2 = c(1-2/p)$ and if also the following inequality is satisfied,

$$c\left(1-\frac{2}{p}-\frac{c_1}{p}\right)\|u\|^p_{L^p} - \frac{c_0c_1}{2}\|u\|^2_{L^p} \ge 0.$$

As before, this is required only when $\mathcal{F}(t,u) \ge |\mathcal{E}u(0)| = \varepsilon$ on $J\times G$, i.e. when $\|u\|_{L^p} \ge \lambda(\varepsilon) = \lambda > 0$. This gives the condition

$$c_1 = \frac{2(p-2)c\lambda^{p-2}}{2c\lambda^{p-2}+pc_0}.$$

Obviously $c_1 > 0$ since $p > 2$ by assumption. Here, as in the applications of Theorem 1, it remains crucial that $c > 0$.

Next consider the problem

$$u_{tt} - \operatorname{div}(|Du|^{s-2}Du) = f(t,x,u), \quad x\in\Omega,\ t\in J,$$

where $s > 1$, $s \ne 2$, and

$$f(t,x,u) = g(t,x)|u|^{s-2}u + c|u|^{p-2}u, \quad c \ge 0,\ p > s.$$

We assume the analogue of (4.2), namely

$$\frac{\partial g}{\partial t} \ge 0 \quad \text{on } J\times\Omega, \qquad \|g(t,\cdot)\|_{L^{p/(p-s)}} \le c_0 \quad \text{for all } t\in J.$$

Here the appropriate spaces are $V = L^2(\Omega)$, $W = W_0^{1,s}(\Omega)$, $X = L^p(\Omega)$ and $G = L^2(\Omega)\cap L^p(\Omega)\cap W_0^{1,s}(\Omega)$. Of course, as before, in applications of Theorem 2 the domain $\Omega$ must be bounded.

Using Theorem 1 when $s > 2$, and choosing $q = s$, we get non–continuation whenever $c \ge 0$. Using Theorem 2 with $q = s$, we get non–continuation for $c > 0$ and $p > \max\{2,\ s\}$.

Hence, when $s > 2$ and $c = 0$ the result given by Theorem 1 is unobtainable from Theorem 2. Conversely, when $1 < s < 2$ and $c > 0$ the result given by Theorem 2 is unobtainable from Theorem 1.

Another function $f$ for which Theorem 1 applies, but not Theorem 2, is given by

$$f(u) = |u|^{p-1}, \quad p > 2.$$

Here

$$\langle F(u), u\rangle_{L^p} - q\mathcal{F}(u) = \left(1-\frac{q}{p}\right)\int_\Omega |u|^{p-1}u\,dx,$$

so we can take $q = p$ for Theorem 1. On the other hand, for Theorem 2 the condition

$$c_2\|u\|_{L^p}^p \leq \left(1 - \frac{q}{p}\right) \int_\Omega |u|^{p-1} u dx$$

cannot in general be satisfied unless one assumes *a priori* that $u \geq 0$.

Theorems 1 and 2 also differ in their requirements concerning the dissipation operator $Q$. We leave such comparisons to the reader.

## References

[1] H.A. Levine, *Some nonexistence and instability theorems for solutions of formally parabolic equations of the form* $Pu_t = Au + \mathcal{F}(u)$, Archive Rational Mech. Anal. **51** (1973), 371–386.

[2] H.A. Levine, *Instability and nonexistence of global solutions of nonlinear wave equations of the form* $Pu_{tt} = Au + \mathcal{F}(u)$, Trans. Amer. Math. Soc. **192** (1974), 1–21.

[3] H.A. Levine, *Some additional remarks on the nonexistence of global solutions to nonlinear wave equations*, SIAM J. Math. Anal. **5** (1974), 138–146.

[4] H.A. Levine, *Nonexistence of global solutions of nonlinear wave equations*, in *Improperly Posed Boundary Value Problems*, Res. Notes Math. **1**, 94–104; Pitman, London, 1975.

[5] H.A. Levine & J. Serrin, *Global nonexistence theorems for quasilinear evolution equations with dissipation*, Archive Rational Mech. Anal., to appear.

[6] P. Pucci & J. Serrin, *Asymptotic stability for non-autonomous dissipative wave systems*, Comm. Pure Appl. Math. **XLIX** (1996), 177–216.

[7] P. Pucci & J. Serrin, *Stability for abstract evolution equations*, in *Partial Differential Equations and Applications – Collected papers in Honor of Carlo Pucci*, P. Marcellini, G. Talenti & E. Vesentini eds., Lecture Notes in Pure and Appl. Math. 1996, M. Dekker, Inc. New York, 279–288.

Nina Ural'tseva.

Hans Weinberger.

# II.5. Wave Operators – Part II

Enzo Vitillaro

**Mathematics Subject Classification (2010).** 34G20, 35L99, 35L70, 35D05.

**Keywords.** Existence, nonexistence, evolution, nonlinear damping, nonlinear wave equation.

The two papers [i, ii] of James Serrin concern evolution equations. The first one (coauthored with Patrizia Pucci) deals with global nonexistence for abstract evolution equations, while the latter (with Grozdena Todorova and the Commentator) deals with existence theory for wave equations with nonlinear damping and source terms.

The paper [i] studies abstract evolution equations of the form

$$Pu_{tt} + Q(t)u_t + A(t,u) = F(t,u), \qquad t \in [0,\infty), \tag{1}$$

where $P : V \to V'$, $Q : [0,\infty) \times Y \to Y'$, $A : [0,\infty) \times W \to W'$, $F : [0,\infty) \times X \to X'$. Here $V$ and $Y$ are Hilbert spaces, and $W$ and $X$ are Banach spaces. Moreover $P$ and $Q(t)$, $t \geq 0$, are linear operators, while $A$ and $F$, respectively, are the Fréchet derivatives, with respect to $u$, of $C^1$ potentials $\mathcal{A} : [0,\infty) \times W \to \mathbb{R}$ and $\mathcal{F} : [0,\infty) \times X \to \mathbb{R}$, with the normalizing condition $\mathcal{A}(t,0) \equiv 0$, $\mathcal{F}(t,0) \equiv 0$. Typically $P$ represents the evolution operator and $Q$ the dissipation operator, both of which are assumed to be symmetric and nonnegative definite. In addition, it is assumed that $Q$ and $Q_t$ are bounded operators which are continuous in $t$, and that $Q_t$ is nonpositive definite for all $t \geq 0$. The cases $P \equiv 0$ and $Q \equiv 0$ are allowed. Moreover $A$ typically represents a divergence operator and $F$ a driving force. When $V$, $Y$, $W$ and $X$ posses a common nontrivial linear manifold $G$ (a condition commonly satisfied in the applications to PDE's, where all spaces are function spaces), a suitable class of solutions of (1) is introduced; namely, *strong solutions*, which are regular enough to define the *energy function*

$$\mathcal{E}u(t) = \frac{1}{2}\langle Pu_t(t), u_t(t)\rangle_V + \mathcal{A}(t,u(t)) - \mathcal{F}(t,u(t)), \qquad t \geq 0.$$

The energy is, in general, undefined. Under the very mild assumptions that there are constants $p \geq q$ such that

$$\mathcal{A}_t(t,u) - \mathcal{F}_t(t,u) \leq 0 \quad \text{and} \quad \langle A(t,u),u\rangle_W - \langle F(t,u),u\rangle_X \leq q\mathcal{A}(t,u) - p\mathcal{F}(t,u)$$

for all $(t,u) \in [0,\infty) \times G$ the authors prove:

(a) when $p > 2$ there is no solution on $[0,\infty)$ such that $\mathcal{A}(t,u(t)) \geq \lambda_0 > 0$ for $t \geq 0$ and $\mathcal{E}u(0) < (1 - q/p)\lambda_0$;
(b) when $q > 2$ there is no solution on $[0,\infty)$ such that $\mathcal{F}(t,u(t)) \geq \lambda_1 > 0$ for $t \geq 0$ and $\mathcal{E}u(0) < (p/q - 1)\lambda_1$.

P. Pucci et al. (eds.), *James Serrin. Selected Papers: Volume 2*, Contemporary Mathematicians,
DOI 10.1007/978-3-0348-0847-7_5, 

Clearly this result deals also with initial data $u_0$ with positive energy $\mathcal{E}u(0)$. Classical papers dealing with positive initial energy at the time were [17], which dealt with wave equations with homogeneous Dirichlet boundary condition (whose proof is not completely clear), [14], which treats classical solutions for wave-type operators, and [16], which concerns wave equation with nonlinear boundary conditions. The paper [i] contains a clear description of the so-called "external potential-well", which is much more general than those contained in the papers above. This description is by now a standard tool. Indeed MathSciNet counts 36 citations of the paper, most of which are related to this fact. In particular, we would like to point out the paper [19] by the Commentator, which extends the above result to nonlinear damping terms and was inspired by Professor Serrin as the natural continuation of [i]. Moreover the idea in [i] was directly adapted to the porous media equation in [20], to linear boundary damping and linear dynamical boundary conditions in [21], and to strongly damped semilinear wave equations in [12]. More generally, a qualitative analysis of the solutions starting from the potential-well was outlined in [10, 11]. This description was also used in [15] to deal with parabolic problems, and it was adapted in [24] to the Dirichlet-to-Neumann operator, which is shown in [9] to be related to the fractional Laplacian. More recently it was a main ingredient of global nonexistence theorems for Kirchhoff systems, under various boundary conditions, static or dynamical, contained in [1, 2, 3], where also several improvements to the theory are given.

The paper [ii] deals with the Cauchy problem

$$\begin{cases} u_{tt} - \Delta u + Q(t,x,u_t) = f(x,u), & (0,T) \times \mathbb{R}^n, \\ u(0) = u_0, \quad u_t(0) = u_1, \end{cases} \tag{2}$$

where $n \geq 1$. Roughly, $Q$ represents a damping term while $f$ represents a source term. Nonlinear terms specifically considered are $Q(t,x,v) = \sigma(x)|v|^{m-2}v$, $m > 1$, $0 < c_1 \leq \sigma(x) \leq c_2 < \infty$ and $f(x,u) = \mu_1(x)|u|^{p-2}u + \mu_2(x)|u|^{q-2}u$, $p > 2$, $1 < q < p$, $\mu_1, \mu_2 \in L^\infty_{\text{loc}}(\mathbb{R}^n)$. The data $u_0 \in H^1(\mathbb{R}^n)$ and $u_1 \in L^2(\mathbb{R}^n)$ are compactly supported. When

$$m > 1, \quad 2 < p < 1 + r/2 \qquad \text{or} \quad m > r/(r+1-p), \quad 1 + r/2 \leq p < r, \tag{3}$$

where $r = 2n/(n-2)$ for $n \geq 3$ and $r = \infty$ for $n = 1, 2$, it is proved in [ii] that problem (2) possesses a weak solution for $T$ small enough; that is, a *local solution*, which is also unique when $p \leq 1+r/2$ and $\mu_2 \equiv 0$. Moreover, this local existence theorem is used to prove that when $p \leq m$, the solution can be extended to a *global solution*, i.e., $T = \infty$. At the time, local existence for problems like (2) was proved only when $p \leq 1 + r/2$ (see [13]); that is when the Nemitskii operator $u \mapsto f(\cdot, u)$ from $H^1(\mathbb{R}^n)$ to $L^2(\mathbb{R}^n)$ is locally Lipschitz continuous, a condition often arising in local existence theory, by energy or semigroup methods. The main idea in [ii], due to Professor Serrin, was to make use of the damping term $Q$ as a regularizing term, which allows one to drop this condition. This allows, when $m \geq 2$, to consider $1 + r/2 < p < 1 + r/m'$. This idea was subsequently adapted to boundary damping and source terms in [6, 7, 8, 22], to the heat equation with nonlinear dynamical boundary conditions in [23], and to degenerate damping terms in a wave equation in [4, 5]. Assumption (3) was improved in [18] by a different technique (which still needs some clarification) which is specific to wave operator in the whole space, while the one used in [ii] has turned out to be a very general one in the presence of nonlinear damping terms. The proof of the local existence result in [ii] is based on a standard two-step procedure; first one studies existence for an auxiliary problem, where the source term is replaced by a given forcing term, then one uses a fixed point argument (Schauder) where through an energy estimate one makes use of damping regularization. To overcome the lack of compactness, one uses the finite speed of propagation property. Here the authors had to face a major difficulty, since when $1 < m < 2$ the damping term is not locally Lipschitz continuous so the proof of that property cannot be completed. Consequently the authors were forced to consider approximating damping terms. However, the time derivatives of the solutions of these approximating problems only live in Orlicz spaces.

**Acknowledgment.** It was a real pleasure and an honor to contribute to this volume. Professor Serrin has been for me an example as well as a master of mathematics. I do not need to emphasize the importance of his contribution to mathematics, which has been clearly shown by Howard Levine in his contribution.

## References

[1] G. Autuori and P. Pucci, *Kirchhoff systems with dynamic boundary conditions*, Nonlinear Anal. **73** (2010), no. 7, 1952–1965.

[2] ———, *Kirchhoff systems with nonlinear source and boundary damping terms*, Commun. Pure Appl. Anal. **9** (2010), no. 5, 1161–1188.

[3] G. Autuori, P. Pucci, and M.C. Salvatori, *Global nonexistence for nonlinear Kirchhoff systems*, Arch. Ration. Mech. Anal. **196** (2010), no. 2, 489–516.

[4] V. Barbu, I. Lasiecka, and M.A. Rammaha, *On nonlinear wave equations with degenerate damping and source terms*, Trans. Amer. Math. Soc. **357** (2005), no. 7, 2571–2611.

[5] ———, *Blow-up of generalized solutions to wave equations with nonlinear degenerate damping and source terms*, Indiana Univ. Math. J. **56** (2007), no. 3, 995–1021.

[6] L. Bociu, *Local and global wellposedness of weak solutions for the wave equation with nonlinear boundary and interior sources of supercritical exponents and damping*, Nonlinear Anal. **71** (2009), no. 12, e560–e575.

[7] L. Bociu and I. Lasiecka, *Uniqueness of weak solutions for the semilinear wave equations with supercritical boundary/interior sources and damping*, Discrete Contin. Dyn. Syst. **22** (2008), no. 4, 835–860.

[8] ———, *Local Hadamard well-posedness for nonlinear wave equations with supercritical sources and damping*, J. Differential Equations **249** (2010), no. 3, 654–683.

[9] L. Caffarelli and L. Silvestre, *An extension problem related to the fractional Laplacian*, Comm. Partial Differential Equations **32** (2007), no. 7-9, 1245–1260.

[10] J.A. Esquivel-Avila, *The dynamics of a nonlinear wave equation*, J. Math. Anal. Appl. **279** (2003), no. 1, 135–150.

[11] ———, *Qualitative analysis of a nonlinear wave equation*, Discrete Contin. Dyn. Syst. **10** (2004), no. 3, 787–804.

[12] F. Gazzola and M. Squassina, *Global solutions and finite time blow up for damped semilinear wave equations*, Ann. Inst. H. Poincaré Anal. Non Linéaire **23** (2006), no. 2, 185–207.

[13] V. Georgiev and G. Todorova, *Existence of a solution of the wave equation with nonlinear damping and source terms*, J. Differential Equations **109** (1994), no. 2, 295–308.

[14] R.J. Knops, H.A. Levine, and L.E. Payne, *Nonexistence, instability, and growth theorems for solutions of a class of abstract nonlinear equations with applications to nonlinear elastodynamic*, Arch. Rational Mech. Anal. **55** (1974), no. 1, 52–72.

[15] H.A. Levine, S.R. Park, and J. Serrin, *Global existence and nonexistence theorems for quasilinear evolution equations of formally parabolic type*, J. Differential Equations **142** (1998), no. 1, 212–229.

[16] H.A. Levine and R.A. Smith, *A potential well theory for the wave equation with a nonlinear boundary condition*, J. Reine Angew. Math. **374** (1987), 1–23.

[17] L.E. Payne and D.H. Sattinger, *Saddle points and instability of nonlinear hyperbolic equations*, Israel J. Math. **22** (1975), 273–303.

[18] P. Radu, *Weak solutions to the Cauchy problem of a semilinear wave equation with damping and source terms*, Adv. Differential Equations **10** (2005), no. 11, 1261–1300.

[19] E. Vitillaro, *Global nonexistence theorems for a class of evolution equations with dissipation and application*, Arch. Rational Mech. Anal. **149** (1999), 155–182.

[20] ———, *Blow-up for the porous media equation with source term and positive initial energy*, J. Math. Anal. Appl. **247** (2000), no. 1, 183–197.

[21] ———, *Some new results on global nonexistence and blow-up for evolution problems with positive initial energy*, Rend. Istit. Mat. Univ. Trieste **31** (2000), no. suppl. 2, 245–275, Workshop on Blow-up and Global Existence of Solutions for Parabolic and Hyperbolic Problems (Trieste, 1999).

[22] ———, *Global existence for the wave equation with nonlinear boundary damping and source terms*, J. Differential Equations **186** (2002), no. 1, 259–298.

[23] ———, *Global existence for the heat equation with nonlinear dynamical boundary condition*, Proc. Roy. Soc. Edinburgh Sect. A **135** (2005), 1–33.

[24] ———, *On the Laplace equation with non-linear dynamical boundary conditions*, Proc. London Math. Soc. (3) **93** (2006), no. 2, 418–446.

Enzo Vitillaro
Dipartimento di Matematica ed Informatica
Università di Perugia, Italy
e-mail: `enzo@dmi.unipg.it`

*J. Differ. Equations* 150 (1998), 203–214.
© Academic Press. Reprinted with permission from Elsevier.

**Reprint II.5.[i]**

# Global Nonexistence for Abstract Evolution Equations with Positive Initial Energy

Patrizia Pucci*

*Dipartimento di Matematica, Università di Perugia, Perugia, Italy*
E-mail: pucci@unipg.it

and

James Serrin

*Department of Mathematics, University of Minnesota, Minneapolis, Minnesota*
E-mail: serrin@math.umn.edu

Received January 12, 1998; revised March 31, 1998

In this paper we consider the problem of noncontinuation of solutions of the initial value problem for abstract evolution equations of the form

$$Pu_{tt} + Q(t)\, u_t + A(t, u) = F(t, u), \qquad t \in J = [0, \infty),$$

where $P$ and $Q$ are linear self-adjoint operators, and $A(t, u)$ and $F(t, u)$ are respectively a linear operator in $u$ (typically of differential type) and a nonlinear driving force. Our principal concern is the noncontinuation (or blow-up) of solutions when the initial energy is positive, but appropriately bounded.
© 1998 Academic Press

## 1. INTRODUCTION

In a recent paper [7] the problem of non–continuation was studied for abstract evolution equations of the type

$$Pu_{tt} + Q(t)\, u_t + A(t, u) = F(t, u), \qquad t \in J = [0, \infty), \tag{1.1}$$

where $P$ and $Q(t)$ are linear self-adjoint operators, and $A(t, u)$ and $F(t, u)$ are typically a divergence operator in $u$ and a nonlinear driving force.

* This work is partially supported by CNR, Progetto Strategico *Modelli e Metodi per la Matematica e l'Ingegneria* and by the Italian *Ministero della Università e della Ricerca Scientifica e Tecnologica* under the auspices of the *Gruppo Nazionale di Analisi Funzionale e sue Applicazioni* of the CNR.



0022-0396/98 $25.00

Other versions of (1.1) were considered earlier by Levine [3–6], for which he introduced the important technique of "concavity" analysis of auxiliary second-order differential inequalities. In all of these papers, the principal mechanism of noncontinuation was the assumption of negative initial energy.

In an interesting paper [10], Ono has also used concavity analysis to study blow-up, but in the more general case when the initial energy is allowed to take appropriately small positive values. His analysis primarily considers linear wave operators, and moreover is restricted to bounded domains in $\mathbb{R}^n$. (It should, however, be added that Ono also allows Kirchhoff type operators, an added generalization but without serious affect on the principal ideas.)

It is the purpose of this paper to extend Ono's analysis to the abstract Eq. (1.1), which we do in Theorem 1. Moreover, in concrete cases, we introduce appropriate methods to treat divergence structure operators in unbounded domains (including but not necessarily restricted to $\mathbb{R}^n$). Our conclusions also yield a larger class of initial data than in [10] for which non-continuation (or blow-up) must occur; see Remark 1 in Section 3.

In the next section we give a precise meaning to Eq. (1.1), and give our main abstract theorem. Section 3 discusses a divergence structure equation in $\mathbb{R}^n$ for which noncontinuation occurs for positive initial energy, even for unbounded domains. Here the primary new idea, in comparison with [7] and [10], is to introduce an appropriate coercive operator associated with the equation.

## 2. THE MAIN THEOREM

Let $X$ be a Banach space, and $X'$ its dual space. If $x \in X$ and $x' \in X'$, we shall write $\langle x', x \rangle_X$ to denote the natural pairing of $x$ and $x'$, that is $\langle x', x \rangle_X = x'(x)$.

Let $V$ be a Hilbert space. An operator $P\colon V \to V'$ will be called *symmetric* if

$$\langle Pv, w \rangle_V = \langle Pw, v \rangle_V \qquad \text{for all} \quad v, w \in V,$$

and *non-negative definite* if

$$\langle Pv, v \rangle_V \geqslant 0 \qquad \text{for all} \quad v \in V.$$

It is easy to check that a symmetric operator must be linear and, moreover, continuous by the uniform boundedness theorem.

We consider the evolution Eq. (1.1), where $P$ is symmetric and non-negative definite from $V$ into $V'$. We suppose that the *dissipation operator*

$Q(t)$ is, for each $t \in J$, symmetric and non-negative definite from an appropriate Hilbert space $Y$ into its dual $Y'$. In addition, assume $Q \in C(J \to B(Y, Y'))$, that is $\langle Q(\cdot) v, w \rangle_Y : J \to \mathbb{R}$ is continuous for each $v, w \in Y$. Note that $P \equiv 0$ and $Q \equiv 0$ are specifically allowed.

Finally, the operators $A$ and $F$ are such that[1]

$$A : J \times W \to W', \qquad F : J \times X \to X',$$

with $W$, $X$ Banach spaces and $W'$, $X'$ their duals. In order to define the energy $\mathscr{E}u$ of a solution of (1.1), see below, it is necessary that there exist $C^1$ potentials

$$\mathscr{A} : J \times W \to \mathbb{R}, \qquad \mathscr{F} : J \times X \to \mathbb{R},$$

such that for each fixed $t$ the operators $A$ and $F$ are the Fréchet derivatives with respect to $u$ of $\mathscr{A}$ and $\mathscr{F}$, respectively; by normalization we can take $\mathscr{A}(t, 0) \equiv 0$, $\mathscr{F}(t, 0) \equiv 0$.

Now suppose that there is given a nontrivial subspace $G$ of $V$, $W$, $X$ and $Y$–not necessarily closed. Let

$$K = \{\varphi : J \to G \mid \varphi \in C(J \to W) \cap C(J \to X) \cap C^1(J \to V) \cap AC(J \to Y)\}.$$

We say that $u$ is a *(strong) solution* of (1.1) if

(a) $u \in K$;

(b) Distribution Identity:

$$\langle Pu_t(\tau), \varphi(\tau) \rangle_V \Big|_0^t = \int_0^t \{\langle Pu_t(\tau), \varphi_t(\tau) \rangle_V - \langle Q(\tau)\, u_t(\tau), \varphi(\tau) \rangle_Y - \langle A(\tau, u(\tau)), \varphi(\tau) \rangle_W + \langle F(\tau, u(\tau)), \varphi(\tau) \rangle_X\} \, d\tau$$

for all $t \in J$ and $\varphi \in K$;

(c) Energy Conservation:

$$\mathscr{E}u(t) - \mathscr{E}u(0) \leqslant - \int_0^t \{\langle Q(\tau)\, u_t(\tau), u_t(\tau) \rangle_Y - \mathscr{A}_t(\tau, u(\tau)) + \mathscr{F}_t(\tau, u(\tau))\} \, d\tau, \tag{2.1}$$

where

$$\mathscr{E}u(t) = \tfrac{1}{2} \langle Pu_t(t), u_t(t) \rangle_V + \mathscr{A}(t, u(t)) - \mathscr{F}(t, u(t)), \qquad t \in J, \tag{2.2}$$

is the total energy of $u$.

[1] Specific examples are given in [4, 7, 8, 11], and also in Section 3 below.

Assume even more that $Q \in C^1(J \to B(Y, Y'))$, with $Q_t(t)\colon Y \to Y'$ being non-positive definite and (necessarily) symmetric for all $t \in J$.

Suppose there are constants $p \geqslant q$ such that, for all $(t, u) \in J \times G$,

$$\langle A(t, u), u \rangle_W - \langle F(t, u), u \rangle_X \leqslant q\mathscr{A}(t, u) - p\mathscr{F}(t, u) \tag{2.3}$$

and

$$\mathscr{A}_t(t, u) - \mathscr{F}_t(t, u) \leqslant 0. \tag{2.4}$$

THEOREM 1. *Assume that* (2.3) *and* (2.4) *hold.*

(i) *Let* $p > 2$. *Then there is no solution* $u = u(t)$ *of* (1.1) *on* $J$ *with*

$$\mathscr{A}(t, u(t)) \geqslant \lambda_0 > 0, \qquad t \in J, \tag{2.5}$$

*and*

$$\mathscr{E}u(0) < \left(1 - \frac{q}{p}\right) \lambda_0 = D_0.$$

(ii) *Let* $q > 2$. *Then there is no solution* $u = u(t)$ *of* (1.1) *on* $J$ *with*

$$\mathscr{F}(t, u(t)) \geqslant \lambda_1 > 0, \qquad t \in J, \tag{2.6}$$

*and*

$$\mathscr{E}u(0) < \left(\frac{p}{q} - 1\right) \lambda_1 = D_1.$$

*Proof.* We define, corresponding to any solution $u$ of (1.1) on $J$,

$$\begin{aligned} \mathscr{I}(t) = \langle Pu(t), u(t) \rangle_V &+ \int_0^t \{\langle Q(\tau)\, u(\tau), u(\tau) \rangle_Y \\ &+ (\tau - t)\langle Q_t(\tau)\, u(\tau), u(\tau) \rangle_Y\}\, d\tau \\ &+ (T_0 - t)\langle Q(0)\, u(0), u(0) \rangle_Y + \beta(t + t_0)^2, \end{aligned} \tag{2.7}$$

where $t_0$, $T_0$, $\beta$ are positive constants which will be fixed later (see Levine [3–5], and also [7]). Then one finds, from the assumption that $P$, $Q(t)$, $Q_t(t)$ are linear, continuous and symmetric,

$$\mathscr{I}'(t) = 2\langle Pu(t), u_t(t)\rangle_V + \langle Q(t)\,u(t), u(t)\rangle_Y - \langle Q(0)\,u(0), u(0)\rangle_Y$$
$$- \int_0^t \langle Q_t(\tau)\,u(\tau), u(\tau)\rangle_Y + 2\beta(t+t_0)$$
$$= 2\langle Pu(t), u_t(t)\rangle_V + 2\int_0^t \langle Q(\tau)\,u(\tau), u_t(\tau)\rangle_Y\,d\tau + 2\beta(t+t_0).$$

From the distribution identity (b), by taking $\varphi = u \in K$ it follows next that

$$\tfrac{1}{2}\mathscr{I}''(t) = \{\langle Pu_t(t), u_t(t)\rangle_V - \langle Q(t)\,u_t(t), u(t)\rangle_Y$$
$$- \langle A(t, u(t)), u(t)\rangle_W + \langle F(t, u(t)), u(t)\rangle_X\}$$
$$+ \langle Q(t)\,u(t), u_t(t)\rangle_Y + \beta$$
$$= \langle Pu_t(t), u_t(t)\rangle_V - \langle A(t, u(t)), u(t)\rangle_W + \langle F(t, u(t)), u(t)\rangle_X + \beta.$$

This may be simplified by using (2.3) and (2.2), namely

$$\frac{1}{2}\mathscr{I}''(t) \geqslant \langle Pu_t(t), u_t(t)\rangle_V - q\mathscr{A}(t, u(t)) + p\mathscr{F}(t, u(t)) + \beta$$
$$= \langle Pu_t(t), u_t(t)\rangle_V - q\mathscr{A}(t, u(t))$$
$$+ p\left(\frac{1}{2}\langle Pu_t(t), u_t(t)\rangle_V + \mathscr{A}(t, u(t)) - \mathscr{E}u(t)\right) + \beta$$
$$= \left(1 + \frac{p}{2}\right)\langle Pu_t(t), u_t(t)\rangle_V + (p-q)\,\mathscr{A}(t, u(t)) - p\mathscr{E}u(t) + \beta.$$

Now $(p-q)\,\mathscr{A}(t, u(t)) \geqslant (p-q)\,\lambda_0 = pD_0$ by the hypothesis (2.5), and also

$$\mathscr{E}u(t) \leqslant \mathscr{E}u(0) - \int_0^t \langle Q(\tau)\,u_t(\tau), u_t(\tau)\rangle_Y\,d\tau$$

by (2.1) and (2.4). Therefore we find

$$\frac{1}{2}\mathscr{I}''(t) \geqslant \left(1 + \frac{p}{2}\right)\langle Pu_t(t), u_t(t)\rangle_V$$
$$+ p\int_0^t \langle Q(\tau)\,u_t(\tau), u_t(\tau)\rangle_Y\,d\tau + p\{D_0 - \mathscr{E}u(0)\} + \beta.$$

Let $\beta = 2\{D_0 - \mathscr{E}u(0)\} > 0$. This gives the main estimate

$$\mathscr{I}''(t) \geqslant (p+2)\{\langle Pu_t(t), u_t(t)\rangle_V + \beta\} + 2p \int_0^t \langle Q(\tau)\, u_t(\tau), u_t(\tau)\rangle_Y \, d\tau. \tag{2.8}$$

The proof of part (i) is now almost exactly the same as in [7], from their formula (2.8) onward. For the convenience of the reader we include it here.

Take $t_0$ so large that $\mathscr{I}'(0) = 2\langle Pu(0), u_t(0)\rangle_V + 2\beta t_0 > 0$. Then, using the fact that $P$, $Q(t)$ are non-negative definite there results

$$\mathscr{I}'', \mathscr{I}', \mathscr{I} > 0 \qquad \text{on } J.$$

We assert that

$$\mathscr{I}\mathscr{I}'' - \alpha \mathscr{I}'^2 \geqslant 0 \qquad \text{on } [0, T_0], \tag{2.9}$$

where $\alpha = (p+2)/4$. Indeed, put

$$\mathbb{A} = \langle Pu(t), u(t)\rangle_V + \int_0^t \langle Q(\tau)\, u(\tau), u(\tau)\rangle_Y \, d\tau + \beta(t+t_0)^2,$$

$$\mathbb{C} = \langle Pu_t(t), u_t(t)\rangle_V + \int_0^t \langle Q(\tau)\, u_t(\tau), u_t(\tau)\rangle_Y \, d\tau + \beta,$$

and $\mathbb{B} = \frac{1}{2}\mathscr{I}'$. Since $Q(t)$ is non-negative definite and $Q_t(t)$ non-positive definite for each $t \in J$, we see that

$$\mathbb{A} \leqslant \mathscr{I} \qquad \text{on } [0, T_0]. \tag{2.10}$$

Moreover, by (2.8) and the fact that $2p > p+2$,

$$\mathbb{C} \leqslant \mathscr{I}''/(p+2) \qquad \text{on } J. \tag{2.11}$$

Now observe that, for all $(\xi, \eta) \in \mathbb{R}^2$ and $t \in J$,

$$\begin{aligned}\mathbb{A}\xi^2 + 2\mathbb{B}\xi\eta + \mathbb{C}\eta^2 &= \langle \xi Pu(t) + \eta Pu_t(t), \xi u(t) + \eta u_t(t)\rangle_V \\ &\quad + \int_0^t \langle \xi Q(\tau)\, u(\tau) + \eta Q(\tau)\, u_t(\tau), \xi u(\tau) + \eta u_t(\tau)\rangle_Y \, d\tau \\ &\quad + \beta\{(t+t_0)\,\xi + \eta\}^2 \geqslant 0,\end{aligned}$$

because $P$, $Q(t)$ are linear, symmetric, and non-negative definite. Thus $\mathbb{A}\mathbb{C} - \mathbb{B}^2 \geqslant 0$. In turn (2.9) is valid by virtue of (2.10), (2.11) and the fact that $\mathbb{A}, \mathbb{C} > 0$.

Of course $\alpha > 1$ since $p > 2$ by assumption. The inequality (2.9) can be written as $(\mathcal{I}^{-\alpha}\mathcal{I}')' \geqslant 0$, so

$$\frac{\mathcal{I}'(t)}{\mathcal{I}^{\alpha}(t)} \geqslant \frac{\mathcal{I}'(0)}{\mathcal{I}^{\alpha}(0)} > 0 \qquad \text{for} \quad t \in [0, T_0].$$

This is a Riccati inequality with blow-up time

$$T < \frac{1}{\alpha - 1} \frac{\mathcal{I}(0)}{\mathcal{I}'(0)}.$$

Consequently, if $T_0$ is chosen as the right hand side of the above inequality, we have a contradiction. In fact, since $\mathcal{I}(0)$ depends on $T_0$, this gives an easily solved linear equation for $T_0$, the solution being *positive* for all $t_0$ large enough, e.g., whenever

$$\beta t_0 > \frac{2}{p-2} \langle Q(0)\, u(0), u(0) \rangle_Y - \langle Pu(0), u_t(0) \rangle_V.$$

(An optimal choice for $t_0$, to minimize $T_0$ and to provide a specific estimate for the blow-up time, is easily determined, but is unnecessary for our purposes.) This completes the proof of part (i) of the theorem.

For the proof of part (ii) we proceed almost exactly as in part (i), obtaining in place of (2.8) the estimate

$$\frac{1}{2}\mathcal{I}''(t) \geqslant \left(1 + \frac{q}{2}\right) \langle Pu_t(t), u_t(t) \rangle_V + q \int_0^t \langle Q(\tau)\, u_t(\tau), u_t(\tau) \rangle_Y \, d\tau$$
$$+ q\{D_1 - \mathscr{E}u(0)\} + \beta.$$

Now set $\beta = 2\{D_1 - \mathscr{E}u(0)\} > 0$, and from here on the proof is the same as in part (i).

*Remarks* 1. If $p = q$ in (2.3), then we see from the proof that the results (i) and (ii) remain valid with $D_0 = D_1 = 0$, but without requiring either (2.5) or (2.6). In other words, noncontinuation holds under the single condition $\mathscr{E}u(0) < 0$, namely, negative initial energy; this is exactly the main result of [7].

2. In the usual applications $\mathscr{A}$ is independent of $t$, in which case (2.4) reduces simply to $\mathscr{F}_t(t, u) \geqslant 0$ on $J \times G$.

3. In interesting recent work [13] Vitillaro has treated positive energy blow-up for abstract evolution equations of the type

$$[P(u_t)]_t + Q(t, u_t) + A(u) = F(u), \qquad t \in J = [0, \infty),$$

where the damping term $Q$ is allowed to be nonlinear in $u_t$. To compensate for this increased generality, however, he requires stronger conditions on $A(u)$ and $F(u)$ than purely the assumption (2.3).

## 3. EXAMPLES

Let $\Omega$ be an open domain in $\mathbb{R}^n$ and consider the model problem

$$u_{tt} - \operatorname{div}(|Du|^{q-2} Du) + \mu |u|^{q-2} u = f(t, x, u), \qquad x \in \Omega, \quad t \in J, \tag{3.1}$$

where

$$f(t, x, u) = g(t, x) |u|^{\sigma-2} u + c |u|^{p-2} u, \tag{3.2}$$

and

$$\mu \geqslant 0, \qquad 1 < q < p; \qquad c > 0, \qquad 1 < \sigma < p. \tag{3.3}$$

For the function $g$ we assume

$$-g, \frac{\partial g}{\partial t} \geqslant 0 \quad \text{on} \quad J \times \Omega, \qquad g(t, \cdot) \in L^{p/(p-\sigma)}(\Omega) \qquad \text{for all} \quad t \in J. \tag{3.4}$$

Here the appropriate spaces are $V = L^2(\Omega)$, $W = W_0^{1,q}(\Omega)$, $X = L^p(\Omega)$, and $G = L^2(\Omega) \cap L^p(\Omega) \cap W_0^{1,q}(\Omega)$. For definiteness the space $W$ will be endowed with the norm

$$\|u\|_W = (\|u\|_{L^q(\Omega)}^q + \|Du\|_{L^q(\Omega)}^q)^{1/q}.$$

(Note that the remaining space $Y$ is unneeded, since for simplicity we have omitted damping terms from the equation. In fact, adding a term $a(t, x)\, u_t$, $a \geqslant 0$, $\partial a/\partial t \leqslant 0$ on $J \times \Omega$, to the left-hand side of (3.1) leaves Theorems 2 and 3 below unchanged.)

The operator $P$ corresponding to (3.1) is given by $\langle Pv, w \rangle_V = (v, w)_{L^2}$; clearly $P$ is symmetric and positive definite. We take $A(u) = -\operatorname{div}(|Du|^{q-2} Du) + \mu |u|^{q-2} u$, so that[2]

$$\langle A(u), u \rangle_W = \|Du\|_{L^q}^q + \mu \|u\|_{L^q}^q = q\mathscr{A}(u).$$

[2] The choice $A(u) = -\operatorname{div}(|Du|^{q-2} Du)$ is inconvenient when $\Omega = \mathbb{R}^n$, or indeed when $\Omega$ has infinite measure, due to the lack of coercivity.

On the other hand, as is easy to see, one must then have

$$\mathscr{F}(t,u)=\frac{1}{\sigma}\int_{\Omega} g(t,x)\,|u|^{\sigma}\,dx+\frac{c}{p}\,\|u\|_{L^p}^{p}. \tag{3.5}$$

By a solution of (3.1) we now mean a solution of the abstract evolution equation (1.1) corresponding to the operators $P$, $A$, and $F$ just defined.

Clearly (2.4) is satisfied on $J\times G$ by $(3.4)_1$—we assume suitable regularity of $g=g(t,x)$ so that $\mathscr{F}_t(t,u)$ can be calculated on $J\times G$ by differentiation under the integral sign in (3.5). Finally,

$$\langle F(t,u),u\rangle_X=\int_{\Omega} g(t,x)\,|u|^{\sigma}\,dx+c\,\|u\|_{L^p}^{p},$$

so $p\mathscr{F}(t,u)\geqslant\langle F(t,u),u\rangle_X$ because $\sigma<p$ and $g\leqslant 0$. Therefore (2.3) is verified on $J\times G$.

In order to apply Theorem 1 we shall verify that condition (2.5) holds for appropriate initial data. The situation is different when $\Omega$ is bounded, and for $\Omega$ unbounded. For definiteness we treat the (somewhat more difficult) unbounded case, assuming the further restrictions

$$\mu>0, \qquad q<p\leqslant r, \tag{3.6}$$

where $r=nq/(n-q)$ is the Sobolev exponent for $W_0^{1,q}(\Omega)$ when $q<n$, or otherwise $q<p<\infty$ if $q\geqslant n$.

It is well known that there is a constant $C>0$ depending only on $n$, $p$, and $q$ such that

$$\|u\|_{L^p}\leqslant C\,\|u\|_W \qquad \text{for all} \quad u\in W,$$

see Adams [1, Theorem 5.4, Part III, pp. 97–98]. It follows that

$$\mathscr{A}(u)\geqslant B\,\|u\|_{L^p}^{q}, \qquad u\in W, \tag{3.7}$$

for $B=\min\{1,\mu\}/qC^q$. For precision, we take $B$ in fact to be the *supremum* of values $B$ for which the coercivity condition (3.7) holds (that (3.7) is actually satisfied for the supremum value of $B$ can be proved, but in any case the final results are independent of this).

Using (3.7), it now follows from (2.2), (2.4) and $(3.4)_1$ that

$$\mathscr{E}u(t)\geqslant\mathscr{A}(u)-\mathscr{F}(t,u)\geqslant B\,\|u\|_{L^p}^{q}-\frac{c}{p}\,\|u\|_{L^p}^{p}=E(\lambda), \tag{3.8}$$

where $E(\lambda)=B\lambda^q-(c/p)\,\lambda^p$ and $\lambda=\lambda(t)=\|u\|_{L^p}$.

The graph of the function $E(\lambda) = B\lambda^q - (c/p)\,\lambda^p$ on $\lambda \geqslant 0$ has a single maximum value $E_0$ at $\lambda = \Lambda_0$, where

$$\Lambda_0^{p-q} = \frac{qB}{c}, \qquad E_0 = \left(\frac{q}{c}\right)^{q/(p-q)} B^{p/(p-q)} \left(1 - \frac{q}{p}\right). \tag{3.9}$$

Suppose the initial data is such that $\mathscr{E}u(0)$, $\lambda(0)$ satisfies

$$\mathscr{E}u(0) < E_0, \qquad \lambda(0) > \Lambda_0. \tag{3.10}$$

Then clearly from (2.1) and (2.4) we have $\mathscr{E}u(t) < E_0$ for all $t > 0$. At the same time, by (3.8) it is evident that there can be no time $t > 0$ for which $\mathscr{E}u(t) < E_0$ and $\lambda(t) = \Lambda_0$. Hence by continuity of $\mathscr{E}u$ and $\lambda$ also we have $\lambda(t) > \Lambda_0$ for all $t > 0$. This in turn implies by (3.7) that

$$\mathscr{A}(u(t)) \geqslant B\,\|u(t)\|_{L^p}^q > B\Lambda_0 \equiv \lambda_0,$$

proving that (2.5) holds when (3.10) is satisfied. This proves our second main theorem.

THEOREM 2. *Under the conditions* (3.3), (3.4), (3.6), *and the further restriction* $p > 2$, *the problem* (3.1), (3.2) *cannot have any global solution* $u$ *corresponding to initial data*

$$\mathscr{E}u(0) < E_0, \qquad \|u(0)\|_{L^p} > \Lambda_0, \tag{3.11}$$

*where* $E_0$, $\Lambda_0$ *are given by* (3.9) *and* $B$ *is the best constant for the coercive potential* $\mathscr{A}(u)$, *see* (3.7).

The corresponding result for bounded domains $\Omega$ is slightly different.

THEOREM 3. *Let the measure of* $\Omega$ *be finite. Then Theorem* 2 *remains valid even when* $\mu = 0$.

To see this, it is enough to note that for domains having finite measure we can use for $W$ the equivalent norm $\|u\|_W = \|Du\|_{L^q}$.

*Remarks* 1. Ono in [10] essentially treats the semilinear case $q = 2$, $\mu = 0$ of Theorem 3, but with the initial data satisfying the somewhat stronger conditions

$$\mathscr{E}u(0) < E_0, \qquad \|Du(0)\|_{L^2}^2 < c\,\|u(0)\|_{L^p}^p.$$

Other work involving positive initial energies appears earlier in [2] and [9], the first however restricted to the wave operator itself (with nonlinear boundary conditions), and the second with less precise bounds for the initial energy.

2. In general, (2.6) cannot be obtained when $\mathscr{E}u(0) > 0$. This can be seen, for example, when the function $-g$ is sufficiently large.

3. The discussion above makes clear the distinction between the case when $\mathscr{E}u(0)$ is taken to be negative, and when it is allowed to be positive, as in (3.11). In particular, in the latter case it is necessary that the potential $\mathscr{A}(u)$ be coercive so that in turn one must assume that $p \leqslant r$, a condition which was not needed in the corresponding examples in Section 4 of [7].

4. In the example concerning the degenerate $s$-Laplacian on p. 262 of [7] the condition $s > 2$ was required for the application of their Theorem 1. Here we assume only that $s = q > 1$. This significant improvement is made possible by the more general form of condition (2.3) here in comparison with the corresponding assumption of [7].

5. The results of Theorems 2 and 3 arise directly from the algebraic behavior of the function $E(\lambda)$ representing the potential well for (3.1). For initial data which lies deep enough in the "well" itself, the corresponding solutions are asymptotically stable. This dichotomy is discussed in detail in [12].

A number of concrete examples relative to linear operators $A$ were given in Section III of [4], to which we refer the reader. Example VI of [4, p. 16] in particular deserves special mention. Here the operator $-Q$ is the Laplacian, so for precision the space $Y$ as well as $W$ must be chosen as $H_0^1(\Omega)$.

Other concrete operators $A(u)$ are given in Section 6 of [11], notably the polyharmonic operator $(-\Delta)^L$, where $L \geqslant 1$ is an integer, and still further examples are given in Section 4 of [8].

All of these examples allow extensions to the time dependent case and also exhibit noncontinuation for positive initial energies, both for bounded and unbounded domains, as discussed above.

## REFERENCES

1. R. Adams, "Sobolev Spaces," Academic Press, San Diego, 1975.
2. R. J. Knops, H. A. Levine and L. E. Payne, Non-existence, instability and growth theorems for solutions of a class of abstract nonlinear equations with applications to nonlinear elastodynamics, *Archive Rational Mech. Anal.* **55** (1974), 52–72.
3. H. A. Levine, Some nonexistence and instability theorems for solutions of formally parabolic equations of the form $Pu_t = Au + \mathscr{F}(u)$, *Arch. Rational Mech. Anal.* **51** (1973), 371–386.
4. H. A. Levine, Instability and nonexistence of global solutions of nonlinear wave equations of the form $Pu_{tt} = Au + \mathscr{F}(u)$, *Trans. Amer. Math. Soc.* **192** (1974), 1–21.
5. H. A. Levine, Some additional remarks on the nonexistence of global solutions to nonlinear wave equations, *SIAM J. Math. Anal.* **5** (1974), 138–146.

6. H. A. Levine, Nonexistence of global weak solutions to nonlinear wave equations, *in* "Improperly Posed Boundary Value Problems," Res. Notes Math., Vol. 1, pp. 94–104, Pitman, London, 1975.
7. H. A. Levine, P. Pucci and J. Serrin, Some remarks on global nonexistence for non-autonomous abstract evolution equations, *Contemp. Math.* **208** (1997), 253–263.
8. H. A. Levine and J. Serrin, Global nonexistence theorems for quasilinear evolution equations with dissipation, *Arch. Rational Mech. Anal.* **137** (1997), 341–361.
9. H. A. Levine and R. A. Smith, A potential well theory for the wave equation with a nonlinear boundary condition, *J. Reine Angew. Math.* **374** (1987), 1–23.
10. K. Ono, Global existence, decay, and blowup of solutions for some mildly degenerate nonlinear Kirchhoff strings, *J. Differential Equations* **137** (1997), 273–301.
11. P. Pucci and J. Serrin, Asymptotic stability for non-autonomous dissipative wave systems, *Comm. Pure Appl. Math.* **XLIX** (1996), 177–216.
12. P. Pucci and J. Serrin, Local asymptotic stability for dissipative wave systems, *Israel J. Math.* **104** (1998), 29–50.
13. E. Vitillaro, Global nonexistence theorems for a class of evolution equations with dissipation, preprint, 1998.

*Differ. Integral Equ.* 16 (2003), 13–50. **Reprint II.5.[ii]**

Pages 35–48 have been removed for the sake of brevity.

©Differential and Integral Equations Volume 16, Number 1, January 2003, Pages 13–50

# EXISTENCE FOR A NONLINEAR WAVE EQUATION WITH DAMPING AND SOURCE TERMS

James Serrin
Department of Mathematics, University of Minnesota
127 Vincent Hall, Minneapolis, MN 55455

Grozdena Todorova
Department of Mathematics, University of Tennessee
Knoxville, Tennessee 37996

Enzo Vitillaro
Dipartimento di Matematica e Informatica, Università di Perugia
Via Vanvitelli,1 06123 Perugia, Italy

## 1. Introduction

We study the Cauchy problem

$$u_{tt} - \Delta u + Q(t, x, u_t) = f(x, u), \quad u(0) = u_0, \; u_t(0) = u_1, \tag{1.1}$$

in $(0, T) \times \mathbb{R}^n$, where $0 < T \leq \infty$, $n \geq 1$. The data $u_0$ and $u_1$ are in the energy space and are compactly supported in $\mathbb{R}^n$. The terms $Q$ and $f$ are nonlinear damping and source terms respectively.

Problems like (1.1) have been studied in many papers. For the case $Q \equiv 0$ see [3], [7], [11], [12], [20], [30]; for the case $f \equiv 0$ see [8], [10]; when both terms $Q$ and $f$ interact, see [6], [9], [13], [14], [15], [18], [19], [25], [26], [27], [31], [32], [33].

For the sake of simplicity we illustrate our results by considering the model power–like equation with damping term

$$Q(x, v) = \sigma(x)|v|^{m-2}v, \tag{1.2}$$

where $m > 1$, $\sigma$ is a measurable function such that $c_1 \leq \sigma(x) \leq c_2$ on $\mathbb{R}^n$, $c_1$, $c_2$ are positive constants, and the source term is in the form

$$f(x, u) = \mu_1(x)|u|^{p-2}u + \mu_2(x)|u|^{q-2}u, \tag{1.3}$$

with $p > 2$, $1 < q < p$, and $\mu_i \in L^\infty_{\text{loc}}(\mathbb{R}^n)$, $i = 1, 2$.

Accepted for publication: June 2002.
AMS Subject Classifications: 34D05, 70D10, 94C99.

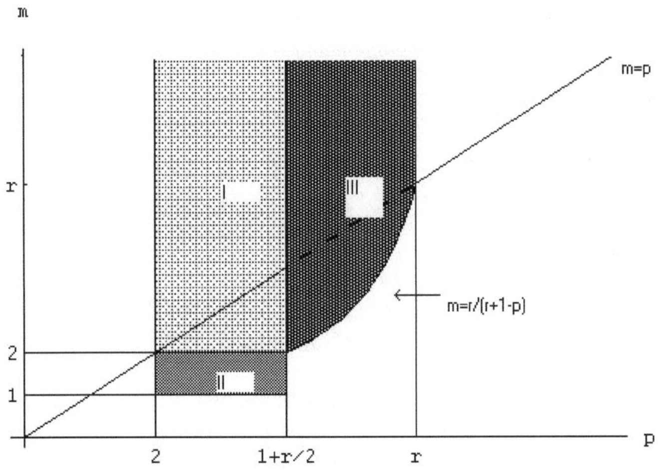


FIGURE 1. The three regions I, II and III in the $(p,m)$ plane, for which we prove local existence and uniqueness in the case $q \geq 2$, given by

$$\begin{aligned} \mathrm{I} &= \{(p,m) \in \mathbb{R}^2 : 2 < p < 1 + r/2, \quad m \geq 2\}, \\ \mathrm{II} &= \{(p,m) \in \mathbb{R}^2 : 2 < p < 1 + r/2, \quad 1 < m < 2\}, \\ \mathrm{III} &= \{(p,m) \in \mathbb{R}^2 : 1 + r/2 \leq p < r, \quad m > r/(r+1-p)\}. \end{aligned}$$

Local existence for solutions of the equation (1.1) in a bounded domain $\Omega$ with zero Dirichlet boundary condition was proved in [6], for $\sigma \equiv \mu_1 \equiv 1$ and $\mu_2 \equiv 0$, where the exponents of nonlinearity $m$ and $p$ satisfy the conditions

$$m \geq 2, \qquad \text{and} \quad 2 < p \leq 1 + r/2, \tag{1.4}$$

with $r = 2n/(n-2)$ for $n \geq 3$ and $r = \infty$ for $n = 1, 2$. The above region for $m$ and $p$ is the right closure of the region I in the plane $(p,m)$ in Figure 1. The particular cut (1.4) for exponents of nonlinearities is assumed in all previously quoted papers where the interaction between source and damping terms is studied. The goal of this paper is to take more careful advantage of the presence of the damping term, namely to clarify the smoothing effect of the damping (which will be manifested in particular in the region III), as well as to consider the influence of sublinear damping terms (region II). We are able to enlarge the local existence region I to the region $\mathrm{I} \cup \mathrm{II} \cup \mathrm{III}$; more precisely, we have the following existence result for the model power–like equation:

**Theorem 1.** *Let $Q$ and $f$ satisfy (1.2) and (1.3), and let $(p,m) \in I \cup II \cup III$, namely*

$$m > 1, \quad 2 < p < 1 + r/2 \quad or \quad m > r/(r+1-p), \quad 1 + r/2 \leq p < r. \quad (1.5)$$

*Moreover, assume that $1 < q < p$, and let the data be compactly supported and in the energy space $u_0 \in H^1(\mathbb{R}^n)$, $u_1 \in L^2(\mathbb{R}^n)$. Then, there is a weak solution of (1.1) with the properties*

$$u \in C([0,T]; H^1(\mathbb{R}^n)), \quad u_t \in C([0,T]; L^2(\mathbb{R}^n)) \quad (1.6)$$

$$u_t \in L^m((0,T) \times \mathbb{R}^n), \quad (1.7)$$

*for $T > 0$ small enough.*

*Moreover, when $\mu_2 \equiv 0$, and exponents $(p,m) \in \overline{I_r} \cup \overline{II_r}$, where*

$$\overline{I_r} = \{(p,m) \in \mathbb{R}^2 : 2 < p \leq 1 + r/2, \quad m \geq 2\},$$

$$\overline{II_r} = \{(p,m) \in \mathbb{R}^2 : 2 < p \leq 1 + r/2, \quad 1 < m < 2\}$$

*respectively are the right closure of I and II, a unique solution of (1.1) exists, with the same regularity.*

**Remark 1.** Theorem 1 is a particular case of our general local existence result (see Theorem 7 below) for the equation (1.1), with damping $Q = Q(t,x,u_t)$ and source $f = f(x,u)$, under mild assumptions which will be presented in the sequel.

As an extension of the above result we have the following global existence result, for the case $p \leq m$.

**Theorem 2.** *Let $Q$ and $f$ satisfy (1.2) and (1.3) and $2 < p < r$, $1 < q < p \leq m$. Then for any compactly supported data in the energy space $u_0 \in H^1(\mathbb{R}^n)$, $u_1 \in L^2(\mathbb{R}^n)$ the Cauchy problem (1.1) has a global solution with the regularity (1.6)–(1.7).*

**Remark 2.** Once again our global existence theorem is for more general nonlinearities $Q$ and $f$ compared with the ones considered in Theorem 2. The general global existence Theorem 8 will be formulated and proved in the sequel. Let us mention that this global existence theorem is a generalization of the global existence theorem in [27].

Actually, with essentially not power like damping and source terms, namely, with $Q(u_t)$ and $f(x,u)$ being derivatives of $N$−functions, then under appropriate conditions an existence – regularity theorem could be proved.

**Remark 3.** Further applications of our general existence–regularity theorem, concerning global existence and energy decay, will be presented in a forthcoming paper [29]. We shall present several blow-up type applications of the general local existence Theorem 7 in another paper [28].

Following the idea of the proof of the general global existence Theorem 8, we can derive a global existence result for any compactly supported data in the energy space, for the Cauchy problem with decaying potential

$$\begin{cases} u_{tt} - \Delta u + \alpha(t)|u_t|^{m-2}u_t = f(x,u), & \text{in } (0,\infty)\times\mathbb{R}^n, \\ u(0) = u_0, \qquad u_t(0) = u_1, \end{cases} \tag{1.8}$$

where $\alpha(t) = 1/(1+t)^\beta$, $\beta > 0$, and $f$ satisfies (1.3). We state this as

**Corollary 1.** *Let $m \geq 2$. Then for any compactly supported data in the energy space the problem* (1.8) *has a global weak solution with the regularity* (1.6)–(1.7).

**Remark 4.** This result does not depends on the range of the potential. Even a very short range potential does not effect the global existence result.

Now we emphasize some of the essential difficulties to be faced and the way to overcome them.

The main difficulty in the region II ($1 < m < 2$) is the lack of an overall Lipschitz behavior for the damping, which badly effects the proof of the finite propagation speed in this region. Indeed a Lipschitz condition is a basic tool for Strauss' argument [25] proving a finite speed of propagation. For the sake of simplicity, we consider the problem when the damping is in the simple form $Q(t,x,v) = |v|^{m-2}v$. We approximate $Q$ uniformly with conveniently chosen Lipschitz functions such that the approximate equation

$$u_{tt} - \Delta u + Q_\varepsilon(u_t) = g(t,x) \tag{1.9}$$

has a finite speed of propagation. Here $g(t,x)$ is a conveniently chosen replacement for the nonlinear source $f(x,u)$, with support in the light cone. To carry out this procedure we need an existence–regularity result for (1.9). Here the main difficulty appears, namely we are no longer able to stay in the Lebesgue spaces used by Strauss. To show that this is the case, we present a list of requirements for $Q_\varepsilon$ which would follow from Lipschitzianity and a finite speed of propagation for (1.9). From the finite speed of propagation it follows that $u = 0$ satisfies (1.9) outside the light cone, i.e., the first requirement is $Q_\varepsilon(0) = 0$. From this and a Lipschitz condition we get the next requirement, namely, the estimate

$$|Q_\varepsilon(v)| \leq c_1^\varepsilon |v| \tag{1.10}$$

for some constant $c_1^\varepsilon$. Then to derive an a priori estimate for $u_t^\varepsilon$, where $u^\varepsilon$ is a solution of the approximate problem (1.9) in some Lebesgue space, we have to ask that

$$Q_\varepsilon(v)v \geq c_2^\varepsilon |v|^\mu \tag{1.11}$$

for some $\mu > 1$. Due to the uniformity of the approximation, $\mu$ cannot be different from $m$, which is clearly impossible because (1.10) and (1.11) lead to a contradiction when $\mu = m < 2$. Thus, we realize that it is not possible to stay within any Lebesgue space. Nevertheless we are still able to derive an estimate

$$Q_\varepsilon(v)v \geq c_2^\varepsilon \Phi_\varepsilon(v),$$

where $\Phi_\varepsilon$ is a conveniently chosen $N$-function. In this way, we naturally come to the idea of embedding $u_t^\varepsilon$ in the corresponding Orlicz space $L^{\Phi_\varepsilon}((0,T) \times \mathbb{R}^n)$. Since $Q_\varepsilon$ is no longer power–like, we need to prove a new existence–regularity result for (1.9), where the damping $Q_\varepsilon$ is essentially the derivative of an $N$ – function (under mild conditions). This result, together with some continuity results for Orlicz spaces, and a density lemma in the intersection of Orlicz and Lebesgue spaces (which we prove in Appendix B) closes the circle.

The smoothness effect of the damping is mainly manifested in the region III. Indeed, due to the presence of the damping term, we are able to obtain a space – time estimate for $u_t$ which plays a basic role in controlling the kinetic energy of the solution. This, together with an appropriate compactness result, allows us to use the Schauder fixed point theorem. On the other hand, for the region $\overline{\mathrm{I}_r} \cup \overline{\mathrm{II}_r}$, we can improve this result by applying the contraction principle and thus obtain uniqueness as well.

## 2. Global existence with a given forcing term

This section is devoted to global existence results for the Cauchy problem

$$\begin{cases} u_{tt} - \Delta u + Q(t,x,u_t) = g(t,x), & \text{in } (0,T) \times \mathbb{R}^n, \\ u(0) = u_0, \qquad u_t(0) = u_1. \end{cases} \tag{2.1}$$

These results will be essentially used in the sequel.

We assume that $Q$ is a Caratheodory function on $(0,T) \times \mathbb{R}^n$, i.e., that $Q = Q(t,x,v)$ is continuous in $v$ for almost all $(t,x) \in (0,T) \times \mathbb{R}^n$ and measurable in $(t,x)$ for all $v \in \mathbb{R}$. Moreover, let the following assumptions hold:

(Q1) $Q$ is increasing in $v$ for almost all $(t,x) \in (0,T) \times \mathbb{R}^n$;

(Q2) there exist $m > 1$ and $c_1 > 0$ such that

$$|Q(t,x,v)| \le c_1|v|^{m-1}$$

for almost all $(t,x) \in (0,T) \times \mathbb{R}^n$ and all $v \in \mathbb{R}$;

(Q3) there is $c_2 > 0$ such that

$$Q(t,x,v)v \ge c_2|v|^m$$

for almost all $(t,x) \in (0,T) \times \mathbb{R}^n$ and all $v \in \mathbb{R}$.

The first result of this section generalizes the Lions and Strauss bounded domain global existence result [16, Theorem 6.1] to the Cauchy problem in $\mathbb{R}^n$ and to nonlinearities satisfying (Q1)–(Q3).

**Theorem 3.** *Let $Q$ satisfy (Q1)–(Q3), let the data satisfy $u_0 \in H^1(\mathbb{R}^n)$, $u_1 \in L^2(\mathbb{R}^n)$, and suppose $g(t,x) \in L^2((0,T) \times \mathbb{R}^n)$. Then there is a unique global weak solution of (2.1) such that*

$$u \in C([0,T]; H^1(\mathbb{R}^n)), \quad u_t \in C([0,T]; L^2(\mathbb{R}^n)) \cap L^m((0,T) \times \mathbb{R}^n).$$

**Remark 5.** The solution also satisfies the energy identity

$$\tfrac{1}{2}\|u_t(s)\|_2^2 + \tfrac{1}{2}\|\nabla u(s)\|_2^2\Big|_0^t + \int_{(0,t)\times\mathbb{R}^n} Q(t,x,u_t)u_t = \int_{(0,t)\times\mathbb{R}^n} g u_t,$$

as follows directly from Strauss' argument [24, Theorem 4.2].

**Proof.** We start by reducing the Cauchy problem to an initial–boundary value problem (IBVP) in $B_\rho = \{x \in \mathbb{R}^n : |x| < \rho\}$. First we approximate the data and the forcing term $g(t,x)$ with conveniently chosen functions in $B_\rho$. Let $\eta \in C^\infty(\mathbb{R})$ be a cut–off function such that $\eta(s) = 0$ for $|s| \ge 1$ and $\eta(s) = 1$ for $|s| \le 1/2$. Put $\eta_\rho(x) = \eta(4|x|^2/\rho^2)$. Then we define

$$u_{0\rho} = \eta_\rho u_0 \in H_0^1(B_\rho), \quad u_{1\rho} = \eta_\rho u_1 \in L^2(B_\rho), \quad g_\rho = \eta_\rho g \in L^2((0,T) \times B_\rho). \tag{2.2}$$

Clearly

$$\nabla u_{0\rho}(x) = \nabla u_0(x)\eta_\rho(x) + u_0(x)\eta'(4|x|^2/\rho^2)8x/\rho^2.$$

Next, by the Lebesgue dominated convergence theorem, we get

$$u_{0\rho} \to u_0 \in H^1(\mathbb{R}^n), \; u_{1\rho} \to u_1 \in L^2(\mathbb{R}^n), \; g_\rho \to g \in L^2((0,T) \times \mathbb{R}^n). \tag{2.3}$$

Now consider the perturbed initial–boundary value problem

$$\begin{cases} u_{tt} - \Delta u + Q(t,x,u_t) = g_\rho, & \text{in } (0,T) \times B_\rho, \\ u = 0 & \text{in } (0,T) \times \partial B_\rho, \\ u(0) = u_{0\rho}, \qquad u_t(0) = u_{1\rho}. & \end{cases} \tag{2.4}$$

Since the nonlinear damping term $Q(t, x, u_t)$ could possibly contain a non–Lipschitz part, we are not able to use the standard Faedo–Galerkin procedure which requires a Lipschitz argument to obtain the solution of the corresponding ordinary differential equation. Instead, we are forced to use Carathéodory's Theorem (see [5]) to insure the existence of a solution of the corresponding ordinary differential systems. We complete the proof of the global existence result for (2.4) in the standard way, obtaining the same regularity for the solution $u^\rho$, i.e.,

$$\begin{aligned} &u^\rho \in L^\infty(0,T;H_0^1(B_\rho)), \quad u_t^\rho \in L^\infty(0,T,L^2(B_\rho)) \cap L^m((0,T)\times B_\rho), \\ &Q(t,x,u_t^\rho) \in L^{m'}((0,T)\times B_\rho). \end{aligned} \tag{2.5}$$

Moreover, the solution $u^\rho$ of (2.4) satisfies the energy identity. As a consequence of this and the convergence (2.3) of $u_{0\rho}$, $u_{1\rho}$ and $g_\rho$, we obtain

$$\sup_{t\in[0,T)} \|u_t^\rho(t)\|_2^2 + \|\nabla u^\rho(t)\|_2^2 \le k_1, \tag{2.6}$$

$$\int_{(0,T)\times\mathbb{R}^n} Q(\cdot,\cdot,u_t^\rho)u_t^\rho \le k_1, \tag{2.7}$$

where the constant $k_1$ is independent of $\rho$. Moreover, since

$$\|u^\rho(t)\|_2 \le \|u_{0\rho}\|_2 + \int_0^T \|u_t^\rho\|_2$$

for all $t \in [0,T]$, from the convergence $(2.3)_1$ and from the estimate (2.6) we get

$$\sup_{t\in[0,T)} \|u^\rho(t)\|_2 \le k_2, \tag{2.8}$$

where the constant $k_2$ is independent of $\rho$. Then, using assumptions (Q1) and (Q2) together with the estimates (2.6)–(2.8), we obtain (up to a subsequence) the convergence

$$\begin{aligned} &u^\rho \to u \;\text{ weakly* in } L^\infty(0,T;H^1(\mathbb{R}^n)), \\ &u_t^\rho \to u_t \;\text{ weakly* in } L^\infty(0,T;L^2(\mathbb{R}^n)) \text{ and weakly in } L^m((0,T)\times\mathbb{R}^n), \\ &Q(\cdot,\cdot,u_t^\rho) \to \chi \;\text{ weakly in } L^{m'}((0,T)\times\mathbb{R}^n), \end{aligned} \tag{2.9}$$

as $\rho \to \infty$. We shall prove that $u$ is a solution in a distribution sense of the equation

$$u_{tt} - \Delta u + \chi - g = 0 \qquad \text{in } H^{-1}(\mathbb{R}^n) + L^{m'}(\mathbb{R}^n), \tag{2.10}$$

where the space $H^{-1}(\mathbb{R}^n)$ is the dual of $H^1(\mathbb{R}^n)$. First, we multiply equation (2.4) by $\phi \in C_c^\infty(0,T)$ and integrate over $[0,T)$, to get

$$\int_0^T -u_t^\rho \phi' + [-\Delta u^\rho + Q(\cdot,\cdot,u_t^\rho) - g_\rho]\phi \, ds = 0, \tag{2.11}$$

where an integration by parts was used. Then we fix $w \in C_c^\infty(\mathbb{R}^n)$ and chose $\rho$ sufficiently large so that supp $w$ is compactly included in $B_\rho$. Multiplying (2.11) by $w$ and integrating over $\mathbb{R}^n$, we derive

$$\int_{(0,T)\times\mathbb{R}^n} -u_t^\rho w\phi' + [\nabla u^\rho \nabla w + Q(t,x,u_t^\rho)w - g_\rho w]\phi = 0. \tag{2.12}$$

Finally, due to the density of $C_c^\infty(\mathbb{R}^n)$ in $H^1(\mathbb{R}^n) \cap L^m(\mathbb{R}^n)$, and using the regularity (2.5), we see that (2.12) is fulfilled for all test functions $w \in H^1(\mathbb{R}^n) \cap L^m(\mathbb{R}^n)$. Passing to the limit as $\rho \to \infty$ in (2.12), and using the convergences (2.9) and $(2.3)_3$, we get

$$\int_0^T -u_t\phi' + [-\Delta u + \chi - g]\phi = 0 \qquad \text{in } H^{-1}(\mathbb{R}^n) + L^{m'}(\mathbb{R}^n).$$

It remain to prove that $\chi = Q(\cdot,\cdot,u_t)$ and $u(0) = u_0$, $u_t(0) = u_1$. Using the assumption (Q1) and the celebrated Lions–Strauss monotonicity argument we obtain that $\chi = Q(\cdot,\cdot,u_t)$. The relations $u(0) = u_0$, $u_t(0) = u_1$, the energy identity and the uniqueness of $u$ are derived exactly as in [16] and [24]. □

We need one more global existence result for equation (2.1) with more regular data. For this we shall require some assumptions on the derivative of the damping term in the case $m \geq 2$. More precisely, we assume that $Q_t$ and that $Q_v$ are Caratheodory functions, and

(Q4) there exists $c_3 \geq 0$ such that

$$|Q_t(t,x,v)| \leq c_3|v|^{m-1}$$

for almost all $(t,x) \in (0,T) \times \mathbb{R}^n$ and all $v \in \mathbb{R}$;

(Q5) there exists $\chi \in L^1_{\text{loc}}(\mathbb{R}^n)$ and $\psi \in L^\infty_{\text{loc}}(\mathbb{R}^n)$ such that

$$|Q_v(t,x,v)| \leq \chi(x)\psi(v)$$

for almost all $(t,x) \in (0,T) \times \mathbb{R}^n$ and all $v \in \mathbb{R}$;

(Q6) there is $c_4 > 0$ such that

$$Q_v(t,x,v) \geq c_4|v|^{m-2}$$

for almost all $(t,x) \in (0,T) \times \mathbb{R}^n$ and all $v \in \mathbb{R}$.

We are now able to state

**Theorem 4.** *Assume that the assumptions* (Q1)*–*(Q6) *hold, that* $m \geq 2$, *and*

$$u_0 \in H^2(\mathbb{R}^n), \quad u_1 \in H^1(\mathbb{R}^n) \cap L^{2(m-1)}(\mathbb{R}^n), \quad g, g_t \in L^2((0,T) \times \mathbb{R}^n). \tag{2.13}$$

*Let* $u$ *be the solution of* (2.1) *given by Theorem* 3. *Then in addition to the regularity assured by Theorem* 3, *we have*

$$u_t \in L^\infty(0,T; H^1(\mathbb{R}^n)), \qquad \text{and} \quad u_{tt} \in L^\infty(0,T; L^2(\mathbb{R}^n)). \tag{2.14}$$

**Remark 6.** The corresponding regularity cannot be obtained for region II, namely when the growth rate $m$ of the damping is between 1 and 2. The main reason for this is the lack of a Lipschitz condition for the damping term $Q$ in this region.

**Proof.** To derive the higher regularity of the solution $u$, we start with the calculation of derivatives for $u_{0\rho}$ and $g_\rho$ (see (2.2)). Namely, we have [1]

$$\begin{aligned} D^2 u_{0\rho}(x) &= \eta_\rho(x) D^2 u_0(x) + 16/\rho^2 \eta' \left(4|x|^2/\rho^2\right) \nabla u_0(x) \otimes x \\ &\quad + 8/\rho^2 u_0(x) \left[\eta' \left(4|x|^2/\rho^2\right) I_n + 8/\rho^2 \eta'' \left(4|x|^2/\rho^2\right) x \otimes x\right] \end{aligned}$$

and $(g_\rho)_t = \eta_\rho g_t$. With the help of (2.13) we can improve the convergence (2.3) in the following way:

$$u_{0\rho} \to u_0 \ \text{ in } H^2(\mathbb{R}^n), \quad u_{1\rho} \to u_1 \quad \text{in } H^1(\mathbb{R}^n) \cap L^{2(m-1)}(\mathbb{R}^n), \tag{2.15}$$

$$g_\rho \to g \ \text{ in } L^2((0,T) \times \mathbb{R}^n), \quad (g_\rho)_t \to g_t \quad \text{in } L^2((0,T) \times \mathbb{R}^n). \tag{2.16}$$

Next we need some estimates for the solution $u_\nu$ in a modified Faedo–Galerkin approximation procedure, namely for the equations

$$\begin{cases} (u_\nu''(t), w_j) + (\nabla u_\nu(t), \nabla w_j) + (Q(t, \cdot, u_\nu'(t)), w_j) = (g_\rho(t), w_j), \\ u_\nu(0) = u_{0\nu}, \qquad u_\nu'(0) = u_{1\nu}, \end{cases} \tag{2.17}$$

where $j = 1, \dots, \nu$, and $w_j \in C^\infty(\overline{B_\rho})$ are the eigenfunctions of the Laplacian on $B_\rho$. In the above, we choose $u_{0\nu}, u_{1\nu} \in W_\nu := \text{span}[w_1, \dots, w_\nu]$ and approximate the initial data in the sense that

$$u_{0\nu} \to u_{0\rho} \quad \text{in } H^2(B_\rho), \quad u_{1\nu} \to u_{1\rho} \quad \text{in } H_0^1(B_\rho) \cap L^{2(m-1)}(B_\rho). \tag{2.18}$$

Following the ideas of [17], we get the estimates

$$\|u_\nu''(0)\|_2 \leq \|\Delta u_{0\nu}\|_2 + \|g_\rho(0)\|_2 + c_1 \|u_{1\nu}\|_{2(m-1)}^{m-1}, \tag{2.19}$$

$$\sup_{t \in (0,T)} \|u_\nu''(t)\|_2^2 + \|\nabla u_\nu'(t)\|_2^2$$

---

[1] $x \otimes y$ denotes the standard tensor product in $\mathbb{R}^n$, and $I_n$ the identity matrix in $\mathbb{R}^n$.

$$\leq \text{Const.} \left( |||u'_\nu|||_m^m + |||(g_\rho)_t|||_2^2 + \|u''_\nu(0)\|_2^2 + \|\nabla u_{1\nu}\|_2^2 \right). \tag{2.20}$$

Here and in the sequel we use the notation $|||h|||_s$ for the time – space norm $(\int_{(0,T)\times\mathbb{R}^n} |h|^s)^{1/s}$. Moreover we have the convergence conditions

$$u''_\nu \to (u^\rho)_{tt} \qquad \text{weakly}^* \text{ in } L^\infty(0,T,L^2(B_\rho)) \tag{2.21}$$

$$u'_\nu \to (u^\rho)_t \qquad \text{weakly}^* \text{ in } L^\infty(0,T,H_0^1(B_\rho)). \tag{2.22}$$

Passing to the limit as $\nu \to \infty$ in (2.19) and using (2.18), we have

$$\limsup_\nu \|u''_\nu(0)\|_2^2 \leq \|\Delta u_{0\rho}\|_2 + \|g_\rho(0)\|_2 + c_1\|u_{1\rho}\|_{2(m-1)}^{m-1}. \tag{2.23}$$

By the energy identity for $u_\nu$, the Gronwall lemma, and passage to the limit as $\nu \to \infty$, we obtain

$$\limsup_\nu \int_{(0,T)\times B_\rho} Q(\cdot,\cdot,u'_\nu)u'_\nu \leq \text{Const.} \left(\|u_{1\rho}\|_2^2 + \|\nabla u_{0\rho}\|_2^2 + |||g_\rho|||_2^2\right). \tag{2.24}$$

Then, using assumption (Q3) we can rewrite the previous estimate in the form

$$\limsup_\nu |||u'_\nu|||_m^m \leq \text{Const.} \left(\|u_{1\rho}\|_2^2 + \|\nabla u_{0\rho}\|_2^2 + |||g_\rho|||_2^2\right). \tag{2.25}$$

Passing to the limit in (2.20), using the convergences (2.16), (2.18), (2.21), (2.22) and the estimates (2.23), (2.25), we obtain

$$\begin{aligned} \sup_{t\in(0,T)} \|u_{tt}^\rho(t)\|_2^2 + \|\nabla u_t^\rho(t)\|_2^2 \leq \text{Const.}\Big( & \|u_{0\rho}\|_{H^2(\mathbb{R}^n)} + \|u_{1\rho}\|_{H^1(\mathbb{R}^n)} \\ & + \|u_{1\rho}\|_{2(m-1)}^{m-1}\|g_\rho(0)\|_2^2 + |||g_\rho|||_2^2 + |||(g_\rho)_t|||_2^2\Big), \end{aligned} \tag{2.26}$$

where we have also used the lower semi–continuity of the norm in the space $L^\infty(0,T;L^2(\mathbb{R}^n))$ with respect to the weak* topology. By the convergence (2.16) together with the inequality (2.26) we get the estimates

$$\sup_{t\in(0,T)} \|u_{tt}^\rho(t)\|_2 \leq \text{Const.}, \qquad \text{and} \qquad \sup_{t\in(0,T)} \|\nabla u_t^\rho(t)\|_2 \leq \text{Const.}, \tag{2.27}$$

where the constants are $\rho$ – independent. Then, passing to the limit as $\rho \to \infty$, we can extend the convergence (2.9) with the further results

$$u_{tt}^\rho \to u_{tt}, \quad \text{and} \quad \nabla u_t^\rho \to \nabla u_t \qquad \text{weakly}^* \text{ in } L^\infty(0,T;L^2(\mathbb{R}^n)). \tag{2.28}$$

This completes the proof. □

**Remark 7.** We can get more regularity for the solution $u$ than is assured by Theorem 4 by assuming additional properties for the damping term and for $g(t,x)$. Namely, provided that

$$|\nabla_x Q(t,x,v)| \le c_5 |v|^{m-1}$$

is fulfilled for almost all $(t,x) \in (0,T) \times \mathbb{R}^n$ and all $v \in \mathbb{R}$, and $g \in L^2(0,T;H^1(\mathbb{R}^n))$, the solution $u$ of equation (2.1) then has the additional regularity $u \in L^\infty(0,T;H^2(\mathbb{R}^n))$. We omit the proof of this remark since it can be performed using arguments similar to those of Theorem 3 and 4.

The above existence–regularity results are basically related with power-type nonlinearities. In Section 3 we shall prove an existence–regularity result for essentially different types of nonlinearities.

## 3. Finite speed of propagation

In this section we prove the finite propagation speed for solutions of equation (2.1) when the growth rate $m$ of the damping term satisfies $1 < m < 2$.

As mentioned in the introduction, since the damping term obeys a Lipschitz condition when $m \ge 2$, the existence of a finite propagation speed for this case is almost a direct consequence of the argument of Strauss (see [25]), adapted for weak solutions.

In the region $1 < m < 2$ the main difficulty is the lack of Lipschitz behavior for the damping. We overcome this difficulty by approximating the damping term with Lipschitz continuous functions $Q_\varepsilon$. This approximation however causes other difficulties since we are not able to prove that the time derivative of the solution of the approximate problems is in the Lebesgue space $L^m$. On the other hand we are able to show that the time derivative is in a conveniently chosen Orlicz space. See Proposition 1 and Theorem 5 below. For this approximating procedure we need to construct sufficiently smooth functions which are quadratic near 0 and behave like $t^m$ for sufficiently large $t$. The construction is given in the following lemma.

**Lemma 1.** *Let $1 < m < 2$. For each $\varepsilon > 0$ there is a strictly convex function $\Phi_\varepsilon \in C^2([0,\infty))$ such that*

$$\Phi_\varepsilon(0) = \Phi'_\varepsilon(0) = 0, \tag{3.1}$$

$$\Phi_\varepsilon(t) = t^m \quad \textit{for all } t \ge \varepsilon, \tag{3.2}$$

$$\Phi_\varepsilon(t) = O(t^2) \quad \textit{as } t \to 0^+. \tag{3.3}$$

*Moreover, $\Phi_\varepsilon$ satisfies the estimates*

$$\Phi_\varepsilon(2t) \le c_{1\varepsilon}\Phi_\varepsilon(t) \quad \textit{for all } t \ge 0, \tag{3.4}$$

$$2l_\varepsilon \Phi_\varepsilon(t) \le \Phi_\varepsilon(l_\varepsilon t) \quad \textit{for all } t \ge 0, \tag{3.5}$$

$$\Phi'_\varepsilon(t) \le c_5 t^{m-1} \quad \textit{for all } t \ge 0 \tag{3.6}$$

$$\Phi''_\varepsilon(t) t \le c_{2\varepsilon} \Phi'_\varepsilon(t) \quad \textit{for all } t \ge 0, \tag{3.7}$$

*where the constants* $c_{1\varepsilon}, c_{2\varepsilon} > 0$, $l_\varepsilon > 1$ *may depend on* $\varepsilon$ *but* $c_5 > 0$ *is independent of* $\varepsilon$.

**Proof.** We define the function $\Phi_\varepsilon$ by

$$\Phi_\varepsilon(t) = \begin{cases} t^2(a_\varepsilon t^2 + b_\varepsilon t + c_\varepsilon), & t < \varepsilon \\ t^m, & t \ge \varepsilon \end{cases} \tag{3.8}$$

where $a_\varepsilon$, $b_\varepsilon$ and $c_\varepsilon$ are conveniently chosen constants in $\mathbb{R}$ such that $\Phi_\varepsilon \in C^2([0,\infty))$, namely $a_\varepsilon = \frac{1}{2}(m-2)(m-3)\varepsilon^{m-4}$, $b_\varepsilon = -(m-2)(m-4)\varepsilon^{m-3}$, and $c_\varepsilon = \frac{1}{2}(m-3)(m-4)\varepsilon^{m-2}$. Elementary calculations using the facts that $a_\varepsilon > 0$ and $1 < m < 2$ show that $\Phi_\varepsilon$ is a strictly convex function, while (3.1)–(3.3) are trivial consequences of the definition (3.8). Properties (3.4) and (3.7) easily follow from (3.8), taking into account the behavior of the functions $\Phi''_\varepsilon(t)t/\Phi'_\varepsilon(t)$ and $\Phi_\varepsilon(2t)/\Phi_\varepsilon(t)$ near 0 and $\infty$. Inequality (3.6) also follows from (3.8), the homogeneity of $a_\varepsilon$, $b_\varepsilon$ and $c_\varepsilon$ and the fact that $1 < m < 2$.

The proof of (3.5) has to be divided in three different cases:

(a) large $t$, i.e., $t \ge \varepsilon$;
(b) intermediate $t$, i.e., $\varepsilon/l_\varepsilon < t < \varepsilon$;
(c) small $t$, i.e., $t \le \varepsilon/l_\varepsilon$.

The case (a) is trivial. For the case (b) it is sufficient to verify

$$2(a_\varepsilon t^2 + b_\varepsilon t + c_\varepsilon) \le l_\varepsilon^{m-1} t^{m-2} \tag{3.9}$$

for $t \in (\varepsilon/l_\varepsilon, \varepsilon)$, which follows from the convexity of the function $a_\varepsilon t^2 + b_\varepsilon t + c_\varepsilon$, namely

$$a_\varepsilon t^2 + b_\varepsilon t + c \le \max\{a_\varepsilon \varepsilon^2 + b_\varepsilon \varepsilon + c_\varepsilon, a_\varepsilon(\varepsilon/l_\varepsilon)^2 + b_\varepsilon(\varepsilon/l_\varepsilon) + c_\varepsilon\},$$

and the inequality $l_\varepsilon^{m-1} t^{m-2} \ge l_\varepsilon^{m-1} \varepsilon^{m-2}$ for $t \in (\varepsilon/l_\varepsilon, \varepsilon)$. Then (3.9) is fulfilled for sufficiently large $l_\varepsilon$. In the third case we have to verify that

$$a_\varepsilon(l_\varepsilon^3 - 2)t^2 + b_\varepsilon(l_\varepsilon^2 - 2)t + c_\varepsilon(l_\varepsilon - 2) \ge 0$$

for $t \le \varepsilon/l_\varepsilon$, which is true for $l_\varepsilon$ sufficiently large. □

For the sake of simplicity we shall prove the main theorem of this section only for damping terms which satisfy the assumption

(Q7) for almost all $x \in \mathbb{R}^n$ and all $v \in \mathbb{R}$,

$$Q(t,x,v) = Q(x,v) = \sigma(x)|v|^{m-2}v$$

where $\sigma \in L^\infty(\mathbb{R}^n)$ is such that $\inf_{\mathbb{R}^n} \sigma > 0$.

It is easy to see that (Q7) implies the previous assumptions (Q1)–(Q3).

Now, using the function $\Phi_\varepsilon$ constructed in Lemma 1, the approximation of the damping term can be done in the following way:

$$Q_\varepsilon(x,v) = \frac{1}{m}\sigma(x)\Phi'_\varepsilon(|v|)\text{sign } v. \tag{3.10}$$

Then $Q_\varepsilon(x,\cdot) \in C(\mathbb{R})$ for almost all $x \in \mathbb{R}^n$ and, more important, $Q_\varepsilon$ is uniformly Lipschitz continuous in $v$, i.e., $(Q_\varepsilon)_v \in L^\infty(\mathbb{R}^{n+1})$. Let us note that

$$Q_\varepsilon(\cdot,v) \equiv Q(\cdot,v), \qquad \text{for } |v| \geq \varepsilon, \tag{3.11}$$

and

$$|Q_\varepsilon(x,v)| \leq |\sigma(x)|\varepsilon^{m-1}, \qquad \text{for } |v| \leq \varepsilon, \tag{3.12}$$

where to get the last inequality we used the convexity of $\Phi_\varepsilon$. We approximate problem (2.1) by

$$\begin{cases} u_{tt} - \Delta u + Q_\varepsilon(x,u_t) = g(t,x), & \text{in } (0,T) \times \mathbb{R}^n, \\ u(0) = u_0, \qquad u_t(0) = u_1. \end{cases} \tag{3.13}$$

Unfortunately, we are not able to assure existence of a regular solution of this approximate problem because all previous existence theorems apply for nonlinearities $Q$ which essentially dominate a power of $v$. Evidently, $Q_\varepsilon$ does not satisfy this requirement, i.e., the assumption (Q3). Thus, we need a specific existence–regularity result for (3.13) (actually this result applies for more general equations, see Theorem 5 below).

The peculiarity of (3.13) is that the time derivative $u_t^\varepsilon$ of its solution lives only in an Orlicz space, and accordingly we are led to prove regularity in Orlicz spaces.

We briefly recall some basic definitions from Orlicz space theory (see [1] or [21]). First we have to define the so–called $N$–functions. We say that a real valued function $\Phi$ on $[0,\infty)$ is an $N$–function if $\Phi(s) = \int_0^s \varphi(\eta)\,d\eta$, where $\varphi : [0,\infty) \to \mathbb{R}$ is such that

(a) $\varphi(0) = 0$, $\varphi(s) > 0$ for $s > 0$, $\lim_{s\to\infty} \varphi(s) = +\infty$;
(b) $\varphi$ is nondecreasing;
(c) $\varphi$ is right continuous.

The simplest example of a $N$–function is a power. The non–trivial example which we shall use in the sequel are functions $\Phi_\varepsilon$ constructed in Lemma 1.

The $N$–function $\Psi(s) = \max_{\eta\geq 0}\{s\eta - \Phi(\eta)\}$ is called the complementary function of $\Phi$ and $(\Phi, \Psi)$ a complimentary pair.

Let $A$ be $\mathbb{R}^n$ or $(0,T)\times\mathbb{R}^N$, and $\Phi$ be a $N$–function. We denote by $L^\Phi(A)$ the Orlicz space on $A$ associated to $\Phi$ in the following way

$$L^\Phi(A)=\{u \text{ is a measurable real function on } A \text{ such that } \textstyle\int_A \Phi(|u|)dx < \infty\}, \tag{3.14}$$

with norm defined by

$$\|u\|_\Phi = \inf\left\{k > 0 : \textstyle\int_A \Phi(|u|/k) \leq 1\right\}. \tag{3.15}$$

We recall the so–called Hölder inequality for Orlicz space, i.e., that given a complimentary pair $(\Phi, \Psi)$ and functions $u \in L^\Phi(A)$ and $v \in L^\Psi(A)$, we can estimate the $L^1$ norm of $uv$ by

$$\left|\int_A uv\right| \leq 2\|u\|_\Phi\|v\|_\Psi. \tag{3.16}$$

Now we are able to formulate our existence and regularity result.

**Proposition 1.** *Assume that the data satisfy the condition* $u_0 \in H^2(\mathbb{R}^n)$, $u_1 \in H^1(\mathbb{R}^n)$, $|u_1|^{m-2}u_1 \in L^2(\mathbb{R}^n)$, *and* $g, g_t \in L^2((0,T)\times\mathbb{R}^n)$. *Then* (3.13) *has a unique weak solution* $u^\varepsilon$ *in the sense that*

$$(u_t^\varepsilon(s), \phi(s))\Big|_0^t = \int_{(0,t)\times\mathbb{R}^n} u_t^\varepsilon\phi_t - \nabla u^\varepsilon\nabla\phi - Q_\varepsilon(\cdot, u_t^\varepsilon)\phi + g\phi \tag{3.17}$$

*for all* $t \in [0,T]$ *and all test functions* $\phi \in C_c^\infty(\mathbb{R}^{n+1})$. *Moreover, we have*

$$u^\varepsilon \in C([0,T]; H^1(\mathbb{R}^n)), \quad u_t^\varepsilon \in L^{\Phi_\varepsilon}((0,T)\times\mathbb{R}^n)\cap L^\infty(0,T;H^1(\mathbb{R}^n)), \tag{3.18}$$

$$u_{tt}^\varepsilon \in L^\infty(0,T;L^2(\mathbb{R}^n)). \tag{3.19}$$

**Remark 8.** In addition there holds

$$Q_\varepsilon(\cdot, u_t^\varepsilon) \in L^{\Psi_\varepsilon}((0,T)\times\mathbb{R}^n), \tag{3.20}$$

where $\Psi_\varepsilon$ is the complementary $N$–function of $\Phi_\varepsilon$.

The proof of Proposition 1 is technical and is postponed to Appendix A, together with the proof of Remark 8. The above existence and regularity result can be extended to more general damping terms, which are essentially not power–like. More precisely, the following result holds.

**Theorem 5.** *Assume that $\tilde{Q} \in C^1(\mathbb{R})$ is a derivative of an $N-$function, i.e., suppose that the $N-$function $\Phi(v)$ and its complimentary function $\Psi(v)$ are $\Delta_2-$regular and moreover the following estimates hold:*

$$\tilde{Q}(v) \leq K_1 v^{m-1}, \qquad \tilde{Q}'(v)v \leq K_2 \tilde{Q}(v)$$

*for all $v \geq 0$, where $K_1$, $K_2$ are positive constants and $m \in (1, \infty)$. Then the statements of Proposition 1 and Remark 8 hold, with $Q_\varepsilon$, $\Phi_\varepsilon$, $\Psi_\varepsilon$ replaced respectively by $\tilde{Q}$, $\Phi$, $\Psi$ and the problem (3.13) by*

$$\begin{cases} u_{tt} - \Delta u + \tilde{Q}(u_t) = g(t,x), & \text{in } (0,T) \times \mathbb{R}^n, \\ u(0) = u_0, \qquad u_t(0) = u_1. \end{cases} \tag{3.21}$$

We omit the proof of Theorem 5 since it is similar to that of Proposition 1 (see Appendix A).

Now we can state the main theorem of this section.

**Theorem 6.** *Suppose that $1 < m < 2$, that (Q7) holds and the data have the following regularity*

$$u_0 \in H^2(\mathbb{R}^n), \quad u_1 \in H^1(\mathbb{R}^n), \quad |u_1|^{m-2}u_1 \in L^2(\mathbb{R}^n), \quad g \in L^2((0,T) \times \mathbb{R}^n).$$

*Moreover, let the data satisfy*

$$\text{supp}\, u_0, \text{supp}\, u_1 \subset B_R, \quad \text{and} \quad \text{supp}\, g(t) \subset B_{R+t} \text{ for almost all } t \in [0,T], \tag{3.22}$$

*for some $R > 0$. Then the solution $u$ of equation (2.1) has finite speed of propagation, i.e.,*

$$\text{supp}\ u(t) \subset B_{R+t} \quad \text{for all } t \in [0,T]. \tag{3.23}$$

**Proof of Theorem 6.** In the proof we essentially use the existence – regularity result for the approximate problem (3.13), namely Proposition 1. Since this proposition is valid when $g$ is more regular, i.e., $g_t \in L^2((0,T) \times \mathbb{R}^n)$, the proof is organized in three steps. First we prove that the solution of (3.13) has finite speed of propagation provided that $g_t \in L^2((0,T) \times \mathbb{R}^n)$. Second, passing to the limit as $\varepsilon \to 0$, we prove that the solution of equation (2.1) has finite speed of propagation provided that $g_t \in L^2((0,T) \times \mathbb{R}^n)$. Third, we remove the extra regularity requirement for $g$ using an additional approximation procedure.

**Step 1.** We apply Proposition 1 to equation (3.13). This gives us the existence and regularity for the solution. We apply (3.17) with test functions of the form $\phi = \alpha u_t^\varepsilon$, where $\alpha \in C_c(\mathbb{R}^{n+1}) \cap W^{1,\infty}(\mathbb{R}^{n+1})$ will be chosen in the sequel. The regularity of $\alpha u_t^\varepsilon$ is sufficient in order to use it as a test function. Indeed, for this purpose we use a density lemma which is formulated and proved in Appendix B. Then we get the needed regularity for $\alpha u_t^\varepsilon$ as a direct consequence of (3.18), of the Hölder inequality in Orlicz spaces, and the density lemma. So, for any fixed $t \in [0,T]$, we obtain

$$\int_{\mathbb{R}^n} \alpha(s,\cdot)(u_t^\varepsilon)^2(s)\Big|_0^t \tag{3.24}$$
$$= \int_{(0,t)\times\mathbb{R}^n} \alpha_t(u_t^\varepsilon)^2 + \alpha u_t^\varepsilon u_{tt}^\varepsilon - \nabla u^\varepsilon \nabla \alpha u_t^\varepsilon - \alpha \nabla u^\varepsilon \nabla u_t^\varepsilon - \alpha Q_\varepsilon(\cdot, u_t^\varepsilon)u_t^\varepsilon + \alpha g u_t^\varepsilon.$$

Before choosing the functions $\alpha$, we adapt a classical argument from [25] to evaluate the weighted energy $\frac{1}{2}\int_{\mathbb{R}^n} \alpha(t,\cdot)((u_t^\varepsilon)^2(t) + |\nabla u^\varepsilon(t)|^2$ over a piece of solid cone $K := \{(\tau,x) \in (0,T)\times\mathbb{R}^n : \tau + |x - x_0| \le r\}$, where $r > t$ and $|x_0| > R + r$. Given $\varepsilon > 0$ such that $R + r + \varepsilon < |x_0|$, we also set

$$K_\varepsilon := \{(\tau,x) \in (0,T)\times\mathbb{R}^n : \tau + |x - x_0| \le r + \varepsilon\},$$

so that $K \subset K_\varepsilon \subset D^c$, [2] where $D := \{(\tau,x) \in (0,T)\times\mathbb{R}^n : |x| \le R + \tau\}$ is the light cone of $B_R$. We define a cut–off function $\sigma \in C^\infty(\mathbb{R})$ such that $\sigma(s) = 1$ for $|s| \le r$, $\sigma(s) = 0$ for $|s| \ge r + \varepsilon$ and $\sigma'(s) \le 0$ for $s \ge 0$. Then we choose $\alpha(\tau,x) = \sigma(\tau + |x - x_0|)$. This choice makes $\alpha$ a function of $K_\varepsilon$. It is easy to verify that $\alpha$ is nonnegative, has the needed regularity and, moreover,

$$\alpha \equiv 1 \text{ on } K, \quad \alpha \equiv 0 \text{ on } K_\varepsilon^c, \quad \alpha_t + |\nabla\alpha| \le 0 \text{ on } (0,T)\times\mathbb{R}^n. \tag{3.25}$$

Next we evaluate the terms in (3.24). Applying $(3.25)_2$ together with

$$B_{r+\varepsilon}(x_0) \cap \operatorname{supp} u_1 = \emptyset$$

gives

$$\int_{\mathbb{R}^n} \alpha(\tau,\cdot)(u_t^\varepsilon)^2(\tau)\Big|_0^t = \int_{\mathbb{R}^n} \alpha(t,\cdot)(u_t^\varepsilon)^2(t). \tag{3.26}$$

Integrating by parts with respect to $t$, we obtain

$$\int_{(0,t)\times\mathbb{R}^n} \alpha u_t^\varepsilon u_{tt}^\varepsilon = \tfrac{1}{2}\int_{\mathbb{R}^n} \alpha(t,\cdot)(u_t^\varepsilon)^2(t)\,dx - \tfrac{1}{2}\int_{(0,t)\times\mathbb{R}^n} \alpha_t(u_t^\varepsilon)^2, \tag{3.27}$$

where (3.26) was used to evaluate the boundary term. One more integration by parts gives

$$\int_{(0,t)\times\mathbb{R}^n} \alpha\nabla u^\varepsilon \nabla u_t^\varepsilon = \tfrac{1}{2}\int_{\mathbb{R}^n} \alpha(t,\cdot)|\nabla u^\varepsilon(t)|^2 - \tfrac{1}{2}\int_{(0,t)\times\mathbb{R}^n} \alpha_t|\nabla u^\varepsilon|^2, \tag{3.28}$$

[2] Here and in the sequel we denote by $S^c$ the complement of $S$ in $(0,T)\times\mathbb{R}^n$.

where the fact that $B_{r+\varepsilon}(x_0) \cap u_0 = \emptyset$ is used. Moreover, since supp $g \subset D$, then supp $g \cap$ supp $\alpha = \emptyset$, which implies

$$\int_{(0,t)\times\mathbb{R}^n} \alpha g u_t^\varepsilon = 0. \tag{3.29}$$

Inserting the identities (3.26)–(3.29) in (3.24), we get the estimate

$$\begin{aligned}
&\frac{1}{2}\int_{\mathbb{R}^n} \alpha(t,\cdot)((u_t^\varepsilon)^2(t) + |\nabla u^\varepsilon(t)|^2) \qquad (3.30)\\
&= \frac{1}{2}\int_{(0,t)\times\mathbb{R}^n} \alpha_t((u_t^\varepsilon)^2 + |\nabla u^\varepsilon|^2) - \int_{(0,t)\times\mathbb{R}^n} \nabla u^\varepsilon \nabla \alpha u_t^\varepsilon - \int_{(0,t)\times\mathbb{R}^n} \alpha Q_\varepsilon(\cdot,\cdot,u_t^\varepsilon) u_t^\varepsilon\\
&\le \frac{1}{2}\int_{(0,t)\times\mathbb{R}^n} (\alpha_t + |\nabla\alpha|)((u_t^\varepsilon)^2 + |\nabla u^\varepsilon|^2) \le 0.
\end{aligned}$$

Here, to estimate the term $\int_{(0,t)\times\mathbb{R}^n} \nabla u^\varepsilon \nabla \alpha u_t^\varepsilon$ we have applied the Cauchy–Schwarz inequality; to estimate $\int_{(0,t)\times\mathbb{R}^n} \alpha Q_\varepsilon(\cdot,\cdot,u_t^\varepsilon)u_t^\varepsilon$ we use the specific form of the damping term (3.10) and the convexity of $\Phi_\varepsilon$; the last estimate of (3.30) follows from $(3.25)_3$.

Since $\alpha \ge 0$, by (3.30) we have $\alpha(t)((u_t^\varepsilon)^2(t) + |\nabla u^\varepsilon(t)|^2) = 0$ a.e. in $\mathbb{R}^n$ and then, using (3.25) we have $(u_t^\varepsilon)^2(t) + |\nabla u^\varepsilon(t)|^2 = 0$ a.e. in $B_{r-t}(x_0)$. Namely, for all $r > t$ and all $|x_0| > R + r$, we obtain $u^\varepsilon(t) =$ Const. for $|x| > R + t$. Then, since $u^\varepsilon(t) \in L^2(\mathbb{R}^n)$, we have

$$\text{supp } u^\varepsilon(t) \subset B_{R+t} \tag{3.31}$$

for all $t \in [0,T]$, which completes the first step in the proof.

**Step 2.** By (3.2) we have $t^m \le \Phi_\varepsilon(t) + \varepsilon^m$ for $t \ge 0$. Moreover, $\int_{(0,T)\times\mathbb{R}^n} \Phi_\varepsilon(|u_t^\varepsilon|) < \infty$ because of the regularity $(3.18)_2$. Using these estimates together with (3.31), we see that $u_t^\varepsilon \in L^m((0,T)\times\mathbb{R}^n)$. Let us emphasize that this last regularity is a consequence of the finite speed of propagation for $u^\varepsilon$, a fact that was impossible to be obtained directly at the beginning of the proof. Moreover, using this regularity together with by (3.6) and (3.10), we get $Q_\varepsilon(\cdot,u_t^\varepsilon) \in L^{m'}((0,T)\times\mathbb{R}^n)$. Let $u$ be the solution of (2.1) given by Theorem 3. Define $w^\varepsilon = u^\varepsilon - u$, which is a solution of

$$w_{tt}^\varepsilon - \Delta w^\varepsilon + Q_\varepsilon(x,u_t^\varepsilon) - Q(x,u_t) = 0, \tag{3.32}$$

with zero initial data. Then we have enough regularity to write the energy identity for equation (3.32),

$$\tfrac{1}{2}\|w_t^\varepsilon(t)\|_2^2 + \tfrac{1}{2}\|\nabla w^\varepsilon(t)\|_2^2 + \int_{(0,t)\times\mathbb{R}^n} (Q_\varepsilon(\cdot,u_t^\varepsilon) - Q(\cdot,u_t))w_t^\varepsilon = 0. \tag{3.33}$$

Using the monotonicity of the damping term $Q$ in $v$, we have

$$\int_{(0,t)\times\mathbb{R}^n}(Q_\varepsilon(\cdot,u_t^\varepsilon)-Q(\cdot,u_t))w_t^\varepsilon \geq \int_{(0,t)\times\mathbb{R}^n}(Q_\varepsilon(\cdot,u_t^\varepsilon)-Q(\cdot,u_t^\varepsilon))w_t^\varepsilon. \tag{3.34}$$

The estimate (3.34) allows us to rewrite (3.33) in the form

$$\tfrac{1}{2}\|w_t^\varepsilon(t)\|_2^2+\tfrac{1}{2}\|\nabla w^\varepsilon(t)\|_2^2 \leq \int_{(0,t)\times\mathbb{R}^n}|Q_\varepsilon(\cdot,u_t^\varepsilon)-Q(\cdot,u_t^\varepsilon)||w_t^\varepsilon|. \tag{3.35}$$

Now, using Hölder's inequality together with (3.11) and the regularity of $Q_\varepsilon(\cdot,u_t^\varepsilon)$ and $Q(\cdot,u_t^\varepsilon)$, namely that they are in $L^m((0,T)\times\mathbb{R}^n)$, we have

$$I_1 := \int_{(0,T)\times\mathbb{R}^n}|Q_\varepsilon(\cdot,u_t^\varepsilon)-Q(\cdot,u_t^\varepsilon)||w_t^\varepsilon| \leq \Big[\int_{A_\varepsilon}|Q_\varepsilon(\cdot,u_t^\varepsilon)-Q(\cdot,u_t^\varepsilon)|^{m'}\Big]^{\frac{1}{m'}}|||w_t^\varepsilon|||_m, \tag{3.36}$$

where $A_\varepsilon=\{(t,x)\in(0,T)\times B_{R+T} : |u_t^\varepsilon|\leq\varepsilon\}$. Here the finite speed of propagation for the approximate problem (3.31) was essentially used. By (3.36) and (3.12) together with the assumption (Q7), we have

$$\begin{aligned} I_1 &\leq 2\varepsilon^{m-1}\Big(\int_{A_\varepsilon}|\sigma|^{m'}\Big)^{1/m'}(|||u_t^\varepsilon|||_m+|||u_t|||_m) \\ &\leq 2\varepsilon^{m-1}\Big(\int_{(0,T)\times B_{R+T}}|\sigma|^{m'}\Big)^{1/m'}(|||u_t^\varepsilon|||_m+|||u_t|||_m) \\ &\leq \text{Const.}\,\varepsilon^{m-1}(|||u_t^\varepsilon|||_m+|||u_t|||_m), \end{aligned} \tag{3.37}$$

where in the last inequality we have used the $L^\infty$ regularity of $\sigma$ and the definition of $A_\varepsilon$. From Theorem 4 we know that $|||u_t|||_m<\infty$. It remains to show $\sup_{\varepsilon>0}|||u_t^\varepsilon|||_m<\infty$, which is easily done by using the Hölder inequality (recall that $m<2$) and the finite speed of propagation for (3.13). Thus we have

$$|||u_t^\varepsilon|||_m \leq \text{Const.}|||u_t^\varepsilon|||_2. \tag{3.38}$$

From the energy inequality for $u^\varepsilon$

$$\tfrac{1}{2}\|u_t^\varepsilon\|_2^2+\tfrac{1}{2}\|\nabla u^\varepsilon\|_2^2 \leq \int_0^t \|g\|_2\|u_t^\varepsilon\|_2.$$

By Gronwall's lemma together with the regularity $g\in L^2((0,T)\times\mathbb{R}^n)$, we get

$$\sup_{t\in[0,T]}\|u_t^\varepsilon(t,\cdot)\|_2 \leq \text{Const.},$$

from which it follows that

$$|||u_t^\varepsilon|||_2 \leq \text{Const.} \tag{3.39}$$

where the constants in above estimates are independent of $\varepsilon$. Inserting (3.39) and (3.38) in (3.37) gives

$$I_1 \leq \text{Const.}\varepsilon^{m-1}.$$

Passing to the limit as $\varepsilon \to 0$ (recall that $m > 1$) we find that $I_1 \to 0$. Then from (3.35) we obtain

$$\tfrac{1}{2}\|w_t^\varepsilon(t)\|_2^2 + \tfrac{1}{2}\|\nabla w^\varepsilon(t)\|_2^2 \to 0, \qquad \text{as } \varepsilon \to 0. \tag{3.40}$$

In view of Cauchy's theorem, and taking into account that $w^\varepsilon(0) = 0$, we derive

$$\|w^\varepsilon(t,\cdot)\|_2 \leq \int_0^T \|w_t^\varepsilon\|_2.$$

Together with (3.40) this implies that $w^\varepsilon \to 0$ in $L^2(\mathbb{R}^n)$. Then (up to a subsequence) $u^\varepsilon(t) \to u(t)$ a.e. on $\mathbb{R}^n$. In view of the finite speed of propagation of $u^\varepsilon$ this completes the proof of Step 2.

**Step 3.** When $g_t \notin L^2((0,T) \times \mathbb{R}^n)$, we approximate $g$ with sufficiently regular functions $g^\nu$ such that $g^\nu, g_t^\nu \in L^2((0,T) \times \mathbb{R}^n)$, supp $g^\nu(t) \subset B_{R+t+1/\nu}$ for all $t \in [0,T]$, and for which $g^\nu \to g$, as $\nu \to \infty$. By the first part of the proof the solution of the approximate problem

$$\begin{cases} u_{tt}^\nu - \Delta u^\nu + Q(x, u_t^\nu) = g^\nu(t,x), & \text{in } (0,T) \times \mathbb{R}^n, \\ u(0) = u_0, \qquad u_t(0) = u_1, \end{cases} \tag{3.41}$$

satisfies

$$\text{supp } u^\nu(t) \subset B_{R+1/\nu+t}, \qquad \text{for all } t \in [0,T].$$

Now we measure the distance between the solution $u$ of (2.1) and $u^\nu$ of equation (3.41). The difference $w^\nu := u - u^\nu$ satisfies the equation

$$\begin{cases} w_{tt}^\nu - \Delta w^\nu + Q(x, u_t^\nu) - Q(x, u_t) = g^\nu(t,x) - g(t,x), & \text{in } (0,T) \times \mathbb{R}^n, \\ u(0) = 0, \qquad u_t(0) = 0. \end{cases} \tag{3.42}$$

Using the energy identity, we have

$$\tfrac{1}{2}\|w^\nu\|_2^2 + \tfrac{1}{2}\|\nabla w^\nu\|_2^2 + \int_{(0,t)\times\mathbb{R}^n} (Q(x,u_t^\nu) - Q(x,u_t))w_t^\nu = \int_{(0,t)\times\mathbb{R}^n} (g^\nu - g)w_t^\nu.$$

By the monotonicity of $Q$ and the Cauchy–Schwarz inequality,

$$\tfrac{1}{2}\|w^\nu\|_2^2 + \tfrac{1}{2}\|\nabla w^\nu\|_2^2 \leq \int_0^t \|g^\nu - g\|_2 \|w_t^\nu\|_2.$$

Then from the convergence of $g^\nu$ to $g$ in $L^2((0,T)\times\mathbb{R}^n)$ and the Gronwall lemma,

$$\sup_{t\in[0,T]}\|w_t^\nu(t)\|_2 \to 0, \text{as } \nu\to\infty.$$

Finally, since

$$\|w^\nu(t)\|_2 \le \int_0^T \|w_t^\nu\|_2,$$

for all $t\in[0,T]$, we conclude that (up to a subsequence) $u^\nu(t)\to u(t)$ a.e. in $\mathbb{R}^n$. This shows that supp $u(t)\subset B_{R+t}$ for all $t\in[0,T]$, as required. □

## 4. Existence for nonlinear sources

In this section we prove our main local and global existence results. We start with our general local existence result for the equation (1.1), with damping term $Q$ satisfying the assumptions (Q1)–(Q6) when the growth rate $m\ge 2$, and the assumption (Q7) when $1<m<2$. Moreover, the assumption (Q1) has to be strengthened to

(Q1)′ there is $c_0>0$ such that

$$(Q(t,x,v)-Q(t,x,w))(v-w)\ge c_0|v-w|^m$$

for almost all $(t,x)\in(0,T)\times\mathbb{R}^n$ and all $v,w\in\mathbb{R}$.

Clearly, because of the elementary inequality

$$(|v|^{m-2}v-|w|^{m-2}w)(v-w)\ge \text{Const.}|v-w|^m$$

for $m\ge 2$, $v,w\in\mathbb{R}$, the damping term in the model power–like equation, i.e., $Q$ given by (1.2), satisfies the assumption (Q1)′.

We consider a nonlinear source $f$ in the form

$$f(x,u)=f_0(x,u)+\mu_2(x)|u|^{q-2}u, \tag{4.1}$$

where $\mu_2\in L^\infty_{\text{loc}}(\mathbb{R}^n)$, $q>1$, and $f_0$ satisfies the assumption:

(F1) $f_0(x,0)\equiv 0$ and

$$|f_0(x,u_1)-f_0(x,u_2)|\le \mu_1(x)|u_1-u_2|(1+|u_1|^{p-2}+|u_2|^{p-2})$$

for almost all $x\in\mathbb{R}^n$ and all $u_1,u_2\in\mathbb{R}$, where $\mu_1\in L^\infty_{\text{loc}}(\mathbb{R}^n)$ and $p>2$.

**Remark 9.** It is easy to check that the source term $f$ in the power–like equation (see (1.3)) satisfies (F1), due to the elementary inequality

$$||u_1|^{p-2}u_1-|u_2|^{p-2}u_2|\le \text{Const.}|u_1-u_2|(|u_1|^{p-2}+|u_2|^{p-2}),\qquad u_1,u_2\in\mathbb{R},$$

for $p>2$.

**Remark 10.** Evidently the essential range for the exponent $q$ is $1 < q < 2$; otherwise for $q \geq 2$ we can combine the second and the first term in (4.1).

**Theorem 7.** *Let the damping $Q$ satisfy the assumptions* (Q1)$'$ *and* (Q2)*–*(Q6) *when $m \geq 2$, and* (Q7) *when $1 < m < 2$. Let the source $f$ obey* (4.1) *and the assumption* (F1)*. Assume also the growth rates $p, m$ of the nonlinearities satisfy $(p,m) \in I \cup II \cup III$, namely that* (1.5) *holds. Finally let the data be compactly supported and be in the energy space*

$$u_0 \in H^1(\mathbb{R}^n), \qquad u_1 \in L^2(\mathbb{R}^n). \tag{4.2}$$

*Then, there is a weak solution of* (1.1) *with the properties*

$$u \in C([0,T]; H^1(\mathbb{R}^n)), \quad u_t \in C([0,T]; L^2(\mathbb{R}^n)) \tag{4.3}$$

$$u_t \in L^m((0,T) \times \mathbb{R}^n), \tag{4.4}$$

*for $T > 0$ small enough.*

*Moreover, for exponents $(p,m) \in \overline{I_r} \cup \overline{II_r}$ , where*

$$\overline{I_r} = \{(p,m) \in \mathbb{R}^2 : 2 < p \leq 1 + r/2, \quad m \geq 2\},$$
$$\overline{II_r} = \{(p,m) \in \mathbb{R}^2 : 2 < p \leq 1 + r/2, \quad 1 < m < 2\}$$

*are the right closure of $I$ and $II$ respectively, and $\mu_2 \equiv 0$, there is a unique weak solution of* (1.1) *with the same regularity.*

We emphasize again that the solution which we obtain in Theorem 7 has finite speed of propagation.

Before proving Theorem 7 we need the following:

**Proposition 2.** *Let $u$ be a function satisfying*

$$supp\ u(t) \subset B_{R+t} \quad for\ all\ t \in [0,T], \tag{4.5}$$

*with smoothness* (1.6)*. Let the data $u_0$ and $u_1$ have support in $B_R$ and smoothness* (4.2)*. Then the problem*

$$\begin{cases} v_{tt} - \Delta v + Q(t,x,v_t) = f(x,u) & in\ (0,T) \times \mathbb{R}^n, \\ v(0) = u_0, \qquad v_t(0) = u_1, \end{cases} \tag{4.6}$$

*has a solution $v$ with regularity* (1.6) *and* (1.7) *which satisfies*

$$supp\ v(t) \subset B_{R+t} \quad for\ all\ t \in [0,T]. \tag{4.7}$$

**Proof.** This follows some ideas from [6], but uses in a more delicate way the gain of regularity due to the presence of the damping term. Moreover, since we are in $\mathbb{R}^n$ we have to pay attention to the change of the support of functions involved in the approximation procedure. We start by approximating the data with more regular data $u_0^\nu$ and $u_1^\nu$ in $C_c^\infty(B_R)$, such that

$$u_0^\nu \to u_0 \quad \text{in } H^1(\mathbb{R}^n), \qquad \text{and} \quad u_1^\nu \to u_1 \quad \text{in } L^2(\mathbb{R}^n). \tag{4.8}$$

Next, we approximate the function $u$ in the topology of the energy space with more regular functions $u^\nu \in C_c^\infty([0,T]\times\mathbb{R}^n)$ such that

$$\operatorname{supp} u^\nu(t) \subset B_{R+t+1/\nu} \tag{4.9}$$

for all $t \in [0,T]$. This can easily be done by using standard convolution arguments (see [4]), which of course only inessentiality increases the support.

Then, the existence of a solution $v^\nu$ of the problem

$$\begin{cases} v_{tt}^\nu - \Delta v^\nu + Q(t,x,v_t^\nu) = f(x,u^\nu) & \text{in } (0,T)\times\mathbb{R}^n, \\ v^\nu(0) = u_0^\nu, \qquad v_t^\nu(0) = u_1^\nu, \end{cases} \tag{4.10}$$

is assured by Theorem 3. In fact, to apply this result it is only necessary to check that $f(\cdot,u^\nu) \in L^2((0,T)\times\mathbb{R}^n)$. Indeed, using the assumption (Q1) for $u_2 = 0$ and $u_1 = u^\nu$ yields

$$|f(x,u^\nu)| \le \mu_1(x)|u^\nu|(1+|u^\nu|^{p-2}) + |\mu_2(x)||u^\nu|^{q-1}.$$

Now, taking into account the regularity of $u^\nu$, $\mu_1(x)$ and $\mu_2(x)$, there holds $f(\cdot,u^\nu) \in L^\infty((0,T)\times\mathbb{R}^n)$. This together with (4.9) gives us the needed regularity $f(\cdot,u^\nu) \in L^2((0,T)\times\mathbb{R}^n)$, since

$$|||f(\cdot,u^\nu)|||_2 \le |||f(\cdot,u^\nu)|||_\infty \int_{(0,T)\times B_{R+T}} 1.$$

Then, due to Theorem 6 for $1 < m < 2$ and Strauss' theorem [3] for $2 \le m < \infty$, one has

$$\operatorname{supp} v^\nu(t) \subset B_{R+t+1/\nu} \qquad \text{for all } t \in [0,T]. \tag{4.11}$$

The key point in the proof is to show that $v^\nu$ is a Cauchy sequence in the topology of the spaces (1.6)–(1.7). This is done in a different way for the three different regions. First of all, the energy identity for $w = v^\nu - v^\mu$ has

[3]Actually for $m \ge 2$, Strauss' argument [25] can be adapted to weak solutions by using the ideas of Steps 1 and 3 in the proof of Theorem 6. In this case Step 2 is no longer needed and can be omitted.

the form

$$\begin{aligned}&\tfrac{1}{2}\|w_t(t)\|_2^2+\tfrac{1}{2}\|\nabla w(t)\|_2^2+\int_{(0,t)\times\mathbb{R}^n}(Q(t,\cdot,u_t^\nu)-Q(t,\cdot,u_t^\mu))w_t\\&=\tfrac{1}{2}\|w_t(0)\|_2^2+\tfrac{1}{2}\|\nabla w(0)\|_2^2+\int_{(0,t)\times\mathbb{R}^n}(f(\cdot,u^\nu)-f(\cdot,u^\mu))w_t.\end{aligned}\tag{4.12}$$

Let us start with the region I. Using (4.1), (F1) and the Hölder continuity of $|u|^{q-2}u$ (here $1<q<2$) and Remark 10, we obtain the estimate

$$\begin{aligned}&|(f(\cdot,u)-f(\cdot,\bar{u}),v-\bar{v})|\\&\le\int_K\left[\|\mu_1\|_{L^\infty(K)}|u-\bar{u}|(1+|u|^{p-2}+|\bar{u}|^{p-2})+\|\mu_2\|_{L^\infty(K)}|u-\bar{u}|^{q-1}\right]|v-\bar{v}|\\&\le\text{Const.}\left[\|u-\bar{u}\|_r(1+\|u\|_{n(p-2)}^{p-2}+\|\bar{u}\|_{n(p-2)}^{p-2})+\|u-\bar{u}\|_2^{q-1}\right]\|v-\bar{v}\|_2\\&\le\text{Const.}\left[\|\nabla u-\nabla\bar{u}\|_2(1+\|\nabla u\|_2^{p-2}+\|\nabla\bar{u}\|_2^{p-2})+\|\nabla u-\nabla\bar{u}\|_2^{q-1}\right]\|v-\bar{v}\|_2,\end{aligned}\tag{4.13}$$

for any $u,\bar{u}\in H^1(\mathbb{R}^n)$ and $v,\bar{v}\in L^2(\mathbb{R}^n)$ with support contained in a fixed compact subset $K$ of $\mathbb{R}^n$. The above estimate requires the restriction $n(p-2)\le r$, which is fulfilled in regions I and II since there $p<1+r/2$. Then using (4.12), (Q1)$'$ together with the property supp $u^\nu(t)\subset K=B_{R+T+1}$ (recall that $1<q<2$) and the fact that $u^\nu$ is a Cauchy sequence in the energy norm, one arrives at the inequality

$$\begin{aligned}&\tfrac{1}{2}\|w_t(t)\|_2^2+\tfrac{1}{2}\|\nabla w(t)\|_2^2+c_0\int_{(0,t)\times\mathbb{R}^n}|v_t^\nu-v_t^\mu|^m\\&\le\tfrac{1}{2}\|w_t(0)\|_2^2+\tfrac{1}{2}\|\nabla w(0)\|_2^2+C(T)\int_0^t\|\nabla u^\nu-\nabla u^\mu\|_2^{q-1}\|w_t\|_2.\end{aligned}\tag{4.14}$$

Here and in the sequel we denote by $C(T)$ some $T$ dependent constants, which decrease when $T\to 0$. Then, from the Gronwall lemma and the fact that $u^\nu$, $u_0^\nu$ and $u_1^\nu$ are Cauchy sequences it follows that $v^\nu$ is a Cauchy sequence in the energy space, uniformly in time. With the help of the estimate (4.14) once again, together with the knowledge that $v^\nu$ is a Cauchy sequence in the energy space, we find that $v_t^\nu$ is a Cauchy sequence in the space – time $L^m$ norm.

In the same way, for region II, $v^\nu$ is again a Cauchy sequence in the energy norm, uniformly in time. Due to the Hölder inequality in space – time together with (4.11) and the fact that $1<m<2$, we have

$$|||w_t|||_m\le C(T)|||w_t|||_2,\le C(T)\sup_{t\in[0,T]}\|w_t(t,\cdot)\|_2.\tag{4.15}$$

Moreover, from the arguments of [4], it follows that $\|\gamma_\varepsilon *_t \phi(t) - \phi(t)\|_2 \to 0$ uniformly in $[0,T]$, as $\varepsilon \to 0$. Hence, for all $t \in [0,T]$,

$$\begin{aligned}\|\rho_\varepsilon * \phi(t) - \phi(t)\|_2 &\le \|\sigma_\varepsilon *_x (\gamma_\varepsilon *_t \phi)(t) - \sigma_\varepsilon *_x \phi(t)\|_2 + \|\sigma_\varepsilon *_x \phi(t) - \phi(t)\|_2 \\ &\le \|\gamma_\varepsilon *_t \phi(t) - \phi(t)\|_2 + \|\sigma_\varepsilon *_x \phi(t) - \phi(t)\|_2 \to 0 \qquad \text{(B.3)}\end{aligned}$$

as $\varepsilon \to 0$. Now, for $s, t \in [0,T]$

$$\|\rho_\varepsilon * \phi(s) - \rho_\varepsilon * \phi(t)\|_2 \le \|\phi(s) - \phi(t)\|_2.$$

In turn, by (B.3) and the Arzelá–Ascoli Theorem one obtains (up to a subsequence) $\|\rho_\varepsilon * \phi(t) - \phi(t)\|_2 \to 0$ uniformly for $t \in [0,T]$. Moreover, $\|\eta_\varepsilon\phi(t) - \phi(t)\|_2 \to 0$ pointwise in $[0,T]$ as $\varepsilon \to 0$. Next, for $s, t \in [0,T]$,

$$\|\eta_\varepsilon\phi(s) - \eta_\varepsilon\phi(t)\|_2 \le \|\phi(s) - \phi(t)\|_2.$$

Consequently, applying the Arzelá–Ascoli Theorem again, it follows that (up to a subsequence), $\|\eta_\varepsilon\phi(t) - \phi(t)\|_2 \to 0$ uniformly in $[0,T]$ as $\varepsilon \to 0$. Since

$$\begin{aligned}\|\phi_\varepsilon(t) - \phi(t)\|_2 &\le \|\eta_\varepsilon(\rho_\varepsilon * \phi)(t) - \eta_\varepsilon\phi(t)\|_2 + \|\eta_\varepsilon\phi(t) - \phi(t)\|_2 \\ &\le \|\rho_\varepsilon * \phi(t) - \phi(t)\|_2 + \|\eta_\varepsilon\phi(t) - \phi(t)\|_2\end{aligned}$$

for all $t \in [0,T]$, we derive $\phi_\varepsilon \to \phi$ in $C([0,T]; L^2(\mathbb{R}^n))$. This completes the proof. □

## References

[1] R.A. Adams, "Sobolev Spaces," Academic Press, New York, 1975.

[2] A. Ambrosetti and G. Prodi, "A Primer of Nonlinear Analysis," Cambridge Univ. Press, 1993.

[3] J. Ball, *Remarks on blow-up and nonexistence theorems for nonlinear evolution equations*, Quart. J. Math. Oxford Ser., **28** (1977), 473–486.

[4] H. Brezis, "Analyse fonctionelle, théorie et applications," Masson, Paris, 1983.

[5] E. A. Coddington and N. Levinson, "Theory of Ordinary Differential Equations," McGraw–Hill, New York–Toronto–London, 1955.

[6] V. Georgiev and G. Todorova, *Existence of a solution of the wave equation with nonlinear damping and source terms*, J. Differential Equations, **109** (1994), 295–308.

[7] R.T. Glassey, *Blow-up theorems for nonlinear wave equations*, Math. Z., **132** (1973), 183–203.

[8] A. Haraux and E. Zuazua, *Decay estimates for some semilinear damped hyperbolic problems*, Arch. Rational Mech. Anal., **100** (1988), 191–206.

[9] R. Ikehata, *Some remarks on the wave equations with nonlinear damping and source terms*, Nonlinear Anal., **27** (1996), 1165–1175.

[10] M. Kopackova, *Remarks on bounded solutions of a semilinear dissipative hyperbolic equation*, Comment. Math. Univ. Carolin., **30** (1989), 713–719.

[11] H.A. Levine, *Instability and nonexistence of global solutions to nonlinear wave equations of the form* $Pu_{tt} = -Au + \mathcal{F}(u)$, Trans. Amer. Math. Soc., **192** (1974), 1–21.

[12] ———, *Some additional remarks on the nonexistence of global solutions to nonlinear wave equations*, SIAM J. Math. Anal., **5** (1974), 138–146.

[13] H.A. Levine, S.R. Park, and J. Serrin, *Global existence and global non existence of solutions of the Cauchy problem for a nonlinearly damped wave equation*, J. Math. Anal. Appl., **228** (1998), 181–205.

[14] H.A. Levine and J. Serrin, *Global nonexistence theorems for quasilinear evolutions equations with dissipation*, Arch. Rational Mech. Anal., **137** (1997), 341–361.

[15] H.A. Levine and G. Todorova, *Blow up of solutions of the Cauchy problem for a wave equation with nonlinear damping and source terms and positive initial energy*, Proc. Amer. Math. Soc., **129** (2001), 693–805.

[16] J.L. Lions, "Quelques méthodes de résolutions des problèmes aux limites non linéaires," Dunod, Paris, 1969.

[17] J.L. Lions and W. A. Strauss, *On some nonlinear evolution equations*, Bull. Soc. Math. France, **93** (1965), 43–96.

[18] M. Ohta, *Remark on blow–uo of solutions for nonlinear evolution equations of second order*, Adv. in Math. Sc. and Appl., **8** (1998), 901–910.

[19] K. Ono, *On global solutions and blow-up solutions of nonlinear Kirchoff string with nonlinear dissipations*, J. Math. Anal. Appl., **216** (1997), 321–342.

[20] L. E. Payne and D. H. Sattinger, *Saddle points and instability of nonlinear hyperbolic equations*, Israel J. Math., **22** (1975), 273–303.

[21] M. M. Rao and Z.D. Ren, *Theory of Orlicz spaces*, Pure and applied mathematics, vol. 146, Marcel Dekker, inc., New York–Basel–Hong Kong, 1991.

[22] H. L. Royden, "Real Analysis," MacMillan, New York, 1963.

[23] I. Segal, *Nonlinear semigroups*, Ann. of Math., **78** (1963), 339–364.

[24] W.A. Strauss, *On continuity of functions with values in various Banach spaces*, Pacific J. Math., **19** (1966), 543–551.

[25] ———, *Nonlinear wave equations*, Regional conference series in mathematics, vol. 73, Amer.Math.Soc., Providence, R.I., 1989.

[26] G. Todorova, *Stable and unstable sets for the Cauchy problem for a nonlinear wave equation with nonlinear damping and source terms*, J. Math. Anal. Appl., **239** (1999), 213–226.

[27] ———, *Cauchy problems for a nonlinear wave equations with nonlinear damping and source terms*, Nonlinear Anal., **41** (2000), 891–905.

[28] G. Todorova and E. Vitillaro, *Blow-up phenomena for the Cauchy problem for a nonlinear wave equation*, in preparation.

[29] ———, *Energy decay for the Cauchy problem for a nonlinear wave equation*, in preparation.

[30] M. Tsutsumi, *On solutions of semilinear differential equations in a Hilbert space*, Math. Japon, **17** (1972), 173–193.

[31] E. Vitillaro, *Global existence for the wave equation with nonlinear boundary damping and source term*, to appear on J. Differential Equations.

[32] ———, *A potential well theory for the wave equation with nonlinear source and boundary damping terms*, to appear on Glasgow Mathematical Journal.

[33] ———, *Global nonexistence theorems for a class of evolution equations with dissipation and application*, Arch. Rational Mech. Anal., **149** (1999), 155–182.

Hans Weinberger, Laura Weinberger, Johannes C. C. Nitsche and James Serrin,
Rocky Mountain National Park, Boulder, CO, 1957.

Donald Aronson and Hans Weinberger, Wisconsin River, WI, 2002.

# III. Real Analysis

# III.1. Differentiation Theory

Giovanni Leoni

**Mathematics Subject Classification (2010).** 26B10, 26B35.

**Keywords.** Real functions, functional analysis, Sobolev spaces, spaces of measurable functions.

Serrin's papers [i], [ii], and [iii] below lie at the heart of modern theory of Sobolev spaces $W^{k,p}(\Omega)$ and spaces of functions of bounded variation $BV(\Omega)$.

It is well known (see, e.g., [AFP00], [EG92], [L09]) that if a function of one variable $u : I \to \mathbb{R}$, where $I \subset \mathbb{R}$ is an interval, has finite pointwise variation, then it can be decomposed into the sum of three functions of finite pointwise variation, $u = u_{\text{ac}} + u_{\text{c}} + u_{\text{j}}$, where $u_{\text{ac}}$ is absolutely continuous, $u_{\text{c}}$ is continuous and has derivative zero $\mathcal{L}^1$ a.e. in $I$ (the Cantor function is the standard example), and $u_{\text{j}}$ is a saltus or jump function. For functions of several variables a similar decomposition holds for the distributional derivative. Indeed, if $u$ belongs to $BV(\Omega)$, then its distributional derivative $Du$, which is a vector-valued Radon measure, can be decomposed into three mutually singular vector-valued Radon measures (see, e.g., [AFP00], [EG92]),

$$Du = D^{\text{ac}}u + D^{\text{c}}u + D^{\text{j}}u,$$

where the measure $D^{\text{ac}}u$ is absolutely continuous with respect to the Lebesgue measure $\mathcal{L}^N$ restricted to $\Omega$, and the Cantor part $D^{\text{c}}u$ and the jump part $D^{\text{j}}u$ of $Du$ are singular with respect to $\mathcal{L}^N$. Since $D^{\text{ac}}u$ is absolutely continuous with respect to $\mathcal{L}^N$, by the Radon–Nikodym theorem there exists an integrable function $f : \Omega \to \mathbb{R}^N$ such that

$$(D^{\text{ac}}u)(B) = \int_B f(x)\,dx \tag{1}$$

for every Borel set $B \subset \Omega$. The function $f$ is the Radon–Nikodym derivative of $Du$ with respect to $\mathcal{L}^N$. In [i, Theorem 1], Serrin identifies the function $f$. Indeed, he shows that there exists a precise representative $v$ of $u$ which admits (classical) partial derivatives $\mathcal{L}^N$ a.e. in $\Omega$ and such that

$$\nabla v(x) = f(x) \quad \text{for } \mathcal{L}^N \text{ a.e. } x \in \Omega. \tag{2}$$

Here, $\nabla v = \left(\frac{\partial v}{\partial x_1}, \ldots, \frac{\partial v}{\partial x_N}\right)$.

The counterpart of Serrin's theorem is a result of Alberti [A91], who proved that for every integrable function $f : \Omega \to \mathbb{R}^N$ there exists a function $u \in BV(\Omega)$ such that the Radon–Nikodym derivative of $Du$ with respect to $\mathcal{L}^N$ is exactly $f$, that is, (1) holds, up to a set of Lebesgue measure zero. By combining both theorems, we obtain that *every* integrable function is, up to a set of Lebesgue measure zero, the (classical) gradient of a function of bounded variation. Note that $f$ does not have to be curl-free. This is in sharp

P. Pucci et al. (eds.), *James Serrin. Selected Papers: Volume 2*, Contemporary Mathematicians,
DOI 10.1007/978-3-0348-0847-7_6, © Springer Basel 2014

contrast with the $C^1$ case, where, in view of Schwarz's theorem, if $f : \Omega \to \mathbb{R}^N$ is of class $C^1$, then a necessary condition for $f$ to be the gradient of a function of class $C^2$ is that $f$ be curl-free.

In the recent paper [MP08] Moonens and Pfeffer have shown that for every measurable function $f : \Omega \to \mathbb{R}^N$ there exists a continuous function $v : \Omega \to \mathbb{R}$, which is differentiable (in the classical sense) $\mathcal{L}^N$ a.e. in $\Omega$ and such that (2) holds.

In the second part of the paper [i], Serrin shows that a function $u$ belongs to the space $BV(\Omega)$ if and only if it admits a representative which has finite pointwise variation on $\mathcal{L}^{N-1}$ almost all lines parallel to the axes (with an appropriate bound on the pointwise variation). This theorem is the extension of a similar result for Sobolev functions, where one has absolute continuity on $\mathcal{L}^{N-1}$ almost all lines parallel to the axes, obtained by Deny and Lions (see Theorem 3.1 in [DL55]). These characterizations are extremely useful in the calculus of variations, where one studies first one-dimensional functionals and then uses a slicing argument to obtain the higher-dimensional case (see, e.g., [AFP00] and the references contained therein).

Finally, in the last part of the paper [i], Serrin proves that when $p > N$ a function $u \in W^{1,p}(\Omega)$ admits a representative which is differentiable (in the classical sense) $\mathcal{L}^N$ a.e. in $\Omega$, and constructs a continuous function $u \in W^{1,N}(\Omega)$, which is nowhere differentiable.

In [ii] Serrin gives a necessary and sufficient condition for strong convergence in $L^p$ in a product space. More precisely, let $M, N \in \mathbb{N}$ and consider the cartesian product $\mathbb{R}^N \times \mathbb{R}^M$. Denote by $x$ a generic point of $\mathbb{R}^N$ and by $y$ a generic point in $\mathbb{R}^M$. The main theorem in [ii] is the following.

**Theorem 1.** *Let $1 \le p < \infty$ and let $\{u_n\} \subset L^p(\mathbb{R}^N \times \mathbb{R}^M)$ be a sequence of functions having joint compact support. Assume that for $\mathcal{L}^M$ a.e. $y \in \mathbb{R}^M$ the sequence $\{u_n(\cdot, y)\}$ is weakly convergent in $L^p(\mathbb{R}^N)$ and that*

$$\sup_n \left( \int_{\mathbb{R}^N} |u_n(x,y)|^p \, dx \right)^{1/p} \le v(y)$$

*for some function $v \in L^p(\mathbb{R}^M)$. Then $\{u_n\}$ converges strongly in $L^p(\mathbb{R}^N \times \mathbb{R}^M)$ if and only if*

$$\lim_{t \to 0} \int_{\mathbb{R}^N \times \mathbb{R}^M} |u_n(x+t, y) - u_n(x,y)|^p \, dx dy = 0 \tag{3}$$

*uniformly in $n$.*

The previous theorem continues to hold in the case $M = 0$, that is, when the functions $u_n$ do not depend on the variable $y$, provided one replaces (3) with

$$\lim_{t \to 0} \int_{\mathbb{R}^N} |u_n(x+t) - u_n(x)|^p \, dx = 0$$

uniformly in $n$. This latter result is essentially due to Riesz [R32].

The proof of Theorem 1 is rather elegant and uses mollifiers in a clever way. As a corollary of this theorem, Serrin gives a simple proof of what is now called the Rellich–Kondrachov compactness theorem (see, e.g., [AF03], [EG92], [L09], [Z89]), namely that a bounded sequence $\{u_n\}$ in $W^{1,p}(\Omega)$, $1 \le p < N$, admits a subsequence converging strongly in $L^q(\Omega)$ for every $1 \le q < \frac{Np}{N-p}$, provided $\Omega$ satisfies appropriate conditions (usually some type of boundedness and boundary regularity).[1]

[1]In [ii] Serrin quoted the result of Rellich, who proved the case $p = 2$, but he did not seem to be aware of Kondrachov's paper [K45].

The full force of Theorem 1 (in the product case) is used to extend to arbitrary $M$ and $p$ a compactness result proved by Hopf (for $M = 1$ and $p = 2$) for Sobolev functions in $W^{1,p}(\mathbb{R}^N \times \mathbb{R}^M)$.

It's easy to comment on the paper of Meyers and Serrin [iii]: It is the paper that every mathematician would have like to have written. Meyers and Serrin proved that $C^\infty(\Omega) \cap W^{k,p}(\Omega)$ is dense in $W^{k,p}(\Omega)$. This result is one of the milestones in the history of Sobolev spaces and it can be found in all textbooks on Sobolev spaces (see, e.g., [AF03], [AFP00], [EG92], [L09], [Z89]) and partial differential equations ([B11], [E98]).

The beauty of the proof lies in its simplicity. Given a function $u \in W^{k,p}(\Omega)$ and a standard mollifier $\varphi_\varepsilon$, the convolution $u_\varepsilon := u * \varphi_\varepsilon$ is of class $C^\infty$ but it is only defined away from the boundary. The idea of Meyers and Serrin is to use a sequence of mollifiers $\varphi_{\varepsilon_n}$, where $\varepsilon_n$ gets smaller and smaller as one approaches the boundary of $\Omega$.

## References


[AF03] R.A. Adams and J.J.F. Fournier, *Sobolev spaces.* Second edition. Pure and Applied Mathematics (Amsterdam), 140. Elsevier/Academic Press, Amsterdam, 2003.

[A91] G. Alberti, *A Lusin type theorem for gradients.* J. Funct. Anal. **100** (1991) 110–118.

[AFP00] L. Ambrosio, N. Fusco, and D. Pallara, *Functions of bounded variation and free discontinuity problems.* Oxford Mathematical Monographs. The Clarendon Press, Oxford University Press, New York, 2000.

[B11] H. Brezis, *Functional analysis, Sobolev spaces and partial differential equations,* Springer, New York, 2011.

[DL55] J. Deny and J.L. Lions, *Les espaces du type de Beppo Levi.* Ann. Inst. Fourier, Grenoble **5** (1955) 305–370.

[EG92] L.C. Evans and R.F. Gariepy, *Measure theory and fine properties of functions.* Studies in Advanced Mathematics. CRC Press, Boca Raton, FL, 1992.

[E98] L.C. Evans, *Partial differential equations.* Second edition. Graduate Studies in Mathematics, 19, American Mathematical Society, Providence, RI, 2010.

[H51] E. Hopf, *Über die Anfangswertaufgabe für die hydrodynamischen Grundgleichungen.* Math. Nachr. **4** (1951) 213–231.

[K45] W. Kondrachov, *Sur certaines propriétés des fonctions dans l'espace.* C. R. (Doklady) Acad. Sci. URSS (N. S.) **48** (1945) 535–538.

[L09] G. Leoni, *A first course in Sobolev spaces.* Graduate Studies in Mathematics, 105. American Mathematical Society, Providence, RI, 2009.

[MP08] L. Moonens and W.F. Pfeffer, *The multidimensional Luzin theorem.* J. Math. Anal. Appl. **339** (2008) 746–752.

[R32] F. Riesz, Sur les ensembles compacts de fonctions sommables. Acta Litt. Sci. Szeged 6 (1932–34) 136–142.

[Z89] W.P. Ziemer, *Weakly differentiable functions. Sobolev spaces and functions of bounded variation.* Graduate Texts in Mathematics, 120. Springer-Verlag, New York, 1989.



Giovanni Leoni
Department of Mathematical Sciences
Carnegie Mellon University
e-mail: giovanni@andrew.cmu.edu

*Arch. Ration. Mech. Anal.* 7 (1961), 359–372.




# *On the Differentiability of Functions of Several Variables*

JAMES SERRIN

Let $u = u(x) = u(x_1, \ldots, x_n)$ be a real-valued locally summable function defined on an open region $R$ of the $n$-dimensional number space $\mathscr{R}^n$. It is our purpose here to investigate the differentiability of $u$ when the gradient of the distribution associated with $u$ is a measure. By this we mean that there exists a Borel measure $\alpha$, finite-valued on compact subsets of $R$, such that the formula

$$-\int u\,\varphi_x\,dx = \int \varphi\,d\alpha \tag{1}$$

holds for any continuously differentiable function $\varphi$ with compact support in $R$. Here $\varphi_x$ denotes the vector $\{\partial\varphi/\partial x_1, \ldots, \partial\varphi/\partial x_n\}$, and the measure $\alpha$ is of course vector-valued and unrestricted as to sign. For convenience, a function satisfying the preceding condition will be called *weakly differentiable.*

In order to discuss the differentiability properties of a summable function, whose values are in the nature of things somewhat arbitrary on a set of measure zero, we make the following definition: A function $u(x)$ will be called *essentially differentiable* if there exists an equivalent function $v(x)$, defined on $R$, which has ordinary partial derivatives $\{\partial v/\partial x_1, \ldots, \partial v/\partial x_n\}$ almost everywhere.

Suppose now that $u$ is essentially differentiable. If $v_1$ and $v_2$ are two functions, each one equivalent to $u$ and having partial derivatives almost everywhere, then obviously the partial derivatives of $v_1$ and $v_2$ agree almost everywhere. We may therefore define the derivative of $u$ with respect to $x_k$, $1 \leq k \leq n$, to be the *class of functions equivalent to* $\partial v/\partial x_k$, where $v$ is any function equivalent to $u$ and having partial derivatives almost everywhere. This derivative will be denoted by $\partial u/\partial x_k$. Although $\partial u/\partial x_k$ is an equivalence class, one naturally thinks of it as a function in the same way as elements of $L_p$ are thought of as functions. Finally, if $u$ is essentially differentiable, we set

$$u_x = \{\partial u/\partial x_1, \ldots, \partial u/\partial x_n\}.$$

The main result of the paper is now the following

**Theorem 1.** *If $u$ is weakly differentiable, then $u$ is essentially differentiable. Moreover $u_x = \alpha'$, where $\alpha'$ is the general derivative of the set function $\alpha$.*

The proof of the first statement in Theorem 1 occupies Sections 2 and 3 of the paper, Section 1 being devoted to a simple but interesting preparatory lemma. At the end of Section 3 we digress from the proof of Theorem 1 to obtain an immediate demonstration of a known theorem for the case when $u$ is continuous. The remaining part of Theorem 1 is then proved in Sections 4 and 5, following which we briefly discuss the relation between weak differentiability and bounded variation. The methods of the paper have their partial genesis in the work of GOFFMAN, KRICKEBERG, and DE VITO, and our main theorem complements the results of these authors[1]. It may be worth adding that the only background necessary for the present work is a knowledge of the elements of the theory of integration, and the Lebesgue decomposition theorem.

*Remarks.* There are several immediate consequences of Theorem 1 which may be conveniently noticed at this point.

1. By use of a well known fact about distributions [*11*, pp. 37–38], it follows from the first part of Theorem 1 that if $T$ is a distribution whose gradient is a measure, then $T$ is defined by a function having partial derivatives almost everywhere.

2. An essentially differentiable function obviously has approximate partial derivatives almost everywhere. Therefore, by Theorem 1 and a result of STEPANOFF [*10*, p. 300], a weakly differentiable function $u$ has an approximate differential almost everywhere. On the other hand, neither $u$ nor any equivalent function need have an ordinary differential at any point. The situation with respect to the existence of a differential is discussed more fully in the concluding sections, 8 and 9, of the paper.

3. Finally, by virtue of the formula $u_x = \alpha'$, one sees that $u_x$ is locally summable and transforms as a Cartesian vector.

## 1. A preparatory lemma

The results of this paper are ultimately based on the well known theorem of Lebesgue which states that any monotone function of a single real variable has a finite derivative almost everywhere. The following lemma generalizes this result to functions of several variables.

**Lemma 1.** *Let $f(x)$ be a measurable function defined on an open interval $I$ in $\mathcal{R}^n$. Suppose that $f(x)$ is monotone on almost all lines parallel to the $x_k$ axis. Then the partial derivative $\partial f/\partial x_k$ exists and is finite almost everywhere in $I$.*

Before proving this result, some comments are in order. Since Lebesgue's theorem for one variable also applies to functions of bounded variation, one might conjecture that the hypothesis of Lemma 1 could be weakened from "monotone on almost all lines" to simply "bounded variation on almost all lines". This change, however, would destroy the lemma: There exist measurable functions of uniformly bounded variation on *all* lines parallel to the $x_1$ axis, say, which do *not* possess a partial derivative $\partial f/\partial x_1$ almost everywhere in $I$. As a matter of interest, such an example is constructed in Section 7.

---

[1] Although our results might seem to follow from Theorems XII and XIV of [*3*], this is not actually the case. In particular, in [*3*] each component of $u_x$ is treated on the basis of a different function $v(x)$, and, more important, the derivatives $\partial u/\partial x_k$ which appear in [*3*] are defined by the special operation $\lim_{n\to\infty} \Delta u/\Delta x_k$, where $\Delta x_k = 1/n$; *cf.* footnote 6 of [*3*].

*Proof of Lemma 1.* It is clear from the hypothesis that $\partial f/\partial x_k$ exists and is finite almost everywhere on almost all lines parallel to the $x_k$ axis. It does not follow from this, of course, that $\partial f/\partial x_k$ exists almost everywhere in $I$. We must therefore resort to further argument.

For convenience, the set of variables $(x_1, \ldots, x_{k-1}, x_{k-1}, \ldots, x_n)$ will be denoted by $y$, and the set $(x_1, \ldots, x_{k-1}, t, x_{k-1}, \ldots, x_n)$ by $(t, y)$. Now for $x=(x_k, y)\in I$ we let

$$E^+f = E^+f(x) = \limsup_{m\to\infty} \frac{f(x_k+1/m, y)-f(x_k, y)}{1/m} \qquad (m = \text{integer})$$

be an upper right "sequential derivate" of $f$. It is clear that $E^+f$ is measurable, and, as one easily deduces, finite almost everywhere in $I$. The same result holds, of course, for the lower right derivate $E_+f$, and the left derivates $E^-f$ and $E_-f$. Now by Fubini's theorem

$$\int_I |E^+f - E_+f|\, dx = \int \left[\int |E^+f - E_+f|\, dx_k\right] dy = 0,$$

since the inner integral vanishes for almost all $y$. It follows at once that both limits

$$\lim_{m\to\infty} \frac{f(x_k+1/m, y)-f(x_k, y)}{1/m}, \qquad \lim_{m\to\infty} \frac{f(x_k-1/m, y)-f(x_k, y)}{-1/m}$$

exist and are equal almost everywhere in $I$. Let $E(x)$ denote the common limit where it exists; otherwise put $E(x)=\infty$. $E(x)$ is obviously finite-valued almost everywhere in $I$.

Now let $T$ denote the set of points $x\in I$ such that $E(x)$ is finite and $f(x)$ monotone in $x_k$. If $x\in T$ and $1/m+1 \leqq \Delta x_k \leqq 1/m$, then

$$\frac{f(x_k+1/m+1, y)-f(x_k, y)}{1/m} \leqq \frac{\Delta f}{\Delta x_k} \leqq \frac{f(x_k+1/m, y)-f(x_k, y)}{1/m+1},$$

provided $f$ is monotone *increasing* at $x$ (when $f$ is *decreasing* the inequalities must be reversed). Since both the right and left sides of the above inequality tend to $E(x)$ as $m\to\infty$, it follows that $\partial f/\partial x_k$ exists and is finite at each point of $T$. But $T$ includes almost all points of $I$, and the proof is thereby completed.

### 2. Proof of Theorem 1. First part

Consider any fixed open interval $I$ in $\mathscr{R}^n$ whose closure is contained in $R$, and let $k$ be a fixed integer, $1\leqq k\leqq n$. In order to prove the first statement of Theorem 1, it will be enough to show that there is a function $v(x)$, equivalent to $u$ *and* independent of $k$, such that $\partial v/\partial x_k$ exists almost everywhere in $I$. The first step in this process is to obtain an estimate of the variation of $u$ along lines parallel to the $x_k$ axis.

Let us associate with each positive number $h$ a corresponding function $K(\xi)\equiv K(\xi, h)$ with the properties:

i) $K$ is non-negative and smooth for all values of $\xi=(\xi_1, \ldots, \xi_n)$.

ii) $K\leqq \text{Const.}/h^n$ for $|\xi|<h$ and $K\equiv 0$ for $|\xi|\geqq h$.

iii) $\int K(\xi)\, d\xi = 1$.

The $\xi$ integration in this and subsequent formulas is assumed to be carried out over the entire $n$-dimensional number space. For $x \in I$ (and for $h$ sufficiently small), we define the function

(2) $$u_h = u_h(x) = \int K(\xi - x)\, u(\xi)\, d\xi .$$

In view of the properties of $K$, it is easy to prove that $\lim u_h$ exists and equals $u$ almost everywhere in $I$.[2] We set $w = \lim u_h$ wherever the limit exists, and otherwise leave the function $w$ undefined. Clearly $w = u$ almost everywhere.

It will be convenient at this stage to introduce some simple notation. We consider the interval $I$ to be the Cartesian product $J \times J'$ of an interval $J$ of the variable $x_k$ and an interval $J'$ of the complementary variable $y = (x_1, \ldots, x_{k-1}, x_{k+1}, \ldots, x_n)$. Correspondingly, the set of variables $(x_1, \ldots, x_{k-1}, t, x_{k+1}, \ldots, x_n)$ will be denoted by $(t, y)$ as in the preceding section. Finally, for any set $B$ in $\mathcal{R}^m$ we let $B_h$ denote the set of points in $B$ whose distance from $\mathcal{R}^m - B$ is greater than $h$.

Now by differentiation of (2) and use of (1), there results

(3) $$\frac{\partial u_h}{\partial x_k} = -\int \frac{\partial K}{\partial \xi_k}(\xi - x)\, u(\xi)\, d\xi = \int K(\xi - x)\, d\alpha_k ,$$

where $\alpha_k$ is the $k^{\text{th}}$ component of the measure $\alpha$. An easy application of Fubini's theorem then yields

$$\int_{I_h} \left| \frac{\partial u_h}{\partial x_k} \right| d x \leqq \int_I d|\alpha_k| .$$

For $y \in J'$ let us define

$$V(y) = \liminf \int_{J_h} \left| \frac{\partial u_h}{\partial x_k}(t, y) \right| dt, \qquad h \in \{1/m\}.$$

Clearly $V(y)$ is measurable on $J'$, whence by Fatou's lemma

$$\int_{J'} V(y)\, d y \leqq \int_I d|\alpha_k|$$

(integrate first over a compact subset $K$ of $J'$, and then let $K \to J'$). Now let $\varphi(x)$ be a real-valued function defined a.e. on $I$. For $y \in J'$, let $\operatorname{Var} \varphi$ be the function defined by

$$[\operatorname{Var} \varphi](y) = \underset{t \in J}{\text{Total variation}}\, \varphi(t, y)$$

with the convention that if $\varphi(x)$ is undefined at a certain set of points, then these points are simply to be disregarded in calculating the total variation. In view of the definition of $V(y)$, it is clear that the variation of the function $w = \lim u_h$ satisfies $\operatorname{Var} w \leqq V(y)$ for $y \in J'$. This completes the first step in the proof.

[2] Indeed, we have

$$|u_h(x) - u(x)| = \left| \int K(\xi - x)\,[u(\xi) - u(x)]\, d\xi \right| \leqq \frac{\text{Const.}}{h^n} \int_{|\xi - x| \leqq h} |u(\xi) - u(x)|\, d\xi ,$$

and the last expression tends to zero at each point of the Lebesgue set of $u$.

### 3. Continuation

The function $w(x)$ constructed in the preceding section has the property

$$\text{(4)} \qquad \operatorname{Var} w \leqq V(y), \quad \int_{J'} V(y)\,dy < \infty.$$

Moreover, $w$ is equal to $u$ almost everywhere in $I$, hence is summable. Such functions allow a canonical decomposition into the difference of two monotone functions, according to the following lemma.

**Lemma 2.** *Let $w(x)$ be a summable function defined a.e. on an open interval $I$, and there satisfying with respect to some variable $x_k$ the condition* (4). *Then there exist two summable functions $w^+$ and $w^-$ on $I$, each monotone increasing in the variable $x_k$, such that almost everywhere in $I$*

$$\text{(5)} \qquad w = w^+ - w^-.$$

*Remark.* A representation of the type (5) is not always possible if the equality is to hold *everywhere* in $I$. Indeed, if it were possible, then by Lemma 1 any function satisfying the hypotheses of Lemma 2 would have a partial derivative $\partial w/\partial x_k$ almost everywhere in $I$. But according to the example of Section 7, this is not the case.

*Proof of Lemma 2.* We shall first construct a function $\omega(x)$, equivalent to $w(x)$ but having "fewer" discontinuities. Let $S$ denote the set of points $y \in J'$, such that

i) $w(x_k, y)$ is defined for almost all $x_k \in J$, and

ii) $V(y) < \infty$.

Obviously $S$ includes almost all points of $J'$. For each $x = (x_k, y) \in I$ let us then define

$$\omega(x) = \begin{cases} w(x_k - 0, y) & y \in S, \\ 0 & y \notin S. \end{cases}$$

$\omega(x)$ is measurable, since it is the limit almost everywhere of the sequence of measurable functions $\{w(x_k - 1/m, y)\}$. Now on almost every line $y = y_0$ the functions $\omega$ and $w$ differ at most at a set of linear measure zero. From this fact, and since $\omega$ and $w$ are both measurable, it follows easily that $\omega = w$ almost everywhere in $I$. The argument above shows not only that $\omega(x)$ is measurable, but also, by a slight variant, that each function $\omega(x_k, y)$ is measurable in $y$ for fixed $x_k$.

Now let $x_k = a$ be a point such that

$$\int_{J'} \omega(a, y)\,dy < \infty$$

(such a point exist, by Fubini's theorem). Also, for fixed $y$, let $\tau(x_k, y)$ be the indefinite total variation of the function $\omega(x_k, y)$, normalized so that $\tau(a, y) = 0$. We assert that the function $\tau(x) = \tau(x_k, y)$ is measurable on $I$.

To prove this, denote by $P_m$ the partition of the interval $J$ by $m$ equally spaced points $t_1 < t_2 < \cdots < t_m$. The *approximate variation* of $\omega(x_k, y)$ corresponding to the partition $P_m$ is then

$$\tau_m(x_k, y) = \sum |\omega(t_i, y) - \omega(t_{i-1}, y)| \cdot \operatorname{sign}(x_k - a),$$

Setting $\varphi = u_h$ in (9), replacing $I$ by $I_h$, and then letting $h \to 0$, we deduce from Fatou's lemma that

$$\int_{I_\delta} F(\Delta u/\Delta x_k)\, dx \leq \liminf \int_{I_h} F(u_{h,k})\, dx.$$

Now let $u$ be replaced by $v$ in the left-hand side of this inequality, and allow $\delta$ to tend to zero. This proves (8).

### 5. Completion of the proof

Here we shall show in contrast to (8) that

$$\int_{I_h} F(u_{h,k})\, dx \leq \int_I F(\alpha_k')\, dx + \Theta(I), \tag{10}$$

where $\Theta$ is a singular measure.

The measure $\alpha(E)$ can be expressed by means of the Lebesgue decomposition theorem in the form

$$\alpha(E) = \int_E \alpha'(x)\, dx + \vartheta(E),$$

where $\vartheta(E)$ is a singular measure. Accordingly, equation (3) can be written

$$u_{h,k} = \int K(\xi - x)\, \alpha_k'(\xi)\, d\xi + \int K(\xi - x)\, d\vartheta_k.$$

Therefore by virtue of (7) and Jensen's inequality

$$F(u_{h,k}) \leq F\left(\int K(\xi - x)\, \alpha_k'(\xi)\, d\xi\right) + \left|\int K(\xi - x)\, d\vartheta_k\right|,$$
$$\leq \int K(\xi - x)\, F(\alpha_k')\, d\xi + \int K(\xi - x)\, d|\vartheta_k|.$$

Integrating over $I_h$ and applying Fubini's theorem yields (10), with $\Theta = |\vartheta_k|$.

We can now prove (6) and thus complete the proof of Theorem 1. By combining (8) and (10) one obtains

$$\int_I F(v_{,k})\, dx \leq \int_I F(\alpha_k')\, dx + \Theta(I).$$

Dividing each side by $|I|$ and letting $I$ tend regularly to $x$ gives, on account of Lebesgue's differentiation theorem [*10*, pp. 118–119],

$$F(v_{,k}) \leq F(\alpha_k') \quad \text{almost everywhere in } R. \tag{11}$$

Now choose

$$F(p) = F_1(p) = \begin{cases} e^p & p < 0, \\ p + 1 & p \geq 0, \end{cases}$$

which certainly satisfies the imposed conditions. Since $F_1(p)$ is monotone increasing, it follows from (11) that $v_{,k} \leq \alpha_k'$ almost everywhere. Similarly, setting $F(p) = F_2(p) = F_1(-p)$, we find in the same way that $v_{,k} \geq \alpha_k'$ almost everywhere. Combining the last two inequalities completes the proof of Theorem 1.

### 6. Bounded variation and weak differentiability

A function $u(x)$ will be called of *bounded variation in the sense of Cesari* if there is an equivalent function $w(x)$ defined a.e. such that, for every open interval $I$ in the interior of the domain of definition of $u$, and for each variable $x_k$,

$1 \leqq k \leqq n$, we have

$$\operatorname{Var} w \leqq V(y), \quad y \in J',$$

for some function $V(y)$ which is summable on the interval $J'$. The function $V(y)$ may of course be different for different directions $x_k$.

**Theorem 3.** *A function $u$ is weakly differentiable if and only if it is of bounded variation in the sense of Cesari*[3].

*Proof.* The "only if" part of this theorem has already been proved in Section 2. Suppose then that $u$ is of bounded variation in the sense of Cesari. By Theorem 5.4 of reference [6], the function $u$ is locally summable.

Now let $C(I)$ denote the linear space whose members are real-valued continuous functions on $I$, with compact support in $I$. Similarly, let $C'(I)$ denote the space of continuously differentiable functions with compact support in $I$. For $\varphi \in C'(I)$ and for any fixed $k$, $1 \leqq k \leqq n$, set

$$A(\varphi) = \int_I u \, \varphi_{,k} \, dx.$$

Evidently $A(\varphi)$ is a linear operator on $C'(I)$. Moreover, $A$ is continuous in the topology of $C(I)$. To see this, let us observe first that $\operatorname{Var} w < \infty$ for almost all $y \in J'$. Thus for almost all $y$ there exists a corresponding measure $\beta_y$ on $J$ such that

$$w(x_k, y) = \int^{x_k} d\beta_y, \quad \text{and} \quad \int_J d|\beta_y| \leqq V(y).$$

It follows that

$$A(\varphi) = \int_{J'} \left[\int_J u \, \varphi_{,k} \, dx_k\right] dy = \int_{J'} \left[\int_J \varphi \, d\beta_y\right] dy,$$

and so

$$|A(\varphi)| \leqq \operatorname{Max}_I |\varphi| \cdot \int_{J'} V(y) \, dy,$$

proving our assertion. Since $A(\varphi)$ is continuous in $C(I)$, it may be extended to a continuous linear functional on $C(I)$. But then by the Riesz theorem there exists a Borel measure $\alpha_k$ such that

$$A(\varphi) = -\int \varphi \, d\alpha_k.$$

This equality, however, is precisely the condition that $u$ should be weakly differentiable on $I$. Since $I$ was any interval in $R$, it follows easily that $u$ is weakly differentiable in $R$.

**Theorem 4.** *If $u$ is of bounded variation in the sense of Cesari, then for any interval $I \subset R$ and for any $k$, $1 \leqq k \leqq n$, we have*

$$\int_{J'} V(y) \, dy \geqq \int_I |\partial u / \partial x_k| \, dx.$$

*Proof.* By the preceding theorem, $u$ is weakly differentiable in $I$, and the measure $\alpha_k$ even satisfies

$$\int_I d|\alpha_k| \leqq \int_{J'} V(y) \, dy.$$

[3] This result is due to Krickeberg; the present proof, which in part parallels that of Krickeberg, is given mainly for the sake of completeness.

On the other hand, it is clear that

$$\int_I d|\alpha_k| \geqq \int_I |\alpha_k'|\, dx = \int_I |u_{,k}|\, dx,$$

and the theorem is proved.

## 7. A counterexample

For the purpose of this section we shall suppose that $I$ is the unit square, $0<x<1$, $0<y<1$, in the plane of the two real variables $x$, $y$. As indicated in Section 1, we intend to construct a function $g(x, y)$ on $I$ with the properties

i) $g=0$ almost everywhere in $I$,

ii) $\operatorname{Var} g = 1$ on each line $y=$ constant,

and nevertheless such that

iii) $\partial g/\partial x$ does not exist at a set of points whose outer measure is one.

The construction is based on the fact that there exists a set $O$ in $I$ which has outer measure one, and yet has only two points in common with each straight line parallel to the $x$-axis (see [12]).

With each point $Q \in O$ let us associate a sequence of points $Q_m$ which lie on the same horizontal line as $Q$, have rational abscissas $x_m$, and tend to $Q$ as $m \to \infty$. By omitting certain of the points $Q_m$ and then renumbering the sequence, we can suppose furthermore that

$$0<\sum |Q-Q_m|<1 \tag{12}$$

and that the sets $\{Q_m\}$ corresponding to the two different points of $O$ on any horizontal line are disjoint. The function $g(x, y)$ can now be defined. First, at any point $(x, y)=Q_m$ associated with a point $Q \in O$, we set

$$g(x, y) = |Q-Q_m|/4\Sigma$$

where $\Sigma = \Sigma(Q)$ denotes the sum of the series (12). Then, at all other points, we put $g(x, y)=0$.

The resulting function is zero so long as $x$ is irrational, hence property i) is assured. Property ii) is likewise immediately clear. Finally, at any point $Q \in O$ we have

$$\liminf \left|\frac{\Delta g}{\Delta x}\right| = 0, \qquad \limsup \left|\frac{\Delta g}{\Delta x}\right| = 1/4\Sigma.$$

Thus $\partial g/\partial x$ does not exist at any point of $O$. Q.E.D.

*Remark.* Although Lemma 1 is thus shown to be incorrect under the hypothesis of bounded variation, a slightly modified version remains true, namely

**Lemma 1′.** *Let $f(x)$ be a measurable function defined on an open interval $I$ in $\mathscr{R}^n$. Suppose that $f$ is of bounded variation on almost all lines parallel to the $x_k$ axis. Then there exists an equivalent function $h(x)$ which has finite partial derivatives almost everywhere in $I$.*

The proof is based on the fact that $f(x)$ can be decomposed into two monotone functions as in the proof of Lemma 2. We omit the details.

## 8. The existence of a differential

We have already remarked in the introduction that a weakly differentiable function need not have an ordinary differential at any point. It is the purpose of this section to investigate the precise situation with respect to the existence of a differential. This can best be done by introducing the concept of strong differentiability. A locally summable function $u(x)$ will be said to be *strongly differentiable* if there exists a vector-valued function $\psi(x)$ such that the formula

$$-\int u\,\varphi_x\,dx = \int \varphi\,\psi\,dx \tag{13}$$

holds for every continuously differential function $\varphi$ with compact support in $R$. Equivalently, $u$ is strongly differentiable if it is weakly differentiable and the associated measure is absolutely continuous. By Theorem 1 a strongly differentiable function is essentially differentiable and has the derivative $u_x=\psi$. The following result now holds.

**Theorem 5.** *Suppose that $u(x)$ is strongly differentiable and that $u_x$ belongs to $L_p$ for some $p>n$. Then $u$ is continuous and has a differential almost everywhere.*

The conclusion of the theorem refers of course only to a particular member of the equivalence class of $u$.

For the case of two independent variables, Theorem 5 is essentially due to CESARI [2]. The proof we give is a modification and simplification of his. Theorem 5 is best possible: If $u$ is strongly differentiable and $u_x$ belongs to $L_n$, then $u$ need not be continuous, and, even if it is continuous, it need not possess a differential at any point. Examples illustrating this behavior are given in the following section.

*Proof of Theorem* 5. It is well known that if $u$ is strongly differentiable and $u_x$ is in $L_p$, then there exists a sequence of continuously differentiable functions $u_m(x)$ such that

$$\int_S |u_m-u|\,dx \to 0, \qquad \int_S |u_{mx}-u_x|^p\,dx \to 0, \tag{14}$$

where $S$ is any compact subregion of $R$.[4] Now let $x$ and $y$ be two points of $R$, and let $G$ denote the sphere whose diameter is the line segment joining $x$ and $y$. If $G$ has compact support in $R$, then

$$|u_m(x)-u_m(y)| \leqq \text{Const.}\,|x-y|^{1-n/p}\Big(\int_G |u_{mx}|^p\,dx\Big)^{1/p} \qquad (p>n), \tag{15}$$

where the constant depends only on $n$ and $p$. A simple proof of (15) may be found in reference [8], pp. 523–524.

From (14) and (15) one easily infers that the sequence $\{u_m(x)\}$ is uniformly bounded and equicontinuous in any compact subregion of $R$. Consequently there exists a subsequence which converges to a continuous function $v(x)$ in $R$. Obviously $v=u$ almost everywhere, and the first part of Theorem 5 is thereby proved.

We may assume from here on that $u$ is a continuous function, and that $u_m\to u$ uniformly on any compact subregion of $R$. Letting $m\to\infty$ in inequality (15) then yields

$$|u(x)-u(y)| \leqq \text{Const.}\,|x-y|^{1-n/p}\Big(\int_G |u_x|^p\,dx\Big)^{1/p}. \tag{16}$$

(Hence $u(x)$ is Hölder continuous in $R$, a result due to MORREY.)

[4] See reference [4]. We can in fact take $u_m=u_h$ with $h=1/m$.

Now let $E$ denote the Lebesgue set of $u_x$ relative to the class $L_p$.[5] We shall show that $u$ has a differential at each point $x_0$ in $E$, and this will prove the theorem. Indeed, consider the function

$$w(x) = u(x) - A \cdot x,$$

where $A$ denotes the constant vector $u_x(x_0)$. Evidently $w(x)$ is strongly differentiable and $w_x = u_x - A$. Therefore, as one sees by comparing with (16),

$$|w(x) - w(x_0)| \leqq \text{Const.}\, |x - x_0|^{1-n/p} \Big( \int_G |u_x - A|^p \Big)^{1/p},$$

it being assumed that $x$ is near enough to $x_0$ so that $G$ is in $R$. Since $x_0$ is in $E$ the integral on the right is $o(|x - x_0|^n)$, whence

$$|w(x) - w(x_0)| = o(|x - x_0|).$$

This is precisely the condition that $u$ have a differential at $x_0$. Q.E.D.

*Remark.* Since a Lipschitz-continuous function is strongly differentiable and its derivative belongs to $L_\infty$, we obtain as a corollary of Theorem 5 that every Lipschitz continuous function has a differential at almost all points of its domain of definition (Rademacher's theorem).

## 9. Two examples

Consider a set of points $Q_k$, $k = 1, 2, 3, \ldots$, dense on some interval $I$. Let $a_k$ be a sequence of positive numbers such that $\sum a_k < \infty$, and define $f_k(x) = \log\log(1 + 1/r_k)$, where $r_k$ denotes the distance from $x$ to $Q_k$. Finally set

$$u(x) = \sum a_k f_k(x). \tag{17}$$

Since $f_k$ is summable, it is clear from Lebesgue's theorem that the partial sums $u_m(x)$ of the series (17) converge in $L_1$ to $u(x)$; this shows, by the way, that $u$ is finite almost everywhere. Moreover, since

$$\|f_k\| \equiv \Big( \int_I |f_{kx}|^n \, dx \Big)^{1/n} < \infty,$$

we see that the sequence $\{u_{mx}\}$ is bounded, and hence weakly compact, in $L_n$. We let $\psi$ denote a weak limit of this sequence. Then from

$$\int u_m \varphi_x \, dx = -\int \varphi \, u_{mx} \, dx,$$

it follows by letting $m \to \infty$ through an appropriate subsequence that

$$\int u \varphi_x \, dx = -\int \varphi \psi \, dx.$$

Thus $u$ is strongly differentiable on $I$, and its derivative is in $L_n$. But $u$ is infinite on a dense set, and so is everywhere discontinuous.

We shall next construct a continuous function $u$ which has strong derivatives in $L_n$, but which nevertheless is nowhere differentiable. (In the paper quoted earlier, CESARI proved the existence of such a function in the case $n = 2$, but

[5] See [*14*, p. 65]. The work here refers only to the one-dimensional case, but carries over to $n$ dimensions in the obvious way.

he did not give an explicit example.) The method is an elaboration and extension of one due to STEPANOFF.

The function to be constructed will again have the basic form $u = \sum a_k f_k$ but will be more complicated in nature. We set

$$a_k = k/2^k, \quad N_k = 2^k, \quad \text{and} \quad \omega(t) = \log\frac{t+\varepsilon}{1+\varepsilon}\Big/\log\frac{\varepsilon}{1+\varepsilon},$$

where $\varepsilon = \varepsilon_k$ is fixed by the condition $\log \varepsilon = -2^k$. Also let $\varrho_k$ be a positive number less than $2^{-2k}$, to be fixed later. In order to define $f_k(x)$ we partition the interval $I$, which may be assumed to be a cube of unit length on each of its sides, into $N_k^n$ subcubes each having side length $2^{-k}$. Let $x_{(k)}$ be a generic symbol for the center of any one of these subcubes. We then set

$$f_k(x) = \omega\left(|x - x_{(k)}|/\varrho_k\right)$$

if $|x - x_{(k)}| \leqq \varrho_k$ for some center $x_{(k)}$, and otherwise put $f_k(x) = 0$. (In geometrical language, the surface $f_k$ consists of a family of "cones" of height one and base $\varrho_k$, centered at the lattice points $x_{(k)}$; in two dimensions this is an Indian mystic's bed of nails.)

Since $0 \leqq f_k \leqq 1$, it is clear that the series $\sum a_k f_k$ converges uniformly to a continuous function $u$. Moreover $u$ is strongly differentiable. Indeed, the partial sums $u_m$ satisfy

$$\|u_m\| = \left\|\sum_1^m a_k f_k\right\| \leqq \sum_1^m a_k \|f_k\|$$

by Minkowski's inequality. But

$$\|f_k\| = N_k \|\omega\| = \text{Const.}\, N_k \left(\int_0^1 |\omega'(t)|^n t^{n-1} dt\right)^{1/n}$$
$$\leqq \text{Const.}\, N_k |\log \varepsilon|^{(1-n)/n},$$

as one easily sees. Consequently $\|u_m\| \leqq \text{Const.} \sum k\, 2^{k(1-n)/n}$, so that $\{u_{m\,x}\}$ forms a uniformly bounded sequence in $L_n$. By our earlier argument, then, $u$ is strongly differentiable with $u_x$ belonging to $L_n$.

It remains to show that $u$ cannot have a differential at any point. Consider first a point $\xi$ where all the functions $f_k$ vanish. This point lies no farther than $\sqrt{n}/2^{k+1}$ from some vertex $x_{(k)}$. Hence

$$\frac{u(x_{(k)}) - u(\xi)}{|x_{(k)} - \xi|} \geqq \frac{a_k f_k(x_{(k)})}{\sqrt{n}/2^{k+1}} = \frac{2k}{\sqrt{n}}.$$

Since the right-hand side can be arbitrarily large for appropriate choice of $x_{(k)}$, it is evident that $u$ cannot have a differential at $\xi$. Next, if only a finite number of functions $f_k$ are non-zero at $\xi$, an easy modification of the above argument again shows that $u$ cannot have a differential at $\xi$.

Finally, suppose that $f_k(\xi) \neq 0$ for an infinite number of values of $k$. Consider a fixed $k$ for which $f_k(\xi) \neq 0$, and form the difference quotient with respect to the nearest of the vertices $x_{(k)}$, or else (if $|x_{(k)} - \xi| < \frac{1}{2}\varrho_k$) with respect to an appropriate point $y_{(k)}$ where $|x_{(k)} - y_{(k)}| = \varrho_k$. Then, provided the radii $\varrho_k$ have been chosen sufficiently small, one sees that the dominant contribution to the difference quotient is due to the function $a_k f_k$. But the slope of this function becomes arbitrarily large as $k$ tends to infinity; hence again $u$ cannot have a differential at $\xi$. Thus $u$ does not have a differential at any point in $I$.

*Note.* After this paper was completed, the author learned that C. GOFFMAN and R. E. HUGHES had independently established that a function of bounded variation in the sense of Cesari is essentially differentiable. Their proof (which will be published elsewhere) and the work of Section 3 of this paper are closely similar.

The above research was supported by the United States Air Force Office of Scientific Research under Contract No. AF 49(638)-262.

## References

[1] CESARI, L.: Sulle funzioni a variazione limitata. Ann. Scuola Normale Superiore Pisa, Series II **5**, 299—316 (1936).
[2] CESARI, L.: Sulle funzioni assolutamente continue di due variabili. Ann. Scuola Normale Superior Pisa, Series II **10**, 91—101 (1941).
[3] DE VITO, L.: Sulle funzioni di più variabili a variazione limitata. Archive for Rational Mechanics and Analysis **3**, 60—81 (1959).
[4] FRIEDRICHS, K. O.: The identity of weak and strong extensions of differential operators. Trans. Amer. Math. Soc **55**, 132—151 (1944).
[5] GOFFMAN, C.: Non-parametric surfaces given by linearly continuous functions. Acta Mathematica **103**, 269—291 (1960).
[6] KRICKEBERG, K.: Distributionen, Funktionen beschränkter Variation, und Lebesguescher Inhalt nichtparametrischer Flächen. Annali di Matematica, Series IV **44**, 105—133 (1957).
[7] MORREY, C. B.: Multiple integral problems in the calculus of variations and related topics. Univ. of California Pub. in Mathematics (new series), **1** (1943); see especially p. 13.
[8] NIRENBERG, L.: Estimates and existence of solutions of elliptic equations. Comm. Pure Applied Mathematics **9**, 509—530 (1956).
[9] RADEMACHER, H.: Über partielle und totale Differenzierbarkeit. Mathematische Annalen **79**, 340—359 (1919).
[10] SAKS, S.: Theory of the Integral, 2nd Ed. New York: Hafner 1937.
[11] SCHWARTZ, L.: Theorie des Distributions, II. Paris: Hermann 1951.
[12] SIERPIŃSKI, W.: Sur un problème concernant les ensembles mesurables superficialement. Fund. Mathematica **1**, 112—115 (1920).
[13] STEPANOFF, W.: Sur les conditions de l'existence de la différentialle totale. Matematicheski Sbornik **32**, 511—526 (1924).
[14] ZYGMUND, A.: Trigonometrical Series, 2nd Ed. Cambridge 1959.

Department of Mathematics
Institute of Technology
University of Minnesota
Minneapolis, Minnesota

*(Received December 28, 1960)*

*Proc. Am. Math. Soc.* 13 (1962), 651–655.




# STRONG CONVERGENCE IN A PRODUCT SPACE

JAMES SERRIN

The purpose of this note is to present a certain necessary and sufficient condition for strong convergence in $L_p$ over a product space. This result is formulated as Theorem 1 below. Theorem 1 includes as a special case a well-known compactness condition of M. Riesz, which we state in slightly different form as Theorem 2. In the final part of the paper we show that Theorem 1 can be used to give very simple proofs of the so-called Rellich lemma of functional analysis.

1. **Weak and strong convergence on product spaces.** Let $S = E^n \times E^m$ be the cartesian product of the Euclidean number spaces $E^n$ and $E^m$. Points in $S$ will be denoted by $z$. Also we shall frequently write $z = (x, y)$, where $x \in E^n$ and $y \in E^m$ are the orthogonal projections of $z$ onto $E^n$ and $E^m$, respectively.

Consider functions $u = u(z)$ of class $L_p$ on $S$, where $1 \leqq p < \infty$. The following norms will be used,

$$\|u\| = \left(\int |u(z)|^p dz\right)^{1/p}, \qquad \|u\|_x = \left(\int |u(x, y)|^p dx\right)^{1/p}.$$

The integrations here are over $S$ and $E^n$ respectively, and $\|u\|_x$ is of course itself a function of class $L_p$ over $E^m$.

THEOREM 1. *Let $\{u_k\}$ be a sequence of functions of class $L_p(S)$, having joint compact support in $S$. Suppose that for almost all fixed $y$ the functions $u_k(x, y)$ converge weakly in $L_p(E^n)$, and that*

$$\|u_k\|_x \leqq v(y), \qquad k = 1, 2, \cdots, \tag{1}$$

*where $v(y)$ is of class $L_p(E^m)$. Then the sequence $\{u_k\}$ converges strongly in $L_p(S)$ if and only if*

$$\lim_{|t| \to 0} \|u_k(x + t, y) - u_k(x, y)\| = 0 \tag{2}$$

*uniformly in $k$.*

PROOF. Let $\tilde{K}(\xi)$ be a non-negative continuous function on $E^n$, with support in $|\xi| \leqq 1$, and such that $\int \tilde{K} d\xi = 1$. For each positive real number $h$ we set

$$K(\xi) = K(\xi, h) = h^{-n}\tilde{K}(\xi/h),$$

Received by the editors April 28, 1961.

and define an integral average of functions $u \in L_p(S)$ by the formula

$$u_h = u_h(z) = \int K(\xi - x)u(\xi, y)d\xi.$$

By Hölder's inequality we have

$$(3)\qquad \begin{aligned} |u_h - u|^p &\leqq \int K(\xi - x)\,|u(\xi, y) - u(x, y)|^p d\xi \\ &\leqq \text{Const. } h^{-n} \int_{|t| \leqq h} |u(x + t, y) - u(x, y)|^p dt. \end{aligned}$$

Now suppose $\{u_k\}$ is a sequence of functions satisfying the hypotheses of Theorem 1, and such that condition (2) holds. Then taking account of (2) and (3) one obtains

$$(4)\qquad \lim_{h \to 0} \|u_{kh} - u_k\| = 0, \qquad \text{uniformly in } k.$$

Let $u_0(x, y)$ denote the weak limit of the functions $u_k(x, y)$. Suppose for the moment that $u_0$ is in $L_p(S)$, which will certainly be the case if, for example, $u_0$ is measurable on $S$. Then we have

$$(5)\qquad \|u_k - u_0\| \leqq \|u_k - u_{kh}\| + \|u_{kh} - u_{0h}\| + \|u_{0h} - u_0\|.$$

Let $\epsilon > 0$ be fixed, and choose $h$ so small that the first and last terms on the right-hand side of (5) are each less than $\epsilon/2$; in view of (4) this choice can clearly be made independent of $k$. Therefore

$$(6)\qquad \lim \|u_k - u_0\| \leqq \epsilon + \lim \|u_{kh} - u_{0h}\|.$$

It follows that $\{u_k\}$ will converge strongly to $u_0$ provided the right-hand limit is zero.

By the hypothesis of weak convergence, $u_{kh} \to u_{0h}$ for almost all $z$. But

$$|u_{kh}|^p \leqq \text{Const. } h^{-n} \int |u_k(x, y)|^p dx \leqq \text{Const. } h^{-n} v^p,$$

so that $|u_{kh} - u_{0h}|^p$ is dominated by a fixed summable function (here we use the fact that the functions $u_k$ have joint compact support). The required conclusion thus follows from Lebesgue's theorem.

If $u_0$ is not originally known to be in $L_p(S)$, then a slightly modified version of the above argument may be used to show that $\{u_k\}$ is Cauchy in $L_p(S)$; we shall omit the details since they add nothing essentially new. The first half of the theorem is therefore proved. To complete the demonstration it is only necessary to show that (2) holds if $\{u_k\}$ is strongly convergent. This is, however, a consequence of the well-known fact that (2) holds *for fixed k*, together with the argument given on p. 100 of [3].

By taking $m=0$ in the preceding theorem we immediately obtain the following corollary result.

THEOREM 2. *Let $\{u_k\}$ be a sequence of functions of class $L_p(E^n)$, having joint compact support. Suppose that $\{u_k\}$ is weakly convergent in $L_p(E^n)$. Then $\{u_k\}$ is strongly convergent if and only if*

$$\lim_{|t|\to 0} \|u_k(x+t) - u_k(x)\| = 0 \qquad \textit{uniformly in } k.$$

Theorem 2 implies a well-known compactness condition of M. Riesz (see [5], or [2, Theorem IV. 8.21]). In fact, for the case of joint compact support and $1<p<\infty$ the latter theorem follows at once from Theorem 2 and the fact that $L_p$ is weakly compact. For $p=1$ a slight additional argument is required which we shall omit. The hypothesis of joint compact support in both Theorems 1 and 2 can easily be replaced by a uniformity condition for large $x$, as in [2] or [5].

2. **Rellich's lemma.** An important application of Theorems 1 and 2 is in showing that certain spaces of differentiable functions are compact. We present two such results here. The first is a slight generalization of Rellich's lemma (cf. [1, p. 489]), and the second is an extension of a similar but lesser known result of E. Hopf.

THEOREM 3. *Let $K$ be a family of differentiable functions $u=u(x)$ with joint compact support. Suppose the norm $\|Du\|_p$ is uniformly bounded in $K$, where $p\leqq n$.*[1] *Then $K$ is compact in $L_q$ for $1\leqq q<np/(n-p)$.*

PROOF. It is enough to consider any fixed $q$ satisfying $1<q<np/(n-p)$. By standard and well-known arguments we obtain the inequality

$$\|u(x+t) - u(x)\|_q \leqq \text{Const. } |t|^{\alpha}\|Du\|_p, \qquad u\in C^1, \tag{7}$$

where $\alpha=1+n/q-n/p>0$, and the Const. depends only on $p$, $q$, and $n$. (In the important case $q=p$ inequality (7) specializes to

$$\|u(x+t) - u(x)\|_p \leqq |t|\cdot\|Du\|_p, \tag{7'}$$

as can be seen at once by integration of the inequality

[1] $\|f\|_p$ denotes the usual $L_p$ norm of $f(x)$ on $E^n$. When $p>n$ the functions $u$ are well known to be uniformly Hölder continuous (C. B. Morrey, Jr., University of California Publ. Math. 1943, p. 13), and Ascoli's compactness theorem applies.

$$|u(x+t) - u(x)| \leqq |t| \cdot \int_0^1 |Du(x+\sigma t)|\, d\sigma.)$$

It now follows from (7) that $\|u\|_q$ is uniformly bounded in $K$, whence $K$ is weakly compact in $L_q$. But then by virtue of (7) and Theorem 2, $K$ is strongly compact in $L_q$. Q.E.D.

Whether $K$ is compact in $L_q$ for $q=np/(n-p)$ is an open question, but the author is inclined to doubt that this is true. The special case $m=1$, $p=2$, of the following theorem is due to E. Hopf.

THEOREM 4. *Let $\{u_k\}$ be a sequence of differentiable functions $u_k(z)$, having joint compact support in S. Suppose that for almost all fixed y the functions $u_k(x, y)$ converge weakly in $L_p(E^n)$ and that* (1) *holds. Finally, assume that the norms $\|Du_k\|$ are uniformly bounded, the differentiations being taken with respect to the space $E^n$. Then $\{u_k\}$ is strongly convergent in $L_p(S)$.*

PROOF. It is enough to show that condition (2) of Theorem 1 holds for the sequence $\{u_k\}$. But by (7′),

$$\|u_k(x+t, y) - u_k(x, y)\|_x \leqq |t| \cdot \|Du_k\|_x,$$

whence integrating over $E_m$ we get $\|u_k(x+t, y) - u_k(x, y)\| \leqq \text{Const. } |t|$, and the proof is complete.

*Added in proof.* Theorems similar to those above are easily proved for functions in the space $L_{p,p'}(S)$, that is, functions for which the norm

$$\left(\int_{E^m}\left(\int_{E^n} |f(x, y)|^p dx\right)^{p'/p} dy\right)^{1/p'} < \infty.$$

In fact, Benedek and Panzone have obtained analogous results even for the class $L^{P,P'}$ (the notation is apparent from their paper, *The spaces $L^P$, with mixed norms*, Duke Math. J. (1961)), and for functions with values in a Banach space.

Benedek and Panzone have also kindly communicated an example showing that the conjecture following the proof of Theorem 3 is correct. Indeed, consider the sequence of functions

$$f_k(x) = \begin{cases} k^{(n-p)/p}(1 - k|x|) & |x| < 1/k, \\ 0 & |x| \geqq 1/k, \end{cases}$$

where $1 \leqq p < n$. Then evidently $\|Df_k\|_p \equiv \text{Constant}$, $k=1, 2, \cdots$. On the other hand, if we put $q=np/(n-p)$, then $\|f_k\|_q \equiv \text{Constant}$. Since $f_k \to 0$ almost everywhere, it follows that the sequence $\{f_k\}$ cannot be compact in $L_q$.

*Note*. The above research was partially supported by the United States Air Force under Contract AF 49(638)-262.

## REFERENCES

1. R. Courant and D. Hilbert, *Methoden der mathematischen Physik*, Vol. II, Springer, Berlin, 1937.

2. N. Dunford and J. Schwartz, *Linear operators*, Interscience, New York, 1959.

3. P. Halmos, *Measure theory*, Van Nostrand, New York, 1950.

4. E. Hopf, *Über die Anfangswertaufgabe für die hydrodynamischen Grundgleichungen*, Math. Nachr. **4** (1951), 213–231.

5. M. Riesz, *Sur les ensembles compacts de fonctions sommables*, Acta. Sci. Math. Szeged **6** (1933), 136–142.

UNIVERSITY OF MINNESOTA

Vladimir G. Maz'ya and James Serrin, Rome, 2008.

Norman Meyers,
University of Minnesota, 2010.

Tatyana O. Shaposhnikova and James Serrin,
Rome, 2008.

# III.2. Integration Theory

Giovanni Leoni

**Mathematics Subject Classification (2010).** 49J45.
**Keywords.** Semicontinuity, relaxation, calculus of variations.

Serrin's papers [i] (together with its sequel [S59]), [ii], and [iii] below are some of the most influential papers in the direct method in the calculus of variations. Indeed, they lay the foundation to the *relaxation method* for variational functionals (see, e.g., [D08], [DM93], [FL07]).

Consider an open set $\Omega \subset \mathbb{R}^N$ and a nonnegative continuous integrand $f : \Omega \times \mathbb{R} \times \mathbb{R}^N \to [0, \infty)$. For $u \in C^1(\Omega)$ (or $u$ in some Sobolev space) we can define the functional

$$I(u, \Omega) := \int_\Omega f(x, u(x), \nabla u(x))\, dx. \tag{1}$$

To extend the definition of $I$ to functions $u$ that are only locally integrable, or, say, that are in the space of functions of bounded variation $BV$, Serrin proposed to consider the "relaxed" functional[1]

$$\mathcal{I}(u, \Omega) := \inf_{\{u_n\}} \left\{ \liminf_{n\to\infty} I(u_n, \Omega) \right\}, \tag{2}$$

where the infimum is taken over all sequences $\{u_n\} \subset C^1(\Omega)$ converging to $u$ in some appropriate sense. The correct type of convergence is usually determined by the coercivity properties of the integrand $f$. In what follows, for simplicity, we will consider local convergence in $L^1$.

The relaxed functional $\mathcal{I}$ is analogous to the generalized area in the Lebesgue's theory of surface area of nonparametric surfaces. In this case $N = 2$ and $f(\xi_1, \xi_2) := \sqrt{1 + \xi_1^2 + \xi_2^2}$.

One of the main goals of [i], [S59], [ii], and [iii] is to study the relation between the two functionals $I$ and $\mathcal{I}$. In [ii, Theorem 12] Serrin gave some simple and elegant conditions on the integrand $f$, which guarantee that $\mathcal{I}(u, \Omega) = I(u, \Omega)$ for functions $u \in C^1(\Omega)$.

**Theorem 1.** *Assume that $f : \Omega \times \mathbb{R} \times \mathbb{R}^N \to [0, \infty)$ is continuous, $f(x, s, \cdot)$ is convex in $\mathbb{R}^N$ for every $(x, s) \in \Omega \times \mathbb{R}$, and that $f$ satisfies any one of the following conditions:*

(i) *$f(x, s, \xi) \to \infty$ as $|\xi| \to \infty$ for every $(x, s) \in \Omega \times \mathbb{R}$.*
(ii) *$f(x, s, \cdot)$ is strictly convex in $\mathbb{R}^N$ for every $(x, s) \in \Omega \times \mathbb{R}$.*

[1] In Serrin's original definition, $u_n$ is allowed to be defined on a different domain $\Omega_n$, so that $I(u_n, \Omega)$ should be replaced by $I(u_n, \Omega_n)$. For simplicity we consider here the case $\Omega_n \equiv \Omega$.

P. Pucci et al. (eds.), *James Serrin. Selected Papers: Volume 2*, Contemporary Mathematicians, DOI 10.1007/978-3-0348-0847-7_7, © Springer Basel 2014

(iii) *The derivatives* $\frac{\partial f}{\partial x_i}$, $\frac{\partial f}{\partial \xi_j}$, *and* $\frac{\partial^2 f}{\partial x_i \partial \xi_j}$ *exist and are continuous,* $i, j = 1, \ldots, N$.

*Then* $I$ *is lower semicontinuous in* $W^{1,1}_{\mathrm{loc}}(\Omega)$ *with respect to convergence in* $L^1_{\mathrm{loc}}(\Omega)$.

Over the years conditions (i)–(iii) of Theorem 1 have been extended in several ways. For the sake of brevity, I will only mention a few. In [M03], [GM04] Gori and Maggi proved that Theorem 1 continues to hold if $f(x, s, \cdot)$ is demicoercive for every $(x, s) \in \Omega \times \mathbb{R}$ (or constantly equal to $\inf f$). A function $g : \mathbb{R}^N \to \mathbb{R}$ is said to be *demicoercive* if there exist $c > 0$, $b \geq 0$, and $v \in \mathbb{R}^N$ such that $g(\xi) + v \cdot \xi + b \geq c|\xi|$ for all $\xi \in \mathbb{R}^N$. Note that strictly convex functions or convex functions $g$ satisfying $g(\xi) \to \infty$ as $|\xi| \to \infty$ are demicoercive, so that demicoercivity provides a geometrical condition weaker than either condition (i) or (ii).

Moreover, in [FGM06] Fusco, Gori, and Maggi proved that if an integrand $f$ satisfies either condition (i) or (ii), or, more generally, if $f(x, s, \cdot)$ is demicoercive, then it can be approximated from below by a sequence of integrands of the type (iii), so that, in a way, part (iii) of Theorem 1 implies parts (i) and (ii).

Another important extension concerns condition (iii). Indeed, Gori and Marcellini in [GM02] showed that the conclusion of Theorem 1 continues to hold if hypothesis (iii) is replaced by the significantly weaker locally Lipschitz condition that for every compact set $K \subset \Omega \times \mathbb{R} \times \mathbb{R}^N$ there exists $L = L(K) > 0$ such that

$$|f(x, s, \xi) - f(y, s, \xi)| \leq L|x - y|$$

for all $(x, s, \xi), (y, s, \xi) \in K$. The main tool in the proof is an approximation result due to De Giorgi (see Theorem 4.79 in [FL07]), which provides an *explicit formula* for the affine functions approximating a given convex function $g$. A major advantage of this characterization is that additional regularity hypotheses on the integrand $f$ are reflected immediately on the approximating affine functions.

Finally, concerning the extension to the vectorial case, that is, to integrands $f : \Omega \times \mathbb{R}^d \times \mathbb{R}^{d \times N} \to [0, \infty)$, Theorem 1 was extended to the case $d > 1$ by Morrey in his book on calculus of variations [M08]. However, several years later Eisen [E78] found a gap in Morrey's proof, thus placing in doubt the validity of Theorem 1 when $d > 1$, and constructed a counterexample to Theorem 1(iii) when $d > 1$.

It took twenty more years to finally solve the question. Indeed, in [FL00], Fonseca and the author proved that Theorem 1(i) continues to hold when $d > 1$ (see also [CM01]), while in [CM04] Černý and Malý constructed a counterexample to Theorem 1(ii) when $d > 1$.

Theorem 1 and its extensions are useful when one is working in the Sobolev setting. However, in minimization problems for integrands with linear growth such as $f(\xi) = \sqrt{1 + |\xi|^2}$, the natural setting is the space $BV(\Omega)$ of functions of bounded variation. In this case, in general, one expects $\mathcal{I}(u, \Omega) > I(u, \Omega)$. Thus, a natural question is to be able to find a representation of the functional $\mathcal{I}$. This problem was first solved by Goffman and Serrin in [iii] in the special case in which $f$ depends only on the gradient variable, that is, $f = f(\xi)$. Indeed, they were able to prove the following *integral representation* result:

**Theorem 2.** *Assume that* $\Omega$ *is bounded and that* $f : \mathbb{R}^N \to [0, \infty)$ *is convex. Then for every* $u \in BV(\Omega)$,

$$\mathcal{I}(u, \Omega) = \int_\Omega f\left(\frac{dDu}{d\mathcal{L}^N}(x)\right) dx + \int_\Omega f^\infty\left(\frac{dD^s u}{d|D^s u|}(x)\right) d|D^s u|.$$

Here $Du$ is the distributional derivative of $u$, which is a vector-valued Radon measure, and according to Lebesgue decomposition and the Radon–Nikodym theorems, for every Borel set $B \subset \Omega$,

$$(Du)(B) = \int_B \frac{dDu}{d\mathcal{L}^N}(x)\, dx + (D^s u)(B),$$

where the integrable function $\frac{dDu}{d\mathcal{L}^N}$ is the Radon–Nikodym derivative of $Du$ with respect the Lebesgue measure $\mathcal{L}^N$ restricted to $\Omega$, and $D^s u$ and $\mathcal{L}^N$ are mutually singular. Finally $f^\infty(\xi) := \lim_{r\to 0} rf(\xi/r)$ is the recession function of $f$.

This pioneering theorem has been extended and generalized in several directions in the past fifty years in order to treat integrands $f = f(x, s, \xi)$ as well as the vectorial setting $d > 1$. We refer to Section 5.5 in [AFP00] and to the references contained therein for a more complete bibliography.

**Acknowledgment**

While this book of selected papers will attest to the enormous impact that James continues to have on mathematics, from personal experience I can say that he also had an equally important influence on the life of his students. I consider myself blessed and I am forever grateful for having him as Ph.D. advisor and for everything he taught me.

## References

[AFP00] L. Ambrosio, N. Fusco, and D. Pallara, *Functions of bounded variation and free discontinuity problems.* Oxford Mathematical Monographs. The Clarendon Press, Oxford University Press, New York, 2000.

[CM01] R. Černý and J. Malý, *Counterexample to lower semicontinuity in calculus of variations.* Math. Z. **238** (2001), 689–694.

[CM04] R. Černý and J. Malý, *Another counterexample to lower semicontinuity in calculus of variations.* J. Convex Anal. **9** (2002), 295–299.

[D08] B. Dacorogna, *Direct methods in the calculus of variations.* Second edition. Applied Mathematical Sciences, 78. Springer, New York, 2008.

[DM93] G. Dal Maso, *An introduction to Γ-convergence.* Progress in Nonlinear Differential Equations and their Applications, 8. Birkhäuser Boston, Inc., Boston, MA, 1993.

[E78] G. Eisen. *A counterexample for some lower semicontinuity results.* Math. Z. **162** (1978), 241–243.

[FL00] I. Fonseca and G. Leoni, *Some remarks on lower semicontinuity.* Indiana Univ. Math. J. **49** (2000), 617–635.

[FL07] I. Fonseca and G. Leoni, *Modern methods in the calculus of variations: $L^p$ spaces.* Springer Monographs in Mathematics. Springer, New York, 2007.

[FGM06] N. Fusco, M. Gori, and F. Maggi, *A remark on Serrin's theorem.* NoDEA Nonlinear Differential Equations Appl. **13** (2006), 425–433.

[GM04] M. Gori and F. Maggi, *The common root of the geometric conditions in Serrin's lower semicontinuity theorem.* Ann. Mat. Pura Appl. (4) **184** (2005), 95–114.

[GM02] M. Gori and P. Marcellini, *An extension of the Serrin's lower semicontinuity theorem.* J. Convex Anal. **9** (2002), 475–502.

[M03] F. Maggi, *On the relaxation on BV of certain non coercive integral functionals.* J. Convex Anal. 10 (2003), 477–489.

[M08] C.B. Morrey, *Multiple integrals in the calculus of variations.* Reprint of the 1966 edition. Classics in Mathematics. Springer-Verlag, Berlin, 2008.

[S59] J. Serrin, *A new definition of the integral for nonparametric problems in the calculus of variations.* Acta Math. 102 (1959), 23–3.

Giovanni Leoni
Department of Mathematical Sciences, Carnegie Mellon University
e-mail: giovanni@andrew.cmu.edu

*Acta Math.* 102 (1959), 1–22.




# ON A FUNDAMENTAL THEOREM OF THE CALCULUS OF VARIATIONS

BY

**JAMES SERRIN**

*University of Minnesota*

Consider the simplest kind of multiple integral variational problem, that of minimizing an integral $I[u]$ of the form

$$I[u]=\int_R F(x, y, u, p, q)\, dA, \tag{1}$$

where the admissible functions $u=u(x, y)$ are continuously differentiable in $R$ and take on given continuous values on the boundary of $R$. Here $R$ denotes an open bounded region of the plane and $dA = dx\, dy$.

It has been known for more than one hundred and fifty years that a *twice differentiable* minimizing function must satisfy the Euler-Lagrange equation

$$\frac{\partial}{\partial x}(F_p)+\frac{\partial}{\partial y}(F_q)=F_u. \tag{2}$$

A function $u(x, y)$ which is continuous in some region $A$ and satisfies (2) at all interior points of $A$ is called an *extremal*. Let us denote by $\mathcal{E}$ an extremal defined over the closure of $R$ which continuously takes on the given boundary values. A major task of the calculus of variations is to give conditions under which $\mathcal{E}$ will minimize $I[u]$. This problem is not completely settled even today, in spite of the efforts of many investigators. We shall consider here one aspect of this problem, having its genesis in the classical field theory of Weierstrass and Hilbert. The latter, in his famous paper "Mathematische Probleme" and in a later paper "Zur Variationsrechnung", proved essentially the following theorem.

*If the extremal $\mathcal{E}$ can be imbedded in a field $\mathfrak{F}$ with slope functions $\mathfrak{p}(x, y, z)$, $\mathfrak{q}(x, y, z)$, such that*

1 – 593804. *Acta mathematica*. 102. Imprimé le 14 septembre 1959

$$E(x, y, z, \mathfrak{p}, \mathfrak{q}; P, Q) \geqslant 0$$

*for each set of values* $(x, y, z)$ *in the field, then* $\mathcal{E}$ *minimizes the integral* $I[u]$ *relative to all admissible functions whose values lie in the field.*(1)

The kind of imbedding used in Hilbert's theorem requires the entire extremal surface corresponding to $\mathcal{E}$ to lie in the field $\mathfrak{V}$; as a consequence $\mathcal{E}$ must be continuously differentiable in the *closure* of $R$. Thus Hilbert's theorem, in spite of its elegance, is significantly restricted in its usefulness and range of application. Indeed, it is a commonplace today that the solution of a boundary value problem in partial differential equations need not be differentiable up to the boundary, even when the boundary values themselves are differentiable. On the other hand, if one considers extremals $\mathcal{E}$ which are not differentiable n the closure of $R$, then the conclusion of Hilbert's theorem comes in doubt, since $\mathcal{E}$ in this case need not give $I[u]$ a finite value.(2) There the matter rests at present, for other authors who have studied the problem from a more or less related point of view have always placed some kind of restriction on the behavior of $\mathcal{E}$ at the boundary (see, for example, the papers of Lichtenstein, Miranda, Karush, and Hestenes listed in the references).

The main purpose of this paper is to give a reformulation of Hilbert's theorem which avoids both the difficulties mentioned above. In the new formulation (Theorem 4) it is assumed that there is at least one admissible function which gives $I$ a finite value, and the field is required to be an extremal field. On the other hand, by virtue of an extension of the notion of imbedding, it is not necessary to make any assumption concerning the behavior of $\mathcal{E}$ at the boundary, beyond simply requiring that it take on the given boundary values. The proof is a modification and extension of Hilbert's classical argument: at no stage is it necessary to resort to deep theorems of integration. Finally the proof applies

(1) The first complete statement of this theorem seems to be due to Bolza ([2], 683); see also [1], § 11. For completeness, we add the definition

$$E(x, y, u, p, q; P, Q) = F(P, Q) - F(p, q) - (P - p)F_p(p, q) - (Q - q)F_q(p, q),$$

in which the arguments $(x, y, u)$ of $F$ have been uniformly suppressed.

(2) This objection (usually attributed to Hadamard) was first pointed out in 1871 by F. Prym. Prym's example is so elegant that I cannot resist reproducting it here: Let $R$ be the circle $|3z - 1| < 1$, and consider the function

$$u = \operatorname{Im}\{\sqrt{\log z}\}, \quad z = x + iy.$$

By direct calculation one finds that the Dirichlet integral of $u$ over $R$ is divergent, while, on the other hand, $u$ is continuous in the closure of $R$. In other words, $u$ is an extremal for a regular variational problem, and at the same time $I[u] = \infty$.

equally in $n$ dimensions and to extremals satisfying only the Haar equations. Further results of the paper are discussed in sections 1 and 2.

The paper is divided into three parts. The first part contains preparatory material and a discussion of the special integrand $F(x, y, p, q)$, the second part treats the full integrand $F(x, y, u, p, q)$, and the final part is devoted to certain subsidiary matters.

## Part I

**1. Preliminaries:** The purpose of this section is to define precisely the class of integrands $F(x, y, u, p, q)$ which we shall treat, and the class of functions which will be admitted to competition in minimizing the integral $I$.

We assume that $F$ is defined and continuous for all values of $u, p, q$ and for all $(x, y) \in R$. Furthermore, we suppose that the partial derivatives $F_p$, $F_q$, and $F_u$ exist and are continuous. Finally, *in all of our results it is supposed that*

$$F(x, y, u, p, q) \geqslant 0, \tag{3}$$

*and that the Weierstrass function is non-negative,*

$$E(x, y, u, p, q; P, Q) \geqslant 0, \quad (p, q) \neq (P, Q). \tag{4}$$

It would be possible to lighten these assumptions somewhat in certain cases, but we leave such refinements to the reader. By a simple change of integrand, condition (3) can be attained for any integrand $F$ which is bounded below by a fixed integrable function of $x$, $y$. Additional requirements will occasionally be placed on $F$, but these will be stated in the hypotheses of the individual theorems.

A real-valued function $v = v(x, y)$ defined on the closure of $R$ will be said to be in the class $\mathfrak{A}$ of admissible functions if and only if it satisfies the following three conditions:

$\mathfrak{A}$ 1. $v$ continuously takes on assigned (continuous) values on the boundary of $R$.

$\mathfrak{A}$ 2. $v$ is continuously differentiable in $R$. The integral $I[v]$ is understood to be the improper Riemann integral obtained by exhaustion of the region $R$.

$\mathfrak{A}$ 3. $I[v]$ is finite.

Various generalizations of $\mathfrak{A}$ 2 can be treated, the simplest being piecewise continuously differentiable functions. The class of admissible functions secured from this particular extension of $\mathfrak{A}$ 2 is not especially interesting, and accordingly we shall set it aside without further discussion. More important is the class $\mathfrak{A}^*$ of admissible functions obtained by replacing $\mathfrak{A}$ 2 by the condition

$\mathfrak{A}$ 2*. $v$ is continuous in $R$. The integral $I[v]$ is to be understood in the following generalized sense: Let $\{v_n\}$ be a sequence of continuously differentiable functions converging uniformly to $v$ in any proper subregion of $R$. Then

$$I[v] = \text{g.l.b.} \liminf_{n\to\infty} I[v_n], \tag{5}$$

where the g.l.b. is taken over all sequences $\{v_n\}$ converging to $v$ as above. It is tacitly assumed that each function $v_n$ is defined in some subregion $R_n$ of $R$, and $I[v_n] = I[v_n, R_n]$.

The generalized integral defined by (5) is somewhat analogous to the generalized area in Lebesgue's theory of surface area, and indeed can be shown equivalent to the latter when $F = \sqrt{1+p^2+q^2}$. In order for (5) to be meaningful and logically consistent, it must still be proved that if $v$ is continuously differentiable in $R$, then $I[v]$, as defined by (5), has the natural value. This we shall do in the following paper in this journal. An important subclass $\mathfrak{L}$ of the admissible functions $\mathfrak{A}^*$ is obtained by restricting $v$ to be Lipschitz continuous in any closed subregion of $R$.

With the preceding definitions in mind, we can now state the problem which will concern us throughout the paper.

*Variational Problem I.* Among all functions satisfying the boundary condition $\mathfrak{A}$ 1 and for which the integral $I$ has a meaning, to determine a function $u$ such that $I[u] = \text{Minimum}$.

Let $u = u(x, y)$ be a continuous function defined in some region $A$. We shall say that $u$ is an *extremal* if and only if

1. $u$ is continuously differentiable in the interior of $A$.
2. For every piecewise smooth function $\zeta$ which vanishes on and near the boundary of $A$, we have

$$\int_A (\zeta_x F_p + \zeta_y F_q + \zeta F_u)\, dA = 0. \tag{6}$$

(A function is said to be *piecewise smooth* in a region $A$ if it is continuous in $A$, and has bounded continuous first derivatives except on a finite number of smooth arcs and at a finite number of isolated points.)

This definition of an extremal, which is equivalent to the Haar equations, will be used henceforth in the paper; it is obviously satisfied by any solution of the Euler-Lagrange equation (2). To simplify the wording of later theorems, we shall say that a function $u$ is an *extremal for the variational problem I* if $u$ is an extremal defined over the closure of $R$ which continuously takes on the given boundary values.

*Remark.* The reader who prefers to deal with integrands and extremals which are of class $C^2$, and who is content to consider only the class $\mathfrak{A}$ of admissible functions, will find the proofs immediately applicable to his case; moreover, he may omit reading sections 4 and 5, the lemma in section 9, the last portion of section 10, and all of Part III.

**2. The special integrand $F(x, y, p, q)$.** In sections 2 through 5 we shall consider integrands which are independent of $u$. Recalling the definitions and terminology of the preceding sections, we can state our first result.

THEOREM 1. *Let $F$ be independent of $u$, and suppose $u = u(x, y)$ is an extremal for the variational problem $I$. Let there exist at least one function $v$ in the class $\mathfrak{A}$ (or $\mathfrak{A}^*$) of admissible functions. Then $I[u]$ exists and*

$$I[u] \leqslant I[v] \tag{7}$$

*for each $v$ in $\mathfrak{A}$ (or $\mathfrak{A}^*$).*

*If the equality sign in (4) is excluded, then $I[u] < I[v]$ for each $v$ different from $u$.*

This result is a corollary of Theorem 4 below. It has an independent proof, however, which is enough different to justify being given separately (section 3). This proof applies equally well in any number of dimensions.

Two special cases of Theorem 1 deserve notice. First, if $F_p(u)$ and $F_q(u)$ are bounded,(1) and if the boundary of $R$ is rectifiable, then (7) follows in a well-known way simply by integration by parts. The merit of Theorem 1 lies precisely in eliminating these hypotheses.

Secondly, for the integrand $F = p^2 + q^2$ (the Dirichlet integral) a result equivalent to Theorem 1 was given by Lebesgue in 1913. It is interesting to examine the connection between this result and what is usually known as Dirichlet's principle. The latter may be interpreted as saying that, under certain circumstances, the Dirichlet integral may be minimized and the minimizing function solves the boundary value (Dirichlet) problem for the Laplace equation. Sufficient conditions for the existence of a minimum were first given by Hilbert, using direct methods of the calculus of variations, and the problem from this point of view has been extensively treated by Courant in a well-known monograph. On the other hand, if the result of Lebesgue is combined with the known fact that a minimizing function must necessarily be harmonic, then we obtain the following result: *a minimizing function exists if and only if* (*i*) *there exists at least one admissible function and* (*ii*) *the Dirichlet problem is solvable for the given boundary values.* This elegant formulation of the Dirichlet principle seems to be due originally to Kamke & Lorentz.

The preceding example has been discussed in some detail because of its close bearing on the problems considered here. Indeed, Theorem 1 shows that the "if" part of the italicized statement holds also for the variational problem $I$. In section 4 we shall prove the "only if"

(1) Boundedness is assured if $u$ has bounded derivatives in $R$ (this is the case usually found in the literature), or it may be an inherent property of the integrand, as in the case $F = \sqrt{1 + p^2 + q^2}$.

part for a large class of integrands. In combination, these results can appropriately be called a generalized Dirichlet principle.

The theorems of section 4 can also be interpreted as conditions for a minimizing function to be of class $C^2$. In particular, Theorems 3.1 and 3.2 are special cases of known results of Morrey. They are included here because they apply to a fairly wide range of integrands and can be proved in a matter of a few lines.

Before proceeding to the proof, I wish to thank Professor Johannes Nitsche for several stimulating and helpful conversations concerning various aspects of the paper.

**3. Proof of Theorem 1.** We recall that the classical "proof" by integration by parts fails because a certain boundary integral may not have meaning. To overcome this difficulty we propose to truncate a given admissible function $v$ by means of surfaces $z = u(x, y) \pm \varepsilon$, and to show that this process actually reduces the integral $I[v]$. It will then follow that $I[u] \leqslant I[v]$ by letting $\varepsilon \to 0$. This process must be handled in a somewhat roundabout fashion so as to make it perfectly rigorous, which accounts for the method finally adopted. (A somewhat similar idea occurs in the note of Lebesgue cited above.)

Now let $v(x, y)$ be any function in $\mathfrak{A}$, or in $\mathfrak{A}^*$. Let $S$ be a fixed closed subregion of $R$ with smooth boundary $\Sigma$, and let $\varepsilon$ be an arbitrary positive number. Since $u$ and $v$ take the same boundary values, we can find a closed region $B$ such that $S \subset B \subset R$, and

$$|u - v| < \varepsilon/3 \quad \text{in } R - B.$$

Furthermore, by virtue of condition $\mathfrak{A}$ 2 or $\mathfrak{A}$ 2* it is clear that there exists a polynomial $w$ such that

$$|v - w| < \varepsilon/3 \quad \text{in } B,$$

and

$$I[w, B] \leqslant I[v] + \varepsilon. \tag{8}$$

Finally, there exists a sequence of polynomials $u_n$ with the property

$$|u - u_n|, \; |\nabla u - \nabla u_n| \to 0 \text{ uniformly in } B,$$

where $\nabla$ denotes the vector gradient. Without loss of generality one can assume that $|u - u_n| < \varepsilon/3$ for all $n$.(1)

We now define the function

(1) By careful use of the theory of measure, one can omit this preliminary approximation. Our preference, however, is for the course of proof actually adopted.

$$\phi = \begin{cases} u_n + \varepsilon & \text{if } w > u_n + \varepsilon \\ u_n - \varepsilon & \text{if } w < u_n - \varepsilon \\ w & \text{otherwise.} \end{cases}$$

Obviously $\phi$ is piecewise differentiable in B, and can be substituted in the integral $I$. Then, in virtue of the definition of the Weierstrass $E$ function and the fact that $F$ is independent of $u$, we have

$$I[w, B] - I[\phi, B] = \int_B [\zeta_x F_p + \zeta_y F_q]_\phi \, dA + K, \tag{9}$$

where $$\zeta = w - \phi, \quad K = \int_B E(x, y, \phi_x, \phi_y; w_x, w_y) \, dA,$$

and the subscript $\phi$ denotes evaluation of $F_p$ and $F_q$ for the arguments $(x, y, \phi_x, \phi_y)$. Now at all points of $B$ at which $\zeta \neq 0$ we have $\phi = u_n \pm \varepsilon$, hence

$$\int_B [\zeta_x F_p + \zeta_y F_q]_\phi \, dA = \int_B [\zeta_x F_p + \zeta_y F_q]_{u_n} \, dA$$

$$= \int_B [\zeta_x F_p + \zeta_y F_q]_u \, dA + \varepsilon_1,$$

where $\varepsilon_1 \to 0$ as $n \to \infty$. In virtue of the preceding construction, $\zeta = 0$ on and near the boundary of $B$, so that by hypothesis the last integral vanishes. Combining this result with (9) yields the formula

$$I[\phi, B] = I[w, B] - K - \varepsilon_1. \tag{10}$$

Turning next to an estimate for the integral $I[u, S]$, we have in the same way as (9),

$$I[\phi, S] - I[u, S] = \int_S [\eta_x F_p + \eta_y F_q]_u \, dA + K_1, \tag{11}$$

where $\eta = \phi - u$ and $K_1 \geqslant 0$.

Using condition (6), the integral appearing on the right in (11) is easily shown to equal

$$\oint_\Sigma \eta \, (F_p \, dy - F_q \, dx).$$

But $|\eta| \leqslant |\phi - u_n| + |u_n - u| \leqslant \frac{4}{3} \varepsilon$, whence from (11),

$$I[u, S] \leqslant I[\phi, S] + \frac{4}{3} \varepsilon \oint_\Sigma |F_p \, dy - F_q \, dx|.$$

Noting that $I[\phi, S] \leqslant I[\phi, B]$, and using (10) and (8) to estimate this last integral, we obtain

$$I[u, S] \leqslant I[v] - K - \varepsilon_1 + \text{Const.}\, \varepsilon.$$

In this inequality we may let $n \to \infty$; setting $\Lambda = \lim_{n\to\infty} K$ there results the main estimate

$$I[u, S] \leqslant I[v] - \Lambda + \text{Const.}\, \varepsilon. \tag{12}$$

Since $\Lambda$ is certainly $\geqslant 0$, we conclude that $I[u, S] \leqslant I[v]$, and (7) follows at once.

In order to prove the remainder of Theorem 1 one can proceed as in sections 9 and 10 below. We shall omit the details, however, and merely accept the final statement of Theorem 1 as a corollary of Theorem 4.

*Remark.* The preceding theorem applies also to certain integrands in which the unknown function appears explicitly. Consider in particular the integrand

$$F(x, y, p, q) + G(x, y, u),$$

in which $G$ and its first partial derivative $G_u$ are continuous functions defined over the *closure* of $R$. *We suppose that $G_u$ is a non-decreasing function of $u$.* Then by carrying out the steps in the preceding proof one finds in place of (10),

$$I[\phi, B] = I[w, B] - K' - \varepsilon_1 - \varepsilon_2, \tag{10$'$}$$

where $\varepsilon_2$ tends to zero with $\varepsilon$, and $K'$ has a slightly different meaning than $K$ but still remains $\geqslant 0$. The main estimate (12) continues to hold, except that the error term Const. $\varepsilon$ must be replaced by a quantity which merely tends to zero as $\varepsilon \to 0$, $S \to R$.

Similar remarks hold for the integrand

$$F(x, y, p, q) + L(x, y, u)p + M(x, y, u)q + N(x, y, u),$$

in which $L$, $M$, $N$, and their partial derivatives $L_x$, $M_y$, and $N_u$ are continuous functions defined in the closure of $R$. The condition on $G_u$ in the preceding example is here replaced by the condition that $N_u - L_x - M_y$ should be a non-decreasing function of $u$. [The reason for this requirement may be seen from the identity

$$Lp + Mq = \left(\int L\,du\right)_x + \left(\int M\,du\right)_y - \int (L_x + M_y)\,du,$$

from which it follows that the function $N - \int (L_x + N_y)\,du$ plays the same role as $G$ did before.] Assuming finally that the total integrand $F + Lp + Mq + N$ is bounded from below, one can again carry out the necessary steps in the proof of Theorem 1. In case the boundary of $R$ is rectifiable we can dispense with the boundedness condition, for in this case the divergence terms above contribute only a fixed boundary integral.

**4. Differentiability of minimizing functions.** By using Theorem 1 one can obtain several simple conditions guaranteeing the differentiability of a minimizing function (even should the minimum be known only in the general class $\mathfrak{A}^*$). For simplicity we shall consider in this respect only integrands $F(x, y, p, q)$ which are of class $C^2$. Corresponding results for more general integrands can, however, easily be formulated.

Let $\mathfrak{K}$ denote a covering of $R$ by open circular disks. We shall say that an integrand $F(x, y, p, q)$ is *tame* if it satisfies conditions (3) and (4), the latter with equality excluded, and if the Dirichlet problem for the Euler-Lagrange equation

$$F_{pp}r + 2F_{pq}s + F_{qq}t + F_{px} + F_{qy} = 0 \tag{13}$$

is solvable for all circles of some covering $\mathfrak{K}$.(1) The functions $p^2 + q^2$ and $\sqrt{1 + p^2 + q^2}$ are simple examples of tame integrands; in Theorems 3.1 and 3.2 we shall note some others.

THEOREM 2. *Let the integrand $F(x, y, p, q)$ be tame, and suppose $u = u(x, y)$ minimizes the integral $I$ among all functions of class $\mathfrak{A}$, or $\mathfrak{L}$, or $\mathfrak{A}^*$. Then $u$ is of class $C^2$ in $R$.*

*Proof.* Let $P$ be any point of $R$, and let $K$ be a circle about $P$ in which the Dirichlet problem is solvable. Suppose $U$ is the solution of (13) which agrees with $u$ on the boundary of $K$. We assert that $U \equiv u$ in $K$. For if not, then we shall be able to define a new admissible function $u^*$ such that $I[u^*] < I[u]$, contradicting the fact that $u$ is a minimizing function. Granting this step for a moment, from the equality $U \equiv u$ it follows that $u$ is of class $C^2$ in the neighborhood of $P$, and the theorem is proved.

It remains therefore to construct the function $u^*$. This will take a slightly different form in each of the three cases of the theorem. Suppose first that $u$ minimizes in the class $\mathfrak{L}$. We define

$$u^* = \begin{cases} U + \delta & \text{if } u > U + \delta \\ U - \delta & \text{if } u < U - \delta \\ u & \text{otherwise.} \end{cases}$$

Then $u^*$ is actually different from $u$, provided that $\delta$ is chosen small enough. Clearly $u^* \in \mathfrak{L}$; moreover, by Theorem 1

(1) There may be some ambiguity in this definition. To be precise, in saying that the Dirichlet problem is solvable for a region $K$ we mean that, given any continuous data on the boundary of $K$, there exists a corresponding solution of (13), of class $C^2$ in the interior of $K$, and continuously taking on the given values on the boundary of $K$.

$$I[u] = I[u, R'] + I[u, R - R'] > I[U \pm \delta, R'] + I[u, R - R'] = I[u^*], \tag{14}$$

where $R'$ denotes the (open) set where $u^* \neq u$. In case $u$ minimizes in the class $\mathfrak{A}$, the function $u^*$ defined above is only piecewise smooth, and therefore not admissible to competition. This difficulty is avoided by smoothing $u^*$ at its ridges, taking care not to destroy the inequality (14).

Finally, if $u$ minimizes in the class $\mathfrak{A}^*$, the preceding argument fails since the two equalities in (14) may no longer hold. In this case we set

$$u^* = \begin{cases} U & \text{in } K \\ u & \text{in } R - K. \end{cases}$$

Using a modified version of the proof of Theorem 1, in which the function $u$ is truncated by means of the surfaces $z = U \pm \varepsilon$, it is easily shown that $u^* \in \mathfrak{A}^*$ and $I[u^*] < I[u]$. This completes the proof of Theorem 2.

While the above result is valid in $n$-dimensional space, the following is so far known to hold *only in the plane*. To extend it to higher dimensions will require a significant advance in the theory of partial differential equations (cf. [22], [5] and [19]).

THEOREM 3.1. *Let the integrand $F(x, y, p, q)$ satisfy the following conditions*

$$k(\xi^2 + \eta^2) \leqslant F_{pp}\xi^2 + 2F_{pq}\xi\eta + F_{qq}\eta^2 \leqslant K(\xi^2 + \eta^2)$$
$$F_{px}^2 + F_{qy}^2 \leqslant K(1 + p^2 + q^2),$$

*where $k$ and $K$ are positive continuous functions of $x, y$. Suppose also that the five partial derivatives of $F$ which appear in the above conditions are Hölder continuous. Let $u = u(x, y)$ minimize the integral $I$ among all functions of class $\mathfrak{A}^*$. Then $u$ is of class $C^2$ in $R$.*

*Proof.* It is clear that $F$ is bounded below by some fixed function of $x, y$, whence by a simple change of integrand we can suppose without loss of generality that $F \geqslant 0$. Also $F_{pp}F_{qq} - F_{pq}^2 > 0$, so that by Taylor's formula $E > 0$ for $(P, Q) \neq (p, q)$. Theorem 3.1 will therefore follow from Theorem 2 if the Dirichlet problem is solvable for equation (13). But the conditions of the theorem are sufficient for the solvability of the Dirichlet problem over circles in $R$ (and in fact over much more general regions) for arbitrary continuous boundary data.(1) This completes the proof.

THEOREM 3.2. *Let the integrand $F = F(p, q)$ be of class $C^3$ and satisfy the conditions*

$$F \geqslant 0, \quad F_{pp}F_{qq} - F_{pq}^2 > 0.$$

(1) This has been proved by P. C. Rosenbloom and the author.

*Let $u = u(x, y)$ minimize the integral $I$ among all functions of class $\mathfrak{L}$. Then $u$ is of class $C^2$ in $R$.*

*Proof.* The conditions placed on $F$ are not enough to conclude that $F$ is tame according to the definition given at the opening of this section. On the other hand, from the proof of Theorem 2 it is apparent that we actually need to solve the Dirichlet problem only for Lipschitzian boundary values. Now the Euler-Lagrange equation takes the form

$$ar + 2bs + ct = 0, \tag{15}$$

where $ac - b^2 > 0$ and $a, b, c$ are Hölder continuous functions of $p$ and $q$. Under these conditions it can be shown(1) that (15) is solvable for convex regions with Lipschitzian boundary data, and the proof is complete.

Theorem 3.1 and 3.2 are contained in known results of Morrey, but the simplicity of their proofs (granted certain existence theorems for quasi-linear partial differential equations) justifies their inclusion here.

**5. Example: the area integral.** Let us set $W = \sqrt{1 + p^2 + q^2}$, and consider the (generalized) integral

$$A[u] = \int_R W\, dA.$$

It is clear that $A[u]$ is exactly the Lebesgue area of the surface $z = u(x, y)$ over $R$. Moreover, since $W \geqslant 0$, $W_{pp} W_{qq} - W_{pq}^2 > 0$, the conclusions of Theorems 1 and 2 apply unaltered in the present case.

Consider, then, the problem of spanning a curve with a surface of least area, area being understood in the sense of Lebesgue. From Theorem 1, *a minimal surface of the form $z = u(x, y)$ which spans a given curve $\Gamma$ has smaller area than any other surface of the same form which spans $\Gamma$*. From Theorem 2, since $W$ is tame (see [21], 101), *any surface of least area which has the form $z = u(x, y)$ is a minimal surface*, and, more generally, *any surface of least area which locally can be represented in the form $z = u(x, y)$, for some orientation of coordinates, is a minimal surface.* It happens that these results are not new, but have already been proved by E. J. McShane (except that in his work $\Gamma$ was required to have a convex projection). On the other hand, it seems that the present proof is both simpler and more general in application than that of McShane.

## PART II

**6. Fields and imbedding.** Before stating our main theorem, it is first necessary to define precisely the nature of the fields which we shall consider.

Consider a family of surfaces, each of which can be represented in the form $z = f(x, y)$.

A region $\mathfrak{V}$ in $(x, y, z)$ space is said to be *smoothly covered* by such a family if (i) through each point of $\mathfrak{V}$ there passes exactly one surface of the family, (ii) each surface is continuously differentiable, and (iii) the slope functions $\mathfrak{p}\ (x, y, z)$ and $\mathfrak{q}\ (x, y, z)$ of the family are continuous in $\mathfrak{V}$ and have continuous first partial derivatives $\mathfrak{p}_z$ and $\mathfrak{q}_z$. We shall say that $\mathfrak{V}$ is $z$-simple if its intersection with any straight line $x = x_0$, $y = y_0$ is a connected set. The principal definitions of field and imbedding can now be stated.

*Definition* 1. A field of extremals is a $z$-simple region $\mathfrak{V}$ of $(x, y, z)$ space together with a family $S$ of extremal surfaces which smoothly covers $\mathfrak{V}$.

*Definition* 2. Suppose that $u = u(x, y)$ is an extremal, and let $D$ denote the interior of its domain of definition. Also let $\mathcal{E}$ denote the extremal surface $z = u(x, y)$ for $(x, y) \in D$. We shall say that $u$ is *weakly imbedded* in a field of extremals $(\mathfrak{V}, S)$ if and only if

1. $\mathcal{E}$ is a member (or a portion of a member) of the family $S$, and
2. If $\mathcal{E}_1$ is any other surface in $S$, then the vertical distance from $\mathcal{E}$ to $\mathcal{E}_1$ is positive.

It is important to observe that the concept of weak imbedding requires no differentiability of $u$ at the boundary of its domain of definition. This is in contrast with the usual situation (Bolza, Bliss) in which $u$ must be continuously differentiable over its entire domain of definition.

**7. Some properties of the Hilbert integral.** Let $S$ be a closed region in the plane with piecewise smooth boundary $\Sigma$. Let $U$ be a piecewise smooth function in $S$, whose values lie in the region $\mathfrak{V}$ of a field of extremals.(¹) Then the Hilbert invariant integral $I^*[U, S]$ is defined by

$$I^*[U, S] = \int_S \{F + (P - \mathfrak{p})\, F_p + (Q - \mathfrak{q})\, F_q\}\, dA,$$

where the arguments of $F$, $F_p$, and $F_q$ are $(x, y, U, \mathfrak{p}, \mathfrak{q})$ and

$$\mathfrak{p} = \mathfrak{p}(x, y, U), \quad \mathfrak{q} = \mathfrak{q}(x, y, U).$$

If $V$ is another function satisfying the same conditions as $U$, then we have the familiar result

$$I^*[U, S] = I^*[V, S] \tag{16}$$

whenever $U = V$ on $\Sigma$. Formula (16) is not difficult to prove when the surfaces of the extremal family $S$ and the integrand $F$ are both of class $C^2$ (cf. [2] and [9]). The general case under consideration here requires a more sophisticated argument which we shall give in § 11. If the condition $U = V$ on $\Sigma$ is not satisfied, then (16) must be replaced by

(¹) By this we mean simply that the points $(x, y, U)$ for $(x, y) \in S$ lie in $\mathfrak{V}$.

$$I^*[U, S] = I^*[V, S] + \oint_{\Sigma} (U - V)(\tilde{F}_p\, dy - \tilde{F}_q\, dx), \tag{17}$$

where the tildes denote evaluation at an intermediate value between $U$ and $V$. Formula (17) is proved in exactly the same way as (16), except that a certain boundary integral which vanishes in the derivation of the former result must now be carried along. Finally, let us note the obvious equality

$$I^*[u, S] = I[u, S] \tag{18}$$

when $u$ is an extremal of the field $(\mathfrak{V}, S)$.

**8. The main theorem.** With the terminology and results of the preceding two sections understood, the main result of the paper may now be stated.

THEOREM 4. *Let $u = u(x, y)$ be an extremal for the variational problem $I$. Suppose that $u$ can be weakly imbedded in an extremal field $(\mathfrak{V}, S)$, and that there exists at least one function $v \in \mathfrak{A}^*$ such that the values of $v$ over $R$ lie in $\mathfrak{V}$. Then $I[u]$ exists, and*

$$I[u] \leqslant I[v] \tag{19}$$

*for each $v \in \mathfrak{A}^*$ whose values over $R$ lie in $\mathfrak{V}$.*

*If the equality sign in* (4) *is excluded, then $I[u] < I[v]$ for each $v$ different from $u$.*

*Proof.* For simplicity we shall assume that the boundary of $\mathfrak{V}$ contains no points of the extremal surface

$$z = u(x, y), \quad (x, y) \in R.$$

It will be clear from the proof that this is no essential restriction.

Now let $v$ be any function in $\mathfrak{A}^*$ whose values over $R$ lie in $\mathfrak{V}$, let $S$ be a fixed closed subregion of $R$ with smooth boundary $\Sigma$, and let $\varepsilon$ be an arbitrary positive number. In virtue of the assumption made in the preceding paragraph, we can find extremals $u'$ and $u''$ in $S$ such that

$$u - \varepsilon < u'' < u < u' < u + \varepsilon \quad \text{in } S.$$

Moreover, by condition 2 in the definition of weak imbedding there exists a number $\delta > 0$ such that

$$|u - u'|, \ |u - u''| > \delta.$$

Next, let $B$ be a closed region such that $S \subset B \subset R$ and

$$|u - v| < \delta/3 \quad \text{in } R - B.$$

By hypothesis $\mathfrak{A}\ 2^*$ one can find a polynomial $w$ with the properties

$$|v-w|<\delta/3 \quad \text{in } B$$

and
$$I[w,B] \leqslant I[v]+\varepsilon. \tag{20}$$

The following two paragraphs are devoted to proving the fundamental inequality (24). In order for the demonstration to be completely rigorous it is necessary to approximate both $u'$ and $u''$ by polynomials. Unfortunately, the idea of the proof is somewhat obscured by this procedure, so we prefer instead to work formally, leaving the details of the approximation process to a later footnote.

We turn now to the computational aspect of the proof. Let the function $\phi$ be defined by

$$\phi=\begin{cases} u' & \text{if } w>u' \\ u'' & \text{if } w<u'' \\ w & \text{otherwise.}\end{cases}$$

Our first task is to find a formula for the quantity

$$\Lambda = I[w,B]-I[\phi,B].$$

To this end, let $C'$ denote the set of points in $B$ where $w>u'$ and $C''$ the set where $w<u''$. Setting $C=C'+C''$ we have

$$\begin{aligned}\Lambda &= I[w,C]-I[u',C']-I[u'',C''] \\ &= I[w,C]-I^*[u',C']-I^*[u'',C''] \\ &= I[w,C]-I^*[w,C].\end{aligned}$$

In the last two equalities we have used properties (18) and (16) of the invariant integral, and the fact that $w=u'$ on the boundary of $C'$, $w=u''$ on the boundary of $C''$. Using the Weierstrass function to represent the last expression yields finally

$$\Lambda=\int_C E(w)\,dA, \tag{21}$$

where $E(w)\equiv E(x,y,w,\mathfrak{p},\mathfrak{q};w_x,w_y)$.

We may now obtain the main estimate for the integral $I[u,S]$. By virtue of (18) and (17) we have

$$I[u,S]=I^*[u,S]=I^*[\phi,S]+\oint_\Sigma (u-\phi)(\tilde F_p\,dy-\tilde F_q\,dx). \tag{22}$$

Moreover, using respectively inequalities (4) and (3),

$$I^*[\phi,S]\leqslant I[\phi,S]\leqslant I[\phi,B]=I[w,B]-\Lambda \tag{23}$$

Thus, since $|u - \phi| \leqslant \varepsilon$ on $\Sigma$, it follows from (22), (23) and (20) that

$$I[u, S] \leqslant I[v] - \Lambda + \text{Const}\cdot\varepsilon, \qquad \Lambda = \int_C E(w)\,dA. \tag{24}$$

This result is fundamental for the further conclusions of the paper.([1])

To establish the first part of Theorem 4 we let $\varepsilon \to 0$ in (24). Using the fact that $\Lambda \geqslant 0$ there results simply

$$I[u, S] \leqslant I[v],$$

and (19) follows at once. The final statement of Theorem 4 requires a fairly elaborate treatment of its own, which will occupy us in the following two sections.

**9. Some inequalities.** We begin by deriving a more explicit form of the fundamental inequality (24). Let $D$ denote an arbitrary closed subregion of $R$ in which $v \neq u$. Choosing $S$ and $\varepsilon$ respectively so that $D \subset S$ and $|u - v| \geqslant 2\varepsilon$ in $D$, it is clear that the set $C$ defined in the preceding section contains $D$. Using the fact that $E \geqslant 0$ then allows us to write in place of (24),

$$I[v] - I[u, S] \geqslant \int_D E(w)\,dA - \text{Const.}\ \varepsilon. \tag{25}$$

Consider now a function $v \in \mathfrak{A}$. The polynomial $w$ can, under these circumstances, be chosen so that $|\nabla w - \nabla v| < \varepsilon$ in $B$. Then letting $\varepsilon \to 0$ in (25), and afterwards $S \to R$, yields

$$I[v] - I[u] \geqslant \int_D E(v)\,dA, \tag{26}$$

where $D$ is an arbitrary subregion of $R$ in which $v \neq u$.

---

([1]) To make the above argument rigorous, we approximate $u'$ and $u''$ by sequences $\{u_n'\}$ and $\{u_n''\}$ of polynomials such that

$$|u' - u_v'|,\ |\nabla u' - \nabla u_n'| \to 0 \quad \text{uniformly in } B,$$

with similar relations for $u_n''$. It may be supposed that $|u' - u_n'|$ and $|u'' - u_n''|$ are less than $\delta/3$ for all $n$.

In the definition of $\phi$ one then replaces $u'$ and $u''$ by their approximants; letting $\Lambda_n$ stand for the resulting difference $I[w, B] - I[\phi, B]$, formula (21) becomes

$$\Lambda_n = \int_{C_n} E(w)\,dA + \varepsilon_1, \tag{21$'$}$$

in which $\varepsilon_1 \to 0$ as $n \to \infty$. Finally, in place of (24) we have

$$I[u, S] \leqslant I[v] - \Lambda_n + \text{Const.}\ \varepsilon. \tag{24$'$}$$

Letting $n \to \infty$ in this formula gives exactly the present estimate (24).

The following lemma, of a kind originally due to E. E. Levi, will be instrumental in the sequel.

*Suppose* $E > 0$ *for all* $(P, Q) \neq (p, q)$. *Let* $\mathfrak{D}$ *denote a bounded set in the five-dimensional space* $(x, y, u, p, q)$. *Then there exists a convex function* $f(t)$ *such that* $f(0) = 0$, $f(t) > 0$ *if* $t > 0$, and

$$E(x, y, u, p, q; P, Q) \geqslant f(X), \quad X = \sqrt{(P-p)^2 + (Q-q)^2}, \tag{27}$$

*for all* $(x, y, u, p, q) \in \mathfrak{D}$.

*Proof*. For fixed $(x, y, u)$ consider the surface $F = F(x, y, u, P, Q)$ over the $P, Q$ plane. Let this surface (the figuratrix) be denoted by S, and let T denote the tangent plane to S at the fixed point $p, q$. Evidently the Weierstrass $E$-function expresses the vertical distance from T to S. Therefore, according to hypothesis, S lies everywhere above T, and since this holds for arbitrary $(p, q)$, S is seen to be a *strictly convex surface*.

This being shown, it is easy to conclude the existence of the required function $f(t)$. For completeness we shall indicate an explicit construction. Set

$$g(t) = \underset{D}{\text{g.l.b.}} \; \underset{X=t}{\text{Min}} \; E(x, y, u, p, q; P, Q).$$

Obviously $g(0) = 0$ and $g(t) > 0$ for $t > 0$; also, because the surface $S$ is convex one sees that

$$g(t) \geqslant g(1)t \quad \text{for } t \geqslant 1.$$

We may now choose for $f(t)$ the function defining the lower boundary of the convex hull of the graph of $g(t)$, $0 \leqslant t < \infty$. That $f(t)$ has the required properties is easily verified, and the lemma is proved.

**10. Completion of the proof.** It must still be shown that $I[u] < I[v]$, provided that $v$ is different from $u$.

For simplicity, consider first the case of a function $v \in \mathfrak{A}$, $v \not\equiv u$. Supposing that $I[u] = I[v]$, then by virtue of (26) and the fact that $E \geqslant 0$ we have

$$E(v) \equiv 0 \quad \text{for } (x, y) \in D. \tag{28}$$

Since (4) is assumed to hold with the equality excluded, this implies

$$v_x = \mathfrak{p}(x, y, v), \quad v_y = \mathfrak{q}(x, y, v).$$

It follows that $z = v(x, y)$ is an integral surface of the slope functions $\mathfrak{p}$ and $\mathfrak{q}$. However, since $\mathfrak{p}_z$ and $\mathfrak{q}_z$ are continuous, there is but one solution surface through each point. Consequently, the surface $z = v(x, y)$ over $D$ coincides with a member of the family $\mathcal{S}$. Since $D$

was an arbitrary set where $v \neq u$, it follows that $v$ is an extremal of $S$ for all $(x, y) \in R$. Finally, by condition 2 in the definition of weak imbedding, the surfaces $z = v(x, y)$ and $z = u(x, y)$ must be a finite distance apart (recall that we have assumed $v \not\equiv u$). But this is impossible, for then $v$ could not satisfy the given boundary condition. This contradiction proves that $I[u] < I[v]$.

Consider next the case when $v$ is contained in $\mathfrak{A}^*$. It is no longer possible to establish (26), of course, because $v$ need not be differentiable. However, if $I[u] = I[v]$ we do have

$$\lim_{\varepsilon \to 0} \int_D E(w)\, dA = 0, \tag{29}$$

as follows at once from (25). We shall suppose in the sequel that $D$ is *convex*.

Now for $(x, y) \in D$ the arguments $(x, y, w, \mathfrak{p}, \mathfrak{q})$ in $E(w)$ are confined to some bounded set $\mathcal{D}$. Therefore, using (29) and (27), together with Jensen's inequality,(¹) we find

$$\lim \int_D X\, dA = 0; \tag{30}$$

here $X = \sqrt{(w_x - \mathfrak{p})^2 + (w_y - \mathfrak{q})^2}$, and the limit is evaluated for $\varepsilon \to 0$.

It is now an easy exercise to show that for any two points $(x_1, y)$ and $(x_0, y)$ in $D$, we have

$$v(x_1, y) - v(x_0, y) = \int_{(x_0, y)}^{(x_1, y)} \mathfrak{p}(x, y, v)\, dx. \tag{31}$$

Consequently, the derivative $v_x$ exists and satisfies the first of the following two equations,

$$v_x = \mathfrak{p}(x, y, v), \quad v_y = \mathfrak{q}(x, y, v); \tag{32}$$

the second is derived by similar reasoning. Equations (32) being verified, the remainder of the proof is exactly the same as in the case $v \in \mathfrak{A}$. This completes the demonstration of Theorem 4.

It remains for us to show that Theorem 1 can be obtained as a corollary of Theorem 4. To do this, it is sufficient to show that the extremal $u(x, y)$ of Theorem 1 can be weakly imbedded in an extremal field for which $\mathcal{V}$ is the entire cylinder $R \times \{z \mid -\infty < z < +\infty\}$. The required field can be obtaining by taking $S$ to be the family of vertical translates of the given surface $z = u(x, y)$. It is clear that $S$ smoothly covers the cylinder in question, and that the imbedding conditions are satisfied. Thus Theorem 1 is proved.

---

(¹) Cf. Polya and Szegö, *Aufgaben und Lehrsätze aus der Analysis*, vol. 1, p. 53. We require only the case $p = 1$.

2 – 593804. *Acta mathematica.* 102. Imprimé le 14 septembre 1959

## PART III

**11. A proof of Hilbert's formula.** It has already been remarked that Hilbert's formula (16) can be proved without difficulty if suitable differentiability conditions are available. In the present circumstances these conditions do not hold, and an alternate proof must be found. We shall follow an argument of Bliss ([1] § 18), except for a few essential changes.

Let the surfaces of $S$ be represented in the form

$$z = u(x, y, a), \tag{33}$$

where $a$ is a real parameter (the surface corresponding to a given value of the parameter may consist of several disjoint parts, but this causes no difficulty). Because of the continuity of $\mathfrak{p}_z$ and $\mathfrak{q}_z$ it can be assumed that $u(x, y, a)$ is of class $C^1$ in all variables, with

$$\frac{\partial u}{\partial a} > 0. \tag{34}$$

An explicit construction of the family (33), so that (34) holds, can be obtained by choosing the parameter $a$ to satisfy

$$u(x_0, y_0, a) = a \tag{35}$$

for some convenient constants $x_0$, $y_0$. Considering $u(x, y, a)$ to be the solution of the system

$$u_x = \mathfrak{p}(x, y, u), \quad u_y = \mathfrak{q}(x, y, u)$$

with (35) as initial condition, it follows from a well-known theorem of ordinary differential equations that $u$ is of class $C^1$ in all variables. Moreover, $\partial u/\partial a > 0$. For otherwise, suppose $\partial u/\partial a = 0$ for some values $x_1$, $y_1$, $a_1$. Then we could reparametrize the surfaces so that $u(x_1, y_1, b) = b$, and it would follow that $\partial u/\partial b = \infty$ for the values $x_0$, $y_0$, $b(a_1)$. This is impossible. Because of (34), the transformation (33) has a *differentiable* inverse $a(x, y, z)$.

There is clearly no loss of generality in supposing that the surfaces $z = U$ and $z = V$ lie entirely interior to $\mathfrak{V}$. This being the case, there then exists a proper subregion $\mathfrak{V}'$ of $\mathfrak{V}$ which contains both surfaces, as well as the volume enclosed between them.

These preliminaries taken care of, let us define a function $\mathcal{A} = \mathcal{A}(x, y, a)$ by means of the integral average

$$\mathcal{A} = \frac{3}{4\pi\delta^3} \iiint F_p(\xi, \eta, u, u_x, u_y)\, d\xi\, d\eta\, d\alpha,$$

in which the arguments of $u$, $u_x$, $u_y$ in the integrand are $(\xi, \eta, \alpha)$, and the domain of integration is a sphere of radius $\delta$ about the point $(x, y, a)$. The constant $\delta$ is chosen so (small) that the integral is well-defined whenever $(x, y, a)$ corresponds to a point in $\mathfrak{V}'$. We define $\mathcal{B}$, $\mathcal{C}$ by similar formulae, but with $F_q$ and $F_u$ respectively replacing $F_p$. Since (6) holds identically in $a$, a well-known argument (cf. [3], or [1], § 15) shows that

$$\mathcal{A}_x + \mathcal{B}_y = \mathcal{C}. \tag{36}$$

This equation may be thought of as a kind of averaged version of the Euler-Lagrange equation.

For convenience in carrying out the remaining steps in the proof we define the two functions

$$\tilde{\mathcal{A}} = \tilde{\mathcal{A}}(x, y, z) \equiv \mathcal{A}(x, y, a(x, y, z)), \qquad \tilde{\mathcal{B}} = \tilde{\mathcal{B}}(x, y, z) \equiv \mathcal{B}(x, y, a(x, y, z)). \tag{37}$$

The functions $\tilde{\mathcal{A}}$ and $\tilde{\mathcal{B}}$ are obviously differentiable, and indeed we have the useful relations

$$\tilde{\mathcal{A}}_x + \tilde{\mathcal{A}}_z \mathfrak{p} = \mathcal{A}_x, \qquad \tilde{\mathcal{B}}_x + \tilde{\mathcal{B}}_z \mathfrak{q} = \mathcal{B}_x, \tag{38}$$

obtained by differentiating (37) while keeping $a$ fixed. Now let us put

$$\mathfrak{J}^* [U, S] = \int_S \{F + (P - \mathfrak{p}) \tilde{\mathcal{A}} + (Q - \mathfrak{q}) \tilde{\mathcal{B}}\} \, dA,$$

where the arguments of $F$, $\mathfrak{p}$, and $\mathfrak{q}$ are the same as in the integral $I^*[U, S]$ of section 7, while the arguments of $\tilde{\mathcal{A}}$ and $\tilde{\mathcal{B}}$ are $(x, y, U)$. Then by virtue of the limit properties of the integral average, it can be seen that $\mathfrak{J}^*[U, S]$ tends to $I^*[U, S]$ as $\delta$ tends to zero. Therefore, in order to prove (16) it is enough to show that

$$\mathfrak{J} = \mathfrak{J}^*[V, S] - \mathfrak{J}^*[U, S] \to 0$$

as $\delta$ tends to zero. Let $W = U + t(V - U)$, $0 \leqslant t \leqslant 1$, and write $\mathfrak{J}^*(t)$ in place of $\mathfrak{J}^*[W, S]$. Then evidently

$$\mathfrak{J} = \int_0^1 \frac{d\mathfrak{J}^*}{dt} \, dt.$$

The completion of the proof rests on a computation of $d\mathfrak{J}^*/dt$. For simplicity in making this computation we write the integrand of $\mathfrak{J}^*(t)$ in the form

$$W_x \tilde{\mathcal{A}}(x, y, W) + W_y \tilde{\mathcal{B}}(x, y, W) + \Phi(x, y, W),$$

where $\Phi = F - \mathfrak{p} \tilde{\mathcal{A}} - \mathfrak{q} \tilde{\mathcal{B}}$. Then, setting $\zeta = V - U$, we have

$\mathcal{I}[u]$ does to $I[u]$. Because of this, the theory of surface area can be used as a guide to the type of results which can be expected in the present more general situation. For example, it is known that $\mathcal{A}[u]=A[u]$ whenever $u$ has strong derivatives, but that equality does not hold beyond this class of functions. A corresponding result for the functionals $I[u]$ and $\mathcal{I}[u]$ will be one of the goals of this paper.

The paper is divided into three parts. The first part contains, in addition to definitions and other preliminary material, a discussion of the special case when the integrand depends only on $p$. The results here are especially simple and elegant, and serve as motivation for much of the later work. §§2 and 4 contain generalizations of the theorem of Tonelli noted in the preceding paragraph. Because of a consistent use of integral averaging the proofs are actually simpler than the corresponding ones given in the theory of surface area (cf., for example, [11, Chapter V]). In §5 we show that $\mathcal{I}[u]$ is a convex functional of $u$, in analogy with the so-called Steiner inequality for surface area.

The results of Part I are extended to a significantly larger class of integrands in Part II. For a detailed idea of the results of Part II the reader is best referred directly to the theorems there. Theorem 10, in particular, deserves special notice as our ultimate generalization of Tonelli's theorem.

There are a number of integrands which do not satisfy the conditions imposed in Parts I and II, but which nevertheless are amenable to an alternate approach. In Part III we investigate the functionals $I[u]$ and $\mathcal{I}[u]$ corresponding to these more difficult integrands. Although the results are no longer so definitive, the greater generality of the integrands lends its own interest. The major effort turns on proving certain lower semicontinuity theorems for the integral $I[u]$, these being used to show, in particular, that $I[u]\leqq\mathcal{I}[u]$ for the corresponding integrands. Since these lower semicontinuity theorems have an independent interest, some of them may be separately noted here. For example, Theorem 11 shows that $I[u]$ is lower semicontinuous with respect to weak convergence in $L_1$ whenever $f$ depends on $p$ alone, or even is of the form $A(x)F(p)$. In Theorem 12 we show that $I[u]$ is lower semicontinuous with respect to (strong) convergence in $L_1$ if $f$ is strictly convex, or if the derivatives $f_x$, $f_p$, and $f_{px}$ exist and are continuous, or finally if only $f\to\infty$ as $|p|\to\infty$. These results, which are probably known for the case of one independent variable, generalize well-known multiple integral semicontinuity theorems of Tonelli and Morrey.

## Part I

1. **Definitions and preliminary results.** Consider a real valued function $\phi=\phi(x)$ defined on an open set $R$ in the $n$-dimensional number space $E^n$. The function $\phi$ is called locally summable if it admits a finite Lebesgue integral over any compact subset of $R$. It is natural to associate a special kind of con-

vergence with locally summable functions; namely, a sequence $\{\phi_m\}$ will be said to *converge locally to $\phi$ in $R$* if for any compact subset $S$ of $R$ the following condition is satisfied:

$\phi_m$ is defined and summable in $S$, at least for all $m$ sufficiently large, and $\int_S |\phi_m - \phi| dx \to 0$ as $m \to \infty$.

To indicate that a sequence $\{\phi_m\}$ converges locally to $\phi$ we shall write $\phi_m \to \phi$.

The function $\phi$ is called *strongly differentiable in $R$* if there exists a vector-valued function $\psi$ such that

$$\int \phi\omega_x dx = -\int \omega\psi dx, \tag{1}$$

for any continuously differentiable function $\omega$ with compact support in $R$. If $\phi$ is strongly differentiable there exists a sequence $\{\phi_m\}$ of continuously differentiable functions such that

$$\phi_m \to \phi, \qquad \phi_{mx} \to \psi. \tag{2}$$

This condition is, in fact, often used to define strong differentiability.

In analogy with the concept of strong differentiation, we say that $\phi$ is *weakly differentiable in $R$* if there exists a vector-valued Borel measure $\alpha$ such that

$$\int \phi\omega_x dx = -\int \omega d\alpha \tag{3}$$

for any continuously differentiable function $\omega$ with compact support in $R$. In general $\alpha$ need not be defined on the set $R$ itself; however, for the results of Theorems 3, 5, and 10 it is convenient to assume that $R$ is measurable and that $\alpha(R)$ is finite. It is evident that the class of weakly differentiable functions *includes* all strongly differentiable functions. In particular, by virtue of the Radon-Nikodym theorem a weakly differentiable function is strongly differentiable if and only if its derivative measure $\alpha$ is absolutely continuous. In this case the relation between $\alpha$ and $\psi$ is simply $\alpha(E) = \int_E \psi dx$.

If $\phi$ is weakly differentiable, its derivative measure is an additive set function. Consequently, we may *define* the derivative $\phi_x$ of $\phi$ by the formula

$$\phi_x = \alpha', \tag{4}$$

where $\alpha'$ denotes the general derivative(1) of $\alpha$. The function $\phi_x$ is then well-defined, equals $\psi$ when $\phi$ is strongly differentiable, and finally agrees with the

(1) Cf. [11, pp. 106, 115]. Equivalently, $\alpha'$ can be defined as the Radon-Nikodym derivative of the absolutely continuous part of $\alpha$, cf. [4, p. 133]. In either case, $\alpha$ and $\alpha'$ are connected by the fundamental relation

$$\alpha(E) = \int_E \alpha'(x)dx + \theta(E), \tag{5}$$

where $\theta(E)$ is a singular measure.

ordinary partial derivatives of $\phi$ when $\phi$ is continuously differentiable. A fuller discussion of the properties of weakly differentiable functions will be found in references [12] and [14].

*Integral averages.* In certain of our results the operation of integral averaging or smoothing can be used to particular advantage. We associate with each positive number $h$ a corresponding function $K(\xi)=K(\xi, h)$ with the properties

(i) $K$ is non-negative and smooth for all values of $\xi=(\xi_1, \cdots, \xi_n)$.

(ii) $K\equiv 0$ for $|\xi|\geqq h$.

(iii) $\int K(\xi)d\xi=1$.

The $\xi$ integration in this and subsequent formulas is assumed to be carried out over the entire $n$-dimensional number space. Now let $\phi$ be a locally summable function in an open region $R$, and let $R_h$ denote the subset of $R$ whose points are farther than $h$ from the boundary of $R$. The integral average $\phi_h$ of $\phi$ is then defined by the formula

$$\phi_h = \phi_h(x) = \int K(\xi - x)\phi(\xi)d\xi, \qquad x \in R_h.$$

It is well known and easy to show that $\phi_h\to\phi$ as $h\to 0$. Furthermore, $\phi_h$ is continuously differentiable and its derivative is given by the formula

$$\phi_{hx} = -\int K_\xi(\xi - x)\phi(\xi)d\xi. \tag{6}$$

From (1) and (6) it follows that if $\phi$ is strongly differentiable, then

$$\phi_{hx} = \phi_{xh}. \tag{7}$$

It is this equation, in fact, which justifies the relations (2); we need only choose $\phi_m=\phi_h$ with $h=m^{-1}$.

*The functionals* $I[u, R]$ *and* $\mathcal{g}[u, R]$. The integrands $f(x, u, p)$ which we shall consider in this paper are assumed to be continuous for all $x$ in some open set $R_0$ and for all values of $u$ and $p$. In addition, without further mention we assume that

$$f \geqq 0, \qquad f \text{ convex in } p. \tag{8}$$

Now let $R$ be an open set, $R\subseteq R_0$, and let $u$ be a weakly differentiable function in $R$. We define

$$I[u, R] = \int_R f(x, u, u_x)dx, \tag{9}$$

the integral being understood in the sense of Lebesgue. Since $f$ is non-negative, the value $+\infty$ is allowed for $I[u, R]$. We shall occasionally write $I[u]$ instead of $I[u, R]$, provided the meaning is clear from the context.

To define the functional $\mathcal{J}[u, R]$ we proceed somewhat analogously to Lebesgue's definition of surface area. Let $R$ be an open set, $R \subseteq R_0$, and let $u$ be a locally summable function defined on $R$. Consider a sequence $\{u_m\}$ of continuously differentiable functions such that $u_m \to u$ in $R$. We set

$$A = \liminf I[u_m, R_m],$$

where $R_m$ denotes the (open) domain of definition of $u_m$. Then $\mathcal{J}[u, R]$ is defined as the infimum of the numbers $A$ corresponding to all possible sequences $\{u_m\}$ converging to $u$ as above. By a diagonal process it is evident that there exists some sequence $\{u'_m\}$ such that $\mathcal{J}[u, R] = \lim I[u'_m, R'_m]$. Hence we can express the above definition in the concise form

$$\mathcal{J}[u, R] \equiv \underset{u_m \to u}{\text{Minimum limit}}\, I[u_m, R_m]. \tag{10}$$

It is an immediate consequence of this definition that $\mathcal{J}[u, R]$ is lower semicontinuous with respect to local convergence. Also, let $u$ be a fixed function defined in $R \subseteq R_0$, and let $Q$ be a variable open subset of $R$. Then $\mathcal{J}[u, Q]$ is a well-defined function of the set $Q$. For later use we observe that this set function is both monotone and countably additive, and in particular

$$\mathcal{J}[u, Q_1] + \mathcal{J}[u, Q_2] \leqq \mathcal{J}[u, R]$$

if $Q_1$ and $Q_2$ are disjoint open subsets of $R$.

Certain slightly different definitions might be advocated for the functional $\mathcal{J}[u, R]$. Some of these are discussed in §13, to which the reader is referred. Another would replace local convergence by uniform convergence, in exact analogy with the procedure for Lebesgue area. It can be shown that the present definition is an extension of the latter one, in that both give the same value for continuous functions $u$ while only the present definition applies at all to discontinuous functions (see reference [13]).

2. **The integrand** $f = f(p)$. For the rest of Part I we shall be concerned with the properties and relationships of the functionals $I[u, R]$ and $\mathcal{J}[u, R]$ when the integrand depends only on $p$. In the following lemma, and throughout the rest of the paper, the region $R_m$ denotes the (open) domain of definition of a function $u_m$. On the other hand, $R_h$ is the set of points of $R$ whose distance from the boundary is greater than $h$.

Lemma 1. *Let $u$ be a locally summable function defined on an open set $R$, and let $\{u_m\}$ be a sequence of strongly differentiable functions converging locally to $u$ in $R$. Then for each $h > 0$, we have*

$$I[u_h, R_h] \leqq \liminf I[u_m, R_m]. \tag{11}$$

**Proof.** Let $S$ denote a fixed compact subset of $R$. Since $u_m \to u$ in $R$, it is clear that the domains $R_m$ eventually contain $S$, say for all $m \geqq m_0$. By virtue of (6) we have then for all $x \in S_h$ and $m \geqq m_0$,

$$|u_{mhx} - u_{hx}| \leqq C(h) \int_S |u_m - u| \, dx.$$

Therefore

$$f(u_{hx}) \leqq f(u_{mhx}) + \epsilon = f(u_{mxh}) + \epsilon \leqq f(u_{mx})_h + \epsilon$$

by Jensen's inequality, where $\epsilon = \epsilon(m, h)$ tends to zero as $m \to \infty$. Integrating this inequality over $S_h$ and applying Fubini's theorem yields

$$I[u_h, S_h] \leqq \int_{S_h} f(u_{mx})_h dx + \epsilon \operatorname{Meas} S \leqq \int_S f(u_{mx}) dx + \epsilon \operatorname{Meas} S;$$

in this transformation we have made essential use of properties (i), (ii) and (iii) of the kernel $K$ and the fact that $f \geqq 0$. Since $S \subset R_m$ the preceding inequality may be written

$$I[u_h, S_h] \leqq I[u_m, R_m] + \epsilon \operatorname{Meas} S.$$

Now let $m \to \infty$ and then $S \to R$, and the lemma is proved.

The following two theorems are simple consequences of Lemma 1. The first gives a constructive procedure for evaluating $\mathcal{J}[u, R]$, analogous to Radó's theorem for the evaluation of Lebesgue area (see reference [11, p. 173]). The second describes the relationship between $I[u]$ and $\mathcal{J}[u]$.

THEOREM 1. *For any locally summable function $u$ in $R$ we have* $\mathcal{J}[u, R] = \lim I[u_h, R_h]$.

**Proof.** From Lemma 1 and the definition of $\mathcal{J}[u, R]$ follows $I[u_h, R_h] \leqq \mathcal{J}[u, R]$. Hence, once more using the definition of $\mathcal{J}[u, R]$, we have

$$\mathcal{J}[u, R] \leqq \liminf I[u_h, R_h] \leqq \limsup I[u_h, R_h] \leqq \mathcal{J}[u, R],$$

and the theorem is proved.

THEOREM 2. *If $u$ is weakly differentiable, then $I[u] \leqq \mathcal{J}[u]$, equality holding whenever $u$ is strongly differentiable.*

**Proof.** We first require some general considerations about weakly differentiable functions $u$. Let $\alpha$ be the derivative measure of $u$. According to formulas (6), (3), and (5) in footnote 1,

$$u_{hx} = \int K(\xi - x) d\alpha(\xi) = \int K(\xi - x) u_x(\xi) d\xi + \int K(\xi - x) d\theta(\xi),$$

where $\theta$ is a singular measure. Now assume that $K$ satisfies the condition $K \leqq \text{Const.}/h^n$ in addition to the properties already listed (this condition is satisfied, for example, by kernels of the form $K(\xi) = K(\xi, h) = h^{-n}\tilde{K}(h^{-1}\xi)$, where $\tilde{K}$ is independent of $h$). Then it is easily shown(²) that $\lim u_{hx} = u_x$ al-

(²) We have $|u_{hx} - u_x| \leqq \int K(\xi - x)|u_x(\xi) - u_x(x)| d\xi + \int K(\xi - x) d|\theta(\xi)| \leqq (\text{Const.}/h^n) \times [\int_X |u_x(\xi) - u_x(x)| d\xi + |\theta|(X)]$, where $X$ denotes the set of points $\xi$ such that $|\xi - x| \leqq h$. The last expression tends to zero almost everywhere as $h \to 0$ [7, p. 189; 11, p. 119].

most everywhere in $R$.

This being the case, we have by virtue of Fatou's lemma

$$I[u, S] \leqq \lim I[u_h, S] \leqq \lim I[u_h, R_h],$$

where $S$ is any compact subset of $R$. But the expression on the right equals $\mathcal{I}[u, R]$, whatever the averaging kernel. This proves the first part of the theorem.

Next suppose that $u$ is strongly differentiable. Setting $u_m \equiv u$ in Lemma 1, there results $I[u_h, R_h] \leqq I[u, R]$. Letting $h \to 0$ yields $\mathcal{I}[u] \leqq I[u]$. In view of the earlier inequality we must therefore have $I[u] = \mathcal{I}[u]$, and the proof is complete.

REMARK. The conclusion of Lemma 1 can now be interpreted as stating that *integral averaging is a norm shrinking operation*. Indeed, since $u_h$ is continuously differentiable, Lemma 1 and Theorem 2 imply

$$\mathcal{I}[u_h, R_h] \leqq \mathcal{I}[u, R].$$

The same result holds for the integral $I[u]$ provided $u$ is strongly differentiable.

**3. Two particular integrals.** The developments of later sections will stand out in sharper relief if we first investigate the particular integrands

$$f(p) = |p| \quad \text{and} \quad f(p) = \Phi(p),$$

where $|p|$ stands for the *sum of the absolute values of the components of* $p$, and $\Phi(p)$ is some fixed function with the property $\Phi(p)/|p| \to \infty$ as $|p| \to \infty$. The corresponding functionals will be denoted by $D[u]$, $\mathfrak{D}[u]$ and $E[u]$, $\mathcal{E}[u]$, respectively.

THEOREM 3. *$\mathfrak{D}[u]$ is finite if and only if $u$ is weakly differentiable. The set function $\mathfrak{D}[u, Q]$ is absolutely continuous if and only if $u$ is strongly differentiable.*

**Proof.** Suppose that $u$ is weakly differentiable, and suppose also (this is tacitly assumed in the statement of the theorem) that the measure $\alpha$ associated with $u$ is finite on $R$. From the formula

$$u_{hx} = \int K(\xi - x)d\alpha, \qquad x \in R_h,$$

one easily infers that $D[u_h, R_h] \leqq |\alpha|(R)$. Hence by Theorem 1,

$$\mathfrak{D}[u, R] \leqq |\alpha|(R) < \infty.$$

Conversely, suppose that $\mathfrak{D}[u, R]$ is finite. We must show that $u$ is weakly differentiable, and that its measure $\alpha$ is finite on $R$. For convenience, let $C(R)$ denote the linear space whose members are real valued continuous functions with compact support in $R$. Similarly, let $C'(R)$ denote the space of continuously differentiable functions with compact support in $R$. For $\omega \in C'(R)$ let us set

$$A(\omega) = \int_R u\omega_x dx.$$

Evidently $A(\omega)$ is a vector-valued linear functional on $C'(R)$. Moreover, $A$ is continuous in the uniform topology of $C(R)$. To see this, we have

$$A(\omega) = \lim \int_{R_h} u_h \omega_x dx = -\lim \int_{R_h} u_{hx}\omega dx,$$

and

$$\left| \int_{R_h} u_{hx}\omega dx \right| \leqq \max |\omega| \cdot \int_{R_h} |u_{hx}| dx.$$

Hence by Theorem 1

$$|A(\omega)| \leqq \max |\omega| \cdot \mathfrak{D}[u, R],$$

proving our assertion. Since $A(\omega)$ is continuous in the topology of $C(R)$, it may evidently be extended to a continuous linear functional defined on all of $C(R)$. But then by the Riesz theorem there exists a (vector-valued) Borel measure $\alpha$ such that

$$A(\omega) = -\int \omega d\alpha, \qquad |\alpha|(R) \leqq n\mathfrak{D}[u, R].$$

This is, however, precisely the condition that $u$ be weakly differentiable, and the first part of the theorem is proved.

The preceding work yields as a byproduct the interesting corollary

$$\mathfrak{D}[u, R] \leqq |\alpha|(R) \leqq n\mathfrak{D}[u, R]. \tag{12}$$

Hence the set function $\mathfrak{D}[u, Q]$ is absolutely continuous if and only if $u$ is weakly differentiable and its derivative measure is absolutely continuous; that is, if and only if $u$ is strongly differentiable.

THEOREM 4. *If $\mathcal{E}[u]$ is finite, then $u$ is strongly differentiable.*

**Proof.** Since $\lim E[u_h, R_h] = \mathcal{E}[u, R]$, the integrals $E[u_h, R_h]$ must be uniformly bounded. But then by a theorem of Nagumo, the integrals $\int |u_{hx}| dx$ are uniformly absolutely continuous. This in turn implies, by Theorem 1, that $\mathfrak{D}[u, Q]$ is an absolutely continuous function of open sets $Q$ in $R$. According to Theorem 3, therefore, $u$ is strongly differentiable in $R$.

Theorems 3 and 4 throw considerable light on the type of functions which are admissible in problems of the calculus of variations. In particular, let us call an integrand of "area" type if it is compatible with $|p|$, and of "non-area" type if it grows faster than any multiple of $|p|$. Then $\mathcal{I}[u]$ will be finite in problems of area type whenever $u$ is weakly differentiable, while in problems of non-area type it is necessary that $u$ be strongly differentiable. The frequently studied quadratic variational problems, for example, are of non-area type, whence for them we should expect the admissible functions to be strongly differentiable (in this connection it is interesting to note that the usual function space approach to these problems also leads to the consideration of strongly differentiable functions as the appropriate admissibility class).

Since for non-area problems the admissible functions are necessarily strongly differentiable, there are reasonable grounds for considering either the Lebesgue integral $I[u]$ or the functional $\mathcal{I}[u]$ as the fundamental quantity. The decision as to which one to use is, however, simplified by the fact that in most cases of practical importance both have the same value. On the other hand, it seems that there are (rather complicated) non-area integrands for which $I[u]$ and $\mathcal{I}[u]$ differ beyond the class of Lipschitz functions. In these cases the choice of the appropriate integral must be based on aesthetic considerations. In any case, we emphasize that for area type problems the functional $\mathcal{I}[u]$ is most appropriate, since it gives a wider class of admissible functions and also because it is closer to Lebesgue's original ideas.

**4. A generalization of Tonelli's theorem.** In 1926 Tonelli discovered a remarkable relation between the Lebesgue area of a surface and the classical surface area integral. His work was later simplified and extended by Saks (see the discussion in Chapter V of reference [11]), and more recently the theory has been further refined by Cesari and Goffman. The following theorem is a generalization of all these results.

THEOREM 5. *Let $R$ be a bounded open set. Suppose that there exist positive constants $a$ and $A$ such that*

$$a|p| \leqq 1 + f(p) \leqq A|p| \qquad \textit{when } |p| \geqq 1. \tag{13}$$

*Then $\mathcal{I}[u]$ is finite if and only if $u$ is weakly differentiable.*

*When this is the case, we have*

$$I[u] \leqq \mathcal{I}[u] \tag{14}$$

*and also $\mathcal{I}'(x) = f(u_x)$ almost everywhere in $R$. In order that $I[u] = \mathcal{I}[u]$ it is necessary and sufficient that $u$ be strongly differentiable; and in order that this be the case it is necessary and sufficient that $\mathcal{I}[u, Q]$ be an absolutely continuous function of open sets $Q$ contained in $R$.*

**Proof.** We may assume without loss of generality that $a \leqq 1$. Then from condition (13) it is clear that

$$a\mathfrak{D}[u] - \text{Meas } R \leqq \mathcal{I}[u] \leqq A(\mathfrak{D}[u] + \text{Meas } R). \tag{15}$$

The first part of the theorem is therefore an immediate consequence of Theorem 3.

Now suppose $\mathcal{g}[u]$ is finite and $u$ is weakly differentiable. Inequality (14) is a consequence of Theorem 2. Moreover if $I[u]=\mathcal{g}[u]$, then using the superadditivity of $\mathcal{g}$ one easily infers that $I[u, Q]=\mathcal{g}[u, Q]$ for each open set $Q$ in $R$ whose boundary is of measure zero. Thus using (12) and (15),

$$|\alpha|(Q) = \mathfrak{D}[u, Q] \leqq a^{-1}(I[u, Q] + \text{Meas } Q).$$

That is, $\alpha$ is an absolutely continuous function of open sets whose boundaries are of measure zero. Since any open set can be approximated from inside by an increasing sequence of open sets whose boundaries are of measure zero, it follows that $\alpha$ is absolutely continuous on all open sets and thence on all Borel sets. The equality $I[u]=\mathcal{g}[u]$ thus implies that $u$ is strongly differentiable; conversely, if $u$ is strongly differentiable, then $I[u]=\mathcal{g}[u]$ by Theorem 2.

The final statement of the theorem is obvious from what has gone before, and a detailed proof is therefore omitted.

It remains only to show that $\mathcal{g}'(x)=f(u_x)$ whenever $u$ is weakly differentiable. Let $J$ denote an arbitrary open interval in $R$. We assert that

$$(16) \qquad I[u, J] \leqq \mathcal{g}[u, J] \leqq I[u, J] + \Theta(J),$$

where $\Theta$ is a singular measure. The first half of (16) is of course obvious. To prove the second, we observe that

$$u_{hx} = u_{xh} + \int K(\xi - x)d\theta(\xi)$$

as in the proof of Theorem 2. Since by hypothesis the partial slopes of $f(p)$ are always less than $A$, we have

$$f(u_{hx}) \leqq f(u_{xh}) + A\left|\int K(\xi - x)d\theta(\xi)\right| \leqq f(u_x)_h + \int K(\xi - x)d\Theta(\xi)$$

by Jensen's inequality, where $\Theta=A|\theta|$ is a singular measure. Integrating over $J_h$ and applying Fubini's theorem, there results

$$I[u_h, J_h] \leqq I[u, J] + \Theta(J).$$

The required inequality is now obtained by letting $h$ go to zero.

Dividing each side of (16) by $|J|$ and letting $J$ tend regularly to $x$ gives, on account of Lebesgue's differentiation theorem [11, pp. 113–119],

$$f(u_x) \leqq \mathcal{g}'(x) \leqq f(u_x)$$

almost everywhere in $E$. This completes the proof.

REMARKS. 1. When $u$ is a continuous function of two real variables $x$ and $y$ and $f=(1+u_x^2+u_y^2)^{1/2}$, Theorem 5 reduces to Tonelli's theorem. To see this, we must verify first that $\mathcal{I}[u]$ represents the Lebesgue area of the surface $z=u(x, y)$; second, that a weakly differentiable continuous function is almost everywhere differentiable in the ordinary sense; and third, that the derivative $u_x=\alpha'$ agrees almost everywhere with the ordinary derivative.

The first result is given as Theorem 2 of reference [3]; a more immediate proof can be based on Theorem 1 and the fact that for a continuous function $u$ the sequence $\{u_h\}$ converges uniformly on compact subsets. The second and third results are special cases of the main theorem of [14]. Indeed, in that paper it is shown that *any* weakly differentiable function is *equivalent* to a function having ordinary partial derivatives almost everywhere, and that these partial derivatives agree almost everywhere with $\alpha'$.

2. The result of Theorem 5 implies that the set function $\mathcal{I}[u, Q]$ can be extended to a (unique) regular Borel measure on $R$(3). Leaving aside for a moment the demonstration of this fact, we obtain from the Lebesgue decomposition of the resulting measure $\mathcal{I}[u, E]$ the formula

$$\mathcal{I}[u, E] = \int_E f(u_x)dx + \Psi(E),$$

where $\Psi$ is a non-negative singular measure, depending of course on the function $u$. This interesting equality may be looked upon as an alternate formulation, and in fact as a generalization, of the fundamental inequality (14).

The extension of the set function $\mathcal{I}[u, Q]$ from open sets $Q$ to Borel sets $E$ is most easily accomplished by the general theory of generation of Borel measures ([4, pp. 231–237]; this theory is phrased so as to apply to set functions defined on compact sets, but the ideas are equally valid for functions defined on open sets). We have already observed in §1 that $\mathcal{I}[u, Q]$ is monotone and additive on open sets. In addition, by virtue of Theorem 1 we have easily

$$\mathcal{I}[u, Q_1 \cup Q_2] \leqq \mathcal{I}[u, Q_1] + \mathcal{I}[u, Q_2],$$

so that $\mathcal{I}[u, Q]$ is subadditive. It only remains to show that $\mathcal{I}[u, Q]$ is "regular," i.e. that

$$\mathcal{I}[u, Q] = \sup \mathcal{I}[u, Q_1],$$

where the supremum is taken over all open sets $Q_1$ properly contained in $Q$. But let $Q_1$ be an open subset of $Q$, and $C$ a compact subset of $Q_1$. Then

$$\begin{aligned}\mathcal{I}[u, Q] &\leqq \mathcal{I}[u, Q_1] + \mathcal{I}[u, Q - C] \\ &\leqq \mathcal{I}[u, Q_1] + I[u, Q - C] + \Theta(Q - C),\end{aligned}$$

by (16). Letting $Q_1 \to Q$, and consequently $C \to Q$, we thus obtain

(3) The corresponding result for Lebesgue area was obtained by Verchenko, and elaborated further by Goffman (Acta Math. vol. 103 (1960)).

$$\mathcal{I}[u, Q] \leqq \lim \mathcal{I}[u, Q_1].$$

On the other hand, by the monotonicity of $\mathcal{I}[u, Q]$ we have $\mathcal{I}[u, Q] \geqq \lim \mathcal{I}[u, Q_1]$, and the required relation is proved.

3. When the integrand $f(p)$ depends on $p$ only through the intermediate variable $(p_1^2 + \cdots + p_n^2)^{1/2} = \|p\|$, it is possible to give an alternate version of Theorem 5. Indeed, in this case the hypothesis (13) can be dropped in favor of the simple condition that $f$ *not be identically constant.* The conclusions are that if $\mathcal{I}[u]$ is finite then $u$ is (at least) weakly differentiable, and that the second part of the theorem remains valid. To see this, we observe that if $f$ depends only on $\|p\|$, and is not identically constant, then either (13) holds, or else $f \geqq \Phi$ for some function $\Phi$ of the type considered in §3. In either case the conclusions noted follow at once.

5. **Further results.** This section is devoted to certain subsidiary results and generalizations of the preceding work.

1. The basic definition of $\mathcal{I}[u]$ can be slightly weakened by allowing sequences $\{u_m\}$ into competition which are only weakly convergent to $u$. That is, in place of the condition $\int_S |u_m - u| dx \to 0$ we could use simply

$$\int (u_m - u)\omega dx \to 0,$$

where $\omega$ is any continuous function with compact support in $R$. Lemma 1 remains true for this kind of convergence after only slight changes in the proof. But then Theorem 1 shows that the resulting functional is exactly the same as before.

2. The so-called inequality of Steiner for surface area has an interesting generalization to the present case. Namely, we assert that $\mathcal{I}[u]$ *is a convex functional over the linear space of locally summable functions on a fixed region* $R$. That is, if $u_1$ and $u_2$ are locally summable in $R$, and $u = tu_1 + (1-t)u_2$, $0 < t < 1$, then

$$\mathcal{I}[u] \leqq t\mathcal{I}[u_1] + (1 - t)\mathcal{I}[u_2]. \tag{17}$$

To prove this, note that since $f$ is convex in $p$,

$$f(u_{hx}) = f[tu_{1hx} + (1 - t)u_{2hx}] \leqq tf(u_{1hx}) + (1 - t)f(u_{2hx}).$$

Integrating over $R_h$ and applying Theorem 1 yields (17).

It may be conjectured that if $u_1$ and $u_2$ are weakly differentiable and if $u_{1x} = u_{2x}$, then equality holds in (17). This conclusion is obviously legitimate if $u_1$ and $u_2$ are strongly differentiable. Moreover, a partial converse is true:

*Suppose that $f$ is strictly convex and satisfies condition* (13). *If for some value of $t$ we have*

$$\mathcal{I}[u] = t\mathcal{I}[u_1] + (1 - t)\mathcal{I}[u_2], \tag{18}$$

*and if $\mathcal{I}[u] < \infty$, then $u_1$ and $u_2$ are weakly differentiable and their derivatives agree almost everywhere.*

**Proof.** Since $\mathcal{I}[u] < \infty$, it is evident that the functions $u$, $u_1$, and $u_2$ are all weakly differentiable. We now assert that the additivity relation

$$\mathcal{I}[u, R] = \mathcal{I}[u, J] + \mathcal{I}[u, \operatorname{Int}(R - J)] \tag{19}$$

holds for any open interval $J$ in $R$ such that $\Theta(\partial J) = 0$, irrespective of whether or not (18) holds. To see this, let $\mathcal{B}$ denote the quantity on the right side of (19), and let $J^*$ denote the set of points in $R$ whose distance from $\partial J$ is less than some number $\epsilon > 0$. Then clearly

$$\mathcal{B} \leqq \mathcal{I}[u, R] \leqq \mathcal{B} + \mathcal{I}[u, J^*].$$

But by the argument proving (16) we have

$$\mathcal{I}[u, J^*] \leqq I[u, J^*] + \Theta(J^*),$$

whence $\lim_{\epsilon \to 0} \mathcal{I}[u, J^*] = \Theta(\partial J) = 0$. This proves our assertion.

It follows from (17), (18), (19), and the superadditivity of $\mathcal{I}$ that

$$\mathcal{I}[u, J] = t\mathcal{I}[u_1, J] + (1 - t)\mathcal{I}[u_2, J] \qquad (\Theta(\partial J) = 0). \tag{20}$$

Also, the set of intervals $J$ such that $\Theta(\partial J) = 0$ is certainly dense in the set of all intervals. We may therefore form the set-theoretic derivative of (20), whence by Theorem 5

$$f(u_x) = tf(u_{1x}) + (1 - t)f(u_{2x})$$

almost everywhere in $R$. Since $u_x = tu_{1x} + (1-t)u_{2x}$ almost everywhere, the strict convexity of $f$ implies $u_{1x} = u_{2x} = u_x$, as required.

The preceding discussion is based upon, and generalizes, a corresponding result of McShane for the surface area functional.

## Part II

6. **Integrands of type** I **and** II. We now turn to a more general class of integrands for which the results of the preceding sections remain valid. An integrand $f(x, u, p)$ will be called of *type* I if for any two points $x$ and $y$ in $R_0$, and for all values of $u$, $v$, and $p$, we have

$$|f(x, u, p) - f(y, v, p)| \leqq \lambda(|x - y|)[1 + f(x, u, p)] + \mu(|u - v|).$$

Here $\lambda(\sigma)$ and $\mu(\sigma)$ are moduli of continuity, with $\mu$ satisfying $\mu(\sigma) \leqq \text{Const.}\ \sigma$ for large values of $\sigma$. An integrand will be called of *type* II if, similarly,

$$|f(x, u, p) - f(y, v, p)| \leqq \lambda(|x - y| + |u - v|)[1 + f(x, u, p)].$$

Important examples of integrands of type I and type II are

$$f = f(x, p) = A(x)F(p), \qquad f = f(x, u, p) = A(x, u)F(p),$$

where $A(x)$ and $A(x, u)$ are continuous functions with some positive lower bound. Finally, an integrand will be called of *type* $\mathrm{II}_B$ if the relevant inequality holds for all $u$, $v$ bounded in absolute value by $B$.

The following two lemmas play the same role for our present class of integrands as Lemma 1 did for the integrand $f(p)$. In these results, and throughout the rest of the paper, it will be tacitly assumed that the regions $R$ under consideration are subsets of the basic domain of continuity $R_0$. We recall that $R_m$ denotes the domain of definition of a function $u_m$.

LEMMA 2. *Suppose $f$ is of type* I, *and let $u$ and $\{u_m\}$ satisfy the hypotheses of Lemma* 1. *Then for every compact subset $S$ of $R$ we have*

$$\limsup I[u_h, S] \leqq \liminf I[u_m, R_m].$$

**Proof.** Let $S_0$ be some fixed compact subset of $R$, containing $S$ in its interior. We suppose that $S_0 \subset R_m$ for all $m \geqq m_0$, and also that $S \subset S_{0h}$ for all $h \leqq h_0$. Then, restricting consideration to such values of $m$ and $h$, we find as in Lemma 1 that for all $x \in S$,

$$\begin{aligned} f(x, u_h, u_{hx}) &\leqq f(x, u_h, u_{mxh}) + \epsilon \\ &\leqq \int K(\xi - x) f[x, u_h(x), u_{mx}(\xi)] d\xi + \epsilon \\ &= \int K(\xi - x) f[\xi, u_m(\xi), u_{mx}(\xi)] d\xi + J + \epsilon, \end{aligned}$$

where $J$ denotes the difference of the integrals on the preceding two lines, and $\epsilon = \epsilon(m, h)$ tends to zero as $m \to \infty$. Integrating over $S$ and applying Fubini's theorem to the first term on the right easily yields

$$I[u_h, S] \leqq I[u_m, R_m] + \int_S J dx + \epsilon \text{ Meas } S. \tag{21}$$

We must now estimate the integral of $J$. Since $f$ is of type I, one sees that

$$\begin{aligned} |J| &\leqq \int K(\xi - x)[\lambda(|\xi - x|)\{1 + f[\xi, u_m(\xi), u_{mx}(\xi)]\} + \mu(|u_m - u_h|)] d\xi \\ &\leqq \lambda(h)\{1 + f(x, u_m, u_{mx})_h\} + \int K(\xi - x)\mu(|u_m - u_h|) d\xi. \end{aligned}$$

In virtue of the hypothesis $\mu \leqq \text{Const. } \sigma$ for large $\sigma$, there is no loss of generality in assuming that $\mu$ is concave (convex downwards). Thus again using Fubini's theorem and Jensen's inequality, we have

$$\left| \int_S J dx \right| \leqq \lambda(h)\{\text{Meas } S + I[u_m, R_m]\} + \mu(P/\text{Meas } S) \cdot \text{Meas } S,$$

where

$$P = \int_S \int K(\xi - x) \mid u_m(\xi) - u_h(x) \mid d\xi dx$$

$$\leq \int_{S_0} \mid u_m - u \mid d\xi + \int_S \int K(\xi - x) \mid u(\xi) - u(x) \mid d\xi dx + \int_S \mid u - u_h \mid dx.$$

The middle integral can be written in the form $\int K(\xi) T(\xi) d\xi$, where

$$T(\xi) = \int_S \mid u(x + \xi) - u(x) \mid dx.$$

It is well known that $T(\xi)$ tends to zero as $\xi \to 0$. Therefore $P \leqq \epsilon_1 + \epsilon_2$, where $\epsilon_1 = \epsilon_1(m)$ and $\epsilon_2 = \epsilon_2(h)$ tend to zero as $m$ and $h$ tend respectively to infinity and zero.

Inserting the estimates of the previous paragraph into inequality (21), there results

$$I[u_h, S] \leqq \{1 + \lambda(h)\} I[u_m, R_m] + \left\{\epsilon + \lambda(h) + \mu\left(\frac{\epsilon_1 + \epsilon_2}{\text{Meas } S}\right)\right\} \text{Meas } S.$$

Letting $m \to \infty$ through an appropriate subsequence now yields

(22) $I[u_h, S] \leqq \{1 + \lambda(h)\} \liminf I[u_m, R_m] + \{\lambda(h) + \mu(\epsilon_2/\text{Meas } S)\} \text{Meas } S.$

To complete the proof of the lemma, it is only necessary to let $h$ tend to zero through an appropriate subsequence.

LEMMA 3. *Let $f$ be of type* II. *Suppose that $u$ is continuous, but otherwise let $u$ and $\{u_m\}$ satisfy the conditions of Lemma* 1. *Then for every compact subset $S$ of $R$ we have*

$$\limsup I[u_h, S] \leqq \liminf I[u_m, R_m].$$

**Proof.** We may assume without loss of generality that the sequence $\{u_m\}$ converges uniformly to $u$ in any compact subset $S$ of $R$. Indeed, if $\{u_m\}$ did not converge uniformly, we could construct for any $\epsilon > 0$ another sequence $\{v_m\}$ which converges uniformly, and such that

$$\lim I[v_m, R_m] \leqq \liminf I[u_m, R_m] + \epsilon.$$

The construction of the alternate sequence $\{v_m\}$ can be carried out as in [13, pp. 30–31], the only difference being that the functions $v_m$ which arise are now strongly differentiable rather than piecewise continuously differentiable.

This much being shown, the rest of the proof is very nearly the same as the proof of Lemma 2, except in some respects simpler. We shall omit the details. The next two results are parallel to Theorems 1 and 2.

THEOREM 6. *Suppose that $R$ is bounded. If $f$ is of type* I *and $u$ is summable over $R$, then $\mathcal{I}[u, R] = \lim I[u_h, R_h]$. If $f$ is of type* II *and $u$ is uniformly con-*

*tinuous over R, then again* $\mathcal{I}[u, R] = \lim I[u_h, R_h]$.

**Proof.** In the first case, by letting $S \to R_h$ in (22) we obtain

(23) $I[u_h, R_h] \leqq \{1 + \lambda(h)\} \liminf I[u_m, R_m] + \{\lambda(h) + \mu(\epsilon_2'/\text{Meas}\, S)\}\, \text{Meas}\, R,$

where

$$\epsilon_2' = \int_{R_h} | u - u_h |\, dx + \operatorname*{Max}_{|\xi| \leqq h} \int_{R_h} | u(x + \xi) - u(x) |\, dx. \tag{24}$$

Since $u$ is summable over $R$, the right hand side of (24) tends to zero as $h \to 0$. Therefore by (23)

$$\limsup I[u_h, R_h] \leqq \liminf I[u_m, R_m].$$

The formula $\mathcal{I}[u, R] = \lim I[u_h, R_h]$ follows at once from this inequality, exactly as in the proof of Theorem 1. The second part of Theorem 6 is treated in much the same way, except that the details are simpler.

THEOREM 7. *Let $f$ be of type* I. *Then for any weakly differentiable function $u$ we have $I[u] \leqq \mathcal{I}[u]$, with equality holding whenever $u$ is strongly differentiable. If $f$ is of type* II, *the same result holds provided $u$ is also continuous.*

**Proof.** By Fatou's lemma (see the proof of Theorem 2), and by either Lemma 2 or Lemma 3, we have

$$I[u, S] \leqq \lim I[u_h, S] \leqq \liminf I[u_m, R_m]. \tag{25}$$

The inequality $I[u] \leqq \mathcal{I}[u]$ is an immediate consequence of (25).

Now suppose that $u$ is strongly differentiable. If we set $u_m \equiv u$ in Lemma 2 or Lemma 3, there results

$$\limsup I[u_h, S] \leqq I[u, R]. \tag{26}$$

Let $\{R_\nu\}$ be a sequence of open sets with compact closure in $R$, such that $R_\nu \to R$ as $\nu \to \infty$. In virtue of (26), there exists for each integer $\nu$ a corresponding number $h = h(\nu) \leqq 1/\nu$ such that

$$I[u_{h(\nu)}, R_\nu] \leqq I[u, R] + 1/\nu.$$

Letting $\nu \to \infty$ and using the definition of $\mathcal{I}$ yields $\mathcal{I}[u] \leqq I[u]$, and the proof is complete.

REMARK. The result of Lemma 3, and the parts of Theorems 6 and 7 which refer to integrands of type II, can be extended to apply to integrands of type $\text{II}_B$. In particular, if $|u| < B$ in $R$ then it is easily seen that all the proofs remain essentially unchanged. We shall make use of this fact in the proof of Theorem 9.

7. **Three lemmas.** The deeper considerations of the following sections depend on several technical approximation lemmas which it is convenient to present at this time.

Let $f(t, p)$ be a continuous function on the product space $\Omega \times E^n$, where $\Omega$ is an open set in the $N$ dimensional space $E^N$. Suppose also that

$$f \geqq 0, \qquad f \text{ convex in } p. \tag{27}$$

If $f$ has continuous partial derivatives $f_p$, we define

$$F(t, p, q) = f(t, q) + (p - q) \cdot f_p(t, q)$$

and, for fixed $L > 0$,

$$f_L(t, p) = \operatorname*{Supremum}_{|q| \leqq L} \{0, F(t, p, q)\}.$$

These functions have a simple geometric interpretation when $t$ is held fixed: In the $(n+1)$-dimensional $(p, z)$ space, the equation $z = F(t, p, q)$ represents the tangent plane to the figurative surface $z = f(t, p)$ at the point $(q, f(t, q))$. Similarly, the equation $z = f_L(t, p)$ represents essentially the envelope of a set of tangent planes of the figurative surface. The following lemma describes certain useful properties of $f_L$.

LEMMA 4. *The function $f_L(t, p)$ defined above has the following properties:*

(i) *$f_L$ is continuous in $\Omega \times E^n$, convex in $p$, and $f_L \geqq 0$.*

(ii) *$f_L \leqq f$, equality holding for all $|p| \leqq L$.*

(iii) *For every compact subset $\Sigma$ of $\Omega$ there exists a constant $A$ and a modulus of continuity $\lambda(\sigma)$ such that*

$$f_L(t, p) \leqq A(1 + |p|),$$
$$|f_L(s, p) - f_L(t, p)| \leqq \lambda(|s - t|) \cdot (1 + |p|).$$

**Proof.** Obviously $f_L \geqq 0$. The convexity of $f_L$ is likewise clear, since $f_L$ is the supremum of a family of linear functions of $p$. Also, $f_L$ is continuous, since $F(t, p, q)$ is uniformly continuous over any compact subset of $\Omega \times E^n \times E^n$. This proves property (i). Condition (ii) is evident from the geometrical nature of the construction, or else follows analytically from the inequality $F(t, p, q) \leqq f(t, p)$. Finally, since uniform inequalities of the form (iii) hold for each function $F(t, p, q)$ when $|q| \leqq L$ and $s$ and $t$ are in $\Sigma$, the same inequalities must hold for the supremum of these functions.

LEMMA 5. *Let $f(t, p)$ be continuous on $\Omega \times E^n$, and satisfy* (27). *Then for every pair of positive numbers $L$ and $\epsilon$ and every compact subset $\Sigma$ of $\Omega$, there exists a function $g(t, p)$ with compact support in $\Omega$, satisfying the same hypotheses as $f$, and such that*

(i) $g \leqq f + \epsilon(1 + |p|)$.

(ii) $|g - f| \leqq \epsilon$ *for* $t \in \Sigma$ *and* $|p| \leqq L$.

(iii) *There exist constants $A$ and $C$ such that*

$$g(t, p) \leqq A(1 + |p|),$$
$$|g(s, p) - g(t, p)| \leqq C|s - t|(1 + |p|).$$

**Proof.** In order to construct the function $g$ we shall first need to introduce a certain integral average of the function $f$. For any real number $h>0$ let us define

$$f_{(h)} = f_{(h)}(t, p) = \int K(\tau - t, \xi - p)f(\tau, \xi)d\tau d\xi,$$

where $K(\tau, \xi)$ is an averaging kernel of the type described in §1. We suppose that $h$ is so small that

$$(28) \qquad | f_{(h)} - f | \leqq \epsilon \qquad \text{for } t \in \Sigma_0 \text{ and } | p | \leqq L + 1,$$

where $\Sigma_0$ is some fixed compact subset of $\Omega$ which contains $\Sigma$ in its interior. Let $\alpha(t)$ be a continuously differentiable function with support in $\Sigma_0$, such that $0 \leqq \alpha \leqq 1$ everywhere and $\alpha \equiv 1$ in $\Sigma$. We then set

$$g(t, p) \equiv \alpha(t)f_{(h)L}(t, p).$$

By Lemma 4, $g$ satisfies the same hypothesis as $f$. Moreover, for $t \in \Sigma$ and $|p| \leqq L$ we have $g = f_{(h)L} = f_{(h)}$, so that condition (ii) holds in virtue of (28). Also by Lemma 4, $f_{(h)L}$ satisfies (iii) when $t \in \Sigma_0$ (recall that $f_{(h)}$ is differentiable with respect to $t$, so that the modulus $\lambda$ can be assumed Lipschitzian). Multiplication by $\alpha(t)$ does not affect the validity of (iii) beyond possibly changing the constants. It remains only to verify condition (i). We have

$$f_{(h)L} \leqq f_{(h)} \leqq f + \epsilon \qquad \text{for } t \in \Sigma_0, \ | p | \leqq L + 1,$$
$$f_{(h)L} = f_{(h)} \geqq f - \epsilon \qquad \text{for } t \in \Sigma_0, \ | p | \leqq L.$$

The first inequality implies (i) in the range $|p| \leqq L+1$. On the other hand, for $|p| \geqq L+1$ we have

$$(29) \qquad \begin{aligned} g &\leqq f_{(h)L} \leqq f + 2\epsilon | p |, & t &\in \Sigma_0, \\ g &= 0, & t &\notin \Sigma_0. \end{aligned}$$

(The first inequality of (29) may be obtained most simply by considering the values of the functions $f_{(h)L}$ and $f$ along a "generator" of $f_{(h)L}$ passing through the point $p$: the values of $f_{(h)L}$ are linear and those of $f$ convex.) Making the inessential replacement of $\epsilon$ by $\epsilon/2$ completes the proof.

REMARKS. For later use we shall need two slightly different versions of Lemma 5.

1. Suppose that the function $f$ of Lemma 5 has the additional property that for every compact subset of $\Omega$ there exists a constant $a$ such that $1+f \geqq a|p|$. Then we can replace condition (i) in Lemma 5 by

$$(\text{i})' \qquad g \leqq f + \epsilon(1 + f),$$

this being evident for $t \in \Sigma_0$, and obviously true elsewhere since $g=0$ outside $\Sigma_0$. Similarly, the second part of (iii) implies the further result

$$(\text{iv}) \qquad | g(s, p) - g(t, p) | \leqq D | s - t | \{1 + g(t, p)\}$$

for all $s$ and $t$ in $\Sigma$, and for $L$ sufficiently large.

2. If $f$ is strictly convex in $p$, then we can derive in place of (29) the stronger conclusion

$$f_{(h)L} \leqq \operatorname{Max}[0, f - \kappa(|p| - L - 1) + 2\epsilon|p|], \qquad t \in \Sigma_0, \ |p| \geqq L + 1,$$

where $\kappa$ is some positive constant. Thus by suitably adjusting the number $\epsilon$, $\epsilon < \kappa/2$, it is possible to replace condition (i) by $g \leqq f + \epsilon$, and a fortiori by (i)′.

The remainder of the section concerns the approximation of $I[u]$ by integrals of continuously differentiable functions. Such an approximation can by no means be taken for granted (cf. [16, p. 212]).

LEMMA 6. *Suppose that there exists a constant $A$ and a function $\gamma(p) \geqq 0$ such that*

$$\gamma(p) \leqq f(x, u, p) \leqq A(1 + |p| + \gamma(p)). \tag{30}$$

*Then for every weakly differentiable function $u$ in $R$, and for every compact subset $S$ of $R$, we have* $\lim \int_S f(x, u_h, u_{xh})dx = I[u, S]$.

**Proof.** If $I[u, S] = \infty$ the conclusion is a consequence of Fatou's lemma. Hence it can be assumed that $I[u, S] < \infty$. By application of Jensen's inequality and the hypothesis (30), one has for $x \in S$,

$$\begin{aligned} f(x, u_h, u_{xh}) &\leqq \int K(\xi - x) f[x, u_h(x), u_x(\xi)]d\xi \\ &\leqq A[1 + |u_x|_h + f(x, u, u_x)_h]. \end{aligned}$$

Since $|u_x|_h \to |u_x|$ and $f_h \to f$, the integrals $\int |u_x|_h dx$, $\int f_h dx$, are uniformly absolutely continuous [4, p. 100]. It follows that the integral $\int f(x, u_h, u_{xh})dx$ is also uniformly absolutely continuous.

We next observe that the functions $u_h$ and $u_{xh}$ converge locally to $u$ and $u_x$, respectively. Consequently, $f(x, u_h, u_{xh})$ converges in measure to $f(x, u, u_x)$. But this together with the result of the preceding paragraph implies

$$\lim \int_S f(x, u_h, u_{xh})dx = \int_S f(x, u, u_x)dx,$$

(cf. [4, p. 108]). This completes the proof. As a corollary of Lemma 6, we have the following important result.

THEOREM 8. *If $u$ is Lipschitzian, or if $u$ is strongly differentiable and $f$ satisfies the hypothesis of Lemma 6, then $\mathcal{I}[u] \leqq I[u]$.*

**Proof.** Consider first the case when $u$ is strongly differentiable. Since $u_{hx} = u_{xh}$, Lemma 6 implies

$$\lim I[u_h, S] = I[u, S] \leqq I[u, R].$$

The conclusion $\mathcal{I}[u, R] \leqq I[u, R]$ follows easily. Next, if $u$ is Lipschitz continuous, it is surely strongly differentiable with *bounded* derivatives in any compact $S$. The conclusion of Lemma 6 can therefore be obtained without auxiliary conditions restricting the behavior of $f$ for large $p$. The remainder of the proof is the same as before.

8. **Normal integrands.** In this section we shall consider the important class of integrands which have the property that for each fixed $(x, u)$

$$f(x, u, p) \to \infty \qquad \text{as } |p| \to \infty.$$

Such integrands will here be called *normal.* Since by assumption $f$ is convex in $p$ and continuous, a necessary and sufficient condition that $f$ be normal is that for every compact set of values of $x$ and $u$, there exists a corresponding positive constant $a$ such that $a|p| \leqq 1+f$.

THEOREM 9. *Let $f$ be normal. Then for every continuous weakly differentiable function $u$ we have $I[u] \leqq \mathcal{I}[u]$, equality holding whenever $u$ is Lipschitzian. If $f$ is independent of $u$, we can drop the hypothesis that $u$ be continuous.*

Though we shall make no use of this fact, it may be remarked here that a function is continuous and weakly differentiable if and only if it is of bounded variation in the sense of Tonelli; cf. [5; 14].

**Proof of Theorem 9.** Let $Q$ be a fixed open set with compact closure in $R$. Let $B$ be a bound on the function $u$ in $Q$, and let $\epsilon$ be a small positive number. By Lemma 5 and by the remark following the proof of that lemma, there exists an integrand $g(x, u, p)$ of type $\mathrm{II}_B$ in $Q$ such that

$$\text{(i)} \qquad g \leqq f + \epsilon(1+f)$$

$$\text{(ii)} \qquad |g - f| \leqq \epsilon \qquad \text{for } |u| \leqq B, \text{ and } |p| \leqq 1/\epsilon.$$

Applying Theorem 7 to the integrand $g$, one finds

$$I_g[u, Q] \leqq \mathcal{I}_g[u, Q] \leqq (1+\epsilon)\mathcal{I}[u, R] + \epsilon \text{ Meas } Q.$$

By property (ii), $\lim_{\epsilon\to 0} g(x, u, u_x) = f(x, u, u_x)$ almost everywhere in $Q$. Therefore by Fatou's lemma $I[u, Q] \leqq \mathcal{I}[u, R]$. Letting $Q \to R$ gives the inequality of Theorem 9. Moreover, in consequence of Theorem 8, $I[u] = \mathcal{I}[u]$ whenever $u$ is Lipschitzian.

When $f = f(x, p)$ the proof is very similar, except that the function $g(x, p)$ is now of type I. The following result represents our ultimate generalization of Tonelli's theorem.

THEOREM 10. *Let $R$ be a bounded open region. Suppose that there exist constants $a$ and $A$ such that*

$$a|p| \leqq 1 + f(x, u, p) \leqq A|p| \qquad \textit{when } |p| \geqq 1.$$

*Then $\mathcal{I}[u]$ is finite if and only if $u$ is weakly differentiable.*

*When this is the case, and when, further, u is continuous if f depends on u, we have*

$$\mathcal{J}[u] \geqq I[u] \tag{31}$$

*and* $\mathcal{J}'(x) = f(x, u, u_x)$ *almost everywhere in R. In order that* $\mathcal{J}[u] = I[u]$ *it is necessary and sufficient that u be strongly differentiable; and in order that this be the case it is necessary and sufficient that* $\mathcal{J}[u, Q]$ *be an absolutely continuous function of open sets Q in R.*

**Proof.** Except for the formula $\mathcal{J}' = f$, everything is just as in the proof of Theorem 5, with the basic inequality (31) now following from Theorems 8 and 9 rather than from Theorem 2.

To prove that $\mathcal{J}' = f$ it is enough (as in the proof of Theorem 5) to show that for every open interval $J$ in $R$

$$\mathcal{J}[u, J] \leqq I[u, J] + \Theta(J). \tag{32}$$

Clearly

$$f(x, u_h, u_{hx}) \leqq f(x, u_h, u_{xh}) + A\left|\int K(\xi - x)d\theta(\xi)\right|.$$

Integrating over $J_h$ easily yields

$$I[u_h, J_h] \leqq \int_J f(x, u_h, u_{xh})dx + \Theta(J), \tag{33}$$

where $\Theta$ is the singular measure $A|\theta|$. Now let $h \to 0$. The left side of (33) certainly tends to a number greater than or equal to $\mathcal{J}[u, J]$. The first term on the right tends, by Lemma 6, to $I[u, J]$. This proves (32), and thus also Theorem 10.

## Part III

9. **General considerations. Lower semicontinuity.** There are many integrands which do not satisfy the conditions of the preceding sections, but which nevertheless are amenable to an alternate approach. It is the purpose of the present part to investigate the functions $I[u]$ and $\mathcal{J}[u]$ corresponding to these more difficult integrands. The major interest will be in determining conditions guaranteeing that $\mathcal{J}[u] = I[u]$ for strongly differentiable functions.

As we have seen, this involves two essentially different considerations. Namely, one must show on the one hand that $\mathcal{J}[u] \leqq I[u]$ and on the other that $I[u] \leqq \mathcal{J}[u]$. A moment's reflection will show that the former inequality is connected with the problem of approximating $I[u]$ by integrals $I[u_m]$ of continuously differentiable functions. The best result available in this respect is Lemma 6, with its attendant corollary, Theorem 8. This being the case, we shall turn our full attention to the latter inequality $I[u] \leqq \mathcal{J}[u]$. The reader has no doubt observed the close relation between this inequality and the traditional lower semicontinuity properties of variational integrals. Indeed, the inequality $I[u] \leqq \mathcal{G}[u]$ is itself a type of lower semicontinuity theorem, while, conversely, certain lower semicontinuity theorems imply the inequality. It thus becomes evident that our goal must be to obtain as strong lower semicontinuity theorems as possible: *These results, in conjunction with Theorem* 8, *will provide us with the required relations between* $I[u]$ *and* $\mathcal{G}[u]$. *In particular, since* $\mathcal{G}[u] \leqq I[u]$ *for any continuously differentiable function, a lower semicontinuity theorem immediately implies the* consistency *of* $\mathcal{G}[u]$ *and* $I[u]$ *for the corresponding integrand.*

Before stating our main results, we shall for completeness first set down the lower semicontinuity theorems implied by the preceding work.

THEOREM 11. *Suppose that f is of type* I. *Let u be weakly differentiable in R, and let* $\{u_m\}$ *be a sequence of strongly differentiable functions which are locally convergent to u in R. Then*

$$I[u, R] \leqq \liminf I[u_m, R_m].$$

*If f is independent of u, we may replace local convergence by weak convergence in the sense of* § 5.

*Suppose that f is of type* II. *Let u be a continuous weakly differentiable function in R, and let* $\{u_m\}$ *be a sequence of strongly differentiable functions which are convergent in measure to u in R. Then*

$$I[u, R] \leqq \liminf I[u_m, R_m].$$

**Proof.** For local convergence, these results follow at once from Lemma 2 and Lemma 3, respectively, and Fatou's lemma. The more general convergence allowed in the statement of the theorem is due to the fact that Lemmas 2 and 3 actually hold for the type of convergence noted here. The main results of Part III may now be stated:

THEOREM 12. *Suppose that* $f=f(x, u, p)$ *satisfies any one of the following conditions*:

A. *f is normal.*

B. *f is strictly convex.*

C. *The derivatives* $f_x$, $f_p$, *and* $f_{px}$ *exist and are continuous.*

*Then on the class of strongly differentiable functions the integral* $I[u]$ *is lower semicontinuous with respect to local convergence in* $L_1$.

THEOREM 13. $I[u]$ *is lower semicontinuous with respect to local convergence in* $L_1$ *on any class of strongly differentiable functions uniformly satisfying the inequality* $D[u] \leqq H < \infty$.

The following three sections of the paper are devoted to the proof of Theorems 12 and 13.

REMARKS. As we have already noted in the introduction, Theorems 11 through 13 are generalizations of well-known lower semicontinuity theorems of Tonelli and Morrey. Other related results are due to S. Cinquini and L. Giuliano.

In consequence of Theorem 12 it is apparent that $I[u] \leqq \mathcal{J}[u]$ for any integrand of type A, B, or C, and any strongly differentiable function $u$. Furthermore, if $u$ is Lipschitzian then $I[u] = \mathcal{J}[u]$.

Finally, because the hypotheses on $f$ in Theorem 12 are very weak, one might suspect that the conclusion holds with no restrictions on $f$ beyond those formulated at the outset of the paper. This is, however, not the case. There do in fact exist non-negative convex integrands for which semicontinuity fails for local convergence. Such an example (necessarily fairly complicated) was constructed by Aronszajn and appears in reference [10]. On the other hand, Theorem 13 shows that if certain restrictions are placed on the functions $u$, then lower semicontinuity does indeed hold for general integrands.

10. **Preliminary lemmas.** It is convenient to begin with some simple lemmas, the first of which generalizes a well-known result of Tonelli.

LEMMA 7. *Let $S$ be a compact set in $E^n$, and define*

$$\mathcal{K}[u] = \int_S Q(x, u) \cdot u_x dx,$$

*where $Q$ is a continuous vector-valued function with support in $S$. Suppose there exist constants $M$ and $M'$ such that*

$$|Q| \leqq M, \qquad |Q(x, u) - Q(y, u)| \leqq M'|x - y|.$$

*Then for every pair of strongly differentiable functions $u$ and $v$ which are defined on some open set containing $S$, we have*

$$|\mathcal{K}[u] - \mathcal{K}[v]| \leqq nM' \int_S |u - v| dx. \tag{34}$$

**Proof.** Suppose first that $Q$ is differentiable with respect to $x$, and that $|Q_x| \leqq M'$. Setting

$$P(x) \equiv \int_{u_h}^{v_h} Q(x, \sigma) d\sigma, \qquad x \in S,$$

we find by the divergence theorem

$$0 = \int_S \nabla \cdot P dx = \mathcal{K}[v_h] - \mathcal{K}[u_h] + \int_S \int_{u_h}^{v_h} \nabla_x \cdot Q dx.$$

Thus

$$|\mathcal{K}[u_h] - \mathcal{K}[v_h]| \leqq nM' \int_S |u_h - v_h| dx.$$

Letting $h \to 0$ and applying Lemma 6 yields (34). The result for nondifferentiable $Q$ is easily reduced to this one by smoothing $Q$ with respect to the variable $x$.

LEMMA 8. *Let $f$ satisfy any one of the hypotheses* A, B, *or* C *of Theorem* 12. *Then for every $\epsilon>0$ and every compact subset $S$ of $R$ there exists a function $g=g(x, u, p)$ with the properties*

(i) $g\leqq f+\epsilon(1+f)$,

(ii) $|g-f|\leqq\epsilon$ *for* $x\in S$, $|u|+|p|\leqq 1/\epsilon$.

(iii) $g$ *is continuous, convex in $p$, and $g\geqq 0$. Furthermore, $g$ has compact support in $x$ and $u$.*

(iv) $g$ *has continuous partial derivatives $g_p$ which are Lipschitz continuous in $x$ and $p$ and satisfy $|g_p|\leqq M$ for some constant $M$.*

**Proof.** Consider first the case when $f$ satisfies hypothesis A or B. By Lemma 5 and the remarks following this lemma, there exists a function $g'$ satisfying conditions (i)–(iii), Lipschitz continuous in $x$, and such that $g'\leqq A(1+|p|)$. Averaging this function with respect to $p$ does not alter the validity of (i)–(iii), while at the same time (iv) becomes true with $M=nA$.

In case C, let $\alpha=\alpha(x, u)$ be a function with compact support in $R\times E^1$, satisfying $0\leqq\alpha\leqq 1$, and such that $\alpha\equiv 1$ when $x\in S$ and $|u|\leqq 1/\epsilon$. We set

$$g'(x, u, p) = \alpha(x, u)f_L(x, u, p), \qquad L = 1/\epsilon.$$

By Lemma 4, conditions (i)–(iii) are satisfied by $g'$. Furthermore, $g'\leqq A(1+|p|)$ and $g'$ is Lipschitz continuous in $x$. Smoothing $g'$ with respect to $p$ completes the proof.

11. **Proof of Theorem 12.** Suppose to begin with that lower semicontinuity has been established for integrands $g$ which satisfy conditions (iii) and (iv) of Lemma 8. We shall show that this implies the truth of Theorem 12.

For let $f$ be an integrand satisfying the hypotheses of Theorem 12, let $S$ be an arbitrary compact subset of $R$, and let $\epsilon>0$ be a small constant. Let $g$ denote the integrand associated to $f$, $S$, and $\epsilon$ by Lemma 8; in view of condition (iii) we may suppose that $g$ vanishes outside some compact subset $S_0$ of $R$. Then by our initial assumption and by property (i) of Lemma 8, if $u_m\to u$,

$$(35)\quad I_g[u, S] \leqq \liminf I_g[u_m, R_m] \leqq (1+\epsilon)\liminf I[u_m, R_m] + \epsilon \operatorname{Meas} S_0.$$

Now let $\epsilon\to 0$. By virtue of (ii) and Fatou's lemma $I[u, S]\leqq\lim I_g[u, S]$, whence we obtain from (35),

$$I[u, S] \leqq \liminf I[u_m, R_m].$$

Letting $S\to R$ completes the proof of Theorem 12, subject of course to the initial assumption concerning $g$.

It remains, then, to show that $I_g[u]$ is lower semicontinuous. Let $S_0$ denote the support of $g$ in $R$, and let $\epsilon$ be a given small number. Let us fix $h$ so small that $u_h$ is defined in $S_0$ and

$$\int_{S_0} | u_x - u_{xh} | \, dx \leq \epsilon. \tag{36}$$

Now the Weierstrass $E$-function associated with the integrand $g$ is defined by

$$E(x, u, p, q) = g(x, u, q) - g(x, u, p) - (q - p) \cdot g_p(x, u, p).$$

Since $g$ is convex and $|g_p| \leq M$, it is apparent that $0 \leq E(x, u, p, q) \leq 2M|p-q|$. Therefore, if we make the abbreviations $u_x = p$, $u_{mx} = p_m$, we have for all $x \in S_0$ and all $m$ suitably large

$$\begin{aligned} &g(x, u_m, p_m) - g(x, u, p) \\ &\quad \geq g(x, u_m, p_h) + (p_m - p_h) \cdot g_p(x, u_m, p_h) \\ &\quad\quad - g(x, u, p_h) - (p - p_h) \cdot g_p(x, u, p_h) - 2M | p - p_h | . \end{aligned} \tag{37}$$

Consider the function $Q(x, u) \equiv g_p[x, u, p_h(x)]$. By hypothesis $g_p(x, u, p)$ is Lipschitz continuous in $x$ and $p$. Moreover, since $g$ has compact support in $x$ and $u$ the Lipschitz coefficient is uniformly bounded over any set of points $(x, u, p)$ for which $|p|$ is uniformly bounded. It follows that there exists a constant $M'$, depending at most on the previously fixed number $h$, such that

$$| Q(x, u) - Q(y, u) | \leq M' | x - y | .$$

In other words, $Q(x, u)$ satisfies precisely the hypotheses of Lemma 7, with $S$ replaced by $S_0$.

The next step is to integrate (37) over $S_0$. Before writing down the result of this integration, it is convenient to estimate the integral of the right hand side. Specifically, since $u_m \rightarrow u$, one has essentially by Lebesgue's bounded convergence theorem

$$\int_{S_0} [g(x, u_m, p_h) - g(x, u, p_h)] dx \rightarrow 0,$$

$$\int_{S_0} p_h \cdot [g_p(x, u_m, p_h) - g_p(x, u, p_h)] dx \rightarrow 0,$$

as $m \rightarrow \infty$. The remaining difference between terms on the second and third lines of (37) can be estimated by Lemma 7. Hence, making use of (36) we obtain

$$I_g[u_m, S_0] - I_g[u, S_0] \geq - nM' \int_{S_0} | u_m - u | \, dx - \epsilon_1 - 2M\epsilon,$$

where $\epsilon_1$ tends to zero as $m \rightarrow \infty$. Now letting $m \rightarrow \infty$ through an appropriate subsequence yields

$$\liminf I_g[u_m, S_0] - I_g[u, S_0] \geq - 2M\epsilon.$$

$$\mathfrak{F}\mu(E) = \sup \sum_1^N \mathfrak{F}(\mu(E_i)),$$

where the supremum is taken over all finite partitions $\cup E_i$ of $E$ with $E_i \in \mathbf{R}$. We assert that $\mathfrak{F}\mu$ is a scalar measure on the ring $\mathbf{R}$. In order to prove this, it is enough to show that $\mathfrak{F}\mu(E) > -\infty$, and that $\mathfrak{F}\mu$ is countably additive on $\mathbf{R}$. Now by the Hahn-Banach theorem there exists a linear functional $\mathfrak{L}(p)$ such that $\mathfrak{L}(p) \leq \mathfrak{F}(p)$ for all $p \in \mathfrak{X}$. If we set $\mathfrak{G}(p) = \mathfrak{F}(p) - \mathfrak{L}(p)$, then evidently $\mathfrak{G}(p)$ is sublinear and

$$\mathfrak{G}\mu \geq 0, \qquad \mathfrak{F}\mu = \mathfrak{G}\mu + \mathfrak{L}\mu.$$

One checks easily that $\mathfrak{L}\mu(E) = \mathfrak{L}(\mu(E))$, and the displayed formulas therefore imply $\mathfrak{F}\mu > -\infty$.

Suppose next that $E = \cup E_j$ is a countable partition of $E$ into sets $E_j \in \mathbf{R}$. Since $\mathfrak{L}(p)$ is continuous (observe that $|\mathfrak{L}(p)| \leq C\,|p|$), it is easy to show that

$$\mathfrak{L}\mu(E) = \sum \mathfrak{L}\mu(E_j).$$

Moreover, in view of the positivity of $\mathfrak{G}$ we have

$$\mathfrak{G}\mu(E) \geq \sum_1^M \mathfrak{G}\mu(E_j)$$

for all integers $M > 0$. Therefore $\mathfrak{G}\mu(E) \geq \sum \mathfrak{G}\mu(E_j)$. To establish the opposite inequality, let $\cup E_i'$ be an arbitrary *finite* partition of $E$ into sets $E_i' \in \mathbf{R}$. Put $E_{ij} = E_i' \cap E_j$. Then $\mu(E_i') = \sum_j \mu(E_{ij})$, and by the continuity and sublinearity of $\mathfrak{G}$,

$$\sum_i \mathfrak{G}(\mu(E_i')) \leq \sum_i \sum_j \mathfrak{G}(\mu(E_{ij}))$$
$$= \sum_j \sum_i \mathfrak{G}(\mu(E_{ij})) \leq \sum_j \mathfrak{G}\mu(E_j).$$

It follows that $\mathfrak{G}\mu(E) \leq \sum \mathfrak{G}\mu(E_j)$, whence $\mathfrak{G}\mu$ as well as $\mathfrak{L}\mu$ is countably additive. This completes the proof of our assertion.

Since $\mathfrak{F}\mu$ is now shown to be a measure on $\mathbf{R}$, it is possible to consider integration with respect to this measure. The following lemmas will be useful in this regard.

LEMMA 1. *Let $\mu$ and $\mu'$ be vector measures. Then*

$$|\mathfrak{F}\mu - \mathfrak{F}\mu'| \leq C\,|\mu - \mu'|.$$

*Proof.* Let $E = \cup E_i$ be a finite partition of $E$ such that, simultaneously,

$$\mathfrak{F}\mu(E) \leq \sum \mathfrak{F}(\mu(E_i)) + \epsilon$$

and

$$\mathfrak{F}\mu'(E) \leq \sum \mathfrak{F}(\mu'(E_i)) + \epsilon, \qquad (\epsilon > 0)$$

(to show that such a partition exists, consider refinements of partitions holding for $\mathfrak{F}\mu$ and $\mathfrak{F}\mu'$ separately). These inequalities yield

$$\begin{aligned} |\mathfrak{F}\mu(E) - \mathfrak{F}\mu'(E)| &\leq |\sum \mathfrak{F}(\mu(E_i)) - \sum \mathfrak{F}(\mu'(E_i))| + 2\epsilon \\ &\leq C \sum |\mu(E_i) - \mu'(E_i)| + 2\epsilon \\ &\leq C\, |\mu - \mu'|\,(E) + 2\epsilon. \end{aligned}$$

Since $\epsilon$ is arbitrary, the required conclusion follows at once.

LEMMA 2. *If $\alpha$ and $\beta$ are vector measures, and if $\mu = \alpha + \beta$, then*

$$\mathfrak{F}\mu \leq \mathfrak{F}\alpha + \mathfrak{F}\beta,$$

*with equality holding whenever $\alpha$ and $\beta$ are mutually singular (i.e. have disjoint supports).*

This is obvious from the definition of $\mathfrak{F}\mu$. The next result shows that $\mathfrak{F}$ has certain properties similar to a norm.

THEOREM 1. *If $\omega$ is a non-negative $\mu$-integrable function on $S$, then*

$$\mathfrak{F}\left(\int \omega\, d\mu\right) \leq \int \omega\, d\mathfrak{F}\mu.$$

*Proof.* If $\mathfrak{L}$ is a bounded linear functional, then

$$\mathfrak{L}\left(\int \omega\, d\mu\right) = \int \omega\, d\mathfrak{L}\mu,$$

cf. [1, Theorem 10.8, Chapter IV]. (The definition of integral in [1, §IV.10] may be given without change for a $\sigma$-ring, and the theory of that section holds in the more general case since the support of a summable function is an element of the ring.) According to the earlier decomposition $\mathfrak{F} = \mathfrak{G} + \mathfrak{L}$, it is therefore enough to prove Theorem 1 for non-negative functionals. We suppose to begin with that $\omega$ is a simple function, with non-negative constant values $x_i$ on the sets $E_i \in \mathbf{R}$. Then

$$\begin{aligned} \mathfrak{F}\left(\int \omega\, d\mu\right) &= \mathfrak{F}(\sum x_i \mu(E_i)) \\ &\leq \sum x_i \mathfrak{F}(\mu(E_i)) \\ &\leq \sum x_i \mathfrak{F}\mu(E_i) = \int \omega\, d\mathfrak{F}\mu. \end{aligned}$$

In the general case, let $\{\omega^k\}$ be a sequence of simple functions such that $\omega^k \leq \omega$ and $\lim \omega^k = \omega$ $\mu$-almost everywhere [1; 322]. Then by the dominated convergence theorem

$$\lim \int \omega^k\, d\mu = \int \omega\, d\mu.$$

Let $K(x) = K(x, k)$ be a continuous, non-negative real function on $R^n$, with support in the sphere $|x| \leq k^{-1}$, and such that

$$\int K \, dm = 1,$$

(here $m$ denotes Lebesgue measure on $S$). Then the following representation theorem holds.

THEOREM 4. *Let $\mu$ be a regular vector measure on the Borel sets of $S$, and let*

$$v^k(x) = \int K(x - \xi) \, d\mu(\xi).$$

*Then for every open set $G$ in $S$,*

$$\mathfrak{F}\mu(G) = \lim \int_{G_k} \mathfrak{F}(v^k) \, dm,$$

*where $G_k$ denotes the open subset of $G$ whose points are further than $k^{-1}$ from the boundary of $G$.*

*Remarks.* It should be observed that $v^k$ is a continuous function defined for all $x \in S_k$. Thus the integral occurring in the representation formula for $\mathfrak{F}\mu$ is well defined. Moreover, the hypothesis that $\mu$ is regular is actually unnecessary, since any bounded measure is regular on a metric space.

*Proof of Theorem* 4. Define a new measure $\mu^k$ by means of

$$\mu^k(E) = \int_{E \cap S_k} v^k \, dm.$$

We assert that $\mu^k$ converges weakly to $\mu$. Indeed, let $\omega$ be any continuous function with compact support in $S$. Then for all sufficiently large $k$

$$\int \omega \, d\mu^k = \int \omega v^k \, dm = \int \omega(x) \left\{ \int K(x - \xi) \, d\mu(\xi) \right\} dm(x)$$

$$= \int \left\{ \int K(x - \xi) \omega(x) \, dm(x) \right\} d\mu(\xi).$$

Since $\omega$ is continuous, the inner integral converges uniformly to $\omega(\xi)$. The right-hand side thus tends to $\int \omega \, d\mu$, and the assertion is proved.

According to the decomposition $\mathfrak{F} = \mathfrak{G} + \mathfrak{L}$, it is sufficient for us to consider only the two special cases (A), when $\mathfrak{F}(p) \geq 0$, and (B), when $\mathfrak{F}(p) = \mathfrak{L}(p)$ is a continuous linear functional.

*Case* (A). By the Corollary to Theorem 3, and by Theorem 2, we have

$$\mathfrak{F}\mu(G) \leq \lim\inf \mathfrak{F}\mu^k(G_k) = \lim\inf \int_{G_k} \mathfrak{F}(v^k) \, dm.$$

On the other hand, using Theorem 1,

$$\begin{aligned}\int_{G_k} \mathfrak{F}(v^k)\, dm &= \int_{G_k} \mathfrak{F}\left(\int K(x-\xi)\, d\mu(\xi)\right) dm(x)\\ &\leq \int_{G_k} \left\{\int K(x-\xi)\, d\mathfrak{F}\mu(\xi)\right\} dm(x)\\ &\leq \int_{G} \left\{\int K(x-\xi)\, dm(x)\right\} d\mathfrak{F}\mu(\xi) = \mathfrak{F}\mu(G).\end{aligned}$$

Consequently lim sup $\int_{G_k} \mathfrak{F}(v^k)\, dm \leq \mathfrak{F}\mu(G)$, and Case (A) of the theorem is proved.

*Case* (B). Let $D_k$ denote the set $G - G_{2k}$. The following calculation is easily checked,

$$\begin{aligned}\int_{G_k} v^k\, dm &= \int_{G_k}\int_G K(x-\xi)\, d\mu(\xi)\, dm(x)\\ &= \int_{G_{2k}}\int_{G_k} K(x-\xi)\, dm(x)\, d\mu(\xi) + \int_{D_k}\int_{G_k} K(x-\xi)\, dm(x)\, d\mu(\xi)\\ &= \mu(G_{2k}) + \int_{D_k}\int_{G_k} K(x-\xi)\, dm(x)\, d\mu(\xi).\end{aligned}$$

It follows that

$$\mathfrak{L}\mu(G) - \int_{G_k} \mathfrak{L}(v^k)\, dm = \mathfrak{L}\left(\mu(G) - \int_{G_k} v^k\, dm\right) = \mathfrak{L}\left(\int_{D_k} \omega\, d\mu\right),$$

where

$$\omega = \omega(x) = 1 - \int_{G_k} K(\xi - x)\, dm(\xi).$$

Since $0 \leq \omega \leq 1$, we have then

$$\left|\mathfrak{L}\mu(G) - \int_{G_k} \mathfrak{L}(v^k)\, dm\right| \leq C \left|\int_{D_k} \omega\, d\mu\right| \leq C\, ||\mu||\, (D_k).$$

Now by the assumed regularity of the measure $\mu$, the expression on the right must tend to zero as $k \to \infty$. Thus

$$\mathfrak{L}\mu(G) = \lim \int_{G_k} \mathfrak{L}(v^k)\, dm,$$

and the theorem is proved.

*Remarks.* Although the fact has little bearing on the present work, it is of some interest that Theorem 4 holds in a more general setting than that of euclidean space. Let $H$ be a locally compact topological group with identity $e$ and, say, left invariant measure $m$. Let $\{U_\alpha\}$ be a directed set of symmetric open neighborhoods of $e$ which forms a base for the topology at $e$. (Then for every $\alpha$, $\beta$ there is $\gamma > \alpha, \beta$ and $U_\gamma \subset U_\alpha \cap U_\beta$ and for every $\alpha$, $U_\alpha^{-1} = U_\alpha$). For every $\alpha$, let $K(x) = K(x, \alpha)$ be a continuous non-negative real function on $H$, with support in $U_\alpha$, such that

$$\int K\, dm = 1.$$

Let $S$ be an open set in $H$ and $\mu$ a regular vector measure on $S$. If $G \subset S$ is open, then for every $\alpha$, let $G_\alpha$ be the interior of the set of $x \in S$ such that $xU_\alpha \subset G$. Then $G = \bigcup G_\alpha$. The functions

$$v^\alpha(x) = K(x\xi^{-1}, \alpha)\, d\mu(\xi)$$

and the measures

$$\mu^\alpha(E) = \int_{E\cap S_\alpha} v^\alpha\, dm$$

remain meaningful, and the proof that $\mu^\alpha$ converges weakly to $\mu$ is virtually unchanged, where the uniform convergence of the directed set

$$\left\{\int K(x\xi^{-1}, \alpha)w(x)\, dm\right\}$$

to $w(\xi)$ means that for every $\epsilon > 0$ there is an $\alpha$ such that $\beta > \alpha$ implies

$$\left|\int K(x\xi^{-1}, \beta)w(x)\, dm(x) - w(\xi^{-1})\right| < \epsilon$$

for every $\xi \in H$.

Regarding the remainder of the proof that

$$\mathfrak{F}\mu(G) = \lim \int_{G_\alpha} \mathfrak{F}(v^\alpha)\, dm,$$

all steps are retained without change. It is only necessary to note that the role of $G_{2k}$ is played by $G_{\beta(\alpha)}$ where $\beta = \beta(\alpha)$ is such that $U_\beta \cdot U_\beta \subset U_\alpha$ and that $D_\alpha = G - G_{\beta(\alpha)}$.

3. **The measure $f\mu$.** From here on we shall suppose that the vector measure $\mu$ takes its values in the euclidean vector space $R^s$. Although this restriction is not essential for most of the results of this section, it greatly simplifies our discussion.

Now let $f(p)$ be a convex function defined at all points $p \in R^s$. With this function we may associate a unique sublinear function $\mathfrak{F}(p, t)$, according to the following simple

LEMMA 3. *There exists a unique lower semicontinuous sublinear function $\mathfrak{F}(p, t)$, defined for all $p \in R^s$ and all $t \geq 0$, such that $\mathfrak{F}(p, 1) = f(p)$.*

*Proof.* For $t > 0$ it is evident that we must have

$$\mathfrak{F}(p, t) = tf(p/t).$$

What must be shown is that $\mathfrak{F}(p, t)$ can be extended lower semicontinuously to the plane $t = 0$, and still remain sublinear. Now for fixed $p$ it is easy to see that $\mathfrak{F}(p, t)$ is a convex function of $t$ in the interval $0 < t < 1$. Consequently

$$\lim_{t \to 0} \mathfrak{F}(p, t)$$

exists (finite or infinite). This limit we accept as the definition of $\mathfrak{F}(p, 0)$.

For $t > 0$, $\mathfrak{F}(p, t)$ is continuous and sublinear. Moreover, with the above definition of $\mathfrak{F}(p, 0)$ it is readily apparent that the sublinearity persists for all $t \geq 0$. Now consider a sequence of points $(p_k, t_k)$ with $t_k \geq 0$, tending to a point $(p_0, 0)$. To complete the proof we must show that

$$\mathfrak{F}(p_0, 0) \leq \liminf \mathfrak{F}(p_k, t_k).$$

Let $\epsilon$ be any positive number. By the definition of $\mathfrak{F}(p, 0)$ we can find a $\tau \leq \epsilon$ such that

$$\mathfrak{F}(p_0, 0) \leq \mathfrak{F}(p_0, \tau) + \epsilon.$$

Also, since $f(p)$ is continuous, there is a neighborhood $N$ of $p_0$ such that

$$|\mathfrak{F}(p, \tau) - \mathfrak{F}(p_0, \tau)| \leq \epsilon, \qquad p \in N.$$

Finally, for $0 \leq t_k < \tau$ and all values of $p$

$$\begin{aligned} \mathfrak{F}(p, \tau) &\leq \mathfrak{F}(p, t_k) + \mathfrak{F}(0, \tau - t_k) \\ &\leq \mathfrak{F}(p, t_k) + |f(0)|\, \epsilon. \end{aligned}$$

Combining the last three displayed inequalities we find

$$\mathfrak{F}(p_0, 0) \leq \mathfrak{F}(p_k, t_k) + \{2 + |f(0)|\}\epsilon$$

for $p_k \in N$ and $0 \leq t_k < \tau$. The required assertion is an immediate consequence, and the lemma is proved. (An essentially similar result was proved by McShane in a study of non-parametric variational problems [3; 188–189].)

In what follows, the letter $\mathfrak{F}$ will always refer to the function $\mathfrak{F}(p, t)$ constructed in Lemma 3. We shall also occasionally find it convenient to denote the point $(p, t)$ by the simple symbol $p$.

Let $\mu$ be a vector measure of the type described in §1, except that its values are now in $R^s$. We denote by $\tilde{\mu}$ the $(s + 1)$-component vector measure $(\mu, \nu)$, where $\nu$ is a given finite-valued positive measure *which will remain fixed during the following discussion.* Finally, for any set $E$ in the ring $\mathsf{R}$ we define

$$f\mu(E) = \mathfrak{F}\tilde{\mu}(E).$$

(In order that this definition be consistent with §1, it must be shown that if $f(p)$ is sublinear (as well as being convex), then the two possible meanings for $f\mu(E)$, namely $\mathfrak{F}\tilde{\mu}(E)$ and sup $\sum_i f(\mu(E_i))$, agree. But this follows at once from the fact that $\mathfrak{F}(p, t) \equiv f(p)$ for sublinear functions $f$.) It is our purpose next to recast the results of §1 in terms of the set function $f\mu$.

To begin with, we observe that by the convexity of $f(p)$ there exists a vector $r \in R^s$ and a number $r'$ such that $f(p) \geq r \cdot p + r'$. The application of the Hahn-Banach theorem in §1 can therefore be avoided by simply defining

$$\mathfrak{L}(\tilde{p}) = \mathfrak{L}(p, t) = r \cdot p + r't.$$

The rest of the proof that $\mathfrak{F}\tilde{\mu}$ is countably additive carries through word for word except that the lower semicontinuity rather than continuity of $\mathcal{G}$ is employed at the final step. Thus $f\mu$ is indeed a scalar measure on the ring $\mathsf{R}$.

The proof of Theorem 1 likewise applies almost unchanged to the measure $\mathfrak{F}\tilde{\mu}$. When translated into a statement involving $f\mu$ we obtain the interesting inequality

$$f\left[\frac{\int \omega \, d\mu}{\int \omega \, d\nu}\right] \leq \frac{\int \omega \, df\mu}{\int \omega \, d\nu},$$

provided that $\int \omega \, d\nu \neq 0$. This result may be considered a generalization of the classical *Jensen inequality*.

Theorem 2 is a more difficult matter because the boundedness condition $\mathfrak{F}(p) \leq C\,|p|$, which was strongly used during the proof, need not hold for the function $\mathfrak{F}(\tilde{p})$. Nevertheless, a corresponding result remains true.

THEOREM 2′. *Let $\mu$ be a vector measure on the ring $\mathsf{R}$, defined by*

$$\mu(E) = \alpha(E) + \beta(E) = \int_E a \, d\nu + \beta(E),$$

*where $a$ is an integrable vector field, and $\beta$ is singular with respect to $\nu$. Then*

$$f\mu(E) = \int_E f(a) \, d\nu + f^*\beta(E),$$

*where*

$$f^*(p) = \mathfrak{F}(p, 0) = \lim_{t \to 0} tf(p/t).$$

*Proof.* The measure $\mu$ can be written in the form

$$\tilde{\mu} = (\mu, \nu) = (\alpha + \beta, \nu) = (\alpha, \nu) + (\beta, 0),$$

so that $f\mu = \mathfrak{F}\tilde{\mu} = f\alpha + f^*\beta$, by Lemma 2. It remains, therefore, to show that

$$f\alpha(E) = \int_E f(a) \, d\nu.$$

Let $\Sigma$ denote the surface $z = f(p)$ in a euclidean space of $s + 1$ dimensions, and set

$$f_k(p) = \sup l(p)$$

where the supremum is taken over all linear functions $l(p)$ defining support planes of $\Sigma$ which touch $\Sigma$ at a point $(z', p')$ with $|p'| \leq k$. Obviously

$$\mathfrak{F}_k(\bar{p}) = \mathfrak{F}(\bar{p}) \quad \text{for} \quad |\mathrm{p}| \leq kt$$

and

$$\mathfrak{F}_k(\bar{p}) \leq \mathfrak{F}(\bar{p}) \quad \text{otherwise.}$$

Moreover, it is geometrically evident that for each $k$ there exists a number $C_k$ with the property $|f_k(p)| \leq C_k \sqrt{1 + |p|^2}$. Hence

$$|\mathfrak{F}_k(\bar{p})| \leq C_k \, |\bar{p}|.$$

Therefore, by Theorem 2,

$$\mathfrak{F}_k\tilde{\alpha}(E) = \int_E \mathfrak{F}_k(a, 1) \, d\nu = \int_E f_k(a) \, d\nu.$$

We may now let $k$ tend to infinity. By the monotone convergence theorem the integral on the right tends to $\int f(a) \, d\nu$. The set function $\mathfrak{F}_k\tilde{\alpha}$ is a bit harder to handle.

Suppose first that $\mathfrak{F}\tilde{\alpha}(E) < \infty$, and let $\cup E_i$ be a finite partition of $E$ into sets $E_i \in \mathbf{R}$ such that

$$\mathfrak{F}\tilde{\alpha}(E) \leq \sum \mathfrak{F}(\tilde{\alpha}(E_i)) + \epsilon \qquad (\epsilon > 0).$$

Since $\nu(E_i) = 0$ implies $\alpha(E_i) = 0$, and since there are only a finite number of sets $E_i$, the numbers $\tilde{\alpha}(E_i)$ all lie in some cone $|p| \leq k't$ in $R^{s+1}$. Consequently, for all $k \geq k'$ we have $\mathfrak{F}(\tilde{\alpha}(E_i)) = \mathfrak{F}_k(\tilde{\alpha}(E_i))$, and so

$$\mathfrak{F}\tilde{\alpha}(E) \leq \sum \mathfrak{F}_k(\tilde{\alpha}(E_i)) + \epsilon \leq \mathfrak{F}_k\tilde{\alpha}(E) + \epsilon.$$

Therefore, $\mathfrak{F}\tilde{\alpha}(E) \leq \liminf \mathfrak{F}_k\tilde{\alpha}(E)$. The opposite inequality being obvious, it follows that

$$\lim \mathfrak{F}_k\tilde{\alpha}(E) = \mathfrak{F}\tilde{\alpha}(E) = f\alpha(E).$$

If $\mathfrak{F}\tilde{\alpha}(E) = \infty$, the argument is essentially the same. We thus have $f\alpha(E) = \int_E f(a) \, d\nu$, and the proof is complete.

Turning next to Theorem 3, we observe that the proof applies almost word for word to the measure $\mathfrak{F}\tilde{\mu}$, except that lower semicontinuity rather than continuity must be used at several intermediate steps. It is also necessary to have $\nu$ a *regular* measure on $\mathbf{R}$.

THEOREM 4′. *Let $S$ be a bounded open set in euclidean space, and $\mu$ a regular vector measure on the Borel sets of $S$. Fix $\nu = m =$ Lebesgue measure, in the* *definition of* $f\mu$. *Then with the remaining notation as in Theorem* 4, *we have*

$$f\mu(G) = \lim \int_{G_k} f(v^k)\, dm.$$

This follows at once by applying Theorem 4 to the measure $\mathfrak{F}\tilde{\mu}$.

*Remark.* Theorem 2′ has a simple application to the case of Orlicz type functions $f(p)$, namely convex functions such that

$$\frac{f(p)}{|p|} \to \infty \quad \text{as} \quad |p| \to \infty.$$

In fact, we can assert that *the measure* $f\mu$ *is finite if and only if* $\mu$ *is absolutely continuous with respect to* $\nu$, *and its Radon-Nikodym derivative* $d\mu/d\nu$ *is in the Orlicz class* $L_f(S)$. The proof is an immediate consequence of the Lebesgue decomposition theorem and Theorem 2′, if we simply observe that $f^*(p) = \infty$ for all $p \neq 0$.

If $f(p) = |p|^r$, $1 < r < \infty$, and we write $\mu^r(E)$ for $f\mu(E)$, the above result can be stated as follows: *the measure* $\mu^r$ *is finite if and only if* $\mu$ *is absolutely continuous with respect to* $\nu$, *and its Radon-Nikodym derivative* $d\mu/d\nu$ *is in the Lebesgue class* $L_r(S)$.

4. **Variational integrals.** Let $S$ be a bounded open set in Euclidean space $R^n$, and let $u = u(x)$ be a real valued locally integrable function on $S$. We say that $u$ is *weakly differentiable* if there exists a (bounded regular) vector measure $\mu$ on the Borel subsets of $S$ such that

$$\int u\phi_x\, dm = -\int \phi\, d\mu$$

for all continuously differentiable functions $\phi = \phi(x)$ with compact support in $S$, (here $\phi_x$ denotes the vector gradient of $\phi$, and $\mu$ of course takes its values in $R^n$). If $u$ is weakly differentiable, and if

$$\mu(E) = \int_E a\, dm + \beta(E)$$

is the Lebesgue decomposition of its derivative measure with respect to Lebesgue measure, we *define* the gradient $u_x$ of $u$ by the formula $u_x = a$, (cf. reference [4]). If the singular measure $\beta$ is identically zero, $u$ is called *strongly differentiable*; in this case the partial integration formula takes the explicit form

$$\int u\phi_x\, dm = -\int \phi u_x\, dm.$$

Consider a variational integral of the form

$$I[u, G] = \int_G f(u_x)\, dm.$$

Here $G$ is an open subset of $S$, and $f = f(p) = f(p_1, \cdots, p_n)$ is a non-negative convex function of $p$, defined for all $p \in R^n$. Evidently, the expression $I[u, G]$ is meaningful for any weakly differentiable function $u$. At the same time, a functional $\mathcal{I}[u, G]$ related to $I[u, G]$ can be defined on the class of locally integrable functions on $S$, cf. [4]. In Part I of that paper, which we assume known to the reader, a number of relations between the functionals $I$ and $\mathcal{I}$ were developed, among them several generalizations of Tonelli's theorem on non-parametric surface area (cf. Theorems 2 and 5 of [4]). Here we shall obtain a significant extension and a new proof of the latter results.

THEOREM 5. *If $u$ is weakly differentiable, then*

$$\mathcal{I}[u, G] = I[u, G] + f^*\beta(G) = f\mu(G),$$

*where $\mu = \int u_x \, dm + \beta$ is the derivative measure of $u$, and $f^*(p) = \lim_{t\to 0} tf(p/t)$. In particular, $\mathcal{I}[u, G]$ can be extended to be a Borel measure on $S$, which we denote by $\mathcal{I}[u, E]$.*

*Proof.* We introduce the auxiliary function

$$u^k = u^k(x) = \int K(x - \xi)u(\xi) \, dm(\xi),$$

and observe that by partial integration

$$(u^k)_x = \int K(x - \xi) \, d\mu(\xi) = v^k(x),$$

in the notation of Theorem 4. Now by Theorem 1 of [4],

$$\mathcal{I}[u, G] = \lim I[u^k, G_k] = \lim \int_{G_k} f(v^k) \, dm.$$

By Theorem 4′ the last limit is precisely $f\mu(G)$. Application of Theorem 2′ completes the proof.

*Remark.* The preceding result obviously includes Theorem 2 of [4]. Those parts of Theorem 5 of [4] which are not directly contained in the above theorem are either quite straightforward in themselves, or are easily proved on the basis of the above result.

COROLLARY 1. *Let $\mathcal{A}[u, G]$ denote the non-parametric Lebesgue area of a surface $z = u(x, y)$ over an open set $G$ in the $(x, y)$ plane. Then $\mathcal{A}$ is finite if and only if $u$ is weakly differentiable, and in this case*

$$\mathcal{A}[u, G] = \int_G \sqrt{1 + u_x^2 + u_y^2} \, dx \, dy + |\beta| \, (G).$$

COROLLARY 2. *Suppose that $u$ is weakly differentiable on $S$. Then $\mathcal{I}[u, S] \geq I[u, S]$, with equality holding if and only if either $I[u, S] = \infty$ or $f^*(\beta(E)) = 0$ for all Borel subsets $E$ of $S$.*

These corollaries are easy consequences of Theorem 5 and of our basic definitions. In Corollary 2, as in the proof of Theorem 5, we make strong use of the fact that $f \geq 0$. The following result is equally simple.

COROLLARY 3. *If $u$ and $v$ are weakly differentiable functions whose derivatives measures agree on all Borel subsets of a Borel set $E$, then $\mathcal{I}[u, E] = \mathcal{I}[v, E]$.*

Verchenko has proved a related (and deeper) result, namely, if $u$ and $v$ are continuous and have finite Lebesgue area over a set $S$, and if $u \equiv v$ on a Borel set $E$ in $S$, then $\mathcal{A}[u, E] = \mathcal{A}[v, E]$. (This is obvious if $E$ is an open set. The point of the result is that the *area measure* $\mathcal{A}[u, E]$ is determined by the values of $u$ on $E$ alone.) The remaining part of the paper has for its purpose the extension of Verchenko's theorem to the functional $\mathcal{I}[u, E]$.

We require some preliminary notation. Let $r$ be a *fixed* integer, $1 \leq r \leq n$. The set of variables $(x_1, \cdots, x_{r-1}, x_{r+1}, \cdots, x_n)$ in $R^{n-1}$ will be denoted by $y$, and the set $(x_2, \cdots, x_{r-1}, t, x_{r+1}, \cdots, x_n)$ by $(t, y)$. If $E$ is a Borel set in $R^n$, and $y$ is fixed in $R^{n-1}$, let $E$ be the (linear) set of values $t$ such that $x = (t, y) \in E$. Now let $u(x) = u(x_r, y)$ be a weakly differentiable function on $S$, with derivative measure $\mu$. Krickeberg has shown that for almost all fixed values of $y$ the function $u(t, y)$ is weakly differentiable with respect to $t$. Let the derivative measure of this function be denoted by $\sigma_y$. Then it was also shown by Krickeberg that

$$\mu_r(E) = \int \sigma_y(E_y)\, dy,$$

expressing the $r$-th component of the vector measure $\mu$ in terms of the measure $\sigma_y$.

We require next a simple fact regarding functions of one variable. Let $u = u(t)$ be weakly differentiable on an open set $T$ in $R^1$, and let its derivative measure be denoted by $\sigma$ (the double use of the letters $u$ and $\sigma$ will cause no confusion). Now $u$ is equivalent to a function $u^*$ of bounded variation on $T$, and $u^*$ has right and left-hand limits at each point of $T$. Let these limits be denoted by $u(t^+)$ and $u(t^-)$. Then for every open interval $J = (a, b)$ in $T$ we have

$$\sigma(J) = u(b^-) - u(a^+).$$

Similarly, let $F$ be compact and interior to some interval $J$ in $T$. Suppose that the decomposition of $J - F$ into open intervals is given by

$$J - F = \bigcup J_i, \qquad J_i = (a_i, b_i),$$

where $J_1$ and $J_2$ are respectively the left and right-hand intervals of the decomposition. Then

$$\begin{aligned}\sigma(F) &= \sigma(J) - \sum \sigma(J_i)\\ &= \{u(a_2^+) - u(b_1^-)\} + \sum_3^\infty \{u(a_i^+) - u(b_i^-)\}.\end{aligned}$$

Observe that the points $a_2$, $b_1$, and $(a_i, b_i)$ for $i \geq 3$, are all contained in $F$. We can now prove

LEMMA 4. *Let $u(t)$ and $v(t)$ be weakly differentiable functions on an open set $T$ in $R^1$, and let their derivative measures be denoted by $\sigma$ and $\tau$, respectively. Suppose that for every point $t$ in some Borel subset $E$ of $T$ we have $u(t^+) = v(t^+)$ and $u(t^-) = v(t^-)$. Then $\sigma(E) = \tau(E)$.*

*Proof.* We may suppose, without loss of generality, that $T$ is connected. Now choose $\epsilon > 0$. There is a compact set $F \subset E$ such that

$$|\sigma|\,(E - F) \leq \epsilon, \qquad |\tau|\,(E - F) \leq \epsilon.$$

Moreover, according to the formulas above

$$\begin{aligned}\sigma(F) &= \{u(a_2^+) - u(b_1^-)\} + \sum_3^\infty \{u(a_i^+) - u(b_i^-)\} \\ &= \{v(a_2^+) - v(a_1^-)\} + \sum_3^\infty \{v(a_i^+) - v(b_i^-)\},\end{aligned}$$

since the partition points are in $F$ and thus *a fortiori* in $E$, it follows that $\sigma(F) = \tau(F)$, whence

$$|\sigma(E) - \tau(E)| \leq 2\epsilon.$$

Since $\epsilon$ is arbitrary, the lemma is proved.

Now consider two weakly differentiable functions $u(x)$ and $v(x)$ defined on $S$. Let $E$ be a Borel subset of $S$. We say that $u$ and $v$ are *r-linearly alike* on $E$ if for almost all fixed values of $y$ we have

$$u(t^+, y) = v(t^+, y) \quad \text{and} \quad u(t^-, y) = v(t^-, y)$$

for all $t \in E_y$. The functions $u$ and $v$ will be called *linearly alike* on $E$ if they are $r$-linearly alike for each $r$, $1 \leq r \leq n$. (The assumption that $u$ and $v$ are weakly differentiable is not essential for the concept of linear alikeness; we have used it only in order to simplify the definition.)

THEOREM 6. *If $u$ and $v$ are weakly differentiable on $S$, and linearly alike on a Borel set $E$ in $S$, then $\mathcal{J}[u, E] = \mathcal{J}[v, E]$.*

By Lemma 4 and Krickeberg's formula the derivative measures of $u$ and $v$ agree on Borel subsets of $E$. The result then follows from Corollary 3 of Theorem 5.

Since continuous functions, or even linearly continuous functions, are linearly alike on any set where they are identical, we have

COROLLARY 1. *If $u$ and $v$ are linearly continuous and weakly differentiable, and if $u = v$ on a Borel set $E$ in $S$, then $\mathcal{J}[u, E] = \mathcal{J}[v, E]$.*

If $f(p)/|p| \to \infty$ as $|p| \to \infty$ and if $\mathcal{J}[u, S]$ is finite, then $u$ must be strongly differentiable. Since a strongly differentiable function is necessarily linearly continuous, this proves

COROLLARY 2. *Suppose that $f(p)$ is an Orlicz type function, and that $u$ and $v$ are locally integrable functions on $S$ such that $\mathcal{I}[u, S]$ and $\mathcal{I}[v, S]$ are finite. Then if $u = v$ on a Borel set $E$ in $S$, we have $\mathcal{I}[u, E] = \mathcal{I}[v, E]$.*

In conclusion it should be noted that the condition of linear alikeness has been defined in terms of a fixed coordinate system, a feature which is not entirely desirable. On the other hand, if this condition is replaced in the hypothesis of Theorem 6 by the simple requirement that $u = v$ on $E$, it is easy to see that the conclusion need not remain true.

5. **Appendix.** We indicate a proof of Krickeberg's theorem. For convenience we confine attention to 2 space; the $n$-dimensional case offers no new difficulties. Let $f$ be such that its weak partial derivatives are measures. Suppose the support of the measure $\alpha$ is contained in a rectangle $R = I \times J$ and that for every $\phi \in C^\infty$ with support in $R$,

$$(*)\qquad \int_R \phi(x, y)\, d\alpha = -\int_J \int_I \phi_x(x, y) f(x, y)\, dx\, dy.$$

For every rational interval $\rho \subset I$ and every Borel set $E \subset J$ consider the number

$$\mu_\rho(E) = \alpha(\rho \times E);$$

then $\mu_\rho$ is evidently a measure on the Borel sets in $J$. For every $\phi \in C^\infty$ let

$$\mu_\rho^{(\phi)}(E) = \int_{\rho \times E} \phi\, dx.$$

Then $\mu_\rho^{(\phi)}$ is a measure on the Borel sets in $J$. It is absolutely continuous with respect to Lebsegue measure on $J$ in view of (*). It then follows that $\mu_\rho$ is absolutely continuous with respect to Lebesgue measure on $J$. There is thus a summable function $m_\rho$ on $J$ such that

$$\mu_\rho(E) = \int_E m_\rho(y)\, dy$$

for every Borel set $E \subset J$.

From the definition of $m_\rho$ it follows that if $\rho = \rho_1 \cup \rho_2$, where $\rho_1 \cap \rho_2$ is empty, then for almost all $y$, $m_\rho(y) = m_{\rho_1}(y) + m_{\rho_2}(y)$. Since there are countably many rational intervals, it follows that, for almost all $y$, $m_\rho(y)$ is an additive function of $\rho$. Moreover, if for every $y$, $V(y)$ is the variation (sets of measure 0 are neglected here in the customary way) of $f(x, y)$ as a function of $x$, then it is known that

$$\int d\,|\alpha| \leq \int V(y)\, dy.$$

It follows that the functions $m_\rho(y)$ are completely additive on the rational intervals in $I$ and so may be extended to a measure $\alpha_y$ for almost all $y$. Moreover since

$$\alpha(\rho \times E) = \int_E m_\rho(y)\, dy = \int_E \alpha_y(\rho)\, dy$$

for every $\rho$ and $E$, we have

$$\alpha(E' \times E) = \int_E \alpha_y(E')\, dy$$

for all Borel sets $E' \subset I$, $E \subset J$.

Now, let $\{\phi_n\}$ be a sequence of functions in $C^\infty$ with compact supports in $R$ such that for every $y \in J$ and every function $\psi$ of one variable in $C^\infty$, with support in $I$, there is an $n$ such that

$$\max_{x \in I} |\phi_n(x, y) - \psi(x)| + \max_{x \in I} \left| \frac{\partial \phi_n(x, y)}{\partial x} - \psi'(x) \right| < \epsilon.$$

Since for every Borel set $E \subset J$ and every $n$,

$$\int_{I \times E} \phi_n\, d\alpha = \int_E dy \int_I \phi_n\, d\alpha_y = -\int_E dy \int_I \frac{\partial \phi_n}{\partial x} f\, dx,$$

it follows that

$$\int_I \phi_n\, d\alpha_y = -\int_I \frac{\partial \phi_n}{\partial x} f\, dx$$

for every $y$, except for a subset $Z_n \subset J$ of Lebesgue measure 0. Let $Z = \cup Z_n$. Then for every $y \in Z$ and every $\psi$ in $C^\infty$ whose support is in $J$,

$$\int \psi\, d\alpha_y = -\int \psi_x f(x, y)\, dx.$$

This proves Krickeberg's theorem.

## References

1. N. Dunford and J. T. Schwartz, *Linear Operators*, Part I, New York, 1958.
2. K. Krickeberg, *Distributionen, Funktionen beschränkter Variation, und Lebesguescher Inhalt nichtparametrischer Flächen*, Annali di Matematica Pura ed Applicata, vol. 44(1957), pp. 105–133.
3. E. J. McShane, *Existence theorems for ordinary problems in calculus of variations*, Annali della Scuola Normale Superiore di Pisa Series II, vol. 3(1934), pp. 183–211.
4. J. Serrin, *On the definition and properties of certain variational integrals*, Transactions of the American Mathematical Society, vol. 101(1961), pp. 139–167.
5. I. Verchenko, *Über des Flächenmass von Mengen*, Mathmaticheskiĭ Sbornik (N.S.), vol. 10(1942), pp. 11–32.

Purdue University
and
University of Minnesota

Conference on Degenerate Diffusions, in honor of James Serrin and the opening of the IMA, Minneapolis, 1991.
Front row: Paul Rabinowitz, Shoshana Kamin, Hans Weinberger, Avner Friedman, Felix Browder, James Serrin, Stuart Antman, Ilya Bakelman, Walter Littman, Gene Fabes, Maura Ughi, Patrizia Pucci, Roberta Dal Passo, Nina Ural'tseva.
Second row includes: Wei-Ming Ni, Lambertus Peletier, Juan-Luis Vazques, Jesus Ildefonso Diaz, Donald Aronson, Joel Smoller, William Ziemer, H.Wilmore, Giovanni Paolo Galdi, Daniel D. Joseph and Nicolai V. Krylov.
Other rows include: David Gilbarg, Hiehiko Yamabe, Howard Levine, Grigory I. Barenblatt, David Owen, John Cannon.

Vicentiu Radulescu, Reggio Calabria, Italy, 2013.

# IV. Critical Point Theory

# IV.1 Variants of the Mountain Pass Theorem

Vicenţiu D. Rădulescu

**Mathematics Subject Classification (2010).** Primary 58E05; Secondary 35B38; 35J60; 49J35.

**Keywords.** Mountain pass, critical point, energy functional.

The solutions of many problems are found to be stationary points of some associated "energy" functionals. Often such a functional is unbounded from above and below, so that it has no maximum or minimum. This forces one to look for saddle points, which are obtained by mini-max arguments. One specifies a functional $I$ on a Banach space $X$, and two points 0 and $e$ of $X$, where 0 is the origin. For each smooth path $g(t)$ which connects 0 with $e$, one defines the functional $\max_t I(g(t))$. One then tries to minimize this functional with respect to the choice of $g(t)$ in the collection $G$ of all such paths. Thus, it is natural to define

$$b = \inf_{g(t)\in G} \sup_{u\in g(t)} I(u) \tag{1}$$

and to prove, under various hypotheses, that $b$ is a critical value of $I$. Indeed, it seems intuitively obvious that $b$ defined in relation (1) is a critical value of $I$. However, this is not true in general, as showed by the following example in the plane: let $I(x,y) = x^2-(x-1)^3y^2$. Because $I$ has a proper local minimum at the origin, then $b>0$ whenever $e\neq 0$. Moreover, choosing $e=(2,2)$, we observe that $I=1$ along the line $x=1$, which separates 0 from $e$, so that $b\geq 1$, hence it is not a critical value. In fact, $b=1$, but there is no path $g(t)$ in $G$ such that $I(g(t))\leq 1$.

One of the most important mini-max properties is the so-called *mountain pass theorem*, which is a deep result in modern nonlinear analysis. It marks the beginning of a new approach to critical point theory. The mountain pass theorem was established by Ambrosetti and Rabinowitz in [1]. Their original proof relies on some deep deformation techniques developed by Palais and Smale [8, 9], who put the main ideas of the Morse theory into the framework of differential topology on infinite-dimensional manifolds.

In the statement of the mountain pass theorem, the strict inequality in the geometric condition

$$\inf_{S(0,R)} I > \max\{I(0), I(e)\}\,, \tag{2}$$

which means that the mountain ridge separating the points 0 and $e$ has an altitude which is strictly higher than those of 0 and $e$, plays an essential role in the proof.

In paper [ii] below P. Pucci and J. Serrin studied what happens when the equality holds in relation (2). In such a case, Pucci and Serrin located a critical point of level $b$ on the sphere $S(0,R)$. Their argument is based on the observation that there is a critical

P. Pucci et al. (eds.), *James Serrin. Selected Papers: Volume 2*, Contemporary Mathematicians, DOI 10.1007/978-3-0348-0847-7_8, © Springer Basel 2014

point in the closure of an open ring $A$ around the sphere such that the distance between the boundary of $A$ and $S(0,R)$ can be taken arbitrarily small. However, in the infinite-dimensional case, they established that the mountain pass theorem still holds true when the strict inequality in (2) is replaced a non-strict inequality. Their result in this case is the following.

**Theorem (P. Pucci & J. Serrin [ii]).** *Let $X$ be a real Banach space and let $I : X \to \mathbb{R}$ be a $C^1$-functional satisfying the following conditions:*

(i) *there exist numbers $a$, $r$, $R$ such that $0 < r < R$ and $I(u) \geq a$ for all $u \in A := \{u \in X : r < \|u\| < R\}$;*

(ii) *$I(0) \leq a$ and $I(e) \leq a$ for some $e \in X$ with $\|e\| \geq R$.*

*Then $I$ has a critical point $x_0$ in $X$, different from 0 and $e$, with critical value $b \geq a$; in addition, $x_0 \in A$ when $b = a$.*

We point out that the restriction of the mountain ridge to an annulus centered at 0 is not necessary and the above result remains true when the annulus is replaced by a topological annulus.

This theorem implies the existence of an infinite number of critical points in the ring $A$, since the preceding theorem applies in any of its sub-rings.

We point out that another extension of the mountain pass theorem is due to Ghoussoub and Preiss [4]. We refer to the monographs [5] and [6] for applications to nonlinear PDEs and to the survey [10] for an historical development of the mountain pass theory.

P. Pucci and J. Serrin [i] have also established two interesting corollaries of their mountain pass theorem in the limiting case. The first one is the *three critical point* theorem, which asserts that a $C^1$ functional with the Palais-Smale property that has two local minimum points has a third critical point. We refer to Ricceri [11] and Bonanno [2] for results concerning the *stability*, resp. the *location* of the critical points in the three critical point theorem. The second relevant application of the limiting case of the mountain pass theorem found in [i,ii] states that a $v$-periodic $C^1$ functional with a local minimum $e$ has a critical point $x_0 \neq e + kv$, $k = 0, \pm 1, \pm 2, \ldots$. This critical point yields a second independent solution of the forced pendulum equation studied by Mawhin and Willem [7].

Brezis and Nirenberg [3] also proved a version of the mountain pass theorem that includes the limiting case corresponding to mountains of zero altitude. Their proof combines a pseudo-gradient lemma, an original perturbation argument and Ekeland's variational principle.

**Acknowledgment**

The author acknowledges the support through Grant CNCSIS PNII–ID–PCE–2011–3–0195.

## References

[1] A. Ambrosetti & P.H. Rabinowitz, *Dual variational methods in critical point theory and applications.* J. Funct. Anal. **14** (1973), 349–381.

[2] G. Bonanno, *Some remarks on a three critical points theorem.* Nonlinear Analysis, TMA **54** (2003), 651–665.

[3] H. Brezis & L. Nirenberg, *Remarks on finding critical points.* Comm. Pure Appl. Math. **44** (1991), 939–963.

[4] N. Ghoussoub & D. Preiss, *A general mountain pass principle for locating and classifying critical points.* Ann. Inst. H. Poincaré Anal. Non Linéaire **6** (1989), 321–330.

[5] Y. Jabri, *The mountain pass theorem. Variants, generalizations and some applications.* Encyclopedia of Mathematics and its Applications, Vol. 95, Cambridge University Press, Cambridge, 2003.

[6] A. Kristály, V. Rădulescu & Cs. Varga, *Variational Principles in Mathematical Physics, Geometry and Economics: Qualitative Analysis of Nonlinear Equations and Unilateral Problems.* Encyclopedia of Mathematics, Vol. 136, Cambridge University Press, Cambridge, 2010.

[7] J. Mawhin & M. Willem, *Variational methods and boundary value problems for vector second order differential equations and applications to the pendulum equation.* Nonlinear analysis and optimization (Bologna, 1982), pp. 181–192, Lecture Notes in Math., Vol. 1107, Springer, Berlin, 1984.

[8] R. Palais, *Ljusternik–Schnirelmann theory on Banach manifolds.* Topology **5** (1960), 115–132.

[9] R. Palais & S. Smale, *A generalized Morse theory.* Bull. Amer. Math. Soc. **70** (1964), 165–171.

[10] P. Pucci & V. Rădulescu, *The impact of the mountain pass theory in nonlinear analysis: a mathematical survey.* Boll. Unione Mat. Ital. **3** (2010), 543–582.

[11] B. Ricceri, *On a three critical points theorem.* Arch. Math. **75** (2000), 220–226.

[i] P. Pucci & J. Serrin, *Extensions of the mountain pass theorem.* J. Funct. Anal. **59** (1984), 185–210.

[ii] P. Pucci & J. Serrin, *A mountain pass theorem.* J. Differential Equations **60** (1985), 142–149.

Vicenţiu D. Rădulescu
"Simion Stoilow" Math. Inst. of the Romanian Academy
RO-014700 Bucharest, Romania

*and*

Department of Mathematics
University of Craiova
RO-200585 Craiova, Romania
e-mail: `vicentiu.radulescu@imar.ro`

*J. Differ. Equations* 60 (1985), 142–149. **Reprint IV.1.[i]**


# A Mountain Pass Theorem

PATRIZIA PUCCI*

*Dipartimento di Matematica, Università degli Studi, 06100 Perugia, Italy*

AND

JAMES SERRIN

*School of Mathematics, University of Minnesota, Minneapolis, Minnesota 55455*

Received March 1, 1984

## 1. INTRODUCTION

The mountain pass theorem of Ambrosetti and Rabinowitz [1] has been applied frequently in order to establish the existence of critical points for functionals. In this lemma one considers a $C^1$ function

$$I: X \to \mathscr{R},$$

where $X$ is a real Banach space, satisfying the conditions

(1) there exist two numbers $R > 0$ and $a$ such that $I(x) \geqslant a$ for every $x \in S_R := \{x \in X: \|x\| = R\}$;

(2) $I(0) < a$ and $I(e) < a$ for some $e$ with $\|e\| > R$.

With an additional compactness condition of Palais–Smale type it then follows that the function $I$ has a critical point $x_0 \neq 0, e$ with critical value $c \geqslant a$.

In essence, this critical value occurs because 0 and $e$ are low points on either side of the "mountain" $S_R$, so that between 0 and $e$ there must be a lowest critical point, or "mountain pass."

In this paper we show that if $X$ has *finite dimension* then condition (2) can be weakened and replaced by

(2′) $I(0) \leqslant a$ and $I(e) \leqslant a$ for some $e$ with $\|e\| > R$.

* Supported by the Consiglio Nazionale delle Ricerche within the framework of the Gruppo Nazionale di Analisi Funzionale e Sue Applicazioni.



0022-0396/85 $3.00

In other words, the mountain separating 0 and $e$ may even be assumed of "zero" altitude and the conclusion is still true; moreover (as we shall show) if $c = a$ the critical point can be chosen with $\|x_0\| = R$.

It is interesting to ask whether this extension of the Ambrosetti–Rabinowitz theorem remains true in the infinite-dimensional case. We prove here that if (1) is strengthened a little, to the form

(1′) there exist real numbers $a, r, R$ such that $0 < r < R$ and $I(x) \geqslant a$ for every $x \in A := \{x \in X: r < \|x\| < R\}$,

then (2′) is sufficient to yield, even in the infinite-dimensional case, the existence of a critical point $x_0$, different from 0 and $e$, with critical value $c \geqslant a$. Moreover, if $c = a$ the critical point can be chosen with $r < \|x_0\| < R$.

Roughly speaking, in brief, the mountain pass theorem continues to hold for a mountain of zero altitude, provided it also has non-zero thickness; in addition, if $c = a$, the "pass" itself occurs precisely on the mountain—i.e., satisfying $r < \|r_0\| < R$.

There are two interesting and immediate corollaries. The first one says that a $C^1$ function which has *two* local minimum points also has a third critical point. The second states that a $v$-periodic $C^1$ function with a local minimum $e$ has a critical point $x_0 \neq e + kv$, $k = 0, \pm 1, \pm 2, \ldots$.

The precise statements of the above results will be given in the next sections. The proof depends upon a lemma of Clark [3], formulated here in a version suitable to our purpose. In Section 3 some applications are presented for the forced pendulum equation.

## 2. Main Results

We shall consider a real Banach space $X$ and a function $I: X \to \mathscr{R}$ of class $C^1 := C^1(X, \mathscr{R})$. It will be convenient to let $I_k$ denote the set $\{x \in X: I(x) < k\}$, where $k$ is a real number.

The main results depend upon the following

Lemma. *Let $E$ be a subset of $X$ and $k$ a positive number such that $I_k \cap E \neq \varnothing$. Define*

$$\tilde{E} := \{x \in X: \operatorname{dist}(x, E) < 1\}.$$

*Then the set $I_K \cap \tilde{E}$ contains a point $y$ where either*

$$I(y) < 0 \qquad \text{or} \qquad \|I'(y)\| < 2k.$$

*Proof.* We follow a method due to Clark [3, pp. 67–68].

By a lemma of Palais there exists a $C^{1-}$ pseudo-gradient vector $v: K \to X$ for the function $I$. Here $K$ denotes the set of points in $X$ such that $I'(x) \neq 0$ and $v$ has the specific properties:

$$\text{(a)}\ \|v(x)\| \leqslant 2\|I'(x)\|, \qquad \text{(b)}\ \langle I'(x), v(x)\rangle \geqslant \|I'(x)\|^2,$$

for all $x \in K$.

We define $V: K \to X$ by $V(x) = -v(x)/\|v(x)\|$. Thus $V \in C^{1-}$ and $\|V(x)\| = 1$, for all $x \in K$.

Now assume that $\|I'(x)\| \geqslant 2k$ for every $x \in I_k \cap \tilde{E}$. Let $z$ be a point in $I_k \cap E$, which we hold fixed in what follows, and let $\sigma = \sigma(t)$ be the solution of the initial value problem

$$d\sigma(t)/dt = V(\sigma(t)), \qquad \sigma(0) = z.$$

We assert that $\sigma$ can be continued at least until $t = 1$ and that $\sigma(t) \in I_k \cap \tilde{E}$ for all $t \in [0, 1)$.

To see this, we first observe that $z$ is in $I_k \cap \tilde{E}$ and hence also in $K$, so that $V(\sigma(0))$ is well defined. Furthermore, since $I_k \cap \tilde{E}$ is open, it is clear that the solution $\sigma$ exists in some interval $[0, T]$, $T > 0$, with $\sigma(t) \in I_k \cap \tilde{E}$ for every $t \in [0, T]$. We now consider the subinterval $\mathscr{T}$ of $(0, 1)$ consisting of all $T$ with this property.

The set $\mathscr{T}$ is both open and closed relative to $(0, 1)$. Indeed, by the standard theory of the initial value problem, $\mathscr{T}$ is obviously open. To see that $\mathscr{T}$ is closed in $(0, 1)$ we observe that for $t \in [0, T]$, $T \in \mathscr{T}$, we have

$$\begin{aligned}\frac{dI}{dt}(\sigma(t)) &= -\langle I'(\sigma(t)), v(\sigma(t))/\|v(\sigma(t))\|\rangle \\ &\leqslant -\|I'(\sigma(t))\|^2/\|v(\sigma(t))\| \leqslant -\|I'(\sigma(t))\|/2 \leqslant -k,\end{aligned}$$

and

$$\frac{d}{dt}\|\sigma(t) - z\| \leqslant \left\|\frac{d}{dt}(\sigma(t) - z)\right\| = \|V(\sigma(t))\| = 1.$$

Hence $I(\sigma(T)) < k - kT$ and $\|\sigma(T) - z\| \leqslant T < 1$.

Now let $(T_n)_{n=1}^{\infty}$ be a sequence in $\mathscr{T}$ converging to some $T_0 \in (0, 1)$. Thus, for each $n = 1, 2, \ldots$ we get $I(\sigma(T_n)) \leqslant k - kl$ and $\|\sigma(T_n) - z\| \leqslant L$, where $l$ and $L$ denote the infimum and the supremum of the sequence, respectively. Clearly $0 < l \leqslant L < 1$. Since $\|\sigma(T_n) - \sigma(T_m)\| \leqslant |T_n - T_m|$ for every $n, m = 1, 2, \ldots$, it follows that $(\sigma(T_n))_{n=1}^{\infty}$ is a Cauchy sequence, with limit $y \in I_k \cap \tilde{E}$. Defining $\sigma(T_0) = y$, we see that $\sigma$ is a solution on the interval $[0, T_0]$. This shows that $\mathscr{T}$ is closed in $(0, 1)$.

It follows that $\mathscr{T} = (0, 1)$ and so the solution $\sigma$ exists at least for $t \in [0, 1)$, with $\sigma(t) \in I_k \cap \tilde{E}$. Clearly

$$I(\sigma(t)) = I(\sigma(0)) + \int_0^t \frac{dI}{dt}(\sigma(t))\, dt \leqslant (k - \varepsilon) - kt,$$

for some $\varepsilon > 0$. Choose $t = 1 - \varepsilon/2k$. Then the corresponding point $y = \sigma(t)$ is in $I_k \cap \tilde{E}$ and satisfies $I(y) \leqslant -\varepsilon/2 < 0$.

Now suppose that $I(x) \geqslant 0$ for all $x \in I_k \cap \tilde{E}$. In this case it is impossible by the above argument to have $\|I'(x)\| \geqslant 2k$ for all $x \in I_k \cap \tilde{E}$. Therefore, there is a point $y$ in $I_k \cap \tilde{E}$ where $\|I'(y)\| < 2k$. This completes the proof.

We now turn to our main result, complementing the Ambrosetti–Rabinowitz theorem for mountains with zero altitude.

THEOREM 1. *Let $I$ satisfy the following conditions*:

($C_1$) *there exist numbers $a, r, R$ such that $0 < r < R$ and $I(x) \geqslant a$ for every $x \in A := \{x \in X \colon r < \|x\| < R\}$;*

($C_2$) *$I(0) \leqslant a$ and $I(e) \leqslant a$ for some $e$ with $\|e\| \geqslant R$;*

($C_3$) *any sequence $(x_n)_{n=1}^{\infty}$ in $X$ such that $I(x_n) \to$ limit $\geqslant a$ and $I'(x_n) \to 0$ possesses a convergent subsequence.*

*Then there exists a critical point $x_0$ in $X$, different from 0 and $e$, with critical value $c \geqslant a$; in addition $x_0 \in A$ when $c = a$.*

*Proof.* Let $G := \{g \in C([0, 1], X) \colon g(0) = 0,\ g(1) = e\}$. For $g \in G$ we have

$$\max_{0 \leqslant t \leqslant 1} I(g(t)) \geqslant a,$$

and thus

$$\inf_{g \in G} \max_{0 \leqslant t \leqslant 1} I(g(t)) = c \geqslant a.$$

There are now two cases to consider: $c > a$ and $c = a$. Without loss of generality we may suppose $a = 0$.

If $c > 0$, then we can use the original proof of Ambrosetti and Rabinowitz [1] to show that $c$ is a critical value of $I$. Indeed, even though their hypotheses are different then ours, it is only the condition $c > 0$ which is required for their proof. Moreover, in this case, it is easily seen that the corresponding critical point $x_0$ must be different from 0 and $e$.

If $c = 0$, we have to apply another method, involving the previous lemma.

Let $A'$ be the annulus $\{x \in X \colon r' < \|x\| < R'\}$, with $0 < r < r' < R' < R$. We may suppose that $R' - r' \geqslant 3$, if necessary by replacing $I(x)$ by $I(\lambda x)$ for some appropriate $\lambda > 0$.

Consider the set

$$E := \{x \in X\colon r' + 1 < \|x\| < R' - 1\}$$

together with a fixed number $k > 0$. We shall show that $I_k \cap E$ is non-empty. With this aim in mind, we note that there exists some path $g \in G$ such that $I(g(t)) < k$ for every $t \in [0, 1]$. By continuity of $t \to \|g(t)\|$, and the fact that $\|g(0)\| = 0$, $\|g(e)\| > R'$ there is a value $t_0 \in [0, 1]$ such that $g(t_0) \in E$. Thus we conclude that $g(t_0) \in I_k \cap E$, that is, $I_k \cap E$ is non-empty.

Therefore, since in the present context we have $\tilde{E} = A'$ and $I(x) \geqslant 0$ for $x \in A'$, the above lemma implies the existence of a point $y$ in the set $I_k \cap A'$ such that $\|I'(y)\| < 2k$.

Because $k > 0$ is arbitrary, choosing the sequence $(2^{-n})_{n=1}^{\infty}$ for the values $k$, we deduce the existence of a sequence $(x_n)_{n=1}^{\infty}$ in $A'$ with

$$0 \leqslant I(x_n) < 2^{-n}, \qquad \|I'(x_n)\| < 2^{1-n}, \qquad n = 1, 2, \ldots.$$

The condition $(C_3)$ allows us to extract from $(x_n)_{n=1}^{\infty}$ a subsequence converging to some $x_0 \in \operatorname{cl} A' \subset A$. By the regularity of $I$ it is easily seen that $I(x_0) = 0$ and $I'(x_0) = 0$.

*Remark.* The compactness condition $(C_3)$ used above is essentially equivalent to the Palais–Smale compactness hypothesis $(I_3)$ in [1].

In the case $c = a$, there are an infinite number of critical points in $A$, since the above arguments apply in any subannulus of $A$.

Moreover, if in condition $(C_1)$ the inequality $I(x) \geqslant a$ is replaced by $I(x) > a$, then the conclusion can be strengthened to say that there is a critical point with critical value $c > a$.

COROLLARY 1. *If condition* $(C_3)$ *holds and if I has two different local minimum points, then I possesses a third critical point.*

*Proof.* By changing variables we may assume that the local minimum point with higher value falls at the origin. If both values are the same, either can be used for this purpose. The other will then be denoted by $e$, so that $\|e\| > 0$ and $I(e) \leqslant I(0)$.

By definition of a local minimum point, we can find an open ball $B_R$ with center 0 and radius $R$, $0 < R < \|e\|$, such that $I(x) \geqslant I(0)$ for $x \in B_R$. The result now follows from Theorem 1, by choosing $r = R/2$. (Note that the critical value at the third critical point is greater than or equal to $I(0)$, that is, at least as large as either local minimum.)

## 3. PERIODIC FUNCTIONS

For $v$-periodic functions the Palais–Smale condition ($C_3$) generally fails, since critical sets are not compact. Accordingly, to study the mountain pass theorem in this case we introduce the following weaker compactness condition.

($C_3'$) *Let* $(x_n)_{n=1}^{\infty}$ *be a sequence in* $X$ *such that*

$$I(x_n) \to \text{limit}, \qquad I'(x_n) \to 0.$$

*Then there exist integers* $(k_n)_{n=1}^{\infty}$ *such that the sequence* $(x_n + k_n v)_{n=1}^{\infty}$ *contains a convergent subsequence.*

It follows at once that the limit value of $I(x_n)$ must be a critical value of $I$. Indeed we can suppose without loss of generality that $x_n + k_n v \to y \in X$ for some sequence of integers $k_n$. But then, since $I$ is assumed to be $v$-periodic and of class $C^1$,

$$I(y) = \lim I(x_n + k_n v) = \lim I(x_n) = b$$

$$I'(y) = \lim I'(x_n + k_n v) = \lim I'(x_n) = 0$$

so that $b$ is a critical value of $I$.

Moreover, if the sequence $(x_n)_{n=1}^{\infty}$ in ($C_3'$) is *bounded*, then it possesses a convergent subsequence, for then the integers $k_n$ such that $x_n + k_n v \to y$ must be bounded. We then have $k_n = k$ for some subsequence, and so $x_n \to y - kv$ on this subsequence.

THEOREM 1′. *Let* $I$ *be* $v$*-periodic in Theorem* 1, *and suppose that conditions* ($C_1$), ($C_2$), ($C_3'$) *hold. Then there exists a critical point* $x_0$ *in* $X$, *different from* 0, $e$ *and their periodic translates, with critical value* $c \geqslant a$. *Moreover if* $c = a$ *we can suppose that* $x_0 \in A$.

*Proof.* We proceed as for Theorem 1. The original proof of Ambrosetti and Rabinowitz applies unchanged when $c > a$, using only the weak compactness condition that when $I(x_n) \to c$, $I'(x_n) \to 0$ then $c$ must be a critical value of $I$ (see [2]). On the other hand, when $c = a$ the sequence $(x_n)_{n=1}^{\infty}$ which we have constructed is bounded, so must contain a convergent subsequence as noted above. Thus the demonstration of Theorem 1 given in the last section carries over immediately to the present case, and Theorem 1′ is proved.

COROLLARY 1′. *If* $I$ *is* $v$*-periodic for some* $v \in X$ *and admits a local* *minimum e, then condition* ($C_3'$) *implies the existence of a critical point* $x_0$ *different from* $e+kv$, $k=0, \pm 1, \pm 2,\ldots$.

*Proof.* Clearly $I$ takes its local minimum at each point $e+kv$, $k=0, \pm 1, \pm 2,\ldots$ Using the proof of Corollary 1 for the two points $e$ and $e+v$, the conclusion follows directly from Theorem 1′.

EXAMPLE. In [4] Mawhin and Willem have studied a variational problem for periodic solutions of the forced pendulum equation. Their formulation involves a $v$-periodic function $I$ defined in the Hilbert space $H^1$ of absolutely continuous functions $x$: $[0, T] \to \mathscr{R}^n$, where $v \neq 0$ is a constant function in $H^1$. They show that $I$ has an absolute minimum $x_0$ and also satisfies the Palais–Smale condition ($C_3'$). It follows immediately from Corollary 1′ that there is another critical point $x_1$ of $I$ different from $x_0+kv$, $k=0, \pm 1, \pm 2,\ldots$ This critical point accordingly yields a second independent solution of the forced pendulum equation in their setting, a result which they obtain by a somewhat less direct method.

## 4. A FINITE-DIMENSIONAL GENERALIZATION

THEOREM 2. *Suppose that X has finite dimension and I verifies the assumptions*:

($A_1$) *there exist two numbers* $r>0$ *and a such that* $I(x) \geqslant a$ *for every* $x \in S_R := \{x \in X: \|x\| = R\}$;

($A_2$) $I(0) \leqslant a$ *and* $I(e) \leqslant a$ *for some e with* $\|e\| > R$.

*Then, under condition* ($C_3$), *there exists a critical point* $x_0 \in X$, *different from* 0 *and e, with critical value* $c \geqslant a$.

*If* $c=a$, *the critical point can be taken with* $\|x_0\| = R$.

*Proof.* We begin as in the proof of Theorem 1, in order to get

$$\inf_{g\in G} \max_{0\leqslant t\leqslant 1} I(g(t)) = c \geqslant a = 0.$$

The case $c>0$ can be treated exactly as before. When $c=0$, we proceed as follows. If there is a point $x_0 \in S_R$ with $I'(x_0)=0$, we are done. Therefore, we assume that $I'(x) \neq 0$ for every $x \in S_R$.

Consider an arbitrary point $y \in S_R$; thus either $I(y)>0$ or $I(y)=0$, by condition ($A_1$). If $I(y)>0$, there exists an open pluri-interval $J_y$ with center $y$ such that $I(x)>0$ for each $x \in J_y$. If $I(y)=0$, then the vector $I'(y)$ has zero projection onto the tangent space to $S_R$ at $y$, for otherwise we contradict the hypothesis that $I$ is non-negative on $S_R$. Hence $I'(y)$ is orthogonal to $S_R$ at $y$; in other words it is either an inner or outer normal. Thus, since $I'$ is continuous and $I$ is non-negative on $S_R$, there is an open pluri-interval $J_y$ with center $y$ such that $I(x) > 0$ and $x \cdot I'(x) \neq 0$ either for each $x \in B_y := J_y \cap \{z \in X\colon \ \|z\| < R\}$ or for each $x \in C_y := J_y \cap \{z \in X\colon \|z\| > R\}$.

From the open covering $\{J_y\colon y \in S_R\}$ of $S_R$ we now extract a finite subcovering $\{J_{y_1}, \ldots, J_{y_M}\}$. It is easily seen that the set $\bigcup_{i=1}^{M} J_{y_i}$ contains an open annulus $A$ around $S_R$. We consider the set $\tilde{E} = \bigcup_{i=1}^{M} (Z_i \cap A)$, where $Z_i$, $i = 1, \ldots, M$, denotes $J_{y_i}$ if $I(y_i) > 0$, or $B_{y_i}$ or $C_{y_i}$ if $I(y_i) = 0$, and $I'(y_i)$ is, respectively, the inner normal or the outer normal to $S_R$. Note, from the method of construction, that if $Z_i$ and $Z_j$ are adjacent, then they cannot be, respectively, of the type $B_{y_i}$ and $C_{y_j}$ or $C_{y_i}$ and $B_{y_j}$. Evidently, $I(x) > 0$ for every $x \in \tilde{E}$ and moreover $S_R \subset \operatorname{cl} \tilde{E}$. Thus, putting

$$E_\delta := \{x \in \tilde{E}\colon \operatorname{dist}(x, X \backslash \tilde{E}) > \delta\},$$

we may choose $\delta > 0$ so small that $E_\delta$ separates 0 and $e$.

The proof is now easily completed. By a change of scale we can suppose $\delta = 1$. Then $E_1$ and $\tilde{E}$ are related as $E$ and $\tilde{E}$ in the above lemma and in Theorem 1. Therefore we are able to apply the final part of the proof of Theorem 1 to get a critical point $w \in \operatorname{cl} \tilde{E}$ with $I(w) = 0$.

In other words, we have proved the existence of a critical point $w$ in the closure of the annulus $A$. Since the distance from the boundary of $A$ to $S_R$ can be taken arbitrarily small, a standard compactness argument shows that there is a critical point $x_0 \in S_R$ with $I(x_0) = 0$.

*Remark.* The assumption $(C_3)$ can be avoided in the case $c = a$. When $c > a$, this condition is necessary as may be shown by easy counter-examples.

## REFERENCES

1. A. AMBROSETTI AND P. H. RABINOWITZ, Dual variational methods in critical point theory and applications, *J. Funct. Anal.* **14** (1973), 349–381.
2. H. BREZIS, J. M. CORON, AND L. NIRENBERG, Free vibrations for a nonlinear wave equation and a theorem of P. Rabinowitz, *Comm. Pure Appl. Math.* **33** (1980), 667–684.
3. D. C. CLARK, A variant of the Lusternik–Schnirelman theory, *Indiana Univ. Math. J.* **22** (1972), 65–74.
4. J. MAWHIN AND M. WILLEM, Variational methods and boundary value problems for vector second order differential equations and applications to the pendulum equation, to appear.
5. L. NIRENBERG, Variational and topological methods in nonlinear problems, *Bull. Amer. Math. Soc.* **4** (1981), 267–302.

*Printed in Belgium*

*J. Funct. Anal.* 59 (1984), 185–210.



# Extensions of the Mountain Pass Theorem

PATRIZIA PUCCI

*Istituto di Matematica, Universitá di Perugia, Perugia, Italy*

AND

JAMES SERRIN

*Department of Mathematics, University of Minnesota, Minneapolis, Minnesota 55455*

*Communicated by H. Brezis*

Received April 10, 1984

## 1

The mountain pass lemma of Ambrosetti and Rabinowitz is a result of great intuitive appeal as well as practical importance in the determination of critical points of functionals, particularly those which occur in the theory of ordinary and partial differential equations. In this paper we wish to point out several further results in the same direction, which are both of geometric interest and practical usefulness.

We consider a $C^1$ functional

$$I: X \to \mathbb{R},$$

where $X$ is a real Banach space and where $I$ satisfies the Palais–Smale compactness condition, namely

*Any sequence* $(x_n)$ *in* $X$ *such that* $I(x_n) \to \text{limit}$, $I'(x_n) \to 0$ *contains a convergent subsequence.*

The mountain pass lemma in its main form, as given originally by Ambrosetti and Rabinowitz [1], can be stated as follows.

*Suppose there exist two real numbers a and R,* $R > 0$, *such that*

$$I(x) \geqslant a \quad \text{on the sphere } S_R = \{x \in X \mid \|x\| = R\},$$

$$I(0) < a \quad \text{and } I(e) < a \quad \text{for some } e \text{ with } \|e\| > R.$$

185

0022-1236/84 $3.00

*Then*

$$\inf_{g\in G}\max_{t\in[0,1]} I(g(t)) = b \geqslant a$$

*is a critical value of I. Here G denotes the class of continuous paths*

$$g\colon [0,1] \to X$$

*joining* 0 *and e, that is, satisfying* $g(0)=0$, $g(1)=e$.

Rabinowitz [8] has raised the question whether critical points of the kind obtained in this result must necessarily be *saddle points*, as one might conjecture from the geometric interpretation of the critical point as the "mountain pass" on the journey from 0 to $e$. He noted, however, that this need not be the case if the "mountain" surrounding 0 everywhere has the same height, for in this case the critical point (actually the critical set) would be a local maximum rather than a saddle. ***We shall nevertheless show that under certain reasonable hypotheses the critical point must indeed be a saddle point (Theorems 6–8).***

In addition, we are able to obtain by the same methods a number of useful extensions of the mountain pass lemma, some of independent geometric interest. These are given in the following set of theorems.

THEOREM 1. *Let* 0 *be a local minimum of I and suppose there exists a point e such that* $I(e) < I(0)$. *Then there exists a critical point z, with* $I(z) \geqslant I(0)$, *which is not a local minimum.*

THEOREM 2. *Let* 0 *be a strict local minimum of I and suppose there is some* $e \neq 0$ *such that* $I(e) \leqslant I(0)$. *Then there exists a critical point z, with* $I(z) > I(0)$, *which is not a local minimum.*

THEOREM 3. *Let* 0 *be a local minimum of I. Then either there exists a second critical point which is not a local minimum, or* 0 *is an absolute minimum and the set of absolute minima is connected.*

THEOREM 4. *Let I have two local minima. Then there is a third critical point.*

THEOREM 5. *Let λ be a fixed real number, and suppose that each critical point with critical value greater than λ is a local minimum. Then each is an absolute minimum and the set of these points is connected.*

An immediate consequence of Theorem 5, which is worth stating because of its minimal hypotheses and equally interesting conclusion, is the following result.

COROLLARY. *If every critical point of $I$ is a local minimum, then every one is an absolute minimum, and the critical set is connected.*

To state our results concerning *existence* of saddle points in the mountain pass lemma, it is convenient first to introduce some standard terminology. For any real number $c$ we define

$$K_c = \{x \in X \mid I(x) = c, I'(x) = 0\},$$

that is, $K_c$ is the set of *critical points* of $I$ with *critical value* $c$. If $K_c \neq \varnothing$ it is called a *critical set*. Note that a critical set $K_c$ is necessarily compact, by the Palais–Smale condition.

A point $x \in X$ is called a *local minimum* of $I$ if there exists a neighborhood $B$ of $x$ such that

$$I(y) \geqslant I(x), \qquad \forall y \in B.$$

A local minimum (or a local maximum, defined similarly) is obviously a critical point. A local minimum (maximum) is called *strict* if the equality sign holds only when $y = x$.

A critical point $x$ is called a *saddle point* of $I$ if any neighborhood $B$ of $x$ contains points $y, z$ such that

$$I(y) > I(x) > I(z).$$

It is evident that a critical point must be either a local maximum or a local minimum or a saddle point (the cases of a local maximum and a local minimum are not mutually exclusive, however, for if $I \equiv$ const. near a point $x$, then $x$ is both a local maximum and a local minimum).

Returning to Rabinowitz's question about whether critical points in the mountain pass lemma are saddle points, let us consider the *set* $K_b$ of critical points with the critical value $b$ defined in the lemma. Of course $K_b \neq \varnothing$. At the same time, it is evident that $K_b$ need not consist only of a single point $\bar{x}$, and, even more, need not be connected.

In view of the remarks made earlier, it might be thought that $K_b$ should contain a saddle point whenever it is *not* composed solely of local maxima. This, however, is not the case, for $K_b$ might consist, for example, of a mountain everywhere of the same height *together with another disconnected part all of whose points are local minima.* In this case, every point on the "mountain top" would be a local maxima and every point on the disconnected part a local minima, and there would be no saddle point whatever in $K_b$.

Even making $K_b$ connected does not improve the situation, as the reader will easily see from other only slightly more complicated examples. Nevertheless we have the following results, listing a number of cases where the critical set $K_b$ *must* contain a saddle point.

THEOREM 6. *Let $K_b$ be the set of critical points in the mountain pass lemma with critical value*

$$b = \inf_{g \in G} \max_{t \in [0,1]} I(g(t)).$$

*Then the following conclusions hold:*

(i) *$K_b$ contains at least one point which is not a local minimum.*

(ii) *If the boundary $\partial K_b$ of $K_b$ is connected, and contains a point which is not a local maximum, then $K_b$ contains a saddle point.*

To state the next two results—our principal theorems concerning the existence of saddle points—we recall that a subset $E$ of $X$ *separates* two points $x$ and $y$ of $X$ if and only if there is no component of $X - E$ containing both $x$ and $y$.

THEOREM 7. *If the critical set $K_b$ in the mountain pass lemma does not separate 0 and e, then $K_b$ contains a saddle point.*

In particular it follows from Theorem 7 that if there exists some path $g \in G$ which contains *no* critical points of $I$, then $K_b$ must possess a saddle point. In descriptive terms, if one can cross a mountain in such a way that at each point of the journey the mountain has a nonzero gradient, then there exists a saddle point somewhere on the mountain.

An especially simple consequence of Theorem 7 is the following interesting

COROLLARY. *If there is only one critical point with the critical value b, then it is a saddle point.*

A slightly stronger result than this corollary was recently found by Hofer [4]; see also [7] and the remarks at the end of Section 5.

In spite of the negative answer which must necessarily be given to Rabinowitz's *general* conjecture, nevertheless *when the underlying Banach space is infinite dimensional the conjecture is in fact true*! Indeed, since a compact set *cannot* separate two points 0 and *e* in an *infinite* dimensional Banach space, the hypothesis of Theorem 7 is automatically satisfied in this case. Thus we have, as a second corollary of Theorem 7,

THEOREM 8. *Let X be an infinite dimensional Banach space. Then the set $K_b$ in the mountain pass lemma always contains at least one saddle point.*

In light of this result, the conclusions of Theorems 6 and 7 (as well as the preceding discussion of counterexamples to Rabinowitz's general conjecture) retain their interest solely for the case of a finite dimensional space.

Surprisingly enough, even in this restricted context the results of Theorems 6 and 7 are new, as far as the authors are aware.

For exactly the same reasons, one can also improve the conclusion of Theorem 2 to state that in the infinite-dimensional case the critical point $z$ can be chosen as a saddle point.

Our proofs of Theorems 1–7 depend crucually on two propositions which extend the result of the mountain pass lemma itself. Because both of these results have independent geometric interest, we state them here rather than in the main body of the paper.

THEOREM A. *Let $K_b$ be the set of critical points in the mountain pass lemma, with critical value*

$$b = \inf_{g \in G} \max_{t \in [0,1]} I(g(t)),$$

*and let $U$ be an arbitrary neighborhood of $K_b$. Then there exists a corresponding real number $\varepsilon > 0$ (depending on $U$) such that the conditions*

$$g \in G, \qquad \max_{t \in [0,1)} I(g(t)) < b + \varepsilon$$

*imply that the path $g$ intersects $U$.*

The essential content of Theorem A is that if a path $g \in G$ joining 0 to $e$ traverses the "mountain" $S_R$ with *very nearly the lowest maximum height* possible, then it must at the same time be very near the mountain pass itself, i.e., the critical set $K_b$. (Here, and in what follows, we understand that a *neighborhood* of a set $S \subset X$ is an open subset of $X$ containing $S$.)

The second proposition concerns the case when the mountain is allowed "zero height" with respect to the altitudes of the initial and final points 0 and $e$, a case not covered in the original results of Ambrosetti and Rabinowitz.

THEOREM B. *Let $N$ be an open set with $0 \in N$ and $e \notin N$. Suppose that for some real numbers $a$ and $l$, $l > 0$, we have*

$$I(x) \geqslant a \quad \textit{for} \quad x \in N_l \equiv \{y \in N \mid \operatorname{dist}(y, \partial N) < l\}, \tag{1}$$

$$I(0) \leqslant a \quad \textit{and} \quad I(e) \leqslant a. \tag{2}$$

*Then*

$$\inf_{g \in G} \max_{t \in [0,1]} I(g(t)) = b \geqslant a$$

*is a critical value of $I$; moreover if $b = a$ the critical set $K_b$ intersects $N_l$.*

Theorem B was obtained by the present writers in 1982; see [6]. A related result was later found by Rabinowitz [9], with (1) replaced by the weaker hypothesis

$$I(x) \geqslant a \qquad \text{for} \quad x \in \partial N, \tag{1'}$$

but with the critical value $b$ defined by a different minimax procedure Whether Theorem B itself holds with hypothesis (1) replaced by (1′) is not known, though in any case this would not affect our present considerations. If $X$ is finite dimensional the result is true, however; see [6].

The essential heuristic content of Theorem B is that should the mountain possibly have zero height, then the mountain pass lemma remains true if the mountain is simultaneously required to have positive width.

In Section 2 we prove Theorem A and establish a generalized version of Theorem B.

Theorems 1–4 are direct consequences of Theorem 5 (see Section 4). Hence it is necessary only to prove Theorems 5–7. Theorem 5 is proved in Section 3, and Theorem 6 follows almost immediately from the same set of ideas (see Section 4). Theorem 7 is proved in Section 5.

Since the corollary to Theorem 7 is of interest independently of the conclusion of Theorem 7 we give a separate and simpler proof at the beginning of Section 5.

For functionals $I$ which are periodic, in the sense that there exists some vector $v \in X$ such that

$$I(x+v) = I(x), \qquad \forall x \in X,$$

the Palais–Smale condition clearly fails. Since such functionals nevertheless are of considerable interest, we introduce in Section 6 a weaker compactness condition under which the mountain pass lemma remains true together with most of our conclusions.

It is not our purpose to describe the possible applications of our results, but in Section 6 we do give one simple example in this direction to indicate their general usefulness. This arises in the case of oscillations of a forced pendulum and shows the existence of oscillations not minimizing the corresponding variational integral.

## 2

We begin with the following variant of a result of Clark [3, Theorem 4]. It will be supposed, as always, that $I \in C^1(X, \mathbb{R})$ satisfies the Palais–Smale compactness condition.

THEOREM A′. *Let $E$ be a closed set in $X$ and $c$ a real number such that $K_c \cap E = \varnothing$. Then for all sufficiently small $d > 0$ there exists a related continuous mapping $\sigma: E \to X$ such that the conditions*

$$x \in E, \qquad I(x) \leqslant c + d$$

*imply*

$$I(\sigma(x)) \leqslant c - d.$$

*Moreover we can suppose that $\sigma(x) = x$ whenever $I(x) \leqslant c - 2d$.*

*Remark.* Clark's theorem is slightly weaker than Theorem A′, but his proof (with one easy modification) yields just the result stated here.

*Proof of Theorem* A. Let $U$ be a given neighborhood of the critical set $K_b$ in the mountain pass lemma. Let $E = X - U$. Then $E$ is closed and $K_b \cap E = \varnothing$. Consequently, by Theorem A′, for all sufficiently small $d$, say $d \leqslant d_0$, there exists a corresponding mapping $\sigma$ such that the conditions

$$x \in E, \qquad I(x) \leqslant b + d \tag{3}$$

imply

$$I(\sigma(x)) \leqslant b - d.$$

We shall show that Theorem A holds with the choice

$$\varepsilon = \tfrac{1}{2} \operatorname{Min}\{2d_0, a - I(0), a - I(e)\}.$$

Indeed suppose for contradiction that, for this $\varepsilon$, there exists a path $g \in G$ such that

(i) $\max_{t \in [0,1]} I(g(t)) < b + \varepsilon$,
(ii) $g$ does not intersect $U$.

Clearly $g([0, 1]) \subset E$. Hence taking $d = \varepsilon$ $(\leqslant d_0)$ in (3) there is a corresponding mapping $\sigma$ such that

$$\max_{t \in [0,1]} I((\sigma \circ g)(t)) \leqslant b - \varepsilon. \tag{4}$$

Moreover, since $a \leqslant b$ we have

$$I(g(0)) = I(0) \leqslant b - 2\varepsilon,$$
$$I(g(1)) = I(e) \leqslant b - 2\varepsilon,$$

so that by the final part of Theorem A′,

$$\sigma(g(0)) = g(0) = 0, \qquad \sigma(g(1)) = g(1) = e.$$

Thus $\sigma \circ g \in G$ and in turn (by the conclusion of the mountain pass lemma)

$$\max_{t\in[0,1]} I((\sigma \circ g)(t)) \geqslant b.$$

This contradicts (4), thus completing the proof.

Theorem B, as we have already noted in the Introduction, is proved in [6]. Actually that paper treats only the case where $N$ is a ball centered at the origin, but the proof carries over immediately to the present more general situation once it is observed that any path $g$ in $G$ must contain at least one point $z$ such that the ball $B_{(1/2)l}(z)$ is in $N_l$. (One can suppose $l < \operatorname{dist}(0, \partial N)$ and take for $z$ the first point of $g$, starting from 0, at which $\operatorname{dist}(z, \partial N) = \frac{1}{2}l$.)

There is a further generalization of Theorem B which we shall also need. We state this as

THEOREM C. *Let $N$ be an open set, and let $O_1$ and $O_2$ be subsets of $X$, such that $O_1 \subset N$, $O_2 \subset X - N$. Suppose for some $l > 0$ that*

$$I(x) \geqslant a \quad \textit{for} \quad x \in N_l,$$
$$I(x) \leqslant a \quad \textit{for} \quad x \in O_1 \cup O_2.$$

*Then*

$$\inf_{h\in\hat{G}} \max_{t\in[0,1]} I(h(t)) = b' \geqslant a$$

*is a critical value of $I$; moreover if $b' = a$, then the critical set $K_{b'}$ intersects $N_l$.*

*Here $\hat{G}$ denotes the set of continous paths $h$: $[0, 1] \to X$ such that $h(0) \in O_1$, $h(1) \in O_2$.*

The proof is essentially the same as the proof of Theorem B.

## 3

We first introduce some notation and terminology which will be useful throughout the paper.

If $d$ is a real number and $E$ is a subset of $X$ we shall say that $I > d$ in $E$ provided that $I(x) > d$ for all $x \in E$. The condition $I \geqslant d$ in $E$ is defined similarly to mean that $I(x) \geqslant d$ for all $x \in E$.

Now let $g\colon [0, 1] \to X$ be a continuous path in $X$, and let $U$ be a subset of $X$. We shall say that the path $g$ *intersects* $U$ whenever the set

$$g([0, 1]) \equiv \{x \in X \mid g(t) = x \text{ for some } t \in [0, 1]\}$$

has a nonempty intersection with $U$. Alternately, $g$ intersects $U$ if and only if there exists $t \in [0, 1]$ such that $g(t) \in U$.

Next let $g$ and $h$ be two continuous paths,

$$g\colon [0, 1] \to X, \qquad h\colon [0, 1] \to X$$

which intersect at some point $y \in X$, that is, have the property $g(t_0) = h(t_1) = y$ for some values $t_0, t_1 \in [0, 1]$. It is evident that there is a continuous path $f$ in $X$, joining $g(0)$ to $h(1)$, obtained by first following $g$ from $g(0)$ to $g(t_0)$ and then following $h$ from $h(t_1)$ to $h(1)$. More precisely, we define

$$f\colon [0, 1] \to X$$

by

$$f(t) = \begin{cases} g(2t_0 t), & 0 \leqslant t \leqslant \frac{1}{2}, \\ h(2(1 - t_1)\, t + (2t_1 - 1)), & \frac{1}{2} \leqslant t \leqslant 1. \end{cases}$$

It is evident that $f$ is continuous and that $f([0, 1]) = g([0, t_0]) \cup h([t_1, 1])$. In the above circumstance, we shall say that the path *f follows g from* $g(0)$ *to* $y$ *and then follows h from* $y$ *to* $h(1)$.

The same idea can be extended in an obvious way to any finite number of paths which have successive intersections.

Finally if $A$ and $B$ are disjoint sets in $X$ we define the *distance* from $A$ to $B$ by

$$\operatorname{dist}(A, B) = \inf_{\substack{x \in A \\ y \in B}} d(x, y).$$

If $A$ is closed and $B$ is compact, then $\operatorname{dist}(A, B) > 0$.

We now turn to the demonstration of our main results, beginning with the

*Proof of Theorem* 5. We first show that when $c > \lambda$ any critical set $K_c$ is connected. If not, we can write

$$K_c = R \cup S \qquad (R, S \text{ nonempty}),$$

where

$$\bar{R} \cap S = \varnothing, \qquad R \cap \bar{S} = \varnothing.$$

Both sets $R$ and $S$ are compact, since $K_c$ is. Indeed, consider a sequence $(x_n)$ contained, say, in $R$. Then there is a subsequence, still denoted the same way, such that $x_n \to x \in K_c$. Naturally $x \in \bar{R}$. Now if $x \notin R$, then $x \in S$. Hence $S \cap \bar{R}$ contains $x$ and is nonempty. Thus $x \in R$ and so $R$ is compact.

Since $R$ and $S$ are compact, we have

$$\operatorname{dist}(R, S) > 0.$$

In addition, every point of $R$ is a local minimum, by hypothesis. Hence there exists a neighborhood $N$ of $R$ which is disjoint from $S$ and in which $I \geqslant c$. We assert that

$$I > c \quad \text{in} \quad N - R.$$

To see this, suppose to the contrary that $z \in N - R$ and $I(z) = c$. Then $I'(z) = 0$, that is $z \in K_c$. But this is impossible because $z \in N - R$ implies that $z$ is neither in $R$ nor in $S$.

We can now apply Theorem B, using the set $N$ just constructed and taking

$$0 \in R, \qquad e \in S,$$

$$a = c, \qquad l = \tfrac{1}{2} \operatorname{dist}(R, \partial N),$$

where $l > 0$ since $R$ is compact, $\partial N$ is closed and $R \cap \partial N = \varnothing$. This yields a critical value

$$b = \inf_{g \in G} \max_{t \in [0,1]} I(g(t)) \geqslant a = c,$$

where $G$ denotes the class of continuous paths joining 0 and $e$. If $b = c$ then the critical set $K_b$ intersects $N_l$, which is impossible since $I > c$ in $N - R$ and hence in $N_l$. Thus $b > c$.

We next consider the critical set $K_b$, consisting of course as before entirely of local minima. Let $\hat{G}$ be the set of all continuous paths $h$ joining some point $y$ of $K_b$ to 0. We assert that

$$d = \inf_{h \in \hat{G}} \max_{t \in [0,1]} I(h(t)) > b.$$

To see this, first observe that there is a neighborhood of $K_b$ in which $I \geqslant b$. The previous argument then yields the required result, provided instead of Theorem B we use Theorem C with $O_1 = K_b$, $O_2 = \{0\}$.

We can now apply Theorem A. For each $x \in K_b$ let $B(x)$ denote an open ball centered at $x$ in which $I < d$. Define

$$U = \bigcup_{x \in K_b} B(x).$$

Clearly $U$ is a neighborhood of $K_b$ in which $I < d$. Let $\varepsilon > 0$ be such that the conditions

$$g \in G, \qquad \max_{t \in [0,1]} I(g(t)) < b + \varepsilon \tag{5}$$

imply that $g$ intersects $U$. We can of course assume also that $\varepsilon \leqslant d - b$.

Now let $g$ be a path satisfying condition (5). Then $g$ intersects $U$, and so in turn there is a point $\bar{x} \in K_b$ such that $g$ intersects the ball $B(\bar{x})$. Consider the combined path $h$ starting at $\bar{x}$ and proceeding (within $B(\bar{x})$) to a point of $g$, after which it follows $g$ back to 0. Along this path we have

$$I(h(t)) < \operatorname{Max}(d, b + \varepsilon) = d. \tag{6}$$

But since $h \in \hat{G}$ we also have

$$\max_{t \in [0,1]} I(h(t)) \geqslant d. \tag{7}$$

This last contradiction proves that every critical set $K_c$ with $c > \lambda$ is connected.

We now show that each critical point with critical value greater than $\lambda$ is an absolute minimum. Indeed, suppose for contradiction that there is some critical point, say 0, with

$$I(0) = c > \lambda$$

which is not an absolute minimum. As in the earlier part of the proof there is a neighborhood $N$ of the critical set $K_c$ such that

$$I > c \qquad \text{in} \quad N - K_c.$$

Also, since 0 is not an absolute minimum there exists a point $e$, with $e \notin N$, such that $I(e) < c$. We can now apply Theorem B as in the first part of the proof, with

$$a = c, \qquad l = \tfrac{1}{2} \operatorname{dist}(K_c, \partial N).$$

Then, exactly as in that proof we are led to the two contradictory inequalities (6) and (7). Hence every critical point with critical value $>\lambda$ is an absolute minimum, and this set of absolute minima is connected.

*Remark.* Theorem 5 holds equally when the condition "greater than $\lambda$" in its statement is replaced by "greater than or equal to $\lambda$." The proof of this version requires only trivial modifications of the one already given.

4

In this section we shall prove Theorems 1–4 as direct consequences of Theorem 5. In addition we shall prove Theorem 6, again using ideas introduced in the proof of Theorem 5.

*Proof of Theorem* 1. Assume for contradiction that all critical points $z$ of $I$ with critical values $I(z) \geqslant I(0)$ are local minima. Then by Theorem 5 with $\lambda = I(0)$ we see that all these critical points are absolute minima. This contradicts the condition $I(e) < I(0)$, and completes the proof.

*Proof of Theorem* 2. We first apply Theorem B, taking for $N$ an open ball centered at 0 with the property that

$$I(x) > I(0) \qquad \text{for} \quad x \in N - \{0\}.$$

It follows that there exists a critical point $\bar{x}$ with $I(\bar{x}) = b > I(0)$. Now suppose for contradiction that all critical points with critical value $>I(0)$ are local minima. Then by Theorem 5 with $\lambda = I(0)$ we see that all critical values are absolute minima. Hence $\bar{x}$ is an absolute minima, which is impossible.

*Proof of Theorem* 3. Let 0 be a local minimum of $I$, and suppose that all critical points are local minima. Then by the corollary to Theorem 5 all the critical points are absolute minima and the set of absolute minima is connected.

*Proof of Theorem* 4. Let the two given local minima be 0 and $e$. If there are no other critical points then each is in fact a strict local minimum. But then by Theorem 2 there exists a critical point $z$ which is not a local minimum. Since $z$ cannot be either 0 or $e$, we have an obvious contradiction.

*Proof of Theorem* 6. (i) Suppose the conclusion is false and that the set $K_b$ consists entirely of local minima. Proceeding as in the proof of Theorem 5, word for word from the fourth paragraph through the sixth paragraph, we obtain a contradiction.

Theorem 6(ii) is a direct consequence of Theorem 6(i). Indeed, suppose for contradiction that $K_b$ satisfies the hypothesis of Theorem 6(ii), but has no saddle points. Let

$$\mathcal{M} = \{x \in \partial K_b \mid x \text{ is a local maximum of } I\}$$

$$\mathcal{N} = \{x \in \partial K_b \mid x \text{ is a local minimum of } I\}.^{1}$$

Clearly $\partial K_b = \mathcal{M} \cup \mathcal{N}$, $\mathcal{N} \neq \varnothing$ by hypothesis, and $\mathcal{M} \neq \varnothing$ by virtue of Theorem 6(i).

[1] When $X$ is infinite dimensional it is clear that $K_b$, being compact, can have no interior points. Therefore in this case $K_b \equiv \partial K_b$. This fact, however, exerts no bearing on the proof.

Moreover both $\mathcal{M}$ and $\mathcal{N}$ are relatively open subsets of $\partial K_b$. For if $x \in \mathcal{M}$ then there exists an open ball $B$ centered at $x$ such that

$$I \leqslant b \quad \text{in} \quad B.$$

If $y \in B \cap \partial K_b$ then clearly $I(y) = b$ and $I \leqslant b$ in some neighborhood of $y$, which we can, of course, take to be $B$. Hence $y \in \mathcal{M}$ so that $\mathcal{M}$ is relatively open. The proof for $\mathcal{N}$ is almost identical.

Since $\partial K_b$ is connected by hypothesis, we must have either

$$\mathcal{M} \cap \bar{\mathcal{N}} \neq \varnothing \qquad \text{or} \qquad \bar{\mathcal{M}} \cap \mathcal{N} \neq \varnothing.$$

Suppose, for example, that the first case holds (the other case is treated in the same way). Because $\mathcal{M}$ is open relative to $\partial K_b$ it follows that there are points of $\mathcal{N}$ also in $\mathcal{M}$. Such a point $y$, however, would be both a local maximum and a local minimum; that is, $I \equiv b$ in some neighborhood of $y$. But then $y$ must be an interior point of $K_b$, not a boundary point. This contradiction proves Theorem 6(ii) and completes the demonstration of Theorem 6.

## 5

The proof which we shall give below for Theorem 7 is delicate in the extreme. It is therefore worthwhile to give an independent proof of the corollary to Theorem 7, both because the corollary is of particular interest in itself and because this will help to illuminate the ideas later to be used in the demonstration of Theorem 7.

*Proof of the Corollary to Theorem* 7. Let $K_b = \{\bar{x}\}$ be the critical set in the mountain pass lemma. By Theorem 6(i) the critical point $\bar{x}$ must be either a saddle point or a local maximum.

In the latter case, namely if $\bar{x}$ is a local maximum, there exists an open ball $U$ centered at $\bar{x}$ such that

$$I \leqslant b \quad \text{in} \quad U.$$

Now consider the number $\varepsilon > 0$ associated to $U$ by Theorem A. Thus if $g$ is a path in $G$ such that

$$\max_{t \in [0,1]} I(g(t)) < b + \varepsilon,$$

then $g$ intersects $U$. Obviously $g$ first enters $\bar{U}$ at some $t_0 > 0$ and last leaves $\bar{U}$ at some $t_1 < 1$. We modify the path $g$ by replacing the part between $g(t_0)$

and $g(t_1)$ by an arc on $\partial U$ joining $g(t_0)$ to $g(t_1)$. Along the resulting path $\tilde{g}$ it is evident that

$$\max_{t\in[0,1]} I(\tilde{g}(t)) < b + \varepsilon.$$

However, $\tilde{g}$ does not intersect $U$; this contradicts Theorem A and shows that $\bar{x}$ is a saddle point.

*Proof of Theorem* 7. Since $K_b$ does not separate 0 and $e$ there exists a component $D$ of the open set $X - K_b$ containing both of these points. Because $D$ is open it is arcwise connected as well as connected. Hence there exists a path $\hat{g} \in G$ joining 0 and $e$ and lying entirely in $D$. In particular

$$\hat{g}([0,1]) \cap K_b = \varnothing$$

and so

$$k = \operatorname{dist}(\hat{g}([0,1]), K_b) > 0.$$

Now suppose for contradiction that $K_b$ contains no saddle points, and therefore consists solely of local maxima and local minima of $I$. Thus we can introduce the sets

$$\mathscr{M} = \{x \in \partial K_b \mid x \text{ is a local maximum of } I\},$$

$$\mathscr{N} = \{x \in \partial K_b \mid x \text{ is a local minimum of } I\},$$

with

$$\mathscr{M} \cup \mathscr{N} = \partial K_b.$$

Evidently $\mathscr{M}$ and $\mathscr{N}$ are disjoint, for otherwise any point $x \in \mathscr{M} \cap \mathscr{N}$ would be such that $I \equiv b$ in some neighborhood of $x$; but then $x$ would not be in $\partial K_b$, a contradiction.[2]

We next observe that $\mathscr{M}$ and $\mathscr{N}$ are compact. In showing this it is enough to consider the case of $\mathscr{M}$, the proof for $\mathscr{N}$ being entirely similar. Thus let $(x_n)$ be a sequence in $\mathscr{M}$, which we can suppose convergent to some $\bar{x} \in \partial K_b$. Suppose for contradiction that $\bar{x} \notin \mathscr{M}$. Then $\bar{x} \in \mathscr{N}$ and there is a neighborhood of $\bar{x}$ where $I \geqslant b$. But each $x_n$ from some $n_0$ onwards is also in this neighborhood, so in fact $I \equiv b$ in a neighborhood of each $x_n$, $n \geqslant n_0$, contradicting (as before) the fact that $x_n \in \partial K_b$.

Since $\mathscr{M}$ and $\mathscr{N}$ are compact and disjoint, we have

$$\operatorname{dist}(\mathscr{M}, \mathscr{N}) = k' > 0.$$

[2] If $\mathscr{N}$ is empty (by Theorem 6 we have $\mathscr{M} \neq \varnothing$) the proof follows the same lines but is much simpler. We therefore suppose throughout that $\mathscr{M} \neq \varnothing$, $\mathscr{N} \neq \varnothing$.

Moreover, each point $x \in \mathscr{M}$ is contained in some open ball $B_r(x)$ with center $x$ and radius $r < \min(k, \frac{1}{2}k')$, such that

$$I \leqslant b \qquad \text{in} \quad B_r(x).$$

Using once more the fact that $\mathscr{M}$ is compact, we can find a *finite* collection of such balls, say

$$B_{r_1}(x_1), B_{r_2}(x_2), \ldots, B_{r_m}(x_m),$$

whose union $\mathscr{M}'$ covers $\mathscr{M}$. In particular

$$I \leqslant b \qquad \text{in} \quad \mathscr{M}'.$$

A similar covering can be found for $\mathscr{N}$, say

$$B_{r_{m+1}}(x_{m+1}), B_{r_{m+2}}(x_{m+2}), \ldots, B_{r_s}(x_s),$$

with $r_i$ again less than $\min(k, \frac{1}{2}k')$ and $x_i \in \mathscr{N}$, $i = m+1, \ldots, s$. Of course in this case the condition $I \leqslant b$ in $\mathscr{M}'$ is replaced by

$$I \geqslant b \qquad \text{in} \quad \mathscr{N}',$$

where $\mathscr{N}'$ denotes the union of the balls $B_{r_{m+1}}(x_{m+1}), \ldots, B_{r_s}(x_s)$. Since $\operatorname{dist}(\mathscr{M}, \mathscr{N}) > 2 \max_{i \leqslant s} r_i$ it is clear that $\operatorname{dist}(\mathscr{M}', \mathscr{N}') > 0$; of course also $\partial K_b \subset \mathscr{M}' \cup \mathscr{N}'$.

Now consider the open set

$$V = \operatorname{Int}(K_b) \cup \mathscr{M}' \cup \mathscr{N}'.$$

Since $\partial K_b \subset \mathscr{M}' \cup \mathscr{N}'$ we see that $V \supset K_b$. Obviously $K_b \cap \partial V = \varnothing$, and, moreover, $V$ has at most a finite number of components. To see this, note that any component of $V$ must surely contain a point of $\partial K_b$, and so some ball $B_{r_k}(x_k)$, while conversely every ball $B_{r_i}(x_i)$ is contained in some component of $V$.

Let $A$ be a fixed component of $V$. If $X$ is infinite dimensional the compact set $K_b$ can have no interior points, so $K_b \subset \mathscr{M}' \cup \mathscr{N}'$ and

$$V = \mathscr{M}' \cup \mathscr{N}',$$

the union of two finite disjoint collections of open balls. Thus we have either $A \subset \mathscr{M}'$ or $A \subset \mathscr{N}'$. In the latter case we say that $A$ is of type $\mathscr{N}$.

A similar, though slightly more complicated situation, occurs when $X$ is finite dimensional (we then assume, for simplicity and without loss of generality, that $X$ is endowed with a Cartesian structure and the Euclidean metric). In this case observe first that $0 \notin V$ by construction, so that

$0 \in X - A$. We let $C$ denote the component of $X - A$ containing 0 ( a better notation might be $C_A$, but this added terminology is not really necessary). By the lemma in the Appendix, $\partial C$ is connected, and because $\partial C \subset \partial V \subset \partial \mathscr{M}' \cup \partial \mathscr{N}'$ it follows that either $\partial C \subset \partial \mathscr{M}'$ or $\partial C \subset \partial \mathscr{N}'$. In the latter case we say that $A$ is of type $\mathscr{N}$.

Now let $N$ be the union of all components of $V$ which are of type $\mathscr{N}$ (if $X$ is infinite dimensional then of course $N = \mathscr{N}'$). Let $\hat{K} = K_b \cap N$ and denote by $\hat{G}$ the set of continous paths joining some point of $\hat{K}$ with 0. We assert that

$$d = \inf_{h \in \hat{G}} \max_{t \in [0,1]} I(h(t)) > b.$$

The proof is slightly different depending on whether or not $X$ is infinite dimensional.

(i) $X$ is infinite dimensional. Let

$$N_l = \{x \in N \mid \operatorname{dist}(x, \partial N) < l\},$$

where $l$ is so small that

$$N_l \cap K_b = \varnothing.$$

Because $I \geqslant b$ in $N$ we obtain in turn that

$$I > b \quad \text{in} \quad N_l.$$

The required assertion now follows from Theorem C exactly as in the proof of Theorem 5.

(ii) $X$ is finite dimensional. Here we let

$$A_{(l)} = \{x \in A \mid \operatorname{dist}(x, \partial C) < l\}.$$

$$N_{(l)} = \bigcup A_{(l)},$$

the union being taken over all components $A$ of type $\mathscr{N}$. We can choose $l$ sufficiently small so that

(a) $N_{(l)} \cap K_b = \varnothing$,

(b) $N_{(l)} \subset \mathscr{N}'$,

(c) $\bar{N}_{(l)} \cap \partial A = \partial C$,

for any component $A$ of type $\mathscr{N}$. [Condition (a) uses the fact that $K_b$ does not intersect $\partial V$, condition (b) that $N_{(l)} \subset \mathscr{M}' \cup \mathscr{N}'$ together with $\operatorname{dist}(\mathscr{M}', \mathscr{N}') > 0$, and condition (c) follows from the local structure of $\partial A$ (see condition (iii) in the proof of Lemma (*) in the Appendix, and note that $\partial C$ is compact when $X$ is finite dimensional).]

The partial boundary neighborhood $N_{(l)}$ of $N$ serves equally as well in the application of Theorem C as the full boundary neighborhood $N_l$, since it is in fact just this boundary strip which actually separates points of $\hat{K}$ from 0. Hence again the required assertion follows.

We next choose, for each $x \in K_b$, an open ball $B(x)$ centered at $x$ in which $I < d$ and also $B(x) \subset V$. Define

$$U = \bigcup_{x \in K_b} B(x).$$

Clearly $U$ is a neighborhood of $K_b$, and $U \subset V$. Now let $\varepsilon > 0$ be the number in Theorem A such that the conditions

$$g \in G, \qquad \max_{t \in [0,1]} I(g(t)) < b + \varepsilon$$

imply that $g$ intersects $U$. We can require as well that $\varepsilon \leqslant d - b$.

Now consider such a path $g$. We assert that $g$ cannot intersect the set $U \cap N$. Indeed, suppose for contradiction that such an intersection took place, say at $z \in U \cap N$. Then there would be a point $\bar{x} \in K_b$ such that $z \in B(\bar{x})$. Since $B(\bar{x}) \subset V$ and since $z \in N$ it then follows that $B(\bar{x}) \subset N$ and in particular that $\bar{x} \in N$. Hence $\bar{x} \in \hat{K}$. We can now apply an argument used in the proof of Theorem 5 to construct a new path $h \in \hat{G}$ satisfying (compare (6)) the condition

$$\max_{t \in [0,1]} I(h(t)) < d.$$

This contradicts the fact that $\max_{t \in [0,1]} I(h(t)) \geqslant d$ for any path $h \in \hat{G}$, and thus proves the assertion. In particular, it follows that $g$ can intersect a set $U \cap A$ only for components $A$ of $V$ which are *not* of type $\mathcal{N}$.

To complete the proof we shall now consider a path $g \in G$ satisfying, as above, the condition

$$\max_{t \in [0,1]} I(g(t)) < b + \varepsilon$$

(and so not intersecting any set $U \cap A$ when $A$ is of type $\mathcal{N}$). We shall then modify the path $g$, as in the proof of the corollary at the beginning of this section, to obtain another path $\tilde{g} \in G$ such that

$$\max_{t \in [0,1]} I(\tilde{g}(t)) < b + \varepsilon$$

but with $\tilde{g}$ *not* intersecting the neighborhood $U$ at all. This will contradict Theorem A and will thus prove the theorem.

To construct $\tilde{g}$ we consider first the *finite dimensional* case. Let $A$ be any component of $V$ not of type $\mathcal{N}$, which is entered by the path $g$. Then because

0 and $e$ are outside $A$ (actually outside $V$) there must exist number $t_0, t_1$ such that

(i) $0 < t_0 < t_1 < 1$,

(ii) $g(t_0), g(t_1) \in \partial A$,

(iii) $g[0, t_0] \cap A = g[t_1, 1] \cap A = \varnothing$.

Consider the path $g_1$ from $g(t_1)$ to $g(1) = e$ defined by

$$g_1(t) = g(t_1 + (1 - t_1)\, t), \qquad 0 \leqslant t \leqslant 1,$$

the path $g_2$ from $\hat{g}(1) = e$ to $\hat{g}(0) = 0$ defined by

$$g_2(t) = \hat{g}(1 - t), \qquad 0 \leqslant t \leqslant 1,$$

and the path $g_3$ from $g(0) = 0$ to $g(t_0)$ defined by

$$g_3(t) = g(t_0 t), \qquad 0 \leqslant t \leqslant 1.$$

If we follow the paths $g_1, g_2, g_3$ in succession we obtain a continuous path which joins $g(t_1)$ and $g(t_0)$, contains 0, and nowhere intersects $A$. Hence $g(t_0)$ and $g(t_1)$ are in the component $C$ of $X - A$ which contains 0. At the same time, $g(t_0)$ and $g(t_1)$ are in $\partial A$, hence also in $\partial C$.

By the lemma in the Appendix, $\partial C$ is arcwise connected. Hence there exists a continuous path $g_A$ joining $g(t_0)$ to $g(t_1)$, with $g_A([0, 1]) \subset \partial C$. Of course

$$\max_{t \in [0,1]} I(g_A(t)) \leqslant \max_{x \in \partial C} I(x) \leqslant b$$

since the balls contributing to $\partial C$ are in $\mathscr{M}'$. We now consider the path $\tilde{g}_A$ consisting of the path $g_3$ followed by $g_A$ followed by $g_1$. Clearly $\tilde{g}_A \in G$ and

$$\max_{t \in [0,1]} I(\tilde{g}_A(t)) < b + \varepsilon.$$

Moreover $\tilde{g}_A$ does not intersect $A$ nor any other component of $V$ previously not intersected by $g$.

Let the components $A$ of $V$ which are *not* of type $\mathscr{N}$ be enumerated $A_1, \ldots, A_j$. A natural recursive procedure, using at each stage the modification just described, enables us to eliminate one by one the intersections of $g$ with the sets $A_1, A_2, \ldots, A_j$. The result of this procedure is ultimately, then, a path $\tilde{g} \in G$ which is disjoint from $U$ and satisfies

$$\max_{t \in [0,1]} I(\tilde{g}(t)) < b + \varepsilon.$$

This contradicts Theorem A, and consequently proves Theorem 7 for the finite dimensional case.

When $X$ is infinite dimensional, the previous idea needs some modification. To begin with, without loss of generality the path $\hat{g}$ joining 0 and $e$ can be assumed *piecewise linear* (since paths are compact and $D$ is open). Moreover, we can modify the path $g$, without changing the condition

$$\max_{t\in[0,1]} I(g(t)) < b + \varepsilon,$$

so that $g$ also becomes piecewise linear (again because paths are compact and $I$ is continuous).

Now let $Y$ be a finite-dimensional subspace of $X$ which contains $\hat{g}([0,1]) \cup g([0,1])$.[3] Let $V' = V \cap Y$, and let the components of $V'$ be denoted by $A'$. As before $g$ cannot intersect any set $U \cap A'$ when $A'$ is of type $\mathcal{N}$ (more precisely, when $A'$ is a subset of some $A$ of type $\mathcal{N}$).

We proceed now in essentially the same way as in the finite dimensional case. First, it may be assumed that $Y$ is endowed with a Cartesian structure and a Euclidean metric. Consider the union of all components $A'$ which are *not* of type $\mathcal{N}$. We can cover this union by a finite collection of open balls (in $Y$). Denote the union of these balls by $V''$. It can be assumed that $V''$ does not intersect $\hat{g}$ (since $\bar{V}'$ does not) and that

$$I \leqslant b + \tfrac{1}{2}\varepsilon \qquad \text{on} \qquad V''$$

(since $I \leqslant b$ on each component $A'$ which is not of type $\mathcal{N}$). Let $A''$ be any component of $V''$.

As in the finite-dimensional case, we can now construct a path $\tilde{g}_{A''} \in G$, with $\tilde{g}_{A''}([0,1]) \subset Y$, for which

$$\max_{t\in[0,1]} I(\tilde{g}_{A''}(t)) < b + \varepsilon$$

and moreover such that $\tilde{g}_{A''}$ does not intersect $A''$ nor any other component of $V''$ previously not intersected by $g$. Of course $\tilde{g}_{A''}$ also cannot intersect any set $U \cap A'$ when $A'$ is of type $\mathcal{N}$.

Since $V''$ has only a finite number of components, we can successively eliminate these exactly as in the finite dimensional case. This leads ultimately to a path $\tilde{g} \in G$, with $\tilde{g}([0,1]) \subset Y$, which is disjoint from $U$ and nevertheless satisfies

$$\max_{t\in[0,1]} I(\tilde{g}(t)) < b + \varepsilon.$$

This contradicts Theorem A, and completes the proof.

[3] This device was suggested to the authors by Donald Kahn and Karel Pikry, to whom we express our indebtedness.

*Remark.* A saddle point as defined in [8] and here is not necessarily the same thing as might be heuristically represented by functions of the type

$$I: \mathbb{R}^2 \to \mathbb{R}^1 \quad \text{with} \quad I(x, y) = x^2 - y^2$$

or

$$I: \mathbb{R}^3 \to \mathbb{R}^1 \quad \text{with} \quad I(x, y, z) = x^2 - y^2 - z^2,$$

for both of which the set of points where $I < 0$ is disconnected. Indeed, examples show that the critical set $K_b$ in the mountain pass lemma need not include any point having such structure (even when the hypothesis of Theorem 7 holds).

If each point of the critical set $K_b$ is isolated then the structure of the critical points can be indicated with more precision, as shown by Hofer [4, Theorem 1]. In fact, in the case of isolated critical points, he has proved that at least one of them, call it $\bar{x}$, has the property that, for any neighborhood $N$ of $\bar{x}$, the set

$$S_N = \{x \in N \mid I(x) < b\}.$$

is nonempty and not path-connected. Such a critical pont $\bar{x}$ is, moreover, a saddle point as well. To see this, suppose for contradiction that $\bar{x}$ is not a saddle point. Then there exists some neighborhood $B$ of $\bar{x}$ such that

$$I(x) \leqslant I(\bar{x}) = b, \qquad \forall x \in B.$$

In particular, from this relation and the fact that $\bar{x}$ is an isolated critical point it follows that there is an open ball $\hat{N}$ centered at $\bar{x}$ such that

$$I(x) \leqslant b, \qquad I'(x) \neq 0$$

for all $x \in \hat{N} \setminus \{\bar{x}\}$. This implies in turn that $I < b$ in $\hat{N} \setminus \{\bar{x}\}$. Consequently

$$S_{\hat{N}} = \{x \in \hat{N} | I(x) < b\} = \hat{N} \setminus \{\bar{x}\},$$

a non-empty, path-connected set, contradicting the fact that for any neighborhood $N$ of $\bar{x}$ the set $S_N$ is not path-connected.

For the critical points illustrated at the beginning of this remark the sets $S_N$ are of course disconnected; it is precisely their isolated character that gives these critical points this special property, and precisely for nonisolated critical points that such behavior can break down.

Finally, we remark that in the direction of Hofer's result one can also prove the following theorem:

*Let $U$ be an arbitrary neighborhood of the critical set $K_b$ in the mountain pass lemma. Then for some component $A$ of $U$ the set*

$$A \cap \{x \in X \mid I(x) < b\}$$

*is nonempty and not path-connected.*

## 6

In this section we shall extend the preceding results to the case of *periodic functions I*. Thus in what follows we suppose that there exists a nonzero vector $v \in X$ such that

$$I(x+v) = I(x) \qquad \text{for all } x \in X.$$

In this circumstance the Palais–Smale condition generally fails (e.g., no critical set $K_c$ could be compact). We therefore adopt the following weaker compactness condition:

(P–S′) *Let* $(x_n)$ *be a sequence in* $X$ *such that* $I(x_n) \to \text{limit}$, $I'(x_n) \to 0$. *Then there exist integers* $k_n$ *such that the sequence* $(x_n + k_n v)$ *contains a convergent subsequence.*

Here the limit value of $I(x_n)$ must be a critical value of $I$. To show this, we can suppose without loss of generality that $x_n + k_n v \to y \in X$ for some sequence of integers $k_n$. But then by periodicity

$$I(y) = \lim I(x_n + k_n v) = \lim I(x_n) = c,$$
$$I'(y) = \lim I'(x_n + k_n v) = \lim I'(x_n) = 0,$$

so that $c$ is a critical value of $I$.

In addition, if the sequence $(x_n)$ in (P–S′) is *bounded*, then it contains a convergent subsequence. To prove this, we start again with the relation $x_n + k_n v \to y$. Since $(x_n)$ is bounded, the integers $k_n$ are also bounded. Hence there is a (further) subsequence on which $k_n = k =$ fixed integer. But then on this subsequence $x_n \to y - kv$. Although a critical set is *not* compact, this result shows at least that bounded closed subsets of critical sets are compact.

Now consider the mountain pass lemma itself. As Brezis, Coron and Nirenberg [2] have pointed out, this result actually requires the Palais–Smale condition only in the following weak form, namely if

$$I(x_n) \to \text{limit}, \qquad I'(x_n) \to 0,$$

then the limit must be a critical value, a conclusion which we have already noted follows directly from (P–S′).

To obtain Theorem A for a periodic function $I$ satisfying (P–S′) it is enough to verify Theorem A′ for this case, since the rest of the proof then applies unchanged. However, *with the additional hypothesis that the set E is periodic*, that is satisfies $x \in E$ if and only if $x + v \in E$, it is easy to extend Clark's argument to the present case. Thus Theorem A holds in the periodic case when $U$ is periodic, this, in fact, being the situation which will be needed in what follows.

Theorem B for the periodic case is proved in [6] when $N$ is an open ball. The proof applies unchanged to any bounded open $N$ and with only slight modifications when $N$ is periodic (these are the only cases needed in what follows). Finally, Theorem C holds equally in the periodic case, provided again that the open set $N$ is either bounded or periodic.

We next consider Theorem 5. The first construction fails since the disjoint sets $R$ and $S$ need not be periodic. On the other hand, the remaining parts of Theorem 5 can be carried through unchanged except to choose the required neighborhoods of $K_b$ to be periodic (which can obviously be done since $K_b$ is periodic). Hence Theorem 6 carries over unchanged, while Theorem 5 remains valid in the slightly weaker form

THEOREM 5′. *Suppose that each critical point with critical value greater than $\lambda$ is a local minimum. Then each is an absolute minimum.*

With this shown, we now obtain Theorems 1 and 2 exactly as before, and Theorem 3 in the weaker form without the final connectedness condition. Theorem 4 likewise continues to hold, even in the stronger form

THEOREM 4′. *Let 0 be a local minimum of $I$. Then there exists a critical point $e \neq kv$ for any integer $k$.*

*Proof.* Clearly all points $kv$ are local minima. If there are no other critical points, then each is in fact a strict local minima. But then by Theorem 2 there exists a critical point $z$ which is not a local minima. Since $z$ cannot equal $kv$ for any $k$, we have an obvious contradiction.

It remains to consider Theorem 7 and 8. The first step is to obtain an appropriate covering for the sets $\mathscr{M}$ and $\mathscr{N}$. To this end we require several preliminary results.

(1) *There exists a ball $\hat{B}$ which together with its translates by $kv$, $k = 0, \pm 1, \pm 2, \ldots$, covers $K_b$.* Let $B_n(0)$ denote the ball of radius $n$ about 0. If the assertion were false, then for every positive integer $n$ and every integer $m$ there would exist an element $x_n \in K_b$ such that $x_n + mv \notin B_n(0)$. By (P–S′) one can suppose $x_n + k_n v \to y$ for some sequence of integers $k_n$. But, taking $m = k_n$, we have $\|x_n + k_n v\| \geqslant n$, an obvious contradiction. It is convenient to suppose $\hat{B}$ closed.

(2) $\operatorname{Dist}(\mathscr{M}, \mathscr{N}) > 0$. This follows from (1) and the fact that $\mathscr{M}, \mathscr{N}$ are disjoint and periodic, and that both $\mathscr{M} \cap \hat{B}$ and $\mathscr{N} \cap \hat{B}$ are compact.

We can cover the set $\mathscr{M} \cap \hat{B}$ by a finite number of balls $B_{r_i}(x_i)$, $i = 1, \ldots, m$, with $I \leqslant b$ in $B_{r_i}(x_i)$, exactly as in the proof of Theorem 7. Similarly we can cover $\mathscr{N} \cap \hat{B}$ by a finite number of balls, this time with

$I \geqslant b$ on each. It can be assumed that the first set of balls is disjoint from the second set, and moreover that if $B$ is one of these balls then each translate of $B$ appears as long it intersects $\hat{B}$.

Let $\mathcal{M}'$ be the union of all translates (by $kv$) of balls of the first set, and $\mathcal{N}'$ the union of all translates of balls of the second set. Then $\mathcal{M}'$ covers $\mathcal{M}$; $\mathcal{N}'$ covers $\mathcal{N}$; $I \leqslant b$ in $\mathcal{M}'$; $I \geqslant b$ in $\mathcal{N}'$; $\operatorname{dist}(\mathcal{M}', \mathcal{N}') > 0$; and $\partial K_b \subset \mathcal{M}' \cup \mathcal{N}'$. Finally each bounded subset of $\mathcal{M}$ or $\mathcal{N}$ is covered by some *finite* union of balls, and both $\mathcal{M}'$ and $\mathcal{N}'$ are periodic sets.

With this covering in hand, and when $X$ is infinite dimensional, the rest of the proof of Theorem 7 for the periodic case proceeds exactly as in the regular case (the union $V''$ at the end of the proof can be assumed finite, since it is necessary to cover only those components $A'$ of $V'$ which intersect $g$). When $X$ is finite dimensional the situation is technically more complicated though the result remains true. We shall omit the details.

Theorem 8 is of course a corollary of Theorem 7 in the periodic case just as in the regular case.

EXAMPLE. In [5] Mawhin and Willem have studied a variational problem for periodic solutions of the forced pendulum equation. Their formulation involves a $v$-periodic function $I$ defined on the Hilbert space $H^1$ of absolutely continuous functions $x$: $[0, T] \to \mathbb{R}^n$, where $v \neq 0$ is a constant function in $H^1$. They show that $I$ satisfies the Palais–Smale-type condition (P–S$'$) and that $I$ has a strict local minimum (actually an absolute minimum) at the point 0 (and hence at each point $kv$, $k = 0, \pm 1, \pm 2, \ldots$). It follows immediately from Theorem 2 that $I$ must have another critical point $\bar{x}$ different from $kv$. This critical point, which of course satisfies $I(\bar{x}) > I(0)$ and is *not* a local minimum, yields a second independent solution of the forced pendulum equation in their setting.

If $\bar{x}$ (and its translates by $kv$) are the *only* other critical points of $I$, then by the corollary of Theorem 7 it is evident that $\bar{x}$ is a saddle point.

## 7. APPENDIX

Here we shall prove the topological results required for the demonstration of Theorem 7.

LEMMA (*). *Let $X$ be $n$-dimensional Euclidean space and let $A$ be a connected set in $X$ which is the union of a finite number of open balls. Let $C$ be any component of $X - A$. Then $\partial C$ is arcwise connected.*

Before turning to the proof, we note that the conclusion of the lemma is in general false if the underlying space $X$ is not simply connected. A simple example of this situation, even with $X$ one dimensional, occurs when $X$ is a circle in the plane. If $A$ is the lower half (connected), then $X-A$ is the upper half, and this has a *disconnected* boundary. This example shows that a proof of the lemma will require some care and cannot be based only on notions of general topology.

It is possible to give a fairly short proof of the lemma using Meyer–Vietoris sequences. For simplicity, however, we shall give a different demonstration which avoids the machinery of algebraic topology.

*Proof of Lemma* (∗). We list several properties of $\partial A$ and of $C$ which hold by virtue of the fact that $A$ consists of a finite number of open balls.

(i) If $z \in \partial A$ there exists an open straight segment $zz_0 \subset A$.

(ii) $\partial C$ is locally arcwise connected, and $C$ is arcwise connected.

(iii) Let $S$ be a component of $\partial C$ and let $z \in S$. Then for all sufficiently small $r>0$ we have

$$B_r(z) \subset A \cup S \cup \operatorname{Int}(C).$$

Let us now suppose for the sake of contradiction that $\partial C$ is not connected. Thus let $S$ be some component of $\partial C$ and suppose that $\partial C - S \neq \emptyset$. Using (ii) and (iii) it is not difficult to construct a piecewise linear path $\gamma$ in $C$ joining a point $z$ of $\partial C - S$ with a point $z'$ of $S$ and intersecting $\partial C$ only at its endpoints.

By (i) there exists an open straight segment $zz_0 \subset A$ and another open straight segment $z'z_0' \subset A$. We can suppose without loss of generality that $z_0$ and $z_0'$ are in $A$. Since $A$ is open and connected (hence arcwise connected) it is clear that $z_0$ and $z_0'$ can be joined by a piecewise linear path in $A$. In turn $z$ and $z'$ can be connected by a piecewise linear path $\gamma'$ in $A \cup \partial A$ which intersects $\partial A$ only at its endpoints.

Let $\Gamma$ be the closed piecewise linear path, going from $z$ to $z'$ along the path $\gamma$ and returning from $z'$ to $z$ along the path $\gamma'$. This path intersects $S$ exactly once, at $z'$.

Now let the path $\Gamma$ be subjected to a similarity contraction, continuously shrinking it toward $z$. Let $\tau \in [0,1)$ be the homotopy parameter defining the corresponding paths (thus $\Gamma(0)=\Gamma$).

Consider any path $\Gamma(\tau)$ for fixed $\tau$. Each line segment of $\Gamma(\tau)$ intersects the boundary of any ball making up $A$ at most twice. Consequently $\Gamma(\tau)$ intersects $S$ at most $2s'$ times, where $s'$ is the number of balls making up $A$.

Let $z_1, \ldots, z_k$ be the points of intersection of $\Gamma(\tau)$ with $S$. By (iii) there exist disjoint open balls $B_{\rho_i}(z_i)$ about the points $z_i$ such that

(a) the part of $\Gamma(\tau)$ inside $B_{\rho_i}(z_i)$ consists either of a single straight segment or two segments joined at $z_i$,

(b) $B_{\rho_i}(z_i) \subset A \cup S \cup \operatorname{Int}(C)$.

We define

$$p_i(\tau)=\begin{cases}0 & \text{if } \Gamma(\tau)\cap B_{\rho_i}(z_i)\subset \mathrm{Int}(C)\cup\{z_i\},\\ 0 & \text{if } \Gamma(\tau)\cap B_{\rho_i}(z_i)\subset A\cup\{z_i\},\\ 1 & \text{otherwise.}\end{cases}$$

That is, $p_i(\tau)=1$ if $\Gamma(\tau)$ crosses $S$ at $z_i$ from Int($C$) to $A$ (or vice versa) and $p_i(\tau)=0$ if $\Gamma(\tau)$ after its intersection with $S$ continues on the same side. Also put

$$\rho(\tau)=\sum_{i=1}^{k} p_i(\tau) \qquad (\text{mod } 2),$$

the total number of times $\Gamma(\tau)$ crosses $S$, mod 2.

We assert that $p(\tau)$ is continuous for $\tau\in[0,1)$. To see this, let $a_i$ and $b_i$ be the two points of intersection of $\Gamma(\tau)$ with $\partial B_{(1/2)\rho_i}(z_i)$, $i=1,\ldots,k$. If $p_i(\tau)=0$ they are both in $A$, or both in Int($C$), while if $p_i(\tau)=1$ then $a_i$, say, is in $A$ and $b_i$ in Int($C$). Then for any $\tau'$ sufficiently near $\tau$ the part of $\Gamma(\tau')$ outside $\bigcup_{i=1}^{k} B_{(1/2)\rho_i}(z_i)$ does not intersect $S$ and the part inside any $B_{(1/2)\rho_i}(z_i)$ joins a point near $a_i$ to a point near $b_i$. The latter part of the path may of course intersect $S$, but by an obvious counting argument it is clear that the number of crossings is even if $p_i(\tau)=0$ and odd if $p_i(\tau)=1$. Thus for $\tau'$ sufficiently near $\tau$

$$p(\tau')=\sum_{i=1}^{k}\{p_i(\tau)+2n_i\} \qquad (\text{mod } 2),$$

where $n_i$ is a positive integer or zero. It follows then that

$$p(\tau')=\sum_{i=1}^{k} p_i(t)\ (\text{mod } 2)=p(\tau)$$

so $p(\cdot)$ is continuous on $[0,1)$, and in fact constant.

For $\tau$ sufficiently near 1, say $\tau\in[\tau_0,1)$, the path $\Gamma(\tau)$ stays in an arbitrarily small neighborhood of $z$ and thus cannot intersect $S$. Hence $p(\tau_0)=0$. On the other hand $p(0)=1$ since $\Gamma(0)=\Gamma$ intersects $S$ exactly once. This contradiction shows that $\partial C$ is connected. That it is arcwise connected now follows from (ii), completing the proof.

ACKNOWLEDGMENTS

This work was partially supported by the Consiglio Nazionale delle Ricerche under the auspices of the Gruppo Nazionale di Analisi Funzionale e Sue Applicazioni, and by the

580/59/2-5

Mathematics Division of the National Science Foundation. The results were presented first in a lecture at the Unione Matematica Italiana held at Perugia, September 6, 1983. We wish to thank Professor Charles Conley and Professor James Milgram for helpful and stimulating conversations during the course of this work.

## References

1. A. Ambrosetti and P. H. Rabinowitz, Dual variational methods in critical point theory and applications, *J. Funct. Anal.* **14** (1973), 349–381.
2. H. Brezis, J. M. Coron, and L. Nirenberg, Free vibrations for a nonlinear wave equation and a theorem of P. Rabinowitz, *Comm. Pure Appl. Math.* **33** (1980), 667–684.
3. D. C. Clark, A variant of the Lusternik–Schnirelman theory, *Indiana Univ. Math. J.* **22** (1972), 65–74.
4. H. Hofer, A note on the topological degree at a critical point of mountain pass type, *Proc. Amer. Math. Soc.* **90** (1984), 309–315.
5. J. Mawhin and M. Willem, Variational methods and boundary value problems for vector second order differential equations and applications to the pendulum equation, *J. Diff. Equations* **52** (1984), 264–287.
6. P. Pucci and J. Serrin, A mountain pass theorem *J. Diff. Equations*, in press.
7. W. M. Ni, Some minimax principles and their applications in nonlinear eliptic equations. *J. Analyse Math.* **37** (1980), 248–278.
8. P. H. Rabinowitz, Variational methods for nonlinear eigenvalue problems, *in* "Lecture Notes C.I.M.E.," Varenna, 1974.
9. P. H. Rabinowitz, "Some Aspects of Critical Point Theory," University of Wisconsin, MRC Technical Report No. 2465, 1983.

# IV.2 Geometric Description of the Mountain Pass Critical Points

Patrizia Pucci

**Mathematics Subject Classification (2010).** 58E05, 35B38, 35J60, 49F15.

**Keywords.** Mountain pass, critical point theory.

In [1] *Ambrosetti* and *Rabinowitz* proved the celebrated *Mountain Pass Theorem* (MPT). The proof used the notion of the critical set $K_b$ of a functional $I$ on a Banach space $X$. Here $b$ is the min-max value defined in (0.1) of the preceding commentary for paper IV.1.[i], and $K_b$ is the set of those critical points $x$ for which $I(x) = b$. *Hofer* in [11] began the investigation of the structure of $K_b$; his work shows for example that the critical set possesses either a point of mountain-pass type, or a local minimum. *Hofer* also gave an example where $K_b$ contains no points of mountain-pass type. Here the set $M$-$P$ of *mountain-pass type critical points* consists of all points $x \in K_b$ such that the set $N \cap I_b$ is non-empty and not path connected, where $N$ is any neighborhood of $x$, and

$$I_b = \{x \in X \,:\, I(x) < b\}.$$

This definition for example implies that all points $x$ in $M$-$P$ are either saddle points or local minima, where a saddle point (at level $b$) is defined to be a point such that both sets $\{I > b\}$ and $\{I < b\}$ are non-empty and intersect any neighborhood of $x$.

Hofer's investigation was carried further in the paper [i] below, which continues the study of the structure of the critical set, being driven as well by the emergence of applications depending on the behavior of $K_b$ rather than on its mere existence. This paper answers such questions as does $K_b$ contain a saddle point, as might be expected from its construction?. (An example of Rabinowitz shows that this need NOT be the case if the "mountain" surrounding 0 everywhere has the same height; in such a case, the critical point (actually the critical set $K_b$) would consist of local maxima and not contain a saddle point.

The following review by *Pierre-Louis Lions* (MR0869402) of paper [i] gives an excellent summary of the main result in [i], see Theorem 2 of Section 3.

"This paper deals with critical points obtained via the mountain pass lemma of A. Ambrosetti and P.H. Rabinowitz. The structure of such critical points is studied with a remarkable precision and several general results are obtained, such as, for example, in the case of an infinite-dimensional problem, a general alternative between saddle points of mountain pass type or a set of local minima whose closure intersects at least two components of the set of saddle points."

P. Pucci et al. (eds.), *James Serrin. Selected Papers: Volume 2*, Contemporary Mathematicians, DOI 10.1007/978-3-0348-0847-7_9, © Springer Basel 2014

As an immediate consequence of this main theorem (see Corollary 1 on page 117 of [i]), it follows that in the infinite-dimensional case $K_b$ *always contains at least one saddle point*, giving a positive answer to the question raised above. Even more, *if there is just one point in* $K_b$, *or even if there is exactly one saddle point in* $K_b$, *then* $K_b$ *is itself a saddle point of mountain-pass type*, thus providing in this important case a geometrically appealing description of the behavior of the function $I$ in the neighborhood of the point in question.

The finite-dimensional case, of particular interest in itself, and also critical in understanding the more important infinite-dimensional case, is treated in detail in Theorem 1 of Section 3. In particular, if one adds the crucial, but very weak, additional assumption that $K_b$ does not separate the base points $O$ and $e$ of the min-max paths of Ambrosetti and Rabinowitz, then the conclusions and the corollaries remain the same for the finite-dimensional as for the infinite-dimensional case.

Of course, the set $K_b$ is compact (Palais–Smale, pages 115–116) so in the infinite-dimensional case $K_b$ *cannot* separate $O$ and $e$. Thus in this case the additional assumption required for the finite-dimensional case is now automatic, and hence not stated in Theorem 2.

The questions which arose in [i] (see also IV.1.[i] and IV.1.[ii]) inspired a number of further research papers. See, e.g., [10] by *Ghoussoub* and *Preiss*, [6, 7] by *Fang*, [2] by *Barletta* and *Marano*, [14] by *Livrea* and *Marano*, [3] by *Buratello* and *Terracini*, [16] by *Ohtsuka* and *Suzuki*, [15] by *Marano*, *Molica Bisci* and *Motreanu*, [4] by *Candela* and *Palmieri*. For further references I refer to the recent surveys [5, 12, 17]. Finally, various applications may be found in [8, 9, 12, 13].

## References

[1] A. Ambrosetti & P.H. Rabinowitz, *Dual variational methods in critical point theory and applications*, J. Funct. Anal. **14** (1973), 349–381.

[2] G. Barletta & S.A. Marano, *The structure of the critical set in the Mountain–Pass Theorem for nondifferentiable functions*, Differential Integral Equations **16** (2003), 1001–1012.

[3] V. Buratello & S. Terracini, *A bisection algorithm for the numerical mountain pass*, NoDEA, Nonlinear Differential Equations Appl. **14** (2007), 527–539.

[4] A.M. Candela & G. Palmieri, *Some abstract critical point theorems and applications*, Discrete and Continuous Dynamical Systems, in Dynamical Systems, Differential Equations and Applications, Supplement to 7th AIMS Conference, (2009), 133–142.

[5] I. Ekeland & N. Ghoussoub, *Selected new aspects of the calculus of variations in the large*, Bull. Amer. Math. Soc. **39** (2002), 207–265.

[6] G. Fang, *The structure of the critical set in the general mountain pass theorem*, Ann. Fac. Sci. Toulouse Math. **3** (1994), 345–362.

[7] G. Fang, *On the existence and classification of critical points for non-smooth functionals*, Canad. J. Math. **47** (1995), 684–717.

[8] M. Ghergu & V. Rădulescu, *Singular elliptic problems. Bifurcation and asymptotic analysis*, Oxford Lecture Series in Mathematics and its Applications, vol. **37**, Oxford University Press, New York, 2008.

[9] N. Ghoussoub, *Duality and perturbation methods in critical point theory*, Cambridge Tracts in Mathematics, Vol. **107**, Cambridge University Press, Cambridge, 1993.

[10] N. Ghoussoub & D. Preiss, *A general mountain pass principle for locating and classifying critical points*, Ann. Inst. H. Poincaré Anal. Non Linéaire **6** (1989), 321–330.

[11] H. Hofer, *A geometric description of the neighborhood of a critical point given by the mountain pass theorem*, J. London Math. Soc. **31** (1985), 566–570.

[12] Y. Jabri, *The mountain pass theorem. Variants, generalizations and some applications*, Encyclopedia of Mathematics and its Applications, Vol. **95**, Cambridge University Press, Cambridge, 2003.

[13] A. Kristály, V. Rădulescu & Cs. Varga, *Variational Principles in Mathematical Physics, Geometry and Economics: Qualitative Analysis of Nonlinear Equations and Unilateral Problems*, Encyclopedia of Mathematics, Vol. **136**, Cambridge University Press, Cambridge, 2010.

[14] R. Livrea & S.A. Marano, *Existence and classification of critical points for non–differentiable functions*, Adv. Differential Equations **9** (2004), 961–978.

[15] S.A. Marano, G. Molica Bisci & D. Motreanu, *Multiple solutions for a class of elliptic hemivariational inequalities*, J. Math. Anal. **337** (2008), 85–97.

[16] H. Ohtsuka & T. Suzuki, *Local property of the mountain pass critical point and the mean field equation.* Differential Integral Equations **21** (2008), 421–432.

[17] P. Pucci & V. Rădulescu, *The impact of the mountain pass theory in nonlinear analysis: a mathematical survey*, Boll. Unione Mat. Ital. **3** (2010), 543–582.

[18] P.H. Rabinowitz, *Some aspects of nonlinear eigenvalue problems*, Rocky Mountain J. Math. **3** (1973), 161–202.

[i] P. Pucci & J. Serrin, *The structure of the critical set in mountain pass theory*, Trans. Am. Math. Soc. **229** (1987), 115–132.

Patrizia Pucci
Dipartimento di Matematica e Informatica
Università degli Studi di Perugia
e-mail: pucci@dmi.unipg.it

*Trans. Am. Math. Soc.* 229 (1987), 115–132. **Reprint IV.2.[i]**


# THE STRUCTURE OF THE CRITICAL SET IN THE MOUNTAIN PASS THEOREM

PATRIZIA PUCCI AND JAMES SERRIN

ABSTRACT. We show that the critical set generated by the Mountain Pass Theorem of Ambrosetti and Rabinowitz must have a well-defined structure. In particular, if the underlying Banach space is infinite dimensional then either the critical set contains a saddle point of mountain-pass type, or the set of local minima intersects at least *two* components of the set of saddle points. Related conclusions are also established for the finite dimensional case, and when other special conditions are assumed. Throughout the paper, no hypotheses of nondegeneracy are required on the critical set.

**1. Introduction.** The existence of critical points of a real-valued $C^1$ functional $I$ defined on a real Banach space $X$ has been studied extensively in recent years. Naturally some kind of compactness is required in such problems, the assumption employed here usually being the Palais-Smale condition or some variation of it. A remarkable result in this setting is the following theorem due to Ambrosetti and Rabinowitz [**1**].

THE MOUNTAIN PASS THEOREM. *Let $I: X \to \mathbf{R}$ be a $C^1$ functional satisfying the Palais-Smale condition:*

> *Any sequence $(x_n)$ in $X$ with the property that $I(x_n) \to$* limit *and $I'(x_n) \to 0$ admits a convergent subsequence.*

*Suppose that there exist two real numbers $a$ and $R$, $R > 0$, such that $I(x) \geq a$ whenever $||x|| = R$, $I(0) < a$, and $I(e) < a$ for some $e \in X$ with $||e|| > R$. Then, with $G$ denoting the class of continuous paths joining 0 and $e$, the number*

$$b = \inf_{g \in G} \max_{t \in [0,1]} I(g(t))$$

*is a critical value of $I$, i.e. the set*

$$K_b = \{x \in X : I(x) = b,\ I'(x) = 0\}$$

*is nonempty.*

The structure of the critical set $K_b$ has been the subject of recent papers by Hofer and by the present authors. This interest stems from the natural desire to understand the structure of the critical set (e.g. does $K_b$ contain a saddle point, as would be expected from its construction), and from the emergence of applications depending on the *nature* of $K_b$ rather than on its mere *existence* (see [**3**]).

Received by the editors November 18, 1985.
1980 *Mathematics Subject Classification* (1985 *Revision*). Primary 58E05, 49F15.
*Key words and phrases.* Critical point theory, nonlinear functional analysis.

In [**4**] Hofer showed that the critical set $K_b$ possesses either a point of mountain-pass type, or a local minimum. On the other hand, in [**7**] we have shown that when $X$ is infinite dimensional there must be a saddle point in $K_b$. Nevertheless the situation still allows considerable refinement, without making any additional hypotheses on the basic functional $I$.

More precisely, the main result of this paper for the infinite dimensional case asserts that either the critical set $K_b$ contains a point which is simultaneously of saddle *and* mountain-pass type, or the set of local minima in $K_b$ is nonempty and its closure intersects at least two components of the set of saddle points.

This result can be viewed as a Morse theorem which is applicable to degenerate critical points as well as to functionals merely of class $C^1$.

Roughly speaking, the first case of the main result says that the "mountain" surrounding 0 possesses a "realistic" mountain pass. That is, at this point the gradient is zero, in each neighborhood of this point there are higher as well as lower points, and the set of lower points is not path-connected. In other words, the pass must be crossed in going from one set of lower values to another, and at the crossing one can see higher points. Of course, this is not the only conceivable situation, and the second alternative of the theorem gives the other possibility, namely, the mountain pass consists of a saddle point, then of a passage at fixed level through a set of local minima and finally of a second saddle point which is disconnected from the first.

The finite dimensional case, rather remarkably, involves a more subtle discussion. Indeed here, without some additional mild assumptions, the main result no longer holds and must be replaced by a weaker version. More precisely, in this case either the critical set contains a point of mountain-pass type or the set of local minima of $I$ in $K_b$ is nonempty *and* its closure intersects at least two components of the set of saddle points and proper local maxima.

To obtain for the finite dimensional case a result corresponding to that stated earlier for the infinite dimensional case it is necessary to add the hypothesis that the critical set $K_b$ does not separate the primary points 0 and $e$ in the Mountain Pass Theorem. That such a hypothesis is not necessary in the infinite dimensional case arises from the underlying strength of the Palais-Smale condition, which implies, among other things, that $K_b$ is compact. Since in an infinite dimensional space a compact set cannot separate two different points in its complement, no matter where they may be, there is accordingly no need to invoke the separation hypothesis required for the finite dimensional case.

These results are best possible. Indeed, Rabinowitz has remarked in [**8**] that, for the finite dimensional case, $K_b$ need not contain any saddle point, while in [**4**] Hofer has given an example where $K_b$ contains no points of mountain-pass type. On the other hand, it might be thought that, at least when $K_b$ consists of a single point or of isolated points, it should contain a saddle point $y$ not only of mountain-pass type but also having the property that for each sufficiently small neighborhood $N$ of $y$ both sets $\{x \in N : I(x) > b\}$ and $\{x \in N : I(x) < b\}$ are nonempty and not path-connected. Such a generalization, however, is *not* valid. For example, when $I(x,y,z) = x^2 + y^2 + z^2(3-z)$ the Mountain Pass Theorem applies with $e = (0,0,3)$ and $a = R = 1$. The critical value $b$ is 4 and the critical set $K_4$ consists only of the point $(0,0,2)$, which is saddle point of mountain-pass type for which the sets

$\{(x, y, z) \in \mathbf{R}^3 : x^2 + y^2 + (z-2)^2 < r^2$ and $I(x, y, z) > 4\}$, $r > 0$, are nonempty and path-connected. Similar examples are possible in any finite dimensional space whose dimension $n$ is greater than two and in suitable infinite dimensional spaces.

When $n = 2$ and the critical set $K_b$ consists of a single point it appears likely that both the sets $\{x \in N : I(x) > b\}$ and $\{x \in N : I(x) < b\}$ are nonempty and not path-connected, but we shall not pursue this.

Precise statements of the results in question, together with their proofs, are given in §3, following an important preliminary lemma which we establish in §2. In §4 we give two further results. The first is essentially a generalization of the theorems of §3, but with an additional and somewhat special hypothesis. The second concerns the structure of the critical set when $K_b$ possibly separates 0 and $e$.

Finally it is worthwhile noting two simple corollaries which follow immediately from the main results.

COROLLARY 1. *In the infinite dimensional case, or in the finite dimensional case when the points* 0 *and* $e$ *are not separated by the critical set* $K_b$, *there exists a saddle point in* $K_b$.

*Moreover, if the set of saddle points in* $K_b$ *is connected, or if the set of local minima in* $K_b$, *is closed, then there exists a saddle point of mountain-pass type in* $K_b$.

In particular, if $K_b$ contains exactly one saddle point, then this point must be of mountain-pass type.

COROLLARY 2. *In the finite dimensional case there exists either a saddle point or a proper local maximum in the critical set* $K_b$.

*In addition, if the set of local minima in* $K_b$ *is closed, then* $K_b$ *contains a point of mountain-pass type.*

*Finally, if* $K_b$ *contains exactly one point which is either a saddle or a proper local maximum, then this point is of mountain-pass type.*

The first part of Corollary 1 was established earlier by the present authors (see Theorem 7 of [**7**]), while Hofer's Theorem given in [**4**] is included in Corollaries 1 and 2.

A final remark may be added here. Our principal result shows that the "typical" critical point obtained from the Mountain Pass Theorem is a saddle point of mountain-pass type, as one would expect from the nature of the construction. For illustrative purposes, the functional $I$ may be thought of as having the asymptotic behavior

$$\sum_{1}^{n-1}(x_i - x_{i0})^2 - (x_n - x_{n0})^2, \qquad X = \mathbf{R}^n,\ n \geq 2,$$

near such a point. On the other hand, additional types of asymptotic behavior are possible near a saddle point, for example

$$\sum_{1}^{k-1}(x_i - x_{i0})^2 - \sum_{k}^{n}(x_i - x_{i0})^2, \qquad 3 \leq k \leq n-1,\ n \geq 4,$$

this corresponding to a point which is *not* of mountain-pass type. Clearly, alternative methods are needed to attain such points.

**2. Preliminaries.** Hereafter $X$ will be a fixed real Banach space, $I: X \to \mathbf{R}$ a $C^1$ functional satisfying the Palais-Smale condition, and $K_c = \{x \in X: I(x) = c, I'(x) = 0\}$, $c \in \mathbf{R}$, the set of critical points of $I$ at the level $c$.

If $A$ is a subset of $X$ we denote its complement by $A^c$, its closure by $\overline{A}$ and its boundary by $\partial A$. If $A$ and $B$ are two nonempty disjoint subsets of $X$ we define the *distance* from $A$ to $B$ by the number

$$\operatorname{dist}(A, B) = \inf\{d(x, y): x \in A,\ y \in B\}.$$

It is evident that if $A$ is closed and $B$ is compact, for instance, then $0 < \operatorname{dist}(A, B)$.

The following lemma is a slight variant of a result due to Clark [**2**] and Rabinowitz (see the proof of Theorem 1.9 in [**8**]). For the convenience of the reader we shall give the main outline of the proof.

LEMMA. *Let $c$ be a fixed real number and $E$ a nonempty closed set in $X$ such that $E \cap K_c = \emptyset$. Then for any positive number $\rho$ and for all sufficiently small $\varepsilon > 0$, say $\varepsilon \le d(c, \rho, E)$, there exists a related continuous mapping $\sigma: X \to X$ such that:*

(i) $\|\sigma(x) - x\| \le \rho$;
(ii) $I(x) \le c - 2\varepsilon$ *implies* $\sigma(x) = x$;
(iii) $x \in E$ *and* $I(x) \le c + \varepsilon$ *implies* $I(\sigma(x)) \le c - \varepsilon$.

PROOF. Assume first that $K_c \neq \emptyset$. Then since $K_c$ is compact and $E$ closed we have $0 < \delta = \operatorname{dist}(K_c, E)$. Put

$$A_\delta = \{x \in X: \operatorname{dist}(x, K_c) \ge \delta\}$$

so that $A_\delta$ is closed and contains $E$.

By the Palais-Smale condition there exists a positive real number $\alpha$ depending on $\delta$ and having the property that

$$\|I'(x)\| \ge \alpha > 0 \quad \text{when } x \in I^{-1}(c-2, c+2) \cap A_{\delta/4}. \tag{1}$$

Given $\rho > 0$, we set

$$d(c, \rho, E) = \min\{1, \alpha\rho/4, \alpha\delta/8\}, \tag{2}$$

and consider any positive number $\varepsilon \le d(c, \rho, E)$.

We now introduce two locally Lipschitz continuous functions $g, \bar{g}: X \to [0, 1]$ such that

$$g \equiv \begin{cases} 0 & \text{outside } I^{-1}(c - 2\varepsilon, c + 2\varepsilon), \\ 1 & \text{in } I^{-1}[c - \varepsilon, c + \varepsilon] \end{cases} \tag{3}$$

and

$$\bar{g} \equiv \begin{cases} 0 & \text{outside } A_{\delta/4}, \\ 1 & \text{in } A_{\delta/2}. \end{cases} \tag{4}$$

By Palais' Lemma (for a proof see, for instance, Lemma 1.6 of [**8**]) there exists a pseudo-gradient vector field $v: \tilde{X} \to X$, $\tilde{X} = \{x \in X: I'(x) \neq 0\}$, of $I$ which is locally Lipschitz continuous. In particular, for $x \in \tilde{X}$

$$\|v(x)\| \le 2\|I'(x)\| \quad \text{and} \quad I'(x)(v(x)) \ge \|I'(x)\|^2.$$

The mapping $V: X \to X$ given by

$$V(x) = -\rho g(x)\overline{g}(x)\,\frac{v(x)}{||v(x)||}$$

is then well defined. Indeed by (1) and (2) it follows that $||v(x)|| \geq ||I'(x)|| \geq \alpha > 0$ when $x \in I^{-1}(c-2\varepsilon, c+2\varepsilon) \cap A_{\delta/4}$, while $V(x) = 0$ when $x \notin I^{-1}(c-2\varepsilon, c+2\varepsilon)$ or $x \notin A_{\delta/4}$ by (3) and (4). By construction $V$ is also bounded and locally Lipschitz continuous. We now consider the flow associated with the ordinary differential equation in $X$

(5) $$\frac{d\sigma_x}{dt} = V(\sigma_x), \qquad \sigma_x(0) = x.$$

By classical existence theorems there exists a unique solution $\sigma_x = \sigma_x(t)$, $t \in \mathbf{R}$, for each $x \in X$. Moreover, $\sigma(x) \equiv \sigma_x(1)$ is a continuous mapping of $X$ into itself. We shall show that $\sigma$ satisfies properties (i), (ii), and (iii).

Since $||V(x)|| \leq \rho$ it follows at once that (i) holds. Also in view of (3) we have $\sigma_x(t) \equiv x$ whenever $x \notin I^{-1}(c-2\varepsilon, c+2\varepsilon)$, proving (ii). It remains to demonstrate (iii).

To this end, put $\omega_x(t) = I(\sigma_x(t))$ and observe that

$$\begin{aligned}\frac{d\omega_x}{dt}(t) &= I'(\sigma_x(t))\left(\frac{d}{dt}\sigma_x(t)\right) = I'(\sigma_x(t))(V(\sigma_x(t)))\\ &= -\frac{\rho g(\sigma_x(t))\overline{g}(\sigma_x(t))}{||v(\sigma_x(t))||}\,I'(\sigma_x(t))(v(\sigma_x(t)))\\ &\leq -\frac{\rho g(\sigma_x(t))\overline{g}(\sigma_x(t))}{||v(\sigma_x(t))||}||I'(\sigma_x(t))||^2\\ &\leq -\tfrac{1}{2}\rho g(\sigma_x(t))\overline{g}(\sigma_x(t))||I'(\sigma_x(t))||\end{aligned}$$

where we have used both properties of the pseudogradient $v$ as well as the fact that $I'(x) \neq 0$ when $g(x)\overline{g}(x) \neq 0$. This inequality entails two conclusions: first, for all $x \in X$, $t \in \mathbf{R}$,

(6) $$\frac{d\omega_x(t)}{dt} \leq 0;$$

and second, for all $x, t$ such that $\sigma_x(t) \in I^{-1}[c-\varepsilon, c+\varepsilon] \cap A_{\delta/2}$,

(7) $$\frac{d\omega_x(t)}{dt} \leq -\frac{1}{2}\rho\alpha$$

by virtue of (1)–(4).

The first conclusion shows that $I(\sigma_x(t)) \leq c-\varepsilon$ when $I(x) \leq c-\varepsilon$. It is thus enough to obtain (iii) when $x \in I^{-1}(c-\varepsilon, c+\varepsilon] \cap A_\delta$.

In this case, suppose for contradiction that $I(\sigma(x)) > c-\varepsilon$. Then by (6) we have $\sigma_x(t) \in I^{-1}(c-\varepsilon, c+\varepsilon]$ for $t \in [0,1]$. There are now two situations to be considered.

1. $\sigma_x(t) \in A_{\delta/2}$ for all $t \in [0,1]$. Then from (7) and (2) we obtain

$$\begin{aligned}c-\varepsilon < I(\sigma(x)) = \omega_x(1) &= \omega_x(0) + \int_0^1 \frac{d\omega_x}{dt}(t)\,dt\\ &\leq (c+\varepsilon) - \tfrac{1}{2}\rho\alpha \leq c-\varepsilon,\end{aligned}$$

a contradiction.

2. $\sigma_x(t) \in A_{\delta/2}$ for $t \in [0,\tau]$, $\tau < 1$, and $\sigma_x(\tau) \in \partial A_{\delta/2}$. Then, as in case 1, we find

$$c - \varepsilon < I(\sigma_x(\tau)) = \omega_x(\tau) \leq (c+\varepsilon) - \tfrac{1}{2}\rho\alpha\tau,$$

so that $\rho\tau < 4\varepsilon/\alpha$. On the other hand, since $||V(x)|| = \rho$ when

$$x \in I^{-1}[c-\varepsilon, c+\varepsilon] \cap A_{\delta/2},$$

we have by (2)

$$||\sigma_x(\tau) - \sigma_x(0)|| \leq \rho\tau < 4\varepsilon/\alpha \leq \delta/2. \tag{8}$$

But $\sigma_x(0) = x \in A_\delta$ and $\sigma_x(\tau) \in \partial A_{\delta/2}$ which gives $||\sigma_x(\tau) - \sigma_x(0)|| \geq \delta/2$, violating (8). This completes the demonstration when $K_c \neq \emptyset$.

If $K_c = \emptyset$ the same proof applies, except that the number $\delta$ need not be introduced, we can simply take $A_\delta = A_{\delta/2} = A_{\delta/4} = X$, and $\overline{g} \equiv 1$. Moreover the second case above does not now occur since obviously $\sigma_x(t) \in X$ for all $t \in \mathbf{R}$.

REMARK. From the proof above it is apparent that for each $x \in X$ there is a continuous map $\sigma_x\colon [0,1] \to X$ such that $\sigma_x(0) = x$, $\sigma_x(1) = \sigma(x)$ and $I(\sigma_x(t)) \leq I(x)$ for every $t \in [0,1]$.

In the remaining part of the paper we shall suppose that $I$ satisfies the hypotheses of the Mountain Pass Theorem, as given in the Introduction. The following notation will be useful throughout the paper:

$$\begin{aligned} I_b &= \{x \in X\colon I(x) < b\},\\ K_b &= \{x \in X\colon I(x) = b, I'(x) = 0\},\\ M_b &= \{x \in K_b\colon x \text{ is a local minimum of } I\},\\ P_b &= \{x \in K_b\colon x \text{ is a proper local maximum of } I, \text{ that is } x\\ &\qquad \text{is a local maximum of } I \text{ and } x \in \overline{I_b}\},\\ S_b &= \{x \in K_b\colon x \text{ is a saddle point of } I, \text{ namely in each}\\ &\qquad \text{neighborhood of } x \text{ there exist two points } y \text{ and } z\\ &\qquad \text{such that } I(y) < I(x) < I(z)\}, \end{aligned}$$

where $b$ denotes the critical value of $I$ given in the Mountain Pass Theorem.

It is easily seen that $K_b$ is the disjoint union of the sets $M_b$, $P_b$ and $S_b$. Moreover, in view of the Palais-Smale condition, $K_b$ as well as $S_b$ and $P_b \cup S_b$ is compact. Finally $P_b \cup S_b \subset \overline{I_b}$.

We say that $K_b$ *does not separate* 0 *and* $e$ if there exists a component $C$ of $(K_b)^c$ such that $0, e \in C$.

PATHS. A path $g$ joining two points $x_0$ and $x_1$ in $X$ is by definition a continuous map $g\colon [0,1] \to X$ such that $g(0) = x_0$ and $g(1) = x_1$.

Let $A$ and $B$ be sets in $X$. We call $A$ a *subcomponent* of $B$ if $A$ is a subset of some component of $B$. In this case, if $B$ is open and $x, y \in A$ then there exists a path joining $x$ and $y$ and contained in $B$.

Now, slightly modifying a terminology introduced by Hofer in [**4**], we say that a point $x \in K_b$ is of *mountain-pass type* (m-p type) if for any neighborhood $N$ of $x$ the set $N \cap I_b$ is not a subcomponent of $I_b$. It follows at once that if $x$ is of mountain-pass type then for any neighborhood $N$ of $x$ the set $N \cap I_b$ must be nonempty and not path-connected (Hofer's definition). Conversely, it is easy to see that if $x$ is not of mountain-pass type then for all suitably small neighborhoods $N$ of $x$ the set $N \cap I_b$ must be a subcomponent of $I_b$.

We say that a path $g$ intersects a nonempty subset $U$ of $X$ if there exists some $t \in [0,1]$ such that $g(t) \in U$.

Two paths $g$ and $h$ with the property that there exist two parameters $t_0$ and $t_1$ in $[0,1]$ such that $g(t_0) = h(t_1)$ determine a new path $f$ joining $g(0)$ to $h(1)$ obtained by following, first, $g$ from $g(0)$ to $g(t_0)$ and then $h$ from $h(t_1)$ to $h(1)$. In other words $f\colon [0,1] \to X$ is defined by

$$f(t) = \begin{cases} g(2t_0t), & 0 \le t \le \frac{1}{2}, \\ h(2(1-t_1)t + 2t_1 - 1), & \frac{1}{2} \le t \le 1, \end{cases}$$

and clearly is continuous with $f([0,1]) = g([0,t_0]) \cup h([t_1,1])$. Obviously the same idea can be used for any finite number of paths having successive intersections.

We finally say that a path $g$ is a *bridge* if $g([0,1]) \subset I_b$. If $x, y$ are two points in a subcomponent of $I_b$ then obviously there is a bridge joining these points.

**3. The main result.** For the sake of simplicity we first prove the main result for the finite dimensional case.

THEOREM 1. *Let $X$ be a finite dimensional Banach space. Assume that $K_b$ in the Mountain Pass Theorem does not separate 0 and $e$. Then at least one of the following two cases occurs:*

($\alpha$) *$K_b$ contains a saddle point of mountain-pass type;*

($\beta$) *$\overline{M_b}$ intersects at least two components of $S_b$.*

PROOF. Without loss of generality we may assume that $X$ is endowed with the Euclidean norm.

Since $K_b$ does not separate 0 and $e$, there exists a component $C$ of the open set $(K_b)^c$ containing 0 and $e$. Because $C$ is open, it is therefore path-connected as well as connected. Hence there is a path $\hat{g} \in G$ joining 0 and $e$ and lying entirely in $C$. In particular $\hat{g}([0,1]) \cap K_b = \varnothing$ and

$$0 < \delta_1 = \operatorname{dist}(\hat{g}([0,1]), K_b).$$

Assume that there are no saddle points in $K_b$ of m-p type, for otherwise ($\alpha$) holds and we are done. Now let us suppose for contradiction that ($\beta$) fails. Then either

$$(*) \qquad \overline{M_b} \cap S_b = \varnothing$$

or

$$(**) \qquad \overline{M_b} \text{ intersects exactly one component, say } S, \text{ of } S_b.$$

Under the assumption $(*)$ we have

$$0 < \delta_2 = \operatorname{dist}(\overline{M_b}, S_b)$$

if $M_b \neq \varnothing$ and $S_b \neq \varnothing$. If either $M_b = \varnothing$ or $S_b = \varnothing$ then we put $\delta_2 = 1$. The case $S_b = \varnothing$ actually does not occur, in view of Theorem 7 of [7], though we make no use of this fact.

When $(**)$ holds, for each point $x \in S$ we choose an open ball $B(x, \rho_x)$ with center at $x$ and radius $\rho_x$ less than $\delta_1$ such that $B(x, \rho_x) \cap I_b$ is a subcomponent of $I_b$, necessarily nonempty since $x$ is a saddle point. This is possible since $(\alpha)$ is assumed false. From the open cover $\{B(x, \frac{1}{4}\rho_x): x \in S\}$ of the compact set $S$ we can extract a finite subcover, namely

$$B(x_1, \tfrac{1}{4}\rho_1), \dots, B(x_k, \tfrac{1}{4}\rho_k), \qquad k \geq 1, \tag{9}$$

whose union will be denoted by $N$. Assumption $(**)$ implies that $N$ is connected, that $\overline{M_b}$ and $S_b \cap N^c$ are disjoint and that

$$0 < \delta_2 = \operatorname{dist}(\overline{M_b}, S_b \cap N^c),$$

if $S_b \cap N^c \neq \emptyset$. When $S_b \cap N^c = \emptyset$, we set $\delta_2 = 1$ as before.

For uniformity of notation, we put $N = \emptyset$ and $k = 0$ also for the simpler case $(*)$. Let $\delta = \min\{\delta_1, \delta_2\}$ and $\hat{K}_b = K_b \cap N^c$.

We now introduce an open cover for $\hat{K}_b$. Since none of the saddle points $x$ of $\hat{K}_b$ is of m-p type, for each of them we again select an open ball $B(x, \rho_x)$ with center at $x$ and radius $\rho_x < \delta$ such that $B(x, \rho_x) \cap I_b$ is a subcomponent of $I_b$. If $y \in \hat{K}_b$ is a local minimum of $I$, we choose an open ball $B(y, r_y)$ such that $I \geq b$ in $B(y, r_y)$. Similarly, if $y \in \hat{K}_b$ is a proper local maximum of $I$, we pick an open ball $B(y, r_y)$ such that $I \leq b$ in $B(y, r_y)$. Again in these two last cases we take the radii $r_y < \delta$. The open cover $\{B(x, \frac{1}{4}\rho_x),\ B(y, \frac{1}{2}r_y): x \in \hat{K}_b \cap S_b,\ y \in \hat{K}_b \cap S_b^c\}$ of the compact set $\hat{K}_b$ has a finite subcover of the form:

$$B(x_{k+1}, \tfrac{1}{4}\rho_{k+1}), \dots, B(x_{k+l}, \tfrac{1}{4}\rho_{k+l}), \tag{10}$$

$$B(y_1, \tfrac{1}{2}r_1), \dots, B(y_m, \tfrac{1}{2}r_m), \tag{11}$$

$$B(y_{m+1}, \tfrac{1}{2}r_{m+1}), \dots, B(y_{m+n}, \tfrac{1}{2}r_{m+n}), \tag{12}$$

whose union will be denoted by $W$. Here $x_{k+1}, \dots, x_{k+l}$ are saddle points of $I$; $y_1, \dots, y_m$ are local minima of $I$; and $y_{m+1}, \dots, y_{m+n}$ are proper local maxima of $I$ in $\hat{K}_b$. If a ball $B(y_i, \frac{1}{2}r_i)$ arises from a flat point $y_i$ of $I$, it will of course appear only in (11). Possibly there are no balls of any particular type (9)–(12), but this will not affect the proof. In the following $U$ will denote the open neighborhood $N \cup W$ of $K_b$. Clearly $U \cap \hat{g}([0,1]) = \emptyset$.

Since $U$ is given by a finite union of open balls, it can be assumed that its components $U_j$, $j = 1, \dots, \mu$, $\mu \leq 1 + l + m + n$, which are of course open, have the additional property that

$$0 < \min\{\operatorname{dist}(\overline{U_i}, \overline{U_j}):\ i, j = 1, \dots, \mu,\ i \neq j\}.$$

In fact, if necessary, we may attain this requirement by slightly reducing each of the numbers $\rho_i$, $i = k+1, \dots, k+l$, and $r_s$, $s = 1, \dots, m+n$.

Since $K_b$ is compact and $U$ is an open neighborhood of $K_b$ it is clear that

$$0 < \rho = \tfrac{1}{2}\operatorname{dist}(K_b, U^c).$$

Define $E = \{x \in X: \operatorname{dist}(x, U^c) \leq \rho\}$, so that $E$ is closed, disjoint from $K_b$ and contains $U^c$. Let $d = d(b, \rho, E)$ be the number given in the lemma of §2 when $K_c = K_b$, and put

$$\varepsilon = \min\{d, \tfrac{1}{2}(b - I(0)), \tfrac{1}{2}(b - I(e))\}.$$

Clearly $\varepsilon > 0$ since $b \geq a > \max\{I(0), I(e)\}$. In view of the definition of $b$ we can fix a path $g$ such that

(13) $$g \in G, \qquad g([0,1]) \subset I_{b+\varepsilon}.$$

Our goal is to modify $g$ to obtain another path $\tilde{g}$ such that

(14) $$\tilde{g} \in G, \qquad \tilde{g}([0,1]) \subset I_b.$$

This will contradict the definition of $b$ and complete the proof.

Consider the mapping $\sigma: X \to X$ given by the lemma in §2, depending on $b, \rho, E$ and $\varepsilon$. By the definition of $\varepsilon$ and property (ii) in the lemma it is evident that the path $\sigma \circ g$ is in $G$. Moreover for all $t$ such that $\sigma \circ g(t) \in U^c$ we have $I(\sigma \circ g(t)) \leq b - \varepsilon$. Indeed the condition $\sigma \circ g(t) \in U^c$ together with property (i) of the lemma shows that $g(t) \in E$. But then $I(\sigma \circ g(t)) \leq b - \varepsilon$ in view of (13), property (iii) of the lemma, and the fact that $\varepsilon \leq d$.

We may assume without loss of generality, therefore, that the path $g$ itself satisfies the condition

$$g(t) \in U^c \quad \text{implies} \quad I(g(t)) \leq b - \varepsilon,$$

and a fortiori that

$$g(t) \in U^c \quad \text{implies} \quad g(t) \in I_b.$$

If $g([0,1]) \subset U^c$, then we reach the required contradiction immediately. Thus, it is enough to consider the case when $g$ intersects $U$. We can suppose that the first component $U_j$ of $U$ intersected by $g$, in the natural order given by $t$, is $U_1$. Let $t_1^0$, $t_1^1$ be the unique pair of parameters such that

$$\begin{aligned} &0 < t_1^0 < t_1^1 < 1; \\ &z_1^0 = g(t_1^0), z_1^1 = g(t_1^1) \in \partial U_1; \\ &g([0, t_1^0]) \cap U = g([t_1^1, 1]) \cap U_1 = \varnothing. \end{aligned}$$

In the same manner, the first component $U_j$ intersected by $g$ for some $t > t_1^1$ can be supposed to be $U_2$. Again there exists a unique pair of parameters $t_2^0$, $t_2^1$ such that

$$\begin{aligned} &t_1^1 < t_2^0 < t_2^1 < 1; \\ &z_2^0 = g(t_2^0), z_2^1 = g(t_2^1) \in \partial U_2; \\ &g([t_1^1, t_2^0]) \cap U = g([t_2^1, 1]) \cap U_2 = \varnothing. \end{aligned}$$

Now, proceeding recursively, we construct $\nu$ pairs of parameters, with $\nu \leq \mu$, such that

(15) $$0 = t_0^1 < t_1^0 < t_1^1 < \cdots < t_\nu^0 < t_\nu^1 < t_{\nu+1}^0 = 1;$$

(16) $$z_j^0 = g(t_j^0), z_j^1 = g(t_j^1) \in \partial U_j, \qquad j = 1, \ldots, \nu;$$

(17) $$\bigcup_{j=1}^{\nu+1} g([t_{j-1}^1, t_j^0]) \subset I_b.$$

For uniformity of notation we write $0 = z_0^1$ and $e = z_{\nu+1}^0$. Note also that each of the points $z_j$, $j = 1, \ldots, \nu$, is in $I_b$, by the result of the previous paragraph.

We now show that there exists a bridge $v_j$ corresponding to each pair of points $z_j^0$ and $z_j^1$, $j = 1, \dots, \nu$. We begin by constructing the bridge $v_1$ from $z_1^0$ to $z_1^1$.

Since by assumption $K_b$ does not separate 0 and $e$, we claim that $z_1^0$ and $z_1^1$ are in the same component $L_1$ of $(U_1)^c$. To show this, we first follow the path $g$ from $z_1^1$ until we reach $e$. Afterwards we proceed along the continuous path $\hat{g}$ in the reverse direction from $e$ to 0, and then along $g$ from 0 to $z_1^0$. The resulting continuous path joins $z_1^1$ to $z_1^0$ and nowhere intersects $U_1$. This proves the claim above.

At the same time, $z_1^0$, $z_1^1$ are in $\partial U_1$ and hence also belong to $\partial L_1$. By the lemma in §7 of [7] we know that $\partial L_1$ is path-connected. Thus, there exists a continuous path $h_1$ joining $z_1^0$ and $z_1^1$ with $h_1([0,1]) \subset \partial L_1$. We recall that $h_1(0) = z_1^0$ and $h_1(1) = z_1^1$ are both in $I_b$. If $h_1([0,1]) \subset I_b$, then the path $v_1 = h_1$ is a bridge from $z_1^0$ to $z_1^1$.

The converse case, $h_1([0,1]) \not\subset I_b$, requires a further construction. Let us write

$$\partial U = \Gamma_1 \cup \Gamma_2 \cup \Gamma_3$$

where

$$\Gamma_1 = \bigcup_{i=1}^{k+l} \left(\partial U \cap \partial B\left(x_i, \tfrac{1}{4}\rho_i\right)\right),$$

$$\Gamma_2 = \bigcup_{s=1}^{m} \left(\partial U \cap \partial B\left(y_s, \tfrac{1}{2}r_s\right)\right),$$

$$\Gamma_3 = \bigcup_{s=m+1}^{m+n} \left(\partial U \cap \partial B\left(y_s, \tfrac{1}{2}r_s\right)\right).$$

In the following, to simplify the notation, we shall denote the closed balls $\overline{B(x_i, \frac{1}{4}\rho_i)}$ by $B_i$ and the sets $B(x_i, \rho_i) \cap I_b$ by $I_b^i$. For later use, recall that each of the sets $I_b^i$ is a nonempty subcomponent of $I_b$. It will be convenient also to refer to the *closure* of any ball listed in (9) as a *closed ball of type* (9), and similarly for the closures of balls listed in (10), (11), and (12).

If $x \in \Gamma_3$, then $x \in \partial B(y_s, \frac{1}{2}r_s)$ for some $s = m+1, \dots, m+n$. Since $I \leq b$ in $B(y_s, r_s)$, the equality $I(x) = b$ implies $I'(x) = 0$. But this is impossible since $x \notin K_b$. Hence $I < b$ in $\Gamma_3$, and by the compactness of $\Gamma_3$ there exists a real number $c$ such that $I \leq c < b$ in $\Gamma_3$. Moreover $\Gamma_2 \cap \Gamma_3 = \emptyset$ and $\Gamma_2 \cap I_b = \emptyset$, since $I \geq b$ in $\Gamma_2$.

We select a parameter $\tau_1^0 \in [0,1)$ as follows:

(a) if $I(h_1(0)) < \frac{1}{2}(b+c)$, then $\tau_1^0$ is the maximal parameter in $(0,1)$ such that $I(h_1(\tau_1^0)) = \frac{1}{2}(b+c)$ and $I(h_1(\tau)) \leq \frac{1}{2}(b+c)$ for $\tau \in [0, \tau_1^0]$;

(b) if $I(h_1(0)) \geq \frac{1}{2}(b+c)$, then $\tau_1^0 = 0$.

The existence of $\tau_1^0$ in (a) follows from the present assumption that $h_1([0,1]) \not\subset I_b$. Here and in the following the parameters $\tau$ which we introduce depend on the path $h_1$, but for simplicity we do not indicate this dependence explicitly. In both cases (a) and (b) we have $c < I(h_1(\tau_1^0)) < b$, and so $h_1(\tau_1^0) \in \Gamma_1$.

Let $C_1$ be the component of the compact set $\Gamma_1 \cup \Gamma_2$ containing $h_1(\tau_1^0)$ and $\tau_1^1$ the maximum value of $\tau$ such that $h_1(\tau) \in C_1$. If $\tau_1^1 < 1$, then $h_1(\tau_1^1) \in \Gamma_3$. Otherwise, there exists $\theta > 0$ such that $h_1([\tau_1^1 - \theta, \tau_1^1 + \theta]) \subset \partial U \backslash \Gamma_3 \subset \Gamma_1 \cup \Gamma_2$. Hence, since $C_1$ is a component of $\Gamma_1 \cup \Gamma_2$ and $h_1(\tau_1^1) \in C_1$, we get $h_1(\tau_1 + \theta) \in C_1$, which is absurd. Thus, if $\tau_1^1 < 1$, then $h_1(\tau_1^1) \in (\Gamma_1 \cup \Gamma_2) \cap \Gamma_3 \subset \Gamma_1 \cap I_b$. Moreover, since $h_1(\tau_1^1) \in \Gamma_3$ we have $I(h_1(\tau_1^1)) \leq c$, while on the other hand $I(h_1(\tau_1^0)) \geq \frac{1}{2}(b+c) > c$. Consequently $\tau_1^1 > \tau_1^0$. If $\tau_1^1 = 1$, then by construction we again have $h_1(\tau_1^1) = h_1(1) = z_1^1 \in C_1 \cap I_b \subset \Gamma_1 \cap I_b$. Finally we observe that $h_1(\tau_1^0)$, $h_1(\tau_1^1) \in C_1 \cap \Gamma_1 \cap I_b$ and hence in particular $C_1 \cap \Gamma_1 \neq \emptyset$.

The next step is to show that there exists a subcomponent $I_b^{(1)}$ of $I_b$ containing both $h_1(\tau_1^0)$ and $h_1(\tau_1^1)$. It is convenient to discuss assumptions (*) and (**) separately.

ASSUMPTION (*). We recall that here $k = 0$. If $x \in \Gamma_1 \cap \Gamma_2$, then there exist two indices $i$ and $s$, $i = 1, \ldots, l$, $s = 1, \ldots, m$, such that $\|x - x_i\| = \frac{1}{4}\rho_i$ and $\|x - y_s\| = \frac{1}{2}r_s$. Therefore $\|y_s - x_i\| \leq \frac{1}{2}r_s + \frac{1}{4}\rho_i < \delta \leq \delta_2 = \operatorname{dist}(\overline{M_b}, S_b)$, which is absurd since $y_s \in M_b$ and $x_i \in S_b$. Thus $\Gamma_1 \cap \Gamma_2 = \emptyset$. Since $C_1 = (C_1 \cap \Gamma_1) \cup (C_1 \cap \Gamma_2)$ and $C_1$ is connected, it now follows that $C_1 \cap \Gamma_2 = \emptyset$ and hence $C_1 \subset \Gamma_1$. Consequently there exist $l'$ saddle points, $l' \leq l$, which without loss of generality we may relabel $x_1, x_2, \ldots, x_{l'}$, such that $\bigcup_{j=1}^{l'} \partial B_j$ is connected and contains $C_1$. In particular, denoting the set $\bigcup_{j=1}^{l'} I_b^j$ by $I_b^{(1)}$, we have $h_1(\tau_1^0)$, $h_1(\tau_1^1) \in C_1 \cap I_b \subset I_b^{(1)}$.

We now show that $I_b^{(1)}$ is a subcomponent of $I_b$. To see this, consider two balls $B_i$, $B_j$, $i, j = 1, \ldots, l'$, $i \neq j$, such that $B_i \cap B_j \neq \emptyset$. Then, either $x_j \in B(x_i, \rho_i)$ or $x_i \in B(x_j, \rho_j)$. In both cases $I_b^i \cap I_b^j \neq \emptyset$ and therefore $I_b^i \cup I_b^j$ is a subcomponent of $I_b$. Next let $x, y$ be two points in $I_b^{(1)}$, say $x \in B(x_1, \rho_1)$ and $y \in B(x_{l'}, \rho_{l'})$. Since $\bigcup_{j=1}^{l'} B_j$ is connected, a standard maximality argument shows that there exists a chain of the balls $B_{j_s}$ in this union joining $B_1$ and $B_{l'}$, namely $B_{j_1} = B_1$, $B_{j_{l''}} = B_{l'}$, $B_{j_s} \cap B_{j_{s+1}} \neq \emptyset$, $s = 1, \ldots, l'' - 1$, $l'' \leq l'$. Hence $I_b^{(1)}$ is a subcomponent of $I_b$, as desired.

ASSUMPTION (**). Let $\hat{N}$ be the component of the set $\bigcup_{i=1}^{k+l} B_i$ which contains the open connected set $N$. Since $h_1(\tau_1^0)$, $h_1(\tau_1^1) \in C_1 \cap \Gamma_1 \cap I_b$ it follows that $h_1(\tau_1^0)$, $h_1(\tau_1^1) \in \bigcup_{i=1}^{k+l} B_i$. There are now two cases to be discussed, that is $\{h_1(\tau_1^0), h_1(\tau_1^1)\} \subset \hat{N}$ and $\{h_1(\tau_1^0), h_1(\tau_1^1)\} \not\subset \hat{N}$. If the first case occurs, then $h_1(\tau_1^0)$, $h_1(\tau_1^1) \in \hat{N} \cap I_b$. This set, which we again call $I_b^{(1)}$, is a subcomponent of $I_b$ by the argument of the previous paragraph.

Next suppose $\{h_1(\tau_1^0), h_1(\tau_1^1)\} \not\subset \hat{N}$, say, for instance, $h_1(\tau_1^0) \notin \hat{N}$. We shall show then that $C_1 \subset \Gamma_1$, so that the situation reduces to the case discussed in (*). Thus assume for contradiction that $C_1 \cap \Gamma_2 \neq \emptyset$. Hence there exists some index $\tilde{s} = 1, \ldots, m$ such that $C_1 \cap \partial B(y_{\tilde{s}}, \frac{1}{2}r_{\tilde{s}}) \neq \emptyset$. Consequently, since $C_1$ is connected, there must be a chain joining $h_1(\tau_1^0)$ and $\overline{B(y_{\tilde{s}}, \frac{1}{2}r_{\tilde{s}})}$ consisting of closed balls of type (9) or (10) or (11). The first ball in the chain must be of type (9) or (10) since $h_1(\tau_1^0) \in I_b$. Hence without loss of generality we may assume that all balls $\tilde{B}$ in the chain before $\overline{B(y_{\tilde{s}}, \frac{1}{2}r_{\tilde{s}})}$ are of type (9) or (10). By the definition of $\hat{N}$ and the fact that $h_1(\tau_1^0) \notin \hat{N}$, it now follows that the balls $\tilde{B}$ cannot intersect $\hat{N}$ and so a fortiori cannot intersect $N$. In particular, each ball $\tilde{B}$ is of type (10). But this cannot occur since closed balls of type (10) do not intersect closed balls of type (11), by construction and the fact that $\delta \leq \delta_2$. Thus, $C_1 \subset \Gamma_1$ as required.

The treatment of the case $h_1(\tau_1^1) \notin \hat{N}$ is exactly the same.

For both assumptions $(*)$ and $(**)$ we stop the construction if $\tau_1^1 = 1$. Otherwise, recalling that $I(h_1(\tau_1^1)) < \frac{1}{2}(b+c)$ and repeating the procedure as often as necessary, we ultimately obtain a finite set of parameters $\tau$ such that

$$0 = \tau_0^1 \le \tau_1^0 < \tau_1^1 < \cdots < \tau_\lambda^0 < \tau_\lambda^1 \le \tau_{\lambda+1}^0 = 1; \tag{18}$$

$$h_1(\tau_s^0), h_1(\tau_s^1) \in I_b^{(s)} \quad \text{for every } s = 1, \ldots, \lambda; \tag{19}$$

$$\bigcup_{s=1}^{\lambda+1} h_1([\tau_{s-1}^1, \tau_s^0]) \subset I_b, \tag{20}$$

where the set $I_b^{(s)}$ is a subcomponent of $I_b$ for every $s = 1, \ldots, \lambda$. That the parameter set (18) is not infinite follows from the fact that the continuous function $I \circ h_1$ on $[0,1]$ cannot have an infinite number of oscillations between the two values $c$ and $\frac{1}{2}(b+c)$. Now by (19) the points $h_1(\tau_s^0)$ and $h_1(\tau_s^1)$ are in the same subcomponent of $I_b$, so there exists a bridge $v_s^1$ joining them. We are thus in position to construct a bridge $v_1$ from $z_1^0$ to $z_1^1$. Indeed, it is enough to follow the path $h_1$ from $z_1^0 = h_1(0)$ to $h_1(\tau_1^0)$, next $v_1^1$ from $h_1(\tau_1^0)$ to $h_1(\tau_1^1)$, subsequently $h_1$ from $h_1(\tau_1^1)$ to $h_1(\tau_2^0)$ and so forth until we reach $z_1^1 = h_1(1)$.

Proceeding in this way we can therefore successively bridge each of the components $U_j$, $j = 1, \ldots, \nu$, by a corresponding path $v_j$ from $z_j^0$ to $z_j^1$.

We can now obtain the required path $\tilde{g}$ satisfying (14). In fact it is sufficient to follow $g$ from $0 = z_0^1$ to $z_1^0$, then $v_1$ from $z_1^0$ to $z_1^1$, next $g$ from $z_1^1$ to $z_2^0$ and so on until reaching $e = z_{\nu+1}^0$. This completes the proof.

Now we state the main result for the infinite dimensional case:

THEOREM 2. *Let $X$ be an infinite dimensional Banach space. Then at least one of the following two cases occurs:*

($\alpha$) *$K_b$ contains a saddle point of mountain-pass type;*

($\beta$) *$\overline{M_b}$ intersects at least two components of $S_b$.*

PROOF. Since $X$ is infinite dimensional and 0 and $e$ do not belong to the compact set $K_b$, they both are in the same component $C$ of $(K_b)^c$. Thus we may follow the main argument introduced for the finite dimensional case. For this reason, there will be no confusion if we do not repeat word for word the previous preliminary construction. Indeed, no changes are necessary in the selection of the number $\delta = \min\{\delta_1, \delta_2\}$ and of the open neighborhood $U = N \cup W$ of $K_b$. It is also worth noting here that $K_b$ contains no flat points. We introduce as before

$$\rho = \tfrac{1}{2}\text{dist}(K_b, U^c), \quad E = \{x \in X: \text{dist}(x, U^c) \le \rho\}, \quad d = d(b, \rho, E),$$

and take

$$\varepsilon = \min\{d, \tfrac{1}{2}(b - I(0)), \tfrac{1}{2}(b - I(e))\}.$$

We now fix a path $g$ satisfying (13) and shall show that it can be modified in order to obtain (14). To see this, we proceed word for word until formula (17).

Thus, it remains to show that there exists a bridge $v_j$ corresponding to each pair of points $z_j^0$, $z_j^1$, $j = 1, \ldots, \nu$. To begin with, we modify the path $\hat{g}$ joining 0 and $e$ into a piecewise linear path, still from 0 to $e$ and nowhere intersecting $\overline{U}$, which we continue to call $\hat{g}$ for simplicity. Then we approximate the parts of $g$ joining $z_j^1$

and $z_{j+1}^0$ by piecewise linear paths $g_j$ from $z_j^1$ to $z_{j+1}^0$, entirely contained in $I_b$ for each $j = 0, 1, \ldots, \nu$.

Let $Y$ be a finite dimensional subspace of $X$ containing the piecewise linear paths $\hat{g}([0,1])$ and $g_j([0,1])$, $j = 0, 1, \ldots, \nu$, as well as the centers of the set of balls listed in (9)–(12). Without loss of generality we may endow $Y$ with the Euclidean norm.

Let $U' = U \cap Y$, $N' = N \cap Y$ and $U_j' = U_j \cap Y$, for $j = 1, \ldots, \mu$. Clearly $N'$ is connected and $U_j'$ are exactly the components of $U'$ in $Y$.

For each point $x \in \overline{U'}$ we choose an open ball $B_Y(x, r_x)$ in $Y$ such that:

(21) $B_Y(x, r_x) \cap \hat{g}([0,1]) = \emptyset$;

(22) if $x$ belongs to some closed ball of type (9)–(12), then $B_Y(x, r_x)$ is entirely contained in the corresponding ball of radius $\frac{1}{3}\rho_i$, $i = 1, \ldots, k+l$, or $\frac{2}{3}r_s$, $s = 1, \ldots, m+n$;

(23) if $x$ and $\tilde{x}$ respectively are in disjoint closed balls of type (9)–(12), then the closures of $B_Y(x, r_x)$ and $B_Y(\tilde{x}, r_{\tilde{x}})$ are disjoint.

From the open cover $\{B_Y(x, r_x): x \in \overline{U'}\}$ of the compact set $\overline{U'}$ of $Y$ we select a finite subcover, whose union will be called $V$. We denote by $V_j$ the component of $V$ containing $U_j'$, $j = 1, \ldots, \mu$, so that by (23) we have $\overline{V_i} \cap \overline{V_j} = \emptyset$ if $i \neq j$.

For uniformity and simplicity of notation we still call $z_j^0$ the first point of intersection of $g_{j-1}$ with $\partial V_j$ and $z_j^1$ the last point of intersection of $g_j$ with $\partial V_j$, $j = 1, \ldots, \mu$. The argument already used for the finite dimensional case implies that $z_j^0$, $z_j^1$ are in the same component $L_j$ of $Y \backslash V_j$. Hence, by the lemma in §7 of [**7**], there is a path $h_j$ from $z_j^0$ to $z_j^1$ entirely contained in $\partial L_j$ for each $j = 1, \ldots, \mu$.

We start by constructing a bridge $v_1$ joining $z_1^0$ to $z_1^1$. If $h_1([0,1]) \subset I_b$, we take $v_1 = h_1$ as before. Otherwise we introduce the following new sets:

$$\Gamma_1' = \partial V \cap \overline{\bigcup_{i=1}^{k+l} B(x_i, \tfrac{1}{3}\rho_i)},$$

$$\Gamma_2' = \partial V \cap \overline{\bigcup_{s=1}^{m} B(y_s, \tfrac{2}{3}r_s)},$$

$$\Gamma_3' = \partial V \cap \overline{\bigcup_{s=m+1}^{l} B(y_s, \tfrac{2}{3}r_s)}.$$

By (22) it is clear that $\partial V = \Gamma_1' \cup \Gamma_2' \cup \Gamma_3'$. Moreover as in the proof of Theorem 1 we have $I(z) < b$ on the closed bounded subset $\Gamma_3'$ of $Y$, so in turn there exists a real number $c$ such that $I \leq c < b$ in $\Gamma_3'$. Since $I \geq b$ on $\Gamma_2'$ it follows also that $\Gamma_2' \cap I_b = \emptyset$ and $\Gamma_2' \cap \Gamma_3' = \emptyset$.

Define $\tau_1^0 \in [0,1)$ as in the proof of Theorem 1. As before, $h_1(\tau_1^0)$ is in $\Gamma_1'$ but not in $\Gamma_2'$ or $\Gamma_3'$. Let $C_1$ be the component of the compact set $\Gamma_1' \cup \Gamma_2'$ containing $h_1(\tau_1^0)$ and $\tau_1^1$ the maximum value of $\tau$ such that $h_1(\tau) \in C_1$. Arguing as in the finite dimensional case, we see at once that $h_1(\tau_1^1) \in \Gamma_3'$ when $\tau_1^1 < 1$. In this case $h_1(\tau_1^1) \in (\Gamma_1' \cup \Gamma_2') \cap \Gamma_3' \subset \Gamma_1' \cap I_b$. Also as before $\tau_1^0 < \tau_1^1$. When $\tau_1^1 = 1$, the definition of $\tau_1^1$ shows that $h_1(\tau_1^1) = h_1(1) = z_1^1 \in C_1 \cap I_b \subset \Gamma_1' \cap I_b$. In both cases $h_1(\tau_1^0)$, $h_1(\tau_1^1) \in C_1 \cap \Gamma_1' \cap I_b$ and so in particular $C_1 \cap \Gamma_1'$ is nonempty.

For convenience at this stage we separate the treatment of assumptions $(*)$ and $(**)$.

ASSUMPTION $(*)$. We recall that $k = 0$ here. Moreover $\Gamma_1' \cap \Gamma_2' = \emptyset$, for if not then there must exist a point $y$ such that $y \in \overline{B(x_i, \frac{1}{3}\rho_i)}$ for some $i = 1, \ldots, l$ and $y \in \overline{B(y_s, \frac{2}{3}r_s)}$ for some $s = 1, \ldots, m$. Hence $\|y_s - x_i\| < \delta \le \delta_2 = \operatorname{dist}(\overline{M_b}, S_b)$, which is absurd. Thus we have $C_1 \cap \Gamma_2' = \emptyset$ by the usual connectedness argument, and so $C_1 \subset \Gamma_1'$. Therefore there exist $l'$ saddle points, $l' \le l$, which we may still call $x_1, x_2, \ldots, x_{l'}$, such that $\bigcup_{j=1}^{l'} \overline{B(x_j, \frac{1}{3}\rho_j)}$ is connected and contains $C_1$. We also have $h_1(\tau_1^0)$, $h_1(\tau_1^1) \in C_1 \cap I_b \subset I_b^{(1)}$ where, as usual, $I_b^{(1)} = \bigcup_{j=1}^{l'} I_b^j$. To show that $I_b^{(1)}$ is a subcomponent of $I_b$ it is sufficient to repeat the argument used in the proof of the finite dimensional case, since if $\overline{B(x_i, \frac{1}{3}\rho_i)} \cap \overline{B(x_j, \frac{1}{3}\rho_j)} \neq \emptyset$, then either $x_i \in B(x_j, \rho_j)$ or $x_j \in B(x_i, \rho_i)$.

ASSUMPTION $(**)$. Let $\hat{N}$ be the component of the set $\bigcup_{i=1}^{k+l} \overline{B(x_i, \frac{1}{3}\rho_i)}$ which contains the open connected set $N$. Since $h_1(\tau_1^0)$, $h_1(\tau_1^1) \in C_1 \cap \Gamma_1' \cap I_b$ it follows that $h_1(\tau_1^0)$, $h_1(\tau_1^1) \in \bigcup_{i=1}^{k+l} \overline{B(x_i, \frac{1}{3}\rho_i)}$. There are two cases to be considered, namely $\{h_1(\tau_1^0), h_1(\tau_1^1)\} \subset \hat{N}$ and $\{h_1(\tau_1^0), h_1(\tau_1^1)\} \not\subset \hat{N}$. In the first case it is clear, as in earlier arguments, that $h_1(\tau_1^0)$ and $h_1(\tau_1^1)$ are in the same subcomponent of $I_b$, and hence in the same component of $I_b$.

In the second case, we may assume for instance that $h_1(\tau_1^0) \notin \hat{N}$. We shall show that $C_1 \cap \Gamma_2' = \emptyset$. Otherwise there exists an index $\tilde{s} = 1, \ldots, m$ such that $C_1 \cap \overline{B(y_{\tilde{s}}, \frac{2}{3}r_{\tilde{s}})} \neq \emptyset$. Hence, since $C_1$ is also connected in $X$, there must be a chain of balls of type

$$\overline{B(x_i, \tfrac{1}{3}\rho_i)}, \quad i = 1, \ldots, k; \qquad \overline{B(x_i, \tfrac{1}{3}\rho_i)}, \quad i = k+1, \ldots, k+l;$$
$$\overline{B(y_s, \tfrac{2}{3}r_s)}, \qquad s = 1, \ldots, m,$$

joining $h_1(\tau_1^0)$ and $\overline{B(y_{\tilde{s}}, \frac{2}{3}r_{\tilde{s}})}$. Now, repeating the argument already used in the finite dimensional case, mutatis mutandis, we reach the same contradiction. The case $h_1(\tau_1^1) \notin \hat{N}$ can be treated similarly.

The remaining part of the proof is exactly as in Theorem 1.

**4. Further related results.** In this section we present two additional theorems whose proofs rely on the main argument introduced in the previous section. The first is a generalization of Theorems 1 and 2, while the second is a result for the case when $K_b$ may possibly separate 0 and $e$.

THEOREM 3. *Suppose that $X$ is infinite dimensional, or that $X$ is finite dimensional and $K_b$ does not separate 0 and $e$. Assume also that $\overline{M_b}$ has only a finite number of components. Then either*

($\alpha$) *$K_b$ contains a saddle point of mountain-pass type, or*

($\beta$) *at least one component of $\overline{M_b}$ intersects two or more components of $S_b$.*

PROOF. For simplicity we shall consider only the finite dimensional case, the treatment of the infinite dimensional case being similar. Since the proof requires only minor modifications of the argument already given for Theorem 1 we shall indicate only the points of difference. In particular assumption $(**)$ must be replaced by a more delicate condition.

Let $M_1, \ldots, M_\lambda$, $\lambda \geq 1$, be the components of the set $\overline{M_b}$. If $(\alpha)$ and $(\beta)$ are assumed false, then either $(*)$ holds or there exists an integer $\lambda'$, with $1 \leq \lambda' \leq \lambda$, such that

$M_i$ intersects exactly one component of $S_b$, say $S_i$, for every $i = 1, \ldots, \lambda'$,

and

$$M_i \cap S_b = \varnothing \text{ for each } i = \lambda' + 1, \ldots, \lambda, \text{ when } \lambda' < \lambda.$$

Since two different components of $\overline{M_b}$ can intersect the same component of $S_b$, it is convenient to group the components of $\overline{M_b}$ in such a way that

$$\overline{M_b} = \bigcup_{i=1}^{\lambda''} \hat{M}_i, \qquad \lambda'' \leq \lambda',$$

where $\hat{M}_i$ are pairwise disjoint compact sets with the property

$(**)$ each $\hat{M}_i$, $i = 1, \ldots, \lambda''$, intersects exactly one component $S_i$ of $S_b$, and $S_i \neq S_j$ when $i \neq j$.

In this case it is clear that

$$0 < \delta_0 = \min\{\operatorname{dist}(\hat{M}_i \cup S_i, \hat{M}_j \cup S_j) : i, j = 1, \ldots, \lambda'', i \neq j\}.$$

We now introduce an open cover of the compact set $S = \bigcup_{i=1}^{\lambda''} S_i$. Since $(\alpha)$ is assumed to be false, for each point $x \in S$ we choose an open ball $B(x, \rho_x)$, with radius $\rho_x < \delta_0$, such that $B(x, \rho_x) \cap I_b$ is a subcomponent of $I_b$. Let

$$B(x_1, \tfrac{1}{4}\rho_1), \ldots, B(x_k, \tfrac{1}{4}\rho_k), \qquad k \geq 1,$$

be a finite subcover of the open cover $\{B(x, \frac{1}{4}\rho_x) : x \in S\}$ of $S$, and denote its union by $N$. Also we put $\delta = \min\{\delta_0, \delta_1, \delta_2\}$.

We now proceed exactly as in the proof of Theorem 1 until the concluding discussion of the assumption $(**)$.

ASSUMPTION $(**)$. Let $\hat{N}_i$ denote the component of $\bigcup_{j=1}^{k+l} B_j$ containing the connected set $S_i$, $i = 1, \ldots, \lambda''$. Since $h_1(\tau_1^0)$, $h_1(\tau_1^1) \in C_1 \cap \Gamma_1 \cap I_b$ it follows that $h_1(\tau_1^0)$, $h_1(\tau_1^1) \in \bigcup_{j=1}^{k+l} B_j$. There are three cases to consider here:

(a) $\{h_1(\tau_1^0), h_1(\tau_1^1)\} \subset \hat{N}_i$ for some $i = 1, \ldots, \lambda''$;

(b) $\{h_1(\tau_1^0), h_1(\tau_1^1)\} \not\subset \bigcup_{i=1}^{\lambda''} \hat{N}_i$;

(c) $h_1(\tau_1^0) \in \hat{N}_i$ and $h_1(\tau_1^1) \in \hat{N}_k$ for some $i, k = 1, \ldots, \lambda''$, with $\hat{N}_i \neq \hat{N}_k$.

Cases (a) and (b) can be treated exactly as in the proof of Theorem 1. The third case (c) cannot occur, as we now show. Indeed, if $h_1(\tau_1^0) \in \hat{N}_i$ and $h_1(\tau_1^1) \in \hat{N}_k$, then, since $C_1$ is connected, there is a chain of closed balls of type (9), (10) or (11) joining $h_1(\tau_1^0)$ and $h_1(\tau_1^1)$. For the remaining part of the argument we note that any consecutive set of balls of type (9) or (10) in this chain must belong to the same component of $\bigcup_{j=1}^{k+l} B_j$. Now consider the first closed ball $\tilde{B}$ of type (11) in the chain. Since $\delta \leq \delta_2$, this ball must be preceded by a closed ball of type (9) whose center is in $S_j \cap \hat{N}_i$ for some $j = 1, \ldots, \lambda''$. Here, if $j \neq i$ then $\hat{N}_j = \hat{N}_i$ even though $S_j \neq S_i$. Since $\delta \leq \delta_0$, it is evident that $\tilde{B}$ has center in $\hat{M}_j$. The next ball in the chain is either of type (9) or type (11). If it is of type (11) then its center is in $\hat{M}_j$ since $\delta \leq \delta_0$, while if it is of type (9) its center is in $S_j$, again because $\delta \leq \delta_0$. In the second case the ball itself is in $\hat{N}_j$, and so finally in $\hat{N}_i$ since $\hat{N}_j = \hat{N}_i$. Continuing in this way we see that every closed ball of type (9) or (10) in the chain is contained in $\hat{N}_i$. Since $h_1(\tau_1^1) \in I_b$, the last ball in the chain cannot be of type (11) and so is contained in $\hat{N}_i$. In particular $h_1(\tau_1^1) \in \hat{N}_i$ and hence $\hat{N}_i \cap \hat{N}_k \neq \emptyset$, which is absurd.

Now following the remaining part of the proof of Theorem 1, we complete the demonstration.

REMARK. From the proof of Theorem 3 it is easily seen that the assumption concerning $\overline{M_b}$ can be replaced by the following:

> *there exists an integer $\lambda$ such that $\overline{M_b} = \bigcup_{i=1}^{\lambda} M_j$ where $M_j$, $j = 1, \ldots, \lambda$, are pairwise disjoint compact sets.*

The appropriate assertion ($\beta$) then becomes

($\beta'$) *there exists at least one set $M_j$, $j = 1, \ldots, \lambda$, which intersects two or more components of $S_b$.*

This version of Theorem 3 contains Theorem 1 when $\lambda = 1$, and reduces exactly to Theorem 3 if each set $M_j$, $j = 1, \ldots, \lambda$, is connected.

In the remaining part of this section we shall prove a further result without invoking the hypothesis that $K_b$ does not separate the points 0 and $e$.

THEOREM 4. *The critical set $K_b$ in the Mountain Pass Theorem is such that either*

($\alpha$) *$K_b$ contains at least one point of mountain-pass type, or*

($\beta$) *$\overline{M_b}$ intersects at least two components of $S_b \cup P_b$.*

PROOF. We shall present the demonstration only for the finite dimensional case. Thus without loss of generality we may assume that $X$ is endowed with the Euclidean norm. Also for convenience in notation we shall put $S_b^* = S_b \cup P_b$.

We start by assuming that there are no points in $S_b^*$ of m-p type, for otherwise ($\alpha$) holds and we are done. Thus we suppose for contradiction that either

$$(*) \qquad \overline{M_b} \cap S_b^* = \emptyset$$

or

$$(**) \qquad \overline{M_b} \text{ intersects exactly one component, say } S, \text{ of } S_b^*.$$

Under assumption ($*$) we have $0 < \delta = \operatorname{dist}(\overline{M_b}, S_b^*)$, whenever the sets $M_b$ and $S_b^*$ are nonempty. If, on the other hand, one or the other of these sets is empty, then we put $\delta = 1$.

Next suppose ($**$) holds. Since ($\alpha$) is assumed to be false, for each point $x \in S$ we can choose an open ball $B(x, \rho_x)$ such that the set $B(x, \rho_x) \cap I_b$ is a subcomponent of $I_b$, which is nonempty since $x$ is either a saddle point or a proper local maximum.

Hence, the open cover $\{B(x, \frac{1}{4}\rho_x): x \in S\}$ of the compact set $S$ possesses a finite subcover, namely

$$(24) \qquad B(x_1, \tfrac{1}{4}\rho_1), \ldots, B(x_k, \tfrac{1}{4}\rho_k), \qquad k \geq 1,$$

whose union will be denoted by $N$. By virtue of assumption $(**)$ the open set $N$ is also connected.

We have $0 < \delta = \operatorname{dist}(\overline{M_b}, S_b^* \cap N^c)$, if $S_b^* \cap N^c \neq \varnothing$. When $S_b^* \cap N^c = \varnothing$ we take $\delta = 1$, as before.

For uniformity of notation we put $N = \varnothing$ and $k = 0$ also for the simpler case $(*)$. In both cases, we denote the set $K_b \cap N^c$ by $\hat{K}_b$.

For each point $x \in \hat{K}_b \cap S_b^*$ we can select an open ball $B(x, \rho_x)$ with radius $\rho_x < \delta$ such that the set $B(x, \rho_x) \cap I_b$ is a subcomponent of $I_b$. If $y \in \hat{K}_b \cap M_b$ we choose an open ball $B(y, r_y)$ such that $I \geq b$ in $B(y, r_y)$. From the open cover $\{B(x, \frac{1}{4}\rho_x),\ B(y, \frac{1}{2}r_y): x \in \hat{K}_b \cap S_b^*,\ y \in \hat{K}_b \cap M_b\}$ of the compact set $\hat{K}_b$ we can extract a finite subcover of the form

$$(25) \qquad B(x_{k+1}, \tfrac{1}{4}\rho_{k+1}), \ldots, B(x_{k+l}, \tfrac{1}{4}\rho_{k+l}),$$

$$(26) \qquad B(y_1, \tfrac{1}{2}r_1), \ldots, B(y_m, \tfrac{1}{2}r_m),$$

whose union will be denoted by $W$.

In the following, $U$ will denote the open neighborhood $N \cup W$ of $K_b$. Since $U$ is given by a finite union of open balls, its components $U_j$, $j = 1, \ldots, \mu$, $\mu \leq 1+l+m$, which are of course open, can be supposed to have the additional property that

$$0 < \min\{\operatorname{dist}(\overline{U_i}, \overline{U_j}):\ i, j = 1, \ldots, \mu, i \neq j\}.$$

This can be done by slightly reducing the numbers $\rho_i$, $i = k+1, \ldots, k+l$, and $r_s$, $s = 1, \ldots, m$.

We now proceed exactly as in the proof of Theorem 1 until formula (17).

The next step is to prove that there exists a bridge $v_j$ corresponding to each pair of points $z_j^0$ and $z_j^1$, $j = 1, \ldots, \nu$. We start by constructing the bridge $v_1$ from $z_1^0$ to $z_1^1$. We recall that $z_1^0$ and $z_1^1$ are both in $I_b$ by (16). For convenience we discuss assumptions $(*)$ and $(**)$ separately.

ASSUMPTION $(*)$. Here $k = 0$ and $\overline{M_b} \cap S_b^* = \varnothing$. Thus $U_1$ consists entirely of balls of type (25), since it is connected and the radii $\rho_i$, $r_s$, $i = 1, \ldots, l$, $s = 1, \ldots, m$, are each less than $\delta$. Moreover $\overline{U_1} \cap I_b = I_b^{(1)}$ is a subcomponent of $I_b$ by virtue of the argument already used to treat the corresponding assumption $(*)$ in the proof of Theorem 1. Hence there exists a bridge $v_1$ joining $z_1^0$ and $z_1^1$, as required.

ASSUMPTION $(**)$. Let $\hat{N}$ be the component of the set $\bigcup_{i=1}^{k+l} B_i$ which contains the open connected set $N$. There are two cases to consider: $\{z_1^0, z_1^1\} \subset \hat{N}$ and $\{z_1^0, z_1^1\} \not\subset \hat{N}$. In the first case $z_1^0, z_1^1 \in \hat{N} \cap I_b$, where $\hat{N} \cap I_b = I_b^{(1)}$ is a subcomponent of $I_b$ by construction. Consequently there is a bridge $v_1$ joining $z_1^0$ and $z_1^1$.

Otherwise we may assume, for instance, that $z_1^0 \notin \hat{N}$. In this case we claim that $U_1$ consists entirely of balls of type (25). Indeed, if there were some ball of type (26) in $U_1$, then there would be a chain consisting of closed balls of type (24) and (25) joining $z_1^0$ to a closed ball of type (26). By definition of $\hat{N}$ and the fact that $z_1^0 \notin \hat{N}$, however, no balls of type (24) can appear in the chain. Moreover, closed balls of type (25) are disjoint from closed balls of type (26). Hence no balls of type (26) can appear in $U_1$ and, by the same argument repeated again, also no balls of type (24). The situation is thus reduced to the case already discussed under assumption (*) and so a bridge $v_1$ joining $z_1^0$ to $z_1^1$ can be found. The case $z_1^1 \notin \hat{N}$ can be handled similarly.

For each $j = 2, \ldots, \nu$ we proceed as above to find a bridge $v_j$ for the component $U_j$.

Finally, denoting by $\tilde{g}$ the path obtained by following, first, $g$ from $0 = z_0^1$ to $z_1^0$, then $v_1$ from $z_1^0$ to $z_1^1$, next $g$ from $z_1^1$ to $z_2^0$ and so forth until reaching $e = z_{\nu+1}^0$, we obtain the required contradiction.

REMARK. In analogy with Theorem 3, it is clear that Theorem 4 can be generalized as follows:

*Assume that* $\overline{M_b} = \bigcup_{j=1}^{\lambda} M_j$, *where* $M_j$, $j = 1, \ldots, \lambda$, *are pairwise disjoint compact sets. Then either*

($\alpha$) $K_b$ *contains a point of mountain-pass type, or*

($\beta'$) *at least one set* $M_j$ *intersects two or more components of* $S_b \cup P_b$.

ACKNOWLEDGMENT. The work of the first author was carried out under the auspices of the Gruppo Nazionale di Analisi Funzionale e sue Applicazioni of the Consiglio Nazionale delle Richerche, Rome, Italy. The paper was written partly at the University of Perugia and partly at the University of Minnesota. The first author wishes to thank the University of Minnesota for its hospitality, and the second author similarly expresses his thanks to the University of Perugia for its support.

We wish finally to thank Professor Helmut Hofer for several helpful comments concerning this work.

## REFERENCES

1. A. Ambrosetti and P. H. Rabinowitz, *Dual variational methods in critical point theory and applications*, J. Funct. Anal. **14** (1973), 349–381.
2. D. C. Clark, *A variant of the Lusternik-Schnirelman theory*, Indiana Univ. Math. J. **22** (1972), 65–74.
3. I. Ekeland and H. Hofer, *Periodic solutions with prescribed minimal period for convex autonomous hamiltonian systems*, Invent. Math. **81** (1985), 155–188.
4. H. Hofer, *A geometric description of the neighbourhood of a critical point given by the Mountain Pass Theorem*, J. London Math. Soc. **31** (1985), 566–570.
5. L. Nirenberg, *Variational and topological methods in nonlinear problems*, Bull. Amer. Math. Soc. (N.S.) **4** (1981), 267–302.
6. P. Pucci and J. Serrin, *A Mountain Pass Theorem*, J. Differential Equations **60** (1985), 142–149.
7. ——, *Extensions of the Mountain Pass Theorem*, J. Funct. Anal. **59** (1984), 185–210.
8. P. H. Rabinowitz, *Variational methods for nonlinear eigenvalue problems*, Eigenvalues of Nonlinear Problems (G. Prodi, ed.), C.I.M.E., Edizioni Cremonese, Rome, 1974, pp. 141–195.
9. ——, *Some aspects of critical point theory*, Univ. of Wisconsin, MRC Tech. Rep. No. 2465, 1983.

DIPARTIMENTO DI MATEMATICA, UNIVERSITÀ DEGLI STUDI, 06100 PERUGIA, ITALY

DEPARTMENT OF MATHEMATICS, UNIVERSITY OF MINNESOTA, MINNEAPOLIS, MINNESOTA

# V. Fluid Mechanics

# V.1. The Navier–Stokes Equation – Theory

Vladimír Šverák

**Mathematics Subject Classification (2010).** 35Q30, 76D05, 76N10.

**Keywords.** Navier–Stokes equations, viscous incompressible fluid, boundary-layer theory.

In a 1911 paper [13] C. W. Oseen proved (among other things) the short-time existence for the classical Cauchy problem for the Navier–Stokes equations in $R^3$, and made the following remark concerning the global existence (p. 254): "According to our theory, it therefore seems likely that singularities can appear in the interior of a viscous incompressible fluid, even in the case when the volume forces and initial velocity are smooth."

At the time of this writing, in 2011, we still do not know whether such singularities can appear, and other fundamental questions remain open, although we do know much more about the Navier–Stokes solutions than what was known in 1911. The papers [i–iv] of J. Serrin re-printed below represent important contributions to the effort towards a better understanding of these issues. The papers appeared between 1959 and 1963. Let us briefly summarize the main results of the PDE theory of the Navier–Stokes theory up to that time.

By the 1950s, Oseen's local-in-time existence theory was generalized in [12, 7].[1] Moreover, in [12] Leray established the existence of a new type of global solutions[2], not known to be smooth, but satisfying the equation in the sense of distributions (to use today's terminology), and proved a number of ground-breaking results about them.[3] In a 1951 paper [5] E. Hopf re-cast Leray's ideas in the by-then-already-common language of functional analysis and imbedding theorems, and re-worked some of the definitions and proofs[4], while also allowing for some generalizations (such as domains with boundaries, as opposed to the whole space used in Leray's paper). A 1957 paper [7] by Kisilev and Ladyzhenskaya emphasizes the problem of finding the class of solutions in which both

[1]For later results, see for example [6, 10].

[2]"Turbulent solutions" in [12], or "Leray–Hopf weak solutions" (today's terminology).

[3]In two dimension he established (in another paper) the full regularity of the weak solutions in all space. In three dimension he established the first "partial regularity" result.

[4]For example, instead of using a regularization of the equation to produce approximate solutions, Hopf used a finite-dimensional approximation (nowadays usually called Galerkin approximation), for which he also established the energy inequality. He also found an elegant simplification of the proof of a crucial compactness property in Leray's paper. See Lemma 5.1, p. 228 in Hopf's paper. In addition, his method works very well in domains with boundaries, where – as he noted – the boundary condition for the weak solution is naturally formulated in terms of the energy space.

P. Pucci et al. (eds.), *James Serrin. Selected Papers: Volume 2*, Contemporary Mathematicians,
DOI 10.1007/978-3-0348-0847-7_10, © Springer Basel 2014

existence and uniqueness can be shown. Several such classes are then established in the paper for the local-in-time problem or a small initial data situation.[5]

Papers [i] and [ii] below represent important contributions to the research on these topics, and had become standard references on the subject soon after their publication. They identify the function spaces $L_t^{s'}L_x^s$ satisfying

$$\frac{2}{s'} + \frac{n}{s} = 1, \tag{1}$$

(where $n$ is the spatial dimension of the domain occupied by the fluid[6] and $s > n$) as good uniqueness and regularity classes for the weak solutions. Strictly speaking, the regularity was proved in [ii] under the condition $\frac{2}{s'} + \frac{n}{s} < 1$, but it was shown later that (1) is still sufficient for regularity.[7] The uniqueness result proved in [i] (Theorem 6, page 22) is in fact stronger than uniqueness in the $L_t^{s'}L_x^s$ class. It only assumes that one of the Leray–Hopf weak solutions in question belongs to this class, and hence it is a weak-strong uniqueness result, generalizing Leray's weak-strong uniqueness theorem in [12], pp. 242–244, and also uniqueness theorems of Prodi [14] and Lions–Prodi [11].

We should emphasize that, due to lack of space, our description here can be only sketchy, and we cannot do justice to some of the subtle points. We recommend that the reader read the original papers in detail, in order to appreciate some of the subtleties and the clever ways in which various technical aspects are dealt with.

Over the years, Serrin's uniqueness and regularity results became one of the centerpieces of the PDE theory of the Navier–Stokes equations, against which all subsequent results have been measured.

The spaces $L_t^{s'}L_x^s$ satisfying condition (1) are invariant under the scaling symmetry

$$u(x,t), p(x,t) \to \lambda u(\lambda x, \lambda^2 t), \lambda^2 p(\lambda x, \lambda^2 t) \tag{2}$$

of the equation, and – in today's terminology – they are "critical spaces" for the perturbation theory.[8] See for example [10, 16]. The original condition $\frac{2}{s'} + \frac{n}{s} < 1$ in Serrin's regularity result is "subcritical", in that the corresponding $L_t^{s'}L_n^s$ space is slightly better than critical, and the problem of extending the same result to the critical case motivated many subsequent developments. It should also be mentioned that Serrin's result is local

[5]The task of finding a solution class for which both global existence and uniqueness can be shown was seen by Ladyzhenskaya as the most basic problem in the PDE theory of the Navier–Stokes equations. A detailed account of her views on the subject can be found in [9]. The classes of solutions for which uniqueness could not be proved were always considered as somewhat deficient by her. (For Leray–Hopf weak solutions in dimension three the uniqueness remains uknown. Hopf's comment on this issue in his 1951 paper, p. 217, is: "It is hard to believe that the initial value problem for the viscous fluid in dimension $n = 3$ could have more than one solution, and more work should be devoted to the study of the uniqueness question.") Ladyzhenskaya's caution in this regard has later received some indirect justification when anomalous weak solutions of inviscid incompressible Euler's equation which defy intuitive expectations (and for which uniqueness fails) were discovered by V. Scheffer, see [17, 18, 3]. Of course, the incompressible Euler's equation is not the same as the Navier–Stokes equation, but the examples still show that some caution about weak solutions may not be out place and the precise class of function where the weak solution lives can be crucial.

[6]In some places a restriction $n \leq 4$ is imposed in the paper, which however is not essential. In any case, the most interesting cases $n = 2$ and $n = 3$ are always included.

[7]This is in fact relatively subtle. At the time when the papers [i] and [ii] were published, analogous "critical space results" were probably just about being clarified for much simpler non-linear equations of elliptic type, such as $-\Delta u = u^a$. The clarification for the Navier–Stokes equation, including the case $s = n$, was achieved only later. See for example [1, 10, 15, 16, 2].

[8]Heuristically this is quite understandable, the behavior of the equation at the borderline should not single out any particular scales.

in space, which makes the proof more difficult, due to the presence of non-smooth solutions of the form $u(x,t) = \nabla h(x,t)$ with $\Delta_x h(x,t) = 0$ and arbitrary dependence on $t$. The difficulty is overcome by working with the vorticity equation, rather than with the equation for $u$. This trick has helped many researches ever since.

Paper [ii] reviews and establishes other notable results. We should at least mention a backward uniqueness result in Theorem 7 (p. 30) and the stability results, first proved in papers [iii] and [iv].

The stability result says that a solution with Reynolds number

$$\mathrm{Re} = \frac{\max|u| \operatorname{diam} \Omega}{\nu} \leq 5.7$$

is stable, in the sense that a perturbation of the solution (with the same lateral boundary condition and volume forces) will approach the given solution as $t \to \infty$. This is shown by an energy method, based on a specific estimate for a linear Stokes problem (inequality (11) in the paper). The method also gives a very specific stability criterion for the Taylor-Couette flow between two rotating cylinders. See condition (1) in the paper. Paper [iv] establishes, under certain assumptions, an existence and stability result for time-periodic solutions (for time-periodic data). For recent developments we refer the reader to [4].

## References

[1] Caffarelli, L., Kohn, R., Nirenberg, L., Partial regularity of suitable weak solutions of the Navier–Stokes equations, Comm. Pure Appl. Math. 35 (1982), no. 6, 771–831.

[2] Eskauriaza, L., Seregin, G.A., Sverak, V., $L^{3,\infty}$-solutions of Navier–Stokes equations and backward uniqueness. (Russian) Uspekhi Mat. Nauk 58 (2003), no. 2(350), 3–44; translation in Russian Math. Surveys 58 (2003), no. 2, 211–250.

[3] De Lellis, C., Szekelyhidi, L., Jr., The Euler equations as a differential inclusion, Ann. of Math. (2) 170 (2009), no. 3, 1417–1436.

[4] Galdi, G., Sohr, H., Existence and uniqueness of time-periodic physically reasonable Navier–Stokes flow past a body, Arch. Ration. Mech. Anal. 172 (2004), no. 3, 363–406.

[5] Hopf, E., Über die Anfangswertaufgabe für die hydrodynamischen Grundgleichungen. Math. Nachr. 4, (1951), 213–231.

[6] Koch, H., Tataru, D., Well-posedness for the Navier–Stokes equations, Adv. Math. 157 (2001), no. 1, 22–35.

[7] Kiselev, A.A., Ladyzhenskaya, O.A., On the existence and uniqueness of the solution of the nonstationary problem for a viscous, incompressible fluid, (Russian), Izv. Akad. Nauk SSSR. Ser. Mat. 21 (1957), 655–680.

[8] Ladyzhenskaja, O.A., Solution "in the large" of the nonstationary boundary value problem for the Navier–Stokes system with two space variables, Comm. Pure Appl. Math. 12, 1959, 427–433.

[9] Ladyzhenskaya, O.A., The sixth millennium problem: Navier–Stokes equations, existence and smoothness. (Russian) Uspekhi Mat. Nauk 58 (2003), no. 2(350), 45–78; translation in Russian Math. Surveys 58 (2003), no. 2, 251–286.

[10] Lemarié-Rieusset, P.G., Recent developments in the Navier–Stokes problem, Research Notes in Mathematics, 431. Chapman & Hall/CRC, Boca Raton, FL, 2002. xiv+395 pp. ISBN: 1-58488-220-4.

[11] Lions, J.-L., Prodi, G., Un théorème d'existence et unicité dans les équations de Navier–Stokes en dimension 2, C. R. Acad. Sci. Paris 248 1959, 3519–3521.

[12] Leray, J., Sur le mouvement d'un liquide visqueux emplissant l'espace. (French) Acta Math. 63 (1934), no. 1, 193–248.

[13] Oseen, C.W., Sur let formules de Green généralisées qui se présentent dans l'hydrodynamique et sur quelques-unes de leurs application, Acta Math. 34 (1911), no. 1, 205–284.

[14] Prodi, G., Un teorema di unicità per le equazioni di Navier–Stokes, Ann. Mat. Pura Appl. (4) 48 1959, 173–182.

[15] Seregin, G.A., Local regularity of suitable weak solutions to the Navier–Stokes equations near the boundary, J. Math. Fluid Mech. 4 (2002), no. 1, 1–29.

[16] Seregin, G.A., On the local regularity of suitable weak solutions of the Navier–Stokes equations. (Russian) Uspekhi Mat. Nauk 62 (2007), no. 3(375), 149–168; translation in Russian Math Surveys 62 (2007), no. 3.

[17] Scheffer, V., An inviscid flow with compact support in space-time, J. Geom. Anal. 3 (1993), no. 4, 343–401.

[18] Shnirelman, A., On the nonuniqueness of weak solution of the Euler equation, Comm. Pure Appl. Math. 50 (1997), no. 12., 1261–1286.

Vladimír Šverák
School of Mathematics
University of Minnesota

*Arch. Ration. Mech. Anal.* 9 (1962), 187–195.


**Reprint V.1.[i]**

# *On the Interior Regularity of Weak Solutions of the Navier-Stokes Equations*

JAMES SERRIN

In this note we intend to show that certain weak solutions of the time dependent Navier-Stokes equations must actually be solutions in an ordinary sense. Though the case of three spatial dimensions is no doubt the most important, we shall consider throughout the paper an arbitrary number $n$ of spatial dimensions.

The Navier-Stokes equations may be written in the form

$$\begin{aligned} u_t - \Delta u + u \cdot \operatorname{grad} u &= f - \operatorname{grad} p, \\ \operatorname{div} u &= 0, \end{aligned} \tag{1}$$

where $u$ is the velocity vector, $p$ is the pressure, and vector notation has been used. Let $D$ denote the test space of $C^\infty$ vector fields which are divergence-free and have compact support in an open region $R$ of space-time. A vector $u$ is then called a ***weak solution*** of (1) in $R$ if it is (weakly) divergence-free and if

$$\int \{(u, \Phi_t) + (u, \Delta \Phi) + (u, u \cdot \operatorname{grad} \Phi)\}\, dt = -\int (f, \Phi)\, dt \tag{2}$$

for all $\Phi \in D$. Here $(u, v)$ denotes the ordinary scalar product $\int u(x) \cdot v(x)\, dx$. A vector $u$ may of course satisfy (2) even if it has no derivatives in the ordinary sense. Our purpose is to prove a differentiability theorem for such weak solutions: assuming to begin with a certain amount of regularity of a solution of (2), we wish to show that this solution actually possesses derivatives in the ordinary sense. The order of differentiability which one can hope to attain turns out to be more or less closely tied to the amount of regularity one assumes at the outset. An example will make this clear. Suppose that $f=0$. Then by direct calculation it is easily verified that

$$u = a(t) \operatorname{grad} \psi(x) \tag{3}$$

is a solution of (2), provided only that $a$ is an integrable function and $\psi$ is harmonic. Now (3) is infinitely differentiable so far as the space variables are concerned but need possess *no* time derivatives whatever. The conclusion which we draw from this example is that the order of time differentiability of a weak solution is intimately tied to the amount of time regularity which is originally assumed[1]. On the other hand, it is reasonable to expect that quite modest assumptions concerning the spatial regularity of a weak solution should be enough to guarantee infinite differentiability with respect to the space variables (at least when $f=0$). The main theorem of the paper, contained in Section 3, shows that this is actually the case.

As an immediate application of the results of Section 3, it is found that the solutions to the initial value problem which were obtained by KISELEV & LADYSHENSKAYA are infinitely differentiable in the space variables, Lipschitz continuous in time, and satisfy (1) almost everywhere. (A similar result, at least so far as the spatial regularity of $u$ is concerned, was obtained recently by OHYAMA [6]: I am indebted to this paper for some of the methods used here. Also, for the case of two space dimensions, GOLOVKIN has proved an interior regularity theorem for the KISELEV-LADYSHENSKAYA solution, using, however, different methods; *cf.* [1], pp. 79–86.) It is not known to the writer whether the KISELEV-LADYSHENSKAYA solution is more than Lipschitz continuous in time, but in view of the above example and remarks it is clear that to prove additional time regularity of this solution, one must make strong use of the fact that the boundary values are zero and probably also of some smoothness of the boundary.

The author wishes to thank Professor DONALD ARONSON for many helpful conversations during the course of this work.

## 1. Preliminary considerations

We shall have frequent occasion to deal with convolution integrals of the type

$$h(x,t)=\iint_S k(x-\xi,t-\tau)\,g(\xi,\tau)\,d\xi\,d\tau,$$

where $S=G\times(T_1,T_2)$ is an open set in space-time. It is convenient to write $h=k\star g$, and to think of this relation as a transformation of functions $g$ on $S$ into functions $h$ also on $S$. In particular, it will be necessary to estimate certain norms of $h$ in terms of corresponding norms of $g$. To this end we define

$$\|g(t)\|_r=\left(\int|g(x,t)|^r\,dx\right)^{1/r}$$

and

$$\|g\|_{r,r'}=\left(\int\|g(t)\|_r^{r'}\,dt\right)^{1/r'},$$

the integrals being respectively over $G$ and $(T_1,T_2)$. The function $\|g(t)\|_r$ is thus the ordinary $L^r$ norm of $g(x,t)$ at the instant $t$. Note also that $r$ and $r'$ are here considered to be *independent* exponents with $r, r'\geqq 1$.

We have the following fundamental inequality connecting the norms of $g$ and $h$,

$$\|h\|_{r,r'}\leqq\|k\|_{p,p'}\|g\|_{q,q'},\tag{4}$$

where the exponents are related by the conditions

$$1\leqq q\leqq r,\quad 1\leqq q'\leqq r',\quad \frac{1}{r}=\frac{1}{p}+\frac{1}{q}-1,\quad \frac{1}{r'}=\frac{1}{p'}+\frac{1}{q'}-1.\tag{5}$$

[1] The physical explanation for this fact can be found in the incompressibility of the fluid; a sudden modification of the boundary values of a motion, say by means of a piston, will cause an immediate percussive change in the fluid velocity throughout the whole flow region.

The proof is simple: by MINKOWSKI's inequality

$$\|h(t)\|_r \leq \int \|\int k(x-\xi, t-\tau)\, g(\xi,\tau)\, d\xi\|_r\, d\tau,$$
$$\leq \int \|k(t-\tau)\|_p \|g(\tau)\|_q\, d\tau$$

according to YOUNG's version of the HÖLDER inequality (*cf.* A. ZYGMUND, *Trigonometrical Series,* Chapter IV). Applying YOUNG's inequality a second time yields (4).

In the applications below the kernel $k(x,t)$ will arise as a space derivative of the fundamental solution of the heat equation

$$K(x,t) = \begin{cases} \text{Const.}\, t^{-\frac{n}{2}} e^{-\frac{|x|^2}{4t}} & t>0, \\ 0 & t \leq 0. \end{cases}$$

For such $k(x,t)$ one has the result:

**Lemma 1.** *Let $h = k \star g$ where $k = \partial K/\partial x$. Then in any region $S = G \times (T_1, T_2)$ we have*

(6) $$\|h\|_{r,r'} \leq \text{Const.} \|g\|_{q,q'}$$

*provided that the exponents satisfy the relations*

(7) $$1 \leq q \leq r, \quad 1 \leq q' \leq r', \quad n\left(\frac{1}{q} - \frac{1}{r}\right) + 2\left(\frac{1}{q'} - \frac{1}{r'}\right) < 1.$$

*The constant depends only on $T = T_2 - T_1$ and upon the exponents.*

**Proof.** By direct calculation one easily verifies that

$$|k(x,t)| \leq \text{Const.} |x|\, t^{-\frac{n}{2}-1} e^{-\frac{|x|^2}{4t}}$$

and

$$\|k\|_{p,p'} \leq \text{Const.}\, T^{-\alpha + \frac{1}{p'}} \leq \text{Const.},$$

where $\alpha = \frac{n}{2}\left(1 - \frac{1}{p}\right) + \frac{1}{2}$ and $\alpha p' < 1$. Inserting this estimate for $\|k\|_{p,p'}$ into (4) we obtain (6), provided only that the exponents satisfy the restriction (5) and $\alpha p' < 1$. Eliminating $p$ and $p'$ from the last named conditions, however, yields precisely the relations (7).

In the sequel we shall require the concept of vorticity of the fluid motion. In a space of 2 or 3 dimensions the vorticity is defined by

(8) $$\omega = \operatorname{curl} u,$$

the derivative being of course interpreted in the generalized or strong sense. For more than three dimensions the appropriate generalization of (8) is

(9) $$\omega = \delta(*u) = *du$$

where the notation is that of exterior calculus: that is, $\omega$ is an $(n-2)$ skew tensor whose components are of the type $(\partial u_i/\partial x_k - \partial u_k/\partial x_i)$. The next lemma will be useful at several points in the proof of our main theorem.

**Lemma 2.** *Suppose that as functions of the space variables, u and ω are locally of class $L^1$ in some open set F in $E^n$. Then in any open region G with compact closure in F we have*

$$(10) \qquad {}^*u = \int_G \operatorname{grad} H(x-\xi) \wedge \omega(\xi)\, d\xi + A(x),$$

*where H(x) is the fundamental solution of Laplace's equation and A(x) is harmonic in G.*

**Proof.** Consider first the case of three dimensions, where (10) may be written

$$u = (4\pi)^{-1} \int_G \operatorname{grad} |x-\xi|^{-1} \times \omega(\xi)\, d\xi + A(x).$$

This formula is well known if $u$ is of class $C^1$ in $F$; *cf.*, for example, H. B. PHILLIPS, *Vector Analysis*, p. 190. The present case follows by standard limiting processes, using of course the facts that $u$ and $\omega$ are integrable.

When $n \neq 3$ the same proof is applicable, the representation formula for $u \in C^1$ being obtained by essentially the same argument as in the case $n=3$. An alternate but perhaps less satisfactory demonstration could be based on HODGE's theorem, $u = d\varphi + \delta\psi + A$.

We conclude this section by introducing certain function spaces which will play a fundamental role in what follows:

**Definition.** *Let $g = g(x,t)$ be defined on an open set R of space-time. Suppose that for every open set $S = G \times (T_1, T_2)$ with compact closure in R the norm $\|g\|_{r,r'}$ over S is finite. Then g is said to be in the class $L^{r,r'}(R)$.*

It should be observed that the norm $\|g\|_{r,r'}$ taken over the *full set* $R$ need not be finite for a function $g$ in $L^{r,r'}(R)$.

## 2. A representation theorem

In this section and throughout the rest of the paper the convolution operator will be used in a somewhat special way. Specifically, if $k$ is a vector, and $g$ is a $\nu^{\text{th}}$ order tensor, we define $k \star g$ to be the $(\nu-1)^{\text{st}}$ order tensor with components

$$(k \star g)_{lm} = \iint_S k_i(x-\xi, t-\tau)\, g_{ilm}(\xi,\tau)\, d\xi\, d\tau,$$

where $i$ is summed from 1 to $n$ and the subscripts $l, m$ denote an arbitrary set of component indices. In addition, the symbol $\operatorname{div} g$ will stand for the $(\nu-1)^{\text{st}}$ order tensor with components $g_{ilm,i}$, summed on $i$.

Now suppose that $u$ is a weak solution of the Navier-Stokes equations in an open region $R$ of space-time, with

$$u \in L^{2,2}(R), \qquad \omega \in L^{2,2}(R).$$

Assume also that the external force $f$ is of class $L^{1,1}$ and conservative, and let $n=3$. Then in any open region $S = G \times (T_1, T_2)$ with compact support in $R$, we have

$$(11) \qquad \omega = \operatorname{grad} K \star g + B$$

where $g = (\omega u - u\omega)$, and $B = B(x,t)$ is a solution of the heat equation in $S$. The remainder of this section is devoted to the proof of (11) and of a corresponding $n$-dimensional analogue.

Let $v_h = v_h(x, t)$ denote an integral average of the function $v(x, t)$ over a sphere of radius $h$ in space-time with center $(x, t)$.[2] By simple arguments it follows from (2) that there exists a differentiable function $p_h$ such that

(12) $$u_{ht} - \Delta u_h = -\operatorname{div}(u\,u)_h + f_h - \operatorname{grad} p_h.$$

Forming the curl of this equation yields

(13) $$\omega_{ht} - \Delta \omega_h = -\operatorname{curl}\operatorname{div}(u\,u)_h = -\operatorname{div}(u\,\omega - \omega\,u)_h,$$

the last equality being clear if $u$ and $\omega$ are of class $C^2$, and therefore true in the present case by standard limiting arguments (observe that by Hölder's inequality $u\,\omega \in L^{1,1}$). Now consider the expression

$$B_h \equiv \omega_h - \operatorname{grad} K \star g_h.$$

It is easily checked that $B_{ht} - \Delta B_h = 0$ and that $B_h$ is contained in $L^{1,1}$ uniformly in $h$. We may therefore let $h \to 0$ in the last display line, thus proving (11).

When $n \neq 3$ the representation (11) is still valid provided

$$g = (n-1)\,\omega \wedge u.$$

This is proved exactly as above, using however the identity

(14) $$\delta \star (\operatorname{div} u\,u) = (n-1)\operatorname{div}(u \wedge \omega).$$

We leave the verification of (14) to the enterprising reader.

## 3. The main theorem

In this section it is assumed that the external force is conservative and at least of class $L^{1,1}$. Our main result follows.

*Let $u$ be a weak solution of the Navier-Stokes equations in some open region $R$ of space-time, with $u \in L^{2,\infty}(R)$, $\omega \in L^{2,2}(R)$. Suppose further that*

(15) $$u \in L^{s,s'}(R)$$

*where the exponents $s$ and $s'$ satisfy the relation*

(16) $$\frac{n}{s} + \frac{2}{s'} < 1.$$

*Then $u$ is of class $C^\infty$ in the space variables, and each derivative is bounded in compact subregions of $R$.*

*Assume in addition that $u_t \in L^{2,p}(R)$, where $p \geqq 1$. Then the space derivatives of $u$ are absolutely continuous functions of time. Moreover, there exists a strongly differentiable function $p = p(x, t)$ such that*

(17) $$u_t - \Delta u + u \cdot \operatorname{grad} u = f - \operatorname{grad} p$$

*almost everywhere in $R$.*

---

[2] The reader is assumed to be familiar with the elementary properties of integral averages, as contained for example in the paper of FRIEDRICHS, Trans. A. M. S. **55**, 523–544 (1944).

*Remarks.* For a space of two or three dimensions, KISELEV & LADYSHENSKAYA have proved the existence of a weak solution of the initial value problem with the properties

$$u \in L^{4,\infty}; \quad u, u_x, u_t \in L^{2,\infty}. \tag{17}$$

The above theorem thus contains as a special case the result that this solution satisfies (17), is of class $C^\infty$ in the space variables, and is Lipschitz continuous in time, at least if the external force is conservative (for the case of non-conservative forces see the following section).

The solution of KISELEV & LADYSHENSKAYA was obtained only for $n=2$ and $n=3$, and moreover may not be "global" if $n=3$, that is, may not exist for all $t>0$. On the other hand, EBERHARD HOPF has constructed a global solution in all dimensions, and even under weaker hypotheses on the initial data than were required by KISELEV & LADYSHENSKAYA. The solution of HOPF, however, is only known to satisfy the conditions

$$u \in L^{2,\infty}, \quad u_x \in L^{2,2},$$

and these are not strong enough to apply the above theorem. We are therefore unable to draw any general conclusions concerning the regularity of HOPF's solution. Nevertheless it is possible to make certain remarks. First, HOPF's solution obeys the representation formula (11) in a space of any number of dimensions. Second, if $n=2$ and the initial data are regular enough for the solution of KISELEV & LADYSHENSKAYA to exist, then by a uniqueness theorem of LIONS & PRODI, HOPF's solution must be identical to that of KISELEV & LADYSHENSKAYA and therefore *a fortiori* must be of class $C^\infty$ in the space variables.

Certain additional conclusions regarding the structure of a weak solution of the Navier-Stokes equations will be found at the end of this section.

**Proof of the main theorem.** First of all, the hypotheses of the theorem are sufficient for the validity of the representation formula (11); that is, in any region $S=G\times(T_1, T_2)$ with compact closure in $R$ we have

$$\omega = \operatorname{grad} K \star g + B \tag{18}$$

and

$$|g| \leqq \text{Const.}\, |u| \cdot |\omega|. \tag{19}$$

The proof now proceeds in three steps:

I. We assert that $\omega \in L^{\infty,\infty}(R)$, that is, $\omega$ is bounded in compact subregions of $R$. Indeed suppose for the moment that $\omega \in L^{\varrho,\varrho'}(R)$ for some exponents $\varrho, \varrho' \geqq 2$. By virtue of (15), (19), and this assumption on $\omega$, it is evident that

$$g \in L^{q,q'}(R),$$

where the exponents $q, q'$ are given by the relations

$$\frac{1}{q} = \frac{1}{s} + \frac{1}{\varrho}, \qquad \frac{1}{q'} = \frac{1}{s'} + \frac{1}{\varrho'}$$

(obviously $q, q' \geqq 1$). Now define a positive constant $\varkappa$ by the equation

$$(n+3)\varkappa = 1 - \left(\frac{n}{s} + \frac{2}{s'}\right)$$

and set

$$r = \frac{\varrho}{1-\varkappa\varrho}, \qquad r' = \frac{\varrho'}{1-\varkappa\varrho'}$$

(put $r=\infty$ if $\varkappa\varrho \geqq 1$, and similarly for $r'$). One easily checks that $1 \leqq q \leqq r$, $1 \leqq q' \leqq r'$ and that

$$n\left(\frac{1}{q}-\frac{1}{r}\right)+2\left(\frac{1}{q'}-\frac{1}{r'}\right)=1-\varkappa<1,$$

whence by Lemma 1 and the representation formula (18) we have $\omega \in L^{r,r'}(R)$. But $r$ and $r'$ are larger than $\varrho$ and $\varrho'$, so that $\omega$ is actually in a higher $L$ class than originally supposed. This process of improving the class of $\omega$ can clearly be repeated as often as one likes, beginning with $\varrho=\varrho'=2$. Obviously after a finite number of steps one will obtain $\omega \in L^{\varrho,\varrho'}(R)$ with $\varrho = \varrho' \geqq \varkappa^{-1}$. At the next step $r=r'=\infty$, and the required assertion is demonstrated.

II. From Lemma 2 we have the representation

$$(20) \qquad {}^*u = \int_G \operatorname{grad} H(x-\xi) \wedge \omega(\xi)\, d\xi + A.$$

But by hypothesis $u \in L^{2,\infty}$, and we have just shown $\omega \in L^{\infty,\infty}$. Thus the harmonic function $A(x,t)$ must be *bounded* on compact subsets of $G$, both as a function of $x$ and as a function of $t$. Hence $u \in L^{\infty,\infty}(R)$. Therefore according to the usual potential theoretic estimates for heat kernel convolutions, $\omega$ is spatially Hölder continuous in any compact subregion of $R$, with arbitrary exponent $\vartheta<1$. Now that $\omega$ is Hölder continuous, $u_x$ is also Hölder continuous by virtue of the representation (20). Thus $g$ is Hölder continuous, and again by the potential theory of the heat convolution $\omega_x$ is Hölder continuous. Repetition of this argument shows successively that $\omega_{xx}$, $\omega_{xxx}$, *etc.*, are Hölder continuous. It follows obviously that $u \in C^\infty$ in the space variables, and the first statement of the theorem is proved.

III. So far the hypothesis $u_t \in L^{2,p}$ has not been invoked. We now investigate the consequences of this assumption. First of all, observe that (18) implies

$$(21) \qquad \omega_t = \Delta\omega + \operatorname{div} g,$$

this relation being meaningful in view of what has already been proved. It follows that $\omega_t$ is of class $C^\infty$ in the space variables, and that its derivatives are bounded in compact subregions of $R$. On the other hand, differentiation of (20) yields

$$(22) \qquad {}^*u_t = \int \operatorname{grad} H(x-\xi) \wedge \omega_t(\xi)\, d\xi + A_t.$$

Thus $u_t$ is of class $C^\infty$ in the spatial variables, and each derivative is of class $L^p$ in time. To complete the proof, we note that (17) follows by letting $h$ tend to zero in (12).

*Corollary remarks.* There are several comments concerning the preceding theorem and proof which are helpful in understanding the general situation. The first three concern the structure of a solution, and the last two deal with various other hypotheses for the theorem.

*1.* From the last part of the theorem it can be seen that the pressure will have exactly as many space derivatives as are permitted by the external force $f$. In particular, if $\Omega$ is the potential of $f$, then the combination $p+\Omega$ will be of class $C^\infty$ in the space variables, and its derivatives will be in $L^p$ with respect to time.

*2.* Even without the assumption $u_t \in L^{2,p}$ the derivative $\omega_t$ still exists, and the vorticity equation is satisfied.

*3.* It should be emphasized that all class inclusion relations in the theorem are actually *a priori* estimates. That is, when we say, for example, that $\omega$ is bounded on compact subsets of $R$, what is meant is that in any compact subset of $R$

$$|\omega| \leqq M,$$

where $M$ depends only of the subset and on the norms appearing (implicitly) in the hypothesis of the theorem.

*4.* If *a priori* estimates are available for the time derivative $u_{tt}$, then this derivative must likewise be of class $C^\infty$ in the space variables. Furthermore, the *first* time derivative is then a continuous function of $x$ and $t$, and the Navier-Stokes equation is satisfied in the classical sense.

*5.* The hypotheses of the theorem are slightly stronger than necessary. Specifically, it would be possible to carry out the same proof under the conditions

$$u \in L^{1,\infty}, \quad u \in L^{s,s'}, \quad \omega \in L^{\sigma,\sigma'},$$

where $\sigma$ and $\sigma'$ are, respectively, the exponents conjugate to $s$ and $s'$. The present statement of the theorem was adopted mostly for simplicity. Finally, we conjecture that in the conclusion of the theorem the words "of class $C^\infty$ in the space variables" can be replaced by "analytic in the space variables". A proof of this fact could possibly be patterned after a rather similar situation discussed by A. FRIEDMAN [Journal of Mathematics and Mechanics **7**, 43—60 (1958)].

## 4. Non-conservative force fields

We consider in this section force fields which do not satisfy the condition $\operatorname{curl} f = 0$. As an alternative to this condition one must require a certain degree of regularity of $f$. Although it would certainly be possible to consider quite general classes of external forces in this regard, nevertheless for our present purposes it will be sufficient to discuss the case when $f \in C^{m+\beta}$ in the space variables, with $m \geqq 0$ and $0 < \beta < 1$.

As a first step in treating this situation, observe that the vorticity equation (13) must be replaced by

$$\omega_{ht} - \Delta \omega_h = \operatorname{div} g_h + \operatorname{curl} f_h. \tag{23}$$

This leads to a modified representation formula, namely

$$\omega = \operatorname{grad} K \star g' + B$$

where $g' = (n-1)(\omega \wedge u + {}^*f)$, and $B$ remains a solution of the heat equation. Since ${}^*f$ is Hölder continuous, the preceding proof applies unchanged up to and including the point where it is stated that $\omega_x$ is Hölder continuous. Further differentiability clearly depends on whether $m=0$ or $m>0$. In particular, we have the result:

*If $f \in C^{m+\beta}$, then under the hypotheses of the first part of the main theorem $u \in C^{m+2+\beta}$ in the space variables, and each derivative up to this order is bounded in compact subregions of $R$.*

Now consider the effect of the assumption $u_t \in L^{2,p}$. If $m \geqq 1$, the preceding proof applies almost unchanged to show that $\omega_t \in C^{m-1+\beta}$, and thus that $u_t \in C^{m+\beta}$. If $m=0$, the same result is true but requires a slightly different argument. It is perhaps easiest to use elementary distribution theory, though this is by no means necessary. In any case, if $m=0$ we have $\omega \in C^{1+\beta}$; hence by (23), $\omega_t$ is the *derivative* of a Hölder-continuous function. Substituting this expression for $\omega_t$ into (22), and then shifting the derivative in question onto the term grad $H$, it is apparent that $u_t$ is represented as the sum of a harmonic function and the second derivative of the potential of a Hölder-continuous function. The first part of the following result is therefore proved.

*If in addition to the previous assumptions we have $u_t \in L^{2,p}$, then actually $u_t$ must be of class $C^{m+\beta}$ in the space variables. Moreover, there exists a function $p \in C^{m+1+\beta}$ such that*

$$u_t - \Delta u + u \cdot \operatorname{grad} u = f - \operatorname{grad} p$$

*almost everywhere in $R$.*

The last statement of the theorem arises simply by letting $h \to 0$ in equation (12). We add the obvious remark that if $f$ is the sum of a conservative force and a smooth force, then the potential of $f$ may be combined with $p$ to obtain a somewhat more general result than above.

*Note.* The above research was supported by the United States Air Force under Contract No. AF 49(638)-262, and also by the Office of Naval Research under Contract No. 225(11) with Stanford University.

## References

[1] Golovkin, K. K.: The plane motion of a viscous incompressible fluid. Trudy Mat. Inst. Steklov. **59**, 37–86 (1960).

[2] Hopf, E.: Über die Anfangswertaufgabe für die hydrodynamischen Grundgleichungen. Math. Nachrichten **4**, 213–231 (1951).

[3] Kiselev, A. A., & O. A. Ladyshenskaya: On existence and uniqueness of the solution of the nonstationary problem for a viscous incompressible fluid. Izvestiya Akad. Nauk SSSR **21**, 655–680 (1957).

[4] Ladyshenskaya, O. A.: Solution "in the large" of the nonstationary boundary value problem for the Navier-Stokes system with two space variables. Comm. Pure Appl. Math. **12**, 427–433 (1959).

[5] Lions, J. L., & G. Prodi: Un théorème d'existence et unicité dans les équations de Navier-Stokes en dimension 2. C. R. Acad. Sci., Paris **248**, 3519–3521 (1959).

[6] Ohyama, T.: Interior regularity of weak solutions of the time dependent Navier-Stokes equation. Proc. Japan Acad. **36**, 273–277 (1960).

Department of Mathematics
Institute of Technology
University of Minnesota
Minneapolis, Minnesota

*(Received September 6, 1961)*

In: R.E. Langer (ed.): *Nonlinear Problems*, 1963, pp 69–98.

**Reprint V.1.[ii]**

# THE INITIAL VALUE PROBLEM FOR THE NAVIER-STOKES EQUATIONS

James Serrin

UNIVERSITY OF MINNESOTA
DEPARTMENT OF MATHEMATICS
Minneapolis, Minnesota

1962

# THE INITIAL VALUE PROBLEM FOR THE NAVIER-STOKES EQUATIONS

James Serrin

The initial value problem for the Navier-Stokes equations has been the object of a remarkable surge of interest in the last five years. Developments have proceeded with such rapidity, and the results have become so specialized, that it is now no mean task to keep abreast of the field. It seems worthwhile, therefore, to attempt a synthesis of some of these results, to see a pattern of sorts emerging, and to point out some directions for future work. In addition to the expository aspects of the paper, it also contains several new results, particularly Theorems 3, 6 and 7, as well as some shortened proofs of known results.

For simplicity of presentation I shall confine myself for the most part to a somewhat restricted version of the problem. Consider a fixed (bounded or unbounded) domain $\Omega$ in $E^n$, where $n \geq 2$. This domain may be considered a rigid vessel filled with incompressible fluid; the fluid is initially set into motion, and we are to determine its subsequent velocity distribution, subject to the Navier-Stokes equation and the condition of adherence at the boundary of $\Omega$. More precisely, it is required to find a velocity field $u = u(x,t)$ and a pressure field $p = p(x,t)$ which, for $x = (x_1, \ldots, x_n) \in \Omega$ and for $t > 0$, satisfy the differential equations

$$u_t + u \cdot \text{grad } u = - \text{grad } p + \Delta u, \tag{1}$$

$$\text{div } u = 0, \tag{2}$$

and obey in some sense the initial condition

$$u(x,0) = u_o(x), \qquad x \in \Omega \tag{3}$$

and the boundary condition

$$u(x,t) = 0, \qquad x \in \dot{\Omega}. \tag{4}$$

(More general formulations of the problem include the effect of an external force in equation (1), non-vanishing boundary conditions, and so forth. On the other hand, the fact that we have set the kinematic viscosity equal to 1 is no restriction of generality.) Historically, the interest of mathematicians in the problem (1) - (4) was aroused by Hadamard, Lichtenstein, Villat, and, most particularly, C. W. Oseen. But it is one man, Jean Leray, who really opened the way. In his three fundamental papers of the years 1933-34 are to be found ideas and problems which have remained valuable to this day. In 1951 Eberhard Hopf discovered that the Navier-Stokes equations could be attacked in a simple, elegant fashion using the techniques of functional analysis. His bold introduction of linear methods into an essentially non-linear problem gave new life to the field, and this trend was further strengthened by the celebrated paper of Kiselev and Ladyshenskaya in 1957. A veritable flood of papers has followed, working out and extending the methods introduced by these various writers. It is to this situation that we now turn.

In the first section below we consider the basic notion of weak solution, which underlies almost all modern work. This somewhat abstract concept has, rather surprisingly, proved itself to be the key to the problem's defenses. Next we quickly review the fundamental existence theorem of Hopf, and present a somewhat streamlined version of the result of Kiselev and Ladyshenskaya. An extension of their work to the case $n = 4$ is also given.

Turning next to the uniqueness and regularity questions which are naturally associated with the existence of weak solutions, we outline in Sections 3 and 4 a new uniqueness theorem, which contains, among other things, the result that in dimension 3 the solutions of Hopf and of Kiselev-Ladyshenskaya are the same, provided the latter exists (cf. the discussion following Theorem 6). The methods employed here are in the main due to Prodi and Lions, see references [25], [29], and [30]. In Section 6 the interior regularity of a weak solution is proved under assumptions rather similar to those required for the uniqueness theorems. In these sections I have also attempted to relate the work discussed to that of various other writers who have treated similar questions.

Section 5 contains a backward uniqueness theorem, applying to classical solutions of the boundary value problem. The final part of the paper is devoted to a resume of Leray's notion of "epochs of irregularity," and to the discussion of a number of unsolved or open problems.

## 1. Preliminaries.

The program outlined in the introduction is relatively sophisticated, requiring a fair amount of mathematical preparation. It is convenient to divide this into three subsections.

1. Let $R$ denote an open region of space-time, which for the purpose of the initial value problem may be considered a cylinder with cross section $\Omega$. Let $u = u(x,t)$ and $v = v(x,t)$ be locally square summable vectors in $R$; the notation

$$(u,v) = \int u \cdot v \, dx = \int u_i v_i \, dx$$

will be used for the spatial inner product, whenever the integral exists. Since in all applications either $u$ or $v$ can be considered identically zero outside $R$, there is no need to specify the domain of integration. The product $(u,v)$ as defined is obviously a function of time; when we wish to make this dependence on $t$ more explicit we shall write the product in the form $(u(t), v(t))$.

The vector $u$ will be called weakly divergence free if

$$(u, \operatorname{grad} \omega)dt = 0 \tag{5}$$

for every $C^1$ function $\omega = \omega(x,t)$ having compact support in $R$. Moreover, $u$ will be called strongly differentiable with respect to $x$ if there exists a locally summable tensor, denoted either by $u_x$ or by $\operatorname{grad} u$, such that

(6) $$\int (\varphi_x, u)\,dt = - \int (u_x, \varphi)\,dt$$

for all $C^1$ vectors $\varphi$ with compact support in $R$. Finally, $u$ is called <u>strongly differentiable with respect to time</u> if there exists a locally summable vector, denoted by $u_t$, such that

$$\int (\varphi_t, u)\,dt = - \int (u_t, \varphi)\,dt$$

for all $C^1$ vectors $\varphi$ with compact support in $R$.

2. Now let $D(R)$ denote the test space of $C^\infty$ vectors which are divergence free and have compact support in $R$. We have the following basic definition.

DEFINITION 1. <u>A vector $u = u(x,t)$ will be called a weak solution of the Navier-Stokes equations in $R$ if it is locally square summable and weakly divergence free in $R$, and if</u>

$$\int \{ (u,\varphi_t) + (u,\Delta\varphi) + (u, u\cdot\operatorname{grad}\varphi) \}\,dt = 0$$

<u>for all $\varphi \in D(R)$</u>.

For smooth vectors $u$ this formulation is obviously equivalent to the original equations (1) and (2) holding in $R$ with some appropriate pressure $p(x,t)$.

3. In order to deal specifically with the initial value problem (1) - (4),* let $D$ denote the test space of $C^\infty$ vector fields $\varphi(x,t)$ which are divergence free and have compact support in $\Omega \times [0,\infty)$, (it should be observed that the vector $\varphi_0 = \varphi(x,0)$ need not vanish identically). With the notations

---

*More correctly we should refer to the <u>initial-boundary value problem</u>, but for the sake of brevity we shall use the shorter terminology of the text.

$$((u,v)) = \int u_x : v_x \, dx = \int u_{i,k} v_{i,k} \, dx$$

and

$$|u|^2 = (u,u), \qquad \|u\|^2 = ((u,u)),$$

the norm

$$\int_0^T \{|\varphi|^2 + \|\varphi\|^2\} dt \tag{7}$$

is then defined and finite for each $\varphi \in D$ and each $T$, $0 < T < \infty$.

DEFINITION 2. <u>A vector $u = u(x,t)$ will be said to be in the class $V$ if</u>

i) <u>for each $T$, $0 < T < \infty$, it is in the closure of $D$ under the norm (7), and</u>

ii) <u>the norm $|u|$ is uniformly bounded in time</u>.

It is evident that any vector $u \in V$ is strongly differentiable with respect to $x$, is weakly divergence free, and for each fixed $T$ has finite norm (7).* Furthermore, $u$ has generalized boundary values zero. (Another way to define the space $V$ would be to consider the vectors $u(x,t)$ as representing a variable point $u(t)$ in an appropriate Hilbert space; cf. [29].)

---

*We can suppose $u$ to be set identically zero outside $\Omega$, in which case the inner product $(u,v)$ may be taken over either $\Omega$ or $E^n$, the choice being immaterial. Moreover, $u$ is then strongly differentiable and divergence free in the entire half space $E^n \times [0,\infty)$.

The initial velocity distributions $u_o = u_o(x)$ which we shall consider are conveniently defined with reference to the test space $D_o$ of $C^\infty$ vectors $\psi = \psi(x)$ which are divergence free and have compact support in $\Omega$. In particular, we let $W$ denote the closure of $D_o$ in the norm $|\psi|$, $W^1$ the closure in $|\psi| + \|\psi\|$, and so forth.

DEFINITION 3. Let $u_o \in W$. A vector $u = u(x,t)$ will be called a weak solution of the initial value problem (1) - (4) if $u \in V$ and if

$$\int_0^\infty \{ (u,\varphi_t) + (u,\Delta\varphi) + (u,u\cdot\text{grad}\,\varphi)\} dt = -(u_o,\varphi_o)$$

for all $\varphi \in D$. Here $\varphi_o$ denotes the vector $\varphi(x,0)$.

As in Definition 1, it is easy to see that for smooth $u$ and $u_o$ this formulation is equivalent to (1) - (4). We add that there are procedures other than the above for taking account of the initial data; the present method, however, is convenient and standard, and will be followed henceforth in the paper.

## 2. Existence Theorems.

In this section we shall consider the existence of weak solutions of the initial value problem (1) - (4), <u>in the sense of Definition 3</u>. Excepting the work of Leray, which is briefly reviewed in Sections 7 and 8, the fundamental result for this problem is the following

THEOREM 1 (Eberhard Hopf). <u>For any initial velocity distribution $u_0 \in W$ there exists a weak solution $u \in V$ of the initial value problem. Moreover</u>

(8) $$|u|^2 + 2\int_0^t \|u\|^2\,dt \le |u_0|^2, \qquad \lim_{t\to 0} |u-u_0| = 0.$$

The proof (cf. [6]) is a beautiful application of the technique of Fourier approximation, unfortunately too long to include here. Nevertheless, we may observe that the process succeeds primarily because the original problem (1) - (4) admits a formal energy identity

(9) $$|u|^2 + 2\int_0^t \|u\|^2\,dt = |u_0|^2.$$

[A formal proof of (9) is easily given: multiply both sides of (1) by $u$ and integrate over $\Omega$, to obtain

$$(u,u_t) + (u,u\cdot\operatorname{grad} u) = -(u,\operatorname{grad} p) + (u,\Delta u),$$

Since $u$ is divergence free and vanishes on the boundary of $\Omega$, we have

$$\begin{aligned}(u,u\cdot\operatorname{grad} u) &= (u, \operatorname{grad}\tfrac{1}{2}u^2) = 0,\\ (u, \operatorname{grad} p) &= 0,\end{aligned}$$

whence it follows that

$$\frac{1}{2}\frac{d}{dt}|u|^2 = (u,u_t) = (u,\Delta u) = -\|u\|^2 .$$

The identity (9) is an immediate consequence of this last relation.] The fact that the formal procedure supplies the identity (9), while the rigorous proof yields only the inequality (8), should be carefully noted.

The solution which Hopf obtained, in spite of its elegance, is of course not a classically differentiable vector field. In an effort to obtain a solution with stronger regularity properties, Kiselev and Ladyshenskaya in the papers [14] and [15] considered initial data in the space $W^2$. Their result is as follows.

THEOREM 2 (Kiselev-Ladyshenskaya). Suppose $n = 2$ or $n = 3$. Then for any initial data $u_o \in W^2$ there exists a weak solution $u \in V$ of the initial value problem, and a positive number $T$ such that $|u_x|$ and $|u_t|$ are uniformly bounded in the interval $0 \leq t < T$.

The number $T$ depends only on bounds for the initial data; in case $n = 2$, or for sufficiently small data if $n = 3$, we can take $T = \infty$.

It almost goes without saying that this solution possesses in addition the properties (8) noted in Theorem 1. A detailed study of further properties of this solution, and its relation to Hopf's solution will be given in the following sections.

The proof of Theorem 2, as of Theorem 1, depends on Fourier approximation and a type of energy estimate. We shall present here only the <u>formal procedure</u> leading to this estimate. By differentiation of (2) with respect to $t$, scalar multiplication by $u_t$, and integration over $\Omega$, there arises

(10) $$\frac{1}{2}\frac{d}{dt}|u_t|^2 = -(u_t, u_t\cdot\operatorname{grad} u) - \|u_t\|^2 ,$$

(we omit the straightforward details of the calculation). The first term on the right can be estimated by Hölder's inequality

(11) $$(u_t, u_t\cdot\operatorname{grad} u) \le |u_t|_4^2 \|u\| ,$$

where $|\ |_4$ denotes the usual norm in $L^4(\Omega)$. Now according to Sobolev's theorem, for any differentiable vector $f$ with compact support we have

(12) $$|f|_4^2 \le C \begin{cases} |f|\cdot\|f\| & n = 2 \\ |f|^{1/2}\|f\|^{3/2} & n = 3 \\ \|f\|^2 & n = 4, \end{cases}$$

where $C$ depends only on the dimension; for $n = 2, 3$, and $4$ it can be taken to be $2^{-1/2}$, $3^{-3/4}$, and $9/16$, respectively (cf. the appendix at the end of this section). Hence

(13) $$(u_t, u_t\cdot\operatorname{grad} u) \le C \begin{cases} |u_t|\cdot\|u_t\|\cdot\|u\| & n = 2 \\ |u_t|^{1/2}\|u_t\|^{3/2}\|u\| & n = 3. \end{cases}$$

A subsequent application of the inequality $ab \le a^p/p + b^q/q$ allows us to conclude from (10) and (13) that

(14) $$\frac{1}{2}\frac{d}{dt}|u_t|^2 \le \begin{cases} 2^{-3}\|u\|^2|u_t|^2 & n = 2 \\ 2^{-8}\|u\|^4|u_t|^2 & n = 3 . \end{cases}$$

In case $n = 2$, integration of this inequality leads to the estimate

$$|u_t| \leq |u_{to}| \exp\left\{\frac{1}{8}\int_0^t \|u\|^2\,dt\right\}.$$

Under the hypothesis $u_o \in W^2$ it is apparent from (1) and (2) that

$$|u_{to}| \leq |\Delta u_o| + |u_o| \cdot |\operatorname{grad} u_o| \leq \text{Const.}$$

But also, by virtue of (9), or (8), the quantity $\int \|u\|^2 dt$ is bounded. Consequently, <u>$|u_t|$ remains uniformly bounded for all $t \geq 0$.</u> In addition, we have formally

(15) $$\|u\|^2 = -\frac{1}{2}\frac{d}{dt}|u|^2 \leq |u| \cdot |u_t| ,$$

so that the known bounds on $|u|$ and $|u_t|$ supply us with an obvious estimate for $\|u\|$, that is, for $|u_x|$.

Consider next the case $n = 3$. We observe that inequality (15) holds also in this case, whence by (14) we have

(16) $$\frac{d}{dt}|u_t| \leq 2^{-8} |u| \cdot \|u\|^2 |u_t|^2 .$$

This leads by integration to the estimate

(17) $$|u_t| \leq \frac{|u_{to}|}{1 - |u_{to}| \cdot A(t)} ,$$

where

$$A(t) = 2^{-8}\int_0^t |u| \cdot \|u\|^2 dt = 2^{-8}3^{-1}\left\{|u_o|^3 - |u|^3\right\}.$$

The estimate (17) obviously remains valid only so long as the denominator is positive. Now if the initial data satisfies the conditio $|u_o|^3 \cdot |u_{to}| < 768$, then clearly $|u_{to}| \cdot A(t) < 1$ for all time. In this case, the norms $|u_t|$ and $|u_x|$ are uniformly bounded for $t \geq 0$. In general, however, we can only assert that $|u_t|$ remains bounded during a certain finite time interval $0 \leq t < T$.

A simple estimate for $T$ can be obtained from the inequality

$$\frac{d}{dt} |u_t| \leq 2^{-8} |u_o|^2 |u_t|^3 ,$$

which follows from (15), (16), and the relation $|u| \leq |u_o|$. By integration of this result, we get

$$|u_t|^2 \leq \frac{|u_{to}|^2}{1 - 2^{-7}|u_o|^2|u_{to}|^2 t} ,$$

again valid whenever the denominator is positive. We have therefore proved the following result. In case $n = 3$, and for initial data in $W^2$, the norms $|u_t|$ and $|u_x|$ are uniformly bounded in the interval

$$0 \leq t \leq 128 \, (|u_o| \cdot |u_{to}|)^{-2} .$$

Moreover, if $|u_o|^3 |u_{to}| < 768$ the above norms are bounded for all $t \geq 0$. (If there is a non-unit kinematic viscosity $\nu$, then the numbers 128 and 768 should be replaced by $128\nu^5$ and $768\nu^5$, respectively, according to dimensional considerations.) This completes the formal proof of Theorem 2. A rigorous proof remains a technical matter which can be handled by Hopf's construction. (It is worth pointing out here that the basic difficulty in the case $n = 3$ as compared to the case $n = 2$ is due to the non-linearity of the differential inequality (16). Unfortunately, there seems to be no direct method for avoiding this difficulty.)

Although the case $n = 4$ was not discussed by Kiselev and Ladyshenskaya, this too can be treated by similar methods. Indeed, from (10), (11), and (12) we obtain

$$\frac{d}{dt}|u_t|^2 + 2(1 - C\|u\|)\|u_t\|^2 \leq 0. \tag{18}$$

Now suppose the initial data satisfies the inequality

$$|u_o| \cdot |u_{to}| < C^{-2} = 256/81 .$$

Then according to (15) one has $\|u_o\| < C^{-1}$. We assert that actuall[y] $\|u\| < C^{-1}$ for all time. Indeed, if this were not so, then there would exist a $\tau > 0$ such that

$$\|u\| < C^{-1} \quad 0 \leq t < \tau, \text{ and } \|u(\tau)\| = C^{-1}.$$

But in this case it follows from (18) that $|u_t| \leq |u_{to}|$ for $0 \leq t$ Therefore at the moment $t = \tau$

$$C^{-2} = \|u\|^2 \leq |u| \cdot |u_t| \leq |u_o| \cdot |u_{to}| < C^{-2},$$

which is a contradiction. This completes the (formal) proof of the following result.

THEOREM 3. Suppose $n = 4$. Then for any initial data $u_o$ with sufficiently small norm there exists a solution of the initial value problem which is strongly differentiable with respect to $x$ a[nd] $t$ and has uniformly bounded norms $|u_x|$ and $|u_t|$.

For $n \geq 5$ I do not know of any way to establish the existence of a solution with stronger regularity properties than that of Hopf. It seems of the greatest interest to know whether this limitation on dimension is a genuine feature of non-linear equations of the type (2), or whether it merely indicates an insufficiently developed mathematical attack. In this regard, the possibility of solutions having fractional order time derivatives should not be overlooked; cf, in particular, the interesting paper of J. L. Lions [22].

Further results concerning the regularity of the solutions determined above, and a discussion of the related existence theorems of Leray, Golovkin, Ito, and Sobolevskii, will be found in Sections 6 and 7.

Appendix. The given values for $C$ in inequality (12) are easily derived from the elementary estimate

$$|\omega|_{\frac{n}{n-1}} \leq \frac{1}{2n} |\omega_x|_1 ,$$

which holds for any differentiable function $\omega$ with compact support in $E^n$ (cf. Nirenberg, Annali Scuola Norm. Sup. Pisa (III), **13** (1959), p. 14). If we set $\omega = (\quad\cdot\quad)^{2(n-1)/n}$ in this inequality, and apply Schwarz's inequality, there results

$$|f|_4^2 \leq \frac{1}{\sqrt{2}} |f| \cdot \|f\| \qquad n = 2,$$

and

$$|f|_4 \leq \frac{3}{4} \|f\| \qquad n = 4.$$

When $n = 3$ the situation is a little more involved. We have in this case

$$|f|_4^4 = \int |f(x_1)|_4^4 \, dx_1,$$

where $|f(x_1)|_4$ is the $L^4$ norm of $f$ in the variables $x_2$ and $x_3$. Now by what has already been shown

$$|f(x_1)|_4^4 \le \frac{1}{2} |f(x_1)|^2 \cdot \|f(x_1)\|^2 ,$$

and using the fact that $f$ has compact support,

$$|f(x_1)|^2 \le |f| \cdot \left| \frac{\partial f}{\partial x_1} \right| .$$

Substituting the last two estimates into the equality for $|f|_4^4$, and performing the $x_1$ integration, then leads to

$$|f|_4^4 \le \frac{1}{2} |f| \cdot \left| \frac{\partial f}{\partial x_1} \right| \cdot \left\{ \left| \frac{\partial f}{\partial x_2} \right|^2 + \left| \frac{\partial f}{\partial x_3} \right|^2 \right\} .$$

With obvious meanings for the letters J, A, B, C, this may be writt

$$J \le \frac{1}{2} |f| \cdot A^{1/2} \{B + C\},$$

whence by permutation there arises

$$J \le \frac{1}{2} |f| \cdot (ABC)^{1/6} \{B+C\}^{1/3} \{A+C\}^{1/3} \{A+B\}^{1/3}$$

$$\le \frac{1}{3\sqrt{3}} |f| \cdot (A+B+C)^{3/2} = \frac{1}{3\sqrt{3}} |f| \cdot \|f\|^3 ,$$

completing the proof of (12).

## 3. The energy equality.

We shall be concerned in this section with certain rather immediate results concerning the continuity in time of a weal solution of the initial value problem, including the important energy equality. Although some of the results are new (especially when $n \geq 4$), the essential ideas are due to Prodi.

THEOREM 4. Suppose that $u \in V$ is a weak solution of the initial value problem. Then after suitable redefinition of $u$ at a set of values of $t$ of measure zero, we have

$$\int_0^t \{(u,\varphi_t) + (u,\Delta\varphi) + (u,u\cdot\text{grad}\,\varphi)\}\,dt = (u,\varphi) - (u_o,\varphi_o) \tag{19}$$

for all $\varphi \in D$ and all $t \geq 0$.

Proof. Let $\theta(t)$ be a smooth function whose graph has the form shown in Figure 1.

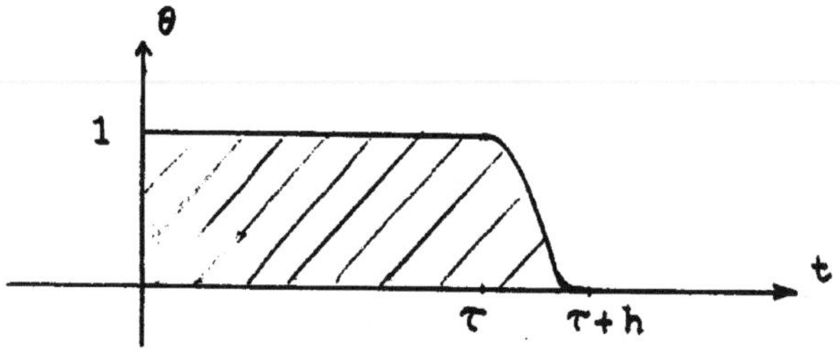


Setting $\Phi(x,t) = \theta(t)\varphi(x,t)$ into the weak form of the Navier-Stokes equations (Definition 3), there results

$$\int_0^\infty \{(u,\varphi_t) + (u,\Delta\varphi) + (u,u\cdot\text{grad}\,\varphi)\}\,\theta\,dt = -\int_0^\infty (u,\varphi)\theta_t dt - (u_o,\varphi_o) \tag{20}$$

As $h \to 0$ the integral on the left tends to

$$\int_0^\tau \{(u,\varphi_t) + (u,\Delta\varphi) + (u,u\cdot\text{grad}\,\varphi)\}\,dt.$$

To investigate the behavior of the integral on the right, observe that $\theta_t$ vanishes outside the interval $(\tau,\tau+h)$, and satisfies $\int \theta_t dt = -1$. Since we may assume further that $\text{Const.}/h \leq \theta_t \leq 0$, it is then a standard procedure to show that this integral tends to $(u(\tau), \varphi(\tau))$ for all $\tau$ belonging to the Lebesgue set $\mathcal{L}$ of $u(t)$. We have thus shown that (19) holds for almost all values of $t$.

Now consider an arbitrary instant of time $\tau \geq 0$. Since $|u(t)|$ is uniformly bounded, it follows from the weak compactness of $L^2$ that there exists a vector $U(x,\tau)$ and a sequence of values $t_j$ in $\mathcal{L}$, $t_j \rightarrow \tau$, such that $u(x,t_j) \rightarrow U(x,\tau)$ weakly in $L^2$. Since (19) holds for $t \in \mathcal{L}$, it is an easy consequence that

$$(21) \quad \int_0^\tau \{(u,\varphi_t) + (u,\Delta\varphi) + (u,u\cdot\text{grad}\,\varphi)\}\, dt = (U(\tau),\varphi(\tau)) - (u_0,\varphi_0)$$

for all $\varphi$ in $D$ and all $\tau \geq 0$. Now if $\tau$ is itself in $\mathcal{L}$ it is not difficult to see that $U(x,\tau)$ must be identical to $u(x,\tau)$, as a function in $L^2$, (if $\tau = 0$, then $U(x,0) = u_0(x)$). Hence the function $U$ is a redefinition of $u$ at most at a set of values of $t$ of measure zero. Replacing $U$ by $u$ and $\tau$ by $t$ in (21) completes the proof of Theorem 4.

We remark that the redefined function $u = u(x,t)$ is weakly continuous in $L^2$ as a function of time. Indeed a necessary and sufficient condition for weak continuity of $u \in V$ is evidently that

$$(u(t),\psi) \rightarrow (u(t_0),\psi) \quad \text{as} \quad t \rightarrow t_0,$$

for all $\psi \in D_0$ and all $t_0 \geq 0$. Since this condition is easily seen to hold by virtue of (19), the assertion is proved.

In what follows we shall regularly assume that all weak solutions under discussion are redefined according to the result of the preceding theorem.

For subsequent work certain spaces $L^{r,r'}$ play a fundamental role. If $g = g(x,t)$ is defined and measurable in a cylindrical domain $S = G \times (T_1, T_2)$ of space time, we set

$$|g(t)|_r = \left( \int |g(x,t)|^r dx \right)^{1/r}$$

and

$$|g|_{r,r'} = \left( \int |g(t)|_r^{r'} \, dt \right)^{1/r'},$$

the integrations being over $G$ and $(T_1, T_2)$, respectively. Here $r$ and $r'$ are considered to be independent exponents with $1 \leq r,r' \leq \infty$. We note that the norm $|u(t)|_2$ may be alternately denoted simply by $|u|$, as in the earlier sections of the paper.

DEFINITION 4. <u>We say that $g = g(x,t)$ is contained in the class $L^{r,r'}(S)$ if $g$ is defined and measurable in $S$, and $|g|_{r,r'} < \infty$.</u>

The following lemma will play a fundamental, though often implicit, role in the proofs of Theorems 5 and 6 below.

LEMMA 1. <u>Suppose that exponents $s$ and $s'$ satisfy the relation</u>

$$\frac{n}{s} + \frac{2}{s'} = 1, \qquad (n \leq s \leq \infty). \tag{22}$$

<u>Let $v$ and $w$ be contained in the space $V$, and suppose also that $u \in L^{s,s'}$ on the cylinder $\Omega \times (0,T)$. Then we have</u>

$$\int_0^T (u, v\cdot\text{grad}\, w)dt \leq \text{Const.}\, |u|_{s,s'} \, |w_x|_{2,2} \cdot \left\{ |v|_{2,\infty}^{2/s'} \, |v_x|_{2,2}^{n/s} \right\},$$

with the exception of the single case $s = n = 2$.

Proof. By Hölder's inequality, if $1/r + 1/s = 1/2$ we have

$$(u, v\cdot \operatorname{grad} w) \leq |u|_s \, |v|_r \, |w_x|_2 \, .$$

Also, by Sobolev's theorem

$$|v|_r \leq \text{Const.}\ |v|^{2/s'} \, |v_x|^{n/s},$$

whence integrating with respect to $t$ and applying Hölder's inequality once more, we obtain the required conclusion.

THEOREM 5 (Energy equality). Let $u \in V$ be a weak solution of the initial value problem, such that $u \in L^{s,s'}$ on the cylinder $\Omega \times (0,T)$ for a pair of exponents $s, s'$ satisfying (22). Then

$$(23) \qquad |u|^2 + 2\int_0^t \|u\|^2 \, dt = |u_0|^2, \qquad 0 \leq t \leq T.$$

Remarks. For the cases $n = 2,3$ this result is essentially due to Prodi. Our proof is based on a technique introduced by Lions and Prodi in a slightly different context.

We observe that in case $n = 2$ the assumption $u \in L^{s,s'}$ is redundant. In fact, by Sobolev's inequality any vector $u \in V$ is automatically in $L^{s,s'}$ for every pair of exponents satisfying (22), except $(\infty, 2)$. A similar result fails for $n \geq 3$.

Proof of Theorem 5. The result is obviously true for $t = 0$, since we recall that $u(x,t)$ has been redefined to equal $u_0(x)$ when $t = 0$. Now let $\tau$ be fixed, $0 < \tau \leq T$. For an even averaging kernel $K(t)$ with compact support in $|t| \leq h$, we set

$$\varphi(x,t) = \int_0^\tau K(t-t')\, u^k(x,t')dt',$$

where $\{u^k\}$ is a sequence of test functions converging to $u$ in the norm (7). The vector $\varphi$ obviously may be substituted into relation (19); if we then let $k \to \infty$ there results

$$(24)\qquad \int_0^\tau \{(u,u_{ht}) - ((u,u_h)) + (u,u\cdot\text{grad } u_h)\}\, dt = (u,u_h) - (u_o,u_{ho}),$$

where

$$u_h = u_h(x,t) = \int_0^\tau K(t-t')\, u(x,t')dt'.$$

(The limiting process involved here is a typical one in what follows; it may be justified by fairly obvious arguments.) Now

$$\int_0^\tau (u,u_{ht})dt = \int_0^\tau\int_0^\tau K_t(t-t')(u(t),\, u(t'))\, dt'\, dt = 0$$

by the symmetry of $K$. We may let $h \to 0$ in the remaining terms of (24). By standard properties of the averaging process, one obtains

$$\int ((u,u_h))dt \to \int\|u\|^2dt,\quad \int(u,u\cdot\text{grad } u_h)dt \to \int(u,u\cdot\text{grad } u)dt.$$

The last integral vanishes since $u$ is weakly divergence free (this is not completely obvious, though we shall omit the demonstration). Finally we have

$$\begin{aligned}(u(\tau),u_h(\tau)) &= (u(\tau), \int_0^\tau K(\eta)\, u(\tau-\eta)d\eta) \\ &= \int_0^h K(\eta)(u(\tau),\, u(\tau-\eta))d\eta \qquad (h < \tau) \\ &= \int_0^h K(\eta)\, \{\, |u(\tau)|^2 + \varepsilon(\eta)\,\}\, d\eta\ ,\end{aligned}$$

the last step in consequence of the weak continuity of $u(t)$. The limit as $h \to 0$ is therefore $\frac{1}{2}|u(\tau)|^2$, and similarly $\lim(u_o,u_{ho}) = \frac{1}{2}|u_o|^2$. Substituting the appropriate limits into (24), we obtain (23), and Theorem 5 is demonstrated.

<u>Concluding remarks</u>. If $n = 2$ the assumption $u \in L^{s,s'}$ is redundant, as has already been remarked. It follows that <u>all</u> weak solutions in $n = 2$ necessarily obey the energy equality, a fact first discovered by Prodi [30].

If $n \geq 3$, it is not known whether every weak solution $u \in V$ obeys an energy equality. The solution of Kiselev and Ladyshenskaya, however, satisfies

$$|u|_4 \leq C\,|u|^{1/4}\,|u_x|^{3/4} \leq \text{Const.}, \qquad 0 \leq t \leq T,$$

so that $u \in L^{4,\infty}$ on the interval $(0,T)$. This solution therefore satisfies the hypotheses of Theorem 5, and consequently <u>does</u> have an energy equality. A similar result is valid for the solution constructed in Theorem 3.

It may be added that any solution $u \in L^{s,s'}$, (and in particular the solution of Kiselev and Ladyshenskaya), is necessarily strongly continuous in time, that is

$$|u(t) - u(t_0)| \to 0$$

as $t \to t_0$. This follows at once from the weak continuity of $u(t)$ together with the continuity of the norm $|u(t)|$. We shall see later that the solutions of Kiselev and Ladyshenskaya are also Lipschitz continuous in $t$ for each fixed $x$.

## 4. Uniqueness and stability.

This section contains a general uniqueness theorem for the initial value problem, and some closely related stability theorems. Roughly speaking, the uniqueness theorem states that if a weak solution has a certain mild degree of regularity, then it is unique among all solutions of class V satisfying (8). This result essentially includes uniqueness theorems of Leray [20, Section 32], Prodi [29], and Lions and Prodi [25], though it should be mentioned that the method of proof derives something from all these papers.

THEOREM 6 (Unicity). Suppose $n \leq 4$. Let $u$ and $v$ be weak solutions of the initial value problem, in the sense of Definition 3. Suppose also that

$$|v|^2 + 2\int_0^t \|v\|^2\,dt \leq |v_o|^2, \qquad 0 \leq t \leq T,$$

and that $u \in L^{s,s'}$ for a pair of exponents satisfying (22) and also $n < s < \infty$. Then

$$|u-v| \leq |u_o - v_o| \exp\left\{\text{Const.} \int_0^t |u|_s^{s'}\,dt\right\}$$

in the interval $0 \leq t \leq T$. In particular, if $u_o = v_o$ then $u \equiv v$.

Proof. (The following demonstration was obtained jointly by myself and Jerome Sather of the University of Minnesota. It may be extended to cover the case of moving boundaries and non-zero boundary values, and, in this generality, an expanded version will be published elsewhere.) By a construction similar to that of Theorem 5, but with some differences in detail, one can show that

$$(25)\quad \int_0^\tau \{(u,v_{ht}) - ((u,v_h)) + (u,u\cdot\text{grad } v_h)\}\, dt = (u,v_h) - (u_o,v_{ho})$$

and

$$(26)\quad \int_0^\tau \{(v,u_{ht}) - ((v,u_h)) - (u_h,v\cdot\text{grad } v)\}\, dt = (v,u_h) - (v_o,u_{ho}) .$$

(The third term in the integral (26) causes the major difficulty, and will be discussed at some length in the paper referred to above.) Now by virtue of Fubini's theorem and the symmetry of the kernel $K$ it is easy to see that

$$\int_0^\tau (u,v_{ht})dt = - \int_0^\tau (u_{ht},v)\, dt.$$

Consequently, addition of (25) and (26) yields

$$- \int_0^\tau \{((u,v_h)) + ((v,u_h)) + (u_h,v\cdot\text{grad } v) - (u,u\cdot\text{grad } v_h)\}\, dt$$
$$= (u,v_h) + (v,u_h) - (u_o,v_{ho}) - (v_o,u_{ho}) .$$

In this identity we may let $h \to 0$. Recalling the calculations of the preceding proof, one sees that the result is

$$- \int_0^\tau \{ 2((u,v)) + (u,w\cdot\text{grad } v) \}\, dt = (u,v) - (u_o,v_o),$$

where $w = v-u$. Now by hypothesis

$$|v|^2 + 2 \int_0^\tau \| v \|^2 dt \le |v_o|^2 ,$$

and by virtue of Theorem 5,

$$|u|^2 + 2 \int_0^\tau \|u\|^2 dt = |u_o|^2 .$$

Adding the last three displayed equations (the first being multiplied by 2) then gives

$$(27)\quad |w|^2 + 2\int_0^\tau \|w\|^2 dt \le |w_0|^2 + 2\int_0^\tau (u, w\cdot\text{grad } v)dt,$$
$$= |w_0|^2 + 2\int_0^\tau (u, w\cdot\text{grad } w)dt.$$

So much holds for $n \le s < \infty$. But by the procedure of Lemma 1, if $s > n$,

$$(28)\quad \int (u, w\cdot\text{grad } w)dt \le \text{Const.}\ |w_x|_{2,2}^{1+n/s} \left\{ \int |u|_s^{s'} |w|^2\, dt \right\}^{1/s'}$$
$$\le |w_x|_{2,2}^2 + \text{Const.} \int |u|_s^{s'}\, |w|^2\, dt.$$

Consequently

$$|w|^2 \le |w_0|^2 + \text{Const.} \int_0^t |u|_s^{s'}\, |w|^2\, dt.$$

This differential inequality is easily integrated (set the right hand side equal to a new function $z$), and one obtains

$$|w| \le |w_0| \exp\left\{ \text{Const.} \int_0^t |u|_s^{s'}\, dt \right\},$$

completing the proof.

A similar result is valid when $s = n$, but requires the additional hypothesis that $|u|_{n,\infty}$ be suitably small. This hypothesis is satisfied by the solution of Theorem 3.

Remarks. Let us consider separately the cases $n = 2, 3,$ and $4$. First, if $n = 2$ the extra hypotheses on $u$ and $v$ are redundant, as follows from Theorem 5 and the accompanying remarks. In other words, for the case of plane flow two solutions $u$ and $v$ in the class $V$ are identical if and only if $u_0 = v_0$. This result was first proved by Lions and Prodi.

If $n \ge 3$, let us denote by $V'$ the subset of vectors in $V$ which satisfy the energy inequality

$$|u|^2 + 2\int_0^t \|u\|^2 \, dt \leq |u_0|^2 .$$

Evidently the solutions constructed in Section 2 are all in $V'$. Theorem 6 may then be rephrased as follows: Suppose that $u$ and $v$ are weak solutions of the initial value problem for the same initial data $u_0$, and let both $u$ and $v$ be in the class $V'$. Suppose also that $u \in L^{s,s'}$ for a pair of exponents satisfying (22) and $n < s < \infty$. Then $v \equiv u$.

The solution of Kiselev and Ladyshenskaya $(n = 3)$ is in the class $L^{4,\infty}$, as has already been noted. Consequently, for initial data in $W^2$ it is the only solution in $V'$, the solution constructed by Eberhard Hopf necessarily being the same.

For $n = 4$ a uniqueness theorem holds for the solution constructed in Theorem 3; if that solution exists for given initial data, then it is the only solution in $V'$.

There is one feature of Theorem 6 which is not quite desirable, namely the restriction to $n \leq 4$. This arises in one of the limiting processes in the proof, and can be removed in any of the following cases: 1) $\Omega \equiv E^n$, 2) $\Omega$ is starlike, and 3) the test functions $u^k$ defining $u$ have uniformly bounded norms $|u^k|_{s,s'}$.

There is, finally, a remarkable uniqueness theorem due to Foias for the case $\Omega \equiv E^n$, which assumes no differentiability of the weak solutions in question. On the other hand, a vector $u \in V'$ does not necessarily satisfy the hypotheses of Foias' theorem, so that even in the case $\Omega \equiv E^n$ the result of Foias does not include Theorem 6, nor vice versa.

<u>Stability</u>. The remaining considerations of this section will be directed to the case where $\Omega$ is a bounded region, say contained within a sphere of diameter $d$. It is well known (Poincare's inequality) that there exists a positive number $\alpha$, depending only on $n$, such that

$$(29) \qquad (\alpha/d)\,|w| \leq \|w\|$$

for all strongly differentiable vectors $w$ in $\Omega$ with generalized boundary values zero.

If we now modify inequality (28) by placing a factor $1/2$ in front of the first term on the right, (this can certainly be done simply by increasing the remaining constant), then from (27) and (28) follows

$$\begin{aligned} |w|^2 &\leq |w_o|^2 + \int_o^t \{\text{Const.}\ |u|_s^{s'}|w|^2 - \|w\|^2\}\ dt \\ &\leq |w_o|^2 + \int_o^t \{\text{Const.}\ |u|_s^{s'} - (\alpha/d)^2\}\ |w|^2\ dt. \end{aligned}$$

Solving this differential inequality yields

$$(30) \qquad |w|^2 \leq |w_o|^2 \exp\left\{\text{Const.} \int_o^t |u|_s^{s'}\ dt - (\alpha/d)^2\, t\right\}.$$

This proof applies, of course, only to the case when both solutions $u$ and $v$ have generalized boundary data zero. However, as we have indicated above, the same methods are valid for the case of non-zero boundary data and moving boundaries. We may assume, therefore, that (30) holds also in this more general case. Perhaps the most important consequence of (30) is the fact that <u>solutions of the initial value problem depend continuously (in</u>

the $L^2$ norm) on the initial data, this condition being one of the requirements of a properly posed problem. Also, from (30) one may derive a number of stability criteria.

I. If for some $s$, $n < s < \infty$, the basic flow $u$ satisfies

$$|u|_{s,\infty} \leq K < \left( \frac{a}{d\sqrt{\text{Const.}}} \right)^{(s-n)/s}, \qquad 0 < t < T,$$

then the perturbation $w = v-u$ corresponding to any solution $v \in V'$ of the initial value problem, having the same boundary values but possibly different initial data, satisfies

$$|w| \leq |w_o| \, e^{-\gamma t} \qquad 0 < t < T$$

for some $\gamma > 0$. The basic solution $u$ is thus exponentially stable.

II. Suppose next the special assumption $u \in L^{\infty,\infty}$ that is, the flow speed itself is uniformly bounded, say $\leq U$. In this case inequality (27) certainly holds, and implies the estimates

$$|w|^2 \leq |w_o|^2 + \int_o^t \{ U^2 - (a/d)^2 \} \, |w|^2 \, dt$$

and

$$|w|^2 \leq |w_o|^2 \exp \{ U^2 - (a/d)^2 \} \, t.$$

This leads to the stability criterion

(31) $$Ud < a,$$

derived in [32] under more restrictive assumptions on the flows than we have assumed here. It should be observed that the quantity Ud (or, if a non-unit kinematic viscosity $\nu$ is present, the quantity $Ud/\nu$) is a Reynolds number for the motion u.

III. The concrete nature of the preceding stability conditions makes it desirable to have a close estimate for the value of the constant $\alpha$ in (29). In [32] it was shown (using the fact that div w = 0) that we can take $\alpha = 5.71$ in case $n = 3$. Weinberger and Payne, making use of eigenvalue theory, have recently proved that the best value for $\alpha$ is 8.98.

For domains which can be enclosed in a cube of edge length d, Velte has found that the best value of $\alpha$ is between 7.70 and 7.91. Finally, for certain particular flows these criteria can be greatly improved by using special techniques (cf. reference [36], and §§ 4-6 of reference [32]).

## 5. Other uniqueness theorems.

Theorem 6 of the preceding section refers to the case of zero boundary data (or, in the case of moving boundaries, to prescribed velocity at the boundary). If the fluid fills the exterior of a bounded region, however, the condition that the motion have prescribed data at infinity is perhaps not an entirely natural assumption.* Investigating this situation for the case of classical solutions, Graffi [5] found that the zero condition at infinity could be replaced by the following assumptions:

i) $u$ and $u_x$ are uniformly bounded in $x,t$;

ii) as $x \to \infty$ the pressure $p$ tends to a limit $p_o$, with $|p-p_o| \leq \text{Const.}\ |x|^{-1}$.

This remarkable result has been further extended by Graffi to include compressible fluid motions [39], [40]. There is not space to give the details here, though we may remark that a corresponding existence theorem would be highly desirable.

Another problem of interest is the question whether a non-constant solution of the initial value problem (1) - (4) can ever come to rest in a finite time. Although a definite result is not known, at least for a classical solution in a bounded domain the answer is no. This fact is implicit in the following somewhat more general theorem.

---

*The situation may be compared with the initial value problem for the heat equation in an exterior domain. For that problem it is well-known that a sufficient condition for uniqueness is boundedness, or even exponential growth, at infinity.

THEOREM 7. Suppose that $u$ and $v$ are (classical) solutions of the Navier-Stokes equations in a bounded region $\Omega$, and let $u = v$ on the boundary of $\Omega$. If $u(x,t_o) = v(x,t_o)$ at some instant $t_o$, then $u$ and $v$ are identical for all $t$.

Proof: The method of proof is a simple modification of one established by Lees and Protter [17] for the study of a single parabolic equation of second order. Let us set

$$L\varphi \equiv \Delta\varphi - \varphi_t \quad \text{and} \quad \lambda(t) = t - \tau$$

where $\tau$ is some fixed constant, $0 < \tau < 1$. Then, essentially by the argument of [17], pp. 371-372, we have the following result.

LEMMA 2. Let $\varphi = \varphi(x,t)$ be a divergence free vector of class $C^2$, vanishing on the complete boundary of the space time domain $\Omega \times (-\tau,0)$. Then for any function $\tilde{p}$ of class $C^1$ and any integer $m > 0$ we have

$$\int \lambda^{-2m} \left\{|\varphi|^2 + \|\varphi\|^2\right\} dt \leq C \int \lambda^{-2m} \, |L\varphi - \operatorname{grad} \tilde{p}|^2 \, dt,$$

where $C = 5(\tau+m^{-1})$. In this inequality it is tacitly understood that the space integration is taken over $\Omega$, and the time integration over the interval $(-\tau,0)$.

This much being shown, it is now easy to complete the proof. Since uniqueness in the forward time direction certainly holds, we have $u \equiv v$ for all $t \geq t_o$. Therefore, if the theorem were not true there would be an instant of time, say $t = 0$, such that $u \equiv v$ for $t \geq 0$, but $u \not\equiv v$ in any interval $-\varepsilon < t < 0$. Let the magnitudes of $u$ and of $\operatorname{grad} v$ be bounded by $M$ in the interval $-1 < t < 0$, so that in this interval

$$|Lw - \mathrm{grad}\,\tilde{p}| \leq 2M^2 \{ |w|^2 + \|w\|^2 \},$$

where $w = v-u$ and $\tilde{p}$ denotes the difference in pressures for the flows $v$ and $u$. Now choose $\tau$ and $m_o$ to satisfy the conditions

$$0 < \tau < 1 \quad \text{and} \quad 4CM^2 < 1 .$$

Finally, let $\theta(t)$ be defined as indicated by the graph in Figure 2.

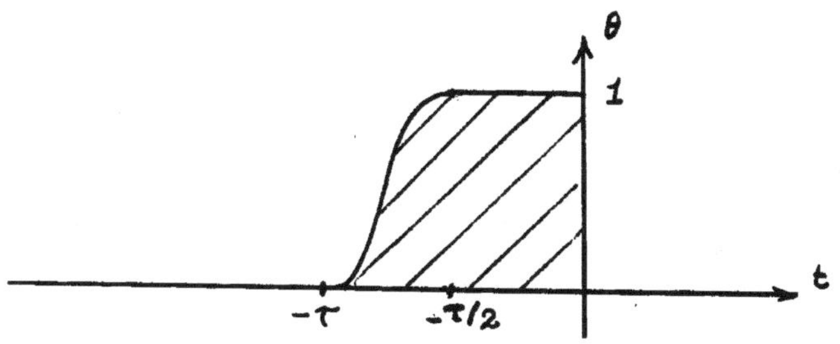


We may set $\varphi = \theta w$ into Lemma 2, with the result

$$\int \lambda^{-2m} \{ |\varphi|^2 + \|\varphi\|^2 \} \, dt$$

$$\leq C \int \lambda^{-2m} |L\varphi - \mathrm{grad}\,\tilde{p}|^2 \, dt$$

$$= C \int_{-\tau}^{-\tau/2} \lambda^{-2m} |L\varphi - \mathrm{grad}\,\tilde{p}|^2 \, dt + \int_{-\tau/2}^{0} \lambda^{-2m} |Lw - \mathrm{grad}\,\tilde{p}|^2 dt$$

$$\leq (2/3\tau)^{2m}\, CD + 2CM^2 \int_{-\tau/2}^{0} \lambda^{-2m} \{ |w|^2 + \|w\|^2 \} \, dt,$$

where $D$ is some appropriate constant, and $m \geq m_o$. Transferring the final term to the left hand side, and then restricting the integration to the interval $(-\tau/4, 0)$ yields after some simple manipulations

$$\int_{-\tau/4}^{0} |w|^2 \, dt \le 2CD \, (5/6)^{2m}, \qquad m \ge m_0.$$

Now if $m \to \infty$ we obtain $w \equiv 0$ for $-\tau/4 < t < 0$, contradicting the fact that $w$ is non-constant on any interval containing the origin. This completes the proof.

## 6. Regularity of weak solutions.

In all treatments of partial differential equations by the methods of functional analysis, it is necessary at some stage to examine the regularity properties of the weak solutions which are obtained. In [34], the author has given a direct and relatively simple proof of the interior regularity of weak solutions of the Navier-Stokes equations, assuming a certain moderate degree of regularity to begin with. We shall review this paper below, but first we must mention the work of various other authors who have also treated the regularity problem, considering it an integral part of their basic existence theorems.

Leray in the important series of papers [18], [19], [20], studied the initial value problem for dimensions $n = 2$ ( $\Omega =$ bounded convex region, and $\Omega \equiv E^2$), and $n = 3$ ( $\Omega \equiv E^3$). The existence of smooth solutions of the initial value problem was there established for finite time intervals (if the data was sufficiently small, also for all time) using the fundamental solution of the equations of slow motion, together with the technique of successive approximations. This work of Leray has recently been extended by Golovkin ($n = 2$) and Ito ($n = 3$), making use of the steady state Green's tensor of F.K.C. Odquist. Also, P. E. Sobolevskii has announced the construction of a classical solution for a bounded domain in $E^3$, valid for some finite time interval. Further details will be found at the conclusion of this section, where we discuss the relation of this work to that of reference [34].

Now let R be an open region of space time. We shall say that a vector u is in $\mathcal{L}^{r,r'}$ over R if u is defined and measurable in R and if $u \in L^{r,r'}(S)$ for every space-time cylinder S with compact support in R. We consider weak solutions u of the Navier stokes equations in R, in the sense of Definition 1. The condition that u be locally square summable in R is alternately expressed, in the notation just introduced, by the statement $u \in \mathcal{L}^{2,2}(R)$. Now suppose that u is strongly differentiable with respect to x, and $u_x \in \mathcal{L}^{2,2}(R)$. Then in any open cylinder $S = G \times (T_1, T_2)$ with compact support in R it can be shown that

$$(32) \qquad \omega(x,t) = \iint k(x-\xi,\ t-\tau) g(\xi,\tau) d\xi\ d\tau + B(x,t)\ .$$

Here $\omega = \text{curl } u$ denotes the vorticity of the fluid motion,* g is the dyadic tensor $\omega u - u\omega$, B is a solution of the heat equation in S, and $k = \text{grad } K$, where

$$K = K(x,t) = \begin{cases} \text{Const. } t^{-n/2} e^{-|x|^2/4t} & t > 0 \\ 0 & t \leq 0. \end{cases}$$

The product kg in the integrand of (32) is to be understood as the tensor sum $k_i g_{ij}$. The proof of (32) is most simply carried out by the process of integral averaging; a <u>formal</u> proof is evident on the basis of the vorticity equation

$$\omega_t - \Delta\omega = \text{div}(\omega u - u\omega), \qquad (n = 3).$$

---

*The definitions of $\omega$ and g are here given only for the case $n = 3$. The modifications necessary for general n are explained in [34].

Making use of the representation formula (32) one can now prove the following interior regularity theorem, (in this result it is worth pointing out the similarity between (33) and the exponent relation (22) which occurred repeatedly in Sections 3 and 4),

THEOREM 8. Let $u$ be a weak solution of the Navier-Stokes equations in an open region $R$ of space time, with $u \in \mathcal{L}^{2,\infty}(R)$, $u_x \in \mathcal{L}^{2,2}(R)$. Suppose further that

$$u \in \mathcal{L}^{s,s'}(R)$$

for a pair of exponents $s$ and $s'$ satisfying the condition

$$\frac{n}{s} + \frac{2}{s'} < 1. \tag{33}$$

Then $u$ is of class $C^{\infty}$ in the space variables $x$, and each (spatial) derivative is bounded in compact subregions of $R$.

Assume in addition that $u$ is strongly differentiable with respect to time. Then $u$ and each of its space derivatives is an absolutely continuous function of time. Moreover, there exists a function $p = p(x,t)$ of class $C^{\infty}$ in $x$ such that (1) holds almost everywhere in $R$.

In essence, the proof depends on the following recursion (or bootstrap) argument. Suppose we know that

$$u \in \mathcal{L}^{s,s'}, \qquad \omega \in \mathcal{L}^{\rho,\rho'}$$

where $\rho, \rho'$ are independent exponents $\geq 2$. Then $g \in \mathcal{L}^{q,q'}$ for

$$\frac{1}{q} = \frac{1}{s} + \frac{1}{\rho}, \qquad \frac{1}{q'} = \frac{1}{s'} + \frac{1}{\rho'} .$$

Hence by virtue of the representation formula (32), and a lemma on the Lebesgue class of convolution integrals, one finds $\omega \in \mathcal{L}^{r,r'}$ where

$$r = \frac{\rho}{1-\varkappa\rho}, \qquad r' = \frac{\rho'}{1-\varkappa\rho'}$$

and

$$\varkappa = \left(1 - \frac{n}{s} - \frac{2}{s'}\right) \Big/ (n+3) > 0.$$

(put $r = \infty$ if $\varkappa\rho \geq 1$, and similarly for $r'$). But $r$ and $r'$ are larger than $\rho, \rho'$, so that $\omega$ is actually in a higher Lebesgue class than originally supposed. This process of improving the class of $\omega$ can clearly be repeated as often as one likes, beginning with $\rho = \rho' = 2$. Obviously, after a finite number of steps one will obtain $\omega \in \mathcal{L}^{\infty,\infty}$. This much being shown, a further and more straightforward recursion procedure establishes that $u$ and $\omega$ are arbitrarily often differentiable with respect to the space variables. The details of this process, and the verification of the final portion of the theorem, will be omitted.

For solutions of the initial value problem in $\Omega \equiv E^n$, the conclusion of Theorem 8 can be considerably strengthened without significantly altering the proof. In fact, we have the result,

THEOREM 8'. Let $u \in V'$ be a weak solution of the initial value problem, where $\Omega \equiv E^n$. Suppose further that $u \in L^{s,s'}$ for a pair of exponents satisfying (33). Then $u$ is of class $C^\infty$ in all variables, and is a classical solution of equations (1) and (2).

The improvement of Theorem 8' over Theorem 8 is essentially due to the fact that in the latter case one can give an explicit formula for the function $B(x,t)$. If we have to deal with a bounded region $\Omega$, however, there is apparently no way to obtain an analogous formula for $B$. Thus, in order to generalize Theorem 8' to this case it seems necessary to use the notion of a Green's function or Green's tensor. This line of attack, if it is successful, should also provide the result that the solution takes on boundary values in the classical sense.

Finally, let us consider Theorems 8 and 8' in connection with the existence theorems of section 2. In both cases $n = 2$ and $n = 3$ the solution of Kiselev and Ladyshenskaya satisfies the hypotheses of these theorems. It follows that this solution is $C^{\infty}$ in the space variables, and (by an obvious extension of the final statement of Theorem 8) Lipschitz continuous in time.* Moreover, if $\Omega \equiv E^n$ these solutions are $C^{\infty}$ in all variables.

The weak solutions of Theorem 1 do not possess sufficient regularity properties to apply the above theorems. We are therefore unable to draw any general conclusions concerning the smoothness of these solutions. Nevertheless, it is possible to make certain remarks. First, Hopf's solution obeys the representation formula (32) in a space of any number of dimensions. Second, if $n = 2$ or $n = 3$ and the initial data are smooth enough for the Kiselev-Ladyshenskaya solution to exist, then according to the remark following Theorem 6, Hopf's solution must be the same and consequently must be of class $C^{\infty}$ in the space variables.

If the boundary of $\Omega$ is smooth, then stronger conclusions are available. In particular, Golovkin and Ito in the

---

*A similar result, at least so far as the spatial regularity of $u$ is concerned, was obtained by Ohyama.

papers referred to at the beginning of this section have proved for $n = 2$ and $n = 3$, respectively, the existence of classical solutions continuously taking on the given boundary data (the solution constructed by Ito exists only for a suitable interval of time). By the uniqueness theorem of Section 4 it is apparent that every solution $u \in V'$ with the same initial data will have the same smoothness properties. This fact can, of course, be regarded as a regularity theorem. It would be worthwhile to have a direct proof along the lines indicated in the remarks following Theorem 8'.

## 7. Open problems.

The most important problem remaining unsolved is without doubt that of constructing a solution of the initial value problem in three dimensions which exists and is regular for all $t \geq 0$, no matter what (smooth) initial data may be assigned. As we have seen above, the corresponding problem has been more or less completely solved in the plane: for data in $W^2$ the solution of Kiselev-Ladyshenskaya exists for all time, is unique in the class $V$ of weak solutions, and by the results of the preceding section is $C^\infty$ in $x$ and Lipschitz continuous in $t$. The entire picture is less clear in three dimensions. If the data is not suitably small the solution of Kiselev-Ladyshenskaya may only exist for a finite time interval $(0,T)$, though certainly a weak solution in the sense of Theorem 1 does persist for all $t \geq T$. There is an important result of Leray which adds a fascinating element of complication to the situation: If $\Omega \equiv E^3$ the weak solution which continues to exist is actually regular except for a set of values of $t$ of measure zero, whose complement is a set of intervals; and the final interval is necessarily of the form $T^* < t < \infty$! In light of this result, one can say in summary that a solution of the initial value problem in $E^3$, for smooth initial data, will be regular for a finite time interval, will then possibly pass through a set of epochs of irregularity where branching (non-uniqueness) occurs, and will eventually settle down again to a regular solution which tends uniformly to zero on compact sets.

Whether the behavior indicated above genuinely occurs, or whether it is only apparent and due to an incomplete theory, remains to be decided. The answer will involve either a new existence theorem in case there are no epochs of irregularity, or else an example to show that breakdown actually occurs. Leray proposed the problem of finding such an example (cf. p. 225 of [20]), and in fact gave a possible method for its construction. In spite of this, I am of the opinion that the more likely alternative is the existence of a regular solution for all time.

The reader will have observed in the course of the paper a number of other places where the theory could be improved. We may mention specifically the possibility of results analogous to Theorems 2 and 3 for dimension $n \geq 5$, an extension of Theorem 7 to weak solutions (or at least to solutions with generalized boundary data), and a proof that classical solutions of the Navier-Stokes equations are analytic in the space variables.

A new and wide field opens if one considers regions with moving boundaries, and non-vanishing boundary data. The existence of periodic solutions has been treated by Prodi [30], Yudovich [38] and Serrin [33], though these results do not yet provide a complete theory. Jerome Sather has studied the extension of Hopf's construction to the case of moving boundaries, and he and I together have proved a uniqueness theorem analogous to Theorem 5 for weak solutions of this problem.

Some significant long range contributions are due to Hopf. In an important series of papers [7], [8], [9], [10], he has raised a number of questions concerning the nature of solutions as the viscosity tends to zero. These questions are also intimately related to the asymptotic behavior of flows as $t \to \infty$ and to the general problem of stability, instability, and turbulence (cf. the opening remarks in [13]). Much work remains before this situation can be clarified.

We may observe finally that the field of classical hydrodynamics blends into more general doctrines, in particular compressible fluid flow and the reçently fashionable subject of magnetohydrodynamics. Some results concerning the initial value problem are known for these fields, and following the main references there is a short list of papers of this type. Needless to say, the problems become more complicated as the physical situation does, and, just as for incompressible fluids, there are a multitude of avenues open to the interested mathematician.

## REFERENCES

[1] C. Foias, Une remarque sur l'unicité des solutions des equations de Navier-Stokes en dimension n, Bull. Soc. Math. France 89 (1961), pp. 1-8.

[2] K. K. Golovkin, The plane motion of a viscous incompressible fluid, Trudy Mat. Inst. Steklov 59 (1960), pp. 37-86.

[3] K. K. Golovkin and O. A. Ladyshenskaya, On the solution of the non-steady boundary problem for the Navier-Stokes equation, Trudy Mat. Inst. Steklov 59 (1960), pp. 100-114.

[4] K. K. Golovkin and V. A. Solonnikov, On the first boundary value problem for the non-stationary Navier-Stokes equations, Dokl Akad. Nauk SSSR 140 (1961), pp. 287-290.

[5] D. Graffi, Sul teorema di unicità nella dinamica dei fluidi, Annali di Mat. 50 (1960), pp. 379-388.

[6] E. Hopf, Über die Anfangswertaufgabe für die hydrodynamischen Grundgleichungen, Math. Nachrichten 4 (1951), pp. 213-231.

[7] -------, Ein allgemeiner Endlichkeitssatz der Hydrodynamik, Math. Annalen 117, pp. 764-775 (1941).

[8] -------, A mathematical example displaying the features of turbulence, Comm. Pure Applied Math. 1, pp. 303-322, (1948).

[9] -------, Statistical hydromechanics and functional calculus, Journ. Rational Mech. Analysis, 1 (1952), pp. 87-124.

[10] -------, Repeated branching through loss of stability, an example Proc. Conference on Differential Equations, Univ. of Maryland (1955).

[11] -------, On non-linear partial differential equations, Lecture Series Symp. P.D.E. (Berkeley, 1955). Univ. of Kansas, 1957.

[12] S. Ito, The existence and uniqueness of regular solution of non-stationary Navier-Stokes equation. Journ. Fac. Science, Univ. of Tokyo, 9 (1961), pp. 103-140.

[13] J. Kampe de Feriet, Statistical fluid mechanics: Two dimensional linear gravity waves. Conference on Partial Differential Equations and Continuum Mechanics, Univ. of Wisconsin, 1961.

[14] A. A. Kiselev and O. A. Ladyshenskaya, On existence and uniqueness of the solution of the nonstationary problem for a viscous incompressible fluid, Izvestiya Akad. Nauk SSSR 21 (1957), pp. 655-680.

[15] O. A. Ladyshenskaya, Solution "in the large" of non-stationary boundary value problem for the Navier-Stokes system with two space variables, Comm. Pure Applied Math. 12 (1959), pp. 427-433.

[16] -------------------, Mathematical problems in the dynamics of viscous incompressible fluids. Moscow 1961.

[17] M. Lees and M. H. Protter, Unique continuation for parabolic differential equations and differential inequalities, Duke Math. Journ. 28 (1961), pp. 369-382.

[18] J. Leray, Étude de diverses equations intégrales non-linéares et de quelques problèmes que pose l'hydrodynamique, Journ. Math. Pures Appl. 12 (1933), pp. 1-82.

[19] --------, Essai sur les mouvements plans d'un liquide visqueux que limitent des parois, Journ. Math. Pures Appl. 13, (1934), pp. 331-418.

[20] --------, Sur le mouvement d'un liquide visqueux emplissant l'espace, Acta Math. 63 (1934), pp. 193-248.

[21] L. Lichtenstein, Grundlagen der Hydromechanik. Berlin 1929.

[22] J. L. Lions, Quelques résultats d'existence dans des équations aux dérivées partielles non linéaires, Bull. Soc. Math. France 87 (1959), pp. 245-273.

[23] -----------, Sur la régularité et l'unicité des solutions turbulentes des équations de Navier-Stokes, Rend. Sem. Mat. Padova 30 (1960), pp. 16-23.

[24] -----------, Equations differentielles - operationnelles, Berlin, 1961. Especially Chapter X.

[25] J. L. Lions et G. Prodi, Un théorème d'existence et unicité dans les équations de Navier-Stokes en dimension 2, C. R. Acad. Sci. Paris 248 (1959), pp. 3519-3521.

[26] F. K. C. Odquist, Über die Randwertaufgaben der Hydrodynamik zäher Flüssigkeiten, Math. Zeitschrift 32 (1930), pp. 325-375.

[27] T. Ohyama, Interior regularity of weak solutions of the time-dependent Navier-Stokes equation, Proc. Japan Acad. 36 (1960), pp. 273-277.

[28] C. W. Oseen, Neuere Methoden und Ergebnisse der Hydrodynamik. Leipzig, 1927.

[29] G. Prodi, Un teorema di unicitá per le equazioni di Navier-Stokes, Annali di Mat. 48 (1959), pp. 173-182.

[30] --------, Qualche risultato riguardo alle equazioni di Navier-Stokes nel caso bidimensionale, Rend. Sem. Mat. Padova 30 (1960), pp. 1-15.

[31] --------, Teoremi ergodici per le equazioni della idrodinamica. Symposium: Sistemi dinamica e teoremi ergodici, Varenna, 1960. CIME, Instituto Mat. dell' Univ. Roma.

[32] J. Serrin, On the stability of viscous fluid motions, Arch. Rational Mech. Analysis 3 (1959), pp. 1-13.

[33] ---------, A note on the existence of periodic solutions of the Navier-Stokes equations, Arch. Rational Mech. Analysis 3 (1959), pp. 120-122.

[34] ---------, On the interior regularity of weak solutions of the Navier-Stokes equations, Arch. Rational Mech. Analysis. 9 (1962), pp. 187-195.

[36] T. Y. Thomas, On the uniform convergence of the solutions of the Navier-Stokes equations, Proc. Nat. Acad. Sci. USA 29 (1943), pp. 243-246.

## SUPPLEMENTARY REFERENCES

[37] W. Velte, Über ein Stabilitäts kriterium der Hydrodynamik, Arch. Rational Mech. Analysis 9 (1962), pp. 9-20.

[38] R. H. Dyer and D. E. Edmunds, A uniqueness theorem in magneto-hydrodynamics, Arch. Rational Mech. Analysis 8 (1961), pp. 254-262.

[39] D. Graffi, Sul teorema di unicità per le equazioni del moto dei fluidi compressibili in un dominio illimitato, Atti Acad. Sci. dell Istituto Bologna, Sci-Fis. Classe, Series XI, 7 (1960), pp. 1-8.

[40] ---------, Ancora sul teorema di unicità per le equazioni del moto dei fluidi, Atti Acad Sci. dell Istituto Bologna, Sci-Fis. Classe, Series XI, 8 (1961), pp. 7-14.

[41] R. P. Kanwal, Uniqueness of magnetohydrodynamic flows, Arch. Rational Mech. Analysis 4 (1960), pp. 335-340.

[42] O. A. Ladyshenskaya and V. A. Solonnikov. On the solution of the non-stationary problem of magneto-hydrodynamics, Dokl. Akad. Nauk SSSR 124 (1959), pp. 26-28. Cf. also Trudy Mat. Inst. Steklov 59 (1960).

[43] B. L. Rozhdestvenskii, Discontinuous solutions of hyperbolic systems of quasi-linear equations, Russian Math. Surveys (trans. of Uspekhi Mat. Nauk), 15 (1960), pp. 53-111 in Number 6.

[44] J. Serrin, On the uniqueness of compressible fluid motions, Arch. Rational Mech. Analysis 3 (1959), pp. 271-288.

*Arch. Ration. Mech. Anal.* 3 (1959), 1–13. Reprint V.1.[iii]


# *On the Stability of Viscous Fluid Motions*

JAMES SERRIN

*Communicated by* C. TRUESDELL

This paper deals with one aspect of the classical problem of hydrodynamical stability, namely, the determination of sufficient conditions for a basic (usually laminar) flow of an incompressible fluid to be stable under arbitrary disturbances. The technique to be applied is the well known method of energy, originated by OSBOURNE REYNOLDS and WILLIAM MCF. ORR and used since that time by many other writers[1]. In spite of this intensive study, it appears that a number of new results can be secured from the method, and it is to these that the paper is devoted.

Our main conclusion is a Reynolds number criterion for the stability of an *arbitrary* fluid motion in a bounded region. In particular, we show that a basic flow in a bounded region $\mathscr{V}$ is stable whenever its Reynolds number $\mathrm{Re} = Vd/\nu$ is less than 5.71; here $V$ is the maximum speed of the basic flow, $d$ is the diameter of $\mathscr{V}$, and $\nu$ is the kinematic viscosity of the fluid. The number 5.71 in itself is neither especially good nor especially bad as a criterion for stability, but what *is* interesting is the fact that it is absolutely rigorous and applies independently of the geometry of the flow region and the particular flow involved. For this reason the result may appropriately be called a Reynolds number for universal stability. We also obtain similar criteria for the stability of flows in unbounded regions, and applications are made to the problem of uniqueness of steady flow.

In the second part of the paper (§ 4) we state a general variational problem connected with the stability of an arbitrary motion. The Euler-Lagrange equations corresponding to this problem bear an interesting and remarkable resemblance to the Navier-Stokes equations, but they are in general too difficult to solve, except for special cases.

The last part of the paper treats a particular example, the laminar Couette flow between rotating coaxial cylinders. The methods of the earlier sections of the paper, when applied to this case, yield the stability criterion

$$\left|\frac{\Omega_2-\Omega_1}{\nu}\right| \leqq (R_2^2 - R_1^2)\left\{\frac{\pi}{R_1 R_2 \log(R_2/R_1)}\right\}^2, \tag{1}$$

in which the notation is practically self-explanatory. The author knows of no other formula of this type which applies to *arbitrary* disturbances of the basic laminar motion. (The criteria of G. I. TAYLOR and others (*cf.* [*14*], Chapter 2)

[1] *Cf.* references [*1*]—[*9*] at the end of the paper.

all refer to infinitesimal disturbances of a special type; thus, although they have indisputable importance if one is considering the breakdown of already established flow, and give quite sharp results in certain limited situations, they do not by any means apply to the whole realm of possible disturbances of the motion.) Finally, the stability of Couette motion is considered from the point of view of the variational method noted above. The results here do not have the finality of criterion (1), but they are nevertheless of interest.

The paper begins with a derivation of a fundamental identity for the rate of change of energy of a perturbation motion. This formula is the basis for all that follows.

## 1. The Reynolds-Orr energy equation

We consider a basic fluid motion occupying a region $\mathscr{V} = \mathscr{V}(t)$ of space and subject to a prescribed velocity distribution on the boundary $\mathscr{S}$ of $\mathscr{V}$. In the cases of greatest interest $\mathscr{V}$ is bounded by material walls and the boundary conditions arise from the motion of these walls, as, for example, in the case of Couette flow. Now suppose the velocity field of the basic flow is altered at some initial instant $t=0$; it is natural to ask whether the subsequent motion, subject to the same boundary conditions, will alter only slightly from what it was, or whether it will change radically in character. To investigate this problem, we shall consider the energy $\mathscr{K}$ of the perturbation (difference) motion. If $\mathscr{K}$ tends to zero as $t \to \infty$, then the basic motion is said to be stable, or, more precisely, stable in the mean.

To be specific, suppose the region $\mathscr{V}$ is bounded, and let $\boldsymbol{v}$ and $\boldsymbol{v}'$ denote, respectively, the velocity vectors of the basic and altered motions. The velocity $\boldsymbol{u} = \boldsymbol{v}' - \boldsymbol{v}$ of the perturbation motion obviously satisfies

$$\boldsymbol{u} = 0 \qquad \text{on } \mathscr{S}, \tag{2}$$

and its kinetic energy is given by

$$\mathscr{K} = \tfrac{1}{2} \int u^2 \tag{3}$$

(in writing integrals, we shall consistently omit the conventional volume infinitesimal; moreover, all integrals are understood to be extended over the entire region $\mathscr{V}$). The rate of change of $\mathscr{K}$ is governed by the fundamental formula

$$\boxed{\frac{d\mathscr{K}}{dt} = -\int (\boldsymbol{u} \cdot \boldsymbol{D} \cdot \boldsymbol{u} + \nu \,\mathrm{grad}\, \boldsymbol{u} : \mathrm{grad}\, \boldsymbol{u})} \tag{4}$$

essentially due to REYNOLDS and ORR. In this equation grad $\boldsymbol{u}$ denotes the matrix with components $(\mathrm{grad}\, \boldsymbol{u})_{ik} = u_{k,i}$ and $\boldsymbol{D}$ denotes the deformation matrix of the basic motion, $D_{ik} = \frac{1}{2}(v_{i,k} + v_{k,i})$.

To prove (4), we begin with the observation that both $\boldsymbol{v}$ and $\boldsymbol{v}'$ satisfy the Navier-Stokes equation

$$\varrho\left(\frac{\partial \boldsymbol{v}}{\partial t} + \boldsymbol{v} \cdot \mathrm{grad}\, \boldsymbol{v}\right) = \varrho \boldsymbol{f} - \mathrm{grad}\, p + \mu \nabla^2 \boldsymbol{v}.$$

By subtraction there arises

$$\frac{\partial \boldsymbol{u}}{\partial t} + \boldsymbol{u} \cdot \mathrm{grad}\, \boldsymbol{v} + \boldsymbol{v}' \cdot \mathrm{grad}\, \boldsymbol{u} = \mathrm{grad}\, \frac{p - p'}{\varrho} + \nu \nabla^2 \boldsymbol{u}.$$

Forming the scalar product of this equation with the vector $\boldsymbol{u}$, and using the incompressibility conditions $\operatorname{div} \boldsymbol{v} = \operatorname{div} \boldsymbol{v}' = 0$, then leads to

$$\frac{\partial}{\partial t}\left(\frac{1}{2} u^2\right) = -\left(\boldsymbol{u}\cdot\boldsymbol{D}\cdot\boldsymbol{u} + \nu \operatorname{grad}\boldsymbol{u} : \operatorname{grad}\boldsymbol{u}\right) + \operatorname{div}\boldsymbol{\Phi}, \tag{5}$$

where

$$\boldsymbol{\Phi} = \frac{p-p'}{\varrho}\boldsymbol{u} + \nu \operatorname{grad}\left(\frac{1}{2} u^2\right) - \frac{1}{2} u^2 \boldsymbol{v}'.$$

On the other hand, we have obviously

$$\frac{d\mathscr{K}}{dt} = \int \frac{\partial}{\partial t}\left(\frac{1}{2} u^2\right) + \oint \frac{1}{2} u^2 \boldsymbol{v}\cdot\boldsymbol{n}, \tag{6}$$

the latter integral being taken over the entire boundary of $\mathscr{V}$. The required formula (4) now follows easily from (5), (6), the divergence theorem, and the fact that $\boldsymbol{u} = \boldsymbol{\Phi} = 0$ on $\mathscr{S}$.[2]

This derivation does not hold when the region $\mathscr{V}$ is unbounded, since both formula (6) and the divergence theorem are applicable only to bounded regions. Under suitable conditions on the asymptotic behavior of $\boldsymbol{v}$, $\boldsymbol{v}'$, however, the above formal steps can still be justified (we shall omit the details). Another justification of (4) for infinite regions is available whenever the flow geometry is such that the disturbances can be assumed spatially periodic at each instant. This will be the case, in particular, for the important Poiseuille and Couette flows. The disturbance $\boldsymbol{u}$ being supposed periodic in the direction of the axis of symmetry (what Lin calls sustained oscillations), the region $\mathscr{V}$ can then be chosen to cover exactly one period. Then the boundary integrals at either end of $\mathscr{V}$, neither of which vanishes separately, just cancel one another. Formula (4) may therefore be assumed to hold in these two important situations.

## 2. Criteria for universal stability

In this section we shall use the method of energy to establish certain criteria for the stability of arbitrary fluid motions. This method is based on the observation that if $\mathscr{K}$ tends to zero, then $\boldsymbol{u}$ must likewise tend to zero almost everywhere. Thus a basic flow will be stable (stable in the mean) provided the energy of any disturbance tends to zero as $t$ increases. To apply the method one considers the right-hand side of (4): if it is negative for arbitrary non-vanishing vectors $\boldsymbol{u}$ satisfying $\operatorname{div}\boldsymbol{u} = 0$, then obviously $d\mathscr{K}/dt < 0$ and there is stability. Since the second term on the right of (4) is always negative, it is seen that viscosity tends to damp out any disturbance. On the other hand, a high rate of shear in the basic flow can cause the first term to be highly positive, thus fostering the growth of a disturbance. The relative importance of these two terms, then, determines the stability of the flow (*cf.* also the discussion in [*14*], § 4.5).

Another criterion of the same sort arises when the right-hand side of (4) is written in slightly different form. Indeed, since $\operatorname{div}\boldsymbol{u} = 0$, we have

$$\boldsymbol{u}\cdot\boldsymbol{D}\cdot\boldsymbol{u} = \operatorname{div}\left[(\boldsymbol{u}\cdot\boldsymbol{v})\,\boldsymbol{u}\right] - \boldsymbol{u}\cdot\operatorname{grad}\boldsymbol{u}\cdot\boldsymbol{v},$$

[2] Several alternate and occasionally useful forms of (4) arise from the simple identity

$$\int \operatorname{grad}\boldsymbol{u} : \operatorname{grad}\boldsymbol{u} = \int (\operatorname{curl}\boldsymbol{u})^2 = 2\int \boldsymbol{D}' : \boldsymbol{D}',$$

where $\boldsymbol{D}'$ denotes the deformation matrix of the perturbation motion.

1*

whence, by application of the divergence theorem, (4) can be written in the form

$$\boxed{\frac{d\mathcal{K}}{dt} = \int (\boldsymbol{u}\cdot \operatorname{grad} \boldsymbol{u}\cdot \boldsymbol{v} - \nu \operatorname{grad} \boldsymbol{u} : \operatorname{grad} \boldsymbol{u}).} \tag{7}$$

If the first term on the right of (7) is less than the second for all admissible vectors $\boldsymbol{u}$, then clearly the basic flow will be stable. But the size of the first term is governed by the magnitude of $\boldsymbol{v}$, from which it follows that high speeds in the basic flow, as well as high rates of shear, tend to cause instability. The qualitative nature of these effects will be investigated in the next several paragraphs.

It is important to note that the energy method cannot provide accurate knowledge of the limits of stability, such as can be gained from the linearized perturbation theory ([*6*], Part II; [*14*], Chap. 1). The reason is that in the energy method one establishes stability relative to arbitrary disturbances, while in reality only those satisfying the hydrodynamical equations need be considered. Nevertheless, because the energy method gives insight into the physical situation, and because the results have the merit of simplicity and complete mathematical rigor, the investigations based on it are both interesting and valuable.

With these preliminary remarks aside, we may now turn to the main result of the paper.

**Theorem 1.** *Let $\mathcal{V} = \mathcal{V}(t)$ be a bounded region of space, which can be included in some cube of edge length $d$. Let $\boldsymbol{v}$ be the velocity vector of any motion in $\mathcal{V}$ satisfying prescribed conditions at the boundary of $\mathcal{V}$. Then the kinetic energy of an arbitrary disturbance motion $\boldsymbol{u} = \boldsymbol{v}' - \boldsymbol{v}$ satisfies the inequalities*

$$\mathcal{K} \leqq \mathcal{K}_0\, e^{2(m - \alpha\nu/d^2)t}, \tag{8}$$

$$\mathcal{K} \leqq \mathcal{K}_0\, e^{(V^2 - \alpha\nu^2/d^2)t/\nu}. \tag{9}$$

*Here $\mathcal{K}_0$ is the initial energy of the disturbance, $-m$ is a lower bound for the characteristic values of the deformation matrix of the basic flow over the time interval 0 to $t$, $V$ is the maximum speed of the basic flow in the same time interval, and $\alpha$ is a pure number,*

$$\alpha = \frac{3+\sqrt{13}}{2}\,\pi^2 \cong 32.6. \tag{10}$$

This theorem is a generalization and improvement of certain results of T. Y. THOMAS and EBERHARD HOPF[3]. Before giving the proof of the theorem, we note two important corollaries, which in a certain sense constitute the heart of the paper. It is to be emphasized that these corollaries are absolutely rigorous, and apply to *all possible motions* in $\mathcal{V}$.

**Universal stability criterion I.** *If the dimensionless "Reynolds number" $md^2/\nu$ of a flow is less than 32.6, then $\mathcal{K} \to 0$ as $t \to \infty$, and the motion is stable.*

[3] THOMAS [*8*] proved that an arbitrary fluid motion is stable if its deformation matrix is sufficiently small, but did not compute numerical values; the present analysis is an extension of THOMAS' work. Similarly, HOPF has shown (essentially) that a fluid motion is stable if its maximum speed is small enough, again with no discussion of numerical values.

**Universal stability criterion II.** *If the Reynolds number $Vd/\nu$ of a flow is less than 5.71, then $\mathcal{K}\to 0$ as $t\to\infty$, and the motion is stable. (Here $V$ is the maximum speed of the flow, and $d$ is the maximum diameter of the flow region.)*

*Proof of Theorem 1.* We begin with an auxiliary computation whose goal is the inequality

$$\alpha d^{-2}\int u^2 \leqq \int \operatorname{grad}\boldsymbol{u} : \operatorname{grad}\boldsymbol{u}, \tag{11}$$

where $\alpha$ is given by (10). Let $\boldsymbol{h}$ be an arbitrary continuously differentiable vector field in $\mathcal{V}$. Then for any value of the constant $\varepsilon$,

$$\begin{aligned} 0 &\leqq (u_i h_k + u_{i,k} + \varepsilon u_{k,i})(u_i h_k + u_{i,k} + \varepsilon u_{k,i}) \\ &= (1+\varepsilon^2) u_{i,k} u_{i,k} + h^2 u^2 + h_i (u^2)_{,i} + 2\varepsilon (u_i h_k + u_{i,k}) u_{k,i}. \end{aligned} \tag{12}$$

(For convenience we use tensor rather than vector notation in carrying out this and several subsequent calculations.) Some of the terms on the right-hand side of (12) can be transformed as follows,

$$\begin{aligned} h_i (u^2)_{,i} &= (u^2 h_i)_{,i} - u^2 h_{i,i} \\ u_i h_k u_{k,i} &= (u_i h_k u_k)_{,i} - u_i h_{k,i} u_k \\ u_{i,k} u_{k,i} &= (u_i u_{k,i})_{,k}. \end{aligned}$$

Here we have made use of the incompressibility condition $u_{i,i}=0$. Making the above changes in (12), integrating the result over $\mathcal{V}$, and using the divergence theorem, we obtain the inequality

$$\int \left[(h_{i,i} - h^2) u^2 + 2\varepsilon u_i h_{i,k} u_k\right] \leqq (1+\varepsilon^2)\int u_{i,k} u_{i,k}. \tag{13}$$

Now the particular vector field $h_i = C\tan C x_i$ is differentiable in a cube of edge length $\pi/C$ centered at the origin. If we set $C=\pi/d$ and suitably locate the origin in $\mathcal{V}$, then this vector can be substituted into (13). An easy computation then yields the inequality

$$(1+\varepsilon^2)^{-1}(3+2\varepsilon)\, C^2 \int u^2 \leqq \int u_{i,k} u_{i,k}.$$

The left-hand side is maximized by choosing $\varepsilon = \frac{1}{2}(\sqrt{13}-3)$, and (11) is thereby proved.

The term $\boldsymbol{u}\cdot\boldsymbol{D}\cdot\boldsymbol{u}$ in (4) satisfies the inequality

$$\boldsymbol{u}\cdot\boldsymbol{D}\cdot\boldsymbol{u} \geqq -m u^2 \tag{14}$$

during the entire time interval 0 to $t$. Therefore, from (4), (11) and (14)

$$\frac{d\mathcal{K}}{dt} \leqq (m - \alpha\nu d^{-2})\int u^2 = 2(m-\alpha\nu d^{-2})\mathcal{K}.$$

Writing this in the form

$$\frac{d}{dt}\{\mathcal{K} e^{-2(m-\alpha\nu d^{-2})t}\} \leqq 0$$

and integrating from 0 to $t$ leads at once to (8).

The proof of (9) is similar but requires in addition the inequality

$$\boldsymbol{u}\cdot\operatorname{grad}\boldsymbol{u}\cdot\boldsymbol{v} \leqq \frac{1}{2}\left(\nu \operatorname{grad}\boldsymbol{u} : \operatorname{grad}\boldsymbol{u} + \frac{u^2 v^2}{\nu}\right), \tag{15}$$

which follows at once from the identity

$$\boldsymbol{A}:\boldsymbol{A} - 2\boldsymbol{u}\cdot\boldsymbol{A}\cdot\boldsymbol{v} + u^2 v^2 \equiv (\boldsymbol{A} - \boldsymbol{u}\boldsymbol{v}):(\boldsymbol{A} - \boldsymbol{u}\boldsymbol{v}) \geqq 0$$

with $\boldsymbol{A} = \nu \operatorname{grad} \boldsymbol{u}$. From (7), (15), and (11) we obtain

$$\frac{d\mathscr{K}}{dt} \leqq \frac{1}{\nu}\left(V^2 - \alpha \nu^2 d^{-2}\right)\mathscr{K},$$

and (9) then follows by integration as in the proof of (8).

### 3. Extensions and applications of Theorem 1

The preceding method can be applied equally well to flows which take place in channels and pipes, and to plane flow problems, so long as the fundamental formula (4) applies. To illustrate the method, consider for example a straight pipe whose cross section has a maximum diameter $d$. If the pipe is directed along the $z$-axis, then the vector

$$\boldsymbol{h} = C(\boldsymbol{i} \tan C x + \boldsymbol{j} \tan C y), \qquad (C = \pi/d),$$

can be used in (13). Choosing $\varepsilon = 0$, we get an inequality of the form (11) with $\alpha = 2\pi^2$. By using this result, the reader can easily obtain estimates for the rate of change of $\mathscr{K}$, and thus formulate analogues of Theorem 1 and the universal stability criteria for flow in a pipe.

Other cases can be treated similarly and the results conveniently grouped in the following table.

| | |
|---|---|
| A. Straight channel, maximum width $d$: | $\alpha = \pi^2$. |
| B. Straight pipe, maximum diameter $d$: | $\alpha = 2\pi^2$. |
| C. Plane flow in a bounded region, maximum diameter $d$, | |
| i) Three dimensional disturbances: | $\alpha = 2\pi^2$, |
| ii) Plane disturbances only: | $\alpha = (1 + \sqrt{2})\,\pi^2$. |
| D. Bounded region, maximum diameter $d$: | $\alpha = \frac{3+\sqrt{13}}{2}\pi^2$. |

(Note: It is only in cases C, ii) and D that we have been able to make use of the incompressibility condition $\operatorname{div} \boldsymbol{u} = 0$.)

We conclude the section with two simple applications of Theorem 1. First, suppose the boundaries of $\mathscr{V}$ consist of *rigid fixed* walls, so that any motion initially present will presumably die out due to lack of an energy source. By choosing $\boldsymbol{v} \equiv 0$ for the basic motion, we see from (8) that the energy $\mathscr{K}$ of an arbitrary motion $\boldsymbol{v}'$ in $\mathscr{V}$ must in fact tend to zero according to the law

$$\mathscr{K} \leqq \mathscr{K}_0 e^{-2\alpha\nu t/d^2}, \qquad \alpha \cong 32.6.$$

Similar estimates, but with smaller values for the coefficient $\alpha$, have been obtained by LERAY (for plane flow) and KAMPÉ DE FERIET[4] and BERKER (for spatial

[4] KAMPÉ DE FÉRIET obtained the value $\alpha = 3\pi^2$, using a method somewhat similar to the one presented here.

flows), and RAYLEIGH in a much earlier paper [4] proved that $\mathscr{K} \to 0$ as $t \to \infty$, though without estimating the rate of convergence. It remains an open question whether the velocity itself must tend to zero as $t \to \infty$; certainly one would expect this, but a strict proof seems to be a matter of more than ordinary difficulty.

As a second application, we have the following uniqueness theorem concerning *steady* motion in a fixed bounded region $\mathscr{V}$.

**Theorem 2.** *Let $\boldsymbol{v}$ and $\boldsymbol{v}'$ be two steady flows in $\mathscr{V}$, subject to a prescribed velocity distribution on the boundary of $\mathscr{V}$. Let $-m$ be a lower bound for the characteristic values of the deformation matrix of the motion $\boldsymbol{v}$, let $V = \max v$, and suppose that either*

$$m\, d^2/\nu \leqq \alpha \quad \text{or} \quad V\, d/\nu \leqq \sqrt{\alpha}. \tag{16}$$

*Then the two flows are identical.*

*Proof.* The kinetic energy of the difference motion $\boldsymbol{v}' - \boldsymbol{v}$ must be constant, since the flows are steady. On the other hand, it must satisfy both (8) and (9). In view of (16) this can happen only if $\mathscr{K} = 0$, which in turn implies $\boldsymbol{v} = \boldsymbol{v}'$.

This theorem depends strongly on the condition (16), but without some such assumption it is extremely unlikely that the conclusion is true.

### 4. Variational techniques

The key step in the proof of Theorem 1 lay in establishing inequality (11). We are concerned, however, not only with the validity of (11), but also with the determination of the largest possible coefficient $\alpha$, for the size of $\alpha$ evidently determines the numerical values in the various stability criteria. Now the method of proof in Theorem 1 clearly gives no guarantee of providing the "best possible" value for $\alpha$; moreover, it is quite crude in its estimate of the term $\boldsymbol{u} \cdot \boldsymbol{D} \cdot \boldsymbol{u}$. For these reasons, it is of great interest to consider an alternate approach which not only supplies a theoretical procedure for determining the best possible $\alpha$, but in addition gives a way to avoid the estimate (14).

Stated in precise terms, our problem is twofold. First, we must determine the greatest coefficient $\alpha$ such that the inequality

$$\alpha\, d^{-2} \int u^2 \leqq \int \operatorname{grad} \boldsymbol{u} : \operatorname{grad} \boldsymbol{u}$$

holds for arbitrary vector fields $\boldsymbol{u}$ satisfying

$$\operatorname{div} \boldsymbol{u} = 0, \qquad \boldsymbol{u} = 0 \text{ on } \mathscr{S}. \tag{17}$$

Second, we must determine the least coefficient $\tilde{\nu}$ such that the inequality

$$\int (\boldsymbol{u} \cdot \boldsymbol{D} \cdot \boldsymbol{u} + \tilde{\nu} \operatorname{grad} \boldsymbol{u} : \operatorname{grad} \boldsymbol{u}) \geqq 0$$

holds, again for arbitrary $\boldsymbol{u}$ satisfying (17). In this case it is clear that the basic flow will be stable provided simply that $\nu > \tilde{\nu}$. The two problems above can be consolidated into the single variational problem:

$$-\int \boldsymbol{u} \cdot \boldsymbol{D} \cdot \boldsymbol{u} = \text{Maximum}, \tag{18}$$

where the vector field $\boldsymbol{u}$ must satisfy the side conditions

$$\int \operatorname{grad} \boldsymbol{u} : \operatorname{grad} \boldsymbol{u} = 1, \qquad \operatorname{div} \boldsymbol{u} = 0, \qquad \boldsymbol{u} = 0 \text{ on } \mathscr{S}. \tag{19}$$

The first problem occurs when $\boldsymbol{D}$ is the negative of the identity matrix.

The variational problem (18)—(19) can be reformulated as a partial differential equation for the extremal function $\boldsymbol{u}$, according to well known procedures of the calculus of variations. Thus, through introduction of Lagrange multipliers $\nu^*$ and $\lambda = \lambda(x, y, z, t)$, the problem becomes

$$\delta \int (\boldsymbol{u} \cdot \boldsymbol{D} \cdot \boldsymbol{u} + \nu^* \operatorname{grad} \boldsymbol{u} : \operatorname{grad} \boldsymbol{u} - 2\lambda \operatorname{div} \boldsymbol{u}) = 0. \tag{20}$$

The Euler-Lagrange equation corresponding to (20) is easily found to be

$$\boldsymbol{u} \cdot \boldsymbol{D} = -\operatorname{grad} \lambda + \nu^* \nabla^2 \boldsymbol{u}, \tag{21}$$

and this is to be solved subject to the side conditions (19). The reader may observe the remarkable similarity between the eigenvalue equation (21) and the equations of hydrodynamics. The equations corresponding to the case $\boldsymbol{D} = -\boldsymbol{I}$ are of sufficient importance to be noted separately, namely

$$\nu^* \nabla^2 \boldsymbol{u} + \boldsymbol{u} = \operatorname{grad} \lambda, \qquad \operatorname{div} \boldsymbol{u} = 0. \tag{22}$$

In spite of the relative simplicity of (22), we have been unable to determine its solution, and in what follows can only offer some general remarks concerning the system (19), (21).

Now for any solution of equation (21) we have

$$-\int \boldsymbol{u} \cdot \boldsymbol{D} \cdot \boldsymbol{u} = \int (\boldsymbol{u} \cdot \operatorname{grad} \lambda - \nu^* \boldsymbol{u} \cdot \nabla^2 \boldsymbol{u}) = \nu^* \int \operatorname{grad} \boldsymbol{u} : \operatorname{grad} \boldsymbol{u} = \nu^*,$$

where use has been made of the divergence theorem and the conditions (19). On the other hand, any vector $\boldsymbol{u}$ which provides the integral (18) with its maximum value must be a solution of (21). It follows that the eigenvalue $\tilde{\nu}$ associated with a maximizing vector is precisely the maximum value of the integral (18). Moreover, no eigenvalue $\nu^*$ can be larger than $\tilde{\nu}$, for if this were the case then the corresponding eigenvector $\boldsymbol{u}^*$ would give to the integral (18) a value larger than $\tilde{\nu}$. This proves the following theorem.

**Theorem 3.** *Suppose there exists a vector* $\boldsymbol{u}$ *which solves the variational problem* (18)—(19). *Then the eigenvalue problem* (19), (21) *has a greatest eigenvalue* $\tilde{\nu}$, *and the basic motion will be stable provided that* $\nu > \tilde{\nu}$.

The difficulty with Theorem 3 is, of course, that there is no direct way of verifying its hypothesis. In circumstances such as these the best that can be hoped for is to determine all the eigenvalues (by Theorem 1 they constitute a bounded set), and to assume that the least upper bound of these eigenvalues is the required maximum of the integral (18). If the eigenvectors are complete in the set of admissible vectors, then this gives another method for proving that the least upper bound of the eigenvalues is the maximum of (18); unfortunately, this appears at least as difficult to verify as the hypothesis of Theorem 3.

## 5. Example. Couette flow

We consider the well known circulatory flow between rotating concentric cylinders. If the inner and outer cylinders, respectively, have radii $R_1$ and $R_2$, and rotate with angular speeds $\Omega_1$ and $\Omega_2$ ($\Omega_1 > 0$), then the velocity field is given by

$$v_\theta = A r + B r^{-1}, \quad v_r = v_z = 0,$$

where

$$A = \frac{R_2^2 \Omega_2 - R_1^2 \Omega_1}{R_2^2 - R_1^2}, \quad B = -\frac{(R_1 R_2)^2 (\Omega_2 - \Omega_1)}{R_2^2 - R_1^2}.$$

One finds easily that the scalar vorticity $\omega$ has the constant value $2A$, while in polar coordinates

$$\boldsymbol{D} = -\frac{B}{r^2}\begin{pmatrix} 0 & 1 & 0 \\ 1 & 0 & 0 \\ 0 & 0 & 0 \end{pmatrix}.$$

The characteristic values of $\boldsymbol{D}$ are $\pm B/r^2$, whence

$$\boldsymbol{u} \cdot \boldsymbol{D} \cdot \boldsymbol{u} \geqq -\frac{|B|}{r^2} u^2. \tag{23}$$

The special form of inequality (23) suggests that the stability estimates of § 2 can be considerably improved. In particular, in place of the inequality (11) we shall look for an estimate of the form

$$\beta \int \frac{u^2}{r^2} \leqq \int \operatorname{grad} \boldsymbol{u} : \operatorname{grad} \boldsymbol{u}. \tag{24}$$

To this end, let us seek a vector field $\boldsymbol{h}$ such that

$$\operatorname{div} \boldsymbol{h} - h^2 = C^2/r^2, \quad R_1 < r < R_2. \tag{25}$$

A radially symmetric solution of (25) is readily found, namely

$$\boldsymbol{h} = \frac{C}{r} \tan(C \log r + D)\, \boldsymbol{i}_r, \tag{26}$$

where the constants $C$ and $D$ are given by

$$C = \frac{\pi}{\log(R_2/R_1)}, \quad D = \frac{\pi}{2} \frac{\log(R_1 R_2)}{\log(R_2/R_1)}.$$

Substituting (26) into (13) and setting $\varepsilon = 0$ then yields an inequality of the form (24), with

$$\beta = C^2 = \left\{\frac{\pi}{\log(R_2/R_1)}\right\}^2.$$

Combining (23) and (24) with the fundamental energy equation (4) yields

$$\frac{d\mathscr{K}}{dt} \leqq (|B| - \beta \nu) \int \frac{u^2}{r^2},$$

and from this it follows that *Couette flow is stable relative to arbitrary disturbances whenever*

$$\left|\frac{\Omega_2 - \Omega_1}{\nu}\right| < (R_2^2 - R_1^2) \left\{\frac{\pi}{R_1 R_2 \log(R_2/R_1)}\right\}^2. \tag{27}$$

The reader should notice that this stability is in the "strong" sense, $\mathscr{K} \to 0$ as $t \to \infty$.

Previously, SYNGE [13] has proved stability relative to infinitesimal disturbances whenever $A$ has the same sign as $\Omega_1$, which in the present case ($\Omega_1 > 0$) gives the stability conditions $\omega = 2A \geqq 0$, or equivalently, $\Omega_2 \geqq (R_2/R_1)^2 \Omega_1$ [5]. The relative zones of stability are shown in Fig. 1. For the celebrated pair of radii $R_1 = 3.55$, $R_2 = 4.03$, inequality (27) reduces to

$$\left| \frac{\Omega_2 - \Omega_1}{\nu} \right| < 10.92. \tag{28}$$

In Fig. 2 we have indicated the stability zones based on (28), on SYNGE'S criterion, and on the calculations and experiments of G. I. TAYLOR. If we suppose TAYLOR'S

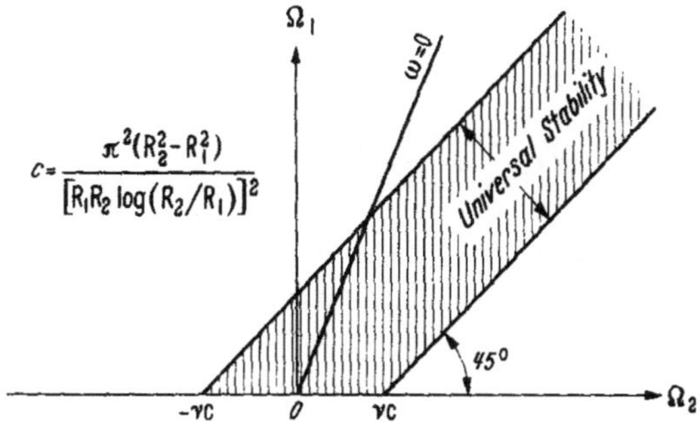


Fig. 1. Stability zones for Couette flow. The zone $\omega \geqq 0$ lies *below* the line $\omega = 0$

experimental data to be an accurate representation of the mathematical situation, it is seen that the right-hand side of (28) cannot possibly be greater than about 50, for otherwise there is an observed secondary motion. Our value is certainly not

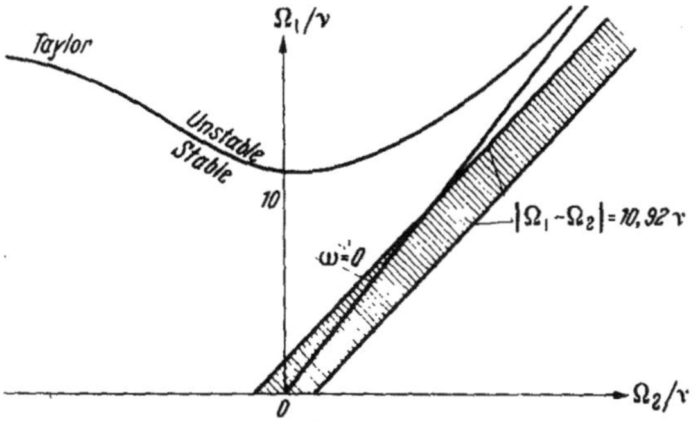


Fig. 2. Stability zones for Couette flow, $R_1 = 3.55$ and $R_2 = 4.03$

very near 50, but considering the difficulty of the problem and the fact that (28) applies to arbitrary disturbances, one does feel that 10.92 is a quite respectable value.

### 6. Couette flow (continued)

In the previous section we studied the stability of Couette flow using the methods of § 2. In this section the same problem will be treated by the variational techniques of § 4. The end result will be seen to have a somewhat tentative

[5] In this connection we may recall some earlier work of RAYLEIGH and SYNGE, in which the same criterion is established for inviscid fluids; *cf.* reference [14], § 4.2.

character, in line with the remarks at the close of § 4, though it still has considerable interest. For Couette flow the differential equation (21) and the side condition div $\boldsymbol{u} = 0$ take the following form (in polar coordinates)

$$\begin{aligned} -\frac{B v}{r^2} &= -\frac{\partial \lambda}{\partial r} + \nu^* \left\{ \Delta u - \frac{2}{r^2} \frac{\partial v}{\partial \vartheta} - \frac{u}{r^2} \right\}, \\ -\frac{B u}{r^2} &= -\frac{1}{r} \frac{\partial \lambda}{\partial \vartheta} + \nu^* \left\{ \Delta v + \frac{2}{r^2} \frac{\partial u}{\partial \vartheta} - \frac{v}{r^2} \right\}, \\ 0 &= -\frac{\partial \lambda}{\partial z} + \nu^* \Delta w, \\ \frac{1}{r} \frac{\partial}{\partial r} (r u) &+ \frac{1}{r} \frac{\partial v}{\partial \vartheta} + \frac{\partial w}{\partial z} = 0, \end{aligned} \tag{29}$$

where

$$\Delta \equiv \frac{1}{r} \frac{\partial}{\partial r} \left( r \frac{\partial}{\partial r} \right) + \frac{1}{r^2} \frac{\partial^2}{\partial \vartheta^2} + \frac{\partial^2}{\partial z^2},$$

and $u, v, w$ denote, respectively, the $r$, $\vartheta$ and $z$ components of the perturbation velocity $\boldsymbol{u}$. Equations (29), as they stand, are too difficult to solve in full generality. We shall therefore seek a particular solution of the form

$$u = \hat{u}(r) \cos k z, \qquad v = \hat{v}(r) \cos k z, \qquad w = \hat{w}(r) \sin k z. \tag{30}$$

Eliminating $\lambda$ and $\hat{w}$ from the system (29), we get

$$\begin{aligned} (L - k^2)^2 \hat{u} &= \frac{B k^2}{\nu^*} \frac{\hat{v}}{r^2}, \\ (L - k^2) \hat{v} &= -\frac{B}{\nu^*} \frac{\hat{u}}{r^2}, \end{aligned} \tag{31}$$

where $L$ denotes the differential operator

$$\frac{1}{r} \frac{d}{dr} \left( r \frac{d}{dr} \right) - \frac{1}{r^2}.$$

(The derivation of (31) is most simply carried out as follows. From the last equation of (29) we have

$$\frac{1}{r} \frac{d}{dr} (r \hat{u}) + k \hat{w} = 0,$$

whence $\boldsymbol{u}$ can be written in the form $\boldsymbol{u} = \text{curl}\, \boldsymbol{\Psi}$, where

$$\boldsymbol{\Psi} = \frac{1}{k} (\hat{v}\, \boldsymbol{i}_r - \hat{u}\, \boldsymbol{i}_\vartheta) \sin k z.$$

Substituting $\boldsymbol{u} = \text{curl}\, \boldsymbol{\Psi}$ directly into (21) and taking the curl of that equation yields

$$B\, \text{curl}\, (v/r^2, u/r^2, 0) = \nu^* \,\text{curl}^4 \boldsymbol{\Psi}, \tag{32}$$

where the identity curl curl = grad div $- \nabla^2$ has been used. The first equation of (31) is now obtained as the $\vartheta$-component of equation (32). Finally, the unkown $\lambda$ must necessarily be independent of $\vartheta$, so that the second equation of (29) reduces immediately to the second equation of (31).)

Equations (31) are formally similar to the classical first order equations for small disturbance in Couette flow. The solution of (31) is a difficult task unless the cylinders are close together, in which case we can suppose (31) to be sufficiently well approximated by the simpler equations

$$\begin{aligned}(D^2 - k^2)^2 \hat{u} &= \Lambda k^2 \hat{v} \\ (D^2 - k^2)\, \hat{v} &= -\Lambda \hat{u},\end{aligned} \tag{33}$$

where $D = d/dr$, $\Lambda = B/\nu^* \bar{R}^2$, and $\bar{R}$ is some appropriate mean radius, say $\bar{R} = \sqrt{R_1 R_2}$. The boundary conditions associated with (33) are

$$\hat{u} = \hat{v} = \frac{d\hat{u}}{dr} = 0 \quad \text{at} \quad r = R_1, R_2. \tag{34}$$

According to the calculations of JEFFREYS and others (*cf.* [*14*], § 2.3) the critical value of $\Lambda$ is $41.2\,(R_2 - R_1)^{-2}$ (that is, this is the smallest value of $\Lambda$ for which a non-trivial solution of (33), (34) exists). We are thus led to the following stability criterion,

$$\left|\frac{\Omega_2 - \Omega_1}{\nu}\right| \leqq 41.2\, \frac{R_1 + R_2}{R_1 R_2 (R_2 - R_1)}\,. \tag{35}$$

For the pair of radii 3.55, 4.03 this gives, in particular,

$$\left|\frac{\Omega_2 - \Omega_1}{\nu}\right| \leqq 45.5\,. \tag{36}$$

(36) is certainly a real improvement over the earlier estimate (28), and, moreover, the right-hand side comes very close to the "best value" 50. It must be borne in mind, however, that (35) and (36) are not yet rigorously established, since it remains to be shown that we have really found the greatest eigenvalue of the system (29). There is some reason to believe that this is so, since hydrodynamical experiments have shown indisputably that the secondary motion occurring in Couette flow is approximately of the form (30), but barring a proof of this fact, we must accept (35)—(36) as only tentatively established, in contrast with the absolute certainty of (27)—(28).

*Note:* The above research was supported by the United States Air Force Office of Scientific Research under Contract No. AF 49(638)-262.

## References

[*1*] REYNOLDS, O.: On the dynamical theory of incompressible viscous fluids and the determination of the criterion. Phil. Trans. Roy. Soc. London (A) **186**, 123—164 (1895); Scientific Papers II, pp. 535—577.

[*2*] ORR, W. MCF.: The stability or instability of the steady motions of a liquid. Part II: A viscous liquid. Proc. Roy. Irish Acad. (A) **27**, 69—138 (1907).

[*3*] HAMEL, G.: Zum Turbulenzproblem. Nachr. Ges. Wiss. Göttingen. **1911**, 261—270.

[*4*] LORD RAYLEIGH: On the motion of a viscous fluid. Phil. Mag. (6) **26**, 776—786 (1913); Papers VI, pp. 187—197.

[*5*] KÁRMÁN, TH. VON: Über die Stabilität der Laminarströmung und die Theorie der Turbulenz. Proc. 1st. Int. Congress Appl. Mech. Delft, 1924, pp. 97—112.

[*6*] SYNGE, J. L.: Hydrodynamical stability. Semi-centennial publications of the Amer. Math. Soc. **2**, 227—269 (1938). Especially Part III.

[7] THOMAS, T. Y.: On the stability of viscous fluids. (1941). Univ. Calif. Publ. Math. New Series **2**, 13–43 (1944).

[8] THOMAS, T. Y.: On the uniform convergence of the solutions of the Navier-Stokes equations. Proc. Nat. Acad. Sci. U.S.A. **29**, 243–246 (1943).

[9] HOPF, E.: On non-linear partial differential equations. Lecture series of the Symposium on partial differential equations. Univ. of Calif. 1955, pp. 7–11.

[10] LERAY, J.: Essai sur les mouvements plan d'un liquide visqueux que limitent des parois. J. Math. Pures. Appl. (9) **13**, 331–418 (1934).

[11] KAMPÉ DE FÉRIET, J.: Sur la decroissance de l'énergie cinétique d'un fluide visqueux incompressible occupant un domaine borné ayant pour frontière des parois solides fixes. Ann. Soc. Sci. Bruxelles **63**, 35–46 (1949).

[12] BERKER, R.: Inégalité vérifiée par l'énergie cinétique d'un fluide visqueux incompressible occupant un domaine spatial borné. Bull. Tech. Univ. Istanbul **2**, 41–50 (1949).

[13] SYNGE, J. L.: On the stability of a viscous liquid between two rotating coaxial cylinders. Proc. Roy. Soc. London (A) **167**, 250–256 (1938).

[14] LIN, C. C.: The theory of hydrodynamic stability. Cambridge 1955. Especially Chaps. 1, 2, 4.

University of Minnesota
Minneapolis, Minnesota

*(Received October 20, 1958)*

*Arch. Ration. Mech. Anal.* 3 (1959), 120–122.


**Reprint V.1.[iv]**

# *A Note on the Exstencie of Periodic Solutions of the Navier-Stokes Equations*

JAMES SERRIN

*Communicated by* C. TRUESDELL

In an earlier paper [*1*] I established the stability of solutions of the Navier-Stokes equations provided that a certain Reynolds number was small enough. This result enables one to deduce from certain plausible hypotheses the existence of stable periodic solutions of the Navier-Stokes equations. In particular we shall prove the following result:

*Let $\mathscr{V}=\mathscr{V}(t)$ be a bounded region in space, and let a flow velocity be prescribed at each point of the boundary of $\mathscr{V}$. Assume furthermore that both $\mathscr{V}$ and the assigned velocities depend periodically on the time $t$. Then, under conditions 1° and 2° stated below, there exists a unique, stable, periodic solution of the Navier-Stokes equations in $\mathscr{V}$ which takes on the prescribed on the boundary of $\mathscr{V}$.*

In view of condition 2° (see below) it is necessary that the boundary conditions be compatible with a flow of Reynolds number less than 5.7. The theorem may therefore be paraphrased by the statement that corresponding to sufficiently low assigned periodic velocities there exists a periodic flow to which every other motion eventually subsides. A simple example would be a fluid enclosed in a fixed container and stirred by a low speed bladed rotor.

Another case of interest occurs when the assigned conditions are *steady*. Then the theorem asserts the existence of a unique, stable, time-independent solution of the Navier-Stokes equations taking on the prescribed velocities on the boundary of $\mathscr{V}$. In several long papers (*e.g.* [*2*]—[*5*]) the existence of steady flows has been proved independently of conditions 1° and 2°, though in these cases a considerable degree of smoothness has been required of the boundary data. Needless to say, the flows thus obtained may be neither stable nor unique, unless their Reynolds numbers are low enough.

We now state conditions 1° and 2° envisaged in the above theorem and remarks.

**1°.** *To every continuous initial distribution of velocities over $\mathscr{V}$ there corresponds a solution of the Navier-Stokes equations satisfying the prescribed boundary conditions.*

**2°.** *There is one solution whose Reynolds number $Re \equiv Vd/\nu$ is less than 5.7. (Here $V$ is the maximum speed of the flow during the whole time interval $0 \leqq t < \infty$, $d$ is the maximum diameter of $\mathscr{V}$, and $\nu$ is the kinematic viscosity.) This solution is equicontinuous in $\boldsymbol{x}=(x, y, z)$ for all $t$.*

Before proceeding to the proof of the theorem there are several remarks to be made. First, it is understood that the flows guaranteed by Condition 1° should be valid for all $t \geqq 0$. Though this is mathematically a very stringent requirement, it is nevertheless quite plausible on the grounds of physical intuition and our knowledge of the behavior of other parabolic systems; moreover, from the proof it will be clear that 1° need hold only for initial velocity distributions whose Reynolds numbers are less than 5.7. Second, the number 5.7 in Condition 2° may possibly be improved, as will be apparent from the proof. Finally, the author wishes to make it clear that he does not consider the above theorem to be equivalent to the usual kind of mathematical existence theorem: it is agreed that for certain types of boundary behavior 1° and 2° might not hold, and that (correspondingly) the limits of application of the theorem are not well-defined. On the other hand, in view of the plausibility of 1° and 2° there is every reason to expect that they will be proved in the future, subject to sufficiently smooth boundary conditions[1], and once such proofs are given the existence of periodic solutions indisputably follows.

Now let $\boldsymbol{v} = \boldsymbol{v}(\boldsymbol{x}, t)$ be the velocity vector of the flow guaranteed by Condition 2°, and suppose the assigned boundary conditions have period 1. We consider the sequence of vector fields

$$\boldsymbol{\varphi}_n(\boldsymbol{x}) = \boldsymbol{v}(\boldsymbol{x}, n), \qquad n = 0, 1, 2, \ldots .$$

This sequence is bounded and equicontinuous by hypothesis, hence by the theorem of Arzela contains a *subsequence* which converges uniformly to a vector $\boldsymbol{\varphi}(\boldsymbol{x})$. We assert that actually the whole sequence converges to $\boldsymbol{\varphi}(\boldsymbol{x})$.

Indeed, if this were not the case we could find another subsequence converging to a vector $\boldsymbol{\psi}(\boldsymbol{x})$ different from $\boldsymbol{\varphi}(\boldsymbol{x})$. But this is impossible, as we shall now show. For positive integers $m$ and $n$, $(m > n)$, set

$$\boldsymbol{v}'(x, t) = \boldsymbol{v}(\boldsymbol{x}, t + m - n), \qquad t \geqq 0.$$

Then obviously $\boldsymbol{v}'$ is a solution of the Navier-Stokes equations, and, according to the periodicity assumption, it satisfies the assigned boundary conditions. Now for any vector $\boldsymbol{f} = \boldsymbol{f}(\boldsymbol{x}, t)$, let $\mathscr{K}(\boldsymbol{f})$ be defined by

$$\mathscr{K}(\boldsymbol{f}) \equiv \frac{1}{2} \int_{\mathscr{V}(t)} |\boldsymbol{f}(\boldsymbol{x}, t)|^2 \, dv.$$

Then by Theorem 1 of [*1*] we have

$$\mathscr{K}(\boldsymbol{v}' - \boldsymbol{v}) \leqq \mathscr{K}_0 e^{-\varepsilon t}, \tag{1}$$

where $\mathscr{K}_0$ denotes the value of $\mathscr{K}(\boldsymbol{v}' - \boldsymbol{v})$ at $t = 0$, and

$$\varepsilon = \frac{1}{\nu} (\alpha \nu^2 / d^2 - V^2) > 0$$

since $\mathrm{Re}[\boldsymbol{v}] \leqq 5.7$. Because $\boldsymbol{v}'$ as well as $\boldsymbol{v}$ satisfies $\mathrm{Re} \leqq 5.7$, it is seen that

$$\mathscr{K}_0 \leqq \frac{1}{2} \int_{\mathscr{V}_0} (2V)^2 \, dv \leqq 2 d^3 V^2 = \text{Const.}$$

Setting $t = n$ in (1) yields

$$\mathscr{K}(\boldsymbol{\varphi}_m - \boldsymbol{\varphi}_n) \leqq \text{Const.}\, e^{-\varepsilon n}, \tag{2}$$

[1] Similar theorems are already known, *cf.* the papers of Leray and Kiselev & Ladyshenskaya cited in the references.

and it follows that

$$\lim_{m,n\to\infty} \mathscr{K}(\varphi_m - \varphi_n) = 0 \tag{3}$$

(the region of integration in (2) and (3) is, of course, $\mathscr{V}(n) \equiv \mathscr{V}_0$). Now letting $m$ and $n$ tend to infinity through sequences of integers such that $\varphi_m \to \psi$, $\varphi_n \to \varphi$ gives an immediate contradiction. This proves that $\lim \varphi_n(x) = \varphi(x)$.

By condition 1° there exists a flow $v^* = v^*(x, t)$ such that $v^*(x, 0) = \varphi(x)$. We assert that $v^*$ is a *periodic* solution of the Navier-Stokes equation. For set $v''(x, t) = v(x, t+n)$. Then we have, in the same way as inequality (1),

$$\mathscr{K}(v^* - v'') \leqq \mathscr{K}_0 e^{-\varepsilon t}, \tag{4}$$

where $\varepsilon$ has the same meaning as before, and $\mathscr{K}_0$ is the value of $\mathscr{K}(v^* - v'')$ when $t = 0$, that is $\mathscr{K}_0 \equiv \mathscr{K}(\varphi - \varphi_n)$. Putting $t = 1$ in (4) yields

$$\mathscr{K}[v^*(x, 1) - \varphi_{n+1}(x)] < \mathscr{K}(\varphi - \varphi_n)$$

and letting $n \to \infty$ gives, finally,

$$\mathscr{K}[v^*(x, 1) - \varphi(x)] = 0.$$

It follows that $v^*(x, 1) = \varphi(x) = v^*(x, 0)$, that is, $v^*$ is periodic.

To complete the proof of the theorem it is enough to show that $\mathrm{Re}[v^*] \leqq 5.7$, for then by [*1*] $v^*$ is stable and unique[2]. But we have

$$\mathscr{K}(v^* - v) \to 0 \quad \text{as} \quad t \to \infty, \tag{5}$$

by virtue of Theorem 1 of [*1*]. Since both $v$ and $v^*$ are equicontinuous, (5) implies $(v^* - v) \to 0$ as $t \to \infty$, and therefore

$$\operatorname*{Max}_{x} v^*(x, t) \leqq V + o(t).$$

$v^*$ being periodic, this in turn proves $V^* \leqq V$ and $\mathrm{Re}[v^*] \leqq 5.7$.

**Note:** This research was supported in part by the United States Air Force Office of Scientific Research under Contract AF 49 (638)—262.

## References

[*1*] SERRIN, J.: On the stability of viscous fluid motions. Arch. Rational Mech. Anal. **3**, 1—13 (1959).
[*2*] LERAY, J.: Étude de diverses équations intégrales nonlinéares et de quelques problèmes que pose l'hydrodynamique. J. Math. Pures Appl. (9) **12**, 1—82 (1933).
[*3*] LERAY, J.: Essai sur les mouvements plan d'un liquide visqueux que limitent des parois. J. Math. Pures Appl. (9) **13**, 331—418 (1934).
[*4*] LERAY, J.: Sur le mouvement d'un liquide visqueux remplissant l'espace. Acta Math. **63**, 193—248 (1934).
[*5*] KISELEV, A. A., & O. A. LADYSHENSKAYA: On the existence and uniqueness of solutions of the initial value problem for viscous incompressible fluids. Izvestia Akad. Nauk SSSR. **21**, 655—670 (1957).

[2] That $v^*$ is stable follows from Theorem 1; that it is unique can be shown by the same argument used to prove Theorem 2.

University of Minnesota
Minneapolis, Minnesota

*(Received December 3, 1958)*

# V.2. The Navier–Stokes Equation – Special Topics

Thomas S. Lundgren

**Mathematics Subject Classification (2010).** 35Q35, 76A15, 76A02.
**Keywords.** Non-Newtonian fluids, Regularity of weak solutions.

### i. The Swirling Vortex

The swirling vortex paper is a mathematical model for tornadoes. Serrin considered the problem of a singular line vortex interacting with a flat plane (the ground) perpendicular to the vortex. While allowing the vortex to be singular he solved the Navier–Stokes equations with a no slip boundary condition at the ground by means of an exact similarity solution: the first complete exact solution of this kind. For a certain range of the parameters of the problem (the strength of the vortex and a constant of integration) the solution has a conical structure, with downflow near the vortex and inflow along the ground. These two streams merge and flow up and out along a conical surface. This gives a theoretical foundation for the observed conical distribution of debris picked up and flung outward by some tornadoes.

There is considerable literature which cites this work, some having to do with uniqueness and with the physical nature of the singularity. I have not found anything which uses the method of matched asymptotic expansions to match the singularity to a finite vortex structure, which seems like an obvious extension.

### iii. Poiseuille and Couette flow of non-Newtonian fluids

This paper is the first theoretical analysis of the observed swelling of jets of non-Newtonian fluids and the tendency of such fluids to climb on a rotating cylinder inserted into the fluid. Serrin studied these problems for a Reiner–Rivlin fluid using Rivlin's exact solutions for Poiseuille and Couette flows.

In its simplest physical aspects the rod climbing phenomenon can be described as stirring a container of liquid by inserting a rotating cylinder. If the liquid is water (a Newtonian fluid) a slight depression of the water surface near the cylinder is observed. However, for certain non-Newtonian fluids the fluid climbs up the cylinder, sometimes to quite spectacular heights. Serrin's seminal analysis was the first in a large number of both theoretical and experimental studies of non-Newtonian fluids with different constitutive equations, mostly polymers but also fluids like pancake batter.

Thomas S. Lundgren
Dept. of Aerospace Engineering and Mechanics
University of Minnesota

P. Pucci et al. (eds.), *James Serrin. Selected Papers: Volume 2*, Contemporary Mathematicians,
DOI 10.1007/978-3-0348-0847-7_11, © Springer Basel 2014

*Philos. Trans. R. Soc. Lond., Ser. A* 271 (1972), 325–360. **Reprint V.2.[i]**


# THE SWIRLING VORTEX

By J. SERRIN
*University of Minnesota*

(*Communicated by L. Howarth, F.R.S.—Received* 9 *December* 1969—*Revised* 27 *July* 1971)

[Plates 2 to 4]

## CONTENTS



We consider an infinite vortex line in a viscous fluid interacting with a plane boundary surface at right angles to the line. If the boundary surface were absent, the vortex would impart to the fluid a circular motion about the vortex line with speed inversely proportional to the distance to the line. The presence of the boundary surface, however, leads to a secondary flow due to the forced adherence of the fluid at the surface.

The purpose of the paper is to describe a family of exact solutions of the Navier–Stokes equations which applies to the above situation. Under quite general hypotheses, it is shown that there can exist only three types of motion compatible with the assumed structure. In the first kind, the radial velocity component (using spherical polar coordinates about the point where the vortex meets the plane surface) is directed inward along the plane surface and upward along the axis of the vortex. In the second type of motion the radial velocity component is directed inward along the plane surface and downward on the axis, with a compensating outflow at an intermediate angle. In the third kind the radial velocity is directed outward near the plane and downward on the central axis. The results can also be used as a basis for numerical calculations of the solutions in question, and several typical flow patterns have been explicitly computed in order to illustrate the theory.

The paper concludes with a discussion of the relation between the theoretical solutions and observed phenomena near the point of contact of tornadoes with the ground; this requires that the flows under discussion be considered as mean motions in a turbulent flow with constant eddy viscosity. The present work adds theoretical weight to the argument that central downdrafts can occur in tornadoes. Moreover, the model provides an explanation, other than centrifugal action, for the frequent appearance of a cascade effect at the foot of both tornadoes and water-spouts; finally it offers a unified point of view from which to consider the diversity of flow patterns observed when vortex fields interact with a boundary surface.

Vol. 271. A. 1214. (Price £1.20; U.S. $3.10) 27 [Published 27 January 1972]

## Introduction

Consider an infinite vortex line in a viscous fluid or gas, interacting with a plane boundary surface at right angles to the line. If the boundary surface were absent, the vortex would impart to the fluid a circular velocity about the vortex line of magnitude $C/r$, where $r$ is the radial distance to the line and the line is assumed for simplicity to be straight. On the other hand, the presence of the boundary surface, which we may suppose to be horizontal and represented by the equation $z = 0$ in rectangular coordinates, leads to a secondary flow due to the forced aherence of the fluid at the surface.

In a general way, under these circumstances we may expect that the loss of circulatory velocity near the plane surface, due to adherence, will produce an unbalanced pressure field near this surface, which falls as one moves towards the axis of the vortex. As a consequence the secondary flow should be directed toward the vortex, and this in turn should create an updraft along its axis. Such a picture is an intriguing one, especially if one is interested in explaining various phenomena of dynamic meteorology. An important and fundamental difficulty arises, however, in attempting to find specific mechanisms (i.e. particular solutions of the Navier–Stokes equations) which account for such swirling or spiral vortex motions in the space $z > 0$.

In this regard, the Russian engineer M. A. Goldstik discovered an interesting exact solution of the Navier–Stokes equations for viscous incompressible flow. His solution partially displays the features required but unfortunately is available only at values of the dimensionless parameter

$$k = C/2\nu$$

so low as to cast doubt on its applicability to many situations of practical significance (here $\nu$ denotes the kinematic viscosity). It turns out, however, that Goldstik's solution is just one of a broader class, and that in this broader class we can find swirling motions which exist at arbitrary values of $k$. Naturally, such vortex motions cannot be entirely transparent and simple to the eye (otherwise surely someone would have long since discovered their structure). Thus it might be expected that the initial picture of a simple secondary motion involving inflow and an updraft needs further modification, and indeed this is the case for the solutions we shall discuss here.

Specifically, we find that there exist three types of swirling vortex motion compatible with the structure assumed here (see § 1). In the first kind, the radial velocity component (in spherical coordinates) is directed inward near the plane $z = 0$ and upward near the axis of the vortex; thus in executing a motion of this kind the fluid near the boundary plane is drawn inward in tightening spirals toward the $z$-axis and is then swirled upwards in helical paths about the axis. This type of motion, similar to that determined by Goldstik and displaying features of the kind indicated in the opening discussion, can exist however only for values of $k$ less than 2.86. In the second kind of motion the velocity component is directed outward near $z = 0$ and downward near the vortex axis. This type of flow exists at arbitrary values of $k$, as does also the third kind in which the radial velocity is directed inward both near the plane $z = 0$ *and* near the vortex filament. For the latter motion there is of course a compensating outflow occurring at a suitable intermediate cone of directions. The downdraft in the second and third types of motion tends to be fairly strong as is the compensating outflow in the third case. Moreover, in the second type of motion, namely when there is a downdraft along the vortex line and an outflow near $z = 0$, the flow pattern develops a boundary layer as $C/2\nu$ becomes large. That is, except in the immediate neighbourhood of the boundary surface the angular velocity about the vortex line becomes nearly $C/r$, as required by the irrotational flow, while in the boundary layer itself the angular velocity rapidly changes from zero at the boundary surface to the main streaming value $C/r$ required by the vortex.

Clearly the variety and complexity of these flows may be of value in explaining the diverse behaviour of naturally occurring vortex phenomena. At the same time, irrespective of such applications, the motions studied here are also interesting for the fact that they are among the more complicated examples in which a boundary value problem for the Navier–Stokes equations can be reduced to the solution of ordinary differential equations (and hence for which a relatively explicit determination of the flow pattern is possible). Considering the difficulty always experienced in finding significant exact solutions of the Navier–Stokes equations, there seems to be sufficient reason on this ground alone to justify the present analysis.

This much being said, I feel compelled to add that the solutions discussed here are not intended as a definitive and final answer to the problem of interaction of a vortex filament with a plane boundary. There is certainly no reason to suppose that some other (possibly non-steady) type of flow might not equally well or better describe this interaction; moreover, the singularity existing along the axis of the line vortex is itself a mathematical idealization of the phenomenon of vortex motion in real fluids. Finally if one is concerned with meteorological phenomena it must surely be decided at some stage whether the Navier–Stokes equations themselves are suitable for the explanation of these phenomena, or alternately whether thermal and compressibility effects play a significant part in the actual motion (besides supplying the ultimate driving energy). On the available evidence, however, there seems to be no conviction that naturally occurring vortices must necessarily and in all cases be fundamentally compressible phenomena. Indeed, the time scales at which tornado phenomena occur do not indicate the action of either buoyancy or compressibility as an important determinant of the local structure of the air motion; and even in cases where compressibility, buoyancy, and other effects are surely present, it is not improbable that further understanding can be gained from an incompressible analogy. Consequently, at the present stage of investigation it remains a reasonable assumption to adopt the Navier–Stokes equations as an appropriate model for fluid motion in a vortex.

Before proceeding to the main considerations, it should perhaps be added that when one is dealing with a solution of the Navier–Stokes equations the actual energy source is irrelevant to the *structure* of the solution: the solution exists as a dynamically possible fluid motion and nothing more is required of it. In the real world the solution expresses a local situation, and the energy to produce and drive it comes from outside. The mode by which this energy is imparted to the flow is of course ultimately of interest; when we deal as here with a solution which idealizes a swirling motion by means of a line vortex, the vortex singularity can serve as a source of momentum, this being entirely consistent with the interpretation of the solution as an asymptotic limit (as the core radii shrinks to zero) of a family of flows not containing a singularity. Morton (1966, p. 186) gives an excellent account both of the relation of the concentrated vortex core to the main motion and of the driving mechanisms likely to be involved in typical tornado and waterspout phenomena.

The paper is divided into four parts. The first chapter is concerned with the basic structure of the vortex and with the derivation of an appropriate system of differential equations governing the motion. The system itself consists of a coupled pair of nonlinear ordinary differential equations, one of the fourth order and the other of second order, subject to five boundary conditions. The value $k = C/2\nu$ appears as a parameter multiplying the highest order derivatives; thus, taking into account that there is one fewer boundary condition than the order of the system allows, it is apparent that we must ultimately deal with a two-parameter family of solutions. Using a technique introduced by Goldstik and in fact appearing in a special form even earlier in the work of Slezkin, the basic equations can be reduced to a first-order integro-differential equation of essentially more manageable proportions. This reduction is carried out in § 2. The second chapter is then concerned with the structure of the solutions of the problem; in particular, it is shown there that solutions have the general form described above.

Chapter III is a general study of the existence and non-existence of motions corresponding to different values of $k$. Because of the inherently complicated nature of the basic equations this work involves some difficulties. We have therefore made every effort to proceed as simply as possible, while at the same time aiming at completeness. The final section of chapter III contains the results of numerical calculations carried out at typical values of $k$.

With the exception of the work of Goldstik, most theoretical models of the interaction of a vortex and a plane surface have not included a singularity at the vortex axis. In the work of Burgers (1948) and Sullivan (1959), maintenance of a continuous velocity field throughout the vortex core is, however, accomplished at the cost of dropping the adherence condition at the boundary surface; moreover, in this work the velocity field does not approach zero as one proceeds radially away from the vortex (in fact, the speed becomes arbitrarily large), a drawback which remains even following boundary-layer analysis of the transition from slip flow to adherence at the boundary plane. Unsteady vortices corresponding to those of Burgers and Sullivan have been studied in interesting papers of Rott (1958), and Bellamy-Knights (1970). In work more closely allied to the present approach, Long (1958, 1961) has carried out a similarity analysis analogous to that given in § 1, and has considered boundary conditions which avoid the singularity at the axis, though again at the expense of dropping the adherence condition at the boundary surface.

It would be quite removed from our purposes to review the literature on boundary layers induced by vortex motion. We do however draw attention to a recent paper of Burggraf, Stewartson & Belcher (1971) in which they present a boundary layer analysis of a potential vortex interacting with a finite disk. In their work it is shown that a finite mass flux occurs at $r = 0$, and they comment that this should cause an eruption at the vortex axis. It is precisely here that a further alteration of their flow field would be necessary; our work bears this out in that we do not find steady motions which exhibit inflow-type boundary layers, but obtain instead the rather more complicated flow patterns described earlier. Further valuable work on the boundary layer induced by a maintained vortex is due to Kuo (1971).

We emphasize that our model obeys both the strict adherence condition at the boundary surface and the requirement that the velocity approach zero as one proceeds radially away from the vortex. On the other hand, because of the singularity at the axis, the physical interpretation of the results must be approached with care. The final part of the paper summarizes our results and contains a discussion of their possible bearing on meteorological phenomena. In particular, the present model adds theoretical weight to the argument (advanced by Oerstad in 1838 and periodically recurring in the literature) that central downdrafts can occur in tornadoes. Moreover, the model provides an explanation, other than centrifugal action, for the frequent appearance of a cascade effect at the foot of both tornadoes and waterspouts; finally it offers a unified point of view from which to consider the diversity of flow patterns observed when vortex fields interact with a boundary surface.

## Chapter I. Formulation of the problem

### 1. *Basic equations*

It is convenient to introduce spherical polar coordinates $(R, \alpha, \theta)$, where $R$ is radial distance from the origin, $\alpha$ is the angle between the radius vector and the positive $z$-axis, and $\theta$ is the meridian angle about the $z$-axis. The positive $z$-axis is then described by $\alpha = 0$, the boundary plane $z = 0$ by $\alpha = \frac{1}{2}\pi$, and the half space $z > 0$ by $R > 0$, $0 \leqslant \alpha < \frac{1}{2}\pi$. The respective physical components of the velocity vector $\boldsymbol{v}$ in this coordinate system will be denoted by

$$v_R, \quad v_\alpha, \quad v_\theta. \tag{1}$$

We shall consider here steady-state fluid motions having the basic structure

$$v_R = \frac{G(x)}{r}, \quad v_\alpha = \frac{F(x)}{r}, \quad v_\theta = \frac{\Omega(x)}{r} \tag{2}$$

where $r = R\sin\alpha$ is the distance to the $z$-axis, and $x = \cos\alpha$. One finds immediately from the equation of continuity that

$$G = F'\sin\alpha,$$

the prime denoting differention with respect to $x$. The functions $F$ and $\Omega$ are of course to be determined so that the flow is dynamically allowable (that is, satisfies the Navier–Stokes equations).

We remark that the particular functions $F \equiv 0$, $\Omega \equiv C$ yield a *line vortex* in space. Our purpose is to find related functions $F$ and $\Omega$ appropriate to a swirling vortex motion adhering to the plane $z = 0$, as described in the introduction.

The first necessity is to derive appropriate differential conditions under which the basic motion (2) satisfies the Navier–Stokes equations. These equations, expressed in spherical polar coordinates, are reproduced in several standard fluid mechanics texts, cf., for example, Berker (1963, p. 6). Using the relation

$$\frac{\partial}{\partial\alpha} = -\sin\alpha\frac{\partial}{\partial x},$$

we find after some calculation that the $R$, $\alpha$, and $\theta$ components of the Navier–Stokes equations become respectively

$$-FF'' - F'^2 - (F^2+\Omega^2)\operatorname{cosec}^2\alpha = -\frac{R^3}{\rho}\frac{\partial p}{\partial R} + \nu(F'''\sin^2\alpha - 2F''\cos\alpha),$$

$$-FF' - (F^2+\Omega^2)\cot\alpha\operatorname{cosec}\alpha = -\frac{R^2\sin\alpha}{\rho}\frac{\partial p}{\partial\alpha} - \nu F''\sin^2\alpha,$$

$$-F\Omega' = -\frac{R^2}{\rho}\frac{\partial p}{\partial\theta} + \nu\Omega''\sin^2\alpha,$$

where $\rho$ is the density, $p$ the pressure, and $\nu$ the kinematic viscosity, and where external forces have been absorbed into the pressure term.

From the last of these equations it is evident that $p$ must be independent of $\theta$ (more precisely, $p$ is at most linear in $\theta$: since $p$ must be periodic in $\theta$, it is thus independent of $\theta$). Now using the first equation, it follows that

$$\frac{p}{\rho} = \frac{A(x)}{R^2} + B(x).$$

Substituting this into the second equation shows that $B(x)$ must be a constant; hence writing $A(x) = \pi(x) \operatorname{cosec}^2 \alpha$ we obtain

$$\frac{p}{\rho} = \frac{\pi(x)}{R^2 \sin^2 \alpha} + \text{constant} = \frac{\pi(x)}{r^2} + \text{constant}. \tag{3}$$

Comparing this with the first equation then gives the pressure formula

$$-2\pi = F^2 + \Omega^2 + \{FF'' + F'^2 + \nu(F''' \sin^2 \alpha - 2F'' \cos \alpha)\} \sin^2 \alpha. \tag{4}$$

With the help of (4) we may eliminate $p$ from the second equation. After some calculation this yields† the following ordinary differential system for the functions $F$ and $\Omega$,

$$\begin{aligned} \nu(1-x^2)\, F^{\mathrm{iv}} - 4\nu x F''' + FF''' + 3F'F'' &= -2\Omega\Omega'/(1-x^2) \\ \nu(1-x^2)\, \Omega'' + F\Omega' &= 0. \end{aligned} \tag{5}$$

Solutions of the system (5) provide a broad class of motions satisfying the Navier–Stokes equations. Our interest here will be in the particular boundary value problem obtained when the $\theta$-component of the velocity approaches the vortex value $C/r$ as $\alpha$ tends to zero and the adherence condition $\mathbf{v} = 0$ is satisfied when $\alpha = \frac{1}{2}\pi$. Thus we shall consider (5) on the interval $0 \leqslant x < 1$ with boundary conditions

$$\left.\begin{aligned} \Omega = F = F' = 0 \quad &\text{when} \quad x = 0 \\ \Omega \to C \quad &\text{as} \quad x \to 1 \end{aligned}\right\}. \tag{6}$$

To these conditions must be added a restriction assuring that the vortex line $\alpha = 0$ is neither a source nor a sink of the fluid motion, namely

$$F \to 0 \quad \text{as} \quad x \to 1. \tag{7}$$

In order to verify the appropriateness of condition (7), it is convenient to introduce the function $\psi = RF(x)$. Let $\mathscr{S}$ be a surface of revolution about the $z$-axis generated by a meridian curve $\mathscr{C}$. Then using the fact that

$$v_R = -\frac{1}{R^2 \sin \alpha} \frac{\partial \psi}{\partial \alpha}, \quad v_\alpha = \frac{1}{R \sin \alpha} \frac{\partial \psi}{\partial R},$$

it is easy to show that the total flux of fluid through $\mathscr{S}$ is $2\pi|\psi_2 - \psi_1|$, where $\psi_1$ and $\psi_2$ denote the respective values of $\psi$ at the end-points of the curve $\mathscr{C}$. If we now suppose that $\mathscr{C}$ joins two distinct points on the $z$-axis, the condition that this axis be neither a source nor a sink of the fluid motion is obviously that the flux through the corresponding closed surface $\mathscr{S}$ be zero. It follows that in this case $\psi_1 = \psi_2$, which in turn holds if and only if $F \to 0$ as $x \to 1$. This proves (7).

For later use, we note that the surfaces $\psi$ = constant contain the streamlines of the fluid motion. Thus these surfaces provide a useful method of flow visualization, in the same way as the streamfunction of an ordinary two-dimensional fluid motion.

*Remarks.* In § 4 we shall show that the reduced pressure $\pi = r^2 p/\rho$ approaches the value $-\frac{1}{2}C^2$ when $\alpha$ tends to zero, in analogy with the case of a line vortex.

The basic separation of variables assumption can be clarified by a simple dimensional analysis. In the interaction problem as posed in the opening paragraphs of the introduction, neither a

† The derivation is facilitated by first writing $\pi$ in the form

$$-2\pi = F^2 + \Omega^2 + \nu(1-x^2)\, S$$

where $S$ denotes the expression enclosed by braces in (4). Equations (5) can also be derived from the last two relations on page 115 of Goldstein's *Modern developments in fluid mechanics*, vol. I (this process does not, however, yield formula (4) for the pressure).

preferred velocity nor a preferred distance is present: we have only the angular momentum $C$ and the kinematic viscosity $\nu$ as basic physical constants, both having the dimensions $L^2/T$. Their ratio is therefore a dimensionless parameter $k$ of the flow. This being the case, if we consider solutions which are independent of time and which are rotationally symmetric (i.e. the velocity components are independent of $\theta$), then necessarily the dimensionless quantity $rv/C$ can depend only on the dimensionless variables $\alpha$ and $k$, unless we are prepared to consider solutions with some preferred distance added in. Since there is no *a priori* reason for including such a distance, it is therefore reasonable to consider solutions of the form (2).

Long (1958) introduced a system equivalent to (5) and gave an asymptotic analysis of the solutions under a set of boundary conditions different from those here. The system (5) also appears in Goldstik's work, though in somewhat disguised form (we note finally that the special case $\Omega \equiv 0$ of (5) was discovered by Slezkin; cf. Berker (1963, p. 59)).

### 2. *Reduction to an integro-differential equation*

In the preceding section we derived for the swirling vortex motion a sixth-order system of ordinary differential equations subject to five boundary conditions. It is therefore apparent that, even when $\nu$ and $C$ are fixed, we must deal with a one-parameter family of fluid motions. This extra degree of freedom is vital to our further considerations.

The system (5) can be considerably simplified, since the first equation can be immediately integrated three times to give the relation

$$2\nu(1-x^2)F'+4\nu xF+F^2 = -\int_0^x dx\int_0^x dx\int_0^x \frac{4\Omega\Omega'}{1-x^2}dx+Px^2+Qx+S, \tag{8}$$

where $P$, $Q$, $S$ are constants of integration.† Setting $x = 0$ and using the boundary conditions (6) yields $S = 0$.

In order to derive a further relation between the constants $P$ and $Q$, note that the triple integral in (8) can be rewritten as

$$2\int_0^x \frac{(x-t)^2}{1-t^2}\Omega\Omega'\,dt;$$

this in turn can be integrated by parts to the form

$$2\int_0^x \frac{(x-t)(1-xt)}{(1-t^2)^2}\Omega^2\,dt$$

by making use of the boundary condition $\Omega(0) = 0$. The function $\Omega$ is bounded on the interval $[0, 1]$. Hence as $x$ tends to 1 the above integral converges to the value

$$2\int_0^1 \frac{\Omega^2\,dt}{(1+t)^2}.$$

Consequently using (7) and (8) we find that

$$(1-x)F' \to \text{constant} \quad \text{as} \quad x \to 1. \tag{9}$$

If the constant is non-zero, say positive, then for $x$ suitably near 1 we have $F' \geqslant \epsilon/(1-x)$. This implies in turn that $F$ becomes infinite as $x$ tends to 1, in contradiction to (7). Since the same argument applies if the constant is assumed to be negative, it follows that $(1-x)F' \to 0$ as $x \to 1$. Consequently from (8) we have

$$P+Q = 2\int_0^1 \frac{\Omega^2\,dt}{(1+t)^2}. \tag{10}$$

† For the special case $\Omega \equiv 0$ this integration was found by Slezkin.

One of the constants $P$ and $Q$ can be assigned arbitrarily, according to the remarks at the beginning of the section. Choosing $P$ as the basic parameter, and using (10) to eliminate $Q$ from (8), then gives the relation

$$2\nu(1-x^2)\,F' + 4\nu x F + F^2 = G(x), \tag{11}$$

where

$$G(x) = 2(1-x)^2 \int_0^x \frac{t\Omega^2}{(1-t^2)^2}\,\mathrm{d}t + 2x \int_x^1 \frac{\Omega^2\,\mathrm{d}t}{(1+t)^2} - P(x-x^2). \tag{12}$$

A final change of variables

$$F = 2\nu(1-x^2)f, \quad k = 1/2\nu \tag{13}$$

yields the integro-differential system

$$\left.\begin{aligned} f' + f^2 &= k^2 \frac{G(x)}{(1-x^2)^2} \\ \Omega'' + 2f\Omega' &= 0 \end{aligned}\right\} \quad (0 \leqslant x < 1), \tag{14}$$

where $G(x)$ is given by (12), and $f$ and $\Omega$ are subject to the boundary conditions

$$f = \Omega = 0 \quad \text{when} \quad x = 0, \qquad \Omega \to C \quad \text{as} \quad x \to 1. \tag{15}$$

In view of the change of variables $\Omega \to C\tilde{\Omega}$, $P \to C^2\tilde{P}$, $k \to \tilde{k}/|C|$, we can suppose without loss of generality that

$$C = 1. \tag{16}$$

*This normalization will be maintained throughout the rest of the paper.* Note that $k$ then becomes the basic dimensionless parameter of the flow (i.e. the normalized value $\tilde{k}$ is $|C|/2\nu$). We note that in laminar flow $k$ may be considered as a *Reynolds number*; if turbulent motion is assumed and $\nu$ correspondingly denotes the kinematic eddy viscosity, then it is more proper to view $k$ as a self-regulating parameter indicating the basic level of turbulence present. Typical values of $k$ are given in the final section of the paper.

So far, it has been shown that (14), (15) is a consequence of (5), (6), (7). In § 4 it will be proved conversely that any solution of (14), (15) is also a solution of (5), (6), (7), so that the two systems are in fact equivalent.

For the special case $P = 1$ the system (14) was derived by Goldstik, by a much more complicated and circuitous procedure.

## Chapter II. Structure of solutions

In this chapter we consider the general qualitative behaviour of solutions of the system (14), (15). The fundamental question of existence of solutions will be taken up in the following chapter.

### 3. *The function* $G(x)$

As a first step in discussing the structure of solutions, it is convenient to consider the behaviour of the function $G(x)$.

Lemma 1. *The angular momentum function* $\Omega(x)$ *increases monotonically from* 0 *to* 1 *as* $x$ *increases from* 0 *to* 1.

*Proof.* From the second equation of (14) we find

$$\Omega' = \text{const.} \exp\left(-2\int_0^x f \,\mathrm{d}x\right). \tag{17}$$

Thus $\Omega'$ always has the same sign. The required conclusion therefore follows immediately from the given boundary conditions. For later use we note that (17) can be integrated once more to yield an explicit formula for $\Omega$ in terms of $f$.

Lemma 2. $G(x)$ *satisfies the conditions*

$$G = 0, \quad G' = 2\int_0^1 \frac{\Omega^2 \mathrm{d}t}{(1+t)^2} - P \quad \textit{when} \quad x = 0;$$

$$G = 0, \quad G' = P - 1 \quad \textit{when} \quad x = 1.$$

*Moreover,* $G'''(x) < 0$ *for* $0 < x < 1$.

*Proof.* We have

$$G(x) = 2(1-x)^2\int_0^x \frac{t\Omega^2}{(1-t^2)^2}\mathrm{d}t + 2x\int_x^1 \frac{\Omega^2 \mathrm{d}t}{(1+t)^2} - P(x-x^2),$$

$$G'(x) = -4(1-x)\int_0^x \frac{t\Omega^2}{(1-t^2)^2}\mathrm{d}t + 2\int_x^1 \frac{\Omega^2 \mathrm{d}t}{(1+t)^2} - P + 2Px.$$

The values of $G$ and $G'$ at $x = 0$ are apparent by inspection. Their values at $x = 1$ follow by application of L'Hôpital's rule for the indeterminate form $0.\infty$. We have also

$$G''(x) = 4\int_0^x \frac{t\Omega^2}{(1-t^2)^2}\mathrm{d}t - \frac{2\Omega^2}{1-x^2} + 2P,$$

$$G'''(x) = -\frac{4\Omega\Omega'}{1-x^2},$$

and the remaining part of the lemma follows at once.

In what follows it is convenient to put $G'(0) = Q$, in agreement with the relation (10). Then the following lemma holds.

Lemma 3. *Let* $Q = G'(0)$. *Then for* $0 < x < 1$ *we have*

$$Q(x-x^2) < G(x) < (1-P)(x-x^2).$$

*Proof.* The function $H(x) = G(x) - Q(x-x^2)$ satisfies $H(0) = H(1) = 0, H'(0) = 0$, and $H''' < 0$. From the last condition it follows that $H''$ has at most a single zero. Hence $H$ is either convex, or first convex and then concave, or everywhere concave in the interval $0 < x < 1$.

The boundary conditions show that $H$ cannot be everywhere convex or concave. In the remaining case (first convex and then concave) the boundary conditions imply without difficulty that $H > 0$. This proves the left hand inequality.

To prove the right hand inequality, consider the function $\hat{H}(x) = G(x) - (1-P)(x-x^2)$. Here $\hat{H}(0) = \hat{H}(1) = \hat{H}'(1) = 0$, and $\hat{H}''' < 0$. The dual of the previous demonstration now yields $\hat{H} < 0$, completing the proof.

As a consequence of lemma 3 it follows that, for $0 < x < 1$,

$$G(x) > 0 \quad \text{when} \quad Q \geqslant 0, \qquad \text{and} \qquad G(x) < 0 \quad \text{when} \quad P \geqslant 1.$$

Moreover, in the remaining case when $Q < 0$ and $P < 1$, the function $G$ is first negative, has a single zero on the interval $0 < x < 1$, and is positive thereafter. In fact, $G''' < 0$ so that $G''$ has at most one zero. Consequently $G$ is either convex, or concave, or first convex and then concave. Evidently only the last case is compatible with the boundary conditions $G(0) = G(1) = 0$ and $G'(0), G'(1) < 0$. But then $G$ must be first negative and then positive.

## 4. *Behaviour of solutions*

In this section we investigate the behaviour of solutions of the system (14), (15). The purpose of the first theorem is to show that this system is equivalent to the original problem posed in § 1. We begin with three important lemmas.

LEMMA 4. *Suppose $P < 1$. Then* $\quad f \leqslant \frac{1}{4}(1-P)\, k^2 \ln \dfrac{1}{1-x}.$

*Furthermore, if $Q \geqslant 0$ then $f$ is positive, while if $Q < 0$ then $f$ is either everywhere negative, or else first negative and then positive thereafter.*†

*Proof.* By lemma 3 we have $G(x) \leqslant (1-P)(x-x^2)$. Thus using (14) and the fact that

$$x \leqslant \tfrac{1}{4}(1+x)^2,$$

we find

$$f' \leqslant \tfrac{1}{4}(1-P)\, k^2/(1-x).$$

Integrating this and making use of the initial condition $f(0) = 0$ proves the first part of the lemma.

Multiplying the first equation of (14) by the integrating factor $\exp\left(\int_0^x f\,dx\right)$ and carrying out the resulting integration yields the relation

$$f = k^2 \int_0^x \frac{G(t)}{(1-t^2)^2} \exp\left(-\int_t^x f\,du\right) dt,$$

where we have used the initial condition $f(0) = 0$. Now if $Q \geqslant 0$, then $G > 0$. Hence $f$ is positive. When $Q < 0$ and $P < 1$ we have seen that $G$ is first negative and then positive. Therefore $f$ is negative for small $x$. At the first zero of $f$ (if any) we surely have $G > 0$. But then $G > 0$ for the remainder of the interval, and $f > 0$ there exactly as before.

LEMMA 5. *If $P \geqslant 1$ then $f$ is negative and decreasing.*

† It is convenient to exclude the points $x = 0, 1$ when referring to the sign of $f$. *This convention, and a similar one for G and $\Omega$, will be adhered to throughout the paper.*

*Proof.* Recalling that $G < 0$ in the present case, we have

$$f' = k^2 \frac{G(x)}{(1-x^2)^2} - f^2 < 0, \tag{18}$$

and the required conclusion follows at once using the fact that $f(0) = 0$.

LEMMA 6. *Let $f$ be a solution of* (14), (15). *Then*

$$f = O\left(\ln \frac{1}{1-x}\right) \quad \text{as} \quad x \to 1.$$

This result is slightly technical, and accordingly its proof will be deferred until the end of the chapter. We may now turn to the main results of the section.

THEOREM 1. *Let $f$, $\Omega$ be a solution of the system* (14), (15). *Then $F$, $\Omega$ is a solution of* (5) *satisfying the boundary conditions* (6), (7). *Moreover the reduced pressure $\pi(x)$ satisfies*

$$\pi(0) = -\tfrac{1}{2}P, \quad \pi(1) = -\tfrac{1}{2}. \tag{19}$$

Here we have assumed the normalization $C = 1$; otherwise the relations (19) would be replaced by $\pi(0) = -\frac{1}{2}PC^2$ and $\pi(1) = -\frac{1}{2}C^2$.

*Proof of Theorem* 1. In view of the derivation of (14), the result is obvious except possibly for the boundary conditions $F'(0) = F(1) = 0$ and the pressure relations (19). We have by (13), however,

$$F' = 2\nu(1-x^2)f' - 4\nu x f.$$

Since $f(0) = G(0) = 0$ it follows from (14) that $f'(0) = 0$. Consequently $F'(0) = 0$. Also by lemma 6

$$F = 2\nu(1-x^2)f \to 0 \quad \text{as} \quad x \to 1.$$

To prove (19), differentiate (11) twice and use the resulting relation to eliminate the quantity $F'''$ from (4). This yields the formula

$$-2\pi = F^2 + \Omega^2 + \tfrac{1}{2}(1-x^2)(G'' - 4\nu F').$$

A further simplification is available if we use the identity $\frac{1}{2}(1-x^2)G'' + xG' - G = P - \Omega^2$; thus finally

$$-2\pi = P + 2F^2 + (4\nu F - G')x.$$

The required result now follows from lemma 2 and the boundary conditions for $F$.

The first condition of (19) provides a physical interpretation for the parameter $P$. Indeed, using (3) we find that

$$\frac{p}{\rho} = \begin{cases} -\dfrac{P}{2r^2} + \text{constant} & \text{when} \quad \alpha \sim 0 \quad (z\text{-axis}), \\[2ex] -\dfrac{1}{2r^2} + \text{constant} & \text{when} \quad \alpha = \tfrac{1}{2}\pi \quad (z = 0). \end{cases}$$

or, without the normalization $C = 1$,

$$\frac{p}{\rho} = \begin{cases} -\dfrac{PC^2}{2r^2} + \text{constant} & \text{when} \quad \alpha \sim 0 \\[2ex] -\dfrac{C^2}{2r^2} + \text{constant} & \text{when} \quad \alpha = \tfrac{1}{2}\pi. \end{cases}$$

Thus at the boundary plane the pressure $p$ is (up to an additive constant) exactly $P$ times the pressure in the corresponding free vortex. In particular if $P > 0$ then the pressure at the boundary

28-2

plane decreases as one approaches the vortex line (as noted heuristically in the introduction), though the rate of decrease will be the same as in a free vortex only when $P = 1$.

Theorem 1 shows that the problem (5), (6), (7) is equivalent to the problem (14), (15). Therefore from here on we may confine our discussion to the latter system. The following theorem is a more precise version of the results of lemmas 1, 4, 5 and 6.

THEOREM 2. *Let $f$, $\Omega$ be a solution of the system* (14), (15). *Then the following results hold.*

(i) *Suppose $P < 1$. If $G'(0) \geqslant 0$ then $f$ is positive, and $\Omega$ is increasing and concave. If $G'(0) < 0$ then $f$ is first negative, has a single zero on the interval $0 < x < 1$, and is positive thereafter; correspondingly $\Omega$ is first convex and then concave.*

(ii) *Suppose $P \geqslant 1$. Then $f$ is negative and decreasing, and $\Omega$ is increasing and convex.*

(iii) *When $P = 1$ the function $f$ tends to a finite negative limit as $x$ tends to 1. In all other cases*

$$f \sim \tfrac{1}{4}(1-P)\,k^2 \ln\frac{1}{1-x}.$$

*Proof.* The first statement of part (i) is simply a rewording of part of lemma 4, together with the observation that $\Omega'' = -2f\Omega' < 0$. To prove the second part of (i), note that from lemma 4 either $f$ is everywhere negative or else first negative and then positive thereafter. By part (iii), which we shall prove below, it is clear that $f > 0$ for $x$ sufficiently near 1. Consequently the only case which can occur is that $f$ is first negative, and then positive thereafter.

Part (ii) follows from lemma 5 and the fact that $\Omega'' = -2f\Omega' > 0$.

It remains only to prove part (iii). Suppose first that $P \neq 1$. Then by lemma 2 and lemma 6

$$(1-x)f' = k^2\frac{G(x)}{(1-x)\,(1+x)^2} - (1-x)f^2 \to \tfrac{1}{4}(1-P)\,k^2,$$

as $x$ tends to 1. Consequently $f$ approaches either $+\infty$ or $-\infty$ and L'Hôpital's rule is applicable: thus

$$\lim\frac{f}{\ln 1/(1-x)} = \lim\,(1-x)f' = \tfrac{1}{4}(1-P)\,k^2,$$

as required.

Suppose next that $P = 1$. We assert that the quantity $G/(1-x)^2$ is integrable. Indeed, since $f = O(\ln 1/(1-x))$ it is clear from (17) that $\Omega'$ has a finite positive limit $B$ as $x$ tends to 1. Thus by the last formula in the proof of lemma 2

$$\lim\,(1-x)\,G''' = -2B.$$

By repeated application of L'Hôpital's rule, therefore,

$$\lim\frac{G}{(1-x)^2\ln\dfrac{1}{1-x}} = \lim\frac{G'}{(1-x)\left(1-2\ln\dfrac{1}{1-x}\right)} = \lim\frac{G''}{2\ln\dfrac{1}{1-x}-3} = \tfrac{1}{2}\lim\,(1-x)\,G''' = -B.$$

Consequently when $P = 1$ we have $G \sim -B(1-x)^2\ln 1/(1-x)$ and the assertion is proved.

Having shown this, it follows with the help of lemma 6 and (18) that $f'$ is negative and integrable. Hence $f$ tends to a finite limit as $x$ tends to 1. This completes the proof of the theorem.

The results of theorem 2 can be used to give a qualitative picture of the flow pattern corresponding to various parameter values. For simplicity we shall confine ourselves to the radial component of the velocity, though the other components can be treated equally well. Now by (2), (13) and (14) have

$$Rv_R = F' = 2\nu(1-x^2)f' - 4\nu xf = k\frac{G(x)}{1-x^2} - \frac{(1-x^2)f^2}{k} - \frac{2xf}{k}.$$

For small values of $x$ the first term clearly dominates, and the sign of $Rv_R$ will correspondingly agree with that of $G$. Moreover, as $x$ tends to 1 we have with the help of lemma 2

$$Rv_R \sim -\tfrac{1}{2}(1-P)\,k\ln\frac{1}{1-x}$$

when $P \neq 1$, while $F'$ tends to a *positive* limit if $P = 1$. In case $P < 1$ and $Q < 0$, the function $f$ has a zero at some point $x = a$. Thus $G(a) > 0$ according to the proof of lemma 4. Therefore at $x = a$

$$Rv_R = k\frac{G(a)}{1-a^2} > 0.$$

Summarizing, we have the following results.

*Let $f$ be a solution of* (14), (15). *Then $f$ is either everywhere positive, everywhere negative, or first negative and then positive. Moreover:*

A. *For positive solutions $f$, the radial velocity is outward near the plane $z = 0$ and downward near the $z$-axis.*

B. *For solutions $f$ which are first negative and then positive, the radial velocity is inward near the plane $z = 0$ and downward near the $z$-axis. This general motion toward the origin is balanced by a compensating outflow near the streamcone $\alpha_0 = \cos^{-1} a$.*

C. *For negative solutions $f$, the radial velocity is inward near the plane $z = 0$ and upward near the $z$-axis (we shall see later that such solutions can exist only at small values of the parameter $k$).*

Typical stream surfaces for these cases are illustrated in § 10. Using theorem 2 we can also obtain some idea of the relative magnitude of the secondary flow in comparison with the free vortex. Writing

$$\boldsymbol{v} = \boldsymbol{V} + \boldsymbol{w},$$

where $\boldsymbol{V}$ has components $(0, 0, 1/r)$, it is easy to check that the quantities $rw_R, rw_\alpha, rw_\theta$ are independent of $R$ and tend uniformly to zero as $\alpha$ tends to zero, indicating clearly that the major effect near the vortex line is precisely the free vortex motion. For positive values of $\alpha$, both $\boldsymbol{v}$ and $\boldsymbol{V}$ tend uniformly to zero as $R$ tends to infinity. Finally, while the energy of the free vortex is infinite in any hemisphere $R \leqslant R_0$, $z \geqslant 0$, it can be shown that the energy of the secondary flow is finite and strictly proportional to $R_0$.

## 5. *Boundary-layer behaviour of positive solutions*

When $f$ is positive, solutions of (14), (15) exhibit a boundary-layer behaviour, in which as $k$ becomes large the function $\Omega(x)$ uniformly approaches 1 on compact subsets of (0, 1]. That is, for suitably large $k$ the velocity component $v_\theta$ is arbitrarily near that of a free vortex, except for a thin boundary layer near the plane $z = 0$. The purpose of this section is to prove this and other related results.

Let $Q = G'(0)$. By lemma 3

$$G(x) \geqslant Q(x-x^2). \tag{20}$$

Here $Q \geqslant 0$ since we are supposing that $f$ is positive (we shall in fact consider only *fixed positive* values of $Q$; the case $Q = 0$ requires a more delicate analysis, which is omitted here). Now using (14) and (20) it is clear that

$$f' \geqslant Qk^2\frac{x}{(1-x)(1+x)^2} - f^2.$$

Setting $c = Qk^2$ for simplicity in printing, and noting that $(1-x)(1+x)^2 < \frac{4}{3}$ for all $x$ in question, we have then

$$f' \geqslant \tfrac{3}{4}cx - f^2. \tag{21}$$

Thus $f \geqslant \sigma$, where $\sigma$ is defined by the differential equation

$$\sigma' = \tfrac{3}{4}cx - \sigma^2, \quad \sigma(0) = 0.$$

[The argument is as follows: put $g = f - \sigma$. Then

$$g' \geqslant \sigma^2 - f^2 = -(\sigma + f)\, g,$$

that is $g' + (\sigma+f)\, g \geqslant 0$. Multiplying by an integrating factor and carrying out the integration then yields (since $g(0) = 0$)

$$g \exp\left(\int_0^x (\sigma + f)\, \mathrm{d}x\right) \geqslant 0$$

as required. A similar comparison method will be used repeatedly in what follows, generally without specific reference.]

The function $\sigma$ is obviously positive. Moreover, $\sigma' \leqslant \frac{3}{4}cx$, so that $\sigma \leqslant \frac{3}{8}cx^2$. Therefore in turn

$$\sigma' \geqslant \tfrac{3}{4}cx - \tfrac{9}{64}c^2x^4,$$

For $0 \leqslant x \leqslant c^{-\frac{1}{3}}$ this yields $\sigma' \geqslant \frac{1}{2}cx$ and $\sigma \geqslant \frac{1}{4}cx^2$. It is apparent that $\sigma$ is an increasing function for $0 < x < 1$.† Thus for $x \geqslant c^{-\frac{1}{3}}$ there results

$$f \geqslant \sigma \geqslant \tfrac{1}{4}c^{\frac{1}{3}}. \tag{22}$$

Let $\delta = c^{-\frac{1}{3}} = (Qk^2)^{-\frac{1}{3}}$. By (17)

$$\Omega'(x) = \Omega'(\delta) \exp\left(-2\int_\delta^x f\,\mathrm{d}x\right).$$

Since $\Omega$ is concave in the present case (see theorem 2) and $0 \leqslant \Omega \leqslant 1$, it is clear geometrically that $\Omega'(\delta) \leqslant 1/\delta$. Hence, making use of (22), we find that for $x \geqslant \delta$

$$\Omega'(x) \leqslant \frac{1}{\delta}\exp\left(\frac{\delta - x}{2\delta}\right) \leqslant \frac{2}{\delta}\exp\left(-\frac{x}{2\delta}\right).$$

Integrating backward from $x = 1$ (where $\Omega = 1$) then gives

$$|\Omega - 1| \leqslant 4\,\mathrm{e}^{-x/2\delta} \quad (x \geqslant \delta).$$

In particular, if $x \geqslant N\delta$ and $N \geqslant 1$, then

$$|\Omega - 1| \leqslant 4\,\mathrm{e}^{-\frac{1}{2}N}. \tag{23}$$

*This establishes the existence of a boundary layer of nominal thickness $\delta \approx k^{-\frac{2}{3}} \approx \nu^{\frac{2}{3}}$ corresponding to fixed $Q > 0$.*

Within the boundary layer the angular velocity component $v_\alpha$ is small, but the radial component $v_R$ is positive and of order $k^{\frac{1}{3}}$. In order to establish this rigorously, we first show that for fixed $Q > 0$,

$$P \to 1 - Q \quad \text{as} \quad k \to \infty. \tag{24}$$

In fact, by lemma 2

$$1 - P - Q = 2\int_0^1 \frac{1-\Omega^2}{(1+t)^2}\,\mathrm{d}t,$$

so that by (23)

$$0 < \tfrac{1}{4}(1 - P - Q) \leqslant \int_0^1 \frac{1-\Omega}{(1+t)^2}\,\mathrm{d}t = \int_0^{N\delta} + \int_{N\delta}^1 < N\delta + 2\,\mathrm{e}^{-\frac{1}{2}N}$$

† The function $\tau = \frac{1}{2}\sqrt{(3cx)}$ satisfies $\tau' > \frac{3}{4}cx - \tau^2$, whence $\tau > \sigma$ and $\sigma' = \tau^2 - \sigma^2 > 0$.

valid for $1 \leqslant N \leqslant 1/\delta$, Now choose $N = 2\ln 1/\delta$, which yields

$$0 < 1-P-Q < 24\delta \ln 1/\delta \tag{25}$$

for $\delta \leqslant 1/\sqrt{e}$, i.e. for $Qk^2 \geqslant e^{\frac{3}{2}}$. The required limit behaviour now follows at once.

By lemma 3 we have $G(x) \leqslant (1-P)(x-x^2)$. Hence for $0 \leqslant x \leqslant \frac{1}{2}$,

$$f' \leqslant (1-P)k^2x, \quad f \leqslant \tfrac{1}{2}(1-P)k^2x^2. \tag{26}$$

Substituting back into (21) gives

$$f' \geqslant \tfrac{3}{4}cx - \tfrac{1}{4}(1-P)^2k^4x^4 \geqslant \tfrac{1}{2}cx \tag{27}$$

for $0 \leqslant x \leqslant \beta\delta$, where $\beta = [Q/(1-P)]^{\frac{1}{3}} < 1$. Thus from (13) and (26)

$$0 \leqslant \frac{F}{\sin\alpha} \leqslant \tfrac{1}{2}(1-P)kx^2 \quad (0 \leqslant x \leqslant \tfrac{1}{2}).$$

Also, since $F' = 2\nu(1-x^2)f' - 4\nu xf$, we have by (26) and (27)

$$\tfrac{1}{2}Qkx\{1-(1+2[(1-P)/Q])\,x^2\} \leqslant F' \leqslant (1-P)kx,$$

valid for $0 \leqslant x \leqslant \beta\delta$ (and of course $0 \leqslant x \leqslant \frac{1}{2}$, as always). In particular, for $k$ sufficiently large

$$|F/\sin\alpha| \leqslant M^2Q^{\frac{1}{3}}k^{-\frac{1}{3}} \quad (x = M\delta,\ M < 1/2\delta)$$

and

$$\frac{M}{3}Q^{\frac{2}{3}}k^{\frac{1}{3}} \leqslant F' \leqslant \frac{4M}{3}Q^{\frac{2}{3}}k^{\frac{1}{3}} \quad (x = M\delta,\ M < 1).$$

*Recalling the relations* (2), *this proves the assertions concerning the asymptotic behaviour of the velocity components* $v_\alpha$ *and* $v_R$ *in the boundary layer.*

It is a remarkable fact that, even though the fluid is slowed near the wall by the adherence condition and the pressure at the wall increases outward (when $P > 0$), nevertheless there is a strong *outward* component of velocity in the boundary layer. The explanation of this phenomenon is that the pressure gradient is for the most part used to oppose the centrifugal force of the vortex flow, leaving a net outward driving force on the radial velocity.

## 6. *Appendix*

Here we shall give the proof of lemma 6. By lemmas 4 and 5, $f$ is either ultimately positive or everywhere negative. In the former case we must have $P < 1$, and the required result then follows from the displayed inequality of lemma 4. We may consequently suppose from there on that $f$ is everywhere negative.

Consider the function $\omega(x) = -1/2(1-x)$. Since $G(x) = O(1-x)$ as $x \to 1$ (see lemma 2) it is easily checked that

$$\omega' < k^2\frac{G(x)}{(1-x^2)^2} - \omega^2$$

for $x$ sufficiently near 1, say $x_0 \leqslant x < 1$. Consequently if $f = \omega$ at some point $x_1 \geqslant x_0$, we have $f > \omega$ thereafter. It follows that either

$$f > \omega \quad \text{or} \quad f < \omega$$

for all $x$ sufficiently near 1. We shall show that only the first case can actually occur.

Thus suppose for contradiction that $f < \omega$ on some interval $x_2 \leqslant x < 1$. Then by (17), for $x \geqslant x_2$,

$$\Omega' = \text{const.}\exp\left(-2\int_{x_2}^{x} f\,dx\right) \geqslant \text{const.}\exp\left(-2\int_{x_2}^{x}\omega\,dx\right) = \text{const.}\frac{1-x_2}{1-x}.$$

Since the function on the right-hand side is not integrable, this contradicts the fact that $\Omega \to 1$ as $x \to 1$.

Having shown that $f > -1/2(1-x)$ for $x$ near 1, let us now make the change of variables

$$\phi = -\sqrt{(1-x)}f$$

in (14). Recalling that $f < 0$, this gives the result

$$\phi' = \{1+2(1-x)f\}\frac{f}{2\sqrt{(1-x)}} - k^2\sqrt{(1-x)}\frac{G(x)}{(1-x^2)^2} \leqslant \frac{\text{const.}}{\sqrt{(1-x)}}$$

for all $x$ near 1. By integration $\phi \leqslant$ constant, and therefore $f \geqslant -\kappa/\sqrt{(1-x)}$ for some positive constant $\kappa$ and for all $x$ near 1. Substituting this in (14) we obtain

$$f' \geqslant k^2\frac{G(x)}{(1-x^2)^2} - \kappa^2\frac{1}{(1-x)} \geqslant -\frac{\text{const.}}{1-x}.$$

Hence by integration $f \geqslant -\text{const.} \ln 1/(1-x)$. Since $f$ was assumed to be negative, this completes the proof of lemma 6.

## Chapter III. General theory

We show in this chapter that the basic set of equations (14), (15) has a solution for certain combinations of the parameter values $k$ and $P$, but not for others.

### 7. *Non-existence of solutions for certain parameter values*

It is a surprising and somewhat paradoxical fact that the problem as posed may not have solutions for high Reynolds numbers and certain fixed values of $P$. This was already pointed out by Goldstik for the special case $P = 1$. Here we shall investigate this phenomenon in more detail (see also the remarks in chapter IV).

Consider first the case $P \geqslant 1$. Then $f$ is negative and $\Omega$ convex, according to theorem 2. Consequently $\Omega(x) \leqslant x$ and from (12) we find after some calculation

$$G(x) \leqslant (1+x)^2 \ln(1+x) + (1-x)^2 \ln(1-x) - 4x \ln 2 + (3-P)(x-x^2) \equiv I(x).$$

By numerical computation† we obtain

$$I(x) \leqslant -\tfrac{1}{2}x(1-x^2)^2 + (1-P)(x-x^2) \leqslant -(\tfrac{1}{2}P)\, x(1-x^2)^2.$$

Consequently if a solution exists, then

$$f' \leqslant -\tfrac{1}{2}Pk^2x - f^2.$$

It is apparent that the solution $f$ cannot exist on the entire interval from 0 to 1 when the coefficient of $x$ is large. More specifically, consider the comparison equation

$$\sigma' = -d^2x - \sigma^2, \quad \sigma(0) = 0.$$

By explicit integration, Goldstik (p. 921) found that $\sigma$ cannot be continued over the interval $[0, 1)$ whenever $d > 3\mu_1/2$, where $\mu_1$ is the first root of the (Bessel) equation $J_{-\frac{1}{3}}(\mu) = 0$. The same conclusion therefore holds for the function $f$ by comparison. Consequently we have proved

Proposition 1. *There can be no solution of* (14), (15) *if* $P \geqslant 1$ *and* $Pk^2 > 9\mu_1^2/2 \approx 16$.

In view of theorem 2 and the result just proved, negative solutions $f$ can exist only when $k$ is less than 4. Therefore the most interesting class of vortex motions of the type (2) will have either $f$ positive or else $f$ first negative and then afterwards positive (such flows correspond to a downward swirling motion in the vortex, as pointed out in § 4).

Consider now solutions $f$ which are first negative and then positive thereafter. Let the zero of $f$ occur at the position $x = a$, $0 < a < 1$. Then the following result holds.

Proposition 2. *For fixed* $a$, $0 < a < 1$, *we must have*

$$P \leqslant \frac{1}{a}\left\{1 + \frac{1-a}{a^2}\ln(1-a^2)\right\}$$

*in order for a solution to exist with* $f(a) = 0$.

† See Goldstik, p. 920. I take it to be legitimate to check an inequality between two elementary functions by a numerical (digital) calculation of the functions involved. In any case, an analytic proof is available for the inequality $I(x) \leqslant -\tfrac{2}{5}Px(1-x^2)^2$, which would lead to essentially the same conclusions.

To prove this, note that by theorem 2 the function $\Omega$ is convex for $0 \leqslant x < a$ and concave for $a < x < 1$. Consequently, taking into account the boundary conditions on $\Omega$, it is easy to see that

$$\Omega(x) \leqslant \begin{cases} x/a, & 0 \leqslant x \leqslant a, \\ 1, & a \leqslant x \leqslant 1. \end{cases}$$

Hence by direct evaluation of the integrals involved in the definition of $G$, we find that

$$G(a) \leqslant (1-a)\left\{1 - aP + \frac{1-a}{a^2}\ln(1-a^2)\right\}. \tag{28}$$

On the other hand, $G(a) > 0$ as noted in the proof of lemma 4. Comparing this inequality with (28) yields the required result.

Proposition 2 implies that motions in which $f$ (and consequently $F$) have both positive and negative values can exist only at certain values of the parameters. It should be noted in particular that the function

$$\frac{1}{a}\left\{1 + \frac{1-a}{a^2}\ln(1-a^2)\right\}$$

is less than 1 for $0 < a < 1$, and tends to 1 as $a \to 0, 1$. We consider finally the case when $f$ is everywhere positive.

PROPOSITION 3. *In order for a positive solution $f$ to exist we must have*

$$P \leqslant K(k),$$

*where $K(k)$ is a positive monotone function of $k$, which increases from $3 - 4\ln 2 (\approx 0.224)$ to 1 as $k$ goes from 0 to $\infty$.*

*Proof.* By integration of (17) from 0 to 1 it is easy to obtain the relation

$$\Omega'(0) = \left\{\int_0^1 \exp\left(-2\int_0^x f\,\mathrm{d}x\right)\mathrm{d}x\right\}^{-1}.$$

Recalling that $P < 1$ in the present case, we see by lemma 4 that the right-hand side is in turn

$$\leqslant \left\{\int_0^1 \exp\left(\tfrac{1}{2}(P-1)\,k^2\int_0^1 \ln\frac{1}{1-x}\,\mathrm{d}x\right)\mathrm{d}x\right\}^{-1} = \exp\left(\tfrac{1}{2}(1-P)\,k^2\right) \equiv A.$$

Therefore since $\Omega$ is concave (cf. theorem 2) it follows that

$$\Omega(x) \leqslant \begin{cases} Ax, & 0 \leqslant x \leqslant 1/A, \\ 1, & 1/A \leqslant x \leqslant 1. \end{cases}$$

Thus by lemma 2 and the fact that $G'(0) \geqslant 0$ for a positive solution $f$, we find

$$P = 2\int_0^1 \frac{\Omega^2\,\mathrm{d}t}{(1+t)^2} - G'(0) \leqslant 4A - 4A^2\ln\frac{1+A}{A} - 1 \equiv \chi(A). \tag{29}$$

Now $$\frac{\mathrm{d}\chi}{\mathrm{d}A} = 4A\left\{\frac{1+2A}{A(1+A)} - 2\ln\frac{1+A}{A}\right\} \quad (A > 1).$$

The quantity in braces is easily found to be a decreasing function, and has the value zero when $A = \infty$. It follows therefore that $\mathrm{d}\chi/\mathrm{d}A > 0$ and that $\chi$ is an increasing function.

If we fix $k$, the function $\chi(A)$ therefore decreases from 1 to $3 - 4\ln 2$ as $P$ increases from $-\infty$ to 1 (and $A$ correspondingly decreases from $\infty$ to 1). Hence for each $k$ inequality (29) places an upper bound $K(k)$ on $P$. It is easy to check that $K(k)$ is increasing, and varies from $3-4\ln 2$ when $k=0$ to 1 when $k=\infty$.

Proposition 3 shows that for $k$ small, $P$ can be little larger than $3-4\ln 2$ in order for a positive solution to exist, while conversely for large $k$ only values of $P$ approaching 1 are ruled out. We shall see later (corollary 4) that the qualitative behaviour indicated by this proposition is essentially best possible, since positive solutions are there proved to exist for all $P$ not exceeding $3-4\ln 2$, and for $P$ arbitrarily near 1 if $k$ is suitably large.

8. *Existence of solutions (I)*

In this section we shall establish the existence of solutions for certain ranges of the parameters $k$ and $P$. Note that according to the work of the preceding section one cannot expect to solve the given problem for arbitrary values of these parameters.

Let $\tilde{k}$, $\tilde{P}$ be fixed parameter values, and let $\tilde{f}$, $\tilde{\Omega}$ be a pair of functions satisfying the conditions $\tilde{f}(0)=\tilde{\Omega}(0)=0$ and such that also

$$\tilde{f}=O\left(\ln\frac{1}{1-x}\right) \quad \text{and} \quad 0\leqslant\tilde{\Omega}\leqslant 1. \tag{30}$$

Then $\tilde{f}$, $\tilde{\Omega}$ will be called a *subsolution* of the system (14), (15) provided

$$\left.\begin{aligned} &\tilde{f}'+\tilde{f}^2\leqslant\tilde{k}^2\frac{\tilde{G}(x)}{(1-x^2)^2}\\ &\tilde{\Omega}''+2\tilde{f}\tilde{\Omega}\geqslant 0\end{aligned}\right\}\quad (0\leqslant x<1), \tag{31}$$

where $\tilde{G}$ denotes the function (12) with $\Omega$ replaced by $\tilde{\Omega}$ and $P$ by $\tilde{P}$. The following main result now holds.

THEOREM 3. *Let $\tilde{f}$, $\tilde{\Omega}$ be a subsolution for given parameter values $\tilde{k}$ and $\tilde{P}$. Then there exists a solution $f$, $\Omega$ of the system* (14), (15) *for the same k and all values of P not exceeding $\tilde{P}$. Moreover*

$$f\geqslant\tilde{f},\quad \Omega\geqslant\tilde{\Omega}.$$

*Proof.* Consider the method of successive approximations given by the scheme

$$\Omega_0\equiv 1$$

and for $n\geqslant 1$,
$$G_n(x)=2(1-x)^2\int_0^x\frac{t\Omega_{n-1}^2}{(1-t^2)^2}\,\mathrm{d}t+2x\int_x^1\frac{\Omega_{n-1}^2\,\mathrm{d}t}{(1+t)^2}-P(x-x^2),$$

$$f_n'+f_n^2=k^2\frac{G_n(x)}{(1-x^2)^2},\quad f_n(0)=0,$$

$$\Omega_n''+2f_n\Omega_n'=0,\quad \Omega_n(0)=0,\quad \Omega_n(1)=1.$$

We shall show that this scheme is well defined, and that for $0\leqslant x<1$

$$f_1\geqslant f_2\geqslant\ldots\geqslant\tilde{f},\quad \Omega_0\geqslant\Omega_1\geqslant\Omega_2\geqslant\ldots\geqslant\tilde{\Omega}. \tag{32}$$

The proof will be by induction. We establish first that $f_1$ and $\Omega_1$ are well defined, and that

$$f_1\geqslant\tilde{f},\quad \Omega_0\geqslant\Omega_1\geqslant\tilde{\Omega}. \tag{33}$$

To begin with, because $0\leqslant\tilde{\Omega}\leqslant 1$ and $\tilde{P}\geqslant P$ it follows that

$$\tilde{G}(x)\leqslant G_1(x)\equiv(1-P)\,(x-x^2).$$

Since
$$f_1' = k^2 \frac{G_1(x)}{(1-x^2)^2} - f_1^2 \tag{34}$$
the function $\tilde{f}$ therefore serves as a minorant for $f_1$, while
$$\max\left(0, \tfrac{1}{4}(1-P)\, k^2 \ln \frac{1}{1-x}\right)$$
serves as a majorant (see lemma 4). Thus according to standard results in the theory of ordinary differential equations, (34) has a solution $f_1$ on the full interval $0 \leqslant x < 1$, and the first inequality of (33) holds.†

Using (30) we see from the above construction that $f_1 = O(\ln 1/(1-x))$. Hence $f_1$ is integrable and $\Omega_1$ can be determined by quadrature as noted in the proof of lemma 1. It remains to prove the second inequality of (33). We require the following comparison lemma.

*Suppose that for* $0 \leqslant x < 1$ *we have*
$$\Omega'' + 2f\Omega' = 0, \quad \bar{\Omega}'' + 2\bar{f}\bar{\Omega}' \geqslant 0$$
*together with the boundary conditions* $\qquad \Omega(0) = \bar{\Omega}(0) = 0$

*and* $\qquad \Omega(1) = 1, \quad \bar{\Omega} \leqslant 1$ *for $x$ near* 1.

*Then if* $f \geqslant \bar{f}$ *we have* $\Omega \geqslant \bar{\Omega}$.

*Proof.* Obviously
$$(\bar{\Omega}-\Omega)'' + 2\bar{f}(\bar{\Omega}-\Omega)' \geqslant 2(f-\bar{f})\,\Omega' \geqslant 0,$$
since $\Omega'$ is certainly positive (see lemma 1). Multiplying through by the integrating factor $\exp\left(2\int_0^x \bar{f}\,dx\right)$ we find
$$\left\{(\bar{\Omega}-\Omega)' \exp\left(2\int_0^x \bar{f}\,dx\right)\right\}' \geqslant 0. \tag{35}$$
Now suppose for contradiction that the function $\bar{\Omega}-\Omega$ takes on positive values. Then it has a positive maximum at some point $x_0 < 1$. Integrating (35) from $x_0$ to $x > x_0$, we get
$$(\bar{\Omega}-\Omega)' \geqslant 0, \quad \bar{\Omega}-\Omega \geqslant \bar{\Omega}(x_0) - \Omega(x_0) > 0 \quad \text{for} \quad x \geqslant x_0,$$
contradicting the boundary conditions at $x = 1$. The lemma is therefore proved.

Applying the lemma with $\Omega = \Omega_1$, $\bar{\Omega} = \tilde{\Omega}$ now establishes $\Omega_1 \geqslant \tilde{\Omega}$. The remaining inequality of (33), namely $\Omega_0 \geqslant \Omega_1$, is obvious since $0 \leqslant \Omega_1 \leqslant 1$ (see lemma 1).

Now consider the induction hypothesis
$$f_1 \geqslant \dots \geqslant f_n \geqslant \tilde{f}, \quad \Omega_0 \geqslant \Omega_1 \geqslant \dots \geqslant \Omega_n \geqslant \tilde{\Omega}, \tag{36}$$
which by (33) surely holds when $n = 1$. Supposing (36) to hold for $n(\geqslant 1)$ we shall verify it for $n+1$. From the fact that $P \leqslant \tilde{P}$ we have easily
$$G_n(x) \geqslant G_{n+1}(x) \geqslant \tilde{G}(x).$$
Using $G_n$ and $\tilde{G}$ as comparison functions, we see that $f_n$ serves as a majorant for $f_{n+1}$ and $\tilde{f}$ serves as a minorant. Thus $f_{n+1}$ is well defined and
$$f_n \geqslant f_{n+1} \geqslant \tilde{f},$$
i.e. the first inequality of (36) holds for $n+1$. Moreover, since $f_{n+1}$ is $O(\ln 1/(1-x))$ it is clear that $\Omega_{n+1}$ is well defined (see the earlier discussion of $\Omega_1$).

By applying the lemma, first with $\Omega = \Omega_n$, $\bar{\Omega} = \Omega_{n+1}$ and then with $\Omega = \Omega_{n+1}$, $\bar{\Omega} = \tilde{\Omega}$, we obtain $\Omega_n \geqslant \Omega_{n+1} \geqslant \tilde{\Omega}$. Hence the second inequality of (36) also holds for $n+1$.

† See Coddington & Levinson (1955) chapter 1.

The validity of (36) is thereby established for all $n$, and accordingly (32) is proved. That is, the sequences $\{\Omega_n\}$, $\{f_n\}$ are monotonically decreasing and bounded below for fixed values of $x$. Consequently they converge to limits:

$$\Omega_n \to \Omega, \quad f_n \to f.$$

Evidently $\Omega \geqslant \tilde{\Omega}$ and $f \geqslant \tilde{f}$. It remains to show that these functions are solutions of (14), (15).

In view of (36) the functions $f'_n$ are uniformly bounded on any closed subinterval of $[0, 1)$. Thus $f_n$ converges uniformly on closed subintervals of $[0, 1)$. Moreover the functions $f_n$ are uniformly integrable according to the construction.

It now follows from the explicit integral formula for $\Omega_n$ in terms of $f_n$ that $\Omega_n$ converges uniformly to $\Omega$ on $0 \leqslant x \leqslant 1$. Furthermore, $\Omega$ is twice differentiable on $0 \leqslant x < 1$ and

$$\Omega'' + 2f\Omega' = 0.$$

Having shown that $\Omega_n$ converges uniformly to $\Omega$, it follows that $G_n$ converges uniformly to $G$; hence letting $n$ tend to infinity in the relation

$$f_n = \int_0^x \left( k^2 \frac{G_n(x)}{(1-x^2)^2} - f_n^2 \right) \mathrm{d}x$$

we find that $f$ is differentiable on $0 \leqslant x < 1$ and is a solution of (14). This completes the proof of theorem 3.

Corollary 1. *If* (14), (15) *is solvable for parameter values* $\tilde{k}$, $\tilde{P}$, *it is solvable for the same* $k$ *and all* $P \leqslant \tilde{P}$. *For the solutions in question, if* $P$ *decreases then* $f$ *increases.*

*If the solution* $\tilde{f}$ *is positive, then the problem is solvable with* $f$ *positive for all* $k \geqslant \tilde{k}$ *and* $P \leqslant \tilde{P}$. *For the solutions in question, if* $P$ *decreases and* $k$ *increases then* $f$ *increases. Finally, if* $\tilde{P} \geqslant 1$ *then the problem is solvable for all* $k \leqslant \tilde{k}$ *and* $P \leqslant \tilde{P}$.

*Proof.* The first statement follows at once from theorem 3, since any solution of (14), (15) is also a subsolution (see lemma 6). Now suppose $\tilde{f}$ is positive. The main proof then applies word for word (with $k \geqslant \tilde{k}$), once we notice that $k^2\tilde{G} \geqslant \tilde{k}^2\tilde{G}$ because $\tilde{G}$ is positive.

When $\tilde{P} \geqslant 1$ we note that $\tilde{G}$ is negative (lemma 3). Thus $k^2\tilde{G} \geqslant \tilde{k}^2\tilde{G}$ when $k \leqslant \tilde{k}$. The main proof then applies word for word.

In the statement of corollary 1, the phrase '(and indeed by successive approximations)' could be added following the words 'it is solvable'. In particular, this means that if a solution (positive solution) exists at all for given $k$, $P$, then the method of successive approximations will converge to a solution (positive solution) for these parameter values. We shall take advantage of this remark in § 10.

Corollary 2. *The system* (14), (15) *is solvable with* $f$ *positive provided that*

$$P \leqslant 3 - 4 \ln 2 \approx 0.224. \tag{37}$$

*Proof.* We shall show that $\tilde{f} \equiv 0$, $\tilde{\Omega} \equiv x$ is a subsolution. Indeed, it is enough to verify that the function $\tilde{G}$ associated with $\tilde{\Omega} = x$ is positive. By lemma 2, however,

$$\begin{aligned} \tilde{G}'(0) &= 2\int_0^1 \frac{t^2\,\mathrm{d}t}{(1+t)^2} - P \\ &= 3 - 4\ln 2 - P \geqslant 0, \end{aligned}$$

and consequently $\tilde{G} > 0$ by lemma 3. This completes the proof.

It should be noted that *condition* (37) *is best possible.* Indeed, if $P > 3-4\ln 2$ then for all sufficiently small $k$ no positive solution can exist, according to the remark following proposition 3 in § 7.

COROLLARY 3. *There exists an absolute constant* $\lambda \approx 2.85$ *such that* (14), (15) *is solvable provided that*

$$Pk^2 < \lambda^2,$$

*but is not solvable when* $(P-1)\,k^2 > \lambda^2$.

*Proof.* We seek a subsolution with $\tilde{\Omega} \equiv 0$. Then $\tilde{G} \equiv P(x^2-x)$, and we can choose $\tilde{f}$ as the solution of the equation

$$\tilde{f}' + \tilde{f}^2 = -Pk^2 \frac{x}{(1-x)\,(1+x)^2} \qquad (0 \leqslant x < 1),$$

with corresponding boundary conditions

$$\tilde{f}(0) = 0, \quad \tilde{f} = O(\ln(1/1-x)).$$

It is evident† that there is a constant $\lambda > 0$ such that this problem is solvable if and only if $Pk^2 < \lambda^2$. The approximate value 2.85 for $\lambda$ was determined by a simple computer program. This proves the first part of the corollary.

Now for any solution of (14), (15), lemma 3 implies

$$f' + f^2 \leqslant (1-P)\,k^2 \frac{x}{(1-x)\,(1+x)^2}.$$

From the remarks above we have $-(1-P)\,k^2 < \lambda^2$, and the remaining part of the corollary follows at once (note that this part of the result overlaps proposition 1 in § 7).

The preceding three corollaries contain the main existence theorem of this section. In the next section we shall use a different method to obtain further results.

## 9. *Existence of solutions (II)*

The results of the previous section establish the existence of two of the three possible types of solutions described by theorem 2. Namely when $P \leqslant 0.224$ we obtain a positive solution $f$, while when $P \geqslant 1$ and $Pk^2 < 8.2$ we find a negative decreasing solution. These results do not, however, shed light one way or the other on the existence of the remaining type of solution, namely one which is first negative, has a single zero, and is positive thereafter. This type of solution seems to be

† To see this, let $S$ be the set of values of $Pk^2$ for which the problem is solvable. Clearly $S$ contains zero, and if $\mu$ is in $S$ then any value $\mu' < \mu$ is also in $S$. Consequently it is only necessary to show that $S$ is open on the right. This being the case, let $\mu$ be in $S$ and let $g$ be the corresponding solution. Put

$$\bar{g} = g - \ln\frac{1}{1-x}.$$

Then an easy calculation shows that

$$\bar{g}' \leqslant -(\mu+1)\,\frac{x}{(1-x)\,(1+x)^2} - \bar{g}^2$$

on some interval $1-\gamma \leqslant x < 1$. Now let $\epsilon(<1)$ be a small positive number such that the equation

$$h' = -(\mu+\epsilon)\,\frac{x}{(1-x)\,(1+x)^2} - h^2$$

has a solution on the interval $0 \leqslant x \leqslant 1-\gamma$, with $h(0) = 0$, $h(1-\gamma) \geqslant \bar{g}(1-\gamma)$ (the existence of such an $\epsilon$ follows from the fact that we can choose $h = g$ when $\epsilon = 0$, together with standard perturbation theory; Coddington & Levinson, p. 29). By the previous construction $h$ can be continued throughout the remaining interval $1-\gamma \leqslant x < 1$, with $\bar{g} \leqslant h \leqslant g$. Hence $\mu+\epsilon$ is in $S$, as required.

of particular interest; our purpose here will be to show that such solutions do in fact exist, and indeed can exist at arbitrarily high values of $k$.

As motivation for our approach, we observe that for such solutions we have necessarily (see lemma 2 and theorem 2)

$$G'(0) = 2\int_0^1 \frac{\Omega^2\,\mathrm{d}t}{(1+t)^2} - P < 0.$$

Let us therefore consider $Q = G'(0)$ as the basic parameter, and correspondingly put

$$P = 2\int_0^1 \frac{\Omega^2}{(1+t)^2}\,\mathrm{d}t - Q. \tag{38}$$

*Our object is now to solve* (14), (15), (38) *for fixed values of* $Q$.

Eliminating $P$ from (12) by using (38), we can write $G(x)$ in the form

$$G(x) = -2\int_0^x \frac{(x-t)(1-tx)}{(1-t^2)^2}\,\Omega^2\,\mathrm{d}t + 2x^2\int_0^1 \frac{\Omega^2\,\mathrm{d}t}{(1+t)^2} + Q(x-x^2). \tag{39}$$

It is apparent from (39) that the method of theorem 3 will not work in the present case, since $G$ is no longer a monotone functional of $\Omega$; that is, when $Q$ remains fixed we no longer have $\bar{G} \geqslant G$ when $\bar{\Omega} \geqslant \Omega$. Nevertheless, a modified approach making use of well-known fixed point theorems can be given.

Consider functions $f$ which are continuous on $0 \leqslant x < 1$ and satisfy the condition

$$f = O(\ln 1/(1-x)). \tag{40}$$

Correspondingly, let us determine $\Omega$ by the equation

$$\Omega'' + 2f\Omega' = 0 \tag{41}$$

together with the usual boundary conditions; finally let $g$ satisfy the differential equation

$$g' + g^2 = k^2 \frac{G(x)}{(1-x^2)^2}, \quad g(0) = 0,$$

where $G(x)$ is the function introduced above. This defines a transformation

$$f \to g = Tf.$$

Note that $g$ need not be determined on the entire interval $0 \leqslant x < 1$; that is, it may diverge to $-\infty$ at some point $x = b < 1$. In any case, our purpose is to find a fixed point $f$ of this transformation, that is, a function $f$ such that $Tf$ exists for $0 \leqslant x < 1$ and satisfies $f = Tf$.

To this end, we shall employ a form of the Schauder fixed point theorem. Let $X$ be the space of continuous functions on the interval $0 \leqslant x < 1$ with the metric

$$\operatorname{dist}(f,g) = \sum_{m=1}^{\infty} \frac{1}{2^m} \frac{M_m(f,g)}{1+M_m(f,g)}, \quad f, g \in X,$$

where

$$M_m(f,g) = \sup_{0 \leqslant x \leqslant 1-\frac{1}{m}} |f(x) - g(x)|.$$

Clearly a sequence of functions $f_n$ converges in $X$ to a function $f$ if and only if $f_n \to f$ uniformly on any compact subinterval of $[0, 1)$. Moreover, under the topology induced by the metric, $X$ is obviously a locally convex linear topological space (see Dunford & Schwartz 1958, for definitions).

This being the case, if $\hat{X}$ is a closed convex subset of $X$ which is continuously mapped by $T$ into a conditionally compact subset of itself, then $T$ has a fixed point in $\hat{X}$ (Dunford & Schwartz, p. 456; the result stated there is slightly different, but is easily seen to imply the present form). It should be noted that since $X$ is a metric space, a set $A$ is conditionally compact in $X$ if every sequence of points in $A$ has a subsequence converging to a point in $X$.

THEOREM 4. *The system* (14), (15) *is solvable, with* $G(x)$ *given by* (39), *provided that*

$$Qk^2 > -\lambda^2,$$

*where* $\lambda$ *is the constant of corollary* 3.

*Proof.* It is necessary to determine an appropriate set $\hat{X}$ for the application of Schauder's theorem.

Let $f$ be a continuous function on $0 \leqslant x < 1$ satisfying condition (40). We examine in detail the transformation $g = Tf$. In particular, with $\Omega$ defined by (41), it is clear that

$$G(x) \geqslant Q(x-x^2)$$

according to lemma 3. Consequently we have

$$g'+g^2 \geqslant Qk^2 \frac{x}{(1-x)(1+x)^2} \quad (g(0) = 0).$$

Let the function $\tau$ be defined by the equation

$$\tau'+\tau^2 = Qk^2 \frac{x}{(1-x)(1+x)^2}.$$

As already remarked in the proof of corollary 3 this equation is solvable for the boundary conditions

$$\tau(0) = 0, \quad \tau = O(\ln 1/(1-x)),$$

provided that $Qk^2 > -\lambda^2$. Hence under the hypothesis of the theorem, comparison of the functions $g$ and $\tau$ yields the relation

$$g(x) \geqslant \tau(x).$$

Also by lemma 3, one has

$$G(x) \leqslant (1-P)(x-x^2) \leqslant (1+Q)(x-x^2)$$

since $P > -Q$ by (38). Consequently, as in the proof of lemmas 4 and 5,

$$g(x) \leqslant \begin{cases} 0, & Q \leqslant -1, \\ \frac{1}{4}(1+Q)k^2 \ln 1/(1-x), & Q > -1. \end{cases}$$

Combining the preceding estimates, it is clear that under the conditions of the theorem $T$ is well-defined over the full interval $0 \leqslant x < 1$.

We now choose $\hat{X}$ to be the set of all functions $f \in X$ such that

$$\tau(x) \leqslant f(x) \leqslant 0 \qquad \text{if} \quad Q \leqslant -1$$

$$\tau(x) \leqslant f(x) \leqslant \tfrac{1}{4}(1+Q)k^2 \ln \frac{1}{1-x} \qquad \text{if} \quad Q > -1.$$

Clearly $Tf$ is well defined for all $f \in \hat{X}$, and moreover we have $T\hat{X} \subset \hat{X}$ according to the first part of the proof. Also using the fact that $g = Tf$ satisfies

$$g' = k^2 \frac{G(x)}{(1-x^2)^2} - g^2,$$

we see easily that $T\hat{X}$ consists of functions which are uniformly continuous on closed subintervals of $[0, 1)$. Hence by Arzela's theorem $T\hat{X}$ is conditionally compact in $X$.

The set $\hat{X}$ is obviously closed and convex in $X$. Finally, the mapping $T$ is continuous on $\hat{X}$. Indeed, using the explicit integral for $\Omega$ in terms of $f$, one sees that if $f$ converges to $f_0$ in $\hat{X}$ then $\Omega$ converges uniformly to $\Omega_0$ on $0 \leqslant x \leqslant 1$. Consequently $G$ converges to $G_0$ uniformly on closed subintervals of $0 \leqslant x < 1$, and $g$ in turn converges uniformly to $g_0$ on such subintervals.

The conditions of Schauder's theorem are therefore satisfied, and accordingly there exists a fixed point $f$ of $T$. It remains only to note that when $\Omega$ is defined by (41) and the usual boundary conditions, then $f$, $\Omega$ satisfy (14), (15). This completes the proof of the theorem.

The preceding theorem obviously establishes the existence of solutions of (14), (15) for all non-negative values of $Q = G'(0)$. These solutions correspond to parameter values $P < 1$, and are of course such that $f$ is positive.

We can in fact be more explicit. Consider a fixed value $k \geqslant e$ and let $Q = k^{-\frac{1}{2}}$. For the corresponding solution we have $P > 1 - 12k^{-\frac{1}{2}}(1 + \ln k)$ by (25). Hence applying corollary 1, there exist positive solutions whenever

$$P \leqslant 1 - 12k^{-\frac{1}{2}}(1 + \ln k), \quad (k \geqslant e)$$

Combining this with the result of corollary 2 now proves

Corollary 4. *There exist positive solutions of* (14), (15) *whenever*

$$P \leqslant \max\left(3 - 4\ln 2,\ 1 - 12k^{-\frac{1}{2}}(1 + |\ln k|)\right).$$

The following result answers the question raised at the beginning of the section.

Corollary 5. *For each* $k > 0$ *there exist solutions of* (14), (15) *such that* $f$ *is first negative, has a single zero on* $0 < x < 1$, *and is positive thereafter.*

*Proof.* By the main theorem of the section, there exists for each fixed $k > 0$ a solution with $Q = -\lambda^2/2k^2$. If the corresponding parameter value $P(k)$ is less than 1, it is clear by theorem 2 that the solution in question is of the required sort and there is nothing more to prove. We may thus suppose that $P(k) \geqslant 1$.

Then by corollary 1 the system (14), (15) is solvable for the fixed value $k$ and for all $P < 1$. In particular, a solution exists when $P = \frac{1}{2}(1 + K(k))$ where the function $K(k)$ is defined in proposition 3. This solution cannot be positive (by proposition 3), and is not negative (since $P < 1$). The only remaining possibility is that the solution has the required behaviour. This completes the proof.

### *Discussion of results of chapter III*

It is convenient to associate a point in the $(P, k^{-2})$ plane with each pair of parameter values $P$ and $k$. Then by corollary 1 (§ 8) positive solutions of the system (14), (15) exist for all parameter values which lie on the left-hand side of a certain fixed monotonically decreasing curve $\Sigma$ in the $(P, k^{-2})$ plane. According to corollary 4 and proposition 3 this curve must satisfy the condition

$$\max\left(3 - 4\ln 2, 1 - 12k^{-\frac{1}{2}}(1 + |\ln k|)\right) \leqslant P \leqslant K(k). \tag{42}$$

Consequently, taking into account proposition 3, the curve approaches the asymptotic value $P = 3-4\ln 2$ as $k^{-2}$ tends to infinity, and tends to 1 as $k^{-2}$ tends to zero.

Similarly, by corollary 1 the zone in which solutions of any kind can exist must be bounded on the right by a second curve $\Sigma'$ which is monotonically increasing when $P \geqslant 1$, and has a single-valued projection on the $k^{-2}$ axis when $P < 1$. This curve obviously lies to the right of the first one. Moreover by corollary 3 and proposition 1 it must satisfy the relation

$$\lambda^2 k^{-2} \leqslant P \leqslant \max\left(1, \tfrac{9}{2}\mu_1^2 k^{-2}\right). \tag{43}$$

These qualitative results are illustrated in figure 1. Specifically, the region A in which positive solutions exist is shown bounded by a monotonically decreasing curve with the properties noted

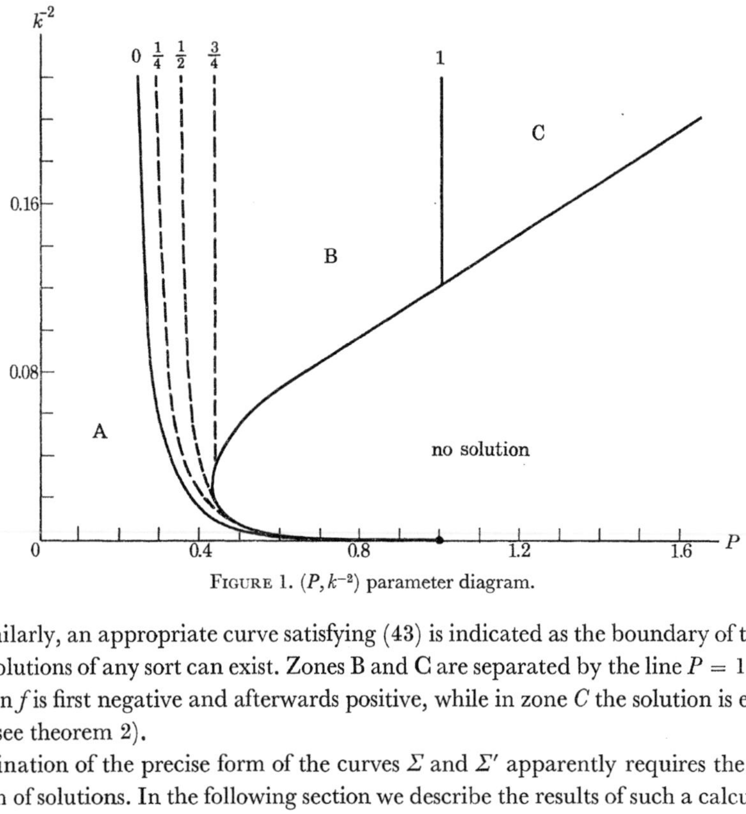


FIGURE 1. $(P, k^{-2})$ parameter diagram.

above; similarly, an appropriate curve satisfying (43) is indicated as the boundary of the domain in which solutions of any sort can exist. Zones B and C are separated by the line $P = 1$. In zone B the solution $f$ is first negative and afterwards positive, while in zone $C$ the solution is everywhere negative (see theorem 2).

Determination of the precise form of the curves $\Sigma$ and $\Sigma'$ apparently requires the numerical calculation of solutions. In the following section we describe the results of such a calculation and present some representative solutions for several typical values of the parameter $k$. The curves in figure 1 are in fact drawn so as to be in agreement with this work (as well as being consistent with the preceding theory). The dotted lines in zone B are curves along which $a =$ constant (where $F(a) = 0$), as interpolated from the numerical results. These curves agree in all respects with the conditions of proposition 2 and corollary 5. Finally, the line $P = 1$ separating zones B and C is found to terminate at the value $k^{-2} \approx 0.122$, that is, when $k \approx 2.86$. Consequently, solutions involving a central updraft can exist only at values of $k$ not greater than this.

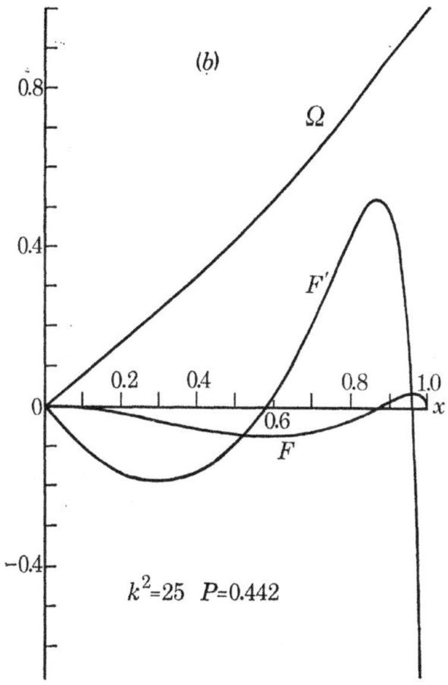


FIGURE 2*b*. Numerical solution when $k = 5$, $P = 0.442$ (zone B).

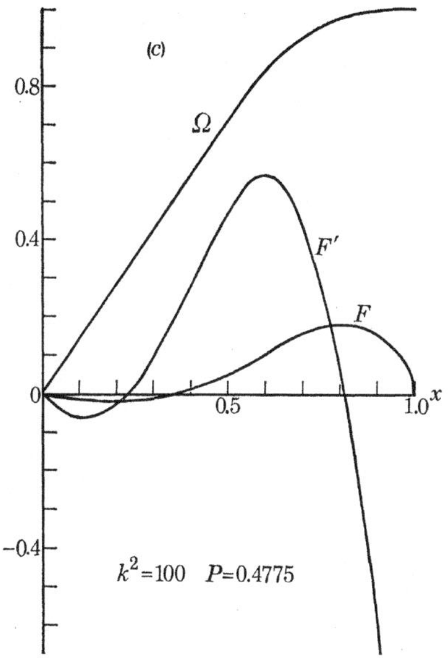


FIGURE 2*c*. Numerical solution when $k = 10$, $P = 0.4775$ (zone B).

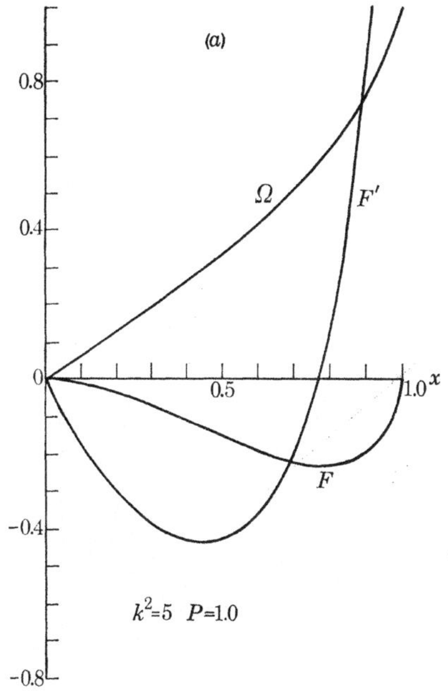


FIGURE 2*a*. Numerical solution when $k = 2.236$, $P = 1$ (zone C).

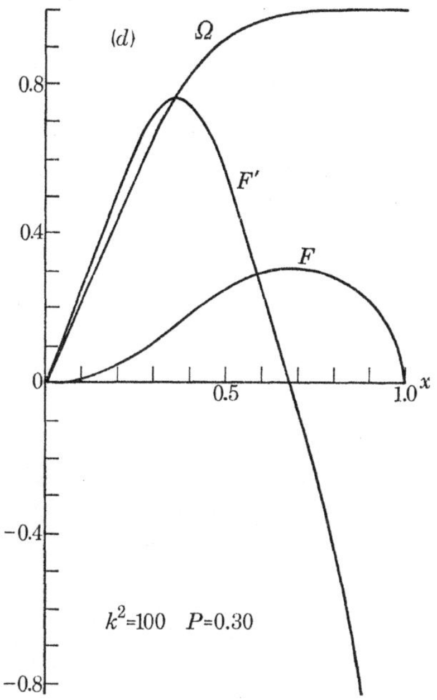


FIGURE 2*d*. Numerical solution when $k = 10$, $P = 0.3$ (zone A).

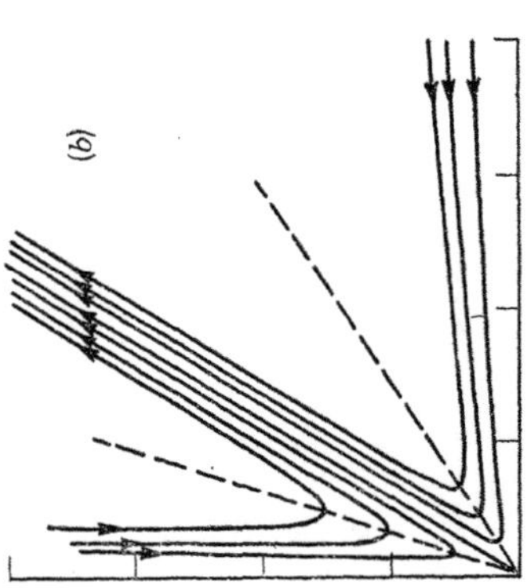


FIGURE 3*b*. Flow pattern when $k = 5$, $P = 0.442$ (zone B).

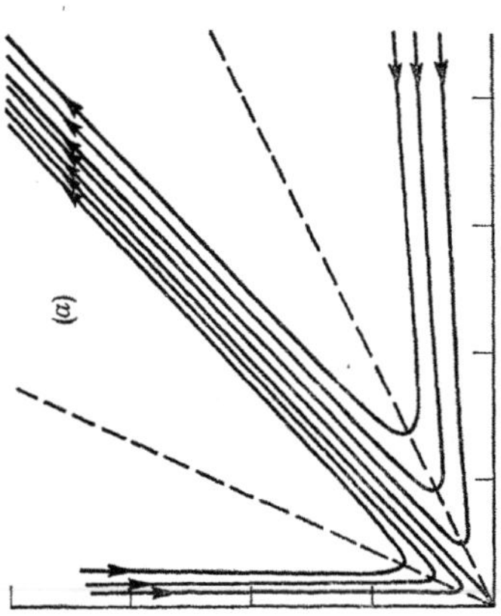


FIGURE 4*a*. Flow pattern when $k = 5.58$, $P = 0.428$ (zone B).

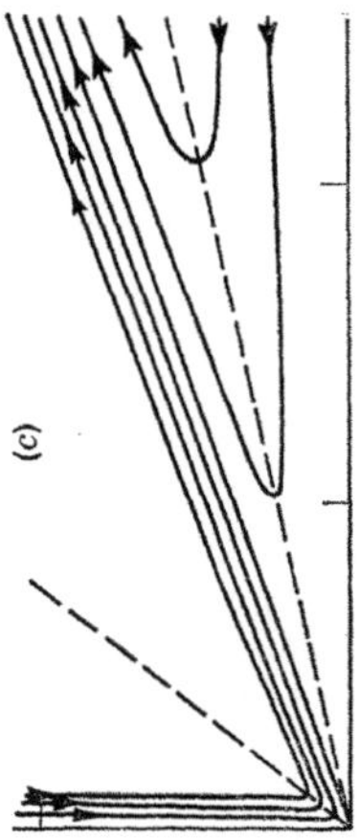


FIGURE 3*c*. Flow pattern when $k = 10$, $P = 0.4775$ (zone B).

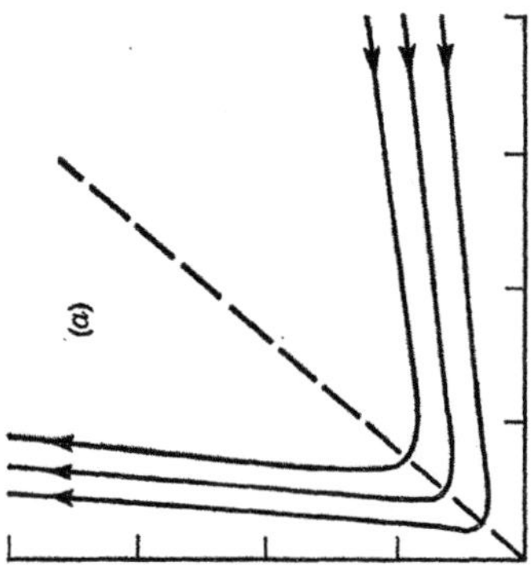


FIGURE 3*a*. Flow pattern when $k = 2.236$, $P = 1$ (zone C).

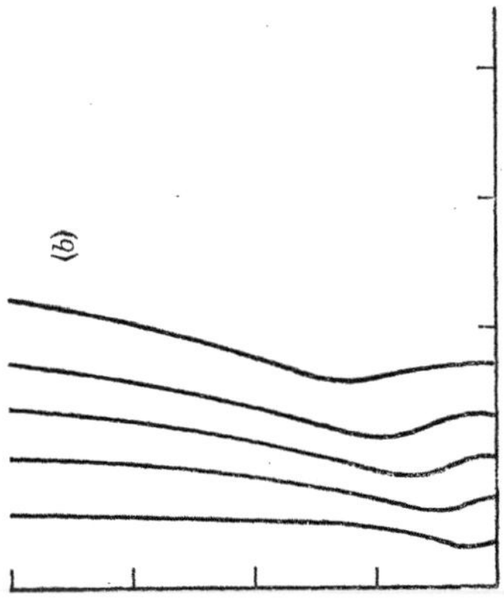


FIGURE 4*b*. Pressure pattern for above flow.

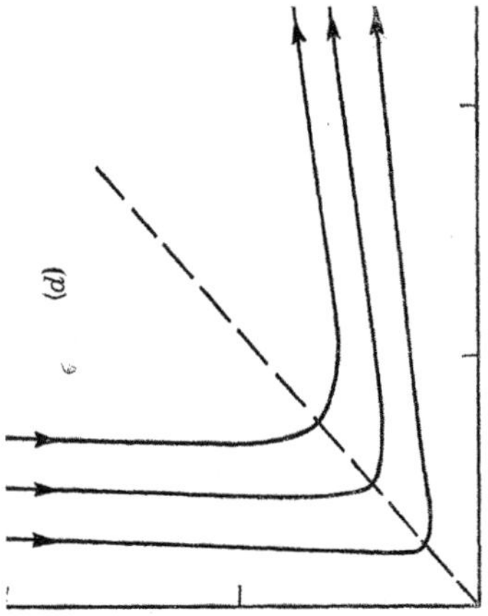


FIGURE 3*d*. Flow pattern when $k = 10$, $P = 0.3$ (zone A).

## 10. *Numerical results*

The discussion at the close of the previous section indicates the need for precise numerical results concerning those parameter values for which solutions exist.

The University of Minnesota CDC 6600 digital computer was programmed to seek solutions according to the successive approximations scheme described in § 8. As we have already observed following corollary 1, this procedure will converge whenever a solution exists. Numerical solutions were obtained for the fifteen values

$$k^2 = 1, 5, 10, 15, 20, 25, 30, 34.2, 40, 60, 100, 200, 625, 1000, 3600,$$

and for corresponding pressure ratios $P$ starting from $P = 0.2$ and continuing at appropriate increments $\Delta P = 0.002$ until breakdown occurred. Figure 1 has been drawn so as to be in agreement with these calculations. In addition, a successive approximation procedure was carried out for the modified function $G(x)$ defined by (39) with $Q = 0$, in order to obtain a more accurate determination of the curve which separates zone A and zone B.

Figure 2*a–d* show the result of the numerical calculations in four typical cases, corresponding respectively to parameter values falling in zones C, B (two of these) and A. Stream surfaces† of the accompanying flows are portrayed in figure 3*a–d*; these graphs illustrate in detail the qualitative behaviour described in the introduction and in chapter II. The stream surfaces in each are drawn for equally spaced values $\psi$ of the streamfunction (see § 1). In consequence, the mass flux of the secondary flow between each of the respective stream surfaces is the same; the particular values of $\psi$ are of course unimportant because of the geometrical similarity of the flow patterns.

The fluid motions shown in figures 3*b* and *c* involve an inflow near the boundary plane and a downdraft along the vortex axis. This is of course balanced by a corresponding outflow which occurs at an intermediate range of angles: in figure 3*b* the outflow occurs between 35° and 73° with an asymptotic direction 60°, angles being measured from the horizontal; for the flow in figure 3*c* the outflow occurs between 13° and 54°, and asymptotically approaches 21°. Because of the interest in motions of this sort, it is valuable to exhibit another example of this type; see figure 4*a*. The asymptotic direction here is precisely 45°. Isobars of the corresponding pressure field were calculated from equations (3) and (4) and are shown in figure 4*b*; the pressure decreases monotonically as one radially approaches the origin.

The arrangement of the figures on p. 352 emphasizes the continuity of the flow patterns when $k$ and $P$ are varied. Reading the figures in order 3*a*, 3*b*, 4*a*, 3*c*, 3*d*, one sees clearly the smooth transition from upflow (figure 3*a*) to downflow (figure 3*d*) through the mechanism of a gradually flattening cascade angle (figures 3*b*, 4*a*, 3*c*).

The particular values of $P$ for the flows shown in figures 3*b*, 3*c* and 4 were chosen so that the point $(P, k^{-2})$ lies on, or nearly on, the right-hand boundary of zone $B$. Thus the value of $P$ in these cases not only makes the pressure distribution on the boundary plane conform as nearly as possible to the corresponding free vortex flow (see the remarks following the proof of theorem 1), but also results in a minimum axial speed near the vortex line (see theorem 2 (iii)). The related choice $P = 1$ for the flow shown in figure 3*a* was made for essentially the same reasons.

Now let $\hat{P}(k)$ denote the value of $P$ associated with the right-hand boundary of zone B. Similarly, let $\beta = \beta(k)$ be the asymptotic direction associated with the parameters $k$, $\hat{P}(k)$, for

† More precisely, their cross-section with planes $\theta$ = constant.

$0 < k^{-2} \leqslant 0.122$. From corollary 1 in § 8, $\beta$ is the *maximum cascade angle* possible for the given value of $k$. The calculated graph of $\beta(k)$ is shown in figure 5. When $k^{-2} > 0.122$ the flow corresponding to $P = 1$ can be considered to have a cascade angle (or asymptotic direction) of 90°. From this point of view we could define $\beta(k) = 90°$ for all $k^{-2} > 0.122$, and continue the graph in figure 5 indefinitely to the right at this constant value. (It also follows from corollary 1 that, for fixed $k$, the asymptotic direction increases monotonically as $P$ goes from the left- to the right-hand boundary of zone B).

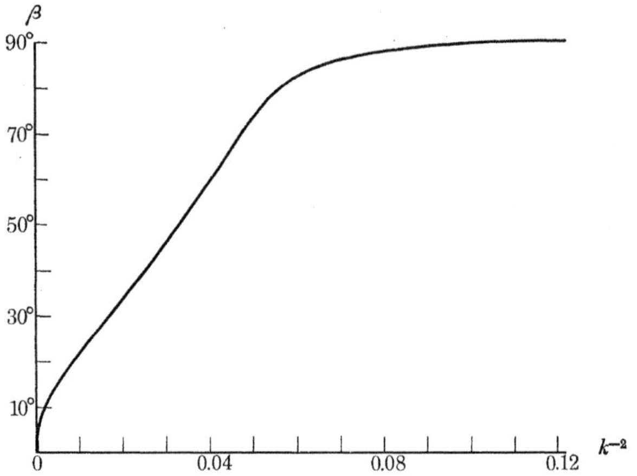


FIGURE 5. Maximum cascade angle $\beta$ against turbulence level $k$.

SUMMARY

In the preceding work we have introduced a class of exact solutions of the Navier–Stokes equations, representing the interaction of an infinite vortex line with a plane boundary surface at right angles to the line. As discussed in the introduction, the presence of the boundary surface induces a secondary flow superimposed on the free vortex motion.

Assuming the basic separation of variables given by equations (2), the differential equations governing the motion then show that the flow can be described in terms of a pair of dimensionless parameters $k$ and $P$. Here $P$ is the ratio of the pressure gradient of the actual flow and of the corresponding free vortex, evaluated on the boundary surface at some fixed radial distance from the vortex line, and $k$ is defined by the ratio $C/2\nu$, where $C = \lim_{r\to 0} rv_\theta$ is the angular momentum of the vortex and $\nu$ is the kinematic viscosity. In laminar motion $k$ may be considered as a Reynolds number of the flow; if turbulent motion is assumed then $k$ is more properly a self-regulating parameter indicating the basic level of turbulence present.

It is found that exactly three types of secondary flow régimes can exist. In particular, let us associate to each pair of parameter values $k$ and $P$ a point in the upper half of the $(P, k^{-2})$ plane, as shown in figure 1. For parameter values falling in zone A the secondary flow is directed downward along, the vortex axis and outward near the boundary surface. For values of $k$ and $P$ falling in zone B the secondary flow again involves a down-draft along the vortex axis, but there is an inflow along the boundary surface. This motion inwards is balanced by a compensating outflow occurring at an intermediate cone of directions around the angle $\alpha_0 = \cos^{-1} a$. Finally, for values of $k$ and $P$ in zone C, the secondary flow consists of an updraft along the vortex axis together with an inflow along the boundary surface. No vortex flow consistent with (2) and the assumed boundary conditions can exist for values of $k$, $P$ which fall in the remaining zone. It is immediately apparent from the diagram that motions of the third kind, involving an updraft and an inflow, can exist only at low values of $k$, less than approximately 2.86. Since the values of $k$ appropriate to meteorological phenomena are in many instances apparently somewhat larger than this (see below) it appears that motions of the third kind need not invariably represent the typical case of vortex phenomena, a point to which we shall return later. Typical streamlines for the secondary flows in the three cases above are shown in figures 3 and 4.

The results described in the previous paragraph are found in § 4 and in chapter III. As far as the mathematical part of the paper is concerned, the main problem which has been left open is the question of uniqueness, namely whether at most one flow consistent with (2) can exist for a given pair of values $k$ and $P$. While this seems highly likely, and while the numerical calculations discussed in § 10 lead one to the belief that solutions are unique, we have been unable to prove this (on page 927, Goldstik claims that solutions are unique for the case $P = 1$, but his remarks are unsupported). In addition to a uniqueness theorem it would also be useful to know that solutions depend continuously on the parameter values. In view of the complicated nature of the relevant differential equations both this problem and the question of uniqueness may be quite difficult.

Some final remarks may be added concerning the implications of the above results for various meteorological phenomena.† Before proceeding with this, however, it is important to make

† Important surveys of the physical and meteorological aspects of tornadoes, dust whirls, and waterspouts have been given by Brooks (1951) and Kessler (1970). Morton (1966) in addition to discussing the behaviour of geophysical vortices also includes a valuable and penetrating discussion of the theoretical aspects of those phenomena. All of these papers have extensive and useful bibliographies.

clear the not inconsiderable problems which attend any such interpretation. We have already argued in the introduction that buoyancy and compressibility effects should not be of major importance in tornado phenomena, for the speeds are apparently neither so high that compressibility will be important nor so low that buoyancy can exert an immediate effect.† More significant is the idealization of a tornado core by a line vortex, for in a real fluid the innermost parts of a vortex surely will not enjoy the inverse distance law of speed which governs a line vortex. Whether such an idealization is justified is, at our present state of knowledge of viscous or turbulent vortex cores, a matter which must remain somewhat intuitive, resting ultimately on the same conviction which justifies the study of line vortices in ideal fluids—the belief that they are reasonably accurate asymptotic models of the physical world. Following this line of thought, one could argue that the velocity singularity at the vortex line (and the corresponding pressure singularity) might reasonably be smoothed out by a matching process analogous to the Rankine combined vortex of ideal fluid theory. While such a process is to be desired, and might produce important further insight into the structure of tornado cores, nevertheless lacking such a desideratum one tends to accept the (admittedly meagre) experimental measurements in tornadoes which point to the validity of the inverse distance law of speed except in the neighbourhood of the core itself (see particularly Lewis & Perkins 1953) and to accept the intuitive picture (Morton 1966) of a tornado or waterspout as a strongly concentrated core of vorticity embedded in a weakly vortical environment, for which the circulation in circular paths outside the core is relatively constant.

In modelling large-scale meteorological phenomena one must assume the flow to be turbulent if any progress is to be made. Following the semi-empirical approach of Boussinesq we thus consider $\nu$ to be a virtual or eddy viscosity, a self-regulating parameter of the flow. There is of course no reason to expect a precise value for $\nu$ nor in fact is one necessary; on the other hand, numerical and empirical evidence given by Schlichting (1960, chapter 23) indicates that a reasonably accurate hypothesis is

$$\nu \approx \sigma q r,$$

where $q$ is the flow speed, $r$ the radial distance from the axis of the vortex, and $\sigma$ a dimensionless factor in the range 0.1 to 0.2. By (2), (6), (7) the quantity $qr$ is constant along radii through the origin, takes the value 0 at the boundary plane, and tends to $C$ as one approaches the vortex axis. One may then put $\nu \approx \frac{3}{4}\sigma C$ throughout the flow field with fair approximation.‡ Thus we are led to consider values of the parameter $k = C/2\nu$ in the approximate range

$$k = 3 \quad \text{to} \quad k = 7.$$

Naturally, rather than taking this *a priori* point of view we could equally well consider $k$ as a parameter to be determined by observational evidence.

With $k$ fixed, numerical values of $P$ presumably should be as near to 1 as possible in order for the pressure distribution to follow most nearly that in a free vortex (see the remarks following theorem 1). For values of $k$ between 3 and 7, this places the parameter point $(P, k^{-2})$ on the curved part of the right-hand boundary of zone B; that is, $P = \hat{P}(k)$ where $\hat{P}(k)$ was defined in the previous section. The associated motions exhibit an inflow along the boundary plane and a

† Vonnegut has argued that electrical activity may be important in certain tornado phenomena; in such cases the present analysis would evidently be insufficient.

‡ Another procedure, computationally and theoretically more difficult, would be to set $\nu = \sigma(\alpha)\, C$ where $\sigma(\alpha)$ is some function deliberately chosen to reflect a dependence of the turbulence level on the angle $\alpha$. This approach would also seem to be in agreement with Turner's view (1966, p. 400) of self-regulating turbulence in a tornado.

descending motion along the axis. The corresponding cascade angle as determined from figure 5 varies from 89° when $k = 3$ to 33° when $k = 7$. Figures 3*b* and 4*a* illustrate two flows from this range of parameter values, those for $k = 5$ and $k = 5.85$. The associated pressure fields are also of interest, since the visible funnel of a tornado or waterspout is caused by the condensation of water vapour when the dynamic pressure falls below the local vapour pressure (Brooks 1951). Consequently the outline of a funnel cloud should approximately represent an isobar of the motion. The general appearance of the equipressure lines in figure 4*b*, and especially their gradual widening at the top, agrees in this respect with the typical configurations of tornado and waterspout funnels.

There is considerable observational evidence concerning the form of dust and mist cascades at the foot of tornadoes and waterspouts, both descriptive and in the form of photographs taken by persons who had, so to speak, the presence of mind to reach for their cameras rather than run for their cellars.

Figure 6*a*, plate 2, is a historic photograph of a waterspout off Cottage City, Martha's Vineyard, 19 August 1896. Though the photograph has suffered from poor development especially at the critical point where the waterspout meets the horizon, one may nevertheless estimate the cascade angle to be about 40° to 45°. The straight funnel and almost stationary aspect of this waterspout are also particularly remarkable. Figure 6*b*, plate 2, shows a tornado at Elbow Lake, Minnesota. The cascade angle averages about 40°; in this case the inclination of the vortex axis explains the fact that the angle is different on the two sides of the tornado. Figure 6*c*, plate 3, shows a similar example, this one from Kansas (the ubiquitous telephone pole dominates the scene as usual). Figure 6*d*, plate 3, is a particularly good picture of the destructive tornado of 11 April 1965 which struck Elkhart, Indiana. Althought the dust funnel is more fully developed here than in the previous pictures, we again see a strong conical outflow, the average angle of the right hand debris cone being about 48°. Figure 6*e*, plate 4, is a beautiful photograph of a tornado whose cascade angle is either 90° (that is, upflow throughout the vortex as in mode C) or very nearly 90°. A number of further photographs can be found on pages 24, 88, and 136 of Flora's monograph (1954).

If air flow in a tornado is assumed to follow the pattern associated with zone B, and if dust or debris accurately traces the motion, then material swept up from the area near the foot of the vortex would become concentrated on the conical stream surface defined by the angle $\beta$ and would appear to the observer as a dark region with a conical lower boundary, beneath which the air is clear. *Therefore to the extent that debris (or mist) in figures 6a to 6d faithfully traces the air motion, we obtain specific observational evidence in favour of mode* B, *and of the associated occurrence of descending air motion in the core of tornadoes whose foot reaches to the ground.* As Professor Lighthill has noted in a private communication, the main problem in identifying the debris shower with the air motion is that debris is subject to gravity and centrifugal force in addition to the swirling motion of the air. Gravitational and inertial effects on dust are doubtless slight over the time scales involved and consequently can be ignored. Centrifugal effects, however, are certainly larger, but whether they can account for cascades in a flow with a central updraft is doubtful. There are a number of reasons for this, but perhaps the strongest is simply that the lower boundaries of observed cascades are relatively straight and well defined. Assuming that such cascades of dusty debris were in fact caused by centrifugal effects in an air motion otherwise involving a simple inflow and updraft, the vertical component of the air velocity would have to fall off like the cube of the radius, which appears unlikely. The random distribution of size among the dust or debris particles would also dictate a variety of trajectories and accordingly a badly defined cascade boundary, accentuated further by what would seem to be an unstable moving front (Taylor instability). The appearance of the cascade itself thus argues against centrifugal force as the generating mechanism, though naturally this feature will have some moderating effect.† The author has seen motion pictures in which large pieces of debris were swirled out at approximately 45° while the dust cascade stood at about 60°. If centrifugal force were the prime mechanism one would expect a very much larger relative effect on the heavier debris; it would thus seem more reasonable to view the dust as a tracer of the air flow and the larger debris alone as significantly affected by centrifugal force.

An extreme example of a dust cloud is shown in figure 6*f*, plate 4. The tornado was 11 000 feet distant and was moving at a speed of about 15 miles/h to the left. It is difficult to explain the immense size of the cloud by centrifugal force; moreover, were there a significant central updraft it is hard to imagine that the funnel could have remained unobscured by debris. One may conjecture that the dust cloud is the result, therefore, of a prolonged cascade effect at a fairly low angle. If we estimate this intuitively from the figure to be about 30°–35°, the corresponding $k$ is about 7, indicating a relatively low level of turbulence. Significantly this tornado not only moved slowly but had a high degree of coherence since it remained actively in contact with the ground and seriously destructive for some $1\frac{1}{2}$ hours.

The possibility of central downflow in a tornado is reinforced by the careful observational evidence of Hoecker (1960), who writes, 'A sequence of photographs of [a] tornado on file [shows] the ground-based debris cloud increasing in diameter as the suspended funnel widened and lowered toward the ground. No debris was seen to rise in the region beneath the cut-off tip of the condensation funnel but all of it ascended exterior to the tapered cylindrical funnel which widened upward. In the pictures and movies of the Dallas tornado no large clouds of dust or spray, or chunks of structures were observed ascending along the trunk when the trunk was touching the ground. Had any large chunks been carried upward inside the funnel, surely a few would have been thrown outward through the funnel wall by centrifugal force much as they were observed to do when the funnel tip had retracted.' The problem of descending flow has also been logically considered by Morton (1966). He concludes that 'The radial flow at ground level is inwards towards the low-pressure core, and hence the net vertical flow must be upward; however, this does not imply that the flow need be upward over the whole core section, and it is quite possible that under appropriate circumstances there may be downflow near the axis'. Following a consideration of Hoecker's work, he states further, 'When the funnel descends to the ground the column of dust and debris often forms a shell which is clearly separated from the funnel within. Such a shell might be formed by particles swept up between an outer flow spiralling in and up, and an inner flow spiralling down the axis and then out and up to form a cellular core; and this seems more plausible than the traditional explanation that the dust column is formed by centrifugal separation of material raised in a strong upflow near the funnel axis, which explains neither the frequent separation between shell and inner core nor the obviously cylindrical shell-form sometimes observed.'

Some direct observations of downflow have also been reported. E. M. Brooks, in the *Compendium of meteorology*, states that 'whirling air coming down at 1800 ft/min has actually been encountered by a glider in a desert dust whirl', and M. Hale, in a private communication to the author, has described the flight of a naval airplane through a mild waterspout, in which the plane

† The fact that figure 6*e* shows no cascade is itself not an argument against centrifugal force, since the dust carried aloft may be relatively fine and dry.

Figure 6*a*. Waterspout off Martha's Vineyard, 19 August 1896. (Courtesy Dukes County Historical Society, Edgartown, Mass.)

Figure 6*b*. Tornado at Elbow Lake, Minnesota, 5 September 1969. (Photograph by Olaf Dybdal.)

(*Facing p.* 358)

FIGURE 6*c*. Tornado in Kansas, 21 April 1967.
FIGURE 6*d*. Tornado at Elkhart, Indiana, 11 April 1965. (Photograph by Paul Huffman.)

FIGURE 6*e*. Tornado at Enid, Oklahoma, 5 June 1966. (Environmental Science Services Administration photograph by Leo Ainsworth.)

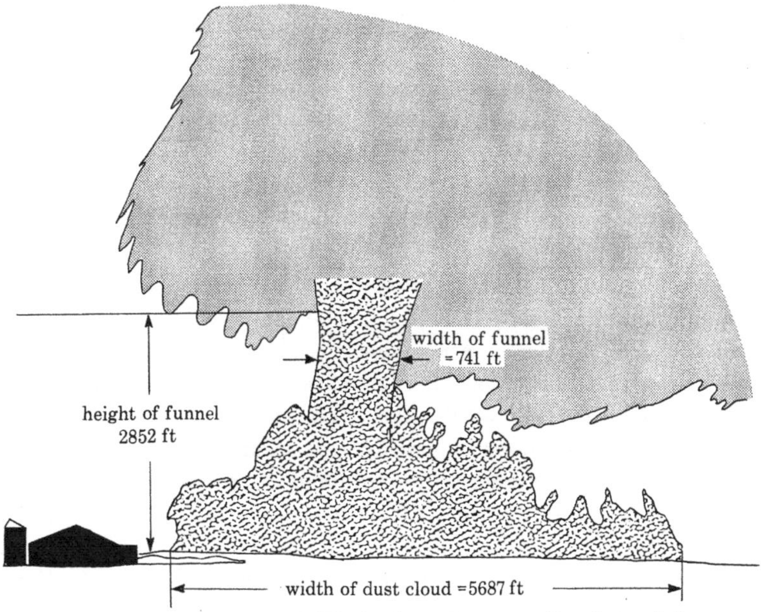


FIGURE 6*f*. Tornado at Scottsbluff, Nebraska, 27 June 1955. (Drawing made from measured photograph, Van Tassel 1955, p. 257.)

was seen to experience a downward turn. In addition to direct reports of this kind, a series of interesting experiments has been conducted by Ward (1956). In laboratory-generated vortices maintained by a relative low-pressure region above the vortex location, Ward has observed combined inflow at the lower boundary and descending air motion in the core at certain critical combinations of the flow parameters when the motion becomes turbulent. The theoretical work of Sullivan (1959) and Bellamy-Knights (1970) likewise admits the possibility of central downdrafts in vortex cores, though as we have remarked in the introduction their models do not effectively reflect either the adherence condition at the boundary surface or the fact that the velocity should approach zero at large radial distances. Kuo (1967) has determined approximate numerical solutions of the equations representing vortex motion in an unstably stratified atmosphere, and finds again the possibility of central downflow. Emphasizing the possible complexity of the flow situation, Turner (1966) has presented approximate solutions of the Navier–Stokes equations with annular downdrafts and a central updraft. His work, like that of Sullivan, assumes solid rotation at large radial distances.

While the preceding discussion has focused particularly on the observational evidence for flow patterns of type B, this was held necessary in view of the widely disseminated opinion that tornadoes and waterspouts are invariably updraft phenomena. It should not be concluded from this particular emphasis, however, that updraft phenomena on the other hand are unlikely or unusual. Besides the observational evidence of figure 6*e*, plate 4, and other similar photographs, Hoecker (1960) specifically notes that 'the air flow pattern in a tornado is complex and changes from one tornado to another and may change from time to time in any one tornado'. Similarly, Morton says 'It may be noted that there is no reason why tornado vortices should all be dynamically similar.' To the extent that the present model represents tornado phenomena, it bears out both statements, since for different values of $k$ the resulting motions are *not* dynamically similar, nor is it necessary that $k$ retain a constant value through the life of a given tornado. On the other hand, one qualitative attribute of the model remains valid irrespective of the choice of $k$, provided only that the parameter $P$ is determined by the physically motivated relation $P = \hat{P}(k)$. Indeed, the corresponding flows are then of type B if $k > 2.86$ and of type C with $P = 1$ when $k \geqslant 2.86$: *in either case the motion near the boundary plane is inwards toward the vortex axis.* Observational evidence (Brooks 1951) on this point is convincingly in agreement.

Although the conditions $k < 2.86$, $P = 1$ are consistent with observational evidence (see particularly figure 6*e*) and with the prevalent inflow-updraft theory, nevertheless, we have earlier concluded that $k$ would tend to fall in the range 3 to 7. The difficulty here is, however, more apparent than real. One must not, to begin with, place excessive confidence in empirical values of $\nu$ carried over from one situation to another, so that we need not automatically eliminate values of $k$ which remain slightly less than 3. Alternately, when $k = 3$ the cascade angle is approximately 89°, and within any reasonable limits of accuracy this amounts to a motion with a central updraft. Finally, the various reservations in the basic model which we have discussed earlier must surely be taken into consideration in applying the conclusions.

In summary, beyond the details of numerical computation and the idealization of the modelling process, the present discussion indicates the possibility of a complete analysis of the steady state interaction of a line vortex and a boundary plane in a viscous fluid. The family of solutions which has been obtained moreover displays a number of the diverse effects observed in both tornadoes and waterspouts, and places particular emphasis on the possibility of centrally descending air motion as a realistic flow pattern in geophysical vortex phenomena.

The author wishes to express his appreciation to Professors Bernard Vonnegut and Thomas Lundgren for their helpful and stimulating conversations during the first drafting of the paper. He also most particularly wishes to thank Dr Edwin Kessler of the National Severe Storms Laboratory in Norman, Oklahoma, for his sustaining interest in the ideas presented here. The results were presented originally at the Instructional Conference on Differential Equations held at Edinburgh, Scotland, in September 1967; the research was partially supported by the Air Force Office of Scientific Research Grant No. AF–AFOSR 883–67.

I am very greatly indebted to Dr Wendell Hung of the Department of Aeronautics and Engineering Mechanics of the University of Minnesota for his skilful and accomplished work in obtaining the numerical results of the paper, and also for his careful preparation of the accompanying figures.

## References

Bellamy-Knights, P. G. 1970 An unsteady two-cell vortex solution of the Navier–Stokes equations. *J. Fluid Mech.* **41**, 673–687.

Berker, R. 1963 Intégration des équations du mouvement d'un fluide visqueux incompressible. *Hand. Phys.* VIII/2. Berlin: Springer-Verlag.

Burgers, J. M. 1948 A mathematical model illustrating the theory of turbulence. *Adv. appl. Mech.* **1**, 197–199.

Burggraf, O., Stewartson, K. & Belcher, R. 1971 Boundary layer induced by a potential vortex. *Phys. Fluids.* **14**, 1821–1833.

Coddington, E. & Levinson, N. 1955 *Ordinary differential equations.* New York: McGraw-Hill.

Dunford, N. & Schwartz, J. 1958 *Linear operators,* **1**. New York: Interscience.

Goldstik, M. A. 1960 A paradoxical solution of the Navier–Stokes equations. *Prik. Mat. Mekh.* **24**, 610–621. Translated in *J. appl. Math. Mech.* **24**, 913–929.

Long, R. R. 1958 Vortex motion in a viscous fluid. *J. Meteor.* **15**, 108–112.

Long, R. R. 1961 A vortex in an infinite viscous fluid. *J. Fluid Mech.* **11**, 611–624.

Rott, N. 1958 On the viscous core of a line vortex. *Z.A.M.P.* **9***b*, 543–553.

Schlichting, H. 1960 *Boundary layer theory,* Fourth ed. New York: McGraw-Hill.

Sullivan, R. D. 1959 A two-cell solution of the Navier–Stokes equations. *J. Aero/Space Sci.* **26**, 767–768.

## Meteorological references

Brooks, E. M. 1951 Tornadoes and related phenomena. *Compendium of meteorology.* Boston: Amer. Met. Society.

Flora, S. D. 1954 *Tornadoes in the United States.* University of Oklahoma Press, Norman.

Hoecker, W. H. 1960 Wind speed and air flow patterns in the Dallas tornado of April 2, 1957. *Mon. Weath. Rev.* **88**, 167–180.

Kessler, E. 1970 Tornadoes. *Bull. Am. Met. Soc.* **51**, 926–936.

Kuo, H. L. 1967 Note on the similarity solutions of the vortex equations in an unstably stratified atmosphere. *J. atmos. Sci.* **24**, 95–97. Cf. also *J. atmos. Sci.* **23**, 25–42 (1966).

Kuo, H. L. 1971 Axisymmetric flows in the boundary layer of a maintained vortex. *J. atmos. Sci.* **28**, 20–41.

Lewis, W. & Perkins, P. J. 1953 Recorded pressure distribution in the outer portion of a tornado vortex. *Mon. Weath. Rev.* **81**, 379–385.

Morton, B. R. 1966 Geophysical vortices. *Prog. aeronaut. Sci.* **7**, 145–193.

Oerstad, H. C. 1839 On water-spouts. *Am. J. Sci. Arts* **37**, 250–267. Translated from *Schumacher's Jahrbuch für* 1838.

Turner, J. S. 1966 The constraints imposed on tornado-like vortices by the top and bottom boundary conditions. *J. Fluid Mech.* **25**, 377–400.

Van Tassel, E. L. 1955 The North Platte Valley tornado outbreak of June 27, 1955. *Mon. Weath. Rev.* **83**, 255–264.

Vonnegut, B. 1960 Electrical theory of tornadoes. *J. geophys. Res.* **65**, 203–212.

Ward, N. B. 1956 Temperature inversion as a factor in the formation of tornadoes. *Bull. Am. Met. Soc.* **37**, 145–151. Cf. also *Nat. Sev. Storms Lab. Tech. Memo.* no. 52 (1970).

Ying, S. J. & Chang, C. C. 1970 Exploratory model study of tornado-like vortex dynamics. *J. atmos. Sci.* **27**, 3–14.

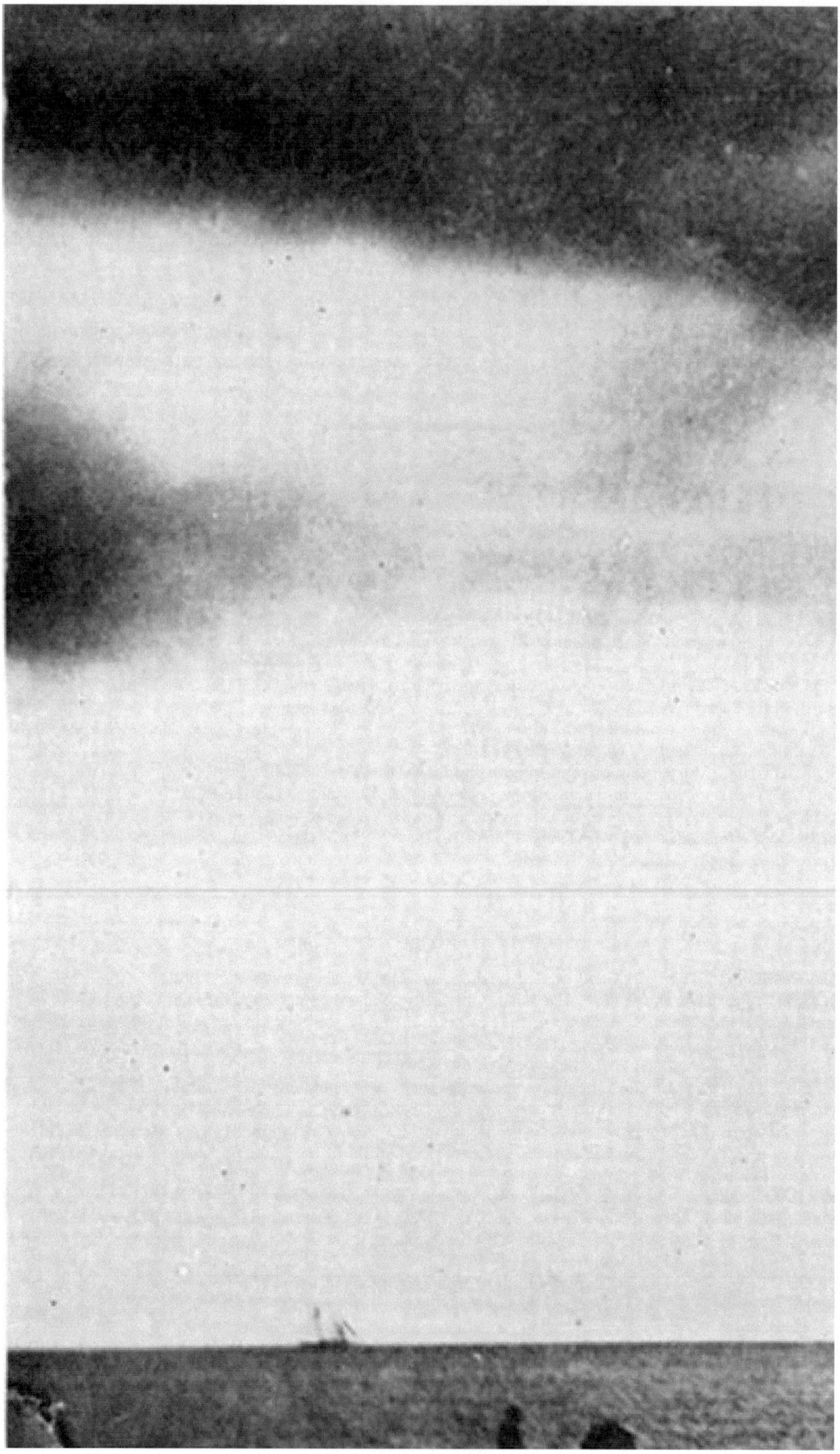

*J. Aerosp. Sci.* 28 (1961), 990–991. **Reprint V.2.[ii]**


## On the Entropy Change Through a Shock Layer

James Serrin and Y. C. Whang
*Department of Mathematics and Department of Aeronautical Engineering, University of Minnesota, Minneapolis, Minn.*
May 11, 1961

When a viscous, heat-conducting gas traverses a shock wave its temperature, density, pressure, and entropy are all increased. It might be supposed, therefore, that during the *process* of shock transition all of these quantities increase monotonically from their initial to their final values. This is indeed true for the temperature, density, and pressure, as we shall show here, but it is not the case for the entropy. Rather surprisingly, the latter quantity first increases to a local maximum in the interior of the shock, and then undergoes a *decrease* to its final value (Fig. 2). These conclusions follow from general thermodynamic assumptions, and therefore apply to all gases of practical interest.

Consider, then, a steady one-dimensional shock transition in an arbitrary viscous, heat-conducting fluid. As is well known, the transition is governed by the equations

$$\begin{aligned} \rho u &= m \qquad (m > 0) \\ \kappa(dT/dx) &= m[E - {}^1/_2(u-a)^2 - b] \qquad (1) \\ (\lambda + 2\mu)(du/dx) &= p + m(u-a) \end{aligned}$$

representing conservation, of mass, energy, and momentum, respectively. Here $a$, $b$, $m$ are constants, $\lambda$, $\mu$, $\kappa$ are constitutive coefficients of the fluid (positive, but not necessarily constant), and the remaining notation is standard.[1] We require the following thermodynamic assumptions concerning the fluid:

*The specific heats are positive, and the following inequalities hold*

$$(\partial p/\partial \rho)_S > 0, \quad (\partial p/\partial S)_\rho > 0, \quad (\partial^2 p/\partial \tau^2)_S > 0$$

*where $\tau = 1/\rho$ is the specific volume, and S is the entropy.*

The first of these inequalities simply states that the speed of sound is real. The second and third are slightly more special assumptions, but are nevertheless satisfied in all practical situations. Now $E$ and $p$ in Eqs. (1) can be expressed as functions of thermodynamic variables $\rho$ and $T$. Thus Eq. (1) can be reduced to the following pair of equations

$$\kappa(dT/dx) = mL(u,T), \quad (\lambda + 2\mu)(du/dx) = mM(u,t) \qquad (2)$$

The curves $L = 0$ and $M = 0$ may be plotted in the $(u,T)$ plane, and have the general form shown in Fig. 1. A *shock layer* or *shock transition* is then a solution of Eq. (2):

$$u = u(x), \quad T = T(x), \quad -\infty < x < +\infty$$

which, considered as a parametric curve in the $(u,T)$ plane, joins the singular points $Z_1$ and $Z_2$. The qualitative form of the curves $L = 0$ and $M = 0$, and the existence of a shock transition

Fig. 1. Shock transition diagram.

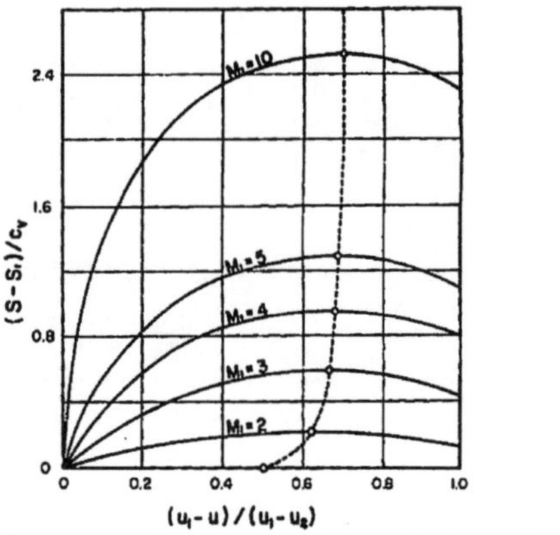


Fig. 2. Entropy variation through shock layer (ideal gas with $\gamma = 7/5$, $(\lambda + 2\mu)c_p/\kappa = 1$).

as shown in Fig. 1, follow from the thermodynamic assumptions noted above; cf. Gilbarg.[2] It should be observed that the shock curve rises to the left as $x$ increases, because $L > 0$ and $M < 0$ throughout the region $R$ between the curves $L = 0$ and $M = 0$.

Consider now the variation of the thermodynamic state as one follows the shock curve from $Z_1$ to $Z_2$. Evidently

$$d\rho = -mu^{-2}du > 0, \quad dT > 0$$

so that both $\rho$ and $T$ increase monotonically from $Z_1$ to $Z_2$. Moreover

$$\begin{aligned} dp &= (\partial p/\partial \rho)_T d\rho + (\partial p/\partial T)_\rho dT \\ &= (1/\gamma T)[T(\partial p/\partial \rho)_S d\rho + c_p(\partial p/\partial S)_\rho dT] > 0 \end{aligned}$$

according to our thermodynamic assumptions. It remains only to investigate the change in entropy. But we have

$$TdS = dE + pd\tau = dL + Mdu$$

Thus at $Z_1$ we have $dS = dL/T > 0$ in the direction of the shock curve, and at $Z_2$ we have $dS = dL/T < 0$ along the shock curve. In other words, although $S$ increases as the shock begins, it must be *decreasing* over the final portion of the transition. This completes the proof of our assertions.

For an ideal gas with constant specific heats and with Prandtl number $(\lambda + 2\mu)c_p/\kappa = 1$, Eq. (2) has the solution

$$c_pT + {}^1/_2u^2 = {}^1/_2u_1^2\{1 + [2/(\gamma - 1)]M_1^{-2}\}$$

as noted by Becker. In this case the entropy variation in the shock transition may be explicitly expressed in the form

$$\begin{aligned} \exp[(S - S_1)/c_v] &= (T/T_1)(\rho_1/\rho)^{\gamma-1} = (T/T_1)(u/u_1)^{\gamma-1} \\ &= \{[(\gamma - 1)/2]M_1^2[1 - (u^2/u_1^2)] + 1\} \times (u/u_1)^{\gamma-1} \end{aligned}$$

which is plotted in Fig. 2 for $\gamma = 7/5$. It may be observed that in the limit $M_1 \to 1$ the maximum entropy occurs exactly when the velocity is half way between its initial and final values.

### References

[1] Serrin, J., *Mathematical Principles of Classical Fluid Mechanics, Handbuch der Physik*, Vol. VIII/1, pp. 226–230; Springer-Verlag, Berlin, 1959.
[2] Gilbarg, D., *The Existence and Limit Behavior of the One-Dimensional Shock Layer*, Am. J. Math., Vol. 73, No. 2, pp., 256–275, April 1951.

*Z. Angew. Math. Mech.* 39 (1959), 295–299.


**Reprint V.2.[iii]**

©Zeitschrift für angewandte Mathematik und Mechanik
Akademie-Verlag
Band 39 Heft 7/8 Juli/August 1959

# Poiseuille and Couette flow of non-Newtonian fluids

By *James Serrin*

*Vor einigen Jahren fand Rivlin einige exakte Lösungen der Bewegungsgleichungen einer nicht-Newtonschen Füssigkeit, die den wohlbekannten Lösungen der Navier-Stokesschen Gleichungen entsprechen, welche von Poiseuille und Couette angegeben wurden. In der vorliegenden Arbeit wird die physikalische Bedeutung der Rivlinschen Lösungen geklärt und eine Methode zur Bestimmung des Koeffizienten der Querviskosität vorgeschlagen. Ferner wird gezeigt, wie sich gewisse Experimente von Merrington und Weissenberg auf Grund der Resultate dieser Arbeit erklären lassen.*

*Some years ago Rivlin discovered several exact solutions of the equations of motion of a non-Newtonian fluid, corresponding to the well-known Poiseuille and Couette solutions of the Navier-Stokes equations. In this paper the physical nature of Rivlin's solutions is clarified, and a method is proposed for determining the value of the cross-viscosity coefficient. In addition, it is shown how certain experiments of Merrington and Weissenberg can be explained by means of the results of the paper.*

*Il y a quelques années que Rivlin a trouvé quelques solutions exactes des équations de mouvement pour un liquide non-Newton, solutions qui correspondent aux solutions bien connues des équations de Navier-Stokes indiquées par Poiseuille et Couette. Dans l'article présent la signification physique de la solution de Rivlin est éclairée, et une méthode pour la détermination du coefficient de la viscosité transversale est proposée. En outre, il est démontré comment certains expériments de Merrington et Weissenberg peuvent etre expliqués sur la base des résultats de cet article.*

Несколько лет тому назад Ривлин нашел некоторые точные решения уравнений движения ненютоновской жидкости, соответсвующие хорошо известным решениям Пуасёля и Куэта для уравнения Навье-Стокса. В настоящей работе выясняется физический характер решений Ривлина и предлогается метод для опрделения значения коэффициента поперечной вязкости. Кроме того показывается, как некоторые опыты Меррингтона и Вейссенберга могут быть обяснены при помощи результатов настоящей работы.

The theory of non-Newtonian fluids is based on a stress-deformation relation of the form

$$\mathsf{T} = \alpha \mathsf{I} + \beta \mathsf{D} + \gamma \mathsf{D}^2 \quad (1),$$

where $\mathsf{T}$ is the stress tensor, $\mathsf{D}$ is the rate-of-deformation tensor, and $\alpha$, $\beta$, $\gamma$ are certain scalar coefficients[1]). We shall be concerned here with incompressible fluids, so that $\alpha$ is identified as the negative of the dynamic pressure, $\alpha = -p$. The coefficients $\beta$ and $\gamma$ are assumed to be known functions of the scalar invariants of $\mathsf{D}$ and of the thermodynamic state of the fluid. That is

$$\beta = \beta(\mathrm{II}, \mathrm{III}; T), \qquad \gamma = \gamma(\mathrm{II}, \mathrm{III}; T),$$

where II, III denote respectively the sum of the principal minors of $\mathsf{D}$, and the determinant of $\mathsf{D}$, and $T$ is the fluid temperature.

The fluid motion is governed by the familiar Cauchy law

$$\varrho \frac{d\mathsf{v}}{dt} = \varrho \mathsf{f} + \operatorname{div} \mathsf{T} \quad (2),$$

in which $\mathsf{v}$ is the velocity vector and $\mathsf{f}$ is the gravitational force, $\mathsf{f} = -g\,\mathsf{k}$. Some years ago Rivlin ([2], [3]) discovered several exact solutions of equations (1), (2), corresponding to certain well-known solutions of the Navier-Stokes equations. We are interested here in two of these exact solutions, namely those corresponding to Poiseuille and Couette flow: our purpose is to clarify the physical nature of these solutions, and especially to point out their similarities and differences with the classical solutions. Some steps in this direction have already been taken by C. Truesdell[2]), but the present analysis seems to be more complete.

[1]) The reader unfamiliar with the subject of non-Newtonian fluids will find adequate references and background material in the extensive monograph of C. Truesdell [5], or in [4], §§ 58—60, 65.

[2]) [5], § 72. This work is, in fact, the inspiration for the present paper.

Finally, we observe that equation (12) offers an excellent opportunity to determine an experimental value for the coefficient $\gamma$ in a second order theory. This statement will be expanded in the work below.

## 1. Poiseuille flow

In cylindrical polar coordinates $(r, \Theta, z)$ we seek, following Rivlin, a solution of equations (1) and (2) of the form

$$v_r = v_\Theta = 0\,, \quad v_z = w = w(r)\,, \quad 0 \leqq r \leqq a \quad \dots\dots\dots \quad (3).$$

This corresponds to steady laminar flow down a straight pipe of diameter $2a$. For such a motion the deformation tensor $\mathsf{D}$ takes the form

$$\mathsf{D} = \frac{1}{2}\begin{pmatrix} 0 & 0 & w' \\ 0 & 0 & 0 \\ w' & 0 & 0 \end{pmatrix},$$

and therefore

$$\mathsf{T} = -p\,\mathsf{I} + \frac{1}{2}\beta w'\begin{pmatrix} 0 & 0 & 1 \\ 0 & 0 & 0 \\ 1 & 0 & 0 \end{pmatrix} + \frac{1}{4}\gamma w'^2\begin{pmatrix} 1 & 0 & 0 \\ 0 & 0 & 0 \\ 0 & 0 & 1 \end{pmatrix},$$

where $\beta = \beta\left(-\frac{1}{4}w'^2, 0; T\right) = \beta(r, T)$, $\gamma = \gamma(r, T)$. Since the motion (3) has zero acceleration, Cauchys law reduces simply to $\varrho\mathsf{f} + \operatorname{div}\mathsf{T} = 0$. Moreover, using the techniques of tensor analysis[3]), one finds that the covariant component of div T is given by

$$(\operatorname{div}\mathsf{T})_i = \frac{1}{r}\left[\frac{\partial}{\partial x^k}(r\,T_i^k) - T_2^2\frac{\partial r}{\partial x^i}\right] \quad \dots\dots\dots \quad (4),$$

where $x_1 = r$, $x_2 = \Theta$, $x_3 = z$. The equations of motion thus become

$$\left.\begin{aligned} -\frac{\partial p}{\partial r} + \frac{1}{r}\left[\frac{\partial}{\partial r}\left(\frac{1}{4}r\gamma w'^2\right) + \frac{\partial}{\partial z}\left(\frac{1}{2}r\beta w'\right)\right] &= 0, \\ -\varrho g - \frac{\partial p}{\partial z} + \frac{1}{r}\left[\frac{\partial}{\partial r}\left(\frac{1}{2}r\beta w'\right) + \frac{\partial}{\partial z}\left(\frac{1}{4}r\gamma w'^2\right)\right] &= 0 \end{aligned}\right\} \quad \dots\dots \quad (5),$$

in which the $\Theta$-dependence of all quantities is taken zero. Equations (5) can be integrated provided that $\beta$ and $\gamma$ do not depend on the temperature, or if the temperature is assumed to be a function of $r$ alone. Indeed in this case one finds the following unique solution of (5),

$$p = (C - \varrho g)z + f(r)\,, \qquad \beta w' = Cr \quad \dots\dots\dots \quad (6),$$

where $C$ is an arbitrary constant, and

$$f(r) = \frac{C^2}{4}\left[\frac{\gamma r^2}{\beta^2} + \int\frac{\gamma r}{\beta^2}\,dr\right].$$

The second equation of (6) is a differential equation for the velocity $w(r)$, to be solved subject to the boundary condition $w(a) = 0$, (i. e., adherence at the pipe wall). Due to the fact that $\beta$ will in general involve $w'$, the solution of this differential equation may be difficult to obtain explicitly.

To complete the solution, we observe that the stress components are given by

$$\widehat{rr} = \widehat{zz} = -p + \frac{1}{4}\gamma w'^2 = (\varrho g - C)z - \frac{C^2}{4}\int\frac{\gamma r}{\beta^2}\,dr\,, \qquad \widehat{rz} = \frac{1}{2}Cr\,,$$

and that the total mass flux $\mathfrak{M}$ is

$$\mathfrak{M} = -\iint \varrho w r\,dr\,d\Theta = -\pi\varrho\int_0^a w\,d(r^2) = \pi\varrho C\int_0^a r^3\beta^{-1}\,dr \quad \dots\dots \quad (7),$$

where in the last equality we have used the boundary condition $w(a) = 0$. Notice, in particular, that the normal stress $\widehat{rr}$ on the pipe wall varies linearly along the length of the pipe, exactly as in the classical solution.

[3]) The reader is assumed to have an elementary knowledge of this subject in order to derive equation (5).

Quadratic dependence. In the special case where the components of T depend at most quadratically on the components of D, it can be seen that $\beta$ and $\gamma$ must depend, at most, on the fluid temperature. Supposing in fact that $\beta$ and $\gamma$ are constant, integration of (6) yields the familiar formula

$$w = \frac{C}{4\mu}(r^2 - a^2), \qquad (\beta = 2\mu) \quad (8)$$

while from (7) follows

$$\mathfrak{M} = \frac{\pi \varrho C}{8\mu} a^4 \quad (9).$$

Since $\gamma$ does not appear in these formulas, it is clear that standard viscocity or velocity measurements will not indicate the presence or absence of non-linear viscosity in a fluid.

The effect of $\gamma$ is felt, however, in a non-uniform distribution of pressure across a section of the pipe. To investigate this effect in detail, we may suppose the fluid to issue from the pipe into an atmosphere at pressure $p_0$, the latter exerting a force on the output cross-section in the amount $\pi a^2 p_0$. Since $\beta$ and $\gamma$ are constant, the stress components are given by

$$\widehat{rr} = \widehat{zz} = (\varrho g - C) z - 2\gamma D^2 r^2 - \text{Const} \quad (10),$$

where $D = C/8\mu$, and the following force balance holds at the exit section (taken to be $z = 0$),

$$\pi a^2 p_0 = -\pi \int_0^a \widehat{zz}\, d(r^2) = \pi \gamma D^2 a^4 + \pi a^2 \text{Const}.$$

Using this equation to eliminate the Const. from (10), we obtain

$$\widehat{rr} = \widehat{zz} = (\varrho g - C) z + \gamma D^2 (a^2 - 2r^2) - p_0 \quad (11).$$

Let $P$ be the force per unit area which the fluid exerts on the pipe walls; that is, $P = -\widehat{rr}$ evaluated at $r = a$. In virtue of (9) and (11) we have then

$$P - p_0^* = \left(\frac{\Gamma}{a}\right)^2 \left(\frac{8\mu}{\Gamma} z + \gamma\right) \quad (12),$$

where $p_0^* = p_0 - \varrho g z$ and $\Gamma = \mathfrak{M}/\pi a^2 \varrho$ = volume of flow per second per unit cross section area.

Formula (12) may serve as the theoretical explanation of an interesting experimental phenomenon discovered by Merrington [1], namely the tendency for a fluid stream to swell at the exit section of a viscometer. Indeed, according to (12) we have

$$P - p_0 = (\Gamma/a)^2 \gamma \quad (13)$$

at the exit section, indicating that excess pressure in the fluid can be caused simply by a positive coefficient $\gamma$. Formula (13) implies, moreover, that the swelling should be emphasized by high flux and small pipe radius, a fact actually observed by Merrington.

An experimental verification of (12) could easily be secured by measuring the variation of $P - p_0$ along the length of the pipe, for various values of $\Gamma$. By plotting $(a/\Gamma)^2 (P - p_0^*)$ against the pipe length $z$, one should then obtain a series of straight lines, each intercepting the axis $z = 0$ at the fixed height $\gamma$; moreover, from the slope of these lines one can read off the coefficient of viscosity $\mu$. The internal consistency of the theory is immediately decidable from the range of values of $\mu$ and $\gamma$ which are experimentally obtained for various values of $\Gamma$. To the author's knowledge, the required experiments have not yet been performed.

### 2. Continuation

The conclusion that the fluid stream should tend to swell at the exit section of a pipe can also be derived without the assumption that $\beta$ and $\gamma$ are constant. Indeed, in the general case let us write

$$\widehat{rr} = \widehat{zz} = (\varrho g - C) z - (C/2)^2 \int_a^r \gamma \beta^{-2} r\, dr - \text{Const.} \quad (10)'.$$

Then the following force balance holds at the exit section,

$$\pi a^2 p_0 = -\pi \int_0^a \widehat{zz}\, d(r^2) = -\pi (C/2)^2 \int_0^a \gamma \beta^{-2} r^3\, dr + \pi a^2 \text{Const}.$$

where a straightforward integration by parts was used in deriving the final equality. Using this equation to eliminate the Const. in (10)′ leads to the formula

$$\widehat{rr} = \widehat{zz} = (\varrho g - C) z - \frac{C^2}{4} \left[ \int_a^r \frac{\gamma r}{\beta^2} dr + \frac{1}{a^2} \int_0^a \frac{\gamma r^3}{\beta^2} dr \right] - p_0 \quad (11)'$$

and finally

$$P - p_0 = \left(\frac{C}{2\,a}\right)^2 \int_0^a \frac{\gamma\, r^3}{\beta^2}\, dr \quad \ldots \ldots \ldots \ldots \quad (13)'$$

at the exit section. Thus whether or not $\beta$ and $\gamma$ are constants, the fluid stream will tend to swell if $\gamma > 0$.

Now let $\Gamma$ denote as before the volume of flow per second per unit cross section area. Then

$$\Gamma = \mathfrak{M}/\pi\, a^2\, \varrho = C\, a^{-2} \int_0^a \beta^{-1}\, r^3\, dr\,,$$

and so by the Cauchy inequality (assuming $\gamma > 0$)

$$\Gamma^2 \leqq \frac{C^2}{a^4} \int_0^a \frac{r^3}{\gamma}\, dr \int_0^a \frac{\gamma\, r^3}{\beta^2}\, dr \leqq \frac{C^2}{4\,\mathrm{Min}\,\gamma} \int_0^a \frac{\gamma\, r^3}{\beta^2}\, dr\,.$$

Combining this inequality with (13)′ yields the interesting result

$$P - p_0 \geqq (\Gamma/a)^2\, \mathrm{Min}\,\gamma \quad \ldots \ldots \ldots \ldots \quad (13)''.$$

Inequality (13)″ not only exhibits a remarkable similarity to formula (13), but also shows as before that the swelling should be emphasized by high flux and small pipe radius.

### 3. Couette flow

Again using cylindrical polar coordinates, we seek a solution of (1), (2) of the form

$$v_r = v_z = 0\,, \qquad v_\Theta = r\,\omega(r) \quad \ldots \ldots \ldots \ldots \quad (14).$$

In this case

$$\mathsf{D} = \frac{1}{2}\begin{pmatrix} 0 & r\,\omega' & 0 \\ r\,\omega' & 0 & 0 \\ 0 & 0 & 0 \end{pmatrix}, \qquad \mathsf{D}^2 = \frac{1}{4}\begin{pmatrix} r^2\,\omega'^2 & 0 & 0 \\ 0 & r^2\,\omega'^2 & 0 \\ 0 & 0 & 0 \end{pmatrix},$$

(using physical components), and $\beta = \beta\left(-\frac{1}{4}\, r^2\,\omega'^2, 0;\, T\right) = \beta(r, T)$, $\gamma = \gamma(r, T)$. With the help of (4) the equations of motion are easily reduced to the following system,

$$-\varrho\,\frac{v_\Theta^2}{r} = -\frac{\partial p}{\partial r} + \frac{\partial}{\partial r}\left(\frac{1}{4}\, r^2\,\gamma\,\omega'^2\right) + \frac{\partial}{\partial \Theta}\left(\frac{1}{2}\,\beta\,\omega'\right),$$

$$0 = -\frac{\partial p}{\partial \Theta} + \frac{1}{r}\left[\frac{\partial}{\partial r}\left(\frac{1}{2}\, r^3\,\beta\,\omega'\right) + \frac{\partial}{\partial \Theta}\left(\frac{1}{4}\, r^3\,\gamma\,\omega'^2\right)\right],$$

$$0 = -\varrho\, g - \frac{\partial p}{\partial z}\,.$$

Assuming that $\beta$ and $\gamma$ do not involve $T$ explicitly, or else that $T$ is a function of $r$ alone, these equations have the integral

$$p = -\varrho\, g\, z + f(r)\,, \qquad r^3\,\beta\,\omega' = C \quad \ldots \ldots \ldots \ldots \quad (15),$$

where $C$ is an arbitrary constant and

$$f(r) = \left(\frac{C}{2\,\beta}\right)^2 \frac{\gamma}{r^4} + \varrho \int r\,\omega^2\, dr\,.$$

The only stress component whose value is affected by $\gamma$ is $\widehat{zz}$, as the reader can easily check.

Quadratic dependence. Setting $\beta$ = constant, and $\gamma$ = constant, as before, integration of (15) yields the velocity distribution

$$v = r\,\omega = A\, r^{-1} + B\, r\,, \qquad (A = -C/2\,\beta) \quad \ldots \ldots \ldots \quad (16),$$

exactly as in the classical Couette flow. A difference arises in the stress component $\widehat{zz}$, which now takes the form

$$\widehat{zz} = -p = \varrho\, g\, z - A^2\,\gamma\, r^{-4} - \varrho \int (A + B\, r^2)^2\, r^{-3}\, dr\,.$$

To investigate the effect of $\gamma$ on the motion, we suppose the flow to take place between two concentric rotating cylinders, with the upper fluid surface being left open to the atmosphere. We shall be interested in the shape of this free surface. Now Couette flow involves a non-vanishing stress component $T_{r\Theta}$, so that (16) can only approximate the actual flow under consideration; nevertheless, as long as the free surface remains relatively horizontal (i. e. at relatively low speeds of rotation) the discrepancy should not be serious. Assuming, then, that the free surface remains nearly horizontal, its locus is given by $\widehat{zz}$ = constant, and its slope satisfies

$$\frac{dz}{dr} = \frac{A^2}{g\, r^5} \left\{ [(1 + \Theta)\, r]^2 - \frac{4\,\gamma}{\varrho} \right\}, \qquad \Theta = \frac{B}{A}\, r^2 .$$

We shall consider two cases in detail: (I) The inner cylinder ($r = r_1$) remains at rest, and (II) the outer cylinder ($r = r_2$) remains at rest. The constants $A$ and $B$ for these cases are easily determined, and in particular we find that

$$\Theta = \begin{cases} -(r/r_1)^2 & \text{in Case I,} \\ -(r/r_2)^2 & \text{in Case II.} \end{cases}$$

When $\gamma = 0$, the slope in Case I varies from 0 at the inner cylinder to a positive value at the outer cylinder, while in Case II it varies from a positive value at the inner cylinder to 0 at the outer cylinder. When $\gamma > 0$ these slopes are obviously decreased somewhat. This leads to the following general conclusions:

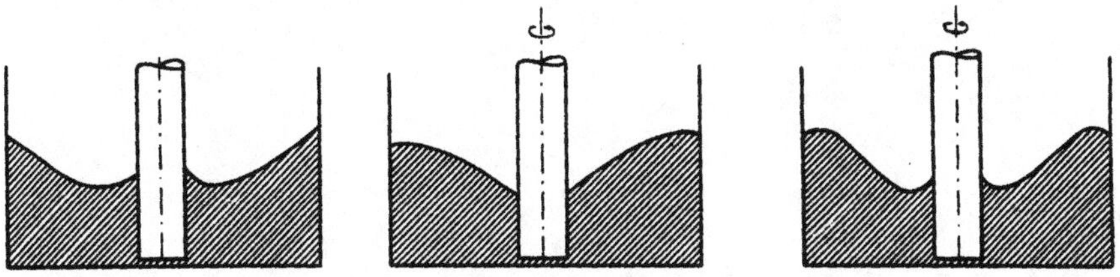

1a 1b 1c

**Fig. 1.** Typical fluid cross sections in Couette flow of a non-Newtonian fluid

1 a. The inner cylinder at rest. 1 b. The outer cylinder at rest, and $r_1 > 2\sqrt{\gamma/\varrho}$. 1 c. The outer cylinder at rest, and $r_1 < 2\sqrt{\gamma/\varrho}$.

Case I. When $\gamma > 0$ the fluid will tend to climb the inner cylinder, and will generally also climb the outer cylinder, unless $\gamma$ is very large (Fig. 1a).

Case II. The fluid will tend to fall slightly at the outer cylinder. At the inner cylinder the fluid will rise or fall according as the quantity

$$4\,\gamma - \varrho\,[(1 + \Theta)\, r_1]^2, \qquad \Theta = -(r_1/r_2)^2,$$

is greater or less than zero (Fig. 1b, c). In the important case ($r_1 \ll r_2$) this leads to the criterion: the fluid will tend to climb the inner cylinder if

$$r_1 < 2\sqrt{\gamma/\varrho} \quad \ldots\ldots\ldots\ldots \quad (17),$$

and to fall away from the inner cylinder otherwise.

The above conclusions are derived under several approximating assumptions, the most important being that the free surface should not greatly vary from the horizontal. This qualification should alway be borne in mind, though in all probability the general conclusions remain valid in a considerably wider range of circumstances. The conclusions above are in general agreement with certain experiments of Weissenberg [6], and very probably furnish the correct theoretical explanation of his results. So far there has been no experimental check of the inequality (17), though this would certainly supply a strong test of the preceding considerations.

This research was supported in part by the United States Air Force Office of Scientific Research under Contract AF 49 (638) —262.

## References

[1] A. C. Merrington, Nature, vol. 152 (1943), p. 663.
[2] R. S. Rivlin, Proc. Royal Society of London, Series A, vol. 193 (1948), p. 260.
[3] R. S. Rivlin, Proc. Cambridge Philosophical Society, vol. 45 (1949), p. 88.
[4] J. Serrin, Mathematical Principles of Classical Fluid Mechanics, Handbuch der Physik, vol. 8. Berlin 1959: Springer.
[5] C. Truesdell, J. Rational Mechanics and Analysis, vol. 1 (1952), p. 125.
[6] K. Weissenberg, Nature, vol. 159 (1947), p. 310.

Manuskripteingang: 13. 8. 1958

*Anschrift:* Prof. Dr. James Serrin,
Institute of Technology, University of Minnesota, Minneapolis 14 / Minn. (USA)

V/12/6 0,075

Barbara Serrin, James Serrin and Hidehiko Matano, home of James Serrin, 1998.
The painting of Amsterdam is by James Serrin's brother Richard Serrin.

James Serrin, Garden at 4422 Dupont Avenue, Minneapolis, MN, 2002.

# V.3. Boundary Layer Theory

John Bryce McLeod

**Mathematics Subject Classification (2010).** 34CXX, 76NXX.
**Keywords.** Fluid mechanics, compressible boundary layer theory.

Jim Serrin's papers on boundary layer theory contain a number of results which were quite fundamental in the subject. In particular, he demonstrated the significance of similarity solutions by proving that they were the limit of more general solutions, and then discussed the existence and behavior of these similarity solutions in situations more general than had been done previously.

Paper [i] below is concerned with the asymptotic behavior of a solution of the Prandtl boundary layer equations as $x \to \infty$, where we are considering the two-dimensional laminar flow of an incompressible viscous fluid past a rigid wall and the coordinates $x, y$ $(> 0)$ measure the length along the wall and perpendicular to it, respectively. The velocity vector $\{u(x,y), v(x,y)\}$ depends on the exterior streaming speed $U(x)$ as well as on the initial data $u(0,y) = \tilde{u}(y)$, and the Prandtl equations take the form (with $\nu$ the kinematic viscosity)

$$u_x + u_y = 0, \quad uu_x + vu_y = UU_x + \nu u_{yy},$$

with the boundary conditions

$$\begin{aligned} u = v = 0 \text{ at } y = 0, \quad & u(x,y) \to U(x) \text{ as } y \to \infty, \\ u(0,y) = \tilde{u}(y) \geq 0, \quad & \tilde{u} \to U(0) \text{ as } y \to \infty. \end{aligned} \tag{1}$$

If, in particular, we take the case

$$U(x) = C(x+d)^m \quad (C > 0, d > 0, m \geq 0) \tag{2}$$

and set

$$u(x,y) = U(x) f'(y/g(x))$$

for a suitable function $g$, then it is found that there is a constant $\lambda$ $(> 0)$ such that $f$ satisfies the Falkner–Skan equation

$$f''' + ff'' + \lambda(1 - f'^2) = 0,$$

with

$$f(0) = 0, \;\; f'(0) = 0, \;\; f'(\infty) = 1.$$

Existence and uniqueness of solutions to this had already been proved by Iglisch and Coppel, and there seemed to have been an assumption on the part of a number of writers that general solutions converged to the Falkner–Skan solution as $x \to \infty$. This is established rigorously by Serrin in two theorems:

P. Pucci et al. (eds.), *James Serrin. Selected Papers: Volume 2*, Contemporary Mathematicians, DOI 10.1007/978-3-0348-0847-7_12, © Springer Basel 2014

I. If $u(x, y)$ is a solution to the Prandtl system satisfying (1), and $U$ satisfies (2), and if $u_y$ is continuous, then

$$\left|\frac{u}{U} - f'\right| \doteq o\left(\frac{1 + m \ln x}{x^m}\right) \quad \text{as } x \to \infty, \text{ uniformly in } y.$$

II. For more general $U$, suppose that $U \in C^2$ and that

$$C_1(x+d)^{2m-1} \le UU_x \le C_2(x+d)^{2m-1},$$

$C_1, C_2 > 0$, $m$ and $n$ satisfying $m \le n < \frac{5}{3}m$. Then if $u, \bar{u}$ are two solutions of the Prandtl system, we have

$$\left|\frac{\bar{u} - u}{U}\right| = o(1) \text{ as } x \to \infty, \text{ uniformly in } y.$$

Serrin modestly says that the proof of these result are not particularly deep, but they are certainly ingenious. There is also a proof that if the pressure gradient is favorable ($dp/dx \le 0$), then the solution is asymptotic to a concave profile.

In paper [ii] below Serrin builds on the work of Paul Fife, who had established a rigorous connection between steady-state solutions of the Navier–Stokes equations and solutions of the Prandtl equations. Serrin looks at the Fife results in the context of a particular example, the radial motion occurring between two straight converging walls, originally treated by Jeffery and Hamel, and obtains sharper estimates. The crucial one is that, if the walls are taken as $\theta = 0$ and $\theta = 2\alpha$, and if the velocity vector in cylindrical polar coordinates $(r, \theta)$ has the form

$$v_r = -\frac{f(\theta)}{r}, \quad v_\theta = 0,$$

where $f = 0$ at the walls and $f > 0$, then

$$|f(\theta) - F(\theta)| \le 12\nu, \quad 0 \le \theta \le \alpha,$$

where $F$ is the corresponding solution for the Prandtl equations, and $\nu$ is the kinematic viscosity. The paper ends with some interesting remarks about directions for future work (existence of "transition zones", for example). It is interesting also that although $F$ is known explicitly in terms of elliptic functions (it is a solution of a Falkner–Skan equation), Serrin's proof does not use the explicit form, although it does use very heavily estimates obtained directly from the Falkner–Skan equation.

Serrin then turns to the existence of similarity solutions of compressible boundary layer problems, where now the effects of temperature are taken into account. Instead of a single ODE (Falkner–Skan), one now gets two linked ODEs, as first shown by Stewarts and by Li and Nagamatsu:

$$f''' + ff'' + \lambda(h - f'^2) = 0, \tag{3}$$

$$h'' + fh' = 0, \tag{4}$$

with $f(0) = f'(0) = 0$, $h(0) = a$; $f'(\infty) = h(\infty) = 1$. Here $f'$ and $h$ are effectively the tangential velocity, and the total energy in the boundary layer, $\lambda$ is a constant, and $a$ measures the ratio of the temperature at the wall to the temperature in the main stream.

Previous work had concentrated on the case $a < 1$ (the wall cooled), where existence had been proved by Ho and Wilson. Paper [iii] assumes existence of solutions (existence is proved in another paper with McLeod, Arch. Rational Mech. Anal. 31 (1968/69), 288–303) but shows that, if $\lambda > 0$ and the solutions do not exhibit back-flow, so that $f' \ge 0$, then, if $0 < a < 1$, we have $h$ and $f'$ monotonically increasing and concave, with $f'^2 < h$, so that there is no velocity overshoot(i.e., $f' < 1$), while if $a > 1$, then $h$ is monotonically decreasing and concave, but $f'$ first increases to a value exceeding 1 before finally decreasing to 1. In addition, in either case, we have $f'^2 < h$, which has the physical interpretation that the Mach number in the boundary layer is less than the Mach number in the mainstream. The proofs are straightforward manipulations on the governing equations.

Existence of solutions is also discussed in paper [iv]. The existence is carried out by a so-called "shooting" argument. Thus we define a solution of (3)–(4) by setting $f''(0) = \beta$, $h'(0) = \alpha$ and show that we can choose $\alpha, \beta$ so that $f'(\infty) = 1$, $h(\infty) = 1$. For the case of the simple Falkner–Skan equation, where $h \equiv 1$, the solution $f$ is determined by $\beta$. It is easy to see that if $\beta(> 0)$ is sufficiently small, then $f'$ (which is initially positive) becomes negative before (if ever) it crosses 1, and if $\beta$ is sufficiently large, $f'$ crosses 1 before (if ever) it becomes negative. A continuity argument in $\beta$ then shows that there must be some $\beta$ for which $0 < f' < 1$, and this leads to $f'(\infty) = 1$. This simple one-dimensional argument in $\beta$ now becomes two-dimensional in $\alpha, \beta$, and the topology somewhat more complicated. We now define sets in the $(\alpha, \beta)$-plane by using the inequalities $f' > 0$, $f'^2 < h$, and show that there is a continuum of values of $(\alpha, \beta)$ such that the limits $f'(\infty)$ and $h(\infty)$ exist with $\{f'(\infty)\}^2 = h(\infty)$. A further continuity argument then ensures that we can find a point on this continuum where in fact $f'(\infty) = h(\infty) = 1$. Details can be found in the paper.

One-dimensional shooting arguments are very common. Two-dimensional ones are much rarer. One example mentioned by Serrin is that of von Karman's swirling flow above an infinite disc, where the equations are

$$f''' + ff'' + \frac{1}{2}\left(h^2 - f'^2\right) = 0, \qquad h'' + fh' - f'h = 0,$$

with the boundary conditions

$$f = f' = 0, \; h = 1 \text{ at } t = 0, \; f' = h = 0 \text{ at } t = \infty.$$

Here two-dimensional shooting does give a solution with $f' > 0$.

There is also a result of Y.-H. Wan which extends the arguments to higher dimensions [S.P. Hastings, C. Lu and Y.H. Wan, Physica D 10 (1996), 301–306] and which was helpful in one three-dimensional setting, and there is a four-dimensional shooting proof by Dunbar [Trans. Amer. Math. Soc. 286 (1984), 557–594]. But the overall conclusion is that shooting in higher dimensions is a difficult sport.

John Bryce McLeod
Mathematical Institute
Oxford University
24–25 St. Giles'
Oxford OX1 3LB, England
e-mail: mcleod@maths.ox.ac.uk

*Proc. R. Soc. Lond., Ser. A* 299 (1967), 491–507. **Reprint V.3.[i]**


# Asymptotic behaviour of velocity profiles in the Prandtl boundary layer theory

BY J. SERRIN
*University of Minnesota*

(*Communicated by L. Howarth, F.R.S.—Received* 19 *Sepetmber* 1966)

Consider the Prandtl boundary layer equation for the steady two-dimensional laminar flow of an incompressible viscous fluid past a rigid wall. On the basis of an arbitrary velocity profile at some initial position on the wall, the analysis presented shows that, for a power law streaming speed $U(x) = C(x+d)^m$ $(m \geqslant 0)$, the velocity profile which develops downstream is asymptotically given by the well known Falkner–Skan similarity solution. Moreover, for a streaming speed satisfying (6), the velocity profile which develops downstream is asymptotically unique, though of course the particular form of the resulting profile depends on the precise nature of the exterior stream. The rate of convergence for this asymptotic behaviour is estimated, as well as corresponding rates for the convergence of the skin friction coefficient. This result verifies the tacit assumption of a number of writers that the downstream velocity profile is essentially independent of the initial profile, and also supplies a theoretical justification for the role of similar solutions in boundary layer theory.

We also prove the *existence* of concave velocity profiles whenever the pressure gradient is favourable. It follows that, for streaming speeds which correspond to a favourable pressure gradient, concave velocity profiles play somewhat the same role as similarity profiles do for a power law streaming speed.

The simplest and earliest example in Prandtl's boundary layer theory concerns the laminar flow of a viscous incompressible fluid past a rigid wall or obstacle. The equations derived by Prandtl to represent this situation constitute an important simplification of the original Navier–Stokes system, and are significant and practically useful in numerous applications. Perhaps the most fruitful source of information in this regard is the body of exact solutions derived under the assumption of similarity, of which the famous Blasius solution is typical. Nevertheless, in spite of the success of these particular solutions in predicting actual motions, a well known and primary problem has been present, namely, in what way are similar solutions unique or special among the totality of solutions of Prandtl's equations? What theoretical justification can be offered for the pre-eminent role of similar solutions in boundary layer theory? This paper presents an answer to these questions, based on the downstream asymptotic behaviour of velocity profiles in the Prandtl theory.

The familiar Prandtl equations for steady laminar two-dimensional flow past a rigid wall are

$$u_x + v_y = 0$$

and

$$uu_x + vu_y = UU_x + \nu u_{yy}.$$

Here $(x, y)$ are orthogonal coordinates in the boundary layer, with $x$ representing arc length along the wall and $y$ perpendicular distance from the wall, $u$ and $v$ are the corresponding velocity components, and $\nu$ is the kinematic viscosity; finally $U = U(x)$ is the assigned exterior streaming speed, which is assumed throughout the paper to be non-negative. Note also that $U$ is related to the pressure $p = p(x)$ in the boundary layer through the Bernoulli relation

$$\mathrm{d}p/\mathrm{d}x + UU_x = 0.$$

The appropriate boundary conditions are

$$u = v = 0 \quad \text{when} \quad y = 0, \tag{1}$$

to which must also be added Prandtl's stream condition,

$$u \to U \quad \text{as} \quad y \to \infty, \quad \text{uniformly in } x. \tag{2}$$

As is well known, though perhaps not sufficiently emphasized, conditions (1) and (2) are in themselves not enough to determine uniquely a solution of Prandtl's equations. In order to obtain a well-set problem, both physically and mathematically, we suppose that at some initial position, say $x = 0$, the velocity component $u$ is assigned, that is

$$u(0, y) = \tilde{u}(y) \quad \text{given}, \quad (0 < y < \infty).$$

Here it is assumed that the function $\tilde{u}(y)$ is non-negative and continuous, and tends to $U(0)$ as $y \to \infty$; moreover, we suppose the initial data is taken on in such a way that $u \geqslant 0$ in some neighbourhood of the initial line. The resulting problem is then that of the *downstream evolution of the velocity component* $u(x, y)$.

It should be emphasized that in general we have no specific information concerning the form of the assigned function $\tilde{u}(y)$. For example, in the flow past a plate one simply does not know the profile to assign at the leading edge, and indeed the very assumptions of boundary layer theory are such that Prandtl's equation should not in any case be used in the neighbourhood of this point. And of course if the initial position $x = 0$ is taken downstream of the leading edge, we remain equally at a loss to assign a definitive form to the velocity profile $\tilde{u}(y)$ there. This being the case, we might for lack of better information simply set $\tilde{u}(y) = U(0)$, or make some other special choice of the initial velocity profile. Without further justification, however, such a procedure is arbitrary and particular. As we shall see, it is more fruitful and to the point to retain the full generality of the initial profile without prejudging its particular form.

The initial profile problem considered here is analogous to the initial value problem in the theory of parabolic partial differential equations, with the coordinate $x$ playing the role of time and the velocity component $u$ the role of a temperature. As in the theory of the heat equation, it has been shown that for appropriate $\tilde{u}(y)$ and $U(x)$ the Prandtl system with boundary conditions (1) and (2) possesses a unique solution (cf. Nickel 1958; Oleinik 1963). We shall therefore turn our attention directly to the downstream asymptotic behaviour of $u(x, y)$.

Suppose first that the exterior streaming speed $U(x)$ is such that a similarity solution of the Prandtl system can exist. We consider particularly the case where $U$ is a non-negative power of the arc length, that is

$$U(x) = C(x + d)^m \quad (0 \leqslant x < \infty), \tag{3}$$

where $C$ and $d$ are constants, with $C > 0$, $d \geqslant 0$. The similar solution corresponding to the streaming speed (3) is given by

$$\bar{u}(x,y) = U(x)f'(\zeta), \tag{4}$$

where $\zeta = y/g(x)$, $g(x) = \sqrt{\{\nu(x+d)/U(x)\}}$, and $f(\zeta)$ is the solution of the Falkner–Skan equation associated with the given value of $m$.† Although the function $f'(\zeta)$ depends on $m$, it is in all cases a monotonically increasing, concave function of $\zeta$ with $f'(0) = 0$, $f'(\infty) = 1$, and $f(0) = 0$.

Now let $\tilde{u}(y)$ be an arbitrary initial profile, and let $u(x,y)$ be a corresponding solution of the Prandtl system, subject to the boundary conditions (1) and the streaming speed (3). Assume that $u_y$ is continuous for $0 < x < \infty$, $0 \leqslant y < \infty$. *Then we have the fundamental asymptotic estimate*

$$\left|\frac{u}{U} - f'\right| = o\left(\frac{1 + m\ln x}{x^m}\right) \quad \textit{as} \quad x \to \infty, \quad \textit{uniformly in } y. \tag{5}$$

*That is, the normalized velocity component $u/U$ tends uniformly to the normalized similarity solution $f'$ as $x$ tends to infinity downstream.*

This result not only provides a well-determined asymptotic behaviour for the velocity profile, but, what is more important, singles out similarity solutions as those which asymptotically develop downstream, whatever may be the state of motion at the initial position $x = 0$, Thus within the limits of the Prandtl system we have a rigorous justification for the central position of similar solutions in boundary layer theory.

In addition to the case where similar solutions exist, it is worth considering more general streaming speeds $U(x)$, for here also one has an interest in the asymptotic behaviour and stability of the velocity profile. Thus let $u$ and $\bar{u}$ be two solutions of the Prandtl system corresponding to the same streaming speed $U(x)$, though with possibly different initial profiles $\tilde{u}(y)$ and $\tilde{\bar{u}}(y)$. Suppose $U$ is twice continuously differentiable and obeys the inequality

$$C_1(x+d)^{2m-1} \leqslant UU_x \leqslant C_2(x+d)^{2n-1} \quad (0 \leqslant x < \infty), \tag{6}$$

where $C_1$ and $C_2$ are positive constants and $m$ and $n$ are exponents satisfying the condition $m \leqslant n < \frac{5}{3}m$. *Then we have the second fundamental asymptotic estimate*

$$\left|\frac{\bar{u} - u}{U}\right| = o(1) \quad \textit{as} \quad x \to \infty, \quad \textit{uniformly in } y. \tag{7}$$

*Thus again the normalized velocity profile is asymptotically unique, whatever may be the state of motion at the initial position $x = 0$.*

The preceding result allows us to conclude, at least when the restriction (6) holds, that the downstream behaviour of the velocity profile is essentially independent of the initial profile. A tacit assumption to this effect seems in fact to have been

† A derivation and discussion of the Falkner–Skan equation may be found in the treatise of Schlichting (1960). The existence and uniqueness of solutions of the Falkner–Skan equation was proved by Iglisch (1953, 1954); cf. also Coppel (1960).

adopted by a number of writers (Hiemenz 1911; Howarth 1935; Goertler 1957); here we have a theoretical justification for their calculations.

The remaining part of the paper is devoted to a proof of the asymptotic estimates (5) and (7). While the argument is not particularly deep, there are, nevertheless, a number of individual parts which are best treated as a series of lemmas.

Before turning to the proofs, we remark that the restriction $m \geqslant 0$ in (3) implies a favourable pressure gradient, $\mathrm{d}p/\mathrm{d}x \leqslant 0$. On the other hand, the case $m < 0$ corresponds to an adverse pressure gradient in which $p(x)$ increases in the downstream direction. Although flows may endure small adverse gradients without backflow or separation, a number of difficult problems intervene. In any case, when $\mathrm{d}p/\mathrm{d}x > 0$ our arguments fail in several crucial places, and it is not at all clear that any worthwhile asymptotic estimates can hold.

The left hand inequality of (6) similarly insures a favourable pressure gradient. However, the physical significance of the right hand inequality and the condition $n < \frac{5}{3}$ is unresolved, beyond the obvious fact that they prevent overlarge oscillations of the streaming speed as we proceed downstream.

## 1. Preliminary results

We recall that $u$ is assumed to be non-negative in some neighbourhood of the initial, line, and continuously differentiable on $y = 0$. Moreover, in the cases under consideration here we have $UU_x \geqslant 0$. Hence, according to a theorem of Velte (1960), we have $u > 0$ in $0 < x < \infty$, $0 < y < \infty$, and $u_y(x, 0) > 0$ for $0 < x < \infty$.

Since our interest is in the downstream asymptotic behaviour of solutions, it is clearly allowable to consider the line $x = 1$, rather than $x = 0$, as the initial position. With this choice, and then a relabelling of coordinates so that the new initial position is again called $x = 0$, we may therefore assume that $u$ and $u_y$ are continuous in $0 \leqslant x < \infty$, $0 \leqslant y < \infty$, and that the initial profile $\tilde{u}(y)$ satisfies

$$\tilde{u}(0) = 0, \quad \tilde{u}_y(0) > 0 \quad \text{and} \quad \tilde{u}(y) > 0 \quad \text{for} \quad y > 0.$$

The relation $u > 0$ also allows us to introduce the so-called von Mises variables $(x, \psi)$ in place of the physical variables $(x, y)$. Indeed, the transformation

$$x = x, \quad \psi = \psi(x, y) = \int_0^y u(x, t)\,\mathrm{d}t$$

is one-to-one between the regions $0 \leqslant x < \infty, 0 \leqslant y < \infty$ and $0 \leqslant x < \infty, 0 \leqslant \psi < \infty$. Moreover, in terms of the new variables the Prandtl system reduces to the single equation (Schlichting 1960, p. 136)

$$(u^2)_x = \nu u (u^2)_{\psi\psi} + (U^2)_x. \tag{8}$$

The corresponding boundary and initial conditions are

$$u = 0 \quad \text{on} \quad \psi = 0,$$

$$u \to U \quad \text{as} \quad \psi \to \infty, \quad \text{uniformly in } x \text{ on any finite interval } 0 \leqslant x \leqslant A,$$

and
$$u(0, \psi) = \tilde{u}(\psi) \quad (0 \leqslant \psi < \infty),$$

where $\tilde{u}(\psi)$ denotes the transformed initial data. The following lemma will be basic in what follows.

LEMMA 1. *Let $u$ and $\bar{u}$ be two solutions of the boundary layer equations as above, corresponding to the respective streaming speeds $U(x)$ and $\bar{U}(x)$. Assume*

$$(U^2)_x \leqslant (\bar{U}^2)_x \quad \text{and} \quad \tilde{u}(\psi)^2 \leqslant \tilde{\bar{u}}(\psi)^2 + a^2$$

*for some constant $a > 0$. Suppose also that either $u_{yy} < 0$ or $\bar{u}_{yy} < 0$. Then we have*

$$u(x, \psi) \leqslant \bar{u}(x, \psi) + a. \tag{9}$$

*Proof.* Suppose first that $u_{yy} < 0$. Since in the $(x, \psi)$ plane we have

$$(\bar{u}^2)_x = \nu\bar{u}(\bar{u}^2)_{\psi\psi} + (\bar{U}^2)_x,$$

the function $\phi(x, \psi) = \bar{u}^2 - u^2$ satisfies the differential equation

$$\begin{aligned}\phi_x &= \nu\bar{u}(\bar{u}^2)_{\psi\psi} - \nu u(u^2)_{\psi\psi} + (\bar{U}^2)_x - (U^2)_x \\ &= \nu\bar{u}\,\phi_{\psi\psi} + \frac{\nu(u^2)_{\psi\psi}}{u+\bar{u}}\phi + (\bar{U} - U^2)_x.\end{aligned} \tag{10}$$

Now $(u^2)_{\psi\psi} = 2u_{yy}/u$, as one sees by comparing the von Mises equation with the Prandtl equations. Thus if we set

$$\alpha = 2\nu u_{yy}/u(u+\bar{u}),$$

equation (10) takes the form

$$\phi_x = \nu\bar{u}\phi_{\psi\psi} + \alpha\phi + (\bar{U}^2 - U^2)_x. \tag{11}$$

Moreover, $\phi$ is subject to the boundary conditions

$$\phi = 0 \quad \text{on} \quad \psi = 0, \quad \phi \geqslant -a^2 \quad \text{on} \quad x = 0,$$

and $\quad \phi \to \bar{U}^2 - U^2 \quad$ as $\quad \psi \to \infty, \quad$ uniformly in $x$ on any finite interval.

Since $(\bar{U}^2 - U^2)_x \geqslant 0$ and $\bar{U}(0)^2 - U(0)^2 \geqslant -a^2$, it is evident that $\bar{U}^2 - U^2 \geqslant -^2$. for all $x$. In view of these conditions, we assert that

$$\phi(x, \psi) \geqslant -a^2$$

for $0 \leqslant x < \infty$, $0 \leqslant \psi < \infty$.

To see this, suppose for contradiction that $\phi < -a^2$ at some point $(x_0, \psi_0)$. Let $R$ denote the set $0 < x \leqslant x_0$, $0 < \psi < \infty$, and consider the function $\phi(x, \psi)$ *restricted* to $R$. Since $\phi < -a^2$ at the point $(x_0, \psi_0)$, it is evident in view of the boundary conditions above that $\phi$ must take on an absolute minimum in $R$. Let this minimum be assumed at the point $(x_1, \psi_1)$. Since at $(x_1, \psi_1)$ we have

$$\phi_x \leqslant 0, \quad \phi_{\psi\psi} \geqslant 0, \quad \phi < 0,$$

while also $\alpha < 0$ and $(\bar{U}^2 - U^2)_x \geqslant 0$, an obvious sign contradiction occurs in (11). This proves the lemma when $u_{yy} < 0$.

When $\overline{u}_{yy} < 0$ we rewrite (10) as

$$\phi_x = \nu u \phi_{\psi\psi} + \frac{\nu(\overline{u}^2)_{\psi\psi}}{u+\overline{u}}\phi + (\overline{U}^2 - U^2)_x$$

and the conclusion follows as before.

Now consider a stream speed $U(x)$ satisfying (6), and let $u$ and $\overline{u}$ be corresponding solutions of the Prandtl system whose downstream behaviour is to be compared. Set

$$U^*(x) = C^*(x+d)^m, \quad U^{**}(x) = C^{**}(x+d)^n, \tag{12}$$

where $C^*$ and $C^{**}$ are appropriate constants which will be determined in a moment. Corresponding to the streaming speeds (12), let $u^*(x,y)$ and $u^{**}(x,y)$ be the respective similar solutions. Thus

$$u^*(x,y) = U^*(x)f'(\zeta),$$

where $\zeta = y/g^*(x)$ and $g^*(x) = \sqrt{\{\nu(x+d)/U^*(x)\}}$, with similar formulae for the solution $u^{**}$. In view of the relabelling of coordinates introduced at the beginning of this section, we note also that the parameter $d$ in the preceding formulae can be assumed $\geqslant 1$.

Lemma 2. *The constants $C^*$ and $C^{**}$ can be chosen so that $u^* \leqslant u \leqslant u^{**}$ and $u^* \leqslant \overline{u} \leqslant u^{**}$ in the von Mises variables.*

*Proof.* Since $(U^{*2})_x = 2mC^{*2}(x+d)^{2m-1}$ and $(U^{**2})_x = 2nC^{**2}(x+d)^{2n-1}$, it is apparent from (6) that through appropriate choice of $C^*$ and $C^{**}$ we can satisfy

$$(U^{*2})_x \leqslant (U^2)_x \leqslant (U^{**2})_x.$$

By making $C^*$ smaller and $C^{**}$ larger, if necessary, we wish to show that

$$u^*(0,\psi) \leqslant \tilde{u}(\psi),\ \tilde{\overline{u}}(\psi) \leqslant u^{**}(0,\psi). \tag{13}$$

Lemma 2 will then follow at once from Lemma 1 with $a = 0$.

It is enough to prove (13) for the function $\tilde{u}(\psi)$. Now for $x = 0$ and small $y$ we have $\tilde{u} \approx by$, $\psi \approx \frac{1}{2}by^2$, and $\tilde{u}^2 \approx 2b\psi$, where $b > 0$. Moreover, $\tilde{u} > 0$ for $\psi > 0$, and $\tilde{u}$ tends to $U(0)$ as $\psi$ tends to infinity; cf. figure 1.

On the other hand, $u^*_y(0,0) = U^*(0)f''(0)/g^*(0) = \text{const. } C^{*\frac{3}{2}}$, so that

$$u^{*2} \approx \text{const. } C^{*\frac{3}{2}}\psi$$

for small $\psi$. Since $u^*$ tends to $C^*d^m$ as $\psi$ tends to infinity, while $(u^{*2})_{\psi\psi} < 0$, it is clear that $C^*$ can be chosen so that $u^*(0,\psi) \leqslant \tilde{u}(\psi)$.

In order to prove the remaining part of (13), it is enough to show that $u^{**}$ at $x = 0$, $\psi = 1$ can be made arbitrarily large by an appropriate choice of $C^{**}$ (recall that $u^{**2}$ is a concave, monotonically increasing function of $\psi$). Now we have for the flow $u^{**}$,

$$\psi = U^{**}g^{**}f(\zeta).$$

Moreover, $f(\zeta) < \zeta$. Hence at the point $x = 0$, $\psi = 1$,

$$\zeta > \{U^{**}g^{**}\}^{-1} = \text{const.}/C^{**\frac{1}{2}},$$

and

$$u^{**} = U^{**}f'(\zeta) \geqslant \text{const. } C^{**\frac{1}{2}}$$

(assuming $C^{**} > 1$). This completes the proof of lemma 2.

LEMMA 3. *Suppose that* $|\tilde{u}(\psi)^2-\tilde{\bar{u}}(\psi)^2| \leqslant a^2$ *and that* $u_{yy} < 0$. *Then*

$$|u(x,\psi)-\bar{u}(x,\psi)| \leqslant a. \tag{14}$$

*Proof*. This is an immediate consequence of lemma 1 with $U \equiv \bar{U}$. (Note that in order to prove the pair of inequalities implicit in (14) it is necessary to use lemma 1 twice.)

Since the flows $u$ and $\bar{u}$ in question satisfy the hypothesis of lemma 3 for *some* number $a$, inequality (14) already yields an asymptotic estimate of $u$ and $\bar{u}$. This comparison occurs at equal values of the stream function $\psi$, however, rather than at equal values of the physical variable $y$. In order to compare $u$ and $\bar{u}$ *as functions of* $x, y$, as required for (5) and (7), further argument is needed. In addition, we shall ultimately have to remove the restriction $u_{yy} < 0$.

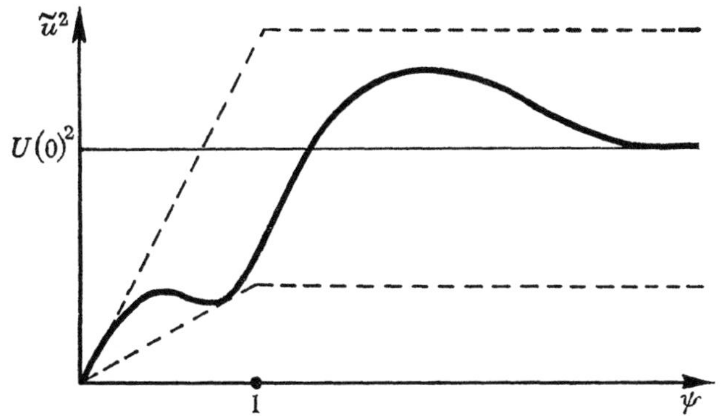


FIGURE 1. General form of the initial function $\tilde{u}^2(\psi)$. The dashed lines describe simple majorant and minorant curves relevant to the proof of lemma 2.

## 2. THE MAIN PROOF (*first stage*)

For a given value of $\psi$, let $y$ and $\bar{y}$ denote the respective physical variables for the flows $u$ and $\bar{u}$, that is

$$y = \int_0^\psi \frac{d\psi}{u(x,\psi)}, \quad \bar{y} = \int_0^\psi \frac{d\psi}{\bar{u}(x,\psi)}.$$

In general $y \neq \bar{y}$.

LEMMA 4. *If* $u_{yy} < 0$ *then we have for* $0 < x < \infty$

$$y-\bar{y} \leqslant \frac{ag^*}{bU^*}\left(1+\frac{a}{bU^*}\right)\ln\left(1+\frac{bU^*}{a}\right)+\frac{a}{bU^*}\bar{y},$$

*where* $b = f'(1)$ *and* $f$ *is the Falkner–Skan function associated with the exponent* $m$.

*Proof*. By lemma 3, since $u_{yy} < 0$,

$$y-\bar{y} = \int\left(\frac{1}{u}-\frac{1}{\bar{u}}\right)d\psi \leqslant \int\left(\frac{1}{u}-\frac{1}{u+a}\right)d\psi = a\int\frac{d\psi}{u(u+a)}.$$

Let $\psi_1 = U^*g^*f(1)$. Then applying lemma 2 yields

$$y-\bar{y} \leqslant a\int_0^{\psi_1}\frac{d\psi}{u^*(u^*+a)}+a\int_{\psi_1}^{\psi}\frac{d\psi}{u(u^*+a)},$$

where the second integral should be omitted if $\psi \leqslant \psi_1$.

Now $d\psi/u^* = g^* d\zeta$ so that

$$\int_0^{\psi_1} \frac{d\psi}{u^*(u^*+a)} = \int_0^1 \frac{g^* d\zeta}{u^*+a} \leqslant \int_0^1 \frac{g^* d\zeta}{b\zeta U^*+a}$$

since for $0 < \zeta < 1$ we have $u^* = U^* f'(\zeta) \geqslant U^* f'(1)\, \zeta$. On the other hand

$$\int_{\psi_1}^{\psi} \frac{d\psi}{u(u^*+a)} \leqslant \int_{\psi_1}^{\psi} \frac{d\psi}{u(bU^*+a)} \leqslant \int_0^y \frac{dy}{bU^*+a}$$

since for $\zeta > 1$ we have $u^* \geqslant U^* f'(1)$. Combining the preceding inequalities there results

$$y - \bar{y} \leqslant \frac{ag^*}{bU^*} \ln\left(1+\frac{bU^*}{a}\right) + \frac{a}{bU^*+a} y,$$

and the lemma follows at once.

Lemma 4′. *If* $u_{yy} < 0$ *then for* $0 < x < \infty$ *we have also*

$$\bar{y} - y \leqslant \frac{ag^*}{bU^*}\left(1+\frac{a}{bU^*}\right) \ln\left(1+\frac{bU^*}{a}\right) + \frac{a}{bU^*} y.$$

*Proof*. This follows exactly as lemma 4 by interchanging $y$ and $\bar{y}$ and $u$ and $\bar{u}$ in the preceding calculation.

Lemma 5. *If* $u_{yy} < 0$ *then for* $0 < x < \infty$

$$|u(x,y) - u(x,\bar{y})| \leqslant \frac{aU^{**}}{bU^*}\left\{1+f''(0)\left(\frac{U^{**}}{U^*}\right)^{\frac{1}{2}}\left(1+\frac{a}{bU^*}\right)\ln\left(1+\frac{bU^*}{a}\right)\right\},$$

*where* $f$ *is the Falkner–Skan function associated with the exponent* $n$.

*Proof*. We suppress the dependence of $u$ on $x$ in the following calculations. By the mean value theorem

$$u(y) - u(\bar{y}) = \hat{u}_y \cdot (y - \bar{y}),$$

where $\hat{u}_y$ denotes an appropriate intermediate value of $u_y$. Since $u$ is assumed to be concave in $y$ and since $u(0) = 0$, $u(\infty) = U$, it is evident that $u_y$ is a positive, decreasing function of $y$. Thus if $y \geqslant \bar{y}$ we have

$$\hat{u}_y \cdot (y-\bar{y}) \leqslant u_y(\bar{y})\,(y-\bar{y}),$$

while if $\bar{y} \geqslant y$,

$$\hat{u}_y \cdot (y-\bar{y}) \leqslant 0.$$

Hence by lemma 4, in both cases,

$$\hat{u}_y \cdot (y-\bar{y}) \leqslant \frac{ag^* u_y(\bar{y})}{bU^*}\left(1+\frac{a}{bU^*}\right)\ln\left(1+\frac{bU^*}{a}\right) + \frac{a}{bU^*}\bar{y} u_y(\bar{y}).$$

Now since $u_{yy} < 0$, it is clear that

$$\bar{y} u_y(\bar{y}) \leqslant U \quad \text{and} \quad u_y(\bar{y}) \leqslant u_y(0).$$

Moreover, by lemma 2, $u \leqslant u^{**}$. Therefore $u_y(0) \leqslant u_y^{**}(0) = U^{**} f''(0)/g^{**}$. Consequently $u(y) - u(\bar{y})$ is bounded by the quantity appearing in lemma 5.

To obtain the reverse inequality, note that

$$u(\bar{y}) - u(y) = \hat{u}_y \cdot (\bar{y} - y) \leqslant \max(0, u_y(y)\,(\bar{y}-y)).$$

Using lemma 4′ rather than lemma 4, we obtain the required inequality exactly as before

LEMMA 6. *If* $u_{yy} < 0$ *then*

$$\left|\frac{u-\bar{u}}{U}\right| = O\left\{\frac{a(1+m\ln x+\ln_{+}1/a)}{x^{\frac{1}{2}(5m-3n)}}\right\} \quad \textit{as} \quad x\to\infty, \quad \textit{uniformly in } y.$$

*Proof.* Clearly for any positive number $\bar{y}$

$$|u(x,\bar{y})-\bar{u}(x,\bar{y})| \leqslant |u(x,\bar{y})-u(x,y)|+|u(x,y)-\bar{u}(x,\bar{y})|.$$

By lemma 3, the second term on the right hand side does not exceed $a$. The first term can be estimated by means of lemma 5. Indeed for any non-negative number $r$ we have $(1+r^{-1})\ln(1+r) \leqslant 2\ln 2+\ln_{+}r$, and the lemma follows at once from a straightforward calculation.

We note that the multiplicative factor implicit in the above order relation depends only on $C^*$, $C^{**}$, and the absolute constants $f'(1)$ and $f''(0)$.

### 3. THE MAIN PROOF (*second stage*)

Here we shall complete the proof of the second asymptotic estimate stated in the introduction. By lemma 6, if $u_{yy} < 0$ and $n < 5m/3$ we have

$$\left|\frac{u-\bar{u}}{U}\right| = o(1) \quad \text{as} \quad x\to\infty, \quad \text{uniformly in } y. \tag{15}$$

In order to complete the proof, it therefore remains only to remove the restriction $u_{yy} < 0$.

LEMMA 7. *Suppose that the streaming speed* $U = U(x)$ *is twice continuously differentiable, and that* $UU_x > 0$. *Then for a suitable choice of the initial data* $\tilde{u}(y)$ *there exists a solution of the initial profile problem such that* $u > 0$ *and* $u_{yy} < 0$ *for* $0 \leqslant x < \infty$, $0 < y < \infty$.

*Proof.* Let $\tilde{u}(y)$ be a four times continuously differentiable function defined on $0 \leqslant y < \infty$ and having the following properties:

(i) At $y = 0$ the derivatives of $\tilde{u}$ satisfy

$$\tilde{u} = 0, \qquad \tilde{u}_y > 0,$$
$$\tilde{u}_{yy} = -U(0)\,U_x(0)/\nu, \quad \tilde{u}_{yyy} = 0.$$

(ii) $\tilde{u} > 0$ and $\tilde{u}_{yy} < 0$ for $0 < y < \infty$.

(iii) $\tilde{u}$ tends to $U(0)$ as $y$ tends to infinity.

(iv) $\tilde{u}$ and its derivatives are uniformly bounded.

Clearly such a function exists. Moreover, under these circumstances Oleinik (1963) has shown that there is a corresponding solution $u(x,y)$ of the Prandtl system for which $u$ and $u_y$ are continuous in $0 \leqslant x < \infty$, $0 \leqslant y < \infty$ and $u > 0$ for $y > 0$. (The condition $\tilde{u}_{yy} < 0$ is not required in Oleinik's treatment, but will be used here to show that $u_{yy} < 0$. The assumption on $UU_x$ ensures that the solution can be continued downstream for all values of $x$.)

In terms of the von Mises variables, $u$ satisfies the equation

$$w_x = \nu u w_{\psi\psi} + 2UU_x \quad (w = u^2). \tag{16}$$

Now in the course of establishing the existence of the solution $u$, Oleinik obtained several properties of the function $w = w(x, \psi)$ which will be useful here. Let $A$ be an arbitrary positive number. Then in the domain $0 \leqslant x \leqslant A$, $0 < \psi < \infty$ the functions $w_\psi$ and $uw_{\psi\psi}$ are bounded and continuous, also

$$|w_x| \leqslant M\psi^{\frac{3}{4}}, \tag{17}$$

and finally $u \leqslant U(A)$.

Now set $z = w_{\psi\psi}$. By differentiating (16) twice with respect to $\psi$ we obtain†

$$z_x = \nu\{uz_{\psi\psi} + 2u_\psi z_\psi + u_{\psi\psi} z\},$$

that is

$$z_x = \nu\left\{uz_{\psi\psi} + \frac{1}{u} w_\psi z_\psi + \frac{1}{4u^3}(2u^2 w_{\psi\psi} - w_\psi^2)\, z\right\}. \tag{18}$$

Using the condition (ii) above and the fact that $2u_{yy} = uz$, it is evident that $z < 0$ on $x = 0$. Furthermore, by (16) and (17)

$$z = \frac{1}{\nu u}(w_x - 2UU_x) \leqslant \frac{1}{\nu u}(M\psi^{\frac{3}{4}} - K),$$

where $K$ is a positive constant. Thus $z < 0$ for $0 < \psi \leqslant \psi_0$, $0 \leqslant x \leqslant A$, provided $\psi_0$ is chosen small enough

The boundary conditions $z < 0$ on $x = 0$ and $z < 0$ on $\psi = \psi_0$ together with the differential equation (18) imply that $z < 0$ throughout $0 \leqslant x \leqslant A$, $\psi_0 \leqslant \psi < \infty$. Indeed, this is an immediate consequence of the following result of Krzyżański (1959).

LEMMA 8. *Let $z(x, y)$ be a bounded solution of the differential equation*

$$z_x = a(x,y)\, z_{yy} + b(x,y)\, z_y + c(x,y)\, z$$

*in $0 < x < A$, $0 < y < \infty$. Suppose that $a(x,y)$, $b(x,y)$, $c(x,y)$ are bounded continuous functions of $(x,y)$ and that $a(x,y) > 0$. Then if $z < 0$ on $x = 0$ and $y = 0$, we have $z < 0$ in the entire domain.*

Actually, Krzyżański proved only that $z \leqslant 0$; in order to obtain the sharper conclusion $z < 0$ we apply Nirenberg's strong maximum principle (1953).

By application of lemma 8 and the results already established we find $z < 0$ for $0 \leqslant x < A$, $0 < \psi < \infty$. Since $A$ is arbitrary and since $2u_{yy} = uz$, the proof of lemma 7 is completed.

Returning to the proof of (7), let $\bar{u}$ and $\bar{\bar{u}}$ be two flows satisfying the given conditions. Moreover, let $u$ now denote the solution constructed in lemma 7. We shall compare $\bar{u}$ and $\bar{\bar{u}}$ each with $u$, using the estimate (15). Clearly we obtain

$$\left|\frac{\bar{u} - \bar{\bar{u}}}{U}\right| \leqslant \left|\frac{\bar{u} - u}{U}\right| + \left|\frac{u - \bar{\bar{u}}}{U}\right| = o(1) \quad \text{as} \quad x \to \infty, \quad \text{uniformly in } y.$$

This proves (7) since $\bar{u}$ and $\bar{\bar{u}}$ are arbitrary solutions of the initial profile problem corresponding to the streaming speed $U(x)$.

† This process is easily justified; in fact, $w$ is arbitrarily often differentiable with respect to $\psi$.

4. The main proof (*conclusion*)

It remains to prove the asymptotic estimate (5). Consider the streaming speed $U(x) = C(x+d)^m$ given by (3), and let $\bar{u}$ denote the corresponding similar solution. When $m > 0$ this function satisfies (6) with $n = m$ and $C_1 = C_2 = 2mC^2$. Consequently by Lemma 6 (the solution $\bar{u}$ rather than $u$ is concave, but this is just a matter of notation),

$$\left|\frac{u}{U} - f'\right| = O\left\{\frac{a(1+m\ln x+\ln_+ 1/a)}{x^m}\right\} \quad \text{as} \quad x \to \infty, \quad \textit{uniformly in } y. \tag{19}$$

When $m = 0$ the speeds $U$, $U^*$, and $U^{**}$ are constant, and the preceding lemmas apply unchanged. Hence (19) holds also in this case.

Suppose $\epsilon > 0$, In order to prove (5), we shall show that suitably far downstream, say at $x = x_0$, there holds

$$|\bar{u}(x_0, \psi)^2 - u(x_0, \psi)^2| \leqslant \epsilon. \tag{20}$$

Assuming (20) for the moment, the proof is then completed by taking $x_0$ as the new initial position. Since the constants $C^*$ and $C^{**}$ need not be changed in this process, we may replace $a$ in (19) by $\surd\epsilon$. Since $\epsilon$ can be made arbitrarily small by taking $x_0$ suitably large, (5) is proved.

The proof of (20) is based on the explicit calculation of a comparison function and a consequent sharpening of the conclusion of lemma 3. Now in terms of the von Mises variables the function $\phi(x, \psi) = \bar{u}^2 - u^2$ satisfies the equation

$$\phi_x = \nu u \phi_{\psi\psi} + \beta(x, \psi)\,\phi, \quad \beta(x, \psi) = \frac{2\nu \bar{u}_{yy}}{\bar{u}(u+\bar{u})}, \tag{21}$$

(cf. the proof of lemma 1). We define

$$h(x, \psi) = \psi / U(x)\, g(x),$$

where $U(x) = C(x+d)^m$ and $g(x) = \surd\{\nu(x+d)/U(x)\}$, and consider the function

$$W(x, \psi) = (x+d)^{-\lambda} \begin{cases} h_0/h & \text{when} \quad h(x, \psi) \geqslant h_0, \\ 1 & \text{when} \quad h(x, \psi) \leqslant h_0, \end{cases}$$

$\lambda$ and $h_0$ being positive constants. We assert that these constants can be chosen so that

$$W_x \geqslant \nu u W_{\psi\psi} + \beta W \quad (h \neq h_0). \tag{22}$$

In order to prove this, introduce the function

$$B(h) = -f'''(\zeta)/\{f'(\zeta)\}^2, \quad \zeta = f^{-1}(h),$$

where $f(\zeta)$ is the similar solution associated with the exponent $m$ in question. Then the following results hold.

Lemma 9. *Let $c = C/C^{**}$. Then*

$$\beta(x, \psi) \leqslant -c\frac{B(h)}{x+d}, \quad h = h(x, \psi).$$

*Proof.* Since $\bar{u}_{yy} < 0$ and since $u, \bar{u} \leqslant u^{**}$ by lemma 2, we have

$$\beta = \frac{2\nu\bar{u}_{yy}}{\bar{u}(u+\bar{u})} \leqslant \frac{\nu\bar{u}_{yy}}{\bar{u}u^{**}} \tag{23}$$

where $\bar{u} = Uf'(\zeta)$ and $u^{**} = U^{**}f'(\zeta^{**})$ in an obvious notation, and

$$U^{**} = C^{**}(x+d)^m.$$

Now in (23) the functions $\bar{u}$ and $u^{**}$ are to be evaluated for the same value of $\psi$. Consequently $\zeta$ and $\zeta^{**}$ are related by

$$\psi = Ugf(\zeta) = U^{**}g^{**}f(\zeta^{**}).$$

Therefore since $Ug \leqslant U^{**}g^{**}$, we have

$$f(\zeta^{**}) \leqslant f(\zeta),$$

which in turn implies that $\zeta^{**} \leqslant \zeta$ and $f'(\zeta^{**}) \leqslant f'(\zeta)$. Hence we have

$$u^{**} \leqslant U^{**}f'(\zeta). \tag{24}$$

From (23) and (24), and the fact that $\bar{u}_{yy} = Ug^{-2}f'''(\zeta)$, there results easily

$$\beta \leqslant \frac{c}{x+d}\frac{f'''(\zeta)}{\{f'(\zeta)\}^2}.$$

On the other hand, according to the definition of $h$ we have $h(x,\psi) = f(\zeta)$. This completes the proof of lemma 9.

Lemma 10. *$B(h) > 0$ for $h > 0$. Moreover, $B(h) \approx$ const.$/h$ for small $h$ when $m > 0$, and $B(0) = \frac{1}{4}$ when $m = 0$.*

*Proof.* The first part of the lemma is obvious, since $f''' < 0$. Now as $h$ tends to zero, clearly $\zeta$ tends to zero. But for small $\zeta$ we have evidently

$$f(\zeta) \approx \tfrac{1}{2}b_0\zeta^2, \quad f'(\zeta) \approx b_0\zeta, \quad f''(0) = b_0, \quad f'''(0) = -m,$$

where $b_0$ is a positive constant depending on $m$ (if $m = 0$ the last relation should be replaced by $f''' = -\frac{1}{2}ff'' \approx -\frac{1}{4}b_0^2\zeta^2$. The first relation implies $\zeta^2 \approx 2h/b_0$, and the lemma follows easily.

Returning to the proof of (22), we have if $h < h_0$

$$W_x - \nu u W_{\psi\psi} - \beta W \geqslant \frac{W}{x+d}\{-\lambda + cB(h)\},$$

where lemma 9 is used to estimate the term $\beta W$. On the other hand, if $h > h_0$ then

$$\begin{aligned} W_x - \nu u W_{\psi\psi} - \beta W &\geqslant W_x - \nu u W_{\psi\psi} \\ &= -\frac{W}{x+d}\left\{\lambda + (x+d)\frac{\partial \ln h}{\partial x} + 2\nu u\frac{x+d}{\psi^2}\right\}. \end{aligned}$$

Now $\partial \ln h/\partial x = -(m+1)/2(x+d)$, $\psi^2 = \nu(x+d)Uh^2$, and $u \leqslant U^{**}$. Hence recalling that $c = C/C^{**}$ we have finally for $h > h_0$

$$W_x - \nu u W_{\psi\psi} - \beta W \geqslant -\frac{W}{x+d}\left\{\lambda - \frac{m+1}{2} + \frac{2}{ch^2}\right\}.$$

In order that (22) be satisfied it is therefore sufficient to require

$$\lambda \leqslant \begin{cases} cB(h) & \text{when} \quad h < h_0, \\ \dfrac{m+1}{2} - \dfrac{2}{ch^2} & \text{when} \quad h > h_0. \end{cases} \tag{25}$$

From lemma 10 it is clear that one can choose $\lambda(<1)$ and $h_0$ so that (25) holds. Thus (22) is proved.

In what follows it is convenient to specify $h_0$ in a definite way, namely

$$h_0^2 = \max\left\{\frac{8}{c(m+1)}, \frac{1}{\nu C d^{m+1}}\right\}. \tag{26}$$

The following result then holds.

LEMMA 11. *Let $\epsilon$ be an arbitrary positive number. Then*

$$|\bar{u}(x,\psi)^2 - u(x,\psi)^2| \leqslant \epsilon + K(\epsilon)\, W(x,\psi),$$

*where $K(\epsilon)$ is an appropriate constant depending only on $\epsilon$ and the initial data.*

*Proof.* Let $\tilde{W}(\psi) = (1+d)^{-1}\min(1,\psi^{-1})$. Since $\lambda < 1$ and $h_0 \geqslant (\nu C d^{1+m})^{-\frac{1}{2}}$ it is clear that $W(0,\psi) \geqslant \tilde{W}(\psi)$. On the initial line $\bar{u}$ and $u$ are bounded functions tending to $U(0)$ as $\psi$ tends to infinity. Hence there exists a constant $K(\epsilon)$ such that

$$|\bar{u}(0,\psi)^2 - u(0,\psi)^2| \leqslant \epsilon + K(\epsilon)\,\tilde{W}(\psi). \tag{27}$$

Put $\qquad \phi = \bar{u}^2 - u^2 \quad \text{and} \quad \Phi = \epsilon + K(\epsilon)\, W - \phi.$

Then by (21), (22), and the fact that $\beta < 0$ we have

$$\Phi_x \geqslant \nu u \Phi_{\psi\psi} + \beta\Phi \quad (h \neq h_0).$$

In addition, since $W > 0$.

$$\Phi \geqslant \epsilon \quad \text{on} \quad \psi = 0, \quad \Phi \geqslant 0 \quad \text{on} \quad x = 0$$

and $\qquad \Phi \to \epsilon \quad \text{as} \quad \psi \to \infty, \quad \text{uniformly in } x.$

This being the case, it follows as in the proof of lemma 1 that $\Phi \geqslant 0$ for $0 \leqslant x < \infty$, $0 \leqslant \psi < \infty$ (observe that $\Phi$ cannot take on a minimum value where $h = h_0$ since the function $\Phi(x,\psi)$ has a ridge, or concave corner, along this curve). We have therefore proved that

$$\bar{u}^2 - u^2 \leqslant \epsilon + K(\epsilon)\, W.$$

The reverse inequality is obtained in the same way, putting $\Phi = \epsilon + K(\epsilon)\, W + \phi$. This completes the proof of lemma 11.

In order to establish (20), choose $x_0$ so that $(x_0+d)^{\lambda} \geqslant K(\epsilon)/\epsilon$. Since $W \leqslant (x+d)^{-\lambda}$, the result follows at once from lemma 11 and the inessential replacement of $2\epsilon$ by $\epsilon$.

## 5. SKIN FRICTION

In addition to the asymptotic behaviour of the velocity profile, it is also of interest to consider the behaviour of the skin friction $\tau$ as $x$ tends downstream to infinity. We consider first the streaming speed $U(x) = C(x+d)^m$. The skin friction of the corresponding similar solution is then given by

$$\bar{\tau} = \mu\bar{u}_y = \mu\frac{U(x)}{g(x)} f''(0).$$

32-2

Now let $\tilde{u}(y)$ be an arbitrary initial profile, and let $u(x, y)$ be the corresponding solution of the Prandtl system. *Then we have the asymptotic estimate*

$$\left|\frac{g(x)}{U(x)}\tau - \mu f''(0)\right| = o\left(\frac{\mu}{x^{2m}}\right) \quad as \quad x \to \infty. \tag{28}$$

*That is, the normalized skin friction $g\tau/U$ tends uniformly to the normalized skin friction of the similar solution as $x$ tends downstream to infinity.*

In order to prove this, we recall from (21) that the difference function $\phi = \bar{u}^2 - u^2$ satisfies the equation

$$\phi_x = \nu u \phi_{\psi\psi} + \beta(x, \psi)\,\phi.$$

Set $V(x, \psi) = (x+d)^{-\lambda}(h - h^2)$, where $\lambda$ is the constant determined in the preceding section and $h = \psi/U(x)\,g(x)$. Then we find after a straightforward calculation (assuming $h \leqslant 1$, so that $V \geqslant 0$)

$$\begin{aligned} V_x - \nu u V_{\psi\psi} - \beta V &\geqslant V_x - \nu u V_{\psi\psi} \\ &= (x+d)^{-\lambda-1}\left\{\frac{2u}{U} - \left(\lambda + \frac{1+m}{2}\right)h + (\lambda + 1 + m)\,h^2\right\}. \end{aligned} \tag{29}$$

Moreover, as in the proof of lemma 9 we have

$$u \geqslant u^* = U^* f'(\zeta^*) \geqslant U^* f'(\zeta).$$

Let $h_1 = f(1)$. Then for $0 \leqslant h \leqslant h_1$ it is evident that

$$f'(\zeta) \geqslant b\zeta, \quad f(\zeta) \leqslant \tfrac{1}{2} b_0 \zeta^2,$$

where
$$b = f'(1) \quad \text{and} \quad b_0 = f''(0).$$

Consequently
$$f'(\zeta) \geqslant b\sqrt{(2h/b_0)} \quad \text{and} \quad u \geqslant U^* b\sqrt{(2h/b_0)}.$$

Inserting this in (29), it is evident that we can choose $h_2$ so small that

$$V_x - \nu u V_{\psi\psi} - \beta V \geqslant 0 \quad (0 \leqslant h \leqslant h_2).$$

Now $V(0, \psi) \geqslant$ const. $\psi$ for $0 \leqslant h \leqslant h_2$. Hence for this interval of the initial line there exists a constant $K_1$ such that

$$|\bar{u}(0, \psi)^2 - u(0, \psi)^2| \leqslant K_1 V(0, \psi).$$

On $\psi = h = 0$ we have $u = \bar{u} = V = 0$. Also on $h = h_2$ we have by lemma 11

$$|\bar{u}(x, \psi)^2 - u(x, \psi)^2| \leqslant \epsilon + K(\epsilon)\,(x+d)^{-\lambda}.$$

Let $x_0$ be an arbitrary positive number. For $0 \leqslant x \leqslant x_0$ and $h = h_2$

$$\begin{aligned} \epsilon + K(\epsilon)\,(x+d)^{-\lambda} &\leqslant \{\epsilon(x_0+d)^{\lambda} + K(\epsilon)\}\,(x+d)^{-\lambda} \\ &= \{\epsilon(x_0+d)^{\lambda} + K(\epsilon)\}\,K_2 V(x, \psi), \end{aligned}$$

where $K_2 = (h_2 - h_2^2)^{-1}$. Consequently, by the maximum principle argument already used several times, we have for $0 \leqslant x \leqslant x_0$, $0 \leqslant h \leqslant h_2$,

$$|\bar{u}(x, \psi)^2 - u(x, \psi)^2| \leqslant L V(x, \psi), \tag{30}$$

where $L$ denotes the maximum of $K_1$ and $\{\epsilon(x_0+d)^{\lambda} + K(\epsilon)\}K_2$.

Since $u$ and $\bar{u}$ are zero when $\psi = 0$, and since $2u_y = (u^2)_\psi$ it follows from (30) that for $x = x_0, y = 0$,

$$|\bar{u}_y - u_y| \leqslant (x_0+d)^{-\lambda} L/2U(x_0)\, g(x_0).$$

Now $(x_0+d)^{-\lambda} L \leqslant \epsilon K_2 + \{K_1 + K_2 K(\epsilon)\}(x_0+d)^{-\lambda} \leqslant 2\epsilon K_2$ if $x_0$ is sufficiently large. This proves (28).

A similar result can be obtained for the asymptotic behaviour of the skin friction when the streaming speed satisfies (6). More precisely, suppose that

$$UU_x \geqslant C_1(x+d)^{2m-1}, \tag{31}$$

where the constants $C_1$ and $m$ are positive and $d \geqslant 0$. Let $u$ and $\bar{u}$ be corresponding solutions of the Prandtl system with respective skin frictions $\tau$ and $\bar{\tau}$. *Then*

$$\frac{g}{U}|\tau - \bar{\tau}| = O\left(\frac{\mu}{x^{2m}}\right) \quad as \quad x \to \infty, \tag{32}$$

*where* $g = \sqrt{\{\nu(x+d)/U\}}$; *that is, the normalized skin friction* $g\tau/U$ *is asymptotically unique as* $x$ *tends to infinity.*

The proof is essentially the same as before. We note first that by lemma 7 one can assume $u_{yy} < 0$ without loss of generality. Then the function $\phi = \bar{u}^2 - u^2$ satisfies the equation

$$\phi_x = \nu\bar{u}\phi_{\psi\psi} + \alpha\phi,$$

where $\alpha > 0$.

Let $u^*$ be the similar solution obtained in lemma 2, and let $U^* = C^*(x+d)^m$ be the corresponding streaming speed. (Note that for the construction of $u^*$ only the inequality (31) is needed.) Consider the function $V(x,\psi) = h - h^2$, where $h = \psi/U^*g^*$. An easy calculation yields, for $0 \leqslant h \leqslant h_1$,

$$V_x - \nu\bar{u}V_{xx} - \alpha V \geqslant \frac{1}{x+d}\left\{2b\sqrt{\left(\frac{2h}{b_0}\right)} - \frac{1+m}{2}h + (1+m)h^2\right\};$$

here we have used the fact that $\bar{u} \geqslant u^*$ as in the previous proof. It follows that for $h_3$ suitably small

$$V_x - \nu\bar{u}V_{xx} - \alpha V \geqslant 0 \quad (0 \leqslant h \leqslant h_3).$$

For $x = 0$ it is clear that $|\phi| \leqslant$ const. $V$. Moreover, $\phi = V = 0$ on $\psi = 0$. By lemma 3 we have $|\phi| \leqslant a^2$, so that $|\phi| \leqslant$ const. $V$ also on $h = h_3$. The usual maximum principle argument therefore implies

$$|\bar{u}^2 - u^2| \leqslant \text{const.}\, V$$

for $0 \leqslant x < \infty$, $0 \leqslant h \leqslant h_3$. It follows as before that

$$|\bar{u}_y - u_y| \leqslant \text{const.}/U^*g^*$$

for $y = 0$, and the required conclusion is then obtained by multiplying both sides by $\mu g/U$.

## 6. Conclusions

The analysis presented here shows that for a power law streaming speed

$$U(x) = C(x+d)^m,$$

the velocity profile which develops downstream is asymptotically the same as the profile of the well known Falkner–Skan similarity solution. Moreover, for a streaming speed satisfying (6) the velocity profile which develops downstream is asymptotically unique, though of course the particular form of the resulting profile depends on the precise nature of the exterior stream. The rate of convergence for this asymptotic behaviour was also estimated, as well as corresponding rates for the convergence of the skin friction coefficient; relevant formulae are given in (5), (7), (28) and (32).

Lemma 2 contains a result of slightly different form, in which direct inequalities for the velocity profile $u(x,y)$ are given. (These inequalities are of course stated in terms of the von Mises variables; an easy calculation shows, however, that the conclusion also holds in the physical variables.) While lemma 2 is perhaps not so immediately useful as the main asymptotic inequalities, it does imply for a power law streaming speed and for a streaming speed satisfying (6) that, for fixed $y$,

$$\lim_{x\to\infty} u(x,y) = \begin{cases} 0 & \text{if} \quad n < \frac{1}{3}, \\ \infty & \text{if} \quad m > \frac{1}{3}; \end{cases}$$

also, in terms of the von Mises variables, for fixed $\psi$,

$$\lim_{x\to\infty} u(x,\psi) = \begin{cases} 0 & \text{if} \quad n < \frac{1}{3}, \\ \infty & \text{if} \quad m > \frac{1}{3}. \end{cases}$$

If $m = n$, as is the case in particular when $U(x) = C(x+d)^m$, the preceding formulae are of course somewhat sharper.

Lemma 7 has an interest in its own right, in showing the existence of concave velocity profiles when the pressure gradient is favourable ($\mathrm{d}p/\mathrm{d}x \leqslant 0$). Although the method of proof clearly indicates that there is no unique concave profile, it is nevertheless the case that no matter what the initial velocity profile may be, its downstream asymptotic development is *towards* a concave profile. Thus for streaming speeds which correspond to a favourable pressure gradient, concave velocity profiles play somewhat the same role as the similarity profiles do for a power law streaming speed. It may also be observed that when the pressure gradient is not favourable, the Prandtl equations automatically require that $u_{yy} > 0$ in the neighbourhood of the wall. Thus in this case concave velocity profiles are impossible. The relation between a favourable pressure gradient and concave velocity profiles has been noted by earlier writers (see, for example, Schlichting (1960, pp. 113–114), though the conclusion drawn by Schlichting that $u_{yy}$ is necessarily less than zero over the whole boundary layer is certainly incorrect and should be replaced by the results which we have just noted.

*Note.* This research was supported by the Air Force Office of Scientific Research under Grant No. AF-AFOSR 883-65 to the University of Minnesota. The results were announced at the sixth meeting of the Society for Natural Philosophy, held at Tucson, Arizona, January 1966.

## References

Coppel, W. 1960 On a differential equation of boundary layer theory. *Phil. Trans.* A **253**, 101–136.

Falkner, V. W. & Skan, S. W. 1931 Solutions of the boundary layer equations. *Phil. Mag.* (7), **12**, 865–896.

Goertler, H. 1957 A new series for the calculation of steady laminar boundary layer flows. *J. Math. Mech.* **6**, 1–166.

Hiemenz, K. 1911 Die Grenzschicht an einem in den gleichformigen Flüssigkeitsstrom eingetauchten geraden Kreiszlinder. *Diss. Gottingen.*

Howarth, L. 1935 Steady flow in the boundary layer near the surface of a cylinder in a stream. *Aero. Res. Comm. R.M.* 1632.

Iglisch, R. 1953 Elementarer Existenzbeweis für die Strömung in der laminaren Grenzschicht zur Potentialströmung $U = u_1 x^m$ mit $m > 0$ bei Absaugen und Ausblasen. *Z. ang. Math. Mech.* **33**, 143–147.

Iglisch, R. 1954 Elementarer beweis für die Eindeutigkeit der Strömung in der laminaren Grenzschicht zur Potentialströmung $U = u_1 x^m$ mit $m > 0$ bei Absaugen und Ausblasen. *Z. ang. Math. Mech.* **34**, 441–443.

Krzyżański, M. 1959 Certaines inégalités relatives aux solutions de l'équation parabolique linéare normale. *Bull. Acad Polon. Sci., Ser. Math. Ast. Phys.* **7**, 131–135.

Nickel, K. 1958 Einzige Eigenschaften von Lösungen der Prandtlschen Grenzschicht-Differentialgleichungen. *Arch. Rat. Mech. Anal.* **2**, 1–31.

Nirenberg, L. 1953 A strong maximum principle for parabolic equations. *Comm. Pure Appl. Math.* **6**, 167–177.

Oleinik, O. A. 1963 On the equations of a boundary layer. *Sem. Ist. Naz. Alta Math.* pp. 372–387.

Oleinik, O. A. 1963 The Prandtl system of equations in boundary layer theory. *Dokl. Akad. Nauk. SSSR* **150**, 28–32 (English trans. *Sov. Math.* **4**, 583–586).

Prandtl, L. 1904 Über Flüssigkeitsbewegung bei sehr kleiner Reibung. *Proc. Third. Int. Congr. Math.* (Heidelberg), 484–491.

Schlichting, H. 1960 *Boundary layer theory.* New York: McGraw-Hill.

Velte, W. 1960 Eine Anwendungen des Nirenbergschen Maximumsprinzips für parabolische Differentialgleichungen in der Grenzschichttheorie. *Arch. Rat. Mech. Anal.* **5**, 420–431.

*Arch. Ration. Mech. Anal.* 28 (1968), 217–225.




# *On the Mathematical Basis for Prandtl's Boundary Layer Theory: an Example*

JAMES SERRIN

In the preceding paper in this journal, FIFE has discussed the foundations of Prandtl's boundary layer theory and has established a firm mathematical connection between the Navier-Stokes equations and the Prandtl equations. The purpose of the present paper is to illustrate his results by studying in detail a specific example of fluid flow where a boundary layer occurs. To some extent we obtain in this way a miniature version of the full theory, which can serve as a concrete introduction to the more abstract and difficult considerations which are necessarily required for the general case. The example can also be used to indicate the way in which an interior flow more and more closely approximates to irrotational motion as the viscosity tends to zero.

The particular flow which we discuss is the well-known radial motion occurring between two straight converging walls, originally treated by JEFFERY and HAMEL in classical papers. We restrict consideration to the case where the flow is directed towards the vertex, since it is only in this situation that a boundary layer develops for small viscosity. The velocity vector in cylindrical polar coordinates $(r, \theta)$ in the plane of flow then has the form

$$v_r = -\frac{f(\theta)}{r}, \qquad v_\theta = 0 \tag{1}$$

where $f=0$ at the walls (taken as $\theta=0$ and $\theta=2\alpha$) and $f>0$ for $0<\theta<2\alpha$. One finds easily that (1) obeys the Navier-Stokes equations provided that

$$\nu(f''+4f)-f^2=\text{Constant}, \tag{2}$$

where $\nu$ denotes the kinematic viscosity and primes indicate differentiation with respect to $\theta$.

We assume the solution normalized so that the speed on the central streamline at the position $r=1$ has the fixed value $a$. That such solutions in fact exist for all sufficiently small values of $\nu$ was shown by HAMEL (cf. also [3], pp. 106 – 109). We shall be interested in the limiting form of these solutions as the viscosity tends to zero. According to boundary layer theory the result "should be" that in the central opening the motion uniformly approaches the irrotational velocity distribution

$$v_r = -\frac{a}{r}, \qquad v_\theta = 0, \tag{3}$$

while in thin boundary layers near the walls the speed rapidly adjusts from zero to its main stream value. We shall show that this is indeed exactly what happens; moreover, inside the boundary layer we shall establish a strict relation between the exact solution and the approximating flow given by boundary layer theory.

It should be remarked that reference [3], pp. 107–109, contains a brief discussion along similar lines. The results there are sketchy, however, and do not include precise asymptotic estimates. Since it is our purpose to present a detailed treatment bearing out FIFE's conclusions, it seems necessary, therefore, to give a fuller discussion.

Before stating the specific conclusions, we introduce the boundary layer approximation for the flow adjacent to the wall $\theta=0$. The natural coordinates for this are the distance $x$ along the wall and the distance $y$ normal to the wall; the appropriate exterior streaming speed is then just $U=-a/x$. A standard similarity solution exists for this streaming speed, with tangential velocity component given by

$$\bar{u}=-\frac{F(y/x)}{x} \tag{4}$$

and with $F$ determined by the Falkner-Skan equation

$$\nu F''+(a^2-F^2)=0 \tag{5}$$

and the boundary conditions

$$F(0)=0, \quad F(\infty)=a, \quad F>0 \quad \text{for } 0<\theta<\infty, \tag{6}$$

cf. [3], p. 144. The system (5), (6) has a unique solution (this is known, but will in any case be proved later), which moreover can be expressed in terms of elementary functions,

$$F(t)=3a\tanh^2\left(\sqrt{\frac{a}{2\nu}}\,t+c\right)-2a$$

where $c=\operatorname{arctanh}\sqrt{2/3}\approx 1.146$. Note that $F$ increases monotonically from zero to its limiting value $a$ as $t$ goes from 0 to $\infty$.

We shall adopt here the conventional boundary layer thickness $\delta=3\sqrt{\nu/a}$; in this distance the function $F(t)$ reaches 98% of its limiting value. (We could also define the width of the boundary layer in terms of displacement thickness, with no essential change required in the general nature of the conclusions. See Remark 3 at the end of the first section.)

## 1.

We may now state our main conclusions, relating the exact solution first with the Prandtl solution inside the boundary layer and then with the irrotational flow outside the boundary layer. For convenience we write $v$ and $V$ for the radial velocity component in (1) and (3), and $u$ for the tangential velocity component in (1).* Then the following results hold for sufficiently large values of the Reynolds number $R=a/\nu$.

(I) *In a boundary layer $0\leqq\theta\leqq 3/\sqrt{R}$ adjacent to the wall $\theta=0$, and similarly in a boundary layer adjacent to the wall $\theta=2\alpha$, we have*

$$\left|\frac{u-\bar{u}}{U}\right|\leqq\frac{32}{R}.$$

* That is, $u=v\cos\theta$.

(II) *Outside the boundary layer, that is, in the central opening* $3/\sqrt{R} \leqq \theta \leqq 2\alpha - 3/\sqrt{R}$, *we have*

$$\left|\frac{v-V}{V}\right| \leqq \frac{1}{50} + \frac{12}{R}.$$

(III) *In the central opening*

$$\sqrt{1/2R}\log R \leqq \theta \leqq 2\alpha - \sqrt{1/2R}\log R$$

*there holds*

$$\left|\frac{v-V}{V}\right| \leqq \frac{24}{R}.$$

**Proof.** Our main goal will be to establish the estimate

$$|f(\theta) - F(\theta)| \leqq 12\nu, \qquad 0 \leqq \theta \leqq \alpha, \tag{7}$$

for sufficiently large $R$. Once this is done the required conclusions will follow easily.

Let $b$ be a positive real parameter, and define

$$F(\theta, b) = 3b\tanh^2\left(\sqrt{\frac{b}{2\nu}}\,\theta + c\right) - 2b, \qquad 0 \leqq \theta < \infty.$$

Clearly, we have

$$\nu F'' + (b^2 - F^2) = 0, \quad F(\theta, b) = 0, \quad F(\infty, b) = b.$$

Now let $b_1^2 = a^2 - 4a\nu$ and $b_2^2 = a^2 + 4a\nu$. Then if $R(\geqq 4)$ is suitably large, we claim that

$$F(\theta, b_1) \leqq f(\theta) \leqq F(\theta, b_2), \qquad 0 \leqq \theta \leqq \alpha. \tag{8}$$

Suppose for contradiction that the left-hand inequality were not true. Then since (in obvious notation)

$$f = F_1 = 0 \text{ at } \theta = 0, \qquad f > F_1 \text{ at } \theta = \alpha,$$

there must exist a point in the interval $0 < \theta < \alpha$ where

$$f < F_1, \quad f' = F_1', \quad f'' \geqq F_1''.$$

This, however, contradicts the relation

$$\nu(f'' - F_1'' + 4f) + F_1^2 - f^2 = \nu f''(\alpha) < 0. \tag{9}$$

[In deriving (9), observe that the constant in (2) has the value $\nu f''(\alpha) + 4a\nu - a^2$, and use the well-known fact that the velocity profile $f(\theta)$ is concave in the interval $0 \leqq \theta \leqq 2\alpha$.]

Before proving the right-hand inequality of (8), we shall show that for suitably large $R$ (depending only on $\alpha$)

$$f < F_2 \text{ and } f'' > -4a \qquad \text{at } \theta = \alpha. \tag{10}$$

The first inequality in fact follows from the relation

$$3\tanh^2\frac{\alpha(R^2+4R)^{\frac{1}{2}}}{\sqrt{2}}>2+\left(1+\frac{4}{R}\right)^{-\frac{1}{2}}, \tag{11}$$

which is obvious for large $R$. To obtain the second, put

$$\gamma=\mathrm{Max}\left(\frac{1}{2},1-\frac{\alpha^2}{2}\right)$$

and assume that

$$3\tanh^2\frac{\alpha(R^2-4R)^{\frac{1}{2}}}{2\sqrt{2}}>2+\gamma\left(1-\frac{4}{R}\right)^{-\frac{1}{2}}, \tag{12}$$

which is also obvious for suitably large $R$. Inequality (12) implies

$$F_1>a\gamma \tag{13}$$

in the interval $\alpha/2\leqq\theta\leqq\alpha$. Consequently, by differentiating (2) and using the first part of (8), we get

$$\nu f'''=2f'(f-2\nu)>2f'(F_1-2\nu)>2\nu f'(\gamma R-2)\geqq 0,$$

valid for $\alpha/2\leqq\theta\leqq\alpha$. Therefore, using TAYLOR's expansion with remainder,

$$f\leqq a+\tfrac{1}{2}f''(\alpha)(\theta-\alpha)^2,\qquad \alpha/2\leqq\theta\leqq\alpha.$$

Evaluating this at $\theta=\alpha/2$, and once more using (13) and the first part of (8), proves the remaining inequality of (10).

With (10) varified, the right-hand inequality of (8) can be proved in the same way as the left-hand inequality. Indeed, suppose it were not true. Then, since

$$f=F_2=0 \quad\text{at}\quad \theta=0,\qquad f<F_2 \quad\text{at}\quad \theta=\alpha,$$

there must exist a point in the interval $0<\theta<\alpha$ where

$$f>F_2,\qquad f'=F_2',\qquad f''\leqq F_2''.$$

This, however, contradicts the relation

$$\nu(f''-F_2'')+F_2^2-f^2=\nu\{f''(\alpha)+4a+4(a-f)\}>0$$

and (8) is therefore established.

Next, using a similar comparison method, we find

$$F(\theta,b_1)\leqq F(\theta)\leqq F(\theta,b_2),\qquad 0\leqq\theta<\infty \tag{14}$$

(since the details are entirely straightforward, we may omit the formal proof).

Combining (8) and (14) gives

$$|f(\theta)-F(\theta)|\leqq F(\theta,b_2)-F(\theta,b_1),\qquad 0\leqq\theta\leqq\alpha. \tag{15}$$

Putting

$$z=\sqrt{\frac{b}{2\nu}}\,\theta+c,$$

we find

$$
\begin{aligned}
(16)\qquad \frac{\partial F}{\partial b} &= 3\tanh^2 z - 2 + 3\theta\sqrt{\frac{b}{2\nu}}\tanh z\,\mathrm{sech}^2 z \\
&\leqq 1 + 3z\tanh z\,\mathrm{sech}^2 z \leqq 2\,{}^{\star}
\end{aligned}
$$

and (assuming always that $R \geqq 4$)

$$(17)\qquad b_2 - b_1 = a\{\sqrt{1+4/R} - \sqrt{1-4/R}\} \leqq 4\sqrt{2}\,\nu .$$

Therefore $|f(\theta) - F(\theta)| \leqq 12\,\nu$ when $0 \leqq \theta \leqq \alpha$, proving (7).

Since $F(3/\sqrt{R}) \geqq 0.98\,a$, it follows that in the central opening $3/\sqrt{R} \leqq \theta \leqq 2\alpha - 3/\sqrt{R}$ we have

$$\left|\frac{v - V}{V}\right| = \frac{|f(\theta) - a|}{a} \leqq \frac{1}{50} + \frac{12}{R},$$

proving (II). To verify (III), we observe that

$$F\left(\frac{\log R}{\sqrt{2R}}\right) \geqq a\left(1 - \frac{12}{R}\right),$$

The result then follows exactly as before.

It remains to prove (I). Here

$$u = \frac{f(\theta)\cos\theta}{r}, \qquad \bar{u} = \frac{F(\tan\theta)}{r\cos\theta}, \qquad U = \frac{a}{r\cos\theta},$$

and

$$\frac{u - \bar{u}}{U} = \frac{1}{a}\{f(\theta)\cos^2\theta - F(\tan\theta)\}.$$

thus

$$a\left|\frac{u - \bar{u}}{U}\right| \leqq \cos^2\theta\,|f(\theta) - F(\theta)| + |F(\theta) - F(\tan\theta)| + F(\theta)\sin^2\theta .$$

By the mean value theorem together with TAYLOR's expansion for $\tan\theta$,

$$(18)\qquad |F(\theta) - F(\tan\theta)| \leqq \frac{2a\sqrt{R}}{\sqrt{3}}\,|\theta - \tan\theta| \leqq \frac{2a\sqrt{R}}{3\sqrt{3}}\,\frac{1 + 2\sin^2\theta}{\cos^4\theta}\,\theta^3 .$$

From (7), (18), and the fact that $\theta \leqq 3/\sqrt{R}$ in the boundary layer, we get then

$$\left|\frac{u - \bar{u}}{U}\right| \leqq \frac{12\cos^2\theta}{R} + \frac{6\sqrt{3}(1 + 2\sin^2\theta)}{R\cos^4\theta} + \frac{9}{R}.$$

If $R$ is suitably large $\cos\theta$ can be taken arbitrarily near 1. Thus in this case

$$\left|\frac{u - \bar{u}}{U}\right| \leqq \frac{32}{R} \qquad \text{for } 0 \leqq \theta \leqq 3/\sqrt{R},$$

completing the proof.

---

* To see this, note that $z$ varies in the range $c \leqq z < \infty$, and that $z\tanh z\,\mathrm{sech}^2 z$ is a decreasing function there.

*Remarks*

*1.* The estimate given by (I) can be improved for small values of $\theta$, a result not without significance since both $u$ and $\bar{u}$ are zero when $\theta=0$. More precisely, for sufficiently large values of the Reynolds number $R$ we have

(I') *In the boundary layer* $0\leqq\theta\leqq 3/\sqrt{R}$, *and similarly in the boundary layer* $2\alpha-3/\sqrt{R}\leqq\theta\leqq 2\alpha$, *not only does the estimate* (I) *hold, but also we have*

$$\left|\frac{u-\bar{u}}{U}\right|\leqq\frac{20}{\sqrt{R}}\,\theta.$$

**Proof.** We first give an alternate estimate to (16), namely

$$\frac{\partial F}{\partial b}=\theta\left\{\frac{3\tanh^2 z-2}{\theta}+3\tanh z\cdot\operatorname{sech}^2 z\right.\left.\sqrt{\frac{b}{2\nu}}\right\}\leqq\theta\sqrt{3R}\left(1+\frac{4}{R}\right)^{\frac{1}{2}}.$$

Combining this with (15) and (17) there results

$$|f(\theta)-F(\theta)|\leqq 12\theta\,\nu\sqrt{R} \tag{19}$$

in place of the previous estimate (7).

Now following the procedure at the end of the previous section, we obtain

$$\left|\frac{u-\bar{u}}{U}\right|\leqq\frac{12\theta\cos^2\theta}{\sqrt{R}}+\frac{2\sqrt{3}\theta(1+2\sin^2\theta)}{\sqrt{R}\cos^4\theta}+\frac{3\theta}{\sqrt{R}}.$$

The required result is an immediate consequence of the last inequality when $R$ is suitably large.

*2.* A result similar to (I) can be given for the normal velocity component $w$ in the boundary layer region; namely, for $0\leqq\theta\leqq 3/\sqrt{R}$ and for sufficiently large $R$

$$\left|\frac{w-\bar{w}}{V}\right|=\tan\theta\left|\frac{u-\bar{u}}{V}\right|\leqq\frac{32}{R}\,\theta.$$

We omit the statement of the result analogous to (I').

*3.* As one can see from the concluding part of the proof of Theorem (II), the following result also holds for sufficiently large $R$.

(II') *For each* $\varepsilon$, $0<\varepsilon<1$, *there exists an* $\eta>0$ *such that in the central opening* $\eta/\sqrt{R}\leqq\theta\leqq 2\alpha-\eta/\sqrt{R}$ *we have*

$$\left|\frac{v-V}{V}\right|\leqq\varepsilon+\frac{12}{R}.$$

## 2.

In addition to the question of the validity of PRANDTL's theory throughout the boundary layer, it is worthwhile discussing its value in predicting stresses at the wall. Let $\tau=\tau(x)$ be the shear stress at the wall $\theta=0$,

$$\tau(x)=\mu\left(\frac{\partial u}{\partial y}+\frac{\partial v}{\partial x}\right)=\mu\frac{\partial u}{\partial y}.$$

Similarly, let $\bar{\tau}=\bar{\tau}(x)$ be the corresponding function calculated from the Prandtl solution (4),

$$\bar{\tau}(x)=\mu\frac{\partial \bar{u}}{\partial y}.$$

We shall show for suitably large values of the Reynolds number $R$ that

$$\text{(IV)}\qquad \left|\frac{\tau-\bar{\tau}}{\bar{\tau}}\right|\leqq\frac{12}{R}.$$

**Proof.** By direct calculation

$$\tau(x)=-\mu\frac{f'(0)}{x^2},\qquad \bar{\tau}(x)=-\mu\frac{F'(0)}{x^2}.$$

Hence, for suitably large $R$

$$|\tau-\bar{\tau}|=\frac{\mu}{x^2}|f'(0)-F'(0)|\leqq\frac{12\mu\nu\sqrt{R}}{x^2}$$

the last step holding in view of (19) and the fact that $f=F=0$ at $\theta=0$.

Since $F'(0)=2a\sqrt{R/3}$, we find also

$$\bar{\tau}(x)=-\frac{2\mu a\sqrt{R/3}}{x^2}$$

and the required result follows at once.

Next consider the relation between the pressure $p$ and its approximation $\bar{p}$ according to boundary layer theory. Now from [3]

$$p=(-4\nu f+\text{Constant})/2r^2,$$

while $\bar{p}$ is given outside the boundary layer by $\bar{p}=\pi=-a^2/2r^2$, and inside the boundary layer by $\bar{p}=-a^2/2x^2$.

Evidently

$$\frac{p-\pi}{\pi}=\frac{1}{R}\left\{4\left(1-\frac{f}{a}\right)+\frac{f''(\alpha)}{a}\right\},$$

whence according to (10),

$$\left|\frac{p-\pi}{\pi}\right|\leqq\frac{4}{R}\qquad\text{for sufficiently large } R.$$

Since $\bar{p}=\pi\sec^2\theta$ in the boundary layer, it follows that for sufficiently large values of $R$ we have

$$\text{(V)}\qquad \left|\frac{p-\bar{p}}{\bar{p}}\right|\leqq\frac{5}{R}$$

throughout the entire opening $0\leqq\theta\leqq 2\alpha$.

## 3.

It was remarked earlier that problem (5), (6) has only one solution. To prove this, it is enough to return to inequality (14),

$$F(\theta,b_1)\leqq F(\theta)\leqq F(\theta,b_2).$$

Note that the explicit formula for $F(\theta)$ was not used in the derivation of this relation, and that it holds whenever $b_1 \leqq a \leqq b_2$. In particular, setting $b_1 = b_2 = a$ we get

$$F(\theta) = F(\theta, a) = 3a \tanh^2\left(\sqrt{\frac{a}{2\nu}}\,\theta + c\right) - 2a,$$

as required.

## 4.

The main Theorems (I) and (IV) bear out the general estimates obtained by FIFE (cf. in particular, his Theorem 5.2), though naturally the specific numerical coefficients here are much sharper. On the other hand, FIFE's approach does not appear to yield conclusions corresponding to (II), (III), and (V). Thus the present paper not only provides a concrete case where FIFE's hypotheses and results apply (as pointed out in the introduction), but also indicates certain directions for further work.

We conclude the paper with a series of somewhat unrelated remarks, grouped under three headings.

*1.* The conventionally defined boundary layer thickness, which we have adopted here, is that in which the similarity solution (4) reaches 98 % of its exterior streaming value, namely

$$\delta = 3/\sqrt{R}.$$

Theorem (II) shows that exterior to this region the motion comes "close" to the irrotational flow (3), though not uniformly so. It is in fact only in a somewhat smaller central opening that the exact solution uniformly approaches the irrotational flow (see Theorem (III), for example). Thus there is a certain *transition zone* outside the boundary layer where neither the Prandtl solution nor the irrotational flow provides an asymptotic approximation valid to order $1/R$. This transition zone lies roughly between $\theta = O(R^{-\frac{1}{2}})$ and $\theta = O(R^{-\frac{1}{2}} \log R)$, and can be conventionally defined by the thicknesses

$$\delta_1 \approx R^{-\frac{1}{2}} \quad \text{and} \quad \delta_2 \approx R^{-\frac{1}{2}} \log R.$$

At the outer edge of this zone the irrotational flow provides a better approximation to the exact solution than does the Prandtl flow, the situation being just reversed at the inner edge. It should be noted that although an approximation to order $1/R$ is not available in this zone, nevertheless the Prandtl solution does provide an approximation uniformly valid to order $(\log R)^3/R$.

It remains an interesting question whether transition zones with similar properties exist for motions more general than the particular one under study.

*2.* Besides the particular flow discussed in this paper, there are a number of other explicit solutions of the Navier-Stokes equation which exhibit boundary layer behaviour as the viscosity goes to zero. So far as I can see, however, none of these supplies as useful an example as HAMEL's flow, since they are either too complicated (as in KÁRMÁN's swirling flow) or too special. In the second category I would include the well-known flow in which fluid impinges from infinity onto a plane wall and exhibits a central stagnation point ([3], pp. 139–143). This example occurs both for plane and axially symmetric flow, and has the feature that the exact solutions also solve the Prandtl equations (the suppressed terms in the Prandtl equations being identically zero). Thus while these flows do serve to justify PRANDTL's theory, they do so in such a strong way that first of all we cannot learn from them the typical sort of error to be expected, and second are left with the unsettling possibility that perhaps PRANDTL's theory is valid *only* when the suppressed terms are actually zero.

*3.* Some comments should be made concerning the proof itself. For one thing, it may be felt that a simpler argument ought to be available in view of the fact that (2) can be integrated explicitly in terms of elliptic functions. While I hesitate to make a strong statement here, it nevertheless seems that the specific integration of (2) would not be immediately helpful in obtaining precise asymptotic estimates (see, however, [2], p. 135).

A related point is that the proofs of *all* the estimates depend on the integration of (5). In other words, in verifying the boundary layer assumptions over the entire flow field it was necessary to use the boundary layer equation itself!

Finally, we should comment on the specific numerical coefficients on the right-hand sides of the main estimates (I) through (IV). These can, of course, be sharpened, both by being somewhat more careful in the preliminary estimates and by dealing with even larger Reynolds numbers. It seems, however, that the slight improvements which could be obtained in this way would not repay the effort involved.

*Added in proof.* Professor W. CHESTER has observed that the conclusion of Theorem (I′) can be rewritten in the form

$$\left|\frac{u-\bar{u}}{\bar{u}}\right| \leqq \frac{60}{R} \qquad (R \text{ sufficiently large}),$$

showing that the difference $u-\bar{u}$ is uniformly small with respect to the actual velocity in the boundary layer.

*Note.* This research was partially supported by the United States Air Force Office of Scientific Research under Grant No. AF-AFOSR-883-67.

## References

1. FIFE, P., Considerations regarding the mathematical basis for Prandtl's boundary layer theory. Preceding in this journal.
2. FRAENKEL, L. E., Laminar flow in symmetrical channels with slightly curved walls. Proc. Roy. Soc. London A **267**, 119–138 (1962).
3. Modern Developments in Fluid Dynamics, edited by S. GOLDSTEIN. Oxford University Press 1938.
4. HAMEL, G., Spiralförmige Bewegungen zäher Flüssigkeiten. Jahresbericht der Deutschen Mathematiker-Vereinigung **25**, 34–60 (1916).
5. JEFFERY, G. B., The two dimensional steady motion of a viscous fluid. Phil. Mag. **29**, 455–465 (1916).
6. ROSENHEAD, A., The steady two-dimensional radial flow of viscous fluid between two inclined plane walls. Proc. Roy. Soc. London A **175**, 436–467 (1940).

University of Minnesota
Minneapolis, Minnesota

*(Received October 10, 1967)*

J. Fluid Mech. 34 (1968), 337–342.

**Reprint V.3.[iii]**

# The behaviour of similar solutions in a compressible boundary layer

**By J. B. McLEOD AND J. SERRIN**

Wadham College, Oxford and University of Sussex

(Received 4 January 1968)

This paper discusses the mathematical properties of similar solutions of the boundary-layer equations in a compressible model fluid, under assumptions first introduced by Stewartson and by Li & Nagamatsu. Assuming a favourable pressure gradient and that backflow is not present, our results include (among other things) a rigorous proof that velocity overshoot occurs in the boundary layer if the wall is heated, and that this is true whether or not suction, blowing or slipping occurs at the wall; while, conversely, velocity overshoot does not occur when the wall is cooled and the amount of slipping at the wall is suitably restricted.

## 1. Introduction

We shall be concerned in this note with the boundary layer of a compressible fluid adjacent to a fixed wall. Under appropriate conditions on the physical properties of the fluid one can look for similar solutions to this problem. As was first shown by Stewartson (1949) and by Li & Nagamatsu (1955) this leads to the linked pair of non-linear differential equations

$$f''' + ff'' + \beta(h - f'^2) = 0, \tag{1}$$

$$h'' + fh' = 0, \tag{2}$$

with the boundary conditions

$$f(0) = f'(0) = 0, \quad h(0) = a; \quad f'(\infty) = h(\infty) = 1. \tag{3}$$

Here $f'$ and $h$ are effectively the tangential velocity and the total energy function in the boundary layer, $\beta$ is a constant dependent on the particular stream conditions, and $a$ measures the ratio of the temperature at the wall to the stagnation temperature in the mainstream. The first two conditions of (3) express the fact that the fluid adheres to the wall without slipping. (It is also possible to consider walls which are permeable to the fluid and even to allow a certain amount of slipping. The changes in the argument which are required to handle these more general cases are discussed briefly at the end of the paper.)

A derivation of the above equations and a discussion of the stream conditions under which they are valid can be found in several standard works, e.g. Schlichting (1960), Rosenhead (1963) and Stewartson (1964). We remark here only that the constant $\beta$ is non-negative when the pressure gradient in the mainstream is favourable.

If $\beta = 0$, equation (1) becomes Blasius's equation, the existence, uniqueness, and behaviour of the solution being well known. Solutions of (1), (2), (3) have been calculated numerically for various values of the parameters $a$ and $\beta$ by Li & Nagamatsu (1955) and by Cohen & Reshotko (1956). In addition, Ho & Wilson (1967) have recently proved the existence of solutions when $\beta > 0$ and $0 < a \leqslant 1$, that is, when the wall is cooled, while in work to appear shortly the present authors have extended this result to the case when the wall is heated ($a > 1$) and slipping, suction and blowing are present.

For the case of a heated wall with $\beta > 0$ numerical work and physical reasoning indicate the existence of velocity overshoot in the boundary layer. (By velocity overshoot we mean the occurence of $f' > 1$ for some range of values of the independent variable $\eta$.) An asymptotic analysis of this phenomenon was given by Cohen & Reshotko (1956). While this analysis is valuable, and suggestive of the general behaviour to be expected of solutions, it nevertheless was based on dropping certain terms from the equations and accordingly is not entirely convincing. This being the case, one of the main purposes of this note is to show that such an overshoot must in fact occur as a direct and mathematically exact consequence of the differential equations and the fact that $a > 1$. At the same time we shall consider the general qualitative behaviour of solutions when $\beta > 0$, both when the wall is heated and when it is cooled. It will be assumed throughout that the solutions under discussion do not exhibit backflow, that is, we assume that $f' \geqslant 0$ for all solutions under consideration.

When $0 < a < 1$ we shall show, specifically, that the functions $h$ and $f'$ are monotonically increasing and concave, with $f'^2 < h$. It follows, in consequence, that similar solutions do not exhibit velocity overshoot when the wall is cooled. (When $a = 1$, see the remark at the end of §3.)

When $a > 1$ we show that $h$ is monotonically decreasing and convex. Moreover, $f'$ first increases to a maximum value greater than 1 and then decreases monotonically to its limit value 1. In addition $f'^2 < h$ everywhere, and the graph of $f'$ is concave in its rising portion, has exactly one inflexion when the parameters $\beta$ and $a$ satisfy the relation

$$2\beta \leqslant \frac{\sqrt{a}}{\sqrt{a}-1},$$

and otherwise has at most a finite (odd) number of inflexions.

The condition $f'^2 < h$, which occurs both for cooled and for heated walls, has the physical interpretation that the local Mach number in the boundary layer is less than the Mach number in the mainstream. To see this, we recall the relationship of $f'$ and $h$ to the physical variables in the flow, namely

$$u = Uf', \quad \tfrac{1}{2}u^2 + c_p T = Hh,$$

where $u$ is the tangential velocity and $T$ the temperature, $c_p$ the specific heat at constant pressure, and $H$ and $U$ mainstream scaling parameters. A simple calculation then shows that $f'^2 < h$ is equivalent to

$$u^2/T < U^2/T_\infty,$$

which establishes the required interpretation (since in the fluid the squared speed of sound is proportional to temperature).

We emphasize finally that what is being proved here is that *if solutions exist which do not exhibit backflow* then they must behave in the manner described. The existence of solutions is another matter, dealt with in the papers noted earlier, while the uniqueness of solutions remains an open question.

## 2. Existence of velocity overshoot

We start by considering the existence of velocity overshoot when $a > 1$. From (2) there follows

$$h'(\eta) = -C\exp\left(-\int_0^\eta f(v)\,dv\right), \tag{4}$$

where the real coefficient $C$ must be positive since $h(0) = a > 1$ and $h(\infty) = 1$. Since $f' \to 1$ as $\eta \to \infty$ we have $f(\eta) \sim \eta$, whence by L'Hôpital's rule

$$\{h-1\}.\left\{\eta\exp\left(\int_0^\eta f\,dv\right)\right\} \to C \quad \text{as } \eta \to \infty. \tag{5}$$

Now set

$$F(\eta) = f''(\eta)\exp\left(\int_0^\eta f(v)\,dv\right)$$

and write (1) in the form

$$F' = -\beta(h-f'^2)\exp\left(\int_0^\eta f(v)\,dv\right). \tag{6}$$

If we suppose, contrary to what has to be proved, that overshoot does not take place, then $f' \leqslant 1$ everywhere. Since $f' \geqslant 0$ by assumption, we find $f'^2 \leqslant 1$ and so, by (5) and (6),

$$F' \leqslant -\beta(h-1)\exp\left(\int_0^\eta f(v)\,dv\right) \leqslant -\beta C/2\eta$$

for $\eta$ sufficiently large. Integrating this yields $F \to -\infty$ as $\eta \to \infty$, whence $f''$ is ultimately negative. This in turn implies (using the condition $f' \to 1$ as $\eta \to \infty$) that $f' > 1$ for all sufficiently large $\eta$, and we have the required contradiction.

## 3. The inequality $f'^2 < h$

We show now that $f'^2 < h$ when $\beta > 0$, irrespective of whether the wall is heated or cooled. Letting $G = h - f'^2$, an easy calculation from (1) and (2) gives

$$G'' + fG' - 2\beta f'G = -2(f'')^2. \tag{7}$$

Suppose for contradiction that $G \leqslant 0$ at some point. Since $G = a > 0$ at $\eta = 0$ and $G = 0$ at $\eta = \infty$, there must then be some point $\eta_0$ at which $G$ takes a negative or zero minimum. At this point

$$G \leqslant 0, \quad G' = 0, \quad G'' \geqslant 0,$$

contradicting (7) unless $f''(\eta_0) = 0$. But then $G' = h' - 2f'f'' = h'$ at $\eta_0$, contradicting the fact that $G' = 0$ at $\eta_0$.

## 4. The cooled wall

Consider now the behaviour of $f'$ and $h$ when $0 < a < 1$. Relation (4) continues to hold, though now the coefficient $C$ must be negative since $h(0) < 1$. Therefore $h' > 0$, and in turn $h'' < 0$ by (2). Thus the graph of $h$ is monotonically increasing and concave.

By (6) the function $F$ is monotonically decreasing. If its limit as $\eta$ tends to infinity is negative, then $f''$ must eventually be negative. But (using the condition $f' \to 1$ as $\eta \to \infty$) this contradicts the fact that $f'^2 < h < 1$. Hence the limit value of $F$ is non-negative and $F$ and $f''$ are positive functions. Finally by (1)

$$f''' = -ff'' - \beta(h - f'^2) < 0.$$

Thus the graph of $f'$ is also monotonically increasing and concave.

*Remark*

When $a = 1$ we are easily led as above to $h' \equiv 0$ and $h \equiv 1$. Equation (1) then reduces to the Falkner–Skan equation; it follows finally by the argument in the preceding paragraph that $f'$ is monotonically increasing and concave.

## 5. The heated wall

We turn lastly to the behaviour of $f'$ and $h$ when $a > 1$. From (4) and (2) it is clear that the graph of $h$ is monotonically decreasing and convex.

Also by (6) the function $F$ is monotonically decreasing. Let its limit value be denoted by $l$ (possibly $-\infty$). The case $l \geqslant 0$ is impossible, since then we always have $f'' \geqslant 0$ and $f' \leqslant 1$, which has already been excluded. Hence $l < 0$ and $f''$ is first positive and then negative (the case where $f''$ takes only non-positive values obviously cannot occur). Consequently $f'$ must first increase to a maximum value greater than 1 and then decrease to its limit value 1 at infinity.

Since $f''' = -ff'' - \beta(h - f'^2)$ it follows that $f''' < 0$ so long as the graph of $f'$ is rising.

It remains to consider the inflexions in the graph. We show first that these are finite in number and that eventually $f''' > 0$. Indeed, once $f'$ becomes greater than 1 we have

$$f''' \geqslant -ff'' - \beta(h-1).$$

For sufficiently large values of $\eta$, however,

$$-ff''\exp\left(\int_0^\eta f(v)\,dv\right) = -fF \geqslant -\tfrac{1}{2}l\eta \qquad (l < 0)$$

and

$$(h-1)\exp\left(\int_0^\eta f(v)\,dv\right) \leqslant 2C/\eta$$

by the results of §2. Thus $f'''$ is ultimately positive. If there were an infinite number of inflexions they would have a finite limit point. But then (since $f$ is an analytic function) it would follow that $f''' \equiv 0$, which is impossible.

Before proving the final result, we show that

$$2F > -C. \tag{8}$$

To begin with, by the definition of $F$ and $G$ and by (4)

$$G' = h' - 2f'f'' = -(C + 2f'F)\exp\left(-\int_0^\eta f(v)\,dv\right).$$

If (8) were not true, then ultimately $G'$ would be positive. Since $G = 0$ at $\eta = \infty$ this would imply that $G$ is ultimately negative, a situation which has already been excluded.

With (8) established, we now observe that

$$f^{\text{iv}} = -ff''' - f'f'' - \beta(h' - 2f'f'').$$

Let $\eta_1$ be an inflexion of $f'$. By (4) and the definition of $F$, we have at $\eta_1$

$$f^{\text{iv}} = \{(2\beta - 1)f'F + \beta C\}\exp\left(-\int_0^{\eta_1} f(v)\,dv\right). \tag{9}$$

We assert that if

$$2\beta \leqslant \frac{\sqrt{a}}{\sqrt{a} - 1} \tag{10}$$

then $f^{\text{iv}} > 0$ at $\eta_1$. Consider separately the two cases

$$0 < 2\beta \leqslant 1, \quad 1 < 2\beta \leqslant \frac{\sqrt{a}}{\sqrt{a} - 1}.$$

For the first one, note that $f'' = -\beta G/f < 0$ at $\eta_1$ according to (1). Hence $F < 0$ at $\eta_1$ and the assertion follows at once from (9). In the second case, using (9), (8) and the fact that $f' < \sqrt{h} < \sqrt{a}$, we find

$$f^{\text{iv}} > \tfrac{1}{2}C\{2\beta - (2\beta - 1)\sqrt{a}\}\exp\left(-\int_0^{\eta_1} f(v)\,dv\right) \geqslant 0.$$

Thus when (10) holds, $f'''$ can only increase as we pass an inflexion point. Consequently there can be no more than one inflexion in the graph of $f'$ under this condition. On the other hand, there must be at least one such point in view of the already established behaviour of $f'$. This completes the proof of the assertions made in the introduction.

## 6. Generalizations

In the preceding sections we restricted ourselves specifically to the situation when $f(0) = f'(0) = 0$, but the analysis easily extends to the more general case $f(0) = \alpha_1, f'(0) = \alpha_2$, where $\alpha_1$ is any real constant and $\alpha_2$ is a non-negative constant satisfying $\alpha_2 \leqslant \sqrt{a}$. (The last condition expresses the fact that the Mach number at the wall should not exceed the Mach number in the mainstream; cf. the remark at the close of the introduction.)

Nothing in §2 is altered by the above changes, and it therefore follows that velocity overshoot continues to hold when $a > 1$. Furthermore, with the help of the condition $\alpha_2 \leqslant \sqrt{a}$ it is not hard to see that the argument of §3 continues to hold; we can therefore conclude that $f'^2 < h$ for all positive values of $\eta$.

When $\alpha_1 > 0$, that is, if fluid is sucked from the boundary layer into the wall, nothing needs to be changed in §§4 and 5. The qualitative behaviour of the graphs of $f'$ and $h$ therefore remains unaltered for this case.

On the other hand, if $\alpha_1 < 0$, that is, if fluid is blown through the wall into the boundary layer, then some significant changes result in the graphs of $f'$ and $h$. In particular, when $\alpha_1 < 0$ the function $f$ will first be negative and then positive, assuming as always that the motion exhibits no backflow. Then, since $h'$ is of one sign, it is apparent from (2) that the graph of $h$ will have exactly one inflexion for $a \neq 1$, being ultimately concave when $a < 1$, and ultimately convex when $a > 1$. Also, following the arguments of §§4 and 5, it is clear that $f'$ will be monotonically increasing when $a \leqslant 1$, while when $a > 1$ it will first increase to a maximum value greater than 1 and then decrease afterwards.

The inflexions of $f'$ are slightly harder to discuss in the present case. We first observe that according to the argument of §5 they must be finite in number whatever the value of $a$, and that $f'$ will be ultimately concave when $a \leqslant 1$ and ultimately convex when $a > 1$.

When $a \leqslant 1$ and $0 < 2\beta \leqslant 1$ one can show as, in the final part of §5, that $f'$ has at most one inflexion. Similarly, when $a > 1$ and $0 < 2\beta \leqslant 1$, the graph of $f'$ has exactly one inflexion on its falling portion (though it may have other inflexions on its rising part), and when $1 \leqslant 2\beta \leqslant \sqrt{a}/(\sqrt{a}-1)$ it has exactly one inflexion on its entire course. There is no need to elaborate the details of the argument.

This work was partially supported by the United States Air Force Office of Scientific Research under Grant No. AF-AFOSR-1301-67.

## REFERENCES

COHEN, C. B. & RESHOTKO, E. 1956 Similar solutions for the compressible laminar boundary layer with heat transfer and pressure gradient. *NACA Report* 1293.

HO, D. & WILSON, H. K. 1967 A boundary value problem arising in boundary layer theory. *Arch. Rational Mech. Analysis* **27**, 165–174.

LI, T. & NAGAMATSU, H. 1955 Similar solutions of compressible boundary layer equations. *J. Aero. Sci.* **22**, 607–616.

MCLEOD, J. B. & SERRIN, J. (To appear.) The existence of similar solutions for some compressible boundary layer problems.

ROSENHEAD, L. (Ed.) 1963 *Laminar Boundary Layers*. Oxford University Press.

SCHLICHTING, H. 1960 *Boundary Layer Theory*, 4th ed. New York: McGraw-Hill.

STEWARTSON, K. 1949 Correlated compressible and incompressible boundary layers. *Proc. Roy. Soc.* A **200**, 84–100.

STEWARTSON, K. 1964 *The Theory of Laminar Boundary Layers in Compressible Fluids*. Oxford University Press.

*SIAM Studies Appl. Math.* 5 (1969), 35–42.

**Reprint V.3.[iv]**



# Existence Theorems for Some Compressible Boundary Layer Problems*

*James Serrin, University of Minnesota*

In the study of the Navier–Stokes equations the position of boundary layer theory is well established, and there are a wide range of situations in which the relevant equations yield successful, rigorous expressions of the asymptotic behavior of fluid flow at small Reynolds numbers. Moreover, the recent mathematical development of the subject has included existence and uniqueness theorems for the well-known initial velocity profile problem [14], [15] as well as a number of results concerning the qualitative behavior of the associated fluid motions [20], [24]. It has also been found that the technique of separation of variables, which leads to the Blasius differential equation and the Falkner–Skan class of similarity solutions, can be equally exploited in many other situations where one wishes to find explicit exact solutions.

Our interest here will be in two examples of this latter sort, which lead to interesting and unexpected problems in the theory of ordinary differential equations. At the same time, we point out some analogous problems which arise in the study of the swirling flow generated by a revolving plane disk, an example of which occurs in the exact theory of the Navier–Stokes equations. Most of the work to be described is the result of joint research with J. B. McLeod of Oxford University.

To introduce the relevant differential equations, we first recall the standard similarity procedure for steady plane flow of an incompressible viscous fluid past a fixed wall. Beginning with the well-known boundary layer equations.

$$u\frac{\partial u}{\partial x} + v\frac{\partial u}{\partial y} = U\frac{dU}{dx} + \nu\frac{\partial^2 u}{\partial y^2},$$

$$\frac{\partial u}{\partial x} + \frac{\partial v}{\partial y} = 0, \qquad x \geqq 0, \quad y \geqq 0,$$

where $U = U(x)$ is the stream speed at the outer edge of the boundary layer, we introduce the separation of variables

$$u = U(x)f'(\eta), \quad \eta = y/g(x),$$

in which the function $g(x)$ must still be determined. A straightforward calculation

* This research was supported in part by the United States Air Force Office of Aerospace Research under Grants AF-AFOSR 883-67 and 1301-67.

first carried out by Falkner and Skan in 1930 shows that this separation of variables is successful if

$$U(x) = cx^m,$$

and that if $c > 0$ and $m > -1$, then the resulting function $f(\eta)$ satisfies the ordinary differential equation[1]

$$f''' + ff'' + \lambda(1 - f'^2) = 0 \tag{1}$$

together with the boundary conditions

$$f = f' = 0 \quad \text{at} \quad \eta = 0, \qquad f' = 1 \quad \text{at} \quad \eta = \infty. \tag{2}$$

The first theoretical study of the singular boundary value problem (1), (2) was due to Herman Weyl in a pioneering paper in 1942. His work, which was based on the Leray–Schauder fixed-point theory, was later simplified and extended by a number of other authors, with the result that problem (1), (2) is now in a relatively complete state [2], [4], [7], [8].

The success of the Falkner–Skan procedure prompted many authors to seek classes of similarity solutions for other flow problems where boundary layer theory is applicable. Here we shall consider the flow of a compressible fluid past a fixed wall and the free thermal convection of a viscous fluid adjacent to a heated vertical plate.

For a convection boundary layer, which is physically the simpler and historically the earlier of the two situations, the appropriate differential equations are

$$\begin{aligned} f''' + ff'' + \tfrac{2}{3}(h - f'^2) &= 0, \\ h'' + \sigma f h' &= 0 \end{aligned} \tag{3}$$

with corresponding boundary conditions

$$f = f' = 0, \quad h = 1 \quad \text{at} \quad t = 0, \qquad f' = h = 0 \quad \text{at} \quad t = \infty. \tag{4}$$

Here primes denote differentiation with respect to $t$, and the dependent variables $f'(t), h(t), 0 \leqq t < \infty$, represent respectively the vertical component of velocity and the temperature in the fluid. The parameter $\sigma$ is the Prandtl number of the fluid, which may take any positive constant value. The preceding equations were first obtained by Pohlhausen in 1930 (see [19, p. 398]) and appear also in a number of standard works on boundary layer theory (cf., for example, [16, p. 543], [18, p. 301]) though with a slightly different normalization of the functions $f$ and $h$. Since the hydrodynamical significance of these equations is fully discussed in the above references, we need only observe here that physically relevant solutions must satisfy the further conditions $f' > 0$ for $t > 0$.

For the case of a compressible fluid flowing past a fixed boundary surface, the partial differential equations of the boundary layer can be specialized to a system of ordinary differential equations provided the fluid has appropriate physical

---

[1] Here $\lambda = 2m/(m + 1)$ and the corresponding function $g(x)$ is proportional to $x^{(1-m)/2}$. If the coefficients $c$ and $m$ do not satisfy the previously noted restriction, then the differential equation takes a slightly different form.

properties and the exterior streaming flow is suitably behaved. The precise equations, obtained first by Stewartson in 1949, are

$$\text{(5)} \qquad \begin{aligned} f''' + ff'' + \lambda(h - f'^2) &= 0, \\ h'' + fh' &= 0 \end{aligned}$$

with boundary conditions

$$\text{(6)} \qquad f = f' = 0, \quad h = a \quad \text{at} \quad t = 0, \quad f' = h = 1 \quad \text{at} \quad t = \infty.$$

Here $f'(t)$ and $h(t)$ represent respectively the tangential component of velocity and the total energy of the fluid, while $\lambda$ and $a$ are certain constants determined by the nature of the exterior streaming flow. For a derivation of these equations the reader is referred either to Stewartson's work or to a paper by Li and Nagamatsu where Stewartson's derivation has been somewhat simplified; the derivation is also given in several recent works on compressible boundary layer theory [18, p. 332], [23, p. 69].

Numerical calculations of the function $f'$ and $h$ in the above problems have been carried out in the past for differing combinations of the assigned parameters $\lambda$, $a$ and $\sigma$, though these calculations have naturally left open the fundamental question whether solutions of the boundary value problems can in fact be proved to exist. In what follows we shall outline a proof that Pohlhausen's equations for the convection problem and Stewartson's equations for the compressible flow problem have solutions satisfying the given boundary conditions when $\lambda \geqq 0$ and $a, \sigma > 0$. Our method of proof, like that used by Iglisch and Coppel for the Falkner-Skan equation and by Ho and Wilson in a previous study of Stewartson's problem, relies on the so-called shooting method, in which one seeks appropriate initial conditions for $f''$ and $h'$ so that the resulting solutions of the initial value problem will have appropriate limiting behavior when $t$ becomes large.

To introduce the method it will be convenient to present it first in the case $\lambda = 0$, where it is extremely simple and shows up to greatest possible effect. Here we are dealing with the Blasius equation

$$\text{(7)} \qquad f''' + ff'' = 0$$

and the boundary conditions

$$\text{(8)} \qquad f = f' = 0 \quad \text{at} \quad t = 0, \qquad f' = 1 \quad \text{at} \quad t = \infty.$$

To prove the existence of a solution of this problem, consider first the *initial value problem*

$$g''' + gg'' = 0$$

with $g = g' = 0, g'' = 1$ at $t = 0$. Of course, we do not know that $g$ can be continued for all positive values of $t$, but it *is* evident that this equation does have a solution on *some* interval $(0, t_0)$. We assert that on any such interval we have

$$g, g', g'' > 0, \quad g'' < 1, \quad g''' < 0.$$

To see this, note that

$$(9) \qquad g'' = \exp\left(-\int_0^t g\,du\right),$$

whence $g'' > 0$. But then by the initial conditions $g, g' > 0$. Finally $g''' = -gg'' < 0$, and since $g''$ is decreasing we have $g'' < 1$.

This being the case, it is clear that the solution $g(t)$ can actually be continued for all values of $t$, since $gg''$ must always lie between 0 and $t^2/2$. Since $g'' > 0$, $g'$ is monotone-increasing and tends to a limit value $k > 0$ as $t$ tends to infinity. We assert that $k$ is finite. To see this, note that $g > 1$ for large values of $t$, whence $g''$ is integrable by (9).

To complete the proof, we set

$$(10) \qquad f(t) = \frac{1}{\sqrt{k}}\, g\left(\frac{t}{\sqrt{k}}\right)$$

and observe that this function now satisfies the required conditions (7) and (8). We have thus proved the first part of the following assertion:

*The Blasius system* (7), (8) *has one and only one solution* $f(t)$. To prove uniqueness, suppose for contradiction that there were two different solutions $f_1$ and $f_2$. Then evidently they have different initial slopes, and accordingly the inverse transformation to (10) would give rise to two functions $g(t)$ having different limiting values $k_1$ and $k_2$. This is impossible, however, since $g(t)$ is unique. *This rapid proof, it may be noted, contains within it a most effective method for the numerical computation of the function* $f(t)$.[2]

Turning to the main problems (3), (4) and (5), (6), these may be conveniently combined into a single more general one, as follows.

*Let* $a$, $c$, $\lambda$ *and* $\sigma$ *be given real constants*:

$$a, \lambda, \sigma > 0, \quad c \geqq 0.$$

*It is required to determine functions* $f(t)$, $h(t)$ *obeying the equations*

$$(11) \qquad \begin{aligned} f''' + ff'' + \lambda(h - f'^2) &= 0, \\ h'' + \sigma f h' &= 0 \end{aligned}$$

*on* $0 \leqq t < \infty$, *together with the boundary conditions*

$$(12) \qquad \begin{aligned} &f = f' = 0, \quad h = a \quad \text{at} \quad t = 0, \\ &\lim f'^2 = \lim h = c \quad \text{as} \quad t \to \infty. \end{aligned}$$

We shall outline a proof that this problem has at least one solution, which moreover satisfies the physical condition $f' > 0$ noted earlier. Let $(\alpha, \beta)$ be a pair of real

---

[2] The first proof that the Blasius problem actually has a solution was given by Weyl. Iglisch later obtained an alternate proof as a special case of his study of the Falkner-Skan system. The present proof, combining appropriate elements from the work of Weyl and Iglisch, has so far as I know not been published previously. I first used it in a course at the University of Minnesota in 1963.

constants and consider the *initial value problem*

$$(13) \qquad f = f' = 0, \quad f'' = \beta, \qquad h = a, \quad h' = \alpha \quad \text{at} \quad t = 0$$

for the system (11). The solution of this problem will for convenience be referred to simply as $(f, h)$. This solution will not, of course, generally satisfy the limiting condition (12), and in fact might not even exist for all values of $t$. What we must show is that there exists some parameter pair $(\alpha, \beta)$ for which the solution $(f, h)$ can be continued for all positive $t$ and such that the limit condition (12) is fulfilled.

To this end we let $k = 2\max(\sqrt{a}, \sqrt{c})$ and subdivide the $(\alpha, \beta)$-plane into five sets, as follows.

$S_+$: Pairs $(\alpha, \beta)$ for which there exists a point $t$ with $f'(t_+) = k$ while $0 < h < k^2$, $f' > 0$ for $0 < t \leqq t_+$.

$S_-$: Pairs $(\alpha, \beta)$ such that either $\beta \leqq 0$ or there exists a point $t_-$ with $f'(t_-) = 0$ while $0 < h < k^2$, $f' < k$ for $0 < t \leqq t_-$.

$A$: Pairs $(\alpha, \beta)$ such that there exists a point $t_A$ with $h(t_A) = 0$ while $0 < f' < k$, $0 < h < k^2$ for $0 < t < t_A$.

$B$: Pairs $(\alpha, \beta)$ such that there exists a point $t_B$ with $h(t_B) = k^2$ while $0 < f' < k$, $0 < h < k^2$ for $0 < t < t_B$.

$C$: Pairs $(\alpha, \beta)$ such that the solution can be continued for all $t > 0$ while $0 < f' < k$, $0 < h < k^2$.

The following results can now be proved (see [13]).

LEMMA 1. *$S_+$ and $S_-$ are open and disjoint.*

LEMMA 2. *Corresponding to any fixed positive constant $\alpha_0$, there exists a constant $\beta_0 > 0$ such that the set $|\alpha| \leqq \alpha_0, \beta \geqq \beta_0$ is contained in $S_+$.*

LEMMA 3. *If $(\alpha, \beta)$ is in $C$, then the limits $f'(\infty)$ and $h(\infty)$ exist and satisfy the relation $\{f'(\infty)\}^2 = h(\infty)$.*

With the help of Lemma 3 we can define a function $K(\alpha, \beta)$ from $D = A \cup B \cup C$ into the reals, by means of the formula

$$K(\alpha, \beta) = \begin{cases} -1/t_A & \text{if} \quad (\alpha, \beta) \in A, \\ k^2 + 1/t_B & \text{if} \quad (\alpha, \beta) \in B, \\ h(\infty) & \text{if} \quad (\alpha, \beta) \in C. \end{cases}$$

We then have the following lemma (see [13]).

LEMMA 4. *$K$ is continuous on $D$ and satisfies the conditions*:

$$K > k^2 \quad \textit{when} \quad \alpha \geqq k^2 e^{ok}, \qquad K < 0 \quad \textit{when} \quad \alpha \leqq -k^2 e^{ok}.$$

In view of Lemma 3 and the definition of the function $K(\alpha, \beta)$, the existence of a solution of the given problem (11), (12) is equivalent to the existence of a point $(\alpha, \beta)$ in $D$ where $K = c$. To establish that there is in fact such a point in $D$, we note that since $D$ is the complement of the sets $S_+$ and $S_-$, then by Lemmas 1 and 2 and by the topology of the Euclidean plane there must be a continuum in $D$ joining the lines $\alpha = \pm k^2 e^{ok}$. But then by the intermediate value property of continuous functions on a continuum, the function $K$ takes all values between 0 and $k^2$ on $D$, completing the proof of existence.

It is highly likely that uniqueness holds for both Pohlhausen's problem and Stewartson's problem when $f' > 0$; a proof of this would be an important addition both to boundary layer theory and to the general study of nonlinear ordinary differential systems.[3]

In reference [12], McLeod and I discussed the qualitative behavior of solutions of Stewartson's problem, and in [13] we similarly considered the behavior of solutions of Pohlhausen's problem. The reader may consult these papers for further details and for a physical interpretation of the results. While some of these matters had been treated previously by other authors using heuristic methods or numerical calculations, it is clearly fundamental to obtain the results directly from a study of the differential equations themselves.

**Swirling flow.** In a classical paper in 1921 von Kármán obtained a pair of ordinary differential equations representing the swirling flow of a viscous incompressible fluid generated by a revolving plane disk of infinite extent. The equations in question have the form

$$(14)\qquad \begin{aligned} f''' + ff'' + \tfrac{1}{2}(h^2 - f'^2) &= 0, \\ h'' + fh' - f'h &= 0 \end{aligned}$$

together with the boundary conditions

$$(15)\qquad f = f' = 0, \quad h = 1 \quad \text{at} \quad t = 0, \qquad f' = h = 0 \quad \text{at} \quad t = \infty$$

(see, for example, [18, p. 105], where, however, the functions $f$ and $h$ are given a slightly different normalization). The similarity of these equations to those already considered strongly suggests that analogous methods can be used to treat this problem, and J. B. McLeod has recently shown in this way that von Kármán's problem does indeed have a solution with $f' > 0$, thus settling a long space problem (see also [10]).

We note that this solution necessarily has the following qualitative behavior: *The function $h$ is monotonically decreasing and convex, while $f'$ first increases to a unique maximum value less than* 1 *and then decreases monotonically to zero, and has only one inflection (occurring at some point on the decreasing course).*

*Proof.* By the second equation of (14), $h$ cannot have a nonpositive minimum. Consequently, according to the boundary conditions we must have $h > 0$. Similarly $h$ cannot have a positive maximum, hence $h$ must be strictly decreasing and convex.

Now observe that

$$f'''' = -ff''' - hh'$$

so that $f'''' > 0$ whenever $f''' = 0$. Therefore the graph of $f'$ can have at most one inflection point. On the other hand, since $f' = 0$ at $t = 0$ and at $t = \infty$ the graph must have at least one inflection. This being shown, it now follows that $f'$ first

[3] When $a = 1$ in Stewartson's problem, then $h \equiv 1$ and (5) reduces to the Falkner-Skan equation for which uniqueness is known to hold (see [8]). Thus in fact Stewartson's problem does admit only a single solution in the two special cases $a = 1$ and $\lambda = 0$.

increases to a unique maximum value and then decreases monotonically to zero with a single inflection on its decreasing course.

At the maximum value of $f'$ we have

$$f'' = 0, \quad f''' \leqq 0,$$

whence by the first equation of (14) one obtains $h^2 - f'^2 \geqq 0$ at this point. Consequently, $f' \leqq h < 1$ there, completing the proof.

Again the major issue which remains is the uniqueness of the solution. Although numerical work seems to indicate that uniqueness does hold, the problem is apparently a difficult one.

Another problem, related to von Kármán's and leading equally to a system of equations of the type already discussed, occurs in determining the swirling flow between two rotating coaxial disks. In this case, the relevant differential equations can be written in the form (see [1])

$$(16) \qquad \begin{aligned} f''' + ff'' + \tfrac{1}{2}(h^2 - f'^2) &= \gamma, \\ h'' + fh' - f'h &= 0 \end{aligned}$$

with boundary conditions

$$(17) \quad f = f' = 0, \quad h = \Omega_0 \quad \text{at} \quad t = 0, \qquad f = f' = 0, \quad h = \Omega_1 \quad \text{at} \quad t = 1,$$

where $\gamma$ is a constant to be determined. Although apparently not greatly different from the preceding problem, this one is in fact of far greater complexity. Preliminary formal and numerical calculations have been carried out by Batchelor, Stewartson, and Pearson in papers noted in the references. This work provides an important challenge to mathematicians interested in the severely nonlinear phenomena which can occur even for rather simple fluid motions. It is not impossible that Herman Weyl's original methods may once more come into use in the study of this problem.

A point of particular interest is that no uniqueness theorem can be expected to hold in general for solutions of the problem (16), (17). Indeed, consider the case $\Omega_1 = -\Omega_0$ and let $(f(t), h(t), \gamma)$ be a corresponding solution. Then also $(-f(1-t), -h(1-t), \gamma)$ is a solution and we must only show that it is not the same as the first one. However, by a numerical calculation Pearson determined a solution with $f(\frac{1}{2}) \neq 0$, so that the two flows just noted are in fact different.[4]

## REFERENCES

[1] G. K. Batchelor, *Note on a class of solutions of the Navier-Stokes equations representing steady rotationally symmetric flow*, Quart. J. Mech. Appl. Math., 4 (1951), pp. 29–41.

[2] W. Coppel, *On a differential equation of boundary layer theory*, Phil. Trans. Royal Soc., 253 (1960), pp. 101–136.

[3] V. W. Falkner and S. W. Skan, *Solutions of the boundary layer equations*, Philos. Mag. (7), 12 (1931), pp. 865–896.

[4] P. Hartman, *Ordinary Differential Equations*, John Wiley, New York, 1964 (especially Chap. XIV).

[4] It would be valuable to make this argument completely rigorous.

[5] D. Ho and H. K. Wilson, *On the existence of a similarity solution for a compressible boundary layer*, Arch. Rational Mech. Anal., 27 (1967), pp. 165–174.

[6] ———, *On the boundary layer near a porous wall*, J. Math. and Phys., 46 (1967), pp. 203–210.

[7] R. Iglisch, *Elementarer Existenzbeweis fur die Stromung in der Laminaren Grenzschicht zur Potentialstromung $U = u_1x^m$ mit $m > 0$ bei Absaugen und Ausblasen*, Z. Angew. Math. Mech., 33 (1953), pp. 143–147.

[8] ———, *Elementarer beweis fur die Eindeutigkeit der Stromung in der laminaren Grenzschicht zur Potentialstromung $U = u_1x^m$ mit $m > 0$ bei Absaugen und Ausblasen*, Ibid., 34 (1954), pp. 441–443.

[9] T. Li and H. I. Nagamatsu, *Similar solutions of compressible boundary layer equations*, J. Aero. Sci., 22 (1955), pp. 607–616.

[10] C. C. Lin and T. von Kármán, *On the existence of an exact solution of the equations of Navier-Stokes*, Comm. Pure Appl. Math., 14 (1961), pp. 645–655.

[11] J. B. McLeod, *Von Kármán's swirling flow problem*, Arch. Rational Mech. Anal., (1969), to appear.

[12] J. B. McLeod and J. Serrin, *The behavior of similar solutions in a compressible boundary layer*, J. Fluid Mech., 24 (1968), to appear.

[13] ———, *The existence of similar solutions for some boundary layer problems*, Arch. Rational Mech. Anal., (1968), to appear.

[14] K. Nickel, *Einzige Eigenschaften von Losungen der Prandtlschen Grenzschicht-Differentialgleichungen*, Ibid., 2 (1958), pp. 1–31.

[15] O. A. Oleinik, *The Prandtl system of equations in boundary layer theory*, Dokl. Akad. Nauk SSSR, 150 (1963), pp. 28–32 (English trans: Soviet Math. Dokl., pp. 583–586.)

[16] S. Ostrach, *Laminar flows with body forces*, High Speed Aerodynamics and Jet Propulsion, vol. IV: Theory of Laminar Flows, F. K. Moore, ed., Princeton University Press, Princeton, 1964.

[17] C. E. Pearson, *Numerical solutions for the time dependent viscous flow between two rotating coaxial disks*, J. Fluid Mech., 21 (1965), pp. 623–633.

[18] H. Schlichting, *Boundary Layer Theory*, 6th ed., McGraw-Hill, New York, 1968.

[19] E. Schmidt and W. Beckman, *Das Temperatur-und Geschwindigkeitsfeld von einer Warme abgebenden, senkrechten Platte bei Naturlicher Konvektion*, Technische Mech. und Thermodynamik (Forsch. Ingenieurwesens), 1 (1930), pp. 391–406.

[20] J. Serrin, *Asymptotic behavior of velocity profiles in the Prandtl boundary layer theory*, Proc. Royal Soc. Ser. A, 299 (1967), pp. 491–507.

[21] K. Stewartson, *Correlated compressible and incompressible boundary layers*, Ibid., 200 (1949), pp. 84–100.

[22] ———, *On the flow between two rotating coaxial disks*, Proc. Cambridge Philos. Soc., 49 (1953), pp. 333–341.

[23] ———, *The Theory of Laminar Boundary Layers in Compressible Fluids*, Oxford University Press, Oxford, 1964.

[24] W. Velte, *Eine Anwendungen des Nirenbergschen Maximumsprinzips fur parabolische Differentialgleichungen in der Grenzschichttheorie*, Arch. Rational Mech. Anal, 5 (1960), pp. 420–431.

[25] T. von Kármán, *Uber laminare und turbulente Reibung*, Z. Angew. Math. Mech., 1 (1921), pp. 233–252.

[26] H. Weyl, *On the differential equations of the simplest boundary layer problems*, Ann. of Math., 43 (1942), pp. 381–407.

*Int. J. Eng. Sci.* 9 (1971), 233–240. **Reprint V.3.[v]**


© *Int. J. Engng Sci.* Vol. 9, pp. 233–240. Pergamon Press 1971. Printed in Great Britain

# RECENT DEVELOPMENTS IN THE MATHEMATICAL ASPECTS OF BOUNDARY LAYER THEORY†

JAMES SERRIN

School of Mathematics, University of Minnesota, Minneapolis.

**Abstract** – This is an expository paper concerning some recent mathematical advances in the study of Prandtl's boundary layer equations. The main topics considered are the rigorous derivation of the equations themselves, the mathematical theory of the associated initial value problem, and the theory of similarity solutions.

THE CONCEPTS of boundary layer theory are of course familiar in the everyday life of the engineering world, the main results being consistent, convincing, and essentially valid in the real world; far from being dead and embalmed in dusty covers, boundary layer theory remains a vital and interesting endeavor. Moreover, it is well known that an enormous range of mathematics has played a part in understanding this phenomenon, from the most elusive of empirical arguments to severely rigorous portions of pure analysis. But this being said, I imagine that many have had the disquieting experience of trying to defend boundary layer theory in the face of some determined mathematician who considers it to be not quite respectable. He will argue that:

I. It is well known that the Prandtl boundary layer scheme is only speculative and approximate, and applies only at small values of viscosity. Indeed, the limits of applicability are not even defined.

II. There is a lack of established and mathematically exact techniques for handling the boundary layer equations.

III. The Prandtl equations, like the Navier–Stokes equations, are nonlinear, and are further approximated in most solution processes, thus further compromising their validity.

I can well sympathize with those in such a situation, but at least they can take leave of the argumentative mathematician. On the other hand, I find that I argue with *myself* about these things and there is unfortunately no way for me to separate the combatants.

Because of this, I have been particularly interested in cases where one can in fact apply rigorous mathematical analysis to the Prandtl equations. To be sure, this means that for the most part it is necessary to consider only the simpler problems of boundary layer theory; but certainly if one finds a solid mathematical basis there, it lends strong support to our belief in the validity of boundary layer theory in other situations. Moreover, the open problems in elementary boundary layer theory constitute an important challenge to our mathematical creativity as well as to our physical insight.

Such an interest in the mathematical problems of boundary layer theory has been shared by others, and a significant degree of progress has been attained in recent years. The purpose of this paper is to outline some of these developments in three distinct directions:

†Talk given at the Conference on Boundary Layer Concepts in Fluid Mechanics, Amherst, Massachusetts, July 30, 1969.

I. The derivation of the Prandtl equations.
II. The initial value problem.
III. The theory of similar solutions.

Except where otherwise noted, we consider steady plane flow of an incompressible fluid of constant viscosity, this case being chosen both for its simplicity and for its representative value.

I. As noted above, Prandtl's equations are speculative and approximate. In the usual treatment, they are derived by a semi-heuristic order of magnitude argument, in which it is assumed that the flow retains certain laminarty properties for large Reynolds numbers. This argument can also be given in terms of stretched variables, but in its essence it requires one to assume a heavy degree of asymptotic smoothness of solutions of the Navier-Stokes equations, an assumption not at all obvious *a priori*. In a recent paper Fife has proved the following much sharper result [3].

*Consider a family of solutions of the Navier–Stokes equations in the domain* $0 \leq x \leq 3$, $0 \leq y \leq 1$, *adhering to the wall* $y = 0$, *and existing for all sufficiently small values of the viscosity* $\nu$.

*Suppose that in the variables* $\xi = x$, $\eta = y/\sqrt{\nu}$ *the family has uniformly bounded derivatives up to order 4, the bound being denoted by M. Suppose also that* $p_x \leq -a < 0$ *(uniformly favorable pressure gradient).*

*Then there exists a constant k, depending only on a and M, such that, given any flow* $(u, p)$ *in the family, there is a solution* $\bar{u}$, $\bar{p}$ *of Prandtl's equations having the properties*

$$\bar{p} = p(x,0), \quad |u - \bar{u}| \leq k\nu \quad \textit{for} \quad 1 \leq x \leq 2, \quad 0 \leq y \leq \nu^{1/2}.$$

Thus under quite minimal laminarity assumptions *every flow in the family can be approximated asymptotically by a solution of Prandtl's equations.* To understand the essence of this result, we may consider the well-known Hamel–Jeffery radial motion between two straight converging walls, shown in the figure. Here the velocity component $v_r$ can be written in the form

$$v_r = -\frac{f(\theta)}{r} \quad (f(0) = a);$$

the Navier–Stokes equations correspondingly yield the relation

$$\nu(f'' + 4f) - f^2 = \text{Const.}$$

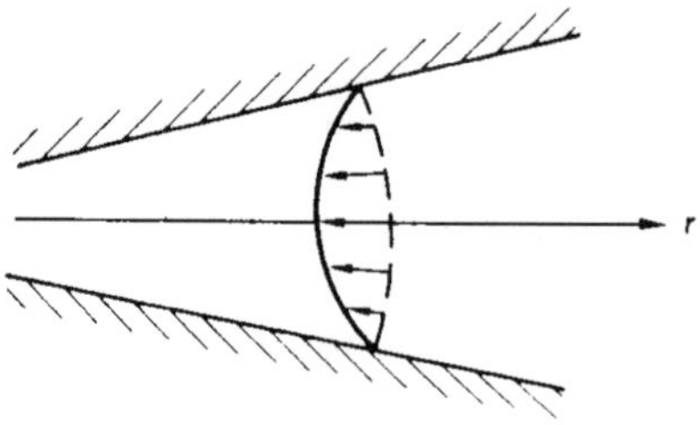


Fig. 1.

A careful analysis shows that as $\nu \to 0$ the solutions rigorously converge as follows: In the central opening, $f(\theta)$ uniformly approaches the irrotational velocity distribution $-a$, while in a thin boundary layer near the wall we have

$$|f(\theta) - F(\theta)| \leq 12\nu$$

where $F$ is the solution of the corresponding Prandtl equation

$$\nu F'' - F^2 = b^2, \quad F(0) = 0, \quad F(\infty) = a.$$

Thus for Hamel's problem, Prandtl's equation occurs naturally and unavoidably in representing the limiting form of the velocity for small $\nu$, without any prior assumptions of order of magnitude. This example, together with Fife's theorem, constitutes a most convincing mathematical argument for the validity of Prandtl's theory in laminar flow situations. (At the same time I must confess a real admiration for the usual heuristic treatments.)

II. *The initial value problem.* The fundamental theoretical problem of elementary boundary layer theory is to solve the Prandtl equations

$$uu_x + vu_y = -UU_x + \nu u_{yy}$$

$$u_x + v_y = 0$$

in the domain $0 \leq x \leq x_0$, $0 \leq y \leq \infty$, subject to the boundary conditions

$$u = v = 0 \quad \text{at} \quad y = 0$$

$$u \to U(x) \quad \text{as} \quad y \to \infty$$

and also satisfying the initial condition

$$u(y, 0) = \bar{u}(y) \quad \text{for} \quad y > 0.$$

Here $U(x) > 0$ is the given streaming speed, and $\bar{u}(y)$ is a given positive function representing the initial velocity profile.

In 1958 Karl Nickel[8] showed that under certain reasonable structure and differentiability assumptions there can be at most one solution to this problem. While Nickel's proof is quite technical, it has since been greatly simplified and can be proved in a page or two. Perhaps the simplest version is as follows.

*Suppose $U_x \geq 0$. If $\bar{u}$ and $u$ are two solutions of the initial value problem, one of which has a concave profile, then $u \equiv \bar{u}$* (Serrin[11]).

More recently, Oleinik has devoted a great deal of attention to the problem of existence of solutions. This work is highly technical, but provides a complete and thorough solution of the problem. The simplest case of her results can be stated in the following way.

*Let $U$ be twice differentiable, and $u$ four times differentiable. Suppose that all derivatives are uniformly bounded, and that the usual compatibility conditions are* *satisfied, i.e.*

$$\tilde{u}(\infty) = U(0)$$

*and*

$$\tilde{u} = 0, \quad \tilde{u}_y > 0, \quad \tilde{u}_{yy} = -U(0)U_x(0)/\nu, \quad \tilde{u}_{yyy} = 0 \quad \text{at} \quad y = 0.$$

*Then a solution exists on some interval* $0 \leqslant x \leqslant x_0$.

*If* $U_x > 0$ *the solution can be extended to arbitrarily large values of x. If, additionally, the initial profile is concave, then the solution will be concave and unique* (Serrin[12]).

The work just described places the mathematical structure of the initial value problem on a firm and rigorous basis. Moreover, the solution which is obtained has a strong degree of regularity—for example, if $\tilde{u} \leqslant U(0)$ at the initial position $x = 0$, then $u < U$ at all downstream points.

As a final example of the type of results available for this problem, consider the situation where

$$U(x) = C(x+d)^m$$

with $C > 0$, $d > 0$, $m > 0$. For such power law behavior it is well known that Prandtl's equations are solvable by the method of separation of variables, i.e. we can find similar solutions in the form

$$\bar{u}(x, y) = U(x)f'(\eta), \quad \eta = y\sqrt{\frac{U(x)}{\nu(x+d)}}$$

Now similar solutions have definite and specific initial values, while according to the remarks above the Prandtl system has solutions corresponding to arbitrary initial values. The question then arises: in what way are similar solutions unique and special among the totality of solutions of Prandtl's equation for the given streaming speed? Why do they play the important role which they do when there are other solutions equally available with different initial values? An answer is provided by the following theorem.

*Under the circumstances above, let u(x, y) be a solution of the initial value problem with arbitrary initial profile* $\tilde{u}(y)$. *Then we have*

$$\left|\frac{u}{U} - f'\right| = 0\left(\frac{1 + m\log x}{x^m}\right) \quad \text{as} \quad x \to \infty;$$

*that is, the normalized velocity u/U tends uniformly to the normalized similarity solution as x tends to infinity downstream* (Serrin, [12]).

This result not only provides a well determined downstream asymptotic behavior for the velocity profile, but, what is more important, singles out similarity solutions as those which asymptotically develop downstream whatever the state of motion at the initial position $x = 0$. This provides a rigorous justification for the central position of similarity solutions in boundary layer theory. An analogous result applies to flows not governed by a power law streaming speed, provided certain other hypotheses are satisfied.

Using an eigenfunction approach, Chen and Libby [1] extended the above result to negative values of $m$ in the range $-m_0 \leq m \leq 0$, where $m_0$ is the well-known Hartree separation exponent. Their analysis applies only in terms of the linearized theory, however, i.e. in the limit of vanishingly small deviations from strict similarity.

Looking back at the results so far described, and the related mathematical effort, one can say with some assurance that, at least for steady plane flow of an incompressible fluid, Prandtl's equations provide a highly successful and mathematically valid model for laminar motion in the neighborhood of a fixed wall. The results noted are all accessible to persons with some elementary experience in partial differential equations, the main tools being the well-known von Mises transformation of Prandtl's equations, and the maximum principles and comparison techniques of the theory of second order parabolic differential equations.

III. *Similarity solutions.* It was remarked earlier that the method of separation of variables can be used in certain important circumstances to reduce the Prandtl equations to ordinary differential equations. Thus the function $f(\eta)$ introduced above satisfies the third order equation

$$f''' + ff'' + \lambda(1 - f'^2) = 0 \tag{1}$$

together with the boundary conditions

$$f = f' = 0 \quad \text{at} \quad \eta = 0; \qquad f' = 1 \quad \text{at} \quad \eta = \infty, \tag{2}$$

as was first shown by Falkner and Skan in 1930 [2].†

The first theoretical study of the singular boundary value problem (1), (2) was given by Herman Weyl in a pioneering paper in 1942 [16]. His work was later simplified and extended by a number of other authors, with the result that the theory of this problem is now in a relatively complete state (cf. [4], Chap. XIV).

The successful reduction of the Prandtl equations to an ordinary differential equation prompted many authors to seek classes of similarity solutions for other problems of boundary layer theory where one wishes to find explicit exact solutions. Two of the most interesting examples of this sort occur in the thermal convection of a viscous fluid adjacent to a heated vertical plate, and the flow of a compressible fluid past a fixed wall. We consider steady plane flow as always, though of course now the basic Navier-Stokes equations need modification for buoyancy and compressibility effects.

For a convection boundary layer, which is physically the simpler and historically the earlier of the two situations, the appropriate differential equations are

$$\begin{aligned} f''' + ff'' + \tfrac{2}{3}(h - f'^2) &= 0, \\ h'' + \sigma f h' &= 0, \end{aligned} \tag{3}$$

with corresponding boundary conditions.

$$f = f' = 0, h = 1 \quad \text{at} \quad t = 0; \qquad f' = h = 0 \quad \text{at} \quad t = \infty, \tag{4}$$

†Here $\lambda = 2m/(m+1)$. If we also admit negative values of $m$, the differential equation takes a slightly different form when $m < -1$.

Here primes denote differentiation with respect to $t$, and the dependent variables $f'(t)$, $h(t)$, $0 \leq t \leq \infty$, represent respectively the vertical component of velocity and the temperature in the fluid. The parameter $\sigma$ is the Prandtl number of the fluid, which may take any positive constant value. The preceding equations were first obtained by Pohlhausen in 1930 (see [10], p. 398). For the case of a compressible fluid flowing past a fixed boundary surface, the partial differential equations of the boundary layer can be specialized to a system of ordinary differential equations provided the fluid has appropriate physical properties and the exterior streaming flow is suitably behaved. The precise equations, obtained first by Stewartson in 1949, are

$$\begin{gathered} f''' + ff'' + \beta(h - f'^2) = 0 \\ h'' + fh' = 0 \end{gathered} \tag{5}$$

with boundary conditions

$$f = f' = 0, h = a \quad \text{at} \quad t = 0; \qquad f' = h = 1 \quad \text{at} \quad t = \infty. \tag{6}$$

Here $f'(t)$ and $h(t)$ represent respectively the tangential component of velocity and the total energy of the fluid, while $\beta$ and $a$ ($> 0$) are certain constants determined by the nature of the exterior streaming flow. For both (3) and (5) we consider only positive solutions $f'$.

In recent work J. B. McLeod and I have shown that these problems have solutions when $\sigma > 0$ and $\beta \geq 0$. In addition to proving existence, we were also able to determine the general qualitative nature of the solutions. Specifically, for the convection problem the graph of the velocity component $f'$ has the form shown in Fig. 2.

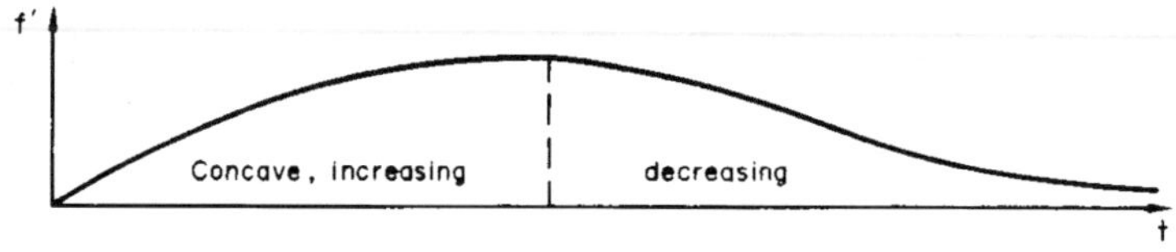


Fig. 2.

While this behavior is also apparent from numerical calculations, it is clearly interesting and valuable to obtain the results directly from a study of the differential equations themselves.

For the compressible flow problem there are two different cases: $a < 1$, corresponding to a cooled wall, and $a > 1$, corresponding to a heated wall (i.e. a wall supplying heat to the fluid). These are illustrated in Fig. 3(a) and 3(b), where it is assumed that $\beta$ is positive.

For the heated wall the solution shows a velocity overshoot, that is, a larger velocity in the boundary layer than in the exterior stream. If we accept the principle that similar solutions are the uniform downstream limits of arbitrary solutions of the initial value problem, then for the heated wall we should always see an eventual velocity overshoot. On the other hand, even though the velocity overshoots, the local Mach number always remains less in the boundary layer than in the exterior stream. To see this, we

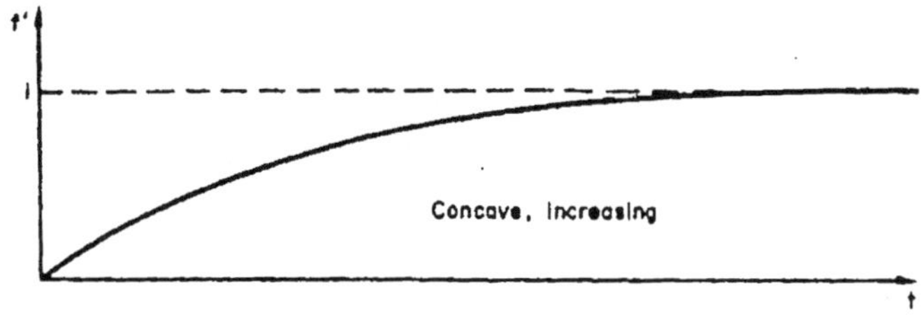


Fig. 3(a). Cooled wall.

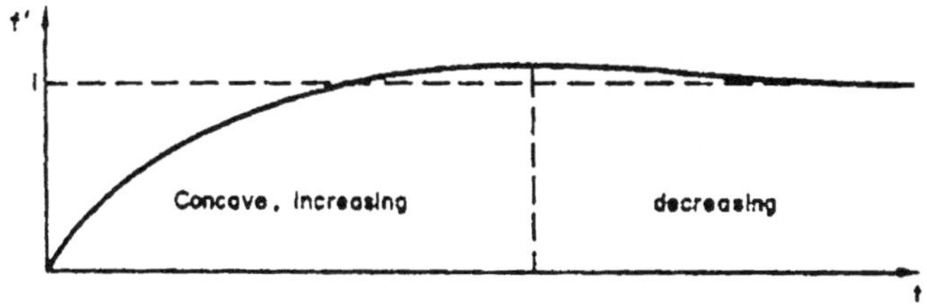


Fig. 3(b). Heated wall.

first observe that the condition for local Mach number overshoot is precisely

$$f'^2 > h \quad \text{at some point } \eta.$$

In order to prove our assertion, therefore, it is enough to show that this inequality is impossible. Setting

$$G = h - f'^2,$$

an easy calculation yields the relation

$$G'' + fG' - 2\beta f' = -2f''^2.$$

If $G < 0$ at some point, then since $G = a > 0$ at $\eta = 0$ and $G = 0$ at $\eta = \infty$, we find that $G$ must have a negative minimum. At this point

$$G < 0, \quad G' = 0, \quad G'' \geq 0, \quad \text{and} \quad f' > 0.$$

These values violate the differential equation for $G$, completing the proof.

I should add that the occurrence of velocity overshoot was observed many years ago on the basis of numerical procedures and a partial asymptotic analysis. Nevertheless, as in the earlier discussion of the convection problem, it is interesting that the result can be obtained as a direct and mathematically exact consequence of the equations.

The existence of solutions of Pohlhausen's problem and Stewartson's problem is proved by the so-called shooting method, in which one seeks appropriate initial conditions for $f''$ and $h'$ so that the resulting solution of the *initial value problem* for the ordinary differential system will have appropriate limiting behavior when $\eta$ becomes large. The technique is one long familiar to numerical analysts, though there it is usually carried out more or less on a trial and error basis. The theoretical approach requires a careful study of the location of the integral curves, and a topological analysis of the initial parameter space. Recently S. Hastings [5] has applied the same techniques to Howarth's differential equations for the boundary layer near a stagnation point in three dimensional flow; he has also presented an alternative approach to the general theory involving application of fixed point methods.

The discussion of the behavior of the solution graphs is based simply on looking at the differential equations in the right way. In particular, the proof of velocity overshoot takes just 12 lines in the *Journal of Fluid Mechanics* (see reference [6]). While one does not usually expect simplicity in the applications of exact mathematics to boundary layer theory, when it does occur it is a justifiable cause of satisfaction.

## REFERENCES

[1] K. CHEN and P. LIBBY, *J. Fluid Mech.* **33**, 273 (1968).
[2] V. W. FALKNER and S. W. SKAN, *Phil. Mag.* **12**, 865 (1931).
[3] P. FIFE, *Arch. ration. Mech. Analysis* **28**, 184 (1968).
[4] P. HARTMAN, *Ordinary Differential Equations*. Wiley, New York (1964).
[5] S. HASTINGS, *Arch. ration. Mech. Analysis* **33**, 103 (1969).
[6] J. B. McLEOD, and J. SERRIN, *J. Fluid. Mech.* **34**, 337 (1968).
[7] J. B. McLEOD, and J. SERRIN, *Arch. ration. Mech. Analysis* **31**, 288 (1968).
[8] K. NICKEL, *Arch. ration. Mech. Analysis* **2**, 1 (1958).
[9] O. OLEINIK, *Uspeki Mat. Nauk* **23**, 3 (1968). English Translation, *Russian Math. Surveys* **23**, (1969).
[10] E. SCHMIDT, and W. BECKMAN, *Forsch. Ingenieurwesens* **1**, 391 (1930).
[11] J. SERRIN, *Mathematical Aspects of Boundary Layer Theory*, Notes, University of Minnesota (1963).
[12] J. SERRIN, *Proc. R. Soc.* **A 299**, 491 (1967).
[13] J. SERRIN, *Arch. ration. Mech. Analysis* **28**, 217 (1968).
[14] K. STEWARTSON, *Proc. R. Soc.* **A 200**, 84 (1949).
[15] K. STEWARTSON, *The Theory of Laminar Boundary Layers in Compressible Fluids*. Oxford University Press (1964).
[16] H. WEYL, *Ann. Math.* **43**, 381 (1942).

(*Received* 17 *December* 1969)

**Résumé** – Ceci est un article de présentation concernant certains progrès mathématiques récents dans l'étude des équations de Prandtl sur les couches limites. Les principaux sujets considérés sont la dérivation rigoureuse des équations elles-mêmes, la théorie mathématique du problème associé de la valeur initiale et la théorie des solutions de similarité.

**Zusammenfassung** – Dieses ist eine erklärende Arbeit betreffend einige jüngere mathematische Fortschritte in der Untersuchung von Prandtl's Grenzschichtgleichungen. Die Hauptgegenstände betrachtet sind die strenge Ableitung der Gleichungen selbst, die mathematische Theorie des assoziierten Anfangswert-problems und die Theorie von Ähnlichkeitslösungen.

**Sommario** – Si tratta di una memoria in cui si espongono alcuni progressi matematici recentemente compiuti nello studio delle equazioni di strato limite de Prandtl. Gli argomenti principali trattati sono la derivazione rigorosa delle equazioni stesse, la teoria matematica del problema di valore iniziale associato e la teoria delle soluzioni di similarità.

**Абстракт**—Эта-изложительная работа о некоторых математических успехах в изучении уравнений граничного слоя Прандтля. Основные рассматрываемые вопросы: строгий вывод уравнений, математическая теория связанной задачи начального значения, и теория подобностных решений.

Reprinted from JOURNAL OF MATHEMATICS AND PHYSICS
Vol. XXXIII, No. 1, April, 1954
*Printed in U.S.A.*

# COMPARISON THEOREMS FOR SUBSONIC FLOWS*

BY J. B. SERRIN

**1. Introduction.** The comparison method of incompressible hydrodynamics, as recently developed for plane and axially symmetric flows, leads to theorems governing the qualitative dependence of flow speed on variations of the streamlines bounding the flow region. The method is to compare the values of the stream function of the original flow and the varied flow, using as a basic tool the fact that the stream function satisfies an *elliptic* differential equation (e.g. Laplace's equation for plane flow) and hence has a maximum principle. Gilbarg [1] has observed that the essential idea of the method is equally valid for irrotational subsonic flows of a compressible fluid, since the stream function for such a flow also satisfies an elliptic differential equation. He obtained in this way the following analogue of a result (Lavrentieff's theorem) already known for incompressible flow:

(*A*) *Let two subsonic flows (plane or axially symmetric) having the same free stream density and velocity be defined in regions $D$ and $\bar{D}$, each bounded by a streamline extending to $\pm\infty$. We suppose that the flow in $D$ or $\bar{D}$ is one of a family of such flows depending continuously on the free stream velocity. Then if $D \subset \bar{D}$ and if the two bounding streamlines have a regular point $P$ in common,*[1] *the respective flow speeds $q$ and $\bar{q}$ satisfy at $P$ the inequality,*

$$q(P) \leqq \bar{q}(P),$$

*equality holding if and only if the flows are identical.*

Gilbarg pointed out that this theorem may be applied to obtain significant results concerning compressible flows with free boundaries. The following comparison theorem which holds for flows of an incompressible fluid, ([2], [3]), would also have important consequences if extended to subsonic flows of a compressible fluid.

(*B*) *Under-Over Theorem. Let two flows (plane or axially symmetric) having well-defined free stream velocities be defined in regions $D$ and $\bar{D}$, bounded respectively by streamlines $S$ and $\bar{S}$ extending to $\infty$. Suppose that $S$ and $\bar{S}$ have a common arc $MN$ which forms part of the boundary of a simply connected region $D'$ interior to both $D$ and $\bar{D}$. Let the remaining boundary of $D'$ consist of an arc $NJ$ of $\bar{S}$ and an arc $JM$ of $S$ meeting at a point $J$ common to $S$ and $\bar{S}$; all points of the arc $NJM$ are presumed finite with the possible exception of $J$. Then*

$$\frac{q(M)}{q(N)} \leqq \frac{\bar{q}(M)}{\bar{q}(N)}, \tag{1}$$

* This work supported in part by the Office of Naval Reseach.

[1] A boundary point of a region is said to be *regular* if it is on the circumference of a circle whose interior falls entirely within the region.

In the original statement of ($A$) the point $P$ was also not allowed to be on the axis in axially symmetric flow, but this restriction is not necessary (cf. the remark following Lemma 2).



P. Pucci et al. (eds.), *James Serrin. Selected Papers: Volume 2*, Contemporary Mathematicians, DOI 10.1007/978-3-0348-0847-7_13, © Springer Basel 2014

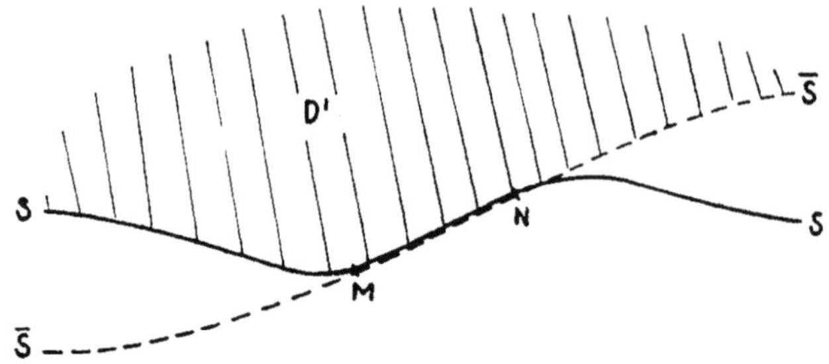


FIG. 1. Configuration illustrating the Under-Over theorem; here $J = \infty$, and both free stream velocities have the same direction.

*equality holding only when the flows are geometrically similar or* $\bar{q}(M) = 0$ *or* $q(N) = 0$.[2]

The purpose of this paper is to make, as far as possible, an extension of this theorem to subsonic flows. Although we do not get a result of equal generality, (even assuming the free stream Mach numbers to be equal), our result does have sufficient validity to be applicable to the compressible free boundary problem, as we shall show in §7, no. 2. Another application concerns the inception of sonic flow on a profile as the free stream Mach number is gradually increased (§7, no. 1).

The basic extension of $(B)$ to subsonic flows is the assertion of its validity *if there exists a point* $P$ *on* $MN$ *where* $q(P) = \bar{q}(P)$. This additional hypothesis could be removed quite simply if, as in the incompressible case, the flows for varying free stream speed (i.e., varying Mach numbers) were dynamically similar, but since this is not true in general, we cannot expect $(B)$ to hold for all subsonic flows. On the other hand, in several special cases the desired similarity does hold: (i) when the flow in $D$ (or $\bar{D}$) is uniform; (ii) when both flows are given by the Prandtl-Glauert approximation

$$(2) \qquad (1 - M^2)\psi_{xx} + \psi_{yy} - py^{-1}\psi_y = 0. \qquad M^2 < 1,$$ [3]

where (as in the rest of the paper) $p = 0$ or 1, depending on whether the flow is plane or axially symmetric. In these two cases of subsonic flow, then, $(B)$ holds without restriction. Theorems 1 and 2 of §4 contain the results described in this paragraph; in §5 corresponding theorems are given for flows in a channel.

An important feature of our results is that we permit the points $M$ and $N$ to be on the axis in axially symmetric flow. This is accomplished by generalizing a lemma of E. Hopf [4] concerning the behavior of a solution of an elliptic differential equation at a boundary point. This generalization also allows us to eliminate the condition in $(A)$ that $P$ be excluded from the axis in axially symmetric

[2] $(B)$ was originally stated only for plane flows with a proof depending strongly on a result of function theory [2]. With the additional hypothesis that $M$ and $N$ be regular points not on the axis it was then proved [3] for axially symmetric flows. The additional hypothesis is unnecessary, however, and for this reason it is omitted from the above formulation (cf. the remark concluding §4).

[3] Here the free stream speed is varied without varying $M$, producing a dynamically similar, but artificial, flow to be used for comparison purposes.

flow (see the remark following Lemma 2). Finally in the cases (i) and (ii) above we can dispense entirely with the requirement that $M$ and $N$ be regular boundary points, a condition which seems to be necessary in the more general case.

In [3] I noted some corollaries of $(B)$ arising when the flow in $D$ or $\bar{D}$ is uniform and giving explicit inequalities for the velocity gradient along straight arcs and at inflection points of a streamline. Completely analogous results hold when the flow is subsonic, as we shall show in §6. The paper concludes with a proof of a comparison theorem of a kind different from $(A)$ or $(B)$.

**2. Preliminary results.** We consider an ideal compressible fluid, and flows which are steady, irrotational, and isentropic. Let $\tilde{p}$, $\rho$, and $q$ be the pressure, density, and flow speed; we may assume $q = 1$ for limit speed, $\rho = 0$ for the corresponding density, and a fixed equation of state $\tilde{p} = \tilde{p}(\rho)$. From Bernoulli's law,

$$\int_0^\rho c^2\, d\rho/\rho = \tfrac{1}{2}(1 - q^2), \qquad c^2 = d\tilde{p}/d\rho > 0, \tag{3}$$

$\rho$ and the sound speed $c$ are obtained as single valued functions of $q$. Introducing the stream function $\psi$ and the mass flow

$$Q = \rho q = y^{-p}\,|\operatorname{grad}\psi| \tag{4}$$

we derive

$$dQ/dq = \rho(1 - m^2), \qquad m^2 = q^2/c^2. \tag{5}$$

Thus as $q$ increases from 0 to sonic speed, $Q$ increases from 0 to some value $b_0$; hence in the totally subsonic range of $Q$, i.e., $0 \leqq Q < b_0$, we have

$$q = q(Q), \qquad \rho = \rho(Q), \qquad c = c(Q). \tag{6}$$

This determined, the stream function of a totally subsonic flow satisfies the quasi-linear elliptic partial differential equation

$$((\rho c)^2 - V^2)\psi_{xx} + 2UV\psi_{xy} + ((\rho c)^2 - U^2)\psi_{yy} - p(\rho c)^2 y^{-1}\psi_y = 0, \tag{7}$$

where $U = y^{-p}\psi_x$, $V = y^{-p}\psi_y$.

If, now, $\psi$ and $\bar{\psi}$ are stream functions for two subsonic flows defined in the same region, say $\Delta$, then in this region $\omega = \bar{\psi} - \psi$ is a solution of a *linear elliptic* partial differential equation *with no term in* $\omega$, namely

$$L(\omega) = A\omega_{xx} + 2B\omega_{xy} + C\omega_{yy} + D\omega_x + (E - pCy^{-1})\omega_y = 0, \tag{8}$$

(see [1], pp. 235–236). Explicit expressions for the coefficients are easily given,

$$A = (\rho c)^2 - V^2 \qquad B = UV, \qquad C = (\rho c)^2 - U^2,$$

$$D = y^{-p}\{(U + \bar{U})(e^2 - \bar{\psi}_{yy}) + (V + \bar{V})\bar{\psi}_{xy}\},$$

$$E = y^{-p}\{(V + \bar{V})(e^2 - \bar{\psi}_{xx}) + (U + \bar{U})\bar{\psi}_{xy}\} - py^{-1}U^2,$$

$$e^2 = \frac{(1 + \bar{\gamma})\bar{\kappa}\bar{m}^2}{2(1 - \bar{m}^2)}\,|\operatorname{grad}\bar{\psi}|, \qquad \gamma = \frac{d\log c^2}{d\log\rho} + 1,$$

where $\kappa$ is the curvature of the streamlines, the bars denote evaluations for the flow given by $\bar{\psi}$, and the tildes denote mean values. The fundamental Lemma 2 below is proved under certain assumptions regarding the boundedness of the coefficients of (8) in the neighborhood of the regular boundary point $P$. In particular, in plane flow, or in axially symmetric flow when $P$ is off the axis, all coefficients must be bounded in the neighborhood of $P$. For this it is sufficient to assume boundedness of the first partial derivatives of the velocity components $\bar{u}$ and $\bar{v}$, or equivalently, boundedness of the acceleration and the curvature (of course, the curvature is bounded *a priori* since $P$ is a regular point). On the other hand, when $P$ is on the axis $A$, $By^{-\frac{1}{2}}$, $C$, $Dy^{\frac{1}{2}}$, and $E$ must be bounded. This may be insured by requiring boundedness of the first partial derivatives of the velocity components $u$, $v$, $\bar{u}$, and $\bar{v}$, as then the quantities

$$y^{-1}\psi_{xx}\,,\ y^{-1}\psi_{xy}\,,\ \psi_{yy}\,,\ \text{and } y^{-1}U$$

will be bounded for both flows. In any event, we assume in all applications of Lemma 2 that the coefficients have the requisite behavior.

**3. Boundary point lemmas.** Since the solutions of an elliptic differential equation of the form (8) satisfy the maximum principle, that is, they cannot have an interior maximum or minimum unless they are constant, we see that the maximum principle holds for the difference $\omega$ of two subsonic stream functions. In addition, under the boundedness conditions just stated, Hopf has shown [4] that if $\omega > 0$ in the neighborhood of a regular point $P$ on the boundary of $\Delta$, while $\omega(P) = 0$, then the inward normal derivative of $\omega$ satisfies the inequality

$$(\partial\omega/\partial n)(P) > 0, \tag{9}$$

provided that $P$ is not on the axis in axially symmetric flow. When $P$ is on the axis we have the corresponding result

$$(\partial\omega/\partial y)(P) = 0, \qquad (\partial^2\omega/\partial y^2)(P) > 0. \tag{9a}$$

This is a consequence of

LEMMA 1. *Let $p$ be a positive constant and let $\omega(x, y)$ be a solution of* (8) *defined in a region $\Delta$ in the upper half plane. Let $P$ be a regular point on the boundary of $\Delta$ with coordinates $(a, 0)$, and suppose that $A$, $By^{-\frac{1}{2}}$, $C$, $Dy^{\frac{1}{2}}$ and $E$ are bounded and $AC - B^2 \geqq \delta$, $\delta = \text{const.} > 0$, in a neighborhood of $P$. If $\omega > 0$ in this neighborhood and $\omega(P) = 0$ then*

$$\lim_{y\to 0} y^{-p}(\partial\omega/\partial y)(a, y) > 0.$$

*Proof.* We shall assume familiarity with [4]. Let the circle $K$ associated with $P$ have radius $R$, and therefore center at $(a, R)$. We define

$$g(x, y) = 1 - e^{-\alpha(R^2-r^2)},$$

where $\alpha$ is a positive constant to be determined later, and $r^2 = (x - a)^2 + (y - R)^2$. One finds easily that

$$0 < g < 2\alpha R y \qquad \text{in } K.$$

Also put

$$h(x, y) = y^p(e^{-\alpha r^2} - e^{-\alpha R^2});$$

obviously

$$h(a, 0) = 0, \qquad h > 0 \qquad \text{in } K.$$

A simple computation shows that

$$\begin{aligned} e^{\alpha r^2}L(h) &= 2\alpha y^p[2\alpha\{A(x-a)^2 + 2B(x-a)(y-R) + C(y-R)^2\} \\ &\quad - \{A + 2p(x-a)y^{-1}B + (p+1)C + (x-a)D \\ &\quad + E[(pg/2\alpha y) + R - y]\} + \{(pC/y)(R - g/2\alpha y)\}] \\ &> 2\alpha y^p[2\alpha\{A(x-a)^2 + 2B(x-a)(y-R) + C(y-R)^2\} \\ &\quad - \{A + 2p\sqrt{2R}y^{-\frac{1}{2}} \mid B \mid + (p+1)C + \sqrt{2R}y^{\frac{1}{2}} \mid D \mid \\ &\quad + (p+1)R \mid E \mid \}], \end{aligned}$$

the last inequality holding in $K$. In view of the ellipticity condition and the boundedness assumptions we may choose $\alpha$ so large that $L(h) > 0$ in the neighborhood of $P$. It follow in the same way as in [4] that, for suitably small positive numbers $\epsilon$,

$$\omega - \epsilon h > 0$$

in the neighborhood of $P$. Hence

$$\lim_{y\to 0} \frac{1}{y^p}\frac{\partial \omega}{\partial y}(a, y) \geq \lim_{y\to 0} \frac{\epsilon}{y^p}\frac{\partial h}{\partial y}(a, y) = 2\epsilon\alpha(p+1)\, Re^{-\alpha R^2} > 0. \quad \text{Q.E.D.}$$

We now specialize the preceding remarks to the situations envisaged in theorems $(A)$ and $(B)$. Let $\psi$ and $\bar{\psi}$ denote the respective stream functions, normalized so that

$$\begin{aligned} \psi &= 0 \text{ on } S, & \bar{\psi} &= 0 \text{ on } \bar{S}, \\ \psi &> 0 \text{ in } D, & \bar{\psi} &> 0 \text{ in } \bar{D}. \end{aligned} \tag{10}$$

The region $\Delta$ becomes the region $D'$ common to $D$ and $\bar{D}$, the boundary of $D'$ then consisting of arcs of $S$ and $\bar{S}$. The function $\omega = \bar{\psi} - \psi$, defined in $D'$, is a solution of (8), and in virtue of the relations (4), (5), (9) and (9a) we get

LEMMA 2. *If $\omega = \bar{\psi} - \psi$ is positive in the neighborhood of a regular boundary point $P$ of $D'$, while $\omega(P) = 0$, then $\bar{q}(P) > q(P)$.*

*Remark.* Since in Lemma 2, $P$ is not excluded from the axis in axially symmetric flow, and since the hypotheses of Lemma 2 are verified in Gilbarg's proof of $(A)$, [1, p. 249], it follows that in the statement of $(A)$ the condition that $P$ be off the axis is unnecessary. For the special cases mentioned in the introduction we have the following improvement of Lemma 2.

LEMMA 3. *If $\omega = \bar{\psi} - \psi$ is positive in the neighborhood of a boundary point $P$ of $D'$, if $\omega(P) = 0$, and if either*

i) *the flow in $\bar{D}$ is one of a family of flows depending continuously on the free stream speed, and $P$ is a regular boundary point of $\bar{D}$, or*

ii) *both flows are given by the Prandtl-Glauert equation* (2),
*then* $\bar{q}(P) \geqq q(P)$, *equality holding only in case* ii) *when* $\bar{q}(P) = 0$.

*Proof*. By hypothesis there exists an arc $P_1PP_2$ of the boundary of $D'$ having the property that $\omega > 0$ at all adjacent points in $D'$. Because $\omega \geqq 0$ on $P_1PP_2$ this arc must be on the curve $S$. If $\omega \equiv 0$ on $P_1P$, let $A$ be a regular point of this arc; otherwise let $A$ be a point on $P_1P$ where $\omega > 0$. In the same way choose a point $B$ on $PP_2$, and join $A$ to $B$ by a curve $T$ lying in $D'$, such that $\omega > 0$ on $T$. Let $\bar{\psi}^*$ be the stream function of a flow in $\bar{D}$ with free stream speed $\bar{q}_\infty^* < \bar{q}_\infty$, and set $\chi = \bar{\psi}^* - \bar{\psi}$. Evidently $\chi$ and $\partial\chi/\partial n$ approach zero as $\bar{q}_\infty^* \to \bar{q}_\infty$. Since $\omega > 0$ on $T$, and since at the endpoints of $T$ either $\omega > 0$ or $\omega = 0$ and $\partial\omega/\partial n > 0$,[4] it follows that, for $\bar{q}_\infty^*$ sufficiently near $\bar{q}_\infty$,

$$\bar{\psi}^* - \psi = \omega + \chi > 0 \qquad \text{on } T.$$

Because of (10) and the fact that $P_1PP_2$ is on $S$,

$$\bar{\psi}^* - \psi = \bar{\psi}^* \geqq 0 \qquad \text{on } P_1PP_2,$$

so that, in virtue of the maximum principle, $\bar{\psi}^* - \psi > 0$ in the region bounded by the arcs $T$ and $P_1PP_2$. Hence

$$(\partial\bar{\psi}^*/\partial n)(P) \geqq (\partial\psi/\partial n)(P),$$

or

$$(\partial^2\bar{\psi}^*/\partial y^2)(P) \geqq (\partial^2\psi/\partial y^2)(P)$$

if $P$ is on the axis in axially symmetric flow. From these inequalities and relations (4) and (5) we get

$$\bar{q}^*(P) \geqq q(P).$$

In case i)

$$\bar{q}(P) > \bar{q}^*(P)$$

since $P$ is a regular point of $\bar{D}$;[5] in case ii)

$$\bar{q}(P) = (\bar{q}_\infty/\bar{q}_\infty^*)\bar{q}_\infty^*(P)$$

and the desired conclusion is a consequence of the last three relations.

The following lemma is similar to the preceding two. Although it will not be needed in the present paper it is of interest because it is applicable in some situations where the other two fail, notably when the curvature at $P$ is positively infinite.

LEMMA 4. *If* $\omega = \bar{\psi} - \psi$ *is positive in the neighborhood of a boundary point* $P$ *of* $D'$, *if* $\omega(P) = 0$, *and if* $\bar{\kappa} \geqq 0$ (*or equally* $\kappa \geqq 0$) *in the neighborhood of* $P$, *then* $\bar{q}(P) \geqq q(P)$, *equality holding if and only if* $\bar{q}(P) = 0$.

[4] If the endpoints are on the axis in axial symmetry, this is replaced by $\partial\omega/\partial y = 0$, $\partial^2\omega/\partial y^2 > 0$, but the conclusion remains the same.

[5] See [1], Theorem 2, or note that $\chi < 0$ in $\bar{D}$ and apply Lemma 2, with $\omega = \bar{\psi} - \bar{\psi}^* = -\chi$.

*Proof.* As in Lemma 3 we find that the function

$$\omega^* = (1 - \epsilon)\bar{\psi} - \psi, \qquad \epsilon > 0,$$

is greater than or equal to zero on an arc $P_1PP_2$ of the boundary of $D'$ and greater than zero on a curve $T$ joining $P_1$ and $P_2$, provided that $\epsilon$ is sufficiently small. We shall show that $\omega^*$ cannot possess an interior minimum in the neighborhood of $P$. For it can be shown that $\omega^*$ satisfies the equation

$$L^*(\omega^*) = -(2 - \epsilon)\epsilon\bar{\kappa}\bar{Q}^2 \,|\,\text{grad}\,\bar{\psi}\,|\left\{1 + \frac{(1+\bar{\gamma})\bar{m}^2}{2(1-\bar{m}^2)}\right\}, \tag{11}$$

where $L^*$ is a linear elliptic operator of the form (8); in fact the coefficients of $L^*$ are obtained from the coefficients of $L$ by replacing $\bar{\psi}$ with $(1 - \epsilon)\bar{\psi}$. Since the right hand side of (11) is negative by hypothesis $\omega^*$ can have no interior minimum. This being shown, it follows that $\omega^* > 0$ in the region bounded by the arcs $T$ and $P_1PP_2$. Hence

$$(1 - \epsilon)\bar{Q}(P) \geqq Q(P),$$

and from this inequality we derive $\bar{q}(P) > q(P)$, unless $\bar{q}(P) = 0$. Q.E.D.

**4. Under-Over theorem for subsonic flows.** We can now state our main result.

THEOREM 1. *(B) holds for subsonic flows of a compressible fluid, provided that $M$ and $N$ are regular boundary points of $D'$ and*

$$q(P) = \bar{q}(P) \tag{12}$$

*at some regular point $P$ of $MN$. If both flows have the same free stream velocity we suppose additionally that the flow in $D$ or $\bar{D}$ is one of a family of flows depending continuously on the free stream velocity.*[6] *Furthermore under these hypotheses equality can hold in* (1) *only when the flows are identical.*

*Note.* Theorem 1 applies also to *linearized subsonic compressible flow at fixed Mach number $M$* (this of course forces both flows to have the same free stream velocity). Indeed the proof below applies almost word for word since the stream functions $\psi$ and $\bar{\psi}$ of a pair of such flows satisfy the same elliptic differential equation (2). Moreover the final hypothesis of Theorem 1 is unnecessary in this case, since constant multiples of the flow in $D$, say, immediately provide the required family.

*Proof.* When the flows are identical the equality in (1) is trivial, and we have therefore to show that inequality holds in (1) when the flows are not identical. Define $\psi$, $\bar{\psi}$ and $\omega$ as in the remarks preceding Lemma 2. Then in virtue of (10) and the geometrical hypothesis of $(B)$ we have

$$\begin{cases} \omega \geqq 0 & \text{on the arc } JMP \text{ of } S, \\ \omega \leqq 0 & \text{on the arc } PNJ \text{ of } \bar{S}. \end{cases} \tag{13}$$

[6] This hypothesis is verified for plane flows if $D$ or $\bar{D}$ is a region admitting a finite Dirichlet integral. See [1], p. 245, and Shiffman, *On the existence of subsonic flows of a compressible fluid*, J. Rational Mech. Anal., 1, (1952), pp. 605–652; for axially symmetric flows a corresponding result is unknown. It may be remarked that the family in question need only contain flows arbitrarily near the given one in $D$ or $\bar{D}$.

Moreover, a level line $C$, $\omega = 0$, issues from $P$ into $D'$. To see this, note that $P$ must be the limit of a set of points where $\omega = 0$; otherwise $\omega \gtrless 0$ near $P$ which, by Lemma 2, contradicts (12). If these points were isolated then the set of level lines through them would tend to a *limit curve*. On the limit curve, $\omega = \partial\omega/\partial n = 0$, which gives a Cauchy problem for $\omega$ having the unique solution $\omega \equiv 0$. This is impossible, so the points may not be isolated: that is, they form a curve $\omega = 0$ entering $D'$ at $P$.

Now consider the case $J = \infty$. Without loss of generality we may assume the free stream velocities to be $(q_\infty, 0)$ and $(\bar{q}_\infty, \bar{\theta})$, $0 \leq \bar{\theta} < \pi$, in polar form.

1) $\theta \neq 0$, occurring only for plane flow. Here $D'$ is a *sector like* region with angular opening from $\bar{\theta}$ to $\pi$. We shall construct a curve in $D'$, joining arc $JM$ of $S$ to arc $NJ$ of $\bar{S}$, and containing exactly one point where $\omega = 0$. Define the vectors

$$\mathbf{Q} = (Q_\infty, \pi/2), \qquad \bar{\mathbf{Q}} = (\bar{Q}_\infty, \pi/2 + \bar{\theta}), \qquad \mathbf{W} = \bar{\mathbf{Q}} - \mathbf{Q} = \lim_{x,y\to\infty} \operatorname{grad} \omega.$$

If $\mathbf{W} = (W, \sigma)$ in polar form then obviously $\bar{\theta} + \pi/2 < \sigma < 3\pi/2$. In virtue of this inequality $\mathbf{W}$ has a non-vanishing component in the direction $\bar{\theta}/2$. Now consider the set of straight lines in $D'$ joining $JM$ to $NJ$ and having inclination $\bar{\theta}/2$. If $L$ is any one of these lines sufficiently far from the origin then $\partial\omega/\partial s \neq 0$ on $L$, where $s =$ arc length on $L$. It follows from (13) that $\omega$ must decrease continually on $L$ as we go from $JM$ to $NJ$. Thus $\omega = 0$ at exactly one point $R$ of $L$, as desired. Counting in $C$ any arcs of $JM$ or $NJ$ along its course, it is evident from elementary continuity considerations and the maximum principle that *$C$ joins $P$ to $R$ and has only the point $P$ in common with the arc $MN$.* This defines two (open) subregions of $D'$, say $D_1$ and $D_2$, (see Figure 2). Using (13), we see

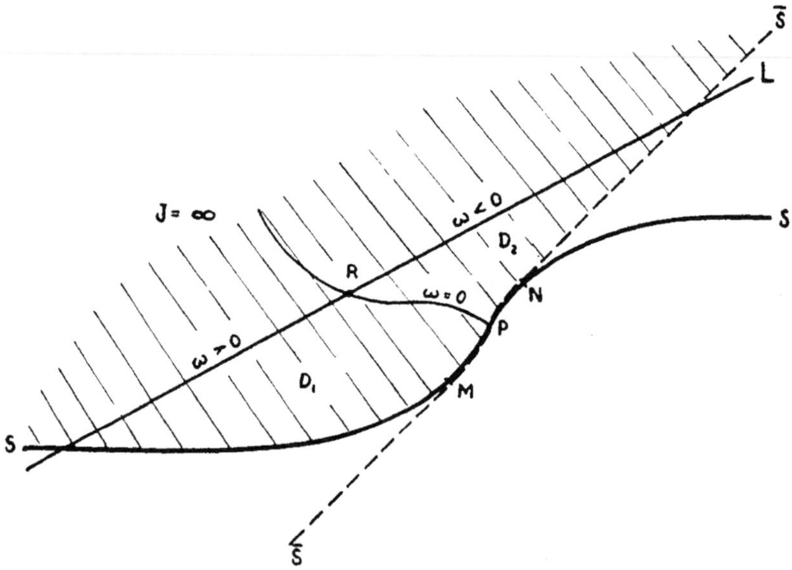


FIG. 2

that $\omega \geqq 0$ on the boundary of $D_1$ and $\omega \leqq 0$ on the boundary of $D_2$, hence according to the maximum principle

$$\omega > 0 \text{ in } D_1, \qquad \omega < 0 \text{ in } D_2.$$

Therefore, by Lemma 2,

(14a) $$\bar{q}(M) > q(M),$$

(14b) $$q(N) > \bar{q}(N),$$

from which we derive

$$\frac{q(M)}{\bar{q}(M)} < 1 < \frac{q(N)}{\bar{q}(N)},$$

or

$$\frac{q(M)}{q(N)} < \frac{\bar{q}(M)}{\bar{q}(N)}$$

as desired.

2) $\bar{\theta} = 0$. Here, when $\bar{q}_\infty \neq q_\infty$ we can easily construct a curve $L$ in $D'$, joining $JM$ to $NJ$, and containing at most one change of sign of $\omega$. Following this construction, which we omit, the remainder of the proof is the same as 1). On the other hand, when $\bar{q}_\infty = q_\infty$, corresponding to equal free stream velocities, it is no longer easy to determine a curve $L$ with the required properties. We resort to a more delicate argument.

As before the level line $C$ extends to infinity, and divides $D'$ into two (open) subregions. Let $D_1$ be the subregion bounded by the arc $JMP$ and the level line $C$. We now make use of the assumption that the flow in $D$ ($\bar{D}$ works equally well) is one of a family of flows depending continuously on the free stream speed; let $\psi^*$ denote the stream function for such a flow with free stream speed $q_\infty^* < q_\infty$, and put $\chi = \psi - \psi^*$. Evidently $\chi$ obeys the maximum principle. Therefore, since $\chi = 0$ on $S$ and

$$\lim_{x,y\to\infty} \partial\chi/\partial y > 0$$

we obtain $\chi > 0$ in $D$. Now $\omega \geqq 0$ on the boundary of $D_1$, for this boundary consists of the arc $JMP$ of $S$ and the level line $C$. In view of the two inequalities just written, $\omega + \chi \geqq 0$ on the boundary of $D_1$; moreover because of the equality

$$\omega + \chi = \bar{\psi} - \psi^*$$

we see that $\omega + \chi$ satisfies the maximum principle and

$$\lim_{x,y\to\infty} \frac{\partial(\omega + \chi)}{\partial y} > 0;$$

hence we conclude

$$\omega + \chi > 0 \qquad \text{in } D_1.$$

By making $q_\infty^* \to q_\infty$, so that $\chi \to 0$, we obtain $\omega \geqq 0$ in $D_1$. But $\omega$ is non constant so that application of the maximum principle in its strong form gives $\omega > 0$ in $D_1$. Therefore

$$\bar{q}(M) > q(M) \tag{15a}$$

by Lemma 2. A similar argument leads to

$$q(N) > \bar{q}(N), \tag{15b}$$

and the desired conclusion is a consequence of the last two inequalities. When $J$ is a finite point, a corresponding proof is very easily given since there are no difficulties with growth conditions. We shall therefore go on to the special cases mentioned in the introduction.

THEOREM 2. *(B) holds for subsonic flows of a compressible fluid if the flow in D (or $\bar{D}$) is uniform, or if both flows are defined by the Prandtl-Glauert equation* (2).

*Proof.* In both cases of this theorem the function $\psi^* = a\psi$ satisfies the same equation as $\psi$. By appropriate choice of the constant $a$ we can make

$$q^*(P) = \bar{q}(P)$$

at a regular point $P$ on $MN$ so that we may apply Theorem 1. If $M$ and $N$ are not regular points, Lemma 3 must be used instead of Lemma 2 in obtaining the inequalities (14) and (15) basic to the proof of Theorem 1. We derive in this way

$$\frac{q(M)}{q(N)} = \frac{q^*(M)}{q^*(N)} \leqq \frac{\bar{q}(M)}{\bar{q}(N)} \tag{16}$$

using the fact that the starred and unstarred flows are similar. When the flow in $D$ is uniform, equality can hold in (16) only when the starred and barred flows are identical, or equivalently when the flows in $D$ and $\bar{D}$ are geometrically similar. When both flows are given by the Prandtl-Glauert equation equality can hold only when the flows are identical (geometrical similarity implying identity in this case) or when $\bar{q}(M)$ or $q(N) = 0$.

*Remark.* Since the Prandtl-Glauert theory reduces, when $M = 0$, to incompressible flow theory, Theorem 2 actually contains as a special case the result that $(B)$ holds for flows of an incompressible fluid.

**5. Flows in a channel.** Analogues of Theorems 1 and 2 are available for flows in a channel. We shall be interested specifically in flows bounded by two stream lines having no points in common other than their extremities, which, in particular, may be at infinity; moreover the flow at an extremity shall be uniform or of simple source (or sink) type. If flows of a real gas only are considered it is impossible to define flows of the type described if either extremity is at a finite point. Since it is useful to allow such flow regions for comparison purposes we arrange for them by the following artifice. Let $d = d(Q)$ be defined for all $Q$ so as to agree with $\rho c$ over the range $0 \leqq Q \leqq b_1$, where $b_1$ is any number less than $b_0$, and so that the equation

$$(d^2 - V^2)\psi_{xx} + 2UV\psi_{xy} + (d^2 - U^2)\psi_{yy} - pd^2\frac{\psi_y}{y} = 0 \tag{17}$$

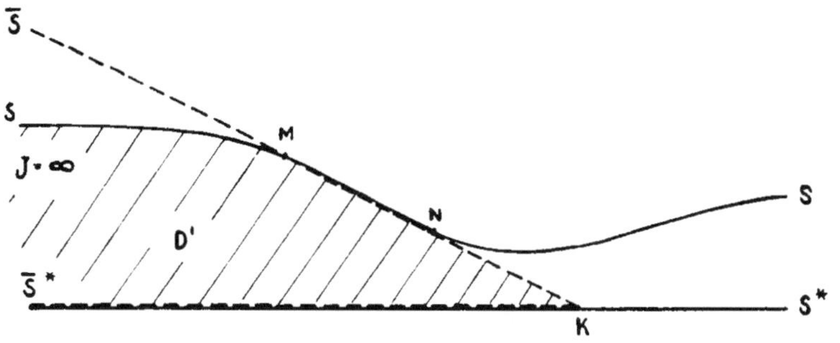


FIG. 3

is elliptic over the whole range $0 \leqq Q < \infty$. Equation (17) can then be used to define a flow in the offending region. Thus, if $\psi$ defines a *bona fide* subsonic flow, if $b_1 = \max Q$, and if $\bar{\psi}$ is the streamfunction of a flow derived from (17), then the developments of §3 hold for the difference $\omega = \bar{\psi} - \psi$ since both $\psi$ and $\bar{\psi}$ satisfy (17). It follows that the methods of §4 may be applied equally well in the envisaged flow situations even though one of the flows may not represent a real gas.

THEOREM 3. *Let two subsonic flows of a compressible fluid be defined in channels $D$ and $\bar{D}$ bounded respectively by streamlines $S$, $S^*$ and $\bar{S}$, $\bar{S}^*$. Let $S$ and $\bar{S}$ have an arc $MN$ in common and suppose the following two conditions are satisfied:*

1. *The curves $S^*$ and $\bar{S}^*$ have a (possibly infinite) arc $JK$ in common, such that $J$ is on $S$ and $K$ is on $\bar{S}$. The arcs $JM$ of $S$ and $NK$ of $\bar{S}$, together with $MN$ and $JK$ bound a simply connected region $JMNK \equiv D'$ interior to $D$ and $\bar{D}$.*[7]

2. *$M$ and $N$ are regular points and there is a regular point $P$ on $MN$ where $q(P) = \bar{q}(P)$. Then*

$$\frac{q(M)}{q(N)} \leqq \frac{\bar{q}(M)}{\bar{q}(N)} \tag{18}$$

*equality holding if and only if the flows are identical.*

THEOREM 4. *If in Theorem 3 the flow in $D$ or $\bar{D}$ is uniform, or if both flows are defined by the Prandtl-Glauert equation, then the restriction 2 need not be made.*

These theorems are the analogues of Theorems 1 and 2 and are proved in the same way (see also [3], p. 568, where some of the details are indicated).

THEOREM 5. *If, in Theorem 3, $\bar{D}$ is a sector and the flow corresponding is of simple source-sink type (Figure 3), then the restriction 2 need not be made, and the conclusion may be written in the form*

$$\frac{\rho q(M)}{\rho q(N)} \leqq \left(\frac{r(N)}{r(M)}\right)^{1+p} \tag{19}$$

*where $r$ is the distance from the vertex $K$ in $\bar{D}$.*

*Proof.* If we write (17) in the form

$$(y^p d)^2(\psi_{xx} + \psi_{yy} - py^{-1}\psi_y) = -\kappa \mid \operatorname{grad} \psi \mid^3$$

[7] It is evident that $J$ is a source (or sink) for the flow in $D$, and $K$ is a source (or sink) for the flow in $\bar{D}$. See Figure 3, where these conditions are fulfilled.

and note that $\kappa \equiv 0$ for the flow postulated in $\bar{D}$, we see that the function $\bar{\psi}^* = a\bar{\psi}$ also solves (17). The inequality (18) then follows as in the proof of Theorem 2 but we prefer to state it in the form

$$\frac{Q(M)}{Q(N)} \leqq \frac{\bar{Q}(M)}{\bar{Q}(N)} \tag{20}$$

which would occur if the results of Lemmas 2 and 3 were stated in terms of $Q$ rather than $q$. Since $\bar{\psi}$ satisfies the equation

$$\psi_{xx} + \psi_{yy} - py^{-1}\psi_y = 0$$

of incompressible fluid theory, we may determine the right hand ratio in (20) from the corresponding incompressible flow. In fact, if $\phi$ is the potential of this flow,

$$\phi = \text{const.} \begin{cases} \log r & \text{plane flow,} \\ 1/r & \text{axially symmetric flow,} \end{cases}$$

then

$$\frac{\bar{Q}(M)}{\bar{Q}(N)} = \frac{|\,\mathrm{grad}\,\phi\,|_M}{|\,\mathrm{grad}\,\phi\,|_N} = \left(\frac{r(N)}{r(M)}\right)^{1+p}. \tag{21}$$

Inequality (19) is now a consequence of (4), (20), and (21).

*Remark.* In this theorem, as well as in any of the preceding ones, the flow in $D$ or $\bar{D}$ need be subsonic only in the subregion $D'$, since the maximum principle has not been applied outside this region. This means that Theorem 5 remains valid in some cases when the flow in the channel $D$ is mixed, as in a transonic nozzle.

**6. Corollaries.** The conclusions of Theorems 2 and 4 may be put in more convenient form. Let $s$ denote arc length along $MN$, increasing from $M$ to $N$. If $P$ is a point on $MN$ with arc length $s$, set $q(P) = q(s)$, $\bar{q}(P) = \bar{q}(s)$. Then for small positive $h$, the conclusions of Theorems 2 or 4 become

$$\frac{q(s)}{q(s+h)} \leqq \frac{\bar{q}(s)}{\bar{q}(s+h)};$$

hence

$$\frac{q(s) - q(s+h)}{hq(s+h)} \leqq \frac{\bar{q}(s) - \bar{q}(s+h)}{h\bar{q}(s+h)},$$

and in the limit $h \to 0$,

$$\frac{d\log q}{ds} \geqq \frac{d\log \bar{q}}{ds}; \tag{22}$$

it follows that

$$\begin{cases} dq/ds \geqq 0 & \text{if the flow in } \bar{D} \text{ is uniform,} \\ d\bar{q}/ds \leqq 0 & \text{if the flow in } D \text{ is uniform.} \end{cases} \tag{23}$$

Similar inequalities may be expected when the arc $MN$ degenerates to a point. At the inflection point thus created we obtain (23) again, where $s$ is now arc length on $S$ or $\bar{S}$.

A consequence of (23) concerns *symmetric subsonic flow past an obstacle* (the obstacle may be in unbounded flow or in a channel): the flow speed on the central streamline decreases monotonically as this streamline is followed from negative infinity to the point on the obstacle where it divides. This result is of course trivial and has been taken for granted in hodograph investigations, but except for convex obstacles in an unbounded plane flow, I do not know of any previous proof.

For Theorem 5 the analogue of (22) is

$$\frac{d \log \rho q}{ds} \geqq \frac{1 + q}{s}$$

where the arc length $s$ is measured from the vertex of $\bar{D}$. If use is made of (3) we get the interesting inequality

$$\frac{d \log q}{ds} \geqq \frac{1 + p}{(1 - m^2)s}.$$

We shall now seek exact solutions of the linearized equation (2) in order to give explicit examples of (22) for this case. Let us consider uniform symmetric flow of Mach number $M < 1$ past a wedge (or cone) of small angle $\alpha$. Taking this for the barred flow will lead to comparison theorems for flows past obstacles like that shaded in Figure 4. Introducing into (2) the new variables

$$\xi = x/\sqrt{1 - M^2}, \qquad \eta = y,$$

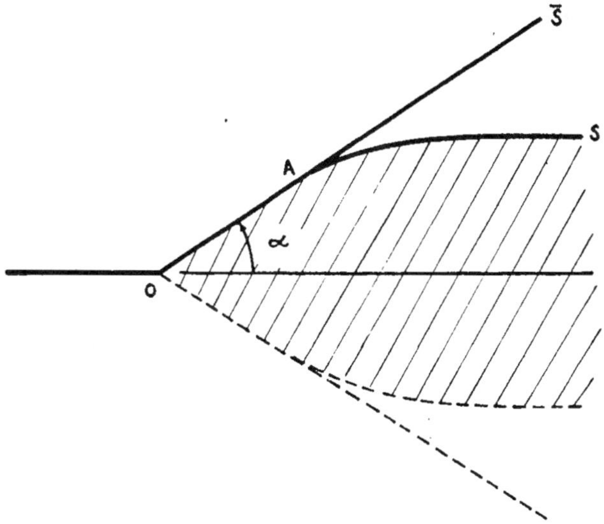


FIG. 4

and polar coordinates $(r, \theta)$ in the $(\xi, \eta)$ plane, (2) becomes

$$(24)\quad \begin{cases} \Delta\psi = \psi_{rr} + r^{-2}\psi_{\theta\theta} + r^{-1}\psi = 0 & \text{(plane flow)}, \\ r^2\psi_{rr} + \psi_{\theta\theta} - \operatorname{ctn}\theta\psi_\theta = 0 & \text{(axially symmetric flow)}; \end{cases}$$

appropriate boundary conditions are

$$(24a)\quad \begin{cases} \psi(r, \pi) = \psi(r, \alpha') = 0, \\ \tan\alpha' = \sqrt{1 - M^2}\tan\alpha. \end{cases}$$

The solution of (24) for plane flow is obviously

$$\psi = Ar^\beta \sin\beta(\theta - \alpha'),$$

where $\beta = \pi/(\pi - \alpha')$ and $A$ is a constant. For $\theta = \alpha'$,

$$\begin{aligned} \bar{q} &= \sqrt{\psi_x^2 + \psi_y^2} \\ &= [(1 - M^2\cos^2\alpha')/(1 - M^2)]^{\frac{1}{2}}(\psi_\theta/r) \\ &= \text{const. } r^{\beta-1}. \end{aligned}$$

Let $s$ denote arc length on $OA$, measured from $O$. Then since

$$r = \text{const. } s,$$

we have

$$\frac{d \log \bar{q}}{ds} = \frac{\beta - 1}{s}.$$

Thus, by (22), for plane flow past the shaded obstacle

$$(25)\quad \frac{d\log q}{ds} \geq \frac{\alpha'}{(\pi - \alpha')s}, \qquad \text{along } OA.$$

For axially symmetric flow, put $\psi = R(r)\Theta(\theta)$. After separation of variables we find that

$$R = r^{n+1},$$

where $n$ is a positive number, while $\Theta$ satisfies the equation

$$(26)\quad \Theta'' - \operatorname{ctn}\theta\Theta' + n(n+1)\Theta = 0.$$

With the substitution

$$\begin{aligned} \Theta &= (1 - z^2)f(z) \\ z &= \cos\theta, \end{aligned}$$

(26) becomes

$$(1 - z^2)f'' - 4zf' + \{n(n+1) - 2\}f = 0,$$

which has the singular points $z = \pm 1$. The solution regular at $-1$ is

$$f(z) = P_{n1}(z) = \frac{dP_n}{dz},$$

TABLE 1

| $\alpha'$ | $n - 1$ | $\alpha'(\pi - \alpha')$ |
|---|---|---|
| 0 | 0 | 0 |
| 1.57 | 1 | 1 |
| 2.04 | 2 | 1.85 |
| 2.27 | 3 | 2.61 |
| 2.44 | 4 | 3.48 |
| 2.56 | 5 | 4.41 |
| 2.62 | 6 | 5.10 |
| 3.14 | $\infty$ | $\infty$ |

where $P_n$ is the Legendre function of order $n$; thus

$$\psi = Ar^{n+1} \sin^2 \theta P_{n1} (\cos \theta).$$

To satisfy the boundary conditions we must have

$$P_{n1} (\cos \alpha') = 0$$

$$P_{n1} (\cos \theta) \neq 0, \qquad \alpha' < \theta < \pi.$$

We shall be mainly interested in the value of $n$ corresponding to a given value of $\alpha'$; table 1 presents values of $n - 1$ for certain values of $\alpha'$.[8]

Now for $\theta = \alpha'$

$$\bar{q} = y^{-1} \sqrt{\psi_x^2 + \psi_y^2} = \text{const. } r^{n-1},$$

hence inequality (22) becomes, for axially symmetric flow past the shaded obstacle,

$$\frac{d \log q}{ds} \geq \frac{n - 1}{s}. \tag{27}$$

Values of $\alpha'(\pi - \alpha')$ are also given in the table in order to compare formulas (25) and (27); the agreement is remarkably close. It would be interesting to see if $n \approx 1 + \alpha'/\pi$ for small $\alpha'$.

**7. Applications.** This section contains several applications of previous results.

*Symmetric subsonic flow past a profile of "airfoil" type.* The profiles we consider are symmetric about the $x$-axis, the upper profile being composed of three arcs $OA$, $AB$, and $BC$ forming a smooth curve. The arcs $OA$ and $AB$ are presumed convex to the flow while $BC$ is concave, so that the trailing edge or tail $C$ is possibly cusped. For the present application it is necessary to assume that $AB$ is symmetric with respect to a line $L$ parallel to the $y$-axis. It is clear, then, that the midpoint $P$ of $AB$ is on $L$, and $L$ cuts the profile in its thickest section. The resemblance of this profile to an airfoil justifies the heading of this section.

[8] Of course, as far as the compressible flow problem is concerned the interest is in small $\alpha'$. On the other hand, for incompressible flow ($M = 0$) the present theory is exact, and the comparison theorem is useful for large $\alpha = \alpha'$. The table is due to Jahnke-Emde, Funktionentafeln.

Now it is well known that there is a critical free stream Mach number $M_0$ such that the flow is totally subsonic for Mach numbers $M < M_0$, while for $M = M_0$ the sonic speed is reached at one or several points on the (upper) profile. We can prove the following result: *if, in the flow at the critical Mach number, the sonic speed is reached at just one point on the profile,*[9] *say $K$, then $K$ is forward of $P$; moreover, if $P$ is on a straight segment $P_1P_2$ of $AB$, then $K$ is forward of $P_1$*.

For this result the flow may be plane or axially symmetric and may occur in a channel or an unbounded domain. We first prove a lemma: if $M$ is a point on $AP$ and if $N$ is its image in $L$, then in any subsonic flow past the profile the speeds $\bar{q}$ at $M$ and $N$ satisfy the inequality

$$\bar{q}(M) > \bar{q}(N). \tag{28}$$

To prove (28) we apply Theorem 1, taking $\bar{D}$ to be the upper half of the flow past the profile. For $D$ we take the reflection of $\bar{D}$ in $L$, and we let the flow in $D$ be the reflection of the given flow in $\bar{D}$. It is evident that $q(P) = \bar{q}(P)$ so that the hypotheses of Theorem 1 are satisfied;[10] hence

$$\frac{q(M)}{q(N)} < \frac{\bar{q}(M)}{\bar{q}(N)}.$$

But $q(M) = \bar{q}(N)$, $q(N) = \bar{q}(M)$, so that

$$[\bar{q}(M)]^2 > [\bar{q}(N)]^2,$$

and (28) follows.

Now suppose $K$ appeared on $BC$. Consider the flow past the profile at Mach number $M_0$. Since the arc $BC$ is concave the well-known formula

$$\partial q/\partial n = \kappa q$$

shows that the speed increases along the normal to $BC$ at $K$, contradicting the fact that the flow is not supersonic. Suppose next that $K$ appeared on the arc $PB$. Then for $M < M_0$ we have from (28),

$$q(K') > q(K)$$

where $K'$ is the image of $K$ in $L$. In the limit $M \to M_0$ we obtain

$$q(K') = q(K) = c,$$

contradicting the assumption that sonic speed occurs at just one point.

To prove the second part it is sufficient to show that $K$ cannot occur on $P_1P$. Again we shall start with a slightly more general result: in any subsonic flow past the profile the speed $\bar{q}$ is monotonically decreasing along $P_1P$. Consider any two points $M$ and $N$ on $P_1P$ with $N$ nearest $P$. Let $P'$ be the midpoint of $MN$, and let $L'$ be a line parallel to the $y$ axis through $P'$. Finally, let $D$ denote

[9] If this assumption is not made, it can only be asserted that the *set* of sonic points is contained in the arc $OAP$.

[10] In the definition of critical Mach number it is tacitly assumed that the flow in $\bar{D}$ is one of a family of flows depending continuously on the free stream speed.

the reflection of $\bar{D}$ in $L'$. In this case $q(P') = \bar{q}(P')$ so that Theorem 1 may be applied as before, with the result that $\bar{q}(M) > \bar{q}(N)$, i.e. the speed is monotonically decreasing on $P_1P$. That $K$ cannot appear on $P_1P$ now follows in the same way as before.

*Flows with free boundaries.* Comparison theorems ($A$) and ($B$) have been used with great success in the study of the classical free boundary problems of hydrodynamics, their application leading specifically to a variety of uniqueness theorems as well as to many qualitative (geometrical) results of importance. The extension of comparison theorems to subsonic flows allows similar conclusions to be drawn for compressible fluids. Indeed, as Gilbarg has observed, the uniqueness of finite cavities and jets carries over to compressible flows with but little alteration in the original proofs, [1]. (The proof of uniqueness for the infinite cavity problem remains incomplete because no formula is available for asymptotic shape of the free boundary, and also for lack of a sufficiently general proof of the comparison theorems.[11])

Likewise Theorems 1 and 2 of reference [5], which are of a geometrical nature, carry over to subsonic flows,[12] although the demonstrations remain incomplete for any axially symmetric problem and for the infinite cavity problem. In the former case this is because the existence of flows in certain auxiliary regions is not known; in the latter, the reason given in footnote 11 stands in the way of a complete proof. Since the required changes in the proofs are not entirely trivial, we shall consider in detail the demonstration of Theorem 1 of [5] for compressible fluids. We shall assume the reader has followed the proof given there through page 565, as no changes are needed up to that point.

The first difficulty encountered is the existence of a flow in the auxiliary region $R_1$. This is assured for plane flow, at least, by the work of Shiffman (see footnote 6); for axially symmetric flow it remains an open question. In any case, for our present purpose we assume that flows in $R_1$ do exist. Let the flow chosen be such that

$$V(P, R_1) = V(P, R) \tag{29}$$

for some point $P$ on $MN$, retaining the notation of [5]. It is convenient to interpret $V(P, R)$ as the value $Q$ at a point $P$ and not as the speed $q$, (thus in the notation of the present paper, (29) would read $Q_1(P) = Q(P)$). By an argument similar to that on page 567, and in virtue of (29), it may be inferred that

$$V(N, R_1) < V(M, R_1). \tag{30}$$

The next step in the proof is to show that $a = (y_m/y_n) < 1$. This result can be obtained as follows. First observe that

$$V(M, R_1) < C/y_m , \tag{31a}$$

[11] That is, the theorems involve embedding a given flow in a family of flows, and it is not known that this is always possible.

[12] Theorem 1 of [5] implies that a starlike obstacle gives rise to a starlike free boundary, a result of importance in the uniqueness problem.

where $C$ is a constant depending only on the strength of the source in $R_1$. This is an immediate consequence of the analogue of $(A)$ for flows in channels, the flow in $R_1$ being compared to a simple source sink flow in a sector. Similarly

$$V(N, R_1) > C/y_N. \tag{31b}$$

Hence

$$V(N, R_1) > a\, V(M, R_1). \tag{32}$$

From (30) and (32) we obtain $a < 1$ as desired. (Note that the original proof does not work since equation (3) is not true for compressible flows.)

The remainder of the proof (p. 568) need not be changed, although it should be pointed out that the existence of a flow in $R_2$ is by no means evident.

*Further applications of comparison methods.* As an application of the methods set out in §4 we consider a type of comparison theorem different from $(A)$ or $(B)$.

$(C)$ *Let a plane subsonic flow having uniform free stream velocity be defined in a region $D$ bounded by a streamline $S$ extending to $\pm\infty$. We suppose that $S$ contains an arc $\sigma$ of a circle whose interior falls entirely outside $D$. Then along the arc $\sigma$ the speed $q$ is first monotonically increasing and then monotonically decreasing.*

*Proof.* To fix the ideas, let the circle have radius $r_0$, let its center be the origin, and take the uniform flow in the positive $x$ direction. Also let the stream function $\psi$ be determined so that $\psi = 0$ on $S$, and $\psi > 0$ in $D$.

Let $r$ designate the radial polar coordinate. For $r \geqq r_0$ define a *circulatory flow* by

$$\bar{q} = kr^{-1}, \qquad \bar{u} = kyr^{-2}, \qquad \bar{v} = -kxr^{-2}, \tag{33}$$

where $k$ is a constant such that $kr_0^{-1} < q^*$, and $q^* = c^* =$ critical speed. The corresponding stream function is given by[13]

$$\bar{\psi} = k\int_{r_0}^{r} (\bar{\rho}/r)\, dr, \tag{34}$$

where $\bar{\rho}$ is given in terms of $r$ by (33) and Bernoulli's law (3). On the boundary $S$ of $D$, $\omega = \bar{\psi} - \psi \geqq 0$, and on $\sigma$, $\omega = 0$. Moreover, $\omega$ has a maximum principle and

$$\lim_{x,y\to\infty} \frac{\partial\omega}{\partial y} < 0. \tag{35}$$

Now let $q_m$ denote the maximum speed on $\sigma$, and let $P$ be any point on $\sigma$ where this speed is attained. If $M$ and $N$ are points of $\sigma$ on one side of $P$, and $N$ is nearest $P$, it is sufficient to show

$$q(M) < q(N) \tag{36}$$

in order to prove $(C)$.

[13] Courant and Friedrichs, *Supersonic Flow and Shock Waves*, N. Y. Interscience (1952), p. 252.

In (33) and (34) put $k = q(N)r_0$; with this choice of $k$ one gets

$$\bar{q} = q(N) = \text{constant} \qquad \text{on } \sigma. \tag{37}$$

It follows as in the proof of Theorem 1 that a level line $C$, $\omega = 0$, enters $D$ at $N$. With the help of (35) it is easy to show that $C$ joins $N$ either to $+\infty$ or to $-\infty$. In either case we can apply the usual arguments based on the maximum principle and the boundary point lemmas; depending on which case actually occurs we obtain

$$\bar{q}(P) > q(P) \tag{37}$$

or

$$\bar{q}(M) > q(M). \tag{38}$$

But $\bar{q}(P) = \bar{q}(M) = q(N)$, so that (37) contradicts the fact that the maximum speed is attained at $P$. The remaining inequality (38) then proves (36).

*When the circle falls entirely within* $D$, a slightly refined version of the above proof shows that *along the arc* $\sigma$ *the speed is first monotonically decreasing and then monotonically increasing*. These theorems remain true in the degenerate case when the circle becomes an entire half plane. In this case the results are also valid for axially symmetric flows.

It is clear that theorems corresponding to (*C*) exist for incompressible fluids. In fact they take a much more general form due to the *linearity* of the basic equations. The analogue of (*C*) is: *Let two flows of an incompressible fluid be defined in regions* $D$ *and* $\bar{D}$, *bounded respectively by streamlines* $S$ *and* $\bar{S}$. *If* $\bar{D} \supset D$ *and if* $S$ *and* $\bar{S}$ *have an arc in common, then along this arc the ratio* $q/\bar{q}$ *of the speeds is first monotonically increasing and then monotonically decreasing.*

## REFERENCES

[1] D. Gilbarg, *Comparison methods in the theory of subsonic flows*, J. Rational Mech. Anal., 2 (1953), pp. 233–251.

[2] J. Serrin, *Uniqueness theorems for two free boundary problems*, Amer. J. Math., 74 (1952), pp. 492–506.

[3] J. Serrin, *Two hydrodynamical comparison theorems*, J. Rational Mech. Anal., 1 (1952), pp. 563–572.

[4] E. Hopf, *A remark on linear elliptic differential equations of the second order*, Proc. Amer. Math. Soc., 3 (1952), pp. 791–793.

[5] J. Serrin, *On plane and axially symmetric free boundary problems*, J. Rational Mech. Anal., 2 (1953), pp. 563–575.

Massachusetts Institute of Technology

(Received Aug. 4, 1953 (Revised Oct. 9, 1953))

*Arch. Ration. Mech. Anal.* 3 (1959), 271–288.


**Reprint V.4.[ii]**

# *On the Uniqueness of Compressible Fluid Motions*

James Serrin

One of the fundamental questions which should be answered concerning any problem of applied mathematics is whether it is well set, that is, whether solutions actually exist and whether they are unique. We shall be concerned here with the initial value problem for compressible fluid flow, and we shall study in particular the uniqueness of its solutions. An exact statement of the problem will be found in § 1, while the main theorems are given in § 3 (viscous fluids) and § 4 (non-viscous fluids). Stated roughly, these theorems assert that fluid motion in a bounded region $\mathscr{V}=\mathscr{V}(t)$ is uniquely determined by its initial velocity, temperature, and density distribution together with certain boundary conditions. The form of the boundary conditions is especially interesting for non-viscous, non-heat conducting fluids, it being found, in fact, that such conditions are superfluous at any point where the fluid is leaving $\mathscr{V}$ at a relatively supersonic speed. This result is intimately associated with the existence of characteristic manifolds in the flow region, and allows us to prove for spatial flows a theorem of the type given by Courant & Friedrichs for one-dimensional motions.

The uniqueness of the initial value problem for incompressible viscous fluids was proved as long ago as 1929 by E. Foà. His proof, which applies equally well to the non-viscous case, was rediscovered and presented more elegantly by Dolidze (see § 72 of reference [*15*] for a resumé of this work). More recently, Graffi showed that the initial value problem for compressible fluids is unique, provided that the pressure and density satisfy a piezotropic relation $p=f(\varrho)$ with $f'(\varrho)>0$. Our work is an extension of Graffi's, in that we make no assumption concerning the equation of state of the fluid, beyond requiring it to satisfy certain universal thermodynamic conditions. This generalization is non-trivial from both the physical and mathematical point of view, for it is well known that a relation between pressure and density is usually incompatible with the motion of a tri-variate fluid, whether or not it is viscous. Moreover, treatment of the general case requires that we use the full set of equations governing the fluid motion, including the complete energy equation.

The proof makes use of the classical energy method, developed originally by Hadamard, Zaremba, and Friedrichs & Lewy in connection with purely mathematical problems, and by Reynolds, Orr and Foà for the study of incompressible fluid motions. Perhaps the most novel feature of our treatment, in comparison with these, is the systematic utilization of the transport equation for the derivation of energy integral identities, thus taking into account in a natural way the special form of the convection terms in the flow equations.

Though a more abstract treatment does seem possible, this has been avoided on the grounds that it would cloud the fundamental issues and make the whole paper more difficult.

In view of the rather strong dependence of the viscosity and heat conduction coefficients on temperature in actual liquids and gases, we have included in the final section of the paper a brief discussion of this case. Also, although we do not treat heterogeneous fluids here, our results remain valid even in this case.

The reader should observe that the uniqueness theorem for non-viscous fluids specifically assumes the flows in question to be continuously differentiable. The proof can be extended without difficulty to cover flows with first order singular surfaces (Mach waves), but in its present form it does not apply to motions admitting shock waves. Such an extension would be very important, since there are numerous cases where the only possible solution involves a shock wave, or in which shock waves occur in the initial data, and in none of these is there reason to expect anomalous behavior so far as uniqueness is concerned. Further discussion of this point will be found in § 4.

## 1. Statement of the problem

We consider a fluid flow occupying a finite region $\mathscr{V}=\mathscr{V}(t)$, and subject to certain prescribed conditions on the boundary $\mathscr{S}$ of $\mathscr{V}$. Our purpose is to answer the following question: given the velocity, temperature and density distribution at some initial instant, does this uniquely determine the subsequent motion, or might it happen that two different motions could be compatible with the same initial conditions? To make the problem precise, we shall now set down the differential equations governing the fluid motion, and the exact boundary conditions which we shall consider. These take a somewhat different form depending on whether the fluid is viscous or not.

### *A. Viscous compressible fluid*

The motion of the fluid is governed by the following standard equations of fluid mechanics,

$$\frac{d\varrho}{dt}+\varrho\operatorname{div}\boldsymbol{v}=0, \tag{1}$$

$$\varrho\frac{d\boldsymbol{v}}{dt}=\varrho\boldsymbol{f}+\operatorname{div}\boldsymbol{T}, \qquad \boldsymbol{T}=(-p+\lambda\operatorname{div}\boldsymbol{v})\boldsymbol{I}+2\mu\boldsymbol{D}, \tag{2}$$

$$\varrho\frac{dE}{dt}=\boldsymbol{T}:\boldsymbol{D}-\operatorname{div}\boldsymbol{q}, \quad \boldsymbol{q}=-\varkappa\operatorname{grad}T. \tag{3}$$

In these equations the symbols have their usual meanings, except perhaps for the stress tensor $\boldsymbol{T}$, the deformation tensor $\boldsymbol{D}$, the specific internal energy $E$, and the heat flux vector $\boldsymbol{q}$. The notation is that of ordinary Cartesian or Gibbsian vector analysis, so that, for example,

$$(\operatorname{div}\boldsymbol{T})_i=T_{ji,j}, \quad \boldsymbol{I}=\text{identity matrix}, \quad \boldsymbol{T}:\boldsymbol{D}=T_{ij}D_{ij}.$$

The material derivative $d/dt$ is of course a differential operator and is defined by

$$\frac{d}{dt}=\frac{\partial}{\partial t}+\boldsymbol{v}\cdot\operatorname{grad}.$$

The reader unfamiliar with equations (1)—(3) will find a complete derivation and explanation in the memoir of C. TRUESDELL, *The mechanical foundations of elasticity and fluid dynamics*, J. Rational Mech. & Analysis, vol. **1**, 1952.

To equations (1)—(3) must be added certain equations of state relating the thermodynamic variables $\varrho, p, E$, and $T$. For this purpose we shall take

$$p = p(\varrho, T) \quad \text{and} \quad E = E(\varrho, T) \tag{4}$$

as the defining relations; here $p(\varrho, T)$ and $E(\varrho, T)$ are *given twice differentiable functions* of the variables $\varrho$ and $T$, whose explicit form depends of course on the particular fluid in question. In order to simplify the proof somewhat, we shall actually treat in detail only the case where the second equation has the special form

$$E = c_v T, \qquad c_v = \text{Const.} > 0; \tag{5}$$

the proof can be carried out equally well in the general case provided that the specific heat $c_v = (\partial E/\partial T)_\varrho$ is positive[1]. It may be added that (5) is quite well satisfied by most real gases and liquids at ordinary conditions.

On the boundary of $\mathscr{V}$ we suppose the velocity distribution to be given (this will be the case in particular if $\mathscr{V}$ is bounded by material walls to which the fluid must adhere). In addition to prescribing the velocity on $\mathscr{S}$, it is necessary to assign certain further conditions in order to determine the fluid motion. Various possibilities present themselves, depending on whether at a particular boundary point the fluid is entering $\mathscr{V}$, or leaving $\mathscr{V}$, or is constrained by a material wall. For convenience in tabulating these cases, let $\boldsymbol{n}$ denote the outer unit normal to $\mathscr{V}$, and $G$ the outward normal velocity of $\mathscr{S}$. Also, let

$$U = \boldsymbol{v} \cdot \boldsymbol{n} - G \tag{6}$$

be the relative normal speed of particles at the boundary. Then we have the following table of

**Boundary Conditions.** *At all points of the boundary $\mathscr{S}$, and at all instants of time, the velocity vector is given. Furthermore,*

*a) At all points where $U < 0$ the absolute temperature and the density are prescribed.*

*b) At all points where $U > 0$ the absolute temperature is given.*

*c) At all points where $U = 0$ one of the following conditions is assumed:*

*c1) The temperature is prescribed.*

*c2) The heat flux $\boldsymbol{q} \cdot \boldsymbol{n}$ is prescribed.*

---

[1] The condition $(\partial E/\partial T)_\varrho > 0$ is a general attribute of all real materials, as shown in textbooks on thermodynamics.

It is also possible to treat heterogeneous fluids which behave as inert mixtures; in this case the defining equations (4) become

$$p = p(\varrho, T, n), \qquad E = E(\varrho, T, n), \qquad \frac{dn}{dt} = 0, \tag{4'}$$

where $n$ is a parameter specifying the constitution of a given particle. The additional details necessary to handle cases (4) and (4') can readily be supplied by the reader.

*c3) The heat flux is assumed to be proportional to the difference between the fluid temperature $T$ and a given wall temperature $T_0$, that is,*

$$\boldsymbol{q} \cdot \boldsymbol{n} = k(T - T_0), \qquad k \geqq 0.$$

Cases *a*), *b*), *c*) above correspond respectively to fluid entering $\mathscr{V}$, fluid leaving $\mathscr{V}$, and fluid constrained to move tangentially to $\mathscr{V}$. Condition *c3*) is the familiar Newtonian law of cooling. We remark that any one of conditions *c1*)−*c3*) could be used at a boundary point where $U > 0$, though only the first one seems to be of any physical interest.

This completes the specification of the differential equations and the boundary conditions for a viscous compressible fluid.

### *B. Non-viscous compressible fluid*

Assuming that the fluid is devoid of heat conduction as well as viscosity, the differential equations (1)—(3) reduce to the familiar system

$$\frac{d\varrho}{dt} + \varrho \operatorname{div} \boldsymbol{v} = 0, \tag{7}$$

$$\varrho \frac{d\boldsymbol{v}}{dt} = \varrho \boldsymbol{f} - \operatorname{grad} p, \tag{8}$$

$$\frac{dS}{dt} = 0, \tag{9}$$

where $S$ denotes the specific entropy. The thermodynamic equation of state is conveniently assumed given in the form $p = p(\varrho, S)$, where the function $p(\varrho, S)$ is continuously differentiable and

$$\left(\frac{\partial p}{\partial \varrho}\right)_S > 0, \tag{10}$$

corresponding to a *real* speed of sound in the fluid.

At each point of the boundary we suppose the normal velocity component $\boldsymbol{v} \cdot \boldsymbol{n}$ to be assigned (this is a weaker condition than was necessary for viscous fluids). The case of most interest is naturally that where the fluid is bounded by material walls and the normal velocity component is determined by the motion of these walls. In this case, in fact, no further boundary conditions should be imposed. The general situation is given in the following table, where $U = \boldsymbol{v} \cdot \boldsymbol{n} - G$.

**Boundary Conditions.** *At all points of the boundary $\mathscr{S}$, and at all instants of time, the normal velocity component $\boldsymbol{v} \cdot \boldsymbol{n}$ is given. Furthermore at all points where $U < 0$ the complete fluid state—velocity, density, and entropy[2]—is assigned.*

It is understood in this statement that no further conditions are to be imposed at points where $U \geqq 0$. On the other hand, the fact that the entire fluid state must be prescribed at points where fluid is entering $\mathscr{V}$ will appear reasonable upon a moment's reflection.

If heat conduction is present, the differential equations and boundary conditions take a somewhat different form. It is convenient to defer this case until a later section.

[2] In place of the entropy one could of course assign the temperature or the pressure and thus indirectly determine the entropy.

### C. *The initial value problem*

For both problems A and B above, *a solution of the initial value problem is defined to be a set of continuously differentiable functions satisfying the given differential equations and the boundary conditions, and reducing when $t=0$ to assigned initial distributions*[3]. These solutions need not be continuable for all times $t>0$: it is not our intention, nor is it necessary in the proofs, to restrict consideration to solutions which exist for all time. Concerning the boundary of $\mathcal{V}$, we assume it smooth enough so that the divergence theorem is applicable at each instant, and also so that the developments of the following section are valid. It would be pedantic to state these agreements with any greater precision.

The solutions which we consider will be understood without further comment to have *positive* density at every point of the closure of $\mathcal{V}$.

## 2. The transport equation

Let $F(x, y, z, t)$ be a continuously differentiable function defined over a region $\mathcal{V}=\mathcal{V}(t)$ occupied by a moving fluid. In the sequel we shall have frequent occasion to use the transport formula

$$\frac{d}{dt}\int_{\mathcal{V}} \varrho F\, dv = \int_{\mathcal{V}} \varrho \frac{dF}{dt}\, dv - \oint_{\mathcal{S}} \varrho U F\, da, \tag{11}$$

where $\varrho$ and $\boldsymbol{v}$ are the density and velocity of the fluid motion, and $U$ denotes as usual the quantity $\boldsymbol{v}\cdot\boldsymbol{n}-G$. To prove (11), we begin with the rather obvious identity[4]

$$\frac{d}{dt}\int \varrho F = \int \frac{\partial}{\partial t}(\varrho F) + \oint \varrho F G, \tag{12}$$

[3] We do not consider "weak solutions" or solutions involving shock waves. On the other hand, with slight modifications it would be possible to deal with flows having first order singular surfaces.

[4] Formula (12) is well known and *physically* obvious: it says that the rate of change of the integral $\int \varrho F$ arises as the sum of internal changes of $\varrho F$ together with changes due to the fact that the boundary of $\mathcal{V}$ is moving with normal speed $G$. A proof may be given as follows:

Suppose the moving region $\mathcal{V}$ consists of those points $(x, y, z)$ such that

$$x = f(a, b, c, t), \qquad y = g(a, b, c, t), \qquad z = h(a, b, c, t),$$

where $a, b, c$ range over some closed region $\mathcal{X}$ in $R^3$, and $f, g, h$ are *continuously differentiable* functions of all their variables; suppose also that the mapping from $\mathcal{X}$ to $\mathcal{V}$ is one-to-one with non-vanishing Jacobian ($\mathcal{V}$ is a differentiable homotope of $\mathcal{X}$), and that the divergence theorem is applicable to $\mathcal{V}$ at each instant. If the "velocity field" $(\dot{x}, \dot{y}, \dot{z})$ is denoted by $\boldsymbol{w}$, then it can be shown that

$$\frac{d}{dt}\int \varphi = \int \frac{\partial \varphi}{\partial t} + \oint \varphi\, \boldsymbol{w}\cdot\boldsymbol{n}, \qquad \varphi = \varphi(x, y, z, t),$$

(*cf.* [*15*], § 4, or [*16*], p. 53). Since $\boldsymbol{w}\cdot\boldsymbol{n}=G$ is the outward normal speed of the boundary of $\mathcal{V}$, this completes the proof of (12).

It is evident that the conditions laid down on $\mathcal{V}$ in the preceding proof embrace all reasonable configurations. In any case, the regions which we shall consider in the rest of the paper are supposed to belong to the class for which (12) holds.

in which (as in the following work) the conventional volume and surface infinitesimals are omitted, and the integrals are understood to be extended over either $\mathscr{V}$ or $\mathscr{S}$ as the context indicates. With the help of the divergence theorem, one sees that the right hand side of (12) is equal to

$$\int \left[\frac{\partial}{\partial t}(\varrho F) + \operatorname{div}(\varrho F \boldsymbol{v})\right] + \oint \varrho F(G - \boldsymbol{v}\cdot\boldsymbol{n}).$$

On the other hand,

$$\frac{\partial}{\partial t}(\varrho F) + \operatorname{div}(\varrho F \boldsymbol{v}) = \varrho\left\{\frac{\partial F}{\partial t} + \boldsymbol{v}\cdot\operatorname{grad} F\right\} + F\left\{\frac{\partial \varrho}{\partial t} + \operatorname{div}(\varrho\boldsymbol{v})\right\} = \varrho\frac{dF}{dt},$$

using the equation of continuity (1) or (6), and the definition of the operator $d/dt$. Combining the results of the last three displayed lines immediately establishes the validity of (11).

### 3. The main uniqueness theorem for viscous fluids

With the conditions and agreements of § 1 understood, we can now state the fundamental uniqueness theorem for compressible viscous fluids.

**Theorem 1.** *Let the coefficients of viscosity and heat conduction be constants satisfying the inequalities*

$$\mu > 0, \quad 3\lambda + 2\mu > 0, \quad \varkappa \geqq 0.$$

*Then there can be at most one solution of the initial value problem described in § 1A. Moreover, if $\varkappa = 0$ no boundary conditions on the temperature (or heat flux) need be imposed at points where $U \geqq 0$.*

The important case where $3\lambda + 2\mu = 0$ is not included in this theorem, but seems to require a more delicate treatment, see § 7. It may be remarked that Theorem 1 applies also to flows occupying unbounded regions, provided that certain boundary integrals at $\infty$ can be assumed to vanish.

*Proof.* Let $(\boldsymbol{v}, \varrho, T)$ and $(\tilde{\boldsymbol{v}}, \tilde{\varrho}, \tilde{T})$ be two solutions of the problem under discussion. We shall show that they are in fact identical. For convenience in notation, a tilde over a flow quantity will mean that the quantity is to be *evaluated* for the flow $(\tilde{\boldsymbol{v}}, \tilde{\varrho}, \tilde{T})$. In addition, for an arbitrary flow quantity $F$ we define

$$F' = \tilde{F} - F;$$

thus in particular, $\boldsymbol{v}' = \tilde{\boldsymbol{v}} - \boldsymbol{v}$, $\varrho' = \tilde{\varrho} - \varrho$, etc. The following identities will be used frequently throughout the paper:

$$\widetilde{AB} - AB = AB' + A'\tilde{B}, \tag{13}$$

$$\frac{d\tilde{F}}{dt} - \frac{dF}{dt} = \frac{dF'}{dt} + \boldsymbol{v}'\cdot\operatorname{grad}\tilde{F}. \tag{14}$$

The first of these is obvious, while the second follows from an easy calculation using the definition of the operator $d/dt$. Note that $dF'/dt$ means

$$\frac{\partial F'}{\partial t} + \boldsymbol{v}\cdot\operatorname{grad} F',$$

and not $(dF/dt)'$.

Since both flows satisfy the equation of continuity, we obtain by subtraction and use of the identities (13) and (14)

$$\frac{d\varrho'}{dt} + \boldsymbol{v}' \cdot \operatorname{grad} \tilde{\varrho} + \varrho \operatorname{div} \boldsymbol{v}' + \varrho' \operatorname{div} \tilde{\boldsymbol{v}} = 0. \tag{15}$$

Multiplication by $\varrho'$ yields

$$\frac{d}{dt}\left(\frac{1}{2}\varrho'^2\right) = -\left(\varrho' \boldsymbol{v}' \cdot \operatorname{grad} \tilde{\varrho} + \varrho \varrho' \operatorname{div} \boldsymbol{v}' + \varrho'^2 \operatorname{div} \tilde{\boldsymbol{v}}\right) = -I,$$

where $I$ is *defined* by the last equality. Now using the transport equation (11) there results

$$\frac{d}{dt}\int \frac{1}{2}\varrho \varrho'^2 = -\int \varrho I - \oint \varrho U \varrho'^2. \tag{16}$$

This formula will be important in the sequel.

Again, since both flows satisfy the equations of motion, we have by subtraction

$$\varrho\left(\frac{d\boldsymbol{v}'}{dt} + \boldsymbol{v}' \cdot \operatorname{grad} \tilde{\boldsymbol{v}}\right) + \varrho' \frac{\widetilde{d\boldsymbol{v}}}{dt} = \varrho' \boldsymbol{f} + \operatorname{div} \boldsymbol{T}'.$$

Forming the scalar product of this equation with the vector $\boldsymbol{v}'$, and rearranging terms, yields

$$\varrho \frac{d}{dt}\left(\frac{1}{2} v'^2\right) = -\left[\varrho' \boldsymbol{v}' \cdot (\tilde{\boldsymbol{a}} - \boldsymbol{f}) + \varrho \boldsymbol{v}' \cdot \tilde{\boldsymbol{D}} \cdot \boldsymbol{v}' + \boldsymbol{T}' : \boldsymbol{D}'\right] + \operatorname{div}(\boldsymbol{T}' \cdot \boldsymbol{v}'),$$

where for brevity we have put $v'^2 = \boldsymbol{v}' \cdot \boldsymbol{v}'$ and $\boldsymbol{a} = d\boldsymbol{v}/dt$. Finally, with the help of the transport equation and the fact that $v' = 0$ on the boundary of $\mathscr{V}$, one obtains

$$\frac{d}{dt}\int \frac{1}{2}\varrho v'^2 = -\int \left[\varrho' \boldsymbol{v}' \cdot (\tilde{\boldsymbol{a}} - \boldsymbol{f}) + \varrho \boldsymbol{v}' \cdot \tilde{\boldsymbol{D}} \cdot \boldsymbol{v}' + \boldsymbol{T}' : \boldsymbol{D}'\right]. \tag{17}$$

Turning next to the energy equation, we observe that in virtue of (5) it can be written in the form

$$\varrho \frac{dT}{dt} = \Psi - \operatorname{div} \boldsymbol{q} \qquad (\Psi \equiv \boldsymbol{T} : \boldsymbol{D}),$$

where without loss of generality we have set $c_v = 1$. Applying the method already described in connection with the previous equations, the reader will have no trouble in proving the identity

$$\frac{d}{dt}\int \frac{1}{2}\varrho T'^2 = \int \left[T' \Psi' + \boldsymbol{q}' \cdot \operatorname{grad} T' - \varrho' T' \frac{\widetilde{dT}}{dt} - \varrho T' \boldsymbol{v}' \cdot \operatorname{grad} \tilde{T}\right] - \\ - \oint \left[T' \boldsymbol{q}' \cdot \boldsymbol{n} + \frac{1}{2}\varrho U T'^2\right]. \tag{18}$$

The next step in the proof is to estimate the various terms which appear on the right hand sides of equations (16), (17) and (18). For this purpose it is convenient to restrict attention to some *fixed* time interval $0 \leqq t \leqq \tau$, where $\tau$ is an arbitrary positive number (if the solutions are known to exist only for a finite time $0 \leqq t \leqq \tau_0$, then $\tau$ is naturally chosen less than $\tau_0$). We shall let $N$ be the generic notation for an upper bound: the number $N$ will then not be the same in each estimate, though it will of course be possible to determine its size at any stage. The expression $\varrho I$ appearing in (16) has the following estimate, valid for $0 \leqq t \leqq \tau$:

$$|\varrho I| = |\varrho \varrho' \boldsymbol{v}' \cdot \operatorname{grad} \tilde{\varrho} + \varrho^2 \varrho' \operatorname{div} \boldsymbol{v}' + \varrho \varrho'^2 \operatorname{div} \tilde{\boldsymbol{v}}|$$
$$\leqq N(\varrho'^2 + v'^2) + \varepsilon \Theta'^2 \qquad (\Theta \equiv \operatorname{div} \boldsymbol{v}).$$

Here the basic tool is CAUCHY's inequality $2ab \leqq a^2 + b^2$, and $\varepsilon$ is an arbitrary positive number to be fixed later; $N$ depends on $\varepsilon$ and on bounds for the magnitude of $(\boldsymbol{v}, \varrho, T)$, $(\tilde{\boldsymbol{v}}, \tilde{\varrho}, \tilde{T})$ and their derivatives during the time interval $0 \leqq t \leqq \tau$. In addition, because of the boundary conditions the term $\varrho U \varrho'^2$ in (16) is $\geqq 0$. Inserting these estimates into (16) yields the inequality

$$\frac{d\mathscr{J}}{dt} \leq \int N(\varrho'^2 + v'^2) + \int \varepsilon \Theta'^2, \qquad 0 \leqq t \leqq \tau,$$

where it is convenient to use the notation

$$\mathscr{J} = \int \tfrac{1}{2} \varrho \varrho'^2, \qquad \mathscr{K} = \int \tfrac{1}{2} \varrho v'^2, \qquad \mathscr{L} = \int \tfrac{1}{2} \varrho T'^2.$$

In considering equation (17) it will be advantageous to write $\boldsymbol{T}$ in the form $\boldsymbol{T} = -p\boldsymbol{I} + \boldsymbol{V}$, where $\boldsymbol{V}$ is the *viscosity tensor*, $\boldsymbol{V} \equiv \lambda \Theta \boldsymbol{I} + 2\mu \boldsymbol{D}$. Then since $\boldsymbol{I}:\boldsymbol{D} = \operatorname{div} \boldsymbol{v} = \Theta$, we have

$$\boldsymbol{T}' : \boldsymbol{D}' = -p' \Theta' + \boldsymbol{V}' : \boldsymbol{D}'.$$

It is now easily verified that

$$\frac{d\mathscr{K}}{dt} \leqq \int N(\varrho'^2 + v'^2 + p'^2) + \int (\varepsilon \Theta'^2 - \boldsymbol{V}' : \boldsymbol{D}'), \qquad 0 \leqq t \leqq \tau.$$

Turning finally to an estimate for the right hand side of (18), we see that

$$\begin{aligned} \Psi' &= \widetilde{\boldsymbol{T} : \boldsymbol{D}} - \boldsymbol{T} : \boldsymbol{D} = \boldsymbol{T} : \boldsymbol{D}' + \boldsymbol{T}' : \tilde{\boldsymbol{D}} \\ &= \boldsymbol{T} : \boldsymbol{D}' + (-p' + \lambda \Theta') \tilde{\Theta} + 2\mu \boldsymbol{D}' : \tilde{\boldsymbol{D}}, \end{aligned} \tag{19}$$

whence for $0 \leqq t \leqq \tau$,

$$|T' \Psi'| = |T' \boldsymbol{T} : \boldsymbol{D}' - p' T' \tilde{\Theta} + \lambda \Theta' T' \tilde{\Theta} + 2\mu T' \boldsymbol{D}' : \tilde{\boldsymbol{D}}|$$
$$\leqq N(T'^2 + p'^2) + \varepsilon (\Theta'^2 + \boldsymbol{D}' : \boldsymbol{D}').$$

Furthermore $\boldsymbol{q}' \cdot \operatorname{grad} T' = -\varkappa \operatorname{grad} T'^2 \leqq 0$, and on the boundary of $\mathscr{V}$, in all cases, $\varrho U T'^2 \geqq 0$ and $T' \boldsymbol{q}' \cdot \boldsymbol{n} \geqq 0$. When the above inequalities are inserted into (18) there results

$$\frac{d\mathscr{L}}{dt} \leq \int N(\varrho'^2 + v'^2 + p'^2 + T'^2) + \int \varepsilon (\Theta'^2 + \boldsymbol{D}' : \boldsymbol{D}').$$

Adding the various estimates for the derivatives of $\mathscr{J}$, $\mathscr{K}$, and $\mathscr{L}$, and using the obvious inequality $p'^2 \leqq N(\varrho'^2 + T'^2)$, gives easily

$$\frac{d}{dt}(\mathscr{J} + \mathscr{K} + \mathscr{L}) \leqq N(\mathscr{J} + \mathscr{K} + \mathscr{L}) + \int [\varepsilon (3\Theta'^2 + \boldsymbol{D}' : \boldsymbol{D}') - \boldsymbol{V}' : \boldsymbol{D}'].$$

This completes the main estimate. It is necessary now to consider the term in braces in the last integral.

The following identity becomes obvious if the matrix $\boldsymbol{D}$ is written in diagonal form,

$$3\boldsymbol{D}:\boldsymbol{D}=\Theta^2+\Delta^2 \qquad \left(\Delta\equiv\sqrt{(d_1-d_2)^2+(d_2-d_3)^2+(d_3-d_1)^2}\right);$$

here $\Theta=\operatorname{div}\boldsymbol{v}=\operatorname{Trace}\boldsymbol{D}$, and $\Delta$ is the octahedral invariant defined in terms of the characteristic values $d_1, d_2, d_3$ of $\boldsymbol{D}$. From this identity follows

$$\boldsymbol{V}':\boldsymbol{D}'=\lambda\Theta'^2+2\mu\boldsymbol{D}':\boldsymbol{D}'=\tfrac{1}{3}\{(3\lambda+2\mu)\Theta'^2+2\mu\Delta'^2\},$$

and also

$$\varepsilon(3\Theta'^2+\boldsymbol{D}':\boldsymbol{D}')=\frac{\varepsilon}{3}(10\Theta'^2+\Delta'^2).$$

Thus if $\varepsilon$ is chosen small enough, we have

$$\varepsilon(3\Theta'^2+\boldsymbol{D}':\boldsymbol{D}')\leqq\boldsymbol{V}':\boldsymbol{D}',$$

and hence

$$\frac{d}{dt}(\mathscr{J}+\mathscr{K}+\mathscr{L})\leqq N(\mathscr{J}+\mathscr{K}+\mathscr{L}), \qquad 0\leqq t\leqq\tau.$$

This differential inequality has the integrating factor $e^{-Nt}$, and the solution

$$\mathscr{J}+\mathscr{K}+\mathscr{L}\leqq(\mathscr{J}+\mathscr{K}+\mathscr{L})_0 e^{Nt}, \qquad 0\leqq t\leqq\tau.$$

Since the integrals $\mathscr{J}$, $\mathscr{K}$, and $\mathscr{L}$ are zero when $t=0$, it follows that they remain zero throughout the whole time interval in question. But this forces $v'=\varrho'=T'=0$, and so the two flows are identical at least until $t=\tau$. Finally, $\tau$ being arbitrary, the flows are identical so long as they exist, and the theorem is completely proved.

Another consequence of the final inequality is that solutions of the initial value problem, so long as they remain uniformly bounded, depend continuously on the initial data.

## 4. The main uniqueness theorem for inviscid fluids

We consider here the most important case $\lambda=\mu=\varkappa=0$, and refer the reader to § 1 for a description of the problem. The principal result is the following

**Theorem 2.** *Let the equation of the state $p=p(\varrho, S)$ be of class $C^2$ for all values of the variables, and suppose that*

$$\left(\frac{\partial p}{\partial\varrho}\right)_S>0. \tag{20}$$

*Then there can be at most one solution of the initial value problem described in § 1B. Moreover, if for some solution of this problem we have*

$$U=\boldsymbol{v}\cdot\boldsymbol{n}-G\geqq c, \qquad c^2=\left(\frac{\partial p}{\partial\varrho}\right)_S$$

*on a subset $\mathscr{S}^*$ of the boundary, then the boundary conditions on $\mathscr{S}^*$ are redundant with respect to the given data; that is, the conditions assigned on the remainder of the boundary, together with the given initial values, uniquely determine the solution.*

(To the author's knowledge, the only previous uniqueness theorem for compressible non-viscous fluids with equation of state $p=p(\varrho, S)$ is that given by Courant & Friedrichs ([1], pp. 82–84 and pp. 48–53, 70–75). There is a large body of literature concerning hyberbolic systems of equations in more than two variables [*5*], but this work does not seem to have immediate application to our problem.)

Before commencing the proof of Theorem 2, there are several points of related interest which we wish to discuss.

To begin with, the second part of the theorem includes as a special case the result that initial values over a region $\mathscr{R}$ uniquely determine the flow (assumed continuously differentiable) throughout a certain "hypercone" in $(x, y, z, t)$ space with its base on $\mathscr{R}$. If for geometric clarity we consider the motion to take place in two spatial dimensions, then this conical *domain of determination* will appear as in Fig. 1. In order to determine the exact shape of the cone it is necessary to know one solution of the initial value problem; the lateral boundary, or

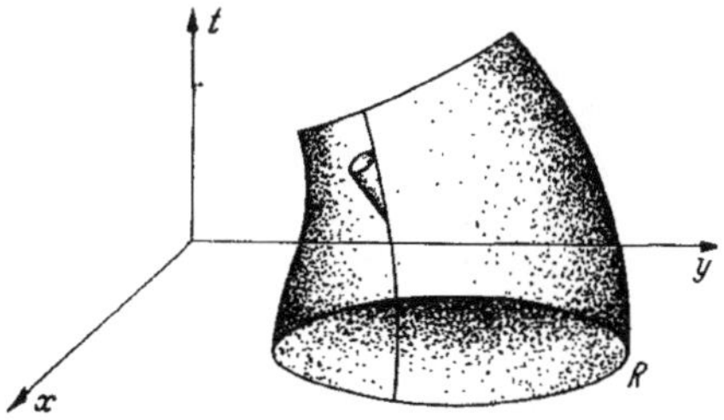

Fig. 1. Domain of determinacy for the region $\mathscr{R}$ (two spatial dimensions)

mantle, of the cone is then characterized by the condition that $U=c$ at each of its regular points[5]. In view of the independent importance of this result, we shall restate it in the following precise form.

**Corollary.** *Consider a solution (with continuous first derivatives) of equations* (6)—(8) *in a region* $\mathscr{R}^*$ *of* $(x, y, z, t)$ *space bounded by the plane* $t=0$ *and a mantle* $\mathscr{S}^*$ *on which* $U \geqq c$. *Suppose another solution with continuous first derivatives is given in* $\mathscr{R}^*$ *which assumes for* $t=0$ *the same values as the first one. Then the two solutions are identical in* $\mathscr{R}^*$.

Another point which should be made in connection with Theorem 2 is that the given initial value problem may easily have no solutions at all, or only solutions which break down after a finite length of time. This occurs most obviously if we simply allow the boundary to penetrate the initial value cone $\mathscr{R}^*$, but it may also occur even when the boundary conditions appear quite innocuous, as has been shown by LUDFORD in a very interesting paper.

---

[5] The equation $U=c$ specifies a *Monge cone* at each point of $(x, y, z, t)$ space where the solution is known. The determination of the lateral boundary thus involves solving a first order partial differential equation; the mantle is then the inner envelope of the family of *Monge conoids* whose vertices lie on the boundary of $R$. Since the relation $U=c$ can be shown to be a characteristic condition of equations (6)—(8), it follows that the lateral boundary is a characteristic manifold of the solution (*cf.* [*2*], pp. 374—376 (plane flow), [*12*], pp. 112—114, or [*15*], § 51; in the last named reference the characteristic condition is derived in exactly the form used here).

The Monge cone noted above has obvious affinities with the Mach cone in $(x, y, z)$ space. Indeed the latter is just the projection of the Monge cone; this projection is real if and only if the flow is supersonic at the point in question.

The fact that initial values uniquely determine the subsequent flow throughout a portion of the flow field, and the fact that other solutions break down after a finite time, arises, of course, because we are considering continuous solutions of a hyperbolic system. If we allow solutions with shock waves, then the initial values no longer dictate the subsequent motion, and the breakdown of a continuous solution no longer disqualifies the future motion from consideration. For this reason, it may be conjectured that the first statement of Theorem 2 remains true for flows with shock waves. The work of LAX, OLEJNIK, and LADYSHENSKAYA (see reference [11] and the accompanying bibliography) may provide the clue to a deeper understanding of the situation.

## 5. Proof of Theorem 2

Let $(\boldsymbol{v}, \varrho, S)$ and $(\tilde{\boldsymbol{v}}, \tilde{\varrho}, \tilde{S})$ be two solutions of the initial value problem under discussion. Also, let $\mathscr{S}^*$ denote the (possibly empty) set of boundary points where the basic solution $(\boldsymbol{v}, \varrho, S)$ satisfies $U \geqq c$. In order to prove Theorem 2, it must be shown that the two solutions are identical, and this must be done without recourse to the assigned boundary conditions on the set $\mathscr{S}^*$.

Retaining the notation introduced in the proof of Theorem 1, and applying the methods developed there, it is clear that from equation (7) follows

$$\frac{d}{dt}\int \frac{1}{2}\varrho v'^2 = -\int [\varrho' \boldsymbol{v}' \cdot (\tilde{\boldsymbol{a}} - \boldsymbol{f}) + \varrho \boldsymbol{v}' \cdot \tilde{\boldsymbol{D}} \cdot \boldsymbol{v}' + \boldsymbol{v}' \cdot \operatorname{grad} p'] - \oint \frac{1}{2}\varrho U v'^2 \tag{21}$$

(this should be compared with the analogous equation (17); note that the boundary term in (21) cannot be assumed to vanish). The energy equation $dS/dt = 0$ likewise yields

$$\frac{dS'}{dt} + \boldsymbol{v}' \cdot \operatorname{grad} \tilde{S} = 0 \tag{22}$$

and

$$\frac{d}{dt}\int \frac{1}{2}\varrho S'^2 = -\int \varrho S' \boldsymbol{v}' \cdot \operatorname{grad} \tilde{S} - \oint \frac{1}{2}\varrho U S'^2. \tag{23}$$

Now let $0 \leqq t \leqq \tau$ be an arbitrary time interval throughout which both solutions $(\boldsymbol{v}, \varrho, S)$ and $(\tilde{\boldsymbol{v}}, \tilde{\varrho}, \tilde{S})$ exist, and let $N$ be the generic notation for an upper bound. Then we have the following estimates valid for $0 \leqq t \leqq \tau$:

$$|\varrho' \boldsymbol{v}' \cdot (\tilde{\boldsymbol{a}} - \boldsymbol{f})| \leqq N(\varrho'^2 + v'^2), \quad |\varrho \boldsymbol{v}' \cdot \tilde{\boldsymbol{D}} \cdot \boldsymbol{v}'| \leqq N v'^2,$$
and
$$|\varrho S' \boldsymbol{v}' \cdot \operatorname{grad} \tilde{S}| \leqq N(v'^2 + S'^2); \tag{24}$$

here $N$ depends on bounds for the magnitude of $(\boldsymbol{v}, \varrho, S)$, $(\tilde{\boldsymbol{v}}, \tilde{\varrho}, \tilde{S})$ and their derivatives during the time interval $0 \leqq t \leqq \tau$.

The remaining term in (21), namely $\boldsymbol{v}' \cdot \operatorname{grad} p'$, requires a separate treatment. The algebraic effort in this estimate can be greatly reduced by agreeing that an equality sign with a dot placed over it means *equality modulo terms which contain products and squares of the quantities* $\boldsymbol{v}'$, $\varrho'$, *and* $S'$. Since such terms can be majorized by inequalities of the form (24), it is clear that this procedure retains at each stage the essential quantities. In any case, setting

$$A = \left(\frac{\partial \varrho}{\partial S}\right)_\varrho$$

we have

$$\begin{aligned} \boldsymbol{v}' \cdot \operatorname{grad} p' &\doteq \boldsymbol{v}' \cdot (c^2 \operatorname{grad} \varrho' + A \operatorname{grad} S') \\ &\doteq -(c^2 \varrho' + A S') \operatorname{div} \boldsymbol{v}' + \operatorname{div}(c^2 \varrho' + A S') \boldsymbol{v}' \\ &\doteq \frac{1}{\varrho}(c^2 \varrho' + A S') \frac{d\varrho'}{dt} + \operatorname{div}(c^2 \varrho' + A S') \boldsymbol{v}', \end{aligned} \tag{25}$$

where the final step is a consequence of equation (15). Now

$$\frac{1}{\varrho} c^2 \varrho' \frac{d\varrho'}{dt} \doteq \varrho \frac{d}{dt}\left(\frac{c^2}{2\varrho^2} \varrho'^2\right), \tag{26}$$

and using equation (22),

$$\frac{1}{\varrho} A S' \frac{d\varrho'}{dt} \doteq \frac{A}{\varrho} \frac{d}{dt}(\varrho' S') \doteq \varrho \frac{d}{dt}\left(\frac{A}{\varrho^2} \varrho' S'\right). \tag{27}$$

When (26) and (27) are inserted into (25) there results

$$\boldsymbol{v}' \cdot \operatorname{grad} p' \doteq \varrho \frac{d}{dt}\left(\frac{c^2}{2\varrho^2} \varrho'^2 + \frac{A}{\varrho^2} \varrho' S'\right) + \operatorname{div}(c^2 \varrho' + A S') \boldsymbol{v}'.$$

Substituting this formula into (21) and then applying the divergence theorem, the transport equation and inequality (24) yields the inequality

$$\begin{aligned} \frac{d}{dt}\int \frac{1}{2}\varrho\left(v'^2 + \frac{c^2}{\varrho^2}\varrho'^2 + \frac{2A}{\varrho^2}\varrho' S'\right) \leq \int N(\varrho'^2 + v'^2 + S'^2) - \\ - \oint \frac{1}{2}\left[\varrho U\left(v'^2 + \frac{c^2}{\varrho^2}\varrho'^2 + \frac{2A}{\varrho^2}\varrho' S'\right) + 2(c^2\varrho' + A S')\boldsymbol{v}' \cdot \boldsymbol{n}\right], \end{aligned} \tag{28}$$

this being valid for $0 \leqq t \leqq \tau$. In addition, from (23) and (24),

$$\frac{d}{dt}\int \frac{1}{2}\varrho S'^2 \leq \int N(v'^2 + S'^2) - \oint \frac{1}{2}\varrho U S'^2. \tag{29}$$

Now let $\vartheta$ be a large positive number, whose value will be fixed in a moment, and define quantities $J$ and $K$ by

$$\begin{aligned} J &\equiv v'^2 + \frac{c^2}{\varrho^2}\varrho'^2 + \frac{2A}{\varrho^2}\varrho' S' + \vartheta S'^2, \\ K &\equiv \varrho U J + 2(c^2 \varrho' + A S') \boldsymbol{v}' \cdot \boldsymbol{n}, \qquad \text{on } \mathscr{S}. \end{aligned}$$

We observe that $\vartheta$ can be chosen so large that

$$J \geqq v'^2 + \frac{c^2}{2\varrho^2}(\varrho'^2 + S'^2) \geqq 0 \tag{30}$$

for $0 \leqq t \leqq \tau$; indeed, recalling that $c > 0$, we may take

$$\vartheta = \operatorname*{Max}_{t \leqq \tau}\left(\frac{2A^2}{\varrho^2 c^2} + \frac{c^2}{2\varrho^2}\right).$$

With this choice of $\vartheta$, it follows from the boundary conditions (p. 274) that $K \geqq 0$ at all points of $\mathscr{S}$. Even more, however, $K \geqq 0$ on the set $\mathscr{S}^*$ *irrespective of the boundary conditions*. For on $\mathscr{S}^*$ we have $U \geqq c$, so that either

$$K \geqq \varrho c J + 2v'(c^2 \varrho' + A S') = Q_1$$

or

$$K \geqq \varrho c J - 2v'(c^2 \varrho' + A S') = Q_2.$$

Since both the quadratic forms $Q$ are easily shown to be non-negative definite, the required inequality follows at once.

With the results of the preceding paragraph in mind, let us multiply inequality (29) by $\vartheta$ and add it to (28). Since $K \geqq 0$, the resulting boundary integral is non-negative, and we obtain

$$\frac{d}{dt}\int \varrho J \leqq \int N(\varrho'^2 + v'^2 + S'^2);$$

therefore, with the help of (30) and the fact that $\varrho$ and $c$ are positive,

$$\frac{d}{dt}\int \varrho J \leqq N \int \varrho J, \qquad 0 \leqq t \leqq \tau.$$

It follows by integration and use of the initial conditions that $\int \varrho J = 0$, and then, using (30) again, that $v' = \varrho' = S' = 0$. This completes the proof of Theorem 2.

In the following section we shall deal with the previously excluded case of an inviscid fluid with non-zero heat conduction.

### 6. The excepted case $\lambda = \mu = 0$, $\varkappa > 0$

The type of fluid considered in this section, having zero viscosity but positive heat conduction, is perhaps rather unlikely to occur in a physical problem. Nevertheless, the results of this section are quite interesting from a mathematical viewpoint, and will also be useful in a later part of the paper

The differential equations for the problem are obtained by specializing (1)—(3) to the forms

$$\frac{d\varrho}{dt} + \varrho \operatorname{div} \boldsymbol{v} = 0, \tag{31}$$

$$\varrho \frac{d\boldsymbol{v}}{dt} = \varrho \boldsymbol{f} - \operatorname{grad} p, \tag{32}$$

$$\varrho \frac{dE}{dt} = -p \operatorname{div} \boldsymbol{v} - \operatorname{div} \boldsymbol{q}. \tag{33}$$

As in the treatment of viscous fluids, we shall consider only the equations of state $\lambda$

$$p = p(\varrho, T), \qquad E = c_v T, \tag{34}$$

where $c_v$ is a positive constant. It will also be assumed that the pressure-density-temperature relation is twice continuously differentiable and satisfies the inequality

$$\left(\frac{\partial p}{\partial \varrho}\right)_T > 0, \tag{35}$$

corresponding to a *real* isothermal speed of sound. Inequality (35), like the earlier conditions $(\partial E/\partial T)_\varrho > 0$ and $(\partial p/\partial \varrho)_S > 0$, is a general attribute of all real materials. Moreover, it is always the case that

$$\left(\frac{\partial p}{\partial \varrho}\right)_S > \left(\frac{\partial p}{\partial \varrho}\right)_T, \tag{36}$$

see [*15*], § 30. The boundary conditions appropriate to equations (31)—(33) are as follows:

*At all points of the boundary $\mathcal{S}$, and at all instants of time, the normal velocity component $\boldsymbol{v}\cdot\boldsymbol{n}$ is given. Furthermore*

*a) At all points where $U<0$ the complete fluid state, velocity, density, and temperature, is assigned.*

*b) At all points where $U\geqq 0$ the temperature is given.*

*c) At all points where $U=0$, one of the conditions c1)—c3) listed on p. 273 is assumed.*

These conditions being understood, we have the following uniqueness theorem for non-viscous, heat-conducting fluids.

**Theorem 3.** *There can be at most one solution of the initial value problem corresponding to the equations and boundary conditions described above. Moreover, if for some solution of this problem we have*

$$U=\boldsymbol{v}\cdot\boldsymbol{n}-G\geqq a,\qquad a^2=\left(\frac{\partial p}{\partial\varrho}\right)_T,$$

*on a subset $\mathcal{S}^*$ of the boundary, then the boundary conditions on $\mathcal{S}^*$ are partly superfluous with respect to the given data: that is, the conditions assigned on the remainder of the boundary, together with the temperature on $\mathcal{S}^*$ and the given initial values, uniquely determine the solution.*

Exactly as in the previous section, this theorem implies the following corollary: Consider a solution (with continuous derivatives) of equations (31)—(33) in a region $\mathcal{R}^*$ of $(x, y, z, t)$ space bounded by the plane $t=0$ and a mantle $\mathcal{S}^*$ on which $U\geqq a$. Suppose another solution with continuous derivatives is given in $\mathcal{R}^*$, which assumes for $t=0$ the same values as the first one, and whose temperature agrees on $\mathcal{S}^*$ with the temperature of the first solution. Then the two solutions are identical.

Since this situation is very much like that of the preceding section, we confine our discussion to the remark that the Monge cone for the present problem lies *inside* the Monge cone $U=c$, in virtue of (36). Thus the region $\mathcal{R}^*$ here is "larger" than the region of determination for a non-viscous, non-heat-conducting fluid, assuming that the two can be compared at all.

*Proof of Theorem 3.* The difference between the present problem and the problem $\mu=\lambda=\varkappa=0$ lies entirely in the energy equation. Equation (21) then holds exactly as before, but in place of (23) we now must deal with

$$\begin{aligned}\frac{d}{dt}\int\frac{1}{2}\varrho T'^2=\int\left[\boldsymbol{q}'\cdot\operatorname{grad}T'-\varrho' T'\frac{d\tilde{T}}{dt}-\varrho T'\boldsymbol{v}'\cdot\operatorname{grad}\tilde{T}-T'\Phi'\right]-\\ -\oint\left[T'\boldsymbol{q}'\cdot\boldsymbol{n}+\frac{1}{2}\varrho U T'^2\right],\end{aligned}\tag{37}$$

where $\Phi\equiv p\operatorname{div}\boldsymbol{v}$ (this identity is proved in the same way as (18)). A perusal of the various terms on the right hand side of equation (21) shows that, as before, the only difficult quantity is $\boldsymbol{v}'\cdot\operatorname{grad}p'$. Using the "dot" notation to denote equality modulo squares and products of $\boldsymbol{v}'$, $\varrho'$, and $T'$, and setting $B=(\partial p/\partial T)_\varrho$, we have (*cf.* formulas (25) and (26))

$$\begin{aligned}\boldsymbol{v}'\cdot\operatorname{grad}p' &\doteq \boldsymbol{v}'\cdot(a^2\operatorname{grad}\varrho' + B\operatorname{grad}T')\\ &\doteq -a^2\varrho'\operatorname{div}\boldsymbol{v}' + B\boldsymbol{v}'\cdot\operatorname{grad}T' + \operatorname{div}(a^2\varrho'\boldsymbol{v}')\\ &\doteq \varrho\frac{d}{dt}\left(\frac{a^2}{2\varrho^2}\varrho'^2\right) + B\boldsymbol{v}'\cdot\operatorname{grad}T' + \operatorname{div}(a^2\varrho'\boldsymbol{v}').\end{aligned}$$

Substituting this formula into (21), and then applying the divergence theorem, the transport equation, and the usual kind of estimates, we obtain

$$\begin{aligned}\frac{d}{dt}\int\frac{1}{2}\varrho\left(v'^2+\frac{a^2}{\varrho^2}\varrho'^2\right) &\leqq \int N(\varrho'^2+v'^2+T'^2)+\int\varepsilon|\operatorname{grad}T'|^2-\\ &-\oint\frac{1}{2}\left[\varrho U\left(v'^2+\frac{a^2}{\varrho^2}\varrho'^2\right)+2a^2\varrho'\boldsymbol{v}'\cdot\boldsymbol{n}\right],\end{aligned}\tag{38}$$

this holding as always for $0\leqq t\leqq\tau$.

Turning next to equation (37), one has the following identity for the term $T'\Phi'$:

$$\begin{aligned}T'\Phi' &= T'(p\operatorname{div}\boldsymbol{v})' = T'(p\operatorname{div}\boldsymbol{v}' + p'\operatorname{div}\tilde{\boldsymbol{v}})\\ &= \operatorname{div}(pT'\boldsymbol{v}') - p\boldsymbol{v}'\cdot\operatorname{grad}T' - T'\boldsymbol{v}'\cdot\operatorname{grad}p + T'p'\operatorname{div}\tilde{\boldsymbol{v}}.\end{aligned}$$

If this formula is inserted into (37), then from the divergence theorem, the boundary conditions, and the usual kind of estimates, we get the inequality

$$\frac{d}{dt}\int\frac{1}{2}\varrho T'^2 \leqq \int N(\varrho'^2+v'^2+T'^2)+\int(\varepsilon-\varkappa)\,|\operatorname{grad}T'|^2.\tag{39}$$

Now define

$$\hat{K}\equiv U(\varrho^2v'^2+a^2\varrho'^2)+2\varrho a^2\varrho'\boldsymbol{v}'\cdot\boldsymbol{n},\qquad\text{on }\mathscr{S}.$$

Clearly $\hat{K}\geqq 0$ in virtue of the given boundary conditions. Moreover

$$\begin{aligned}|2\varrho a^2\varrho'\boldsymbol{v}'\cdot\boldsymbol{n}| &\leqq 2a\,|a\varrho'|\cdot|\varrho\boldsymbol{v}'|\\ &\leqq a(\varrho^2v'^2+a^2\varrho'^2),\end{aligned}$$

so that $\hat{K}\geqq 0$ on $\mathscr{S}^*$ *irrespective of the boundary conditions.* Adding inequalities (38) and (39), and using the fact that $\hat{K}\geqq 0$, gives without difficulty

$$\frac{d\mathscr{M}}{dt}\leqq N\mathscr{M}+\int(2\varepsilon-\varkappa)\,|\operatorname{grad}T'|^2,\qquad 0\leqq t\leqq\tau,$$

where

$$\mathscr{M}\equiv\int\frac{1}{2}\varrho\left(v'^2+\frac{a^2}{\varrho^2}\varrho'^2+T'^2\right).\tag{40}$$

Setting $\varepsilon=\frac{1}{2}\varkappa$ and solving the resulting differential inequality leads easily to the identical vanishing of $v'$, $\varrho'$, and $T'$.

### 7. The excepted case $3\lambda+2\mu=0$

As we have already remarked, Theorem 1 does not apply in the important case when $3\lambda+2\mu=0$, that is, when the Stokes relation holds. However, this case is covered by the following

**Theorem 4.** *Let the coefficients of viscosity and heat conduction be constants satisfying*

$$\mu > 0, \quad \varkappa > 0, \quad 3\lambda + 2\mu = 0,$$

*and suppose also that* $(\partial p/\partial \varrho)_T > 0$. *Then there can be at most one solution of the initial value problem described in § 1A.*

*Proof.* Because of the similarity between Theorem 4 and Theorems 1 and 3, we can make considerable use of formulas already developed in the preceding sections. In fact, since the only difference between equation (2) and equation (32) is that the former has an extra term div $\boldsymbol{V}$ on the right, we have, by applying to equation (2) the same development which led from (32) to (38),

$$\frac{d}{dt}\int \frac{1}{2}\varrho\left(v'^2 + \frac{a^2}{\varrho^2}\varrho'^2\right) \leqq \int N(\varrho'^2 + v'^2 + T'^2) + \int \varepsilon\,|\operatorname{grad} T'|^2;$$

in this inequality, the boundary conditions and the fact that $\boldsymbol{V}':\boldsymbol{D}' \geqq 0$ have been employed. Besides this estimate, which takes the place of the inequality for $d\mathscr{K}/dt$ in the proof of Theorem 1, it is necessary to find an estimate for $\int \frac{1}{2}\varrho T'^2$ corresponding to the earlier inequality for $d\mathscr{L}/dt$. To this end, we observe that from (19),

$$\begin{aligned} T'\Psi' &= T'\boldsymbol{T}:\boldsymbol{D}' - p'\,T'\tilde{\Theta} + \lambda\Theta' T'\tilde{\Theta} + 2\mu\, T'\boldsymbol{D}':\tilde{\boldsymbol{D}} \\ &= \boldsymbol{v}'\cdot\boldsymbol{T}\cdot\operatorname{grad} T' + \lambda\tilde{\Theta}\,\boldsymbol{v}'\cdot\operatorname{grad} T' + 2\mu\,\boldsymbol{v}'\cdot\tilde{\boldsymbol{D}}\cdot\operatorname{grad} T' \\ &\quad + \{\text{squares and products of } \boldsymbol{v}', p', \text{ and } T'\} + \{\text{divergence terms}\}. \end{aligned}$$

Inserting this expression for $T'\Psi'$ into (18), and applying the divergence theorem, the boundary conditions, and the usual sort of estimates, yields

$$\frac{d}{dt}\int \frac{1}{2}\varrho T'^2 \leqq \int N(\varrho'^2 + v'^2 + p'^2 + T'^2) + \int (\varepsilon - \varkappa)\,|\operatorname{grad} T'|^2.$$

Now adding the two preceding inequalities, there results

$$\frac{d\mathscr{M}}{dt} \leqq N\mathscr{M} + \int (2\varepsilon - \varkappa)\,|\operatorname{grad} T'|^2, \qquad 0 \leqq t \leqq \tau,$$

where $\mathscr{M}$ is defined by (40). The rest of the proof is standard.

*Remarks.* In addition to the various combinations of inequalities restricting $\lambda$, $\mu$, and $\varkappa$ which have already been treated, there are two others which deserve notice, namely

$$\mu = 0, \quad \lambda > 0, \quad \varkappa \geqq 0 \tag{41}$$

and

$$\mu > 0, \quad 3\lambda + 2\mu = 0, \quad \varkappa = 0. \tag{42}$$

Case (41) can be discussed exactly as in Theorem 1; when $\varkappa = 0$ the appropriate boundary conditions are those in § 1B, and when $\varkappa > 0$ the appropriate boundary conditions are those of § 6. The remaining case (42) does not seem to be accessible by our methods.

Cases (41) and (42) exhaust all the situations for which the constitutive equations $(2)_2$ and $(3)_2$ satisfy the thermodynamic restrictions

$$\boldsymbol{V}:\boldsymbol{D} \geqq 0, \quad \boldsymbol{q}\cdot\operatorname{grad} T \leqq 0. \tag{43}$$

It would be interesting to consider the initial value problem for more general constitutive equations obeying (43) or some modification of (43).

## 8. Temperature dependent viscosity and heat conduction

As mentioned in the introduction, the coefficients $\lambda, \mu$, and $\varkappa$ are rather strongly dependent on the temperature in actual liquids and gases. For example, in the case of air

$$\mu = \frac{k T^{\frac{3}{2}}}{T + \text{const}}, \qquad \text{const} \cong 200,$$

while for water $\mu(100°)/\mu(0°) = 6.314$[6]. The following theorem shows that no significant qualitative changes are brought about by treating non-constant viscosity and heat conduction.

**Theorem 5.** *If the coefficients of viscosity and heat conduction in Theorem 1 are assumed to be differentiable functions of $\varrho$ and $T$, satisfying*

$$\mu > 0, \quad 3\lambda + 2\mu > 0, \quad \varkappa > 0,$$

*then the conclusion of Theorem 1 remains valid.*

*Proof.* If one repeats the steps in the proof of Theorem 1, he is led with only a few modifications (specifically, in the estimation of $T'\Psi'$) to the inequality

$$\frac{d}{dt}(\mathcal{J} + \mathcal{K} + \mathcal{L}) \leqq N(\mathcal{J} + \mathcal{K} + \mathcal{L}) + \int [\varepsilon(3\Theta'^2 + \boldsymbol{D}' : \boldsymbol{D}') - \boldsymbol{V}' : \boldsymbol{D}' + \boldsymbol{q}' \cdot \operatorname{grad} T'].$$

On the other hand,

$$\begin{aligned} \boldsymbol{V}' : \boldsymbol{D}' &= (\lambda\Theta' + \lambda'\tilde{\Theta})\Theta' + (2\mu \boldsymbol{D}' + 2\mu'\tilde{\boldsymbol{D}}) : \boldsymbol{D}' \\ &= \lambda\Theta'^2 + 2\mu \boldsymbol{D}' : \boldsymbol{D}' + R_1, \end{aligned}$$

where $|R_1| \leqq N(\varrho'^2 + T'^2) + \varepsilon(\Theta'^2 + \boldsymbol{D}' : \boldsymbol{D}')$. Similarly,

$$\begin{aligned} \boldsymbol{q}' \cdot \operatorname{grad} T' &= (-\varkappa \operatorname{grad} T' - \varkappa' \operatorname{grad} \tilde{T}) \cdot \operatorname{grad} T' \\ &= -\varkappa |\operatorname{grad} T'|^2 + R_2, \end{aligned}$$

where $R_2 \leqq N(\varrho'^2 + T'^2) + \varepsilon |\operatorname{grad} T'|^2$. It is evident then (by the argument used near the end of the proof of Theorem 1) that if $\varepsilon$ is chosen small enough,

$$\varepsilon(3\Theta'^2 + \boldsymbol{D}' : \boldsymbol{D}') - \boldsymbol{V}' : \boldsymbol{D}' + \boldsymbol{q}' \cdot \operatorname{grad} T' \leqq N(\varrho'^2 + T'^2).$$

Thus we have

$$\frac{d}{dt}(\mathcal{J} + \mathcal{K} + \mathcal{L}) \leqq N(\mathcal{J} + \mathcal{K} + \mathcal{L}),$$

and the theorem follows at once.

Note: The above research was supported in part by the United States Air Force Office of Scientific Research under Contract No. AF 49 (638)-262.

[6] Water, being an incompressible fluid insofar as its usual hydrodynamic properties are concerned, does not properly come within the scope of this paper. Nevertheless, by slight modifications Theorem 5 can be shown to hold also for incompressible fluids.

## References

[1] COURANT, R., & K. FRIEDRICHS: Supersonic Flow and Shock Waves. New York: Interscience 1948.

[2] COURANT, R., & D. HILBERT: Methoden der Mathematischen Physik, Bd. 2. Berlin: Springer 1936.

[3] DOLIDZE, D. E.: Doklady Akademii Nauk SSSR. **96**, 437—439 (1954).

[4] FOÁ, E.: Sull'impiego dell'analisi dimensionale nello studio del moto turbolento. L'industria (Milano) **43**, 426—429 (1929).

[5] FRIEDRICHS, K.: Symmetric hyperbolic linear differential equations. Communications on Pure and Applied Math. **7**, 345—392 (1954).

[6] FRIEDRICHS, K., & H. LEWY: Über die Eindeutigkeit und das Abhängigkeitsgebiet der Lösungen beim Anfangswertproblem linearer hyperbolischer Differentialgleichungen. Math. Annalen **98**, 192—204 (1927).

[7] GRAFFI, D.: Il teorema di unicità nella dinamica dei fluidi compressibili. J. Rational Mechanics and Analysis **2**, 99—106 (1953).

[8] GRAFFI, D.: Il teorema di unicità per i fluidi incompressibili, perfetti, eterogenei. Revista de la Unión Matemática Argentina **17**, 73—77 (1956).

[9] HADAMARD, J.: Sur l'intégrale residuelle. Bull. Société Mathématique de France **28**, 69—90 (1900).

[10] KISELEV, A. A., & O. A. LADYSHENSKAYA: On the existence and uniqueness of solutions of the initial value problem for viscous incompressible fluids. Izvestia Akademii Nauk SSSR. **21**, 655—680 (1957).

[11] LAX, P.: Hyperbolic systems of conservation laws II. Communications on Pure and Applied Math. **10**, 537—566 (1957).

[12] MISES, R. v.: Mathematical theory of compressible fluid flow. New York: Academic Press 1958.

[13] ORR, W. McF.: The stability or instability of the steady motions of a liquid. Part II. A viscous liquid. Proc. Royal Irish Academy (A) **27**, 69—138 (1907).

[14] REYNOLDS, O.: On the dynamical theory of incompressible viscous fluids and the determination of the criterion. Philosophical Transactions Royal Society of London (A) **186**, 123—164 (1895).

[15] SERRIN, J.: Mathematical principles of classical fluid mechanics. Handbuch der Physik, Bd. 8. Berlin-Göttingen-Heidelberg: Springer. In press.

[16] TRUESDELL, C.: The Kinematics of Vorticity. Indiana Univ. Press 1954.

[17] ZAREMBA, S.: Sopra un teorema d'unicita relativo alla equazione delle onde spheriche. Rendiconti, Accademia Nazionale dei Lincei (5) **24**, 904—908 (1915).

University of Minnesota
Minneapolis, Minnesota

*(Received March 6, 1959)*

James Serrin, Patrizia Pucci and Lucio Boccardo, Tempio di Sant'Angelo, Perugia, Italy, 1994.

Bert Peletier and James Serrin, Minneapolis, MN, 2000.

# VI. Thermodynamics

# VI.1. Thermodynamic Theory

Roger Fosdick

**Mathematics Subject Classification (2010).** 49XX.

**Keywords.** Commentary, continuum Mechanics, thermodynamics, invariance, objectivity.

James Serrin has expressed through his writings over many years a strong interest in the fundamentals and foundations of continuum mechanics. Continuum mechanics is a thread woven throughout his papers since his earliest works in the 1950s on free-boundaries and jets with David Gilbarg. An engagement with continuum thermodynamics, the mathematical science of its foundation and development, was a natural extension within the mechanics setting that combined his keen mathematical skills with his strong appreciation for physics and his desire to form a foundational link between the mathematical and physical sciences. This commentary concerns a contribution that has enhanced our understanding and teaching of continuum thermodynamics.

The paper [i] below considers a deformable heat-conducting body $\mathcal{B}$, immersed in an environment or *reservoir* and subject to loading and heating due to interaction with the reservoir. A thermodynamic process is envisioned to take place during an interval of time, and the interest here is in describing the relationship between continuum thermodynamics for ponderable bodies and well-known classical statements regarding such fundamental concepts as efficiency, cycles, and the second law of thermodynamics due to Carnot, Clausius, Kelvin, and Planck. The Clausius–Duhem inequality, which is a fundamental element of continuum thermodynamics, is shown to be intimately related to the classical expressions of the second law of thermodynamics.

During a time interval $I := [t_0, t_1] \subset \mathbb{R}^+$, a body, which occupies the configuration $\mathcal{B}_t$, is immersed in a reservoir $\mathcal{R}$ which maintains the absolute temperature $\tau(t) > 0$. The first law and the second law of thermodynamics for the body $\mathcal{B}$ in the form of the *balance of energy* and the *Clausius–Duhem inequality* are assumed to hold for the thermodynamic process that takes place. Thus,

$$\frac{d}{dt}E(\mathcal{B}_t) + \frac{d}{dt}K(\mathcal{B}_t) = P(\mathcal{B}_t) + Q(\mathcal{B}_t), \quad \frac{d}{dt}H(\mathcal{B}_t) \geq S(\mathcal{B}_t) \quad \forall t \in I, \tag{1}$$

where $E, K, P, Q, H$ and $S$ denote, respectively, the *internal energy*, *kinetic energy*, *power supply* (*to* $\mathcal{B}_t$ *if positive*), *heating supply* (*to* $\mathcal{B}_t$ *if positive*), *internal entropy* and *entropy flux* for $\mathcal{B}_t$. The *work done by* (*if positive*) *the body* $\mathcal{B}$ during $I$ and the *efficiency* of the

P. Pucci et al. (eds.), *James Serrin. Selected Papers: Volume 2*, Contemporary Mathematicians,
DOI 10.1007/978-3-0348-0847-7_14, © Springer Basel 2014

thermodynamic process are given by

$$W := \int_I -P(\mathcal{B}_t)\,dt, \quad \text{eff} := \frac{W}{C^+}, \tag{2}$$

where $C^+ := \int_{I^+} Q(\mathcal{B}_t)dt$, $I^+ := \{t \in I : Q(\mathcal{B}_t) > 0\}$.

In the early developments of thermodynamics, inequalities were most often expressed in words rather than mathematical expressions. In this paper, one that can be traced back to Clausius (1850) and Maxwell (1872) is converted into a mathematical inequality which gives rise to the *global heat transfer inequality*

$$S(\mathcal{B}_t) \geq \frac{Q(\mathcal{B}_t)}{\tau(t)} \quad \forall t \in I. \tag{3}$$

This paper then goes on to develop the elementary efficiency estimates and to prove the classical verbal statements of the second law of thermodynamics due to Carnot, Clausius, Kelvin and Planck for any *global thermodynamic process*, i.e., for any set of functions

$$\Pi(t) := \{E(\mathcal{B}_t), K(\mathcal{B}_t), P(\mathcal{B}_t), Q(\mathcal{B}_t), H(\mathcal{B}_t), S(\mathcal{B}_t), \tau(t)\}, \quad t \in I$$

that satisfy the balance of energy $(1)_1$, the Clausius–Duhem inequality $(1)_2$ and the global heat transfer inequality (3).[1] Such a process is envisioned to occupy a path $\{E,\ K,\ H,\ \tau\}(t)$, $t \in I$ in a four-dimensional *state space* and at each time $t$ the state is subject to the *actions* $\{P,\ Q,\ S\}(t)$. The *reversal* of a process is simply the path traced in reverse direction with actions of the opposite sign and it may not satisfy the three requirements for a global thermodynamic process. However, $\Pi(t)$ is said to be a *reversible global thermodynamic process* if the reversal is indeed a process. In that case, it follows that equality holds in all three of (1) and (3), and, in particular

$$\frac{d}{dt}H(\mathcal{B}_t) = S(\mathcal{B}_t) = \frac{Q(\mathcal{B}_t)}{\tau(t)} \quad \forall t \in I. \tag{4}$$

$\Pi(t)$ is *cyclic* during $I$ if it begins and ends at the same state. Classical elementary thermodynamics is restricted to homogeneous processes wherein the relevant fields are uniform in space throughout the body at each instant of time. This work broadly generalizes the classical viewpoint.

A major theorem of this work shows that *if $\Pi(t)$ is a cyclic global thermodynamic process during $I$ for which $\tau_{\min}$ and $\tau_{\max}$ correspond, respectively, to the minimum and maximum reservoir temperatures reached during $I$, then*

$$\text{eff} \leq 1 - \frac{\tau_{\min}}{\tau_{\max}},$$

*and equality holds if and only if $\Pi(t)$ is a reversible, cyclic Carnot process.* This theorem emphasizes the importance of the ideal *Carnot process* in thermodynamics and it may be considered to be at the root of the definition of such a process.

This paper establishes a complete framework in which various classical statements of the second law of thermodynamics are proved for global thermodynamic processes.

Finally, this paper closes with a 'curious generalization' that is repeated in a constructive way in the later works of Serrin that are dedicated to the foundations of thermodynamics. In the present paper, it is observed that *within a theory of thermodynamics*

[1] For convenience, and with little compromise, this definition of a global thermodynamic process is slightly more compact than that used in the original paper.

*for which the balance of energy* $(1)_1$ *is replaced by the energy inequality*

$$\frac{d}{dt}E(\mathcal{B}_t) + \frac{d}{dt}K(\mathcal{B}_t) \leq P(\mathcal{B}_t) + Q(\mathcal{B}_t), \tag{5}$$

*all the classical statements of the second law continue to be consequences of the fundamental equations.* It is remarked that a thermodynamic theory based on the imbalance (5) would in many ways be no different than one based on the balance of energy $(1)_1$. This insightful idea is pursued by Serrin [1] and Šilhavý [2] in later important publications on the subject.

## References

[1] Serrin, J., An outline of thermodynamical structure. pp. 3–32 in "New Perspectives in Thermodynamics". J. Serrin (ed.) Springer, Berlin, Heidelberg, New York, 1986.

[2] Šilhavý, M., On measures, convex cones, and foundations of thermodynamics. I. Systems with vector-valued actions; II. Thermodynamic systems. *Czech. J. Phys.* **B30**, 841–861, 961–991 (1980).

Roger Fosdick
Department of Aerospace Engineering and Mechanics
University of Minnesota
Minneapolis, MN 55455, USA
e-mail: fosdick@aem.umn.edu

*Arch. Ration. Mech. Anal.* 59 (1975), 97–109.


**Reprint VI.1.[i]**

# *Global Properties of Continuum Thermodynamic Processes*

ROGER L. FOSDICK & JAMES SERRIN

**Contents**



## Introduction

This paper is concerned with the physical situation of a deformable heat conducting body immersed in an environment (or reservoir) having uniform, but not necessarily constant, temperature, and otherwise subject to loading and heating.

Specifically, our interest is centered on the relation between rational *continuum* thermodynamics, which accepts the Clausius-Duhem inequality as a primitive assertion, and the well-known classical statements of the second law of thermodynamics due to CARNOT, CLAUSIUS, KELVIN and PLANCK. The classical versions of the second law are global in character in that they refer to cyclic processes of bodies; correspondingly our attention will, for the most part, similarly be directed to such cyclic situations.

The global structure of continuum thermodynamics is outlined in the first two sections of this work. Section 1 refers mainly to the balance of energy and to the related notion of efficiency. Two fundamental identities for the efficiency are recorded which will form the basis for several theorems given later on. In Section 2 we discuss the crucial inequalities of rational thermodynamics which restrict the interaction of a body and its environment, and which are of fundamental importance to everything which follows. First we introduce two heat transfer inequalities which we take to be the expression of what PLANCK and others refer to as the "self-evident" condition that heat does not flow directly from a cold to a hot body. Then we introduce the global Clausius-Duhem inequality and finally conclude with some remarks on reversible and irreversible processes.

In Section 3 we introduce the idea of a Carnot process. Here the environmental temperature $\tau$ plays an important role since the body temperature cannot be considered uniform throughout the material. (A similar motivation for the use of environmental temperature was indirectly in the minds of some of the early writers on thermodynamics; not having available any continuum structure within which to discuss material bodies, however, they were generally unable to clarify their ideas.)

A main efficiency estimate for any global thermodynamic process is given in Section 4. In particular this estimate shows that for a cyclic process in which the total internal energy, the total kinetic energy and the total internal entropy undergo zero net change the efficiency is never greater than $1-\tau_{\min}/\tau_{\max}$; moreover this value is assumed when and only when a quasi-reversible Carnot process has been performed. We also show that, for a certain restricted class of Carnot processes over a time interval $\mathscr{I}$, if our main efficiency estimate is assumed in place of the Clausius-Duhem inequality, then the integrated form of the Clausius-Duhem inequality over $\mathscr{I}$ must follow as a consequence.

In Section 5 we show that the usual structure of continuum thermodynamics, as introduced in Sections 1 and 2, leads to the following classical versions of the second law of thermodynamics:

(i) (CARNOT-KELVIN-PLANCK). It is impossible for a body to undergo a cyclic process during which a net positive amount of work is performed but no heat is emitted.

(ii) (CARNOT-CLAUSIUS). It is impossible for a body to undergo a cyclic process during which heat is absorbed but none is emitted.

(iii) (KELVIN). It is impossible for a body to undergo a cyclic process during which a net positive amount of work is performed, a net amount of heat is absorbed from an environment at some given temperature, heat is absorbed and emitted at other fixed environmental temperatures (the net heat transfer, however, being zero at any of these other temperatures) and which otherwise proceeds adiabatically.

(iv) (CLAUSIUS). It is impossible for a body to undergo a cyclic process during which a net positive amount of work is performed, heat is absorbed from an environment which is colder than a given temperature, heat is emitted to an environment which is hotter than that given temperature, and which otherwise proceeds adiabatically.

Stated in other terms, our proof shows that the above statements are necessary consequences of the structure of rational continuum thermodynamics. That is, the Clausius-Duhem inequality turns out to be *consistent* with these classical statements, a fact which seems not to have been noted previously. This observation clearly adds a further degree of credibility to the Clausius-Duhem inequality beyond its degree of successful application in other directions of continuum mechanics.

In Section 6 we discuss the rarity of reversible processes in continuum thermodynamics and indicate that for cyclic processes the classical maximum efficiency value $1-\tau_{\min}/\tau_{\max}$ is essentially never achieved. While it is possible to obtain a sharper estimate of the maximum efficiency for certain materials, this estimate will depend upon particular constitutive assumptions; we shall not pursue this matter further here.

In the classical thermodynamics of homogeneous phases the temperature field is required to be uniform over the body. In Section 7 we briefly consider this special situation in relation to our earlier conclusions. In this case the efficiency estimates of Section 4 reduce to those obtained earlier by TRUESDELL [10] using a rather different approach.

In the last section of this work we observe as a curious generalization that the balance of energy is not essential for any of our conclusions to hold, but that we merely need to postulate an energy *inequality*.

## 1. Thermodynamic Preliminaries; Efficiency

Consider a body $\mathscr{B}$ consisting of particles $X$ and endowed with a continuous mass measure $m$. We let $\mathscr{B}_t$ denote the configuration of $\mathscr{B}$ at time $t$ and use $\partial\mathscr{B}_t$ to denote the boundary of this configuration; $\boldsymbol{n}$ is the outer unit normal on $\partial\mathscr{B}_t$.

We assume throughout that $\mathscr{B}$ is immersed in an environment or heat reservoir $\mathscr{R}$ which supports a spatially homogeneous but time dependent absolute temperature $\tau=\tau(t)>0$. We shall be concerned with various manifestations of the second law of thermodynamics, including the efficiency of piecewise differentiable processes undergone by $\mathscr{B}$ during a time interval $\mathscr{I}=[t_1,t_2]$.

The global balance of energy for $\mathscr{B}$ requires that the total internal energy $E(\mathscr{B}_t)$, kinetic energy $K(\mathscr{B}_t)$, rate of mechanical working $W(\mathscr{B}_t)$ on $\mathscr{B}_t$, and rate of heat working $Q(\mathscr{B}_t)$ on $\mathscr{B}_t$, satisfy the relation

$$\dot{E}(\mathscr{B}_t)+\dot{K}(\mathscr{B}_t)=W(\mathscr{B}_t)+Q(\mathscr{B}_t), \tag{1}$$

where superposed dots denote time differentiation. We let $e(X,t)$, $\boldsymbol{v}(X,t)$, $\boldsymbol{t}(X,t)$, $\boldsymbol{b}(X,t)$, $\boldsymbol{q}(X,t)$, and $r(X,t)$ denote, respectively, the internal energy per unit mass, the velocity, the traction on $\partial\mathscr{B}_t$, the body force density per unit mass, the heat flux on $\partial\mathscr{B}_t$, and the body heat supply per unit mass passing from $\mathscr{R}$ to $\mathscr{B}_t$. The four energetic quantities introduced in (1) have the usual representations

$$\begin{aligned}
E(\mathscr{B}_t)&=\int_{\mathscr{B}} e\,dm, \qquad K(\mathscr{B}_t)=\tfrac{1}{2}\int_{\mathscr{B}} |\boldsymbol{v}|^2\,dm,\\
W(\mathscr{B}_t)&=\int_{\mathscr{B}} \boldsymbol{b}\cdot\boldsymbol{v}\,dm+\int_{\partial\mathscr{B}_t} \boldsymbol{t}\cdot\boldsymbol{v}\,da,\\
Q(\mathscr{B}_t)&=\int_{\mathscr{B}} r\,dm-\int_{\partial\mathscr{B}_t} \boldsymbol{q}\cdot\boldsymbol{n}\,da.
\end{aligned} \tag{2}$$

For the most part we shall need only the representation $(2)_4$ for $Q(\mathscr{B}_t)$. In fact, it is not necessary to introduce the remaining fields mentioned in (2) until Section 6, where we discuss the paucity of reversible processes.

The body $\mathscr{B}$ is said to *absorb* heat or *emit* heat at time $t$ according to whether $Q(\mathscr{B}_t)>0$ or $Q(\mathscr{B}_t)<0$. We denote by $\mathscr{I}^+\subset\mathscr{I}$ and $\mathscr{I}^-\subset\mathscr{I}$ those intervals of time in $\mathscr{I}$ during which heat is absorbed and emitted, respectively, and write

$$C^+\equiv\int_{\mathscr{I}^+} Q(\mathscr{B}_t)\,dt, \qquad C^-\equiv-\int_{\mathscr{I}^-} Q(\mathscr{B}_t)\,dt$$

for the corresponding total heat absorbed and emitted. Clearly, when $\mathscr{I}^+$ and $\mathscr{I}^-$ are non-null we have $C^+>0$ and $C^->0$. *We shall assume throughout this work that* $\mathscr{I}^+$ *is non-null.* During the remainder of the time interval, clearly $Q(\mathscr{B}_t)=0$. At these moments the body is proceeding *adiabatically*.

The net work done *by* $\mathscr{B}$ during the time interval $\mathscr{I}$ is given by

$$U=-\int_{\mathscr{I}} W(\mathscr{B}_t)\,dt, \tag{4}$$

and the efficiency over the time interval $\mathscr{I}$ is then

$$\mathrm{Eff}=\frac{U}{C^+}. \tag{5}$$

With the aid of (1), (3) and (4) this formula can be rewritten in this form:

$$\mathrm{Eff}=1-\frac{C^-}{C^+}-\frac{\Delta E+\Delta K}{C^+}, \tag{6}$$

where the prefix $\Delta$ denotes the difference operation on functions over the interval $\mathscr{I}$.

While the absolute temperature of the heat reservoir $\tau(t)$ is spatially homogeneous, the absolute temperature of the body $\theta(X,t)$ may depend on the particular particle as well as on time. We understand $\tau_{\max}$ and $\tau_{\min}$ to be the supremum and infimum of $\tau(t)$ over $\mathscr{I}$.

As a last primitive notion, we introduce the total internal entropy $H(\mathscr{B}_t)$, with its density per unit mass denoted by $\eta(X,t)$; thus

$$H(\mathscr{B}_t)=\int_{\mathscr{B}}\eta\,dm. \tag{7}$$

The following two main efficiency formulae are easily checked to be equivalent to (6), and are, in fact, valid for *arbitrary* functions $\tau(t)>0$, $\theta(X,t)>0$, and $H(\mathscr{B}_t)$:

$$\begin{aligned}\mathrm{Eff}=1&-\frac{\tau_{\min}}{\tau_{\max}}-\frac{\Delta E+\Delta K-\tau_{\min}\,\Delta H}{C^+}\\&-\frac{\tau_{\min}}{C^+}\left\{\int_{\mathscr{I}^+}\left(\frac{1}{\tau}-\frac{1}{\tau_{\max}}\right)Q(\mathscr{B}_t)\,dt+\int_{\mathscr{I}^-}\left(\frac{1}{\tau_{\min}}-\frac{1}{\tau}\right)(-Q(\mathscr{B}_t))\,dt\right.\\&+\int_{\mathscr{I}}\left[\dot H(\mathscr{B}_t)+\left(\int_{\partial\mathscr{B}_t}\frac{\boldsymbol{q}\cdot\boldsymbol{n}}{\theta}\,da-\int_{\mathscr{B}}\frac{r}{\theta}\,dm\right)\right]dt\\&\left.+\int_{\mathscr{I}}\left[\int_{\partial\mathscr{B}_t}\left(\frac{1}{\tau}-\frac{1}{\theta}\right)\boldsymbol{q}\cdot\boldsymbol{n}\,da-\int_{\mathscr{B}}\left(\frac{1}{\tau}-\frac{1}{\theta}\right)r\,dm\right]dt\right\},\end{aligned} \tag{8a}$$

and (with $\tau^*$ any non-zero constant)

$$\begin{aligned}\mathrm{Eff}=&-\frac{\Delta E+\Delta K-\tau^*\,\Delta H}{C^+}\\&-\frac{\tau^*}{C^+}\left\{\int_{\mathscr{I}^+}\left(\frac{1}{\tau}-\frac{1}{\tau^*}\right)Q(\mathscr{B}_t)\,dt+\int_{\mathscr{I}^-}\left(\frac{1}{\tau^*}-\frac{1}{\tau}\right)(-Q(\mathscr{B}_t))\,dt\right.\\&+\int_{\mathscr{I}}\left[\dot H(\mathscr{B}_t)+\left(\int_{\partial\mathscr{B}_t}\frac{\boldsymbol{q}\cdot\boldsymbol{n}}{\theta}\,da-\int_{\mathscr{B}}\frac{r}{\theta}\,dm\right)\right]dt\\&\left.+\int_{\mathscr{I}}\left[\int_{\partial\mathscr{B}_t}\left(\frac{1}{\tau}-\frac{1}{\theta}\right)\boldsymbol{q}\cdot\boldsymbol{n}\,da-\int_{\mathscr{B}}\left(\frac{1}{\tau}-\frac{1}{\theta}\right)r\,dm\right]dt\right\}.\end{aligned} \tag{8b}$$

## 2. Fundamental Thermodynamic Inequalities; Reversibility

Inequalities which are sometimes tacit and often considered self-evident in the thermodynamics literature concern the conduction and radiation of heat between the body $\mathscr{B}$ and the surrounding heat reservoir $\mathscr{R}$. MAXWELL [3, p. 154] writes that "heat ... never passes out of a body except when it flows by conduction or radiation into a colder body". PLANCK [4, p. 85] has a similar statement, and he also remarks that CLAUSIUS[1] repeatedly stressed this position so as to prevent misunderstanding of his fundamental principle.[2] Apropos these remarks we record the following *heat transfer inequalities*, which we take to be the mathematical expression of the ideas just described:

$$\begin{aligned} (\theta-\tau)\,\boldsymbol{q}\cdot\boldsymbol{n} &\geqq 0 \quad \text{on } \partial\mathscr{B}_t, \\ (\theta-\tau)\,r &\leqq 0 \quad \text{on } \mathscr{B}_t. \end{aligned} \tag{9}$$

Since $\boldsymbol{n}$ is the outer unit normal on $\partial\mathscr{B}_t$, these inequalities give pointwise expression to the idea that heat transfer cannot occur $\left\{\begin{matrix}\text{from}\\ \text{to}\end{matrix}\right\}$ $\mathscr{B}$ $\left\{\begin{matrix}\text{to}\\ \text{from}\end{matrix}\right\}$ a heat reservoir $\mathscr{R}$ which is at a $\left\{\begin{matrix}\text{higher}\\ \text{lower}\end{matrix}\right\}$ temperature. Although $\boldsymbol{q}\cdot\boldsymbol{n}\equiv 0$ and $r\equiv 0$ satisfy (9), our earlier assumption that $\mathscr{I}^+$ is non-null requires the body to absorb *some* heat during the time interval $\mathscr{I}$. Thus $\boldsymbol{q}\cdot\boldsymbol{n}$ and $r$ cannot both be identically zero during $\mathscr{I}$.

Clearly the heat transfer inequalities (9) imply

$$\int_{\partial\mathscr{B}_t}\left(\frac{1}{\tau}-\frac{1}{\theta}\right)\boldsymbol{q}\cdot\boldsymbol{n}\,da-\int_{\mathscr{B}}\left(\frac{1}{\tau}-\frac{1}{\theta}\right)r\,dm\geqq 0, \tag{10}$$

and this shows that the last integral inside the brackets in (8) is non-negative.

We define the entropy flux from $\mathscr{R}$ into $\mathscr{B}_t$ by

$$S(\mathscr{B}_t)=\int_{\mathscr{B}}\frac{r}{\theta}\,dm-\int_{\partial\mathscr{B}_t}\frac{\boldsymbol{q}\cdot\boldsymbol{n}}{\theta}\,da. \tag{11}$$

In a sense, this quantity can be thought of as the rate at which energy per unit temperature is introduced into $\mathscr{B}_t$ by means of heat conduction and radiation. The global *Clausius-Duhem inequality* then asserts that

$$\dot{H}(\mathscr{B}_t)\geqq S(\mathscr{B}_t). \tag{12}$$

This relation is the second of the crucial inequalities of rational continuum thermodynamics, and because of it the third integral inside the brackets in (8) is non-negative.

Inequality (12) applies globally to $\mathscr{B}_t$, not to the constituent parts of $\mathscr{B}$. As an axiom of rational thermodynamics it is usually justified as a generalization of the classical conclusions of the thermodynamics of elementary material bodies (see, for example, [5], [8], [9]).

[1] See *e.g.* [2, p. 11].

[2] *I.e.*, heat cannot pass from a cold to a hot body without compensation.

In order to qualify as a *global* thermodynamic *process* for $\mathscr{B}$ during the time interval $\mathscr{I}=[t_1, t_2]$ the eight quantities

$$\{E(\mathscr{B}_t), H(\mathscr{B}_t), K(\mathscr{B}_t), W(\mathscr{B}_t), \boldsymbol{q}(X,t), r(X,t), \theta(X,t), \tau(t)\}$$

are required to satisfy the balance of energy (1), the heat transfer inequalities (9), and the Clausius-Duhem inequality (12). The thermomechanic *path* of a process for $\mathscr{B}$ is the location-temperature versus time graph of all of its particles during $\mathscr{I}$; its reversal is constructed by the mapping $t \to t_1+t_2-t$. While in general a path reversal is associated with an entirely independent global thermodynamic process, if this so-called "reversal process" is of the related form

$$\{E(\mathscr{B}_t), H(\mathscr{B}_t), K(\mathscr{B}_t), -W(\mathscr{B}_t), -\boldsymbol{q}(X,t), -r(X,t), \theta(X,t), \tau(t)\}$$

then the original process is said to be *reversible*. Obviously in this case equality holds in both (9) and (12) with the exception of a finite set of times in $\mathscr{I}$; otherwise $\mathscr{B}$ undergoes an *irreversible process*. In particular by (9), *during a reversible process* $\theta(X,t)$ *must coincide with* $\tau(t)$ *at those times in* $\mathscr{I}$ *and at all points where* $\boldsymbol{q}\cdot\boldsymbol{n}$ *and* $r$ *are non-null*. Moreover, using this conclusion, (12) with equality, (11), and $(2)_4$, we obtain for a reversible process

$$\dot{H}(\mathscr{B}_t)=\frac{Q(\mathscr{B}_t)}{\tau}. \tag{13}$$

Conversely, if we assume (13), then with the aid of (11), the relation (12) reduces to (10) with the inequality reversed. This together with (10), which also holds, shows that in fact (10) must be true with equality. The inequalities in (9) are then forced to hold as equalities. Thus if (13) holds, then also $\theta=\tau$ at all places where $\boldsymbol{q}\cdot\boldsymbol{n}\neq 0$ or $r\neq 0$, the same conclusion noted above for reversible processes.

A process in which (13) holds will be called *quasi-reversible*. Note that a reversible process is necessarily quasi-reversible, but that a quasi-reversible process need *not* be reversible.

## 3. Carnot Processes

As in the classical thermodynamics of homogeneous phases, here also we say that the body $\mathscr{B}$ undergoes a *Carnot process* during the time interval $\mathscr{I}$ if the heat rate $Q(\mathscr{B}_t)$ and the reservoir[3] temperature $\tau(t)$ are arranged so that for all but a finite set of $t\in\mathscr{I}$ only the following three alternatives are possible:

(i) $\mathscr{B}$ absorbs heat at a maximum reservoir temperature (more specifically, $Q(\mathscr{B}_t)>0$ when $\tau(t)=\tau_{\max}$),

(ii) $\mathscr{B}$ emits heat at a minimum reservoir temperature (more specifically, $Q(\mathscr{B}_t)<0$ when $\tau(t)=\tau_{\min}$),

(iii) $\mathscr{B}$ proceeds adiabatically (*i.e.*, $Q(\mathscr{B}_t)=0$).

[3] Use of the reservoir temperature in defining a CARNOT process is deep rooted in the development of thermodynamics [1, 2]. It is of particular advantage in the present context of continuum thermodynamics since here $\mathscr{B}$ is not necessarily a homogeneous phase and thus generally does not have a single spatially independent temperature.

We note that heat can be transmitted into or out of $\mathscr{B}$ both by conduction and radiation; it is possible that both of these methods of heat transfer take place simultaneously even during the adiabatic portion of a Carnot process.

For a quasi-reversible Carnot process, let $\mathscr{C}_t$ denote the set of points of $\mathscr{B}_t \cup \partial\mathscr{B}_t$ where $r \neq 0$ or $\boldsymbol{q} \cdot \boldsymbol{n} \neq 0$. Then from the remarks at the end of Section 2 it follows that during the time interval $\mathscr{I}^+$ when $\mathscr{B}$ absorbs heat $\theta(X, t)$ must coincide with $\tau_{\max}$ at all points of $\mathscr{C}_t$ and similarly that during the time interval $\mathscr{I}^-$ when $\mathscr{B}$ emits heat $\theta(X, t)$ must coincide with $\tau_{\min}$ on $\mathscr{C}_t$; in the remaining time interval $\mathscr{I} - (\mathscr{I}^+ \cup \mathscr{I}^-)$, when $\mathscr{B}$ proceeds adiabatically, $\theta(X, t)$ and $\tau(t)$ must coincide on $\mathscr{C}_t$.

## 4. Estimates of Efficiency

From the efficiency formula (8a), the definition of $\mathscr{I}^+$ and $\mathscr{I}^-$, and the thermodynamic inequalities (9) and (12), it readily follows that

$$\mathrm{Eff} \leqq 1 - \frac{\tau_{\min}}{\tau_{\max}} - \frac{\Delta E + \Delta K - \tau_{\min} \Delta H}{C^+}, \tag{14}$$

where *equality holds if and only if $\mathscr{B}$ undergoes a quasi-reversible Carnot process.*

Moreover, if

$$\Delta E + \Delta K - \tau_{\min} \Delta H \geqq 0, \tag{15}$$

then

$$\mathrm{Eff} \leqq 1 - \frac{\tau_{\min}}{\tau_{\max}}, \tag{16}$$

where *equality holds if and only if $\mathscr{B}$ undergoes a quasi-reversible Carnot process for which $\Delta E + \Delta K - \tau_{\min} \Delta H = 0$.*

It is of interest to observe that if the efficiency estimate (14) is postulated as a thermodynamic inequality, then for any Carnot process of $\mathscr{B}$ we must have

$$\Delta H \geqq \int_{\mathscr{I}} \frac{Q(\mathscr{B}_t)}{\tau}\, dt$$

which follows directly from (8a) and $(2)_4$, independently of the heat transfer inequalities (9) and the Clausius-Duhem inequality (12). If this Carnot process is, in addition, executed so that the body and its environment undergo a perfect global heat exchange in the sense that equality holds in (10) during $\mathscr{I}$, then we must have

$$\Delta H \geqq \int_{\mathscr{I}} S(\mathscr{B}_t)\, dt$$

which is the integrated form of the Clausius-Duhem inequality.

We wish to thank Professor TRUESDELL for pointing out an error in our original statement of this result and for noting the appropriate conclusion to be drawn from the hypothesis (14).

## 5. Classical Statements of the Second Law

In classical thermodynamics the various accepted, but not necessarily equivalent, versions of the so-called second law invariably are phrased in terms of a *cyclic process;* we follow most classical interpretations and by such a process for $\mathscr{B}$ understand one for which $\Delta E = \Delta H = \Delta K = 0$, *i.e.*, $E$, $H$ and $K$ return at the end of the process to their initial values. One popular description of the second law of thermodynamics, which is essentially due to CARNOT, KELVIN, and PLANCK, states that *it is impossible for a body to undergo a cyclic process*[4] *in which* $U > 0$ *but in which no heat is emitted.* To see that this follows from our hypotheses, suppose that $\mathscr{I}^-$ is null so that $C^- = 0$. Then (6) shows that for a cyclic process the efficiency is 1. Hence by (8a) and (5) it follows that

$$0 = \frac{U}{\tau_{\max}} + \int_{\mathscr{I}^+} \left(\frac{1}{\tau} - \frac{1}{\tau_{\max}}\right) Q(\mathscr{B}_t)\, dt + \int_{\mathscr{I}} \left[\dot{H}(\mathscr{B}_t) + \left(\int_{\partial \mathscr{B}_t} \frac{\boldsymbol{q}\cdot\boldsymbol{n}}{\theta}\, da - \int_{\mathscr{B}} \frac{r}{\theta}\, dm\right)\right] dt + \int_{\mathscr{I}} \left[\int_{\partial \mathscr{B}_t} \left(\frac{1}{\tau} - \frac{1}{\theta}\right) \boldsymbol{q}\cdot\boldsymbol{n}\, da - \int_{\mathscr{B}} \left(\frac{1}{\tau} - \frac{1}{\theta}\right) r\, dm\right] dt.$$

However, the fact that this conclusion cannot hold is immediate from the definition of $\mathscr{I}^+$, (10), (11), (12), and the hypothesis that $U > 0$. In a similar fashion we have the following slightly different version of the second law of thermodynamics which is associated with CARNOT and CLAUSIUS: *It is impossible for a body to undergo a cyclic process, having absorbed some heat and emitted none.* The essential difference between these two statements is that in the former work is assumed to be accomplished by $\mathscr{B}$ while in the latter heat is assumed to be absorbed. The latter statement remains valid for (non-cyclic) processes which merely satisfy $\Delta H \leqq 0$, as the reader can easily check.

If we now use the efficiency formula (8b) in combination with the thermodynamic inequalities (9) and (12), it follows that for any constant $\tau^* > 0$,

$$\mathrm{Eff} \leqq -\frac{\Delta E + \Delta K - \tau^* \Delta H}{C^+} - \frac{\tau^*}{C^+}\left\{\int_{\mathscr{I}^+} \left(\frac{1}{\tau} - \frac{1}{\tau^*}\right) Q(\mathscr{B}_t)\, dt + \int_{\mathscr{I}^-} \left(\frac{1}{\tau^*} - \frac{1}{\tau}\right)(-Q(\mathscr{B}_t))\, dt\right\}. \tag{17}$$

Now consider a process for $\mathscr{B}$ with the time interval $\mathscr{I}^+ \cup \mathscr{I}^-$ of heat absorption and emission partitioned into $k+1$ sets each having a constant distinct environmental temperature $\tau^*, \tau_1, \tau_2, \ldots, \tau_k$ and such that $\mathscr{B}$ absorbs a (positive) net amount of heat when the environmental temperature is $\tau^*$ while the net heats absorbed and emitted at *each* of the remaining environmental temperatures is zero. Then it follows from (4), (5) and (17) that $U \leqq 0$ provided $\Delta E + \Delta K - \tau^* \Delta H \geqq 0$, and this yields a version of the second law of thermodynamics which usually is attributed to KELVIN: *It is impossible for a body to accomplish work in a cyclic process*[5]*, the body having absorbed a (positive) net heat at the constant environmental temperature* $\tau^*$*, while having absorbed and emitted zero net heat at each of the*

[4] The assumption of a cyclic process is not necessary for the truth of this statement; in fact it is valid for any process which merely satisfies $\Delta E + \Delta K - \tau_{\max} \Delta H \geqq 0$, as one can easily establish.

[5] This statement remains valid if "cyclic process" is replaced by "process with $\Delta E + \Delta K - \tau^* \Delta H \geqq 0$".

*distinct constant environmental temperatures* $\{\tau_1, \tau_2, \ldots, \tau_k\}$, *all different from* $\tau^*$, *and otherwise having proceeded adiabatically.*

Now suppose that $\mathscr{B}$ undergoes a process during which heat is absorbed only when the reservoir temperature is lower than or equal to some fixed number $\tau^* > 0$, say, and emits heat only when the reservoir temperature is higher than or equal to $\tau^*$. Then the two integral terms in (17) are non-negative and again from (4), (5) and (17) we see that $U \leqq 0$ if $\Delta E + \Delta K - \tau^* \Delta H \geqq 0$. Thus we obtain a form of the second law of thermodynamics which is due essentially to CLAUSIUS: *It is impossible for a body to accomplish work in a cyclic process* [5], *the body having absorbed heat from an environment at temperature* $\tau(t) \leqq \tau^*$ *and emitted heat to an environment at temperature* $\tau(t) \geqq \tau^*$, *and otherwise having proceeded adiabatically.*

Finally, we may restate the result of Section 4 in terms of cycles: *the efficiency of a cyclic process conducted between the temperatures* $\tau_{\min}$ *and* $\tau_{\max}$ *does not exceed*

$$1 - \frac{\tau_{\min}}{\tau_{\max}},$$

*and can reach this value only for quasi-reversible Carnot cycles.*

## 6. The Paucity of Reversible Processes

In continuum thermodynamics a *thermodynamic process* is characterized by the following prescribed set of nine sufficiently smooth functions of $(X, t)$ on $\mathscr{B} \times \mathscr{I}$: $\{\boldsymbol{\chi}, \theta, e, \boldsymbol{T}, \eta, \boldsymbol{q}, \boldsymbol{b}, r, \rho\}$. The first, the fourth, and the last members of this set have not been introduced previously in this work. They correspond to the motion, the Cauchy stress tensor (related to $\boldsymbol{t}$ and $\boldsymbol{n}$ through $\boldsymbol{t} = \boldsymbol{T}^T \boldsymbol{n}$), and the mass density, respectively. These nine functions are, of course, not arbitrarily prescribed, but are required to satisfy the local balance of mass [6], the local balance of linear and angular momentum [7], the local balance of energy [8], and the local Clausius-Duhem inequality [9] throughout $\mathscr{B}_t$ for all $t \in \mathscr{I}$. In addition, if $X$ is identified with the points occupied by the particles of $\mathscr{B}$ in the reference configuration $\mathscr{B}_0$, say, at $t = 0$, then $\det \nabla \boldsymbol{\chi}(X, t) > 0$, where $\nabla$ denotes the spatial gradient operation in $\mathscr{B}_0$.

In a theory which is broad enough to include both elastic solid and classical frictional fluid effects [10], constitutive relations for $e$, $\eta$, $\boldsymbol{T}$ and $\boldsymbol{q}$ of the following kind are assumed:

$$\begin{aligned} e &= e(\theta, D\theta, \boldsymbol{F}, D\boldsymbol{v}), \\ \eta &= \eta(\theta, D\theta, \boldsymbol{F}, D\boldsymbol{v}), \\ \boldsymbol{T} &= \boldsymbol{T}(\theta, D\theta, \boldsymbol{F}, D\boldsymbol{v}), \\ \boldsymbol{q} &= \boldsymbol{q}(\theta, D\theta, \boldsymbol{F}, D\boldsymbol{v}), \end{aligned} \tag{18}$$

where $D$ denotes the spatial gradient operator with respect to points which belong to $\mathscr{B}_t$, and where $\boldsymbol{F}$ denotes the deformation gradient tensor $\nabla \boldsymbol{\chi}$. In addition,

[6] See (156.2), (156.5) of [5] or (15.14), (21.18) of [6].
[7] See (205.2), (205.11) of [5] or (16.6), (16.7) of [6].
[8] See (241.4) of [5] restricted to the non-polar case or (70.3) of [6].
[9] Set $\gamma \geqq 0$ in (79.8) of [6] which is equivalent to (96b.00).
[10] See COLEMAN & MIZEL [7].

the reference mass density in $\mathscr{B}_0$ is prescribed. These constitutive assumptions are said to be *compatible with thermodynamics* if for every motion $\chi$ and temperature field $\theta$ they induce a thermodynamic process for some choice of the body force density $\boldsymbol{b}$ and the body heat supply density $r$. Not all forms of the constitutive functions in (18) will be compatible with thermodynamics. COLEMAN & MIZEL [7] have found the following set of necessary and sufficient conditions for such compatibility[11]:

(i) The internal energy density and the entropy density per unit mass are independent of $D\theta$ and $D\boldsymbol{v}$. Moreover the Gibbs relation

$$\dot{e}=\theta\dot{\eta}+\frac{1}{\rho}\boldsymbol{T}_0^T\cdot D\boldsymbol{v}$$

holds, where the *equilibrium stress* $\boldsymbol{T}_0$ is defined by

$$\boldsymbol{T}_0\equiv\boldsymbol{T}(\theta,0,\boldsymbol{F},0).$$

(ii) The constitutive function $\boldsymbol{T}$ is symmetric.

(iii) The constitutive functions $\boldsymbol{T}$ and $\boldsymbol{q}$ obey the *general dissipation inequality*

$$\boldsymbol{T}_e^T\cdot D\boldsymbol{v}-\frac{1}{\theta}\boldsymbol{q}\cdot D\theta\geqq 0$$

for all $\theta>0$, $D\theta$, and $D\boldsymbol{v}$, and for all $\boldsymbol{F}$ with $\det\boldsymbol{F}>0$, where the *extra stress* $\boldsymbol{T}_e$ is defined by

$$\boldsymbol{T}_e\equiv\boldsymbol{T}-\boldsymbol{T}_0.$$

Now from (i) and the local balance of energy and mass we are led to the identity

$$\dot{H}(\mathscr{B}_t)+\left(\int_{\partial\mathscr{B}_t}\frac{\boldsymbol{q}\cdot\boldsymbol{n}}{\theta}da-\int_{\mathscr{B}}\frac{r}{\theta}dm\right)=\int_{\mathscr{B}_t}\frac{1}{\theta}\left(\boldsymbol{T}_e^T\cdot D\boldsymbol{v}-\frac{1}{\theta}\boldsymbol{q}\cdot D\theta\right)dv, \tag{19}$$

the right hand side being non-negative by (iii). Thus for a reversible process (or even for a quasi-reversible process) it is necessary that the right hand side of (19) vanish for all but a possible finite set of times in $\mathscr{I}$.

For general *thermo-elastic materials* the constitutive relations (18) should, by definition, omit any dependence upon the velocity gradient $D\boldsymbol{v}$. In this case the theorem of COLEMAN & MIZEL is valid with $\boldsymbol{T}_e\equiv\boldsymbol{0}$. Then if $\mathscr{B}$ is a *strong conductor* in the sense that there exists a number $\alpha>0$ such that

$$\boldsymbol{q}(\theta,D\theta,\boldsymbol{F})\cdot D\theta\leqq-\alpha\,|D\theta|^2, \tag{20}$$

we see that (iii) is satisfied. Moreover the right hand side of (19) is strictly positive unless $\theta(X,t)$ is uniform throughout $\mathscr{B}_t$, in which case it vanishes. Thus, in addition to the earlier conclusions in Sections 2, 3 and 4 concerning reversible processes, reversible processes for thermo-elastic strongly conductive materials are to be found, if at all, among those which have spatially uniform temperature fields over $\mathscr{B}_t$ for almost all $t\in\mathscr{I}$.

[11] This set is not exactly that of COLEMAN & MIZEL inasmuch as they expressly include the symmetry of the Cauchy stress tensor (*i.e.*, condition (ii)) in their *definition* of a thermodynamic process whereas we have not.

For *compressible linearly viscous* fluids the dependence on the deformation gradient $\boldsymbol{F}$ in the constitutive relations (18) appears only through its determinant which, by the balance of mass, is equivalent to the mass density $\rho$. Moreover the stress response function $\boldsymbol{T}$ is assumed to be independent of $D\theta$. The extra stress $\boldsymbol{T}_e$ then has the isotropic linear form

$$\boldsymbol{T}_e = \lambda \mathbf{1}\,\mathrm{tr}\,\boldsymbol{D} + 2\mu \boldsymbol{D}. \tag{21}$$

Here, the stretching tensor $\boldsymbol{D}$ is the symmetric part of the velocity gradient $D\boldsymbol{v}$, and $\lambda$ and $\mu$ are scalar material properties which may depend upon the temperature and the mass density. Since $\boldsymbol{T}_e$ is independent of $D\theta$ it follows from (iii) that the quadratic form $\boldsymbol{T}_e \cdot \boldsymbol{D}$ associated with (21) is positive semi-definite. However, if we assume that this quadratic form is positive definite, as is customary, then for a strong conductor the right hand side of (19) is either positive or zero, and moreover, vanishes only if the motion of $\mathscr{B}$ is rigid and the temperature field is spatially uniform over $\mathscr{B}_t$. Therefore, reversible processes for compressible strongly conductive linearly viscous fluids are restricted to this very special class.

In general, a process for either a strongly conductive thermo-elastic material or a strongly conductive compressible linearly viscous fluid has neither the temperature field uniform nor the motion rigid over every sub-interval of the time interval $\mathscr{I}$ of the process. Hence, by the remarks in the previous two paragraphs together with (8a), (9) and (19), we see that for such a process for either of these materials the efficiency estimate (14) will apply with *strict* inequality. In fact, this remains so if only the temperature field is non-uniform in the case of the former material, while either a non-uniform temperature field or a non-rigid motion is present in the case of the latter. *A fortiori* in such *cyclic* processes of these materials we see that the classical maximum efficiency $1-\tau_{\min}/\tau_{\max}$ can never be achieved.

Finally for a compressible elastic fluid neither $D\theta$ nor $D\boldsymbol{v}$ occur in the functional relation (18), while $\boldsymbol{F}$ appears only through the density $\rho$. Following the above arguments, we see that both $\boldsymbol{T}_e$ and $\boldsymbol{q}$ are zero. Thus reversible processes are precisely those for which $(9)_2$ holds. In particular, if we set $r\equiv 0$ then all piecewise differentiable processes (*i.e.* those not involving shock waves) are reversible, and the usual formulations of the second law in which $C^+$ is assumed non-vanishing are in operative. This is the case encountered in the subject of classical gas dynamics: In this subject, in fact, the set of solutions is invariant under time reversal.

## 7. Spatially Uniform Temperature

When the body temperature field $\theta(X,t)$ is spatially uniform over the configurations $\mathscr{B}_t$ during $\mathscr{I}$, as is assumed in the classical thermodynamics of homogeneous phases, or, as is the case for "most" reversible processes, then the last two $\mathscr{I}$-integrals inside the brackets in the efficiency formulae (8a, b) reduce to the form

$$\int_{\mathscr{I}} \left[\dot{H}(\mathscr{B}_t) - \frac{Q(\mathscr{B}_t)}{\theta}\right] dt - \int_{\mathscr{I}} \left(\frac{1}{\tau} - \frac{1}{\theta}\right) Q(\mathscr{B}_t)\, dt. \tag{22}$$

In addition, the Clausius-Duhem inequality (12) correspondingly reduces to

$$\dot{H}(\mathscr{B}_t) \geqq \frac{Q(\mathscr{B}_t)}{\theta}, \tag{23}$$

and the integrated heat transfer inequality (11) may be written as

$$(\theta-\tau)\,Q(\mathscr{B}_t)\leqq 0. \tag{24}$$

Thus the total rate of heat working $Q(\mathscr{B}_t)$ now appears in both the main efficiency formulae and the Clausius-Duhem inequality as the *sole effect* of heating, without regard to the specific detail (conduction and/or radiation) of heat transmission. While the heat transfer inequalities (9) contain reference to these details, if in place of (9) either the integrated form (10) or its equivalent (24) is now taken as fundamental, as is, in fact, the case for the thermodynamics of homogeneous phases, and if a reversible process is now characterized by equality in (23) and (24) for all but a finite set of times in $\mathscr{I}$, rather than as previously given, *then in any reversible process $\theta$ and $\tau$ must coincide whenever $Q(\mathscr{B}_t)\neq 0$.* A fortiori *for a reversible Carnot process, $\mathscr{B}$ must absorb heat at a temperature $\theta=\tau_{\max}$ and emit heat at $\theta=\tau_{\min}$.*

The theorem associated with (13) and those theorems which are given in Sections 4 and 5 all remain intact within the present structure. In fact, TRUESDELL [10] has previously obtained the efficiency results of Section 4 for a *homogeneous* heat engine within the context of classical thermodynamics. However, his approach is somewhat different from that suggested here in that he does not introduce the environment[12] of $\mathscr{B}$, and neither the efficiency formulae (8a, b) reduced with the aid of (22) nor the heat transfer inequality (24) appears in his work.[13]

## 8. A Curious Generalization

We observe that *every one of the results of the paper remains true even in a more general thermodynamical setting, namely one in which the balance of energy* (1) *is replaced by an energy* **inequality**

$$\dot{E}(\mathscr{B}_t)+\dot{K}(\mathscr{B}_t)\leqq W(\mathscr{B}_t)+Q(\mathscr{B}_t). \tag{1'}$$

(This is formally reminiscent of the Clausius-Duhem inequality (12)). Whether a thermodynamics based on (1′) is true or false in the real world is not our contention, but only that *within such a theory all the classical formulations of the second law continue to be consequences of the fundamental equations, as well as* **all** *the conclusions of Section* 6.

The reader who desires to check these statements need only make the following changes in the preceeding development:

In equations (6), (8 a), (8 b) and (19) the equality sign (=) should be replaced by inequality (≦), where in each case the actual equality will hold if and only if it holds in (1′) above.

In the definition of a quasi-reversible process in Section 2, moreover, condition (13) should be augmented by the energy *equality*

$$\dot{E}(\mathscr{B}_t)+\dot{K}(\mathscr{B}_t)=W(\mathscr{B}_t)+Q(\mathscr{B}_t).$$

---

[12] This is tantamount to setting $\theta=\tau$ in our present context.

[13] In a recent unpublished manuscript TRUESDELL [11] has used the efficiency formula (8a), reduced by (22) and with $\theta\equiv\tau$, in order to streamline his earlier [10] proofs.

Finally in Section 6 the local balance of energy (see footnote 8) should of course be read with the inequality sign rather than the equality sign.

We may remark in conclusion that a thermodynamics based on (1′) would in many ways be no different than one based on (1). In particular, all reversible processes would be the same as before, and all purely dynamical results (those which can be based solely on the laws governing linear and angular momentum balance) would likewise continue to hold. On the other hand, irreversible processes would become indeterminate in the resulting structure, and further constitutive relations would be needed to specify the actually occuring motions.

*Acknowledgement.* This work was supported by the U.S. National Science Foundation under Grant No. ENG 74-04815. Partial support by the U.S. Air Force Office of Scientific Research under Grant AFOSR 883-67 is also acknowledged. We thank Professor C. TRUESDELL for his thoughtful consideration of our first rough draft.

## References

1. CARNOT, SADI, Reflections on the Motive Power of Fire, and on Machines Fitted to Develop that Power. Translated and edited by R. H. THURSTON. New York: Dover, 1960.
2. CLAUSIUS, R., On the Motive Power of Heat, and on the Laws which can be Deduced from it for the Theory of Heat. Translated by W. F. MAGIE. New York: Dover, 1960.
3. MAXWELL, J. CLERK, Theory of Heat. Second Edition. London: Longmans, Green and Co. 1872.
4. PLANCK, MAX, Treatise on Thermodynamics. Third Edition. Translated from the seventh German edition by ALEXANDER OGG. New York: Dover 1945.
5. TRUESDELL, C., & R. TOUPIN, The Classical Field Theories. FLÜGGE's Handbuch der Physik, III/1. Berlin-Göttingen-Heidelberg: Springer 1960.
6. TRUESDELL, C., & W. NOLL, The Non-Linear Field Theories of Mechanics. FLÜGGE's Handbuch der Physik, III/3. Berlin-Heidelberg-New York: Springer 1965.
7. COLEMAN, BERNARD D., & VICTOR J. MIZEL, Existence of caloric equations of state in thermodynamics. J. Chem. Phys. **40**, 1116–1125 (1964).
8. SERRIN, J., Foundations of Classical Thermodynamics. Lecture Notes, Department of Mathematics, University of Chicago, 1975.
9. ZEMANSKY, MARK W., Heat and Thermodynamics. Fifth Edition. New York, Toronto, London: McGraw-Hill 1968.
10. TRUESDELL, C., The efficiency of a homogeneous heat engine. J. Math. and Phys. Sci. **7**, 349–371 (1973).
11. TRUESDELL, C., Improved estimates of the efficiency of irreversible heat engines, Annali di Matematica, in press.

Department of Aerospace Engineering & Mechanics
and
Department of Mathematics
University of Minnesota
Minneapolis

*(Received June 20, 1975)*

# VI.2 Thermodynamics – Continuum Mechanics

Chi-Sing Man and David R. Owen

**Mathematics Subject Classification (2010).** 80A05, 80A10, 80A17, 80A22, 80A32.

**Keywords.** Neoclassical thermodynamics, foundations of thermodynamics, thermodynamics of continua, Clausius–Duhem inequality, Maxwell's equal area rule, chemical mixture dynamics.

Commenting on the axioms of continuum physics, Truesdell and Noll, in the introduction to their treatise *The non-linear field theories of mechanics* (Springer, Berlin, 1965), wrote about a general "principle of irreversibility", the "true form" of which "is not yet known", but they qualified their statement with a "note added in proof" that "[m]ajor progress toward finding this law has been made by Coleman". The general principle of irreversibility that Coleman and his coworkers adopted in their work was the Clausius–Duhem inequality in the form written down by Truesdell and Toupin in their 1960 *Handbuch* article, where the entropy and the absolute temperature are taken as primitives given a priori. By hindsight, one major progress made in the rational thermodynamics of Coleman and coworkers was the development of what is now known as the Coleman–Noll procedure, by which the power of the Clausius–Duhem inequality in imposing constitutive restrictions can be harnessed.

In the classical thermodynamics of reversible systems, the existence of an absolute temperature scale and of entropy as a function of state are consequences that follow from the laws of thermodynamics. This stands in stark contrast with the starting point of rational thermodynamics, which became in the late 1960s a source of criticisms from proponents of other approaches. Perhaps arising partly as attempts to allay such criticisms, soon after the early successes scored by Coleman and coworkers, there followed a stream of foundational studies (the efforts of W.A. Day and of Coleman and Owen being good representatives) examining the general validity of the Clausius–Duhem inequality and investigating the existence and uniqueness of entropy, or the lack thereof, for various material systems. These efforts set the stage for the work of Serrin, Šilhavý, and others in the late 1970s and early 1980s. The article "The laws of thermodynamics" is an outline of some of the main findings, particularly those by Serrin himself and by his student M. Ricou, during this period.

Following the pioneers of classical thermodynamics, Serrin and independently Šilhavý take the concepts of hotness, heat and work as primitives. As was intuitively clear to Carnot and was explicitly pointed out by Gibbs, "in thermodynamic problems, heat received at one temperature is by no means the equivalent of the same amount of heat received at another temperature ... This is a result of the general law, that heat can only pass from a hotter to a colder body ..." In other words, we should consider

P. Pucci et al. (eds.), *James Serrin. Selected Papers: Volume 2*, Contemporary Mathematicians,
DOI 10.1007/978-3-0348-0847-7_15, © Springer Basel 2014

in thermodynamics not only the quantity but also the quality of heat as characterized by the hotness at which the heat is received or given out. If a body receives a quantity of heat over a range of temperatures, just knowing the total amount received will not suffice for us to carry out a full thermodynamic analysis; in addition, the quality of the heat received must be specified by a detailed breakdown of the total according to the temperatures of receipt.

It will be a problem for future historians of thermodynamics to explain why a suitable mathematical expression to capture both the quantitative and qualitative aspects of heat was not created until Serrin and Šilhavý, around 1978, independently proposed their "accumulation function" and "heat distribution measure", respectively, so that once and for all classical thermodynamics could be set on a firm mathematical foundation. A general theory of thermodynamics, where the First and the Second Law refer to work, heat, and the accumulation function for systems undergoing cycles (i.e., cyclic processes) or approximate cycles, was subsequently developed. We would call this theory neoclassical thermodynamics, for it is best described as the classical thermodynamics of Clausius and Kelvin put in the most general setting, plus a careful delineation of the required axioms and logical arguments. In neoclassical thermodynamics, the existence of the mechanical equivalent of heat (even under the "Weak First Law" – a weakened form of the First Law – that would lead to an energy inequality rather than energy balance) and that of an absolute temperature scale follow as general theorems. Furthermore, whenever the existence of a local entropy function can be demonstrated for a classical continuous body, the Clausius–Duhem inequality will indeed be the "true form" of the principle of irreversibility. Sections 1 through 3 of "The laws of thermodynamics" give a brief account of this "clean and precise theory".

Section 4 concerns internal energy and entropy for systems with state structure. Here only the barest outline of a general theoretical framework is sketched, with the remark (and supporting references) that "for simple reversible systems the existence of internal energy and entropy is equivalent to the First and Second Laws." On the other hand, as pointed out by Serrin himself, the real problem at issue is the existence of internal energy and entropy for more general thermodynamic systems such as continuous material bodies.

The subject of internal energy and entropy for systems with state structure is the focus of a frequently cited memoir of Coleman and Owen, published in 1974, on mathematical foundations for thermodynamics. There Coleman and Owen present a general theory on the existence, uniqueness, and regularity of internal energy and entropy for systems with state structure. They apply their theory to material elements (i.e., material points of continuous bodies with specific constitutive structure), rather than to material bodies themselves. The connection of Coleman and Owen's thermodynamics of material elements to neoclassical thermodynamics, however, is tenuous, because in neoclassical thermodynamics the presence of the accumulation function makes it more natural to treat at the outset material bodies, rather than material elements, as thermodynamical systems. Subsequent research by Coleman, Owen, and Serrin, and by Owen recast the general thermodynamics of systems with state structure and approximate accessibility into the framework of neoclassical thermodynamics, but without treating material bodies as examples.

The paper "The equations of continuum mechanics and the laws of thermodynamics" represents the first successful attempt, in the context of neoclassical thermodynamics, to prove the existence of internal energy and entropy for a class of material bodies that are not classical reversible systems. Here Serrin considers a class of compressible fluids of complexity 1, which includes the (compressible) Navier–Stokes fluids as a special case. Under mild constitutive and mathematical assumptions, he derives for the fluid media in question the existence of an internal energy and entropy which observe the Gibbs relation and satisfy the equation of energy balance and the Clausius–Duhem inequality, respectively. Serrin's derivation is based on the First and the Second Law for cyclic processes. A crucial (and perhaps the most innovative) step in his argument is the introduction of a new term, which he calls "frictional heating", to the classical formula for the heating in what Truesdell called the "doctrine of latent and specific heats". Without making any a priori assumption on the frictional heating except that it be a function of various constitutive variables, he appeals to the First Law to prove that it is equal to the stress working for the class of fluids in question. The last section of this paper is devoted to the interesting scenario where the First Law is replaced by the Weak First Law. Serrin shows that most of the conclusions mentioned above remain valid. The exceptions are: the equation of energy balance is replaced by an energy inequality, and the frictional heating now obeys two inequalities – it is bounded above by the stress working and bounded below by the negative of the product of the absolute temperature and the rate of entropy production due to heat conduction. Foundational studies in thermodynamics, however, seem to have become out of fashion among researchers since the late 1980s. As far as we are aware, there is only one subsequent article (by Man and Massoudi) which takes up Serrin's ideas and develops neoclassical thermodynamics for a class of generalized second-grade fluids.

A theory of thermodynamics based on laws written as statements on cycles or approximate cycles has its intrinsic limitations because there are material systems which admit few or none such processes. A lawful assertion on cycles or approximate cycles will provide little or no information on those systems. Beginning with his doctoral dissertation at the University of Minnesota, M. Ricou set his aim to divorce the laws of thermodynamics from the concept of cycle. Section 5 of the article "The laws of thermodynamics" gives an outline of Ricou's work on the First Law. See Ricou's paper "A general first law for thermodynamics" (*ARMA*, vol. 101, pp. 365–385, 1988) for a full account of this work. Section 6 of "The laws of thermodynamics" provides a brief glimpse of the joint work of Ricou and Serrin on a general second law of thermodynamics along the same lines. Unfortunately no full paper of this joint work seems to have appeared in print, although Ricou gave another description of this work in the collection of essays titled *New Perspectives in Thermodynamics* and edited by J. Serrin (Springer, Berlin, 1986).

The array of Serrin's contributions outlined in these two articles has been pivotal for the emergence of neoclassical thermodynamics as a firmly established component of thermomechanics, and future researchers will find within these developments a rich source of issues remaining to be examined critically when foundational studies again gain the support of the research community.

In "The area rule for simple fluid phase transitions" Serrin treats a topic within the realm of classical thermodynamics. He provides a significant contribution to the analysis of phase transitions in a broad class of homogeneous fluid bodies and to the identification of a special subclass of these bodies for which the results of the broader analysis agree remarkably well with experimental observations. At issue is the relation between the pressure and temperature at which a liquid-to-vapor transition may occur and, more specifically, the applicability of the Maxwell "equal area rule" for determining that relation. The classical argument that leads to the equal area rule ignores the fact that the fluid body no longer is homogeneous during the phase transition, and appropriate modifications of that argument in the case of the van der Waals fluid had led earlier researchers to a modified area rule. Serrin's analysis shows that their modified area rule applies in a far broader setting and provides two classes of models for which his derivation is valid. The modified rule yields necessary and sufficient conditions for the validity of Maxwell's equal area rule that, for the case of the van der Waals fluid, reduce to the equalitiy of the specific heats in the vapor and liquid phases, a condition not confirmed by experimental data on real fluids. Serrin identifies a multi-parameter family of equations of state that includes the van der Waals fluid and for which the parameters can be calculated from available data. This, in turn, permits the calculation of the boiling pressure by means of the modified area rule for a broad range of temperatures and the successful comparison of calculated and measured values for several substances. Serrin's analysis is cited in a few subsequent articles, one of which provides an alternative derivation of the modified area rule and shows the disparity between boiling pressure values calculated with the equal area rule and those calculated with the modified area rule.

The paper "A continuum model for chemical mixture dynamics" appeared in the proceedings volume *Developments in Mechanics*, Volume 13 – Midwestern Mechanics Conference, Columbus, OH, 1985 and, to our knowledge, has not been reviewed or cited in the subsequent literature. Moreover, Serrin does not seem to have undertaken additional research on this topic. The rather condensed format of the exposition contains only brief references to the large body of related works, many of which appeared in the period 1950–1980 and belong to one or both of the categories "rational thermodynamics" and "mixture theory". Within the context of mixture theory, Serrin has in mind mixtures of fluids and treats such a mixture as a continuum, while formulating balance laws and a dissipation inequality for the mixture, supplemented by a law of balance of mass for each constituent. Serrin's formulation leads to the appearance of an "interstitial working" term, both in the equation of energy balance and in the dissipation inequality, and to the appearance in all of the basic relations of terms associated with external supplies of the constituents. Within the context of rational thermodynamics, Serrin emphasizes the use of external supplies of the constituents in the application of the Coleman–Noll procedure for determining the restrictions placed on constitutive relations of the mixture by the Second Law of Thermodynamics. This feature of Serrin's approach obviates the need for assumptions of limited solvability of the governing equations of the mixture that appear frequently in the literature. An alternative approach available at the time of publication avoids the underlying assumptions required in treating the mixture, itself, as a continuous body and emphasizes the treatment of each constituent within the mixture as a continuous body that experiences additional interactions due to its intimate contact with other constituents of the mixture. More recently, an assumed submacroscopic structure of each constituent as a system of particles has been employed to formulate and analyze governing relations for mixtures; alternatively, multiscale geometry can be used to approximate mixing constituents by "intermingling" (i.e., non-interpenetrating) constituents in order to clarify the nature of interactions due to intimate contact experienced by each constituent.

Chi-Sing Man
Department of Mathematics
University of Kentucky

David R. Owen
Department of Mathematical Sciences
Carnegie Mellon University

*Res Mechanica* 21 (1987), 361–379. **Reprint VI.2.[i]**


# The Laws of Thermodynamics

James Serrin

**Mathematics Department, University of Minnesota,
Minneapolis, Minnesota 55455, USA**

*ABSTRACT*

***The goals of much modern work in the foundations of thermodynamics can be succinctly expressed: to find an appropriate general structure for the fundamental concepts of the subject; to formulate credible laws of thermodynamics in a clear, concise, reasonable and useful way; to prove the existence of the mechanical equivalent of heat and the absolute temperature scale; to define the concepts of internal energy and entropy in concise and reasonable terms.***

***The purpose of this paper is to outline some recent results which have been obtained in these directions of inquiry. Using the concepts of work, heat and hotness as basic primitives of the subject, a precise theory has emerged in connection with the first three goals, as well as an important understanding of the crucial issues involved in the fourth.***

## INTRODUCTION

In the light of past experience as well as recent research, it has become increasingly clear that for the study of phenomenological thermodynamics—namely, the overarching non-statistical subject—the appropriate primitive variables are *work, heat,* and *hotness.* A physical system of whatever sort—for example, a body of gas on a viscous fluid, an elastic solid or a chemically reacting mixture—whose interactions with its exterior are reflected by various transfers of work and heat is thus called a *thermodynamical system.*

Notwithstanding the broad range of intended applications as well as



*Res Mechanica* 0143-0084/87/$03·50 © Elsevier Applied Science Publishers Ltd, England, 1987. Printed in Northern Ireland

its fundamental position in science, the study of thermodynamics has always been impeded by an inadequate presentation of its foundational aspects. Thus Kelvin could write 'A mere quicksand has been given as a foundation for thermometry' and Cardwell could add only recently 'The student is usually introduced to the concepts of thermodynamics · · · in a way which does violence to credibility.' While there have been occasional attempts to clarify the situation—a notable but not entirely successful one being due to Carathéodory (see Ref. 1)—it has only been in the past several years that a concerted effort has been undertaken to rigorize the subject, involving the work of a number of mathematicians in various different centers. The goals of this research fall into four related categories:

(I) To find an appropriate general *structure* in which to express the fundamental concepts of the subject.
(II) To formulate the *laws of thermodynamics* in ways which are clear and concise, physically reasonable and useful.
(III) To prove existence of the *mechanical equivalent of heat* and the *absolute temperature scale,* in analytically precise terms.
(IV) To define the concepts of *internal energy* and *entropy* for material systems, and to characterize those systems for which these functions can be shown to exist.

The purpose of this paper is to outline some recent results which have been obtained in these directions. In particular, for the first three categories a clean and precise theory has emerged, while for the fourth there is important new understanding of the crucial issues and how to treat them.

It should finally be emphasized that the discussion is not directed toward specific problems for particular materials systems, important as these may be. Rather, we are concerned with providing a *general structure* within which each individual physical system appears as a *special* case. This point of view clearly reflects the beliefs of the founders of thermodynamics (however much they may have limited themselves in practice to the treatment of special systems) and moreover closely parallels modern approaches to continuum physics.

## SECTION 1

The formal structure presented here was first developed during the period 1977–1979 in papers by the author.[2–4] A similar structure was found slightly later but entirely independently by M. Silhavy.[5–7]

Silhavy's development requires considerably deeper topological and measure theoretic considerations, and accordingly we follow the approach in Refs 2–4. Another concurrently developed and philosophically related approach to the foundations of thermodynamics is due to Feinberg and Lavine. In their treatment, beginning in 1978[8] and ultimately published as Ref. 9, it is not necessary to include the notion of hotness as a primitive concept, though this gain is balanced by certain alternate difficulties.

It is convenient to begin our treatment by introducing the concept of hotness, which will here be considered a primitive notion within the theory. It is represented by a *thermal manifold* $\mathscr{H}$ consisting of the set of *hotness levels* $L$ open to material systems. In accord with experience we assume that $\mathscr{H}$ is a totally ordered set, with order relation $>$. The sentence $L_2 > L_1$ will then be read '$L_2$ is hotter than $L_1$.'

A *temperature scale* is a strictly monotone increasing map from $\mathscr{H}$ into the reals $\mathbb{R}$. If $\psi$ is a temperature scale then $\psi(L)$ is called the *temperature* of $L$ in the scale $\psi$.

Fundamental to thermodynamical structure is the concept of a *thermodynamical system,* examples of which might be a body of gas or an elastic solid, to name two particularly simple cases. *Every* thermodynamical system $\mathscr{S}$ comes endowed with a set $\mathbb{P}(\mathscr{S})$ of *processes* which the system may undergo, together with a subset $\mathbb{P}_{\text{cyc}}(\mathscr{S})$ of *cyclic processes* of the system. To every process $P \in \mathbb{P}(\mathscr{S})$ there correspond real numbers $\bar{W}(P)$ and $\bar{Q}(P)$, respectively the *total work* done by the process $P$ and the *total heat* used by the process $P$. Formally

$$\bar{W}:\mathbb{P}(\mathscr{S})\to\mathbb{R}, \qquad \bar{Q}:\mathbb{P}(\mathscr{S})\to\mathbb{R}$$

We adopt the standard sign convention that $\bar{W}(P) > 0$ if work is done by the system on the exterior environment and $\bar{W}(P) < 0$ if the exterior environment does work on the system. Similarly $\bar{Q}(P) > 0$ if heat is supplied to the system, while $\bar{Q}(P) < 0$ means that the system has supplied heat to the environment.

We require one more primitive concept, namely a more refined and subtle measure of the heat used by a process than is directly given by the total heat $\bar{Q}(P)$. The reason for this, of course, is that heat supplied at one temperature is very different from heat supplied at another. To this end, we suppose (in accord with intuition) that to every process $P \in \mathbb{P}(\mathscr{S})$ and every hotness level $L \in \mathscr{H}$ there is associated a real number $Q(P, L)$ representing the total or net heat transferred to the system during the process $P$ *at hotness levels $L'$ not exceeding $L$.* Formally we have

$$Q:\mathbb{P}(\mathscr{S})\times\mathscr{H}\to\mathbb{R}$$

The function $Q(P, \cdot)$ is called the *accumulation function* of the process $P$.

The accumulation function expresses analytically the essential properties of the relation between heat and hotness for a given process $P$. For example, during a process $P$ the total heat added *between* the hotness levels $L_1$ and $L_2$ (with $L_1 < L_2$ say) is given by $Q(P, L_2) - Q(P, L_1)$. It follows in particular that the accumulation function of an *isothermal process* $P$ (operating at a single hotness level $L_0$) is constant except for a single jump at $L_0$, the jump being positive if $\bar{Q}(P) > 0$ and negative if $\bar{Q}(P) < 0$. Similarly if the system only absorbs heat during a process $P$—but never emits heat—then $Q(P, \cdot)$ is monotonically increasing. In the same way, if $P$ is *adiabatic*—that is, exchanges no heat whatsoever with its environment—then $Q(P, \cdot) \equiv 0$.

For any given process $P$ of a system $\mathscr{S}$, one may suppose that heat is supplied to or emitted from the system only on some bounded range of hotnesses.† Reflecting this fact, we assume that the accumulation function has the following property.

(a) For every $P \in \mathbb{P}(\mathscr{S})$ there exists a *lower* hotness level, denoted by $L_l$, such that

$$Q(P, L) = 0 \quad \text{when} \quad L < L_l$$

and an *upper* hotness level, denoted by $L_u$, such that

$$Q(P, L) = \bar{Q}(P) \quad \text{when} \quad L > L_u$$

In addition we suppose a minimal degree of regularity for the accumulation function, namely

(b) For every $P \in \mathbb{P}(\mathscr{S})$ the function $Q(P, \cdot)$ is bounded and has at most a denumerable number of discontinuities.

A final necessary concept is the idea of *products* of thermodynamical systems. The well-known heuristic arguments presented in standard treatments of thermodynamics to justify the classical efficiency theorem, arguments which ultimately go back to Carnot, involve comparing Carnot cycles for two different systems by forming a third (union) system for which the heat and work are found by adding the corresponding quantities for the original systems. In effect, the union idea involves taking the heat emitted by one body and transferring it to a second body, with a corresponding reduction of the heat supplied to the second system from its other surroundings. These well-known but nevertheless somewhat vague ideas require a formal description.

† In more picturesque terms, no process taking place on a finite time interval can ever attain either absolute zero or absolute infinity in temperature.

Let $\mathscr{S}_1$ and $\mathscr{S}_2$ be a pair of physical systems. The *product system*, $\mathscr{S}_1 \oplus \mathscr{S}_2$, is characterized by its processes and their work and heat functions, which are required to satisfy the following conditions:

(i) $\mathbb{P}(\mathscr{S}_1 \oplus \mathscr{S}_2) = \mathbb{P}(\mathscr{S}_1) \times \mathbb{P}(\mathscr{S}_2)$
(ii) $\mathbb{P}_{\text{cyc}}(\mathscr{S}_1 \oplus \mathscr{S}_2) = \mathbb{P}_{\text{cyc}}(\mathscr{S}_1) \times \mathbb{P}_{\text{cyc}}(\mathscr{S}_2)$
(iii) $\bar{W}(P_1 \oplus P_2) = \bar{W}(P_1) + \bar{W}(P_2)$
(iv) $\bar{Q}(P_1 \oplus P_2) = \bar{Q}(P_1) + \bar{Q}(P_2)$
(v) $Q(P_1 \oplus P_2, \cdot) \geq 0 \quad \text{if} \quad Q(P, \cdot) + Q(P, \cdot) \geq 0$

Here $P_1 \oplus P_2$ denotes the union process (in $\mathbb{P}(\mathscr{S}_1 \oplus \mathscr{S}_2)$) corresponding to the pair of processes $P_1 \in \mathbb{P}(\mathscr{S}_1)$, $P_2 \in \mathbb{P}(\mathscr{S}_2)$.

It is open to question whether the concept of a product system should be meaningful for all conceivable pairs of thermodynamical systems. To avoid such metaphysical points, we shall henceforth restrict the formation of product systems only to special and distinguished pairs of systems, which will be called *thermodynamically compatible systems* (or simply *compatible* systems). Thus if $\mathscr{S}_1$ and $\mathscr{S}_2$ are a pair of compatible systems, then the product system $\mathscr{S}_1 \oplus \mathscr{S}_2$ is itself assumed to be a meaningful thermodynamical system satisfying the laws of thermodynamics.

## SECTION 2

Since the mid 19th century, a first principle of thermodynamics has been the basic interconvertibility of work and heat. Stated without recourse to special assumptions regarding state spaces and internal energy, this principle asserts that there exists a universal constant $\mathscr{J} > 0$ such that $\bar{W}(P) = \mathscr{J}\bar{Q}(P)$ for any cyclic process $P$ of any physical system.

A particular feature of this formulation which may strike one as unusual is the appearance of the universal constant $\mathscr{J}$. In developing the theory of absolute temperature, for example, the existence of this canonical scale is *not* postulated, but rather is *derived* from more basic laws. It would therefore seem more appropriate to state the first law *without reference to an absolute equivalent of work and heat,* and to demonstrate (within the theory) that such an absolute equivalent must exist. Once one turns in this direction, however, a number of alternatives present themselves, and it is not immediately clear which of these should be taken as the fundamental expression of the relation between work and heat.

The first important step in this direction was made by Truesdell,[10] who set up such an axiomatic structure for *reversible systems* and who proved in this framework the existence of internal energy, entropy, and an absolute equivalent of work and heat for certain two-variable systems. Truesdell and Bharatha[11] then derived stronger statements from somewhat weaker assumptions of the same kind (see also Ref. 12). The extension to $n$ variables is fairly easy if $n-1$ of them belong to a flat space; the difficulties when that space is a more general manifold were overcome by Pitteri[13] in the course of a memoir on several aspects of the foundations.

Though limited to reversible systems, Truesdell's work nevertheless strikingly shows that a fundamental axiom structure for thermodynamics need not involve a direct postulation that heat and work are interconvertible. The axioms used by Truesdell derive from Carnot's ideas; whether these can lead to a satisfactory theory of *irreversible* systems is uncertain.

In formulating an appropriate and general set of laws for thermodynamics we shall seek here statements which (following the above discussion) do not give *a priori* significance either to a mechanical equivalent of heat or to the existence of a canonical absolute temperature scale. Moreover, the statements should express only the most certain and secure of our intuitive beliefs about heat and work, should apply to general thermodynamical systems without prior restriction, and above all should yield a comprehensive and satisfying theory. In particular if we give some thought to the gist of the first law, namely that work can only be produced at the expense of heat energy, we are led to the following formulation.

*Weak First Law. If $\bar{W}(P)>0$ for a cyclic process $P$ of a thermodynamical system $\mathscr{S}$, then also $\bar{Q}(P)>0$.*

This version of the first law has been noted by Šilhavý in his fundamental paper,[5] though in his context some additional topological considerations appear which are extraneous to our purposes. Moreover Šilhavý does not emphasize this statement as an independent expression of the law (see Ref. 5, Part II, Theorem 2.2.1; Ref. 6, Part 1, Sections 4.6, 4.7; and Ref. 6, Part II, Section 4.12) as we prefer to do here.

The Weak First Law formalizes the idea that positive work can be obtained from a cyclically operating process only when a positive total amount of heat is supplied to the system during the process. While representing a generally weaker requirement than the strict interconvertibility of heat and work, it nevertheless carries great conviction and provides all the normal conclusions drawn from the stronger statement. Of course the Weak First Law (as stated above) is logically consistent with the strict interconvertibility of work and heat in the sense that if the latter is asserted to hold, then the Weak First Law is an obvious consequence.

To obtain strict interconvertibility, certainly a desideratum, we shall also consider a stronger version of the first law, again due to Šilhavý.

*Strong First Law. For a cyclic process P of a thermodynamical system* $\mathcal{S}$, *the conditions* $\bar{W}(P) > 0$ *and* $\bar{Q}(P) > 0$ *are equivalent.*

The principal goal of elementary thermodynamics is to provide analytic tools for studying thermal systems. The following result, essentially due to Šilhavý (see Ref. 5, Part I, Section 2.1), is a consequence of the Weak First Law.

*The Energy Inequality. Let* $\mathcal{U}$ *be a collection of thermodynamic systems containing a perfect gas* $\mathcal{G}$, *and suppose each system* $\mathcal{S}$ *in* $\mathcal{U}$ *is compatible with* $\mathcal{G}$. *Then there exists a unique (universal) constant* $\mathcal{J} > 0$ *such that for every cyclic process P of every system* $\mathcal{S}$ *in* $\mathcal{U}$ *we have* $\bar{W}(P) \leq \mathcal{J}\bar{Q}(P)$.

The energy inequality, as an axiom, was first stated by R. L. Fosdick and the author;[14] for a demonstration of the present form, see Ref. 15. The energy inequality of course applies only to systems $\mathcal{S}$ in the collection $\mathcal{U}$. Since we may assume, realistically, that any system $\mathcal{S}$ of interest belongs to such a collection it follows that the relation $\bar{W}(P) \leq \mathcal{J}\bar{Q}(P)$ can be presumed to hold for cyclic processes of arbitrary thermodynamic systems. The constant $\mathcal{J}$ is called the *mechanical equivalent of heat.*

When the *Strong First Law* is posited instead of the *Weak Law,* a similar proof yields the conclusion

$$\bar{W}(P) = \mathcal{J}\bar{Q}(P)$$

for all cyclic processes $P$ of systems $\mathcal{S}$ in the universe $\mathcal{U}$. That is, the Strong First Law is equivalent to the interconvertibility of heat and work for arbitrary cyclic processes.

In what follows we can assume without loss of generality that units have been chosen so that $\mathcal{J} = 1$.

## SECTION 3

The second law of thermodynamics involves more subtle ideas than the first since it deals with the quality of heat at different hotness levels. Moreover, the physical notions which originally motivated the various

19th century statements of the second law are fairly obscure, requiring some effort to phrase clearly.

The first essentially correct formulation of the second law is due to Rudolf Clausius, namely

> A passage of heat from a colder to a hotter body cannot take place without compensation.[16]

While this is not at all precisely stated, we may consider it to mean that if a cyclic process absorbs heat at a hotness level $L_0$ and emits heat at a hotness level $L_1 > L_0$ then necessarily $\bar{W}(P) < 0$. That is, the accumulation function of a cyclic process $P$ with $\bar{W}(P) \geq 0$ cannot have either of the forms shown:

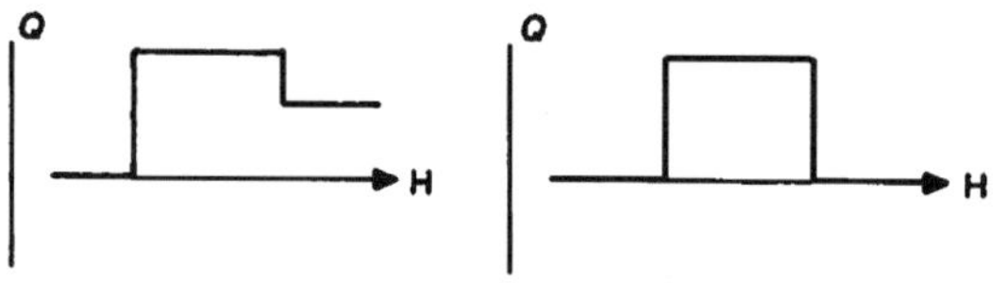


This version of the second law is not easily applicable to the case of general thermal processes without the further intervention of sophisticated topological notions. On the other hand, a slightly stronger formulation of essentially the same idea can be given which avoids this difficulty. In particular, since graphs of the types illustrated above are disallowed by Clausius' version of the second law, it seems equally the case that no *linear combination* of such graphs could occur as the accumulation function of a cyclic process with $\bar{W}(P) \geq 0$. Indeed, such an accumulation function would represent a process which raises various low temperature heat supplies to various higher temperatures, without the need of doing work on the system. In the same way, any *closure* of such linear combinations would also appear impossible for cyclic processes (by continuity considerations) at least if $\bar{W}(P) > 0$. But the set of such closures coincides with the set of *non-negative* accumulation functions. We are thus led to the following general version of the second law (see Serrin[2,3]).

*Second Law. If $\bar{W}(P) > 0$ for a cyclic process $P$ of a thermodynamical system $\mathcal{S}$, then there is some hotness level $L_0$ for which $Q(P, L_0) < 0$.*

The reader will surely have noticed the interesting duality between the (weak) First Law and the Second Law, namely that when $\bar{W}(P) > 0$ the former requires a *positive* value for $\bar{Q}(P)$ while the latter implies a *negative* value for $Q(P, \cdot)$ at some hotness level. This allows the two laws to be stated in a simple *combined form,* a form whose impressive elegance cannot help but be remarked.

*Combined Laws. For any cyclic process P with* $\bar{W}(P) > 0$ *there holds*

$$Q(P, L) > 0 \quad \text{for some} \quad L > L_u$$
$$Q(P, L) < 0 \quad \text{for some} \quad L < L_u.$$

The Second Law is an *intrinsic* statement about the relation between work and heat in cyclic processes. In parallel with the discussion of the First Law in the previous section, the Second Law also has an equivalent *analytical* formulation of great usefulness. We state this as follows (see Refs 2, 3 and, from other standpoints, also 5–7, 9).

*The Accumulation Inequality. Let the hypotheses of the energy inequality hold. Then there exists an (absolute) temperature scale* $\hat{T}$ *on the hotness manifold* $\mathcal{H}$, *with* $\hat{T}(\mathcal{H}) = \mathbb{R}^+$, *such that for every cyclic process P of every thermodynamic system* $\mathcal{S}$ *in* $\mathcal{U}$ *we have*

$$\int_0^\infty \frac{Q(P, L)}{T^2} \, \mathrm{d}T \le 0$$

*where* $L = \hat{L}(T)$ *is the hotness level associated with the temperature T in the scale* $\hat{T}$. *Any temperature scale* $\hat{T}$ *with the above property either agrees with the ideal gas scale of* $\mathcal{G}$ *or is a positive constant multiple of this scale.*

The accumulation theorem immediately accomplishes two purposes: *it establishes the concept of absolute temperature without ambiguity, and it characterizes once and for all the allowable behavior of the accumulation function of any cyclic process.* Indeed the accumulation theorem implies the Second Law, for if $Q(P, \cdot) \ge 0$ and $\bar{W}(P) > 0$ then by the First Law $\bar{Q}(P) > 0$ so that the accumulation integral must be positive, and the process hence cannot be cyclic.

The reader should observe that the accumulation integral is well-defined and finite, as follows easily from properties (1) and (2) of the accumulation function given in Section 1.

The accumulation inequality is a generalization of the Clausius inequality. In particular, should $Q(P, \cdot)$ be of bounded variation then we can obviously write

$$\int_0^\infty \frac{Q(P, L)}{T^2} \, \mathrm{d}T = \int_0^\infty \frac{\mathrm{d}Q(P, L)}{T}$$

the latter integral representing the 'sum' of the heats added divided by their absolute temperatures. The advantage of the accumulation integral compared to the Clausius integral is that it can be expressed analytically in terms of clearly formulated primitive concepts, and at the same time is applicable to a broader class of processes since its existence relies only on the structural properties (1) and (2) of the accumulation function.

For a proof of the accumulation inequality see Ref. 4 or, in a more general setting, Ref. 17. In these papers it is also shown how to replace the perfect gas $\mathcal{G}$ by less special model materials—essentially those with a suitably rich supply of Carnot cycles.

## SECTION 4

In order to provide a concrete framework for the notions of internal energy and entropy it is necessary to introduce the idea of a state space, and an associated state structure. As the simplest (and most general) level this may be defined as follows.

A *state structure* for a system $\mathcal{S}$ consists of a set $\Sigma$, whose elements are called *states* of the system, and a corresponding family of processes $\mathbb{P}_\Sigma(\mathcal{S}) \subset \mathbb{P}(\mathcal{S})$, with each process $P \in \mathbb{P}_\Sigma(\mathcal{S})$ having a well-defined *initial* state $P_i \in \Sigma$ and *final* state $P_f \in \Sigma$.† Moreover if $P$ is a cyclic process in $\mathbb{P}_\Sigma(\mathcal{S})$ then $P_i = P_f$. (In practice, a state structure should also be compatible with the notion of a process $P$ *following* another process $P'$, and should include in this case the axiom $P_i = P'_f$.)

We shall say that a system $\mathcal{S}$ has an *internal energy* corresponding to the state structure $(\Sigma, \mathbb{P}_\Sigma)$ if there exists a function

$$U : \Sigma \to \mathbb{R}$$

such that

$$\Delta U \leq \bar{Q}(P) - \bar{W}(P)$$

for each $P \in \mathbb{P}_\Sigma(\mathcal{S})$. Here $\Delta U$ denotes the difference between $U$ evaluated at the final state and the initial state of $P$, that is

$$\Delta U \equiv U(P_f) - U(P_i)$$

Roughly speaking, then, a function $U$ is an internal energy for a system if it is a lower potential for the difference $\bar{Q}(P) - \bar{W}(P)$.

† Formally, the assignment of initial and final states can be considered as a pair of mappings

$$i : \mathbb{P}_\Sigma(\mathcal{S}) \to \Sigma, \qquad i(P) = P_i$$
$$f : \mathbb{P}_\Sigma(\mathcal{S}) \to \Sigma, \qquad f(P) = P_f$$

If $P \in \mathbb{P}_{cyc}(\mathscr{S}) \cap \mathbb{P}_\Sigma(\mathscr{S})$ then necessarily $P_i = P_f$ and in turn $\Delta U = 0$. Consequently we recover the energy inequality

$$\bar{W}(P) \leq \bar{Q}(\bar{P})$$

from the above formula (this in fact being the *motivation* for the definition of internal energy).

In parallel with the Strong First Law, we may also introduce the idea of a (strong) internal energy, in which the relation $\Delta U \leq \bar{Q}(P) - \bar{W}(P)$ is replaced by the stronger requirement

$$\Delta U = \bar{Q}(P) - \bar{W}(P)$$

Turning to the Second Law it is natural to proceed in a similar way, but now based on the accumulation inequality. For convenience in formulation, we introduce the abbreviation $\bar{A}(P)$ for the integral appearing in the accumulation theorem; thus

$$\bar{A}(P) \equiv \int_0^\infty \frac{Q(P, L)}{T^2} \mathrm{d}T$$

Naturally once one has a (definite) absolute temperature scale in hand one can define $\bar{A}(P)$ whether or not the process $P$ is cyclic. This being understood, we shall say that a system $\mathscr{S}$ has an *entropy* corresponding to the state structure $(\Sigma, \mathbb{P}_\Sigma)$ if there exists a function

$$S : \Sigma \to \mathbb{R}$$

such that

$$\Delta S \geq \bar{A}(P)$$

for each $P \in \mathbb{P}_\Sigma(\mathscr{S})$. Again roughly speaking, the entropy is an upper potential for the accumulation integral $\bar{A}(P)$.

If $P \in \mathbb{P}_{cyc}(\mathscr{S}) \cap \mathbb{P}_\Sigma(\mathscr{S})$ then of course $\Delta S = 0$ so that we recover the cyclic condition

$$\bar{A}(P) \leq 0$$

stated in the accumulation theorem. Another case of interest is that of an *adiabatic process,* for which $Q(P, \cdot) = 0$. In this situation one has $\bar{A}(P) = 0$, whence in turn the entropy hypothesis yields

$$\Delta S \geq 0$$

the celebrated condition of spontaneous entropy increase.

It is one of the principal conclusions of classical thermodynamics that simple reversible systems necessarily possess both an internal energy and an entropy. This result can be obtained within the present structure as a direct and simple consequence of the energy inequality (Weak First Law) and the accumulation inequality (Second Law), cf. Refs 3, 15. Thus *for simple reversible systems the existence of internal energy and entropy is equivalent to the First and Second Laws.*

A main problem of modern thermodynamics is to characterize the general class of thermodynamic systems for which the same conclusion holds (Goal IV in the introduction). Important work in this direction was initiated by Coleman and Owen[18] and has been continued in a series of more recent papers.[2,17,19–21] A far-reaching result in this direction was reached by M. Ricou in his doctoral dissertation,[22] and later in joint work with the present author.[23] We outline these ideas in the next Section.

## SECTION 5

In order to give the greatest generality to the laws of thermodynamics it was observed by Coleman and Owen that they should be formulated not simply as statements about cycles, but rather as conditions on much larger classes of processes. The ultimate motivation for this train of thought lies in the fact that there exist physical materials which admit few, if any, cycles, so that a merely cyclic statement of the laws will provide little or no information concerning these systems. Exploiting fully the power of this idea, Ricou developed a form of the First Law which applies to processes quite independently of whether or not they are cyclic.

To formulate his approach we first formalize the important concept of processes that can *follow* one another.

Thus let $\mathscr{S}$ be a thermodynamic system and let $\mathbb{P}(\mathscr{S})$ be the set of processes available to $\mathscr{S}$. To each $P \in \mathbb{P}(\mathscr{S})$ we associate a corresponding family of processes, called $\operatorname{Foll}(P)$, consisting of all processes $R \in \mathbb{P}(\mathscr{S})$ which can *follow* $P$. Intuitively speaking, this means that the system $\mathscr{S}$, after first executing the process $P$ can then immediately (in temporal juxtaposition) continue on with the process $R$, with no break or discontinuity in its configuration or motion.

Formally, if $P \in \mathbb{P}(\mathscr{S})$ and $R \in \operatorname{Foll}(P)$ then there exists a corresponding process $Q \in \mathbb{P}(\mathscr{S})$, *called $P$ followed by $R$ (and hereafter denoted by $PR$) such that*

(i) $\bar{W}(PR) = \bar{W}(P) + \bar{W}(R)$
$\bar{Q}(PR) = \bar{Q}(P) + \bar{Q}(R)$
$Q(PR, L) = Q(P, L) + Q(R, L)$ for all $L \in \mathscr{H}$

(ii) If $P \in \text{Foll}(N)$ and $R \in \text{Foll}(P)$, then $R \in \text{Foll}(NP)$, $PR \in \text{Foll}(N)$. If $P \in \text{Foll}(N)$ and $R \in \text{Foll}(NP)$, then $R \in \text{Foll}(P)$.

(iii) Let $\mathcal{S}_1$ and $\mathcal{S}_2$ be thermodynamically compatible systems and suppose $P_1 \in \mathbb{P}(\mathcal{S}_1)$ and $P_2 \in \mathbb{P}(\mathcal{S}_2)$. Then

$$\text{Foll}(P_1 \oplus P_2) = \text{Foll}(P_1) \times \text{Foll}(P_2)$$

We note from (i) and (ii) that

$$\bar{W}(N(PR)) = \bar{W}((NP)R) = \bar{W}(N) + \bar{W}(P) + \bar{W}(R)$$

with analogous relations holding also for the total heat $\bar{Q}$ and the accumulation function $Q$.

The notion of followers allows us to introduce the class of cyclic processes by means of a definition rather than as a separate primitive concept. In particular we adopt the following:

*Definition.* A process $P \in \mathbb{P}(\mathcal{S})$ is called *cyclic* if $P \in \text{Foll}(P)$, that is if $P$ can follow itself.

The principal operative feature of cyclic processes is condition (ii) of Section 1. To see that this condition continues to apply in the present circumstance, suppose $P_1 \in \mathbb{P}(\mathcal{S}_1)$ can follow itself and that $P_2 \in \mathbb{P}(\mathcal{S}_2)$ can follow itself. Then by (iii) it is clear that $P_1 \oplus P_2$ can follow itself. More generally, if $P_1 \in \mathbb{P}(\mathcal{S}_1)$ is cyclic and $R \in \mathbb{P}(\mathcal{S}_2)$ can follow $P_2$ then $P_1 \oplus R$ can follow $P_1 \oplus P_2$.

We can now state the non-cyclic general version of the First Law, as formulated by Ricou.[22]

*Non-cyclic First Law. Let $\mathcal{S}$ be a thermodynamic system. For each process $R$ in $\mathbb{P}(\mathcal{S})$ there exists a number $\Phi = \Phi(R)$ such that if $P$ follows $R$ then*

$$\bar{Q}(P) \leq 0 \Rightarrow \bar{W}(P) \leq \Phi(R)$$

In words, this law asserts that *from a given starting point R* there is a *maximum* amount of work that a system can accomplish when no further total heat is added to it. Just as for the cyclic versions of the First Law, this statement serves to reflect some of our most deeply held beliefs about the nature of work and energy.

It is interesting that the first law in this non-cyclic form implies the weak (cyclic) First Law. To prove this fact, let $P$ be cyclic (so that $P$ follows itself) and define for each positive integer $n$,

$$P^n = \underbrace{PP \cdots P}_{n \text{ times}}$$

Then by (i) above

$$\bar{W}(P^n) = n\bar{W}(P), \qquad \bar{Q}(P^n) = n\bar{Q}(P)$$

Moreover $P^n$ can follow $P$, as is obvious from (ii).

Now suppose $\bar{Q}(P) \leq 0$. Then $\bar{Q}(P^n) = n\bar{Q}(P) \leq 0$, and so by the non-cyclic First Law

$$\bar{W}(P^n) \leq \Phi(P)$$

Thus in turn

$$\bar{W}(P) = \frac{1}{n}\bar{W}(P^n) \leq \frac{1}{n}\Phi(P)$$

Letting $n \to \infty$ we get $\bar{W}(P) \leq 0$.

This proves the required result, for if $\bar{W}(P) > 0$ it is necessary that $\bar{Q}(P) > 0$.

The following major result can now be obtained.[22]

*Theorem. Let the hypotheses of the energy inequality (Section 2) hold. Then for every system $\mathscr{S}$ in $\mathscr{U}$ there exists a function*

$$E : \mathbb{P}(\mathscr{S}) \to \mathbb{R}$$

*such that for every process $P$ in $\mathbb{P}(\mathscr{S})$*

$$\Delta E \leq \bar{Q}(P) - \bar{W}(P)$$

*where $\Delta E = E(P) - E(R)$ and $R$ is any process which precedes $P$.*†

The function $E$ is not an internal energy in the sense of Section 4, though naturally there are close parallels. If however $\mathscr{S}$ is endowed with a state structure, then under certain natural circumstances one can demonstrate that an ordinary internal energy function must exist.

We shall say that a state structure $(\Sigma, \mathbb{P}_\Sigma)$ is *deterministic* if the condition

$$P_f = R_i$$

(where $P, R \in \mathbb{P}_\Sigma(\mathscr{S})$) implies $R \in \text{Foll}\,(P)$.

*Theorem. Suppose the hypotheses of the energy inequality hold. Then for every system $\mathscr{S}$ in $\mathscr{U}$ which has a deterministic state structure, there exists an internal energy function*

$$U : \Sigma \to \mathbb{R}$$

*such that for every process $P$ in $\mathbb{P}_\Sigma(\mathscr{S})$ we have*

$$\Delta U \leq \bar{Q}(P) - \bar{W}(P)$$

† Naturally we say that a process $P$ can precede a process $R$ if and only if $R$ can follow $P$.

*where* $\Delta U = U(P_f) - U(P_i)$.

To obtain strict equality in the above theorems one cannot proceed as in the cyclic case, since there appears not to be any *non-cyclic* equivalent for Šilhavý's strong version of the cyclic First Law.

Nevertheless, equality can be proved in some cases. We shall say that a process $P$ is strongly reversible if there exists a second process $P'$ such that

$$\bar{W}(P') = -\bar{W}(P), \qquad \bar{Q}(P') = -\bar{Q}(P)$$
$$Q(P', L) = -Q(P, L)$$

and moreover if

$$P' \in \mathrm{Foll}\,(P), \qquad P \in \mathrm{Foll}\,(P')$$

It is then easy to see that

$$E(P) - E(P') \leq \bar{Q}(P) - \bar{W}(P)$$
$$E(P') - E(P) \leq \bar{Q}(P') - \bar{W}(P')$$

so in fact

$$E(P) - E(P') = \bar{Q}(P) - \bar{W}(P)$$

Similarly, when there is a deterministic state structure we obtain the equality $\Delta U = \bar{Q}(P) - \bar{W}(P)$ when $P$ is strongly reversible.

## SECTION 6

The ideas of Section 5 can also be extended to the Second Law. Following earlier studies of Day and Owen, the present author together with Ricou has proposed a general non-cyclic version of the Second Law:

*For each process $R$ in $\mathbb{P}(\mathcal{S})$ and each $L \in \mathcal{H}$ there exists a number $\Psi(R, L)$ such that if $P$ precedes $R$ then*

$$Q(P, L) \geq 0 \Rightarrow \bar{Q}(P) \leq \Psi(R, L_u(P))$$

(*for the meaning of* $L_u(P)$, *see condition* (1) *in Section* 1).

This formulation is not quite so transparent as was the case for the first law, due to the intervention of the hotness level $L_u(P)$. Nevertheless, one can understand the principal idea, namely if during a process heat is being raised from lower temperatures to higher temperatures (that is, $Q(P, L) \geq 0$) then there is an upper bound to the total amount of heat which can be absorbed by the system if it is to reach a definite target point $R$. The purpose of the temperature $L_u(P)$ is to guarantee that the absorbed heat is not all taken on at arbitrarily high temperatures.

Again the cyclic second law is an immediate consequence. To see this, let $P$ be cyclic and suppose $\bar{W}(P)>0$. Then $\bar{Q}(P)>0$ by the cyclic First Law. If there is no hotness level $L_0$ such that $Q(P, L_0)<0$ we would have $Q(P, L)\geq 0$.

Now consider the process $P^n$. Clearly

$$\bar{Q}(P^n)>0, \qquad Q(P^n, L)\geq 0$$

and

$$L_u(P^n)=L_u(P)$$

Consequently by the non-cyclic law (since $P^n$ can precede $P$)

$$\bar{Q}(P^n)\leq \Psi(P, L_u(P))$$

Letting $n\to\infty$ we obtain $\bar{Q}(P)\leq 0$, a contradiction.

In parallel with the results of the preceding section for internal energy, the following conclusions also hold.

*Theorem. Let the hypotheses of the energy inequality hold. Then for every system $\mathcal{S}$ in $\mathcal{U}$ there exists a function*

$$H:\mathbb{P}(\mathcal{S})\to\mathbb{R}$$

*such that for every process $P$ in $\mathbb{P}(\mathcal{S})$ we have*

$$\Delta H\geq \bar{A}(P)$$

*where $\Delta H=H(R)-H(P)$ and $R$ is any process which can follow $P$.*

As in the case of the First Law, the addition of a deterministic state structure yields a standard entropy function.

*Theorem. Let the hypotheses of the energy inequality hold. Then for every system $\mathcal{S}$ in $\mathcal{U}$ which has a deterministic state structure, there exists an entropy function*

$$S:\Sigma\to\mathbb{R}$$

*such that for every process $P\in\mathbb{P}_\Sigma(\mathcal{S})$ we have*

$$\Delta S\geq \bar{A}(P)$$

*where $\Delta S=S(P_f)-S(P_i)$*

Just as was the case for internal energy, so also in the case of entropy the condition of a deterministic-type state space is crucial for the standard results. Indeed, from the vantage point of hindsight *the*

***well-known results of classical thermodynamics can now be seen to hinge on the classical notion of reversibility and on a tacitly built-in determinism of the state structure.***

In some cases it is possible to obtain the existence of a standard internal energy density and a standard entropy density. This is the case, for example, for elastic solids and heat conducting viscous fluids, but the general question of energy and entropy is still a challenging open problem.

## SECTION 7

Phenomenological thermodynamics studies and interrelates two basic physical quantities, *heat* and *work*. There are two intrinsic principles governing this interrelation—the first and second laws of thermodynamics. Each states a reasonable, even if somewhat pessimistic, conviction about the physical world. There are, next, analytical formulations of these laws—first, the energy inequality (or the interconvertibility of heat and work for cycles, if the Strong First Law is used), and second, the accumulation inequality. These analytical formulations pave the way to all direct applications of the theory. They rely in turn on two derived *scale* concepts—the Joule mechanical equivalent of heat and the Kelvin absolute temperature scale, each among the greatest conceptions of 19th century physics. Finally, there are two fundamental potentials, or more accurately semi-potentials—the internal energy and the entropy—which extend the direct cyclic principles to much broader classes of thermal processes. On this structure hangs the science of heat, from the elementary theory of reversible systems, to Gibbs' magnificent conception of thermal and chemical equilibrium, to sophisticated theories of material dynamics.

The most far-reaching implication of the structure, however, is the fact that it is *not* limited to equilibrium. In fact, a dogmatism which would lay this restriction on thermodynamics would in turn invalidate a massive sector of thermal physics, including heat transfer theory, compressible fluid mechanics, shock wave theory, and combustion theory. Conversely, allowing thermodynamics a natural scope beyond equilibrium yields a powerful and far-reaching theory with which to attack those *dynamical* problems where hotness and heat play a crucial role.

## REFERENCES

1. Truesdell, C., What did Gibbs and Carathéodory leave us in thermo-

dynamics? In *New Perspectives in Thermodynamics*, Springer-Verlag, Berlin, 1986, pp. 101–24.
2. Serrin, J., The concepts of thermodynamics. *Contemporary Developments in Continuum Mechanics and Partial Differential Equations*, (Proceedings of a conference held in Rio de Janeiro, August, 1977), North-Holland, Amsterdam, 1978, pp. 411–51.
3. Serrin, J., Conceptual analysis of the classical second laws of thermodynamics, *Archive for Rational Mechanics and Analysis*, **70** (1979) 355–71.
4. Serrin, J., *Lectures on Thermodynamics*, University of Naples, 1979.
5. Šilhavý, M., On measures, convex cones, and foundations of thermodynamics. I. Systems with vector-valued actions; II. Thermodynamic systems. *Czech. J. Phys. B*, **30** (1980) 841–61, 961–91.
6. Šilhavý, M., On the second law of thermodynamics for cyclic processes. I. General framework; II. Inequalities for cyclic processes. *Czech. J. Phys. B* **32** (1982) 987–1010, 1073–99.
7. Šilhavý, M., On the Clausius inequality. *Archive for Rational Mechanics and Analysis*, **81** (1983) 221–43. (Work originally presented at Euromech 111 Symposium, September 1978.)
8. Feinberg, M., Private communication, 1978.
9. Feinberg, M. and Lavine, R., Thermodynamics based on the Hahn–Banach theorem: the Clausius inequality. *Archive for Rational Mechanics and Analysis*, **82** (1983) 203–93. Cf. also *Foundations of the Clausius–Duhem Inequality*. Appendix of Truesdell, C., *Rational Thermodynamics*, 2nd edition, Springer-Verlag, Berlin, 1984.
10. Truesdell, C., How to understand and teach the logical structure and the history of classical thermodynamics. Proceedings International Congress of Mathematicians, Vancouver, 1974, pp. 577–86.
11. Truesdell, C. and Bharatha, S., *The Concepts and Logic of Classical Thermodynamics*, Springer-Verlag, Berlin, 1977.
12. Truesdell, C., Absolute temperatures as a consequence of Carnot's general axiom. *Archive for History of Exact Sciences*, **20** (1980) 357–80.
13. Pitteri, M., Classical thermodynamics of homogeneous systems based upon Carnot's general axiom. *Archive for Rational Mechanics and Analysis*, **80** (1982) 333–85.
14. Fosdick, R. L. and Serrin, J., Global properties of continuum thermodynamical processes. *Archive for Rational Mechanics and Analysis*, **59** (1975) 108–9.
15. Serrin, J., An outline of thermodynamical structure In *New Perspectives in Thermodynamics*, Springer-Verlag, Berlin, 1986, pp. 3–32.
16. Clausius, R., *Abhandlungen über die mechanischen Wärmetheorie*, Braunschweig 1864, 1967. Translated by W. R. Browne as *The Mechanical Theory of Heat*, London, 1879 (see especially pp. 76–9).
17. Coleman, B., Owen, D. and Serrin, J., The second law of thermodynamics for systems with approximate cycles. *Archive for Rational Mechanics and Analysis*, **77** (1981) 103–42.
18. Coleman, B. and Owen, D., A mathematical foundation for thermodynamics. *Archive for Rational Mechanics and Analysis*, **54** (1974) 1–104.
19. Coleman, B. and Owen, D., On the thermodynamics of semi-systems with restrictions on the accessibility of states. *Archive for Rational Mechanics and Analysis*, **66** (1977) 173–81.

20. Owen, D., The second law of thermodynamics for semi-systems with few approximate cycles. *Archive for Rational Mechanics and Analysis,* **80** (1982) 39–55.
21. Šilhavý, M., Foundations of continuum thermodynamics. In *New Perspectives in Thermodynamics,* Springer-Verlag, Berlin, 1986, pp. 33–48.
22. Ricou, M., Energy, entropy, and the laws of thermodynamics, Thesis, University of Minnesota, 1983.
23. Ricou, M. and Serrin, J., A general second law of thermodynamics (to appear).

**James Serrin** was born in Chicago in 1926 and pursued his undergraduate education at *Northwestern University* and *Western Michigan College* where he obtained a B.S. degree in Mathematics in 1947. He received a M.A. and a Ph.D. in Mathematics from *Indiana University* in 1949 and 1951 respectively. After a short stay at *Princeton University* 1951–52 and *Massachusetts Institute of Technology* 1952–54, he joined the Faculty of Mathematics at the *University of Minnesota* where he became professor of Mathematics in 1959. He has been President of the *Society for Natural Philosophy* 1969–70, and a Member of the Council of the *American Mathematical Society* 1972–74. In 1980 he was elected to the *National Academy of Sciences* for his work in Partial Differential Equations and Continuum Mechanics, and became a Fellow of the *American Academy for the Advancement of Science.* In 1969 he was awarded the *Regents Professorship of Mathematics* of the University of Minnesota; in 1972 an honorary degree of Doctor of Science from the *University of Sussex*; in 1973 the *G. D. Birkhoff Prize* in Applied Mathematics; and in 1979 the *Distinguished Alumni Award* of Indiana University. He was a co-editor of *Archive for Rational Mechanics and Analysis,* and serves on the Editorial Boards of *Journal of Differential Equations, Communications in Partial Differential Equations* and *Rendiconti del Circolo Matematico di Palermo.*

Professor Serrin is universally recognized as one of today's greatest contributors in the subjects of Partial Differential Equations, Continuum Mechanics and Phenomenological Thermodynamics. His work (in depth, variety, and endurance) can be compared to Caratheodory's, the great Greek mathematician of the past century. His two-page article (co-authored with N. Meyers) '$H = W$' published in 1964 in the *Proceedings of the National Academy of Sciences* derived a major result speculated but not proven by mathematicians. Two more mathematical articles of great impact are 'On the definition and properties of certain variational integrals' published in the *Transactions of the American Mathematical Society* in 1961, and 'The problem of Dirichlet for quasi-linear elliptic differential equations with many independent variables' published in *Philosophical Transactions of the Royal Society* in 1969. His classic 'Mathematical principles of classical fluid mechanics' published in the *Handbuch der Physik* in 1959 has had a strong effect on contributors of the quality of Daniel Joseph and Roger Fosdick and also exerted an influence on Rutherford Aris, another giant of chemical engineering science today. Two more fluid mechanics articles of substantial influence are 'On the asymptotic behavior of velocity profiles in the Prandtl boundary layer theory' published in the *Proceedings of the Royal Society* in 1967 and 'The swirling vortex' published in the *Philosophical Transactions of the Royal Society of London* in 1972. His recent article 'Conceptual analysis of the classical second laws of thermodynamics' published in the *Archive for Rational Mechanics and Analysis* in 1979 goes beyond Caratheodory's papers in rigor, clarity and generality.

Serrin's personality has the mark of a prince.

E.C.A.

*Meccanica* 31 (1996), 547–563.




# The Equations of Continuum Mechanics and the Laws of Thermodynamics

JAMES SERRIN
*University of Minnesota, School of Mathematics, Minneapolis, MN, USA*

(Received: 29 May 1995)

**Abstract.** With appropriate constitutive assumptions on the stress tensor, the heat flux vector, and the frictional heating associated to a process, we derive for a fluid media the existence of internal energy and entropy as well as the classical energy balance equation and the Clausius–Duhem inequality.

**Sommario.** Con riferimento ad un continuo fluido, si dimostra – sotto appropriate ipotesi costitutive sul tensore delle tensioni, il vettore del flusso termico e il riscaldamento per altrito associato a un processo – l'esistenza della energia interna e dell'entropia di un continuo fluido. Si derivano inoltre la classica equazione di bilancio dell'energia e la diseguaglianza di Clausius–Duhem.

**Key words:** Internal energy, Entropy, Clausius–Duhem inequality, Heat balance, Thermomechanics of continua.

## 1. Introduction

In classical thermodynamics, the concepts of internal energy and entropy are principal pillars of the subject. The familiar demonstrations which have been given for the existence of these functions depend on various assumptions, particularly on the structure of the quasi-static processes which a system may undergo. In a recent paper [8], the author has given an account of the arguments and hypotheses involved; at the same time, he has noted that for a thermodynamic system endowed with dynamical structure, as is the case for continuum mechanics, it is not at all obvious that there exist suitable families of quasi-static processes of the required type.

Notwithstanding this difficulty, the equations of thermomechanics are generally formulated with an internal energy as an integral part of the Energy Balance axiom, and an entropy as part of the Clausius–Duhem inequality. This produces a partially circular logic, although one that has been accepted from the beginnings of the subject (see [11], page 28).

On the other hand, Coleman and Owen in their well-known paper of 1974 initiated a program in which internal energy and entropy functions could be obtained from basic thermodynamic considerations, at least in the context of material elements of continuous systems; see the penetrating discussion of this matter by Man [4]. This being the case, it seems worthwhile to consider the corresponding situation for *continuous systems* themselves, our goal being to obtain analogous results in this context by a direct appeal to thermodynamical arguments. (Earlier versions of the results presented here were given in [6] and [8].)

The idea behind our approach is formulated in detail in the next section, including the introduction of a new heat balance axiom, and need not be restated here. Nevertheless, several points should be emphasized.

First, the thermal structure which is introduced here is based on the idea that a material media is a thermodynamical system, with total heat and total work defined in natural ways.

Second, with appropriate constitutive assumptions on the stress tensor, the heat flux vector, and the frictional heating associated to a process, we derive for a fluid media both the existence of internal energy and entropy as well as the classical energy balance equation and the Clausius–Duhem inequality. This is of course a different approach than the standard one, in which the energy equation and the Clausius–Duhem inequality are deduced from the *a priori* existence of internal energy and entropy.

As noted above, in order to carry out the alternative derivation it is necessary at the outset to introduce 'appropriate constitutive assumptions', delimiting the nature of the material systems which are treated. This may actually be considered an advantage of the derivation, in that the equations are thus obtained in a form directly allied to the assigned structure. One might even imagine that if this structure is taken to be too general, depending, say, on an inappropriately large set of state variables, then not only might the derivation of the energy balance equation fail, but, even more, there might be no appropriate internal energy or entropy functions at all.

Third, even though the derivations in Section 3 involve the notion of quasi-static (retarded) processes as part of the argument, this alone is not sufficient for the crucial proof of the main energy balance equation itself. That the demonstrations are not entirely short is due to the nature of the conclusions, both in a mathematical and a physical sense: indeed if one pauses a moment for reflection, it must surely seem remarkable, and a fact worthy of deductive verification, that 'energy' can be stored, its essence retained, through the artifice of an internal state function in a balance equation.

The final section in the paper is devoted to the case when the thermodynamic laws are weakened, replacing the standard First Law condition $\bar{W}(P) = \bar{Q}(P)$ for cyclic processes $P$ by the inequality $\bar{W}(P) \leqslant \bar{Q}(P)$, a generalization which has been suggested by several authors [2, 7, 9]. At first sight, this weakening apparently precludes the existence of internal energy as a function of state, and even more seems to rule out a strict governing equation of energy. In fact, the same arguments which provide an internal energy when the standard First Law is invoked, also apply when only the weaker version of the law is assumed! Moreover, an energy balance equation continues to be valid, even if it is modified by the addition of non-positive higher order terms representing *energy loss*; these terms play the same role for the energy equation as *entropy production* does for the entropy balance.

Our results for the weak First Law are summarized in Section 4 by the dual conclusions (i), (ii)″, (iii)″ and (iv), conclusions which can be compared directly with the main results (i)–(iii) of Section 3.

Because the equations of continuum thermodynamics are so well-known, we do not always define the notation which we use, though it generally conforms with classical terminology [5].

## 2. Thermomechanics

The equations of continuum mechanics, that is, the continuity equation, the motion equation, and the energy equation

$$\dot{\rho} + \rho \operatorname{div} v = 0, \tag{2.1}$$

$$\rho \dot{v} = \rho b + \operatorname{div} T, \tag{2.2}$$

$$\rho \dot{u} = \rho r - \operatorname{div} h + T : D, \tag{2.3}$$

where $u$ denotes the specific internal energy function, $T$ the symmetric stress tensor, $D$ the symmetric part of the velocity gradient tensor and $h$ the heat flux vector, are well-known, and their derivation has for many years been completely standard (see for example [5], Sections 33–34, and for a further discussion, [11], Chapter 2). To these equations are appended the Gibbs relation

$$\theta \dot{s} = \dot{u} + p(1/\rho)^{\cdot}, \tag{2.4}$$

and the Clausius–Duhem inequality (in differential form)

$$\rho\theta\dot{s} \geqslant \rho r - \operatorname{div} h + \frac{h \cdot \operatorname{grad}\theta}{\theta}. \tag{2.5}$$

In spite of the standard nature of the derivations, two crucial points are nevertheless taken for granted.

(1) The a priori *existence* of an internal energy $u$ and entropy $s$, satisfying the Gibbs relation, and
(2) The *assumed* validity of the Clausius–Duhem inequality, independent of any separate thermodynamic argument.

The first of these points is usually allowed by analogy with classical (textbook) thermodynamics, while the second is more obscure, as is apparent from the fact that when it was originally proposed by Truesdell, and by Truesdell and Toupin, as the appropriate version of the Second Law of Thermodyanamics in a continuum setting, it was considered to be an innovative and daring idea. Thus it is that one would wish for an alternate approach to the energy equation and the Clausius–Duhem inequality, avoiding these difficulties.

Our purpose here is, then, to present a new derivation of the energy equation and the Clausius–Duhem inequality in which it is not necessary to assume either of the points (1) or (2). To accomplish this, we shall begin with nothing more than the standard cyclic laws of thermodynamics, which themselves involve *neither* the concept of energy nor entropy, *nor* any a priori belief that these entities should exist at all in a continuum setting, much less obey the Gibbs relation.

A program of this kind was first envisioned in a brilliant and seminal paper of Coleman and Owen in 1974. In addition to the positive aspects of their work, they also found, as might be guessed, that actually carrying out the program was not entirely simple, and, even more, that restrictions were necessary on the type of material systems for which the derivations could proceed.

Our present point of view will be clearer if we first introduce the *specific heat and work rates*, $\mathcal{Q}$ and $\mathcal{W}$, for a process or motion of the material, themselves constitutive functions for the material. To this end, with the help of (A5) we first observe that the motion equations can be rewritten in the form[1]

$$\rho b \cdot v + v \cdot \operatorname{div} T + T : D = (p/\rho)\dot{\rho} + \Psi, \tag{2.6}$$

where $p$ is the pressure and

$$\Psi = \rho v \cdot \dot{v} + V : D, \qquad T = -pI + V. \tag{2.7}$$

[1] Some standard but useful identities are placed in an Appendix at the end of the paper, with numbering (A1)–(A6).

The relation (2.6) in turn gives, after integration,

$$-\int_I \left\{ \int_{S(t)} \rho b \cdot v \,\mathrm{d}x + \int_{\partial S(t)} Tn \cdot v \,\mathrm{d}x \right\} \mathrm{d}t = \int_I \int_{S(t)} \rho \mathcal{W} \,\mathrm{d}x \,\mathrm{d}t. \tag{2.8}$$

Here $(t, \mathcal{S}(t), t \in I,\ \mathcal{S}(t) \subset \mathbf{R}^3$, is a *configuration tube* of a motion of the continuum, and

$$\rho \mathcal{W} = -(p/\rho)\dot{\rho} - \Psi, \tag{2.9}$$

with $\Psi$ given by (2.7).

The left-hand side of (2.8) is the *total work* which is done by the body during the time interval $I$. It is clear therefore that the function $\mathcal{W}$ represents the *specific work rate* for the medium, and that (2.8) plays the role of a Work Balance Axiom.

Next, we rewrite the energy equation in the form

$$\rho r + \operatorname{div} h = \rho \dot{u} - T:D,$$

or, with the help of the Gibbs relation and (A5),

$$\rho r + \operatorname{div} h = \rho \theta \dot{s} - V:D. \tag{2.10}$$

By integration, there results

$$\int_I \left\{ \int_{S(t)} \rho r \,\mathrm{d}x - \int_{\partial S(t)} h \cdot n \,\mathrm{d}s \right\} \mathrm{d}t = \int_I \int_{S(t)} \rho \mathcal{Q} \,\mathrm{d}x \,\mathrm{d}t, \tag{2.11}$$

where

$$\rho \mathcal{Q} = \rho \theta \dot{s} - V:D. \tag{2.12}$$

Since the left-hand side of (2.11) is precisely the *total heat* which is supplied to the body during the time interval $I$ it is clear that the function $\mathcal{Q}$ represents the *specific heat rate* for the medium. In turn (2.11) represents a Heat Balance Axiom for thermomechanics.

We emphasize that $\mathcal{Q}$ *and* $\mathcal{W}$ *are both constitutive quantities*, depending explicitly on the material variables

$$\rho,\ \theta,\ v,\ \dot{s},\ \dot{v},\ \operatorname{grad} v,\ V.$$

Note finally that the integral balance (2.11), or its differential form (2.10), is essentially equivalent to the energy equation (2.3), provided we use the Gibbs relation and the equation of continuity (see the Remark at the end of Section 3). This suggests that we take, as the fundamental integral equations of thermomechanics, the balances (2.8) and (2.11), and the constitutive rules (2.9) and (2.12).

At the same time, since our purpose is to obtain the equation of thermomechanics *without* the a priori assumption of the existence of entropy (or of internal energy), the condition (2.12) must still be modified to avoid its explicit appeal to entropy. To this end, we take the final and crucial step of replacing (2.12) by the generic relation

$$\rho \mathcal{Q} = q_1 \dot{\theta} + q_2 \dot{\rho} - \Phi, \tag{2.13}$$

it being assumed that $\Phi$, as well as $V$ and $h$, depend continuously on the material variables

$$\rho, \theta; \dot{\rho}, \dot{\theta};\ \operatorname{grad}\theta,\ \operatorname{grad}\rho,\ \operatorname{grad}v \tag{2.14}$$

and satisfy as well

$$\Phi,\ V\!:\!D,\ h\cdot\operatorname{grad}\theta \;=\; o(|\dot{\rho}|,\ |\dot{\theta}|,\ |\operatorname{grad}\theta|,\ |\operatorname{grad}v|) \tag{2.15}$$

for small values of $\dot{\rho}$, $\dot{\theta}$, grad $\theta$ and grad $v$. The quantity $\Phi$ may be thought of as the *frictional heating* associated with the process; we do not assume a priori that it necessarily coincides with the stress work $V:D$, though this will be proved as part of our main result in the next section.

With the preceding discussion in mind, we now take *as the fundamental equations of thermomechanics* the balances (2.8), (2.11) and the (new) pair of constitutive formulas (2.9) and (2.13). Our purpose is then to derive from these relations the energy equation (2.3) and the Clausius–Duhem inequality (2.5).

To this end we shall rely on the well-known cyclic laws of general thermodynamics. These take the analytic form (see [4, 7, 9])

$$\bar{W}(P) = \bar{Q}(P),$$

$$\bar{A}(P) = \int_0^\infty \frac{\bar{Q}(P,\theta)}{\theta^2}\,\mathrm{d}\theta \leqslant 0$$

*for cyclic processes* $P$ of thermodynamic systems.[2] In the present context, the appropriate forms for $\bar{W}(P)$ and $\bar{Q}(P)$ are exactly the expressions which appear on the left-hand sides of (2.8) and (2.11), respectively. Moreover, as has been noted in [3, 7], we have obviously

$$Q(P,\theta) = \int_I \left\{ \int_{S_\theta(t)} \rho r\,\mathrm{d}x - \int_{\partial_\theta S(t)} h\cdot n\,\mathrm{d}s \right\} dt,$$

where $S_\theta(t)$ and $\partial_\theta S(t)$ are respectively the parts of $S(t)$ and of $\partial S(t)$ where the absolute temperature does not exceed $\theta$. Thus using Fubini's theorem, there results

$$\begin{aligned}\bar{A}(P) &= \int_I \left\{ \int_{S(t)} \frac{\rho r}{\theta}\,\mathrm{d}x - \int_{\partial S(t)} \frac{h\cdot n}{\theta}\,\mathrm{d}s \right\} \mathrm{d}t \\ &= \int_I \int_{S(t)} \frac{1}{\theta} \left\{ \rho r - \operatorname{div} h + \frac{h\cdot\operatorname{grad}\theta}{\theta} \right\} \mathrm{d}x\,\mathrm{d}t.\end{aligned} \tag{2.16}$$

Here, a process $P$ for the material system is a $C^2$ motion, and a corresponding temperature distribution, given by

$$x = \varphi(X,t), \qquad \theta = \hat{\theta}(X,t), \quad t \in I = [a,b], \tag{2.17}$$

[2] Units are chosen so that the mechanical equivalent of heat is 1.

with $\varphi(X,0) = X \in \mathcal{S}_0$, the initial configuration (body location at $t = 0$); correspondingly the density $\rho = \rho(X,t)$ is determined by

$$\rho \det\left(\frac{\partial\varphi}{\partial X}\right) = \rho_0(X), \tag{2.18}$$

where $\rho_0(X)$ is the initial density distribution. Note that the *configuration of the body* at time $t$ is given by $\mathcal{S}(t) = \varphi(\mathcal{S}_0, t)$.

Finally, we adopt the (by now) conventional agreement that all kinematically possible motions and temperature distributions are also dynamically possible for some force field and heat supply.

### 3. Main Results

Suppose the cyclic laws of thermodynamics hold, where $\bar{W}(P), \bar{Q}(P)$ are expressed by the left sides of (2.8), (2.11) and $\bar{A}(P)$ by (2.16). Let the specific heat and work rates be given by (2.9) and (2.13), the constitutive coefficients $p, q_1, q_2$ being continuous functions of $\rho, \theta$ in some domain $\Omega$ in $\mathbf{R}^+ \times \mathbf{R}^+$ (state space), and where the further constitutive functions $\Phi, V, h$ depend continuously on the variables (2.14). Finally we suppose that (2.15) is satisfied, this being a natural condition, for example, in the case of a Navier–Stokes fluid, where $V = \lambda \operatorname{div} v\, I + 2\mu\, D$, $h = \kappa \operatorname{grad}\theta$, and $V\!:\!D = \lambda|\operatorname{div} v|^2 + 2\mu D\!:\!D$, $h\cdot\operatorname{grad}\theta = \kappa|\operatorname{grad}\theta|^2$. Our main result is now the following.

*Under the conditions just given, there exist on $\Omega$ two continuously differentiable functions, the internal energy $u(\rho,\theta)$ and the entropy $s(\rho,\theta)$, having the properties that*

(i) $\rho\, \mathrm{d}s = q_1\, \mathrm{d}\theta + q_2\, \mathrm{d}\rho, \qquad \theta\, \mathrm{d}s = \mathrm{d}u - (p/\rho^2)\, \mathrm{d}\rho,$

(ii) *the frictional heating $\Phi$ is equal to $V\!:\!D$, and*

(iii) *the energy equation (2.3) and the Clausius–Duhem inequality (2.5) hold.*

*Remarks.* The list of constitutive variables (2.14) clearly limits the validity of the derivation to special materials; of course, all classical fluid materials, including in particular Navier–Stokes fluids, have exactly these material variables. For more general materials it is an open problem whether corresponding derivations are available – this seems to the author a matter of considerable interest. Nevertheless some comments can be made. First, the formula (2.13) for the specific heat rate can be written in the generic form

$$\mathcal{Q} = \hat{\mathcal{Q}}(\rho,\theta;\ \dot{\rho},\dot{\theta};\ \operatorname{grad}\theta,\ \operatorname{grad} v),$$

(with grad $\rho$ dropped).

If it is now assumed that $\hat{\mathcal{Q}}$ is continuously differentiable and that the natural condition

$$\hat{\mathcal{Q}}(\rho,\theta;\ 0,0;\ 0,0) = 0$$

is satisfied, it is easy to see from Taylor's expansion that (2.13) holds (linear first order terms in grad $\theta$ and in grad $v - \frac{1}{3}\operatorname{div} v\, I$ cannot occur in view of elementary invariance arguments) and moreover that $\Phi$ satisfies (2.15).

A slightly more general case which can be treated is that of viscoelasticity, when

$$\mathcal{Q} = \hat{\mathcal{Q}}(\rho,\theta,F;\ \dot{\rho},\dot{\theta},\dot{F};\ \operatorname{grad}\theta).$$

(here $F = \partial\phi/\partial X$; and grad $\rho$, grad $F$ are dropped), where $\hat{Q}$ is continuously differentiable and

$$\hat{Q}(\rho, \theta, F;\ 0, 0, 0;\ 0) = 0.$$

We omit the details since reference configuration arguments now play a role.

We emphasize a final time that in the main result both energy and entropy are now *derived quantities*, appearing nowhere in the given conditions (2.8), (2.9), (2.11), (2.13) or (2.15), nor in the statement of the general cyclic laws above.

*Derivation of the main result.* We proceed in several steps, obtaining in order (i), (ii) and finally (iii). The procedure in each case is to construct a suitable *cyclic process* $P$ (that is, a process of the form (2.17) which is periodic in $t$ and has $I$ as its fundamental interval), then to apply the analytical laws of thermodynamics to this process, and finally to conclude with an appropriate limit procedure. In all cases, the motion will have the general form

$$x = \varphi(X, t) = X \exp\{C(t)A\}, \tag{3.1}$$

where $A$ is a fixed $3 \times 3$ matrix, $X \in \mathcal{S}_0$ and $t \in I = [0, b]$. Clearly $C(0) = 0$ in order to have $x = X$ when $t = 0$.

Putting $c(t) = C'(t)$, it is easy to see that

$$v = \frac{\partial \varphi}{\partial t} = c(t)Ax,$$

$$\operatorname{grad} v = c(t)A, \qquad \operatorname{div} v = c(t)\operatorname{Trace} A.$$

To find the corresponding density, we have from (2.18), or equivalently from the equation of continuity,

$$\dot{\rho} = -\rho \operatorname{div} v = -c(t)\rho \operatorname{Trace} A,$$

that is $(\log \rho)^{\cdot} = -c(t)\operatorname{Trace} A$. Hence by integration and use of the initial condition $\rho(X, 0) = \rho_0(X)$, there results

$$\rho(X, t) = \rho_0(X) \exp\{-C(t)\operatorname{Trace} A\}.$$

To insure that the motion is cyclic and of class $C^2$, it is enough that $C(\cdot)$ be of class $C^2$ with $C(b) = C(0)$, $c(b) = c(0)$, $c'(b) = c'(0)$.

For parts (ii), (iii) below, we shall also need the formula

$$\operatorname{grad} \rho = \frac{\rho}{\rho_0} \frac{\partial \rho_0}{\partial X} \cdot \exp(-C(t)A),$$

where $x, X$ are related by (3.1); this is derived by the calculation

$$\begin{aligned} \operatorname{grad} \rho &= \frac{\partial \rho}{\partial X} \cdot \frac{\partial X}{\partial x} = \exp(-C(t)\operatorname{Trace} A)\frac{\partial \rho_0}{\partial X} \cdot \{\partial\varphi/\partial X\}^{-1} \\ &= \frac{\rho}{\rho_0} \frac{\partial \rho_0}{\partial X} \cdot \exp(-C(t)A). \end{aligned}$$

*Proof of* (i). Take $A = I$ and $\rho_0(X) = \text{Const}$, so in particular

$$\rho = \hat{\rho}(t) = \hat{\rho}(0)\exp\{-3C(t)\},$$

and choose also $\theta = \hat{\theta}(t)$. By defining $C(t)$, $\hat{\theta}(t)$ appropriately,[3] the pair $\{\hat{\rho}(t), \hat{\theta}(t)\}$ can be taken as the parametrization of an arbitrary $C^2$ closed curve $\Gamma$ in the state space $\Omega$. Clearly $\operatorname{grad}\rho = \operatorname{grad}\theta = 0$ for the corresponding process, while $\operatorname{grad} v = c(t)I$.

Then from $\bar{W}(P) = \bar{Q}(P)$ follows

$$\int_I \int_{S(t)} \rho(\mathcal{Q} - \mathcal{W})\,\mathrm{d}x\,\mathrm{d}t = 0, \tag{3.2}$$

that is, by (2.9) and (2.13),

$$\int_I \int_{S(t)} \left\{ q_1\dot{\theta} + q_2\dot{\rho} + \frac{p}{\rho}\dot{\rho} + \rho v\cdot\dot{v} \right\} \mathrm{d}x\,\mathrm{d}t$$
$$+ \int_I \int_{S(t)} \{V : D - \Phi\}\,\mathrm{d}x\,\mathrm{d}t = 0.$$

Let $J_1$ denote the first integral on the left-hand side. By (A3) in the Appendix, together with the fact that $\rho$, $\theta$, $p$, $q_1$, $q_2$ depend only on $t$, it is easy to see that

$$\begin{aligned} J_1 &= \int_I \left\{ \frac{1}{\rho}\left(q_1\dot{\theta} + q_2\dot{\rho} + \frac{p}{\rho}\dot{\rho}\right) \int_{S(t)} \rho\,\mathrm{d}x \right\} \mathrm{d}t \\ &= M \int_I \frac{1}{\rho}\left(q_1\dot{\theta} + q_2\dot{\rho} + \frac{p}{\rho}\dot{\rho}\right) \mathrm{d}t \;=\; M \int_\Gamma \frac{1}{\rho}(q - w), \end{aligned}$$

where $M = \int_{S(t)} \rho\,\mathrm{d}x = \int_{S_0} \hat{\rho}(0)\,\mathrm{d}x = \hat{\rho}(0)|S_0|$ is the total mass corresponding to the motion, and where

$$q = q_1\,\mathrm{d}\theta + q_2\,\mathrm{d}\rho, \quad w = -(p/\rho)\,\mathrm{d}\rho, \tag{3.3}$$

are differential forms on the state space. Similarly, by (2.15),

$$\begin{aligned} |J_2| &\leqslant \int_I \int_{S(t)} |V : D - \Phi|\,\mathrm{d}x\,\mathrm{d}t \\ &= \int_I \int_{S(t)} o(|\dot{\theta}| + |c(t)|)\,\mathrm{d}x\,\mathrm{d}t = \int_I o(|\dot{\theta}| + |c(t)|)\,\mathrm{d}t. \end{aligned}$$

We now change the time parametrization of the motion, writing $\tau/n$ in place of $t$, and $I_n = [0, bn]$ in place of $I$. The line integral for $J_1$ is of course unaltered. On the other hand,

$$\frac{\mathrm{d}\theta}{\mathrm{d}\tau} = \frac{1}{n}\dot{\theta}(t) \quad \text{and} \quad \frac{\mathrm{d}C}{\mathrm{d}\tau} = \frac{1}{n}c(t).$$

[3] That is, $C(t) = -\frac{1}{3}\log\{\hat{\rho}(t)/\hat{\rho}(0)\}$

Hence

$$|J_2| \leqslant \int_{I_n} o(1/n)\,\mathrm{d}\tau = o(1),$$

as $n \to \infty$.

Since $J_1 + J_2 = 0$, we find at once that $J_1 = 0$, that is

$$\int_\Gamma (1/\rho)(q - w) = 0.$$

Since $\Gamma$ is an arbitrary $C^2$ closed curve in the $(\rho, \theta)$ plane, or by approximation (smoothing the corners) an arbitrary piecewise smooth closed curve, it follows from the standard theory of line integrals that there exists a $C^1$ function $u(\rho, \theta)$ in $\Omega$ such that

$$\rho\,\mathrm{d}u = q - w. \tag{3.4}$$

(Here $\mathrm{d}u$ denotes the differential $(\partial u/\partial\theta)\,\mathrm{d}\theta + (\partial u/\partial\rho)\,\mathrm{d}\rho$.)

Now we turn to the second law condition $\bar{A}(P) \leqslant 0$. By (2.11) and a usual limiting technique (see [5]), we obtain

$$\rho r - \operatorname{div} h = \rho Q = q_1\dot\theta + q_2\dot\rho - \Phi. \tag{3.5}$$

Hence from (2.16)

$$\bar{A}(P) = \int_I \int_{S(t)} \frac{1}{\theta}\left\{q_1\dot\theta + q_2\dot\rho - \Phi + \frac{h\cdot\operatorname{grad}\theta}{\theta}\right\}\,\mathrm{d}x\,\mathrm{d}t. \tag{3.6}$$

In turn, exactly as in the calculation just concluded, we find that

$$\int_\Gamma \frac{q}{\rho\theta} \leqslant 0.$$

But the closed path $\Gamma$ can be traversed in either direction, so in fact

$$\int_\Gamma \frac{q}{\rho\theta} = 0.$$

It now follows that there exists a $C^1$ function $s(\rho, \theta)$ in $\Omega$ such that

$$\rho\theta\,\mathrm{d}s = q, \tag{3.7}$$

which by (3.2) is just the first half of (i). Combining (3.4) and (3.5) yields the Gibbs relation

$$\theta\,\mathrm{d}s = \mathrm{d}u + w/\rho, \tag{3.8}$$

completing the proof of (i).

The functions $u$ and $s$ are easily seen to be uniquely determined (up to additive constants) in terms of the pressure $p(\rho, \theta)$ and the constitutive coefficients $q_1$ and $q_2$.

*Proof of* (ii). For the second part of the proof the function $C(t)$ and the temperature distribution $\hat{\theta}$ must be chosen with considerably more care. In fact, rather than making the parameter change $\tau = t/n$, we shall instead introduce a family of processes $P_\alpha$, $0 < \alpha < 1$, in which $C(t)$ is replaced by functions

$$C_\alpha(t) = \int_0^t c_\alpha(t)\,dt,$$

where $t \in I = [0, \beta]$, $\beta = 1 + \alpha + \alpha^2$, and

$$c_\alpha(t) = \begin{cases} 1, & t \in [0, \alpha], \\ 1 - \dfrac{\beta}{\alpha^2}(t - \alpha), & t \in [\alpha, \alpha + \alpha^2], \\ -\alpha - \alpha^2, & t \in [\alpha + \alpha^2,\ 1 + \alpha], \\ 1 + \dfrac{\beta}{\alpha^2}(t - \beta), & t \in [1 + \alpha, \beta]. \end{cases}$$

It is easy to check that $c_\alpha(\cdot)$ is continuous, with $c_\alpha(\beta) = c_\alpha(0) = 1$, and that $\int_0^\beta c_\alpha(t)\,dt = 0$. Hence $C_\alpha(\cdot)$ is piecewise of class $C^2[0,\beta]$ with $C_\alpha(0) = C_\alpha(\beta) = 0$, $C'_\alpha(0) = C'_\alpha(\beta) = 1$. Thus $C_\alpha$ has a periodic piecewise $C^2$ extension. (It has previously been required that $C(t)$ should be of class $C^2$. By a simple smoothing procedure we could also arrange this here, but in the interests of simplicity we leave the easily carried out argument to the reader.)

Next, we take

$$\rho_0(X) = a_1 + B_1 \cdot X,$$

where $a_1 \in \mathbf{R}^+$ and $B_1 \in \mathbf{R}^3$ are arbitrary, and

$$\hat{\theta}(X, t) = a_2 + c_\alpha(t) B_2 \cdot x + C_\alpha(t) d,$$

where $a_2 \in \mathbf{R}^+$, $d \in \mathbf{R}$ and $B_2 \in \mathbf{R}^3$ are arbitrary.

The set $\mathcal{S}_0$ $(= \mathcal{S}(0))$ will be chosen as a small ball $|X| \leqslant \alpha^2$ in $\mathbf{R}^3$. With these choices, we observe that

$$\dot{\theta} = c_\alpha(t) d + c'_\alpha(t) B_2 \cdot x, \qquad \operatorname{grad}\theta = c_\alpha(t) B_2,$$

and

$$\operatorname{grad} v = c_\alpha(t) A.$$

Moreover,

$$\dot{\theta} = d, \qquad \operatorname{grad}\theta = B_2, \qquad \operatorname{grad} v = A \qquad \text{when } 0 \leqslant t \leqslant \alpha. \tag{3.9}$$

Finally, for all $t \in I$ we have

$$|C_\alpha(t)| < \alpha + \alpha^2 < 2\alpha, \qquad |c_\alpha(t)| \leqslant 1,$$

$$|x| < a^2 \exp\{2\alpha\|A\|\} \leqslant \text{Const. } \alpha^2,$$

and

$$|c'_\alpha(t)| = (1+\alpha+\alpha^2)/\alpha^2 \qquad \text{for } t \in [\alpha, \alpha+\alpha^2] \cup [1+\alpha, 1+\alpha+\alpha^2].$$

Using these estimates, we also see that

$$\rho = a_1 + O(\alpha), \qquad \operatorname{grad}\rho = B_1 + O(\alpha), \qquad \theta = a_2 + O(\alpha), \tag{3.10}$$

as $\alpha \to 0$.

Now return to the First Law condition $\bar{W}(P) = \bar{Q}(P)$ for the cyclic processes $P_\alpha$ just described. Then by (3.3) and (3.4)

$$\begin{aligned} 0 &= \int_I \int_{S(t)} \left\{ q_1\dot{\theta} + q_2\dot{\rho} + \frac{p}{\rho}\dot{\rho} + \rho v \cdot \dot{v} \right\} \mathrm{d}x\,\mathrm{d}t + \int_I \int_{S(t)} (V\!:\!D - \Phi)\,\mathrm{d}x\,\mathrm{d}t \\ &= \int_I \int_{S(t)} \rho(u + \tfrac{1}{2}|v|^2)^{\cdot}\,\mathrm{d}x\,\mathrm{d}t + \int_I \int_{S(t)} (V\!:\!D - \Phi)\,\mathrm{d}x\,\mathrm{d}t \\ &= \int_I \int_{S(t)} (V\!:\!D - \Phi)\,\mathrm{d}x\,\mathrm{d}t, \end{aligned} \tag{3.11}$$

as noted in the appendix, condition (A3).

Let $\Sigma = V\!:\!D - \Phi$. For the four intervals $I_1, I_2, I_3, I_4$ comprising $I$, there holds (by (3.9), (3.10) and the continuity of $\Sigma$)

$$\Sigma = \widehat{\Sigma}(a_1, a_2; d; B_1, B_2, A) + o(1) \quad \text{on } I_1,$$

the entry for $\dot{\rho}$ being omitted, since this term is dependent on grad $v$;

$$\Sigma = O(1) \quad \text{on } I_2, I_4,$$

since for these intervals

$$|c_\alpha(t)d| \leqslant \text{Const}, \qquad |c'_\alpha(t)B_2 \cdot x| \leqslant |B_2|\,\frac{1+\alpha+\alpha^2}{\alpha^2}\cdot\alpha^2 \leqslant \text{Const};$$

and finally

$$\Sigma = o(|\dot{\theta}| + |\operatorname{grad}\theta| + |\operatorname{grad} v|) = o(|c(t)|) = o(\alpha) \quad \text{on } I_3.$$

Consequently, from (3.11)

$$0 = \frac{\alpha M}{a_1}\widehat{\Sigma}(a_1, a_2; d; B_1, B_2, A) + o(\alpha^3) + O(\alpha^4) + o(\alpha^3).$$

Since also $M = \int_S \rho_0\,\mathrm{d}x \geqslant \text{Const. } a_1\alpha^2$, we can divide both sides by $\alpha M/a_1$ to get

$$0 = \widehat{\Sigma}(a_1, a_2; d; B_1, B_2, A) + o(1).$$

Let $\alpha \to 0$ to conclude that

$$\widehat{\Sigma}(a_1, a_2; d; B_1, B_2, A) = 0.$$

Since all the arguments are arbitrary, this gives $\Phi = V\!:\!D$, as required.

*Proof if* (iii) We apply the argument of the previous part to the thermodynamic condition $\bar{A}(P) \leqslant 0$ on cycles. To begin with, using (3.6), (3.3), and (3.7) we get

$$\begin{aligned}\bar{A}(P) &= \int_I \int_{S(t)} \frac{1}{\theta}\left\{\rho\theta\dot{s} - \Phi + \frac{h\cdot \operatorname{grad}\theta}{\theta}\right\} dx\, dt \\ &= \int_I \int_{S(t)} \left\{\frac{h\cdot \operatorname{grad}\theta}{\theta} - \Phi\right\} dx\, dt,\end{aligned}$$

by (A3). The preceding cyclic calculation now yields, exactly as in (ii),

$$\frac{h\cdot \operatorname{grad}\theta}{\theta} \leqslant \Phi. \tag{3.12}$$

We can now complete the proof of (iii), namely the derivation of the energy equation and the Clausius–Duhem inequality. Indeed from (3.5) and (3.7) we have

$$\rho r - \operatorname{div} h = \rho\theta\dot{s} - \Phi. \tag{3.13}$$

But

$$\begin{aligned}\rho\theta\dot{s} - \Phi &= \rho\theta\dot{s} - V\!:\!D = \rho\theta\dot{s} - T\!:\!D + (p/\rho)\dot{\rho} \\ &= \rho\dot{u} - T\!:\!D, \end{aligned} \tag{3.14}$$

respectively using (ii), (A5) and the Gibbs relation (3.8). Combining (3.13) and (3.14) gives the energy equation (2.3).

On the other hand, from (3.13) and (3.12),

$$\rho r - \operatorname{div} h \leqslant \rho\theta\dot{s} - \frac{h\cdot \operatorname{grad}\theta}{\theta},$$

which is the Clausius–Duhem inequality (2.5). This completes the proof of (iii), and so of our main result.

*Remark.* The same procedure does *not* yield the equations of motion; in fact from (2.7)–(2.9) one gets only

$$v\cdot(\rho\dot{v} - \rho b - \operatorname{div} T) = 0. \tag{3.15}$$

Hence we must either obtain the motion equations independently, as in classical derivations (see [5]), or postulate that the formula $\bar{W}(P) = \int\int \rho\mathcal{W}\, dx\, dt$ is Galilean invariant, i.e. that it is satisfied equally for $v$ and for $v + v_0$, with no change in the external force $b$. In this last case, we get (in place of (3.15))

$$(v + v_0)\cdot(\rho\dot{v} - \rho b - \operatorname{div} T) = 0, \tag{3.16}$$

(recall that $T$ depends on $v$ only through grad $v$). Using the fact that $v_0$ is arbitrary then yields the equations of motion (2.2).

## 4. The Weak First Law

A weaker version of the first law has been proposed by R.L. Fosdick and the present author [2], and by Miroslav Šilhavý [9], namely that

$$\bar{W}(P) \leqslant \bar{Q}(P) \tag{4.1}$$

for every cyclic process $P$ of a thermodynamic system (see also [7]). Though relation (4.1) contains less information than the usual First Law, we shall see that in many ways it implies equivalent or only modestly different results. For this reason one may wish to consider (4.1) as a serious alternative to the standard First Law.

In fact, if merely (4.1) is assumed, then the main results (i)–(iii) of the previous section are altered *only* in parts (ii) and (iii), these now taking the related forms

(ii)′ *the frictional heating* $\Phi$ *satisfies the inequality*

$$\frac{h \cdot \operatorname{grad} \theta}{\theta} \leqslant \Phi \leqslant V : D,$$

(iii)′ *the Clausius–Duhem inequality continues to hold, while the energy equation is replaced by an energy inequality*

$$\rho \dot{u} \leqslant \rho r - \operatorname{div} h + T : D.$$

The conclusions (ii)′, (iii)′ can be written more succinctly if we introduce the *entropy production* $\mathcal{R}$ and *energy loss* $\mathcal{S}$ by the definitions

$$\mathcal{R} = \Phi - \frac{h \cdot \operatorname{grad} \theta}{\theta}, \qquad \mathcal{S} = V : D - \Phi.$$

Then we have

(ii)″ $\mathcal{R} \geqslant 0, \qquad \mathcal{S} \geqslant 0$

*and*

(iii)″ $\rho \dot{u} = \rho r - \operatorname{div} h + T : D - \mathcal{S},$

$$\rho \dot{s} = \rho r - \operatorname{div} h + \frac{h \cdot \operatorname{grad} \theta}{\theta} + \mathcal{R}.$$

The derivation is almost identical to that in the previous section. First, in proving that (i) continues to hold, the only change arises from the fact that the equality (3.2) must be replaced by

$$\int_I \int_{S(t)} \rho(\mathcal{Q} - \mathcal{W})\, \mathrm{d}x\, \mathrm{d}t \geqslant 0.$$

This leads in turn to the conditions

$$J_1 + J_2 \geqslant 0 \quad \text{and } J_1 \geqslant 0,$$

that is

$$\int_\Gamma \frac{1}{\rho}(q - w) \geqslant 0.$$

But then, exactly as in the treatment of entropy, we get

$$\int_\Gamma \frac{1}{\rho}(q - w) = 0$$

since $\Gamma$ can equally be traversed in the reverse direction. The existence of an internal energy function satisfying (3.4) thus arises as before, and (i) is proved.

The proof of (ii)$'$ is also essentially unchanged. The only difference stems from the fact that (3.11) now becomes the inequality

$$0 \leqslant \int_I \int_{S(t)} (V:D - \Phi)\,\mathrm{d}x\,\mathrm{d}t.$$

This leads in turn to the condition $\Phi \leqslant V:D$, which in combination with (3.12) proves (ii)$'$.

Finally (iii)$'$ follows at once, since (3.14) is now replaced by

$$\rho\theta\dot{s} - \Phi \geqslant \rho\dot{u} - T:D,$$

in view of the second inequality of (ii)$'$. Using (3.13) then completes the derivation of (iii)$'$.

Conditions (ii)$''$ are of course nothing but a restatement of (ii)$'$, while (iii)$''$ arises directly from (3.12) and the defintions of $\mathcal{R}$ and $\mathcal{S}$.

As a complement to (ii)$''$ and (iii)$''$, we have

(iv) *the frictional heating* $\Phi$ *is given by*

$$\Phi = \frac{h \cdot \operatorname{grad} \theta}{\theta} + \mathcal{R} = V:D - \mathcal{S}$$

*and the sum of the entropy production and energy loss satisfies the identity*

$$\mathcal{R} + \mathcal{S} = V:D - \frac{h \cdot \operatorname{grad} \theta}{\theta}.$$

The elegant duality of the results (ii)$''$, (iii)$''$ and (iv) should not be overlooked.

It remains to consider the specific example of a Navier–Stokes fluid. Here

$$T = -pI + \lambda \operatorname{div} v\, I + 2\mu D, \quad h = -\kappa \operatorname{grad} \theta, \tag{4.2}$$

where $\lambda, \mu, \kappa$ are constants, necessarily satisfying the conditions

$$\kappa \geqslant 0, \quad \mu \geqslant 0, \quad 3\lambda + 2\mu \geqslant 0. \tag{4.3}$$

By (ii)$'$, the frictional heating then obeys

$$-(\kappa/\theta)|\operatorname{grad}\theta|^2 \leqslant \Phi \leqslant \lambda|\operatorname{div} v|^2 + 2\mu D:D.$$

Accordingly, we may introduce the constitutive hypothesis

$$\Phi = -(\kappa'/\theta)|\operatorname{grad}\theta|^2 + \lambda'|\operatorname{div} v|^2 + 2\mu' D:D, \tag{4.4}$$

where

$$0 \leqslant \kappa' \leqslant \kappa, \quad 0 \leqslant \mu' \leqslant \mu, \quad 0 \leqslant 3\lambda' + 2\mu' \leqslant 3\lambda + 2\mu. \tag{4.5}$$

Then (iii)$''$ takes the form (we need only give the entropy equation, since the energy equation is equivalent)

$$\rho\dot{s} = \rho r - \operatorname{div} h + \Phi. \tag{2.3)$'$}$$

A *full set of governing equations* is now given by (2.1), (2.2), (2, 3)$'$, with $T, h, \Phi$ defined by (4.3) and (4.4), and with the coefficients $\kappa, \lambda, \mu$ and $\kappa', \lambda', \mu'$ satisfying (4.5).

Note that (2.3) appears as the special case of (2.3)$'$ when $\kappa' = 0,\ \lambda' = \lambda$ and $\mu' = \mu$. We then have $\mathcal{S} = 0$ and $\mathcal{R} = \kappa|\operatorname{grad}\theta|^2 + \lambda|\operatorname{div} v|^2 + 2\mu D : D$, rather than the more general relations

$$\mathcal{R} \geqslant 0, \quad \mathcal{S} \geqslant 0,$$
$$\mathcal{R} + \mathcal{S} = \kappa|\operatorname{grad}\theta|^2 + \lambda|\operatorname{div} v|^2 + 2\mu D:D.$$

A more unusual set of assumptions for a Navier–Stokes type fluid, still however compatible with (ii)$'$, is given by

$$T = -pI + (\lambda \operatorname{div} v + \delta|\operatorname{grad}\theta|^2)I + 2\mu D,$$
$$h = -\kappa \operatorname{grad}\theta + \delta\theta\,(\operatorname{grad}\theta) \operatorname{div} v, \tag{4.6}$$
$$\Phi = -(\kappa'/\theta)|\operatorname{grad}\theta|^2 + \lambda'|\operatorname{div} v|^2 + 2\mu' D:D + \delta|\operatorname{grad}\theta|^2 \operatorname{div} v,$$

where $\kappa, \lambda, \mu$ and $\kappa', \lambda', \mu'$ satisfy (4.5) and $\delta$ is an arbitrary coefficient (positive *or* negative). These conditions show that, for a fluid, *it is thermodynamically possible for the stress tensor to depend explicitly on the temperature gradient.*

Perhaps unexpectedly, the frictional heating $\Phi$ given by (4.6) can be *negative* for large values of $\operatorname{grad}\theta$ if either $\kappa' > 0$ or $\delta \neq 0$. Adding the (extraneous) thermodynamic condition $\Phi \geqslant 0$, moreover, does not imply that the energy loss $\mathcal{S}$ vanishes, since even when $\kappa' = \delta = 0$ one still has $\mathcal{S} > 0$ when $\lambda' < \lambda$ or $\mu' < \mu$. That is, in the context of the weak first law the condition $\Phi \geqslant 0$ is neither necessary nor sufficient to imply $\mathcal{S} = 0$.

## Appendix

At a number of places in the paper we have used standard calculations of continuum mechanics. First, we have the well-known formula (see [5], Section 4)

$$\frac{\mathrm{d}}{\mathrm{d}t}\int_{S(t)} \rho F\,\mathrm{d}x = \int_{S(t)} \rho\dot{F}\,\mathrm{d}x, \tag{A1}$$

where the dot denotes the partial derivative $\partial F/\partial t$ with $X$ constant, in other words

$$\dot{F} = \frac{\partial F(x,t)}{\partial t} + v \cdot \operatorname{grad} F.$$

From (A1) we easily derive

$$\int_I \int_{S(t)} \rho \dot{F} \, \mathrm{d}x \, \mathrm{d}t = \int_I \left\{ \frac{\mathrm{d}}{\mathrm{d}t} \int_{S(t)} \rho F \, \mathrm{d}x \right\} \mathrm{d}t = \left[ \int_{S(t)} \rho F \, \mathrm{d}x \right]_{t=0}^{t=b}. \tag{A2}$$

In particular, for a *cyclic process* we have $\mathcal{S}(b) = \mathcal{S}(0)$ and $\rho(X,b) = \rho(X,0)$. Thus if $F$ is a constitutive function depending on the material variables $(\rho, \theta, v;\ \dot{\rho}, \dot{\theta}, \dot{v};\ \operatorname{grad}\rho,\ \operatorname{grad}\theta,\ \operatorname{grad} v)$, then in the context of the cyclic processes used in the main result in Section 3,

$$F(x,b) \equiv F(x,0), \quad x \in \mathcal{S}_0 = \mathcal{S}(0).$$

Hence for such cycles,

$$\int_I \int_{S(t)} \rho \dot{F} \, \mathrm{d}x \, \mathrm{d}t = 0. \tag{A3}$$

This formula is the crucial underpinning for the derivations in (ii), (iii) of Section 3.

Recall from (2.7) that

$$V = pI + T, \tag{A4}$$

where $T$ is the stress tensor, $p$ the (scalar) pressure and $V$ the viscosity tensor. We assert that

$$T\!:\!D = (p/\rho)\dot{\rho} + V\!:\!D, \tag{A5}$$

where $D$ is the symmetric part of the velocity gradient tensor grad $v$.

To prove this, note from the equation of continuity that

$$(p/\rho)\dot{\rho} = -p \operatorname{div} v = -pI\!:\!D. \tag{A6}$$

Relation (A5) now follows by forming the inner product of both sides of (A4) with $D$, and then applying (A6).

## References

1. Coleman, B.D. and Owen, D.R., 'A mathematical foundation for thermodynamics', *Arch. Rational Mech. Anal.*, **54** (1974), 1–104.
2. Fosdick, R.L. and Serrin, J., 'Global properties of continuum thermodynamic processes' *Arch. Rational Mech. Anal.* **59** (1975), 108–109.
3. Man, C.-S., 'Thermodynamics based on a work inequality', *Arch. Rational Mech. Anal.*, **106** (1989), 1–61.
4. Man, C.-S., 'Remarks on global and local versions of the second law of thermodynamics', In: *Thermoelastic Problems and the Thermodynamics of Continua*, Amer. Soc. Mech. Eng., Appl. Mech. Div., vol. **198** (1995), pp. 33–39.
5. Serrin, J., 'Mathematical principles of classical fluid mechanics', In: *Handbuch der Physik*, vol. VIII/1, Springer, Berlin, 1959.
6. Serrin, J., 'The concepts of thermodynamics', In: *Contemporary Developments in Continuum Mechanics and Partial Differential Equations*, North-Holland, Amsterdam 1978, pp. 411–451.

7. Serrin, J., 'An outline of thermodynamical structure', In: *New Perspectives in Thermodynamics*, Springer, Berlin, 1986, pp. 3–32.
8. Serrin, J., 'On the elementary thermodynamics of quasi-static systems and other remarks', In: *Thermoelastic Problems and the Thermodynamics of Continua*, Amer. Soc. Mech. Eng., Appl. Mech. Div., vol. **198** (1995), 53–62.
9. Šilhavý, M., 'The foundations of thermodynamics', In: *New Perspectives in Thermodynamics*, Springer, New York, 1986, pp. 34–48.
10. Truesdell, C., and Toupin, R., 'The classical field theories', In: *Handbuch der Physik*, vol. III/1, Springer, Berlin, 1960.
11. Truesdell, C., *Rational Thermodynamics*, McGraw-Hill, New York 1969.

*J. Elasticity* 90 (2008),129–159. **Reprint VI.2.[iii]**


# The Area Rule for Simple Fluid Phase Transitions

**James Serrin**

Received: 1 June 2007 / Accepted: 11 July 2007 / Published online: 13 September 2007


**Abstract** We propose a new version of the classical van der Waals equation of state. When this is combined with a modified Maxwell construction (replacing the logically inconsistent equal-area rule), it yields remarkably accurate boiling curves for real fluids over large temperature ranges. For the water-steam phase transition, for example, we find a calculated boiling curve which is accurate to within one percent of tabulated values over the entire temperature range 20 to 374°C, the latter being the critical temperature for water, beyond which a phase transition does not occur.

**Keywords** Phase transitions · Boiling pressure · Maxwell construction

**Mathematics Subject Classifications (2000)** 80A10 · 80A22

## 1 Introduction

The first theoretical approach to simple liquid-vapor transitions was given by the Dutch physicist J.D. van der Waals in his celebrated thesis of 1873, when he introduced the famous equation of state associated with his name,

$$p = \frac{RT}{v-b} - \frac{a}{v^2}, \tag{1.1}$$

where $p$ denotes pressure, $T$ absolute temperature, $v$ specific volume, $R$ the gas constant, and $a$ and $b$ positive constants. Van der Waals' purpose was to explain the phenomena of phase transitions, in which empirical isotherms have the typical form shown in Fig. 1, e.g., the carbon dioxide transition. The flat isotherms in the shaded region occur at constant pressure, and are interpreted as a varying mixture of

J. Serrin (✉)
Department of Mathematics, University of Minnesota, Minneapolis, USA

**Fig. 1** Empirical isotherms for carbon dioxide. The critical pressure is $p_c = 71.3$ atm, the critical molal volume is $v_c = 94.86\ \mathrm{cm}^3/\mathrm{mol}$

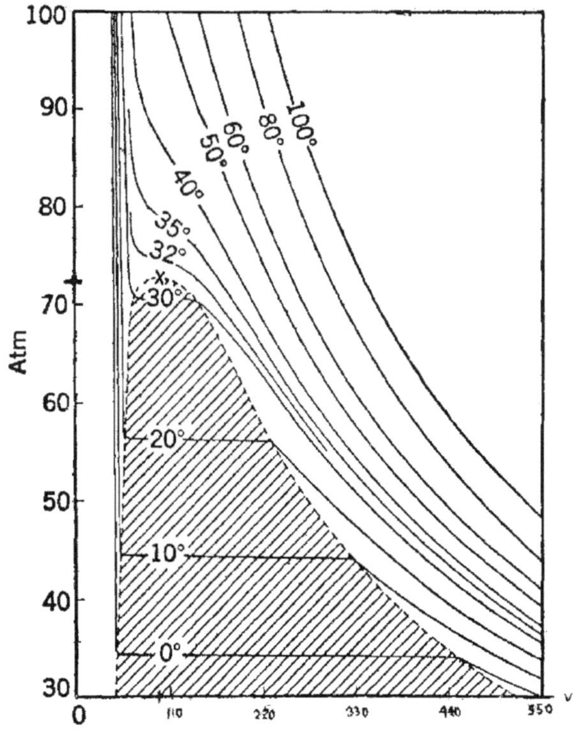


liquid and vapor phases coexisting at the temperature and pressure in question; the isotherms above and to the right of the shaded region display a typical "hyperbolic" form corresponding to the elementary ideal gas relation $pv = RT$. The point $\times$ is called the critical point, the corresponding pressure, temperature and specific volume being the critical parameters of the transition.

In Fig. 2 the isotherms for the van der Waals equation of state (1.1) are shown as solid lines. A correspondence between Figs. 1 and 2 is then obtained by introducing into Fig. 2 straight segments of the form $BE$, representing the phase transition. The state $B$ represents pure liquid and the state $E$ pure (saturated) vapor, with all the liquid being evaporated. The temperature $T$ of the flat isothermal segment $BE$ is called the boiling temperature associated with the pressure $\bar{p}$; it is taken to be the same as the temperature of the full isotherm $ABCDEF$. Three salient points immediately follow:

(1) Only the arcs $AB$ and $EF$ of the isotherm $ABCDEF$ define actual fluid states.

(2) The portion $CD$ of the isotherm where $\partial p/\partial v > 0$ can have no physical meaning – it corresponds to dynamically unstable states which would immediately degenerate into a mixture of fluid and vapor phases. (States on the arcs $BC$ and $DE$ are called metastable.)

(3) The question, which specific segment $BE$ should be associated with a given isotherm is left open for the moment – that is, given the temperature $T$ of an isotherm, the corresponding pressure level $\bar{p}$ must still be determined by further argument; this amounts in essence to a free boundary problem.

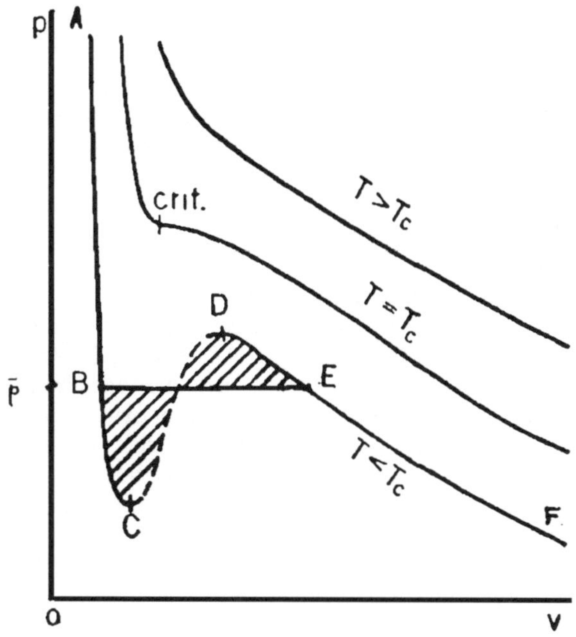


**Fig. 2** Isotherms for the van der Waals equation of state

These points are amplified in most thermodynamics textbooks (e.g., Epstein [3], Wilson [11], Pippard [6], to name three classical works where the ideas are carefully presented). Even more, it is understood that they apply not only to the van der Waals equation but equally to more general equations of state.

In accordance with this point of view, in the sequel we shall consider general equations of state

$$p = \rho(v, T) \tag{1.2}$$

for which the isothermals behave similarly to those shown in Fig. 2; we shall make the context precise in the next section. (The use of the letter $\rho$ to represent the pressure function should not be confused with the frequent use of $\rho$ to represent fluid density).

As is well-known, the classical method for assigning an appropriate boiling pressure $\bar{p} = \bar{p}(T)$ to a given temperature $T$ is the famous equal-area rule (Maxwell construction): the boiling pressure $\bar{p}(T)$ should be such that the two shaded areas in Fig. 2, determined by the segment $BE$, have equal measure. The rule is based on a cyclic thermodynamic argument in which the fluid is taken around the reversible cycle of states $BCDEB$, the final part $EB$ being at constant pressure (see [2, 3, 6, 11]). We shall return to this argument later, but for the present note that it is open to the serious objection that states on the arc $CD$ can have no meaning for a fluid (at least a fluid in equilibrium, as required for the cyclic argument).

Pippard in fact writes "From a logical point of view the cyclic thermodynamic arguments leading to the equal-area rule are worthless" [6], p. 117. After offering the above criticism, Pippard however goes on to say that "it is probable that (the equal-area rule) is correct, though this would require a more detailed investigation of the model to justify the result completely".

On the other hand, especially after Pippard's criticism appeared, further writers have claimed proof based on properties of the thermodynamic potential of the fluid, see, e.g., [1, 2, 10]. As we shall see, this second approach to the equal-area rule involves a crucial, though hidden assumption concerning the "constant of integration" associated to the thermodynamic potential. This assumption, however, is justified neither by logic nor by physical reality (Theorem 4.3, and the remarks after Theorem 3.1), so that once again the equal-area rule faces objection.

(In an isolated and unremarked paper devoted specifically to the van der Waals equation, G.D. Kaul in 1967 raised the same difficulties and went on to resolve the issue by means of a modified Maxwell construction. The point was also noted a little later by R.B.Griffiths, again only for the van der Waals model and again with no attempt to develop a general theory applicable to other equations of state. It is possibly for these reasons that, interesting and logically impeccable as the comments of Kaul and Griffiths may be, they seem to have had no effect on later literature).

With this discussion in mind, it is then the specific purpose of this paper to give a detailed thermodynamic investigation of the liquid–vapor transition process, *avoiding both Pippard's objection and the questionable "constant of integration" assumption noted above,* so going some way to settle the status of the equal-area rule. In Section 2 we introduce two explicit phase transition models based on a given state equation (1.2) together with an appropriate transition topology which reflects the transition structure illustrated in Figs. 1 and 2. The models moreover take into account a precisely defined heat form $q$, as is required in applying the principal laws of classical thermodynamics, that is

$$\mathrm{d}e = T\,\mathrm{d}s - p\mathrm{d}v, \qquad T\,\mathrm{d}s = q,$$

where $e$ is the specific internal energy and $s$ the specific entropy of the fluid. We then obtain explicit expressions for $e$ and $s$ in terms of the pressure and the heat form $q$; Theorems 3.1 and 3.2. The results seem to be new, as they explicitly, and in an unexpected way, take into account the topology of the phase diagram.

We then prove for model 1, independently of the usual thermodynamical equilibrium arguments, that the thermodynamic potential $g = e + pv - Ts$ at the states $B$ and $E$ must be equal, this being the basis for determining the boiling pressure and for resolving the free boundary problem for the segment $BE$.

A short calculation then yields an area (but *not* an equal-area) rule for $\bar{p}$, expressed in the following governing formula, see (4.1), (4.2):

$$\oint_{BCDEB} p\mathrm{d}v = \int_T^{T_c} (1 - T/\tau)E(\tau)\mathrm{d}\tau,$$

where

$$E(T) = c_B - c_E + \int_{BCDE} T\frac{\partial^2 p}{\partial T^2}\mathrm{d}v$$

and $c_B, c_E$ are the specific heats at constant volume of the fluid at the liquid phase $B$ and vapor phase $E$ at the boiling temperature $T$.

The formula can be interpreted as saying that the difference of the upper and lower areas in Fig. 2, namely

$$\oint_{BCDEB} p\mathrm{d}v,$$

is given by the integral on the right side, that is, a modified version of the Maxwell construction.

A first main consequence of the area rule is the following theorem:

*The equal-area rule is valid for a fluid obeying the van der Waals equation of state if and only if the specific heats of the liquid phase B and the vapor phase E are equal along the complete transition boundary.*

A second consequence is that *the boiling pressure curve $\bar{p} = \bar{p}(T)$ is flat at its endpoint, that is,*

$$\frac{\mathrm{d}\bar{p}}{\mathrm{d}T}(T_c) = \frac{\partial p}{\partial T}(v_c, T_c) \text{ (finite)}, \qquad \frac{d^2\bar{p}}{\mathrm{d}T^2}(T_c) = 0,$$

where $v_c$ and $T_c$ are respectively the critical volume and critical temperature; see Proposition 2.

Not surprisingly, the results also supply a proof of the well-known Clayperon–Clausius formula (Theorem 3.8). Finally, a second order ordinary differential equation for $\bar{p}(T)$ is developed, which the author believes can have further significant use.

Though of interest in itself, particularly as a clarification of the Maxwell construction, in the final analysis the area rule depends on the availability of appropriate equations of states to which it can be applied. That is to say, if one is concerned with a particular material, say carbon dioxide as in the case of Fig. 1, a definite equation of state (1.2) must be specified. This is not a trivial concern, since an area rule (or an equal-area rule) can scarcely even be stated if there is no formulated equation of state.

In the years following van der Waals' original paper a number of such equations were suggested, though so far as the author knows none of them have proved to be both relatively accurate and relatively simple. For example, the equation of Bertholot

$$p = \frac{RT}{v-b} - \frac{a}{Tv^2} \tag{1.3}$$

is only slightly more useful than van der Waals' equation, while the equations of Kamarlingh–Onnes and of Beattie and Bridgeman involve at least six empirical constants. I hope therefore to be forgiven for proposing still another candidate for the state equation (1.2), which as far as I know has not been suggested before, namely

$$p = \frac{RT}{v-b} - \frac{a}{T^s v^r}; \tag{1.4}$$

here in addition to the positive scaling constants $a$ and $b$, there are just two further parameters $r > 1$ and $s > -1$.[1]

Equation (1.4), first of all, allows a general law of corresponding states, which simplifies it to the form

$$\pi = \frac{4r}{r-1} \cdot \frac{\theta}{(r+1)\nu - (r-1)} - \frac{r+1}{r-1} \cdot \frac{1}{\theta^s \nu^r}, \tag{1.5}$$

where $\pi, \theta$ and $\nu$ are respectively the reduced pressure $p/p_c$, the reduced temperature $T/T_c$ and the reduced specific volume $v/v_c$. Only the two central parameters $r$ and $s$ remain; these in turn can be immediately determined in terms of the critical parameters of the fluid, that is, as we shall show,

$$r = \frac{2}{K} + \sqrt{\frac{4}{K^2} + 1}, \qquad \text{where } K = \frac{RT_c}{p_c v_c}, \tag{1.6}$$

with a similar formula for $s$. [(1.5) corresponds to the well-known reduced form of the van der Waals equation

$$\pi = \frac{8\theta}{3v - 1} - \frac{3}{v^2},$$

where $r = 2$ and $K = 8/3$.]

The theory of (1.4) is carried out in Section 5, while in Section 6 we develop an analytical program (algorithm) for applying the area rule to determine the boiling pressure. Remarkably, the boiling curve for (1.4) depends, up to obvious scaling of the temperature and pressure, only on the two parameters $r$ and $s$, together with a single additional value $\gamma$ whose significance will emerge later.

When applied to real fluids (here we consider water, carbon dioxide, nitrogen and Freon), (1.4) together with the area rule provides a boiling pressure function $\bar{p}(T)$ which is remarkably accurate over large ranges of the temperature, see Figs. 5, 6, 7, 8, 9, 10. The case of water is illustrative: from 150 to 374°C it over-represents the boiling pressure by less than 1%, and from 20 to 120°C it over-represents the pressure by less than 1/2%.

It is worth adding that the parameters $r$ and $s$ for water are appropriately taken as $r = 8/5, s = 0.45$, for carbon dioxide $r = 1.7, s = 0.44$, for nitrogen $r = 9/5, s = 0.33$ and for Freon $r = 12/9$, $s = -0.23$. The latter value for $s$ shows that the "van der Waals" term $a/T^s v^r$ in (1.4) can *increase* with increasing temperature, in contrast to the case for water, nitrogen and carbon dioxide, and moreover in contradiction to Epstein's statement that this term decreases as the temperature increases ([3], p.15).

[1]The nearest model to (1.4) which I can find in the literature is

$$p = \frac{RT}{v-b} \exp\left(-\frac{a}{RT^{s+1}v}\right)$$

[Wilson (8.23.5)], whose expansion for small values of $a/RT^{s+1}v$ takes the form

$$p = \frac{RT}{v-b} - \frac{a}{T^s v(v-b)}.$$

Wilson's equation yields the fixed ratio $RT_c/p_c v_c = \frac{1}{2}e^2$ for the critical parameters, in conflict with experimental values (again see Wilson, pages 212–213, or Section 6 below) and is thus open to the same objection as the van der Waals equation, for which the ratio takes the fixed value 8/3.

That $r$ does not have the "mutual attraction" value of 2 as in van der Waals' model should not be taken as contradicting fundamental physics, since for fluids one must deal with short range molecular forces which may be neither "hard" nor "soft."

The striking agreement between tabulated and calculated boiling pressures for the equation of state (1.4) needs some further comment. First, there is the elegant relation between the dimensionless parameter $K$ and the exponent $r$, which allows one to model the critical parameters $p_c$, $v_c$, $T_c$ with accuracy and simplicity, something generally not available for other models. The second exponent $s$ adds further to this accuracy. The parameter $r$ of course may be different than 2, though since $r > 1$ the perturbation term $a/T^s v^r$ continues to represent molecular attraction as in the van der Waals model itself. What is interesting is that the attraction exponent $r$ keeps the same value throughout the state space, and that this seems the ultimate reason behind the agreement. In other terms, the (1.4) supplies a rational generalization of van der Waal's original idea, at essentially the simplest level consistent with agreement at the critical point, while also allowing a Law of Corresponding States.

*Other approaches to the boiling pressure function.* Wilson (Section 9.3) has suggested another method for the determination of the boiling curve, based again on use of the thermodynamic potential but involving the specific heat difference $c_{p,B} - c_{p,E}$, where $c_p$ denotes specific heat at constant pressure. His analysis may be considered in some ways a precursor of the area rule; at the same time, he leaves the situation in a nebulous and unfinished state, and gives no indication of further work.

Numerous empirical expressions for the boiling curve have been proposed, going back to van der Waals' original paper, see Wilson, Section 9.3. All these formulas beg the question of any predictive or explanatory usefulness.

Use of the equation of state (1.4) for purposes beyond the determination of the boiling curve has not been explored. Its application to the Joule–Thomson experiment seems a particularly interesting avenue for further study.

*The status of the equal-area rule.* On the basis of the results of this paper I cannot say that this is an easy question. It depends first of all on whether one is prepared to believe in equations of state of the form (1.2), which are defined and continuously differentiable for full domains of values $T$, $v$ including the transition zone. If one does not accept even this first step, then any area rule becomes meaningless, or else simply tautological ($0 = 0$) if (1.2) *defines* straight line isotherms in the transition zone.

Thus let us accept the classical situation in which (1.2) has no straight isotherms, and contains an unstable region where $\partial p/\partial v > 0$. Then based on the transition models of Section 2 *the equal-area rule governs the phase transition if and only if*

$$E(T) = c_B - c_E + T \int_{v_B}^{v_E} \frac{\partial^2 p}{\partial T^2} \mathrm{d}v \equiv 0, \tag{1.7}$$

with $v_B, v_E$ the specific volumes of the liquid and vapor phases at transition. Of course we understand that this criterion concerns particular substances for which the values $c_B, c_E$ are available.

That is, when the application of thermodynamic theory is restricted to those points which represent either *stable* or *neutrally stable* states, as in the two models of Section 2 and as seems entirely natural, then we are logically led to the generalized area rule (4.1). This procedure is not inconsistent with the standard one, in that the possibility $E \equiv 0$ *is* indeed allowed, but at the same time is more general, since this condition is not forced to hold.

The reader may argue that the unstable region can be traversed by paths of quasi-static processes for which the state pressure is given by (1.2), *provided appropriate exterior constraints are applied*, and therefore that such regions are in fact open to thermodynamic argument - thus yielding the equal-area rule as a consequence. This is a difficult position to hold, in view of the remarks of Pippard [6] and Epstein [3], but in this direction we can say no more. The reader may alternately hold that the equal-area rule is correct, irrespective of thermodynamic arguments. This is also a difficult position to maintain, because of the experimental situation. The alternative simply seems a far more attractive point of view.

The question then is, whether for a given state equation the transition pressure $\bar{p}$ obtained from the area rule agrees, over some reasonably large range of temperatures, with the boiling pressure as observed or tabulated. If it does not, then the state equation in question cannot be acceptable.[2] That is, ultimately, the problem is one of finding (or deriving) an equation of state for which the area rule (not the equal-area rule) provides a boiling pressure $\bar{p}$ which agrees acceptably with observed values. This is the point of view we have adopted in Section 6, the state equation (5.1) being *postulated* while the parameters $r$, $s$ are then *derived* from knowledge of the critical parameters of the fluid.

In the final analysis, the phase transition model in Section 2 together with an appropriate equation of state (1.2) has its value in the predictive and explanatory theory of phase transitions. Whether the equal-area condition holds is itself immaterial, its validity essentially being a fortuitous matter for a particular fluid and a particular equation of state.

## 2 The Transition Model

In this section we introduce an unambiguous thermodynamical model of the transition situation depicted in Fig. 1 and informally discussed in the introduction.

I. *State Space*. The physical state of the fluid is represented by the variables $T$ and $v$, namely the absolute temperature and specific molal volume. These take their values in the open set

$$\Omega = \{(v, T) \in \mathbb{R}^2 : v > b,\ T > 0\}, \tag{2.1}$$

where $b$ is a given positive value representing the smallest molal volume the substance can occupy.[3]

II. *The Pressure Function* $\rho(v, T) : \Omega \to \mathbb{R}$, a twice continuously differentiable function of its variables. For a van der Waals fluid, for example, $\rho(v, T)$ is the

[2] An example here is the attempt to model the water-steam transition by the van der Waals equation (1.1). With $a$ and $b$ chosen so that the critical pressure and temperature are correct, we obtain the boiling curve in Fig. 12: tabulated values are shown as the dotted points, the agreement not being particularly good. One may surmise that it is for such a reason that Kaul did not consider his results of sufficient interest to justify application to other equations of state.

[3] Allowing for the full range of temperature values down to absolute zero is a formal convenience, which in practical circumstances is easily avoided.

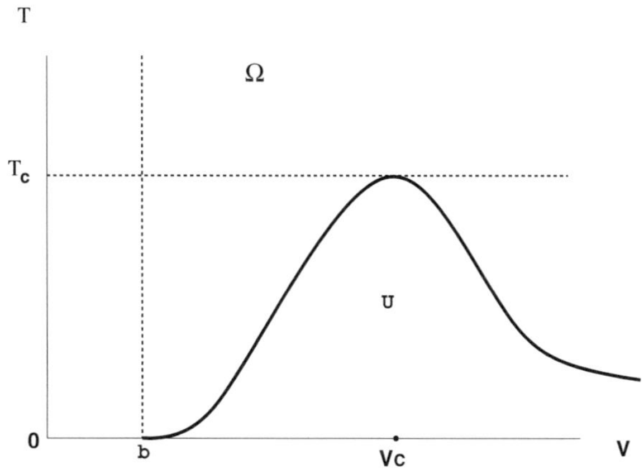


**Fig. 3** State domain for phase transitions. The unstable set $U$ is shaded

function on the right hand side of (1.1). As in the earlier informal discussion, the pressure function is not intended to represent the "state pressure" of the fluid interior to the transition region.

The pressure function is assumed to have the following properties for fixed $T > 0$:

$$\frac{\partial \rho}{\partial v} > 0 \quad \text{in } U, \quad \frac{\partial \rho}{\partial v} \leq 0 \quad \text{in } \Omega \backslash U; \quad \rho(v, T) \to 0 \quad \text{as } v \to \infty. \tag{2.2}$$

Here $U$ is the *unstable region of the model*, assumed to have the form shown in Fig. 3. In particular, the boundary of $U$ consists of two parts, *a strictly increasing curve from the initial point* $(b, 0)$ *to a maximum point* $(v_c, T_c)$*, and a strictly decreasing curve from* $(v_c, T_c)$ *to* $(\infty, 0)$. The values $v_c$, $T_c$ are the critical volume and the critical temperature. The corresponding value $p_c = \rho(v_c, T_c)$ is the critical pressure.

III. *Transition Region.* We suppose that there is a subdomain $\Omega_4$, the transition region, with the properties

$$\begin{cases} \Omega_4 \supset U \\ \text{Each isothermal } T < T_c \text{ in } \Omega \text{ intersects } \Omega_4 \text{ in a line segment} \\ \qquad (v_B, v_E), \quad b < v_B < v_E < \infty. \end{cases} \tag{2.3}$$

See Fig. 4: the points $B = (v_B, T)$, $E = (v_E, T)$ on $\Omega_4$ correspond to the transition endpoints $B$, $E$ in Fig. 2; of course $v_B$ and $v_E$ depend on the choice of $T$, as similarly the points $B$ and $E$ in Fig. 2 depend on the corresponding value $T < T_c$.

IV. *The State Pressure* $p(T, v)$. We give two different models for the state pressure,[4] the first a phenomenological one based on empirical isothermal diagrams, and the second Gibbsian, based on thermodynamical equilibrium principles.

[4]The state pressure is not to be confused with the pressure function $\rho$; they are the same only in the domain $\Omega \backslash \Omega_4$.

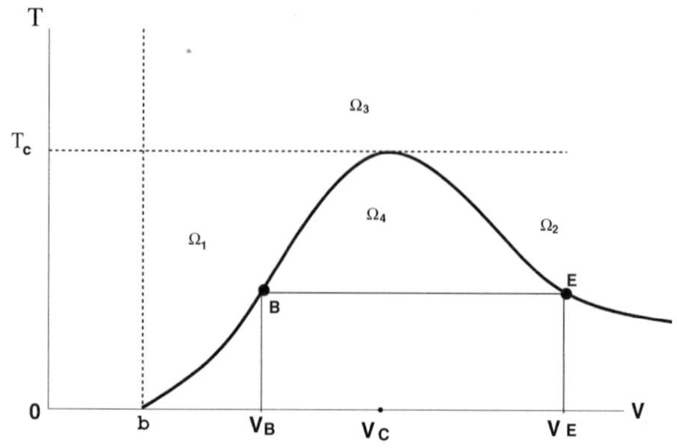


**Fig. 4** The transition domain $\Omega_4$. The points $B$ and $E$ correspond to the analogous points $B$ and $E$ in Fig. 2

*Model 1* (*Phenomenological*). The *state pressure* $p = p(v, T)$ is defined by

$$p = \begin{cases} \rho(v, T) & \text{when } (v, T) \in \Omega \backslash \Omega_4, \text{ and} \\ \bar{p}(T) & \text{when } (v, T) \in \Omega_4 \end{cases} \tag{2.4}$$

where $\bar{p}(T)$, $0 < T < T_c$, is a twice continuously differentiable function, the boiling pressure.

**Axiom 1** $p$ is continuous function over the full state space $\Omega$.

It is convenient, finally, to add the condition that $\partial \rho / \partial T > 0$ in $\Omega \backslash \Omega_4$.

The definition (2.4) of the state pressure reflects exactly the informal transition structure exhibited in Fig. 1. In particular, it is clear to begin with that

$$p(B) = p(E). \tag{2.5}$$

The conditions (2.2)–(2.4) moreover imply that $p(E) > 0$. Hence $p(B) > 0$, so without difficulty one finds that $p > 0$ in $\Omega$ and $\bar{p} > 0$. Finally it is easy to see that any isotherm with temperature $T \geq T_c$ cannot intersect $\Omega_4$; thus as in Fig. 4 the transition region is bounded above by the critical line $T = T_c$.

(It does not seem possible to prove that the boundary of $\Omega_4$ has the monotonicity properties shown in Fig. 4, though in practical cases this always seems true.)

Referring to Fig. 2, it is an easy consequence that the value $\bar{p}(T)$ must satisfy the inequality

$$\min_{ABCD} \rho(v, T) \leq \bar{p} \leq \max_{CDEF} \rho(v, T).$$

This condition, together with the fact that $\bar{p} > 0$, is the most that can be stated at this time concerning the boiling pressure function $\bar{p}$. We emphasize that the location of the transition domain $\Omega_4$ is *not* known a priori (free boundary problem), but depends on thermodynamic theory.

For the sequel, it is convenient to divide the set $\Omega \backslash \Omega_4$ into three subregions $\Omega_1, \Omega_2, \Omega_3$ as shown in Fig. 4 (with $\Omega_3$ including the line $T = T_c$). These regions are respectively called *liquid, vapor* and *superheated*.

The model above might be considered by some thermodynamicists as overly ambitious (or, for purists, even speculative), in that it ascribes a *state pressure* to liquid–vapor mixtures in the transition region. We can avoid this by introducing a second weaker model.

*Model 2 (Gibbsian).* The *state pressure* $p = p(v, T)$ is defined solely on the subset $\Omega\backslash\Omega_4$ of $\Omega$, by

$$p = \rho(v, T) \quad \text{when } (v, T) \in \Omega\backslash\Omega_4, \tag{2.6}$$

Here there is no function $\bar{p}$ and no need for the Axiom 1. In their stead, the following axiom is postulated.

**Axiom 2** $p(B) = p(E)$.

This axiom is a natural expression of the transition phenomena illustrated in Fig. 1. Alternately it is a standard consequence of equilibrium thermodynamics, see, e.g., [3], (6.26), or [10], (19.4.5). The *boiling pressure* can then be introduced by the definition

$$\bar{p}(T) = p(B) = p(E).$$

In turn $\bar{p}(T)$ can be considered the mixture pressure during the transition process; the comments after (2.5) now apply unchanged.

V. *The Laws of Thermodynamics.* The further elucidation of the liquid-vapor phase transition depends on applying the laws of equilibrium thermodynamics. Here we take these to be the relations

$$de = T ds - p\,dv, \qquad T ds = q, \tag{2.7}$$

where $e$ and $s$ are respectively the specific internal energy and the specific entropy (per mole), and $q$ is the heat form for the material in question.
Because of the different assumptions ascribed to the state pressure in Models 1 and 2, the laws (2.7) must be interpreted differently in these cases.

*Model 1. Phenomenological.* The heat form is taken to be

$$q = \ell dv + c dT,$$

where $\ell = \ell(v, T)$ and $c = c(v, T)$ are the latent heat, and the specific heat at constant volume of the system (per mole). The functions $\ell(v, T)$, $c(v, T)$ are however *not* continuous on the state space $\Omega$, as is the case for the pressure. (The reader may verify this for himself, e.g., by consulting tables of specific heats for various fluid substances.) In particular, we shall assume (1) that $\ell$ and $c$, when they are restricted to $\Omega\backslash\Omega_4$ are continuously differentiable, while (2) in $\Omega_4$ they are the restriction of functions which are continuously differentiable in the closure of $\Omega_4$. Thus $\ell$ and $c$ can be discontinuous across the boundary of $\Omega_4$. This being understood, the classical laws (2.7) then apply separately in $\Omega\backslash\Omega_4$ and in $\Omega_4$, but not on the boundary of $\Omega_4$. Necessarily, however,

*The internal energy $e$ and entropy $s$ are continuous over the full state space $\Omega$.*

We shall exploit these basic structure conditions in the next section.

*Model 2. Gibbsian.* The heat form (as well as the state pressure $p$) is defined solely in the region $\Omega\backslash\Omega_4$. In consequence, (2.7) together with $e$ and $s$ then apply only in the region $\Omega\backslash\Omega_4$.[5] In turn, a further axiom is required.

**Axiom 3** Let $g = e + pv - Ts$ be the thermodynamic potential (defined in $\Omega\backslash\Omega_4$). Then

$$g(B) = g(E).$$

As was the case for Axiom 2, this condition equally is a consequence of well-known equilibrium reasoning, see, e.g., [6], (7.11); [10], (19.4.5) or [11], (3.61.7).

For our present purposes, the particular justification offered for Axiom 3 is not an issue, though it is worth adding that in applying equilibrium theory to obtain the condition $g(B) = g(E)$ one does *not* require the free energy $e - Ts$ to exist or even be defined in the unstable set $U$, that is, Gibbsian minimization procedures are equally valid whether or not one can adduce a value for the free energy in $U$. We note finally that Theorems 3.3 and 3.4 below are inoperable and irrelevant in the case of Model 2. In all other respects, the developments in the sequel, and particularly the area rule, apply equally to both models.

*Remark* Both of the models 1 and 2 are closely related to the (free energy) principles C1–C4 of Griffiths [4]. Griffiths also adds two further analyticity assumptions, C5–C6, which the present writer considers too restrictive (and even Griffiths considers "very strong" and not thermodynamically motivated). He further treats a "double-tangent" construction in his Fig. 2, where the relevant function is also analytic, again a much too restrictive condition. Later in the paper however he remarks that analyticity conditions need not in fact be assumed, thus coming in essence to the models proposed here.

## 3 Energy and Entropy for Fluid Transitions

Our first results give explicit expressions for the functions $e$ and $s$ in $\Omega\backslash\Omega_4$. We assume without loss of generality that $e$ and $s$ vanish at the critical point $(v_c, T_c)$.

**Theorem 3.1** *For* $(v, T) \in \Omega\backslash\Omega_4$ *we have*

$$e = e(v, T) = \int_{v_c}^{v} \left( T\frac{\mathrm{d}\rho}{\partial T}(y, T) - \rho(y, T) \right) \mathrm{d}y + k(v, T) \tag{3.1}$$

$$s = s(v, T) = \int_{v_c}^{v} \frac{\partial \rho}{\partial T}(y, T)\,\mathrm{d}y + \int_{T_c}^{T} \frac{1}{\tau}\frac{\partial}{\partial \tau} k(v, \tau)\,\mathrm{d}\tau, \tag{3.2}$$

*where* $k(v, T)$ *is a continuously differentiable function in* $\Omega\backslash\Omega_4$ *satisfying* $k(v, T_c) = 0$ *and* $\partial k/\partial v = 0$.

[5]To study the question of metastable states the definitions of $\ell$ and $c$ can be extended to the domain $\Omega\backslash U$, and corresponding functions $e$, $s$ then introduced by means of (2.7). The transition models 1 and 2 would however remain unaffected.

*Remark* We emphasize that the formulas for the functions $e$, $s$ in Theorem 3.1 *do not apply when* $(v, T)$ *is in the transition region* $\Omega_4$. For Model 1, when $(v, T) \in \Omega_4$ the appropriate expressions for $e$ and $s$ are given in Theorem 3.3.

The function $k$ is of prime importance to the computation of $e$ and $s$. Since $\partial k/\partial v = 0$ in its domain of definition $(\Omega\backslash\Omega_4)$, we see that $k$ must be constant on horizontal lines in the liquid region $\Omega_1$ and also in the vapor region $\Omega_2$, though the constant values *need not be the same*. This fact is expressed in the following key result, in which we recall that $\Omega_3$ denotes the superheated region in $\Omega$.

**Theorem 3.2** *There are continuously differentiable functions* $k_1(T)$, $k_2(T)$, $k_3(T)$ *such that*

$$k(v, T) = \begin{cases} k_1(T) & \text{when } (v, T) \in \Omega_1, \\ k_2(T) & \text{when } (v, T) \in \Omega_2, \\ k_3(T) & \text{when } (v, T) \in \Omega_3. \end{cases}$$

*Moreover* $k_1 = k_2 = k_3 = 0$ *when* $T = T_c$, *while when* $T < T_c$ *we have*

$$k_1'(T) = c_B + T\int_{v_B}^{v_c} \partial^2\rho/\partial T^2 \mathrm{d}y, \qquad k_2'(T) = c_E - T\int_{v_c}^{v_E} \partial^2\rho/\partial T^2 \mathrm{d}y;$$

*finally*

$$c(v, T) = k_3'(T) \quad \text{when} \quad T \geq T_c.$$

*Here* $c_B$, $c_E$ *are the specific heats of the fluid at the states B and E.*

*Proof of Theorem 3.1* Throughout the proof we restrict the variables $(v, T)$ to the set $\Omega\backslash\Omega_4$. Using the facts that $q = \ell\,\mathrm{d}v + c\,\mathrm{d}T$ and $p = \rho$ in $\Omega\backslash\Omega_4$, we can write (2.7) in the form

$$\mathrm{d}e = (\ell - \rho)\mathrm{d}v + c\,\mathrm{d}T, \qquad T\mathrm{d}s = \ell\,\mathrm{d}v + c\,\mathrm{d}T. \tag{3.3}$$

It then follows that

$$\frac{\partial^2 e}{\partial v\partial T} = \frac{\partial c}{\partial v} = \frac{\partial(\ell - \rho)}{\partial T}, \qquad \frac{\partial^2 s}{\partial v\partial T} = \frac{\partial}{\partial v}\left(\frac{c}{T}\right) = \frac{\partial}{\partial T}\left(\frac{\ell}{T}\right).$$

Carrying out the differentiations on the right sides, an algebraic reduction then gives the Maxwell relations

$$\ell = T\frac{\partial\rho}{\partial T}, \qquad \frac{\partial c}{\partial v} = T\,\frac{\partial^2\rho}{\partial T^2}. \tag{3.4}$$

Hence from (3.3)

$$\frac{\partial e}{\partial v} = \ell - \rho = T\frac{\partial\rho}{\partial T} - \rho.$$

Integration now yields the first conclusion of the theorem. [In (3.1), the variable $v$ remains in $\Omega\backslash\Omega_4$, although the integration interval extends into $\Omega_4$ and even into $U$. The integral of course is meaningful since the pressure function $\rho$ is itself of class $C^2$ in all of $\Omega$.]

To obtain the second part we observe that

$$\ell = T\frac{\partial s}{\partial v} = T\frac{\partial \rho}{\partial T}. \tag{3.5}$$

Also by (2.7) and (3.1)

$$c = T\frac{\partial s}{\partial T} = \frac{\partial e}{\partial T} = \int_{v_c}^{v} T\frac{\partial^2 \rho}{\partial T^2}(y, T)\mathrm{d}y + \frac{\partial k}{\partial T}. \tag{3.6}$$

Consider the function

$$\hat{s}(t, v) = \int_{v_c}^{v} \frac{\partial \rho}{\partial T}(y, T)\mathrm{d}y + \int_{T_c}^{T} \frac{1}{\tau}\frac{\partial k}{\partial \tau}\mathrm{d}\tau;$$

then by (3.5) and (3.6)

$$\frac{\partial \hat{s}}{\partial v} = \frac{\partial \rho}{\partial T} = \frac{\partial s}{\partial v}, \qquad \frac{\partial \hat{s}}{\partial T} = \int_{v_c}^{v} \frac{\partial^2 \rho}{\partial T^2}\mathrm{d}v + \frac{1}{T}\frac{\partial k}{\partial T} = \frac{\partial s}{\partial T},$$

so that $d(s - \hat{s}) = 0$, as required. □

*Proof of Theorem 3.2.* When $T < T_c$ the sets $\Omega_1$ and $\Omega_2$ are disconnected. Because $\partial k/\partial v = 0$ it follows as already noted above that $k$ must be constant on horizontal lines in both $\Omega_1$ and $\Omega_2$, but the constant values *need not be equal.* On the other hand, $\Omega_3$ is connected, so that the first part of the theorem is proved.

Next, by (3.6) we get

$$k_1'(T) = c(v, T) - \int_{v_c}^{v} T\frac{\partial^2 \rho}{\partial T^2}(y, T)\mathrm{d}y, \qquad b < v \le v_B.$$

Putting $v = v_B$ gives the required expression for $k_1'(T)$, while the formula for $k_2'(T)$ is obtained similarly. To get the final result we put $v = v_c$, which can be done since $T \ge T_c$. □

The following Theorems 3.3 and 3.4 apply only to Model 1.

**Theorem 3.3** *For* $(v,\ T)$ *in* $\Omega_4$ *we have*

$$e = e(v, T) = \left(T\frac{\mathrm{d}\bar{p}}{\mathrm{d}T} - \bar{p}\right)(v - v_c) + \bar{k}(T)$$

$$s = s(v, T) = \frac{\mathrm{d}\bar{p}}{\mathrm{d}T}(v - v_c) + \int_{T_c}^{T} \frac{1}{\tau}\frac{\mathrm{d}\bar{k}}{\mathrm{d}\tau}\mathrm{d}\tau$$

*where* $\bar{k}(T)$ *is a continuously differentiable function in* $0 < T < T_c$.

This is obtained exactly as Theorem 3.1, with the single exception that the condition $\partial\bar{k}/\partial v = 0$ in $\Omega_4$ implies at once that $\bar{k} = \bar{k}(T)$; this deduction is not available for Theorem 3.1 since the region $\Omega\backslash\Omega_4$ is *not connected in the* $T$ *variable when* $T < T_c$.

**Theorem 3.4** *Let $g = e + pv - Ts$ be the thermodynamic potential of the transition system. Then for $0 < T < T_c$ we have*

$$g(B) = g(E),$$

*where $B = (v_B, T)$ and $E = (v_E, T)$ are the intersection points of the boundary of $\Omega_4$ with the isotherm in $\Omega$ at temperature $T$.*

*Proof* Since $g$ is continuous function in all of $\Omega$, we can evaluate $g(B)$ and $g(E)$ using Theorem 3.3. In particular in $\Omega_4$ we have

$$g(v,T) = e + \bar{p}(T)v - Ts = \bar{k}(T) - \int_{T_c}^{T} \frac{T}{\tau}\frac{d\bar{k}}{\mathrm{d}\tau}\mathrm{d}\tau + \bar{p}(T)v_c.$$

But the right side *depends only on* $T$ so it follows that $g(E) = g(B)$, as required. □

## 4 The Area Rule

The main result of the paper can now be stated.

**Proposition 1** (*General area rule*). *For $0 < T < T_c$ we have*

$$\int_{v_B}^{v_E} \rho(y,T)\,\mathrm{d}y - \bar{p}(T)[v_E - v_B] = \int_{T}^{T_c} \left(1 - \frac{T}{\tau}\right) E(\tau)\mathrm{d}\tau, \tag{4.1}$$

*where*

$$E(T) = c_B - c_E + \int_{v_B}^{v_E} T\frac{\partial^2 \rho}{\partial T^2}(y,T)\,\mathrm{d}y. \tag{4.2}$$

*Proposition* 1 *can be written alternatively as*

$$\oint_{BCDEB} p\mathrm{d}v = \int_{T_c}^{T} (1 - T/\tau)E(\tau)\mathrm{d}\tau,$$

*a result which takes the place of the standard Maxwell construction in determining the boiling pressure $\bar{p}$.*

Before proving Proposition 1 we note two immediate corollaries.

**Corollary 4.1** *The equal-area rule holds for a simple fluid transition if and only if*

$$c_B - c_E + \int_{v_B}^{v_E} T\frac{\partial^2 \rho}{\partial T^2}\mathrm{d}v = 0, \quad 0 < T < T_c.$$

**Corollary 4.2** *For a van der Waals fluid the equal-area rule holds if and only if*

$$c_B - c_E = 0, \quad 0 < T < T_c.$$

Since fluid tables give $c_B - c_E \neq 0$, it follows that the equal-area rule cannot generally hold for a van der Waals fluid.

*Proof of Proposition 1* This follows by evaluation of the quantities $g(B)$ and $g(E)$, using the expressions for $e$ and $s$ given in Theorem 3.1 together with the key relation $p(B) = p(E) = \bar{p}(T)$, that is

$$\begin{aligned} g(B) - g(E) = & \int_{v_E}^{v_B} \left( T\frac{\partial \rho}{\partial T}(y,T) - \rho)(y,T) \right) \mathrm{d}y + k(v_B,T) - k(v_E,T) \\ & + p(B)v_B - p(E)v_E \\ & - T\int_{v_E}^{v_B} \frac{\partial \rho}{\partial T}(y,T)\,\mathrm{d}y - T\int_{T_c}^{T} \left[ \frac{1}{\tau}\frac{\partial k}{\partial \tau}(v_B,\tau) - \frac{1}{\tau}\frac{\partial k}{\partial \tau}(v_E,\tau) \right] \mathrm{d}\tau \\ = & \int_{v_B}^{v_E} \rho \mathrm{d}v - \bar{p}(t)[v_E - v_B] \\ & + k_1(T) - k_2(T) - \int_{T_c}^{T} \frac{T}{\tau}\big[k_1'(\tau) - k_2'(\tau)\big]\,\mathrm{d}\tau. \end{aligned}$$

Thus, since $k(v_c, T_c) = 0$ and $g(B) = g(E)$ – for Model 1 see Theorem 3.4, and for Model 2 see Axiom 3 – we get

$$\int_{v_B}^{v_E} \rho \mathrm{d}v - \bar{p}(t)[v_E - v_B] = \int_{T}^{T_c} (1 - T/\tau)\big[k_1'(\tau) - k_2'(\tau)\big]\,\mathrm{d}\tau. \tag{4.3}$$

On the other hand, from Theorem 3.2

$$k_1'(T) - k_2'(T) = c_B - c_E + \int_{v_B}^{v_E} T\frac{\partial^2 \rho}{\partial T^2}(y,T)\mathrm{d}y. \tag{4.4}$$

The area rule now follows from (4.3) and (4.4). □

The next result shows the relation between the "constants of integration" $k_1$ and $k_2$ and the equal-area rule. Its proof follows from the calculations in the demonstration of the area rule.

**Theorem 4.3** *The following statements are equivalent*:

(1) $k_1(T) \equiv k_2(T)$,

(2) $c_B - c_E + \int_{v_B}^{v_E} T\frac{\partial^2 \rho}{\partial T^2}(y,T)\mathrm{d}y \equiv 0$,

(3) *The equal-area rule holds.*

We emphasize again that there is no logical necessity for any of these conditions to hold.[6] Even more, condition (2) is highly unlikely to occur in practice, since it relates the physical values $c_B$, $c_E$ with the assigned pressure function $\rho$.

[6]Griffiths ([4], page 187) puts it similarly, "[T]here is no thermodynamic inconsistency in using a rule other than Maxwell's for modifying the ... equation of state." In the same way, Kaul [5] writes "[The] Maxwell method is not unique, and a generalization of it permits a more realistic description of a two-phase fluid."

**Theorem 4.4** *The function $\bar{p}(T)$ satisfies*

$$\int_{v_B}^{v_E} \frac{\partial \rho}{\partial T}(y,\tau)\mathrm{d}y - \frac{\mathrm{d}\bar{p}}{\mathrm{d}T}[v_E - v_B] = -\int_T^{T_c} \frac{E(\tau)}{\tau}\mathrm{d}\tau,$$

$$\frac{d^2\bar{p}}{\mathrm{d}T^2}[v_E - v_B] + \frac{\partial \rho}{\partial v}(v_E, T)\left(\frac{\mathrm{d}v_E}{\mathrm{d}T}\right)^2 - \frac{\partial \rho}{\partial v}(v_B, T)\left(\frac{\mathrm{d}v_B}{\mathrm{d}T}\right)^2 = -\frac{1}{T}(c_B - c_E).$$

*Proof* Differentiating the area rule with respect to $T$ yields the first result, after noting that the derivatives of $v_E$ and $v_B$ cancel.

Differentiating again with respect to $T$ we get (using primes to indicate differentiation with respect to $T$)

$$\int_{v_B}^{v_E} \rho''(y,T)\,\mathrm{d}y + \rho'(v_E,T)v_E' - \rho'(v_B,T)v_B' - \bar{p}''(T)[v_E - v_B] - \bar{p}'(T)[v_E' - v_B']$$

$$= \frac{E(T)}{T} = \frac{c_B - c_E}{T} + \int_{v_B}^{v_E} \rho''(y,T)\,\mathrm{d}y \tag{4.5}$$

by (4.2). But $\bar{p}(T) = \rho(v_B, T) = \rho(v_E, T)$, so differentiation of this relation gives

$$\bar{p}'(T) = \frac{\partial \rho}{\partial v}(v_B, T)v_B' + \frac{\partial \rho}{\partial T}(v_B, T) = \frac{\partial \rho}{\partial v}(v_E, T)v_E' + \frac{\partial \rho}{\partial T}(v_E, T). \tag{4.6}$$

The second conclusion now follows by combining (4.5) and (4.6).

Theorem 4.4 gives a complicated second order ordinary differential equation for $\bar{p}(T)$, of the form

$$\frac{d^2\bar{p}}{\mathrm{d}T^2} = \mathcal{F}\left(T, \bar{p}, \frac{\mathrm{d}\bar{p}}{\mathrm{d}T}\right) + C(T).$$

The author has been unable to see how to use this equation to obtain $\bar{p}$ independently of the area rule. At the same time it shows that the form of the function $\rho$ in the unstable region $U$ cannot affect the boiling pressure. □

**Proposition 2** *We have*

$$\bar{p}(T_c) = p_c, \qquad \frac{\mathrm{d}\bar{p}}{\mathrm{d}T}(T_c) = \frac{\partial \rho}{\partial T}(v_c, T_c), \qquad \frac{d^2\bar{p}}{\mathrm{d}T^2}(T_c) = 0.$$

The second part of Proposition 2 has been noted by Rowlinson ([8], p. 63), while the third part seems to be new. It is interesting that the latter result agrees with the observed behavior of $\bar{p}$ near the critical point.

*Proof* Divide both sides of (4.1) by $v_E - v_B$ to obtain

$$\frac{1}{v_E - v_B}\int_{v_B}^{v_E} \rho(y,T)\mathrm{d}y - \bar{p}(T) = \frac{1}{v_B - v_E}\int_T^{T_c} (1 - T/\tau)E(\tau)\mathrm{d}\tau.$$

Letting $T \to T_c$, we see that the left side approaches $\rho(v_c, T_c) - \bar{p}(T_c)$, that is $p_c - \bar{p}(T_c)$. The limit of the right side is somewhat more difficult. But from Fig. 2 one deduces that $v_E - v_B = o(1)$ as $T \to T_c$, while also obviously $v_E - v_B \geq T_c - T$, from which it follows that the right side tends to zero (use (4.2) and the fact that $c_B - c_E = o(1)$ as $T \to T_c$). This proves the first claim.

Next, divide the first result of Theorem 4.4 by $v_E - v_B$ to obtain

$$\frac{1}{v_E - v_B}\int_{v_B}^{v_E} \frac{\partial \rho}{\partial T}\mathrm{d}y - \frac{\mathrm{d}\bar{p}}{\mathrm{d}T} = \frac{1}{v_B - v_E}\int_T^{T_c} \frac{E(\tau)}{\tau}\mathrm{d}\tau.$$

Letting $T \to T_c$, the left side obviously approaches

$$\frac{\partial \rho}{\partial T}(v_c, T_c) - \frac{\mathrm{d}\bar{p}}{\mathrm{d}T}(T_c),$$

while as before the right side tends to zero.

The third result requires an extended proof, which we defer to Section 4.7. □

**Theorem 4.5** *We have*

$$T\frac{\mathrm{d}\bar{p}}{\mathrm{d}T} = \frac{L}{v_E - v_B}, \qquad L = \int_{v_B}^{v_E} \ell(y, T)\mathrm{d}y$$

*Proof* From Theorem 3.2 there follows

$$\frac{\mathrm{d}\bar{p}}{\mathrm{d}T}[v_E - v_B] = s(v_E, T) - s(v_B, T) = \int_{v_B}^{v_E} \mathrm{d}s,$$

and the conclusion follows since $\mathrm{d}s = (\ell/T)\,\mathrm{d}v$ when $T$ is constant.

The quantity $L$ is the *heat of vaporization* in the transition from liquid to vapor. Theorem 3 is known as the Clayperon–Clausius formula. It should be observed that the Clayperon–Clausius formula is independent of the area rule. □

## 5 A Model Equation of State

Beyond the specific van der Waals equation of state, many further equations have been suggested, as discussed in the introduction. I have found that the following equation is useful over a large temperature range, while moreover allowing the hypotheses of the transition model of Section 2 to be verified by fairly direct computation. We take in particular

$$\rho(v, T) = \frac{RT}{v - b} - \frac{a}{T^s v^r}, \tag{5.1}$$

where $a, b$ are positive constants, $R$ is the gas constant, and $r, s$ are structure constants,

$$-1 < s < 1, \qquad r > 1.$$

As convenient units, for definiteness we take bars (bar) for pressure, $^o$K for temperature, cm$^3$/mol for specific molal volume, in which case $R = 83.144$ bars cm$^3$/mol $^o$K. (*Note.* 1 atm = 1.0131 bars.)

We prove next that the pressure function $\rho$ given by (5.1) has the properties required in Part II of Section 2.

**Theorem 5.1** (1) *Let*

$$v_c = \frac{r+1}{r-1}b, \qquad T_c^{s+1} = \frac{4r}{(r+1)^2}\cdot\frac{a}{Rv_c^{r-1}},$$

*The boundaries of the unstable set $U$ are monotone, rising from* 0 *to* $T_c$ *as* $v$ *goes from* $b$ *to* $v_c$ *and decreasing from* $T_c$ *to* 0 *as* $v$ *goes from* $v_c$ *to* $\infty$.

(2) $\partial p/\partial v < 0$ *for all* $v > b$ *if and only if* $T > T_c$.

To prove (1), we observe that both the left and right boundaries of $U$ are defined by $\partial p/\partial v = 0$, that is

$$T^{s+1} = \frac{ar}{R}\frac{(v-b)^2}{v^{r+1}}.$$

Hence along these boundaries we have

$$(s+1)T^s\frac{\mathrm{d}T}{\mathrm{d}v} = \frac{ar}{R}\left\{\frac{2(v-b)}{v^{r+1}} - \frac{(r+1)(v-b)^2}{v^{r+2}}\right\},$$

which gives

$$\begin{aligned}\frac{\mathrm{d}T}{\mathrm{d}v} &= \frac{ar}{R}\frac{1}{(s+1)T^s}\frac{v-b}{v^{r+2}}\{2v-(r+1)(v-b)\}\\ &= \frac{r-1}{s+1}\frac{ar}{R\,T^s\,v^r}\frac{v-b}{v}\frac{v_c-v}{v}.\end{aligned}$$

This is positive for $v < v_c$, negative for $v > v_c$ and zero for $v = v_c$, proving (1). Note also that, along the boundary of $U$,

$$T,\ \frac{\mathrm{d}T}{\mathrm{d}v} \to 0 \quad \text{as } v \to b \text{ and } v \to \infty.$$

Condition (2) is an immediate consequence of (1) and the fact that $\partial p/\partial v < 0$ when $v = v_c$, $T > T_c$ and $\partial p/\partial v > 0$ when $v = v_c$, $T < T_c$.

Finally $dp/\partial T > 0$ in $\Omega\backslash\Omega_4$. Indeed, if $0 \le s < 1$ then

$$\frac{\partial p}{\partial T} = \frac{R}{v-b} + \frac{as}{T^{s+1}v^r} > 0$$

and the conclusion is obvious. When $-1 < s < 0$ the situation is more complicated. First, clearly $\partial p/\partial T > 0$ above the curve $\Gamma$ and $\partial p/\partial T < 0$ below $\Gamma$, where $\Gamma$ is defined by

$$RT^{s+1} = a|s|\frac{v-b}{v^r}.$$

Recalling the equation for the boundary of $U$:

$$RT^{s+1} = ar\frac{(v-b)^2}{v^{r+1}},$$

we see that the $\Gamma$ crosses the boundary of $U$ when $v$ satisfies

$$\frac{v-b}{v} = \frac{|s|}{r}.$$

Since $|s|/r < 2/(r+1)$, it follows that the crossing point is on the left of $v_c$.

To continue we shall add the further assumption that $\bar{p}(T)$ is an increasing function, as in the case of real fluids. It is then not difficult to show that if $\partial\Omega_4$ dips below $\Gamma$ then it necessarily moves downward to the right as $T$ decreases. But this is impossible since $\partial\Omega_4$ is always above $U$.

In summary, the critical values $v_c$, $T_c$ for (5.1) are

$$v_c = \frac{r+1}{r-1}b, \qquad T_c^{s+1} = \frac{4r}{(r+1)^2} \cdot \frac{a}{Rv_c^{r-1}}, \tag{5.2}$$

with corresponding critical pressure

$$p_c = \frac{RT_c}{v_c - b} - \frac{a}{T_c^s v_c^r} = \frac{r-1}{r+1} \cdot \frac{a}{T_c^s v_c^r}. \tag{5.3}$$

Moreover

$$K = \frac{RT_c}{p_c v_c} = \frac{4r}{r^2-1}, \qquad \left(\text{inversely } r = \frac{2}{K} + \sqrt{\frac{4}{K^2}+1}\right). \tag{5.4}$$

Finally,

$$\frac{\mathrm{d}\bar{p}}{\mathrm{d}T}(T_c) = \frac{2r + (r+1)s}{r-1} \cdot \frac{p_c}{T_c}. \tag{5.5}$$

Only (5.5) requires further proof. But at the critical point we have

$$\frac{T}{p}\frac{\mathrm{d}p}{\mathrm{d}T} = \frac{T}{p}\left\{\frac{R}{v-b} + \frac{as}{T^{s+1}v^r}\right\} = \frac{4r}{r^2-1} \cdot \frac{r+1}{2} + \frac{r+1}{r-1}s,$$

and (5.5) is now a consequence of Proposition 2. The importance of equation (5.1) in modeling real fluids lies in the fact that the parameters $r$, $s$ can be chosen so that *calculated values* for the critical parameters will agree with their observed values. That is, to use (5.1) to model a particular real fluid, one can first determine an appropriate value for the parameter $r$ by applying the dimensionless formula (5.4), with $T_c$, $v_c$, $p_c$ being the observed critical values for the fluid; the parameter $s$ can similarly be found from the relation (5.5). In turn, $b$ can be found from the first equation of (5.2), and then $a$ from the second (though these constants are seldom needed).

*The reduced state equation* The model (5.1) allows a law of corresponding states. That is, when expressed in the reduced variables

$$\nu = \frac{v}{v_c}, \qquad \theta = \frac{T}{T_c}, \qquad \pi = \frac{p}{p_c}$$

it becomes, with the help of Proposition 3,

$$\pi = \frac{4r}{r^2-1} \cdot \frac{\theta}{\nu - t} - \frac{1}{t\theta^s \nu^r}, \qquad t = \frac{r-1}{r+1}, \tag{5.6}$$

in which the parameters $R, a, b$ disappear, and only the two determining parameters $r$ and $s$ remain. The state relation (5.6) is called the reduced equation for (5.1).

[For a van der Waals fluid, where $r = 2$. $s = 0$, the reduced equation takes the familiar form

$$\pi = \frac{8\theta}{3\nu - 1} - \frac{3}{\nu^2} \tag{5.7}$$

see [2], (11.2.17); [3], (1.26); [11], (8.211.2).]

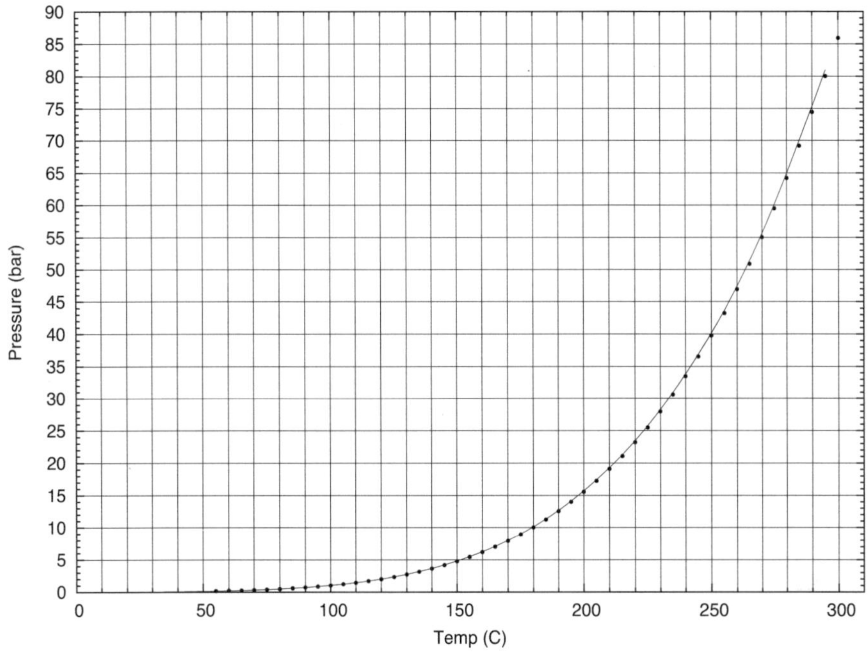


**Fig. 5** Water–steam boiling curve. The *solid line* is calculated from the area rule with $r = 8/5$, $s = 0.45$, $\gamma = 8$. The *dots* are tabulated values ([7, 9]). (The remaining temperature range $T > 290°\mathrm{C}$ [$T > 560\mathrm{K}$] is shown in Fig. 7)

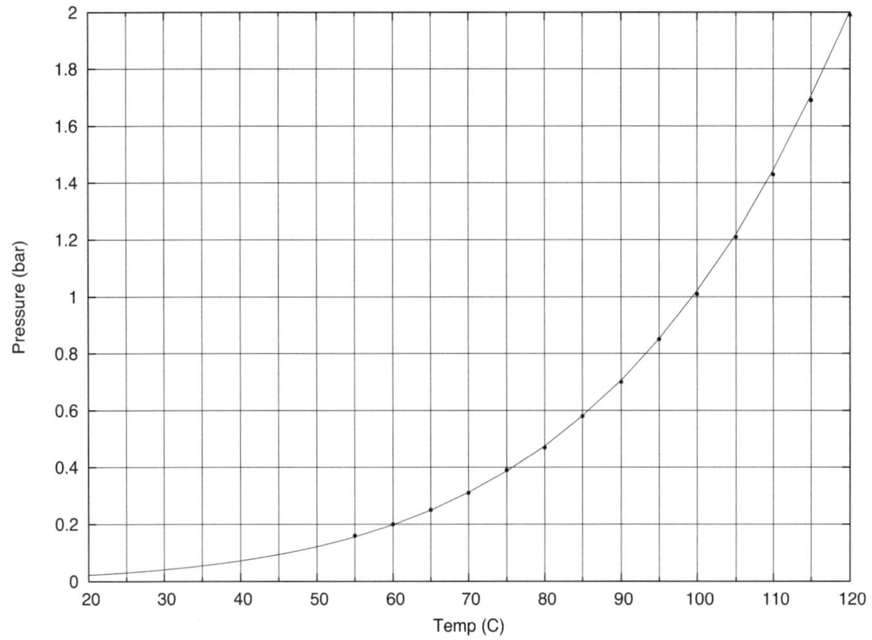


**Fig. 6** Enlarged water–steam boiling curve

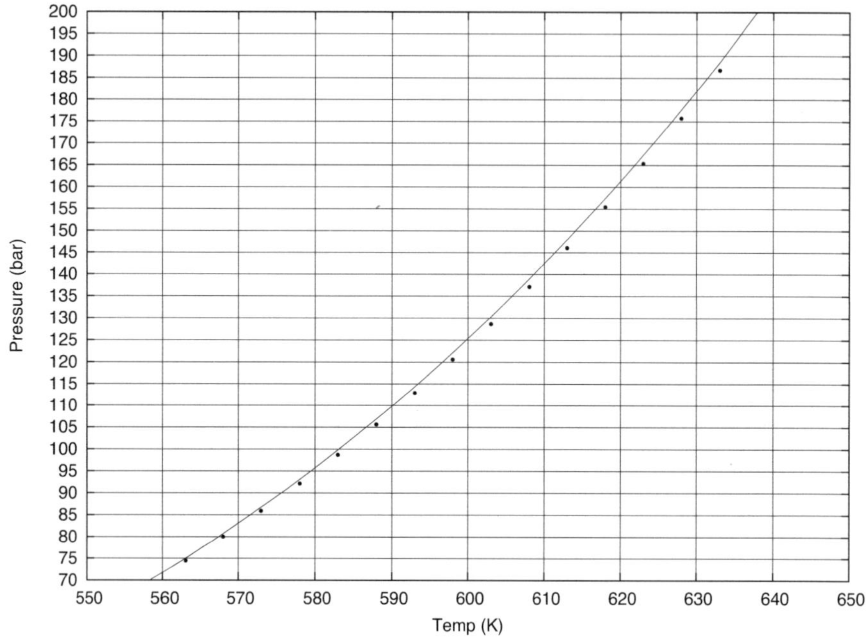


**Fig. 7** Water–steam boiling curve

## 6 Real Fluids

Here we shall apply the area rule together with the equation of state (5.1) to obtain the boiling curve for various fluids. First, using tabulated values of $v_c$, $T_c$, $p_c$ together with $R = 83.144$ one can calculate by means of (5.4) the corresponding values for the dimensionless parameters $K$ and $r$. The values of $K$ in fact vary quite widely, from 3.17 for helium to 6.69 for the refrigerant gas Freon 13 ($CF_3Cl$), with a great number in the range 3.4 to 3.9.

For definiteness we shall deal here with four representative fluids:

$$H_2O: \quad T_c = 647\,\text{K}, \; p_c = 221\,\text{bars}, \; v_c = 58.6\,\text{cm}^3/\text{mol}; \; \text{d}\bar{p}/\text{d}T_{[\text{crit}]} = 2.65\,\text{bars}/^o\text{K}.$$

Thus from Proposition 3,

$$K = 4.15, \quad r = 8/5, \quad s = 0.45.$$

$$CO_2: T_c = 304\text{K}, \; p_c = 73.5\,\text{bars}, \; v_c = 94.86\,\text{cm}^3/\text{mol}; \text{d}\bar{p}/\text{d}T|_{[\text{crit}]} = 1.59.\,\text{bars}/^o\text{K}.$$

Hence,

$$K = 3.626, \quad r = 1.7, \quad s = 0.44.$$

$$N_2: \; T_c{=}126.25\text{K}, \; p_c{=}33.96\,\text{bars}, \; v_c{=}94.92\,\text{cm}^3/\text{mol}; \; \text{d}\bar{p}/\text{d}T|_{[\text{crit}]}{=}1.55.\,\text{bars}/^o\text{K}.$$

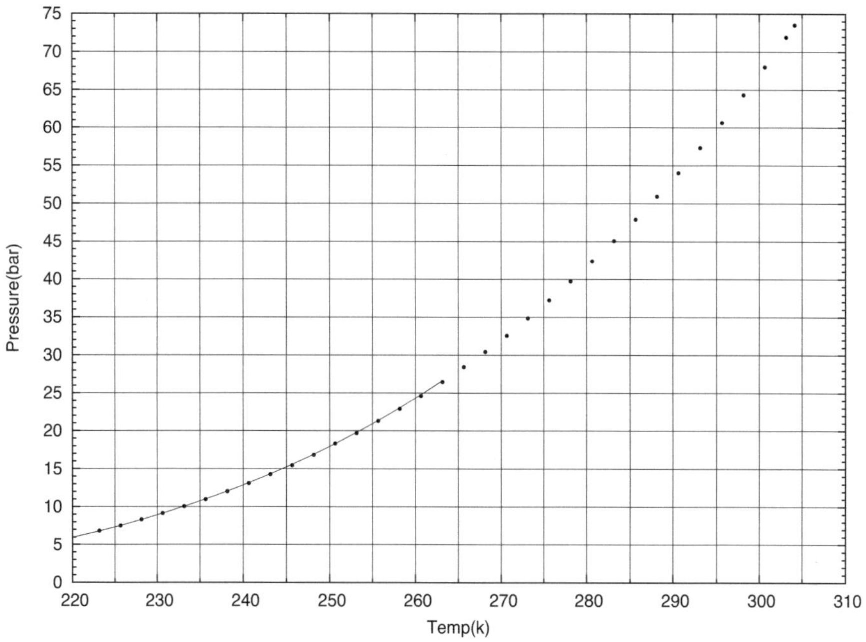


**Fig. 8** Carbon dioxide boiling curve, with $r = 1.7, s = 0.4, \gamma = 12$. *Solid line*, calculated values; *dots*, tabulated values

Hence,

$$K = 3.256, \quad r = 9/5, \quad s = 0.33.$$

Freon : $T_c = 302\text{K}, \; p_c = 38.6\,\text{bars}, \; v_c = 95.9\,\text{cm}^3/\text{mol}; \; \text{d}\bar{p}/\text{d}T_{[crit]} = 0.82.\,\text{bars}/^o\text{K}.$

Hence,

$$K = 6.694, \quad r = 4/3, \quad s = -0.23.$$

[Values of $r$ are chosen to be fractions with small denominators. Note also that in each case $r > 1$ and $|s| < 1$.]

For these materials we wish to use the general area rule of Proposition 1 to determine the boiling pressure $\bar{p}(T)$. It is then necessary to investigate the quantity $c_B - c_E$ in the function $E(T)$. Fairly good values for this quantity, i.e., the difference in specific molal heats (in kJ/kmol $^o$K) for the liquid phase $B$ and the saturated vapor phase $E$, can be found for water by using a Mollier chart for isochors (constant volume curves) in an entropy–temperature diagram, see, e.g., [3], Appendix B, and then noting that $c_v = T\text{d}s/\text{d}T$ at constant volume. Thus, as a reasonable approximation to this quantity, we find the relation

$$c_B - c_E \approx \gamma R(1 - \theta), \qquad \gamma = 8. \tag{6.1}$$

(See also *Note added in proof* at the end of Section 6) I have not located good tabulations of the specific heats $c_B$, $c_E$ for carbon dioxide, nitrogen and Freon.

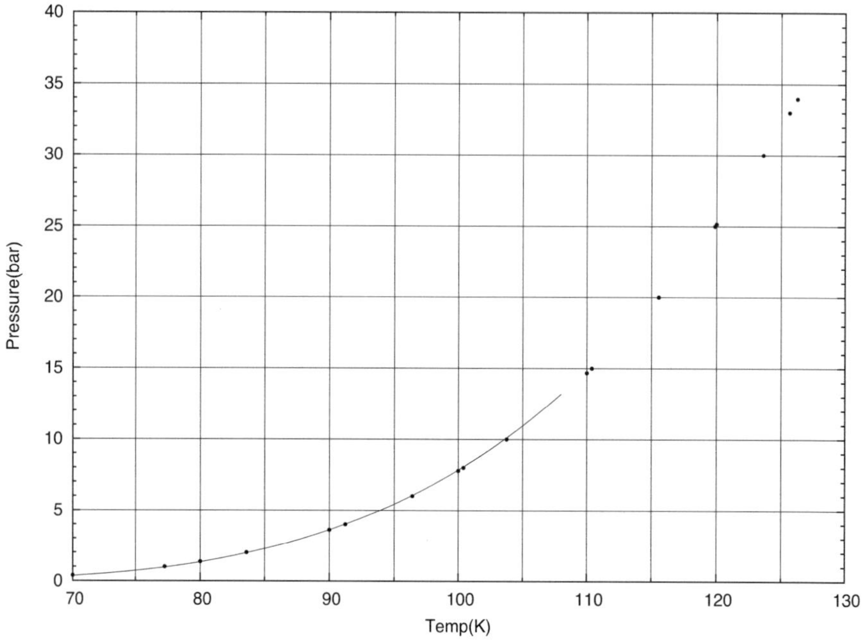


**Fig. 9** Nitrogen boiling curve, with $r = 9/5$, $s = 0.33$, $\gamma = 6$. *Solid line*, calculated values; *dots*, tabulated values

Consequently, for lack of good values, one may take the following relations as temporarily valid working approximations:

$$\begin{aligned} CO_2: \; c_B - c_E &\approx \gamma\, R(1-\theta), & \gamma &= 12, \\ N_2: \; c_B - c_E &\approx \gamma\, R(1-\theta), & \gamma &= 6, \\ CF_3Cl: \; c_B - c_E &\approx \gamma\, R(1-\theta). & \gamma &= 12. \end{aligned}$$

It now falls upon us to set up an appropriate method for obtaining $\bar{p}(T)$ from the area rule (Proposition 1). For temperatures not too near critical, we use the following program, which gives $\bar{p}$ in terms of the reduced temperature $\theta = T/T_c$ with high accuracy in the range $\theta < \theta_0 = 0.24\,[(r+1)/r]^{r+1}$:

Step 1: Put

$$\begin{aligned} A = A(\theta) &= \theta^s - s\theta^s(1-\theta) + 2r - 1, \\ B = B(\theta) &= \theta^s - s\theta(1-\theta) + 1, \\ C = C(\theta) &= \tfrac{1}{2}(1/\theta - \theta) + \theta \log\theta, \\ D = D(\theta) &= \frac{(r+1)^2}{8r}\,\frac{A}{\theta^{s+1}}. \end{aligned}$$

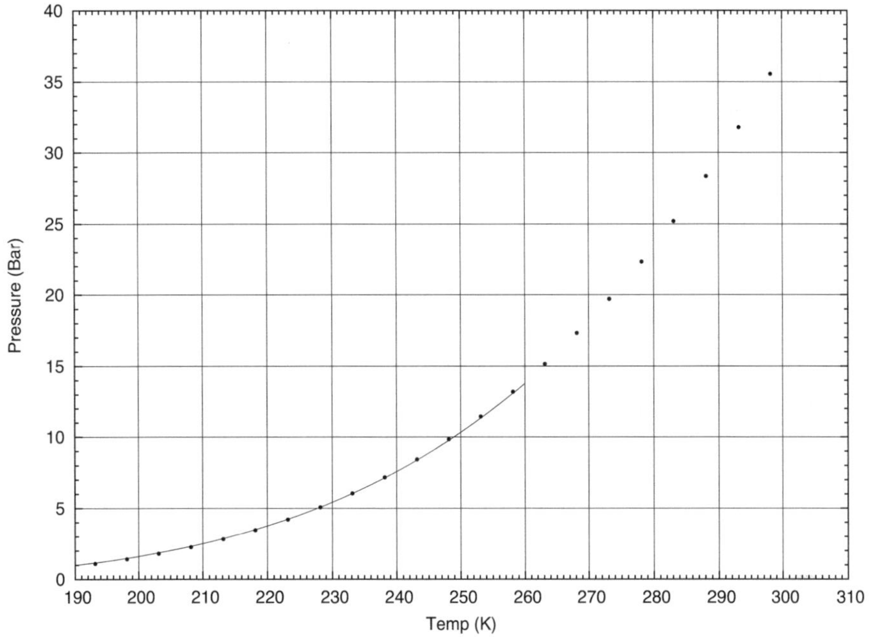


**Fig. 10** Freon 13 boiling curve, with $r = 4/3$, $s = -0.23$, $\gamma = 12$. *Solid line*, calculated values; *dots*, tabulated values

Step 2: Find $x < r/(r+1)$ satisfying the equation

$$\frac{4r}{(r+1)^2}\,\theta^{s+1}x^r - x + t = 0, \qquad t = \frac{r-1}{r+1}.$$

Step 3: Find $w < 1$ such that

$$w\,e^{-w} = \frac{D}{(x-t)^{r-1}} \cdot \exp\left\{-\frac{B}{2}\frac{x}{x-t} - (r+1)(C\gamma+1)\right\},$$

where $\gamma$ is the leading coefficient in the expressions (6.1), et al, for the specific heat differences $c_B - c_E$.

Step 4: Then $\bar{p}(\theta) = p_c\bar{\pi}(\theta)$, where

$$\bar{\pi}(\theta) = K\theta\left\{1 - \frac{2w}{A}\right\}\left(\frac{w}{D}\right)^{1/(r-1)}.$$

The specific molal volumes $v_B$ and $v_E$ at the transition temperature can also be obtained by a further calculation,

Step 5: At the temperature $\theta = T/T_c$ we have

$$v_B = xv_c, \qquad v_E = (D/w)^{1/(r-1)}v_c.$$

(The complication of the program is a consequence of the complicated nature of the area rule itself, which as far as the writer knows cannot be avoided. A derivation of the program is available from the author by writing to his email address.)

When applied to water, carbon dioxide, nitrogen and Freon the results are shown in Figs. 5, 6, 7, 8, 9, 10. Tabulated values from [7] and [9] are shown by dots, with the agreement being remarkably close, as discussed in the introduction.

For the further temperature range $(\theta_0,\ 1)$ the above program is inoperable, and an alternate procedure must be adopted. For this, we observe that for temperatures near critical the integral on the right side of (4.1) is extremely small, of order $O(T_c - T)^{5/2}$, whereas the shaded regions of Fig. 2 have measures of order $O(T_c - T)^{3/2}$. When $\theta > \theta_0$ it is therefore reasonable to approximate the boiling pressure $\bar{p}(T)$ by the simple equal-area relation

$$\bar{p}(T) = \frac{1}{v_E - v_B} \int_{v_B}^{v_E} \rho(y, T)\, dy.$$

This can be solved by the following iterative procedure.

Step 1: Choose $\bar{p}_0 > 0$ and find $x_1$ and $z_1$, the least and greatest roots $\nu$ of the equation

$$\frac{4r}{r^2 - 1} \cdot \frac{\theta}{\nu - t} - \frac{1}{t\,\theta^s\,\nu^r} = \frac{\bar{p}_0}{p_c},$$

where $\theta = T/T_c$ and $t = (r - 1)/(r + 1)$.

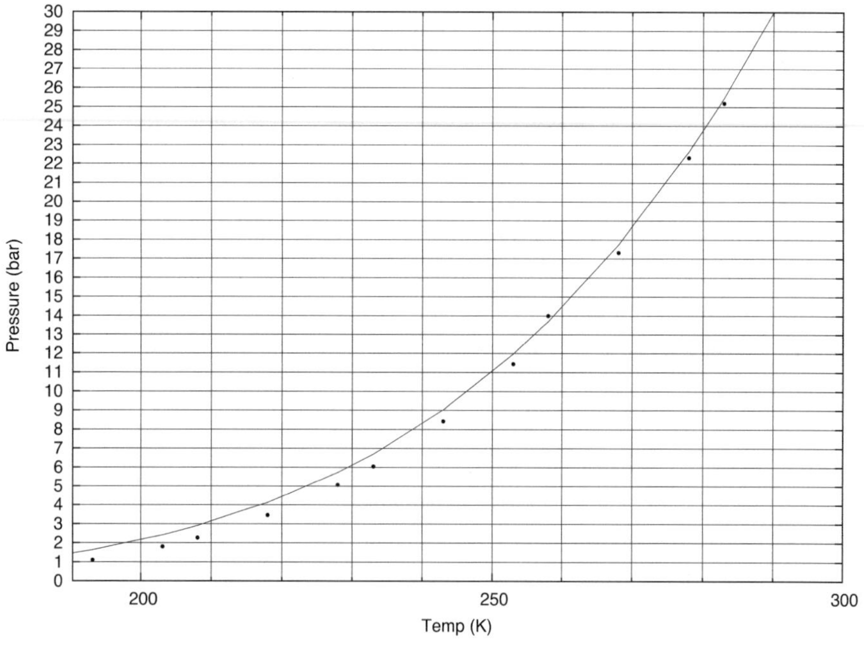


**Fig. 11** Freon 13 boiling curve, according to the equal-area rule. *Solid line*, calculated values; *dots*, tabulated values: note the accuracy above 260 K and the inaccuracy below this temperature

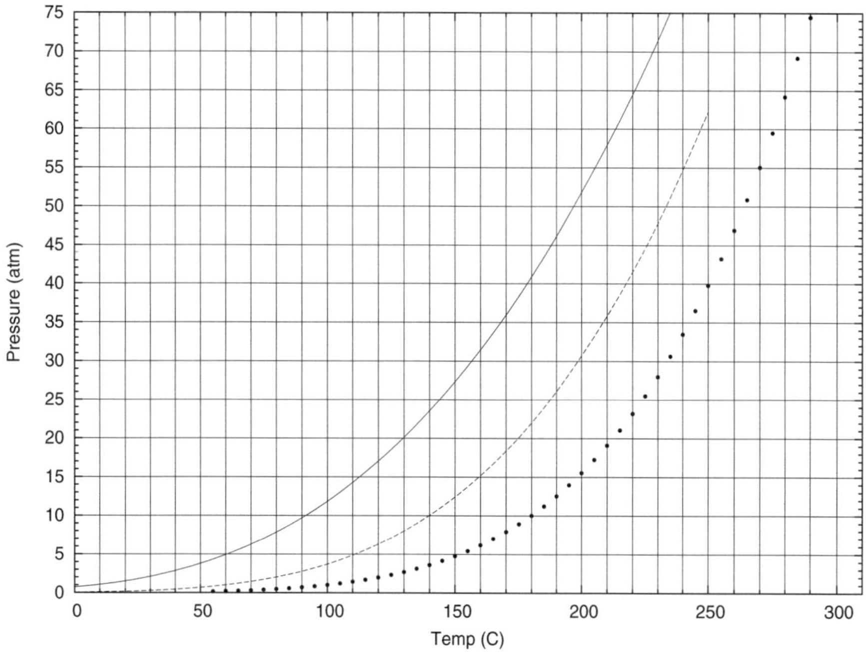


**Fig. 12** Boiling curve for water–steam transition according to the van der Waals equation of state: *solid line* for $r = 2, s = 0, \gamma = 8$; *dashed line* for $r = 2, s = 0.7, \gamma = 8$. *Dots* give the tabulated values

Step 2: Find

$$\bar{p}_1 = \frac{p_c}{z_1 - x_1} \int_{x_1}^{z_1} \pi(y, \theta)\, \mathrm{d}y,$$

where $\pi(v, \theta)$ is given by (5.6).

Step 3: Find $x_2$ and $z_2$, the least and greatest roots $v$ of the equation

$$\frac{4r}{r^2 - 1} \cdot \frac{\theta}{v - t} - \frac{1}{t\theta^s v^r} = \frac{\bar{p}_1}{p_c}, \qquad \text{etc.}$$

Step 4: Then $\bar{p}(T) = \lim_{i\to\infty} \bar{p}_i$.

In fact, if $\bar{p}_0$ is relatively close to the tabulated value for $\bar{p}(T)$, the convergence is rapid, requiring only one or two iterations (up to four significant figures).

Results for the water–steam transition are shown in Fig. 7, the calculated boiling pressures over-representing the tabulated pressures by less than 1%. For carbon dioxide, nitrogen and Freon the results confidently can be expected to be equally close.

It is interesting that the first procedure is inoperable for $\theta > \theta_0$, while the second is inaccurate for $\theta < \theta_0$; see Fig. 11. In fact the inaccuracy of the second procedure

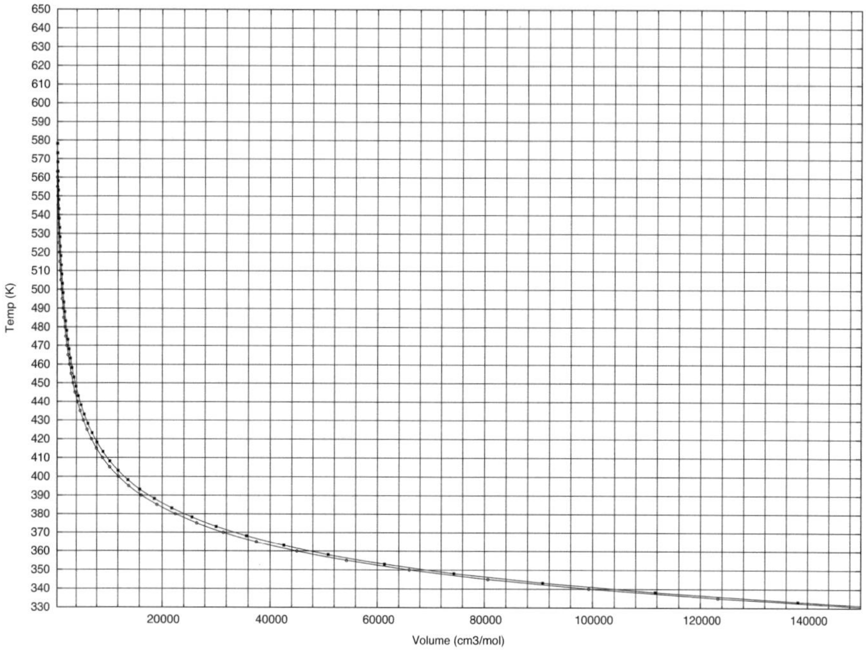


**Fig. 13** Specific molal volume of the vapor phase at saturation for the water–steam transition. Lower curve, calculated; upper curve, observed

when $\theta < \theta_0$ is in itself an argument against the equal-area rule, though, remarkably, the accuracy of the procedure when $\theta > \theta_0$ shows that *the equal-area rule is acceptable near the critical temperature,* $\frac{9}{10}T_c < T < T_c$.

The final Fig. 13 shows calculated and tabulated values for the specific molal volume of the vapor (steam) phase of water at given boiling temperatures, again the accuracy being in the range 1–2% for temperatures $T < 560\,\mathrm{K}$.

**Acknowledgements** The author wishes to thank Professors Roger Fosdick and Grozdena Todorova for their help in improving the presentation of the paper, and also Ms Lynn Tran of the Mathematics Library of the University of Minnesota for valuable bibliographic assistance in locating older references. Finally, he wishes to acknowledge the invaluable help of Benjamin van der Sluis in carrying out the numerical work of Section 6.

*Note added in proof* Measurement of the specific heat difference $\gamma$ in (6.1) is a delicate matter, as noted above. Another calculation of this parameter shows that the value $\gamma = 11$ may be an improvement over the indicated value $\gamma = 8$. At the same time, the calculated boiling curve using $r = 1.6$, $\gamma = 11$ and the slightly modified parameter $s = 0.43$ again is not more than 1% to 2% different from tabulated values for the boiling pressure over the range 70 C to 374 C.

## 7 Appendix: Behavior at the Critical Point

The behavior of the transition region near the critical point is well-known in the case of the equal area rule, yielding a parabolic asymptotic form, which in fact is not as flat at the critical point as observations indicate. While of great interest in itself, this fact alone clearly does not invalidate analytic investigation of the transition region *away from criticality*, as in the main body of the paper.

For completeness, we show here that the same parabolic phenomenon occurs also for the general area rule, at least when the specific heat $c_v$ is continuous at the critical point. (Again this does not invalidate investigation of the transition region away from criticality). The calculations also are useful in proving the third part of Proposition 2.

**Lemma 7.1** *Near the critical point,*

$$(v_c - v_B)^2 = (6\alpha/\beta)(T_c - T) + O\big((T_c - T)^{3/2}\big),$$
$$(v_E - v_c)^2 = (6\alpha/\beta)(T_c - T) + O\big((T_c - T)^{3/2}\big) \tag{7.1}$$

*and*

$$v_B' = \sqrt{\frac{3\alpha}{2\beta}}\,(T_c - T)^{-1/2} + O(1), \qquad v_E' = -\sqrt{\frac{3\alpha}{2\beta}}\,(T_c - T)^{-1/2} + O(1), \tag{7.2}$$

*where*[7]

$$\alpha = -\left(\frac{\partial^2 p}{\partial v\,\partial T}\right)_{crit} > 0, \qquad \beta = -\left(\frac{\partial^3 p}{\partial v^3}\right)_{crit} > 0.$$

*Proof* We assume that the pressure function $p$ is *four times continuously differentiable.* Then by Taylor's expansion with remainder about the critical point, we find (since $\partial p/\partial v = \partial^2 p/\partial v^2 = 0$ at critical)

$$\begin{aligned} p(v, T) =& p_c + \left(\frac{\partial p}{\partial T}\right)_{\text{crit}} (T - T_c) \\ &+ \frac{1}{2}\left(\frac{\partial^2 p}{\partial T^2}\right)_{\text{crit}} (T - T_c)^2 + \left(\frac{\partial^2 p}{\partial v\,\partial T}\right)_{\text{crit}} (T - T_c)(v - v_c) \\ &+ \frac{1}{6}\left(\frac{\partial^3 p}{\partial v^3}\right)_{\text{crit}} (v - v_c)^3 + \frac{1}{2}\left(\frac{\partial^3 p}{\partial v^2\,\partial T}\right)_{\text{crit}} (T - T_c)(v - v_c)^2 \\ &+ \frac{1}{2}\left(\frac{\partial^3 p}{\partial v\,\partial T^2}\right)_{\text{crit}} (T - T_c)^2(v - v_c) + \frac{1}{6}\left(\frac{\partial^3 p}{\partial T^3}\right)_{\text{crit}} (T - T_c)^3 \\ &+ \text{terms of order four.} \end{aligned}$$

At the same time $p(v_B, T) = p(v_E, T) = \bar{p}(T)$, while from the first two parts of Proposition 2,

$$\bar{p}(T) = p_c + (\partial p/\partial T)_{\text{crit}}\,(T - T_c) + O((T - T_c)^2).$$

Putting

$$x = v_c - v_B, \qquad y = T_c - T,$$

[7]One checks easily that $\alpha$ and $\beta$ are positive at the critical point for the model (5.1).

then at $v = v_B$ we have (since $x^2y \le (y^2 + x^4)/2$)

$$-\alpha xy + (\beta/6)x^3 + O(y^2 + x^4) = 0.$$

To investigate this relation, we claim first that $y = O(x^2)$. Assuming this for the moment, it follows that

$$\alpha xy - (\beta/6)x^3 + O(x^4) = 0.$$

Solving this for $y$ gives

$$y = (\beta/6\alpha)x^2 + O(x^3).$$

But then $x = O(y^{1/2})$ and we obtain the first line of (7.1). The second follows similarly.

To prove the claim, we have, again by Taylor's expansion,

$$\begin{aligned}\frac{\partial p}{\partial v}(v, T) = {} & \alpha(T_c - T) - \frac{1}{2}\beta(v - v_c)^2 \\ & + O\big(|v - v_c|(T_c - T) + (T_c - T)^2 + |v - v_c|^3\big). \qquad (7.3)\end{aligned}$$

Since $\partial p/\partial v = 0$ on $\partial U$ this gives $T_c - T = O\big((v - v_c)^2\big)$ on $\partial U$. However $\Omega_4 \supset U$ so $y = T_c - T = O\big((v_c - v_B)^2\big)$ proving the claim, and thus also (7.1).

Next, by (7.3) with $v = v_B$, together with (7.1), we find

$$\frac{\partial p}{\partial v}(v_B, T) = \alpha y - \frac{1}{2}\beta \cdot \frac{6\alpha}{\beta}\, y + O\big(y^{3/2}\big) = -2\alpha y + O\big(y^{3/2}\big). \qquad (7.4)$$

Moreover, by two further Taylor expansions,

$$\frac{\partial p}{\partial T}(v_B, T) = \left(\frac{\partial p}{\partial T}\right)_{\text{crit}} + \alpha x + O(y + x^2) = \left(\frac{\partial p}{\partial T}\right)_{\text{crit}} + \alpha\sqrt{\frac{6\alpha}{\beta}}\, y^{1/2} + O(y)$$

and

$$\bar{p}'(T) = \bar{p}'(T_c) + O(y) = \left(\frac{\partial p}{\partial T}\right)_{\text{crit}} + O(y).$$

Hence, from (4.6) and (7.4) we get

$$[-2\alpha y + O(y^{3/2})]v_B' + \alpha\sqrt{6\alpha/\beta}\, y^{1/2} + O(y) = 0;$$

the first result of (7.2) now follows at once. The second is obtained similarly, proving the lemma. □

[The asymptotic behavior (7.1), though without the explicit exponent for the higher order terms, is well-known in the literature of phase transitions.]

We can now complete the proof of the proposition. Using (7.2) and (7.4) we find

$$\frac{\partial p}{\partial v}(v_E, T)(v_B')^2 = [-2\alpha y + O(y^{3/2})] \cdot \left(\frac{3\alpha}{2\beta} y^{-1}\big(1 + O(y^{1/2})\big)\right) = -\frac{3\alpha^2}{\beta} + O\big(y^{1/2}\big),$$

with a similar expression for $\partial p/\partial v(v_E, T)(v_E')^2$. Hence, the difference quantity

$$\frac{\partial p}{\partial v}(v_E, T)\left(\frac{dv_E}{dT}\right)^2 - \frac{\partial p}{\partial v}(v_B, T)\left(\frac{dv_B}{dT}\right)^2$$

in Theorem 4.4 is of order $O\big(y^{1/2}\big) = O\big((T_c - T)^{1/2}\big)$ as $T \to T_c$. This completes the proof of Proposition 2.

*Remark* In view of the earlier expressions for $v_B - v_c$ and $v_E - v_c$, it follows that the boundary of $\Omega_4$ near the critical point has the asymptotic form

$$T_c - T = \frac{\beta}{6\alpha}(v - v_c)^2,$$

namely, a parabola with explicit coefficient $\beta/6\alpha$ depending only on the pressure function $\rho$. Experimentally, the phase boundary is actually flatter than parabolic at the critical point ([8], p. 64). Thus a full treatment of behavior near the critical point requires further care, complementary to the treatment of the transition process away from criticality, as in the main part of the paper.

## References

1. Callen, H.B.: Thermodynamics and an Introduction to Thermostatics (2nd edn), pp. 235–241. Wiley, New York (1985)
2. Carrington, G.: Basic Thermodynamics, pp. 261–163, 315–316. Oxford Univ. Press (1994)
3. Epstein, P.S.: Textbook of Thermodynamics, pp. 9–12, 150. Wiley, New York (1937)
4. Griffiths, R.B.: Thermodynamic functions for fluids and ferromagnets near the critical point. Phys. Rev. **158**, 176–187 (1967)
5. Kaul, G.D.: Generalization of the Maxwell criterion for van der Waals equation. Phys. Rev. **155**, 78–80 (1967)
6. Pippard, A.B.: The Elements of Classical Thermodynamics, pp. 116–119. Cambridge University Press (1964)
7. Raznjevic, K.: Handbook of Thermodynamic Tables and Charts. McGraw-Hill, New York (1976)
8. Rowlandson, J.S., Swinton, F.I.: Liquids and Liquid Mixtures (3rd edn). Butterworth Scientific, pp. 235–241. London (1981)
9. Schmidt, E.: Properties of water and steam in SI-units. Grigoli, U. (ed.) Springer-Verlag, Berlin, and R. Oldenbourg, Munich (1982)
10. Šilhavý, M.: The Mechanics and Thermodynamics of Continuous Media, pp. 319–320. Springer, Berlin (1997)
11. Wilson, A.H.: Thermodynamics and Statistical Mechanics, pp. 207–215. Cambridge University Press (1957)

In: H.A. Sunder, A.W. Leisse, and C.H. Popelar (eds.): *Proceedings of the 19th Midwestern Mechanics Conference, Columbus, Ohio, Sept. 9–11, 1985.*
Developments in Mechanics 13 (1985), 59–64. © Reprinted with permission.

**Reprint VI.2.[iv]**

# A CONTINUUM MODEL FOR CHEMICAL MIXTURE DYNAMICS

James Serrin
Department of Mathematics, University of Minnesota

Models for the dynamics of a moving chemically reacting fluid mixture have been proposed by many investigators (see particularly Gurtin & Vargas (1971) and the bibliography quoted therein). Moreover such models in more or less <u>ad hoc</u> form have also been widespread in studies of combustion theories and flame fronts (see, for example, Williams (1965)). These models, which certainly contain the essential ideas, at the same time exhibit various failures or difficulties. For example, standard combustion models do not systematically take into account restrictions on the fluid behavior arising from the second law of thermodynamics, while the model of Gurtin & Vargas on the other hand, in dealing with just this matter, requires a delicate and not particularly obvious physical presumption (namely their Axiom B) which one would otherwise not wish to make. Finally, in all the models so far known to the author, the role of the chemical potential in the formulation of the fundamental dynamical equations either does not appear at all, or cames only as an afterthought. The purpose of this paper is to present a new generalized continuum model which avoids the difficulties mentioned above.

The fluid mixtures which we envisage involve N different chemical species, which interact chemically and diffusively, and are subject as well to boundary stresses, external body forces and external body heat supplies. We suppose the mixture obeys the usual rules of mass continuity and momentum balance, and that the first and second laws of thermodynamics are respectively modelled by means of an energy balance identity and the Clausius - Duhem inequality (see Truesdell (1957) and Truesdell & Toupin (1960)).

An important feature of the approach is that we allow for an <u>open chemical system - the external addition of mass to each individual species being admitted throughout</u>. This is standard in classical reversible and quasi-static thermodynamics, but seems to be new in the context of continuum mechanics. One may in fact envisage mass addition in complete parallel with the notion of external body forces and external heat supplies. Indeed <u>it is this concept, and precisely this concept, which allows one to avoid the delicate axiom of Gurtin & Vargas as well as to formalize the notion of chemical potential</u>.

Since the derivations are not entirely transparent we state the final equations here to indicate their essential simplicity, namely

$$
\begin{array}{lll}
\dot{\rho}_i + \rho_i \operatorname{div} \mathbf{v} & = & m_i + a_i + \operatorname{div} \boldsymbol{\sigma}_i \\
\rho \dot{\mathbf{v}} + m \mathbf{v} & = & m_i v_i + \rho \mathbf{b} + \operatorname{div} \mathbf{T} \\
\rho \dot{\epsilon} + m \epsilon & = & m_i \mu_i + \rho r + \operatorname{div}(\mathbf{k} + \mathbf{q}) + \mathbf{T} : \operatorname{grad} \mathbf{v} \\
\rho \dot{\eta} + m \eta & \geq & \eta \quad \frac{\rho r}{\theta} + \operatorname{div} \frac{\mathbf{q}}{\theta}
\end{array}
$$

The notation is standard (and in any case will be restated later), with the exception that $m_i$ denotes the <u>external mass supply rate</u> for the ith species, $m$ the total external mass supply rate (sum of the $m_i$), $a_i$ the <u>mass rate due to chemical reactions</u> between the species, $\mu_i$ the <u>chemical potential</u> of the ith species, and $\mathbf{k}$ the vectorial <u>interstitial working</u>. Of course $i$ runs from 1 to N ; we also sum over repeated indices unless otherwise specifically indicated. When the supplies $m_i$ all vanish these equations are due to Gurtin & Vargas, though their interpretation of the vector $\mathbf{k}$ is somewhat different than here and they do not include a full derivation.

Three points may be made before we turn to the main derivations. First, the equations contain only a small set of response functions. This point is not as obvious as it might seem at first glance, for in the derivations one must allow as well for <u>individual</u> stresses for <u>each species</u>, <u>individual</u> internal energy and entropy transfer for <u>each species</u>, and <u>individual</u> diffusive velocities for <u>each species</u>.

Second, classical reversible and quasi-static thermodynamics finds a logical and well-defined position within the theory; indeed certain paradoxes of this subject can then be satisfactorily resolved. [I refer particularly to the problem that, outside equilibrium, chemical kinetics apparently do not allow arbitrary cyclic processes, which in turn casts some doubt on the proofs of the fundamental differential relations of the subject, i.e. the Gibbs relation and its various permutations and combinations.]

The inclusion of the second law in the form of a generalized Clausius - Duhem inequality not only presents an invaluable tool for the discovery of underlying relations which must necessarily hold among the response functions, but also provides (as we shall see) a canonical <u>Liapunov function</u> for any travelling wave solution of the equations. This fact cannot but have far-reaching implications for the study of combustion fronts, not only in theoretical studies of existence but also in the practical calculation of solutions. At the level of generality here intended, it almost goes without saying that no <u>a priori</u> constitutive relations can be assumed between diffusions, chemical potentials, reaction rates, and the entropy and energy, so the universal availability of a Liapunov function is all the more surprising.

We emphasize again that our goal is to formalize the basic dynamical equations of a chemical mixture, considered as a single continuous media. We assume that the reader is familiar with standard derivations of the equations of fluid dynamics, as presented for example by Serrin (1957) or Aris (1962).

The author also wishes to express his sincere thanks to Professor Morton Gurtin and Dr. J.E. Dunn for a number of helpful conversations.

## 1. Derivation of the Equations

We consider a fluid mixture involving $N$ different chemical species, which may diffuse and interact. Letting the index $i = 1, \ldots, N$ denote the individual species, we put

$\rho_i$ = density of the ith species
$v_i$ = velocity vector of the ith species
$a_i$ = mass rate generation of the ith species through chemical reactions
$m_i$ = external mass supply rate,

all these being functions of the variables $x, t$ defining position and time during the motion. The <u>center of mass velocity</u> $v$ is given by the relation $\rho v = \rho_i v_i$, where

$$\rho = \Sigma \rho_i \qquad (\Sigma \equiv \Sigma_1^N)$$

is the density of the mixture. Similarly we define the <u>diffusive mass rate</u> $\sigma_i$ of the ith species by the relation

$$\sigma_i = \rho_i(v - v_i) \qquad (\text{ i unsummed}).$$

For an arbitrary smooth control volume <u>fixed in space</u>, conservation of mass then requires for each species, $i = 1, \ldots, N$,

$$\frac{d}{dt}\int \rho_i \, dx = \int (m_i + a_i) dx - \oint \rho_i (v_i \cdot n) ds ,$$

where the symbol $\oint$ denotes integration over the full boundary of the control volume and $n$ is the outer normal (again the final term is unsummed). Standard use of the divergence theorem and the fact that the control volume is arbitrary then leads to the <u>equation of continuity for the ith species</u>

$$\dot{\rho}_i + \rho_i \operatorname{div} v = m_i + a_i + \operatorname{div} \sigma_i$$

where $\dot{}$ denotes the material derivative $\partial/\partial t + v \cdot \text{grad}$. Since $\Sigma \sigma_i = \Sigma a_i = 0$ (no mass is generated in a reaction) we obtain also, by summation on the index $i$,

$$\dot{\rho} + \rho \operatorname{div} v = m \qquad (m = \Sigma m_i).$$

Turning to the momentum equation, let $\tilde{\tau}_i$ be the <u>putative stress vector</u> for the ith species. We assume $\tilde{\tau}_i = n \cdot \tilde{T}_i$, where $\tilde{T}_i$ is the <u>putative stress tensor</u> (for simplicity assumed to be symmetric). Balance of momentum then requires

$$\frac{d}{dt}\int \rho_i v_i \, dx = \int (m_i v_i + \rho b) dx + \oint ( \tilde{\tau} - \rho_i v_i (v_i \cdot n)) ds ,$$

where $b$ is the body force per unit mass and $\tilde{\tau} = \Sigma \tilde{\tau}_i$. Here the term $m_i v_i$ represents the momentum rate generated by the external mass supply rates $m_i$. The final summation can be simplifed by writing

$$\rho_i v_i (v_i \cdot n) = \rho_i v_i ((v - \sigma_i/\rho_i)\cdot n)$$
$$= \rho v(v\cdot n) - v_i(\sigma_i \cdot n) = \rho v(v\cdot n) + (\sigma_i/\rho_i)(\sigma_i \cdot n)$$

(since $\Sigma\, \sigma_i = 0$) . Then by standard procedures we get the equation of motion

$$\rho \dot{v} + mv = m_i v_i + \rho b + \operatorname{div}(\tilde{T} - \sigma_i \sigma_i/\rho_i) \ ,$$

where the equation of continuity has been used at the last stage, where $\tilde{T} = \Sigma\, \tilde{T}_i$ and where the expression $\sigma_i\sigma_i$ denotes the (sum of the) dyadic matrices based on the vector $\sigma_i$ .

Energy balance requires considerable additional care, and the introduction of further thermodynamical variables. We let

$\tilde{\varepsilon}$ = putative specific internal energy
$\tilde{\mu}_i$ = putative chemical potential of the ith species
$\tilde{q}$ = putative heat flux vector
$\tilde{k}$ = boundary work flux vector
$r_i$ = heat supply per unit mass to the ith species

(The vector $\tilde{k}$ accounts for short range working additional to stress work; as shown in Dunn & Serrin (1985) and in the references quoted therein, this term is necessary if all working sources are to be included.) Balance of energy is then expressed through the relation

$$\frac{d}{dt}\int\left(\rho\tilde{\varepsilon} + \frac{1}{2}\ \rho_i v_i^2\right)dx = \int\left(m_i\left(\tilde{\mu}_i + \frac{1}{2}\ v_i^2\right) + \rho b\cdot v + \rho_i\ r_i\right)dx$$
$$+ \oint\left(\tilde{T}_i \cdot v_i + (\tilde{q} + \tilde{k})\cdot n - \rho_i\left(\ell_i + \frac{1}{2}\ v_i^2\right)(v_i\cdot n)\right) ds \ ,$$

where $v_i^2 = v_i \cdot v_i = |\,v_i|^2$ . The terms $\ell_i$ account for convective internal energy transfer to the control volume through the boundary. Since the total internal energy is given by the quantity $\rho\tilde{\varepsilon}$ on the left-hand side of the balance, we assume for compatibility that

$$\rho_i\ \ell_i = \rho\tilde{\varepsilon} \ .$$

(This relation can also be justified by applying the principle of frame indifference; for simplicity we shall not carry out the argument here.) Straightforward (though somewhat lengthy) reductions now yield the equation

$$\rho\left(\tilde{\varepsilon} + \frac{1}{2}\ \sigma_i^2/\rho\ \rho_i\right)^{\bullet} + m\left(\tilde{\varepsilon} + \frac{1}{2}\,\sigma_i^2/\rho\,\rho_i\right)$$
$$= m_i\left(\tilde{\mu}_i + \frac{1}{2}\,\sigma_i^2/\rho_i^2\right) + \rho_i r_i + (\tilde{T} - \sigma_i\sigma_i/\rho_i)\ :\ \operatorname{grad} v$$
$$+ \operatorname{div}\left(\tilde{q} + \tilde{k} + \left(\ell_i + \frac{1}{2}\,\sigma_i^2/\rho_i^2\right)\sigma_i - \tilde{T}_i\cdot\sigma_i/\rho_i\right)$$

where the equation of motion has been used to eliminate the term $\rho b\cdot v$ .

We turn finally to the generalized <u>Clausius-Duhem inequality</u>,

$$\frac{d}{dt}\int\rho\eta\, dx \geq \int\frac{\rho_i r_i}{\theta}\ dx + \oint\left(\frac{\tilde{q}\cdot n}{\theta} - \rho_i\ \ell_i'(v_i\cdot n)\right) ds$$

where $\eta$ is the specific entropy, $\theta$ the absolute temperature, and $\ell_i'$ the convective entropy transfers across the boundary. In analogy with the previous assumption concerning $\ell_i$ we here require that $\rho_i \ell_i' = \rho\eta$. We now obtain easily

$$\rho\dot{\eta} + m\eta \geq \rho_i r_i/\theta + \operatorname{div}(\tilde{q}/\theta + \ell_i' \sigma_i) .$$

Let us define

$$\epsilon = \tilde{\epsilon} + \frac{1}{2}\sigma_i^2/\rho_i\rho, \qquad \mu_i = \tilde{\mu}_i + \frac{1}{2}\sigma_i^2/\rho_i^2$$

$$T = \tilde{T} - \sigma_i\sigma_i/\rho_i , \qquad q = \tilde{q} + \theta\, \ell_i'\sigma_i$$

$$k = \tilde{k} + \left(\ell_i - \theta\, \ell_i' + \frac{1}{2}\sigma_i^2/\rho_i^2\right)\sigma_i - \tilde{T}_i \cdot \sigma_i/\rho_i$$

and $r = \rho_i r_i/\rho$. The previous equations then take the simple forms already stated in the introduction. Since $\epsilon$, $\mu_i$, $T$, $k$ and $q$ are in fact the operative response functions we call these quantities respectively the internal energy, chemical potential, stress, interstitial working and heat flux. It is the anticipation of this state of affairs that prompted the earlier use of the word "putative" to label the variables $\tilde{\epsilon}$, $\tilde{\mu}_i$, $\tilde{T}$ and $\tilde{q}$.

## 2. The Reduced Clausius-Duhem Inequality

In this section we obtain the reduced Clausius-Duhem equation for the mixture, an inequality which must be satisfied by any kinematically allowed motion. This inequality in fact expresses the ultimate content of the second law of thermodynamics for the mixture. The idea is simply to eliminate the external supplies $m_i$, $b$, and $r_i$ from the previously derived equations.

The body force $b$ has already been eliminated from the energy balance equation. Next, eliminating $\rho r$ between the energy balance equation and the Clausius-Duhem inequality yields

$$\rho(\dot{\epsilon} - \theta\dot{\eta}) + m(\epsilon - \theta\eta) \leq m_i \mu_i + \operatorname{div} k + T : \operatorname{grad} v + q \cdot \operatorname{grad} \theta/\theta .$$

Finally, using the equations of continuity to eliminate $m_i$ and $m$, we reach the principal result

$$(\rho\psi)^{\bullet} \leq (\rho_i \mu_i - \rho\psi)\operatorname{div} v + \mu_i \dot{\rho}_i - \rho\eta\dot{\theta} - a_i \mu_i + \operatorname{div}(k - \mu_i \sigma_i) + \sigma_i \cdot \operatorname{grad} \mu_i + T : \operatorname{grad} v + q \cdot \operatorname{grad} \theta/\theta$$

where $\psi = \epsilon - \theta\eta$ is the specific free energy. This relation reduces to the standard Clausius-Duhem inequality obtained by Gurtin & Vargas when the mass supplies $m_i$ all vanish, but otherwise is more general.

At this stage we have gone as far as possible without introducing response (i.e. constitutive) assumptions on the mixture, that is, statements asserting the particular functional dependence of the fields $\epsilon$, $\eta$, $\mu$, $T$, $q$,

$\mathbf{k}$, and $a_i$ on the kinematical arguments $\rho_i$, $\theta$ and $\mathbf{v}$.

Gurtin & Vargas have shown, using their Axiom B and well-known arguments of Coleman and Noll (1963), that the reduced Clausius-Duhem inequality yields crucial restrictions on the responses. The same arguments can be carried out in the present context; moreover Axiom B is no longer necessary, since the availability of mass supply rates as well as external body forces and external heat supplies shows that every kinematically allowable notion is also dynamically allowable. For lack of space we shall not reproduce these arguments here, nor indicate various generalizations which are needed to model Korteweg fluids and Cahn-Hilliard mixtures. Nevertheless, the simple but important special case of reversible systems can be indicated briefly. Here the values of the response variables at any point $(\mathbf{x},t)$ in the motion are **assumed** to depend on the densities $\rho_i$ and the temperature $\theta$ at the point. Using the procedures of Coleman and Noll, one then finds that (Theorem) the thermodynamics variables $\epsilon$, $\eta$, $\mu_i$ and $p$ necessarily satisfy the classical relations

$$\theta\, d\eta = d\epsilon + p\, d\upsilon - \mu_i\, d(\rho_i/\rho)\,, \qquad \rho g = \rho_i\, \mu_i\,,$$

while $\mathbf{T} = -p\mathbf{I}$ and $\mathbf{q} = \mathbf{k} = \boldsymbol{\sigma}_i = 0$ (here $\upsilon = 1/\rho$ is the specific volume and $g = \epsilon + p\upsilon - \theta\eta$ the thermodynamic potential of the mixture). The reduced Clausius-Duhem inequality finally simplifies to the well-known condition $\mu_i\, a_i \leq 0$.

We close by noting that for one-dimensional travelling waves in the absence of external forces and supplies the summed equation of continuity may be written in the form

$$\frac{d}{dx}(\rho v) = 0\,.$$

Thus $\rho v = \ell =$ constant. The Clausius-Duhem inequality then becomes

$$\ell \frac{d\eta}{dx} \geq \frac{d}{dx}\left(\frac{q}{\theta}\right).$$

The quantity $\varphi = \ell\eta - q/\theta$ accordingly increases with $x$ in the direction of the wave, and so serves (ultimately because the reduced Clausius-Duhem inequality must be valid) as a universal Liapunov function.

## References

M.E. Gurtin & A.S. Vargas, Arch. Rational Mech. Anal. 43 (1971), 179.

F.A. Williams, Combustion Theory. Addison-Wesley, 1965.

C. Truesdell, Rend. Acad. Lincei 22 (1957), 33, 158.

C. Truesdell & R. Toupin, Encylopedia of Physics, Vol. III/1 (1960). Sect.258.

J. Serrin, Encyclopedia of Physics, Vol. VIII/1 (1957).

R. Aris, Vectors, Tensors and the Basic Equations of Fluid Mechanics. Prentice-Hall (1962).

B.D. Coleman & W. Noll, Arch. Rational Mech. Anal. 13 (1963), 167.

J.E. Dunn & J. Serrin, Arch. Rational Mech. Anal. 89 (1985).

David Owen, Carnegie-Mellon University, 1983.

Roger Fosdick, Minneapolis, MN, 2010.

Chi-Sing Man, University of Kentucky, 1996.

# VI.3. Thermomechanics and Invariance

Roger Fosdick

**Mathematics Subject Classification (2010).** 74AXX.

**Keywords.** Commentary, continuum Mechanics, thermodynamics, invariance, objectivity.

A high regard for the observational space in which physical events take place is essential for developing objective theories that are compatible with what can be measured. Thus, James Serrin began early to wonder about the connections between motion, force, energy, mass, heat and the various balance laws that govern these physical entities. With paper [i] below, he has improved our understanding of the invariant structure of the physical laws that describe the thermomechanical behavior of bodies in a non-relativistic environment.

In [i], Serrin takes up his fascination with the invariant structure of the classical balance laws of continuum mechanics. He was intrigued by the works of Noll, Green and Rivlin, and Šilhavý who showed that within classical continuum mechanics the balance laws are a result of a natural invariance condition that the balance of energy must satisfy. Moreover, he was most interested in the idea put forward by Šilhavý in 1989 that because of this invariance the total intrinsic energy of a body must split into the sum of the internal energy and the kinetic energy.[1] This, and his awareness that a concept of Euclidean invariance would express the special group properties commonly postulated for the event world of non-relativistic classical mechanics, is the basis of his paper. Serrin follows the path of Leibniz, by not introducing a special preferred (inertial) frame, and his aim is to expose in an elementary way how the subtle notion of invariance structures the classical balance laws of continuum mechanics.

For the purposes of this commentary, $\mathsf{F}$ denotes a *frame* in a three-dimensional vector space $\mathbb{E}^3$. The points of $\mathbb{E}^3$ in $\mathsf{F}$ are represented by $\mathbf{x}$. An isometrically equivalent, time-dependent re-framing of $\mathbb{E}^3$ results in a frame $\mathsf{F}^*(t)$ which allows at each time $t$ the points of $\mathbb{E}^3$ to be represented by $\mathbf{x}^* = \mathbf{Q}(t)\mathbf{x} + \mathbf{c}(t), \forall\, \mathbf{x} \in \mathbb{E}^3$, where $\mathbf{Q}(t) \in \text{Orth}$ and $\mathbf{c}(t) \in \mathbb{E}^3$. This may be called a *change of frame* for Euclidean space-time. Thus, under a change of frame the motion $\mathbf{x} = \boldsymbol{\chi}(\mathbf{X}, t)$ of a body $\mathcal{B} \subset \mathbb{E}^3$ with elements $\mathbf{X} \in \mathcal{B}$ in frame $\mathsf{F}$ is represented as $\mathbf{x}^* = \boldsymbol{\chi}^*(\mathbf{X}, t) = \mathbf{Q}(t)\boldsymbol{\chi}(\mathbf{X}, t) + \mathbf{c}(t)$ relative to frame $\mathsf{F}^*(t)$.

The first of the two main concerns in Serrin's paper describes how the balance equations of classical continuum mechanics emerge from minimal invariance hypotheses.

[1] J.E. Dunn and R. Fosdick began working on this question of splitting in the late 1960's, and in December, 1969 R.F. presented invited colloquium talks at the University of Illinois and the University of Kentucky titled "A Causality Approach to Continuum Mechanics". In these talks, a strategy for splitting and the existence of mass and its balance was presented. Serrin was not made aware of this work until January, 1993, after I read a draft of his wonderful Lectio Doctoralis, "Space, Time and Energy", given at the University of Ferrara in October, 1992.

P. Pucci et al. (eds.), *James Serrin. Selected Papers: Volume 2*, Contemporary Mathematicians,
DOI 10.1007/978-3-0348-0847-7_16, © Springer Basel 2014 

Serrin's specific result is stated in his Theorem I, which supposes that the mass density $\rho$, the internal energy $\varepsilon$, the radiant heating $r$, the stress $\mathbf{T}$ and the heat flux $\boldsymbol{q}$ are *objective* fields: $\mathsf{F} \mapsto \mathsf{F}^*(t) \Longrightarrow \rho \to \rho^* = \rho$, $\varepsilon \to \varepsilon^* = \varepsilon$, $r \to r^* = r$, $\mathbf{T} \to \mathbf{T}^* = \mathbf{QTQ}^{\mathrm{T}}$, $\boldsymbol{q} \to \boldsymbol{q}^* = \mathbf{Q}\boldsymbol{q}$. In addition, the classical *balance of energy* for arbitrary parts $\mathcal{P} \subset \mathcal{B}$ is assumed and it is required to be *form invariant*, i.e.,

$$\frac{d}{dt}E(\mathcal{P}_t) + \frac{d}{dt}K(\mathcal{P}_t) = P(\mathcal{P}_t) + Q(\mathcal{P}_t) \quad \forall \mathcal{P} \subset \mathcal{B},\ \forall t, \tag{1}$$

holds with identical structure regardless of the frame in which the various entries of this equation are represented. Here, for the frame $\mathsf{F}$, $\mathcal{P}_t := \chi(\mathcal{P}, t)$, and the entries correspond to the internal and kinetic energies

$$E(\mathcal{P}_t) := \int_{\mathcal{P}_t} \rho\varepsilon\, dv(\mathbf{x}), \quad K(\mathcal{P}_t) := \frac{1}{2}\int_{\mathcal{P}_t} \rho|\mathbf{v}|^2\, dv(\mathbf{x}), \tag{2a}$$

and the mechanical power and heating supplies

$$P(\mathcal{P}_t) := \int_{\partial\mathcal{P}_t} \mathbf{Tn}\cdot\mathbf{v}\, da(\mathbf{x}) + \int_{\mathcal{P}_t} \mathbf{b}\cdot\mathbf{v}\, dv(\mathbf{x}), \quad Q(\mathcal{P}_t) := \int_{\partial\mathcal{P}_t} \boldsymbol{q}\cdot\mathbf{n}\, da(\mathbf{x}) + \int_{\mathcal{P}_t} r\, dv(\mathbf{x}). \tag{2b}$$

The form invariance requirement asserts that for the frame $\mathsf{F}^*(t)$, the balance law (1) remains the same except for the replacement $\mathcal{P}_t \to \mathcal{P}_t^* := \chi^*(\mathcal{P}, t)$ (which implies that $\mathbf{n} \to \mathbf{n}^* = \mathbf{Qn}$), and the corresponding entries in (2) are replaced according to $(\rho, \varepsilon, r, \mathbf{T}, \boldsymbol{q}) \to (\rho, \varepsilon, r, \mathbf{QTQ}^{\mathrm{T}}, \mathbf{Q}\boldsymbol{q})$. The frame-dependent body force field $\mathbf{b} = \mathbf{b}_{\mathsf{F}}(\mathbf{x}, t)$ is replaced by $\mathbf{b}^* = \mathbf{b}_{\mathsf{F}^*(t)}(\mathbf{x}^*, t)$. Of course, $\mathbf{x} \to \mathbf{x}^* = \mathbf{Qx} + \mathbf{c}$ and $\mathbf{v} \to \mathbf{v}^* = \dot{\chi}^* = \mathbf{Qv} + \dot{\mathbf{Q}}\chi + \dot{\mathbf{c}}$.

A *Galilean change of frame*, for which $\mathbf{Q}(t) = \mathbf{1}$ and $\mathbf{c}(t) = t\mathbf{a}$, $\mathbf{a} =$ const., and a subtraction of the balance of energy laws as expressed in the two frames, readily gives

$$\begin{aligned} &\frac{1}{2}|\mathbf{a}|^2 \frac{d}{dt}\int_{\mathcal{P}_t} \rho\, dv(\mathbf{x}) + \mathbf{a}\cdot\Big(\frac{d}{dt}\int_{\mathcal{P}_t} \rho\mathbf{v}\, dv(\mathbf{x}) - \int_{\mathcal{P}_t} \operatorname{div}\mathbf{T}\, dv(\mathbf{x})\Big) \\ &\quad = \int_{\mathcal{P}_t} \big(\mathbf{b_a}\cdot(\mathbf{v}+\mathbf{a}) - \mathbf{b}\cdot\mathbf{v}\big)\, dv(\mathbf{x}) \quad \forall \mathcal{P} \subset \mathcal{B},\ \forall t, \end{aligned} \tag{3}$$

where, for convenience, $\mathbf{b_a}$ has been written in place of $\mathbf{b}^* = \mathbf{b}_{\mathsf{F}^*(t)}$. When the *balance of mass* is assumed, the first term on the left-hand side of (3) vanishes, so the left-hand side and, consequently, the right-hand side must be a linear function of $\mathbf{a}$. Thus, by localizing the right-hand side of (3) Serrin concluded that, given $\mathbf{b}$ and $\mathbf{b_a}$, there exists a unique $\mathbf{f} = \mathbf{f}(\mathbf{x}, t) = \mathbf{f}_{\mathsf{F}}(\mathbf{x}, t) \in \mathbb{E}^3$, independent of $\mathbf{a}$, such that

$$\mathbf{b_a}\cdot(\mathbf{v}+\mathbf{a}) - \mathbf{b}\cdot\mathbf{v} = \mathbf{f}\cdot\mathbf{a} \quad \forall \mathbf{a} \in \mathbb{E}^3 \tag{4}$$

in $\mathcal{B}_t$, where it is clear that $\mathbf{f}$ is a affine function of $\mathbf{v} \in \mathbb{E}^3$. In avoiding the common requirement that the body force is Galilean invariant, Serrin then introduced the idea that the frame-dependent body force field $\mathbf{b} = \mathbf{b}_{\mathsf{F}}(\mathbf{x}, t)$ is a 'work effective field', and he proposed

**Axiom 1.** *Fixing the body supply system at one point*[2] *does not restrict the value of* $\mathbf{v}$ *which can be attained at the point by some motion of* $\mathcal{B}$.

Then, using the fact that $\mathbf{f}$ is unique, independent of $\mathbf{a}$ and affine in its dependence on $\mathbf{v}$, he noted that $\mathbf{f} = \mathbf{f}_0 + \mathbf{Mv}$, where $\mathbf{f}_0 \in \mathbb{E}^3$ and $\mathbf{M}$ is a linear transformation, and then set $\mathbf{v} = -\mathbf{a}$ in (4) and used the arbitrariness of $\mathbf{a}$ to determine $\mathbf{f}_0$ and $\mathbf{M}$ and to

[2]Here, Serrin means that $\mathbf{v}(\mathbf{x}_0, t_0)$ can be varied arbitrarily even though $\mathbf{b}_{\mathsf{F}}(\mathbf{x}, t)$ and $r(\mathbf{x}, t)$, evaluated at the arbitrary point $(\mathbf{x}_0, t_0)$, are fixed.

conclude the existence of a unique *Coriolis supply system* $\boldsymbol{\Omega} = \boldsymbol{\Omega}_{\mathsf{F}}(\mathbf{x}, t) \in \mathbb{E}^3$ such that

$$\mathbf{f} = \mathbf{f}_{\mathsf{F}}(\mathbf{x}, t) = \mathbf{b}_{\mathsf{F}}(\mathbf{x}, t) + \boldsymbol{\Omega}_{\mathsf{F}}(\mathbf{x}, t) \times \mathbf{v}(\mathbf{x}, t) \quad \forall \mathbf{v} \in \mathbb{E}^3. \tag{5}$$

Consequently, (3) and (4) yield the *balance of linear momentum*:

$$\frac{d}{dt} \int_{\mathcal{P}_t} \rho \mathbf{v} \, dv(\mathbf{x}) = \int_{\mathcal{P}_t} \big(\operatorname{div} \mathbf{T} + \mathbf{f}\big) \, dv(\mathbf{x}) \quad \forall \mathcal{P} \subset \mathcal{B}, \, \forall t. \tag{6}$$

Because $\mathbf{b}$ may be replaced by $\mathbf{f}$ in $(2b)_1$, the balance of energy (1) and the balance of linear momentum (6) contain the same dynamic body force field $\mathbf{f}$.

Further analysis of (1) under the change of frame $\mathsf{F} \mapsto \mathsf{F}^*(t)$ with $\mathbf{c}(t) = \mathbf{0}$ and $\mathbf{Q}(t)$ such that $\mathbf{Q}(t_0) = \mathbf{1}$ and $\dot{\mathbf{Q}}(t_0) = \mathbf{W} \in \mathrm{Skew}$ at an arbitrary time $t_0$, led Serrin to the *balance of angular momentum*, $\mathbf{T}(\mathbf{x}, t) \in \mathrm{Sym}$. Thus, he completed his derivation of the classical balance equations without invoking the idea of an inertial frame and without requiring the body force to be Galilean invariant.

The second of the two main concerns of Serrin's paper describes the splitting of the *total intrinsic energy* $\Phi(\mathcal{P}_t)$ of $\mathcal{P} \subset \mathcal{B}$ into the sum of an internal energy plus a kinetic energy, i.e., $\Phi(\mathcal{P}_t) = E(\mathcal{P}_t) + K(\mathcal{P}_t)$. This represents a much deeper and more fundamental concern. His Theorem II announces such a decomposition, for which he provides only a partial proof. A full derivation is the subject of [1] and only a brief commentary on the issue of splitting will be included here.

In paper [i], Serrin introduces the mass density as a primitive field variable, but it is clear that he had some fundamental concerns about introducing mass and its balance as the last paragraph of [i] shows. Here, I purposely avoid this issue and, for the sake of clarity, do not introduce mass density as a primitive, recognizing that it is neither an essential element of Serrin's energy splitting theorem nor is it an essential ingredient in the idea he presented. Rather, with that one exception, I follow Serrin's idea and introduce the primitive form of the balance of energy as

$$\frac{d}{dt} \Phi(\mathcal{P}_t) = P(\mathcal{P}_t) + Q(\mathcal{P}_t) \quad \forall \mathcal{P} \subset \mathcal{B}, \quad \Phi(\mathcal{P}_t) := \int_{\mathcal{P}} \varphi \, dv(\mathbf{X}) \quad \forall t, \tag{7}$$

wherein it is assumed that $\varphi = \hat{\varphi}(\mathbf{X}, t) = \bar{\varphi}(\mathbf{X}; \mathsf{u}, \mathbf{v})$. Here, $\mathsf{u} = \hat{\mathsf{u}}(\mathbf{X}, t) \in \mathbb{E}^{\mathrm{N}}$ denotes a *state*[3] at $(\mathbf{X}, t)$, $\mathbf{v} = \hat{\mathbf{v}}(\mathbf{X}, t)$ is the particle velocity, and the *local intrinsic energy* $\bar{\varphi}(\mathbf{X}; \cdot, \cdot)$ is measured *per unit volume* of $\mathcal{B}$. The presence of $\mathbf{X} \in \mathcal{B}$ in $\bar{\varphi}(\mathbf{X}; \cdot, \cdot)$ allows $\mathcal{B}$ to be inhomogeneous. Serrin supposes that *the balance of energy* (7) *is form invariant under a change of frame* $\mathsf{F} \mapsto \mathsf{F}^*(t)$. Of course, this includes the form invariance[4] of $\bar{\varphi}(\mathbf{X}; \cdot, \cdot)$, so that $\bar{\varphi}(\mathbf{X}; \mathsf{u}, \mathbf{v}) \to \bar{\varphi}(\mathbf{X}; \mathsf{u}^*, \mathbf{v}^*)$. Serrin also requires that the state $\mathsf{u}$ be invariant under a change of frame (i.e., $\mathsf{u} \to \mathsf{u}^* = \mathsf{u}$), but much less is needed in his argument[5]

Now, in the same spirit as in the development of (3), Serrin introduced a Galilean change of frame and (7) was written for this second frame. Subtraction of the two balance

---

[3]A *state* was not defined by Serrin, but my interpretation is that it could refer to a list such as the deformation gradient $\mathbf{F}$ and the absolute temperature $\theta$, or, more generally, to $\{\mathbf{F}, \dot{\mathbf{F}}, \ddot{\mathbf{F}}, \dots, \theta\}$. But, see footnote 5, below.

[4]*Form invariance* is not the same as *frame indifference*. While the form of the function $\bar{\varphi}(\mathbf{X}; \cdot, \cdot)$ is supposed to remain the same under a change of frame, the arguments of the function are different and the corresponding values of the function may not be equal. For scalar-valued functions, the condition of form invariance with this added feature of equality of values defines frame indifference.

[5]Serrin used only that $u$ is *invariant under Galilean and constant acceleration changes of frame*. Even so, this shows that a list, as in footnote 3, defining a state, may be replaced by $\{\mathbf{C}, \dot{\mathbf{C}}, \ddot{\mathbf{C}}, \dots, \theta\}$, where $\mathbf{C} := \mathbf{F}^{\mathrm{T}}\mathbf{F}$ is the right Cauchy–Green strain.

laws corresponding to the two frames and localization of the result then gives

$$\begin{aligned}&\big(\partial_{\mathsf{u}}\bar{\varphi}(\mathbf{X};\mathsf{u},\mathbf{v}+\mathbf{a})-\partial_{\mathsf{u}}\bar{\varphi}(\mathbf{X};\mathsf{u},\mathbf{v})\big)\cdot\dot{\mathsf{u}}+\big(\partial_{\mathbf{v}}\bar{\varphi}(\mathbf{X};\mathsf{u},\mathbf{v}+\mathbf{a})-\partial_{\mathbf{v}}\bar{\varphi}(\mathbf{X};\mathsf{u},\mathbf{v})\big)\cdot\dot{\mathbf{v}}\\&\quad=|\det\nabla\boldsymbol{\chi}|\big(\mathbf{a}\cdot\operatorname{div}\mathbf{T}+\mathbf{b}_{\mathbf{a}}\cdot(\mathbf{v}+\mathbf{a})-\mathbf{b}\cdot\mathbf{v}\big).\end{aligned}\tag{8}$$

In a similar way, Serrin considered the constant acceleration change of frame $\mathsf{F}\mapsto\mathsf{F}^*(t)$ defined by $\mathbf{Q}(t)=\mathbf{1}$ and $\mathbf{c}(t)=\frac{1}{2}\mathbf{e}(t-t_0)^2$, $\mathbf{e}=$ const., and he concluded that

$$\partial_{\mathbf{v}}\bar{\varphi}(\mathbf{X};\mathsf{u},\mathbf{0})=\mathbf{0}.\tag{9}$$

The equations (8) and (9) form the basis for the splitting theorem[6] of Serrin. To go further, Serrin introduced

**Axiom 2.** *Fixing the body supply system at one point*[7] *does not restrict the value of* $\mathsf{u},\dot{\mathsf{u}},\mathbf{v},\dot{\mathbf{v}}$ *which can be attained at the point by some motion of* $\mathcal{B}$.

With this background, Serrin announced that the local intrinsic energy $\varphi=\bar{\varphi}(\mathbf{X};\mathsf{u},\mathbf{v})$ in the balance law (7) must split into the sum of two terms: Aside from dependencies on $\mathbf{X}$, one term is dependent only upon the state $\mathsf{u}$ and the other dependent only upon $\mathbf{v}$, and is proportional to $|\mathbf{v}|^2$. A sketchy argument, which introduces as primitive the mass density $\rho$ of $\mathcal{B}_t$ and its balance, is given in [i] for reaching the split form

$$\varphi=\rho_0\varepsilon+\tfrac{1}{2}\rho_0|\mathbf{v}|^2,\quad \varepsilon:=\bar{\varepsilon}(\mathbf{X};\mathsf{u}),\quad \rho_0:=\rho|\det\nabla\boldsymbol{\chi}|,\tag{10}$$

but a proof is not provided. With this, (7) is reduced to the classical form of the balance of energy, (1) and (2a). Serrin further used (8) to see that (5) again holds and that $\rho$ and $\mathbf{f}$ of (5) enter the equation

$$\rho\dot{\mathbf{v}}=\operatorname{div}\mathbf{T}+\mathbf{f}\ \ \forall\mathbf{x}\in\mathcal{B}_t,\ \forall t,\tag{11}$$

which is the local form of the *balance of linear momentum* in $\mathcal{B}_t$. Finally, he noted that the symmetry of $\mathbf{T}$, and consequently the balance of angular momentum, follow as earlier.

Motivated by Serrin's paper [i], in [1] we introduce as fundamental the concepts of motion, force, power, heating, intrinsic energy, and the form invariant energy balance law[8] (7). Based on non-relativistic space-time invariance considerations, we then show that there exists a field $\rho_0(\cdot)$, commonly called the *referential mass density* measured per unit volume of $\mathcal{B}$, such that

$$\rho:=\frac{\rho_0}{|\det\nabla\boldsymbol{\chi}|}\quad\text{satisfies}\quad\frac{d}{dt}\int_{\mathcal{P}_t}\rho\,dv(\mathbf{x})=0\quad\forall\mathcal{P}\subset\mathcal{B},\ \forall t.\tag{12}$$

Here, $\rho$ is commonly referred to as the *present mass density* measured per unit volume of $\mathcal{B}_t$, and $(12)_2$ represents a statement of the *balance of mass*. We prove the splitting theorem and recover (10) and (11).

## Reference

[1] Roger Fosdick & James Serrin, The splitting of intrinsic energy and the origin of mass density in continuum mechanics. *Continuum Mech. Thermodyn.*, 2013, DOI:org/10.1007/s00161-013-0301-1.

Roger Fosdick, Department of Aerospace Engineering and Mechanics
University of Minnesota, Minneapolis, MN 55455, USA
e-mail: fosdick@aem.umn.edu

[6] They also form the basis for the existence of mass density and its balance, which Serrin did not observe.
[7] See footnote 2.
[8] In the early developments of mechanics, this was sometimes referred to as a 'law of causality', i.e., a statement about the evolutionary behavior of a system.

In: G. Ferrarese (ed.): *Advances in Continuum Mechanics* 1993, 217–225. **Reprint VI.3.[i]**


# The equations of continuum mechanics as a consequence of group invariance

J. Serrin

School of Mathematics
University of Minnesota
Minneapolis

The classical balance equations of continuum mechanics, that is, the Cauchy laws of motion and the equation of energy balance, are well-known and familiar objects. This paper concerns a related issue that has fascinated me since my earliest acquaintance with the subject, namely the remarkable degree of internal consistency which these equations show. The first indication that there may be an underlying structure which gives rise to this consistency came in a short and relatively unnoticed paper of Walter Noll in 1959, in which he showed that an abstract power balance implied an abstract version of the Cauchy equations (with the force and acceleration lumped together as a single quantity). Shortly afterward, this idea was extended by Green and Rivlin, but then lay dormant until an important paper of Miroslav Šilhavý appeared in 1989. Šilhavý's main, and major, contribution was to show that the *total energy* of a continuum system *must* split into an internal and a kinetic part, the demonstration, moreover, being carried out without invoking Cauchy's motion equations. To see how remarkable this is, let me emphasize that in this approach the kinetic energy $\frac{1}{2}v^2$ arises without any of the usual preparation from dynamics—it cannot be replaced by $|v|^4$, for example — not because this would contradict a Newtonian world-view and give instead some unusual or new kind of dynamics, but rather because classical space-time forbids it! I can recall my astonishment, and disbelief, when Šilhavý first told me of this result, while we were walking together on a promenade in Valencia. The puzzle had come together at last, though I refused to believe it — until his paper appeared.

The purpose of the present note is to give a somewhat different approach to this idea then that of Šilhavý, based on what seems to me a simpler and more direct version of the notion of force and its relation to work, and also to motivate the ideas behind the result.

It is convenient to begin with Cauchy's equations for the motion of a continuous media $S$, namely

$$\rho\dot{v} = b + s \quad (s = \operatorname{div} T), \qquad T \text{ symmetric},$$

the notation being standard. Similarly the standard energy balance equation has the form

$$\rho(\varepsilon + \tfrac{1}{2}v^2)^{\cdot} = q + w,$$

where

$$q = \operatorname{div} h + r \qquad (h := \text{heat flux}, \ r = \text{radiative energy supply})$$
$$w = T : \operatorname{grad} v + v \cdot (b + s).$$

With the definitions

$$D = (\operatorname{grad} v)^{sym}, \qquad \omega = \operatorname{curl} v, \qquad T_x = \operatorname{skew} T,$$

this can be rewritten

$$\rho\dot{\varepsilon} = q + T : D + \omega \cdot T_x - v \cdot (\rho\dot{v} - b - s).$$

In turn, using the Cauchy equations, we get the *reduced* form of the energy balance equation

$$\rho\dot{\varepsilon} = q + T : D.$$

Note that the Cauchy equations seem almost ready-made for the purpose of obtaining this beautiful final form of the energy equation.

There remains another pillar of continuum mechanics, in a way more subtle, and certainly less emphasized, than the equations themselves, namely the idea of frame indifference which is used crucially in the determination of actual formulas for $h$ and $T$. This concept is itself intimately connected to the group properties which are postulated for the event world of mechanics. Fundamentally and essentially, classical mechanics is non-relativistic. Space and time are separate: at each instant there is a Cartesian-Euclidean simultaneity space, with a time value attached. However, *different coordinate frames are allowed in the simultaneity space at each instant, moving relative to each other.*

It is how we view these different coordinate frames which is critical. The notion of moving frames of reference can be traced back to Galilei

Galileo. Its modern form stems from ideas due to Gottfried Wilheln Leibniz, who emphasized that, among all moving frames, no particula one should be specifically preferred — all should stand with equal im portance as far as dynamics is concerned. We shall gradually see thi more clearly, but for now I simply point out some natural examples o coordinate frames in common use.

There are coordinates fixed in the room where this paper is bein read (on the rotating earth); or fixed in the sun and its backgroun of so-called "fixed stars"; or fixed within a satellite in orbit abou the earth. Whatever else, however, Leibniz denied that any preferre "inertial coordinate system" should be required for mechanics. Th idea of the "fixed stars" was simply not relevant.

The Cartesian coordinates $x$ and $\overline{x}$ in any pair of (relatively moving frames $F$ and $\overline{F}$ are related at time $t$ by the formula

$$\overline{x} = Q(t)x + \ell(t),$$

where

$$Q = Q(t), \text{ an orthogonal transformation matrix at each time } t$$
$$\ell = \ell(t), \text{ a translation vector at each time } t.$$

The following relations hold between the velocities and the accelera tions of a moving particle in the two frames,

$$\overline{v} = Qv + a, \qquad a = a(t),$$
$$\dot{\overline{v}} = Q\dot{v} + c, \qquad c = c(t),$$

where $c$ denotes the combination of Coriolis, centrifugal, etc., acceler ations which appear due to the relative motion of the frames. For ou purposes it is not necessary to express the specific dependence of $a$ an $c$ on $Q$ and $\ell$.

The appropriate transformation conditions for the physical quantitie in the earlier equations is expressed in the following

**Invariance Axiom.** *$\rho$ and $\epsilon$ are scalar invariants, while $h$ and T transform (through $Q$) as natural tensorial quantities.*

**Agreement.** *For simplicity, the radiative energy supply $r$ is also as sumed invariant.*

Let me return to the energy balance equation, but no longer invok the Cauchy equations. Thus we have simply

$$\rho(\epsilon + \tfrac{1}{2}v^2)^{\cdot} = r + \operatorname{div} h + T : D + \omega \cdot T_x + v \cdot (b + s).$$

219

Following Leibniz, this relation is naturally assumed to hold *whatever the coordinate frame $F$ may be.* We can now prove

**Theorem A.** *If the body force $b$ is a Galilean invariant, then the Cauchy equations are a consequence of the energy balance equation.*

**Proof:** If $F$ and $\overline{F}$ are Galilean related, then for some constant vector $a$, we have

$$\overline{v} = v + a, \qquad \dot{\overline{v}} = \dot{v}, \qquad \overline{D} = D, \qquad \overline{\omega} = \omega.$$

Subtracting the energy equation in $F$ from the equation in $\overline{F}$ and using the preceding invariance axiom and the Galilean invariance of $b$ then gives

$$a \cdot (\rho\dot{v} - b - s) = 0.$$

The Cauchy equations now follow immediately since $a$ is arbitrary.

To show that $T_x = 0$, consider the frame $F$ as before and a second frame $\overline{F}$ rotating with constant angular velocity $\alpha$ relative to $F$. At an arbitrary instant when $Q = I$, we have

$$\overline{D} = D, \qquad \overline{\omega} = \omega + 2\alpha$$

and so by subtraction as before,

$$2\alpha \cdot T_x = 0,$$

hence $T_x = 0$. □

We may now ask two fundamental questions.

I. Can Theorem A be proved *without* using the Galilean invariance of the body force?
II. Why does the total energy $\varepsilon + \frac{1}{2}v^2$ consist of two additive parts, with $\varepsilon$ invariant?

An answer to these questions can be provided on the basis of a proper axiomatization of the notion of force, delineating those properties which are, ultimately, basic and fundamental. Of course this axiomatization should be independent of any notion of inertial coordinate frame.

A *body supply system* $B$ is an assignment, to *each* frame $F$ in the Leibnizian group, of a vector field $\beta_F(x,t)$, called the *work effective force field,* namely

$$F \mapsto \beta_F(x,t) \quad (= \beta(x,t))$$

where in the obvious notation

$$\beta_F : \mathcal{E}_F \times I \mapsto \mathbf{R}^3.$$

Any motion of $S$ is associated to a body supply system $B$ which is presumed to govern, or cause, the motion. There is of course also a radiative energy supply $r$ associated with the motion; for convenience we agree to include this as an integral part of the body supply system. To express this fundamental relation between a motion $P$ and its associated supplies, we use the symbolic notation

$$P \mapsto B,$$

that is, $B$ can be considered a function of $P$. We can leave undetermined the class of motions under consideration, though this set must at the very least be so large as not to limit the scope of Axioms 1 and 2 below.

THEOREM I. *Assume that the energy balance equation*

$$\begin{aligned} \rho(\varepsilon + \tfrac{1}{2}v^2)^{\cdot} &= q + w \\ &= r + \operatorname{div} h + T : D + \omega \cdot T_x + v \cdot (\beta + s) \end{aligned}$$

*is valid in every frame $F$. Then there exists a Coriolis supply system $C$*

$$F \mapsto \gamma_F(x,t), \qquad \gamma_F : \mathcal{E}_F \times I \to \mathbf{R}^3$$

*such that the Cauchy equations*

$$\rho\dot{v} = b + s, \qquad T_x = 0$$

*hold (in every frame $F$) with*

$$b = \beta + \gamma_\wedge v$$

*(that is, $b_F = \beta_F + \gamma_{F\wedge} v$).*

Let me carefully emphasize that in the energy equation as stated here it is precisely *the work effective force field* $\beta$ which appears, not the dynamic force $b$. Of course the relation $b = \beta + \gamma_\wedge v$ yields the identity $\beta \cdot v = b \cdot v$, so that $b$ is also a *work effective force field* (the Coriolis force $\gamma_\wedge v$ does no work).

**Proof of Theorem I:** Let $(x_0, t_0)$ be a fixed point in a frame $F$ and let $\overline{F}$ be a Galilean related frame, with $Q \equiv I$ and

$$\overline{v} = v + a, \qquad a \in \mathbf{R}^3.$$

By subtracting the energy balance equation for $F$ from that for $\overline{F}$, and evaluating at $(x_0, t_0)$, we get

$$\rho a \cdot \dot{v} = (v + a) \cdot \beta(a) - v \cdot \beta(0) + a \cdot s$$

where $\beta(a) = \beta_{\overline{F}}(x_0, t_0)$ and $\beta(0) = \beta_F(x_0, t_0)$.

Define $I(a, v) = (v + a) \cdot \beta(a) - v \cdot \beta(0)$. Then $I(a, v)$ is linear in $a$ for fixed $v \in \mathbf{R}^3$ and affine in $v$ for fixed $a \in \mathbf{R}^3$ (see remark following the proof), that is

$$I(a, v) = a \cdot (A + \Gamma v),$$

where $A$ is a constant vector and $\Gamma$ a constant matrix. Set $v = -a$. Then

$$a \cdot \beta(0) = a \cdot (A + \Gamma a).$$

Since $a$ is arbitrary in $\mathbf{R}^3$ we get

$$A = \beta(0), \qquad a \cdot \Gamma a = 0.$$

Thus $\Gamma$ is skew, with axial vector $\gamma$ (the Coriolis force!), and

$$I(a, v) = a \cdot (\beta(0) + \gamma_\wedge v).$$

Hence

$$a \cdot \rho \dot{v} = a \cdot (\beta(0) + \gamma_\wedge v) + a \cdot s$$

and finally, again since $a$ is arbitrary,

$$\rho \dot{v} = b + s, \qquad b = \beta + \gamma_\wedge v; \qquad T_x = 0. \quad \square$$

*Comment on the Proof.* We have allowed $v$ to vary independently of the supply system $B$. This is in fact an assumption concerning the structure of body supply systems, which we state as

*J. Serrin*

AXIOM 1. *Fixing the body supply system at one point does not restrict the values of $v$ which can be attained at the point by some motion of $S$.*

Of course, this axiom is always satisfied in classical mechanics—i.e., we can vary initial velocities arbitrarily.

The next theorem provides an answer to the second question, why does the total energy consist of an internal energy $\varepsilon$ and a kinetic energy $\frac{1}{2}v^2$? Since this result in fact proves the existence of an internal energy as a consequence of more fundamental considerations, the reference to internal energy in the basic Invariance Axiom should therefore be deleted with regard to the remaining parts of the paper.

THEOREM II. *Let the energy balance equation have the form*

$$\rho\dot{U} = q + w \qquad \text{(valid in each frame)}.$$

*Assume that the total energy $U$ depends on a state $\sigma = \sigma(x,t)$ and on the velocity $v$, that is*

$$U = \hat{U}(\sigma, v)$$

*where $\hat{U}$ is a function independent of frame. Suppose also that the states $\sigma$ are points in an open subset of $\mathbf{R}^K$ and are invariant under change of frame. Then*

$$\hat{U}(\sigma, v) = \hat{\varepsilon}(\sigma) + \text{Const. } v^2$$

SKETCH OF THE PROOF: Consider the energy balance equation in a frame $F$, and a Galilean related frame $\overline{F}$. By subtraction and evaluation at $(x_0, t_0)$ we obtain

$$\rho\{[U_\sigma(v+a) - U_\sigma(v)]\cdot\dot{\sigma} + [U_v(v+a) - U_v(v)]\cdot\dot{v}\} = I(a,v) + a\cdot s,$$

where the argument $\sigma$ is suppressed. Set $a = -v$ to get

$$\begin{aligned}\rho\{U_\sigma(0)\cdot\dot{\sigma} + U_v(0)\cdot\dot{v}\} &= \rho\{U_\sigma(v)\cdot\dot{\sigma} + U_v(v)\cdot\dot{v}\} - v\cdot\beta(0) - v\cdot s\\ &= \rho\dot{U} - v\cdot(\beta + s)\\ &= r + \operatorname{div} h + T : D + \omega\cdot T_x\end{aligned}$$

which now holds for every frame $F$.

Next let $\overline{F}$ be related to $F$ so that

$$Q \equiv I, \qquad \ell(t) = \tfrac{1}{2}c(t-t_0)^2, \qquad c \in \mathbf{R}^3.$$

Then $\bar{v} = v$, $\dot{\bar{v}} = \dot{v} + c$, $\bar{\sigma} = \sigma$, $\bar{\omega} = \omega$, $\bar{D} = D$, $\bar{T} = T$ at the point $(x_0, t_0)$, so by subtraction

$$\rho U_v(0) \cdot c = 0, \qquad U_v(0) = 0.$$

Define $\varepsilon = \hat{\varepsilon}(\sigma) = \hat{U}(\sigma, 0)$. Then

$$\rho\dot{\varepsilon} = \rho\hat{U}_\sigma(0) \cdot \dot{\sigma} = r + \text{ div } h + T : D + \omega \cdot T_x.$$

The skew part of the stress tensor vanishes by the argument of Theorem A, leading therefore to

$$\rho\dot{\varepsilon} = r + \text{ div } h + T : D,$$

the reduced energy equation.

The remaining discussion is more elaborate, and will be omitted. However, as the conclusion of Theorem II goes well beyond the result of Theorem I, it is not surprising that at least some (mild) further structure must be imposed on the force system:

AXIOM 2. *Fixing a supply system at one point does not restrict the values of $\sigma$, $\dot{\sigma}$, $v$, $\dot{v}$ which can be attained at the point by some motion of $S$.*

This assumption may seem self-evident in view of the wide range of motions which normally can be presumed compatible with a given force structure. Nevertheless in the present context of axiomatic proof, this requirement is indeed necessary, and without its specific availability very little can otherwise be concluded.

Theorem II yields the splitting

$$U = \varepsilon + \text{Const. } v^2, \quad \textit{not} \quad U = \varepsilon + \tfrac{1}{2}v^2,$$

which of course is all that can be expected. In turn one obtains the following equations of dynamics,

$$d\rho\dot{v} = b + s, \qquad T_x = 0,$$

where $d$ is a constant (we may assume $d \neq 0$ as the natural case).

This may actually be treated as a *derivation of the law conservation of mass*: in fact, throughout the previous work replace the density

$\rho$ by the Jacobian determinant $J = \det F$. Then, at the end of th derivation, *define*

$$\rho = dJ$$

so that $\dot{\rho} + \rho \operatorname{div} v = 0$ and $\rho\dot{v} = b + s$. Finally, redefining $\varepsilon$ by divisio by $d$ to obtain the internal energy *per unit mass*, there results

$$\rho\dot{\varepsilon} = r + \operatorname{div} h + T : D,$$

the appropriate balance of energy. The idea that mass density as we as the equations of motion should arise from a balance of energy is du to Šilhavý (cf. also Green and Rivlin); the present derivation gives a alternative and mathematically simpler treatment of these ideas, base on the physically natural notion of a work effective force structure.

*Note.* Theorem I appeared first in a limited circulation unpublishe manuscript, 1974; cf. Šilhavý (1989). Theorem II was announced at conference held at Rutgers University, May 1990.

## References

[1] Noll, W., La méthode axiomatique dans les mécaniques classiqu et nouvelles, Colloque International, 1959. Gauthier – Villars, Par (1963).

[2] Green A.E. and Rivlin R.S., Z. Angew. Math. Phys. *15* (1964).

[3] Šilhavý, M., Arch. Rational Mech. Anal. *107* (1989).

# VII. Miscellaneous Topics

# VII.1 Surfaces of Constant Mean Curvature

Haim Brezis

**Mathematics Subject Classification (2010).** 49Q05, 53A10, 49F10.

**Keywords.** Plateau problem, minimal surfaces, surfaces with prescribed mean curvature.

In the early thirties J. Douglas and T. Rado independently solved the *Plateau problem* concerning the existence of a minimal surface which spans a given Jordan curve $\Gamma$ in $\mathbb{R}^3$. The same question for surfaces of constant mean curvature $H > 0$, in brief $H$-surfaces, was tackled twenty years later. A well-known result of S. Hildebrandt [4] (which improves earlier contributions by E. Heinz and H. Werner) asserts that if $\Gamma \subset B_R = B_R(0)$ and $0 < H \leq 1/R$, there exists an $H$-surface $S$ contained in $B_R$ which spans $\Gamma$.

In the paper [i] below, J. Serrin established three remarkable additions:

**A)** Any $H$-surface $S$ which spans $\Gamma$ and is contained in $B_{1/H}$ must be contained in $B_R$.

**B)** Any $H$-surface $S$ without self-intersections which spans $\Gamma$ must be contained in $B_{R+2/H}$.

**C)** Under some additional geometric conditions on $\Gamma$, there are exactly two $H$-surfaces without self-intersections, contained in $B_R$, which span $\Gamma$.

If $\Gamma$ is a circle of radius $R$ centered at 0, and $0 < H < 1/R$, there are, of course, two *small* spherical caps, $\underline{S}_1$ and $\underline{S}_2$, of curvature $H$ which span $\Gamma$. Moreover there are two *large* spherical caps $\overline{S}_1$ and $\overline{S}_2$ of curvature $H$ which span $\Gamma$. The surfaces $\underline{S}_1$ and $\underline{S}_2$ are precisely the two $H$-surfaces which appear in assertion C), while $\overline{S}_1$ and $\overline{S}_2$ are *not* contained in $B_R$.

Rellich's conjecture concerning the existence of a *large* $H$-surface for any Jordan curve $\Gamma \subset B_R$ and any $H \in (0, 1/R)$ was settled in 1982 by H. Brezis and J.M. Coron [2], (see also related works by K. Steffen and M. Struwe). A year later I visited the U. of Minnesota and had a very stimulating conversation with J. Serrin about $H$-surfaces during a cruise over the Mississippi river. In the course of our discussion he raised the following:

**Question (J. Serrin, 1983).** *Let $(\Gamma_n)$ be a sequence of Jordan curves such that $\Gamma_n \subset B_{R_n}$ with $R_n \to 0$. Assume that, for each $n$, $S_n$ is a large $H$-surface with $H$=1, which spans $\Gamma_n$. Does $S_n$ converge (modulo a subsequence) to a sphere of radius 1 containing* 0*?*

In a joint work with J.M. Coron [3] we gave two partial answers:

**Theorem 1** ([3]). *Assume that the area of the surface $S_n$ remains bounded as $n \to \infty$. Then a subsequence of $(S_n)$ converges to a "bouquet" of spheres of radius* 1 *containg* 0*,*

P. Pucci et al. (eds.), *James Serrin. Selected Papers: Volume 2*, Contemporary Mathematicians,
DOI 10.1007/978-3-0348-0847-7_17, © Springer Basel 2014

*i.e., a finite connected union of spheres of radius* 1 *such that, at least, one of them contains* 0.

In general we do not have more precise information about the limiting configuration. Indeed, it would be interesting to determine whether an arbitray bouquet may be achieved as a limit of some sequence $(S_n)$ corresponding to appropriate $\Gamma_n$'s. However we do have a positive answer to Serrin's question when the $S_n$'s are chosen in a *special way.*

**Theorem 2** ([3]). *Assume that $S_n$ is a large 1-surface obtained through the construction of Brezis-Coron [2]. Then a subsequence $S_{n_k}$ converges to a sphere of radius* 1 *containing* 0.

A key ingredient in the proofs is the following sharp description of the blow-up mechanism for a sequence $u^n \in H_0^1(\Omega;\mathbb{R}^3)$, ($\Omega =$ unit disc in $\mathbb{R}^2$), of solutions of

$$\begin{aligned} &\Delta u^n = 2u_x^n \wedge u_y^n + f^n \text{ in } \Omega, \quad \text{where} \\ &\|f^n\|_{H^{-1}} \to 0 \text{ and } \|u^n\|_{H^1} \text{ remains bounded.} \end{aligned} \tag{1}$$

**Theorem 3** ([3]). *There exist an integer $k$, functions $\omega_1, \omega_2, \dots, \omega_k : \mathbb{R}^2 \to \mathbb{R}^3$, sequences $a_1^n, a_2^n, \dots, a_k^n$ in $\Omega$, sequences of positive numbers $\varepsilon_1^n, \varepsilon_2^n, \dots, \varepsilon_k^n$ with $\lim_{n\to\infty} \varepsilon_i^n = 0\,(\forall i)$, such that, (modulo a subsequence),*

$$\lim_{n\to\infty} \left\| u^n(\cdot) - \sum_{i=1}^k \omega_i\left(\frac{\cdot - a_i^n}{\varepsilon_i^n}\right) \right\|_{H^1} = 0, \tag{2}$$

$$\lim_{n\to\infty} \frac{1}{\varepsilon_i^n} \operatorname{dist}(a_i^n, \partial\Omega) = \infty \quad \forall i, \tag{3}$$

*and*

$$\lim_{n\to\infty} \left\{ \frac{\varepsilon_i^n}{\varepsilon_j^n} + \frac{\varepsilon_j^n}{\varepsilon_i^n} + \frac{|a_i^n - a_j^n|}{\varepsilon_j^n + \varepsilon_j^n} \right\} = \infty \quad \forall i \neq j.$$

The functions $\omega_i$ are of the form $\omega(z) = \Pi(R(z))$+Const, where $R$ is a rational function and $\Pi : \mathbb{C} \to S^2$ is a stereographic projection. Properties (2) and (3) imply that $(u^n)$ behaves like a superposition of well-differentiated peaking waves. As pointed out in [1] the functions $\omega_i\left(\frac{\cdot - a_i^n}{\varepsilon_i^n}\right)$ are almost orthogonal (in $H^1$). In particular $\int_\Omega |\nabla u^n|^2$ tends to $\sum_i \int_{\mathbb{R}^2} |\nabla \omega_i|^2$ which is an integer multiple of $8\pi$. A similar conclusion (without (3)) was obtained independently by M. Struwe [5] for an elliptic equation with critical exponent. Such a technique, with roots in the 1981 paper by J. Sacks and K. Uhlenbeck, is a refinement of the concentration-compactness principle of P.L. Lions. Subsequently, it has been widely used under the name "profile decomposition" in a large variety of problems; in particular, Yamabe-type equations (e.g., by A. Bahri, J.M. Coron, O. Druet, E. Hebey), nonlinear wave and nonlinear dispersive equations, such as nonlinear Schrödinger equations (e.g., by H. Bahouri, P. Gérard, C. Kenig, F. Merle, R. Killip, M. Visan), the Navier–Stokes equation (e.g., by I. Gallagher, G. Koch).

## References

[1] H. Brezis, Problèmes de convergence dans certaines EDP non linéaires et applications géométriques, in *Séminaire Goulaouic–Meyer–Schwartz*, 1983–84, 11p.

[2] H. Brezis and J.M. Coron, *Multiple solutions of H-systems and Rellich's conjecture*, Comm. Pure Appl. Math **37** (1984), 149–187; see also C. R. Acad. Sci. **295** (1982), 615–618.

[3] H. Brezis and J.M. Coron, *Convergence of solutions of H-systems or how to blow bubbles*, Archive Rat. Mech. Anal. **89** (1985), 21–56; see also C. R. Acad. Sci. **298** (1984), 389–392.

[4] S. Hildebrandt, *On the Plateau problem for surfaces of constant mean curvature*, Comm. Pure Appl. Math. **23** (1970), 97–114.

[5] M. Struwe, *A global compactness result for elliptic boundary value problems involving limiting nonlinearities*, Math. Z. **187** (1984), 511–517.

Haim Brezis
Rutgers, The State University
Department of Mathematics
Piscataway, NJ 08854, USA
e-mail: brezis@math.rutgers.edu

*Math. Z.* 112 (1969), 77–88.




# On Surfaces of Constant Mean Curvature which Span a Given Space Curve

James Serrin

Let $\Gamma$ be a Jordan curve in $E^3$, which is contained in the closed unit ball $B$ about the origin. Recently it has been shown by Heinz [6] and Werner [18], and by Hildebrandt [7], that there exists a surface of constant mean curvature which spans the curve $\Gamma$ and lies inside $B$, provided the mean curvature in question does not exceed 1. The surface is of course defined parametrically as in the case of Plateau's problem for minimal surfaces, and accordingly may have self-intersections, multiple coverings, and branch points.

The purpose of this note is to describe a sufficient condition for this surface to be unique, and to obtain an *a priori* limitation on the form of the region in which it is contained. As a corollary, the results include a well-known criterion due to Radó for a solution of Plateau's problem to be unique and regular.

Besides the results just noted, Theorems 1 and 1′ contain a general limitation on the diameter of the region in which a surface of constant mean curvature must lie, assuming that it spans a curve $\Gamma$ contained in $B$. Finally, Theorem 4 gives an interesting maximum principle for non-parametric surfaces of constant mean curvature.

The methods by which we arrive at our conclusions belong to the realm of differential geometry. Nevertheless the ultimate basis of the results comes from an analysis of the partial differential equations obeyed by the surfaces in question. To illustrate, we require at one stage the construction of a non-parametric surface of constant mean curvature spanning a given space curve. This surface, once its existence is established, can be treated as a geometric structure or entity; but the basic existence theorem is a matter depending heavily on the theory of *a priori* estimates of non-linear partial differential equations and the delicate fixed point analysis of Leray and Schauder (see [16]). Again, we make use of the fact that two surfaces of constant mean curvature which touch at a point, but do not cross one another, must in fact coincide. This is a geometric statement, but its proof relies on the analytical maximum principle of E. Hopf (see footnote 4). The result just noted is a fundamental one in surface theory, and might in fact appropriately be called the "touching principle".

Another construction, used here as a geometrical tool apparently for the first time, is due to Bonnet, and states that parallel to any regular surface of constant mean curvature there is a second surface also having constant mean curvature. In point of fact, to apply Bonnet's construction requires an analysis of the behavior of the second surface in the neighborhood of points which are images of umbilical points of the original surface, an analysis which can be carried through only after a careful discussion of the local behavior of surfaces of constant mean curvature near an umbilic.

We note that a similar interplay between analytic and geometrical concepts also occurs in other work in the theory of surfaces of constant mean curvature, notably that of H. Hopf and A. D. Alexandrov.

## 1. A Priori Limitations

We shall retain the notations of the introduction throughout the paper. Thus $\Gamma$ will denote a Jordan curve (not necessarily rectifiable) contained in the closed unit ball $B$ about the origin. Let $D$ denote the parameter domain $u^2+v^2<1$ and let $\bar{D}$ be its closure. A surface $S$ will be called a solution of Plateau's problem for constant mean curvature $H$ provided that $S$ can be represented by a vector function $\vec{x}(u,v)\in C(\bar{D})\cap C^2(D)$ such that both

$$\Delta\vec{x}=2H\,\vec{x}_u\times\vec{x}_v$$

and

$$|\vec{x}_u|=|\vec{x}_v|,\qquad x_u\cdot x_v=0$$

in $D$, while the boundary of $D$ is topologically mapped onto $\Gamma$. We may obviously assume that $H>0$ without loss generality.

Before turning to the main *a priori* limitations, we note a few simple results concerning the local behavior of solutions of Plateau's problem. A *branch point* $\vec{x}_0$ of $S$ is a point where the surface has the local representation

$$\vec{x}=\vec{x}_0+\sum_{i+j\geqq\mu}\vec{a}_{ij}(u-u_0)^i(v-v_0)^j,$$

where $\mu\geqq 2$ is the order of the branch point. We can assume without loss of generality that $\vec{x}_0=0$ and $u_0=v_0=0$. Then exactly as in the case of Plateau's problem for minimal surfaces, it follows that

$$\begin{aligned} x+iy&=A(u+iv)^\mu+O(|u+iv|^{\mu+1})\\ z&=\qquad\qquad\quad\; O(|u+iv|^{\mu+1})\end{aligned}$$

for appropriately oriented coordinates $\vec{x}=(x,y,z)$, where $A$ is a suitable positive constant[1]. It is an immediate consequence that the surface $S$ has a unique tangent plane at each point not on $\Gamma$ (points of $S$ which arise from different values $(u,v)$ are considered to be different). Moreover, it is clear that branch points of $S$ must necessarily be isolated[2].

1 The result for minimal surfaces is due to Chen [4]; cf. also [12], p. 235. I first learned of its generalization to surfaces of constant mean curvature in a communication of Hildebrandt.

2 Hildebrandt, ibid.

Somewhat less immediate, but still a consequence of the development above, the surface $S$ must have self-intersections in the neighborhood of any branch point [3]. Indeed, it is easily verified that any suitably small neighborhood of (0, 0) in the $(u, v)$ plane is mapped by

$$x+iy=A(u+iv)^{\mu}+O(|u+iv|)^{\mu}$$

onto a neighborhood of the origin of the Riemann surface for $(u+iv)^{\mu}$. Consequently we can find a curve $\gamma$ surrounding the origin in the $(u, v)$ plane, whose image in the $(x, y)$ plane is the circle $x^2+y^2=\delta^2$ covered $\mu$ times. Consider the values of $(x, y, z)$ as $(u, v)$ traverses $\gamma$ once. If $S$ has no self-intersections, then each time the point $(x, y)$ circles the origin once the value of $z$ must increase (or decrease). But this is impossible because after $\mu$ circuits $z$ returns to its initial value. Consequently $S$ must have self-intersections in the neighborhood of a branch point, as asserted.

We may now turn to our main results. Recall here that $S$ denotes a solution of the Plateau problem spanning a Jordan curve $\Gamma$ contained in the unit ball $B$.

**Theorem 1.** *Suppose that $H \leqq 1$ and that $S$ is contained in the closed ball of radius $1/H$ about the origin. Then $S$ is contained in $B$.*

Before proving this result, we observe that the conclusion becomes false without the hypothesis that $S$ is contained in the closed ball of radius $1/H$ about the origin, as is apparent by considering a sphere of radius $1/H$ which intersects $B$. Whether the radius $1/H$ in the theorem is best possible is not known, but the same example shows that the conclusion is definitely untrue if $1/H$ is replaced by any number greater than $(2/H)-1$. (We note that if the hypothesis $H \leqq 1$ is dropped the theorem remains true, but becomes trivial.)

*Proof of Theorem* 1. Let us suppose for contradiction that $S$ contains points outside $B$. Let $K$ denote the smallest closed ball with center at the origin which contains $S$, and let $\Sigma$ be the spherical boundary of $K$. Evidently $\Sigma$ has mean curvature $\geqq H$ and is tangent to $S$ at some point $P \notin \Gamma$. We now consider two cases.

1. $P$ is a regular point of $S$. Then, in a rectangular coordinate system with the $z$-axis oriented along the exterior normal to $\Sigma$ at $P$, the surfaces $\Sigma$ and $S$ can be represented in the neighborhood of $P$ by

$$z=U(x, y) \quad \text{and} \quad z=u(x, y)$$

respectively, where the functions $U(x, y)$ and $u(x, y)$ satisfy the non-parametric equations

$$(1+U_y^2)\,U_{xx}-2U_x\,U_y\,U_{xy}+(1+U_x^2)\,U_{yy}=2\Lambda(1+U_x^2+U_y^2)^{\frac{3}{2}}$$
$$(1+u_y^2)\,u_{xx}-2u_x\,u_y\,u_{xy}+(1+u_x^2)\,u_{yy}=2\lambda(1+u_x^2+u_y^2)^{\frac{3}{2}}$$

3 The corresponding result for the case of minimal surfaces is due to Chen [4]; see also Nitsche [13]. The present proof is similar to that for minimal surfaces.

By saying that $S$ has *self-intersections* we mean here that there are at least two distinct points $(u_1, v_1)$, $(u_2, v_2)$ for which $\mathring{x}(u_1, v_1)=\mathring{x}(u_2, v_2)$.

6*

with $\Lambda \leqq -H$ and $\lambda = H$ or $-H$. Since $S$ is internally tangent to $\Sigma$ at $P$, we have

$$u \leqq U \text{ near } P, \quad u = U \text{ at } P.$$

This immediately implies $\Lambda \geqq \lambda$, whence the cases $\Lambda < -H$ and $\lambda = H$ are ruled out. It follows finally from E. Hopf's maximum principle that $u = U$ near $P$.[4] Clearly then $S$ coincides with $\Sigma$ and consequently cannot span $\Gamma$.

We note that the same argument applies even if two or more branches of $S$ should intersect at $P$, provided $P$ is regular for at least one of the branches.

2. $P$ is a branch point of $S$. Since branch points are isolated, it is evident from the previous case that $\Sigma$ cannot touch $S$ anywhere in the neighborhood of $P$ except for the point $P$ itself. Consider then a small spherical cap of $\Sigma$ about $P$. If this cap is tilted slightly its boundary will still not touch $S$. Then letting the cap retreat from $S$ (in the direction normal to $S$ at $P$) it is clear that the cap will have a last position where it touches $S$, say at $P'$. Evidently $P'$ will be a regular point of $S$, or at worst a point where regular branches intersect. But then we obtain a contradiction exactly as in case 1, and this completes the proof. (The same proof in fact shows that $S$ can intersect the boundary of $B$ only at points of $\Gamma$, interior points of $S$ being strictly contained in the interior of $B$.)

Theorem 1 yields an interesting qualitative addition to the result of Heinz, Werner, and Hildebrandt noted at the beginning of the paper. In particular, while they prove that there exists a solution of Plateau's problem which is contained in $B$, Theorem 1 shows that there can be no solution which extends outside $B$ and yet remains in the closed ball of radius $1/H$ about the origin. We remark that the same result can be obtained by considering the differential equation

$$\Delta |\vec{x}|^2 = 2(|\vec{x}_u|^2 + |\vec{x}_v|^2) + 4H\vec{x} \cdot (\vec{x}_u \times \vec{x}_v)$$

---

4 Cf. [8], p. 152. To give the argument in detail, let the nonlinear operator $\mathscr{L}v$ be defined by

$$\mathscr{L}v = (1+v_y^2)\,v_{xx} - 2v_x v_y v_{xy} + (1+v_x^2)\,v_{yy} + 2H(1+v_x^2+v_y^2)^{\frac{3}{2}}.$$

Then we have $\mathscr{L}u = \mathscr{L}U = 0$. The difference $w = U - u$ therefore satisfies a certain linear elliptic differential equation

$$a\,w_{xx} + 2b\,w_{xy} + c\,w_{yy} + d\,w_x + e\,w_y = 0,$$

and consequently cannot have an interior minimum unless it is constant. [The argument can also be handled in an alternate way to avoid use of the maximum principle. In particular, if $u = u(x, y)$ and $v = v(x, y)$ are solutions of the equation of constant mean curvature, such that $u \not\equiv v$ and

$$u = v \quad \text{and} \quad \nabla u = \nabla v = 0 \qquad \text{at } 0,$$

then there is an integer $\nu \geqq 2$ such that

$$u - v = H(x, y) + O(r^{\nu+1})$$

where $H(x, y)$ is a harmonic polynomial of degree $\nu$ (see [17], Lemma 1). In consequence, it is again evident that a one-sided tangency of two solutions is possible only if the solutions are identical.

The main advantage of the argument as given is that it applies to quite general partial differential equations and does not make use of the analyticity of solutions.]

satisfied by the function $|\vec{x}|^2$, but we have given the present proof because of its separate geometric interest and relation to later results in the paper.

When the spanning surface $S$ has no self-intersections an overall *a priori* limitation on its spatial location can be given.

**Theorem 1′.** *Suppose that $S$ is a solution of Plateau's problem for constant mean curvature $H>0$. Then if $S$ has no self-intersections it must be contained in the open ball of radius $1+2H^{-1}$ about the origin. The number $1+2H^{-1}$ is best possible.*

*Proof.* Recall that $S$ spans a curve $\Gamma$ located in the unit ball $B$. We show first that no radius smaller than $1+2H^{-1}$ suffices for the conclusion. For let $\tilde{S}$ be a sphere of radius $1/H$ which intersects the boundary of $B$ along a circle $\Gamma$. The portion $S$ of $\tilde{S}$ which lies outside $B$ and spans $\Gamma$ then satisfies the hypotheses of the theorem. Clearly the distance from the outermost point of $S$ to the origin can be made arbitrarity near $1+2H^{-1}$ by a suitable choice of $\tilde{S}$, proving that the radius $1+2H^{-1}$ cannot be improved.

Now consider a solution $S$ of the Plateau problem which extends outside $B$, and has no self-intersections. It is clear from the remarks preceding Theorem 1 that $S$ can have no branch points and hence is regular. Denote by $\rho$ the maximum distance from the origin to points of $S$. We must show that $\rho<1+2H^{-1}$.

Let $P$ be a point of $S$ whose distance from the origin is $\rho$, and consider the tangent plane $T_0$ to $S$ at $P$. We now suppose this plane to be continuously moved normal to itself to positions nearer the origin. At each stage of the motion the resulting plane $T$ will cut off from $S$ a (closed) cap $\Sigma(T)$. We assert that this cap has a single-valued projection onto $T$ as long as the distance from $T$ to the origin is not less than $(1+\rho)/2$; and even more, that no point of the cap has a tangent plane orthogonal to $T$.

This result is clear for all positions of $T$ sufficiently near $T_0$. Moreover (by continuity) the set of positions of $T$ where it holds is certainly open and connected. It remains to show that the set is also closed, provided the distance from $T$ to the origin is at least $(1+\rho)/2$.

Thus let $T_1$ denote the limiting position of a set of planes $T$ for which the result holds, the distance from $T_1$ to the origin being at least $(1+\rho)/2$. It must be shown that $\Sigma(T_1)$ has a single-valued projection onto $T$, and that no tangent plane of this cap is orthogonal to $T_1$.

For any cap $\Sigma(T)$ let $\Sigma'(T)$ be the reflection of this cap in $T$. We claim to begin with that $\Sigma'(T_1)$ lies in the region bounded by $S$ and the surface of the ball $B$. Indeed if this were not true, it is evident that by continuously withdrawing the plane $T$ from a position coincident with $T_1$ we would arrive at a position $T_2$ intermediate between $T_1$ and $T_0$ for which the cap $\Sigma'(T_2)$ is internally tangent to $S$ (here we use the fact that for all planes $T$ between $T_1$ and $T_0$ the cap $\Sigma'(T)$ lies outside $B$ and has a single-valued projection on $T$). The maximum principle of Hopf then shows that the part of $S$ on the same side of $T_2$ as $\Sigma'(T_2)$ must coincide with $\Sigma'(T_2)$, a patent absurdity. We have thus shown that $\Sigma'(T_1)$ is interior to $S$, as claimed.

Now suppose that there were a point $P$ on $\Sigma(T_1)$ with a tangent plane $V$ orthogonal to $T_1$. Clearly $P$ must lie on $T_1$, by the assumption that $T_1$ is the limit of planes for which our result holds. Let us introduce rectangular coordinates with origin at $P$, with the $z$-axis orthogonal to $V$, and with the $x$-axis along the intersection of $V$ and $T_1$. Then locally $S$ and $\Sigma'(T_1)$ have representations $z=u(x,y)$ and $z=U(x,y)$ for $y\leqq 0$ (say), and either $u\leqq U$ or $u\geqq U$ since $\Sigma'(T_1)$ is interior to $S$. Clearly $\partial u/\partial y=\partial U/\partial y=0$ at $(0,0)$ since both $\Sigma'(T_1)$ and $S$ have $V$ as tangent plane at $P$. By Hopf's boundary point lemma [9], it follows that the part of $S$ below $T_1$ must coincide with $\Sigma'(T_1)$, which is certainly impossible. Thus no tangent plane of $\Sigma(T_1)$ is orthogonal to $T_1$.

We must still show that $\Sigma(T_1)$ has a single-valued projection onto $T_1$. From what has gone before, it is clear that $\Sigma(T_1)$ consists of one or more components each of which *separately* has a single valued projection. The only overlapping of these projections which could occur would be due to new components appearing at the instant $T$ reaches coincidence with $T_1$. But these latter components must also have projections which are distinct from other components, in view of the fact that the reflected cap $\Sigma'(T_1)$ is interior to $S$. The original assertion is thereby proved; that is, for any plane $T$ whose distance from the origin is at least $(1+\rho)/2$ the corresponding cap $\Sigma(T)$ has a single-valued projection onto $T$, and no point of $\Sigma(T)$ has a tangent plane orthogonal to $T$.

Consider the cap $\tilde{\Sigma}$ cut off from $S$ when the distance from $T$ to the origin is exactly $(1+\rho)/2$. By what we have just shown, $\tilde{\Sigma}$ can be considered a non-parametric solution of the equation of constant mean curvature over a portion $\tilde{T}$ of $T$, which takes on the value zero on the boundary of $\tilde{T}$. According to the maximum principle proved in the final section of this paper, the distance from points of $\tilde{\Sigma}$ to the set $\tilde{T}$ cannot exceed $1/H$. Consequently, by considering the point $P$ on $\tilde{\Sigma}$ we obtain the relation

$$\rho-\frac{\rho+1}{2}\leqq\frac{1}{H},$$

that is $\rho\leqq 1+2H^{-1}$. If equality holds, then by Theorem 4 the cap $\tilde{\Sigma}$ must be a hemisphere, whence $S$ is a portion of a sphere. But then $S$ does not after all extend the full distance $1+2H^{-1}$ from the origin. This completes the proof of the theorem [5].

We note that the assumption that $S$ has no self-intersections is similar in both concept and use with the corresponding assumption in Alexandrov's proof that a closed surface of constant mean curvature having no self-intersections is a sphere.

We suppose from here on that the curve $\Gamma$ has a single-valued parallel projection $\Gamma^*$ onto some plane $L$ in $E^3$, and that each point $P$ of $\Gamma^*$ (viewed as a plane curve in $L$) admits a supporting circle of radius $\frac{1}{2}$ which surrounds $\Gamma^*$.

5 *Note added in proof.* A generalization of Theorem 1′ to apply to elliptic Weingarten surfaces has been obtained by A. Aeppli and the author.

(If $\Gamma^*$ is of class $C^2$ this condition is equivalent to the requirement that the curvature of $\Gamma^*$ is everywhere greater than or equal to $\frac{1}{2}$.) A surface $\bar{S}$ which spans $\Gamma$ and has a single-valued projection on $L$ will be called a non-parametric spanning surface. The following result then holds.

**Lemma.** *For any constant $\Lambda$ satisfying $0<\Lambda\leqq 1$ there exist exactly two regular non-parametric spanning surfaces of $\Gamma$ having (unsigned) constant mean curvature $\Lambda$.*

To see this, we first introduce a fixed rectangular coordinate system with the $z$-axis normal to $L$. Then any regular non-parametric spanning surface may be represented in the form $z=U(x,y)$ where $U(x,y)$ must satisfy either

$$(1+U_y^2)\,U_{xx}-2\,U_x\,U_y\,U_{xy}+(1+U_x^2)\,U_{yy}=2\Lambda(1+U_x^2+U_y^2)^{\frac{3}{2}}$$

or

$$(1+U_y^2)\,U_{xx}-2\,U_x\,U_y\,U_{xy}+(1+U_x^2)\,U_{yy}=-2\Lambda(1+U_x^2+U_y^2)^{\frac{3}{2}},$$

and where $U(x,y)$ takes on appropriately defined values on the curve $\Gamma^*$. The assertion is therefore equivalent to the existence of a unique solution of the Dirichlet problem for each equation. In view of the results of [16], both problems are solvable provided that the curve $\Gamma$ is of class $C^2$. An additional limiting argument based on an interior estimate for the gradient of solutions then shows that the problems are solvable even for continuous $\Gamma$, cf. [17]. Uniqueness follows at once from the maximum principle for the difference of solutions.

Let the two surfaces described above be denoted by $\bar{S}_+$ and $\bar{S}_-$ respectively, and let $R(\Lambda)$ be the closed region bounded by these surfaces [note that $\bar{S}_+$ lies below $\bar{S}_-$ if the upward direction is associated with the positive $z$-axis].

**Theorem 2.** *Suppose $0<H\leqq 1$. Let $\Gamma$ have the properties noted above and let $S$ be a solution of Plateau's problem for constant mean curvature $H$, contained in the closed ball of radius $1/H$ about the origin. Then $S$ is contained in $R(H)$.*

*Proof.* We show first that $S$ must be contained in the right cylinder with cross section $\Gamma^*+\text{Interior}(\Gamma^*)$, and can intersect the boundary of this cylinder only along the curve $\Gamma$. Suppose in fact that this were not the case. Then, recalling that $S$ is contained in $B$ according to Theorem 1, it is evident that there would exist a right circular cylinder of radius $\frac{1}{2}$ which is tangent to $S$ at some point $P\notin\Gamma$. (To see this, begin with the cylinder in a position tangent to $B$ and translate it in an appropriate direction parallel to $L$ until it first touches $S$.) Since the cylinder has mean curvature 1, our assertion now follows as in the concluding argument of the proof of Theorem 1.

Next suppose that $S$ is not contained in $R(H)$. Let $\bar{S}_+$ and $\bar{S}_-$ be the lower and upper bounding surfaces of this region, as before. Then we can translate $\bar{S}_+$ downwards (or $\bar{S}_-$ upwards) so that it reaches a final position tangent to $S$ at some point $P\notin\Gamma$. This being done, we again reach a contradiction as in the proof of Theorem 1, and the demonstration of Theorem 2 is completed.

The following result is due to Radó [14], though the proof we give is quite different and independent of the monodromy theorem which was required by Radó.

**Theorem 2′.** *Let $S$ be a minimal surface spanning a Jordan curve $C$ in $E^3$, where $C$ has a single-valued convex projection $C^*$ onto a plane $L$ in $E^3$. Then $S$ is a regular non-parametric spanning surface, and is unique.*

*Proof.* By replacing the right circular cylinders used in the preceding proof by planes normal to $L$, it is easy to see that $S$ must be contained in the right cylinder with cross section $C^*+\text{Interior}\,(C^*)$, and can intersect the boundary of this cylinder only along $C$.

We now observe that there is exactly one regular non-parametric minimal surface $\bar{S}_0$ spanning $\Gamma$.[6] By the argument of Theorem 2, it follows that $S$ lies both above and below $\bar{S}_0$, and hence coincides with $\bar{S}_0$. This completes the proof.

*Remark.* Theorems 1 and 2 have been stated in terms of a Jordan curve $\Gamma$ contained in the unit ball. This has been done in order to facilitate comparison with the existence theorem of Heinz, Werner, and Hildebrandt, but it is quite apparent that the results would apply equally after an arbitrary change of scale, that is, with $\Gamma$ contained in a ball of radius $1/\rho$ and with $H \leqq \rho$ where $0<\rho<\infty$. In this way, it becomes clear that Theorem 2′ is precisely the limiting case $H\to 0$ of Theorem 2.

## 2. A Uniqueness Theorem

The method of proof introduced in Section 1 allows us to obtain a uniqueness theorem for spanning surfaces of constant mean curvature, which is a partial analogue of Radó's theorem for minimal surfaces (Theorem 2′ of Section 1).

We suppose as before that $\Gamma$ is contained in the unit ball $B$, has a single valued parallel projection $\Gamma^*$ onto a plane $L$ in $E^3$, and that each point of $\Gamma^*$ admits a supporting circle of radius $\frac{1}{2}$ which surrounds $\Gamma^*$. Then we have the following theorem.

**Theorem 3.** *Suppose that $0<H\leqq 1$. Then there are exactly two solutions of the Plateau problem for constant mean curvature $H$ which are contained in $B$ and have no self-intersections. Moreover, each of these surfaces is regular and has a single-valued parallel projection on $L$.*

6 This is of course well-known and was proved first by Radó, *making use of the result of Theorem 2′*. Since we are here proving Theorem 2′ on the basis of the *existence* of $\bar{S}_0$ it is necessary to cite an independent proof of the latter result in order to avoid circularity. Should $C$ be smooth and $C^*$ have positive curvature this goes back to Bernstein [2]; see also [11] and [15] for simpler proofs of the same result. For the general case the first independent proof was given by Finn, [5], p. 415. Another demonstration was given by Jenkins and Serrin, [10], pp. 323–324, and an outline proof appears in the article of Nitsche, [12], p. 201. From a more general point of view the result can be considered as the special case $\Lambda=0$ of the existence theorem which was used in the proof of the lemma preceding Theorem 2.

*Proof.* Let $S$ be a solution of the Plateau problem which is contained in $B$ and has no self-intersections. As in the proof of Theorem 1′ it follows that $S$ can have no branch points and hence is regular. We shall show that $S$ must coincide with one of the surfaces $\bar{S}_+$ or $\bar{S}_-$ defined previously, which will complete the proof.

Thus suppose for contradiction that $S \neq \bar{S}_+, \bar{S}_-$. Then since $S$ is contained in $R(H)$ according to Theorem 2, it is apparent that $S$ lies strictly between $\bar{S}_+$ and $\bar{S}_-$. This being the case, we can translate the (upper) boundary surface $\bar{S}_-$ of $R(H)$ downwards until it reaches a final position tangent to $S$ at some point $P \notin \Gamma$. Then by the argument of Theorem 1 the signed mean curvature of $S$ at $P$ with respect to the upward directed normal must be $+H$ (if it were $-H$ then $S$ would coincide with the translation of $\bar{S}_-$ and would not span $\Gamma$).

In the same way, we can translate $\bar{S}_+$ upwards until it reaches a final position tangent to $S$ at some point $Q \notin \Gamma$. The mean curvature of $S$ at $Q$ would then be $-H$.

Now the surface $S$ must be oriented, and, moreover, the upward normals to $S$ at both $P$ and $Q$ must point to the same side of $S$. (This can be seen most easily by considering the two relatively open subsets of $R(H)$ which are separated from each other by $S$.) Accordingly the signed mean curvatures at $P$ and $Q$ must be equal, and this contradicts the first part of the proof. Consequently $S$ must coincide with $\bar{S}_+$ or $\bar{S}_-$ and the proof is completed.

## 3. A Maximum Principle

In this section we prove a maximum principle for non-parametric solutions of the equation of constant mean curvature. Besides being required for the proof of the earlier Theorem 1′, this result has a basic interest in its own right.

**Theorem 4.** *Let $\Lambda$ be a positive constant, and suppose $u = u(x, y)$ is a solution of the equation*

$$(1+u_y^2)\, u_{xx} - 2u_x\, u_y\, u_{xy} + (1+u_x^2)\, u_{yy} = 2\Lambda (1+u_x^2+u_y^2)^{\frac{3}{2}}$$

*in a bounded domain $\Omega$. If $u$ is continuous up to the boundary of $\Omega$, then we have*

$$m - 1/\Lambda \leqq u < M \qquad \text{in } \Omega$$

*where $m$ and $M$ are respectively the minimum and maximum boundary values of $u$. Equality is attained on the left only if $u$ represents a hemisphere of radius $1/\Lambda$.*

*Proof.* The right hand inequality is obvious, since a surface of positive mean curvature surely cannot posses an interior maximum point.

In order to prove the left hand inequality, we shall use of a theorem of Bonnet ([3], p. 119) which states that if $S$ is a surface of constant mean curvature $\Lambda$, then the parallel surface to $S$ at a distance $1/\Lambda$ in the direction of the preferred normal to $S$ has mean curvature $-\Lambda$. In point of fact, this applies only at points of the parallel surface which are regular, since focal points of $S$ must obviously be excluded. Such focal points can arise only as images of umbilics of $S$, for at a non-umbilical point the principal curvatures $k_1$ and $k_2$ (being unequal and summing to the value $2\Lambda$) could not have the value $\Lambda$. More precisely, then, Bonnet's theorem asserts that the parallel surface (or set) is regular and has constant mean curvature $-\Lambda$ at all points which are not the image of an umbilical point of $S$.

Now let $S$ denote specifically the surface $z=u(x, y), (x, y)\in\Omega$. If $S$ is a hemisphere, then the left hand inequality is trivial, and we must therefore show, when $S$ is *not* a hemisphere, that

$$u>m-\frac{1}{\Lambda} \quad \text{in } \Omega.$$

Let $S^*$ be the parallel surface associated to $S$ by Bonnet's theorem. The required inequality will follow at once if we can prove that $S^*$ lies entirely above the plane $z=m$.

Assume first for contradiction that $S^*$ contains some point below this plane. Since the image of points of $S$ which are sufficiently near its boundary clearly lies above any plane $z=m-\varepsilon$, it follows that $S^*$ contains some point $P$ with minimum ordinate $m'<m$. We consider two cases.

1. $P$ is a regular point of $S^*$. Then in the neighborhood of $P$ the set $S^*$ is a regular surface having constant mean curvature $-\Lambda$. Such a surface obviously cannot possess an interior minimum point, whence we obtain a contradiction and need consider only case

2. $P$ is the image of an umbilical point $Q$ of $S$. Introduce new rectangular coordinates $(x', y', z')$ with origin at $Q$ and with the $z'$-axis directed along the (upper) normal to $S$. The spherical bowl of radius $1/\Lambda$ which is tangent to $S$ at $Q$ is represented in this coordinate system by the relation

$$z'=v(x', y')=\frac{1}{\Lambda}\left(1-\sqrt{1-\Lambda^2 r'^2}\right), \qquad r'^2=x'^2+y'^2.$$

Similarly let $S$ be locally represented in the new coordinates by $z'=u(x', y')$. Then since the normal curvatures of $S$ at $Q$ are all equal to $\Lambda$, the first differing terms in the Taylor expansions of $u$ and $v$ about $(0, 0)$ will be of degree $\nu\geqq 3$. Using Lemma 1 of [17], we find therefore

$$u(x', y')=v(x', y')+H(x', y')+O(r'^{\nu+1})$$

where $H(x', y')$ is a harmonic polynomial of degree $\nu$.

Now using $(x', y')$ for local coordinates in the set $S^*$ near $P$, the points $(x^*, y^*, z^*)$ of this set are described parametrically by the relations

$$x^*=x'-\frac{u_{x'}}{\Lambda(1+u_{x'}^2+u_{y'}^2)^{\frac{1}{2}}}, \qquad y^*=y'-\frac{u_{y'}}{\Lambda(1+u_{x'}^2+u_{y'}^2)^{\frac{1}{2}}}$$

$$z^*=u(x', y')+\frac{1}{\Lambda(1+u_{x'}^2+u_{y'}^2)^{\frac{1}{2}}},$$

as one sees easily. By expansion around the function $v$ we obtain without difficulty

$$(1+u_{x'}^2+u_{y'}^2)^{-\frac{1}{2}}=(1+v_{x'}^2+v_{y'}^2)^{-\frac{1}{2}}\{1-v_{x'}H_{x'}-v_{y'}H_{y'}+O(r'^{\nu+1})\}$$

(this calculation is made under the assumption $\nu\geqq 3$; if $\nu=2$ an extra term would appear). Hence by a straightforward evaluation we get

$$x^*-iy^*=c(x'+iy')^{\nu-1}+O(r'^{\nu})$$

$$z^*=\frac{1}{\Lambda}+\Lambda\left(1-\frac{1}{\nu}\right)\text{Real}\{c(x'+iy')^{\nu}\}+O(r'^{\nu+1}),$$

where $c$ is a complex constant. In accordance with these formulae, it is apparent that the set $S^*$ must have a unique tangent plane at $P$ and must furthermore extend both above and below the horizontal plane $z=m'$ through $P$, which contradicts the fact that $P$ is a point of $S^*$ with minimum ordinate. Thus $S^*$ cannot extend below the plane $z=m$.

Next suppose that $S^*$ contained a point of the plane $z=m$. This point would have a minimum ordinate, and the same proof as before then leads to a contradiction.

Having shown that $S^*$ lies entirely above $z=m$, we note that each point $Q$ of $S$ is associated with a point $Q^*$ of $S^*$ such that $|QQ^*|=1/\Lambda$. Hence $S$ must lie everywhere above the plane $z=m-1/\Lambda$, completing the proof of the theorem.

It is remarkable that the bound given by the left hand inequality of Theorem 4 tends to minus infinity as the value of $\Lambda$ tends to zero, even though in this limit the equation of constant mean curvature comes more and more nearly to resemble the minimal surface equation, for which the maximum principle holds in the form

$$m\leqq u\leqq M \qquad \text{in } \Omega.$$

This anomaly is partly explained by the fact that the domain must become increasingly large if the lower bound $m-1/\Lambda$ is to be attained. For a *fixed* domain, and for suitally small values of $\Lambda$, an alternate lower bound is available which (while depending on the size of the domain) does at least tend to the value $m$ as $\Lambda$ tends to zero. More precisely, we have the following result complementing the conclusion of Theorem 4.

**Theorem 4'.** *Suppose that the hypotheses of Theorem 4 are satisfied, and that in addition $\Omega$ is contained in a disk of radius $a\leqq 1/\Lambda$. Then we have*

$$m-\Lambda a^2\leqq u<M \qquad \text{in } \Omega.$$

*The equality can hold on the left hand side only if $u$ represents a hemisphere.*

*Proof.* With the help of Hopf's maximum principle it is evident that in $\Omega$

$$M>u\geqq m-\frac{1}{\Lambda}\{\sqrt{1-\Lambda^2(x^2+y^2)}-\sqrt{1-\Lambda^2a^2}\}$$

$$\geqq m-\frac{1}{\Lambda}\{1-\sqrt{1-\Lambda^2a^2}\}\geqq m-\Lambda a^2$$

as required.

In this form of the maximum principle, it is clear that one recovers the conclusion $m \leqq u \leqq M$ in the limit as $\Lambda$ tends to zero. We note finally that Theorems 4 and 4′ can be combined into a single statement, as follows.

*Let the hypotheses of Theorem 4 be satisfied, and suppose that $\Omega$ is contained in a disk of radius a. Then we have*

$$m - \mathrm{Min}(\Lambda^{-1}, \Lambda a^2) \leqq u < M \qquad \text{in } \Omega,$$

*where the equality can hold on the left hand side only when u represents a hemisphere.*

## References

1. Alexandrov, A. D.: Uniqueness theorems for surfaces in the large, V. Vestnik Leningrad. Univ. **13** (No. 19) (1958), 5–8. Translated in Amer. Math. Soc. Translations (Series 2) **21**, 412–416.
2. Bernstein, S. N.: Sur la généralisation du problème de Dirichlet, II. Math. Ann. **69**, 82–136 (1910).
3. Blaschke, W.: Differentialgeometrie, Bd. I. Berlin: Springer 1930.
4. Chen, Y. W.: Branch points, poles, and planar points of minimal surfaces in $R^3$. Ann. of Math. **49**, 790–806 (1948).
5. Finn, R.: On equations of minimal surface type. Ann. of Math. **60**, 397–416 (1954).
6. Heinz, E.: Über die Existenz einer Fläche konstanter mittlerer Krümmung bei vorgegebener Berandung. Math. Ann. **127**, 258–287 (1954).
7. Hildebrandt, S.: Über Flächen konstanter mittlerer Krümmung. (To appear.)
8. Hopf, E.: Elementare Bemerkungen über die Lösungen partieller Differentialgleichungen zweiter Ordnung von Elliptischen Typen. S.-ber., Preuss. Akad. Wiss. **19**, 147–152 (1927).
9. – A remark on linear elliptic differential equations of the second order. Proc. Amer. Math. Soc. **3**, 791–793 (1952).
10. Jenkins, H., Serrin, J.: Variational problems of minimal surface type, II. Arch. Rat. Mech. Analysis **21**, 321–342 (1966).
11. Nirenberg, L.: On nonlinear elliptic partial differential equations and Hölder continuity. Commun. Pure Appl. Math. **6**, 103–156 (1952).
12. Nitsche, J. C. C.: On new results in the theory of minimal surfaces. Bull. Amer. Math. Soc. **71**, 195–270 (1965).
13. – Vorlesung über Minimalflächen. Berlin-Heidelberg-New York: Springer, to appear.
14. Radó, T.: The problem of least area and the problem of Plateau. Math. Z. **32**, 763–796 (1930).
15. – The problem of Plateau. Berlin: Springer 1932. Especially Chapter IV.
16. Serrin, J.: The problem of Dirichlet for quasilinear elliptic equations with many independent variables. Philos. Trans. Roy. Soc. London A **264**, 413–496 (1969).
17. – The Dirichlet problem for surfaces of constant mean curvature. (To appear, Proc. London Math. Soc.)
18. Werner, H.: The existence of surfaces of constant mean curvature with arbitrary assigned Jordan curves as assigned boundary. Proc. Amer. Math. Soc. **11**, 63–70 (1960).

Prof. James Serrin
School of Mathematics
University of Minnesota
Minneapolis, Minnesota

*(Received May 13, 1969)*

# VII.2 Positive Definite Matrices

Charles H. Conley

**Mathematics Subject Classification (2010).** 35J15, 15A48, 35B50, 35J70.

**Keywords.** Quasilinear singular elliptic inequalities, strong maximum principle, products of Hermitian matrices.

This paper represents something of a departure for Serrin, in that it is concerned mainly with algebra. It grew from a question arising in the study of the maximum principle: if $A$ and $B$ are both positive definite symmetric $n \times n$ matrices, is their symmetric product $\{AB\} := \frac{1}{2}(AB + BA)$ necessarily positive definite? The answer is no, and so one is led to ask for conditions on $A$ and $B$ under which $\{AB\}$ *is* positive definite. This is a very natural question which arises in many branches of mathematics.

In fact, one asks for *coordinate-free* conditions on $A$ and $B$, by which we mean conditions invariant under conjugation by orthogonal matrices, *i.e.*, conditions on the eigenvalues of $A$ and $B$. An elegant geometric solution was given by T. Tao (private communication). Define $\phi_A$ to be the maximal angle between $x$ and $Ax$ for any vector $x$:

$$\phi_A := \sup_{x \in \mathbb{R}^n \setminus \text{kernel}\, A} \text{Angle}(x, Ax).$$

(The computation of $\phi_A$ is an exercise in Lagrange multipliers.) Tao observes that there exists an orthogonal matrix $T$ such that $\{A(TBT^{-1})\}$ is indefinite if and only if $\phi_A + \phi_B$ is obtuse. His result is presented in Part I of the paper [i] below, as Theorem 2.1 (with Tao's observation in hand, the proof of this theorem amounts to factoring (2.3) via (2.4)). In Part III, Theorem 2.1 is used to obtain results on a certain class of quasilinear partial differential equations.

The answer can also be deduced as a corollary of a beautiful paper of W.G. Strang [2] giving tight coordinate-free bounds on the least and greatest eigenvalues of $\{AB\}$ for any two real symmetric matrices $A$ and $B$, not necessarily positive definite. By "tight coordinate-free bounds", we mean the two scalars

$$\Lambda_{\min}(A, B) := \inf \big\{\text{minimum eigenvalue of } \{A(TBT^{-1})\} : T \text{ orthogonal}\big\},$$

$$\Lambda_{\max}(A, B) := \sup \big\{\text{maximum eigenvalue of } \{A(TBT^{-1})\} : T \text{ orthogonal}\big\}.$$

The paper [i] extends Strang's work, so we first describe the latter. Strang begins with a subtle argument proving that $\Lambda_{\min}(A, B)$ and $\Lambda_{\max}(A, B)$ depend only on the least and greatest eigenvalues of $A$ and $B$ (see (2) in [2]). This means that the problem is essentially two-dimensional: $\Lambda_{\min}$ and $\Lambda_{\max}$ are unchanged if $A$ and $B$ are replaced by the matrices

$$\begin{pmatrix} A_{\min} & 0 \\ 0 & A_{\max} \end{pmatrix} \quad \text{and} \quad \begin{pmatrix} B_{\min} & 0 \\ 0 & B_{\max} \end{pmatrix},$$

P. Pucci et al. (eds.), *James Serrin. Selected Papers: Volume 2*, Contemporary Mathematicians,
DOI 10.1007/978-3-0348-0847-7_18, 

where $A_{\min}$ and $A_{\max}$ are the least and greatest eigenvalues of $A$, respectively: its *extremal eigenvalues.*

Even in the two-dimensional case the computation of $\Lambda_{\min}$ and $\Lambda_{\max}$ is involved. Observe that if $A$ and $B$ are diagonal (or simultaneously diagonalizable), then the extremal eigenvalues of $\{AB\}$ are the least and greatest of the four numbers $A_{\min}B_{\min}$, $A_{\min}B_{\max}$, $A_{\max}B_{\min}$, and $A_{\max}B_{\max}$ (keep in mind that Strang allows $A$ and $B$ to have negative eigenvalues). In particular, the interval $[\Lambda_{\min}, \Lambda_{\max}]$ must contain these four numbers. However, Strang shows that it need not be the shortest such interval.

To explain, consider the ratios $k := A_{\max}/A_{\min}$ and $\ell := B_{\max}/B_{\min}$. Strang's main result is that there is a region $\mathcal{R}$ in the $k\ell$-plane, defined by a certain pair of inequalities, with the following properties: if $(k, \ell)$ is not in $\mathcal{R}$, then in fact $\Lambda_{\min}$ and $\Lambda_{\max}$ *are* simply the least and greatest of the above four obvious candidates for bounds, but if $(k, \ell)$ is in $\mathcal{R}$, then $\Lambda_{\min}$ and $\Lambda_{\max}$ are the least and greatest of these four *and* a surprising fifth candidate,

$$E_5 := \frac{16A_{\max}A_{\min}B_{\max}B_{\min} - (A_{\max} - A_{\min})^2(B_{\max} - B_{\min})^2}{8(A_{\max} + A_{\min})(B_{\max} + B_{\min})}.$$

We can now describe the main results of [i], which are contained in Part II. The inequalities by which Strang defined $\mathcal{R}$ are rather complicated: in [i] they are analyzed and graphed, and it is proven that at all points of $\mathcal{R}$, $E_5$ actually is either $\Lambda_{\min}$ or $\Lambda_{\max}$ (see Figures 1–3). It turns out that $\mathcal{R}$ is quite pretty: it is bounded by four hyperbolas, which are permuted by the Klein 4-group generated by $k \mapsto 1/k$ and $\ell \mapsto 1/\ell$. These hyperbolas have exactly three intersections, two simple and one triple (see Figure 3).

Finally, let us return to the original question, rephrased: if $A$ and $B$ are positive definite, when is $\{AB\}$ indefinite? It is immediate from Strang's results that this cannot happen unless $E_5$ is negative, and conversely, if $E_5$ is negative, then there does exist some orthogonal matrix $T$ such that $\{A(TBT^{-1})\}$ is indefinite. Of course, $\{AB\}$ itself can be positive definite even if $E_5$ is negative, for example if $A$ and $B$ are simultaneously diagonalizable.

This leads to an interesting open question: if $A$ and $B$ are positive definite but $E_5 < 0$, what is the least "angle between the eigenbases of $A$ and $B$" such that $\{AB\}$ is indefinite? In other words, if $A$ and $B$ are positive definite, diagonal, and satisfy $E_5 < 0$, what is the smallest special orthogonal matrix $T$ such that $\{A(TBT^{-1})\}$ is indefinite? Here "smallest" means of minimal $\phi_T$.

In two dimensions this question is answered in [i] (see the end of Section 6). However, in general there is no analog of Strang's initial argument: the higher-dimensional problem cannot be reduced to two dimensions. Already in three dimensions it seems to be quite difficult. Preliminary steps are taken in N. Bajracharya's thesis [1].

## References

[1] N. Bajracharya, The level curves of the angle function of a positive definite symmetric matrix, *Int. J. Algebra,* **5** (2011), no. 9, 429–449.

[2] W.G. Strang, Eigenvalues of Jordan products, *Amer. Math. Monthly,* **63** (1962), 37–40.

Charles H. Conley
Department of Mathematics
University of North Texas
e-mail: conley@unt.edu

*Math. Nachr.* 278 (2005), 1490–1508. **Reprint VII.2.[i]**

Pages 1491–1497 have been removed for the sake of brevity.

# Elliptic equations and products of positive definite matrices

**Charles H. Conley**[*1], **Patrizia Pucci**[**2], and **James Serrin**[***3]

[1] Department of Mathematics, College of Arts and Sciences, University of North Texas, Denton TX, 76203-1430, USA
[2] Dipartimento di Matematica e Informatica, Università degli Studi di Perugia, Via Vanvitelli 1, 06123 Perugia, Italy
[3] Department of Mathematics, University of Minnesota, Minneapolis, Minnesota, USA

Received 15 November 2004, revised 10 February 2005, accepted 15 March 2005
Published online 8 September 2005

**Key words** Quasilinear singular elliptic inequalities, strong maximum principle, products of Hermitian matrices
**MSC (2000)** Primary: 35J15, 15A18; Secondary: 35J70, 15A23, 15A57

*Dedicated to the memory of Professor F. V. Atkinson*

We present necessary and sufficient conditions under which the symmetrized product of two $n \times n$ positive definite Hermitian matrices is still a positive definite matrix (Part I, Sections 2 and 3). These results are then applied to prove the validity of the strong maximum principle, as well as of the compact support principle, for nonnegative $C^1$ distribution solutions of general quasilinear inequalities, possibly not elliptic at points where the gradient variable is either zero or large (Part III, Sections 9 and 10).

In Part II (Sections 4–8) we consider the general problem of finding bounds for the least and greatest eigenvalues of the product of two (not necessarily definite) Hermitian matrices. In particular, we refine earlier results of Strang for this problem.



## 1 Introduction

In this note we present sufficient and (essentially) necessary conditions under which the symmetrized product of two $n \times n$ positive definite complex valued Hermitian matrices is still positive definite (see Theorem 2.1). These results are then applied to prove the validity of a strong maximum principle for nonnegative $C^1$ distribution solutions of quasilinear differential inequalities of the general form

$$D_i\{a_{ij}(x,u)A(|Du|)D_ju\} - B(x,u,Du) \le 0, \quad u \ge 0 \text{ in } \Omega, \tag{1.1}$$

where $[a_{ij}(x,u)]$, $i,j = 1,\dots,n$, is a continuously differentiable, real symmetric coefficient matrix defined in $\Omega \times \mathbb{R}_0^+$, which is positive definite when $u = 0$, that is

$$a_{ij}(x,0)\eta_i\eta_j > 0 \quad \text{for all} \quad x \in \Omega, \quad \boldsymbol{\eta} \in \mathbb{R}^n \setminus \{\mathbf{0}\}. \tag{1.2}$$

We suppose furthermore that the principal operator $A = A(\varrho)$ and the main nonlinearity $B = B(x,u,\boldsymbol{\xi})$ satisfy the following conditions

(A1) $A \in C^1(\mathbb{R}^+)$, $\mathbb{R}^+ = (0,\infty)$;

(A2) $\Phi'(\varrho) > 0$ for $\varrho > 0$, and $\Phi(\varrho) \to 0$ as $\varrho \to 0$, where $\Phi(\varrho) := \varrho A(\varrho)$;

(B1) $B \in C(\Omega \times \mathbb{R}_0^+ \times \mathbb{R}^n)$ and

$$B(x,u,\boldsymbol{\xi}) \le \kappa\Phi(|\boldsymbol{\xi}|) + f(u)$$

for $x \in \Omega$, $u \ge 0$, and all $\boldsymbol{\xi} \in \mathbb{R}^n$ with $|\boldsymbol{\xi}| \le 1$, where $\kappa > 0$, and the nonlinearity $f$ is continuous in $\mathbb{R}_0^+$ and nondecreasing on some interval $(0,\delta)$, $\delta > 0$, with $f(0) = 0$.

In fact, the strong maximum principle for nonnegative $C^1$ distribution solutions of the general quasilinear differential inequality (1.1) was already stated in [7, Theorem 8.1], but with inadequate conditions on the matrix $[a_{ij}]$. This is because the assertion on [7, p. 42], that the product matrix $[a_{ik}b_{kj}]$ is positive definite,

* e-mail: conley@unt.edu, Phone: +00 (940) 565 3326, Fax: +00 (940) 565 4805
** e-mail pucci@dipmat.unipg.it, Phone: +39 075 585 5038 Fax: +39 075 585 5024
*** Corresponding author: e-mail: serrin@math.umn.edu, Phone: +00 (612) 624 9530, Fax: +00 (612) 626 2017

**Theorem 5.9** *A transition occurs between the value* $\lambda_{\min} = \widehat{E}_i$ *and* $\lambda_{\min} = \widehat{E}_5$ *when and only when one crosses the upper branch of the hyperbola* $H_i$, $i = 1, 2, 4$, *or the lower branch of the hyperbola* $H_3$, *with the exception of the three points* $(0,1)$, $(1,0)$ *and* $(-1,-1)$ *where the hyperbolas intersect.*

*In addition* $\lambda_{\min}$ *is continuous in* $\mathbb{R}^2$.

Proof. We shall actually prove somewhat more. Denote by $\widehat{H}_i$, $i = 1,2,4$, the upper branches of $H_1$, $H_2$ and $H_4$, and similarly by $\widehat{H}_3$ the lower branch of $H_3$, as in the statement of the theorem. Divide the $(k,\ell)$-plane into eight (closed) regions $R$ by means of the following five lines and half lines:

$$k = 0,\ \ell \geq 0; \quad k \geq 0,\ \ell = 0; \quad k = 1; \quad \ell = 1; \quad k = \ell < 0.$$

In each one of the eight regions $R$ there is contained a unique maximal arc $\sigma = \sigma(R)$ of one of the branches $\widehat{H}_1$, $\widehat{H}_2$, $\widehat{H}_3$ and $\widehat{H}_4$. We assert that in each such region $R$ there holds

$$\lambda_{\min} = \begin{cases} E & \text{on one side of the arc } \sigma, \\ \widehat{E}_5 < E & \text{on the other side,} \end{cases}$$

and $\widehat{E}_5 = E$ exactly on the arc $\sigma$. (Thus $\lambda_{\min}$ is continuous in $R$.)

To prove this assertion we shall consider in detail only the regions

$$R_1 = \{(k,\ell) \in \mathbb{R}^2 : k \geq 1,\ \ell \geq 1\}$$

and

$$R_2 = \{(k,\ell) \in \mathbb{R}^2 : 0 \leq k \leq 1,\ \ell \geq 1\},$$

the corresponding arcs $\sigma$ being contained in the branches $\widehat{H}_1$ and $\widehat{H}_2$, respectively, as one sees from Figure 1.

In $R_1$, by Lemma 5.3 we have

$$\lambda(k,\ell;0) = \min\{k,\ell\} \geq 1 \quad \text{and} \quad \lambda(k,\ell;1) = 1 = \widehat{E}_1 = E,$$

so that also $\lambda(k,\ell;1) \leq \lambda(k,\ell;0)$. Moreover in the interior of $R_1$ there holds $k\ell > 1$ and $(k-1)(\ell-1) > 0$. Hence by (5.7),

$$\begin{cases} \dfrac{\partial\lambda}{\partial t}(k,\ell;1) < 0 & \text{if } (k,\ell) \text{ is below } \widehat{H}_1, \\[2ex] \dfrac{\partial\lambda}{\partial t}(k,\ell;1) > 0 & \text{if } (k,\ell) \text{ is above } \widehat{H}_1. \end{cases}$$

Finally, the graph of $\lambda(k,\ell;\cdot)$ is convex by Corollary 5.5. From the above considerations, it now follows easily that

$$\begin{aligned} \lambda_{\min}(k,\ell) &= \lambda(k,\ell;1) = E & &\text{if } (k,\ell) \text{ is below } \widehat{H}_1, \\ \lambda_{\min}(k,\ell) &= \lambda(k,\ell;t_0) = \widehat{E}_5 < E & &\text{if } (k,\ell) \text{ is above } \widehat{H}_1. \end{aligned} \tag{5.10}$$

By Lemma 5.8 we have $\widehat{E}_5 = \widehat{E}_1$ exactly on $\widehat{H}_1 \cap R_1$. This proves the assertion for the region $R_1$.

In $R_2$ we have

$$\lambda(k,\ell;0) = k = \widehat{E}_2 = E \quad \text{and} \quad \lambda(k,\ell;1) = \min\{1, k\ell\} \geq \min\{1,k\} = k,$$

so that now $\lambda(k,\ell;0) \leq \lambda(k,\ell;1)$. Moreover in the interior of $R_2$ there holds $k\ell < 1$ and $(k-1)(\ell-1) < 0$. Hence by (5.6),

$$\begin{cases} \dfrac{\partial\lambda}{\partial t}(k,\ell;0) > 0 & \text{if } (k,\ell) \text{ is below } \widehat{H}_2, \\[2ex] \dfrac{\partial\lambda}{\partial t}(k,\ell;0) < 0 & \text{if } (k,\ell) \text{ is above } \widehat{H}_2. \end{cases}$$

From Corollary 5.5 and the above considerations, it again follows easily that

$$\begin{aligned} \lambda_{\min}(k,\ell) &= \lambda(k,\ell;0) = k = E & &\text{if } (k,\ell) \text{ is below } \widehat{H}_2, \\ \lambda_{\min}(k,\ell) &= \lambda(k,\ell;t_0) = \widehat{E}_5 < E & &\text{if } (k,\ell) \text{ is above } \widehat{H}_2. \end{aligned} \tag{5.11}$$

By Lemma 5.8 we have $\widehat{E}_5 = \widehat{E}_2$ exactly on $\widehat{H}_2 \cap R_2$. This proves the assertion for the region $R_2$. The remaining six regions $R_3, \ldots, R_8$ are treated in the same way.

It remains to be shown that $\lambda_{\min}$ is continuous in $\mathbb{R}^2$. In view of what has been shown so far, it is enough to prove that $\lambda_{\min}$ is continuous at the common boundary of any two adjacent regions $R$. For the common boundary $k = 1$ of the regions $R_1$ and $R_2$, we have by (5.10) and (5.11) that $\lambda_{\min} = \widehat{E}_2$ immediately to the left of $k = 1$ and $\lambda_{\min} = \widehat{E}_1$ immediately to the right. But $\widehat{E}_1 = \widehat{E}_2 = 1$ when $k = 1$, proving continuity at the common boundary of $R_1$ and $R_2$. The remaining common boundaries can obviously be treated in the same way. □

**Remark 5.10** It is worth noting also that by Lemma 5.8 the function $\lambda_{\min}$ is of class $C^1(\mathbb{R}^2)$ except where $k = 1$ or $\ell = 1$.

## 6 The general case $\{\mathbb{AB}\}$

It remains to apply the special case of the matrices $\mathbb{K}$, $\mathbb{L}$, considered above, to obtain a lower bound for the least eigenvalue of $\{\mathbb{AB}\}$.

We assume as in Section 4 that the eigenvalues of $\mathbb{A}$ satisfy $a_1 \le \ldots \le a_n$ and those of $\mathbb{B}$ obey $b_1 \le \ldots \le b_n$, where now we suppose, without loss of generality, that $a_n, b_n > 0$ but of course *no longer* that $a_1, b_1 > 0$. Let $\Lambda_{\min}$ be the best lower bound for the least eigenvalue of $\{\mathbb{AB}_{\text{rot}}\}$, in terms of $a_1, \ldots, a_n$ and $b_1, \ldots, b_n$. Since it is enough to treat the $2 \times 2$ real case (see Strang [11]), we have

$$\Lambda_{\min} = C\lambda_{\min}(k, \ell) = \Lambda_{\min}(a_1, a_n, b_1, b_n), \tag{6.1}$$

where $k$, $\ell$, $C$ are given in the following table:

| | | | |
|---|---|---|---|
| $a_1, b_1 \ge 1$ : | $k = a_n/a_1$, | $\ell = b_n/b_1$, | $C = a_1 b_1$, |
| $a_1 < 1, \quad b_1 \ge 1$ : | $k = a_1/a_n$, | $\ell = b_n/b_1$, | $C = a_n b_1$, |
| $a_1 \ge 1, \quad b_1 < 1$ : | $k = a_n/a_1$, | $\ell = b_1/b_n$, | $C = a_1 b_n$, |
| $a_1, b_1 < 1$ : | $k = a_1/a_n$, | $\ell = b_1/b_n$, | $C = a_n b_n$. |

The numbers $k, \ell$ are sometimes called the *condition numbers* of the matrices $\mathbb{A}$, $\mathbb{B}$. From the table we also get

$$C\widehat{E}_5(k, \ell) = \frac{16 a_1 b_1 a_n b_n - (a_n - a_1)^2 (b_n - b_1)^2}{8(a_n + a_1)(b_n + b_1)} \equiv E_5(a_1, a_n, b_1, b_n) \tag{6.2}$$

(notation of Strang), while moreover the quantities $C\widehat{E}_i$, $i = 1, \ldots, 4$, have one of the four values $a_1 b_1$, $a_1 b_n$, $a_n b_1$, $a_n b_n$.

We can now obtain the diagram of Figure 1, showing the values of $\Lambda_{\min}$ over the whole $(k, \ell)$-plane. First note by Theorem 5.9 that there is a transition each time one crosses an upper branch of the hyperbola $H_i$, $i = 1, 2, 4$, or the lower branch of $H_3$. If we consider the transition across $H_1$, for example, we have seen previously that $\lambda_{\min} = \widehat{E}_1 = 1$ below $H_1$, while $\lambda_{\min} = \widehat{E}_5$ above $H_1$. In particular, one then finds that $\Lambda_{\min} = C\lambda_{\min} = a_1 b_1$ below $H_1$, while $\Lambda_{\min} = E_5(a_1, a_n, b_1, b_n)$ above $H_1$.

All the transitions can be treated the same way, but (as follows from the proof of the Theorem 5.9) it is easier simply to alternate between the values $E_5$ and $\min\{a_1 b_1, a_1 b_n, a_n b_1\}$ each time one crosses the upper branch of a hyperbola $H_i$ (lower branch if $i = 3$).

The diagram of Figure 1 is the first main result of Part II of the paper, significantly refining earlier results of Strang. More particularly, in [11, Theorem 1] there is no attempt to locate the subset of points where the condition $\Lambda_{\min} = E_5$ applies. Here it is shown that this region is precisely the shaded region bounded by branches of the four hyperbolas $H_1, \ldots, H_4$, to us an unexpected and pretty result.

By Remark 5.10 and by (6.1) also $\Lambda_{\min}$ is a continuously differentiable function of the variables $(a_1, \ldots, a_n; b_1, \ldots, b_n)$.

**Remark 6.1** When $\mathbb{A}$ is not positive definite, then $\Lambda_{\min} \le 0$ for all matrices $\mathbb{B}$. Equality can occur only when $\mathbb{A}$ is positive semi-definite and $\mathbb{B} = \mathbb{I}$.

### 6.1 The rotation required for indefiniteness

Let $\mathbb{K}$, $\mathbb{L}$ both be positive definite $2 \times 2$ real matrices. If $E_5 < 0$, where $k = a_n/a_1 > 0$ and $\ell = b_n/b_1 > 0$, then $\{\mathbb{KL}\}$ will be indefinite, either by Theorem 3.1 or by Figure 1. Our interest is in the rotation angles $\vartheta$ in (5.1) which produce $\lambda_{\min} < 0$. In fact, when $\lambda_{\min} < 0$ we have $k + \ell + (k-1)(\ell-1)t < S$ by (5.1), that is

$$[k + \ell + (k-1)(\ell-1)t]^2 \; < \; (k^2-1)(\ell^2-1)t + (\ell-k)^2 .$$

This is a quadratic inequality for $t$ whose solutions

$$t_\pm \; = \; \frac{1}{2}\left[1 \pm \frac{\sqrt{-8(k+1)(\ell+1)\widehat{E}_5}}{(k-1)(\ell-1)}\right]$$

give in turn the decisive angles $\vartheta_\pm$, i.e., $\widehat{E}_5 < 0$ exactly when

$$t_- \; < \; t \; < \; t_+ \tag{6.3}$$

(clearly $0 < t_- < t_+ < 1$). Of course (6.3) can be rewritten as

$$\vartheta_+ \; < \; \vartheta \; < \; \pi/2 - \vartheta_+ ,$$

where $\cos^2\vartheta_+ = t_+$, a sector centered on either side of $\pi/4$. (Other sectors symmetrically related to (6.3) clearly also produce indefiniteness).

As an immediate corollary it is obvious that at the onset of indefiniteness ($E_5 = 0$) one requires the rotation angle $\vartheta = \pi/4$.

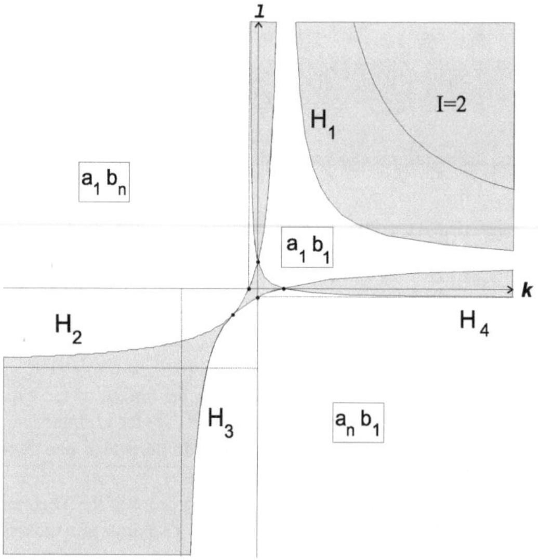


**Fig. 1** The eigenvalue function $\Lambda_{\min}$

The diagram is to be read as follows: given $a_1, \ldots, a_n; b_1, \ldots, b_n$, one first computes $(k, \ell)$ from the table preceding (6.2). If this point occurs in one of the unshaded regions, the value of $\Lambda_{\min}$ is shown in a corresponding box directly on the diagram. If it is in a shaded region, then $\Lambda_{\min} = E_5$ (see (6.2)).

The hyperbolas $H_1, H_2, H_3, H_4$ are defined preceding Theorem 5.9 (see also (8.1)). Their vertical asymptotes are respectively $k = 1,\ 1,\ -3,\ -1/3$ and their horizontal asymptotes are similarly $\ell = 1,\ -3,\ 1,\ -1/3$.

The curve $I = 2$ is the borderline case in Nicholson's theorem (Theorem 2.1).

**Remark 6.2** The region $\widehat{E}_5(k,\ell) > 0$ is of course given by (2.1). We can also obtain this result directly from $\widehat{E}_5 > 0$ by the factorization

$$8(k+1)(\ell+1)\widehat{E}_5 = \left[4\sqrt{k\ell} - (k-1)(\ell-1)\right] \cdot \left[4\sqrt{k\ell} + (k-1)(\ell-1)\right].$$

Hence in the set $\{(k,\ell) \in \mathbb{R}^2 : k > 1,\ \ell > 1\}$, $\widehat{E}_5 > 0$ is equivalent to $4\sqrt{k\ell} > (k-1)(\ell-1)$, which can be rewritten as $\left(\sqrt{k}+\sqrt{\ell}\right)^2 > \left(\sqrt{k\ell}-1\right)^2$, or again

$$\left(\sqrt{k}+\sqrt{\ell}-\sqrt{k\ell}+1\right)\left(\sqrt{k}+\sqrt{\ell}+\sqrt{k\ell}-1\right) > 0.$$

But this immediately gives (2.1).

The same calculation also gives the remarkable identity

$$\widehat{E}_5(k,\ell) = -\frac{1}{8(k+1)(\ell+1)} \prod_{\varepsilon_k=\pm1,\ \varepsilon_\ell=\pm1} \left[\left(\sqrt{k}+\varepsilon_k\right)\left(\sqrt{\ell}+\varepsilon_\ell\right) - 2\varepsilon_k\varepsilon_\ell\right]. \tag{6.4}$$

## 7 Maximum eigenvalues

The maximum eigenvalue problem is treated in almost exactly the same way. The following obvious changes occur:

$$\begin{aligned}
\lambda(k,\ell;t) &= \frac{1}{2}\left[k+\ell+(k-1)(\ell-1)t+S\right],\\
\frac{\partial\lambda}{\partial t}(k,\ell;t) &= \frac{(k-1)(\ell-1)}{4S}\left[2S+(k+1)(\ell+1)\right],\\
\frac{\partial^2\lambda}{\partial t^2}(k,\ell;t) &= -\frac{(k^2-1)^2(\ell^2-1)^2}{8S^3} \le 0,\\
\frac{\partial\lambda}{\partial t}(k,\ell;0) &= \frac{(k-1)(\ell-1)}{4(\ell-k)} \cdot \begin{cases}\gamma_2(k,\ell), & \text{if } \ell < k,\\ \gamma_3(k,\ell), & \text{if } \ell > k,\end{cases}\\
\frac{\partial\lambda}{\partial t}(k,\ell;1) &= \frac{(k-1)(\ell-1)}{4(k\ell-1)} \cdot \begin{cases}\gamma_1(k,\ell), & \text{if } k\ell < 1,\\ \gamma_4(k,\ell), & \text{if } k\ell > 1.\end{cases}
\end{aligned}$$

The remaining lemmas of Section 5 are similarly changed, giving the main

**Theorem 7.1** *A transition occurs between values $\lambda_{\max} = \widehat{E}_i$ and $\lambda_{\max} = \widehat{E}_5$ when and only when one crosses the lower branch of the hyperbola $H_i$, $i = 1, 2, 4$, or the upper branch of the hyperbola $H_3$, with the exception of the point $(-1,-1)$ where the hyperbolas intersect. In addition $\lambda_{\max}$ is continuous in $\mathbb{R}^2$.*

A discussion similar to that given earlier for $\Lambda_{\min}$ applies to the corresponding *upper bound function* $\Lambda_{\max}$, giving the diagram of Figure 2, this being the second main result of Part II.

**Remark 7.2** The pair of matrices

$$\mathbb{A} = \begin{bmatrix}2 & 1\\ 1 & 2\end{bmatrix} \quad \mathbb{B} = \begin{bmatrix}14 & 0\\ 0 & 1\end{bmatrix} \tag{7.1}$$

provides an interesting example.[2]

Here $n = 2$ and the eigenvalues of $\mathbb{A}$ are 1 and 3, and of $\mathbb{B}$ are 1 and 14, so that

$$a_1 = 1, \quad a_n = 3, \quad b_1 = 1 \quad \text{and} \quad b_n = 14.$$

Hence according to the table in Section 6,

$$k = \frac{a_n}{a_1} = 3 \quad \text{and} \quad \ell = \frac{b_n}{b_1} = 14.$$

By Theorem 2.1 the least eigenvalue of $\{\mathbb{AB}\}$ need not be positive, and in fact by direct computation is $-0.008331$. On the other hand, from Figure 1 we have (since the point $(3,14)$ lies in the shaded region)

$$\Lambda_{\min} = E_5(1,3,1,14) = \widehat{E}_5(3,14) = -0.008333.$$

[2] A slightly more complicated example is given in [4, p. 464].

Thus the particular matrices (7.1) give the value of $\Lambda_{\min}(3,14)$ to within the error $2 \times 10^{-6}$. At the same time, of course,

$$\Lambda_{\max}(3,14) = 42$$

(the greatest eigenvalue of $\{\mathbb{AB}\}$ in fact is 30.008331).

It is interesting to note further that the rotation angle from $\hat{\mathbb{A}}$ to $\mathbb{A}$ in this example is just the borderline case $\vartheta = \pi/4$.

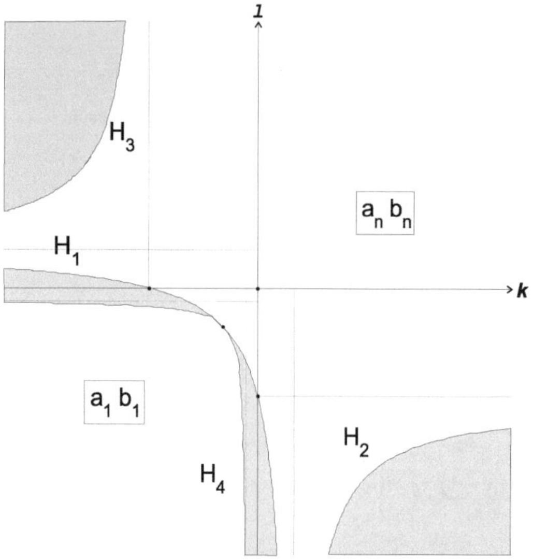


**Fig. 2** The eigenvalue function $\Lambda_{\max}$

The diagram is to be read in the same way as Figure 1.

## 8 Strang's approach

Here we present proofs of Theorems 5.9 (in the $n \times n$ case) and 7.1 based on Strang's method. Let $\mathbb{A}$ be an $n \times n$ Hermitian matrix with smallest and largest eigenvalues $a_1$ and $a_n$. However, we do not specify which of $a_1$ and $a_n$ is larger. Similarly, let $\mathbb{B}$ be an $n \times n$ Hermitian matrix with extremal eigenvalues $b_1$ and $b_n$.

Following Strang, we define

$$E_1 = a_1 b_1 \,, \quad E_2 = a_n b_1 \,, \quad E_3 = a_1 b_n \,, \quad E_4 = a_n b_n \,,$$

and $E_5$ by Eq. (6.2). As in Section 5 of this paper, we define

$$k = a_n/a_1 \,, \quad \ell = b_n/b_1 \quad \text{and} \quad \widehat{E}_i = E_i/a_1 b_1 \,, \quad 1 \le i \le 5 \,.$$

This notation is consistent with (5.8) and (5.5). We emphasize that since we do not specify which of the $a_i$ is larger, $\mathbb{A}$ only determines $k$ up to one of two reciprocal values.

The paper [11] is concerned with the following quantities:

$$\Lambda_{\min} = \inf \left\{ \text{minimum eigenvalue of } \left\{\mathbb{A}\mathbb{B}^{\mathbb{T}}\right\} : \mathbb{T} \text{ unitary} \right\} ,$$

$$\Lambda_{\max} = \sup \left\{ \text{maximum eigenvalue of } \left\{\mathbb{A}\mathbb{B}^{\mathbb{T}}\right\} : \mathbb{T} \text{ unitary} \right\} .$$

Let us define $\mathcal{R}$ to be the region in the $(k,\ell)$-plane where the following two inequalities hold:

$$\left|L(k) + L(k)^{-1} - L(k)L(\ell)^{-2}\right| < 2 \quad \text{and} \quad \left|L(\ell) + L(\ell)^{-1} - L(\ell)L(k)^{-2}\right| < 2 \,,$$

where $L$ is the involutive linear fractional transformation $z \mapsto (z+1)/(z-1)$. Clearly $\mathcal{R}$ is symmetric across $k = \ell$, and we will see during the proof of Theorem 8.2 that it is invariant under the transformations $k \mapsto k^{-1}$, $\ell \mapsto \ell^{-1}$. Therefore the matrices $\mathbb{A}$ and $\mathbb{B}$ determine whether or not $(k,\ell)$ is in $\mathcal{R}$, even though they only determine $k$ and $\ell$ up to reciprocals. Strang's main result can be formulated as follows:

**Theorem 8.1** (Strang [11]). *If $(k,\ell) \in \mathcal{R}$, then $\Lambda_{\min}$ and $\Lambda_{\max}$ are the least and greatest of $E_1, \ldots, E_5$, respectively. If $(k,\ell) \notin \mathcal{R}$, then $\Lambda_{\min}$ and $\Lambda_{\max}$ are the least and greatest of $E_1, \ldots, E_4$, respectively.*

Our Theorems 5.9 and 7.1 extend this theorem by graphing $\mathcal{R}$ and determining explicitly at each point of $\mathcal{R}$ whether $E_5$ is $\Lambda_{\min}$, $\Lambda_{\max}$, or neither. Of course, this is elementary, but it requires a significant calculation and the results are appealing.

We now restate our two theorems as a single theorem and give an alternate proof. Write $\partial\mathcal{R}$ for the boundary of $\mathcal{R}$, and recall that the zero-loci of the conics $\gamma_i$ are the hyperbolas $H_i$, $1 \le i \le 4$.

**Theorem 8.2** *$\mathcal{R}$ is the union of the open shaded areas (both dark and light) in Figure 3. If $a_1 b_1 > 0$, then $E_5 = \Lambda_{\max}$ in the darkly shaded areas and $E_5 = \Lambda_{\min}$ in the lightly shaded areas. This is reversed if $a_1 b_1 < 0$.*

*To put it in words, $(0,0)$ is in $\mathcal{R}$ and $\partial\mathcal{R} = \bigcup_1^4 H_i$, so that a path either enters or leaves $\mathcal{R}$ every time it crosses either branch of any of the $H_i$. Moreover, $E_5$ is either $\Lambda_{\min}$ or $\Lambda_{\max}$ on all of $\mathcal{R}$. If $a_1 b_1 > 0$, then $E_5 = \Lambda_{\max}$ for $(k+1)(\ell+1) > 0$ and $E_5 = \Lambda_{\min}$ for $(k+1)(\ell+1) < 0$. This is reversed if $a_1 b_1 < 0$.*

*Finally, $\Lambda_{\min}$ and $\Lambda_{\max}$ are of class $C(\mathbb{R}^2)$, and they are $C^1$ except on $(k-1)(\ell-1) = 0$. In particular, they are of class $C^1$ across $\partial\mathcal{R}$.*

Proof. By rescaling, we may assume without loss of generality that $a_1 = b_1 = 1$, so that $a_n = k$, $b_n = \ell$, and $E_i = \widehat{E}_i$. Some algebra shows that the inequalities defining $\mathcal{R}$ are equivalent to

$$(1 + |L(k)|) \cdot (1 - |L(\ell)|) < 1, \quad |1 - |L(k)|| \cdot |L(\ell)| < |L(k)|,$$

and the two inequalities obtained from these by exchanging $k$ and $\ell$. Since $L(1/z) = -L(z)$, this makes it clear that $\mathcal{R}$ is invariant under both $k \mapsto k^{-1}$ and $\ell \mapsto \ell^{-1}$.

First we claim that $\partial\mathcal{R} \subset \bigcup_1^4 H_i$. For this, recall that $\gamma_1(k,\ell) = (k-1)(\ell-1) - 4$, and check that

$$\begin{aligned} -\gamma_2(k,\ell) &= -k\gamma_1(k^{-1},\ell) = (k-1)(\ell+3) + 4, \\ \gamma_3(k,\ell) &= -\ell\gamma_1(k,\ell^{-1}) = (k+3)(\ell-1) + 4, \\ \gamma_4(k,\ell) &= -k\ell\gamma_1(k^{-1},\ell^{-1}) = \frac{1}{3}[(3k+1)(3\ell+1) - 4]. \end{aligned} \tag{8.1}$$

Thus it is clear that $\bigcup_1^4 H_i$ is invariant under $k \mapsto k^{-1}$ and $\ell \mapsto \ell^{-1}$. This is also true of $\mathcal{R}$ and hence $\partial\mathcal{R}$, and so to prove the claim it suffices to prove that $\partial\mathcal{R} \cap \{(k,\ell) : |k|, |\ell| > 1\}$ is in $\bigcup_1^4 H_i$.

Now $\{(k,\ell) : |k|, |\ell| > 1\}$ is equal to $\{(k,\ell) : L(k), L(\ell) > 0\}$. Therefore on this set our restatement of the inequalities defining $\mathcal{R}$ becomes

$$[1 + L(k)] \cdot [1 - L(\ell)] < 1, \quad |1 - L(k)| \cdot L(\ell) < L(k),$$

and the two inequalities obtained by exchanging $k$ and $\ell$. It is easy to check that the four $H_i$ are defined by the four equations $(1 \pm L(k))(1 \pm L(\ell)) = 1$, whence the claim follows.

The complement of $\bigcup_1^4 H_i$ in the $(k,\ell)$-plane consists of 14 connected open components (see Figure 3). Since $\partial\mathcal{R} \subset \bigcup_1^4 H_i$, each of these regions is either in $\mathcal{R}$ or its complement, so it is only necessary to check one point from each region to decide. We can reduce to checking only $(0,0)$, $(2,2)$, and $(0,-2)$ by noting that the orbits of the three components containing these points under the group generated by $k \leftrightarrow \ell$, $k \mapsto k^{-1}$, and $\ell \mapsto \ell^{-1}$ intersect all 14 regions. The first sentence of the theorem follows.

For the second sentence, observe in (6.2) that $E_5(a_1, a_n, b_1, b_n)$ is invariant under the exchanges $a_1 \leftrightarrow a_n$ and $b_1 \leftrightarrow b_n$. Deduce from this that

$$\widehat{E}_5(k,\ell) = k\widehat{E}_5(k^{-1},\ell) = \ell\widehat{E}_5(k,\ell^{-1}) = k\ell\widehat{E}_5(k^{-1},\ell^{-1}).$$

Now by Lemma 5.8 we know that $\widehat{E}_5(k,\ell) = 1 - \gamma_1^2/Q$, where we define $Q = 8(k+1)(\ell+1)$ for brevity. Combine this with the preceding identities to obtain

$$\widehat{E}_5 = 1 - \gamma_1^2/Q = k - \gamma_2^2/Q = \ell - \gamma_3^2/Q = k\ell - \gamma_4^2/Q,$$

as already noted in Lemma 5.8.

These four forms of $\widehat{E}_5$ show that for $i = 1, 2, 3, 4$, we have $\widehat{E}_5 = \widehat{E}_i$ precisely on $H_i$. Now it is not hard to determine which of the $\widehat{E}_1, \ldots, \widehat{E}_4$ is least and which is greatest at any point $(k, \ell)$. It turns out that near $H_i$, $\widehat{E}_i$ is always either least or greatest, depending on the branch of $H_i$. Moreover, using the four forms of $\widehat{E}_5$ we easily see that in each shaded region, $\widehat{E}_5$ is either greater or less than the greatest or least of $\widehat{E}_1, \ldots, \widehat{E}_4$, depending on the sign of $Q$. This completes the proof of the first two paragraphs.

The key to the proof of the third paragraph is again the four forms of $\widehat{E}_5$, which, coupled with the above discussion, immediately show that $\Lambda_{\min}$ and $\Lambda_{\max}$ are $C^1$ across the $H_i$. For the rest, draw a picture dividing the $(k, \ell)$-plane into zones according to which of $\widehat{E}_1, \ldots, \widehat{E}_4$ are least and greatest, and superimpose it on Figure 3. □

**Corollary 8.3** *For $a_1 b_1 > 0$, $\{\mathbb{A}\mathbb{B}^{\mathbb{T}}\}$ is positive definite for all unitary $\mathbb{T}$ if and only if $(k, \ell)$ satisfies $\widehat{E}_5(k, \ell) > 0$, i.e., $16kl > (k-1)^2(\ell-1)^2$, which defines the open region between the two dashed curves on Figure 3 (the upper right dashed curve corresponds to Theorem 2.1, and the other curve completes its orbit under $k \mapsto k^{-1}$, $\ell \mapsto \ell^{-1}$). If $a_1 b_1 \leq 0$, then there exists a unitary $\mathbb{T}$ such that $\{\mathbb{A}\mathbb{B}^{\mathbb{T}}\}$ is not positive definite.*

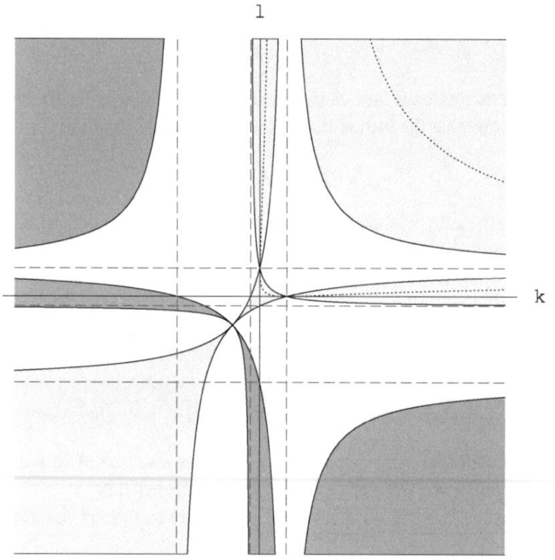


**Fig. 3** The region $\mathcal{R}$ is the union of the shaded areas. $E_5$ is $\Lambda_{\min}$ in the lightly shaded areas and $\Lambda_{\max}$ in the darkly shaded areas. The two dotted curves graph $E_5 = 0$; they bound the positive definite region (see Corollary 8.3). The dashed lines are the asymptotes of the hyperbolas: the asymptotes of $H_1$, $H_2$, $H_3$, and $H_4$ are the pairs $(k = 1,\ \ell = 1)$, $(k = 1,\ \ell = -3)$, $(k = -3,\ \ell = 1)$, and $(k = -1/3,\ \ell = -1/3)$, respectively.

Proof of Corollary 8.3. The second sentence is obvious. For the first, note that $\{\mathbb{A}\mathbb{B}^{\mathbb{T}}\}$ is positive definite for all $\mathbb{T}$ if and only if $\Lambda_{\min} > 0$, and apply Theorem 8.2 and Eq. (5.5). We remark that one can use (6.4) to see that the curves where $\widehat{E}_5 = 0$ are $\ell = (\sqrt{k} - \varepsilon)^2 / (\sqrt{k} + \varepsilon)^2$ for $\varepsilon = \pm 1$. □

# Part III

## 9 The strong maximum principle

Consider the differential inequality

$$D_j\{a_{ij}(x,u)A(|Du|)D_j u\} - B(x,u,Du) \leq 0, \quad u \geq 0, \tag{9.1}$$

Math. Nachr. **278**, No. 12–13 (2005) / www.mn-journal.com 1505

in the domain $\Omega \subset \mathbb{R}^n$, $n \geq 2$, where $[a_{ij}(x,u)]$, $i,j = 1,\ldots,n$, is a continuously differentiable, symmetric coefficient matrix, verifying condition (1.2). We also assume the validity of (A1), (A2) and (B1), given in the introduction.

By the strong maximum principle for (9.1) we mean the statement that if $u$ is a $C^1$ solution of (9.1) with $u(x_0) = 0$ for some $x_0 \in \Omega$ then $u \equiv 0$ in $\Omega$.

**Theorem 8.1′** (Strong maximum principle—Modified version of the sufficiency condition of [7, Theorem 8.1]) *Suppose that*

$$\lim_{\varrho \downarrow 0} \frac{\varrho A'(\varrho)}{A(\varrho)} = c > -1 \tag{9.2}$$

*and when* $c \neq 0$ *assume also that the positive definite matrix* $[a_{ij}]$ *in* (1.2) *satisfies*

$$s_0 < \frac{2 + c + 2\sqrt{1+c}}{|c|}, \tag{9.3}$$

*where*

$$s_0 = \sup_{x \in \Omega} \sqrt{\Lambda(x,0)/\lambda(x,0)} \quad (< \infty) \tag{9.4}$$

*and*

$$\Lambda(x,u) = \max\{\textit{eigenvalues of } [a_{ij}] \textit{ at the point } (x,u)\},$$
$$\lambda(x,u) = \min\{\textit{eigenvalues of } [a_{ij}] \textit{ at the point } (x,u)\}.$$

*For the strong maximum principle to be valid for* (9.1) *it is sufficient that either* $f \equiv 0$ *in* $[0,\mu)$, $\mu > 0$, *or that*

$$\int_{0+} \frac{ds}{H^{-1}(F(s))} = \infty, \tag{9.5}$$

*is satisfied, where* $F(u) = \int_0^u f(s)\,ds$.

Proof. This is almost exactly the same as the proof already given for the sufficiency condition of [7, Theorem 8.1].[3] In fact, the proof is word-for-word the same until the assertion at the end of the proof of the comparison [7, Lemma 8.2], that the matrix

$$\{\mathbb{A}\mathbb{B}\}, \quad \text{where} \quad \mathbb{A} = [a_{ij}(x,u(x))] \quad \text{and} \quad \mathbb{B} = [b_{ij}(x)],$$

*is positive definite*, where (in the notation used earlier in this paper)

$$\mathbb{B} = A(|\boldsymbol{\xi}|)\mathbb{I} + \frac{A'(|\boldsymbol{\xi}|)}{|\boldsymbol{\xi}|}\boldsymbol{\xi} \otimes \boldsymbol{\xi}, \quad \boldsymbol{\xi} \neq \mathbf{0}.$$

As noted already in the introduction this assertion is in general incorrect. To continue correctly, rewrite $\mathbb{B}$ in the form

$$\mathbb{B} = A(|\boldsymbol{\xi}|)\mathbb{C}_{\boldsymbol{\eta}},$$

where

$$\mathbb{C}_{\boldsymbol{\eta}} = \mathbb{I} + \hat{c}\,\boldsymbol{\eta} \otimes \boldsymbol{\eta}, \quad \hat{c} = \hat{c}(|\boldsymbol{\xi}|) = \frac{|\boldsymbol{\xi}|A'(|\boldsymbol{\xi}|)}{A(|\boldsymbol{\xi}|)}, \quad \boldsymbol{\eta} = \frac{\boldsymbol{\xi}}{|\boldsymbol{\xi}|},$$

so that

[3] It is assumed that the reader is familiar with the proof of this result. For convenience in this connection, we also supply several errata and improvements to [7]:

Errata:
Middle of page 33: "Theorem 17" → "Theorem 5.4".
Line 6 from top of page 42: "identifications $a = 0$," → "identification".
Line 7 from foot of page 47: "Theorem 8.1" → "Theorem 10.1".
Line 6 from foot of page 47: "$M = -a \geq 0$" → "$M = a$", and "$u - v \geq M$" → "$u - v \geq -M$".
Line 6 from foot of page 55: "$\Sigma_a$" → "$\Sigma$".

Improvements:
Page 39: "$\mathcal{D}$" → "$\Omega$".
Delete paragraph before Theorem 8.5.
Line 8 from top of page 54: "$\boldsymbol{\xi} \neq \mathbf{0}$ in $\mathbb{R}^n$." → "$\boldsymbol{\xi} \neq \mathbf{0}$ such that $|\boldsymbol{\xi}| < b$."

$$\{\mathbb{AB}\} = A(|\boldsymbol{\xi}|)\{\mathbb{A}\mathbb{C}_{\boldsymbol{\eta}}\}, \quad \mathbb{A} = \mathbb{A}(x,u) = [a_{ij}(x,u)]. \tag{9.6}$$

From Theorem 3.2 it then follows that $\{\mathbb{AB}\}$ will be positive definite if either $\hat{c} = 0$ or if $\hat{c} \neq 0$ and

$$s < \frac{2+\hat{c}+2\sqrt{1+\hat{c}}}{|\hat{c}|}, \quad \text{where} \quad s = s(x,u) = \sqrt{\frac{\Lambda(x,u)}{\lambda(x,u)}} \geq 1. \tag{9.7}$$

In turn, by (9.3) the condition (9.7) holds when $u$ and $\boldsymbol{\xi}$ are sufficiently small. Consequently, under the assumption (9.3) we see that $D_{\xi_j}\hat{A}^i(x,\boldsymbol{\xi})$ is *uniformly* positive definite when $u$ and $\boldsymbol{\xi}$ are sufficiently small, say $u \leq b_0$ and $|\boldsymbol{\xi}| \leq b_0$. (Notation as on [7, p. 42], and positive definiteness being understood in the sense of the corresponding quadratic form.)

Accordingly, using [7, Theorem 10.1] with the parameter $b = b_0$ as above, we can state a *weak version* of [7, Lemma 8.2], as follows.

**Lemma 8.2′** (Comparison lemma) *Assume* (B1). *Let $u$ and $v$ be respectively solutions of* (9.1) *and of*

$$D_i\{a_{ij}(x,u(x))A(|Dv|)D_jv\} - \kappa\Phi(|Dv|) - f(v) \geq 0$$

*in a bounded domain $\Omega$. Suppose that $u$ and $v$ are continuous in $\overline{\Omega}$, and that*

$$0 \leq u \leq b_0, \qquad 0 < |Dv| \leq b_0 \quad in \quad \Omega, \qquad and \qquad u \geq v - M \quad on \quad \partial\Omega,$$

*where $M$ is constant. Then $u \geq v - M$ in $\Omega$.*

The final step in the proof of Theorem 8.1′, as in the proof of [7, Theorem 1.1], is to apply the standard Hopf comparison technique. This technique depends on (i) the construction of a tangent ball $B$ to the set where $u \equiv 0$, together with (ii) an appropriate choice of the value $m$ for the comparison function $v$. The ball $B$ can however be chosen arbitrarily small without affecting the proof. Then $u$ will be arbitrarily small in $B$ (where the ultimate comparison takes place), while simultaneously forcing $m$ ($< u$) to be small (see [7, page 33]). Then in turn $|Dv|$ will be arbitrarily small (see [7, Proposition 4.1]).

The remainder of the proof of sufficiency is then exactly as on [7, p. 42], except one now relies on Lemma 8.2′ (with $M = 0$) instead of on the originally given form of this lemma. This completes the proof of the strong maximum principle for (9.1). □

**Remarks 9.1** When $c \neq 0$ in (9.2) and

$$s_0 > \frac{2+c+2\sqrt{1+c}}{|c|}, \tag{9.8}$$

with $c$ and $s_0$ given as in Theorem 8.1′, the proof given above fails, since by Theorem 3.1 the matrix $\{\mathbb{AB}\}$ is then indefinite for some directions $\boldsymbol{\nu}$ of the vector $Du$ and for some points $x \in \Omega$. Of course exactly such points and directions occur when the normal at the tangent point $x = x_0$ is a direction $\boldsymbol{\nu}$. Thus the proof of Theorem 8.1′ fails in this case, since $x_0$ could be any point in $\Omega$ and the normal could have *any* direction $\boldsymbol{\nu}$, depending on the particular outcome of the Hopf construction.

It is still *an open question* whether [7, Theorem 8.1′] fails for values $s_0$ satisfying (9.8). We have not been able to find a counterexample for such cases, though it may be conjectured that the condition (9.3) is in fact necessary.

If $a_{ij}(x,u) = a(x,u)\delta_{ij}$, where $a : \Omega \times \mathbb{R}_0^+ \to \mathbb{R}^+$ is of class $C^1$, then the differential operator in (9.1) has the variational form

$$\operatorname{div}\{a(x,u)A(|Du|)Du\}.$$

For this special case, [7, Theorem 8.1] continues to hold without the help of (9.3), since then $[a_{ij}b_{jk}] = a(x,u)[b_{ik}]$ is of course positive definite without further argument.

Condition (9.2) applies to the $p$-Laplace operator $A(\varrho) = \varrho^{p-2}$, $p > 1$, with $c = p-2$. In this case, when $c \neq 0$, namely when $p \neq 2$, the condition (9.3) takes the explicit form

$$s_0 < \frac{p+\sqrt{p-1}}{|p-2|}. \tag{9.9}$$

Moreover, if $c = 0$ in (9.2), as occurs for example when $A(\varrho) = 1$, i.e., for the Laplace operator, or when $A(\varrho) = 1/\sqrt{1+\varrho^2}$, i.e., the mean curvature operator, then condition (9.3) is empty and so Theorem 8.1′, and in turn [7, Theorem 8.1], is correct even with no additional conditions on $[a_{ij}]$ outside of positive definiteness and regularity! For the mean curvature type inequality we have the following formal statement.

**Theorem 9.2** *Let* (9.5) *be satisfied. Then the strong maximum principle is valid for the mean curvature type differential inequality*

$$D_j\left\{\frac{a_{ij}(x,u)D_ju}{\sqrt{1+|Du|^2}}\right\} - B(x,u,Du) \le 0\,, \quad u \ge 0\,. \tag{9.10}$$

Here it is worth noting explicitly that (9.10) *need not* be elliptic when

$$|Du| > 2\frac{r}{r^2-1}\,, \quad r = \sqrt[4]{s}\,,$$

where $s = s(x,u)$ is given by (9.7).

Finally, the validity of Theorem 8.1′ can of course also be asserted *if the differential inequality* (9.1) *is assumed to be elliptic for all arguments* $(x,u,Du) \in \Omega \times \mathbb{R}_0^+ \times \mathbb{R}^n$ such that $0 < u < \varepsilon$, $0 < |Du| < \varepsilon$ for some $\varepsilon > 0$.

## 10 The compact support principle

The compact support principle for the differential inequality

$$D_j\{a_{ij}(x,u)A(|Du|)D_ju\} - B(x,u,Du) \ge 0\,, \quad u \ge 0 \quad \text{in} \quad \Omega\,, \tag{10.1}$$

has been treated earlier in [7, Theorem 8.5]. However, just as was the case for the strong maximum principle, the validity of Theorem 8.5 requires further hypotheses on the matrix $[a_{ij}]$, analogous to the conditions (9.3)–(9.4) already given for Theorem 8.1′. As in [7] we suppose that $\Omega$ is an exterior domain, with $\Omega \supset \Omega_R = \mathbb{R}^n \setminus \Omega$; that (A1), (A2) hold; while (B1) is replaced by

(B2) $B \in C(\Omega \times \mathbb{R}_0^+ \times \mathbb{R}^n)$ and

$$B(x,u,\boldsymbol{\xi}) \ge -\kappa\Phi(|\boldsymbol{\xi}|) + g(u)$$

for $x \in \Omega$, $u \ge 0$, and all $\boldsymbol{\xi} \in \mathbb{R}^n$ with $|\boldsymbol{\xi}| \le 1$, where $\kappa > 0$, and the nonlinearity $g$ is continuous in $\mathbb{R}_0^+$ and nondecreasing on some interval $(0,\delta)$, $\delta > 0$, with $g(0) = 0$.

By the compact support principle for (10.1) we mean the statement that *if $u$ is a nonnegative $C^1$ distribution solution of* (10.1) *in an exterior domain $\Omega$, with $u(x) \to 0$ as $|x| \to \infty$, then $u$ has compact support in $\Omega$.*

**Theorem 8.5′** (Compact principle— Modified version of the sufficiency condition of [7, Theorem 8.5]) *Assume that*

$$\limsup_{|x|\to\infty,\ u\to 0} \Lambda(x,u) < \infty\,, \tag{10.2}$$

$$s_1 = \limsup_{|x|\to\infty,\ u\to 0} \sqrt{\frac{\Lambda(x,u)}{\lambda(x,u)}} < \infty\,, \tag{10.3}$$

*where $\Lambda(x,u)$, $\lambda(x,u)$ are given in Theorem* 8.1′.

*Finally suppose that* (9.2) *holds and also* (*when* $c \neq 0$) *that*

$$s_1 < \frac{2+c+2\sqrt{1+c}}{|c|}\,. \tag{10.4}$$

*Then for the compact support principle to hold for* (10.1) *it is sufficient that* (B2) *be satisfied with* $g(u) > 0$ *for* $u > 0$ *and with*

$$\int_{0+} \frac{ds}{H^{-1}(G(s))} < \infty\,, \tag{10.5}$$

*where* $G(u) = \int_0^u g(s)\,ds$.

Proof. This is again essentially the same as the proof of the sufficiency part of [7, Theorem 8.5], of course with modifications due to the lack of global ellipticity.

First, at the foot of [7, p. 45] we replace $C$ by an arbitrarily small value $\widehat{C} < C$. Then by the dead core [7, Lemma 7.1], with however $f(u)$ replaced by $g(u)$, it follows that $\hat{\gamma} = w(\widehat{C}) \to 0$ as $\widehat{C} \to 0$, while also for the central comparison function $v$ we have (see the top of page 46)

$$v(R) = w(\widehat{C}) = \hat{\gamma}, \quad |Dv| \le H^{-1}(G(\hat{\gamma})),$$
$$D_i\{a_{ij}(x, u(x))A(|Dv|)D_jv\} + \kappa\Phi(|Dv|) - g(v) \le 0 \tag{10.6}$$

(here one uses (10.2)).

Moreover, we can assume that $u \le \hat{\gamma}$ in $\Omega$, see the paragraph following (8.16) in [7]. The function $z = v - u$ then satisfies, as in [7], the condition $|z| \le \hat{\gamma}$ in $\Omega$, and correspondingly $0 < \hat{\varepsilon} \le \hat{\gamma}$. Of course, also $|x|$ can be made as large as we wish by taking the radius $R_0$ suitably large.

Proceeding to Case 1, and noting that $\sqrt{\Lambda(x,u)/\lambda(x,u)} \approx s_1$ because $\hat{\gamma}$ can be arbitrarily small, one can then argue exactly as in [7], but using now the new version Theorem 8.1′ of the strong maximum principle, to show that in fact this case cannot occur.

In Case 2 the boundary point lemma still applies either when $c = 0$ in (9.2), or when $c \neq 0$ and (10.3)–(10.4) are satisfied. Again this case cannot occur.

Finally for the crucial Case 3, by using (10.3) and (10.4) it follows as in Section 9 that $D_{\xi_j}\hat{A}^i(x, \boldsymbol{\xi})$ is uniformly positive definite when $u$ and $|Dv|$ are smaller than $b_0$, and $x$ is suitably large.

Next, exactly as in [7], we can apply [7, Theorem 10.1] to get a variant of Lemma 8.2′ in which $u$ and $v$ respectively satisfy the reverse inequalities (10.1) and (10.6), while the inequality $u \ge v - M$ is replaced by $v \ge u - M$.

To complete the argument for Case 3, it is now enough to recall that $u < \hat{\gamma}$ and $0 < |Dv| \le H^{-1}(G(\hat{\gamma}))$, and to take $\hat{\gamma}$ suitably small and $|x|$ suitably large so that the above variant of Lemma 8.2′ applies. This being done, the remaining part of Case 3 is then exactly as in [7], and accordingly this case also cannot occur.

It is thus proved that $z \ge 0$ in $\Omega$, and in turn $u \equiv 0$ for $|x| > R_1$, completing the proof. □

The case $c = 0$ in (9.2) can be treated exactly as in Section 9, leading to the following result for the mean curvature type inequality.

**Theorem 10.1** *Let* (B2) *hold, and assume* (10.5) *is satisfied with* $g(u) > 0$ *for* $u > 0$. *Then the compact support principle is valid for the mean curvature type differential inequality*

$$D_j\left\{\frac{a_{ij}(x,u)D_ju}{\sqrt{1+|Du|^2}}\right\} - B(x,u,Du) \ge 0, \quad u \ge 0 \ in\ \Omega. \tag{10.7}$$

**Acknowledgements** We wish to particularly thank Roger Horn for his valuable advice in writing the paper.

## References

[1] N. Alikakos and P. W. Bates, Estimates for the eigenvalues of the Jordan product of Hermitian matrices, Linear Algebra Appl. **57**, 41–56 (1984).
[2] N. Alikakos and P. W. Bates, Estimates for the eigenvalues of the Jordan product of Hermitian matrices. Erratum, Linear Algebra Appl. **65**, 282 (1985).
[3] K. Gustafson and D. K. M. Rao, Numerical Range: The Field of Values of Linear Operators and Matrices (Springer-Verlag, New York, 1997).
[4] R. A. Horn and C. R. Johnson, Matrix Analysis (Cambridge University Press, Cambridge, U.K., 1985).
[5] D. W. Nicholson, Eigenvalue bounds for $\mathbb{AB} + \mathbb{BA}$ with $\mathbb{A}$, $\mathbb{B}$ positive definite matrices, Linear Algebra Appl. **24**, 173–183 (1979).
[6] P. Pucci and J. Serrin, A note on the strong maximum principle for elliptic differential inequalities, J. Math. Pures Appl. **79**, 57–71 (2000).
[7] P. Pucci and J. Serrin, The strong maximum principle revisited, J. Differential Equations **196**, 1–66 (2004).
[8] P. Pucci and J. Serrin, The strong maximum principle revisited. Erratum, J. Differential Equations **207**, 226–227 (2004).
[9] P. Pucci, J. Serrin, and H. Zou, A strong maximum principle and a compact support principle for singular elliptic inequalities, J. Math. Pures Appl. **78**, 769–789 (1999).
[10] V. Ptak, The Kantorevich inequality, Amer. Math. Monthly **102**, 820–821 (1995).
[11] W. G. Strang, Eigenvalues of Jordan products, Amer. Math. Monthly **63**, 37–40 (1962).
[12] O. Taussky (Todd), Research Problem 2, Bull. Amer. Math. Soc. (N. S.) **66**, 275 (1960).

*Transl., Ser. 2, Am. Math. Soc* 104 (1976), 223–231. **Reprint VII.3.[i]**

First published in *Some problems of Mathematics and Mechanics – M.A. Lavrentieff Anniversary Volume*, Nauka Leningrad 1970, pp.242–250.

# CURVATURE INEQUALITIES FOR SURFACES OVER A DISK

UDC 517.947.42

JAMES SERRIN AND HANS F. WEINBERGER

Serge Bernstein showed in 1910 that a surface of constant mean curvature $k > 0$ could not be defined over any domain containing in its interior a closed disk of radius $1/k$. Recently Robert Finn established a stronger version of Bernstein's theorem, namely that if a surface having mean curvature greater than or equal to $k$ is defined over an open disk of radius $1/k$, then the surface is necessarily a hemisphere. The surfaces in question are assumed to be regular 2-manifolds of the form $z = u(x, y)$ embedded in 3-space, though in fact the results apply also in higher dimensions (see also [2], [4]).

In the present paper we shall generalize these results to a larger class of surfaces, including in particular surfaces whose principal curvatures are functionally related (Weingarten surfaces). For simplicity we shall only treat 2-surfaces in 3-spaces, though it will be clear thorughout that the results can be extended at once to higher dimensions.

Before stating our conclusions we make some standard agreements. We consider only surfaces $S$ which are oriented and of class $C^2$, and we let $k_1$ and $k_2$ denote the principal curvatures of $S$ at each point, with $k_1 \leqslant k_2$. Throughout the paper $k$ will denote a positive real constant. Our main results can now be stated.

THEOREM 1. *Let $S$ be a nonparametric surface of class $C^2$ defined over a domain $D$, with the property that $k_2 \geqslant k$ at each point of $S$. Then $D$ cannot contain a closed disk of radius $1/k$.*

Note that if $S$ is a surface of constant mean curvature $k$, then for an appropriate orientation of the principal normal on $S$ we have $(k_1 + k_2)/2 = k$, so that $k_2 \geqslant k$. Thus Bernstein's theorem is an obvious corollary of our result. Moreover, Theorem 1 applies equally well to surfaces of constant Gaussian curvature $k^2$, since again the condition $k_1 k_2 = k^2$ implies $k_2 \geqslant k$ (the orientation being chosen so that the curvatures are positive).

These examples show that an appropriate orientation of $S$ may be needed to apply Theorem 1 to best effect. The same objective can be reached in an

*AMS (MOS) subject classifications* (1970). Primary 53A05.

alternate way by noting that the conclusion of Theorem 1 also holds if $k_1 < -k$. Similar remarks apply also to the next result.

THEOREM 2. *Let* $S$ *be a nonparametric surface of class* $C^2$ *defined over an open disk of radius* $1/k$. *Suppose that the principal curvatures of* $S$ *satisfy the conditions*

(i) $k_2 > k$ *and*

(ii) $k_1 > k - f(k_2 - k)$,

*where* $f(t)$, $t > 0$, *is a continuous nondecreasing function satisfying*(1)

(iii) $\int_0 dt/(t + f(t)) = \infty$.

*Then* $S$ *is a hemisphere of radius* $1/k$.

For a surface having mean curavture greater than or equal to $k$ we see easily that conditions (i)–(iii) are satisfied with $f(t) = t$. Hence Finn's theorem is an immedaite consequence of Theorem 2. Similarly, a surface having Gaussian curvature everywhere greater than or equal to $k^2$ satisfies conditions (i)–(iii) with $f(t) = kt/(k + t)$. Thus we obtain the following conclusion, paralleling Finn's theorem.

*Let* $S$ *be a nonparametric surface defined over a disk of radius* $1/k$, *whose Gaussian curvature is everywhere greater than or equal to* $k^2$. *Then* $S$ *is a hemisphere.*

The conclusion of Theorem 2 is evidently stronger than that of Theorem 1, but this is bought at the price of an additional restriction placed on the curvature $k_1$. That such an additional condition is necessary will be shown in the final section of the paper, where we construct a surface over the disk of radius $1/k$ which satisfies the conditions

$$k_2 \geqslant k, \; k_1 = k - f(k_2 - k),$$

where $f$ is any continuous functions such that

$$0 \leqslant f(t) < k \quad \text{and} \quad \int_0 \frac{dt}{t + f(t)} < \infty,$$

but which is *not* a hemisphere. Thus condition (iii) is necessary.

To illustrate further the range of application of Theorems 1 and 2, we recall that a Weingarten surface is an oriented $C^2$ surface whose principal curvatures are functionally related:

(iv) $W(k_1, k_2) = 0$ $(k_2 > k_1)$.

The surface is called *elliptic* if (iv) implies that $k_2$ increases as $k_1$ decreases.

(1) Note that these conditions imply $f(0) = 0$ and $f(t) > 0$.

Clearly in this case the graph of (iv) intersects the line $k_1 = k_2$ in a unique point $(l, l)$. By appropriate orientation of the principal normal direction we can (and will) assume that $l \geq 0$. The number $l$ thus determined is called the *umbilical constant* of the surface.

A surface will be called a *regular Weingarten surface* if it is an elliptic Weingarten surface for which relation (iv) can be written in the form

$$k_1 - l - f(k_2 - l) = 0,$$

where $f(t)$, $t \geq 0$, is a continuous nondecreasing function satisfying the integral condition (iii). We remark that surfaces of constant mean curvature $k$ and constant Gaussian curvature $k^2$ are both regular Weingarten surfaces with umbilical constant $k$. The reader can easily supply further examples, and in fact a simple sufficient condition for an elliptic Weingarten surface to be regular is that $W(k_1, k_2)$ be differentiable with $W_{k_1} W_{k_2} < 0$.

Applying Theorems 1 and 2 to Weingarten surfaces now yields the following corollaries.

COROLLARY 1. *A nonparametric elliptic Weingarten surface with umbilical constant* $k$ *cannot be defined over any domain which contains a closed disk of radius* $1/k$.

COROLLARY 2. *Let* $S$ *be a nonparametric regular Weingarten surface with umbilical constant* $k$, *defined over an open disk of radius* $1/k$. *Then* $S$ *is a hemisphere.*

The first section of the paper is devoted to a preliminary lemma of some interest in itself, which establishes certain curvature restrictions for surfaces defined over a disk. Theorem 1 is an immediate consequence of this lemma. In §2 we prove Theorem 2, and finally in the closing section we establish the counterexample noted previously.

## §1. Proof of Theorem 1

We begin with the following lemma, which will serve to establish Theorem 1.

LEMMA. *Let* $S$ *be a nonparametric surface of class* $C^2$ *defined over an open disk of radius* $1/k$. *Then* $\inf k_2 \leq k$.

PROOF. We introduce Cartesian coordiantes $(x, y, z)$ in the ambient 3-space, such that the disk lies in the $(x, y)$-plane, with its center at the origin. We choose the positive $z$ direction so that it agrees with the orientation of $S$, and call this the upward direction.

Now choose any $\epsilon > 0$ and let $\bar{S}(c)$ be the sphere of radius $(k+\epsilon)^{-1}$ centered at $(0, 0, c)$. If $c$ is sufficiently large, this sphere lies above $S$, while if $c$ is sufficiently small, it lies below $S$. We let $c_0$ be the infimum of the values of $c$ for which $\bar{S}(c)$ lies above $S$. Then no point of $\bar{S}(c_0)$ lies below $S$ and $\bar{S}(c_0)$ is tangent to $S$ at least at one point. Evidently $k_2 \leq k+\epsilon$ at this point. Since there, is such a point for every $\epsilon > 0$, the lemma follows.

To obtain Theorem 1, we observe that if the domain $D$ contained a closed disk of radius $1/k$, it would also contain a closed disk of slightly larger radius, say $(k-\epsilon)^{-1}$ for some $\epsilon > 0$. Applying the lemma in this latter disk, we obtain $\inf k_2 \leq k - \epsilon$, contradicting the hypothesis $k_2 \geq k$.

The example of a sequence of hemispheres of radii $(k - n^{-1})^{-1}$ shows that the conclusion of the lemma is best possible, in the sense that it could not be replaced by $\inf k_2 \leq l$ for any real constant $l < k$.

## §2. Proof of Theorem 2

We adopt the conventions introduced in the proof of the preceding lemma. It will also be convenient to suppose that $S$ has the representation

$$z = u(x, y) \quad (x^2 + y^2 \leqslant k^{-2}).$$

We shall show by a modification of the previous proof that $S$ is tangent to the sphere of radius $1/k$ centered at the point $(0, 0, u(0, 0) + k^{-1})$, and that the interior of this sphere lies entirely above $S$. We shall then show that $S$ coincides with the lower hemisphere of the sphere.

Let $\epsilon$ be a small positive constant, $0 < \epsilon < \max(k, 1/4k)$, and let $c$ be a real parameter. We consider the family $\Sigma$ of surfaces of revolution

$$z = h(r) = c - \int_r^{(k+\delta)^{-1}} \frac{(k+\eta)\rho\, d\rho}{\sqrt{1-(k+\eta)^2\rho^2}}, \quad \epsilon \leq r \leqslant (k+\delta)^{-1},$$

where $\delta$ ($< \epsilon$) is a positive constant whose value will be assigned later, $r = (x^2 + y^2)^{1/2}$, and $\eta = \eta(\rho)$, $\epsilon \leq \rho \leq (k+\delta)^{-1}$, is a positive decreasing function defined implicitly by the relation

$$\int_\eta^{\epsilon} \frac{dt}{f(t) + t + \delta} = \log \frac{\rho}{\epsilon}.$$

It must first be established that the integral defining $h(r)$ exists. To this end, we show that $\delta$ can be chosen (in terms of $\epsilon$) so that $\eta = \delta$ when $\rho = (k+\delta)^{-1}$. This evidently requires that the following condition hold:

$$\int_{\delta}^{\epsilon} \frac{dt}{f(t)+t+\delta} = \log \frac{1}{(k+\delta)\epsilon}.$$

Now as $\delta$ decreases from $\epsilon$ to 0, the left-hand side of this relation increases from 0 to $\infty$, because of hypothesis (iii). On the other hand, the right-hand side always takes its value in the interval from $\log(1/(k+\epsilon)\epsilon)$ to $\log(1/k\epsilon)$. Because the restriction $0 < \epsilon < \max(k, 1/4k)$ implies that both these numbers are positive, it is evident that there exists at least one value $\delta = \delta(\epsilon)$ satisfying the preceding conditions, as requried.

Now we have

$$\frac{d}{d\rho}[(k+\eta)\rho] = k - f(\eta) - \delta \geqslant k - f(\epsilon) - \epsilon.$$

Therefore, since $(k+\eta)\rho = 1$ when $\rho = (k+\delta)^{-1}$,

$$1 - (k+\eta)\rho \geqslant \{k - f(\epsilon) - \epsilon\}\{(k+\delta)^{-1} - \rho\}.$$

Thus the family $\Sigma$ is well defined if $\epsilon$ is further assumed to be less than $\bar{\epsilon}$, where $f(\bar{\epsilon}) + \bar{\epsilon} = k$. Indeed, since $(k+\eta)\rho > (k+\delta)\rho$, it follows from the above inequality that

$$1 - (k+\eta)^2\rho^2 \geqslant \frac{k - f(\epsilon) - \epsilon}{k+\delta}\{1 - (k+\delta)^2\rho^2\},$$

and therefore by a simple computation we obtain

$$h(r) - c \geqslant -\sqrt{\frac{k+\epsilon}{k - f(\epsilon) - \epsilon}}\sqrt{(k+\delta)^{-2} - r^2}.$$

On the other hand, again since $(k+\eta)\rho > (k+\delta)\rho$, we have

$$h(r) - c \leqslant -\sqrt{(k+\delta)^{-2} - r^2}.$$

The preceding two estimates show that as $\epsilon \to 0$ the surfaces in the family $\Sigma$ approach the lower half of the sphere of radius $1/k$ centered at $(0, 0, c)$.

The principal curvatures of a surface of revolution $z = h(r)$ are given by the formulae

$$\frac{h'}{r(1+h'^2)^{1/2}} \quad \text{and} \quad \frac{h''}{(1+h'^2)^{3/2}}.$$

Thus, by a short computation, the principal curvatures $\tilde{k}_1$ and $\tilde{k}_2$ of the family $\Sigma$ are found to be

$$\tilde{k}_1 = k - f(\eta) - \delta, \quad \tilde{k}_2 = k + \eta,$$

where $\eta = \eta(r)$.

We now let $c$ decrease from $+\infty$ until the corresponding surface in $\Sigma$ first contacts $S$. If this first contact occurs at some radius in the interval $\epsilon < r < (k + \delta)^{-1}$ then we must have $k_1 \leq \tilde{k}_1$, $k_2 \leq \tilde{k}_2$ at the point of tangency. Using the fact that $f(t)$ is nondecreasing, we find

$$k_1 \leqslant \tilde{k}_1 = k - f(\eta) - \delta = k - f(\tilde{k}_2 - k) - \delta < k - f(k_2 - k)$$

at the point of tangency. This contradicts condition (ii). We conclude that the first point of contact occurs either for $r = \epsilon$ or $r = (k + \delta)^{-1}$. At the latter radius, however, $dh/dr = +\infty$, as one easily checks. Consequently, such points of contact are also points of tangency, which, as we showed above, are disallowed. Therefore the first point of contact occurs at the radius $r = \epsilon$.

Letting $\epsilon \to 0$ we see at once that $S$ lies below the interior of the sphere $\bar{S}$ of radius $1/k$ centered at $(0, 0, u(0, 0) + k^{-1})$ and that it touches this sphere at $(0, 0, u(0, 0))$. This completes the first part of the proof. (We note that a special case of the above argument was given by Finn [3].)

We now show by a variation of the proof of E. Hopf's maximum principle for elliptic differential inequalities that $S$ must agree with the lower hemisphere of $\bar{S}$, which will complete the proof.

Thus suppose for contradiction that $S$ does not agree with this hemisphere. Let $D$ denote the set of points on the hemisphere which are not also points of $S$. Then $D$ is nonempty and lies above $S$, and its complement contains the point of tangency $(0, 0, u(0, 0))$. Let $P$ be a point of $D$ which is nearer to the complement of $D$ than to the boundary of the hemisphere. We choose a new coordinate system with the positive $z$ axis directed from $P$ towards the center of $\bar{S}$. In this system (which we retain henceforth) the hemisphere has the representation $z = \sigma(r)$ and there exists a maximal value $R > 0$ such that the cap $z = \sigma(r)$, $r < R$, lies entirely above $S$. Note particularly that $S$ is tangent to this cap at some point $Q$ on the boundary $r = R$. For simplicity in what follows we shall finally translate the coordinates in the $z$-direction so that $\sigma(R) = 0$. It should also be noted that $R$ is less than the radius $1/k$ of $\bar{S}$.

Now let $\epsilon$ be a small constant, $0 < \epsilon < R^{-1} - k$, and consider the surface of revolution

$$z = \tilde{h}(r) = -\int_r^R \frac{(k + \eta)\rho\, d\rho}{\sqrt{1 - (k + \eta)^2\rho^2}}, \quad \tfrac{1}{2}R \leqslant r \leqslant R,$$

where $\eta = \eta(\rho)$ is defined implicitly by

$$\int_{\eta}^{\epsilon} \frac{dt}{f(t)+t+\delta} = \log\frac{2\rho}{R}, \quad \frac{1}{2}R \leqslant \rho \leqslant R,$$

and where $\delta$ is chosen (in terms of $\epsilon$) so that

$$\int_{0}^{\epsilon} \frac{dt}{f(t)+t+\delta} = \log 3.$$

(This can be done by virtue of hypothesis (iii).) It is apparent that $0 < \eta \leqslant \epsilon$ for the range of values $\rho$ in question; hence we easily find that

$$\sqrt{1-(k+\epsilon)^2R^2} - \sqrt{1-(k+\epsilon)^2r^2} \leqslant \bar{h}(r) \leqslant \sqrt{1-k^2R^2} - \sqrt{1-k^2r^2} = \sigma(r).$$

Since $S$ lies strictly below the sphere $\sigma(r)$ for $r < R$, there is an $\epsilon > 0$ such that

$$T\colon\ z = \bar{h}(r), \quad \frac{1}{2}R \leqslant r \leqslant R,$$

lies strictly above $S$ at $r = \frac{1}{2}R$. On the other hand, $\bar{h}(R) = \sigma(R) = 0$, so that $T$ does not go below $S$ at $r = R$. Finally, $S$ intersects $T$ at the point $Q$.

Now at $Q$,

$$\frac{d\bar{h}}{dr} = \frac{(k+\eta)R}{\sqrt{1-(k+\eta)^2R^2}} > \frac{kR}{\sqrt{1-k^2R^2}} = \frac{d\sigma}{dr}.$$

Hence, recalling that $S$ is tangent to $z = \sigma(r)$ at $Q$, we see that there are surely points of $T$ near $Q$, which lie below $S$. We may therefore translate $S$ downward to a position of last contact with $T$; at the point of final contact we have obviously $k_1 \leqslant \bar{k}_1$ and $k_2 \leqslant \bar{k}_2$, where $\bar{k}_1$ and $\bar{k}_2$ are the principal curvatures of the surface $z = \bar{h}(r)$. Since $\bar{k}_1 = k - f(\eta) - \delta$ and $\bar{k}_2 = k + \eta$, this yields a contradiction as before, completing the proof of the theorem.

## §3. A counterexample

Let $f(t)$, $t \geqslant 0$, be any nonnegative continuous function such that $f(t) < k$ and

$$\int_{0}^{\epsilon} \frac{dt}{t+f(t)} < \infty.$$

We shall construct a surface $y = m(x, z)$ over the disk $k^2(x^2 + z^2) < 1$, which satisfies the conditions

$$k_2 \geqslant k, \quad k_1 = k - f(k_2 - k),$$

but which is not a hemisphere. This will show that condition (iii) is necessary to prove Theorem 2.

As in the preceding section we set

$$h(r) = -\int_r^{k^{-1}} \frac{(k+\eta)\rho\, d\rho}{\sqrt{1-(k+\eta)^2\rho^2}},$$

where $\eta = \eta(\rho)$ is defined implicitly by the relation

$$\int_0^{\eta} \frac{dt}{f(t)+t} = \log \frac{1}{k\rho}.$$

A simple computation shows that

$$\frac{d}{d\rho}[\rho(k+\eta)] = k - f(\eta) > 0$$

and that $\rho(k+\eta) = 1$ at $\rho = k^{-1}$. It follows easily that $h(r)$ is defined for all $0 \leqslant r \leqslant k^{-1}$. Another computation shows that the symmetric surface of revolution $z = \pm h(r)$ has the continuous curvatures

$$k_1 = k - f(\eta), \quad k_2 = k + \eta,$$

except at the tips where $r = 0$. Since $k_1 < k$ is the curvature of the profile and since $h(k^{-1}) = 0$, it follows that the sphere $k^2(x^2 + y^2 + z^2) = 1$ lies inside this surface except on the meridian plane $z = 0$ where it is tangent to the surface. Consequently, the part $y \leqslant 0$ of the surface has a *single-valued projection* on the disk $\{k^2(x^2 + z^2) \leqslant 1,\ y = 0\}$ and may be written in the form $y = m(x, z)$, where the function $m$ is the class $C^2$. This surface is clearly not a hemisphere. Consequently, Theorem 2 does not hold when condition (iii) is removed.

We remark that the surface $y = m(x, z)$ is tangent to the hemisphere $y = \sqrt{k^{-2} - x^2 - z^2}$ on the whole meridian $z = 0$. Consequently, condition (iii) is also necessary in order that the second part of the proof of Theorem 2 (that is, the generalized Hopf lemma) should be correct.

An extension of the above results to the case of parametric surfaces will be presented in a forthcoming paper.

*Note.* The preceding work was partially supported by the U. S. Air Force Office of Scientific Research and the U. S. National Science Foundation.

BIBLIOGRAPHY

1. S. N. Bernšteĭn, *Sur les surfaces définies au moyen de leur courbure moyenne ou totale*, Ann. Sci. École Norm. Sup. (3) 27 (1910), 233–256.

2. S.-S. Chern, *On the curvatures of a piece of hypersurface in Euclidean space*, Abh. Math. Sem. Univ. Hamburg 29 (1965), 77–91. MR 32 #6376.

3. R. Finn, *Remarks relevant to minimal surfaces and to surfaces of prescribed mean curvature*, J. Analyse Math. 14 (1965), 139–160. MR 32 #6337.

4. E. Heinz, *Über Flächen mit eineindeutiger Projektion auf eine Ebene, deren Krümmung durch Ungleichungen eingeschränkt sind*, Math. Ann. 129 (1955), 451–454. MR 17, 189.

Translation provided by the authors

Barbara Serrin, Clifford Truesdell and James Serrin, Piazza San Marco, Venice, 1998.

James Serrin, Minneapolis, MN, 1980. Lecturing on thermodynamics.

# VIII. Essays

# JEAN LERAY'S CONTRIBUTIONS TO NONLINEAR DIFFERENTIAL EQUATIONS, FIXED POINT THEORY AND FLUID MECHANICS

James SERRIN
*University of Minnesota*

Between 1931 to 1939 Jean Leray wrote a series of papers on differential equations, fixed point theory and fluid mechanics which changed the face of each of these subjects, and initiated countless other works based on his ideas. Not often in this century has one man's work had a similarly broad effect.

Jean Leray's work was in every case first presented in concise and even telegraphic style in papers (13 in all) for the *Comptes rendus* de l'Académie des sciences de Paris, and then later published as major works in the *Journal de mathématiques pures et appliquées*, *Acta Mathematica*, *Annales de l'École normale supérieure*, and *Commentarii Mathematici Helvetici* (7 papers in all). Several smaller expository works also appeared at this time, of which we mention chiefly the paper *Les problèmes non linéaires* in l'*Enseignement mathématique*, 1936.

Leray's first interests were in integral equations and in fluid mechanics. His celebrated papers in the latter subject comprise two separate and essentially unrelated topics:

(i) viscous fluid motion, either steady flow past a fixed obstacle or the initial value problem for unsteady flow in a given domain and

(ii) the Helmholtz conformal representation problem in the theory of cavitating wakes (*problème des sillages* and *problème des proues* in Leray's terminology).

The work on viscous fluids and the Navier-Stokes equations has been discussed separately in this journal by M. Chemin and will not concern us here. The writer wishes to add however that Leray's magnificent paper *Sur le mouvement d'un fluide visqueux remplisant l'espace* was the principal stimulation for several of his own papers on the uniqueness, stability and regularity of such motions, and indeed was the inspiration for a vast modern literature on the subject. We shall discuss Leray's contributions to the Helmholtz problem in the final part of this article.

Leray's second major interest lay in fixed point theory. Here, in the great joint paper with Jules Schauder *Topologie et équations fonctionnelles*, the Leray-Schauder fixed point theory first appears. Both Leray and Schauder had been searching for a version of fixed point theory which

P. Pucci et al. (eds.), *James Serrin. Selected Papers: Volume 2*, Contemporary Mathematicians, DOI 10.1007/978-3-0348-0847-7_19, © Springer Basel 2014 

could be applied to compact mappings, but without the restrictive conditions of previous fixed point theorems of Banach and Schauder himself. In those results, the mapping $\mathcal{F}$ had been assumed to satisfy either a contraction condition,

$$\|\mathcal{F}(x) - \mathcal{F}(y)\| \leqslant k\|x - y\|,$$

or the requirement that a convex set in the underlying Banach space be mapped into itself. Neither condition applies to the quasilinear elliptic problems which were of interest to Schauder, or to the Helmholtz problem to which Leray had been devoting his thought.

The idea behind the Leray-Schauder theorem is essentially to generalize L.E.J. Brouwer's theory of degree of a mapping to compact mappings of a Banach space into itself. The difficulty in accomplishing this purpose clearly lies in defining an appropriate concept of degree of a mapping. Here the required idea is the assumption that the map should be compact (*completely continuous* in their terminology, that is, continuous in norm and mapping bounded sets into precompact sets). They proceed thereby to obtain a theory in which the degree is an appropriate homotopic invariant of the mapping, as desired.

For simplicity in presentation we shall confine ourselves to a principal corollary of their theory, in fact one which applies in exactly the contexts to which Leray and Schauder made explicit application.

*Consider a Banach space $\mathcal{B}$ and a compact mapping $\mathcal{F}(x,\sigma)$ of $\mathcal{B} \times [0,1]$ into $\mathcal{B}$. Suppose that there exists an open bounded set $\Omega$ in $\mathcal{B}$ and a point $x_0$ in $\Omega$ such that*

*(i) $\mathcal{F}(x,0) = x_0$ for all $x$ in $\Omega$, and*

*(ii) for any $\sigma$ in $[0,1]$, there is no fixed point of $\mathcal{F}(x,\sigma)$ on the boundary $\partial\Omega$.*

*Then the mapping $\mathcal{F}(x,1)$ has a fixed point $x$ in $\Omega$.*

[The direct connection with degree theory is that degree$(\Phi, \Omega, 0) = 1$ for all $\sigma$ in $(0,1)$, where $\Phi(x) = \Phi(x,\sigma) = x - \mathcal{F}(x,\sigma)$.]

This theorem in turn yields the important principle that when one can find *a priori bounds* for the norm of fixed points, then existence theorems can be obtained. To be specific, *suppose that in the main result we have $\mathcal{F}(x,0) = 0$ while also $\|x\| < M$ for any fixed point $x$ of $\mathcal{F}(x,\sigma)$ with $\sigma$ in $[0,1]$. Then $\mathcal{F}(x,1)$ has a fixed point in the ball $\|x\| < M$.* An elegant and independentproof of the Leray-Schauder theorem in this form was given by F. Browder (1966).

As Leray and Schauder remark, "Le champ d'applications de ce théorème d'existence est vaste", a statement which has been fulfilled one hundred times over in modern analysis. Leray himself applied the result to the two main problems noted above: existence of solutions of second order quasilinear equations (together with Schauder in the paper *Topologie et équations fonctionnelles*, and also separately in his 1939 paper on the Dirichlet problem) and existence of solutions of the Helmholtz problem.

In the first instance, consider the quasilinear elliptic equation

$$\mathcal{L}u \equiv a^{ij}(x,u,Du)D_{ij}u + b(x,u,Du) = 0 \tag{1}$$

in a bounded domain $\Sigma$ of $\mathbf{R}^n$, where $i,j = 1,...,n$ and $Du = (\partial u/\partial x_1,...,\partial u/\partial x_n)$, with Dirichlet boundary conditions being prescribed, that is,

$$u(x) = \varphi(x), \qquad x \in \partial\Sigma. \tag{2}$$

If for any eventual solution $u$ of this problem one can obtain the a priori estimate

$$\|u\|_{C^{1+\alpha}(\Sigma)} < M, \tag{3}$$

then existence follows when appropriate smoothness conditions are imposed on the coefficients $a^{ij}$, on $b$, and on $\partial\Sigma$.

In particular, one introduces the Banach space $\mathcal{B} = C^{1+\alpha}(\Sigma)$ and the mapping $\mathcal{F}(x,\sigma)$ defined by $(u, \sigma) \to \sigma v$, where $v$ solves the Dirichlet problem

$$a^{ij}(x,u,Du)D_{ij}v + b(x,u,Du) = 0 \quad \text{in } \Sigma$$

with $v = \varphi$ on $\partial\Sigma$. That the mapping is compact is not difficult to show by means of Schauder estimates. The uniform boundedness of the $C^{1+\alpha}$ norm of any eventual solution, condition (3), remains as the main problem. For $n = 2$ the required estimates go back to the work of S. Bernstein and to the famous "saddle point"lemma of T. Radó. With the aid of these results, Leray and Schauder then obtained the beautiful and elegant theorem that the two dimensional Dirichlet problem (1), (2) has at least one solution whenever $b = 0$ and the domain $\Sigma$ is convex. Their argument was simplified and the required smoothness conditions were weakened in later work of C.B. Morrey and L. Nirenberg.

Leray and Schauder remark that when the coefficients $a^{ij}$ depend explicitly on $u$, then uniqueness of the solution need not hold even when $b = 0$; that is, the degree of a mapping being 1 does not itself guarantee a simple covering. An explicit example of non-uniqueness was later found by N. Meyers, with however a nonconvex domain $\Sigma$.

It is interesting that Leray and Schauder did not consider the case of uniformly elliptic operators $\mathcal{L}$, for which existence of solutions of the Dirichlet problem holds *without assuming the domain $\Sigma$ to be convex.* Later, in sections 27, 28 of his 1939 paper on the Dirichlet problem, Leray returned to the full equation (1) and gave a more complete analysis of the existence and non-existence of solutions, moreover including uniformly elliptic equations as a special case. Of course, when $\mathcal{L}$ is not uniformly elliptic, convexity of the domain may be required: indeed Leray found even further relations which must be satisfied between the second derivatives of the boundary curve and norms of the inhomogeneous term $b$.

The Dirichlet problem for quasilinear elliptic equations in higher dimensions ($n > 2$) remained open for some 25 years after the work of Leray and Schauder, the required estimate (3) then being far more difficult to obtain. The development of a priori bounds for this case is originally due to De Giorgi and Nash (in 1957), work which was later expanded and generalized by Moser, Ladyshenskaya and Uraltseva, Serrin and others (cf. Gilbarg and Trudinger, Chapters 11-14). In higher dimensions, moreover, even further restrictions on the curvature of the boundary are crucial for the existence of solutions of the Dirichlet problem when the operator is not uniformly elliptic, as shown originally by Jenkins and Serrin and later in full generality by the writer in his monograph in the *Philosophical Transactions of the Royal Society*.

Leray was aware that one could apply fixed point theory even to fully nonlinear elliptic equations. This direction was foreshadowed in the final sections of the paper with Schauder, though a fuller treatment waited until the two papers in the *Journal de mathématiques pures et appliquées* in 1939. The equations considered there have the form

$$f\left(x, y, u, \frac{\partial u}{\partial x}, \frac{\partial u}{\partial y}, \frac{\partial u^2}{\partial x^2}, \frac{\partial u^2}{\partial x \partial y}, \frac{\partial u^2}{\partial y^2}\right) = 0, \tag{4}$$

where $(x, y) \in \Sigma \subset \mathbf{R}^2$, and where (4) is elliptic in the sense that

$$4f_r f_t - f_s^2 > 0, \qquad f_u > 0$$

when $f = f(x, y, u, p, q, r, s, t) = 0$. Leray considers the Dirichlet problem with given boundary values

$$u(x, y) = \varphi(x, y), \qquad (x, y) \in \partial\Sigma,$$

and restricts the problem further by considering solutions which additionally satisfy the condition

$$z_1(x, y) \leqslant u(x, y) \leqslant z_2(x, y), \qquad (x, y) \in \Sigma. \tag{5}$$

Here $z_1$, $z_2$ are given functions satisfying $z_1 < z_2$ in $\Sigma$, while $f$ evaluated for $z_1$ and $z_2$ is respectively $\geqslant 0$ and $\leqslant 0$. Naturally one must have

$$z_1(x, y) \leqslant \varphi(x, y) \leqslant z_2(x, y), \qquad (x, y) \in \partial\Sigma. \tag{6}$$

It is evident that for the existence of a solution of the Dirichlet problem one must provide a priori bounds for any eventual solution $u$ together with its first and second derivatives. This is the program which Leray undertakes. At the initial level one requires a maximum principle which entails (5) whenever $u$ is a solution of the Dirichlet problem and $\varphi$ satisfies (6). The idea for this principle can be expressed in a simple lemma (Lemma 1 of Leray), which we give here in the special quasilinear case (see Serrin, Lectures, Berkeley, 1971):

*Consider the equation* (1) *in* $n$ *independent variables. Let* $v_\lambda = v(x, \lambda)$ *be a family of functions which is increasing in* $\lambda$, *for all* $x \in \Sigma$ *and* $\lambda \in [0, 1]$. *Assume for each fixed* $\lambda$ *that* $v_\lambda$ *satisfies the quasilinear differential inequality*

$$\mathcal{L}v_\lambda \leqslant 0 \quad \text{in } \Sigma,$$

*where* $\mathcal{L}$ *is defined in (1) and* $n \geqslant 2$. *Suppose also that* $\mathcal{L}u \geqslant 0$ *and that* $\mathcal{L}$ *is elliptic either for* $u$ *or for the family* $v_\lambda$. *Then if* $u \leqslant v_1$ *in* $\Sigma$ *and* $u \leqslant v_0$ *on* $\partial\Sigma$, *it follows that either* $u \equiv v_0$ *or* $u < v_0$ *in* $\Sigma$.

The proof is easily accomplished by means of the classical strong maximum principle. We note particularly that if $v_\lambda$ is an elliptic family for the operator $\mathcal{L}$, then no corresponding assumption needs to be made on $u$. Moreover no assumption need be made on the signs of either $\partial a^{ij}/\partial u$ or $\partial b/\partial u$.

The main problem in the development of Leray's existence program for (4) thus turns on estimates for the first and second derivatives of solutions. (For quasilinear equations, estimates for *second* derivatives are a consequence of the Schauder estimates and so are obtained easily and directly).

Leray writes concerning the fully nonlinear problem: "Malheureusement la majoration des dérivées secondes est à l'heure actuelle extrêmement compliquée". In particular, he begins by introducing carefully chosen functions $w$, whose form is far from simple. If it is then supposed that $w$ attains its maximum at a point interior to $\Sigma$, then at this point $dw = 0$, $d^2w \leqslant 0$. These relations, combined with previously obtained limitations on $u$, $\partial u/\partial x$, $\partial/\partial y$, together with the equation $f = 0$ itself (and with various first order derivatives of $f = 0$) then allow one to bound the second order derivatives of $u$ in $\Sigma$. It remains finally to derive bounds for the second derivatives at the boundary $\partial\Sigma$. This step is accomplished by means of a further change of unknown function, together with a difficult geometric analysis. In this anaysis, Leray also finds various conditions for the non-solvability of the Dirichlet problem. For quasilinear equations this work was refined (and generalized to higher dimensions) in work of the writer, *Philosophical Transactions*, Sections 12, 14 and 18.

Leray's analysis moreover brings out and distinguishes two fundamentally different cases, requiring for their treatment separate choices of the function $w$: first, when the relation $f(x, y, u, p, q, r, s, t) = 0$ implies the inequality

$$s^2 - rt > \text{const.}(r^2 + s^2 + t^2) - \text{const.},$$

and second, when conversely

$$f(x, y, u, p, q, r, s, t) = rt - s^2 + g(x, y, u, p, q, r, s, t),$$

where $g$ consists of lower order terms in $r$, $s$, $t$ as these variables approach infinity. The latter situation of course includes the Monge-Ampere equation $rt - s^2 = g(u)$.

The striking difficulty of these papers has led to their neglect in more recent times; though much has been rediscovered and much has since been generalized, nevertheless the question of boundary restrictions for fully nonlinear equations still seems not to have been fully explored since the time of his papers.

As we have already noted, Leray's motivation for developing homotopic fixed point theory lay in the Helmholtz wake problem for general curved obstacles. He attacked this subject two years later in a pair of papers in the *Commentarii Mathematici Helvetici* (preliminary versions in the *Comptes rendus* in 1934 and 1935).

Leray's interest in the problem arose from a lecture course given by H. Villat in 1933. In fact Villat had been a major figure in the development of the equations of Helmholtz flow for curved obstacles, along with T. Levi-Civita, G. Hamel, H. Weyl, K. Friedrichs and A. Weinstein. (It may be noted as well that Leray had written a note in the *Comptes rendus* concerning fluid jets issuing from nozzles, in collaboration with Weinstein). The high interest at that time in the existence problem for Helmholtz flow past a curved obstacle was clearly a great attraction to a young man at the beginning of his mathematical career.

The historical impetus was the celebrated Helmholtz description of plane irrotational flow past a flat plate (held perpendicular to the flow) in two dimensions, when a cavitating wake forms behind the plate, the wake based itself being bounded by two fluid streamlines on which the flow speed is constant (classically, these are the so-called "free boundaries" of the flow, to be determined as part of the solution of the problem). In this form, the Helmholtz flow is a standard part of elementary fluid mechanics, solved by conformal mapping and the Schwarz-Christoffel theorem (see for example Milne-Thomsen's classical treatise *Theoretical Hydrodynamics*).

Replacing the flat plate by a curved obstacle (everything else still being based on irrotational plane flow) drastically complicates the existence and uniqueness problem, as well as the explicit computation of the flow. In order to make the problem tractable, Leray considered various special assumptions on the shape of the curved obstacles under study, which we need not detail here except to note that combined concave-convex obstacles, and also obstacles symmetric with respect to the horizontal $x$-axis, were of particular importance.

The corresponding existence problem leads *à la* Villat to a complicated intego-differential equation. Leray's idea was to homotopically deform the obstacle from its original curved form to a flat plate, for which, as noted just above, one knows the existence of an explicit unique solution. Correspondingly of course the Villat equation also has a unique solution. This being the case, the Leray-Schauder theorem then yields existence of the free boundary problem for the original curved obstacle, provided solutions of the Villat equation are uniformly bounded during the homotopy. The supreme accomplishment of Leray consists exactly in carrying out the (not easy) details of this program.

In particular, the Villat equation can be written in the form

$$x = \mathcal{F}(x),$$

where the function $\mathcal{F}(x)$ depends explicitly on the form of the obstacle. Letting the deformation depend on a parameter $\sigma$, with $\sigma = 0$ and $\sigma = 1$ corresponding respectively to the flat plate and the original obstacle, we then find the equation $x = \mathcal{F}(x, \sigma)$, with $\mathcal{F}(x, 0) =$ constant. The last relation of course guarantees condition (i) of the fixed point theorem.

The present writer's doctoral thesis was strongly based on Leray's work; it may be worth adding that it contains among other things an alternate proof of the crucial uniform boundedness criteria required for the application of the Leray-Schauder theory.

In addition to studying the existence of solutions of the Helmholtz problem, Leray in his second *Commentarii* paper also considered the corresponding uniqueness problem: whether there can be at most one solution for a given obstacle. In a remarkable work, memorable for its subtlety, he first considered local uniqueness (following ideas initiated by Weinstein and Friedrichs). By a further continuation process, global uniqueness then followed for symmetric problems and convex problems.

The difficulty of Leray's uniqueness proof was circumvented in the symmetric case by a remarkable comparison argument devised by

M. Lavrentieff. In this respect, a paper by the present writer in the *American Journal of Mathematics* intersected again with Leray's work. In this work, as well as in several further papers elaborating the concept of hydrodynamical comparison theorems, the uniqueness results for symmetric obstacles are generalized and other qualitative results given, including a discussion of the problem of least wake drag.

The work of Jean Leray everywhere bears the mark of a great mathematician. His results for the non-symmetric Helmholtz problem should never be superseded or improved; his work on the Navier-Stokes equations is still a source of inspiration and open problems; his discussion of the Dirichlet problem is remarkable for its outlook and depth; and the Leray-Schauder fixed point theorem will always be a shining jewel in the crown of mathematics.

## Bibliography

Following Leray's practice, we have not attached reference numbers to the papers cited in the text. The references are gathered below. grouped under three headings: Leray's work; quasilinear equations; and the Helmholtz problem.

**Jean Leray's work, 1931-1939**

*In Comptes rendus, Académie des sciences, Paris*

Sur le système d'équations aux dérivées partielles qui régit l'écoulement permanent des fluides visqueux, **192** (1931), pp. 1180-1182.

Mouvement d'un fluide visqueux à deux dimensions limite par des parois fixes, **193** (1931), pp. 1165-1167.

Sur certaines classes d'équations intégrales non linéaires, **194** (1932), pp. 1627-1629.

Sur les mouvements de liquides illimités, **194** (1932), pp. 1892-1894.

Sur le mouvement d'un liquide visqueux emplissant l'espace, **196** (1933), pp. 527-529.

Topologie et équations fonctionelles, (WITH J. SCHAUDER), **197** (1933), pp. 115-117.

Sur un problème de représentation conforme posé par la théorie de Helmholtz, (WITH A. WEINSTEIN), **198** (1934), pp. 430-432.

Les problèmes de représentation conforme de Helmholtz; théorie des sillages et des proues, **198** (1934), pp. 1282-1284.

Topologie des espaces abstraits de M. Banach, **200** (1935), pp. 1082-1084.

Les problèmes de la représentation conforme de Helmholtz; théorie des sillages et des proues, **200** (1935), pp. 2007-2008.

Compléments à l'étude des mouvements d'un liquide visqueux illimité, (WITH L. ROBIN), **205** (1937), pp. 18-20.

Discussion du problème de Dirichlet, **205** (1937), pp. 269-271.

Sur la résolution du problème de Dirichlet, **205** (1937), pp. 785-787.

Sur le problème de Dirichlet quasilinéaire d'ordre 2, (WITH Y. CHOQUET-BRUHAT), **274** (1972), pp. 81-85.

*Main papers*

Étude de diverses équations intégrales non linéaires et de quelques problèmes que pose l'hydrodynamique (Thesis), *Journ. de math. pures et appl.* **12** (1933), pp. 1-82.

Essai sur les mouvements plans d'un fluide visqueux que limite des parois, *Journ. de math. pures et appl.* **13** (1934), pp. 331-418.

Sur le mouvement d'un fluide visqueux remplissant l'espace, *Acta Math.* **63** (1934), pp. 193-248.

Topologie et équations fonctionnelles, (WITH J. SCHAUDER), *Ann. École norm. sup.* **51** (1934), pp. 45-78.

Sur la validité des solutions du problème de la proue, in: *Livre jubilaire de M. Marcel Brillouin*, Gauthier-Villars, Paris, 1935, pp. 1-12.

Les problèmes de représentation conforme de Helmholtz; théorie des sillages et des proues, *Comm. Math. Helv.* **8** (1936), pp. 149-180 and pp. 250-263.

Les problèmes non linéaires, *l'Enseign. Math.* **35** (1936), pp. 139-151.

Majoration des dérivées secondes des solutions d'un problème de Dirichlet, *Journ. de math. pures et appl.* **17** (1938), pp. 89-104.

Discussion d'un problème de Dirichlet, *Journ. de math. pures et appl.* **18** (1939), pp. 249-284.

**Quasilinear second order elliptic equations**

BERNSTEIN, S., Méthode générale pour la résolution du problème de Dirichlet, *C. R. Acad. Sci. Paris* 144, 1907, pp. 1025-1027.

BERNSTEIN, S., Conditions nécessaires et suffisantes pour la possibilité du problème de Dirichlet, *C. R. Acad. Sci. Paris* 150, 1910, pp. 514-515.

BERNSTEIN, S., Sur la généralisation du problème de Dirichlet, *Math. Ann.*, **69**, pp. 82-126.

BERNSTEIN, S., Sur les équations du calcul de variations, *Ann. Sci. École norm. sup.*, **29** (1912), pp. 431-485.

BROWDER, F., Problèmes Non-linéaires. *Séminaire de Mathématiques Supérieure*, University of Montreal Press, 1966.

DE GIORGI, E., Sulla differenziabilità e l'analiticità della estremali degli integrali multipli regolari, *Mem. Acad. Sci. Torino, Cl. Sci. Fis. Mat.*, **3** (1957), pp. 25-43.

GILBARG, D., Boundary value problems for nonlinear ellipic equations in n variables. *Nonlinear problems*, University of Wisconsin Press, Madison, 1963, pp. 151-159.

GILBARG, D., TRUDINGER, N., *Elliptic Partial Differential Equations of Second Order*, Springer-Verlag, 1977.

JENKINS, H., SERRIN, J., The Dirichlet problem for the minimal surface equation in higher dimensions, *J. Reine Ang. Math.* **229** (1968), pp. 170-187.

LADYZHENSKAYA, O.A., URALTSEVA, N.N., *Linear and Quasilinear Elliptic Equations*, Moscow, 1964.

MEYERS, N.G., An example of non-uniqueness in the theory of quasilinear elliptic equations of second order, *Arch. Rational Mech. Anal.* **14** (1963), pp. 177-179.

MORREY, C.B., JR., On the solution of quasi-linear elliptic partial differential equations, *Trans. Amer. Math. Soc.* **43** (1934), p. 126-166.

MOSER, J., A new proof of de Giorgi's theorem concerning the regularity problem for elliptic differential equations, *Comm. Pure Appl. Math.* **13** (1960), pp. 456-468.

NASH, J., Continuity of solutions of parabolic and elliptic equations, *Amer. Journ. Math.* **80** (1958), pp. 931-954.

NIRENBERG, L., On nonlinear elliptic partial differential equations and Holder continuity, *Comm. Pure Appl. Math.* **6** (1952), pp. 103-156.

RADÓ, T., Contributions to the theory of minimal sutfaces, *Acta Szeged*, **6** (1932), p. 1-20.

SERRIN, J., The Dirichlet problem for quasilinear equations with many independent variables, *Phil. Trans. Royal Soc., London* **264** (1969), p. 413-496.

SERRIN, J., Nonlinear elliptic equations of second order, *Lectures, Symp. on Partial Diff. Eqs.*, Berkeley, 1971.

**The Helmholtz problem**

FRIEDRICHS, K., Ueber ein Minimumproblem mit freiem Rande, *Math. Ann.* **109** (1933), p. 64.

HAMEL, G., Ein hydrodynamischer Unitatssatz, *Proceedings of the Second International Congress of Applied Mechanics*, Zurich, 1926, p. 489.

HELMHOLTZ, H., Uber diskontinuierliche Flussigkeitsbewegungen, *Monatsberichte Berlin Akad.*, 1868, p. 215.

LAVRENTIEFF, M., On certain properties of univalent functions and their application to wake theory (in Russian), *Math. Sbornik*, **46** (1938), pp. 331-458.

LEVI-CIVITA, T., Scie e leggi di resistenza, *Rend. Circolo Mat. Palermo*, 1907, p. 1.

MILNE-THOMPSON, L.M., *Theoretical Hydrodynamics*, Second edition, Macmillan, 1950, p. 291.

SERRIN, J., Existence theorems for some hydrodynamical free boundary problems (Thesis), *Journ. Rational Mech. Anal.* **1** (1952), pp. 1-48.

SERRIN, J., Uniqueness theorems for two free boundary problems, *Amer. Journ. Math.* **74** (1952), pp. 492-506.

VILLAT, H., *Mécanique des fluides*, Seconde édition, Paris, 1938.

WEINSTEIN, A., Les conditions aux limites introduites par l'hydrodynamique, *Enseign. Math.* **35** (1936), p. 107 (contains an extensive bibliography).

WEINSTEIN, A., Non-linear problems in the theory of fluid motion with free boundaries, *Proc. Symposia in Applied Math.*, vol. 1, American Math. Soc., 1949, pp. 1-19.

WEYL, H., Strahlbildung, nach der Kontinuitatsmethode behandelt, *Nachr. Ges. der Wiss. Gottingen*, 1927, p. 227.

In: C.S. Morawetz el al. (eds.): *Selected Works of Eberhard Hopf.* **Reprint VIII.[ii]**
Amer. Math. Soc. (2002), pp. 213–219.


**Über die Anfangswertaufgabe für die hydrodynamischen Grundgleichungen,** *Math. Nachr.* **4 (1950), 213-231.**

# Commentary

James B. Serrin

In 1951 Eberhard Hopf reopened the existence question for the initial-boundary value problem for incompressible Navier-Stokes equations, a problem left essentially untouched since the pioneering papers of Jean Leray in 1932-1934. Leray had established the existence and uniqueness of regular solutions *in space dimension 2 when the domain of the solution is the whole space*; and had also found regular solutions *in a finite time interval* $[0, T)$ for bounded domains in dimension $n = 3$, but there the problem had rested until Hopf's work.

Hopf's paper "Über die Anfangswertaufgabe für die hydrodynamischen Grundgleichungen" brought new life and cohesiveness to the problem, and has been a source of stimulation to authors ever since that time. In this commentary I wish to consider Hopf's work in some detail, especially since the paper itself, as a *mathematical contribution*, has been unduly neglected. The reasons for this lie, I believe, in the difficult German text, in the lengthy elaboration of differentiation theory in the first part of the paper (material now well known in the theory of partial differential equations), and finally in some troubles in reading the proof, which I shall return to later. At the same time, one cannot over-emphasize the far reaching nature of Hopf's fundamental result, that it applies in an arbitrary number of space dimensions, and the fact that the proof is carried out by simpler means than Leray's, simpler also than the later proofs which I know. In this respect, Hopf makes the following interesting comment (page 217): "Der Hauptpunkt der Arbeit ist, dass die Konstruktion der angenäherten Lösungen die in Lerays Arbeiten einen so breiten Raum einnimmt, hier durch einen einfachen Prozess ersetzt wird, des sich auch auf viel weitere Klassen partieller Differentialprobleme anwenden lässt."

The commentary will consist of three sections. First, the statement of Hopf's result and a short discussion of the proof, including an elegant but little known compactness lemma. I also emphasize several places which Hopf left without sufficient discussion. The second section concerns the exceptional case of two spatial dimenions, where a remarkable degree of completeness has subsequently been obtained through work of Prodi, Lions and Ladyzhenskaya.

The third section discusses various refinements of Hopf's work in the space dimensions $n = 3$ and $n = 4$ (and some additional considerations for $n \geq 5$), largely due to Ladyzhenskaya (1957).

There are any number of monographs describing the Navier-Stokes equations and the associated initial-boumdary value problem, excellent sources being Ladyzhenskaya's book (1967), Temam's treatise (2000) and the monograph of Sell and You (2002). Historically minded readers will also be rewarded by consulting Leray's original papers.

**1. Hopf's fundamental result.** Hopf's main existence theorem for the initial-boundary value problem is formulated for weak solutions

$$u = u(x,t) = (u_1(x,t), u_2(x,t), \cdots, u_n(x,t))$$

in a strict distributional sense. For the convenience of the reader we restate the result (Existenzsatz, page 223) in a more compact form. Here it is useful first to introduce the function space $V$:

*Let $G$ be an arbitrary domain of $\mathbf{R}^n$. By $V$ we denote the closure of smooth divergence-free vector fields $\phi(x,t)$ in $\mathbf{R}^n$ with compact support in $\hat{G} = G \times (0,\infty)$, under the space-time norm*

$$\int_0^\infty (|\phi|^2 + |D\phi|^2)\, dt,$$

*where $|\phi|^2 = |\phi|^2(t) = \int_G \phi^2(x,t)\, dx$ and $D\phi$ denotes the gradient matrix of $\phi$ (notation obvious).*

[The space $V$ is essentially Hopf's space $H'(N)$. The definition of $V$ encloses the notion of weakly divergence-free vector fields as well as the weak satisfaction of the imposed boundary condition $u = 0$. In his definitions Hopf uses weak rather than strong closures, although these are equivalent in the present case, as Hopf notices in the footnote on page 223.]

**Main Existence Theorem** (page 223). *Let $G$ be an arbitrary domain in $\mathbf{R}^n$, and consider initial data $u_0(x)$ which is in the $L^2(G)$ closure of smooth divergence-free vector fields in $\mathbf{R}^n$, with compact support in $G$.*

*Then there exists a velocity vector field $u = u(x,t)$ in $V$, such that the distribution identity* (1.4) (*page* 215) *for the Navier-Stokes equations is satisfied. Moreover the initial data $u_0$ is attained in the sense that $|u(\cdot,t) - u_0(\cdot)| \to 0$ as $t \to 0$.*

The vector field $u$ in the main theorem has two important properties not included in the statement, ($i$) the identity (1.4) can be replaced by (4.1), and ($ii$) the solution obeys an *energy inequality*

$$\frac{1}{2}|u|^2 + \nu \int_0^t |Du|^2 dt \le \frac{1}{2}|u_0|^2$$

at each time $t \ge 0$, where $\nu$ denotes the kinematic viscosity of the fluid in question. In fact, it is precisely (4.1) which Hopf actually proves in the paper; see the third part (Dritter Schritt) of the the demonstration, page 228. The energy inequality, finally, is included (in passing) at the last paragraph of the proof, page 229.

Turning to the main proof of the theorem, the application of the Galerkin method is particularly to be emphasized. Hopf in fact describes the technique as "dieses einfachsten und nächstliegenden Annäherungsverfahrens" (page 225). Hopf's preparation of the Galerkin method in Section 4 is exemplary: we would only add that the orthonormal construction preceding Lemma 4.3 on page 225 is in fact the well-known Gram-Schmidt procedure. [Two years before Hopf's work, Faedo had suggested application of the Galerkin method to the simpler case of linear evolutionary equations, but it was Hopf's use of the technique for *nonlinear* problems that cemented its importance in the context of partial differential equations.]

The proof itself (Section 5, pages 226–229) rests squarely on the basic (pre-energy) identity (4.10) and the subsequent relation

$$|u^{(k)}|^2 = \sum_{i=1}^{k} (\lambda_i^{(k)})^2$$

which appears a few lines later.

The whole treatment is masterful. It is worth noting that the ultimate step in the demonstration necessarily involves proving the (strong) compactness of the approximating family $u^{(k)}$; here Hopf proceeds in a beautiful way through Lemma 5.1 (page 228). [In fact, the present writer, in admiration of Hopf's lemma, gave a generalization of the result (1962), with perhaps a slightly more transparent proof.]

Altogether, Hopf's result and its proof are remarkable, though a small final point should be noted: the proof of the last part of the theorem, namely that the initial data $u_0$ is strongly attained in $L^2(G)$, requires for the validity of the argument (foot of page 229), the assertion of *weak continuity in time of the vector* $u(x,t)$. This however follows immediately from (4.1).

**2. The special dimension** $n = 2$. Prodi (1960) showed by not entirely obvious means that the energy inequality for $n = 2$ can be improved to an equality

$$|u|^2 = |u_0|^2 - \nu \int_0^t |Du|^2 dt.$$

Even more, from this it follows that $|u|^2$ is weakly differentiable (as a function of time) with $L^1$ derivative $-\nu|Du|^2$; hence, for *bounded* domains contained in some ball of radius $R$, one gets from Poincaré's inequality

$$\frac{d|u|^2}{dt} \leq -\frac{2\nu}{R^2}|u|^2.$$

By integration we find that $|u|^2 e^{2\nu t/R^2}$ is a decreasing function, and in particular

$$|u|^2 \leq |u_0|^2 e^{-2\nu t/R^2}.$$

Thus in turn $|u| \to 0$ as $t \to \infty$. This calculation of course applies in any case where the energy equality holds and the domain is bounded.

It is surprising that the exponential decay condition continues to be valid for arbitray dimensions even though the energy equality itself may no longer hold. In fact by using Hopf's pre-energy equality (4.10) together with the above calculation, one finds at once that the approximating energy $|u^k|$, see above, also exhibits the exponential decay found before. But then, since $|u|^2 \leq |u^k|^2$ as in Hopf's proof, the desired conclusion follows at once.

It was also shown by Lions and Prodi (1959) that *Hopf's weak solution (for* $n = 2$*) is uniquely determined by its initial data* $u_0$. In their proof, the inequality

$$\int_0^t (u, v\cdot Dw)dt \leq \text{Const.}\, |u|^{1/2}|v|^{1/2} \left(\int_0^t |Du|^2 \int_0^t |Dv|^2\right)^{1/4} \left(\int_0^t |Dw|^2\right)^{1/2} \quad (1)$$

plays a crucial role: (1) is in fact an easy consequence of Hölder's inequality together with the Sobolev interpolation inequality

$$|u|_4 \leq c|u|^{1/2}|Du|^{1/2}; \tag{2}$$

here $(\cdot,\cdot)$ denotes the $L^2$ inner product on $G$, and $|u|_s$ the standard $L^s(G)$ norm of $u$.

It seems highly unlikely that Hopf's weak solution should be a regular function of the spatial variable $x$; on the other hand, if the data $u_0$ is somewhat more regular, then this will in fact be the case (precisely, $u$ will then be analytic in the spatial variable $x$ at each time $t > 0$). This result is due to the combined work of Ladyzhenskaya (1957, 1959), Golovkin (1960), Ohyama (1960), Serrin (1962) and Kahane (1969). Ladyzhenskaya, first of all, showed that *if the initial data $u_0$ is in $H^2$, namely the closure under the norm*

$$|\varphi|^2 + |D\varphi|^2 + |D^2\varphi|^2 \tag{3}$$

*of smooth divergence-free vector fields $\varphi$ with compact support in $G$, then there are weak solutions $u(x,t)$ of the Navier-Stokes equatiions such that $|Du(\cdot,t)|$ and $|u_t(\cdot,t)|$ exist and are uniformly bounded, at least for initial time intervals $[0,\tau]$.*

Here $\tau$ is any fixed time, $0 < \tau < T$, while $T < \infty$ depends only on the norm $\|u_0\|_H$. Serrin (1963) showed that one can in fact take $T = \infty$ here, and the demonstration is then completed through the work of the other authors cited. It is interesting that the inequality (1) for the triple product $(u, v\cdot Dw)$ is closely related to this proof.

The preceding results, that is, the energy equality and uniqueness and regularity, provide a satisfactory resolution of the initial-boundary value problem for $n = 2$ dimensions, in the classical sense of Hadamard. The situation when $n \geq 3$ is less complete, as we shall see, though important conclusions can still be drawn.

**3. Dimensions $n \geq 3$.** Here very little additional information can be deduced without first establishing further regularity of the solution vector $u(x,t)$. To unify the presentation, we begin by presenting a general result applying for all dimensions $n \geq 3$. In stating this result, we shall denote by $L^{s,s'}(\hat{G})$ $(\hat{G} = G \times (0,\tau))$ the function space defined by the condition

$$\int_0^\tau |u(\cdot,t)|_s^{s'}\,dt < \infty,$$

where $1 \leq s, s' < \infty$, with natural extensions to the values $s,\ s' = \infty$.

**Theorem 1.** *Let $u$ be a weak solution of the Navier-Stokes equations in the sense of the Main Existence Theorem of Section 1. Suppose that $u \in L^{s,s'}(\hat{G})$ $(\hat{G} = G \times (0,\tau))$, where*

$$\frac{n}{s} + \frac{2}{s'} = 1. \tag{4}$$

*Then the energy equality holds for all $t \leq \tau$. Moreover, if $v$ is any solution satisfying the conditions of Hopf's theorem, and $u_0 = v_0$, then $u \equiv v$ for all $t \leq \tau$.* (Serrin (1963, Theorems 5, 6).

The basic inequality

$$\int_0^\tau (u, v\cdot Dw)dt \leq \text{Const.}\, |u|_{s,s'}|Dw|_{2,2}\left(|v|_{2,\infty}^{2/s'}\,|Dv|_{2,2}^{n/s}\right)$$

plays a key role in the proofs when $n \geq 3$. In the same way as (1), this inequality arises from a straightforward application of the Hölder and Sobolev inequalities. (The space $L^{s,s'}$ was also used later by J.-L. Lions (1969).)

A direct connection with the two dimensional case is also worth noting: that is, for the Hopf solution in $n = 2$ one finds from (2) and the energy inequality that

$$|u|_{4,4} \leq \text{Const.}\, |u_0|^{1/2} \cdot |u_0|^{1/2}/\nu^{1/4} < \infty$$

on $G \times (0, \infty)$; the pair $(s, s') = (4, 4)$ moreover satisfies (4) when $n = 2$, thus providing another motivation for the uniqueness of weak solutions in this case.

The following general regularity result also holds.

**Theorem 2.** *Let the hypotheses of Theorem 1 be satisfied, except that* (4) *is replaced by*

$$\frac{n}{s} + \frac{2}{s'} < 1. \tag{5}$$

*Then $u(x,t)$ is analytic in the space variable $x$ for each $t$ in $(0, \tau)$, and each spatial derivative is bounded in compact subregions of $\hat{G}$.*

This result is due to Kahane (1969). The condition (4) precludes direct application of Theorem 2 to Hopf's weak solutions in any dimension $n \geq 3$, while the stronger condition (5) even disallows direct application when $n = 2$. It is consequently necessary once more to have stronger existence theorems.

**Theorem 3.** (Ladyzhenskaya (1957, 1959). *Let $n = 3$. Then for initial data $u_0 \in H^2$, there exists a time $T$, $0 < T < \infty$, and weak solutions $u(x,t)$ of the Navier-Stokes equations, such that $|Du(\cdot,t)|$ and $|u_t(\cdot,t)|$ are uniformly bounded for all times $t$ in $[0, \tau]$ with $\tau < T$.*

This result is crucially different than the corresponding one in two dimensions, in that the solution may not satisfy the boundedness conclusions for all times $\tau < \infty$. A refinement of Ladyzhenskaya's result does show however that if $\|u_0\|_H$ is sufficiently small, then one can take $T = \infty$; see Serrin (1963, Theorem 2), and Fujita and Kato (1964).

With the help of Theorem 3 and Sobolev's inequality, one obtains for the solutions in question that, for $n = 3$,

$$|u|_6 \leq \text{Const.}\, |Du| \leq \text{Const.},$$

for $0 \leq t \leq \tau$, $\tau < T$. This in turn gives

$$|u|_{6,4} \leq \text{Const.}\, t^{1/4}, \quad 0 \leq t \leq \tau,$$

where the pair $(s, s') = (6, 4)$ now obeys (4). In consequence of Theorem 1, Ladyzhenskaya 's solution $u(x,t)$ then obeys the *energy equality* and is *unique* – for $0 \leq t < T$ – in the class of weak solutions obtained by Hopf; see Serrin (1963, Theorem 6). It should be noted as well that Ladyzhenskaya (1957) had obtained a preliminary version of the uniqueness of solutions, this being the foundation for the more general result stated here.

Also one has $|u|_{6,\infty} \leq$ Const., so Theorem 2 shows that Ladyzhenskaya's solution is analytic in the space variables at each fixed time $0 < t < T$.

Turning to higher dimensions, one can in fact extend Theorem 3 to the case $n = 4$, the conclusion being that *if the initial data $u_0$ is sufficiently small in the* *norm* $H^2$, *then* $|Du(\cdot,t)|$ *and* $|u_t(0,t)|$ *are uniformly bounded for all times* $t \geq 0$. (Serrin (1963); Theorem 3); see also Temam (2000, Chapter 3, Theorem 3.1) and J.-L. Lions (1959).)

Again with the help of the Sobolev inequality, one has for $n = 4$,

$$|u|_4 \leq \text{Const.}\, |Du| \leq \text{Const.}$$

for all $t \geq 0$, that is $|u|_{4,\infty} \leq$ Const. for $t \geq 0$. The pair $(s, s') = (4, \infty)$ satisfies (4) when $n = 4$. Thus by Theorem 1 the solution again obeys the *energy equality* and is *unique* in the class of weak solutions obtained by Hopf.

The pair $(4, \infty)$ on the other hand does *not* satisfy (5) when $n = 4$ (though it comes close). Consequently, when $n = 4$ one does not have a regularity theorem *for solutions which are known to exist.*

For $n \geq 5$ the result of Theorem 1 is of course still valid, but a corresponding existence theorem is unavailable; in other words, we know the existence but *not* the uniqueness and regularity of weak solutions, and we know uniqueness and regularity of suitably restricted solutions, but *not* the existence of such solutions.

A pivotal question of intense interest, finally, is the behavior of solutions for $n = 3$ when the initial data is not small enough to allow $T = \infty$; see Theorem 3 and the remark following it (explicit bounds are stated in Serrin (1963)). In such cases, even for smooth data, a regular solution may exist only for some sufficiently small time interval, after which it loses both its regularity and uniqueness. It would be fruitless to speculate here on whether such situations can actually occur, but the reader can be referred to Temam's monograph (2000), page 345, and to the recent review article of J.-Y. Chemin (2000). At ths same time, one may express the belief that irregularity and non-uniqueness do in fact occur for the three dimensional Navier-Stokes equations for appropriate initial data $u_0$. On the other hand, Hopf writes on page 217: "Es ist schwer zu glauben, dass die Anfangswertaufgabe zäher Flussigkeiten für $n = 3$ mehr als eine Lösung haben könnte, und der Erledigung der Eindeutigkeitsfrage sollte mehr Aufmerksamkeit geschenkt werden."

Of course, the validity of the Navier-Stokes equations themselves may be questioned in cases where velocity gradients become excessively large, a point raised by Temam in the comments noted above. Thus if one assumes that stresses grow superlinearly with respect to velocity gradients (not just linearly as for the Navier-Stokes equations), it becomes believable that uniqueness and regularity will persist for all time even in dimension 3, or for that matter, in arbitrary dimensions. A result to this effect has in fact been established in a little known but perceptive paper of Ladyzhenskaya!

*Acknowledgement.* The author wishes to thank Prof. Norman Meyers for a number of helpful conversations concerning this work.

## References

CHEMIN, J.-Y., "Jean Leray and Navier-Stokes." *Gazette des Mathématiciens* **84** (Supplement) (2000), pp. 71-82.

FAEDO, S., "Un nuove metodo per l'analisi esistenziale e quantitativa dei problemi di propagazione." *Annali Scuola Norm. Sup. Pisa*, Ser 3 **1** (1949), pp. 1-41.

FOIAS, C., "Une remarque sur l'unicité des solution des équations de Navier-Stokes en dimension n." *Bull. Soc. Math. France* **89** (1961), pp. 1-8.

FUJITA, H., AND T. KATO, "On the Navier-Stokes initial value problem." *Archive Rational Mech. Analysis* **16** (1964), pp. 269-315.

GALDL, G. P., *An Introduction to the Mathematical Theory of the Navier-Stokes Equations.* New York: Springer, 1994.

ITO, S., "The existence and uniqueness of regular solution of non-stationary Navier-Stokes equation." *J. Faculty Science, Univ. Tokyo,* **9** (1961), pp. 103-140.

KAHANE, C., "On the spatial analyticity of solutions of the Navier-Stokes equations." *Archive Rational Mech. Analysis* **33** (1969), pp.386-405.

KISELEV, A.A., AND O. A. LADYZHENSKAYA, "On existence and uniqueness of the solution of the non-stationary problem for a viscous incompressible fluid." *Izvestiya Akad. Nauk SSSR* **21** (1957), pp. 655-680.

LADYZHENSKAYA, O.A., "Solution "in the large" of non-stationary boundary value problem for the Navier-Stokes system with two space variables." *Comm. Pure Applied Math.* **12** (1959), pp. 427-433.

LADYZHENSKAYA, O. A., *Mathematical problems in the dynamics of viscous incompressible fluids.* Moscow 1961.

LERAY, J., "Etude de diverses équations intégrales non-linéares et de quelques problèmes que pose l'hydrodynamique." *Journ. Math. Pures Appl.* **12** (1933), pp.1-82.

LERAY, J., "Essai sur les mouvements plans d'un liquide visqueux que limitent des parois." *J. Math. Pures Appl.* **13** (1934), pp. 331-418.

LERAY, J., "Sur le mouvement d'un liquide visqueux emplissant l'espace." Acta Math. **63** (1934), pp. 193-248.

LION, J.-L., "Quelques résultats d'existence dans les équations aux dérivées partielles nonlinéaires." *Bull. Soc. Math. France* **87** (1959), pp. 245-273.

LIONS, J.-L., *Quelques methodes de resolution des problèmes aux limites non linéaires* Paris: Dunod 1969.

LIONS, J. L., AND G. PRODI, "Un theoreme d'existence et unicite dans les equations de Navier-Stokes en dimension 2." *C. R. Acad. Sci. Paris* **248** (1959), pp. 3519-3521.

PRODI, G., *Un teorema di unicitá per le equazioni di Navier-Stokes.*

PRODI, G., "Qualche risultato riguardo alle equazioni di Navier-Stokes nel caso bidimensionable." *Rend. Sem. Mat. Padova* **30** (1960), pp. 1-15.

SELL, G, AND Y. YOU, *Dynamics of Evolutionary Equations.* New York: Springer 2002. See especially Chapter 6.

SERRIN, J., "Strong convergence in a product space." *Proc. Amer. Math. Soc.* **13** (1962). pp. 651-655.

SERRIN, J., "On the interior regularity of weak solutions of the Navier-Stokes equations." *Arch. Rational Mech. Analysis.* **9** (1962), pp. 187-195.

SERRIN, J., "The initial value problem for the Navier-Stokes equations." In *Nonlinear Problems,* ed. R. E. Langer, pp. 69-98. Univ. Wisconsin Press, 1963.

TEMAM, R., *Navier-Stokes Equations.* Third edition, North-Holland, 1984; revised and with extensive additional comments, Providence: AMS Chelsea Publ., 2000.

*J. Elasticity* 70 (2003), 31–38.


**Reprint VIII.[iii]**

# An Appreciation of Clifford Truesdell

JAMES SERRIN
*School of Mathematics, University of Minnesota, Minneapolis, MN 55455, U.S.A.*

*The following comments are a slightly emended version of a talk delivered at a meeting of the Society for Natural Philosophy in December 1985, together with additional remarks appended in May 2003.*

This meeting of the Society for Natural Philosophy has been arranged as a special tribute to its principal founding member, Clifford Truesdell, in recognition of his many contributions to the Society since its beginning at The Johns Hopkins University in March 1963. In this group picture of the participants of the first meeting (see Figure 1) you will see Clifford Truesdell at next to the left in the front row, and no doubt you will also recognize a number of other faces, all 22 years younger and perhaps 22 years wiser, than today.

In the intervening years the Society has sponsored 27 additional meetings, holding fast throughout its existence to the original form, to small size, no publications, and the encouragement of quality of research – primarily in rational continuum mechanics and its foundations, though with occasional forays into related mathematical disciplines. And outstanding menus, as well.

Consistent maintainance of the original principles of the Society is an achievement due more to Clifford Truesdell than to any other single individual. It is impossible in a few short minutes to do justice to his remarkable and myriad accomplishments, in papers, monographs, memoirs, and books ranging from mathematical sciences, to rational mechanics, natural philosophy, and the history of science. Already in the forties there was clear evidence of this future in several unusual publications, perhaps not today known to everyone here.

As a student at Indiana University I was privileged early to see the power of his thought in exceptional lectures which he offered in elasticity and kinetic theory – perfect in content and stately in development – thus, the appearance in 1952 of the seminal paper "Mechanical Foundations of Elasticity and Fluid Mechanics" came to me not so much as a surprise but rather as a remarkable opportunity to learn more from a great master. Several years before, he had initiated research on the kinematics of vorticity, together with the interrelations of vorticity to thermodynamics. This culminated in his next work, the elegant monograph "The Kinematics of Vorticity", published by the Indiana University Press but now out of print, and, following without pause, his great paper on the absorption and dispersion

*Figure 1.* First Meeting of the Society for Natural Philosophy, March 25–26, 1963. The front row left-to-right: L.M. Milne-Thomson, C. Truesdell, B. Coleman, J. Serrin, J.L. Ericksen, H. Markovitz, E. Sternberg, W. Noll, H. Grad, R. Toupin, R. Rivlin. Back row second from the right: Charlotte Truesdell.

of sound according to the viscosity formulae of Navier and Stokes. Altogether one of the most audacious first appearances of a new star in the firmament of mechanics!

While time makes it impossible to dwell upon still another quiver in his bow, I must add that it was exactly following these works that he began the three celebrated prefaces for the Opera Omnia of Leonhard Euler – one each on fluid mechanics, acoustics, and elasticity. These established his fame as the leading modern interpreter of Eighteenth Century Mechanics, and of course brought to our attention, as nothing else, the contributions of Euler to our science.

In 1960 Truesdell and Richard Toupin discovered the important entropy relations now known as the Clausius–Duhem inequality, specifically including in it the crucial external heat supply term. This development made possible the later theory of Coleman and Noll, providing thereby, for the first time, a rational and rigorous approach to the foundational aspect of constitutive restrictions. Moreover, as appeared in retrospect many years following, the radiation supply term provided a vital link between classical thermodynamics and modern theory.

Truesdell's work on the Clausius–Duhem inequality was severly criticized by those whose understanding of thermodynamics had been formed without benefit from study of the masters of the subject. The criticisms could not be fully answered, ironically, because the foundations of thermodynamics were themselves open to question. In any case, this neglected and nettle-choked thicket of physics became one of the principal areas of Truesdell's later work: of course his detractors never dreamed that anyone would set foot in those Augean stables, nor of course did they have the courage to do so themselves. Nevertheless, the difficulties of thermodynamics were enough to require many further years of thought before the ideas were ripe for presentation, and other projects loomed immediately.

Two monumental volumes for the Handbuch der Physik appeared in the early sixties, "The Classical Field Theories" with Richard Toupin, and "Non-linear Field Theories of Mechanics" with Walter Noll. It is almost impossible to know how to approach these volumes. Their weight of scholarship and perception, their very completeness, daunt the mere mortal who opens them. Even so, the rich veins of ore were, individually and internationally, mined by the many. Here one may also recall the impress which Truesdell's views had on the seven companion volumes of the Handbuch der Physik; for the most part he selected not only the general subject matter, but also the corresponding authors. Even now, thirty years later, many of these volumes are as fresh as ever – required reading for the acolyte.

After 1964 thermodynamics and kinetic theory began to occupy Clifford more heavily than ever before, as if only the most monumental problems were now sufficient of challenge. In 1957 he had published a first paper on chemical mixtures, followed by a second paper in 1968 which inspired much further work on this most difficult of subjects. Then came in 1969 the polemical masterpiece "Rational Thermodynamics", summarizing his thoughts to that time and forming a springboard (I almost wrote "launching pad" but rejected this) for future work.

*Figure 2.* Journal of Rational Mechanics and Analysis: Volume 1, Number 1.

If I may digress a moment, let me recall that this scientific activity was paralleled by continuing effort to re-establish both Rational Mechanics and the methods of Natural Philosophy as integral parts of physical science. Beyond the formation of the Society for Natural Philosophy, there is the founding of the Journal of Rational Mechanics and Analysis in 1952 (see Figure 2), its reappearance as the Archive for Rational Mechanics and Analysis in 1957, and the inauguration of the Archive for History of Exact Sciences in 1960, to name his major and most longstanding contributions in this direction. Galileo had said "Nature first made things in her own way, and then made human reason skillful enough to understand – but only by hard work – some parts of her secrets." Clifford saw well that a combination of mathematical analysis and historical sense was the key to this understanding, and he emphasized always, as Dedekind put it, "What is provable should not be believed in science without proof." In these endeavors his practice emulated his preachment to the Society at an earlier meeting. "The aim of rational mechanics is to provide a sound conceptual framework for description of nature as human senses perceive it and to create patterns of systematic inquiry and inference such as to order and interrelate the phenomena thereby conceived." Deeper understanding was to come through the successive works of generations, together with our experience of materials themselves.

I would like to bring you the message that Clifford's evangelism succeeded beyond his expectations, but must sadly report that the world is still peopled by pagans (and worse) – a theme which Clifford of course has voiced frequently in his scientific essays. In any case I herewith dedicate for his perusal two sentences culled at random from today's scientific journalese – the first "If space has nine dimensions and matter is strings then the mysteries of the universe may soon come clear" (so help me God), and the second, "Among enhancement techniques the National Research Council will investigate are sleep learning, group cohesion technologies, accelerated intelligencing and parapsychological stress." But one must put aside the ridiculous in the study of nature.

Returning then to the question of classical thermodynamics, one observes from the outset that this is the sole discipline of classical physics where serious argument about fundamentals remains. Even while moguls of science grant entropy and energy all-pervasive influence, it is impossible to obtain agreement from these grandees on the two laws of nature from which these concepts derive. As recently as 1970 the subject was innocent of mathematical structure, and the lack of precision showed throughout. The introductory essay in "Rational Thermodynamics", written in 1968, tells this story faultlessly.

Truesdell attacked this miasmic swamp with vital energy, and brought others as well to the fray. The outcome is now clear; there is no longer doubt that energy and entropy can be set upon sure and credible foundations,* vindicating Clifford's intuition when he proposed the Clausius–Duhem inequality in 1960. His own part of this research, carried out from 1974–1984, included in particular a rigorous axiomatization of reversible thermodynamics. There is also in this work a corollary result, a beautiful pearl, which I think deserves to be better known than it is. This is his discovery that, for reversible systems, the law of thermodynamics can be stated without recourse to the mechanical equivalent of heat, indeed in such a way that the existence of this mechanical equivalent can be deduced, much as the existence of absolute temperature can be inferred from the second law. Generalizations of this to arbitrary thermal systems have been given by later writers, leading me to predict that ultimately this method of viewing the First Law will become a new paradigm of physics.

Books issue in continuing stream from "Il Palazzetto": "Introduction to Rational Elasticity" (with Wang), "The Tragicomical History of Thermodynamics, 1822–1854", "A First Course in Rational Mechanics", "An Idiot's Fugitive Essays on Science", "Fundamentals of Maxwell's Kinetic Theory of a Simple Monatomic Gas" (with Muncaster), the last his magnus opus on a subject which has been with him for 40 years, and even papers on antique furniture and the routes to Hell. What

* Note added, March 2003. This statement needs some clarification: The context should be restricted to phenomenological thermodynamics, in which heat and work are the primitive elements, and where also there are given differential forms defining the heat and work associated to processes in question. Beyond this, there are systems where one may credibly define energy and entropy, though such definitions may restrict the set of available processes of the systems.

more proof is needed for the converse of the famous epigram of Leonardo da Vinci: "Quiet water becomes stagnant. Iron rusts from disuse. So doth inactivity sap the vigor of the mind."

Clifford Truesdell inspired a generation to study Natural Philosophy, and persuaded the learned that Athena is not dead. He has repeatedly and deservedly been honored for his many accomplishments. There is the Bingham Medal of the Society of Rheology which he received in 1963, the Modesto Panetti Gold Medal and Prize from the Turin Academy of Sciences in 1967, and the George David Birkhoff Award for Applied Mathematics from the American Mathematical Society in 1978, as well as Foreign Membership in the Accademia Nazionale dei Lincei. I dare say that the Society for Natural Philosophy as well would wish to present formally to him its highest constitutional award: the great regard in which he is held, the high esteem he is owed, and our continuing best wishes for the furtherance of the scientific program in rational mechanics which he initiated at the beginning of his career.

*The following additional words were appended in May, 2003, drawn from a lecture given in November, 2000, at Pisa at the meeting in memory of Clifford Truesdell.*

Generous in praise, criticism, judgement and friendship, none bestowed lightly: ordinary words do not suffice for Clifford Truesdell. He had great vision accompanied by exceptional accomplishments, at the same time supremely gifted, prodigiously learned and with almost overwhelming energy. He lived life not just to the full, but to overflowing in every activity he undertook: his prodigal lifestyle, his many volumes of correspondence, his historical research and essays, his many books, both scientific and philosophical, and his exceptional editorial activities.

He was lavish in his help for young researchers and was happy with the success of those who followed in the directions he pioneered. He chose his friends well, and from the group around him and around the Archive some of the greatest work of mechanics and thermomechanics in the last century resulted.

Theodore Roosevelt's old fashioned words apply still to Clifford and to his goals, "Far better it is to dare mighty things, to win glorious triumphs, even though checkered by failure, than to take rank with these poor spirits who neither enjoy nor suffer much, because they live in the gray twilight that knows neither victory nor defeat."

All the same, it is necessary to add that Clifford was not without the faults of the gods – perhaps he was aware of this – even allowing this aspect of his character to be a full part of his personality. Clifford could become upset on occasion, not always fairly; there were exasperating times when he became all too human.

Let me mention one moment engraved on my memory. On one occasion when my wife Barbara and I were visiting Clifford and Charlotte, after the dinner, while he and I were standing at the top of the magnificent divided stairway at the Palazzetto, the subject of absolute temperature in classical thermodynamics came up. After all at that time we were both violently infected by the thermodynamic plague. He would not accept my view of the "hotness manifold", which perhaps unduly boldly I had ventured to discuss. While always expressed with politeness, the differing views clearly upset him – he was finally led to utter the ultimate put down: "The problem with you, dear James, is that you never understood Bernard Coleman." Of course, that is no doubt true, . . . .

Let me add something about the Archive (Archive for Rational Mechanics and Analysis). I will ramble a bit, because the main aspect of the Archive – what it is and what it has meant is familiar.

In the years between 1946 to 1960 there was a wonderful period of optimistic activity in the United States, based on the successful ending of the Great War against Germany and Japan, and also on the need to supply many of the requirements of life which had not been met since the beginning of the Great Depression in 1930.

Thus a period of optimism was just beginning and American mathematics was coming to maturity. Before that time, American higher mathematics had been concentrated almost entirely in the universities of the East Coast – Harvard, MIT, Yale, Columbia and Princeton; and otherwise only at the University of Chicago and at Berkeley. There had been only 500 members of the American Mathematical Society, when now there are 10,000. A year's collection of mathematical review was less than half what we now receive in one month. This small world was to change dramatically after the war, with the coming of European mathematicians, mostly German, who had escaped from Hitler's tyranny or from the devastation of Europe. They were welcomed in America, to aid in the war as scientists, or as part of the generosity towards the world which Americans in those days so strongly showed.

It was my good fortune to go to Indiana University at just this time, to study with a newly formed but impressive faculty, including among others Max Zorn, Eberhard Hopf and David Gilbarg. Clifford Truesdell arrived in Bloomington in 1950, one year before I left, brought there to solidify the study of continuum mechanics. The students were completely in awe of him, and indeed his lectures were amazing tours de force, which those who knew him later can easily imagine. That was the setting.

Now let me turn specifically to the Archive, at first in Clifford's words.

"A mathematical journal of a different kind had to arise. It was only a question of persons, place, and time. In those circumstances, T.Y. Thomas, the Head of the Department, asked me to join him in founding a journal to serve the then growing fields of mathematical continuum mechanics and the analysis of nonlinear partial differential equations."

"Beyond the scope and editorial policy, there was some discussion about the unusual title 'Rational Mechanics', but in the end it was adopted because Newton had introduced it in his Principia and had not only exemplified it but defined it."

Even the form of the cover of the new journal occupied his mind – for what it is worth, my small contribution in this direction was to choose the color, red instead of blue, to be used as the contrasting ink on the cover. Thus the JOURNAL of Rational Mechanics and Analysis was born, complete with Latin inscription.

Number 1 of Volume 1 of the Journal of Rational Mechanics and Analysis appeared in January, 1952.

Clifford left Indiana in 1957 for Johns Hopkins, and the old Journal became the new Journal of Mathematics and Mechanics (since that time renamed again as the Indiana University Mathematics Journal). At the same time, a new periodical emerged from the ashes, the ARCHIVE for Rational Mechanics and Analysis. This time by the way, there was no problem in choosing a color for the cover, the publisher Springer-Verlag being inextricably committed to yellow, and yellow only. Both journals have of course flourished and are still here, a tribute to the far-seeing eye and organizational ability of Truesdell.

The early years of the Archive saw a great revival of mechanics as a rational doctrine, with the contributions of Stuart Antman, Bernard Coleman, Jerald Ericksen, Roger Fosdick, Morton Gurtin, Daniel Joseph, Victor Mizel, Walter Noll, David Owen, Richard Toupin, as well as the celebrated analysts, Antonio Ambrosetti, Haïm Brezis, Constantine Dafermos, P.L. Lions, J.B. McLeod, Paul Rabinowitz, and others. The Archive had become necessary for every fine scientific library.

It was thus fitting and natural that when Clifford Truesdell retired from the editorship in 1989, Stuart Antman was chosen to follow in the same tradition. The Archive requires special qualities of mind of its main editors – taste in subject and style, judgement of scientific merit, dedication to an ideal, sheer plodding effort – together with the ability to see a union between mathematical analysis and the physical or rational world. These are talents which Stu had in abundance – and, during the time I was co-editor of the Archive, talents which I could draw on from Clifford Truesdell when required.

The focus of the Archive has turned gradually in recent years, in part due to changed research directions in rational mechanics and analysis, a change which affected both contributors and Members of the Editorial Board as well as the response of the chief Editors.

Clifford viewed mathematics and mechanics as a type of art, as part of our living culture. Here are the words of a famous Finnish composer – which seem equally relevant to the life of mathematics:

"It is my belief that art is great if, at some moment, it catches 'a glimpse of eternity through the window of time' – if the experience is one which we might call 'the oceanic feeling'. This to my mind, is the only true justification for all art. All else is of secondary importance." (Einojuhani Rautavaara)

In: C. Vinti (ed.): *Nonlinear Analysis and Optimization (Bologna, 1982).*
Lecture Notes in Math. 1107 (1984), pp 19–27.




# APPLIED MATHEMATICS AND SCIENTIFIC THOUGHT

J. Serrin

University of Minnesota

The mathematical work of Lamberto Cesari has always had a strong element of concrete reality running throught it, from his earliest work on surface theory to his study of the asymptotic behavior of solutions of ordinary differential equations to his modern research on control theory. In this note I wish to pay tribute to this side of Professor Cesari's career, by setting down several thoughts on the nature of mathematics as it is related to the study of the physical world. At the same time I would particularly offer encouragement to those who see the application of mathematics as a central part of our scientific culture, holding the same possibility of achievement in the future that it has always held in the past.

These remarks will display certain biases, since the time is short and the subject immense. But in addition to noting that personal experience and prejudice cannot be avoided when discussing mathematical applications, let me add one overarching principle: "applied" mathematics should be "good" mathematics, and should be marked by the same clarity which all mathematicians necessarily strive for.

Necessarily, applied mathematics cannot be separated from the physical sciences since the goals of both are essentially similar,but at the same time applied mathematics has a distinguished and special role to play because of its logical imperatives. These work in two principal directions, which are quite distinct from each other. The first involves <u>the reductions of various aspects of the physical world to mathematical formalism</u>. The catch word to describe this direction of inquiry is "modelling", and certainly this terminology is currently fashionable enough to serve as a first approximation.

Nevertheless it carries a certain pedestrian tone, subtly lowering the intellectual level. Indeed, would one describe Newton's laws as mere modelling, or Euler's and Cauchy's discovery of the field theories of deformable media, or the invention of non-Euclidean geometries, or the theory of special relativity? These discoveries rather are paramount applications of mathematical thought, turned onto physical problems. They should serve continually to remind us of the power of disciplined logic in formulating paradigms for the world, and they clearly form a central part of our mathematical heritage. Accordingly I am not wholly in accord with the phrase "model-building", even though I have no simple alternative and will even use it occasionally in what follows. But its casual use, without qualification, can only detract from the position which mathematics should occupy in the physical sciences.

In addition to the development of major concepts, this first direction of inquiry proceeds also through axiomatization of existing theories, the purpose being to clarify and simplify their underlying concepts. One frequently hears that the pursuit of axiomatization is "intellectually sterile" and uninteresting. Whatever may be meant by this, surely it cannot have mathematical content, since mathematics itself is based on coherence, rigor, and care. Indeed, to the extent the mathematical community agrees with critics of axiomatization it weakens its case for centrality in all directions. It is the clarity of thought involved in axiomatization which is the essence; and surely thoroughness and rigour are fundamental not only in pure mathematics but also in the foundations of science itself. Without a clear statement of fundamental principles, it is hardly possible to define alternate positions or to determine where some particular theory requires modifications. Indeed it is precisely the clarity of Newtonian mechanics which presaged both its triumphs and its eventual relativistic modifications.

There is still a third focus of inquiry within the broad content of "modelling", namely the specialization and refinement of existing theories, that is, research directed not at the primary level of raw physical experience but at organizing an already available structure for some new application. Such a process occurs for example when one studies boundary layer theory as a substructure within viscous flow theory, itself a substructure within continuum mechanics. Subsuming all cases, the result of the model-building endeavor is a logically articulated structure of justifiable beliefs about nature.

A second general focus of applied mathematical work, and one no less significant than the first, is the creation and the use of abstract mathematics to treat the analytical problems which arise from the modelling of physical structure. The temptation to label this aspect of applied mathematical thought as "routine problem solving" must be resisted vigorously. Naturally, once this is said it becomes obvious, but it is surprisingly common as an immediate reaction. The antidote is a moment's reflection on the origin of the ideas which are today the heritage of pure mathematics. One might conceivably enumerate and catalogue thos ideas whose genesis lay in some application, but discretion forbids and no doubt space precludes. To begin such a list would be too easy; to end it, too difficult. It is the glory of our subject and the justification for our endeavors. And, I cannot help adding, this aspect of applied mathematics clearly shows how fruitless it must ever be to separate "pure" and "applied" research: even if the extremes are distinct the spectrum includes all shades,as is clearly evident in the work of Lamberto Cesari. Obvious as the point is, I will have to return to it again.

Simply pointing out the contributions to mathematics which have applied mathematical origins is only half the story, however. There is an impressive list of achievement also in the problems which have successfully been treated. From the early developments of celestial mechanics, conformal mapping, and calculus of variations to the more recent focus on control theory, quasi-variational inequalities, singular perturbation problems, regularity of solutions of partial differential equations, scattering theory, Fourier analysis, bifurcation theory, turbulence, the criteria for well-posed problems, and phase transition theory, to form only the most minimal list, the quality and breadth of applied mathematical research becomes evident.

To study a physical problem with the help of mathematics requires critical reflection, an unwillingness to acquiesce in logically imprecise development. Even with purposeful direction, views may change due to accumulated experience, as we criticize, discard, or refine what we have developed in order better to tell the story of reality. Overall, there is simply more to it than guessing that A→B and then proving the conjecture: one must also realize the formulation of A itself, entailing thereby subtle questions of proper concepts and definitions.

In this regard, it is remarkable that the sole discipline of classical physics where serious argument about fundamentals still remains is that of thermodynamics. Even though entropy and energy are now granted almost all pervasive influence, it is nearly impossible to obtain agreement on the two laws of thermodynamics from which these concepts derive. Truesdell considered this a subject which cannot be understood as it is taught, Kac called it a "dehydrated elephant", and Bridgman wrote that thermodynamics is "a branch of physics different from all others, having a palpably verbal feel". Here is a subject essentially untouched by mathematical structure, and the lack of precision shows throughout: after 150 years of development it stands in undeniable contrast to other classical fields of mathematical physics, on whose foundations rested the hands of great mathematicians like Newton, Euler, Cauchy, Hamilton, Maxwell and Hilbert. (The work of Carathéodory in thermodynamics, for all its formal and abstract elegance, does not constitute a complete structure for the subject, nor I think would Carathéodory ever have so claimed). While it is no doubt too soon to claim a realistic part for mathematics in delineating the logical structure of classical thermodynamics - it is conceivable for example that the subject is simply not amenable to the logic which pervades other fields of physics and that the failure to agree is a consequence simply of the complications of material structure - there have nevertheless been recent attempts to formulate the principles of thermodynamic structure in a precise way. Some degree of success has been attained, including at least the understanding of the role of joint systems for the concepts of absolute temperature and the mechanical equivalent of heat, as well as in axiomatizing the notion of reversible and quasi-static processes, and in formulating the energy principles of classical continuum mechanics. These advances not only point to the need for further mathematical study, but show again the important role which applied mathematics can play in the foundations of physics.

I have emphasized the crucial and catalytic nature of applied mathematical thought, but for all this the broad public (and even many scientists) knows little of this role and imagines that the major contribution of mathematics to modern life lies in the ability to perform rapid calculation by computer. For this reason it seems particularly important today that mathematicians make a strong case for the relevance of their subject - in both its pure and applied aspects. I am not thinking only in political terms, though it does appear that eschewing the political arena will court serious losses later. Rather it appears that we should examine the philosophical and empirical roots of our subject as evidence for the centrality which I noted earlier. As a case in point, the very success of mathematical thought in explaining the physical world has prompted several writers, notably E. Wigner and P.C. Hamming to examine the reason for what they consider an unreasonable degree of success. It almost goes without saying, in spite of the subtle arguments advanced, that it will be impossible ever to encompass this metaphysical question. On the other hand, one may argue differently. We need not ask why logic is sufficient for physical understanding, but rather agree that progress of any kind is difficult if not impossible without a logical approach: in other words the questions asked by physics require mathematics as a necessary handmaiden of the advances which are achieved.

Whenever great mathematicians faced the problems of physics they kept the experience of nature before them, though always with the goal of rigorous understanding. Daniel Bernoulli in the eighteenth century stated that "to get profit from physical knowledge it is absolutely necessary to be a mathematician", and Euler has written in his paper on the foundations of fluid mechanics that the results he was uncovering "reveal to us the true laws of nature in all their brilliance". In summary, mathematics can be both necessary and sufficient, bringing order, elegance and beauty to parts of science which otherwise can seem complex, disjoint and confining. And surely it goes without saying that the development of mathematical coherence cannot rely solely on collections of fact, but rather on insight, experience, and the development of appropriate abstraction.

The examples I have noted above, and others equally well-known, should make the point that logical consistency and accuracy are marks equally of pure and applied mathematics, that it is not merely expedient ingenuity which alone holds power. Nevertheless, the impression is widespread that the mathematician, whether in his research or in his teaching, will accept without question a given set of equations describing a physical phenomenon without in the slightest inquiring into their meaning and derivation. On the contrary, he should always clarify for himself, and all the more for his students, the purpose and reality of the model he is studying. In this regard, almost at random I found in Lamberto Cesari's treatise on asymptotic behavior a nice phrase clarifying the physical meaning of linear modelling, and indicative of the sort of things teachers should say to their students: "In a deep sense, all mechanical and physical systems are nonlinear, but they are more and more similar to linear systems the smaller are their displacements from their positions of equilibrium. As soon as amplitudes increase systems become essentially nonlinear". Moreover, he goes on to say, "In a wide range of applications it is in fact only asymptotic stability which is of interest. For plain stability is very labile, and may change into instability by the slightest variation of the physical system, of which the differential problem under discussion is, often, only an approximate representation".

That mathematicians should not be separated from the physical world by artificially erected barriers of "pure" and "applied" might seem a truism. Even so, especially in the last 40-50 years, there has been a tendency to claim particular virtue for "purity" in mathematics, and to hold that research is strongest and best when it is most abstract. This belief has its mirror image in the scientist's view that mathematicians themselves are superfluous, even as he continues to use mathematics as a "tool". Though a scientist or an engineer may personally enjoy pursuing some aspects of mathematical culture, the deductive "proof" method may still seem to him of value mainly for cosmetic purposes, adding little of lasting depth. An interesting passage in a recent paper of William Pohl underlines both the problem and the more significant reality. He writes :

> "The physicist C.N. Yang, in a recent lecture, expressed his surprise that the notion of a fibre bundle, which the quantum field theorists had discovered in a primitive form and with which they had dealt with such great difficulty - it is basic to gauge theories, and hence to the new theories of the electromagnetic force and the weak force - had been discovered and developed previously by the mathematicians. How could mathematicians, who have nothing to do with experiments, discover something about the world? This cannot be explained by chance, since such events have happened often in the history of science. Nor can it be explained according to the view of mathematics held by many contemporary philosophers that mathematics is identical with logic and consists of all hypothetical statements, since of the infinity of possible hypothetical statements, mathematicians have chosen those which are later seen to apply to the world. Rather, it can only be explained, it seems to me, on the assumption that mathematics studies the world. It is true that we look at the world at a higher level of abstraction than other natural scientists and we study various possible worlds; but our possible worlds, sooner or later, seem to reflect aspects of the actual world. Often, of course, we look at aspects of the world discovered by other natural sciences, so that the other sciences enrich and aid mathematics, just as mathematics, in turn, provides the basis of the other sciences."

26

The crucial point is the unity of mathematical thought, which requires not only an appreciation of the power of abstraction but equally an understanding of the relation of mathematical thought and the physical world.

In concluding these brief remarks, I can do no better than a quotation from Hilbert's famous list of problems for the twentieth century. "While I insist upon rigor in proofs as a requirement for a perfect solution of a problem, I should like, on the other hand, to oppose the opinion that only the concepts of analysis, or even those of arithmetic alone, are susceptible of a fully rigorous treatment. This opinion .....I consider entirely mistaken. Such a one-sided interpretation of the requirement of rigor would soon lead us to ignore all concepts that derive from geometry, mechanics, and physics, to shut off the flow of new material from the outside world... What an important, vital nerve would be cut, were we to root out geometry and mathematical physics!".

In addition Hilbert pointedly discusses the role of axioms in model-building, in a passage which deserves more attention: "...whenever mathematical ideas come up, whether from the theory of knowledge or in geometry, or from the theories of natural science, the task is set for mathematics to investigate the principles underlying these ideas and establish them upon a simple and complete system of axioms in such a way that in exactness and in application to proof the new ideas shall be no whit inferior to the old arithmetical concepts". These words have had more impact than I think is credited to them by pbysicists.

For our part, if we stay alive to the sources of new concepts, if we continue to be interested in the genuine application of genuine mathematical ideas to significant problems, and if we maintain an environment in which interchange of ideas is possible, then we can expect others to remain interested in what we have to say.

Bibliography

[1] KAC M., Birkhoff Prize Talk, 1978. Mathematical Intelligencer, I, 1978, pp. 97-98.

[2] BONSALL F.F., "A down-to-earth view of mathematics", American Mathematical Monthly, 89, 1982, pp. 8-14.

[3] BROWDER F., Does pure mathematics have a relation to the sciences? (unpublished lecture), pp. 1-29.

[4] CESARI L., Asymptotic behavior and stability problems in ordinary differential equations. Ergebnisse Math. No. 16, Springer-Verlag, Berlin, 1959.

[5] HAMMING R.W., The unreasonable effectiveness of mathematics, American Mathematical Monthly, 87, 1980, pp. 81-90.

[6] JAFFE A., The importance of mathematics, unpublished,1983,pp.1-47.

[7] MURRAY F.J., "Applied Mathematics, An Intellectual Orientation", Plenum Press, New-York-London, 1978. (cf. also Everett Hafner, Bulletin American Mathematical Society, 3, 1980, pp. 735-739.

[8] POHL W., Applications of global differential geometry, Rheinisch-Westfälische Akademie der Wissenschaften, N. 308, 1982, pp. 7-24.

[9] SERRIN J., Conceptual analysis of the classical second laws of thermodynamics, Archive for Rational Mechanics and Analysis, 70, 1979, pp. 355-371.

[10] TRUESDELL C., Is there a philosophy of science?, Centaurus, 17, 1973, pp. 147-161.

[11] TRUESDELL C., The role of mathematics in science as exemplified by the work of the Bernoullis and Euler, Verhandl. Naturf. Ger. Basel, 91, 1980.

[12] TSILIKIS J.D., Simplicity and elegance in theoretical physics, American Scientist, 47, 1959, pp. 87-96.

[13] WIGNER E.P., The unreasonable effectiveness of mathematics in the natural sciences, Comm. Pure Appl. Math., 13, 1960.

In: Giuseppe Buttazzo et al. (eds.): *Nonlinear Analysis and Continuum Mechanics (Ferrara, 1992),* (1998), pp 139–144.


**Reprint VIII.[v]**

# SPACE, TIME AND ENERGY

JAMES SERRIN

A not inaccurate paraphrase of modern cosmological opinion is that our universe initiated in, and expanded from, an incendiary moment some twenty billion years ago, and is governed by the resulting space-time structure and the induced curvature of space, as seen through the eyes of Albert Einstein. Moreover it is held that Isaac Newton's classical world view–though complete and perfect, arguably the pinnacle of man's intellectual achievement–nevertheless does not meet the requirements of modern cosmology and can be confidently discarded as a guide.

These views are persuasively presented in numberless popular expositions of science, a recent and notable example being the best-selling book "A Brief History of Time"; its catchy and marvelously paradoxical title says everything: it is an invitation to the painless acquisition of wisdom. I would, however, advance a contrary case, that the last cosmological word is not in, and may never be delivered to us. If scholarship and ability are not snuffed out by ideology and conformity, then the mythologies of our time should have no better chance of survival than those of earlier days.

In a letter to his patron Chantelou, the artist Nicholas Poussin wrote in 1642:

"Things which partake of perfection should not be looked at in haste, but call for time, judgment and intelligence. The means employed in their appraisal must be the same as those used in their making."

These are remarkable words, their originality no doubt even more striking today then when they were first written. They supply the motif for this lecture, my thesis being that we can indeed look with deeper understanding at Newtonian mechanics, that a sharper perception of its structure emerges when we take into account its underlying relation with the geometry of space and time.

I suppose no apology need be given for engaging your interest in such classical matters, especially in a University where Copernicus and Galileo are honored. The science of our time, just as the science of past centuries, depends crucially and precisely on the mechanics of Newton. It forms the basis for our technological society, for the curriculum in every engineering school in the world, for the successful exploration of space. Dare I say that every gain which we achieve in understanding the Newtonian world should be reflected later in the understanding of the vaster cosmological universe.

Newton's ideas and his life work, the *Philosophiae Naturalis Principia Mathematica*, originally captured the public imagination with its revelation, explanation and prediction of the phenomena of celestial mechanics. Our science and our world has not since been the same. At the heart of the issue is the power that arises from the interrelation between acceleration, as a second derivative of position, and the notion of universal forces.

I have introduced the three threads of the lectures space, time and energy. How are they to be tied together, placed in a mathematical context so that their interrelation can teach us something new about the world? Here we tread on difficult, though hallowed ground. The great physicists of the past already treated these granitic concepts in their struggle to understand the physical world: Newton, and before him Galileo; Leibniz; Sadi Carnot, Joule, Lord Kelvin and Rudolph Clausius–the founders of thermodynamics; George Gabriel Stokes–who saw at least dimly the central role of geometry in mechanics; Leonhard Euler and Augustin Louis Cauchy, who found the key ideas of internally organized forces.

The public barely knows these pioneers of physical science, yet they are the peers of Palestrina, Bach, Mozart, Handel and Verdi in the musical world. The musical analogy is apt. What we may add to the original discoveries of natural philosophy cannot vie with the original conceptions, but as Berlioz, Brahms, and Respighi contributed still to music's power to move us, so we may hope to supply something new as well.

Returning to the threads of space, time and energy, let me explain as simply as I can, how these threads are already woven into mechanics. For a moment, let us go back to Galileo. Galileo lived, as he thought, in a Euclidean world; he saw time as a uniform stream divided only by the markings of simple clocks, and space is Euclidean. We would say that his world is a classical space-time manifold of 4 dimensions, having associated to each instant of time a 3 dimensional Euclidean spatial simultaneity sheet.

How do these simultaneity sheets join together as time moves forward? That is, if we imagine that each observer of the world sees himself in some preferred Cartesian frame in the manifold, and watches events transpire in that frame, how are the frames (charts) of different observers related? For Galileo they simply stand in uniform motion; in any simultaneity sheet they move relative to each other, with a uniform and fixed translational velocity.

Galileo's structure cannot, however, satisfy the needs of Newtonian mechanics, nor the philosophical penetration of Liebniz. For them, relatively rotating and moving frames are surely required–the rotating earth we live on, or, for that matter, satellite technology, immediately forces this issue on us. More fundamentally, Newton's work leads to a coherent unification of mechanical and inertial effects, and externally operating forces. For this purpose, Newton had a felt need for one ultimately preferred frame, the absolute fixed space of God, an unmoving so-called inertial frame set into the distant fixed stars and in which all forces arise by material interaction. Leibniz saw the hand of God differently–that each frame, in the family of all possible relatively rotating and moving frames, was the equal of any other. Thus, in accordance with his doctrine of monads, there could not be individually preferred frames, and a fortiori no special inertial frame. The idea of the "fixed stars" was simply not relevant.

Historically, Leibniz lost the battle, partly through a change in philosophical focus, partly because mechanics became ever more the domain of the scientist rather than the natural philosopher. There is more to say, but first let me turn to the third thread, energy.

This is woven into the tapestry of mechanics through the First Law of Thermodynamics. Let us imagine a given material object, or physical system, which we shall call $S$, and focus our interest on it. Associated to $S$ is its *total energy* - whatever else, a scalar time dependent attribute of

$\mathcal{S}$, calculated in terms of the state variables which delineate $\mathcal{S}$. That it is not an easy idea should be manifest from the fact that the concept did not occur in science for almost 200 years following Newton's initial discoveries. The physical chemist Peter Atkins has commented that:

"The Second Law of Thermodynamics [is] in my opinion much easier to understand than the [law of] conservation of energy. Energy is a word so familiar to us today that we can hardly grasp either the intellectual Everest it represents or the conceptual difficulty we face in saying exactly what it is."

We may recall C.P. Snow's famous vignette, making the point that educated people should be able to discuss the second law of thermodynamics as intelligently as they would a work of Shakespeare. While I have grave doubts that what passes for an education today would equip anyone to discuss sensibly either the second law or Shakespeare, it does seem that Lord Snow might have added the concept of energy to his list.

The First Law grants the primacy of energy, and states that the time rate of change of the total energy of a system is caused by, or more technically, balanced by the *rate of working* of exterior forces on system together with the *rate of heat addition* to the system, i.e. in my only equation

$$dE/dt = W + Q.$$

This relation, originally discovered in the mid nineteenth century for quite restricted circumstances, was given broader application by C. Neumann in 1894, and finally codified in the encyclopedic and celebrated treatise on the Mechanics of Continua by Clifford Truesdell and Richard Toupin. Along with the momentum balance equations of Newton, Euler and Cauchy, it constitutes the heart of classical mechanics.

It is well-known to every student of natural philosophy, though not particularly emphasized in textbooks, that this structure exhibits a remarkable degree of internal consistency the more closely it is examined, an issue which has fascinated me ever since my earliest acquaintance with the subject.

The first indication that there may be an underlying structure which gives rise to this consistency came in a short and relatively unnoticed paper of Walter Noll in 1959, in which he showed that an abstract power balance implied an abstract version of the Cauchy equation (with the force and acceleration lumped together as a single quantity). Shortly afterward, this idea was extended by Albert Green and Ronald Rivlin, but then lay dormant until an important paper of Miroslav Silhavy appeared in 1989. Silhavy's main, and major, contribution was to show that *the total energy of a continuum system must split into an internal and a kinetic part*, the demonstration, moreover, being carried out without invoking Cauchy's motion equations. To see how remarkable this is, let me emphasize that in this approach the kinetic energy $|v|^2/2$ arises without any of the usual preparation from dynamics–it cannot be replaced by $|v|^4/4$ , for example–not because this would give some unusual or new kind of dynamics, but rather because classical space-time forbids it! I can recall my astonishment, and disbelief, when Silhavy first told me of this result, while we were walking together on a promenade in Valencia. The threads had come together at last, though I refused to believe it–until his paper appeared.

The explanation for Silhavy's remarkable result, and indeed for Walter Noll's original idea, lies in the invariance of the energy balance equation in the Leibnizian group of frames, that is, the family of all relatively moving Cartesian charts on the simultaneity sheets. When one says that the energy balance equation is invariant, what is meant is that different observers occupying different frames, but each watching a given motion of the system $\mathcal{S}$, each interpret the energy balance equation for the motion of $\mathcal{S}$ in their own frames and carry out *(in their frames)* the calculation for the energy, the work and the heat according to *the same* strictly given formulas. The idea then is to transform each of these interpretations into a common frame, and then, *since all interpretations ultimately represent the same motion,* to glean from this fact further conditions which must be satisfied by the assigned structure. The surprising thing is the really great amount of information which can be obtained this way, by appropriately choosing relations between the frames. To these comments, one must add that the whole program owes its success also to the major contributions in the 1960's of Walter Noll and Bernard Coleman in their study of the effects of frame invariance on the constitutive equations of materials.

A main mathematical and physical difficulty in carrying out Silhavy's program lies in the force structure which is allowed. Since we cannot make circular use of ultimate conclusions in order to obtain those same conclusions, there is no *a priori* reason why the forces in one frame should have any definite relation to those in another, beyond what is determined from the methodology of invariance. Indeed, even in the context of Newtonian mechanics, the transformation rules for forces from one frame to another are quite complicated, involving Coriolis and centrifugal effects, and others even more subtle. From the Leibnizian point of view these forces retain a full reality, even as their form and their chart transformations must be a priori unknown.

To circumvent this difficulty requires a careful examination of the amount of freedom of reaction which a fixed force structure will allow to a physical system; the needed Freedom Axiom must simultaneously be strong enough to lead to Silhavy's splitting theorem, and at the same time one which is satisfied for ordinary dynamical systems.

Leaving aside these technically subtle matters, as we surely must for now, let me pass to further consequences of Silhavy's theorem for the structure of mechanics. In fact, once it has been proved that the total energy splits into kinetic and internal parts, it is well-known that the primary equations a momentum balance, the Cauchy equations themselves, are a consequence of, and are equivalent to, the group invariance of the energy balance equation. This is no trivial exercise: nonetheless, after the splitting theorem is proved there is a sufficient mathematical structure which one can exploit so that the equivalence becomes a straightforward matter.

In conclusion, in brief summary, the invariance of energy balance within the Leibnizian group of frames in classical space-time contains within it the *necessity* that the total energy of a physical system (of the type studied in continuum mechanics) must split into the sum of a kinetic energy and an internal energy; the coefficient of the squared velocity in the kinetic energy can be interpreted as the mass density of the system; and the Cauchy momentum balance equations are derived consequences of the structure.

Where does this leave the disagreement between Newton and Leibniz about the place of inertial frames in the foundations of mechanics? From what we have seen, certainly inertial frames are not *needed* for mechanics:

everything in the above argument and in the final, striking conclusion that the Cauchy equations must hold because of space-time invariance, can be carried out without the intervention of an inertial frame. Invoking Occam's Razor as a philosophical tool, we shall conclude in favor of Leibniz: what is not needed should not be presumed.

There is concealed in this summary an almost microscopic qualification, a parenthetical comment added almost in passing: the discussion applies to physical systems "of the type studied in continuum mechanics" and, it almost goes without saying, "to the systems of analytical dynamics". Is this qualification only a happenstance of the methodology? Are the Cauchy equations derivable from space, time and energy considerations for every possible physical system? It would seem almost beyond conception that this could be true–the energy structures of general systems can be very intricate and indeed may even be unknowable–for instance, can one assign such a structure to the incredible complications of living things? But if this is the case, what then should we say about Newtonian mechanics as a general principle for non-relativistic experience? I cannot dare to suggest an answer, but if Silhavy's splitting theorem is not a universal attribute of materials then surely the world can turn out to be far more complicated even at the non-relativistic level than we might presently imagine.

I have recently taken up with Professor Lawrence Markus the cosmologically complementary question of classifying 4 dimensional space-time manifolds which are locally, though not globally, Leibnizian. This is motivated by the observation that if space-time is locally Leibnizian then the local equations of motion still remain those of Newton, Cauchy and Euler–that is, local mechanics is entirely unchanged. It is thus conceivable that what occurs at cosmological distances is influenced not only by mechanical principles but also by the fact that local Leibnizian charts may have no global extensions.

We are still left with an unexplained and inexplicable fact. If in practice the solar system apparently floats in a genuine inertial frame, why is it that this inertial frame is anchored in the fixed stars, which after all are not really fixed in the first place? Moreover, the vanishing force structure which characterizes our local inertial frame may not have a global inertial extension. Indeed, if an inertial extension were not available, or if it did not exist, then there would arise the possibility that the cosmological data could be interpreted in alternative and different ways, depending on whatever actual extension might be chosen. Due to the high level of speculation already present in the subject, it might not be too much to add that even further speculation is supportable. It is certainly not my intention to expand further on these elusive ideas, but let me at least add that the nature of any postulated space-time structure, whether it be Newtonian or relativistic, could deeply influence cosmological speculation. For me the hand of God continues to move mysteriously, and the paradigms of modern science are not yet impressed on the tablets of eternity. It is left to reiterate that the more closely one probes the classical mechanics of Newton, the more remarkable is the vista which opens to us; the questions which remain should all the more make us beware of mechanical and cosmological certitude.

[An earlier version of the lecture, in Italian, was published by the Università degli Studi di Ferrara in 1995.]

In: *L'Anno Galileiano,* Vol. I, Edizioni Lint, Trieste, 1995, 120–124. **Reprint VIII.[vi]**


JAMES SERRIN

# REFLECTIONS ON MATHEMATICS

On the occasion of this Galileo celebration year in Padova a mathematician is tempted to reflect upon the changing role of his subject from Galileo's time until the present. Galileo died the year Isaac Newton was born, in a sense at the very moment modern mathematics started. Before this cultural divide, and with only small exceptions, the science of mathematics was intimately connected with axiomatic geometry, and with practical questions arising in measurement, navigation, astronomy and elementary mechanics. Newton of course followed directly in this path, building on the work of Copernicus, Galileo and Kepler, but he also turned of necessity to subtle questions of calculus, approximation theory, and differential equations. In this, he was both precursor and catalyst for the great changes introduced into mathematics at the beginning of the Eighteenth Century.

Mathematical investigation at that time rapidly broadened in scope, not only with the important contributions of the Bernoullis, but primarily under the influence and stimulation of the unparalleled work of Leonhard Euler in Basel and St. Petersburg. Euler wrote seriously in nearly every field of mathematics then known, as well as initiating the study of modern continuum mechanics, topology and the calculus of variations. His work included, in the best sense, both pure and applied mathematics, and his name is everywhere remembered for innumerable fundamental contributions.

With mathematics thus having emerged on its own merits, and with its tremendous expansion in range and depth throughout the nineteenth and twentieth centuries, we may try to see how it has come to eminence as queen of sciences, and to see its place in general culture — even as it has become ever more abstract and beyond the reach of the ordinarily educated person.

The most immediate response to the question, what is the real position of mathematics in the world, is that it is a primary foundation for physical science. Yet this answer is inadequate or even misleading. Indeed, physics today, though using a range of mathematical tools, does not particularly share the goals of mathematics — on the other hand, the logical imperatives of mathematics have become central to the world in quite a different direction — through the revolution wrought by the modern computer. This revolution is not simply one in which we can accomplish more and more quickly various scientific calculations, but, much more fundamentally, one in which the rapid storage, transmission and retrieval of information changes the very way in which we live, so far apparently for the good of mankind, though like all technologies having also the potential for unscrupulous use in the wrong hands.

But beyond these practical uses of mathematical science, and even the great range of industrial and engineering problems which are now treated by concrete applications of mathematics, and which will be increasingly vital to our society in the future, there are other issues which play a major role in the way in which mathematicians conceive their science.

These directions, it seems to me, arise directly from the same cultural, intrinsically humanistic drives found in other avenues of artistic or intellectual life: precision and economy of thought; a desire for beauty, for clarity, for certainty; a pleasure in creativity and elegance; and an intensely human satisfaction, which can hardly be motivated from outside ourselves, in logical and unassailable argument. In fact, just as the ultimate philosophical foundations of mathematical logic remains elusive, mathematics may well be, in the words of Herman Weyl, «like language or music, of primary originality, defy[ing] complete objective rationalization».

If we can offer an «apology» for mathematics today, for its persistence in scientific life, and its flourishing study in universities throughout the world, it must certainly be in all of the above terms. I emphasize this all the more strongly because the professional mathematician is frequently quite divorced from particular uses of his work. He may even take pride, as did the celebrated English mathematician G.H. Hardy, in the non-applicability of his research, and write and speak in a vocabulary which can easily be unintelligible outside of his own small research circle.

Perhaps this is farther from the world of Galileo, Newton and Euler than one would like to admit; nevertheless some fraction of modern discoveries will retain eternal permanence. This permanence must be viewed exactly as the permanence of great works of musical composition and pictorial art, of which there are such wonderful examples in Padova. Mathematicians naturally pay a price for the abstract nature of their thought — mathematics is indeed a quiet, solitary subject, for the most part without those aspects of «popularity» associated with medicine, physics, astronomy and cosmology, or engineering. The rewards of beauty, elegance, certainty and permanence, and the pleasure of creativity are the other side of the coin. Newton, Leibniz, Euler, Lagrange, Gauss, Riemann, Poincaré and Hilbert are the mathematical peers of Palestrina, Scarlatti, Bach, Mozart and Beethoven; even though their names and accomplishments are barely known to the general public, their discoveries will live for as long as men value mathematical thought.

I have so far avoided any personal comment, feeling honored enough to play a part in this 400th anniversary of Galileo's life and discoveries in Padova that I find no need to add words about my own work. But, to give only the briefest indication, I would find my mathematical home in the calculus of variations, in the mechanics of deformable bodies, in the science of heat, in unraveling various mysteries in the purely abstract study of differential equations, and in recent collaborative work on the stability of oscillators. These are subjects with a history stretching back to the Seventeenth and Eighteenth Centuries — and thus intellectually not far removed even from Galileo. If I have any apology at all, it is that I have pursued mathematics from a determinedly classical point of view, though I should add that this has not been without the excitement of theoretical discovery nor without the belief that I have contributed something of lasting permanence to the theory of heat and energy, by clarifying the mathematical principles behind the first and second laws of thermodynamics.

From my point of view, in looking to the future, there is a real danger to be found in the baroque refinement of structure for its own sake; and in an exploding diversity of fields of study, which leads one of necessity to pursue an ever smaller fraction of the totality of mathematics. I would also condemn a growing anti-intellectualism in the world, which threatens the suppression of our transcendent cultural heritage at the hands of cant and ideology, and brings in its train both divisiveness and violence. Equally disturbing are the modern religions of utopianism and populism, whether of the right or of the left, whether isolationist or separatist, whether nationalistic or fundamentalist, since all of these movements exacerbate political instability, this being above all the most serious threat to the world today — though it is closely followed by the grave problem of rapid population increase, which provides ever ready fuel for mass conflagration.

But setting these dark thoughts aside, mathematics still presents a marvellous vista. We have lived in a golden age of discovery, achievement and insights — and this applies with full justice even in the classical fields originally developed just after the time of Galileo. The art of mathematics is useful and beautiful beyond any of our dreams and expectations — and it has given many a wonderful life, a hope I would have for all the young people who are now intending to make their careers in mathematics.

# James Serrin's Publications


[**1**] *Free boundaries and jets in the theory of cavitation* (with D. Gilbarg), J. Math. and Physics, **29** (1950), pp. 1–12.

[**2**] *Commentary on Eberhard Hopf's hydrodynamic work*, Math. Nachr., **4** (1950), pp. 213–231.

[3] *Uniqueness theorems for two free boundary problems*, Amer. J. Math., **74** (1952), pp. 492–506.

[4] *Existence theorems for some hydrodynamical free boundary problems*, J. Rational Mech. Anal., **1** (1952), pp. 1–48.

[5] *Two hydrodynamical comparison theorems*, J. Rational Mech. Anal., **1** (1952), pp. 563–572.

[6] *Notes on hydrodynamics*, Photoprinted by Princeton University, 1952, 218 pages.

[7] *On plane and axially symmetric free boundary problems*, J. Rational Mech. Anal., **2** (1953), pp. 563–575.

[**8**] *A note on the wave equation*, Proc. Amer. Math. Soc., **5** (1954), pp. 307–308.

[**9**] *Comparison theorem for subsonic flows*, J. Math. and Physics, **23** (1954), pp. 27–45.

[10] *On the Phragmen–Lindelőf theorem for elliptic partial differential equations*, J. Rational Mech. Anal., **3** (1954), pp. 395–413.

[11] *A uniqueness theorem for the parabolic equation* $u_t = a(x)u_{xx} + b(x)u_x + c(x)u$, Bull. Amer. Math. Soc., **60** (1954), p. 344.

[12] *Uniqueness of axially symmetric subsonic flow past a finite body* (with D. Gilbarg), J. Rational Mech. Anal., **4** (1955), pp. 169–175.

[13] *A characterization of regular boundary points for second order linear differential equations*, Bull. Amer. Math. Soc., **61** (1955), p. 224.

[**14**] *On the Harnack inequality for linear elliptic equations*, J. d'Analyse Math., **4** (1956), pp. 292–308.

[**15**] *On isolated singularities of solutions of second order linear elliptic equations* (with D. Gilbarg), J. d'Analyse Math., **4** (1956), pp. 309–340.

[16] *A note on harmonic functions defined in a half-plane*, Duke Math. J., **24** (1956), pp. 523–526.

[17] *On the Hölder continuity of quasi-conformal and elliptic mappings* (with R. Finn), Trans. Amer. Math. Soc., **89** (1958), pp. 1–15.

[18] *Mathematical principles of classical fluid mechanics* (Monograph), Handbuch der Physik, **VIII/1** (1959), pp. 125–263; Russian translation: Foreign Literature Publ. House, Moscow 1963, 265 pages.

[**19**] *On the stability of viscous fluid motion*, Archive Rational Mech. Anal., **3** (1959), pp. 1–13.



P. Pucci et al. (eds.), *James Serrin. Selected Papers: Volume 2*, Contemporary Mathematicians, DOI 10.1007/978-3-0348-0847-7, © Springer Basel 2014

**[20]** *A note on the existence of periodic solutions of the Navier–Stokes equations*, Archive Rational Mech. Anal., **3** (1959), pp. 120–122.

**[21]** *On the uniqueness of compressible fluid motions*, Archive Rational Mech. Anal., **3** (1959), pp. 271–288.

[22] *On the derivation of stress-deformation relations for a Stokesian fluid*, J. Math. and Mech., **8** (1959), pp. 459–470.

**[23]** *Poiseuille and Couette flow of non-Newtonian fluids*, Zeit. Ang. Math. Mech., **39** (1959), pp. 295–299.

**[24]** *On a fundamental theorem of the calculus of variations*, Acta Math., **102** (1959), pp. 1–22.

[25] *A new definition of the integral for non-parametric problems in the calculus of variations*, Acta Math., **102** (1959), pp. 23–32.

**[26]** *The exterior Dirichlet problem for second order elliptic partial differential equations* (with N. Meyers), J. Math. Mech., **9** (1960), pp. 513–538.

[27] *On the area of curved surfaces*, Amer. Math. Monthly, **68** (1961), pp. 435–440.

**[28]** *On the differentiability of functions of several variables*, Archive Rational Mech. Anal., **7** (1961), pp. 359–372.

**[29]** *On the definition and properties of certain variational integrals*, Trans. Amer. Math. Soc., **101** (1961), pp. 139–167.

**[30]** *On the entropy change through a shock layer* (with Y.C. Whang), J. Aerospace Sciences, **28** (1961), pp. 990–991.

[31] *Dirichlet's principle in the calculus of variations*, Proc. Symposia in Pure Math., vol. 4, Providence, 1961, pp. 17–22.

[32] *Interior estimates for solutions of the Navier–Stokes equations*, in Partial Differential Equations and Continuum Mechanics, edited by R. Langer, University of Wisconsin Press, 1961, pp. 376–378.

**[33]** *On the interior regularity of weak solutions of the Navier–Stokes equations*, Archive Rational Mech. Anal., **9** (1962), pp. 187–195.

**[34]** *Strong convergence in a product space*, Proc. Amer. Math. Soc., **13** (1962), pp. 651–655.

**[35]** *The initial value problem for the Navier–Stokes equations*, in Nonlinear problems, edited by R.E. Langer, University of Wisconsin Press, 1963, pp. 69–98.

[36] *Variational problems of minimal surface type, I* (with H. Jenkins), Archive Rational Mech. Anal., **12** (1963), pp. 185–212. See also entries 51, 60.

[37] *A Harnack inequality for non-linear equations*, Bull. Amer. Math. Soc., **69** (1963), pp. 481–486.

[38] *Comparison and averaging methods in mathematical physics*, in *Proprietà di Media e Teoremi di Confronto in Fisica Matematica*, Centro Internazionale Matematico Estivo, Rome, Edizioni Cremonese, 1965, pp. 1–87.

[39] *A priori estimates for solutions of the minimal surface equation*, Archive Rational Mech. Anal., **14** (1963), pp. 376–383. See also entry 58.

[40] *Mathematical Aspects of Boundary Layer Theory*, notes taken by H.K. Wilson, Department of Mathematics, University of Minnesota, 1963 (131 pages).

**[41]** *Sublinear functions of measures and variational integrals* (with C. Goffman), Duke Math. J., **31** (1964), pp. 159–178.

[**42**] *Local behavior of solutions of quasi-linear equations*, Acta Math., **111** (1964), pp. 247–302.

[**43**] $H = W$ (with Norman Meyers), Proc. Nat. Acad. Sciences, **51** (1964), pp. 1055–1056.

[44] *Removable singularities of solutions of elliptic equations*, Archive Rational Mech. Anal., **17** (1964), pp. 67–78. See also entry 49.

[**45**] *Pathological solutions of elliptic differential equations*, Ann. Sc. Normale Sup. Pisa, Sci. Fis. Math., (3) **18** (1964), pp. 385–387.

[46] *Singularities of solutions of nonlinear equations*, Proc. Symposia in Pure Math., **17**, pp. 68–88, Providence, 1965.

[**47**] *Isolated singularities of solutions of quasi-linear equations*, Acta Math., **113** (1965), pp. 219–240.

[48] *Theory of differentiation*, notes taken by T. Hatcher, Department of Mathematics, University of Minnesota, 1965 (135 pages).

[**49**] *Removable singularities of solutions of elliptic equations, II*, Archive Rational Mech. Anal., **20** (1965), pp. 163–169.

[50] *The Dirichlet problem for the minimal surface equation with infinite data* (with H. Jenkins), Bull. Amer. Math. Soc., **72** (1966), pp. 102–106. See also entry 60.

[51] *Variational problems of minimal surface type, II: Boundary value problems for the minimal surface equation* (with H. Jenkins), Archive Rational Mech. Anal., **21** (1966), pp. 321–342.

[**52**] *Isolated singularities of solutions of linear elliptic equations* (with H. Weinberger), Amer. J. Math., **88** (1966), pp. 258–272.

[**53**] *Local behavior of solutions of quasi-linear parabolic equations* (with D.G. Aronson), Archive Rational Mech. Anal., **25** (1967), pp. 81–122.

[54] *A maximum principle for nonlinear parabolic equations* (with D.G. Aronson), Ann. Sc. Normale Sup. Pisa, Sci. Fis. Mat., **21** (1967), pp. 291–305.

[**55**] *The Dirichlet problem for quasi-linear elliptic equations with many independent variables*, Proc. Nat. Acad. Science, **58** (1967), pp. 1829–1835. See also entry 65.

[**56**] *On the asymptotic behavior of velocity profiles in the Prandtl boundary layer theory*, Proc. Royal Soc., Ser. A, **299** (1967), pp. 491–507.

[**57**] *The Dirichlet problem for the minimal surface equation in higher dimensions* (with H. Jenkins), J. Reine Ang. Math., **223** (1968), pp. 170–187.

[58] *Addendum to A priori estimates for solutions of the minimal surface equation*, Archive Rational Mech. Anal., **28** (1968), pp. 149–154.

[**59**] *On the mathematical basis of Prandtl's boundary layer theory: an example*, Archive Rational Mech. Anal., **28** (1968), pp. 217–225.

[**60**] *Variational problems of minimal surface type, III: The Dirichlet problem with infinite data* (with H. Jenkins), Archive Rational Mech. Anal., **29** (1968), pp. 304–322.

[**61**] *The behavior of similar solutions in a compressible boundary layer* (with J.B. McLeod), J. Fluid Mech., **34** (1968), pp. 337–342.

[62] *A new proof in differentiation theory*, Notices Amer. Math. Soc., **15** (1968), p. 1036, Abstract; see also W. Rudin, *Real and Complex Analysis*, Second Edition, McGraw–Hill, 1974, pp. xii, 435, and 162–167.

[63] *A general chain rule for derivatives and the change of variable formula for the Lebesgue integral* (with D.E. Varberg), Amer. Math. Monthly, **76** (1969), pp. 514–520.

[64] *The existence of similar solutions for some laminar boundary layer problems* (with J.B. McLeod), Archive Rational Mech. Anal., **31** (1969), pp. 288–303.

[**65**] *The problem of Dirichlet for quasilinear elliptic differential equations with many independent variables*, Philosophical Trans. Royal Soc. London, **A 264** (1969), pp. 413–496.

[**66**] *On surfaces of constant mean curvature which span a given space curve*, Math. Zeit., **112** (1969), pp. 77–88.

[**67**] *Existence theorems for some compressible boundary layer problems*, in *Qualitative Theory of Nonlinear Differential and Integral Equations*, SIAM Studies in Appl. Math., **5** (1969), pp. 35–42.

[68] *The Dirichlet problem for surfaces of constant mean curvature*, Proc. London Math. Soc., **21** (1970), pp. 361–384.

[69] *On the strong maximum principle for nonlinear second order differential inequalities*, J. Funct. Anal., **5** (1970), pp. 184–193.

[70] *Curvature inequalities for surfaces over a disk* (with H.F. Weinberger), in *Some problems of Mathematics and Mechanics – M.A. Lavrentieff Anniversary Volume*, "Nauka" Leningrad, 1970, pp. 242–250 (in Russian).

[**71**] *Boundary curvatures and the solvability of Dirichlet's Problem*, Actes du Congrès International des Mathématiciens (Nice, 1970), Tome **2**, pp. 867–875, Gauthier-Villars, Paris, 1971.

[**72**] *Recent developments in the mathematical aspects of boundary layer theory*, International J. Engineering Sciences, **9** (1971), pp. 233–240.

[73] *Uniqueness and comparison theorems for nonlinear elliptic equations in divergence form* (with J. Douglas, Jr., and T. Dupont), Archive Rational Mech. Anal., **42** (1971), pp. 157–168.

[**74**] *A symmetry problem in potential theory*, Archive Rational Mech. Anal., **43** (1971), pp. 304–318.

[75] *Gradient estimates for solutions of nonlinear elliptic and parabolic equations*, in *Contributions to Nonlinear Functional Analysis*, edited by E. Zarantonello, Univ. of Wisconsin Press, 1971, pp. 565–601.

[76] *Nonlinear elliptic equations of second order*, Lectures at *Symposium on Partial Differential Equations*, Berkeley, 1971, Mimeographed notes (57 pages).

[**77**] *The swirling vortex*, Philosophical Trans. Royal Soc. London, **A271** (1972), pp. 325–360.

[**78**] *Entire solutions of nonlinear Poisson equations*, Proc. London Math. Soc., **24** (1972), pp. 348–366.

[79] *A note on the preceding paper of Amann*, Archive Rational Mech. Anal., **44** (1972), pp. 182–186.

[**80**] *Rectilinear steady flow of simple fluids* (with R.L. Fosdick), Proc. Royal Soc. London, **332** (1973), pp. 311–333.

[81] *Global properties of continuum thermodynamic processes* (with R.L. Fosdick), Archive Rational Mech. Anal., **59** (1975), pp. 97–109.

[82] *On the axioms of classical mechanics*, Department of Mathematics, University of Minnesota, 1974 (74 pages).

[83] *Liouville theorems for nonlinear Poisson equations*, in *Convegno Internazionale sui Metodi Valutativi nella Fisica-Matematica*, Accad. Naz. Lincei, Problemi Attuali di Scienza e di Cultura, **217** (1975), pp. 207–215.

[84] *Doomsday: on target?*, Science Magazine, **189**, July 11, 1975, pp. 86–87.

[**85**] *The solvability of boundary value problems*, in *Mathematical Developments arising from Hilbert Problems* (Proc. Sympos. Pure Math., Northern Illinois Univ., De Kalb, Ill., 1974), Proc. Sympos. Pure Math., **28**, Amer. Math. Soc., Providence, R.I., 1976, pp. 507–524.

[**86**] *Curvature inequalities for surfaces over a disk* (with H.F. Weinberger), Amer. Math. Soc. Translation, (2) **104** (1976), pp. 223–231 (English translation of item [70]).

[87] *Liouville theorems and gradient bounds for quasilinear elliptic systems*, Archive Rational Mech. Anal., **66** (1977), pp. 295–310.

[88] *Optimal shapes for brittle beams under torsion* (with H.F. Weinberger), in *Complex Analysis and its Applications – jubilee volume for I.Y. Vekua*, "Nauka" Moscow, 1978, pp. 88–91.

[89] *Gradient estimates and Liouville theorems for quasilinear elliptic equations* (with L.A. Peletier), Ann. Sc. Normale Sup. Pisa, **5**, Ser. IV (1978), pp. 65–104.

[90] *The concepts of thermodynamics*, in *Continuum Mechanics and Partial Differential Equations*, edited by G.M. de la Penha et al., North–Holland Publ., Amsterdam, 1978, pp. 411–451.

[91] *On the impossibility of linear Cauchy and Piola–Kirchhoff constitutive theories for stress in solids* (with R.L. Fosdick), J. Elasticity, **9** (1979), pp. 83–89.

[92] *Conceptual analysis of the classical second laws of thermodynamics*, Archive Rational Mech. Anal., **70** (1979), pp. 254–272.

[93] *Foundations of thermodynamics*, Lecture Notes, University of Naples, 1979 (150 pages).

[94] *Phase transitions and interfacial layers for van der Waals fluids*, in *Recent Methods in Nonlinear Analysis and Applications* (Proc. Fourth International Meeting of SAFA), edited by A. Canfora et al., Liguori Editore, Naples, 1981, pp. 169–175.

[95] *The second law of thermodynamics for systems with approximate cycles* (with B.D. Coleman and D.R. Owen), Archive Rational Mech. Anal., **77** (1981), pp. 103–142.

[96] *Uniqueness of solutions of semilinear Poisson equations* (with K. McLeod), Proc. Nat. Acad. Sciences USA, **78** (1981), pp. 6592–6598. See also entry 116.

[97] *Uniqueness of positive solutions of semilinear equations in* $\mathbb{R}^n$ (with L.A. Peletier), Archive Rational Mech. Anal., **81** (1983), pp. 181–197. See also entry 112.

[98] *The form of interfacial surfaces in Korteweg's theory of phase equilibria*, Quart. Appl. Math., **41** (1983), pp. 357–364.

[99] *The mechanical theory of fluid interfaces and Maxwell's rule* (with E. Aifantis), J. Coll. and Interface Sci., **96** (1983), pp. 519–547.

[100] *The structure and laws of thermodynamics*, Proc. International Congress of Mathematicians (Warsaw 1983), pp. 1717–1728.

[**101**] *Applied mathematics and scientific thought*, in *Nonlinear analysis and optimization* (Bologna, 1982), Lecture Notes in Math. **1107**, Springer, 1984, pp. 19–27.

[102] *One-dimensional shock layers in Korteweg fluids* (with R. Hagan), in *Phase Transformations and Material Instabilities in Solids*, Academic Press, 1984, pp. 113–128.

[**103**] *Extensions of the mountain pass theorem* (with P. Pucci), J. Funct. Anal., **59** (1984), pp. 185–210.

[**104**] *A mountain pass theorem* (with P. Pucci), J. Diff. Equations, **60** (1985), pp. 142–149.

[105] *Esistenza ed unicità degli stati fondamentali per equazioni ellittiche quasilineari* (with B. Franchi and E. Lanconelli), Rend. Accad. Naz. Lincei, Serie 8, **79** (1985), pp 121–126. See also entries 121, 151.

[**106**] *A continuum model for chemical mixture dynamics*, (Proc. Nineteenth Midwestern Mechanics Conference, Columbus, Ohio) *Developments in Mechanics*, **13** (1985), pp. 59–64.

[107] *Nonexistence theorems for quasilinear partial differential equations* (with W.-M. Ni), Suppl. Rend. Circolo Mat. Palermo, **8** (1985), pp. 171–185.

[108] *On the thermomechanics of interstitial working* (with J.E. Dunn), Archive Rational Mech. Anal., **88** (1985), pp. 95–133.

[109] *An outline of thermodynamical structure*, in *New Perspectives in Thermodynamics*, edited by J. Serrin, Springer-Verlag, 1986, pp. 3–32.

[110] *Dynamic changes of phase in a van der Waals fluids* (with R. Hagan), in *New Perspectives in Thermodynamics*, edited by J. Serrin, Springer-Verlag, 1986, pp. 241–260.

[**111**] *Existence and non-existence theorems for ground states of quasilinear partial differential equations. The anomalous case* (with W.-M. Ni), Accad. Naz. Lincei, Atti dei Convegni, **77** (1986), pp. 231–257.

[**112**] *Uniqueness of nonnegative solutions of semilinear equations in* $\mathbb{R}^n$ (with L.A. Peletier), J. Diff. Equations, **61** (1986), pp. 380–397.

[**113**] *Nonexistence theorems of singular solutions for quasilinear partial differential equations* (with W.-M. Ni), Comm. Pure Appl. Math., **39** (1986), pp. 379–399.

[**114**] *A general variational identity* (with P. Pucci), Indiana Univ. Math. J., **35** (1986), pp. 681–703.

[**115**] *The laws of thermodynamics*, Res Mechanica, **21** (1987), pp. 361–379.

[116] *Uniqueness of positive radial solutions of* $\Delta u + f(u) = 0$ *in* $\mathbb{R}^n$ (with K. McLeod), Archive Rational Mech. Anal., **99** (1987), pp. 115–145.

[**117**] *The structure of the critical set in the mountain pass theory* (with P. Pucci), Trans. Amer. Math. Soc., **229** (1987), pp. 115–132.

[118] *Ground states for a prescribed mean curvature equation* (with L.A. Peletier), Proc. Amer. Math. Soc., **100** (1987), pp. 694–700.

[119] *Positive solutions of a prescribed mean curvature equation*, in *Calculus of Variations and Partial Differential Equations*, edited by S. Hildebrandt et al., Springer Lecture Notes in Math. **1340** (1988), pp. 248–255.

[120] *Ground states for the prescribed mean curvature equation: the supercritical case* (with F.V. Atkinson and L.A. Peletier), in *Nonlinear Diffusion Equations and their*

*Equilibrium States (I)*, edited by W.-M. Ni et al., MSRI Publ., **12**, Springer-Verlag, 1988, pp. 51–74.

[121] *Existence and uniqueness of ground states of quasilinear elliptic equations* (with B. Franchi and E. Lanconelli), in *Nonlinear Diffusion Equations and their Equilibrium States (I)*, MSRI Publ., **12**, Springer-Verlag, 1988, pp. 293–300. See also entry 151.

[122] *Remarks on the first eigenspace for polyharmonic operators* (with P. Pucci), Atti Sem. Mat. Fis. Univ. Modena, **36** (1988), pp. 107–118.

[123] *Asymptotic properties for solutions of strongly nonlinear ordinary differential equations* (with P. Pucci), in *Conference on Differential Equations and Geometry*, Rend. Sem. Mat. Univ. Politec. Torino, Fascicolo Speciale (1989), pp. 121–129.

[**124**] *Critical exponents and critical dimensions for polyharmonic operators* (with P. Pucci), J. Math. Pures Appl., **69** (1990), pp. 55–83.

[**125**] *Continuation and limit properties for solutions of strongly nonlinear second order differential equations* (with P. Pucci), Asymptotic Anal., **4** (1991), pp. 97–160.

[**126**] *Estimates for vertical points of solutions of prescribed mean curvature type equations, II* (with F.V. Atkinson and L.A. Peletier), Asymptotic Anal., **5** (1992), pp. 283–310.

[127] *The nature of thermodynamics*, Atti Sem. Mat. Fis. Univ. Modena, **39** (1991), pp. 445–472; Italian translation: *Sulla natura della termodinamica*, Atti e Memorie Accad. Naz. Scienze Lettere e Arti di Modena, Ser. VII, **7** (1992), pp. 185–202.

[128] *Global asymptotic stability for strongly nonlinear second order systems* (with P. Pucci), in *Nonlinear Diffusion Equations and their Equilibrium States (III)*, edited by N.G. Lloyd et al., Birkhäuser, 1992, pp. 437–449.

[129] *Existence and non-existence for ground states of quasilinear elliptic equations* (with H. Zou), Archive Rational Mech. Anal., **121** (1992), pp. 101–130.

[**130**] *Precise damping conditions for global asymptotic stability for nonlinear second order systems* (with P. Pucci), Acta Math., **170** (1993), pp. 275–307. See also entry 135.

[131] *Continuation and limit behavior for damped quasi-variational systems* (with P. Pucci), in *Degenerate Diffusions*, edited by W.-M. Ni et al., IMA Vols. in Math. Appl., **47**, Springer-Verlag, 1993, pp. 157–174.

[**132**] *The equations of continuum mechanics as a consequence of group invariance*, in *Advances in Continuum Mechanics*, ed. by G. Ferrarese, Pitagora Editrice, Bologna, 1993, pp. 217–225.

[133] *Asymptotic behavior of solutions of a nonstandard second order differential equation* (with W.A. Harris and P. Pucci), Diff. Integral Equations, **6** (1993), pp. 1201–1215.

[134] *Existence and non-existence results for ground states of degenerate Laplace equations* (with H. Zou), in *Differential Equations with Applications to Mathematical Physics*, **192**, Academic Press, 1993, pp. 287–305.

[135] *Precise damping conditions for global asymptotic stability of nonlinear second order systems, II* (with P. Pucci), J. Diff. Equations, **113** (1994), pp. 505–534.

[**136**] *Asymptotic stability for intermittently controlled nonlinear oscillators* (with P. Pucci), SIAM J. Math. Anal., **25** (1994), pp. 815–835.

[137] *On the derivation of Hamilton's equations* (with P. Pucci), Archive Rational Mech. Anal., **125** (1994), pp. 297–310.

[**138**] *Ground states of a quasilinear equation* (with L.A. Peletier and H. Zou), Diff. Integral Equations, **7** (1994), pp. 1063–1082.

[139] *Asymptotic estimates for a nonstandard second order differential equation* (with W.A. Harris and P. Pucci), in *Differential Equations, Dynamical Systems, and Control Science: A Festschrift in Honor of Lawrence Markus*, edited by K.D. Elworthy et al., Lecture Notes in Pure and Appl. Math., M. Dekker, New York, 1994, pp. 75–85.

[140] *Classification of positive solutions of quasilinear elliptic equations* (with H. Zou), Topological Methods in Nonlinear Analysis, **3** (1994), pp. 1–26.

[**141**] *Asymptotic integration of second order systems* (with H. Zou), Amer. J. Math., **116** (1994), no. 5, pp. 1241–1264.

[**142**] *A bifurcation problem at a singular limit* (with F. Merle and L.A. Peletier), Indiana Univ. Math. J., **43** (1994), pp. 585–609.

[143] *Non-existence of positive solutions of semilinear elliptic systems* (with H. Zou), in *A Tribute to Ilya Bakelman*, edited by S.A. Fulling et al., Discourses in Math. and its Applications, **3**, Department of Mathematics, Texas A&M University, 1994, pp. 56–69.

[144] *A numerical study of the existence and non-existence of ground states and their bifurcation for the equations of Chipot and Weissler* (with Y. Yan and H. Zou), unpublished manuscript, 1994.

[**145**] *Reflections on mathematics*, Lectio Doctoralis at the University of Padova. In "*L'Anno Galileiano*", Vol. I, Ed. Lint, Trieste, 1995, pp. 120–124.

[146] *Remarks on Lyapunov stability* (with P. Pucci), Differential and Integral Equations, **8** (1995), pp. 1265–1278.

[147] *On the elementary thermodynamics of quasistatic systems and other remarks*, in *Thermoelastic Problems and the Thermodynamics of Continua*, AMD – Vol. **198**, pp. 53–62, Amer. Soc. Mech. Eng., Applied Mech. Div., 1995.

[**148**] *Asymptotic stability for ordinary differential systems with time dependent restoring potentials* (with P. Pucci), Arch. Rational Mech. Anal., **132** (1995), pp. 207–232.

[**149**] *Asymptotic stability for nonautonomous dissipative wave systems* (with P. Pucci), Comm. Pure Applied Math., **49** (1996), pp. 177–216.

[150] *Quasi-variational systems*, in *First World Congress of Nonlinear Analysts*, ed. V Lakshmikantham, De Gruyter, Berlin, 1996, pp. 2055–2068.

[**151**] *Existence and uniqueness of ground states of quasilinear elliptic equations* (with B. Franchi and E. Lanconelli), Advances in Math., **118** (1996), pp. 177–243.

[**152**] *The equations of continuum mechanics and the laws of thermodynamics*, Meccanica, **31** (1996), pp. 547–563.

[153] *Stability for abstract evolution equations* (with P. Pucci), in *Partial Differential Equations and Applications – Collected papers in Honor of Carlo Pucci*, P. Marcellini, G. Talenti and E. Vesentini eds., Lecture Notes in Pure and Appl. Math. **177**, 1996, M. Dekker, Inc. New York, pp. 279–288.

[**154**] *Non-existence of positive solutions of Lane–Emden systems* (with H. Zou), Differential and Integral Equations, **9** (1996), pp. 635–653.

[155] *Asymptotic stability for nonlinear parabolic systems* (with P. Pucci), in *Energy Methods in Continuum Mechanics*, S.N. Antonsev, J.L. Diaz and S.I. Shmarev, eds., Kluwer Acad. Publ., Dordrecht, 1996, pp. 66–74.

[156] *Some remarks on global nonexistence for nonautonomous abstract evolution equations*, in *Harmonic analysis and nonlinear differential equations*, (Riverside, CA, 1995), pp. 253–263, *Contemp. Math.*, **208**, Amer. Math. Soc., Providence, RI, 1997.

[157] *Existence, stability and blow-up for dissipative evolution equations* (with P. Pucci), *Partial Differential Equations and Applications*, ed. G. Caristi and E. Mitidieri, Lecture Notes in Pure and Applied Math., **194**, Marcel Dekker, New York 1997, pp. 299–317.

[**158**] *Some remarks on global nonexistence for nonautonomous abstract evolution equations* (with H. Levine and P. Pucci), in *Harmonic analysis and nonlinear differential equations* (Riverside, CA, 1995), pp. 253–263, *Contemporary Math.*, **208**, Amer. Math. Soc., Providence, RI, 1997.

[159] *Global non-existence theorems for quasilinear evolution equations with dissipation* (with H. Levine), Arch. Rational Mech. Anal., **137** (1997), pp. 341–361.

[**160**] *Space, time and energy*, Lectio Doctoralis at the University of Ferrara, in *Nonlinear Analysis and Continuum Mechanics, Papers for the 65th Birthday of James Serrin*, G. Buttazzo, G.P. Galdi, E. Lanconelli and P. Pucci, eds., Springer Verlag, 1998, pp. 195–208.

[161] *Global existence and global nonexistence of solutions of the Cauchy problem for a nonlinearly damped wave equation* (with H.A. Levine and S.R. Park), J. Math. Anal. Appl., **228** (1998), pp. 181–205.

[162] *Global existence and nonexistence theorems for quasilinear evolution equations of formally parabolic type* (with H.A. Levine and S.R. Park), J. Differential Equations, **142** (1998), pp. 212–229.

[163] *Existence of positive solutions of the Lane–Emden system* (with H. Zou), Atti Sem. Mat. Fis. Univ. Modena, supplemento, **46** (1998), pp. 368–380.

[164] *Local asymptotic stability for nonlinear evolution equations with amplification forces* (with P. Pucci), Israel J. Math., **104** (1998), pp. 29–50.

[165] *Uniqueness of ground states for quasilinear elliptic operators* (with P. Pucci), Indiana Univ. Math. J., **47** (1998), pp. 501–525.

[166] *Uniqueness of ground states for quasilinear elliptic equations in the exponential case* (with P. Pucci), Indiana Univ. Math. J., **47** (1998), pp. 523–539.

[**167**] *Existence of positive entire solutions of elliptic Hamiltonian systems* (with H. Zou), Comm. Partial Differential Equations, **23** (1998), pp. 577–599.

[**168**] *Global nonexistence for abstract evolution equations with positive initial energy* (with P. Pucci), J. Differ. Equations, **150** (1998), pp. 203–214.

[**169**] *Symmetry of ground states of quasilinear elliptic equations* (with H. Zou), Archive Rational Mech. Anal., **148** (1999), pp. 265–290.

[170] *A strong maximum principle and a compact support principle for singular elliptic inequalities* (with P. Pucci and H. Zou), *Papers in memory of Ennio De Giorgi*, Ricerche Mat., **48** (1999), supplemento, pp. 373–398.

[**171**] *A strong maximum principle and a compact support principle for singular elliptic inequalities* (with P. Pucci and H. Zou), J. Math. Pures Appl., **78** (1999), pp. 769–789.

[172] *Existence of ground states and free boundary problems for quasilinear elliptic operators*(with F. Gazzola and M. Tang), Adv. Differential Equations, **5** (2000), pp. 1–30.

[173] *A note on the strong maximum principle for elliptic differential inequalities* (with P. Pucci), J. Math. Pure Appl., **79** (2000), pp. 57–71.

[174] *Ground states and free boundary problems for the n-Laplacian in n dimensional space* (with M. García–Huidobro, R. Manásevich, M. Tang and C. Yarur), J. Funct. Anal., **172** (2000), pp. 177–201.

**[175]** *Uniqueness of ground states for quasilinear elliptic equations* (with M. Tang), Indiana Univ. Math. J., **49** (2000), pp. 897–923.

**[176]** *Jean Leray's contributions to nonlinear differential equations, fixed point theory and fluid mechanics*, Gaz. des Mathématiciens, Supplement to volume **84** (2000), pp. 61–70.

**[177]** *The Harnack inequality for quasilinear elliptic equations in* $\mathbb{R}^2$ (with P. Pucci), J. d'Anal. Math., **85** (2001), pp. 307–321.

**[178]** *Cauchy–Liouville and universal boundedness theorems for quasilinear elliptic equations and inequalities* (with H. Zou), Acta Math., **189**, (2002), pp. 79–142.

[179] *Asymptotic behavior of ground states of quasilinear elliptic equations with two vanishing parameters* (with F. Gazzola), Ann. Inst. H. Poincaré Anal. Non Linéaire, **19** (2002), pp. 477–504.

[180] *Commentary on the Hopf strong maximum principle*, in *Selected Works of Eberhard Hopf with commentaries*, ed. by C.S. Morawetz, J.B. Serrin and Y.G. Sinai, *Amer. Math. Soc.*, Providence, 2002.

**[181]** *Existence for a nonlinear wave equation with damping and source terms* (with G. Todorova and E. Vitillaro), Differential and Integral Equations, **16** (2003), pp. 13–50.

[182] *Asymptotic behavior of ground states of quasilinear elliptic equations with two vanishing parameters, II* (with F. Gazzola, L.A. Peletier and P. Pucci), Ann. Inst. H. Poincaré Anal. Non Linéaire, **20** (2003), pp. 947–974.

**[183]** *An appreciation of Clifford Truesdell*, in *Essays and papers dedicated to the memory of Clifford Ambrose Truesdell III*, Vol. I, J. Elasticity, **70** (2003), pp. 31–38.

[184] *The strong maximum principle revisited* (with P. Pucci), J. Differ. Equations, **196** (2004), pp. 1–66; *Erratum*, J. Differ. Equations **207** (2004), pp. 226–227.

**[185]** *Elliptic equations and products of positive definite matrices* (with C. H. Conley and P. Pucci), Math. Nachrichten, **278** (2005), pp. 1490–1508.

[186] *Qualitative properties of ground states for singular elliptic equations with weights* (with P. Pucci, M. García–Huidobro and R. Manásevich), Annali Mat. Pura Appl., **185** (2006), pp. 205–243.

**[187]** *Dead cores and bursts for quasilinear singular elliptic equations* (with P. Pucci), SIAM J. Math. Anal., **38** (2006), pp. 259–278.

**[188]** *A maximum principle for mean-curvature type elliptic inequalities*, J. Eur. Math. Soc. (JEMS), **8** (2006), pp. 389–398.

[189] *Qualitative properties for solutions of singular elliptic inequalities on complete manifolds* (with P. Pucci and M. Rigoli), J. Diff. Equations, **234** (2007), pp. 507–543.

[190] *A remark on the Morrey potential*, Contemp. Math., Special Volume "Control methods in PDE-dynamical systems", **426** (2007), pp. 307–315.

[191] *Maximum Principles for Elliptic Partial Differential Equations* (with P. Pucci), in *Handbook of Differential Equations – Stationary Partial Differential Equations*, Ed. M. Chipot, Elsevier BV, **4** (2007), pp. 355–483.

[192] *Partial and full symmetry of solutions of quasilinear elliptic equations, via the comparison principle* (with P. Pucci and B. Sciunzi), Contemp. Math., special volume dedicated to H. Brezis "*Perspectives in nonlinear partial differential equations*", **446** (2007), 437–444.

[193] *The Maximum Principle* (with P. Pucci), in "*Progress in Nonlinear Differential Equations and their Applications*", Vol. **73**, Birkhäuser Publ., Switzerland, 2007, X, 234 pages.

[**194**] *The area rule for simple fluid phase transitions*, J. Elasticity, **90** (2008), pp. 129–159.

[195] *Entire solutions of quasilinear elliptic equations*, J. Math. Anal. Appl., **352** (2009), pp. 3–14.

[196] *Entire solutions of completely coercive quasilinear elliptic equations* (with A. Farina), J. Differential Equations, **250** (2011), pp. 4367–4408.

[197] *Entire solutions of completely coercive quasilinear elliptic equations, II* (with A. Farina), J. Differential Equations, **250** (2011), pp. 4409–4436.

[198] *A remark on entire solutions of completely coercive quasilinear elliptic equations* (with P. Pucci), J. Differential Equations, **250** (2011), pp. 675–689.

[199] *The Liouville theorem for homogenous elliptic differential inequalities*, J. Math. Sciences (St. Petersberg, Russia), **179** (2011), pp. 176–185.

[200] *Weakly subharmonic functions*, Boll. Unione Mat. Ital., (9) **4** (2011), pp. 347–361.

[201] *Comparison principles, uniqueness and symmetry results of solutions of quasilinear elliptic equations and inequalities* (with L. D'Ambrosio, A. Farina and E. Mitidieri), Nonlinear Analysis, **90** (2013), pp. 135–158.

[202] *The splitting of intrinsic energy and the origin of mass density in continuum mechanics* (with R. Fosdick), Contin. Mech. Thermodyn., DOI 10.1007/s00161-013-0301-1.

[203] *Foundations of Thermodynamics* (with C.-S. Man), monograph in preparation.